College Algebra and Trigonometry with Applications

THIRD EDITION

College Algebra and Trigonometry with Applications

Cheryl Cleaves
Margie Hobbs
Paul Dudenhefer

State Technical Institute at Memphis
Memphis, Tennessee

Prentice Hall Career & Technology
Englewood Cliffs, New Jersey 07632

Library of Congress Cataloging-in-Publication Data

Cleaves, Cheryl S.
College Algebra and Trigonometry with Applications / Cheryl Cleaves, Margie Hobbs, Paul Dudenhefer. — 3rd ed.
p. cm.
Rev. ed. of: Basic mathematics for trades and technologies. 2nd ed.
Includes index.
ISBN 0-13-318131-6
1. Mathematics. I. Hobbs, Margie II. Dudenhefer, Paul. III. Cleaves, Cheryl S. Basic mathematics for trades and technologies. IV. Title.
QA39.2.C58 1994
512′.1—dc20 95-5641
CIP

Editorial/production supervision: *Julie Boddorf*
Interior design: *Kenny Beck*
Cover design: *Kenny Beck*
Cover photography: *George Mattei Photography Inc.*
Cover prop sourcing: *Irene Bolm/Wendy Helft*
Color retouching: *Mark Lasalle*
Creative director: *Paula Maylahn*
Director of production and manufacturing: *David Riccardi*
Production coordinator: *Ed O'Dougherty*
Acquisitions editor: *Frank Burrows*
Editorial assistant: *Scott Montgomery*
Supplements editor: *Judith Casillo*

Prentice Hall Career & Technology

A Paramount Communications Company
Englewood Cliffs, NJ 07632

Portions of this book previously published as *Basic Mathematics for Trades and Technologies* by Cleaves, Hobbs, and Dudenhefer (© 1990), *Introduction to Technical Mathematics* by Cleaves, Hobbs, and Dudenhefer (© 1988), and *Vocational-Technical Mathematics Simplified* by Cleaves, Hobbs, and Dudenhefer (© 1987).

Printed in the United States of America

10 9 8 7 6 5

ISBN 0-13-318131-6

Prentice-Hall International (UK) Limited, *London*
Prentice-Hall of Australia Pty. Limited, *Sydney*
Prentice-Hall Canada Inc., *Toronto*
Prentice-Hall Hispanoamericana, S.A., *Mexico*
Prentice-Hall of India Private Limited, *New Delhi*
Prentice-Hall of Japan, Inc., *Tokyo*
Simon & Schuster Asia Pte. Ltd., *Singapore*
Editora Prentice-Hall do Brasil, Ltda., *Rio de Janeiro*

Dedicated To

Wilmer Crockett Smith
Robert Siscero Johnson
Haley Nicole Fink

Contents

CHAPTER 8

PART TWO

CHAPTER 9

CHAPTER 10

CHAPTER 11

CHAPTER 12

CHAPTER 17

Inequalities and Absolute Values 637

CHAPTER 18

Graphing 673

CHAPTER 19

Slope and Distance 718

CHAPTER 20

Systems of Equations and Inequalities 752

CHAPTER 21

Selected Concepts of Geometry 781

Preface

In *College Algebra and Trigonometry with Applications*, Third Edition, we have preserved all the features that made the First and Second Editions the most appropriate text on the market for a comprehensive review of mathematics in post-secondary and college technical programs. No other comprehensive text we know of presents such a mathematically sound coverage of arithmetic, algebra, geometry, and trigonometry topics as the Third Edition. It remains one of the few such texts to include geometry. Our emphasis for geometry remains on practical applications and not proofs.

Readability Level

As in the Second Edition, which was titled *Basic Mathematics for Trades and Technologies*, we have attempted to keep the language simple but yet not talk down to the student. We applied two different readability formulas on selected 100-word passages and averaged the results. The readability level generally compares to that of daily newspapers and popular magazines, making our text suitable for a wide range of instructional uses. Of course, when mathematical vocabulary is taken into account, some passages may prove more challenging to read until the definitions are understood. Along these lines, we have re-examined our definitions carefully and have taken all possible steps to ensure their mathematical soundness as well as their readability.

Content of the Third Edition

The Third Edition, we feel, is a substantially improved text because we have been able to incorporate many valuable suggestions made by users of the Second Edition and reviewers of the Third Edition manuscript. For example, although we continue to use a step-by-step approach, we introduce many topics earlier so that a spiraling development of these topics is possible in later chapters. We have also given greater emphasis to the problem sets and problem solving and have included new material on such topics as graphing calculators, polynomials, exponential expressions, and logarithms. In addition, we have integrated much of the geometry into various arithmetic and algebra chapters throughout the text, an integration suggested in the reform movements of the American Mathematical Association of Two-Year Colleges (AMATYC), the National Council of Teachers of Mathematics (NCTM), and the Mathematical Association of America (MAA).

A number of other changes have been made in the content of *College Algebra and Trigonometry with Applications*, Third Edition, which are highlighted below.

We begin with a thorough review of arithmetic, but we have included in Chapter 1, Whole Numbers, a basic introduction to exponents, roots, and powers of 10 and increased the emphasis on estimation. We also moved order of operations to this chapter and added the concepts of prime and composite numbers, factorization and divisibility, least common multiples, and greatest common factors. Basic, scientific, and graphing calculators are likewise introduced here.

We have moved signed numbers to Chapter 2, which includes integers only. Many technical programs early on introduce signed numbers, so an early introduction of signed numbers in the Third Edition complements cross-instructional learning. Of course, this chapter can be postponed if desired, as well as sections of subsequent chapters dealing with signed numbers. This chapter is followed by chapters on fractions and decimal fractions, including sections on signed fractions and signed decimal fractions.

In addition to integrating geometry throughout the text, we have also combined some material. The separate chapters on the U.S. customary and metric measurements have been combined into one chapter and the separate chapters on percents and their applications have been combined into one chapter.

A new chapter, Interpreting and Analyzing Data, has been added to introduce basic statistics and probability. These topics are included in curriculum guidelines and standards; and, with the availability of calculators and computer software, these topics are receiving increased emphasis in business and industry.

When we introduce algebra, we briefly review signed numbers. We do this to capitalize on the spiraling effect for learning and because students usually need the review. Also, some users may prefer to start with algebra and not cover the arithmetic chapters in any depth.

The Third Edition reflects a condensed presentation of equations containing fractions and decimal fractions and includes significant new material. For example, we have added material on imaginary and complex numbers, factoring the sum and difference of cubes, selecting efficient methods of solving quadratic equations, solving equations with rational expressions, solving inequalities with absolute values, evaluating logarithms with bases other than 10 or e, and simplifying logarithmic expressions by applying the laws of exponents.

As indicated earlier, we integrated geometry throughout the text. However, Chapter 21 collects the geometry concepts that have not been integrated in earlier chapters. We have designed this chapter to emphasize multi-stepped application problems that are both challenging and interesting. This chapter offers an excellent opportunity for integrating numerous algebra skills and critical-thinking skills.

Our introduction to trigonometry retains such topics as radians, trigonometric functions, and the trigonometry of right and oblique triangles. But we have also added material on topics like vectors and graphing trigonometric functions on a calculator. The calculator is used quite extensively in the trigonometry chapters.

Use of Calculators

After introducing basic, scientific, and graphing calculators in Chapter 1, we continue with calculator tips throughout the text. Calculator tips are presented each time a new key is introduced or a new application is encountered. Of course, calculator usage remains optional with the instructor throughout the text, except for those sections where the calculator is the norm rather than the exception, such as for finding square root values and trigonometric values and working with exponential and logarithmic expressions. Our feeling is that today most of these calculations are made with a scientific or graphing calculator.

Flexibility of Use

College Algebra and Trigonometry with Applications, Third Edition, is designed for use in classrooms, business and industrial training programs, or learning laboratories. It is easily adapted to a variety of instructional delivery systems.

We have attempted to combine the best of the old with the new, so we continue with an abundance of numbered examples showing solutions step by step with explanatory marginal notes. The Third Edition also features all rules, formulas, and definitions highlighted in a second color. New with this edition are Tips and Traps boxes which draw attention to special cautions and procedures.

We also continue with the popular Self-Study Exercises but have placed them at the end of each section of each chapter. The Self-Study Exercises are coded by number to a set of Learning Objectives that begins each section of the Third Edition. The answers to the Self-Study Exercises are located at the back of the text.

We have also retained the Assignment Exercises but have relocated them to the end of each chapter, where they are identified by section number. The answers for the odd-numbered Assignment Exercises are located at the back of the text. We have also retained the end-of-chapter Trial Tests and have placed the odd-numbered answers at the back of the book. A new feature of the Third Edition is a chapter summary in grid format at the end of each chapter. The summary lists objectives, rules, and examples.

Also available for the Third Edition is a *Student's Solutions Manual* that shows detailed solutions to every odd-numbered problem of the Assignment Exercises and Trial Tests, as well as an *Instructor's Manual* with the solutions to all even-numbered problems of the Assignment Exercises and Trial Tests. The *Instructor's Manual* also includes projects and activities that can be used as classroom or out-of-class assignments. A computerized test bank is available for the instructor and transparency masters are also available for most figures.

The material may be covered in three semesters or quarters if a thorough presentation is desired. However, the material may be covered in one semester or two quarters if only a review or survey is desired or if chapters or sections less relevant to specific programs are omitted or treated in less depth.

Applications

In *College Algebra and Trigonometry with Applications*, Third Edition, we have added to the number of problems in many of the exercise sets, responding to comments from the many users of the Second Edition as well as from the reviewers of the Third Edition. This edition contains well over 600 numbered examples with step-by-step explanations, showing nearly 1000 different problems and their solutions. There are over 100 sets of Self-Study Exercises and 24 sets of Assignment Exercises, representing well over 12,000 problems. The 24 chapter Trial Tests present well over 600 additional problems.

We include an abundance of technical and other job-related applications. Included are applications in electronics, electricity, robotics, computer-aided design, computer-aided manufacturing, microprocessing, welding, bookkeeping, pipefitting, carpentry, finance, mechanics, air conditioning, refrigeration, economics, TV service, drafting, building construction, surveying, masonry, business operations, machinery, manufacturing, automotive technology, and other areas.

We continue to use a practical, learn-by-doing approach to technical mathematics without frills and with a minimum of theory. Yet we have added higher-order thinking skills and "writing to learn" exercises in the Concepts Analysis problem sets which follow the Assignment Exercises in each chapter. Additionally, each chapter now contains two or three open-ended problems designed as Challenge Problems.

Field Testing

College Algebra and Trigonometry with Applications, Third Edition, is the result of extensive field testing of both the First and Second Editions and the revisions that are included in this Third Edition. These materials have been field tested at State Technical Institute at Memphis, Richards Manufacturing Company, and Ripley Industries, Inc., all located in or near Memphis, Tennessee. We wish to thank these institutions and the instructors who have used the original text and its revisions and who have made valuable suggestions for the Third Edition.

Acknowledgments

We wish to thank those who reviewed the manuscript and offered valuable suggestions in developing this project:

Max Cisneros
Developmental Studies
Albuquerque T-VI
Albuquerque, NM

Don Garmer
ITT Technical Institute
Virginia Beach, VA

Jerry Hartman, Chief Instructor
ITT Technical Institute
Portland, OR

Abdul Kazi, Director of Education
ITT Technical Institute
Omaha, NE

Vrej Khachikian, Associate Professor
DeVry
Industry, CA

Donald E. King
ITT Technical Institute
Youngstown, OH

Don Montgomery, Chief Instructor
Electronic Engineering Department
ITT Technical Institute
Houston, TX

Kevin O'Neill
Math Science & Engineering Div.
Kankakee Community College
Kankakee, IL

John A. Weese, Ph.D., P.E.
Professor & Head
Engineering Technology Department
Texas A&M
College Station, TX

We also wish to thank the many users of our text for their helpful comments and suggestions. In addition, the staff at Prentice Hall Career and Technology, especially Frank Burrows and Julie Boddorf, has been invaluable throughout this process.

Special thanks are given to Kim Collier for the solutions for the *Student's Solutions Manual* and the *Instructor's Manual.* We also give special recognition to Barbara Bode-Snyder who provided the electronics applications, exercises, and solutions and to John DeCoursey and Brian Randall for the computerized test bank. Ted Davis, John Paszel, Jim Smith, and Renee Smith are recognized for accuracy checking many of the exercises. We appreciate the many suggestions from users of previous editions, especially Frank Caldwell, Paul Calter, and Sue Stokley. The authors take full responsibility for all misprints or errors.

We also thank the administration and staff of State Technical Institute at Memphis for their encouragement and support of our efforts. We thank also our coworkers in the computer, business, and engineering technologies and in general studies for their many suggestions and helpful comments during the development of the original manuscript as well as the Third Edition. A special thanks goes to the faculty in the mathematics program at State Tech for their continual suggestions and comments.

Finally, we recognize the support and strength we draw from our families: Charles Cleaves, Sarah Smith, Shirley Riddle, Jim and Renee Smith, Susan and Ben Duke, Tosha Riddle, Jim Riddle, Jr., Brienne Smith, and Keaton Smith; Allen Hobbs, Holly Hobbs, Norman Johnson, Rubin Johnson, Martha Dawson, Byron Johnson, Savannah Byrd, Bill Johnson, and Docie Johnson; and Gaynell Dudenhefer, Paul and Donna Dudenhefer, Paulette and Terry Fink, David and Deanne Dudenhefer, Diane Dudenhefer, and Haley Fink.

Cheryl Cleaves
Margie Hobbs
Paul Dudenhefer

CHAPTER

1 Whole Numbers

When we study a subject for the first time or review in some detail a subject we studied some time ago, we need to begin with the basics. Often, as we examine the basics of a subject, we discover—or rediscover—many bits and pieces of useful information. In this sense, mathematics is no different from any other subject. We begin with a study of whole numbers and the basic operations that we perform with them. Whole numbers and their operations form the foundation of our study of mathematics.

1-1 READING, WRITING, AND ROUNDING WHOLE NUMBERS

Learning Objectives

L.O.1. Identify place values in whole numbers.
L.O.2. Write whole numbers in words and standard notation.
L.O.3. Round a whole number to a place value.
L.O.4. Round a whole number to a number with one nonzero digit.

Our system of numbers, which is called the *decimal number system*, uses 10 individual figures called *digits*: 0, 1, 2, 3, 4, 5, 6, 7, 8, 9. A *whole number* is made up of one or more digits. When a number contains two or more digits, each digit must be in the correct place for the number to have the value we want it to have. If we mean ''ninety-eight,'' we must place the 9 first and the 8 second to represent 98. If we change the places of these two digits by putting the 8 first and the 9 second, we get a new value (eighty-nine) and a new number (89).

L.O.1. Identify place values in whole numbers.

Place Value

Each place a digit can occupy in a number has a value called a *place value*. If we know the place value of each digit in a number, we can read the number and understand how much it means. Look at the chart of place values in Fig. 1-1. Notice that each place value *increases* as we move from *right to left* and that each increase is *10 times* the value of the place to the right. For example, the tens place is 10 times the ones place, the hundreds place is 10 times the tens place, and so on.

Billions			Millions			Thousands			Units		
Hundred billions (100,000,000,000's)	Ten billions (10,000,000,000's)	Billions (1,000,000,000's)	Hundred millions (100,000,000's)	Ten millions (10,000,000's)	Millions (1,000,000's)	Hundred thousands (100,000's)	Ten thousands (10,000's)	Thousands (1,000's)	Hundreds (100's)	Tens (10's)	Ones (1's)

Figure 1-1

The place values are arranged in *periods* or groups of three to make numbers easier to read. The first group of three is called *units*, the second group of three is called *thousands*, the third group is *millions*, and the fourth group is *billions*. Commas are used to mark off these groups in a number. In four-digit numbers the comma separating the units group from the thousands group is optional. Thus, both 4,575 and 4575 are correct.

> **Rule 1-1.1** ***To identify place value of digits:***
>
> **1.** Mentally position the number on the place-value chart so that the last digit on the right aligns under the ones place.
> **2.** Identify the place value of each digit according to its position on the chart.

EXAMPLE 1-1.1 In the number 2,472,694,500, identify the place value of the digit 7.

Solution: To solve this problem, we align the number on the place-value chart as shown in Fig. 1-2.

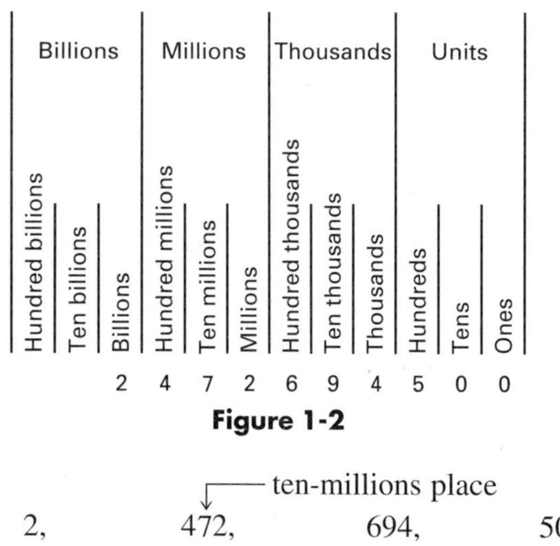

Figure 1-2

ten-millions place

2,	472,	694,	500
Billions	Millions	Thousands	Units

7 is in the ten-millions place. ■

L.O.2. Write whole numbers in words and standard notation.

Writing Numbers in Words and Standard Notation

Now that we have reviewed place values, we can read numbers with little difficulty. All we need to do is use the following rule.

Rule 1-1.2 *To read numbers:*

1. Mentally position the number on the place-value chart so that the last digit on the right lines up under the ones place.
2. Examine the number from right to left, separating each group with commas.
3. Identify the leftmost group.
4. From the left, read the numbers in each group and the group name. (The group name *units* is not usually read. A group containing all zeros is not usually read.)

EXAMPLE 1-1.2 Read 7543026129 by writing it in words.

Solution: Mentally align the digits on the place-value chart (Fig. 1-3). Starting at the right, separate each group of three digits with commas. Identify the leftmost group (billions). The number is seven *billion*, five hundred forty-three *million*, twenty-six *thousand*, one hundred twenty-nine. ■

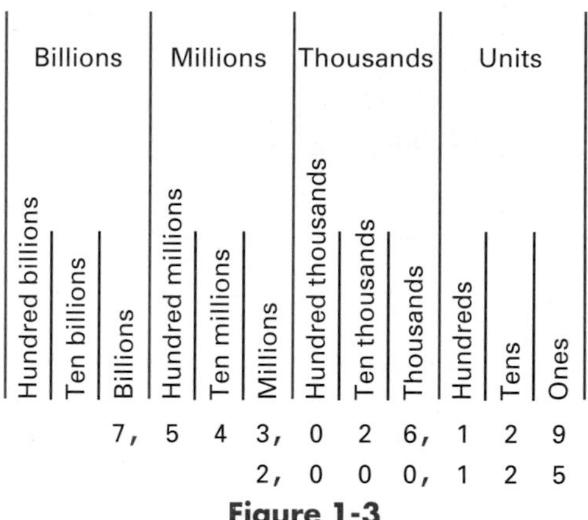

Figure 1-3

Tips and Traps

1. A group name is inserted at each comma.
2. The word "and" should not be used when reading whole numbers.
3. The numbers from 21 to 99 (except 30, 40, 50, and so on) use a hyphen when they are written (forty-three, twenty-six, and so on).

EXAMPLE 1-1.3 Show how 2000125 would be read by writing it in words.

Solution: See Fig. 1-3. The number is two *million*, one hundred twenty-five.

Note: Since the thousands group contains all zeros, it is *not* read. ■

Of course, there are times we have to write down spoken numbers on paper. For example, a contractor may give us a price estimate on the telephone and we need to write it down to compare it with other estimates. Knowing how many places the number contains will help us write the number correctly. We can tell the number of places in a number by the number of digits in each group and the number of groups in the number.

For example, if the number is "twenty-seven thousand, four hundred twenty-one," we know that there are two digits in the thousands group (27) and three digits in the units group (421). We can tell that the number contains only two groups because the *first* group is thousands, meaning that units must follow it. The first group determines how many groups are in the number. Therefore, by a combination of digits in each group and the first group name in the number, we can figure how many places are in the number and can correctly write the words as numbers. Numbers written with digits in the appropriate place-value positions are numbers in *standard notation*.

Rule 1-1.3 ***To write numbers in standard notation:***

1. Write the group names in order from left to right, starting with the first group in the number.
2. Fill in the digits in each group. Leave blanks for zeros if necessary.
3. Supply zeros as needed for each group. Each group except the leftmost group must have three digits.
4. Starting at the right, separate the groups with commas as needed.

EXAMPLE 1-1.4 The exact cost of a group of airplanes is eight million, two hundred four thousand, twelve dollars. Write this amount as a number in standard notation.

Solution

	Millions	Thousands	Units
1.	Millions	Thousands	Units
2.	Millions	Thousands	Units
	8	204	_12
3.	Millions	Thousands	Units
	8	204	012
4.	Millions	Thousands	Units
	8,	204,	012

The cost is $8,204,012. ■

EXAMPLE 1-1.5 One hundred six thousand, three bolts are required for the assembly of a group of light standards. Write this amount as a number.

Solution

	Thousands	Units
1.	Thousands	Units
2.	Thousands	Units
	106	__3
3.	Thousands	Units
	106	003
4.	Thousands	Units
	106,	003

106,003 bolts are required. ■

L.O.3. Round a whole number to a place value.

Rounding a Whole Number to a Place Value

To check calculations made with a calculator or to make mental calculations, it is useful to use approximate numbers instead of exact numbers. To *round* a number is to express it as an approximation.

Rule 1-1.4 *To round a whole number to a certain place value:*

1. Locate the digit occupying the desired place value.
2. Examine the digit to the right of the digit in the desired place value.
3. If the digit to the right is less than 5, round to a smaller approximate value by replacing this digit and any digits to its right with zeros.
4. If the digit to the right is 5 or greater, round to a larger approximate value by adding 1 to the digit in the desired place value and replacing any digits to its right with zeros.

EXAMPLE 1-1.6 Round 3,743 to the nearest hundreds place.

Solution

3,7̲43 Identify the hundreds place and examine the digit to its right. The hundreds digit is 7 and the digit to the right, 4, is less than 5. Replace the digit 4 and any digits to its right with 0. The rounded number is 3,700. ■

L.O.4. Round a whole number to a number with one nonzero digit.

Rounding a Whole Number to a Number with One Nonzero Digit

In estimating sums, it is sometimes necessary to round a number so that only one of its digits is *not* zero. The process of rounding a number so that only one digit is not zero and all other digits are zero is called *rounding to one nonzero digit.*

Rule 1-1.5 *To round a whole number to a number with one nonzero digit:*

1. Identify the first digit on the left that is not zero.
2. Examine the next digit on the right.
3. If the next digit on the right is less than 5, round to a smaller approximate number by replacing the digit and any digits to its right with 0.
4. If the digit to the right is 5 or greater, round to a larger approximate number by adding 1 to the first nonzero digit and replacing any digits to its right with 0.

EXAMPLE 1-1.7 Round 6,415 to a number with one nonzero digit.

Solution

6̲,415 Identify the first nonzero digit on the left and examine the digit to its right. The first nonzero digit on the left is 6 and the digit to its right is 4. The 4 is less than 5, so round to a smaller approximate number by replacing the 4 and all digits to its right with 0. The rounded answer is 6,000. ■

SELF-STUDY EXERCISES 1-1

L.O.1. In the number 2,304,976,186, identify the place value of the following digits.

1. 3	**2.** 7	**3.** 1
4. 0	**5.** 2	

In the number 8,972,069,143, identify the place value of the following digits.

6. 0 **7.** 4 **8.** 7
9. 8 **10.** 6

L.O.2. Show how these numbers would be read by writing them out as words.

11. 6704 **12.** 89021 **13.** 662900714
14. 3000101 **15.** 15407294376 **16.** 150

Write these words as numbers in standard notation. Use commas when necessary.

17. Seven billion, four hundred

18. One million, six hundred twenty-seven thousand, one hundred six

19. Fifty-eight thousand, two hundred one

20. In a telephone conversation a contractor submitted the following bid for a job: "one thousand six dollars." Write this bid as a number.

L.O.3 Round the following whole numbers to the indicated place value.

21. 593 (tens) **22.** 8,934 (hundreds) **23.** 9253 (tens)
24. 82,746 (thousands) **25.** 197,305 (ten-thousands)

L.O.4. Round the following whole numbers to a number with one nonzero digit.

26. 52 **27.** 896 **28.** 7,523
29. 4,007 **30.** 530

1-2 ADDING WHOLE NUMBERS

Learning Objectives

L.O.1. Add single-digit numbers.
L.O.2. Add numbers of two or more digits.
L.O.3. Estimate and check addition.
L.O.4. Add using a calculator.

Addition of whole numbers is used on all jobs and occupations at one time or another. Our purpose in this section is to review some ways to gain speed and make fewer mistakes when adding whole numbers.

L.O.1. Add single-digit numbers.

Commutative and Associative Properties of Addition

It is sometimes easier to discuss the procedures required to work a problem if we name the parts of the problem. In an addition problem, the numbers being added are called *addends* and the answer is called the *sum* or *total.*

Two useful things to know about addition are that it is *commutative* and *associative.* By *commutative* we mean that it does not matter in what *order* we add numbers. We can add 7, 4, and 6 in any order and still get the same answer:

$$7 + 4 + 6 = 17 \qquad 4 + 7 + 6 = 17 \qquad 6 + 7 + 4 = 17$$

■ **Definition 1-2.1 Commutative Property.** The *commutative property* means that values being added (or multiplied) may be added (or multiplied) in any order.

By *associative,* we mean that we can *group* numbers together any way we want when we add and still get the same answer. To add 7 + 4 + 6, we can group the 4 + 6 to get 10; then we add the 10 to the 7:

$$7 + (4 + 6) = 7 + 10 = 17$$

Or we can group 7 + 4 to get 11; then we add the 6 to the 11:

$$(7 + 4) + 6 = 11 + 6 = 17$$

Addition is a *binary operation*. That is, the rules of addition apply to adding *two* numbers at a time. The associative property of addition shows how addition is extended to more than two numbers.

■ **DEFINITION 1-2.2 Associative Property.** The *associative property* means that values being added (or multiplied) may be grouped in any manner.

The commutative and associative properties of addition can help us add whole numbers quickly and accurately. Look for these two aids in the examples that follow.

Tips and Traps

Many rules and definitions can be written symbolically. Symbolic representation allows a quick recall of the rule or definition.

Commutative Property of Addition:

$$a + b = b + a$$

where a and b are numbers.

Associative Property of Addition:

$$a + (b + c) = (a + b) + c$$

where a, b, and c are numbers.

The associative property of addition also allows other possible groupings and extends to more than three numbers.

$$7 + 4 + 6 = 13 + 4 = 17$$

$$3 + 5 + 7 + 9 = 8 + 16 = 24$$

EXAMPLE 1-2.1 Add 7 + 3 + 6 + 8 + 2 + 4 using the commutative and associative properties.

Solution

$$\begin{array}{r} 7 \\ 3 \\ 6 \\ 8 \\ 2 \\ +\ 4 \\ \hline 30 \end{array} \quad \begin{array}{l} 7 + 3 = 10 \\ 8 + 2 = 10 \\ 6 + 4 = 10 \end{array}$$

[*Group 7 + 3, 8 + 2, and 6 + 4. Then add the three groups (10 + 10 + 10 = 30).*] ■

Another important property of addition is the *zero property of addition.*

> **Rule 1-2.1** ***Zero property of addition:***
>
> Zero added to any number results in the same number.
>
> $$n + 0 = n \quad \text{or} \quad 0 + n = n$$
>
> $$5 + 0 = 5 \quad \text{or} \quad 0 + 5 = 5$$

L.O.2. Add numbers of two or more digits.

Adding in Columns

All the numbers in the preceding section consisted of one digit. When adding numbers of two or more digits, such as 238 + 456, we must make sure that the numbers are placed in columns properly. This means that the same place values must be aligned under one another so that all the ones are in the far right column, all the tens in the next column, and so on.

Rule 1-2.2 ***To add numbers of two or more digits:***

1. Arrange the numbers in columns so that the ones place values are in the same column.
2. Add the ones column, then the tens column, then the hundreds column, and so on, until all the columns have been added. *Carry* whenever the sum of any column is more than one digit.

Example 1-2.2 Add 3785 + 5497.

Solution

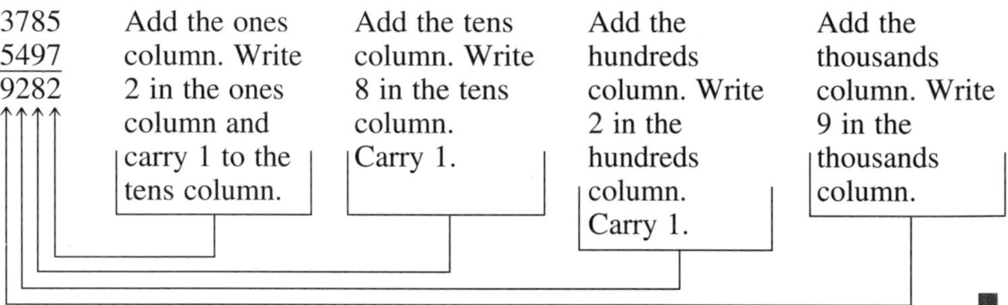

Example 1-2.3 Shipping fees are sometimes charged by the total weight of the shipment. Find the total weight of the following order: nails, 250 pounds (lb); tacks, 75 lb; brackets, 12 lb; and screws, 8 lb. Arrange in columns and add.

Solution

$$\begin{array}{r} 250 \\ 75 \\ 12 \\ +\quad 8 \\ \hline 345 \end{array}$$

The total weight is 345 lb.

L.O.3. Estimate and check addition.

Estimating and Checking Addition

Estimating a sum before performing the actual addition gives us an approximate answer. Estimating is a process that is usually done mentally. Estimating sums is important when performing addition mentally or with a calculator.

When the difference between the exact sum and the estimated sum is large, the exact sum may be incorrect and we should add the numbers again.

Rule 1-2.3 ***To estimate the answer to an addition problem:***

1. Round each addend to the indicated place value or to a number with one nonzero digit.
2. Add the rounded addends.

Rule 1-2.4 *To check an addition problem:*

1. Add the numbers a second time and compare to the first sum.
2. Use a different order or grouping if convenient.

EXAMPLE 1-2.4 Acme Contractor spent the following amounts on a job: $16,466, $23,963, and $5,855. Estimate the total amount by rounding to thousands. Then find the exact amount.

Thousands place	Estimate	Exact	Check
$16,466	$16,000	$16,466	$16,466
23,963	24,000	23,963	23,963
5,855	6,000	5,855	5,855
	$46,000	$46,284	$46,284

■

Estimate and exact answer are close. The exact answer is reasonable.

The level of accuracy of estimates will vary depending on what place value the addends are rounded to.

L.O.4. Add using a calculator.

Adding Using a Calculator

It is important to know the properties of addition so that we can estimate the sums mentally or perform the addition of small numbers mentally. However, in many cases a calculator is used when adding large numbers or when adding several numbers. Several different types of calculators may be used, but we will focus on the most common types of hand-held calculators.

The three most common types of hand-held calculators are the basic, scientific, and graphing calculator. Each type may perform even the simplest calculations differently.

General Tips for Using the Calculator

Addition on the basic and scientific calculators.

1. Enter the first addend.
2. Press the addition operation key (+).
3. Enter the next addend and press the addition operation key.
4. Continue until all addends have been entered.
5. After entering the final addend, pressing the addition operation key or the equal key (=) will place the sum in the display window.

Both basic and scientific calculators will accumulate the sum of entered addends each time the addition operation key or the equal key is pressed.

General Tips for Using the Calculator

Addition on the graphing calculator:

1. Enter the first addend.
2. Press the addition operation key (+).
3. Enter the next addend and press the addition operation key.
4. Continue until all addends have been entered.
5. Press the appropriate key to instruct the calculator to perform the calculations that are in the display window. This key is labeled EXE for execute on some graphing calculators. Another common label is ENTER.

The graphing calculator does not accumulate the sum of entered addeneds as operation keys are pressed. All indicated calculations are performed after the execute key is pressed.

EXAMPLE 1-2.5 Find the sum of the following numbers using a handheld calculator.

$$2345 + 3894 + 758 =$$

Solution: Basic or scientific:

2345 [+] 3894 [+] 758 [+] *(= could be pressed instead of +)*

Basic or scientific display: 6997.

Graphing:

2345 [+] 3894 [+] 758 [EXE]

Graphing display: 2345 + 3894 + 758

6997 ■

SELF-STUDY EXERCISES 1-2

L.O.1. Add.

1. 4 + 1 + 5 + 3 + 2

$$\begin{array}{r} 4 \\ 1 \\ 5 \\ 3 \\ +\ 2 \\ \hline \end{array}$$

2.

$$\begin{array}{r} 6 \\ 9 \\ 7 \\ 4 \\ +\ 1 \\ \hline \end{array}$$

3.

$$\begin{array}{r} 3 \\ 5 \\ 2 \\ 4 \\ 7 \\ +\ 1 \\ \hline \end{array}$$

4.

$$\begin{array}{r} 6 \\ 4 \\ 5 \\ 6 \\ 9 \\ +\ 6 \\ \hline \end{array}$$

5. In taking inventory, the following $\frac{1}{4}$-inch (in.) hex bolts of various lengths were counted: nine 1 in. long, four $1\frac{1}{4}$ in., seven $1\frac{1}{2}$ in., six 2 in., two $2\frac{1}{4}$ in., and nine $2\frac{1}{2}$ in. How many $\frac{1}{4}$-in. hex bolts were there in all?

L.O.2. Add.

6.

$$\begin{array}{r} 1072 \\ 6710 \\ +\ 1410 \\ \hline \end{array}$$

7.

$$\begin{array}{r} 5273 \\ 4001 \\ +\ 7682 \\ \hline \end{array}$$

8.

$$\begin{array}{r} 59{,}718 \\ +\ 46{,}567 \\ \hline \end{array}$$

Write in columns and add.

9. 36 + 482 + 961 + 27 + 804

10. 4582 + 86,724 + 482 + 5826

11. A mechanic was paid the following for 5 days of work: $86, $124, $67, $85, and $94. How much was the mechanic paid for the 5 days?

12. A truck driver had the following weight slips on three loads of gravel: 8114 lb, 8027 lb, and 8208 lb. What was the total weight of the gravel hauled by the driver?

13. During inventory, the following numbers of slotted-head screws were counted in four different boxes: 84, 63, 72, and 79. How many slotted-head screws were there all together?

14. Four bricklayers were working on the same job. In one day the bricklayers laid the following number of bricks: 1217, 1103, 1039, and 1194. How many bricks did all four lay that day?

15. Canty O'Neal, a purchasing clerk placed an order for 15 gallons (gal) of white paint, 27 gal of black paint, 5 gal of crimson red paint, and 3 gal of canary yellow paint. How many gallons of paint were ordered?

L.O.3. Estimate the sum by rounding to hundreds. Then find the exact sum and check.

16.
$$\begin{array}{r} 4{,}256 \\ 3{,}892 \\ 576 \\ \underline{8{,}293} \end{array}$$

17.
$$\begin{array}{r} 52{,}843 \\ 17{,}497 \\ 13{,}052 \\ \underline{821} \end{array}$$

18.
$$\begin{array}{r} 24{,}003 \\ 5{,}874 \\ 319{,}467 \\ \underline{52{,}855} \end{array}$$

19. Palmer Associates provided the following prices for items needed to build a sidewalk: concrete, \$2,583; wire, \$43; frame material, \$18; labor, \$798. Estimate the cost by rounding each amount to the nearest ten. Find the exact amount.

20. Antonio Juarez expects to spend the following amounts for one semester of college: food, \$395; lodging, \$1,285; books, \$288; supplies, \$130; transportation, \$162. Estimate the expenditures for one semester by rounding each amount to the nearest hundred. Calculate the exact amount.

Estimate the sums by rounding addends to numbers with one nonzero digit. Then find the exact sum. Finally, check your answer.

21. $\begin{array}{r} 940 \\ \underline{+\ 8299} \end{array}$

22. $\begin{array}{r} 478 \\ \underline{+\ 146} \end{array}$

23. $\begin{array}{r} 1901 \\ \underline{+\ 6548} \end{array}$

24. $\begin{array}{r} 149 \\ \underline{+\ 652} \end{array}$

25. $\begin{array}{r} 16{,}259 \\ \underline{+\ 36{,}542} \end{array}$

26. $\begin{array}{r} 32{,}501 \\ \underline{+\ 16{,}740} \end{array}$

27. A hardware store filled the following order of nails: 25 lb, $2\frac{1}{2}$-in. common; 16 lb, 4-in. common; 12 lb, 2-in. siding; 24 lb, $2\frac{1}{2}$-in. floor brads; 48 lb, 2-in. roofing, and 34 lb, $2\frac{1}{2}$-in. finish. What is the total weight of the order?

28. If four containers have a capacity of 12 gal, 27 gal, 55 gal, and 21 gal, can 100 gal of fuel be stored in these containers? (Find the total capacity of the containers first.)

29. A printer has three printing jobs that require the following numbers of sheets of paper: 185, 83, and 211. Will one ream of paper (500 sheets) be enough to finish the three jobs?

30. How many feet of fencing are needed to enclose the area illustrated in Fig. 1-4?

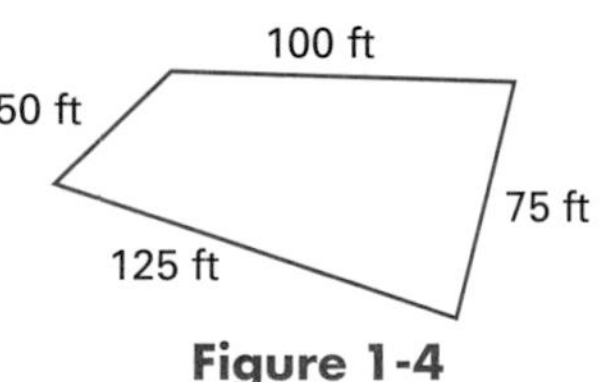

Figure 1-4

L.O.4. Estimate the sums by rounding addends to numbers with one nonzero digit. Then, use a calculator to find the exact sum.

31. 47,287 + 33,409 + 81,496 + 28,594

32. 387,483 + 879,583 + 592,801

33. 31,592 + 8,584 + 13,215 + 968

34. 1,328,591 + 35,803,502 + 10,387,921

35. Mario's Restaurant had the following sales daily: Sunday, \$3,842; Monday, \$1,285; Tuesday, \$1,195; Wednesday, \$1,843; Thursday, \$1,526; Friday, \$2,984; and Saturday, \$4,359. Find the total sales for the week.

1-3 SUBTRACTING WHOLE NUMBERS

Learning Objectives

L.O.1. Subtract single-digit numbers.
L.O.2. Subtract numbers with two or more digits.
L.O.3. Estimate and check subtraction.
L.O.4. Subtract using a calculator.

Subtraction of whole numbers is another basic skill required in most occupations. Subtraction is the *inverse* of addition. In addition, we add numbers to get their total (such as $5 + 4 = 9$), but to solve the subtraction problem $9 - 5 = ?$, we ask ourselves what number must be added to 5 to give us 9. The answer is 4, since 4 must be added to 5 to give a total of 9. When we subtract two numbers, the answer is called the *difference* or *remainder*. The initial quantity is the *minuend*. The amount being subtracted from the initial quantity is the *subtrahend*.

L.O.1. Subtract single-digit numbers.

Subtracting Single-digit Numbers

In addition, it does not matter in which order numbers are added. But *in subtraction order is important:* $8 - 3 = 5$, but $3 - 8$ does not equal 5; that is, $3 - 8 \neq 5$. Subtraction is *not* commutative, whereas addition is. In addition, grouping does not matter when a problem has three or more numbers. But *in subtracting, grouping is important.*

EXAMPLE 1-3.1 Show that $9 - (5 - 1)$ does not equal $(9 - 5) - 1$.

Solution

$$9 - \underbrace{(5 - 1)}_{4} = 9 - 4 = 5, \quad \text{but} \quad \underbrace{(9 - 5)}_{4} - 1 = 4 - 1 = 3$$

■

If a subtraction problem contains no parentheses to show the grouping, subtract the first two numbers at the left. Then subtract from that difference the next number in the problem.

EXAMPLE 1-3.2 Subtract $8 - 3 - 1$.

Solution

$$\underbrace{8 - 3}_{5} - 1 = 5 - 1 = 4$$

■

Rule 1-3.1 ***Zero subtracted from a number results in the same number:***

$$n - 0 = n, \qquad 7 - 0 = 7$$

Tips and Traps

Subtracting a number from zero is not the same as subtracting zero from a number. That is $7 - 0 = 7$, but $0 - 7$ does not equal seven.

L.O.2. Subtract numbers with two or more digits.

Subtracting Numbers with Two or More Digits

> **Rule 1-3.2** *To subtract numbers of two or more digits:*
>
> 1. Arrange the numbers in columns with the minuend at the top and the subtrahend at the bottom.
> 2. Make sure the ones digits are in a vertical line on the right.
> 3. Subtract the ones column first, then the tens column, the hundreds column, and so on.
> 4. When subtracting a larger digit from a smaller digit in any given column, borrow 1 from the column to the left. To borrow 1 from the next column to the left is the same as borrowing *one* group of 10; thus, place the borrowed 1 in front of the digit in the minuend.

EXAMPLE 1-3.3 Subtract 5327 − 3514.

Solution: Arrange in columns.

$$\begin{array}{r} {\scriptstyle 4\,13} \\ 5{,}327 \\ -\ 3{,}514 \\ \hline 1{,}813 \end{array}$$

Subtract the ones column.	Subtract the tens column.	Subtract the hundreds column.	Subtract the thousands column.
7 − 4 = 3	2 − 1 = 1	Borrow. 13 − 5 = 8	4 − 3 = 1

■

Here are some typical phrases that indicate subtraction in applied problems: how many are left, how many more, how much less, how much larger, and how much smaller.

EXAMPLE 1-3.4 Abdullah Samardar had a box containing 144 light switches. If 47 switches were used for a job, how many were left?

Solution: The phrase, *how many were left*, indicates subtraction.

$$144 - 47 = 97$$

There were 97 switches left. ■

Some applied problems require more than one operation. Many problems requiring more than one operation involve parts and a total. If we know a total and all the parts but one, we must add all the known parts and subtract the result from the total to find the missing part.

EXAMPLE 1-3.5 The Randles left Memphis and drove 356 miles on the first day of their vacation. They drove 426 miles on the second day. If they are traveling to Albuquerque, which is 1,050 miles from Memphis, how many more miles do they have to drive?

Solution:

$$1050 \text{ miles} = \text{total miles}$$

$$356 + 426 + \text{miles left to drive} = 1050 \text{ miles}$$

To find the miles left to drive, add 356 + 426 and subtract the result from 1050.

$$356 + 426 = 782, \qquad 1050 - 782 = 268$$ ■

L.O.3. Estimate and check subtraction.

Estimating and Checking Subtraction

Estimating a subtraction problem is similar to estimating an addition problem. The numbers in the problem are rounded before the subtraction is performed. The accuracy of the estimate depends on the method used for rounding.

> **Rule 1-3.3** *To estimate the difference:*
>
> **1.** Round each number to the indicated place value or to a number with one nonzero digit.
> **2.** Subtract the rounded numbers.

> **Rule 1-3.4** *To check a subtraction problem:*
>
> **1.** Add the subtrahend and difference.
> **2.** Compare the result of Step 1 with the minuend. If the two numbers are equal the subtraction is correct.

EXAMPLE 1-3.6 Estimate, find the exact difference, and check. Round to hundreds.

$$427 - 125$$

Solution

	Estimate	Exact	Check
427	400	427	125
− 125	− 100	− 125	+ 302
	300	302	427

■

L.O.4. Subtract using a calculator.

Subtraction Using a Calculator

General Tips for Using the Calculator

Subtraction on the basic and scientific calculators:

1. Enter the minuend.
2. Press the subtraction operation key (−).
3. Enter the subtrahend.
4. After entering the subtrahend, pressing the subtraction operation key or the equal key (=) will place the difference in the display window.

General Tips for Using the Calculator

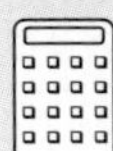

Subtraction on the graphing calculator:

1. Enter the minuend.
2. Press the subtraction operation key (−).
3. Enter the subtrahend.
4. Press the appropriate key to instruct the calculator to perform the calculations that are in the display window. This key is labeled EXE for execute on some graphing calculators.

EXAMPLE 1-3.7 Find the difference between 53,943 and 34,256.

Solution: Basic or scientific:

53943 [−] 34256 [=]

Basic or scientific display: 19687.

Graphing:

53943 [−] 34256 [EXE]

Graphing display: 53943 − 34256

19687 ■

Commas are not entered into the calculator, and most calculator displays do not place commas in the answer.

EXAMPLE 1-3.8 Two cuts were made from a 72-in. pipe. The two lengths cut from the pipe were 28 in. and 15 in. How much of the pipe was left after these cuts were made?

Solution

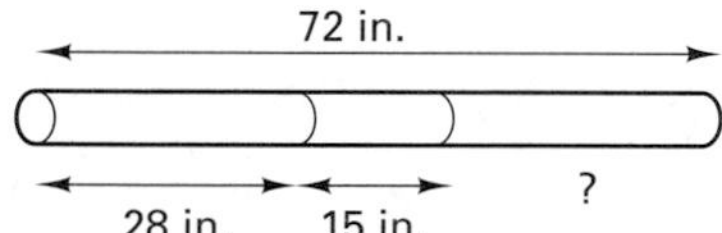

72 [−] 28 [−] 15 [=]

Basic or scientific display: 29.

There were 29 in. of pipe left. ■

SELF-STUDY EXERCISES 1-3

L.O.1. Do the following subtraction problems.

1. 7 − 2
2. (8 − 3) − 4
3. 8 − (5 − 4)
4. 9 − 5 − 2
5. (7 − 1) − 3
6. 8 − (2 − 1)
7. (8 − 2) − 1
8. 9 − 3 − 4
9. 8 − (5 − 1)
10. 6 − 3 − 2

L.O.2. Solve the following subtraction problems.

11. 47 − 23
12. 427 − 26
13. 3672 − 2652
14. 946 − 831
15. 53,867 − 831

16. If a mason ordered 75 bags of cement for a job and used only 53, how many bags were left?

17. An inventory sheet shows that 468 outlet boxes were in stock on March 1. Sales during March were 127. How many outlet boxes were left at the end of the month?

18. A reel of cable contained 575 ft. The following amounts were used on three separate jobs: 112 ft, 101 ft, and 41 ft. How much cable was left on the reel?

L.O.3. Subtract the following problems. Check your answers.

19. 82 − 37
20. 961 − 353
21. 4070 − 2497
22. 30,021 − 7,816

Estimate differences by rounding numbers to one nonzero digit. Then, find the exact answer.

23. $\begin{array}{r} 589{,}760 \\ -\ 498{,}726 \\ \hline \end{array}$

24. $\begin{array}{r} 4903 \\ -\ 2814 \\ \hline \end{array}$

25. $\begin{array}{r} 3070 \\ -\ 2896 \\ \hline \end{array}$

26. $\begin{array}{r} 1401 \\ -\ 802 \\ \hline \end{array}$

27. $\begin{array}{r} 80{,}096 \\ -\ 9{,}714 \\ \hline \end{array}$

28. $\begin{array}{r} 1801 \\ -\ 704 \\ \hline \end{array}$

L.O.4. Estimate differences by rounding numbers with one nonzero digit. Then, find the exact answer using your calculator.

29. $\begin{array}{r} 8001 \\ -\ 3604 \\ \hline \end{array}$

30. $\begin{array}{r} 3804 \\ -\ 2617 \\ \hline \end{array}$

31. $\begin{array}{r} 895{,}740 \\ -\ 387{,}465 \\ \hline \end{array}$

32. A fuel tank on one model automobile has a capacity of 75 liters (L). The tank on a different model has a capacity of 92 L. How much larger is the second tank?

33. A bricklayer laid 1283 bricks on one day. A second bricklayer laid 1097 bricks. How many more bricks did the first bricklayer lay?

34. In a stockroom there are 285 in. of bar stock and 173 in. of round stock. How many inches of bar stock remain after the object in Fig. 1-5 is made from this stock?

35. Referring to Exercise 34, how many inches of round stock remain after the object in Fig. 1-5 is made?

36. Find the missing dimension in Fig. 1-6.

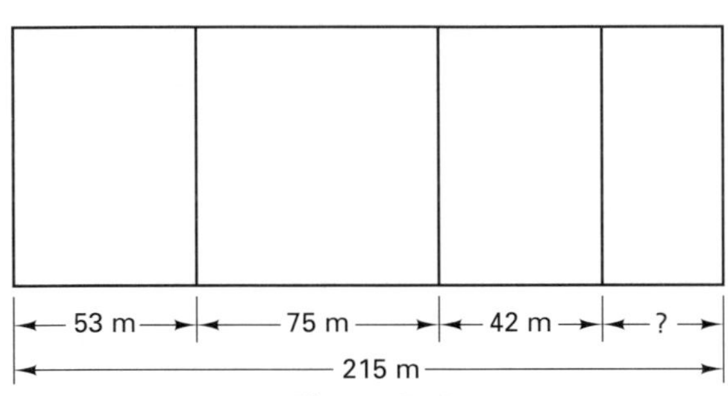

Figure 1-6

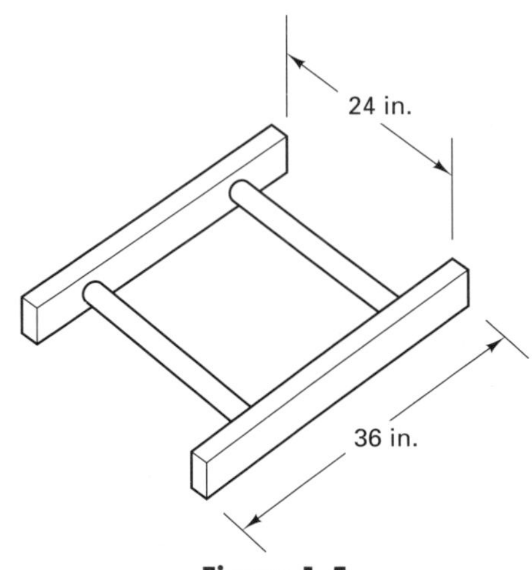

Figure 1-5

1-4 MULTIPLYING WHOLE NUMBERS

Learning Objectives

L.O.1. Multiply single-digit factors.
L.O.2. Multiply factors of more than one digit.
L.O.3. Apply the distributive property.
L.O.4. Estimate and check multiplication.
L.O.5. Multiply using the calculator.

L.O.1. Multiply single-digit factors.

Multiplying Single-Digit Factors

Multiplication is repeated addition. If we have 3 ten-dollar bills, we have \$10 + \$10 + \$10, or \$30. Using multiplication, this is the same as 3 times \$10, or \$30.

When we multiply two numbers, the first number is called the *multiplicand*, and the number we multiply by is called the *multiplier*. Either of the numbers may be referred to as *factors*. The answer or result of multiplication is called the *product*.

2	$\times$	3	=	6
multiplicand or factor		multiplier or factor		product

Tips and Traps

Besides the familiar $\times$ or "times" sign, the raised dot $(\cdot)$, the asterisk $(*)$, and parentheses ( ) may be used to show multiplication.

$$2 \cdot 3 = 6, \quad 2 * 3 = 6, \quad 2(3) = 6, \quad (2)(3) = 6$$

Multiplication is commutative and associative, just like addition.

The *commutative property* of multiplication permits two numbers to be multiplied in any order.

$$4 \times 5 = 20, \qquad 5 \times 4 = 20$$

When multiplying more than two numbers, the numbers must be grouped, and the *associative property* of multiplication permits the numbers to be grouped in any way.

$2 \times (3 \times 5)$	or	$(2 \times 3) \times 5$
2×15		6×5
30		30

EXAMPLE 1-4.1 Multiply $3 \times 2 \times 9$.

Solution

$3 \times 2 \times 9$	
$(3 \times 2) \times 9$	*(Group any two factors.)*
6×9	*(Multiply grouped factors.)*
54	*(Multiply the factors: 6 and 9.)*

■

Another property is the *zero property of multiplication.*

Rule 1-4.1 ***The product of a number and zero is zero:***

$$n \times 0 = 0, \quad 0 \times n = 0, \quad 4 \times 0 = 0, \quad 0 \times 4 = 0$$

EXAMPLE 1-4.2 Multiply $2 \times 5 \times 0 \times 7$.

Solution

$2 \times 5 \times 0 \times 7$	
$(2 \times 5) \times (0 \times 7)$	*(Group factors.)*
10×0	*(Multiply each group of factors.)*
0	*(Multiply products.)*

■

L.O.2. Multiply factors of more than one digit.

Multiplying Factors with More Than One Digit

When either factor contains two or more digits, such as 45×3 or 204×103, the factors can be arranged one under the other. For convenience, we place the larger number at the top and the smaller number at the bottom.

> **Rule 1-4.2** ***To multiply a factor of two or more digits by a factor of one digit:***
>
> **1.** Arrange the factors one under the other.
> **2.** Multiply each digit in the multiplicand (beginning with the ones place) by each digit in the multiplier.

EXAMPLE 1-4.3 Multiply 45×3.

Solution

$$\begin{array}{r} {}^{1} \\ 45 \\ \times\ \ 3 \\ \hline 135 \end{array}$$

(Multiply ones digit of multiplicand by 3.)
($3 \times 5 = 15$. Place 5 in the ones place of the product and carry 1. Then multiply the tens digit by 3. That is, $3 \times 4 = 12$. Add the carried 1. $12 + 1 = 13$.) ■

> **Rule 1-4.3** ***To multiply factors of two or more digits:***
>
> **1.** Arrange the factors one under the other.
> **2.** Multiply each digit in the multiplicand by each digit in the multiplier. The product of the multiplicand and each digit in the multiplier gives a *partial product*.
>   **a.** Start with the ones digit in the multiplicand and multiply from right to left.
>   **b.** Start with the ones digit in the multiplier and multiply from right to left.
>   **c.** Align each partial product with its first digit directly under its multiplier digit.
> **3.** Add the partial products.

EXAMPLE 1-4.4 Multiply 204×103.

Solution

$$\begin{array}{r} {}^{1} \\ 204 \\ \times\ 103 \\ \hline 612 \\ 000 \\ 204 \\ \hline 21012 \end{array}$$

(Multiply: $3 \times 204 = 612$. Align 612 under 3 in the multiplier.)
(Multiply: $0 \times 204 = 000$. Align 000 under 0 in the multiplier.)
(Multiply: $1 \times 204 = 204$. Align 204 under 1 in the multiplier.)
(Add the partial products as they are aligned.)

Partial products 000 and 204 could be combined on a single line to be 2040, with the rightmost 0 aligned under the 0 in the multiplier and the 4 aligned under the 1 in the multiplier. ■

Multiplication is used in an applied problem for repeated addition. The following example shows the use of multiplication.

EXAMPLE 1-4.5 Marjorie Young ordered three gross of pencils. If a gross is 144, how many pencils did she order?

Solution: Marjorie ordered 3 boxes of 144 pencils each. That is, $144 + 144 + 144$, or 3×144.

$$3 \times 144 = 432$$

Marjorie ordered 432 pencils in all. ■

EXAMPLE 1-4.6 Anfernee Hardaway scored 27 points in the first basketball game of the season. Anfernee is expected to play 32 games this season. If he continues to score 27 points each game, how many total points will he score for the season?

Solution: The total points could be found by adding 27 thirty-two times. However, this is the same as 27×32.

$$27 \times 32 = 864$$

Anfernee will score 864 points for the season if he continues to score 27 points per game. ■

Some applied problems require more than one operation to answer the question. The following example requires both multiplication and addition.

EXAMPLE 1-4.7 John Paszel is taking inventory and finds 27 unopened boxes of candy. Each of the boxes contain 48 candy bars. In another part of the warehouse, John counts 33 unopened cases of the same candy. These cases each contain 288 bars of candy. How many candy bars should John show on his inventory report?

Solution: John must first calculate the number of candy bars in the 27 unopened boxes.

$$27 \times 48 = 1296$$

Next, the number of candy bars in the cases must be calculated.

$$33 \times 288 = 9504$$

Finally, the total number of candy bars is $1296 + 9504$, or 10,800. ■

L.O.3. Apply the distributive property.

Applying the Distributive Property

Another property of multiplication is called the distributive property. The *distributive property of multiplication* means that multiplying a sum or difference by a factor is equivalent to multiplying each term of the sum or difference by the factor.

Rule 1-4.4 ***Distributive Property of Multiplication:***

1. Add or subtract the numbers within the grouping.
2. Multiply the result of step 1 by the factor outside the grouping.

or

1. Multiply each number inside the grouping by the factor outside the grouping.
2. Add or subtract the products from step 1.

Symbolically,

$$a \times (b + c) = a \times b + a \times c \quad \text{or} \quad a(b + c) = ab + ac$$

$$a \times (b - c) = a \times b - a \times c \quad \text{or} \quad a(b - c) = ab - ac$$

Tips and Traps

Parentheses show multiplication when the distributive property is used. $a(b + c)$ means $a \times (b + c)$. Letters can be used to represent numbers. Letters that are written together with no operation sign between them show multiplication. ab means $a \times b$; ac means $a \times c$.

Example 1-4.8 Multiply $3(2 + 4)$.

Solution

Multiplying first gives:	Adding first gives:
$3(2 + 4) =$	$3(2 + 4) =$
$3(2) + 3(4) =$	$3(6) = 18$
$6 + 12 = 18$	

Example 1-4.9 Multiply $2(6 - 5)$.

Solution

Multiplying first gives:	Subtracting first gives:
$2(6 - 5) =$	$2(6 - 5) =$
$2(6) - 2(5) =$	$2(1) = 2$
$12 - 10 = 2$	

The distributive principle is found in many applied formulas. One example is the formula for the perimeter of a rectangle. A *rectangle* is a four-sided geometric shape whose opposite sides are equal in length, and each corner makes a square corner. The *perimeter* of a rectangle is the distance around the figure. We would find the perimeter if we were fencing a rectangular yard, installing baseboard in a rectangular room, framing a picture, or outlining a flower bed with landscaping timbers.

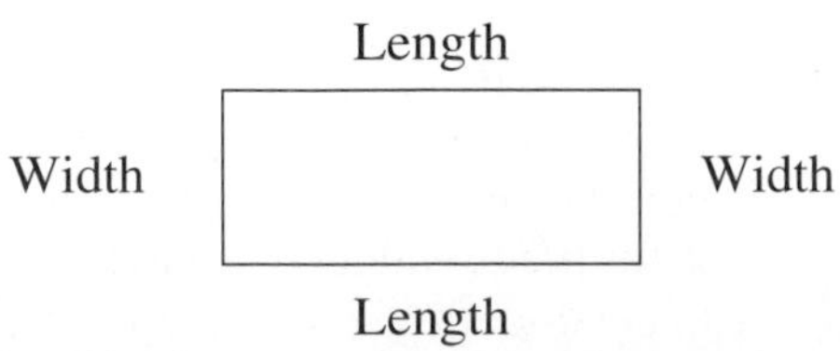

The formula for finding the perimeter of a rectangle is

$$P = 2(l + w) \quad \text{or} \quad P = 2l + 2w$$

Example 1-4.10 Find the number of feet of fencing needed to enclose a rectangular pasture that is 1,784 feet long and 847 feet wide.

Solution

$$\begin{aligned} P &= 2(l + w) & \text{or} \quad P &= 2l + 2w \\ P &= 2(1{,}784 + 847) & P &= 2(1{,}784) + 2(847) \\ P &= 2(2631) & P &= 3568 + 1694 \\ P &= 5{,}262 & P &= 5{,}262 \end{aligned}$$

The amount of fencing needed is 5,262 feet. ■

L.O.4. Estimate and check multiplication.

Estimating and Checking Multiplication

Rule 1-4.5 ***To estimate the answer for a multiplication problem:***

1. Round both factors to a certain place value or to one nonzero digit.
2. Then multiply the rounded numbers.

Rule 1-4.6 ***To check a multiplication problem:***

1. Multiply the numbers a second time and check the product.
2. Interchange the factors if convenient.

Example 1-4.11 Find the approximate and exact cost of 22 books if each book costs \$29. Estimate the answer by rounding each factor to a number with one nonzero digit. Then find the exact answer and check your work.

Solution: 22 rounds to 20 and \$29 rounds to \$30.

$$20 \times \$30 = \$600$$

The approximate cost of the books is \$600.

$$\begin{array}{r} 22 \\ \times\ 29 \\ \hline 198 \\ 44 \\ \hline 638 \end{array} \qquad \text{Check} \qquad \begin{array}{r} 29 \\ \times\ 22 \\ \hline 58 \\ 58 \\ \hline 638 \end{array}$$

The exact cost of the books is \$638. ■

Estimating is used in many do-it-yourself projects before exact calculations are made. One such example involves finding the area of a city house lot in the shape of a rectangle. The *area* of a geometric shape is the number of square units needed to cover the shape. Area is used when we are finding the amount of carpet or floor covering needed for a rectangular room, the amount of paint needed to paint a wall, the amount of asphalt needed to pave a parking lot, the amount of fertilizer needed to treat a yard, or the amount of material needed to produce a rectangular sign. Area is measured in *square measures*. One square foot (1 ft^2) indicates a square that measures 1 foot on each side.

5 feet long by 3 feet wide = 15 square feet

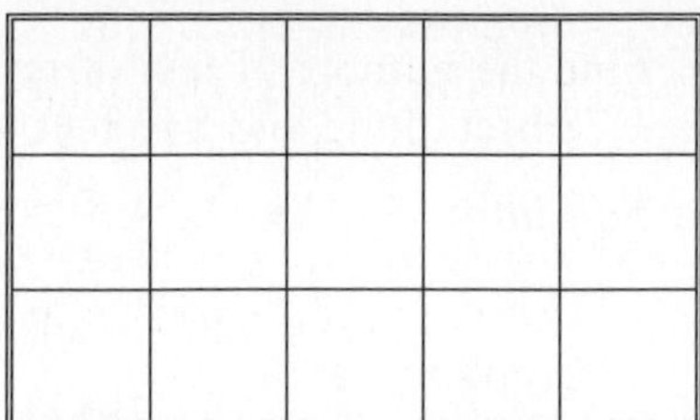

To find the area of a rectangle, we multiply the length by the width. This rule is abbreviated by using a *formula*.

$A = lw$ *(When two letters are written side-by-side with no operation sign, multiplication is implied.)*

EXAMPLE 1-4.12 Maintenance Consultants needs to apply fertilizer to a customer's lawn. The lawn measures 150 feet long and 125 feet wide. How many square feet are in the yard?

Solution: To estimate the number of square feet in the yard, round the length and width each to one nonzero number. 150 rounds to 200 and 125 rounds to 100.

$$\begin{aligned} A &= lw \\ &= 200(100) \\ &= 20{,}000 \end{aligned}$$

The lawn has approximately 20,000 ft^2 of ground to be covered.

To find the exact number of square feet in the yard, use the same formula with the exact measurements.

$$\begin{aligned} A &= lw \\ &= (150)(125) \\ &= 18{,}750 \text{ ft}^2 \end{aligned}$$

■

L.O.5. Multiplying using a calculator.

Multiplying Using a Calculator

General Tips for Using the Calculator

Multiplication on the basic and scientific calculators:

1. Enter the first factor.
2. Press the multiplication operation key (×).
3. Enter the next factor and press the multiplication operation key.
4. Continue until all factors have been entered.
5. After entering the final factor, pressing the multiplication operation key or the equal key (=) will place the product in the display window.

Both basic and scientific calculators will accumulate the product of entered factors each time the multiplication operation key or the equal key is pressed.

General Tips for Using the Calculator

Multiplication on the graphing calculator:

1. Enter the first factor.
2. Press the multiplication operation key (×).
3. Enter the next factor and press the multiplication operation key.
4. Continue until all factors have been entered.
5. Press the appropriate key to instruct the calculator to perform the calculations that are in the display window. This key is labeled EXE for execute on some graphing calculators.

The graphing calculator does not accumulate the product of entered factors as operation keys are pressed. All indicated calculations are performed after the execute key is pressed.

EXAMPLE 1-4.13 Find the product of the following numbers using a handheld calculator.

$$3{,}283 \times 346 \times 34 =$$

Solution: Basic or scientific:

3283 [×] 346 [×] 34 [=]

Basic or scientific display: 38621212.

Graphing:

3283 [×] 346 [×] 34 [EXE]

Graphing display: 3283 × 346 × 34

38621212. ■

EXAMPLE 1-4.14 Solve the problem 2600 × 70.

Solution

1.
```
   26|00
 ×  7|0
```
(Separate the zeros at the end of the factors from the other digits.)

2.
```
   26|00
 ×  7|0
   182|
```
(Multiply the other digits as if the zeros were not there (26 × 7 = 182).)

3.
```
   26|00
 ×  7|0
   182|000
```
(Affix the zeros to the basic product. Note that the number of zeros affixed to the basic product is the same as the number of zeros at the end of the factors.) ■

This process is sometimes necessary whenever a multiplication problem is too long to fit into a calculator. Many handheld calculators only have an eight-digit display window for calculations. The problem 26,000,000 × 300 would not fit many basic calculators. This shortcut allows us to work the problem quickly, with or without a calculator.

EXAMPLE 1-4.15 Multiply 26,000,000 × 300

Solution

$$\begin{array}{r} 26{,}000{,}000 \\ \times \quad 3\ 00 \\ \hline 78\ 00\ 000\ 000 \end{array}$$ *(Separate ending zeros and multiply 26 × 3.)* ■

SELF-STUDY EXERCISES 1-4

L.O.1.

1. Find the following products.
(a) 5 × 3 **(b)** 7 * 8
(c) (9)(7) **(d)** 4 · 6

2. Jaime Oxnard has 9 wood boxes that he intends to sell for $7 each. If Jaime sells 6 of the boxes, how much money will he get?

3. What property of multiplication justifies the statement "5(3) = 3(5)"?

4. Explain the associative property of multiplication and give an example to illustrate the property.

Multiply.

5. 5 · 3 · 0 · 6

6. 3 * 7 * 9 * 2

7. 7 × 3 × 4

8. (3)(2)(0)(8)

L.O.2. Multiply.

9. $\begin{array}{r} 32 \\ \times \ 7 \\ \hline \end{array}$

10. $\begin{array}{r} 83 \\ \times \ 7 \\ \hline \end{array}$

11. $\begin{array}{r} 90 \\ \times \ 7 \\ \hline \end{array}$

12. 503 × 204

13. Margaret Johnston purchased 3 cases of candy bars. Each case contained 12 bags and each bag contained 24 pieces of individually wrapped candies. How many individually wrapped candies did Margaret purchase?

14. Chuckee Wright counted 8 unopened boxes of washers. Each box contained 512 washers. What is the total number of washers to be shown on the inventory sheet?

15. Tracie Burke reported reading 6 books a week over a period of 14 weeks. How many books did Tracie read during the time period?

16. Each officer in the Public Safety office wrote, on the average, 25 tickets a week. If there are 7 officers, how many tickets were written over a 4-week period?

17. Jossie Moore is planning to sell candy bars in her Smart Shop. She receives 12 boxes and each box contains 24 candy bars. If Jossie sells the bars for $1 each, how much money will she get for all the bars?

18. Terry Kelly teaches 5 classes and each class has 46 students. She teaches a sixth class that has 19 students enrolled. How many students is Terry teaching?

L.O.3. Find the product in each of the following by multiplying first, and check your answers by adding first.

19. 5(7 + 4) **20.** 3(9 + 11) **21.** 14(3 + 11)

22. 4(5 − 2) **23.** 12(16 − 3) **24.** 7(18 − 1)

Find the perimeter of the following rectangles.

25.

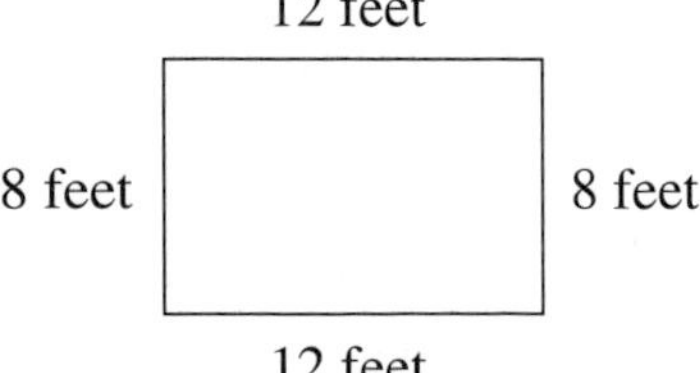

26.

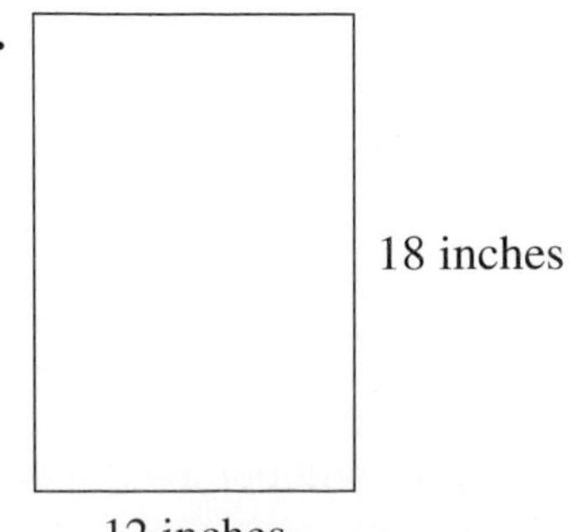

27. How much baseboard is required to surround a kitchen if the floor measures 15 feet by 13 feet and no allowance is made for door openings?

28. A dog pen is staked out in the form of a rectangle so that the width is 38 feet and the length is 47 feet. How much fencing must be purchased to build the pen?

L.O.4.

29. Find the approximate cost of 24 shirts if each shirt costs $32. What is the exact cost of the shirts?

Find the number of square feet in each of the illustrated rectangles. Check your work.

30.

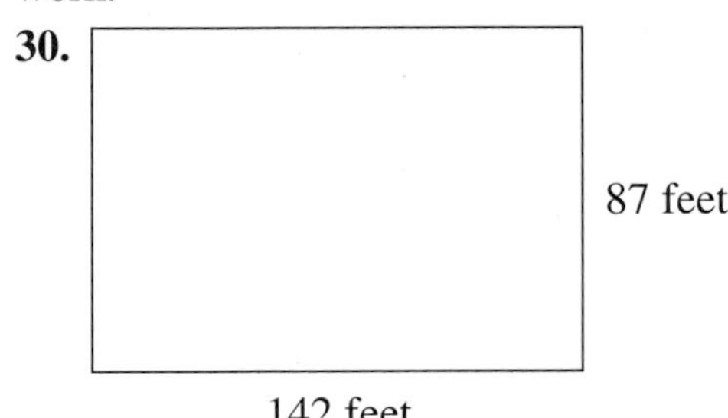

31.

12 centimeters

97 centimeters

32. How many square feet of carpet should be purchased to cover a rectangular room that measures 42 feet by 32 feet?

33. A rectangular table top is 12 feet long and 8 feet wide. How many square feet of glass are needed to cover it?

L.O.5. Use the calculator to find the following products.

34. $5{,}896 \times 473 \times 12$

35. $52{,}834 * 15 * 295$

36. $832 \cdot 703 \cdot 960$

37. $583(294)(627)(0)(8)$

38. Paul Thomas earns $3,015 per month. What is Paul's annual salary?

39. Ertha Brower earns $423 per week. Calculate her annual salary.

Perform the following multiplications.

40. $3{,}500 \times 42{,}000{,}000$

41. $30{,}800 * 544{,}000$

42. $70{,}000(6{,}000)$

1-5 DIVIDING WHOLE NUMBERS AND ESTIMATING

Learning Objectives

L.O.1. Use various symbols to indicate division.
L.O.2. Divide whole numbers.
L.O.3. Estimate and check division.
L.O.4. Divide using a calculator.

L.O.1. Use various symbols to indicate division.

Using Symbols to Indicate Division

Division is the inverse operation for multiplication. If we are trying to determine what number we should multiply 7 by to get a product of 28, that would be equivalent to dividing 28 by 7. The result is 4. In division the number we divide by is the *divisor*. The number being divided is the *dividend*. The answer is the *quotient*.

$$28 \div 7 = 4 \longleftarrow \text{quotient}$$

dividend ↑ (28) ↑ (7) divisor

Besides the "divided by" sign (÷), there is also the symbol $\overline{)}$ in such problems as $7\overline{)28}$. The problem $7\overline{)28}$ means the same as 28 ÷ 7.

$$\begin{array}{c} 4 \longleftarrow \text{quotient} \\ 7\overline{)28} \\ \text{divisor} \uparrow \quad \uparrow \text{dividend} \end{array}$$

Another division symbol is a bar, such as $\frac{28}{7}$, which also means the same as 28 ÷ 7.

$$\begin{array}{l} \text{dividend} \longrightarrow 28 \\ \text{divisor} \longrightarrow 7 \end{array} \quad \frac{28}{7} = 4 \longleftarrow \text{quotient}$$

Words indicating division include:

"divided by" ⟶ dividend divided by divisor
"goes into" ⟶ divisor goes into dividend
"divided into" ⟶ divisor divided into dividend

EXAMPLE 1-5.1 Express 2 goes into 8 as a division using the symbol ÷. Then solve the problem.

Solution

$8 \div 2 = ?$ Since $4 \times 2 = 8$, $8 \div 2 = 4$. ■

EXAMPLE 1-5.2 Express 18 divided by 3 as a division using the symbol $\overline{)}$. Then solve the problem.

Solution

$3\overline{)18}$ Since $6 \times 3 = 18$, $\begin{array}{r} 6 \\ 3\overline{)18} \end{array}$. ■

EXAMPLE 1-5.3 Express 5 divided into 15 using the bar symbol —. Then solve the problem.

Solution

$\frac{15}{5} = ?$ Since $3 \times 5 = 15$, $\frac{15}{5} = 3$. ■

L.O.2. Divide whole numbers.

Dividing Whole Numbers

Some of the problems that arise in division involve *long division.* Study the following example. Long division, like long multiplication, involves using the simple one-digit multiplication facts repeatedly to find the quotient.

EXAMPLE 1-5.4 Divide 975 ÷ 12.

Solution

$12\overline{)975}$ (Use long-division symbol.)

Tips and Traps

The correct placement of the first digit in the quotient is critical. If out of place, all digits that follow will be out of place, giving the quotient too few or too many digits.

```
   8
12)975
```
(12 does not divide into 9, but it does go 8 times into 97. Put the 8 over the 7, the last digit of 97.)

```
   8
12)975
   96
    15
```
(Multiply 8 × 12 = 96. Put the 96 under the 97 of the dividend.
Subtract 97 − 96 = 1. This difference must be less than 12, the divisor.
Bring down the next digit in the dividend, 5.)

```
   81
12)975
   96
    15
    12
     3
```
(12 divides into 15 one time. Put the 1 over the 5 of the dividend.
Multiply 1 × 12 = 12. Put the 12 under the 15.)
(Subtract 15 − 12 = 3. This difference must be less than 12.)

```
   81  R 3
12)975  ↑
   96   |
    15  |
    12  |
     3--┘
```
(The 3 is the remainder.)

Tips and Traps

Keep digits in line.

Thus, 975 ÷ 12 is 81 with 3 left over. ■

At times, after subtracting and bringing down the next digit, the divisor will not divide into the resulting number at least one time. In this case, we place a zero in the quotient.

EXAMPLE 1-5.5 Divide $5\overline{)2535}$.

Solution

```
   5
5)2535
  25
  03
```
(5 divides into 25 five times. Put 5 over the last digit of the 25. Subtract and bring down the 3.)

```
  507
5)2535
  25
  035
   35
    0
```
(5 divides into 3 zero times, so put 0 over the 3 of the dividend. Bring down next digit, which is 5. Divide 5 into the 35. 35 ÷ 5 = 7. Put the 7 over the 5 of the dividend. Continue as usual.) ■

EXAMPLE 1-5.6 A dealer has 58,284 lb of scrap metal that must be transferred to another site. If a truck will hold a maximum of 8000 lb, what is the least number of loads required to transfer the metal?

Solution

```
           7
8000)58,284
     56 000
      2 284
```

After 7 full loads are hauled, there will still be 2284 lb of scrap metal left. Therefore, another trip will have to be made even though the truck will not have a full load. 8 loads or trips will be required to transfer the metal. ■

EXAMPLE 1-5.7 After a tune-up, a mechanic wanted to know if the engine was running more efficiently. When mileage was checked on a tankful of gasoline, the automobile traveled 414 miles on 18 gal of gasoline. What was the average number of miles per gallon?

Solution

$$\begin{array}{r} 23 \\ 18\overline{)414} \\ \underline{36} \\ 54 \\ \underline{54} \end{array}$$

The automobile traveled an average of 23 miles per gallon of gasoline. ■

When dividing by a one-digit divisor, we can perform the division mentally. This process is usually referred to as *short division.*

EXAMPLE 1-5.8 Divide $7\overline{)180}$.

$$\begin{array}{r} 2 \\ 7\overline{)180} \end{array} \quad 18 \div 7 = 2 \text{ with 4 left over.}$$

$$\begin{array}{r} 2\,5 \\ 7\overline{)18^40} \end{array} \quad 40 \div 7 = 5 \text{ with 5 left over}$$

$$\begin{array}{r} 2\,5 \;\text{R}\,5 \\ 7\overline{)18^40} \end{array}$$

■

EXAMPLE 1-5.9 Heritage House transports its residents in vans that will accommodate 18 passengers. If 96 residents plan to attend a picnic, how many vans will be needed for transportation?

Solution: We need to know how many 18's are in 96. Division will give us the answer.

$$\begin{array}{r} 5 \\ 18\overline{)96} \\ \underline{90} \\ 6 \end{array} \quad \textit{(There is a remainder of 6 residents.)}$$

■

Thus, one additional van or a total of 6 vans is needed to transport the residents.

Tips and Traps

In applied problems involving division, it is often necessary to use critical thinking skills to determine how a remainder should be handled. In Example 1-5.9, for instance, the additional van is needed to transport the remaining residents.

Special properties are associated with division involving zeros, division by 1, and division of a number by itself.

Rule 1-5.1 ***The quotient of zero divided by a nonzero number is zero:***

$$0 \div n = 0, \qquad n\overline{)0}\,\text{ with quotient } 0, \qquad \frac{0}{n} = 0$$

$$0 \div 5 = 0, \qquad 5\overline{)0}\,\text{ with quotient } 0, \qquad \frac{0}{5} = 0$$

If a number divided by zero is to have an answer, then that answer times zero should equal the number. However, we learned in multiplication that zero times any number is zero. Therefore, a number divided by zero has *no* answer.

$8 \div 0 = ?$ What number $\times\ 0 = 8$? Since any number times $0 = 0$, $8 \div 0$ has no answer.

The quotient of a division problem can be only *one* number. Consider the problem $0 \div 0$.

$0 \div 0 = ?$ What number $\times\ 0 = 0$? Since *any number* times $0 = 0$, $0 \div 0$ does not have just *one* number as its quotient. That is, the quotient is not *unique*. For this reason, we say that division of zero by zero has no answer or division of zero by zero is impossible.

Rule 1-5.2 ***The quotient of a number divided by zero is impossible* (undefined):**

$n \div 0$ is impossible, $\quad 0\overline{)n}$ with quotient impossible, $\quad \frac{n}{0}$ is impossible

$12 \div 0$ is undefined or impossible

$0 \div 0$ is undefined or impossible

Rule 1-5.3 ***To divide any nonzero number by itself is 1:***

$n \div n = 1$, if n is not equal to zero; $\quad 12 \div 12 = 1$

Rule 1-5.4 ***To divide any number by 1 yields the same number:***

$n \div 1 = n$, $\quad 5 \div 1 = 5$

In division, as in subtraction, order is important. For example, $12 \div 4 = 3$; however, $4 \div 12$ is not 3. Therefore, division is *not* commutative. When dividing more than two numbers, again as in subtraction, the way the numbers are grouped is important. For example, $(16 \div 4) \div 2 = 4 \div 2 = 2$. If we change the grouping to $16 \div (4 \div 2)$, we have $16 \div 2 = 8$. Notice the different answers, 2 and 8. Therefore, division is *not* associative. If no grouping symbols are included, we divide from left to right.

L.O.3. Estimate and check division.

Estimating and Checking Division

Rule 1-5.5 ***To estimate division:***

1. Round the divisor and dividend to one nonzero digit.
2. Find the first digit of the quotient.
3. Attach a zero in the quotient for each remaining digit in the dividend.

Since division and multiplication are inverse operations, division may be checked by multiplication.

Rule 1-5.6 ***To check division:***

1. Multiply the divisor by the quotient.
2. Add any remainder to the product in step 1.
3. The result of step 2 should equal the dividend.

EXAMPLE 1-5.10 Estimate, find the exact answer, and check 947 ÷ 23.

Solution

Estimate: $20\overline{)900}$ with quotient 40 *(20 divides into 90 four whole times. Attach zero after 4.)*

Exact: $23\overline{)947}$ with quotient 41 R 4

```
      41  R 4
23 | 947
     92
      27
      23
       4
```

Check:

```
   41
 × 23
  123
  82
  943 + 4 = 947
```

(The answer checks.) ■

EXAMPLE 1-5.11 If 23 office managers contributed a total of $575 for an earthquake relief program, how much did each manager contribute (assuming each gave the same amount)? Estimate, find the exact quotient, and check answer.

Solution

Estimate: $20\overline{)600}$ with quotient 30 — 20 goes into 60 three whole times. Write the 3 over the 60. Attach 0.

Exact:

```
     25
23 | 575
     46
     115
     115
       0
```

Check:

```
   25
 × 23
   75
  50
  575
```

Answer checks. ■

L.O.4. Divide using a calculator.

Dividing Using the Calculator

General Tips for Using the Calculator

Division on the basic and scientific calculators:

1. Enter the dividend.
2. Press the division operation key (÷).
3. Enter the divisor.
4. After entering the divisor, pressing the equal key (=) will place the quotient in the display window.

General Tips for Using the Calculator

Division on the graphing calculator:

1. Enter the dividend.
2. Press the division operation key (÷).
3. Enter the divisor.
4. Press the appropriate key to instruct the calculator to perform the calculation that is in the display window. This key is labeled EXE for execute on some graphing calculators.

EXAMPLE 1-5.12 Find the quotient of 9,264 and 24.

Solution: Basic or scientific:

9264 [÷] 24 [=]

Basic or scientific display: 386.

Graphing:

9264 [÷] 24 [EXE]

Graphing display: 9264 ÷ 24

386. ∎

Tips and Traps

The display on some graphing calculators will use a slash (/) instead of the division symbol (÷).

EXAMPLE 1-5.13 How many times will 37 divide into 7,392?

Solution: Basic or scientific:

7392 [÷] 37 [=]

Basic or scientific display: 199.7837838

Graphing:

7392 [÷] 37 [EXE]

Graphing display: 7392 ÷ 37
199.7837838 ■

Divisions that have remainders will be expressed in decimal form. Decimals will be discussed in Chapter 4. However, if the context of the problem makes knowing the exact amount of the remainder important, you can find the remainder by using the following process.

General Tips for Using the Calculator

To find the remainder of a calculator division:

1. While the quotient is displayed, subtract the digits to the left of the decimal point (the whole-number part).
2. Multiply the result of the subtraction by the divisor in the original division.
3. The resulting display will either be a whole number or it will round to a whole number that represents the remainder. If the display is not a whole number, the first digit to the right of the decimal will be 9. Using the rounding process described earlier, since 9 is 5 or greater, add one to the digit in the ones place.

EXAMPLE 1-5.14 Find the remainder of the division in Example 11.

Solution: Basic or scientific display: 199.7837838

Enter: [−] 199 [=]

Next display: 0.783783783

Enter: [×] 37 [=]

Next display: 28.99999997 (Remainder rounds to 29.)

On the graphing calculator:

Enter: 7392 [÷] 37 [EXE]

Graphing display: 7392 ÷ 37
199.7837838

Enter: [Ans] [−] 199 [EXE]

Next display: 7392 ÷ 37
199.7837838

Ans − 199
0.7837837837

Enter: [Ans] [×] 37 [EXE]

Next display: 7392 ÷ 37
199.7837838

Ans − 199
0.7837837837

Ans × 37
29. ■

SELF-STUDY EXERCISES 1-5

L.O.1. Write the following as divisions using the symbol ÷.

1. 4 into 8
2. 3 into 9
3. 6 divided into 24
4. 7 divided into 30
5. 6 divided by 2
6. 8 divided by 4

Write the following as divisions using the symbol $\overline{)\quad}$.

7. 9 into 45
8. 6 into 24
9. 3 divided into 21
10. 8 divided into 56
11. 8 divided by 2
12. 12 divided by 6

Write the following as divisions using the bar symbol —.

13. 2 into 14
14. 2 into 16

L.O.2. Perform the following long-division problems.

15. $8\overline{)600}$
16. $9\overline{)207}$
17. $6\overline{)120}$
18. $7\overline{)21}$
19. $3\overline{)48}$
20. $8\overline{)96}$
21. $5\overline{)215}$
22. $37\overline{)1739}$
23. $26\overline{)312}$
24. $4\overline{)93}$
25. $48\overline{)5982}$
26. $133\overline{)7528}$

27. In a building where 46 outlets are installed, 1472 ft of cable are used. What is the average number of feet of cable used per outlet?

28. Twelve water tanks are constructed in a welding shop at a total contract price of $14,940. What is the price per tank?

29. Five holes that are equally spaced are drilled in a piece of $\frac{1}{4}$-in. flat metal stock. The centers of the first and last holes are 2 in. from the end (see Fig. 1-7). What is the distance between the centers of any two adjacent holes? (*Caution:* How many equal center-to-center distances are there?)

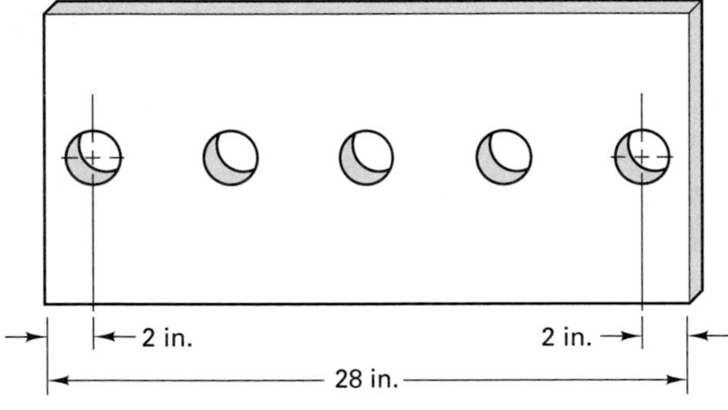

Figure 1-7

L.O.3. Estimate by rounding to one nonzero digit; then find the exact answer. Check your answers.

30. $72\overline{)9240}$
31. $122\overline{)384{,}512}$
32. $221\overline{)824{,}604}$
33. $7\overline{)5901}$

34. A land fill job needs 2294 cubic yards of dirt to be delivered. If one truck can carry 18 cubic yards, how many loads will have to be made?

35. If a reel of wire contains 6362 ft of wire, how many 50-ft extension cords can be made?

L.O.4. Work the following division problems using your calculator. Round the answer to the nearest whole number.

36. $58\overline{)85{,}297}$ **37.** $4\overline{)1203}$ **38.** $56\overline{)224{,}178}$

39. $45\overline{)4{,}027{,}500}$ **40.** $32\overline{)1{,}286{,}400}$ **41.** $71\overline{)440{,}271}$

42. Three bricklayers laid 3,210 bricks on a job in one day. What was the average number of bricks laid by each bricklayer that day?

43. A developer divided a tract of land into 14 equally valued parcels. If the tract is valued at $147,000, what is the value of each parcel?

44. A shipment of 150 machine parts costs $15,775. If this includes a $25 shipping charge, how much does each part cost if shipping charges are not included?

45. If a shop foreman earns $23,400 annually, what is the monthly salary?

46. A machine operator earns a monthly salary of $2465. What is the operator's annual salary?

47. A shipment of 288 headlights is billed out at $1200, which includes a $48 freight charge. What is the cost per headlight excluding freight costs?

48. An invoice showing the total cost of refrigerator replacement parts indicated $2118 which included a $30 freight charge. The catalog price of the parts is $87 each. How many parts were ordered?

49. A job requires 95 feet of pipe which will be cut into 9-ft lengths. How many 9-ft lengths of pipe will there be?

50. A wiring job requires 587 feet of wire to install 37 receptacles. If the receptacles are equally spaced, approximately how far apart are they?

51. Find the missing measure in Fig. 1-8.

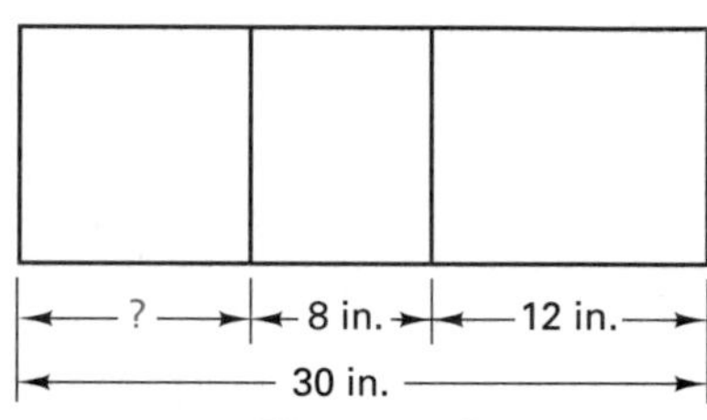

Figure 1-8

1-6 EXPONENTS, ROOTS, AND POWERS OF 10

Learning Objectives

L.O.1. Simplify expressions containing exponents.
L.O.2. Square numbers and find square roots of numbers.
L.O.3. Use a calculator to find powers and roots.
L.O.4. Use powers of 10 to multiply and divide.

L.O.1. Simplify expressions containing exponents.

Simplifying Expressions with Exponents

The product of repeated factors can be written in shorter form with the use of whole-number exponents. For example, $4 \times 4 \times 4 = 4^3$. The number 4 is called the *base* and is the repeated factor. The number 3 is called the *exponent* and indicates how many times the repeated factor is used. The expression 4^3 is read "four *cubed*" or "four to the third *power*" or "four raised to the third *power*." In the example $2 \times 2 \times 2 \times 2 \times 2 = 2^5$, the number 2 is the base and the number 5 is the exponent. The expression 2^5 is read "two to the fifth power." In expressions that have 2 as an exponent, such as 4^2, the expression is usually read as "four *squared*"; however, it may also be read as "four to the second power."

To perform the multiplicaton of 2^5 (or $2 \times 2 \times 2 \times 2 \times 2$) = 32 is to *simplify* the expression. The expression 2^5 is written in *exponential notation*. The number 32 is written in *standard notation*.

$$\text{base} \longrightarrow 2^5 \longleftarrow \text{exponent}$$

EXAMPLE 1-6.1 Identify the base and exponent of the following expressions. Then simplify each expression.

(a) 5^3 5 is the base. 3 is the exponent.
$5^3 = 5 \times 5 \times 5 = 125$

(b) 3^2 3 is the base. 2 is the exponent.
$3^2 = 3 \times 3 = 9$

(c) 2^4 2 is the base. 4 is the exponent.
$2^4 = 2 \times 2 \times 2 \times 2 = 16$ ■

When a number has 1 for an exponent, the 1 indicates that the number is used as a factor 1 time. For example, $5^1 = 5$. When the number does not have an exponent, the exponent is understood to be 1. For example, the number 8 equals 8^1.

An exponent can be any number, including 1 and 0. Exponents that are not whole numbers will be discussed later. Examine the following expressions in exponential notation.

$4^3 = 4 \times 4 \times 4$	4 used as a factor 3 times
$4^2 = 4 \times 4$	4 used as a factor 2 times
$4^1 = 4$	4 used as a factor 1 time

> **Rule 1-6.1** *Any number with an exponent of 1 is that number itself:*
>
> $a^1 = a$, for any base a

EXAMPLE 1-6.2 Write the following exponential expressions as standard numbers.

(a) $3^1 = 3$ **(b)** $8^1 = 8$ **(c)** $5^1 = 5$ ■

EXAMPLE 1-6.3 Express the following standard numbers in exponential notation.

(a) $2 = 2^1$ **(b)** $7 = 7^1$ **(c)** $12 = 12^1$ ■

> **Rule 1-6.2** ***Any number (except zero) with an exponent for 0 equals 1. The expression 0^0 is undefined.***
>
> $a^0 = 1$, for any nonzero base a

EXAMPLE 1-6.4 Simplify the following exponential expressions.

(a) $9^0 = 1$ **(b)** $6^0 = 1$ **(c)** $27^0 = 1$ ■

L.O.2. Square numbers and find square roots of numbers.

Finding Squares and Square Roots

The standard number that results from squaring any number is called a *perfect square* or *square*. To square a number is to find the result of using that number as a factor twice. If asked to square 3, we would say $3^2 = 3 \times 3 = 9$. We say 9 is the square of 3.

Example 1-6.5 Find the square of each of the following.

(a) 2; $2^2 = 2 \times 2 = 4$ **(b)** 7; $7^2 = 7 \times 7 = 49$ ■

The process of squaring has an inverse operation, just as addition and multiplication have the inverse operations of subtraction and division, respectively. The inverse operation of squaring is taking the square root of a number. To take the square root of a number (a perfect square) is to determine the number (called the *square root*) which was used as a factor twice to equal the perfect square. The square root of 9 is 3, since 3^2 or $3 \times 3 = 9$.

The *radical sign* $\sqrt{}$ is an operational symbol that indicates that the square root is to be taken of the number under the bar. This bar serves as a grouping symbol just like parentheses. The number under the radical sign is called the *radicand*. The entire expression is called a *radical expression.*

radical sign → $\sqrt{25} = 5 \longleftarrow$ square root
radicand → (25)

Rule 1-6.3 ***To estimate the square root of a number:***

1. Select the trial estimate of the square root.
2. Square the estimated answer.
3. If the square of the estimated answer is less than the original number, adjust the estimated answer to a larger number. If the square of the estimated answer is more than the original number, adjust the estimated answer to a smaller number.
4. Square the adjusted answer from step 3.
5. Continue the adjusting process until the square of the trial estimate is the original number.

Square roots of whole numbers that are perfect squares will be whole numbers. The calculator display for the square roots of other numbers will be shown as decimal numbers. When estimating the square root of other numbers, the estimate can show the two whole numbers that the square root is between.

Example 1-6.6 Find $\sqrt{256}$.

Solution: Select 15 as the estimated square root. 15^2 or $15 \times 15 = 225$. The number 225 is less than 256, so the square root of 256 must be larger than 15. We adjust the estimate to 17. 17^2 or $17 \times 17 = 289$. The number 289 is more than 256, so the square root of 256 must be smaller than 17. Adjust the estimate to 16. Since $16^2 = 256$, 16 is the square root of 256. ■

L.O.3. Use a calculator to find powers and roots.

Using the Calculator to Find Powers and Square Roots

To raise a number to a power using a basic calculator requires repeated multiplication. However, scientific and graphing calculators have keys designed to raise numbers to any power so that the repeated multiplication is done internally. Since scientific or graphing calculators can be used to raise numbers to powers more efficiently, as well as perform other mathematical operations needed in algebra and other mathematics, further discussion of calculators will focus on these two types.

Scientific and graphing calculators have special power keys, such as keys for squares and cubes. They also have a "general power" key that can be used for all powers. Even though the "general power" key can be used to square and cube numbers, fewer keystrokes are needed when using the "square" and "cube" keys for these operations.

General Tips for Using the Calculator

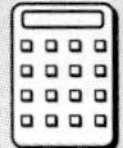

Squaring or cubing on the scientific calculator:

1. Enter the base.
2. Press the square key (x^2) or cube key (x^3).

Raising a number to a power on the scientific calculator:

1. Enter the base.
2. Press the general power key (x^y).
3. Enter the exponent.
4. Press the equal key (=).

General Tips for Using the Calculator

Squaring or cubing on the graphing calculator:

1. Enter the base.
2. Press the square key (x^2) or cube key (x^3).
3. Press the EXE key.

Raising a number to a power on the graphing calculator:

1. Enter the base.
2. Press the general power key (x^y).
3. Enter the exponent.
4. Press the EXE key.

Tips and Traps

To make calculator steps easier to follow, we will identify a common label for a function in a box. The exact label will vary with the specific calculator and model.

EXAMPLE 1-6.7 Use the calculator to find 35^2.

Solution

Scientific: 35 $\boxed{x^2}$

Display: 1225.

Graphing: 35 $\boxed{x^2}$ $\boxed{\text{EXE}}$

Display: 35^2

1225. ■

EXAMPLE 1-6.8 Use the calculator to find 8^5.

Solution

Scientific: 8 $\boxed{x^y}$ 5 $\boxed{=}$

Display: 32768.

Graphing: 8 $\boxed{x^y}$ 5 $\boxed{\text{EXE}}$

Display: 8 x^y 5

32768. ■

General Tips for Using the Calculator

Finding the square root of a number on the scientific calculator:

1. Enter the radicand.
2. Press the square root key $\boxed{\sqrt{\ }}$.

Finding the square root of a number on the graphing calculator:

1. Press the square root key $\boxed{\sqrt{\ }}$.
2. Enter the radicand.
3. Press EXE key.

EXAMPLE 1-6.9 Evaluate $\sqrt{529}$.

Solution

Scientific: 529 $\boxed{\sqrt{\ }}$

Display: 23.

Graphing: $\boxed{\sqrt{\ }}$ 529 $\boxed{\text{EXE}}$

Display: $\sqrt{\ }$ 529

23. ■

L.O.4. Use powers of 10 to multiply and divide.

Using Powers of 10

Powers of 10 are numbers whose only nonzero digit is 1. Thus, 10, 100, 1000, and so on, are powers of 10.

In the following compare the number of zeros in standard notation with the exponent of 10 in exponential notation.

$$\begin{aligned} 1{,}000{,}000 &= 10^6 && \text{6 zeros} \\ 100{,}000 &= 10^5 && \text{5 zeros} \\ 10{,}000 &= 10^4 && \text{4 zeros} \\ 1000 &= 10^3 && \text{3 zeros} \\ 100 &= 10^2 && \text{2 zeros} \\ 10 &= 10^1 && \text{1 zero} \\ 1 &= 10^0 && \text{0 zeros} \end{aligned}$$

The exponents in powers of 10 indicate the number of zeros used in standard notation.

EXAMPLE 1-6.10 Express the following as powers of 10.

(a) $10{,}000{,}000 = 10^7$
(b) $100{,}000{,}000 = 10^8$
(c) $100{,}000{,}000{,}000 = 10^{11}$ ■

EXAMPLE 1-6.11 Express the following in standard notation.

(a) $10^5 = 100{,}000$
(b) $10^{13} = 10{,}000{,}000{,}000{,}000$ ■

In some applications, powers of 10 are used in multiplication and division problems.

Compare each example in the left column with the corresponding example in the right column to find the pattern for multiplying a whole number by a power of 10.

$5 \times 100 = 500$	$5 \times 10^2 = 500$	*(2 zeros attached)*
$27 \times 1000 = 27{,}000$	$27 \times 10^3 = 27{,}000$	*(3 zeros attached)*
$30 \times 10{,}000 = 300{,}000$	$30 \times 10^4 = 300{,}000$	*(4 zeros attached)*

Rule 1-6.7 ***To multiply a* whole *number by a power of 10:***

1. Write the whole number being multiplied by the power of 10.
2. After the whole number, attach as many zeros as the exponent of 10 indicates if the power of 10 is in exponential notation. Attach as many zeros as the number of zeros in the power of 10 if it is in standard notation.

EXAMPLE 1-6.12 Multiply the following whole numbers by using powers of 10.

(a) $2 \times 1000 = 2000$ (Attach 3 zeros after 2.)
(b) $75 \times 10^2 = 7500$ (Attach 2 zeros after 75.) ■

Compare each example in the left column with the corresponding example in the right column to find the pattern for dividing a whole number by a power of 10.

$300 \div 100 = 3$	$300 \div 10^2 = 3$	*(last 2 digits dropped)*
$\frac{90{,}000}{1000} = 90$	$\frac{90{,}000}{10^3} = 90$	*(last 3 digits dropped)*
$5360 \div 100 = 53$ R60	$5360 \div 10^2 = 53$ R60	*(last 2 digits dropped, dropped digits expressed as a remainder)*
$7045 \div 1000 = 7$ R45	$7045 \div 10^3 = 7$ R45	*(last 3 digits dropped, dropped digits expressed as a remainder)*

> **Rule 1-6.8** *To divide a* **whole** *number by a power of 10:*
>
> 1. Drop from the right end of the whole number as many digits as the exponent of 10 indicates if the power of 10 is in exponential notation. Drop from the right end of the whole number as many digits as the number of zeros in the power of 10 if it is in standard notation.
> 2. If the dropped digits are zeros, write the whole number without the dropped zeros.
> 3. If the dropped digits contain one or more nonzero digits, write the whole number without the dropped digits and express the dropped digits as a remainder, beginning with the first nonzero digit of the remainder.

EXAMPLE 1-6.13 Divide the following whole numbers by using powers of 10.

(a) $600 \div 100 = 6$ (Drop last 2 digits. No remainder.)
(b) $\frac{7580}{10^2} = 75$ R80 (Drop last 2 digits. 80 is the remainder.)
(c) $9025 \div 1000 = 9$ R25 (Drop last 3 digits. 25 is the remainder.) ■

Multiplication and division by powers of 10 will be expanded after we study fractions and decimal numbers.

EXAMPLE 1-6.14 Universal Landscaping needs to apply fertilizer to a customer's lawn. The lawn measures 150 feet long and 125 feet wide. Each bag of fertilizer will cover 5,000 ft^2. Estimate the number of bags of fertilizer needed to treat the yard.

Solution: To estimate the number of square feet in the yard, round the length and width to one nonzero number. 150 rounds to 200 and 125 rounds to 100.

$$\begin{aligned} A &= lw \\ &= 200(100) \quad \textit{(Use rule for multiplying by a power of 10.)} \\ &= 20{,}000 \end{aligned}$$

The lawn has approximately 20,000 ft^2 of ground to be covered. Since one bag of fertilizer will cover 5000 ft^2, we must divide 20,000 by 5000 to determine the number of bags needed.

$$20{,}000 \div 5000 = 4$$

Approximately 4 bags of fertilizer are needed to treat the lawn. ■

SELF-STUDY EXERCISES 1-6

L.O.1. Give the base and exponent of the following expressions.

1. 4^3 **2.** 9^4 **3.** 2^9

Simplify the following exponential expressions.

4. 1^3 **5.** 7^2 **6.** 10^3
7. 2^4 **8.** 3^2 **9.** 15^1
10. 8^1 **11.** 9^0

Express the following with an exponent of 1.

12. 8 **13.** 145 **14.** 12
15. 23

16. Explain what an exponent of 1 does to the base. **17.** Explain what an exponent of 0 does to the base.

L.O.2. Write in standard notation.

18. 8^2 **19.** 2^2 **20.** 25^2
21. 14^2 **22.** 13^2 **23.** 1^2
24. 100^2

Square the following numbers.

25. 8 **26.** 9 **27.** 17
28. 18 **29.** 101 **30.** 22

Perform the following operations.

31. $\sqrt{25}$ **32.** $\sqrt{49}$ **33.** $\sqrt{81}$
34. $\sqrt{196}$

35. What does it mean to square a number?

36. What does it mean to take the square root of a number?

L.O.3. Use a calculator to evaluate the following.

37. 15^2 **38.** 7^3 **39.** 5^7
40. $\sqrt{324}$ **41.** $\sqrt{784}$ **42.** $\sqrt{1,089}$

L.O.4. Multiply the following whole numbers by using powers of 10.

43. 10×10^2 **44.** 32×1000 **45.** $3 * 10^4$
46. $2 * 10^4$ **47.** $102 * 100$ **48.** 22×10^3

Divide the following whole numbers by using powers of 10.

49. $250 \div 10$ **50.** $210 \div 10$ **51.** $\frac{300}{10^2}$
52. $\frac{900}{10^2}$ **53.** $2500 \div 10$ **54.** $120 \div 10^2$

55. What does the exponent of a power of 10 mean when multiplying?

56. What does the exponent of a power of 10 mean when dividing?

1-7 ORDER OF OPERATIONS

Learning Objectives

L.O.1. Apply the order of operations to problems containing several operations.
L.O.2. Use the calculator to perform a series of operations.

L.O.1. Apply the order of operations to problems containing several operations.

Applying the Order of Operations

Whenever several mathematical operations are to be performed, they must follow the proper *order of operations*.

Rule 1-7.1 *The order of operations follows:*

1. Perform operations within parentheses (or other grouping symbols) beginning with the innermost set of parentheses.
2. Evaluate exponential operations and find square roots in order from left to right.
3. Multiply and divide in order from left to right.
4. Add and subtract in order from left to right.

To summarize, use the following key words:

Parentheses (grouping), Exponents (and roots), Multiplication and Division, Addition and Subtraction

Tips and Traps

A memory tip for remembering the order of operations is the sentence "Please Excuse My Dear Aunt Sally."

Example 1-7.1 Evaluate $3(2 + 3) - 7$.

Solution

$3(2 + 3) \quad -7$

$3(5) \quad -7$ *(Do operations within parentheses first. Then multiply.)*

$15 \quad -7$

$15 - 7 = 8$ *(Subtract last.)* ■

Example 1-7.2 Simplify $4^2 - 5(2) \div (4 + 6)$.

Solution

$4^2 - 5(2) \div (4 + 6)$

$4^2 - 5(2) \div 10$ *(Do operations within parentheses first. $4 + 6 = 10$.)*

$16 - 5(2) \div 10$ *(Evaluate exponentiation. $4^2 = 16$.)*

$16 - 10 \div 10$ *(Then multiply. $5(2) = 10$.)*

$16 - 1$ *(And divide. $10 \div 10 = 1$.)*

$16 - 1 = 15$ *(Subtract last.)* ■

Tips and Traps

As in Example 1-7.2, parentheses can be used to indicate multiplication or to indicate an operation that should be done first. If the parentheses contain an operation, the parentheses indicate a grouping. Otherwise, the parentheses indicate multiplication. The expression 5(2) indicates multiplication while (4 + 6) indicates a grouping.

Example 1-7.3 Evaluate $5 \times \sqrt{16} - 5 + [15 - (3 \times 2)]$.

Solution

$5 \times \sqrt{16} - 5 + [15 - (3 \times 2)]$

$5 \times \sqrt{16} - 5 + [15 - 6]$ *(Work inner most grouping. 3×2.)*

$5 \times \sqrt{16} - 5 + 9$ *(Work remaining grouping. $15 - 6$.)*

$5 \times 4 - 5 + 9$ *(Find square root. $\sqrt{16} = 4$.)*

$20 - 5 + 9$ *(Multiply. 5×4.)*

$15 + 9 = 24$ *(Add and subtract from left to right.)* ■

Example 1-7.4 Evaluate $3^2 + \sqrt{21 - 5} \times 2$.

Solution

$3^2 + \sqrt{21 - 5} \times 2$

$3^2 + \sqrt{16} \times 2$ *(Do operations within grouping first; the bar part of radical symbol is a grouping symbol. $21 - 5 = 16$.)*

$9 + 4 \times 2$	*(Then do exponentiation and square root from left to right.) ($3^2 = 9$. $\sqrt{16} = 4$.)*
$9 + 8$	*(Then multiply. $4 \times 2 = 8$.)*
$9 + 8 = 17$	*(Add last.)* ■

L.O.2. Use the calculator to perform a series of operations.

Using the Calculator to Perform a Series of Operations

Both the scientific and graphing calculators have been programmed to perform the order of operations in the proper order.

EXAMPLE 1-7.5 Evaluate $4^2 - 5(2) \div (4 + 6)$.

Solution

Scientific:

4 [x^2] [−] 5 [×] 2 [÷] [(]4 [+] 6[)] [=]

Display: 15.

Graphing:

4[x^2] [−] 5[(]2[)] [÷] [(]4 [+] 6[)] [EXE]

Display: $4^2 - 5(2) \div (4 + 6)$

15 ■

General Tips for Using the Calculator

The scientific calculator requires that the multiplication operation key be used to indicate the product of 5 and 2. When using the graphing calculator, the expression is entered exactly as it appears in the example or the multiplication operation key can be used. Parentheses keys, [(] and [)], are used to show groupings.

EXAMPLE 1-7.6 Evaluate $3^4 + \sqrt{77 - 13} \times 2[26 - (45 - 38)]$.

Solution

Scientific:

3 [x^y] 4 [+] [(]77 [−] 13[)] [√] [×] 2 [×] [(]26 [−] [(]45 [−] 38[)][)] [=]

Display: 385.

Graphing:

3 [x^y] 4 [+] [√] [(]77 [−] 13[)] [×] 2[(]26 [−] [(]45 [−] 38[)][)] [EXE]

Display: $3x^y4 + \sqrt{\ }\ (77 - 13) \times 2(26 - (45 - 38))$

385 ■

SELF-STUDY EXERCISES 1-7

L.O.1. Evaluate each problem following the order of operations.

1. $5^2 + 4 - 3$	**2.** $4^2 + 6 - 4$	**3.** $4 \times 3 - 9 \div 3$
4. $5 \times 2 - 4 \div 2$	**5.** $25 \div 5 \times 4$	**6.** $64 \div 4 \times 2$

7. $6 \times \sqrt{36} - 2 \times 3$

8. $3 \times \sqrt{81} - 3 \times 4$

9. $4^2 \times 3^2 + (4 + 2) \times 2$

10. $2^2 \times 5^2 + (2 + 1) \times 3$

11. $54 - 3^3 - \frac{8}{2}$

12. $56 - 2^3 - \frac{9}{3}$

13. $32 - 2 \times 4^2 \div 2$

14. $5 - 3 \times 2^2 \div 6$

15. $4 + 15 \div 3 - \frac{0}{7}$

16. $3 + 8 \div 4 - \frac{0}{5}$

17. $3 - 2 + 3 \times 3 - \sqrt{9}$

18. $2 - 1 + 4 \times 4 - \sqrt{4}$

19. $2^4 \times (7 - 2) \times 2$

20. $3^4 \times (9 - 3) \times 3$

L.O.2. Use the calculator to perform the following operations.

21. $6 \times 9^2 - \frac{12}{4}$

22. $7 \times 8^2 - \frac{10}{2}$

23. $4^3 + 14 - 8$

24. $2^3 + 12 - 7$

25. $2 \times \sqrt{16} + (8 - \sqrt{25})$

26. $4 \times \sqrt{49} + (9 - \sqrt{64})$

27. $3(2^2 + 1) - 30 \div 3$

28. $4(3^2 + 2) - 60 \div 12$

29. $3 + 10 \div 5 + 2$

30. $4 + 12 \div 4 + 7$

31. $25 - 5^2 \div (7 - 2)$

32. $36 - 6^2 \div (8 - 2)$

33. State the first operation in the order of opertions.

34. State the last operation in the order of operations.

1-8 FACTORIZATION, MULTIPLES, AND DIVISIBILITY

Learning Objectives

L.O.1. Find all factorizations of a natural number.
L.O.2. Find multiples of a natural number.
L.O.3. Determine the divisibility of a number.

L.O.1. Find all factorizations of a natural number.

Factoring Natural Numbers

The *natural numbers* are the counting numbers, such as 1, 2, 3, 4, 5, 6, 7, 8, 9, 10, 11, 12, and so forth. The natural numbers can also be described as the whole numbers excluding zero.

To factor a natural number is to express the number as a multiplication of two or more natural numbers. These expressions are called *factorizations.* The numbers multiplied are called *factors.*

One and the number itself are factors of every number, such as $3 = 1 \times 3$.

Example 1-8.1 List all the pairs of factors of 18; then write the factors in order from smallest to largest.

Solution

Factorizations: $18 = 1 \times 18$, $18 = 2 \times 9$, $18 = 3 \times 6$

Factors: 1, 2, 3, 6, 9, 18 ■

Example 1-8.2 List all the pairs of factors of 48; then write the factors in order from smallest to largest.

Solution

Factorizations: $48 = 1 \times 48$, $48 = 2 \times 24$, $48 = 3 \times 16$, $48 = 4 \times 12$, $48 = 6 \times 8$

Factors: 1, 2, 3, 4, 6, 8, 12, 16, 24, 48 ■

L.O.2. Find multiples of a natural number.

Multiples

If we count by threes, such as 3, 6, 9, 12, 15, 18, we obtain natural numbers that are *multiples* of 3. Each number above is a multiple of 3 because each is the product of 3 and a natural number. That is,

$$3 = 3 \times 1, \qquad 6 = 3 \times 2, \qquad 9 = 3 \times 3$$
$$12 = 3 \times 4, \qquad 15 = 3 \times 5, \qquad 18 = 3 \times 6$$

A *multiple* of a natural number is the product of that number and a natural number.

EXAMPLE 1-8.3 Show that 2, 4, 6, 8, and 10 are multiples of 2 by listing their factorizations with 2.

Solution

$$2 = 2 \times 1, \qquad 4 = 2 \times 2, \qquad 6 = 2 \times 3$$
$$8 = 2 \times 4, \qquad 10 = 2 \times 5$$

■

Natural numbers that are multiples of 2 are called *even numbers*. Numbers that are not multiples of 2 are called *odd numbers*.

L.O.3. Determine the divisibility of a number.

Determining the Divisibility of a Number

A number is *divisible* by another number if the quotient has no remainder or if the dividend is a multiple of the divisor.

EXAMPLE 1-8.4 Is 35 divisible by 7?

Solution: 35 is divisible by 7 if 35 ÷ 7 has no remainder or if 35 is a multiple of 7.

$$35 \div 7 = 5 \quad \text{or} \quad 35 = 5 \times 7$$

Yes, 35 is divisible by 7. ■

EXAMPLE 1-8.5 A holiday party is planned for 47 first-grade students. If the teachers have a gross (144) of assorted party favors, can each child receive an equal number of favors with none left over so that no child will receive more than another?

Solution: This problem is really asking whether 144 is divisible by 47.

$$144 \div 47 = 3.063829787 \quad \text{(using a calculator)}$$

The calculator answer is not a natural number; therefore, 144 is not divisible by 47. To determine the remainder, use the procedure found in Section 1-5.

(*calculator steps:* 3.063829787 [−] 3 [=] [×] 47 [=] 2.999999998, which rounds to 3).

No, there will be 3 party favors left over. ■

A number of rules or tests can help us decide if certain numbers are divisible by certain other numbers.

Rule 1-8.1 ***Tests for Divisibility:***

1. A number is divisible by 2 if the last digit is 0, 2, 4, 6, or 8, that is, if the last digit is an even digit.
2. A number is divisible by 3 if the sum of its digits is divisible by 3.
3. A number is divisible by 4 if the last two digits form a number that is divisible by 4.
4. A number is divisible by 5 if the last digit is 0 or 5.
5. A number is divisible by 6 if the number is divisible by *both* 2 *and* 3.
6. A number is divisible by 7 if the division has no remainder.
7. A number is divisible by 8 if the last three digits form a number divisible by 8.
8. A number is divisible by 9 if the sum of its digits is divisible by 9.
9. A number is divisible by 10 if the last digit is 0.

EXAMPLE 1-8.6 Use the tests for divisibility to identify which number in each pair of numbers is divisible by the given number.

Numbers	Divisor	Answer
874 or 873	2	874; the last digit is an even digit (4).
275 or 270	2	270; the last digit is an even digit (0).
427 or 423	3	423; the sum of the digits is divisible by 3. $4 + 2 + 3 = 9$.
5912 or 5913	4	5912; the last two digits form a number divisible by 4. $12 \div 4 = 3$.
80 or 82	5	80; the last digit is 0.
56 or 65	5	65; the last digit is 5.
804 or 802	6	804; the last digit is even and the sum of the digits is divisible by 3.
58 or 56	7	56; it divides by 7 with no remainder.
3160 or 3162	8	3160; the last three digits form a number divisible by 8. $160 \div 8 = 20$.
477 or 475	9	477; the sum of the digits is divisible by 9. $4 + 7 + 7 = 18$.
182 or 180	10	180; the last digit is 0.

SELF-STUDY EXERCISES 1-8

L.O.1. Use factorization to find all the factors for each of the following numbers.

1. 24 **2.** 36 **3.** 45
4. 32 **5.** 16 **6.** 27
7. 20 **8.** 30 **9.** 12
10. 8 **11.** 4 **12.** 15
13. 81 **14.** 64 **15.** 38
16. 46

L.O.2. Show that the following numbers are multiples of the first number by listing their factorizations with the first number.

17. 5, 10, 15, 20, 25, 30 **18.** 6, 12, 18, 24, 30, 36 **19.** 8, 16, 24, 32, 40, 48
20. 9, 18, 27, 36, 45, 54 **21.** 10, 20, 30, 40, 50, 60 **22.** 30, 60, 90, 120, 150, 180

Find 5 multiples of the given numbers by listing their factorizations, starting with the given number times 1.

23. 5 **24.** 12 **25.** 7
26. 3 **27.** 50 **28.** 4

L.O.3. Are the following numbers divisible by the given number? Explain your answer.

29. 2434 by 6
30. 230 by 5
31. 2434 by 4
32. 1221 by 3
33. 756 by 7
34. 920 by 8
35. 621 by 3
36. 426 by 6

1-9 PRIME FACTORIZATION

Learning Objectives

L.O.1. Factor prime and composite numbers.
L.O.2. Determine the prime factorization of composite numbers.

L.O.1. Factor prime and composite numbers.

Factoring Prime and Composite Numbers

When listing all the factors of natural numbers, some numbers may have only two factors, the number itself and 1. These numbers form a special group of numbers called prime numbers. A *prime number* is a whole number greater than 1 that has only two factors, the number itself and 1. One is not a prime number since it has only one factor.

EXAMPLE 1-9.1 Identify the prime numbers by factorization.

(a) 8 **(b)** 1 **(c)** 3 **(d)** 9 **(e)** 7

Solution

(a) 8 is *not* a prime number, since its factors are 1, 2, 4, and 8.
(b) 1 is *not* a prime number, since a prime must be greater than 1 or since it does not have two *different* factors.
(c) 3 is a prime number, since it has only two factors, 1 and 3.
(d) 9 is *not* a prime number, since its factors are 1, 3, and 9.
(e) 7 is a prime number, since it has only two factors, 1 and 7. ■

A *composite number* is a whole number greater than 1 that is not a prime number. Therefore, 8 and 9 in Example 1-9.1 are composite numbers. A composite number has factors other than itself and 1.

EXAMPLE 1-9.2 Identify the composite numbers by factorization.

(a) 4 **(b)** 10 **(c)** 13 **(d)** 12 **(e)** 5

Solution

(a) 4 is a composite number, since its factors are 1, 2, and 4.
(b) 10 is a composite number, since its factors are 1, 2, 5, and 10.
(c) 13 is *not* a composite number, since its only factors are 1 and 13. It is a prime number.
(d) 12 is a composite number, since its factors are 1, 2, 3, 4, 6, and 12.
(e) 5 is *not* a composite number, since its only factors are 1 and 5. It is a prime number. ■

L.O.2. Determine the prime factorization of composite numbers.

Determining the Prime Factorization of Composite Numbers

A composite number can be expressed as a product of prime numbers. *Prime factorization* refers to writing a composite number as the product of *only* prime numbers. In this case, the factors are called *prime factors*.

Example 1-9.3 Find the prime factorization of 30.

Solution

$30 = 2 \times 15$ *(Factor 30 into two factors using its smallest prime factor.)*

first prime ↑ (the 2)

$30 = 2 \times 3 \times 5$ *(Factor the composite 15 using its smallest prime factor. Since 5 is also prime, the factoring is complete.)*

last primes ↑ ↑ (the 3 and 5)

■

The prime factorization of 30 is $2 \times 3 \times 5$.

Example 1-9.4 Find the prime factorization of 16.

Solution

$16 = 2 \times 8$ *(Factor 16 into two factors using its smallest prime factor.)*

first prime ↑

$16 = 2 \times 2 \times 4$ *(Factor 8 into two factors using its smallest prime factor.)*

second prime ↑

$16 = 2 \times 2 \times 2 \times 2$ *(Factor 4 into two factors using its smallest prime factor.)*

last two primes ↑ ↑

■

The prime factorization of 16 is $2 \times 2 \times 2 \times 2$. We can write this expression in exponential notation as 2^4.

SELF-STUDY EXERCISE 1.9

L.O.1. List the pairs of factors of each number. Then write all the factors in order from smallest to largest.

1. 14	**2.** 22	**3.** 11
4. 17	**5.** 18	**6.** 24

L.O.2. Find the prime factorization of the following numbers.

7. 50	**8.** 52	**9.** 225
10. 125	**11.** 100	**12.** 200
13. 65	**14.** 75	**15.** 121
16. 144		

Write the prime factorization of the following numbers using exponential notation.

17. 568	**18.** 112	**19.** 124
20. 164	**21.** 72	**22.** 900

1-10 LEAST COMMON MULTIPLE AND GREATEST COMMON FACTOR

Learning Objectives

L.O.1. Find the least common multiple of two or more numbers.
L.O.2. Find the greatest common factor of two or more numbers.

L.O.1. Find the least common multiple of two or more numbers.

Finding the Least Common Multiple of Two or More Numbers

Multiples, especially the *least common multiple*, will be useful when adding and subtracting fractions with different denominators. The least common multiple will be used as the lowest common denominator. Recall from Section 1-8 that the multiple of a natural number is the product of that number and a natural number.

If we count by threes, we obtain natural numbers that are multiples of 3 ($3 \times 1 = 3$, $3 \times 2 = 6$, and so on). If we count by fives, we obtain natural numbers that are multiples of 5 ($5 \times 1 = 5$, $5 \times 2 = 10$, and so on).

Multiples of 3: 3, 6, 9, 12, 15, 18, 21, 24, . . .

Multiples of 5: 5, 10, 15, 20, 25, 30, 35, 40, . . .

The smallest multiple that is common to both sets of numbers above is 15. So 15 is the *least common multiple* of 3 and 5. It is the smallest number divisible by both 3 and 5.

The *least common multiple* (*LCM*) of two or more natural numbers is the smallest number that is a multiple of each number. It is divisible by each number.

Prime factorization can be used to find the least common multiple of two or more numbers.

Rule 1-10.1 ***The least common multiple of numbers may be found by using the prime factorization of the numbers as follows:***

1. List the prime factorization of each number using exponential notation.
2. List the prime factorization of the least common multiple by including each prime factor appearing in *each* number. If a prime factor appears in more than one number, use the factor with the *largest* exponent.
3. Write the resulting expression in standard notation.

EXAMPLE 1-10.1 Find the least common multiple of 12 and 40 by prime factorization.

Solution

$$12 = 2 \times 2 \times 3 = 2^2 \times 3 \quad \textit{(Prime factorization of 12)}$$

$$40 = 2 \times 2 \times 2 \times 5 = 2^3 \times 5 \quad \textit{(Prime factorization of 40)}$$

$$\text{LCM} = 2^3 \times 3 \times 5 \quad \textit{(Prime factorization of LCM)}$$

$$\text{LCM} = 120 \quad \textit{(LCM in standard notation)}$$

■

L.O.2. Find the greatest common factor of two or more numbers.

Finding the Greatest Common Factor of Two or More Numbers

A *common factor* is a factor common to two or more numbers or products. Common factors, especially the *greatest common factor*, will be useful later in simplifying or reducing fractions.

The *greatest common factor* (*GCF*) of two or more numbers is the largest factor common to each number. Each number is divisible by the GCF.

Let's take the numbers 30 and 42. The prime factors are

$$30 = 2 \times 3 \times 5, \qquad 42 = 2 \times 3 \times 7$$

The *common* prime factors of both 30 and 42 are 2 and 3, which represent the composite factor 6. The *greatest* common factor is the product of the common prime factors, $2 \times 3 = 6$. So 6 is the GCF of 30 and 42. Each number is divisible by the 6.

Rule 1-10.2 ***To find the greatest common factor (GCF) of two or more natural numbers:***

1. List the prime factorization of each number using exponential notation.
2. List the prime factorization of the greatest common factor by including each prime factor appearing in *every* number. If a prime factor appears in more than one number, use the factor with the *smallest* exponent. If there are no common prime factors, the GCF is 1.
3. Write the resulting expression in standard notation.

EXAMPLE 1-10.2 Find the greatest common factor of 15, 30, and 45.

Solution

$$15 = 3 \times 5 = 3 \times 5 \quad \textit{(Prime factorization of 15)}$$

$$30 = 2 \times 3 \times 5 = 2 \times 3 \times 5 \quad \textit{(Prime factorization of 30)}$$

$$45 = 3 \times 3 \times 5 = 3^2 \times 5 \quad \textit{(Prime factorization of 45)}$$

$$\text{GCF} = 3 \times 5 \quad \textit{(Common prime factors)}$$

$$\text{GCF} = 15 \quad \textit{(GCF in standard notation)}$$ ■

EXAMPLE 1-10.3 Find the greatest common factor of 10, 12, and 13.

Solution

$$10 = 2 \times 5 = 2 \times 5 \quad \textit{(Prime factorization of 10)}$$

$$12 = 2 \times 2 \times 3 = 2^2 \times 3 \quad \textit{(Prime factorization of 12)}$$

$$13 = 13 = 13 \quad \textit{(Prime factorization of 13)}$$

$$\text{GCF} = 1 \quad \textit{(No common prime factors)}$$ ■

SELF-STUDY EXERCISES 1-10

L.O.1. Find the least common multiple of the following numbers.

1. 2 and 3
2. 5 and 6
3. 7 and 8
4. 3 and 4
5. 18 and 60
6. 10 and 12
7. 12 and 24
8. 9 and 18
9. 4, 8, and 12
10. 20, 25, and 35
11. 3, 9, and 27
12. 2, 8, and 16
13. 6, 15, and 18
14. 20, 24, and 30
15. 12, 18, and 20
16. 6, 10, and 12
17. 8, 12, and 32
18. 8, 12, and 18
19. 10, 15, and 20
20. 30, 50, and 60
21. 6, 11, and 33
22. 8, 13, and 39

L.O.2. Find the greatest common factor of the following numbers.

23. 18 and 24
24. 15 and 25
25. 10, 11, and 14
26. 12, 13, and 16
27. 6, 8, and 14
28. 4, 10, and 18
29. 30 and 45
30. 40 and 55
31. 36, 60, and 216
32. 18, 30, and 108

ASSIGNMENT EXERCISES

1-1

1. Identify the place value of the digit 3 in each of the following:
(a) 430 **(b)** 34,789 **(c)** 3,456,321

2. Identify the place value of the digit 2 in the following:
(a) 2,785,901 **(b)** 45,923

3. Write 56,109,110 in words.

4. Write 61,201 in words.

5. Write one million, two hundred sixty-five thousand, four hundred one in standard notation.

6. Write thirty-two thousand, three hundred twenty-one in standard notation.

7. Round the following to the nearest tens place.
(a) 36 **(b)** 74 **(c)** 23

8. Round the following to the indicated places.
(a) 65,763 to nearest thousands
(b) 7432 to nearest hundreds
(c) 4499 to nearest thousands

9. Round the following to the indicated places.
(a) 29,656 to nearest thousands
(b) 4945 to nearest hundreds
(c) 9652 to nearest thousands
(d) 49,345 to nearest thousands
(e) 194,745 to nearest ten thousands
(f) 9443 to nearest thousands

10. Round the whole number to a number with one nonzero digit.
(a) 98 **(b)** 94 **(c)** 25,786
(d) 34,786

1-2

11. Add the following. Use a calculator as directed by your instructor.
(a) $6 + 9 + 3 + 5$
(b) $5 + 1 + 6 + 3 + 3$

(c)
$$\begin{array}{r} 8 \\ 5 \\ 3 \\ 6 \\ 2 \\ +\ 4 \\ \hline \end{array}$$

(d)
$$\begin{array}{r} 7 \\ 4 \\ 3 \\ 2 \\ 5 \\ +\ 4 \\ \hline \end{array}$$

12. A collector has 3 African coins, 7 Canadian coins, 4 German coins, and 6 U.S. coins to sell. How many coins does the collector have for sale?

13. A shopper checked out of a grocery store with 2 cans of soup, 4 cans of green beans, 6 cans of carrots, and 8 cans of mixed vegetables. How many cans of food did the shopper buy?

14. A vendor at a flea market sold a lamp for \$6, an old camera for \$9, and a framed picture for \$3. How much did the vendor receive for the items sold?

15. If a card player made 5 points for a king, 9 points for an ace, 4 points for a jack, 1 point for a trey, and 1 point for a deuce, how many points did the player make in all?

16. Add the following.

(a)
$$\begin{array}{r} 3456 \\ +\ 2147 \\ \hline \end{array}$$

(b)
$$\begin{array}{r} 12,467 \\ +\ 24,378 \\ \hline \end{array}$$

(c)
$$\begin{array}{r} 23,609 \\ 2,200 \\ 76 \\ +\quad 124 \\ \hline \end{array}$$

(d)
$$\begin{array}{r} 43,045 \\ 5,047 \\ 87 \\ +\quad 213 \\ \hline \end{array}$$

17. Estimate by rounding to thousands. Then find the exact answer. Check the answer.

(a)
$$\begin{array}{r} \$16,742 \\ +\ 12,349 \\ \hline \end{array}$$

(b)
$$\begin{array}{r} 17,402 \\ +\ 18,646 \\ \hline \end{array}$$

18. A do-it-yourself project requires \$57 for concrete, \$74 for fence posts, and \$174 for fence boards. Estimate the cost by rounding to numbers with one nonzero digit. Then find the exact cost. Check the answer.

1-3

19. Subtract the following.
(a) $8 - 5$ (b) $4 - 2$
(c) $9 - 7$ (d) $5 - 0$
(e) $8 - 3 - 2 - 3$ (f) $8 - 3 - 2 - 1$
(g) $345 - 201$ (h) $13{,}342 - 1{,}202$

20. Estimate by rounding to hundreds. Then find the exact answer. Check the answer.
(a) $12{,}346 - 4{,}468$
(b) $3495 - 3090$
(c) $6767 - 478$

21. A door-to-door salesperson started out with 24 sets of hair brushes. By the afternoon the salesperson had 7 sets left. How many sets were sold?

22. For a moving sale, a family sold a sofa for $75 and a table for $25. If a newspaper advertisement announcing the sale cost $12, how much did the family clear on the two items sold?

23. Planning a vacation, a family selected a scenic route covering 653 miles and a direct route covering 463 miles. Estimate the difference by converting to round numbers. Then find the exact answer. Check the answer.

24. A new foreign car costs $12,677 and a comparable American car costs $11,859. Estimate the difference by rounding to thousands. Then find the exact answer. Check the answer.

1-4

Multiply the following:

25. $2 \times 6 \times 7$
26. $6 \times 3 \times 2 \times 4$
27. 127×9
28. 76×5
29. 305×45
30. $236 \cdot 244$
31. $12{,}407(270)$
32. $527 * 342$
33. $56{,}002 \times 7040$

34. A college bookstore sold 327 American history textbooks for $39 each. How much did the bookstore receive for the books?

35. Enhanced 101-key computer keyboards sell through a mail-order supplier for $67 each. How much would a business pay for 21 keyboards?

36. An automotive tire dealer runs a special on heavy-duty, deluxe whitewall truck tires. If the dealer sold 105 tires for $112 each, how much did the dealer take in on the sale?

37. If a wholesaler ordered 144 bare-bones computers for $305 each, how much did the dealer pay for the order?

Perform the indicated operations.

38. $4(3 + 1)$
39. $5(6 - 2)$
40. $2(9 - 4)$
41. $6(3 + 7)$
42. $3(5 - 3)$
43. $8(7 + 3)$

44. What number is obtained if the sum of 5 and 2 is multiplied by 6? Use the distributive property.

45. A VGA monitor for a computer sells for $687, and an EGA monitor for a computer sells for $523. An RGB monitor for a computer sells for two times the difference in cost between the VGA and EGA monitors. How much does an RGB monitor cost? Use the distributive property.

46. A luxury automotive dealer pays a sound system installer $33 per hour. Estimate by rounding to a number with one nonzero digit how much the sound system installer was paid for 37 hours of work. Then find the exact answer. Check the answer.

47. A worker was offered a job paying $365 per week. If the worker takes the job for 36 weeks, how much will the worker earn? Estimate by rounding to tens. Then find the exact answer. Check the answer.

48. Find the area of a field that is 234 feet long by 123 feet wide. Estimate the area by rounding to one nonzero digit. Then find the exact answer. Check the answer. Express the area in square feet ($A = l \times w$).

49. A parcel of land measures 1940 feet by 620 feet. Estimate the area by rounding to hundreds. Then find the exact area. Check the answer. Express the area in square feet ($A = l \times w$).

1-5

50. Express 3 divides into 4 using the following symbols.
(a) ÷ (b) $\overline{)\quad}$ (c) ——

51. Express 5 divided by 3 using the following symbols.
(a) ÷ (b) $\overline{)\quad}$ (c) ——

Divide the following:

52. $8 \div 8$ **53.** $1 \div 1$ **54.** $0 \div 3$
55. $7 \div 0$ **56.** $4 \div 1$ **57.** $843 \div 16$
58. $2925 \div 36$ **59.** $325 \div 25$ **60.** $3648 \div 6$
61. $30{,}126 \div 15$ **62.** $10{,}160 \div 20$

63. A group of 27 volunteers will seek contributions to send a first-grade class to the circus. The group leader has 632 envelopes for the collection. If the envelopes are divided equally among the 27 volunteers, how many will each receive? How many will be left over?

64. A school marching band is in a formation of 7 rows, each with the same number of students. If the band has 56 members, how many are in each row?

65. Divide by short division: $467 \div 8$.

66. Divide by short division: $7032 \div 9$.

Estimate and check the exact answer.

67. $475 \div 86$ **68.** $3516 \div 104$ **69.** $23{,}405 \div 185$

70. If 27 volunteers collected \$287 to send a first-grade class to the circus, how much was collected per volunteer? Estimate the answer, then find and check exact answer.

71. A class of 16 students made 768 cookies for a school open house. How many cookies did each student make, assuming each made the same number? Estimate the answer, and then find and check the exact answer.

1-6

72. Give the base and exponent of each expression. Then simplify each.
(a) 7^3 **(b)** 2^4 **(c)** 8^4

73. Give the base and exponent of each expression. Then simplify each.
(a) 5^6 **(b)** 12^2 **(c)** 10^6

74. Simplify the following expressions:
(a) 9^1 **(b)** 35^1 **(c)** 1^1

75. Express the following with an exponent of 1.
(a) 904 **(b)** 76 **(c)** 3

76. Simplify the following:
(a) 17^0 **(b)** 1^0 **(c)** 8^0 **(d)** 149^0

77. Find the value of the following:
(a) 2^2 **(b)** 13^2 **(c)** 7^2 **(d)** 12^2

78. Evaluate the following:
(a) 1^2 **(b)** 125^2 **(c)** 56^2 **(d)** 21^2

79. Find the square root of the following:
(a) $\sqrt{2500}$ **(b)** $\sqrt{144}$ **(c)** $\sqrt{289}$
(d) $\sqrt{81}$

80. Express the following as powers of 10.
(a) 10 **(b)** 1000 **(c)** 10,000
(d) 100,000

81. Multiply the following by using powers of 10.
(a) 3×100 **(b)** $75 \times 10{,}000$
(c) 22×1000 **(d)** 5×100

82. Divide the following by using powers of 10.
(a) $700 \div 100$ (b) $4056 \div 1000$
(c) $605 \div 100$ (d) $23{,}079 \div 10{,}000$
(e) $44{,}582 \div 1000$

1-7

Evaluate the following:

83. $2 + (3 + 6) \div 3$ **84.** $4^2 \times (12 - 7) - 8 + 3$ **85.** $8^2 - (3 - 1) \times 5$

86. $4 + 5 - 2 \times 3$ **87.** $4 + \frac{8}{2} \times 2$ **88.** $3 \times 4 \times \sqrt{16} - 6^2$

89. $\sqrt{100} \times (4 - 2) + 8$ **90.** $5^3 - \sqrt{81} \times (2 + 1)$. **91.** $2^4 \div 2 - \sqrt{9}$

92. $\frac{\sqrt{36}}{3} + 9 - 2$

1-8

Find all the factors for each of the following numbers:

93. 72 **94.** 98
95. 21 **96.** 40

Find five multiples of the given numbers.

97. 15 **98.** 9

Are the following numbers divisible by the given number? Explain.

99. 153 by 3

100. 8234 by 4

1-9

Find the prime factorization of each number. Write in factored form and then in exponential notation.

101. 44

102. 128

103. 216

104. 98

List all pairs of factors for each number. Write in order from smallest to largest.

105. 48

106. 50

107. 51

108. 63

1-10

Find the least common multiple of the following numbers:

109. 18 and 40

110. 12 and 18

111. 12, 18, and 30

112. 6, 10, and 12

Find the greatest common factor of the following numbers:

113. 10 and 12

114. 12 and 18

115. 12, 18, and 30

116. 4, 9, and 16

CHALLENGE PROBLEMS

117. You are charged with designing a playground that has 2,160 square yards of space. If the playground is to be rectangular, determine how long and how wide it should be if one of your pieces of equipment requires a space at least 15 yards long.

118. Of all the possible ways to design the playground in Problem 117, which rectangle would require the least amount of fencing?

CONCEPTS ANALYSIS

1. Addition and subtraction are inverse operations. Write the following addition problem as a subtraction problem and find the value of the number represented by the letter n. $12 + n = 17$.

2. Multiplication and division are inverse operations. Write the following multiplication problem as a division problem and find the value of the number represented by the letter n. $5 \times n = 45$.

3. Squaring and square roots are inverse operations. Write the following square root as a squaring problem and find the value of the number represented by the letter n. $\sqrt{n} = 6$.

4. Give an example illustrating that subtraction is not associative.

5. Give an example illustrating that division is not commutative.

6. What is the difference between the set of whole numbers and the set of natural numbers? Is every natural number also a whole number? Is every whole number also a natural number?

7. Give the steps in the order of operations.

8. Explain the difference between a prime number and a composite number.

9. Explain the difference between the least common multiple (LCM) and the greatest common factor (GCF).

10. Expanding your knowledge of divisibility, write a rule to determine if a number is divisible by 12.

Find and explain the mistake in the following. Rework each problem correctly.

11. $25 + 49 =$

$$\begin{array}{r} 2\ 5 \\ +\ 4\ 9 \\ \hline 614 \end{array}$$

12. $2 + 5(4) =$

$7(4) = 28$

13. $\sqrt{9} = 81$

14. Prime factors of 40 are $2^2 \times 10$.

15. The GCF of 12 and 15 is 60.

CHAPTER SUMMARY

Objectives	What to Remember	Examples
Section 1-1		
1. Identify place values in whole numbers.	Each digit has a place value. Place values are grouped in periods of three digits. Place value names can be found in Figure 1-1.	Identify the place value of each of the digits in the number 3628: 3 thousands 6 hundreds 2 tens 8 ones
2. Write whole numbers in words and standard notation.	Standard notation is the usual form of a number or the number written in digits using place value.	The number 345,230 is written in standard form.
	Read from the left the numbers in each period and the period name. The period name *units* is not read.	Write 3,462 in words. Three thousand, four hundred sixty-two.
3. Round a whole number to a place value.	**1.** Locate rounding place. **2.** Examine digit to the right. **3.** Round down if digit is less than 5. **4.** Round up if digit is 5 or more.	Round 3,624 to the nearest tens place. 2 is in the tens place, 4 is the digit to the right, 4 is less than 5, so round down. The rounded value is 3,620.
4. Round a whole number to a number with one nonzero digit.	**1.** Locate first nonzero digit on the left. **2.** Examine the digit to the right. **3.** Round down if digit is less than 5. **4.** Round up if digit is 5 or more.	Round 4,789 to one nonzero digit. 4 is the first nonzero digit on the left, 7 is the digit to the right, 7 is 5 or more, so round up. The rounded value is 5,000.
Section 1-2		
1. Add single-digit numbers.	Addition is a binary operation that is commutative and associative.	$5 + 7 = 7 + 5 = 12$; $(3 + 2) + 6 = 3 + (2 + 6) = 11$
	Zero property of addition: $0 + n = n + 0 = n$.	$0 + 3 = 3 + 0 = 3$
2. Add numbers of two or more digits.	Arrange numbers in columns of like places. Add each column beginning with the ones place. Carry when necessary.	$\begin{array}{r} 4824 \\ +\ 745 \\ \hline 5569 \end{array}$
3. Estimate and check addition.	Estimate by rounding the addends before finding the sum. Check addition by adding a second time.	Estimate the sum by rounding to the nearest hundred: $483 + 723$ $500 + 700 = 1200$. Exact sum: 1206.
4. Add using a calculator.	Enter each addend, followed by the addition operation key. After the last addend, press = or EXE to display the sum.	Add: $483 + 8492 + 3823$. Basic or scientific calculator steps: 483 + 8492 + 3823 = Graphing calculator steps: 483 + 8492 + 3823 EXE. Sum = 12,798.
Section 1-3		
1. Subtract single-digit numbers.	Subtraction is a binary operation that is *not* commutative or associative. Addition and subtraction are inverse operations.	If $5 + 4 = 9$, then $9 - 5 = 4$ or $9 - 4 = 5$.
	Zero property of subtraction: $n - 0 = n$.	$7 - 0 = 7$
2. Subtract numbers with two or more digits.	Arrange numbers in columns of like places. Subtract each column beginning with the ones place. Borrow when necessary.	$\begin{array}{r} 4227 \\ -\ 745 \\ \hline 3482 \end{array}$

3. Estimate and check subtraction.	Estimate by rounding the minuend and subtrahend before finding the difference. Check subtraction by adding the difference and the subtrahend. The check should equal the minuend.	Estimate the difference by rounding to the nearest hundred: 783 − 423 800 − 400 = 400. Exact difference: 783 − 423 = 360. Check: 360 + 423 = 783.
4. Subtract using a calculator.	Enter the minuend, followed by the subtraction operation key. Enter the subtrahend and press = or EXE.	Subtract: 2592 − 1474. Basic or scientific calculator steps: 2592 − 1474 = Graphing calculator steps: 2592 − 1474 EXE. Difference = 1118.
Section 1-4		
1. Multiply single-digit factors.	Multiplication is a binary operation that is commutative and associative. Zero property of multiplication: $n \times 0 = 0 \times n = 0$.	$5 \times 7 = 7 \times 5 = 35$; $(3 \times 2) \times 6 = 3 \times (2 \times 6) = 36$: $5 \times 0 = 0 \times 5 = 0$
2. Multiply factors of more than one digit.	Arrange numbers in columns of like places. Multiply the multiplicand by each digit in the multiplier. Add the partial products.	$\begin{array}{r} 259 \\ \times\ 23 \\ \hline 777 \\ 518 \\ \hline 5957 \end{array}$
3. Apply the distributive property.	$a(b + c) = ab + ac$ $a(b - c) = ab - ac$	$3(5 + 6) = 3(5) + 3(6)$ $3(11) = 15 + 18$ $33 = 33$
4. Estimate and check multiplication.	Estimate by rounding the factors before finding the product. Check multiplication by multiplying a second time.	Estimate the product by rounding to one nonzero digit: 483 × 72; 500 × 70 = 35,000
5. Multiply using a calculator.	Enter each factor followed by the multiplication operation key. After the last factor, press = or EXE to display the product.	Multiply: 48 × 42 × 23. Basic or scientific calculator steps: 48 × 42 × 23 = Graphing calculator steps: 48 × 42 × 23 EXE. Product = 46,368.
Section 1-5		
1. Use various symbols to indicate division.	The division a divided by b can be written as $a \div b$, $b\overline{)a}$, $\frac{a}{b}$, or a/b The divisor, b, cannot be zero.	Write 12 divided by 4 in four ways. $12 \div 4$, $4\overline{)12}$, $\frac{12}{4}$, $12/4$
2. Divide whole numbers.	Align the numbers properly in long division.	$\begin{array}{r} 20\ \text{R}15 \\ 23\overline{)475} \\ 46 \\ \hline 15 \\ 0 \\ \hline 15 \end{array}$
3. Estimate and check division.	To estimate division, round the dividend and divisor before finding the quotient. Find the first digit of the quotient and add a zero for each remaining digit in the dividend. To check division, multiply the quotient by the divisor and add the remainder.	Estimate 2934 ÷ 42 $\begin{array}{r} 70 \\ 40\overline{)3000} \end{array}$ Exact quotient = 69 R36 Check: 69 × 42 = 2898 2898 + 36 = 2934

4. Divide using a calculator.	Enter the dividend followed by the division operation key. Enter the divisor and press = or EXE. To find remainder, subtract part of quotient left of decimal. Multiply times divisor.	Divide: 295 ÷ 15. Basic or scientific calculator steps: 295 ÷ 15 = Graphing calculator steps: 295 ÷ 15 EXE. Quotient = 19.66666667. 19.66666667 − 19 = 0.666666667; 0.666666667 × 15 = 9.999999999 or 10; 295 ÷ 15 = 19 R10
Section 1-6		
1. Simplify expressions containing exponents.	To simplify an exponential expression, use the base as a factor the number of times indicated by the exponent.	$5^3 = 5 \times 5 \times 5 = 125$
	$a^1 = a$, for any base a	$7^1 = 7$
	$a^0 = 1$, for any nonzero base	$9^0 = 1$
2. Square numbers and find square roots of numbers.	Squaring and finding square roots are inverse operations.	$7^2 = 49$ $\sqrt{49} = 7$
3. Use a calculator to find powers and roots.	The special operation keys x^2, x^3, and $\sqrt{\ }$ perform the respective operations with a minimum of keystrokes.	7^2, 6^5 Scientific: 7 x^2; Display: 49
	The general power key x^y can be used to raise a number to any power.	6 x^y 5 = Display: 7776. Graphing: 7 x^2 EXE. Display: 7^2 49. 6 x^y 5 EXE. Display: 6 x^y 5 7776.
4. Use powers of 10 to multiply and divide.	To multiply by a power of 10, add to the whole number as many zeros as indicated by the exponent of 10 or as many zeros as in the standard notation of the power of 10.	$27 \times 10^3 = 27{,}000$ $18 \times 100 = 1800$
	To divide by a power of 10, drop from the whole number as many zeros as indicated by the exponent of 10 or as many zeros as in the standard notation of the power of 10.	$400 \div 10^2 = 4$ $23{,}000 \div 10^3 = 23$
Section 1-7		
1. Apply the order of operations to problems containing several operations.	Order of operations: **1.** Parentheses or groupings, innermost first. **2.** Exponential operations and roots from left to right. **3.** Multiplication and division from left to right. **4.** Addition and subtraction from left to right.	$3^2 + 5(6 - 4) \div 2$ $3^2 + 5(2) \div 2$ $9 + 5(2) \div 2$ $9 + 10 \div 2$ $9 + 5$ 14

2. Use the calculator to perform a series of operations.	Both the scientific and graphing calculators will perform the proper order of operations. Using the scientific calculator, the times sign must be used with multiplication. With the graphing calculator, the times sign is not required if a parenthesis follows.	$3^2 + 5(6 - 4) \div 2$. Scientific: 3 x^2 + 5 × (6 − 4) ÷ 2 = . Display: 14. Graphing: 3 x^2 + 5(6 − 4) ÷ 2 EXE. Display: $3^2 + 5(6 - 4) \div 2$ 14.
Section 1-8		
1. Find all factorizations of a natural number.	Start with the pair of factors, using the number and 1. Examine other natural numbers in order to find other pairs of factors, if they exist. Continue until the factors begin to repeat, but are in the opposite order.	Find all the factors of 24: 1×24, 2×12, 3×8, 4×6; 5 is not a factor, 6×4 is a repeat. Factors are 1, 2, 3, 4, 6, 8, 12, 24.
2. Find multiples of a natural number.	Multiply a number by a natural number to find a multiple.	Find the first five multiples of 7: $7 \times 1 = 7$, $7 \times 2 = 14$, $7 \times 3 = 21$, $7 \times 4 = 28$, $7 \times 5 = 35$. The first five multiples of 7 are 7, 14, 21, 28, and 35.
3. Determine the divisibility of a number.	Apply the appropriate divisibility test or perform the division to see if the division has no remainder.	Is 2195 divisible by 6? No, the number should be even and the sum of the digits should be divisible by 3.
Section 1-9		
1. Factor prime and composite numbers and identify factors of composite numbers.	Identify the factors of the number. Numbers whose only factors are the number and 1 are prime. Other numbers are composite.	Is 18 prime or composite? Factors of 18 are 1, 2, 3, 6, 9, 18; therefore, 18 is composite.
2. Determine the prime factorization of composite numbers.	Find the smallest prime factor of the composite number. Retain prime factors and continue factoring composite factors until all factors are prime.	Find the prime factorization of 28: $28 = 2 \times 14$ $= 2 \times 2 \times 7$ or $2^2 \times 7$
Section 1-10		
1. Find the least common multiple of two or more numbers.	**1.** Find the prime factorization of each number in exponential notation. **2.** The LCM includes each prime factor appearing in any of the numbers the largest number of times that it appears in any factor. **3.** Write the result in standard notation.	Find the LCM for 12, 15, and 30: $12 = 2^2 \times 3$, $15 = 3 \times 5$, $30 = 2 \times 3 \times 5$ LCM $= 2^2 \times 3 \times 5$ or 60
2. Find the greatest common factor of two or more numbers.	**1.** List the prime factorization of each number in exponential notation. **2.** The GCF includes factors that are common to each number. **3.** Write the result in standard notation.	Find the GCF of 12, 15, 30: $12 = 2^2 \times 3$, $15 = 3 \times 5$, $30 = 2 \times 3 \times 5$ GCF = 3

WORDS TO KNOW

place value (p. 1)
whole number digits (p. 1)
decimal number system (p. 2)
periods (p. 2)
standard notation (p. 4)
round (p. 4)

nonzero digit (p. 5)
sum (p. 6)
addition (p. 6)
addends (p. 6)
total (p. 6)
commutative property of addition (p. 6)
associative property of addition (p. 7)
binary operation (p. 7)
carry (p. 8)
subtraction (p. 12)
inverse operations (p. 12)
inverse (p. 12)
minuend (p. 12)
subtrahend (p. 12)
difference (p. 12)
multiplication (p. 16)
multiplicand (p. 16)
multiplier (p. 16)
factors (p. 17)
product (p. 17)
associative property of multiplication (p. 17)
commutative property of multiplication (p. 17)
partial product (p. 18)
distributive property (p. 19)
area (p. 21)
formula (p. 22)
square measures (p. 22)
division (p. 25)
divisor (p. 25)
dividend (p. 25)
quotient (p. 25)
exponent (p. 34)
simplify (p. 34)
exponential notation (p. 34)
standard notation (p. 34)
power (p. 34)
square (p. 35)
squared (p. 35)
perfect square (p. 35)
radical sign (p. 36)
square root (p. 36)
radical expression (p. 36)
cube (p. 37)
cubed (p. 37)
order of operations (p. 41)
natural numbers (p. 44)
factorizations (p. 44)
multiple (p. 45)
even numbers (p. 45)
odd numbers (p. 45)
divisible (p. 45)
prime number (p. 47)
composite number (p. 47)
prime factorization (p. 48)
prime factors (p. 48)
least common multiple (p. 49)
common factor (p. 49)
greatest common factor (p. 50)

TRIAL TEST

1. Write 5,030,102 in words.
2. Write three hundred twenty-four thousand, five hundred twenty in standard notation.
3. Round 2,743 to hundreds.
4. Round 34,988 to a number with one nonzero digit.

Perform the indicated operations.

5. $37 + 158 + 764 + 48$
6. \$61,532 − \$47,245
7. \$13,207 × 702
8. \$25,600 ÷ 12
9. $3^2 + 5^3$
10. 46×10^3
11. $3 \times 6^2 - 4 \div 2$
12. $5^3 - (3 + 2) \times \sqrt{9}$
13. If a stereo costs \$495 and a stereo cabinet costs \$269, use numbers with one nonzero digit to estimate the total cost. Find the exact cost.
14. If a man has a bill for \$165 and his pay check is \$475, estimate how much of the pay check is left after paying the bill by rounding to tens. Find the exact amount left.
15. A corporation wants to buy 45 VCRs for \$335 each. Use numbers with one nonzero digit to estimate the total cost. Find the exact cost.
16. A softball coach paid \$126 for 9 pizzas for a party after a successful season. Estimate the cost of each pizza by using numbers with one nonzero digit. Find the exact cost per pizza.
17. How many books did a bookstore have in stock if it had 36 dictionaries, 19 children's books, and 127 novels?
18. A mathematics professor promised to give 2 extra points for every time a student worked a set of self-study exercises. If one student worked 17 sets of self-study exercises, how many points should have been given?
19. Give all factorizations of 42.
20. Which of the numbers 2572, 3522, and 5493 are divisible by 6?
21. Give the prime factorization of 84.
22. Select the prime numbers from the following: 2, 5, 8, 15, 17, 27, and 39.
23. Find the least common multiple for 12, 14, and 15.
24. Find the greatest common factor for 12, 14, and 15.
25. Explain the difference between the commutative property of addition and the associative property of addition. Given a numerical illustration to support your explanation.

CHAPTER

2 Integers

In our study of mathematics we will extend our numbers to include integers. For example, if we tried to subtract a larger number from a smaller number, such as $8 - 10$, we would need a new type of number to express the result.

2-1 INTEGERS

Learning Objectives

L.O.1. Place integers on the number line.
L.O.2. Compare integers.
L.O.3. Find the absolute value of integers.
L.O.4. Find the opposites of integers.

Integers can be placed on a horizontal line, much like degrees on the Fahrenheit thermometer scale. On the thermometer, all degree readings above zero are "*positive*." A temperature of 72 is read as "72 degrees" and represents 72 units above zero. All degree readings below zero are "*negative*." A temperature of 8 degrees below zero is read as "8 degrees below zero" or "minus 8 degrees" and represents 8 units below zero. Zero is neutral, neither "positive" nor "negative." When numbers are placed on a horizontal line with zero at the origin, that line is called a *number line*.

L.O.1. Place integers on the number line.

The Number Line

All numbers extending to the right of zero on the number line are positive. All numbers extending to the left of zero are negative.

While the number line includes types of numbers such as fractions and their decimal equivalents, for now we will discuss natural numbers, their negative counterparts, and zero.

The space or distance between each number illustrated in Fig. 2-1 is the same and represents one unit.

■ **DEFINITION 2-1.1 Integers.** The set of **integers** includes the natural numbers, their negative counterparts, and zero. The natural numbers are located to the right of zero and are positive. Their opposites are located to the left of zero and are negative.

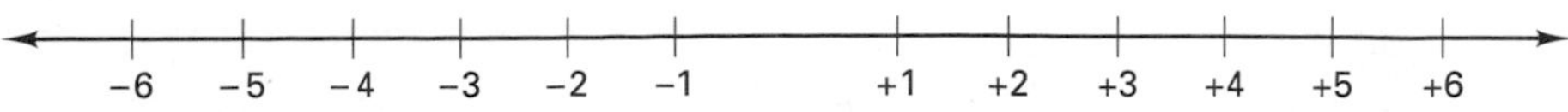

The integral number line

Figure 2-1

Normally, the positive sign (+) is not written before a positive number, but it can be written for emphasis or clarity. Thus, +2 and 2 mean the same. They may be read as "positive 2" or "2." On the other hand, the negative numbers to the left of zero are always written with a negative sign (−). Thus, −2 is read as "negative 2" or "minus 2."

L.O.2. Compare integers.

Comparing Integers

As we move from *left to right* on the number line (Fig. 2-2), the numbers *increase* in value. As we move from *right to left*, the numbers *decrease* in value. Therefore, −5 is less than −4 because −5 is to the *left* of −4 on the number line. Likewise, 0 is greater than −1 because 0 is to the *right* of −1. Also, −3 is less than 5 because −3 is to the *left* of 5. In the same way, 2 is greater than −3 because 2 is to the *right* of −3.

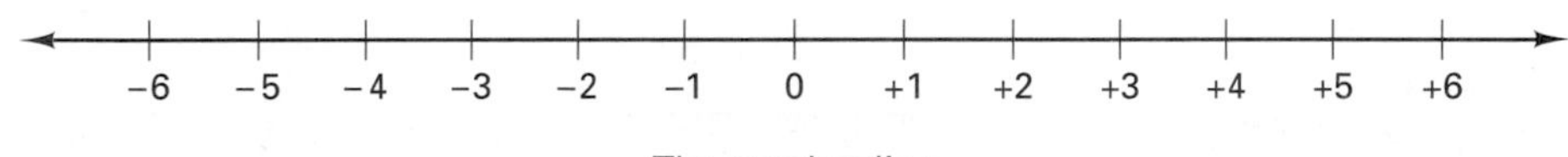

The number line

Figure 2-2

The symbol $>$ means *is greater than.* That is, if we say that 2 is greater than 1, we can represent this fact by writing $2 > 1$ (read "2 is greater than 1"). To express the idea that −1 is less than 0, we can use the symbol $<$ for *is less than.* Therefore, we can write $-1 < 0$. The expression is read "−1 is less than 0." The use of $>$ and $<$ saves time in expressing these relationships in writing.

Tips and Traps

One way to remember that $>$ means "is greater than" is to remember that numbers get larger as we move to the right on the number line. The arrow on the right end of the number line points in the same direction as the "is greater than" symbol.

The left arrow on the number line points in the same direction as the "is less than" symbol.

L.O.3. Find the absolute value of integers.

Absolute Value

Even though −3 and 3 are not equal, the two numbers are both the same distance from zero. That is, both are 3 spaces or units away from zero and are said to have the same absolute value.

■ **DEFINITION 2-1.2 Absolute Value.** *Absolute value* is the distance a number is from zero without regard to direction. The symbol | | is used to indicate absolute value.

$$|5| = 5, \qquad |-5| = 5, \qquad |5| = |-5|$$

L.O.4. Find the opposites of integers.

Opposites

Every number, except zero, has an opposite number.

■ **DEFINITION 2-1.3 Opposites.** Opposites are two numbers that have the same absolute value (distance from zero), but have opposite signs.

Tips and Traps

Integers and other signed numbers tell us two things. In the integer -8, the "$-$" sign tells us the number is on the *left* of zero and it is referred to as the *directional sign* of the number. The 8 tells us how many units the number is from zero and is referred to as the *absolute value* of the number.

The opposite of $+5$ is -5.

SELF-STUDY EXERCISES 2-1

L.O.1. Place the following integers on the number line of Fig. 2-3.

1. 1 **2.** -1 **3.** 3
4. -6 **5.** 2 **6.** 4
7. -2 **8.** 6 **9.** -5
10. -3 **11.** -4 **12.** 5
13. 8 **14.** -11 **15.** 12

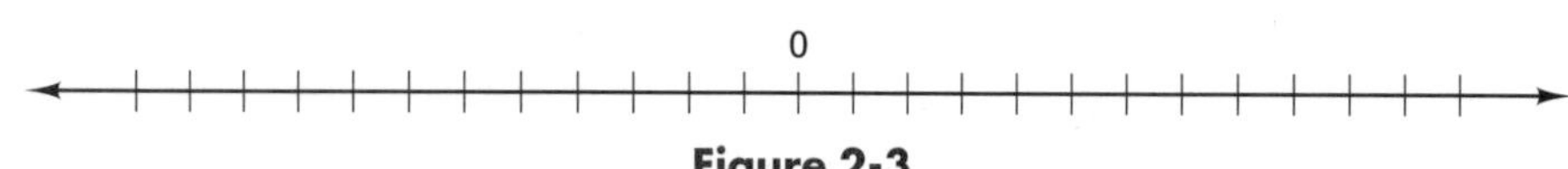

Figure 2-3

L.O.2. Use the symbols $>$ and $<$ to indicate which number is greater than or less than the other.

16. 0 ____ 1 **17.** -1 ____ 1 **18.** 4 ____ -3
19. 2 ____ -2 **20.** 60 ____ -60 **21.** -9 ____ -12
22. -8 ____ -9 **23.** 2 ____ 3 **24.** -16 ____ -15
25. -20 ____ 0

L.O.3. Give the value of each of the following numbers.

26. $|-8|$ **27.** 0 **28.** $|86|$
29. $|-32|$ **30.** $|-74|$ **31.** -1
32. $|32|$ **33.** -74 **34.** $|-2|$
35. 49

L.O.4. Give the opposite of the following numbers:

36. -12 **37.** 58 **38.** 0
39. 529 **40.** $-3,825$

2-2 ADDING INTEGERS

Learning Objectives

L.O.1. Add integers with like signs.
L.O.2. Add integers with unlike signs.
L.O.3. Add numbers that are opposites.
L.O.4. Add zero to an integer.

So far we have looked at integers and how they compare in value to one another. Now we will add integers. As for whole numbers, the numbers being added are *addends* or *terms*, and the answer is the *sum*.

L.O.1. Add integers with like signs.

Adding Integers with Like Signs

We will use the number line to illustrate addition of integers. Suppose that we wanted to add 2 plus 3 (two positive numbers) on the number line. To do this, we would start at zero and move two spaces to the *right;* then we would move three more spaces to the *right*. Notice the arrows in Fig. 2-4.

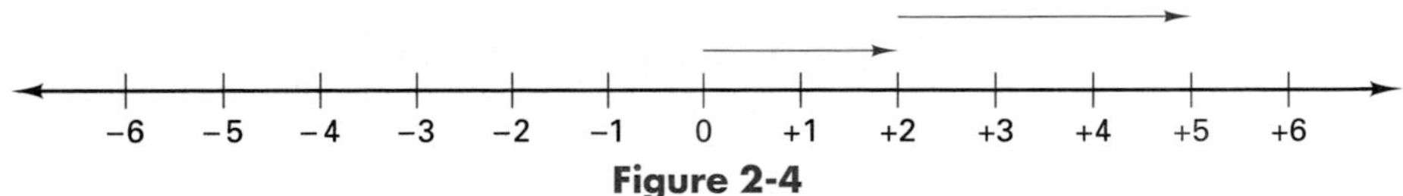

Figure 2-4

The first move (two spaces to the right) takes us to $+2$. The second move (three more spaces to the right) takes us to $+5$. Therefore, $2 + 3 = 5$.

On the other hand, to add negative numbers, such as -3 plus -1, we would start at zero and move three spaces to the *left*. Then we would move one more space to the *left*. Notice the arrows in Fig. 2-5.

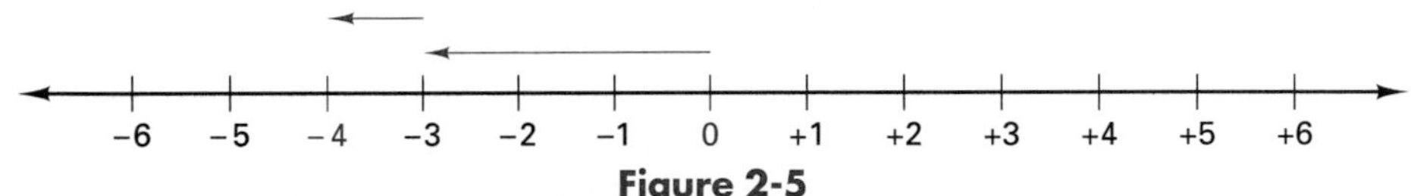

Figure 2-5

The first move (three spaces to the left) takes us to -3. The second move (one more space to the left) takes us to -4. Therefore, $-3 + (-1) = -4$.

Notice the parentheses enclosing the negative number. Parentheses may be used around an integer when its sign is expressed and an operational sign meaning add $(+)$ or subtract $(-)$ precedes the signed number. To avoid writing $-3 + -1$, where the $+$ and $-$ signs can cause confusion, the parentheses separate the signs for clarity, such as $6 + (-2)$ and $-1 + (-3)$.

Rather than use the number line to add the two positive numbers or the two negative numbers, we can add their absolute values and keep the same signs. In $2 + 3 = 5$, the answer is positive, just as the 2 and 3 are positive. And in $-3 + (-1) = -4$, the answer is negative, just as the -3 and -1 are negative.

Rule 2-2.1 ***To add integers having* like *(the same) signs:***

1. Add their absolute values.
2. Give the answer the common sign.

Tips and Traps

In this chapter we will give rules for adding, subtracting, multiplying, and dividing integers. These same rules can be extended to other *signed numbers*, such as fractions and decimals, as illustrated in Chapters 3 and 4.

EXAMPLE 2-2.1 Add $4 + 6 + 8$.

Solution

$$\underbrace{4 + 6}_{10} + 8 = 18$$

(All numbers have same sign, so add absolute values and keep the common sign.) ■

EXAMPLE 2-2.2 Add $-8 + (-7) + (-1)$.

Solution

$$-8 + \underbrace{(-7) + (-1)}$$
$$-8 + \quad (-8) \quad = -16$$

(All numbers have same sign, so add absolute values and keep the common sign.) ■

L.O.2. Add integers with unlike signs.

Adding Integers with Unlike Signs

Let's return to the number line to illustrate addition of integers with *different* or *unlike* signs. Let's add $4 + (-3)$. We start at zero on the number line and move four spaces to the *right*, since 4 is a positive number. Then we move from that point on the number line three spaces to the *left*, since -3 is a negative number. Notice the arrows in Fig. 2-6.

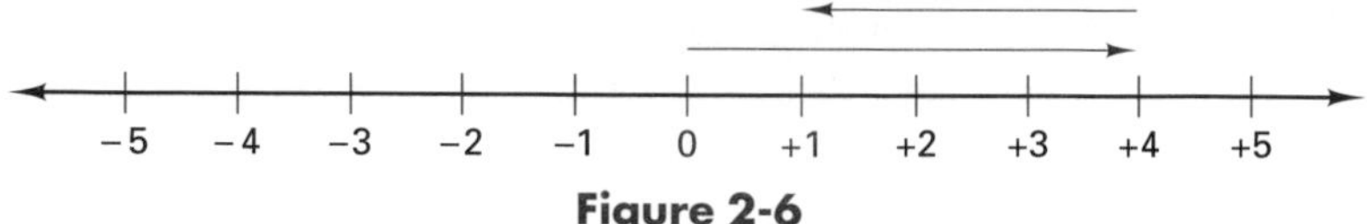

Figure 2-6

The first move (four spaces to the right) takes us to $+4$. The next move (three spaces to the left) takes us to $+1$. Therefore, $4 + (-3) = +1$.

On the other hand, let's add $-5 + 2$ on the number line. We start at zero and move five spaces to the *left*, since -5 is negative. Then from that point we move two spaces to the *right*, since 2 is positive. Notice the arrows in Fig. 2-7.

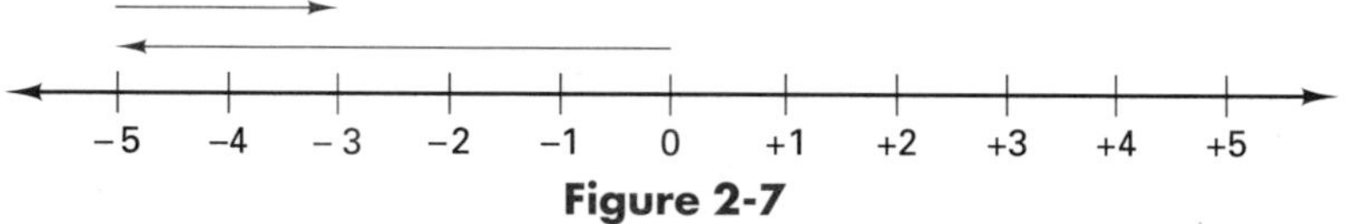

Figure 2-7

The first move (five spaces to the left) takes us to -5. The second move (two spaces to the right) takes us to -3. Therefore, $-5 + 2 = -3$.

In both problems, the absolute value of the answer is the *difference* between the absolute values of the numbers being added. In other words, we can *subtract* the smaller absolute value from the larger absolute value to get the absolute value of the answer. Thus, to add $4 + (-3)$, we subtract 3 from 4 to get 1, which is the absolute value of the answer. To add $(-5) + (2)$, we subtract 2 from 5 to get 3, which is the absolute value of the answer.

We give the answer the *sign of the number with the larger absolute value.* Therefore, in the problem $4 + (-3)$, the 4 has a larger absolute value than -3. Since its sign is positive, the sign of the answer is positive. In the problem $-5 + 2$, the -5 has a larger absolute value than 2. Since its sign is negative, the sign of the answer is negative.

Rule 2-2.2 ***To add integers having* unlike *(different) signs:***

1. *Subtract* the smaller absolute value from the larger absolute value.
2. Give the answer the sign of the number with the larger absolute value.

EXAMPLE 2-2.3 Add $14 + (-6)$.

Solution

$14 + (-6)$ *(The signs are different.)*

$14 - 6 = 8$ *(Subtract smaller absolute value from larger absolute value.)*

$14 + (-6) = +8$ *(Give answer the sign of number with larger absolute value.)* ■

EXAMPLE 2-2.4 Add $20 + (-32)$.

Solution

$20 + (-32)$ *(The signs are different.)*

$32 - 20 = 12$ *(Subtract smaller absolute value from larger absolute value.)*

$20 + (-32) = -12$ *(Give answer the sign of number with larger absolute value.)* ■

When more than two numbers with different signs are to be added, as for whole numbers, they may be added in any order and grouped in any way convenient (*commutative* and *associative properties*). Refer to Definitions 1-2.1 and 1-2.2 for a review of these properties.

The rules for adding integers apply to each step in the problem. Example 2-2.5 illustrates the commutative property of addition. Regardless of what order we add two numbers, we get the same result. Example 2-2.6 illustrates the associative property of addition. Even though terms may be grouped differently, they result in the same sum.

EXAMPLE 2-2.5 Add $3 + (-2)$.

Solution

$3 + (-2) = 1$ *(Signs are different.)*

or

$(-2) + 3 = 1$ *(Change order of addends or terms.)* ■

EXAMPLE 2-2.6 Add $8 + (-7) + 2$.

Solution

$\underbrace{8 + (-7)}_{1} + 2 = 3$ *(Signs are different.)* *(8 + (−7) = 1; 1 + 2 = 3)*

or

$8 + \underbrace{(-7) + 2}_{(-5)} = 3$ *(Regroup addends or terms.)* *(−7 + 2 = −5; 8 + (−5) = 3)* ■

EXAMPLE 2-2.7 Add $-7 + 4 + (-1) + 2$.

Solution

$\underbrace{-7 + 4}_{-3} + \underbrace{(-1) + 2}_{1} = -2$ *(Group addends or terms.)*

or

$\underbrace{-7 + (-1)}_{-8} + \underbrace{4 + 2}_{6} = -2$ *(Rearrange addends; add all negatives, then all positives, then all sums.)* ■

L.O.3. Add numbers that are opposites.

Adding Opposites

Every integer has an *opposite* except zero. Opposites, also called *additive inverses*, are illustrated on the number line in Fig. 2-8.

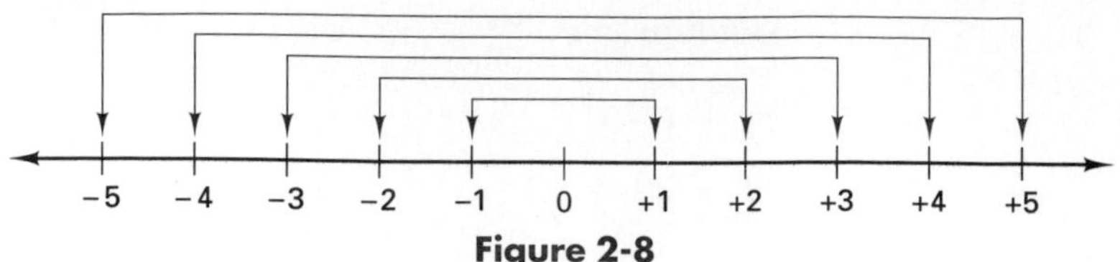

Figure 2-8

> **Rule 2-2.3** ***To add opposites, subtract their absolute values to get zero.***

EXAMPLE 2-2.8 Add $8 + (-8)$.

Solution

$8 + (-8)$ *(These are opposites.)*

$|8| - |-8|$ *(Subtract absolute values.)*

$8 - 8 = 0$ *(Zero has no sign.)* ■

L.O.4. Add zero to an integer.

Adding Zero to an Integer

Zero, since it has no sign, is also a special case in addition problems with integers. As for whole numbers, zero added to a number gives the number itself.

Thus, $12 + 0 = 12$, $-4 + 0 = -4$, and $0 + (-2) = -2$.

> **Rule 2-2.4** ***Adding zero to an integer does not change that number or its sign.***

EXAMPLE 2-2.9 Add $0 + (-6)$.

Solution

$0 + (-6) = -6$ *(Zero added to a number does not change that number.)* ■

SELF-STUDY EXERCISES 2-2

L.O.1. Add the following:

1. $7 + 10$
2. $-5 + (-8)$
3. $12 + 87$
4. $-21 + (-38)$
5. $-32 + -16$
6. $(-58) + (-103)$

L.O.2. Add the following.

7. $-4 + (+2)$
8. $-3 + 7$
9. $4 + (-6)$
10. $-4 + 6$
11. $-18 + 8$
12. $32 + (-72)$
13. $21 + (-14)$
14. $17 + (-4) + 3 + (-1)$
15. $-3 + 2 + (-7)$
16. $47 + (-82) + 2$
17. $14 + (-6) + 1$
18. $-7 + (-3) + (-1)$
19. $4 + 2 + (-3) + 10$
20. $2 + (-1) + 8$
21. $-2 + 1 + (-8) + 12$

L.O.3. Find the sum.

22. $15 + (-15)$ **23.** $-7 + 7$ **24.** $92 + (-92)$
25. $(-396) + (396)$ **26.** $503 + (-503)$

L.O.4. Find the sum.

27. $0 + 9$ **28.** $-3 + 0$ **29.** $0 + (-7)$
30. $18 + 0$ **31.** $(-8) + 0$ **32.** $0 + (-28)$

2-3 SUBTRACTING INTEGERS

Learning Objectives

L.O.1. Subtract integers.
L.O.2. Combine addition and subtraction.
L.O.3. Subtract with zero.

L.O.1. Subtract integers.

Subtracting Integers

As for whole numbers, addition and subtraction of integers are inverse operations. To illustrate the relationship between adding integers and subtracting integers, look at the following examples.

Subtracting Whole Numbers	Subtracting Integers	Adding Integers
$7 - 3 = 4$	$+7 - (+3) = 4$	$+7 + (-3) = 4$
$8 - 5 = 3$	$+8 - (+5) = 3$	$+8 + (-5) = 3$
$12 - 3 = 9$	$+12 - (+3) = 9$	$+12 + (-3) = 9$
$25 - 18 = 7$	$+25 - (+18) = 7$	$+25 + (-18) = 7$

By examining each row, we can see that each subtraction of integers can be written as an equivalent addition of the minuend and the opposite of the subtrahend.

Rule 2-3.1 *To subtract integers:*

1. Change the subtraction sign $(-)$ to an addition sign $(+)$.
2. Change the sign of the *subtrahend* (number being subtracted) to its opposite.
3. Use the appropriate rule for adding integers.

EXAMPLE 2-3.1 Subtract $8 - 6$.

Solution

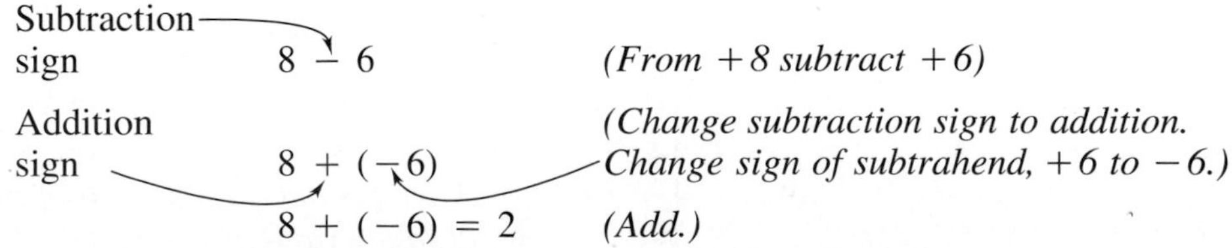

■

EXAMPLE 2-3.2 Subtract $9 - (-2)$.

Solution

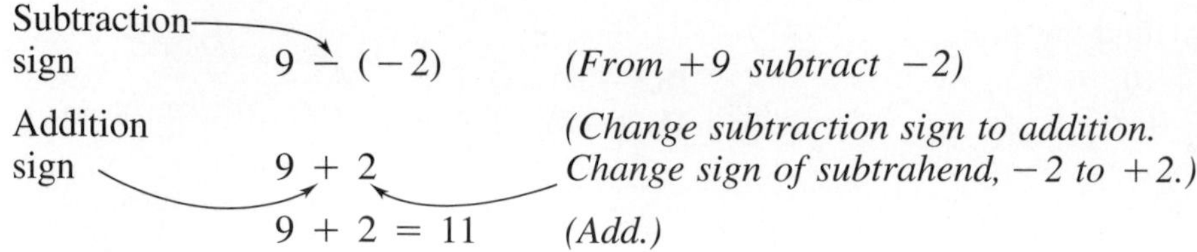

■

EXAMPLE 2-3.3 Subtract $-4 - 7$.

Solution

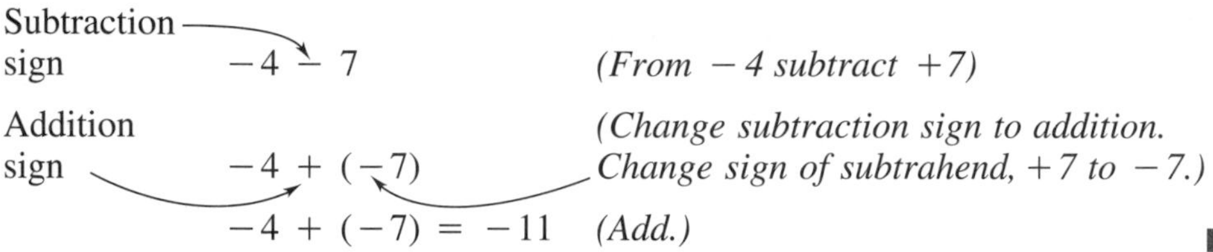

■

EXAMPLE 2-3.4 Subtract $-10 - (-3)$.

Solution

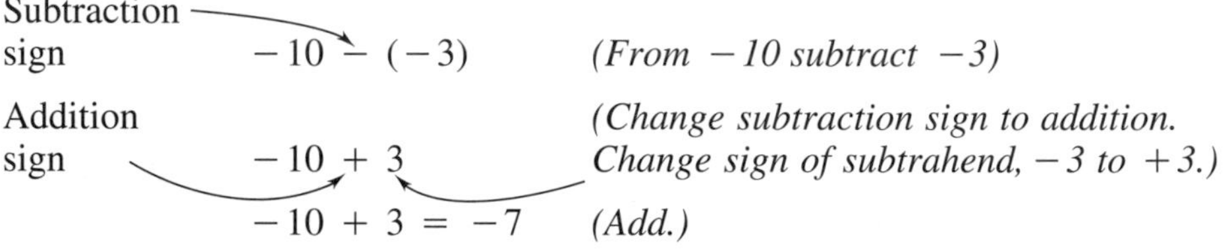

■

The sign of the first number, the *minuend*, does *not* change. The only signs to change are the subtraction sign and the sign of the subtrahend.

Since all subtractions are converted to additions, we want to develop the habit of mentally converting subtractions to additions and of interpreting problems combining subtractions and additions as only additions. In both addition and subtraction, two signs sometimes fall between two numbers. When two signs (one operational and one directional) fall between two numbers, we can combine the two signs into one sign, which will be the directional sign of the number that follows. The operation will be understood to be addition, and we apply the appropriate rule according to the signs of the numbers.

EXAMPLE 2-3.5 Subtract $7 - (+9)$.

Solution

$7 - 9$	*(Think of adding +7 and −9).*
7 plus -9	*(The signs are different.)*
$\lvert -9 \rvert - \lvert 7 \rvert = \quad 2$	*(Subtract smaller absolute value from larger absolute value.)*
7 plus $-9 = -2$	*(Give the answer the sign of the number with larger absolute value.)*

■

EXAMPLE 2-3.6 Subtract $-5 - (+7)$.

Solution

$-5 - 7$	*(Think of adding -5 and -7.)*
-5 plus -7	*(The signs are the same.)*
$\lvert -5\rvert$ plus $\lvert -7\rvert = 12$	*(Add absolute values.)*
-5 plus $-7 = -12$	*(Keep common sign.)* ■

EXAMPLE 2-3.7 Subtract $-2 - (-5)$

Solution

$-2 + 5$	*(Think of adding -2 and $+5$.)*
-2 plus 5	*(The signs are different.)*
$\lvert 5\rvert - \lvert -2\rvert = 3$	*(Subtract smaller absolute value from larger absolute value.)*
$-2 + 5 = 3$	*(Give the answer the sign of the number with larger absolute value.)* ■

EXAMPLE 2-3.8 Add $13 + (-6)$.

Solution

$13 - 6$	*(Think of adding $+13$ and -6.)*
13 plus -6	*(The signs are different.)*
$\lvert 13\rvert + \lvert -6\rvert = 7$	*(Subtract smaller absolute value from larger absolute value.)*
13 plus $-6 = 7$	*(Give the answer the sign of the number with larger absolute value.)* ■

L.O.2. Combine addition and subtraction.

Subtraction and Addition in Same Problem

Whenever numbers are written with one "+" or "−" between them, we will consider the "+" or "−" as the sign of the number that follows and apply the appropriate rule for adding signed numbers.

Thus, from now on, we will usually express addition and subtraction of signed numbers with one sign between the numbers as follows:

$$7 + 3 - 5, \quad \text{meaning} \quad +7 \text{ plus } +3 \text{ plus } -5$$

Tips and Traps

To add or subtract integers by the shortcut:

1. If two signs fall between two numbers, convert to the appropriate single sign.
2. Think of an addition or subtraction sign as the sign of the number that follows the sign.
3. Add the numbers using the appropriate rule for adding signed numbers.

EXAMPLE 2-3.9 Evaluate $8 - 2 + 1 - 9$.

Solution

$8 - 2 + 1 - 9$ *(Think of adding 8, −2, +1, and −9.)*

$\underbrace{8 \text{ plus } -2}_{6} \text{ plus } \underbrace{1 \text{ plus } -9}_{-8} = -2$ *(Group and add.)*

■

Because addition is commutative and associative we could work this problem by grouping numbers with like signs.

$$\underbrace{8 + 1}_{9} + \underbrace{-2 + (-9)}_{-11} = -2$$

L.O.3. Subtract with zero.

Zero in Subtraction

When subtracting zero from a signed number, the subtraction of zero does not change the original number or its sign. This is similar to addition of zero and a signed number.

> **Rule 2-3.2** ***Subtracting zero* from *an integer results in the same number with the same sign.***

EXAMPLE 2-3.10 Subtract $-4 - 0$.

Solution

$-4 - 0$ *(−4 minus 0)*

$-4 - 0 = -4$ *(no change)*

■

On the other hand, when a signed number is subtracted from zero, we must change subtraction to addition and then perform the addition.

> **Rule 2-3.3** ***Subtracting an integer* from *zero results in the opposite of the number.***

EXAMPLE 2-3.11 **(a)** Subtract $0 - 8$.

Solution

$0 - 8$ *(Use the shortcut.)*

0 plus -8 *(Think of adding 0 and −8.)*

-8 *(0 plus −8 = −8.)*

(b) Subtract $0 - (-7)$.

Solution

$0 - (-7)$ *(Change subtraction to addition.)*

$0 + 7$ *(The problem is now addition.)*

7 *(Add 0 + 7 = 7.)*

■

SELF-STUDY EXERCISES 2-3

L.O.1. Subtract.

1. $-3 - 9$
2. $8 - 2$
3. $9 - 15$
4. $-3 - (-7)$
5. $-11 - 14$
6. $-6 - (-3)$
7. $5 - (-3)$
8. $8 - 11$
9. $-8 - 1$
10. $11 - (-2)$
11. $(-8) - (-7)$
12. $(-15) - (-7)$

L.O.2. Evaluate the following:

13. $-1 + 1 - 4$
14. $5 + 3 - 7$
15. $7 + 3 - (-4)$
16. $-8 - 2 - (-7)$
17. $-3 + 4 - 7 - 3$
18. $2 - 4 - 5 - 6 + 8$
19. $8 - 3 + 2 - 1 + 7$
20. $-5 - 3 + 8 - 2 + 4$

L.O.3. Subtract.

21. $15 - 0$
22. $0 - 8$
23. $-12 - 0$
24. $0 - (-8)$
25. $0 - (-7)$
26. $10 - 0$

2-4 MULTIPLYING INTEGERS

Learning Objectives

L.O.1. Multiply integers with like signs.
L.O.2. Multiply integers with unlike signs.
L.O.3. Multiply by zero.
L.O.4. Multiply several integers.
L.O.5. Evaluate powers of integers.

The same methods of expressing multiplication of whole numbers can also be used to express multiplication of integers.

$$5 \times 7 \qquad (5)7 \qquad 5 \cdot 7$$
$$5(7) \qquad (5)(7) \qquad 5 * 7$$

As for whole numbers, multiplication of integers is commutative and associative. That is, the factors may be multiplied in any order and grouped in any manner (see Definitions 1-1.1 and 1-1.2).

L.O.1. Multiply integers with like signs.

Multiplying Numbers with Like Signs

In multiplication of integers, multiply the absolute values as for whole numbers. The result is called the *product*. Examine these illustrations to see how the signs are handled.

$$4(6) = 24, \qquad (-3)(-7) = 21, \qquad -10(-2) = 20, \qquad 8(9) = 72$$

Rule 2-4.1 *To multiply two integers having* **like** *signs:*

1. Multiply their absolute values.
2. Make the sign of the product *positive*.

Example 2-4.1 Evaluate $(-2)(-9)$.

Solution

$(-2)(-9)$

$(-2)(-9) = 18$ *(Same signs give positive product.)* ■

L.O.2. Multiply integers with unlike signs.

Multiplying Numbers with Unlike Signs

When multiplying two numbers with unlike signs, multiply their absolute values. Study the illustrations that follow to see how the signs are handled.

$7(-12) = -84, \quad -5(3) = -15, \quad (-2)(4) = -8, \quad (3)(-1) = -3$

> **Rule 2-4.2** *To multiply two integers having* **unlike** *signs:*
>
> **1.** Multiply their absolute values.
> **2.** Make the sign of the product *negative.*

Example 2-4.2 Evaluate $4(-5)$.

Solution

$4(-5)$

$4(-5) = -20$ *(Different signs give negative product.)* ■

Now let's apply both rules in evaluating multiplication problems with signed numbers.

Example 2-4.3 Evaluate $4(5)(-3)(-2)$.

Solution

$\underbrace{4(5)}_{(20)}\underbrace{(-3)(-2)}_{(6)} = 120$ *(Multiply in pairs, following rules for signs.)* ■

Example 2-4.4 Evaluate $-3(2)(-4)(-1)$.

Solution

$\underbrace{-3(2)}_{(-6)}\underbrace{(-4)(-1)}_{(4)} = -24$ *(Multiply in pairs, following rules for signs.)* ■

L.O.3. Multiply by zero.

Multiplication of Integers by Zero

> **Rule 2-4.3** ***Multiplying an integer by zero results in zero, which has no sign.***

Example 2-4.5 Evaluate $4(2)(0)$.

Solution

$4(2)(0)$

$(8)(0) = 0$ *(Multiply in pairs, following rules for signs.)* ■

Tips and Traps

As a shortcut in Example 2-4.5, we can immediately write the answer of zero if we notice that *all* numbers in the problem are factors; that is, all are multiplied together. In this special case the product is zero because zero times any number is zero.

EXAMPLE 2-4.6 Evaluate $-1(21)(-6)(0)(2)$.

Solution

$-1(21)(-6)(0)(2) = 0$ *(All numbers are factors; one factor is zero.)* ■

Regardless of where zero appears, the product of the above example is zero. Therefore, we can go *directly* from $-1(21)(-6)(0)(2)$ to the product, 0.

L.O.4. Multiply several integers.

Multiplying Several Integers

When we encounter multiplication problems with several factors having different or unlike signs, the product will be *negative* if the number of *negative* factors is *odd*. The product will be *positive* if the number of *negative* factors is *even*.

EXAMPLE 2-4.7 Evaluate $6(-2)(1)(-3)$.

Solution

$6(-2)(1)(-3)$ *(Since number of negative factors is even, product will be positive.)*

$(6)(2)(1)(3) = 36$ *(Multiply absolute values.)*

$6(-2)(1)(-3) = 36$ *(Answer is positive.)* ■

EXAMPLE 2-4.8 Evaluate $-8(2)(-3)(-7)$.

Solution

$-8(2)(-3)(-7)$ *(Since number of negative factors is odd, product will be negative.)*

$(8)(2)(3)(7) = 336$ *(Multiply absolute values.)*

$(-8)(2)(-3)(-7) = -336$ *(Answer is negative.)* ■

L.O.5. Evaluate powers of integers.

Powers of Integers

Since raising a number to a natural-number power is an extension of multiplication, determining the sign of the result will be similar to multiplying several integers.

Examine the following illustrations to observe the pattern for determining the sign of the result.

$(+5)^2 = (+5)(+5) = +25,$ $(-5)^2 = (-5)(-5) = +25$

$(+5)^3 = (+5)(+5)(+5) = +125,$ $(-5)^3 = (-5)(-5)(-5) = -125$

Rule 2-4.4 *Use the following patterns for powers of integers with natural-number exponents:*

1. A positive number raised to a power is positive.
2. Zero raised to a power is zero.
3. A negative number raised to an even-numbered power is positive.
4. A negative number raised to an odd-numbered power is negative.

EXAMPLE 2-4.9 **(a)** $(+4)^3 = (+4)(+4)(+4) = +64$
(b) $(0)^8 = 0$
(c) $(-2)^4 = (-2)(-2)(-2)(-2) = +16$
(d) $(-3)^5 = (-3)(-3)(-3)(-3)(-3) = -243$ ■

SELF-STUDY EXERCISES 2-4

L.O.1. Multiply.

1. $5 \cdot 8$	**2.** $-4(-3)$	**3.** $7 * 5$
4. $(-3)(-7)$	**5.** $-8(-3)$	**6.** $(-2)(-3)$

L.O.2.

7. $5(-3)$	**8.** $(-2)(5)$	**9.** $-4 * 8$
10. $-3 \cdot 4$	**11.** $-7 * 8$	**12.** $6(-4)$

L.O.3.

13. $-8(0)$	**14.** $5(0)$	**15.** $0(-12)$
16. $0 \cdot 3$	**17.** $0(-15)$	**18.** $0(-17)$

L.O.4.

19. $5(-2)(3)(2)$	**20.** $6(1)(-3)(-2)$	**21.** $4(0)(-12)(3)$
22. $5(2)(-3)(0)$	**23.** $-3(2)(-7)(-1)$	**24.** $9(-1)(3)(-2)$

L.O.5. Evaluate.

25. $(-3)^2$	**26.** $(-2)^3$	**27.** $(-5)^2$
28. $(-8)^0$	**29.** $(5)^4$	**30.** 3^4

2-5 DIVIDING INTEGERS

Learning Objectives

L.O.1. Divide integers.
L.O.2. Divide with zero.

L.O.1. Divide integers.

Dividing Integers

The same methods of expressing division of whole numbers can also be used to express division of integers.

$$6\overline{)12}, \qquad 12 \div 6, \qquad \frac{12}{6}$$

Division of integers is most often written using the bar notation.

Study the following illustrations to see how the signs are handled. The *quotient* is the result of division.

$$\frac{10}{2} = 5, \qquad \frac{8}{-4} = -2, \qquad \frac{-14}{2} = -7, \qquad \frac{-15}{-5} = 3$$

As we observe, the signs of the quotients are determined just as they are in multiplication.

Rule 2-5.1 *To divide integers having* **like** *signs:*

1. Divide their absolute values.
2. Make the quotient *positive*.

Rule 2-5.2 *To divide integers having* **unlike** *signs:*

1. Divide their absolute values.
2. Make the quotient *negative*.

EXAMPLE 2-5.1 Evaluate $\frac{18}{9}$.

Solution

$$\frac{18}{9}$$

$$\frac{18}{9} = 2 \qquad \textit{(Same signs give positive quotient.)}$$ ■

EXAMPLE 2-5.2 Evaluate $\frac{20}{-4}$.

Solution

$$\frac{20}{-4}$$

$$\frac{20}{-4} = -5 \qquad \textit{(Different signs give negative quotient.)}$$ ■

EXAMPLE 2-5.3 Evaluate $\frac{-25}{5}$.

Solution

$$\frac{-25}{5}$$

$$\frac{-25}{5} = -5 \qquad \textit{(Different signs give negative quotient.)}$$ ■

L.O.2. Divide with zero.

Zero in Division

As with whole numbers, we cannot divide integers by zero.

Rule 2-5.3 ***Division of an integer by zero is undefined or impossible.***

On the other hand, we *can* divide zero by a nonzero number. For example:

$$\frac{0}{15} = 0, \quad \frac{0}{7} = 0, \quad \frac{0}{-6} = 0$$

Rule 2-5.4 *Zero divided by any nonzero number is zero.*

SELF-STUDY EXERCISES 2-5

L.O.1. Divide.

1. $\frac{-12}{-6}$ **2.** $-15 \div 5$ **3.** $\frac{8}{-2}$

4. $\frac{-24}{6}$ **5.** $-28 \div 7$ **6.** $\frac{-32}{-4}$

7. $\frac{-50}{-10}$ **8.** $\frac{36}{-6}$ **9.** $\frac{-48}{-6}$

10. $\frac{-25}{-5}$

L.O.2. Divide.

11. $\frac{12}{0}$ **12.** $\frac{-15}{0}$ **13.** $\frac{0}{+3}$

14. $\frac{0}{-12}$ **15.** $\frac{0}{-8}$ **16.** $\frac{-20}{0}$

17. $\frac{-100}{0}$

2-6 ORDER OF OPERATIONS

Learning Objectives

L.O.1. Use the order of operations for integers.
L.O.2. Use calculators to evaluate operations with integers.

L.O.1. Use the order of operations for integers.

Order of Operations for Integers

Integers follow the same order of operations as whole numbers.

Rule 2-6.1 *Perform operations in the following order as they appear from left to right:*

1. Parentheses used as groupings and other grouping symbols.
2. Exponents (powers and roots).
3. Multiplications and divisions.
4. Additions and subtractions.

EXAMPLE 2-6.1 Evaluate $(4 + 3) - (3 + 1)$.

Solution

$$\underbrace{(4 + 3)}_{(7)} - \underbrace{(3 + 1)}_{(4)}$$ *(Perform operations inside parentheses.)*

$7 - 4 = 3$ *(Then add or subtract last.)* ■

EXAMPLE 2-6.2 Evaluate $8(4 + 6)$.

Solution: Evaluate by working within the grouping symbols first.

$8(4 + 6)$ *(Add $4 + 6 = 10$.)*

$8(10)$ *(Multiply 8 by 10.)*

80

Or, evaluate using the distributive principle.

$8(4 + 6)$ *(Multiply each term by the factor 8.)*

$8(4) + 8(6)$

$32 + 48$ *(Add $32 + 48$.)*

80 ■

Another common grouping symbol is the bar showing division.

EXAMPLE 2-6.3 Evaluate $\frac{3 - 5}{2} - \frac{9}{2 + 1} + 4(5)$.

Solution

$\frac{3 - 5}{2} - \frac{9}{2 + 1} + 4(5)$ *(Perform operations grouped by bar or fraction line.)*

$\frac{-2}{2} - \frac{9}{3} + 4(5)$ *(Work multiplications and divisions.)*

$-1 - 3 + 20$ *(Work additions and subtractions last.)*

$-4 + 20 = 16$ ■

EXAMPLE 2-6.4 Evaluate $-12 \div 3 - (2)(-5)$.

Solution

$-12 \div 3 - (2)(-5)$

$$\underbrace{-12 \div 3}_{-4} - \underbrace{(2)(-5)}_{(-10)}$$ *(Work multiplications and divisions first.)*

$-4 + 10$ *(Change to a single sign between the numbers.)*

$-4 + 10 = 6$ *(Add integers.)* ■

The problem in Example 2-6.4 could have been written without parentheses around the 2. The problem could have been expressed as follows:

$$-12 \div 3 - 2(-5)$$

Again, the multiplications and divisions are done first. But notice how the multiplication can be performed.

$\underbrace{-12 \div 3}_{-4} \quad \underbrace{- 2(-5)}_{+10}$ *(We can consider the minus sign as the sign of the 2, and we are adding the result of the division and the result of multiplication.)*

$-4 + 10 = 6$ *(Add integers.)*

EXAMPLE 2-6.5 Evaluate $10 - 3(-2)$.

Solution

$\underbrace{10}_{10} \quad \underbrace{-3(-2)}_{+6}$ *(Think of the multiplication as -3 times -2.)*

$10 + 6 = 16$ *(Add integers.)* ■

Only after parentheses and all multiplications and/or divisions have been taken care of do we perform the final additions and/or subtractions from left to right.

EXAMPLE 2-6.6 Evaluate $4 + 5(2 - 8)$.

Solution

$4 + 5(2 - 8)$ *(First, perform operations in parentheses.)*

$4 + 5(-6)$ *(Then work multiplications and/or divisions.)*

$4 - 30$ *(Finally, perform remaining additions and/or subtractions.)*

$4 - 30 = -26$ ■

Tips and Traps

Note what would happen if we proceeded *out of order* in Example 2-6.6!

Incorrectly Worked

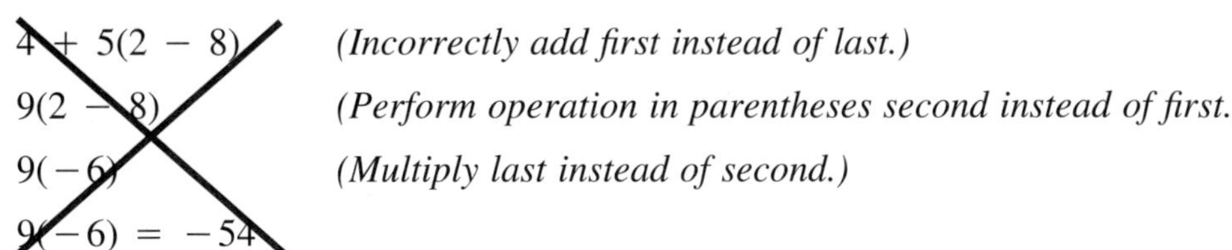

$4 + 5(2 - 8)$ *(Incorrectly add first instead of last.)*

$9(2 - 8)$ *(Perform operation in parentheses second instead of first.)*

$9(-6)$ *(Multiply last instead of second.)*

$9(-6) = -54$

Incorrect Solution

We get two different answers. *The order of operations must be followed to arrive at a correct solution.*

EXAMPLE 2-6.7 Evaluate $5 + (-2)^3 - 3(4 + 1)$.

Solution

$5 + (-2)^3 - 3(4 + 1)$ *(Raise to a power.)*

$5 + (-8) - 3(4 + 1)$ *(Change to single sign between numbers.)*

$5 - 8 - 3(4 + 1)$ *(Perform operation in parentheses.)*

$5 - 8 - 3(5)$ *(Perform multiplication.)*

$5 - 8 - 15$ *(Perform first addition of integers.)*

$-3 - 15$ *(Perform remaining addition of integers.)*

-18 ■

L.O.2. Use calculators to evaluate operations with integers.

Use of Calculators

Calculators using algebraic logic may be used to perform operations with signed numbers by using the [+], [−], [×], [÷], and [+/−] keys. With a graphing calculator, numbers and signs are entered in the order they are written.

With a scientific calculator, the absolute value of the number is entered first. The [+/−] key, called the "*sign change key*," is used to make the number negative. This key is a *toggle* key. That means it will change the sign of whatever number is showing in the display. For example, if 3 is showing in the display, press [+/−] to show −3 in the display. Press [+/−] again to show 3 in the display.

Tips and Traps

The symbol ⇒ indicates final answer in calculator display.

EXAMPLE 2-6.8 Evaluate the following.

(a) $2 - 7$ **(b)** $-7 + 2$

(c) $(-7)(2)$ **(d)** $\frac{-4}{2}$

Solution: Scientific calculator steps:

(a) $2 - 7$ [2] [−] [7] [=] ⇒ −5

(b) $-7 + 2$ [7] [+/−] [+] [2] [=] ⇒ − 5

(c) $(-7)(2)$ [7] [+/−] [×] [2] [=] ⇒ −14

(d) $\frac{-4}{2}$ [4] [+/−] [÷] [2] [=] ⇒ −2

Graphing calculator steps:

(a) $2 - 7$ [2] [−] [7] [EXE] ⇒ −5

(b) $-7 + 2$ [−] [7] [+] [2] [EXE] ⇒ −5

(c) $(-7)(2)$ [−] [7] [×] [2] [EXE] ⇒ − 14

or

[−] [7] [(] [2] [)] [EXE] ⇒ −14

(d) $\frac{-4}{2}$ [−] [4] [÷] [2] [EXE] ⇒ −2

■

Tips and Traps

Although most scientific and graphing calculators follow the order of operations given in Section 2-6, it is a good idea to check the operation of any given calculator to make sure. The owner's manual will be a useful guide to the operation of a particular calculator.

Following the order of operations in this text, a scientific or graphing calculator will allow us to enter many operations as they occur from *left to right*. The calculator will impose the correct order of operations for us.

When a bar is used as a grouping symbol and division, we must instruct the calculator to work the grouping first by enclosing the grouping in parentheses.

EXAMPLE 2-6.9 Evaluate the following.

(a) $\frac{5 + 1}{3} + 2$ **(b)** $2(4 - 1) + 3$

Solution: Scientific calculator steps:

(a) $\frac{5 + 1}{3} + 2$ [(] [5] [+] [1] [)] [÷] [3] [+] [2] [=] $\Rightarrow 4$

(b) $2(4 - 1) + 3$ [2] [×] [(] [4] [−] [1] [)] [+] [3] [=] $\Rightarrow 9$

Graphing calculator steps:

(a) $\frac{5 + 1}{3} + 2$ [(] [5] [+] [1] [)] [÷] [3] [+] [2] [EXE] $\Rightarrow 4$

(b) $2(4 - 1) + 3$ [2] [(] [4] [−] [1] [)] [+] [3] [EXE] $\Rightarrow 9$ ■

SELF-STUDY EXERCISES 2-6

L.O.1. Evaluate the following problems, paying careful attention to order of operations.

1. $-12 + 8$

2. $6(-5)$

3. $\frac{-10}{5}$

4. $8 - (-9)$

5. $\frac{-18}{9}$

6. $3(7) + 9 \div 3$

7. $-6(-2) - 4$

8. $-6(-4)$

9. $7(1 - 4) \div 3 + 2$

10. $\frac{2 - 8}{3} + 5(-2) \div 2$

11. $\frac{5 - 9}{4} + 4(-3) \div 6$

12. $4 \times 8 - 7 \times 3 + 18 \div 6$

L.O.2. Use a scientific or graphing calculator to evaluate the following.

13. $5(3 - 4) - 7(2 - 5) \div (-3)$

14. $142(3 - 21) + 48(27)$

15. $24 - 3(2 + 7) \div 3 + 12$

16. $\frac{12 - 18}{3} + 15 - 81 \div 3$

17. $14 - \frac{3 + 7}{5} \div 2 + 5(3)$

18. $2(3 - 12) \div 4 \times 6 - 8$

19. $2(4 - 3) - 7 + 4(2 - 8)$

20. $7 - 21 + 138 - 256$

CHAPTER 2 ASSIGNMENT EXERCISES

The number line in Fig. 2-9 contains 14 positions lettered *A* through *N*. Give the letter that corresponds to each of the following numbers.

2-1

1. -1
2. $+1$
3. -2
4. $+2$
5. -3
6. $+3$
7. -7
8. 4
9. 5
10. -5
11. 6
12. -6

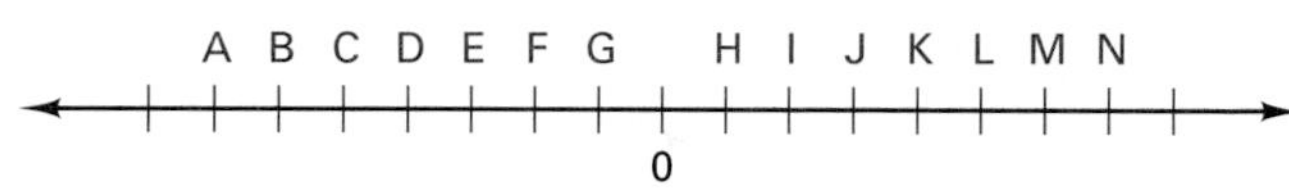

Figure 2-9

Use the symbols $>$ and $<$ to show the relationships of the numbers.

13. Is 72°F more than or less than -80°F?
14. Is 0 more than or less than -3?
15. Is 7 more than or less than -5?
16. Is 5 more than or less than 8?
17. Is -9 more than or less than 5?
18. Is -12 more than or less than -8?

Give the value of each of the following numbers:

19. $|5|$
20. $|-8|$
21. $|-3|$
22. $|+7|$
23. $|+8|$
24. $|-11|$

Give the opposite of each of the following numbers:

25. -12
26. 8
27. 15
28. -13

2-2

Add.

29. $-3 + (-8)$
30. $5 + 12$
31. $-7 + 12$
32. $(-15) + 8$
33. $7 + (-11)$
34. $-5 + (-8)$
35. $-6 + 6$
36. $8 + (-8)$
37. $7 + 0$
38. $0 + (-8)$

2-3

Evaluate.

39. $8 - 5$
40. $-9 - 4$
41. $-7 - (-2)$
42. $11 - (-3)$
43. $12 + 3 + (-8) - 5$
44. $-6 + 3 - 5 - 7$
45. $8 - 0$
46. $-5 - 0$
47. $0 - 3$
48. $0 - (-2)$

2-4

Evaluate.

49. $5(8)$
50. -3×-7
51. 7×-2
52. $(-3)(+2)$
53. $6(-2)$
54. $-7(3)$
55. $-8(0)$
56. $0(5)$
57. $2(3)(-7)(0)$
58. $5(-2)(-1)(-3)$
59. $4(3)(-2)(7)$
60. $(-3)^2$
61. $(7)^3$
62. $(-4)^3$

2-5

Divide.

63. $-8 \div (-4)$
64. $12 \div 3$
65. $\frac{18}{9}$

66. $\frac{-20}{-5}$

67. $\frac{16}{-4}$

68. $\frac{-24}{6}$

69. $\frac{0}{-8}$

70. $\frac{-7}{0}$

2-6

Use the order of operations to evaluate the following. Verify the results with a calculator.

71. $7(3 + 5)$

72. $-2(3 - 1)$

73. $\frac{15 - 7}{8}$

74. $-20 \div 4 - 3(-2)$

75. $12 - 8(-3)$

76. $7 + 3(4 - 6)$

77. $4 + (-3)^4 - 2(5 + 1)$

78. $(-3)^3 + 1 - 8$

79. $-3 + 2^3 - 7$

80. $4(-6 - 2) - \frac{8 + 2}{7 - 5}$

81. $296 - 382(-4)^5$

82. $-71 + 3(-19)$

CHALLENGE PROBLEM

83. The temperature at 8:00 A.M. is recorded as -3°C. Calculate the temperature at each hour as recorded by increases and decreases.

9:00 A.M.: increase 2°
10:00 A.M.: increase 1°
11:00 A.M.: increase 0°
12:00 P.M.: increase 1°
1:00 P.M.: no change
2:00 P.M.: increase 3°
3:00 P.M.: decrease 4°
4:00 P.M.: decrease 7°
5:00 P.M.: decrease 8°
6:00 P.M.: decrease 12°

CONCEPTS ANALYSIS

1. What two operations for integers use similar rules for handling the signs? Explain the rules for these operations.
2. Explain what is meant by "the absolute value of a number." Illustrate your explanation with an example.
3. What operation with 0 is not defined?
4. Describe the process of adding two integers that have different signs.
5. Illustrate the symbol used for "is greater than" by writing a true statement using the symbol.
6. Write the words that describe the correct order for operations with integers.
7. Explain how to determine the sign of a power if the base is a negative integer. Give an example for an even power and for an odd power.
8. Give an example of multiplying two negative integers and give the product.
9. Draw a number line that shows positive and negative integers and zero and place the following integers on the number line: $-3, 8, -2, 0, 3, 5$.
10. Find and correct the mistakes in the following problem.

$$(-8)^2 - 3(2)$$
$$16 - 3(2)$$
$$13(2)$$
$$26$$

CHAPTER SUMMARY

Objectives	What to Remember	Examples
Section 2-1		
1. Place integers on the number line.	Positive numbers are on the right of zero and negative numbers are on the left of zero on the number line.	Arrange from smallest to largest: $5, -3, 0, 8, -5$ $-5, -3, 0, 5, 8$
2. Compare integers.	The "greater than" symbol is $>$. The "less than" symbol is $<$.	Use $>$ or $<$ to make a true statement. $5 \, ? \, -3$: $5 > -3$
3. Find the absolute value of integers.	The absolute value of a number is its *distance* from zero without regard to direction.	Evaluate the following absolute values: $\|-3\|, \|5\|$ $\|-3\| = 3, \quad \|5\| = 5$
4. Find the opposites of integers.	Opposites are numbers that have the same absolute value, but opposite signs.	Give the opposite of the following: $8, -4, -2, +4$ $-8, 4, 2, -4$
Section 2-2		
1. Add integers with like signs.	To add integers with like signs, add the absolute values and give the sum the common sign.	Add: $13 + 7 = 20$ $-35 + (-13) = -48$
2. Add integers with unlike signs.	To add integers with unlike signs, subtract the smaller absolute value from the larger absolute value. The sign of the sum will be the sign of the number with the larger absolute value.	Add: $-12 + 7$. Subtract absolute values: $12 - 7 = 5$. Give the 5 the $-$ sign since -12 has the larger absolute value. $-12 + 7 = -5$
3. Add numbers that are opposites.	The result of adding numbers that are opposites is zero.	Add: $3 + (-3) = 0$ $-23 + 23 = 0$ $7 + (-7) = 0$
4. Add zero to an integer.	When zero is added to a signed number, the result is unchanged: $0 + a = a + 0 = a$.	Add: $-42 + 0 = -42$ $0 + 38 = 38$
Section 2-3		
1. Subtract integers.	To subtract integers, change the sign of the subtrahend, change subtraction to addition, and use the appropriate rule for adding signed numbers.	Subtract: $-32 - (-28)$ $-32 - (-28) = -32 + (+28)$ $= -4$
2. Combine addition and subtraction.	Change all subtractions to additions and work the additions from left to right.	Simplify: $5 - 3 + 2 - (-4)$ $5 + (-3) + 2 + 4 =$ $2 + 2 + 4 = 4 + 4 = 8$
3. Subtract with zero.	To subtract with zero, change the subtraction to addition and use the appropriate rule for addition.	$-8 - 0 = -8 + (-0) = -8$ $0 - (-4) = 0 + (+4) = 4$
Section 2-4		
1. Multiply integers with like signs.	To multiply integers with like signs, multiply the absolute values and make the sign of the product positive.	$-6(-7) = +42$ $8 \times 6 = 48$
2. Multiply integers with unlike signs.	Multiply the absolute values and make the sign of the product negative.	$-7(2) = -14$ $8(-3) = -24$

3. Multiply by zero.	Any number (including signed numbers) multiplied by zero results in zero: $a \times 0 = 0 \times a = a$	$0 \times 3 = 0$ $-5 \times 0 = 0$ $0(-7) = 0$
4. Multiply several integers.	To multiply several integers, perform the multiplication from left to right.	$5(-2)(3) = -10(3) = -30$
5. Evaluate powers of integers.	A positive integer raised to a natural-number power will result in a positive integer. A negative integer raised to an even power will result in a positive integer. A negative integer raised to an odd power will result in a negative integer.	$(3)^4 = 81$ $(-2)^4 = 16$ $(-2)^3 = -8$
Section 2-5		
1. Divide integers.	To divide integers, divide the absolute values of the dividend and the divisor. If the dividend and divisor have like signs, the sign of the quotient will be positive. If the dividend and divisor have unlike signs, the sign of the quotient will be negative.	$-12 \div (-4) = 3$ $15 \div (-3) = -5$
2. Divide with zero.	Zero divided by any nonzero signed number is zero. A nonzero number *cannot* be divided by zero. $0 \div a = 0$ (if a is not equal to zero); $a \div 0$ is not possible.	$0 \div 12 = 0$ $-7 \div 0$ is not possible or not defined.
Section 2-6		
1. Use the order of operations for integers.	Expressions are evaluated using the following order of operations: parentheses, powers (and roots), multiplication and division, addition and subtraction.	Evaluate: $5 - 4(7 + 2)^2 - 15 \div 5$ $5 - 4(9)^2 - 15 \div 5 =$ $5 - 4(81) - 15 \div 5 =$ $5 - 324 - 15 \div 5 =$ $5 - 324 - 3 =$ $-319 - 3 =$ -322
2. Use calculators to evaluate operations with integers.	Use of the scientific and graphing calculators to evaluate expressions is different.	Use the scientific or graphing calculator to evaluate $5(-3) + 3 - (-2)$. Scientific steps: 5 [×] 3 [+/−] [+] 3 [−] 2 [+/−] [=] ⇒ −10 Graphing steps: 5 [×] [−]3 [+] 3 [−] [−] 2 EXE ⇒ −10

WORDS TO KNOW

positive (p. 60)
negative (p. 60)
number line (p. 60)
integers (p. 60)
greater than (>) (p. 61)
less than (<) (p. 61)
absolute value (p. 61)
opposites (p. 62)
addends or terms (p. 62)
like signs (p. 63)
signed numbers (p. 63)
different or unlike signs (p. 64)
additive inverse (p. 66)
sign-change key (p. 79)
toggle key (p. 79)

CHAPTER TRIAL TEST

Use the sumbols > and < to answer the following questions.

1. Is -8 more than or less than 0?

2. Is 2 more than or less than 3?

3. Is -5 more than or less than -1?

Answer the following questions.

4. What is the value of $|-12|$?

5. What is the opposite of 8?

Perform the following operations.

6. $-8 - 2$

7. $-3 + 7$

8. $\dfrac{8}{-2}$

9. $2(6)(-4)$

10. $\dfrac{0}{0}$

11. $-5(-2)$

12. $\dfrac{-6}{-2}$

13. $8 + 4 + (-2) + (-7)$

14. $7 - (-3)$

15. $2(-1)(-4)$

16. $875 - 428$

17. $-7 + (-3)$

18. $(-8)(3)(0)(-1)$

19. $\dfrac{0}{3}$

20. $2 + (-18)$

21. $0 - 7$

22. $-3 + 5 + 0 + 2 + (-5)$

23. $\dfrac{-7}{0}$

24. $4(-13)$

25. $5 + (-7) - (-3) + 2 - 6$

26. $\dfrac{6}{-1}$

27. $\dfrac{4}{-2}$

28. $8(3) - 2 + 6(7)$

29. $2(3 - 9) \div 2 + 7$

30. $\dfrac{10 - 4}{3} - 3$

Perform the following operations. Use a scientific or graphing calculator *if your instructor permits use of a calculator.*

31. $5(6) - 2 + 10 \div 2$

32. $5(2 - 3) + 16 \div 4$

33. $-2 - 3(-4) + (4)(3)$

34. $\dfrac{-18}{3}$

35. $\dfrac{4 - 2}{2} - 6$

CHAPTER

3 Fractions

Thus far we have reviewed the mathematics of whole numbers and integers. However, we often must work with numbers that are not whole numbers or integers (such as $\frac{1}{2}$, $\frac{1}{4}$, $1\frac{1}{4}$, or $1\frac{1}{2}$). In this chapter we study how fractions are used to express numbers and how we add, subtract, multiply, and divide fractions quickly and efficiently.

3-1 FRACTION TERMINOLOGY

Learning Objective

L.O.1. Identify fraction terminology.

L.O.1. Identify fraction terminology.

Fraction Terminology

If one unit (or amount) is divided into four parts, we can write the fraction $\frac{4}{4}$ to represent this single unit. Figure 3-1 illustrates a unit divided into four equal parts.

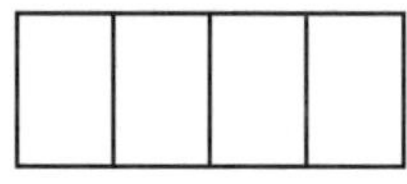

Figure 3-1

The fraction $\frac{4}{4}$ is an example of a *common fraction.* A common fraction consists of two whole numbers. The bottom number, called the *denominator*, indicates *how many equal parts* one whole unit has been divided into. The top number, called the *numerator*, tells *how many of these parts* are being considered. Thus, in Fig. 3-2, one of the four equal parts has been shaded and is represented by the fraction $\frac{1}{4}$. Similarly, the shaded part in Fig. 3-3 is written as the fraction $\frac{3}{7}$.

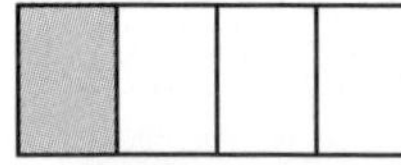

Figure 3-2

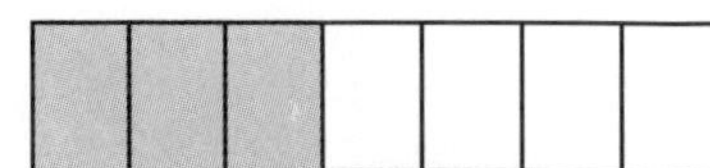

Figure 3-3

The unit is always our *standard* amount when writing fractions. Thus, $\frac{4}{4}$ and $\frac{3}{3}$ represent one unit each, or simply the number 1. Some fractions represent less than one unit (less than 1), for example, $\frac{1}{4}$ or $\frac{3}{4}$. Other fractions, however, represent more than one unit (more than 1), for example, $\frac{7}{5}$.

Suppose that three units are each divided into five equal parts. If we have seven of these parts, the fraction $\frac{7}{5}$ represents this amount. This fraction is larger than one unit (Fig. 3-4).

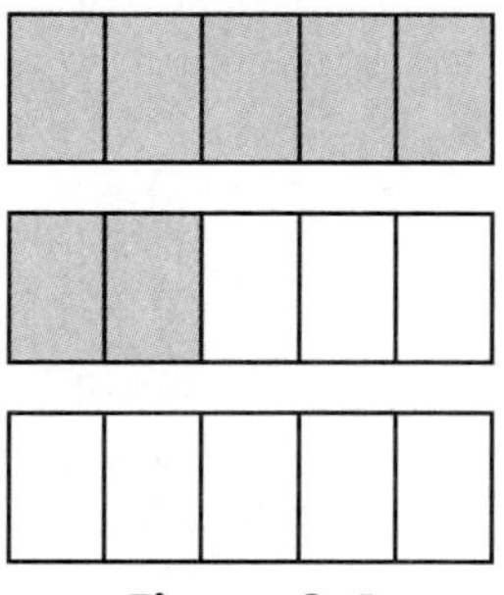

Figure 3-4

Similarly, if we have 10 of these parts, the fraction $\frac{10}{5}$ will represent this amount. If five parts equal one unit, then 10 parts will equal two units (Fig. 3-5). Fifteen parts, $\frac{15}{5}$, will equal three units (Fig. 3-6).

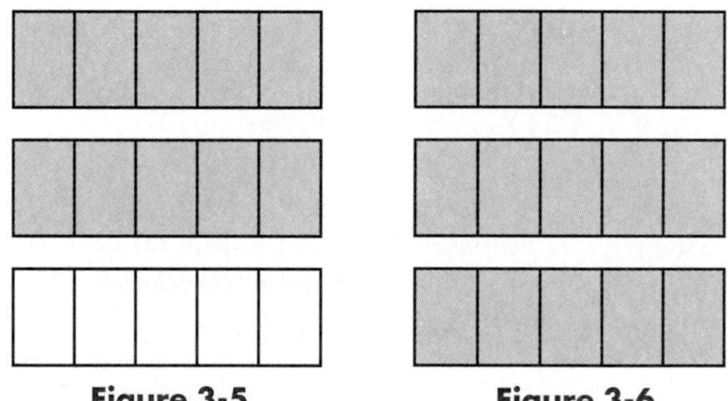

Figure 3-5 **Figure 3-6**

There is a relationship between fractions and division. We can see this in the fractions $\frac{10}{5}$ (two units) and $\frac{15}{5}$ (three units), since $10 \div 5 = 2$ and $15 \div 5 = 3$. This relationship to division is important to our understanding of fractions.

Now we are ready to learn some important definitions concerning fractions.

■ **DEFINITION 3-1.1 Common Fraction.** A *common fraction* represents the division of one whole number by another.

■ **DEFINITION 3-1.2 Denominator.** The *denominator* of a fraction tells how many parts one unit has been divided into. It will be the *bottom* number of a fraction or the *divisor* of the indicated division.

■ **DEFINITION 3-1.3 Numerator.** The *numerator* of a fraction tells how many of the parts are being considered. It will be the *top* number of a fraction or the *dividend* of the indicated division.

The numerator and denominator of a fraction are usually separated by a horizontal line, although sometimes a slash is used. We will call this line the *fraction line* and it will also serve as a division symbol. The fraction can be read from top to bottom as the numerator *divided by* the denominator. Six divided by two can be written as $2\overline{)6}$, $6 \div 2$, or $\frac{6}{2}$ or $6/2$.

There are two basic types of common fractions: those whose value is less than 1 and those whose value is equal to or greater than 1.

■ **Definition 3-1.4 Proper Fraction.** A *proper fraction* is a common fraction whose value is less than one unit. That is, the numerator is less than the denominator. *Examples:*

$$\frac{2}{5}, \quad \frac{3}{7}, \quad \frac{15}{16}, \quad \frac{1}{3}, \quad \frac{7}{10}$$

■ **Definition 3-1.5 Improper Fraction.** An *improper fraction* is a common fraction whose value is equal to or greater than one unit. That is, the numerator is equal to or greater than the denominator. *Examples:*

$$\frac{4}{4}, \quad \frac{7}{3}, \quad \frac{8}{4}, \quad \frac{100}{10}, \quad \frac{17}{5}$$

When the denominator divides evenly into the numerator, the improper fraction can be written as a whole number, such as $\frac{8}{4} = 2$. But when the value is more than one unit and the denominator cannot divide evenly into the numerator, the improper fraction can be written as a combination of a whole number and a fractional part, such as $\frac{9}{4} = 2\frac{1}{4}$. Such a combination is called a *mixed number.*

■ **Definition 3-1.6 Mixed Number.** A *mixed number* is a number that consists of both a whole number and a fraction. The whole number and fraction are added together. *Example:*

$3\frac{2}{5}$ means three whole units and $\frac{2}{5}$ of another unit.

Sometimes mathematical computations in technical applications become more complex than the fractions and mixed numbers we have looked at so far. We have seen that a fraction like $\frac{2}{7}$ means that a unit has been divided into seven equal parts and we are considering only two of those seven parts. But suppose that our job required us to consider $2\frac{1}{3}$ of those seven parts instead of just two. In this instance our fraction would be $\dfrac{2\frac{1}{3}}{7}$. A fraction like $\dfrac{2\frac{1}{3}}{7}$ is a *complex fraction.*

■ **Definition 3-1.7 Complex Fraction.** A *complex fraction* is a fraction that has a fraction or mixed number in its numerator or denominator or both.

Examples:

$$\frac{\frac{1}{2}}{2}, \quad \frac{7}{\frac{3}{4}}, \quad \frac{3\frac{1}{3}}{7}, \quad \frac{5}{4\frac{1}{2}}, \quad \frac{\frac{1}{8}}{\frac{5}{16}}, \quad \frac{6\frac{3}{8}}{4\frac{1}{2}}$$

It is very important to understand that fractions indicate division. The fraction is read from *top to bottom*, with the fraction line read as "divided by." $\frac{\frac{1}{2}}{2}$ is read as $\frac{1}{2}$ divided by 2. $\frac{7}{\frac{3}{4}}$ is read as 7 divided by $\frac{3}{4}$.

Many job-related applications make use of another kind of fraction, whose denominator is 10 or a power of 10, such as 100, 1000, and so on. Instead of writing $\frac{3}{10}$, we may write 0.3 (with the zero indicating no whole number, the period or *decimal point* indicating the beginning of the fraction, and the 3 indicating how many $\frac{1}{10}$'s, in this case 3). The mixed number $2\frac{17}{100}$ can be written as 2.17 (with the 2 indicating the whole number, the decimal point indicating the beginning of the fraction, and the 17 indicating how many $\frac{1}{100}$'s, in this case 17).

Tips and Traps

The number of places after the decimal point indicates the number of zeros in the denominator of the power of 10. Note also that the number after the decimal point indicates the numerator of the fraction.

$$0.015 = \frac{15}{1{,}000}, \qquad 2.43 = 2\frac{43}{100}$$

■ **DEFINITION 3-1.8 Decimal Fraction.** A *decimal fraction* is a fractional notation that uses the decimal point and the place values to its right to represent a fraction whose denominator is 10 or some power of 10, such as 100, 1000, and so on.

■ **DEFINITION 3-1.9 Mixed-decimal Fraction.** A *mixed-decimal fraction* is a notation used to represent a decimal fraction that contains a whole number part as well as a fractional part.

EXAMPLE 3-1.1 Rewrite the decimal fractions and mixed-decimal fractions below as proper fractions or mixed numbers:

0.4, 0.18, 0.014, 3.7, 4.103

Solution: The number of places to the right of the decimal indicates the number of zeros in the denominator of the fraction.

$$0.4 = \frac{4}{10} \qquad 3.7 = 3\frac{7}{10}$$

$$0.18 = \frac{18}{100} \qquad 4.103 = 4\frac{103}{1000}$$

$$0.014 = \frac{14}{1000}$$

■

See Chapter 4 for more information on decimal fractions, which are, as a matter of convenience, referred to as *decimals* or *decimal numbers.*

To summarize, let's place the various fractions on a chart (Fig. 3-7) to show the whole picture at a glance.

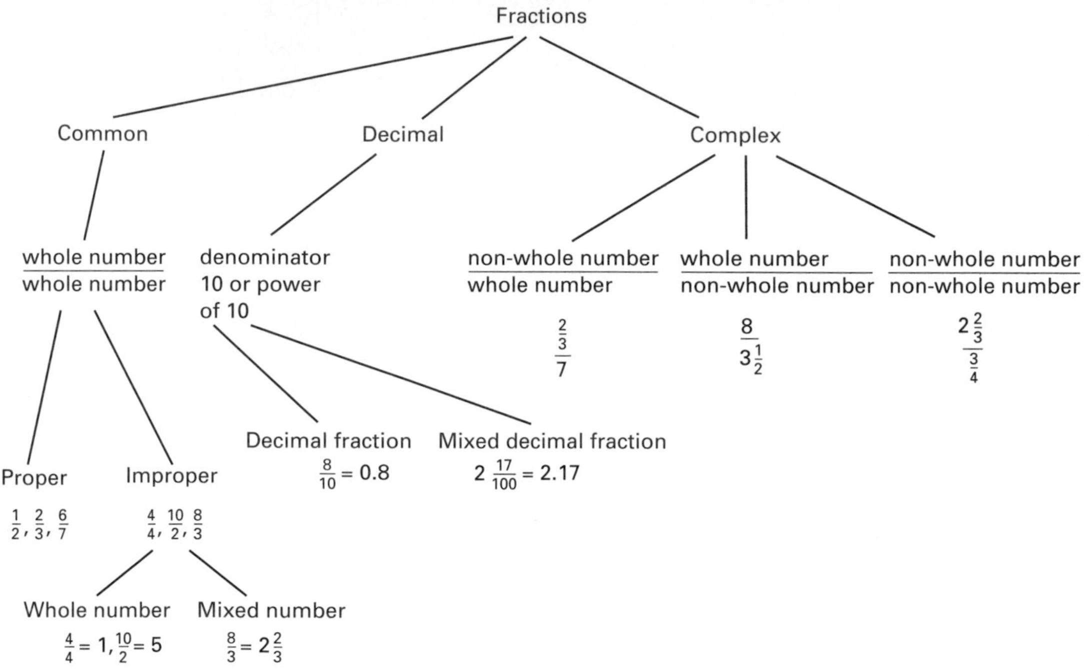

Figure 3-7

Fractions extend our types of numbers to another level called *rational* numbers.

■ **DEFINITION 3-1.10 Rational Number.** Any number that can be written in the form of a fraction, with an integer for a numerator and a nonzero integer for a denominator, is a **rational number**.

Rational numbers include whole numbers, natural numbers, integers, fractions, and decimal numbers because all these types of numbers can be expressed in fraction form. Other types of rational numbers will be introduced in later chapters, as well as numbers that are not rational numbers (irrational numbers).

SELF-STUDY EXERCISES 3-1

L.O.1. Write a fraction to represent the shaded portion of Exercises 1 through 10.

1. ▮▯▯ **2.** ▮▮▯▯▯ **3.** ▮▯

4. 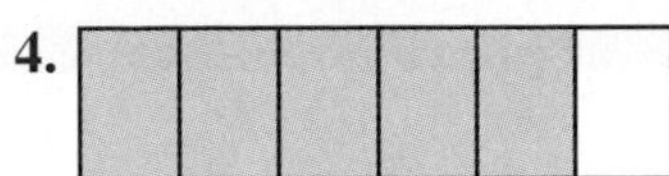

5. 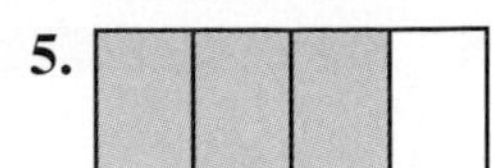

6. 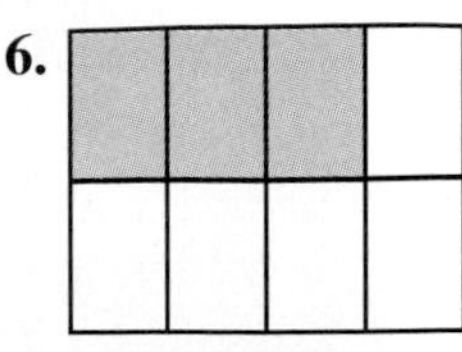

7. 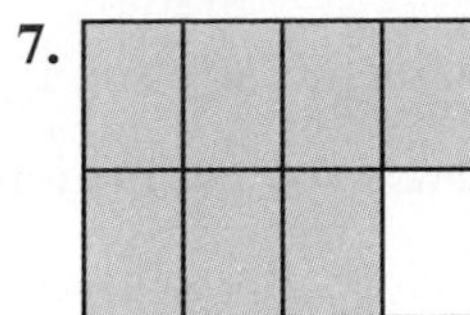

8. 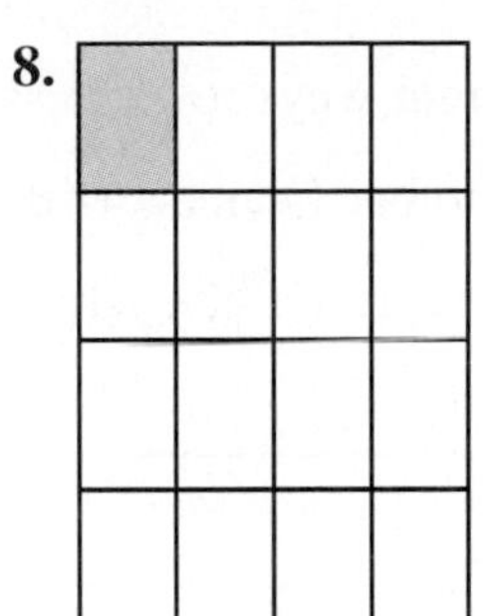

9. 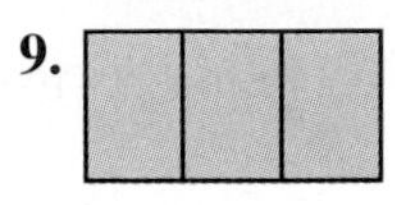

10. 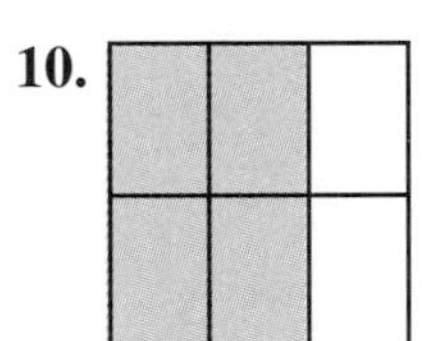

Examine each common fraction in Exercises 11 through 14 and answer the following questions.

11. $\frac{5}{6}$

12. $\frac{8}{8}$

13. $\frac{11}{5}$

14. $\frac{12}{3}$

(a) One unit has been divided into how many parts?
(b) How many of these parts are used?
(c) The numerator of the fraction is ____.
(d) The denominator of the fraction is ____.
(e) The fraction can be read as ____ divided by ____.
(f) ____ is the divisor of the division.
(g) ____ is the dividend of the division.
(h) Is this common fraction proper or improper?
(i) This fraction represents ____ out of ____ parts.
(j) Does this fraction have a value less than, more than, or equal to 1?

Match the best description with each number given.

____ **15.** $2\frac{3}{7}$

____ **16.** $\frac{8}{3}$

____ **17.** 6.27

____ **18.** 0.3

____ **19.** $\frac{3}{\frac{1}{2}}$

____ **20.** $\frac{\frac{3}{4}}{\frac{2}{7}}$

____ **21.** 0.689

____ **22.** $14\frac{3}{16}$

____ **23.** 3.7

____ **24.** $\frac{3}{7}$

____ **25.** $\frac{1\frac{1}{2}}{3}$

(a) Decimal fraction
(b) Mixed number
(c) Complex fraction
(d) Mixed-decimal fraction
(e) Proper fraction
(f) Improper fraction

Rewrite the fractions below as proper fractions or mixed numbers.

26. 6.27 **27.** 0.3 **28.** 0.689
29. 3.7 **30.** 1.53

3-2 EQUIVALENT FRACTIONS

Learning Objectives

L.O.1. Write equivalent fractions.
L.O.2. Reduce fractions.

There are many different ways to express the same value in fractional form. For example, the whole number 1 can be written as $1, \frac{1}{1}, \frac{4}{4}, \frac{7}{7}, \frac{15}{15}$, and so on. Fractions are equivalent if they represent the same value. Let's compare the illustrations in Fig. 3-8.

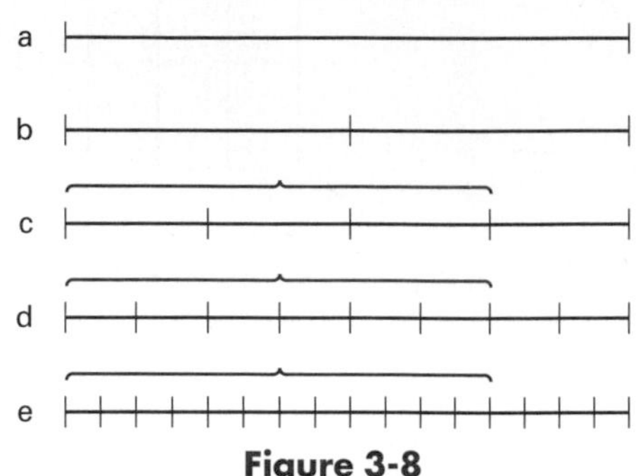

Figure 3-8

In Fig. 3-8, line *a* is one whole unit divided into only one part $\left(\frac{1}{1}\right)$. Line *b* is one unit divided into two parts $\left(\frac{2}{2}\right)$. Line *c* is divided into 4 parts $\left(\frac{4}{4}\right)$, line *d* is divided into eight parts $\left(\frac{8}{8}\right)$, and line *e* is divided into 16 parts $\left(\frac{16}{16}\right)$.

Look at the dimensions indicated on lines *c*, *d*, and *e*. What fraction represents this dimension on line *c*? $\left(\text{Three out of four parts, or } \frac{3}{4}.\right)$ What fraction represents this dimension on line *d*? $\left(\text{Six out of eight parts, or } \frac{6}{8}.\right)$ What fraction represents this dimension on line *e*? $\left(\text{Twelve out of sixteen parts, or } \frac{12}{16}.\right)$

$\frac{3}{4}, \frac{6}{8}$, and $\frac{12}{16}$ are equivalent fractions. That is,

$$\frac{3}{4} = \frac{6}{8} = \frac{12}{16}$$

L.O.1. Write equivalent fractions.

Writing Equivalent Fractions

Fractions that are equivalent are in the same "family of fractions." The first member of the "family" is the fraction in lowest terms; that is, no number divides evenly into *both* the numerator and denominator except the number 1. Other "family members" can be found by multiplying both the numerator and denominator by the same number.

EXAMPLE 3-2.1 Find five fractions that are equivalent to $\frac{1}{2}$.

Solution

$$\frac{1 \times 2}{2 \times 2} \text{ or } \frac{1}{2} \times \frac{2}{2} = \frac{2}{4}$$

$$\frac{1 \times 3}{2 \times 3} \text{ or } \frac{1}{2} \times \frac{3}{3} = \frac{3}{6}$$

$$\frac{1 \times 4}{2 \times 4} \text{ or } \frac{1}{2} \times \frac{4}{4} = \frac{4}{8}$$

$$\frac{1 \times 5}{2 \times 5} \text{ or } \frac{1}{2} \times \frac{5}{5} = \frac{5}{10}$$

$$\frac{1 \times 6}{2 \times 6} \text{ or } \frac{1}{2} \times \frac{6}{6} = \frac{6}{12}$$

■

Tips and Traps

In each case, $\frac{1}{2}$ was multiplied by a fraction whose value is 1, and 1 times any number does not change the value of that number.

$$1 \times n = n$$

Many times we will need to find a fraction with a specific denominator that is equivalent to a given fraction. In this case, we first decide what number should be multiplied by the given denominator to give the specified denominator. Then we multiply the numerator and denominator by that number. This is allowed because of the **fundamental principle of fractions**.

■ **DEFINITION 3-2.1 Fundamental Principle of Fractions.** If the numerator and denominator of a fraction are multiplied by the same nonzero number, the value of the fraction remains unchanged.

EXAMPLE 3-2.2 Find a fraction equivalent to $\frac{2}{3}$ that has a denominator of 12.

Solution

3 times what number is 12?

or

3 divides into 12 how many times?

The answer is 4. Then

$$\frac{2 \times 4}{3 \times 4} \text{ or } \frac{2}{3} \times \frac{4}{4} = \frac{8}{12}$$

■

Rule 3-2.1 ***To change a fraction to an equivalent fraction with a larger specified denominator:***

1. Divide the original denominator into the desired denominator.
2. Multiply the original numerator and denominator by the quotient found in step 1.

EXAMPLE 3-2.3 Change $\frac{5}{8}$ to an equivalent fraction whose denominator is 32.

Solution

$$\frac{5}{8} = \frac{?}{32}, \qquad 8\overline{)32}^{\,4}, \quad 4 \times 5 = 20$$

$$\frac{5}{8} = \frac{20}{32} \quad \text{or} \quad \frac{5}{8} \times \frac{4}{4} = \frac{20}{32}$$

■

L.O.2. Reduce fractions.

Reducing Fractions

Since any fraction will have an unlimited number of equivalent fractions, we will want to work with fractions in *lowest terms.* This gives us smaller numbers to work with.

By *lowest terms*, we mean that there is no *whole* number other than 1 that will divide evenly into both the numerator and denominator. Another way of saying this is that the numerator and denominator have no common factors other than 1.

Rule 3-2.2 ***To change a fraction to an equivalent fraction with a smaller denominator or to reduce a fraction to lowest terms:***

1. Find a number other than 1 that will divide evenly into both the numerator and denominator. This number is a common divisor of the numerator and denominator.
2. Divide both the numerator and denominator by the common factor.
3. Continue until the fraction is in lowest terms or has the desired smaller denominator.

With many fractions, the greatest common divisor (GCD) can be found by *inspection*. That is, *visually* examine the numerator and denominator and *mentally* determine the GCD. For larger numerators or denominators, you may need to use the procedure for finding the GCD discussed in Chapter 1.

EXAMPLE 3-2.4 Reduce $\frac{8}{10}$ to lowest terms.

Solution: What number other than 1 will divide evenly into *both* 8 and 10? 2 and 4 will divide evenly into 8, and 2 and 5 will divide evenly into 10. But only the number 2 will divide evenly into *both* 8 and 10.

$$\frac{8 \div 2}{10 \div 2} \quad \text{or} \quad \frac{8}{10} \div \frac{2}{2} = \frac{4}{5}$$

■

Tips and Traps

When reducing the fraction $\frac{8}{10}$, we are dividing by the whole number 1 in the form of $\frac{2}{2}$. To divide a number by 1 does not change the value of the number.

$$\frac{n}{n} = 1 \quad \text{and} \quad n \div 1 = n$$

EXAMPLE 3-2.5 Reduce $\frac{18}{24}$ to lowest terms.

Solution

What numbers other than 1 will divide evenly into 18? 2, 3, 6, 9, 18

What numbers other than 1 will divide evenly into 24? 2, 3, 4, 6, 8, 12, 24

What numbers other than 1 will divide evenly into *both* 18 and 24? 2, 3, and 6

To reduce a fraction to *lowest terms*, in the fewest number of steps, use the *greatest* common divisor or greatest common factor.

$$\frac{18 \div 6}{24 \div 6} \quad \text{or} \quad \frac{18}{24} \div \frac{6}{6} = \frac{3}{4}$$

■

A fraction can be reduced to lowest terms in the fewest number of steps by using the *greatest common divisor*; however, it can still be reduced correctly in more steps by using any common divisor.

EXAMPLE 3-2.6 Reduce $\frac{72}{100}$ to lowest terms.

Solution

$$\frac{72 \div 4}{100 \div 4} = \frac{18}{25} \quad \text{or} \quad \frac{72 \div 2}{100 \div 2} = \frac{36}{50}$$

$$\frac{36 \div 2}{50 \div 2} = \frac{18}{25}$$

■

SELF-STUDY EXERCISES 3-2

L.O.1.

1. Find five fractions that are equivalent to $\frac{4}{5}$.
2. Find five fractions that are equivalent to $\frac{7}{10}$.
3. Find a fraction that is equivalent to $\frac{3}{4}$ and has a denominator of 24.

Find the equivalent fractions having the indicated denominators.

4. $\frac{3}{8} = \frac{?}{16}$
5. $\frac{4}{7} = \frac{?}{21}$
6. $\frac{9}{11} = \frac{?}{44}$
7. $\frac{1}{3} = \frac{?}{15}$
8. $\frac{5}{6} = \frac{?}{24}$
9. $\frac{7}{8} = \frac{?}{24}$
10. $\frac{2}{5} = \frac{?}{30}$

L.O.2. Reduce the following fractions to lowest terms.

11. $\frac{4}{8}$
12. $\frac{6}{10}$
13. $\frac{12}{16}$
14. $\frac{10}{32}$
15. $\frac{16}{32}$
16. $\frac{28}{32}$
17. $\frac{20}{64}$
18. $\frac{2}{8}$
19. $\frac{8}{32}$

20. $\frac{12}{50}$ 21. $\frac{10}{16}$ 22. $\frac{4}{16}$

23. $\frac{24}{32}$ 24. $\frac{12}{64}$ 25. $\frac{14}{64}$

3-3 IMPROPER FRACTIONS AND MIXED NUMBERS

Learning Objectives

L.O.1. Convert improper fractions to whole or mixed numbers.
L.O.2. Convert mixed numbers and whole numbers to improper fractions.

L.O.1. Convert improper fractions to whole or mixed numbers.

Converting Improper Fractions to Whole or Mixed Numbers

As we recall from Section 3-1, an improper fraction is a fraction whose value is equal to or greater than one unit, such as $\frac{6}{3}$, $\frac{15}{7}$, or $\frac{5}{5}$. Problems involving improper fractions are sometimes easier to solve if we change the improper fractions into whole numbers or mixed numbers.

> **Rule 3-3.1** ***To convert an improper fraction to a whole or mixed number:***
>
> Perform the division indicated. Any remainder of the division is expressed as a fraction.

Example 3-3.1 Convert $\frac{15}{3}$ to a whole or mixed number.

Solution

$$\frac{15}{3} \quad \text{means} \quad 15 \div 3 \quad \text{or} \quad \begin{array}{r} 5 \\ 3\overline{)15} \\ \underline{15} \\ 0 \end{array}$$

$$\frac{15}{3} = 5$$

If there is no remainder from the division, the improper fraction converts to a whole number. ■

Example 3-3.2 Convert $\frac{23}{4}$ to a whole or mixed number.

Solution

$$\frac{23}{4} \quad \text{means} \quad 23 \div 4 \quad \text{or} \quad \begin{array}{r} 5 \\ 4\overline{)23} \\ \underline{20} \\ 3 \end{array}$$

Thus, $\frac{23}{4}$ represents five whole units with a remainder of three parts of another unit. This is expressed as the mixed number $5\frac{3}{4}$. ■

Tips and Traps

Converting an improper fraction to a whole or mixed number should not be confused with reducing to lowest terms. An improper fraction is in lowest terms if its numerator and denominator have no common factor. Therefore, the improper fraction $\frac{10}{7}$ is in lowest terms. The improper fraction $\frac{10}{4}$ is not in lowest terms. $\frac{10}{4}$ will reduce to $\frac{5}{2}$, which is in lowest terms.

When converting an improper fraction to a whole or mixed number, the fractional part should be in lowest terms. We can convert the fractional part in lowest terms either before dividing or after dividing.

EXAMPLE 3-3.3 Convert $\frac{28}{8}$ to a mixed number.

Solution

$$\frac{28}{8} = \frac{7}{2} \qquad \begin{array}{r} 3 \\ 2\overline{)7} \\ \underline{6} \\ 1 \end{array} = 3\frac{1}{2}$$

(Fraction is reduced before dividing.)

or

$$\frac{28}{8} \qquad \begin{array}{r} 3 \\ 8\overline{)28} \\ \underline{24} \\ 4 \end{array} = 3\frac{4}{8} = 3\frac{1}{2}$$

(Fraction is reduced after dividing.) ■

L.O.2. Convert mixed numbers and whole numbers to improper fractions.

Converting Mixed Numbers and Whole Numbers to Improper Fractions

On the other hand, it is sometimes useful to convert a mixed number to an improper fraction when solving certain problems. In the improper fraction, the numerator will indicate the total number of parts considered. The denominator still tells how many parts one unit has been divided into.

Rule 3-3.2 *To convert a mixed number to an improper fraction:*

1. Multiply the denominator of the fractional part by the whole number.
2. Add the numerator of the fractional part to the product; this sum becomes the numerator of the improper fraction.
3. The denominator of the improper fraction is the same as the denominator of the fractional part of the mixed number.

Example 3-3.4 Change $6\frac{2}{3}$ to an improper fraction.

Solution

$$6\frac{2}{3} = \frac{(3 \times 6) + 2}{3} = \frac{20}{3}$$ ■

Whole numbers can also be written as improper fractions. Any whole number can be written as a fraction with a denominator of 1. Then the whole number divided by 1 can be changed to an improper fraction having any denominator by using the procedures described in Section 3-2.

Example 3-3.5 Change 8 to fifths.

Solution

$$8 = \frac{8}{1} = \frac{8 \times 5}{1 \times 5} = \frac{40}{5}$$ ■

Example 3-3.6 How many sixteenths are in three units?

Solution

$$3 = \frac{3}{1} = \frac{3 \times 16}{1 \times 16} = \frac{48}{16}$$ ■

SELF-STUDY EXERCISES 3-3

L.O.1. Convert the following improper fractions to whole or mixed numbers.

1. $\frac{12}{5}$ **2.** $\frac{10}{7}$ **3.** $\frac{12}{12}$

4. $\frac{32}{7}$ **5.** $\frac{24}{6}$ **6.** $\frac{15}{7}$

7. $\frac{23}{9}$ **8.** $\frac{47}{5}$ **9.** $\frac{86}{9}$

10. $\frac{38}{21}$ **11.** $\frac{57}{15}$ **12.** $\frac{64}{4}$

13. $\frac{72}{10}$ **14.** $\frac{19}{2}$ **15.** $\frac{36}{4}$

L.O.2. Change the following whole or mixed numbers to improper fractions.

16. $2\frac{1}{3}$ **17.** $3\frac{1}{8}$ **18.** $1\frac{7}{8}$

19. $6\frac{5}{12}$ **20.** 9 **21.** $3\frac{7}{8}$

22. $7\frac{5}{12}$ **23.** $6\frac{7}{16}$ **24.** $8\frac{1}{32}$

25. $1\frac{5}{64}$ **26.** $7\frac{3}{10}$ **27.** $8\frac{2}{3}$

28. $33\frac{1}{3}$ **29.** $66\frac{2}{3}$ **30.** $12\frac{1}{2}$

Change each of the following whole numbers to an equivalent fraction having the indicated denominator.

31. $5 = \frac{?}{3}$ **32.** $9 = \frac{?}{2}$ **33.** $7 = \frac{?}{8}$

34. $8 = \frac{?}{4}$ **35.** $3 = \frac{?}{16}$

3-4 FINDING COMMON DENOMINATORS AND COMPARING FRACTIONS

Learning Objectives

L.O.1. Find common denominators.
L.O.2. Compare fractions.

L.O.1. Find common denominators.

Finding Common Denominators

Is is possible to have a pipe with an outside diameter of $\frac{5}{8}$-in. and an inside diameter of $\frac{21}{32}$ in. (Fig. 3-9)?

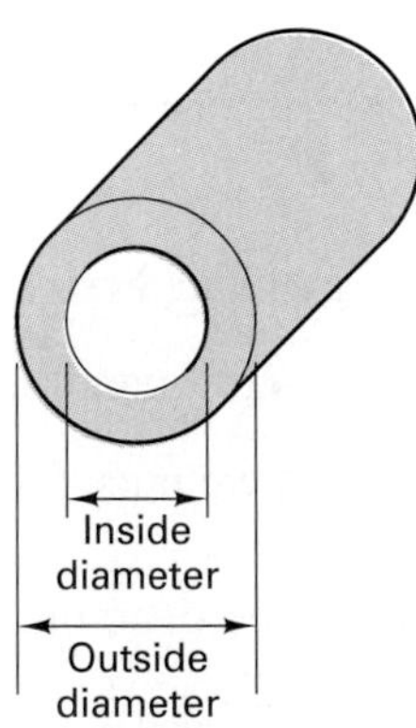

Figure 3-9

To answer this question, it is necessary to compare the size of two fractions. To do this, we first select a common denominator. That is, we find a denominator that each denominator will divide into evenly. A common denominator can always be found by multiplying the two denominators together. However, many times a smaller number can be used as a common denominator. We should always try to find the *least common denominator* (LCD). Since 8 divides evenly into 32, we can use 32 for our common denominator. Next, we change $\frac{5}{8}$ to an equivalent fraction with a denominator of 32.

$$\frac{5}{8} = \frac{5 \times 4}{8 \times 4} = \frac{20}{32}$$

Now, is $\frac{20}{32}$, which is equivalent to $\frac{5}{8}$, larger than $\frac{21}{32}$? No. Then $\frac{21}{32}$ in. cannot be the inside diameter of a $\frac{5}{8}$-in. pipe.

The *least common denominator* is also the *least common multiple* (LCM) of the denominators of the fractions. With many fractions, the least common denominator can be found by inspection. For fractions with larger denominators, you may need to use the procedure for finding the LCM discussed in Chapter 1.

Rule 3-4.1 ***To find the least common denominator (LCD) of two or more fractions, find the least common multiple (LCM) of the denominators of each fraction.***

EXAMPLE 3-4.1 Find the least common denominator for the following fractions.

$$\frac{5}{12}, \quad \frac{4}{15}, \quad \frac{3}{8}$$

Write the prime factorization of each denominator.

12	15	8
2×6	3×5	2×4
$2 \times 2 \times 3$		$2 \times 2 \times 2$
$2^2 \times 3$		2^3

$$\begin{aligned} \text{LCM or LCD} &= 2^3 \times 3 \times 5 \\ &= 8 \times 3 \times 5 \\ &= 120 \end{aligned}$$

■

L.O.2. Compare fractions.

Comparing Fractions

To compare quantities, you must compare like amounts. Thus, when comparing fractions, we must have fractions with the same denominator. In comparing $\frac{3}{7}$ and $\frac{5}{7}$, since the denominators are the same, we compare numerators. Therefore, $\frac{3}{7}$ is smaller than $\frac{5}{7}$.

To compare fractions with different denominators, we must first change the fractions to equivalent fractions having a common denominator.

Rule 3-4.2 ***To compare fractions:***

1. Find the least common denominator (the *smallest* number each denominator will divide into evenly), which is also the least common multiple.
2. Change each fraction to an equivalent fraction having the least common denominator.
3. Compare the numerators.

Tips and Traps

Fractions can be compared by changing them to equivalent fractions having *any* common denominator, but using the least common denominator gives us smaller numbers to work with.

EXAMPLE 3-4.2 Two drill bits have diameters of $\frac{3}{8}$ in. and $\frac{5}{16}$ in., respectively. Which drill bit will make the larger hole?

Solution: Finding the least common denominator, we get 16.

$$\frac{3}{8} = \frac{6}{16}, \qquad \frac{5}{16} = \frac{5}{16}$$

Now that each fraction has been changed to an equivalent fraction with the same denominator, we compare the numerators: $\frac{6}{16}$ is larger than $\frac{5}{16}$ since 6 is larger than 5, so $\frac{3}{8}$, which is equivalent to $\frac{6}{16}$, is larger than $\frac{5}{16}$. The drill bit with a $\frac{3}{8}$-in. diameter will drill the larger hole. ■

EXAMPLE 3-4.3 One brand of insecticide requires $\frac{3}{4}$ qt of insecticide mixed with 1 gal of water, while another brand requires $\frac{4}{5}$ qt of insecticide mixed with 1 gal of water. Which brand of insecticide requires more insecticide to be mixed with 1 gal of water?

Solution: Again we are comparing fractions. So we begin by finding the least common denominator.

$$\frac{3}{4} = \frac{15}{20}$$

$$\frac{4}{5} = \frac{16}{20}$$

Comparing the numerators, we find that $\frac{16}{20}$ is larger than $\frac{15}{20}$, so the brand requiring $\frac{4}{5}$ qt of insecticide requires more insecticide.

■

SELF-STUDY EXERCISES 3-4

L.O.1. Find the least common denominator for the following fractions.

1. $\frac{5}{8}, \frac{4}{9}$
2. $\frac{3}{10}, \frac{4}{15}$
3. $\frac{9}{10}, \frac{4}{25}$
4. $\frac{7}{12}, \frac{9}{16}, \frac{5}{8}$
5. $\frac{2}{3}, \frac{5}{12}, \frac{7}{8}$

L.O.2. Show which fraction is larger.

6. $\frac{2}{3}, \frac{3}{5}$
7. $\frac{5}{12}, \frac{7}{16}$
8. $\frac{8}{9}, \frac{7}{8}$
9. $\frac{5}{8}, \frac{11}{16}$
10. $\frac{15}{32}, \frac{29}{64}$
11. $\frac{7}{12}, \frac{9}{16}$

12. $\frac{3}{8}, \frac{4}{5}$

13. $\frac{7}{11}, \frac{9}{10}$

14. $\frac{4}{15}, \frac{3}{16}$

15. $\frac{1}{2}, \frac{7}{16}$

16. Is the thickness of a $\frac{3}{16}$-in. sheet of metal greater than the length of a $\frac{15}{64}$-in. sheet metal screw?

17. Will a pipe with $\frac{5}{16}$-in. outside diameter fit inside a pipe with a $\frac{3}{8}$-in. inside diameter?

18. A hollow-wall fastener has a grip range up to $\frac{7}{16}$ in. Is it long enough to fasten a thin sheet metal strip to a plywood wall if the combined thickness is $\frac{3}{8}$ in.?

19. A range top is $29\frac{1}{8}$ in. long by $19\frac{1}{2}$ in. wide. Is it smaller than an existing opening $29\frac{3}{16}$ in. long by $19\frac{9}{16}$ in. wide?

20. A wrench is marked $\frac{5}{8}$ at one end and $\frac{19}{32}$ at the other. Which end is larger?

21. Charles Bryant has a wrench marked $\frac{25}{32}$, but it is too large. Would a $\frac{7}{8}$-in. wrench be smaller?

22. For a do-it-yourself project, Brenda Jinkins needs to cut a piece of sheet metal slightly longer than the required $10\frac{21}{32}$ in. of the plans and trim it down to size. Brenda cut the piece $10\frac{3}{4}$ in. Was it cut too short?

23. A plastic anchor for a No. 6 × 1-in. screw requires that at least a $\frac{3}{16}$-in. diameter hole be drilled. Will a $\frac{1}{4}$-in. drill bit be large enough?

24. The plastic anchor in Exercise 23 requires a minimum hole depth of $\frac{7}{8}$ in. Is a $\frac{3}{4}$-in. hole deep enough?

25. Is a $\frac{3}{8}$-in. wrench too large or too small for a $\frac{1}{2}$-in. bolt?

3-5 ADDING FRACTIONS AND MIXED NUMBERS

Learning Objectives

L.O.1. Add fractions.
L.O.2. Add mixed numbers.

So far we have been studying what fractions and mixed numbers represent and how to convert them to their equivalents. Now it is time to put that knowledge to use to solve the kinds of problems commonly found on the job in the various technologies. Let's look first at problems using addition.

L.O.1. Add fractions.

Adding Fractions

The most important rule to remember in adding fractions is that all fractions to be added must have the same denominator.

Trying to add unlike fractions is like trying to add unlike objects or measures, However, when unlike fractions are to be added, we can change the fractions to equivalent fractions having a common denominator and then add the fractions.

The following rule is used when adding fractions.

Rule 3-5.1 *To add fractions:*

1. If the denominators are not the same, find the least common denominator.
2. Change each fraction not already expressed in terms of the common denominator to an equivalent fraction having the common denominator.
3. Add the numerators only.
4. The common denominator will be the denominator of the sum.
5. Reduce the answer to lowest terms and change improper fractions to whole or mixed numbers.

EXAMPLE 3-5.1 Find the sum of $\frac{3}{8} + \frac{1}{8}$.

Solution: Since the denominators are the same, start with step 3 of Rule 3-5.1.

$$\frac{3}{8} + \frac{1}{8} = \frac{4}{8} = \frac{1}{2}$$

■

Tips and Traps

We are beginning to work some of the basic steps mentally. For example, $\frac{4}{8} = \frac{4 \div 4}{8 \div 4} = \frac{1}{2}$ was written as $\frac{4}{8} = \frac{1}{2}$. The understanding of basic principles is important; however, we do not want to burden ourselves with unnecessary written procedures.

EXAMPLE 3-5.2 Add $\frac{5}{32} + \frac{3}{16} + \frac{7}{8}$.

Solution: The least common denominator may be found by inspection. Both 8 and 16 divide evenly into 32, so 32 may be used as the common denominator. Next, change each fraction to an equivalent fraction whose denominator is 32.

$$\frac{5}{32} = \frac{5}{32}, \quad \frac{3}{16} = \frac{6}{32}, \quad \frac{7}{8} = \frac{28}{32}$$

Then add the numerators.

$$\frac{5}{32} + \frac{6}{32} + \frac{28}{32} = \frac{39}{32}$$

Finally, change to a mixed number.

$$\frac{39}{32} = 1\frac{7}{32}$$

■

Notice that in a problem of this type nearly all our fraction skills are used.

EXAMPLE 3-5.3 Find the total of $\frac{3}{4} + \frac{2}{3}$.

Solution: The least common denominator may be found by inspection. The smallest number that both 3 and 4 will divide into evenly is 12. The common denominator is 12. Some students prefer writing the problem vertically.

$$\begin{aligned} \frac{3}{4} &= \frac{9}{12} \\ + \frac{2}{3} &= \frac{8}{12} \\ \hline & \frac{17}{12} = 1\frac{5}{12} \end{aligned}$$

■

EXAMPLE 3-5.4 A plumber uses a $\frac{9}{16}$-in.-diameter copper tube wrapped with $\frac{5}{8}$-in. insulation. What size hole must be bored in the stud for the insulated pipe to be installed?

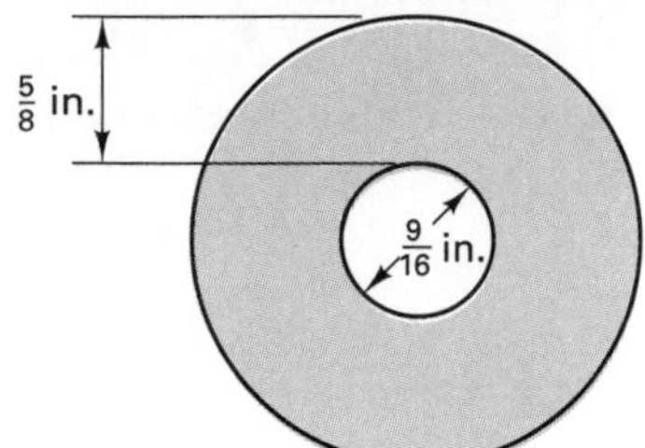

Figure 3-10

Solution: From Fig. 3-10, we can see that the $\frac{5}{8}$-in. insulation increases the diameter on each side of the pipe. To get the total diameter of the pipe and insulation, we add $\frac{9}{16} + \frac{5}{8} + \frac{5}{8}$. The thickness of the insulation is added twice since it appears on both sides of the original diameter as it wraps around the pipe.

$$\begin{aligned} \frac{9}{16} &= \frac{9}{16} \qquad \textit{(The LCD is 16.)} \\ \frac{5}{8} &= \frac{10}{16} \\ + \frac{5}{8} &= \frac{10}{16} \\ \hline & \frac{29}{16} = 1\frac{13}{16} \end{aligned}$$

The total diameter is $1\frac{13}{16}$ in. and the diameter of the hole must be at least this measure. ■

L.O.2. Add mixed numbers.

Adding Mixed Numbers

There are two popular methods for adding mixed numbers. One way is to add the whole numbers and fractions separately and then combine the sums. The other way is to change each mixed number to an improper fraction and then follow the rules for adding fractions. The first method is more appropriate for most situations.

EXAMPLE 3-5.5 Add $5\frac{2}{3} + 7\frac{3}{8} + 4\frac{1}{2}$.

Solution: It is usually more convenient to arrange the problem vertically when adding.

Using the first method, we have

$$\begin{array}{rcl} 5\frac{2}{3} & = & 5\frac{16}{24} \\ 7\frac{3}{8} & = & 7\frac{9}{24} \\ +\ 4\frac{1}{2} & = & 4\frac{12}{24} \\ \hline & & 16\,\frac{37}{24} \end{array} \qquad \begin{array}{l} \textit{(The LCD is 24.)} \\ \\ \\ \left(\frac{37}{24} = 1\frac{13}{24}\right) \end{array}$$

$$16 + 1\frac{13}{24} = 17\frac{13}{24}$$

Using the second method, we have

$$\begin{array}{rcccl} 5\frac{2}{3} & = & \frac{17}{3} & = & \frac{136}{24} \\ 7\frac{3}{8} & = & \frac{59}{8} & = & \frac{177}{24} \\ +\ 4\frac{1}{2} & = & \frac{9}{2} & = & \frac{108}{24} \\ \hline & & & & \frac{421}{24} = 17\frac{13}{24} \end{array}$$

■

EXAMPLE 3-5.6 Find the sum of $32\frac{5}{8} + 15\frac{1}{2}$.

Solution: Using the first method, we have

$$\begin{array}{rcl} 32\frac{5}{8} & = & 32\frac{5}{8} \\ +\ 15\frac{1}{2} & = & 15\frac{4}{8} \\ \hline & & 47\frac{9}{8} = 48\frac{1}{8} \end{array} \qquad \begin{array}{l} \textit{(The LCD is 8.)} \\ \\ \left(\frac{9}{8} = 1\frac{1}{8},\ 47 + 1\frac{1}{8} = 48\frac{1}{8}\right) \end{array}$$

■

EXAMPLE 3-5.7 A carpenter cut the following pieces from a metal rod: $3\frac{3}{8}$ in., $9\frac{5}{16}$ in., and $7\frac{3}{4}$ in. What was the total length cut from the rod?

Solution: To find the total length cut from the rod, we add the mixed numbers.

$$
\begin{array}{rcll}
3\frac{3}{8} & = & 3\frac{6}{16} & \textit{(The LCD is 16.)} \\
9\frac{5}{16} & = & 9\frac{5}{16} & \\
+\ \ 7\frac{3}{4} & = & 7\frac{12}{16} & \\
\hline
19\frac{23}{16} & = & 20\frac{7}{16} & \left(\frac{23}{16} = 1\frac{7}{16},\ 19 + 1\frac{7}{16} = 20\frac{7}{16}\right)
\end{array}
$$

$20\frac{7}{16}$ in. were cut from the rod. ■

Tips and Traps

When adding mixed numbers and whole numbers, the whole number can be thought of as a mixed number with zero as the numerator of the fraction.

$$
5 + 3\frac{1}{3} = \begin{array}{r} 5\frac{0}{3} \\ +\ 3\frac{1}{3} \\ \hline \end{array}
$$

Recall the addition property of zero. Zero added to any number does not change the value of the number. Thus, $5 + 3\frac{1}{3} = 8\frac{1}{3}$.

EXAMPLE 3-5.8 Add $15\frac{1}{2} + 27 + 35\frac{3}{4}$.

Solution

$$
\begin{array}{rcll}
15\frac{1}{2} & = & 15\frac{2}{4} & \textit{(The LCD is 4.)} \\
27 & = & 27 & \\
+\ \ 35\frac{3}{4} & = & 35\frac{3}{4} & \\
\hline
77\frac{5}{4} & = & 78\frac{1}{4} & \left(\frac{5}{4} = 1\frac{1}{4},\ 77 + 1\frac{1}{4} = 78\frac{1}{4}\right)
\end{array}
$$

■

EXAMPLE 3-5.9 Find the largest permissible measurement of a part if the blueprint calls for the part to be 2 in. long and the tolerance is $\pm\frac{1}{8}$ in.

Solution: Tolerance is the amount a part can vary from the blueprint specification. To find the largest permissible measure, we add.

$$2 + \frac{1}{8} = 2\frac{1}{8}$$

The largest permissible measure is $2\frac{1}{8}$ in. ■

Tips and Traps

Tolerance is an often used concept in technical applications. The "plus or minus" symbol ($\pm$) indicates that the actual measure of an object can vary by being no more than a specified amount larger (plus) or no less than a specified amount smaller (minus). If you are asked to find the largest possible acceptable measure, *add* the specified amount. If you are asked to find the smallest possible acceptable measure, *subtract* the specified amount. If you are asked to find the *limits* or *limit dimensions*, find *both* the largest and smallest acceptable measures.

SELF-STUDY EXERCISES 3-5

L.O.1. Add; reduce answers to lowest terms and convert any improper fractions to whole or mixed numbers.

1. $\frac{5}{16} + \frac{1}{16}$

2. $\frac{1}{2} + \frac{1}{8} + \frac{3}{4}$

3. $\frac{1}{8} + \frac{1}{2}$

4. $\frac{3}{8} + \frac{5}{32} + \frac{1}{4}$

5. $\frac{5}{16} + \frac{1}{4}$

6. $\frac{15}{16} + \frac{1}{2}$

7. $\frac{3}{32} + \frac{5}{64}$

8. $\frac{7}{8} + \frac{3}{5}$

9. $\frac{3}{4} + \frac{8}{9}$

10. $\frac{7}{8} + \frac{5}{24}$

11. What is the thickness of a countertop made of $\frac{7}{8}$-in. plywood and $\frac{1}{16}$-in. Formica?

12. Three pieces of steel are joined together. What is the total thickness if the pieces are $\frac{1}{2}$ in., $\frac{7}{16}$ in., and $\frac{29}{32}$ in.?

13. Three books are placed side by side. They are $\frac{5}{16}$ in., $\frac{7}{8}$ in., and $\frac{3}{4}$ in. wide. What is the total width of the books if they are polywrapped in one package?

14. Find the outside diameter of a pipe whose wall is $\frac{1}{2}$ in. thick if its inside diameter is $\frac{7}{8}$ in.

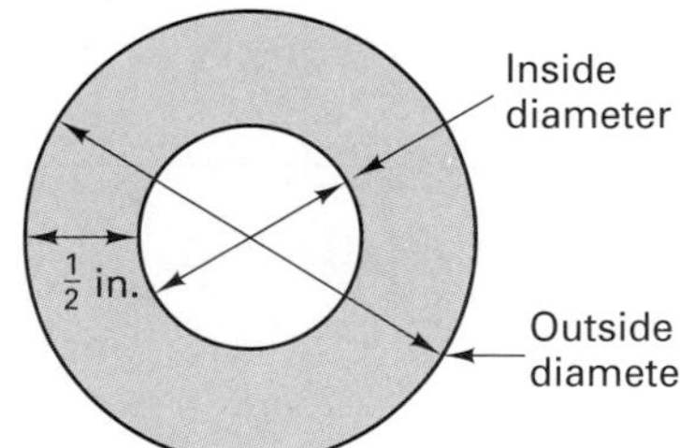

15. How long a bolt is needed to fasten two pieces of metal each $\frac{7}{16}$ in. thick if a $\frac{1}{8}$-in. lockwasher is used and a $\frac{1}{4}$-in. nut is used?

L.O.2. Add; reduce answers to lowest terms and convert any improper fractions to whole or mixed numbers.

16. $2\frac{3}{5} + 4\frac{1}{5}$

17. $1\frac{5}{8} + 2\frac{1}{2}$

18. $3\frac{3}{4} + 7\frac{3}{16} + 5\frac{7}{8}$

19. $\frac{1}{6} + \frac{7}{9} + \frac{2}{3}$

20. $2\frac{1}{4} + 2\frac{9}{16}$

21. $1\frac{5}{16} + 4\frac{7}{32}$

22. $3\frac{1}{4} + 1\frac{7}{16}$

23. $4\frac{1}{2} + 9$

24. $\frac{2}{3} + 3\frac{4}{5}$

25. $5\frac{1}{8} + 3 + 4\frac{9}{16}$

26. The studs of an outside wall are $5\frac{3}{4}$ in. thick. The inside wall board is $\frac{7}{8}$ in. thick and the outside covering is $2\frac{3}{16}$ in. thick. What is the total thickness of the wall?

27. A blueprint calls for a piece of bar stock $3\frac{7}{8}$ in. long. If a tolerance of $\pm\frac{1}{16}$ in. is allowed, what is the longest permissible measurement for the bar stock?

28. If $4\frac{3}{8}$ gal of water are used to dilute $7\frac{1}{4}$ gal of acid, how many gallons are in the mixture?

29. How much bar stock is needed to make bars of the following lengths? $10\frac{1}{4}$ in., $8\frac{7}{16}$ in., $5\frac{15}{32}$ in. Disregard waste.

30. Three pieces of I-beam each measuring $7\frac{5}{8}$ in. are needed to complete a job. How much I-beam is needed?

3-6 SUBTRACTING FRACTIONS AND MIXED NUMBERS

Learning Objectives

L.O.1. Subtract fractions.
L.O.2. Subtract mixed numbers.

L.O.1. Subtract fractions.

Subtracting Fractions

Knowing how to add fractions and mixed numbers helps us when we need to subtract these same numbers. The rules for subtracting fractions and mixed numbers are very similar to the rules for adding fractions and mixed numbers.

Rule 3-6.1 *To subtract fractions:*

1. If the denominators are not the same, find the least common denominator.
2. Change each fraction not expressed in terms of the common denominator to an equivalent fraction having the common denominator.
3. Subtract the numerators.
4. The common denominator will be the denominator of the difference.
5. Reduce the answer to lowest terms.

Fractions must have the same denominator before they can be subtracted. Chapter 1 emphasized a very important contrast between addition and subtraction. In addition, the order of the terms (addends) does not matter. That is, 3 + 4 is the same as 4 + 3. However, in subtraction, the order of the terms is very important, for subtraction is not commutative. We must be careful in arranging a subtraction problem.

EXAMPLE 3-6.1 Subtract $\frac{3}{8} - \frac{7}{32}$.

Solution

$$\frac{3}{8} = \frac{12}{32}$$

Therefore,

$$\frac{12}{32} - \frac{7}{32} = \frac{5}{32}$$

■

EXAMPLE 3-6.2 Find the difference between $\frac{2}{3}$ and $\frac{5}{8}$.

Solution

$$\begin{array}{rcr} \frac{2}{3} & = & \frac{16}{24} \\ -\ \frac{5}{8} & = & \frac{15}{24} \\ \hline & & \frac{1}{24} \end{array}$$

■

L.O.2. Subtract mixed numbers.

Subtracting Mixed Numbers

As in addition, there is more than one method for subtracting mixed numbers. Because the method that requires changing each mixed number to an improper fraction is not the most practical method, the examples below use the method that considers the whole numbers and fractional parts separately.

EXAMPLE 3-6.3 Subtract $15\frac{7}{8} - 4\frac{1}{2}$.

Solution

$$\begin{array}{rcr} 15\frac{7}{8} & = & 15\frac{7}{8} \\ -\ 4\frac{1}{2} & = & 4\frac{4}{8} \\ \hline & & 11\frac{3}{8} \end{array}$$

■

An additional procedure is used when the fractional part of the subtrahend (number being subtracted) is larger than the fractional part of the minuend (first number). This causes us to have to borrow from the whole-number part of the minuend.

Rule 3-6.2 *To borrow in subtracting mixed numbers:*

1. If the fractional parts of the mixed numbers do not have the same denominator, change them to equivalent fractions having a common denominator.
2. When the fraction in the subtrahend is larger than the fraction in the minuend, borrow one whole number from the whole-number part of the minuend. That makes the whole number 1 less.
3. Change the one whole number borrowed to an improper fraction having the common denominator. For example, $1 = \frac{3}{3}$, $1 = \frac{8}{8}$, $1 = \frac{n}{n}$, where n is the common denominator.
4. Add the borrowed fraction $\left(\frac{n}{n}\right)$ to the fraction already in the minuend.
5. Subtract the fractional parts and the whole-number parts.
6. Reduce the answer to lowest terms.

Example 3-6.4 Subtract $15\frac{3}{4}$ from $18\frac{1}{2}$.

Solution

$$\begin{array}{rcrcrcr} 18\frac{1}{2} & = & 18\frac{2}{4} & = & 17\frac{4}{4} + \frac{2}{4} & = & 17\frac{6}{4} \\ -\ 15\frac{3}{4} & = & 15\frac{3}{4} & = & 15\frac{3}{4} & = & 15\frac{3}{4} \\ \hline & & & & & & 2\frac{3}{4} \end{array} \qquad \left(18 - 1 = 17,\ 1 = \frac{4}{4}, \frac{4}{4} + \frac{2}{4} = \frac{6}{4}\right)$$

■

Example 3-6.5 Find the difference between $237\frac{3}{8}$ and $197\frac{17}{32}$.

Solution

$$\begin{array}{rcrcr} 237\frac{3}{8} & = & 237\frac{12}{32} & = & 236\frac{44}{32} \\ -\ 197\frac{17}{32} & = & 197\frac{17}{32} & = & 197\frac{17}{32} \\ \hline & & & & 39\frac{27}{32} \end{array} \qquad \left(237 - 1 = 236,\ 1 = \frac{32}{32}, \frac{32}{32} + \frac{12}{32} = \frac{44}{32}\right)$$

■

Example 3-6.6 A metal block weighing $127\frac{1}{2}$ lb is removed from a flatbed truck that carried a payload of $433\frac{3}{8}$ lb. How many pounds remain on the truck?

Solution

$$\begin{array}{rcccl} & 433\frac{3}{8} & = 433\frac{3}{8} & = 432\frac{11}{8} & \quad \left(433 - 1 = 432,\ 1 = \frac{8}{8}, \frac{8}{8} + \frac{3}{8} = \frac{11}{8}\right) \\ - & 127\frac{1}{2} & = 127\frac{4}{8} & = 127\frac{4}{8} & \\ \hline & & & 305\frac{7}{8} & \end{array}$$

$305\frac{7}{8}$ lb remain on the truck. ■

Example 3-6.7 A plastic pipe $21\frac{7}{8}$ ft long is cut from a piece of pipe measuring $32\frac{5}{16}$ ft long. How much of the original pipe remains?

Solution: To find the length of the remaining pipe, we subtract.

$$\begin{array}{rcccl} & 32\frac{5}{16} & = 32\frac{5}{16} & = 31\frac{21}{16} & \quad \left(32 - 1 = 31,\ 1 = \frac{16}{16}, \frac{16}{16} + \frac{5}{16} = \frac{21}{16}\right) \\ - & 21\frac{7}{8} & = 21\frac{14}{16} & = 21\frac{14}{16} & \\ \hline & & & 10\frac{7}{16} & \end{array}$$

$10\frac{7}{16}$ ft of pipe remain. ■

Tips and Traps

As in addition, when subtracting whole numbers and mixed numbers, consider the whole number to have zero fractional parts. Then follow the same procedures as before. Borrow when necessary.

Example 3-6.8 Subtract 27 from $45\frac{1}{3}$.

Solution

$$\begin{array}{rcc} & 45\frac{1}{3} & = 45\frac{1}{3} \\ - & 27 & = 27\frac{0}{3} \\ \hline & & 18\frac{1}{3} \end{array}$$

■

Example 3-6.9 Find the difference between 138 and $104\frac{11}{16}$.

Solution

$$\begin{array}{rcccl} 138 & = 138\frac{0}{16} & = & 137\frac{16}{16} & \left(138 - 1 = 137,\ 1 = \frac{16}{16}, \frac{16}{16} + \frac{0}{16} = \frac{16}{16}\right) \\ -\ 104\frac{11}{16} & = 104\frac{11}{16} & = & 104\frac{11}{16} & \\ \hline & & & 33\frac{5}{16} & \end{array}$$

■

Example 3-6.10 How many feet of wire are left on a 100-ft roll if $27\frac{1}{4}$ ft are used from the roll?

Solution

$$\begin{array}{rcl} 100 & = & 99\frac{4}{4} \\ -\ 27\frac{1}{4} & = & 27\frac{1}{4} \\ \hline & & 72\frac{3}{4} \end{array}$$

There are $72\frac{3}{4}$ ft left on the roll. ■

Example 3-6.11 Three lengths measuring $5\frac{1}{4}$ in., $7\frac{3}{8}$ in., and $6\frac{1}{2}$ in. are cut from a 64-in. bar of angle iron. If $\frac{3}{16}$in. is wasted on each cut, how many inches of angle iron remain?

Solution: We will first find the total amount of angle iron used. This will include the three lengths and the waste for three cuts.

$$\begin{array}{rcl} 5\frac{1}{4} & = & 5\frac{4}{16} \\ 7\frac{3}{8} & = & 7\frac{6}{16} \\ 6\frac{1}{2} & = & 6\frac{8}{16} \\ \frac{3}{16} & = & \frac{3}{16} \\ \frac{3}{16} & = & \frac{3}{16} \\ +\ \frac{3}{16} & = & \frac{3}{16} \\ \hline & & 18\frac{27}{16} = 18 + \frac{27}{16} = 18 + 1\frac{11}{16} = 19\frac{11}{16} \end{array}$$

Next we will subtract to find the amount of angle iron remaining.

$$\begin{array}{rcr} 64 & = & 63\frac{16}{16} \\ -19\frac{11}{16} & = & 19\frac{11}{16} \\ \hline & & 44\frac{5}{16} \end{array}$$

There are $44\frac{5}{16}$ in. of angle iron remaining. ■

SELF-STUDY EXERCISES 3-6

L.O.1. Subtract; reduce when necessary.

1. $\frac{7}{8} - \frac{5}{8}$ **2.** $\frac{9}{16} - \frac{3}{8}$ **3.** $\frac{7}{16} - \frac{3}{8}$

4. $\frac{5}{8} - \frac{1}{2}$ **5.** $\frac{5}{32} - \frac{1}{64}$ **6.** $\frac{7}{8} - \frac{3}{4}$

L.O.2.

7. $9\frac{11}{16} - 5$ **8.** $23\frac{3}{16} - 5\frac{7}{16}$ **9.** $9\frac{1}{4} - 4\frac{5}{16}$

10. $9\frac{1}{32} - 3\frac{3}{8}$

11. A length of bar stock $16\frac{3}{8}$ in. long is cut so that a piece only $7\frac{9}{16}$ in. long remains. What is the length of the cutoff piece? Disregard waste.

12. A concrete foundation includes $7\frac{7}{8}$-in. of base fill. If the foundation is to be 18 in. thick, how thick must the concrete be?

13. A casting is machined so that $22\frac{1}{5}$ lb of metal remain. If the casting weighed $25\frac{3}{10}$ lb, how many pounds were removed by machine?

14. A bolt $2\frac{5}{8}$ in. long fastens two pieces of steel 1 in. and $1\frac{7}{32}$ in. thick. If a $\frac{3}{32}$-in.-thick lock-washer and a $\frac{1}{8}$-in.-thick washer are used, what thickness is the nut if it is flush with the bolt? When giving the measure of a bolt, the length does not include the bolt head.

15. Find the missing length.

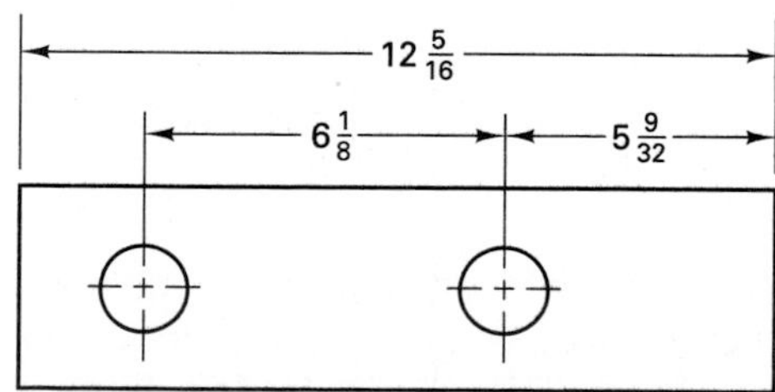

3-7 MULTIPLYING FRACTIONS AND MIXED NUMBERS

Learning Objectives

L.O.1. Multiply fractions.
L.O.2. Multiply mixed numbers.

We saw in Chapter 1 that the multiplication of whole numbers is a shortcut for addition and therefore a time saver on the job. We saw that multiplying 2×4 to get 8 was quicker and more efficient than adding $2 + 2 + 2 + 2$ to get the same number 8. When multiplying a whole number by a whole number, we are increasing a given number of units by a certain number of times. Thus, in $2 \times 7 = 14$, we are increasing 2 by seven times. However, when multiplying a fraction by a fraction, we are doing something different.

When multiplying a fraction by a fraction, we are trying to find *a part of a part*. For instance, $\frac{1}{2} \times \frac{1}{2}$ is $\frac{1}{2}$ of $\frac{1}{2}$. The word "of" is the clue that we must multiply to find the part we are looking for. The examples of Fig. 3-11 illustrate what is meant by finding a part of a part.

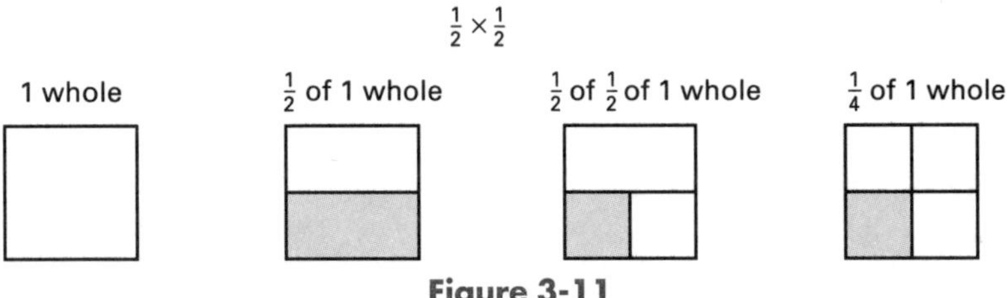

Figure 3-11

When adding or subtracting fractions and mixed numbers, it is necessary to have a common denominator. Otherwise, the fractional parts cannot be added or subtracted. However, this is not the case when multiplying. In multiplying fractions, we do *not* change fractions to equivalent fractions having a common denominator. Look at two more examples of taking a part of a part (Figs. 3-12 and 3-13).

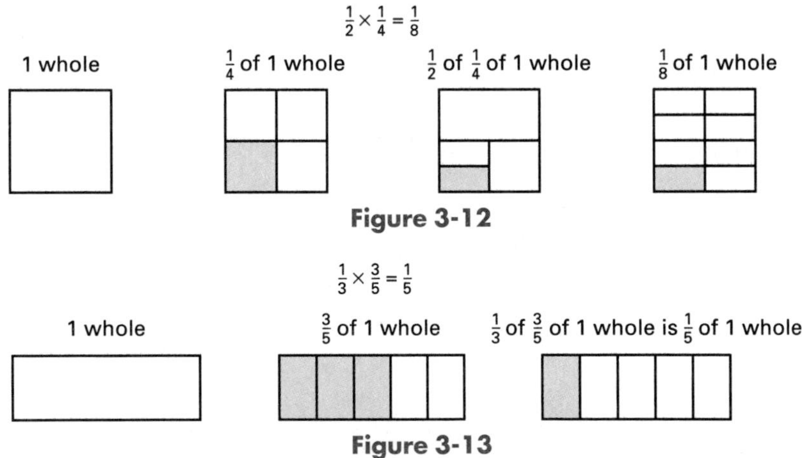

Figure 3-12

Figure 3-13

L.O.1. Multiply fractions.

Multiplying Fractions

The preceding examples graphically show how we can multiply a fraction by a fraction, that is, take a part of a part. On the job, however, we cannot stop and draw pictures or figures like the examples. Instead, we need rules to follow so that we can quickly make our computations.

Rule 3-7.1 *To multiply fractions:*

1. Multiply the numerators of the fractions for the numerator of the answer.
2. Multiply the denominators for the denominator of the answer.
3. Reduce the answer to lowest terms.

EXAMPLE 3-7.1 Find $\frac{1}{2}$ of $\frac{1}{2}$.

Solution

$$\frac{1}{2} \times \frac{1}{2} = \frac{1}{4}$$ ■

EXAMPLE 3-7.2 Find $\frac{1}{2}$ of $\frac{1}{4}$.

Solution

$$\frac{1}{2} \times \frac{1}{4} = \frac{1}{8}$$ ■

EXAMPLE 3-7.3 Find $\frac{1}{3}$ of $\frac{3}{5}$.

Solution

$$\frac{1}{3} \times \frac{3}{5} = \frac{3}{15} = \frac{1}{5}$$ ■

Tips and Traps

In Example 3-7.3, reducing was necessary. In multiplying fractions, it is much easier to reduce common factors before they are multiplied.

$$\frac{1}{3} \times \frac{3}{5} = \frac{1 \times \overset{1}{\cancel{3}}}{\underset{1}{\cancel{3}} \times 5} = \frac{1}{5}$$

Since both the numerator and denominator have a common factor of 3, the factor can be reduced before multiplying. Remember from Section 3-2 that reducing applies the principles $\frac{n}{n} = 1$ and $1 \times n = n$. This process is also referred to as *cancelling*.

EXAMPLE 3-7.4 Find $\frac{3}{8}$ of $\frac{4}{5}$.

Solution

$\frac{3}{8} \times \frac{4}{5}$ *(4 is a common factor of both a numerator and a denominator. Reduce before multiplying.)*

$$\frac{3}{\cancel{8}_{2}} \times \frac{\cancel{4}^{1}}{5} = \frac{3}{10}$$

■

Example 3-7.5 Find $\frac{3}{9}$ of $\frac{2}{7}$.

Solution

$\frac{3}{9} \times \frac{2}{7}$ *(3 is a common factor of both a numerator and a denominator. Reduce before multiplying.)*

$$\frac{\cancel{3}^{1}}{\cancel{9}_{3}} \times \frac{2}{7} = \frac{2}{21}$$

■

Example 3-7.6 Multiply $\frac{2}{3} \times \frac{5}{9} \times \frac{1}{6}$.

Solution

$\frac{2}{3} \times \frac{5}{9} \times \frac{1}{6}$ *(2 is a common factor of both a numerator and a denominator. Reduce before multiplying.)*

$$\frac{\cancel{2}^{1}}{3} \times \frac{5}{9} \times \frac{1}{\cancel{6}_{3}} = \frac{5}{81}$$

■

Tips and Traps

Common factors that are reduced can be diagonal to each other, one above the other, or separated by another fraction, but one factor *must* be in the numerator and the other in the denominator.

Example 3-7.7 Multiply $\frac{2}{5} \times \frac{10}{21} \times \frac{6}{12}$.

Solution

$$\frac{\cancel{2}^{1}}{\cancel{5}_{1}} \times \frac{\cancel{10}^{2}}{21} \times \frac{\cancel{6}^{1}}{\cancel{12}_{\cancel{2}_{1}}} = \frac{2}{21}$$

(5 and 10 are diagonal to each other. 6 and 12 are above each other. 2 and 2 are separated by another fraction.)

Other patterns of reducing could have been done. ■

L.O.2. Multiply mixed numbers.

Multiplying Mixed Numbers

Multiplying mixed numbers or a combination of whole numbers, fractions, and mixed numbers is based on the same rules as for multiplying fractions. When multiplying mixed numbers or any combination of whole numbers, fractions, and mixed numbers, we first change each whole number or mixed number to an improper fraction. Then we can proceed as in multiplying fractions.

Rules 3-7.2 ***To multiply mixed numbers, fractions, and whole numbers:***

1. Change each mixed number or whole number to an improper fraction.
2. Reduce as much as possible.
3. Multiply numerators.
4. Multiply denominators.
5. Convert the answer to a whole or mixed number if possible.

EXAMPLE 3-7.8 Multiply $2\frac{1}{2} \times 5\frac{1}{3}$.

Solution

$$2\frac{1}{2} \times 5\frac{1}{3} = \frac{5}{2} \times \frac{16}{3} = \frac{5}{\cancel{2}_{1}} \times \frac{\cancel{16}^{8}}{3} = \frac{40}{3} = 13\frac{1}{3}$$

■

EXAMPLE 3-7.9 Multiply $5\frac{3}{8} \times 6$.

Solution

$$5\frac{3}{8} \times 6 = \frac{43}{8} \times \frac{6}{1} = \frac{43}{\cancel{8}_{4}} \times \frac{\cancel{6}^{3}}{1} = \frac{129}{4} = 32\frac{1}{4}$$

■

EXAMPLE 3-7.10 An engineering technician needed 10 pieces of wire each $3\frac{5}{8}$ in. long. What length will be needed for the wire pieces?

Solution

$$3\frac{5}{8} \times 10 = \frac{29}{\cancel{8}_{4}} \times \frac{\cancel{10}^{5}}{1} = \frac{145}{4} = 36\frac{1}{4}$$

■

$36\frac{1}{4}$ in. of wire will be needed. ■

EXAMPLE 3-7.11 Bricks that are $2\frac{1}{4}$ in. thick are used to lay a brick wall with $\frac{3}{8}$-in. mortar joints. What will be the height of the wall after nine courses?

Solution: A *course* of bricks means a horizontal row (see Fig. 3-14). If nine horizontal rows are laid, how many mortar joints are included? There will be eight mortar joints between nine rows of bricks and one mortar joint between the first row of bricks and the foundation. The total height of the wall after nine courses of bricks will include nine rows of bricks at $2\frac{1}{4}$ in. per row and nine mortar joints at $\frac{3}{8}$ in. per mortar joint.

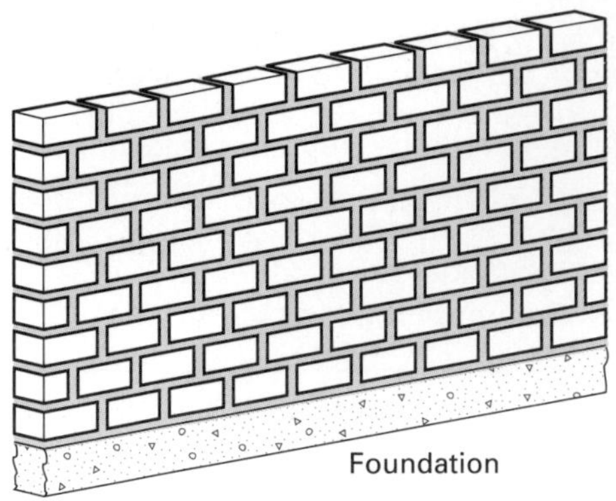

Figure 3-14

$$\left(9 \times 2\frac{1}{4}\right) + \left(9 \times \frac{3}{8}\right)$$

$$\left(\frac{9}{1} \times \frac{9}{4}\right) + \left(\frac{9}{1} \times \frac{3}{8}\right)$$

$$\frac{81}{4} + \frac{27}{8}$$

$$20\frac{1}{4} + 3\frac{3}{8}$$

$$20\frac{2}{8} + 3\frac{3}{8} = 23\frac{5}{8}$$

The wall will be $23\frac{5}{8}$ in. high. ■

SELF-STUDY EXERCISES 3-7

L.O.1. Multiply and reduce answers to lowest terms.

1. $\frac{3}{4} \times \frac{1}{8}$

2. $\frac{1}{2} \times \frac{7}{16}$

3. $\frac{5}{8} \times \frac{7}{10}$

4. $\frac{2}{3} \times \frac{7}{8}$

5. $\frac{1}{2} \times \frac{3}{4} \times \frac{8}{9}$

6. $\frac{3}{8} \times \frac{5}{6} \times \frac{1}{2}$

7. $\frac{7}{8} \times \frac{2}{5} \times \frac{4}{21}$

8. $\frac{11}{12} \times \frac{9}{10} \times \frac{8}{15}$

9. $\frac{3}{10} \times \frac{4}{15}$

10. $\frac{15}{16} \times \frac{7}{10}$

L.O.2. Multiply and reduce answers to lowest terms. Convert improper fractions to whole or mixed numbers.

11. $7 \times 3\frac{1}{8}$

12. $\frac{3}{5} \times 125$

13. $2\frac{3}{4} \times 1\frac{1}{2}$

14. $9\frac{1}{2} \times 3\frac{4}{5}$

15. $\frac{1}{5} \times 7\frac{5}{8}$

16. $\frac{2}{3} \times 3\frac{1}{4}$

17. A fuel tank that holds 75 liters (L) of fuel is $\frac{1}{4}$ full. How many liters of fuel are in the tank?

18. If steps are 12 risers high and each riser is $7\frac{1}{2}$ in. high, what is the total rise of the steps?

19. A piece of sheet metal is $\frac{1}{16}$ in. thick. How thick would a stack of 250 pieces be?

20. An alloy, which is a substance composed of two or more metals, is $\frac{11}{16}$ copper, $\frac{7}{32}$ tin, and $\frac{3}{32}$ zinc. How many kilograms (kg) of each metal are needed to make 384 kg of alloy?

3-8 DIVIDING FRACTIONS AND MIXED NUMBERS

Learning Objectives

L.O.1. Find reciprocals.
L.O.2. Divide fractions.
L.O.3. Divide mixed numbers.

Once we can multiply fractions, we can easily handle the division of fractions and mixed numbers. The reason for this is that multiplication and division are inverse operations.

Let's make a comparison. If six units are divided into two equal parts, how many units will be in each part? $6 \div 2 = ?$ The answer is 3. But, at the same time, how many units are $\frac{1}{2}$ of 6 units? $\frac{1}{2} \times 6 = ?$ Again, the answer is 3.

In the two examples above, $6 \div 2$ and $\frac{1}{2} \times 6$ represent three units. To divide by two, then, is the same as taking half of a quantity.

L.O.1. Find reciprocals.

Finding Reciprocals

If we compare $12 \div 3$ and $\frac{1}{3} \times 12$, we would find that both answers are the same, 4. This means that $12 \div 3 = \frac{1}{3} \times 12$ or $12 \times \frac{1}{3}$. So, not only is there a relationship between multiplication and division, there is also a relationship between 3 and $\frac{1}{3}$. Pairs of numbers like $\frac{1}{2}$ and 2 or $\frac{1}{3}$ and 3 are called *reciprocals.*

■ **DEFINITION 3-8.1 Reciprocals.** Two numbers are *reciprocals* if their product is 1. Thus, $\frac{1}{2}$ and 2 are reciprocals because $\frac{1}{2} \times 2 = 1$, and $\frac{1}{3}$ and 3 are reciprocals because $\frac{1}{3} \times 3 = 1$.

Rule 3-8.1 ***To find the reciprocal of a number:***

1. Write the number in fractional form.
2. Interchange the numerator and denominator so that the numerator becomes the denominator and the denominator becomes the numerator.

Tips and Traps

Interchanging the numerator and denominator of a fraction is commonly referred to as *inverting* the fraction.

EXAMPLE 3-8.1 Find the reciprocals of $\frac{2}{3}, \frac{4}{7}, \frac{1}{5}$, 3, 1, $2\frac{1}{2}$, and 0.

Solution

The reciprocal of $\frac{2}{3}$ is $\frac{3}{2}$.

The reciprocal of $\frac{4}{7}$ is $\frac{7}{4}$.

The reciprocal of $\frac{1}{5}$ is $\frac{5}{1}$ or 5.

The reciprocal of 3 is $\frac{1}{3}$. $(3 = \frac{3}{1})$

The reciprocal of 1 is 1. $(1 = \frac{1}{1})$

The reciprocal of $2\frac{1}{2}$ is $\frac{2}{5}$. $(2\frac{1}{2} = \frac{5}{2})$

0 has no reciprocal since division by 0 is impossible.

$\left(0 = \frac{0}{1}, \frac{1}{0} \text{ is undefined.}\right)$ ■

L.O.2. Divide fractions.

Dividing Fractions

Let's review the parts of a division problem.

15	÷	3	=	5
dividend		divisor		quotient

To help identify the divisor, remember that the symbol ÷ is always read ''divided by.''

Rule 3-8.2 ***To divide fractions:***

1. Change the division to an equivalent multiplication by replacing the divisor with its reciprocal and replacing the division sign (÷) with a multiplication sign (×).
2. Perform the resulting multiplication.

EXAMPLE 3-8.2 Find $\frac{5}{8} \div \frac{2}{3}$.

Solution

$$\frac{5}{8} \div \frac{2}{3} = \frac{5}{8} \times \frac{3}{2} = \frac{15}{16}$$ ■

Tips and Traps

A more common way of stating the division of fractions rule is ''invert the divisor and multiply.''

EXAMPLE 3-8.3 Divide $\frac{7}{8}$ by $\frac{14}{15}$.

Solution

$$\frac{7}{8} \div \frac{14}{15} = \frac{\overset{1}{\cancel{7}}}{8} \times \frac{15}{\underset{2}{\cancel{14}}} = \frac{15}{16}$$

■

EXAMPLE 3-8.4 An auger bit will advance $\frac{1}{16}$ in. for each turn (see Fig. 3-15). How many turns are needed to drill a hole $\frac{5}{8}$ in. deep? $\left(\frac{5}{8}\right.$ in. can be divided into how many $\left.\frac{1}{16}\text{-in. parts?}\right)$

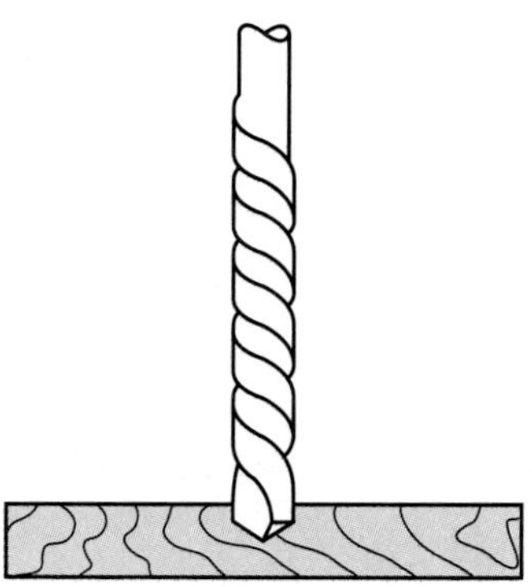

Figure 3-15

Solution

$$\frac{5}{8} \div \frac{1}{16} = \frac{5}{\underset{1}{\cancel{8}}} \times \frac{\overset{2}{\cancel{16}}}{1} = 10$$

Ten turns are needed. ■

L.O.3. Divide mixed numbers.

Dividing Mixed Numbers

Mixed numbers and whole numbers can be divided by first writing the mixed numbers or whole numbers as fractions and then following the rules for dividing fractions.

> **Rule 3-8.3** ***To divide mixed numbers, fractions, and whole numbers:***
>
> 1. Change each mixed number or whole number to an improper fraction.
> 2. Convert to an equivalent multiplication problem by using the reciprocal of the divisor.
> 3. Multiply according to the rule for multiplying fractions.

EXAMPLE 3-8.5 Find $2\frac{1}{2} \div 3\frac{1}{3}$.

Solution

$$2\frac{1}{2} \div 3\frac{1}{3} = \frac{5}{2} \div \frac{10}{3} = \frac{\overset{1}{\cancel{5}}}{2} \times \frac{3}{\underset{2}{\cancel{10}}} = \frac{3}{4}$$ ■

Example 3-8.6 Find $10 \div 1\frac{1}{4}$.

Solution

$$10 \div 1\frac{1}{4} = \frac{10}{1} \div \frac{5}{4} = \frac{\overset{2}{\cancel{10}}}{1} \times \frac{4}{\underset{1}{\cancel{5}}} = 8$$ ■

Example 3-8.7 Find $5\frac{3}{8} \div 3$.

Solution

$$5\frac{3}{8} \div 3 = \frac{43}{8} \div \frac{3}{1} = \frac{43}{8} \times \frac{1}{3} = \frac{43}{24} = 1\frac{19}{24}$$ ■

Example 3-8.8 A developer subdivided $5\frac{1}{4}$ acres into lots, each containing $\frac{7}{10}$ of an acre. How many lots were made?

Solution

$$5\frac{1}{4} \div \frac{7}{10} = \frac{21}{4} \div \frac{7}{10} = \frac{\overset{3}{\cancel{21}}}{\underset{2}{\cancel{4}}} \times \frac{\overset{5}{\cancel{10}}}{\underset{1}{\cancel{7}}} = \frac{15}{2} = 7\frac{1}{2}$$

Seven lots can be made so that each is $\frac{7}{10}$ of an acre. The $\frac{1}{2}$ lot is left over. ■

Example 3-8.9 How many pieces of cable, each $1\frac{3}{4}$ ft long, can be cut from a reel containing 100 ft of cable? See Fig. 3-16.

Figure 3-16

Solution

$$100 \div 1\frac{3}{4} = \frac{100}{1} \div \frac{7}{4} = \frac{100}{1} \times \frac{4}{7} = \frac{400}{7} = 57\frac{1}{7}$$

Fifty-seven pieces of cable can be cut to the desired length. The $\frac{1}{7}$ of the desired length is considered waste. ■

EXAMPLE 3-8.10 A pipe that is $21\frac{1}{2}$ in. long is cut into four equal pieces. If $\frac{3}{16}$ in. is wasted on each cut, how long will each piece be?

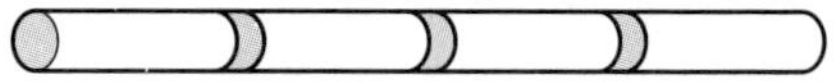

Figure 3-17

Solution: How many cuts will need to be made? (See Fig. 3-17.) Since three cuts are to be made and each cut wastes $\frac{3}{16}$ in., first find the amount wasted.

$$\frac{3}{16} \times 3 = \frac{9}{16}$$

Next, find how much pipe will be left to divide equally into four parts.

$$21\frac{1}{2} - \frac{9}{16}$$

$$\begin{array}{rcrcr} 21\frac{1}{2} & = & 21\frac{8}{16} & = & 20\frac{24}{16} \\ -\quad \frac{9}{16} & = & \frac{9}{16} & = & \frac{9}{16} \\ \hline & & & & 20\frac{15}{16} \text{ in.} \end{array} \qquad \textit{(Align vertically.)}$$

Now find the length of each part.

$$20\frac{15}{16} \div 4 = \frac{335}{16} \div \frac{4}{1} = \frac{335}{16} \times \frac{1}{4} = \frac{335}{64} = 5\frac{15}{64}$$

Each part will be $5\frac{15}{64}$ in. long ■

SELF-STUDY EXERCISES 3-8

L.O.1. Give the reciprocal of the following:

1. $\frac{3}{7}$ **2.** 8 **3.** $2\frac{1}{5}$

4. $\frac{1}{5}$ **5.** 7

L.O.2. Divide and reduce answers to lowest terms. Convert improper fractions to whole or mixed numbers.

6. $\frac{4}{5} \div \frac{8}{9}$

7. $\frac{11}{32} \div \frac{3}{8}$

8. $\frac{3}{4} \div \frac{3}{8}$

9. $\frac{7}{10} \div \frac{2}{3}$

10. $\frac{5}{6} \div \frac{2}{5}$

11. $\frac{7}{8} \div \frac{3}{16}$

12. $\frac{1}{5} \div \frac{1}{2}$

13. One-half inch is divided into $\frac{1}{16}$-in. segments. How many $\frac{1}{16}$-in. segments are in $\frac{1}{2}$ in.?

L.O.3. Divide and reduce answers to lowest terms. Convert improper fractions to whole or mixed numbers.

14. $10 \div \frac{3}{4}$

15. $8 \div \frac{1}{4}$

16. $2\frac{1}{2} \div 4$

17. $7\frac{3}{8} \div \frac{1}{2}$

18. $30\frac{1}{3} \div 4\frac{1}{3}$

19. Lumber is given in rough dimensions. Rough lumber that is 2 in. thick will dress out to $1\frac{5}{8}$ in. How many dressed 2 × 4's are in a stack that is $29\frac{1}{4}$ in. high?

20. How many $4\frac{5}{8}$-ft lengths can be cut from a 50-ft length of conduit?

21. A truck will hold 21 yd^3 (cubic yards) of gravel. If an earth mover has a shovel capacity of $1\frac{3}{4}$ yd^3, how many shovels full will be needed to fill the truck?

22. Three shelves of equal length are to be cut from a 72-in. board. If $\frac{1}{8}$ in. is wasted on each cut, what is the maximum length each shelf can be? (Two cuts will be made to divide the entire board into three equal lengths.)

23. If $\frac{1}{8}$ in. represents 1 ft on a drawing, find the dimensions of a room that measures $2\frac{1}{2}$ in. by $1\frac{7}{8}$ in. on the drawing. (How many $\frac{1}{8}$'s are there in $2\frac{1}{2}$; how many $\frac{1}{8}$'s are there in $1\frac{7}{8}$?)

24. A segment of I-beam is $10\frac{1}{2}$ ft long. Into how many whole $2\frac{1}{4}$-ft pieces can it be divided? Disregard waste.

25. How many $17\frac{5}{8}$-in. strips of quarter-round molding can be cut from a piece $132\frac{3}{4}$ in. long? Disregard waste.

26. How many $9\frac{1}{4}$-in. drinking straws can be cut from a $216\frac{1}{2}$-in. length of stock? How much stock will be left over?

27. It takes $12\frac{1}{2}$ minutes to sand a piece of oak stock. If Florence Randle worked 100 minutes, how many pieces did she sand?

3-9 COMPLEX FRACTIONS

Learning Objective

L.O.1. Simplify complex fractions.

As we recall from Section 3-1, a complex fraction is one in which either the numerator or the denominator or both contain a number other than a whole number. Examples of complex fractions are

$$\frac{\frac{3}{4}}{2}, \quad \frac{6}{\frac{1}{2}}, \quad \frac{\frac{2}{3}}{\frac{1}{5}}, \quad \frac{2\frac{1}{2}}{7}, \quad \frac{1\frac{2}{3}}{3\frac{5}{8}}, \quad \frac{4}{1\frac{1}{2}}$$

Recall also that fractions indicate division as we read from top to bottom. The large fraction line is read as "divided by." In the complex fraction $\frac{2\frac{1}{2}}{7}$, for example, we would read "$2\frac{1}{2}$ divided by 7." The complex fraction $\frac{4}{1\frac{1}{2}}$ would be read "4 divided by $1\frac{1}{2}$."

L.O.1. Simplify complex fractions.

Simplifying Complex Fractions

If we think of fractions as divisions, then complex fractions are merely another way of writing problems like the ones we worked in Section 3-8.

> **Rule 3-9.1** *To simplify a complex fraction:*
>
> Perform the indicated division.

Example 3-9.1 Simplify $\frac{6\frac{3}{8}}{4\frac{1}{2}}$.

Solution

$$\frac{6\frac{3}{8}}{4\frac{1}{2}} = 6\frac{3}{8} \div 4\frac{1}{2} = \frac{51}{8} \div \frac{9}{2} = \frac{\overset{17}{\cancel{51}}}{\underset{4}{\cancel{8}}} \times \frac{\overset{1}{\cancel{2}}}{\underset{3}{\cancel{9}}} = \frac{17}{12} = 1\frac{5}{12}$$

■

Example 3-9.2 Simplify $\frac{\frac{4}{5}}{6}$.

Solution

$$\frac{\frac{4}{5}}{6} = \frac{4}{5} \div 6 = \frac{4}{5} \div \frac{6}{1} = \frac{\overset{2}{\cancel{4}}}{5} \times \frac{1}{\underset{3}{\cancel{6}}} = \frac{2}{15}$$

■

SELF-STUDY EXERCISES 3-9

L.O.1. Divide and reduce answers to lowest terms. Convert improper fractions to whole or mixed numbers.

1. $\dfrac{\frac{1}{2}}{5}$ **2.** $\dfrac{2}{\frac{2}{3}}$ **3.** $\dfrac{7}{1\frac{1}{4}}$

4. $\dfrac{2\frac{2}{5}}{5}$ **5.** $\dfrac{3\frac{1}{4}}{9\frac{3}{4}}$ **6.** $\dfrac{12\frac{1}{2}}{2\frac{1}{2}}$

7. $\dfrac{7}{\frac{1}{4}}$ **8.** $\dfrac{1\frac{1}{3}}{6}$ **9.** $\dfrac{10}{3\frac{1}{3}}$

10. $\dfrac{33\frac{1}{3}}{100}$

3-10 SIGNED FRACTIONS

Learning Objectives

L.O.1. Change a signed fraction to an equivalent signed fraction.
L.O.2. Perform basic operations with signed fractions.

L.O.1. Change a signed fraction to an equivalent signed fraction.

Signs of a Fraction

A fraction has three basic signs, the sign of the fraction, the sign of the numerator, and the sign of the denominator. The fraction $\frac{2}{3}$ expressed as a *signed fraction* is $+\frac{+2}{+3}$. When a signed fraction has negative signs, it is sometimes convenient to change the signed fraction to an equivalent signed fraction.

The rules for operating with integers can be extended to apply to signed fractions. Applying the rules of subtraction and division allows us to manipulate the signs of a fraction.

> **Rule 3-10.1** ***To find an equivalent signed fraction:***
>
> Change any two of the three signs to the opposite sign.

EXAMPLE 3-10.1 Change $+\frac{+2}{+3}$ to three equivalent signed fractions.

Solution

$+\frac{+2}{+3} = -\frac{-2}{+3}$ *(Change the sign of the fraction and the sign of the numerator.)*

$+\frac{+2}{+3} = -\frac{+2}{-3}$ *(Change the sign of the fraction and the sign of the denominator.)*

$+\frac{+2}{+3} = +\frac{-2}{-3}$ *(Change the sign of the numerator and the sign of the denominator.)* ■

Changing the signs of a fraction is a manipulation tool that will allow us to simplify our work when performing basic operations with signed fractions.

L.O.2. Perform basic operations with signed fractions.

Basic Operations with Signed Fractions

The same rules for performing the basic operations with integers can be used with signed fractions. In applying these rules, we will also change the signs of a fraction whenever it will give us a simpler problem to work.

EXAMPLE 3-10.2 Add $\frac{-3}{4} + \frac{5}{-8}$.

Solution

$\frac{-3}{4} + \frac{-5}{8}$ *(Change the signs of the second fraction so that both denominators are positive.)*

$\frac{-6}{8} + \frac{-5}{8}$ *(Change to equivalent fractions with a common denominator.)*

$\frac{-11}{8}$ *(Add numerators, applying the rule for adding numbers with like signs.)*

$-1\frac{3}{8}$ *(Change to mixed number. The sign of the mixed number is determined by the rule for dividing numbers with unlike signs.)* ■

EXAMPLE 3-10.3 Subtract $\frac{-3}{7} - \frac{-5}{7}$.

Solution

$\frac{-3}{7} + \frac{5}{7}$ *(Change the signs of the second fraction by changing the sign of the fraction and the numerator.)*

$\frac{2}{7}$ *(Apply the rule for adding numbers with unlike signs.)* ■

Tips and Traps

Manipulating the signs of a fraction is used to simplify the steps of a problem. Generally, we use this tool to create an equivalent problem with fewer negative signs.

Example 3-10.4 Multiply $\left(\frac{-4}{5}\right)\left(\frac{3}{-7}\right)$.

Solution

$\frac{-4}{5} \cdot \frac{3}{-7}$ *(Multiply numerators and denominators.)*

$\frac{-12}{-35}$ *(Apply rule for dividing numbers with like signs.)*

$\frac{12}{35}$ ■

Example 3-10.5 Simplify $\left(\frac{-2}{3}\right)^3$.

Solution

$\left(\frac{-2}{3}\right)^3$ *(Cube numerator and cube denominator.)*

$\frac{(-2)^3}{3^3}$ *(Apply rule for raising a negative number to an odd power.)*

$\frac{-8}{27}$ or $-\frac{8}{27}$ ■

SELF-STUDY EXERCISES 3-10

L.O.1. Change the following fractions to three equivalent signed fractions.

1. $+\frac{+5}{+8}$ **2.** $-\frac{3}{4}$ **3.** $\frac{-2}{-5}$

4. $-\frac{-7}{-8}$ **5.** $\frac{7}{8}$

L.O.2. Perform the indicated operations.

6. $\frac{-7}{8} + \frac{5}{8}$ **7.** $\frac{-4}{5} + \frac{-3}{10}$ **8.** $\frac{1}{2} - \frac{-3}{5}$

9. $\frac{-3}{5} \times \frac{10}{-11}$ **10.** $-\frac{5}{8} \div \frac{4}{5}$

3-11 CALCULATORS WITH FRACTION KEY

Learning Objective

L.O.1. Use the fraction key on a calculator to perform operations with fractions.

L.O.1. Use the fraction key on a calculator to perform operations with fractions.

Calculators with Fraction Key

Some scientific and graphing calculators have a special key for making calculations with fractions. On these calculators, numbers can be entered as fractions and results can be displayed either in fraction or decimal form (or mixed number form for improper fractions). The key is generally labeled [a b/c]. The numerator and denominator are separated with a special symbol. The fraction $\frac{2}{3}$ appears as 2 ⌟ 3. The mixed number $3\frac{1}{5}$ appears as 3 ⌟ 1 ⌟ 5.

> **Rule 3-11.1**
>
> To reduce the fraction, enter the numerator, press the fraction key, and then enter the denominator. To display the fraction in lowest terms, press the equal key (on scientific calculator) or the EXE key (on graphing calculator).

EXAMPLE 3-11.1 Reduce $\frac{30}{36}$.

Scientific: [30] [a b/c] [36] [=] ⇒ 5 ⌟ 6

Graphing: [30] [a b/c] [36] [EXE] ⇒ 5 ⌟ 6 ■

Although the process for finding the decimal equivalent of a fraction is introduced in Chapter 4, with the fraction key you can see these equivalents now.

> **Rule 3-11.2**
>
> To display the decimal equivalent, press the fraction key following the equal key or EXE key.

EXAMPLE 3-11.2 Use the fraction key to show the decimal equivalent of $\frac{5}{6}$.

Solution

[5] [a b/c] [6] [=] [a b/c] ⇒ 0.833333333 *(or EXE for =)* ■

Improper fractions can be found by accessing the function [d/c] above the fraction key which appears on the calculator as d/c over [a b/c].

Tips and Traps

A special key is used to access calculator functions that are written above the keys. This key is labeled differently on various calculators. Frequently used labels are [Shift] [INV] and [2nd]. Many times, these keys and the labels above the keys are color coded to make the functions and the access key easier to locate.

EXAMPLE 3-11.3 Change $2\frac{7}{8}$ to an improper fraction using the improper fraction key.

Solution

[2] [a b/c] [7] [a b/c] [8] [Shift] [d/c] *(Shift key may be labeled differently.)*

$\frac{23}{8}$ ■

Tips and Traps

Some keys on a calculator act as a toggle in certain situations. With $\frac{23}{8}$ in the display, press the fraction key several times. The display alternates among the improper fraction form, the decimal equivalent, and the mixed number. The [a b/c] key is a toggle key in this instance.

The fraction key can also be used to add, subtract, multiply, or divide fractions.

Rule 3-11.3

To perform calculations of fractions on the calculator, enter the fraction or mixed number using the fraction key and use the operation keys as usual.

EXAMPLE 3-11.4 Simplify $\frac{3}{4} - \frac{2}{3} + 3\frac{1}{2}$.

Solution

[3] [a b/c] [4] [−] [2] [a b/c] [3] [+]
[3] [a b/c] [1] [a b/c] [2] [=] *(or EXE)*

$3\frac{7}{12}$ ■

SELF-STUDY EXERCISES 3-11

L.O.1. Use a calculator with a fraction key to perform the following operations.

1. $\frac{5}{8} + \frac{7}{12}$ **2.** $3\frac{1}{7} - 5\frac{3}{5}$ **3.** $2\frac{7}{8} \times 5\frac{1}{12}$

4. $\frac{3}{5} \div \frac{12}{35}$ **5.** $3\frac{1}{5} \div 4$ **6.** $\frac{-1}{4} + \frac{3}{4}$

7. $\frac{-5}{8} + \frac{7}{12}$ **8.** $15\frac{3}{5} \times 18\frac{1}{12}$

ASSIGNMENT EXERCISES

3-1

Write a fraction to represent the shaded portion of Figs. 3-18 to 3-22.

1. 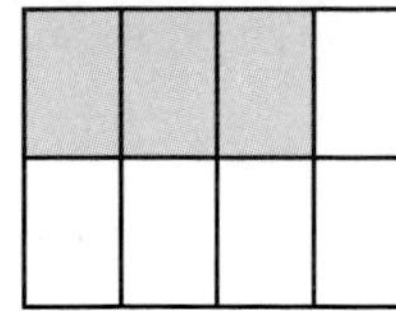

Figure 3-18

2. 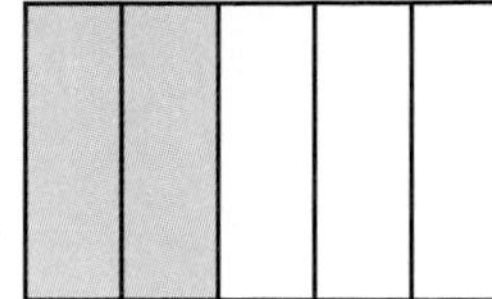

Figure 3-19

3. 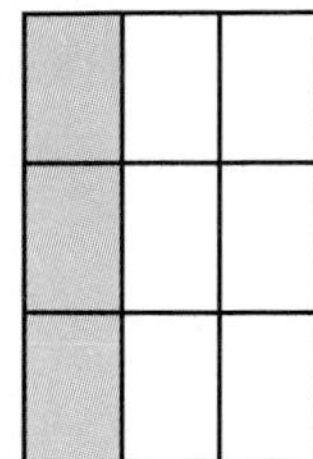

Figure 3-20

4. 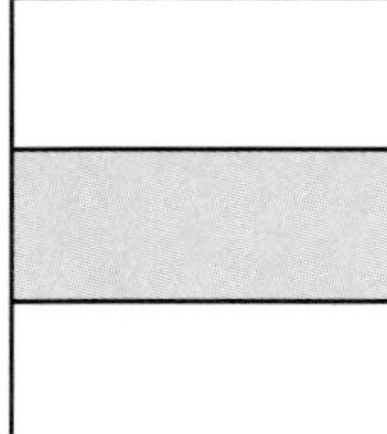

Figure 3-21

5. 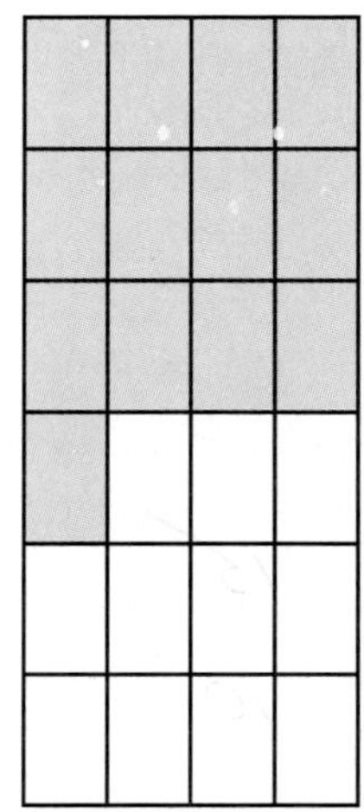

Figure 3-22

Answer the following questions for the numbers in Exercises 6 through 8.

6. $\frac{3}{7}$

7. $\frac{9}{4}$

8. $\frac{2}{2}$

(a) What is the numerator of the fraction?
(b) What is the denominator of the fraction?
(c) One unit has been divided into how many parts?
(d) How many of these parts are used?
(e) The fraction can be read as _____ divided by _____.
(f) What is the divisor of the division?
(g) What is the dividend of the division?
(h) Is this fraction proper or improper?
(i) Does this fraction have a value less than, more than, or equal to 1?
(j) This fraction represents _____ out of _____ parts.

Match the best description with each fraction given.

____ **9.** $\frac{5}{2}$	**(a)** Mixed-decimal fraction
____ **10.** 0.172	**(b)** Decimal fraction
____ **11.** $6\frac{1}{7}$	**(c)** Complex fraction
____ **12.** 4.59	**(d)** Mixed number
____ **13.** $\frac{7}{11}$	**(e)** Proper fraction
____ **14.** $\frac{\frac{2}{3}}{\frac{1}{4}}$	**(f)** Improper fraction
____ **15.** $33\frac{1}{3}$	

Rewrite the decimals below as proper fractions or mixed numbers.

16. 4.273 **17.** 0.87 **18.** 1.002

19. 2.03 **20.** 1.1

3-2

Find the equivalent fractions having the denominators indicated.

21. $\frac{5}{8} = \frac{?}{24}$ **22.** $\frac{3}{7} = \frac{?}{35}$ **23.** $\frac{5}{12} = \frac{?}{60}$

24. $\frac{4}{5} = \frac{?}{40}$ **25.** $\frac{2}{3} = \frac{?}{15}$ **26.** $\frac{4}{9} = \frac{?}{18}$

27. $\frac{3}{4} = \frac{?}{32}$ **28.** $\frac{1}{6} = \frac{?}{30}$ **29.** $\frac{1}{5} = \frac{?}{55}$

30. $\frac{7}{8} = \frac{?}{64}$

Reduce the following fractions to lowest terms.

31. $\frac{6}{12}$ **32.** $\frac{8}{10}$ **33.** $\frac{4}{32}$

34. $\frac{26}{64}$ **35.** $\frac{2}{8}$ **36.** $\frac{8}{32}$

37. $\frac{34}{64}$ **38.** $\frac{16}{64}$ **39.** $\frac{12}{32}$

40. $\frac{45}{90}$ **41.** $\frac{6}{8}$ **42.** $\frac{75}{100}$

3-3

Write the following improper fractions as whole or mixed numbers.

43. $\frac{9}{2}$ **44.** $\frac{35}{7}$ **45.** $\frac{18}{5}$

46. $\frac{27}{6}$ **47.** $\frac{39}{8}$ **48.** $\frac{21}{15}$

49. $\frac{43}{8}$ 50. $\frac{22}{7}$ 51. $\frac{175}{2}$

52. $\frac{135}{3}$

Write the following whole or mixed numbers as improper fractions.

53. 8 54. $10\frac{1}{2}$ 55. $7\frac{1}{8}$

56. $5\frac{7}{12}$ 57. $9\frac{3}{16}$ 58. $7\frac{8}{17}$

59. $4\frac{3}{5}$ 60. $9\frac{1}{9}$ 61. 12

62. $16\frac{2}{3}$

Change each of the following whole numbers to an equivalent fraction having the denominator indicated.

63. $2 = \frac{?}{10}$ 64. $6 = \frac{?}{4}$ 65. $11 = \frac{?}{3}$

66. $7 = \frac{?}{5}$

3-4

Determine which fraction is smaller.

67. $\frac{3}{8}, \frac{4}{8}$ 68. $\frac{5}{9}, \frac{4}{9}$ 69. $\frac{1}{4}, \frac{3}{16}$

70. $\frac{5}{8}, \frac{11}{16}$ 71. $\frac{7}{8}, \frac{27}{32}$ 72. $\frac{7}{64}, \frac{1}{4}$

73. $\frac{1}{2}, \frac{9}{19}$ 74. $\frac{3}{7}, \frac{1}{9}$

Find the least common denominator for the fractions.

75. $\frac{1}{4}, \frac{1}{3}, \frac{1}{5}$ 76. $\frac{7}{8}, \frac{2}{3}$ 77. $\frac{3}{4}, \frac{1}{16}$

78. $\frac{1}{12}, \frac{3}{4}$ 79. $\frac{5}{12}, \frac{3}{10}, \frac{13}{15}$ 80. $\frac{1}{12}, \frac{3}{8}, \frac{15}{16}$

81. Is a $\frac{5}{8}$-in. wrench larger or smaller than a $\frac{9}{16}$-in. wrench?

82. An alloy contains $\frac{2}{3}$ metal A and the same quantity of another alloy contains $\frac{3}{5}$ metal A. Which alloy contains more metal A?

83. Is a $\frac{3}{8}$-in.-thick piece of plaster board thicker than a $\frac{1}{2}$-in.-thick piece?

84. A $\frac{9}{16}$-in. tube must pass through a wall. Is a $\frac{3}{4}$-in.-diameter hole large enough?

85. Is a $\frac{19}{32}$-in. wrench larger or smaller than a $\frac{7}{8}$-in. bolt head?

3-5

Add; reduce sums to lowest terms and convert any improper fractions to mixed numbers or whole numbers.

86. $\frac{1}{8} + \frac{5}{16}$ **87.** $\frac{3}{16} + \frac{9}{64}$ **88.** $\frac{3}{14} + \frac{5}{7}$

89. $\frac{3}{5} + \frac{5}{6}$ **90.** $3\frac{7}{8} + 7 + 5\frac{1}{2}$ **91.** $2\frac{7}{16} + 6\frac{5}{32}$

92. $9\frac{7}{8} + 5\frac{3}{4}$ **93.** $3\frac{7}{8} + 5\frac{3}{16} + 1\frac{7}{32}$ **94.** $2\frac{1}{4} + 3\frac{7}{8}$

95. A pipe is cut into two pieces measuring $7\frac{5}{8}$ in. and $10\frac{7}{16}$ in. How long was the pipe before it was cut? Disregard waste.

96. Find the total thickness of a wall if the outside covering is $3\frac{7}{8}$ in. thick, the studs are $3\frac{7}{8}$ in., and the inside covering is $\frac{5}{16}$-in. paneling.

97. If $7\frac{5}{16}$ in. of a piece of square bar stock is turned so that it is cylindrical and $5\frac{9}{32}$ in. remains square, what is the total length of the bar stock?

98. Two castings weigh $27\frac{1}{2}$ lb and $20\frac{3}{4}$ lb. What is the total weight of the two castings?

99. Three metal rods $3\frac{1}{8}$ in., $5\frac{3}{32}$ in., and $7\frac{9}{16}$ in. were welded together end to end. How long is the welded rod?

100. In Fig. 3-23, what is the length of side A?

101. In Fig. 3-23, what is the length of side B?

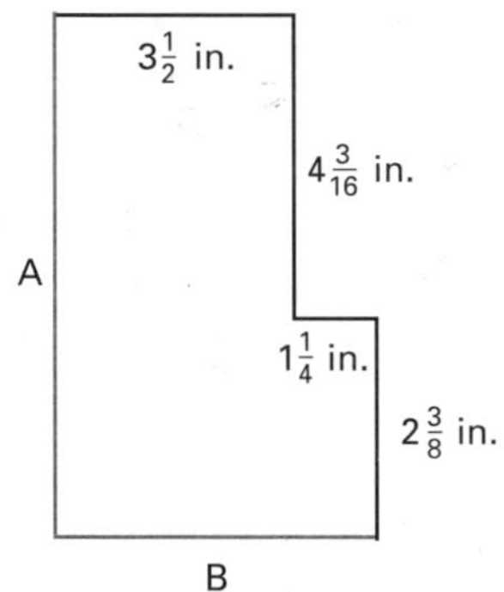

Figure 3-23

102. A hollow-wall fastener has a grip range up to $\frac{3}{4}$ in. Is it long enough to fasten three sheets of metal $\frac{5}{16}$ in. thick, $\frac{3}{8}$ in. thick, and $\frac{1}{16}$ in. thick?

103. Figure 3-24 shows $\frac{1}{2}$-in. copper tubing wrapped in insulation. What is the distance across the tubing and insulation?

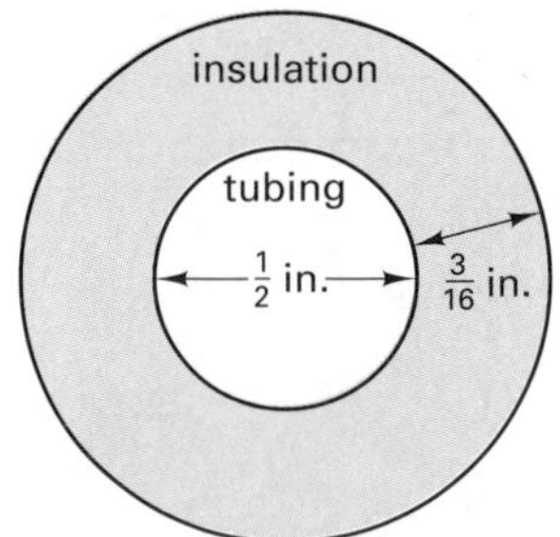

Figure 3-24

104. In Exercise 103, what would be the overall distance across the tubing and insulation if $\frac{3}{8}$-in.-ID tubing were used?

3-6

Subtract; reduce to lowest terms when necessary.

105. $\frac{5}{9} - \frac{2}{9}$ **106.** $\frac{11}{32} - \frac{5}{64}$ **107.** $3\frac{5}{8} - 2$

108. $7 - 4\frac{3}{8}$ **109.** $8\frac{7}{8} - 2\frac{29}{32}$ **110.** $7 - 2\frac{9}{16}$

111. $12\frac{11}{16} - 5$ **112.** $48\frac{5}{12} - 12\frac{11}{15}$

113. A bolt 2 in. long fastens a piece of $\frac{7}{8}$-in.-thick wood to a piece of metal. If a $\frac{3}{16}$-in.-thick lock washer, a $\frac{1}{16}$-in. washer, and a $\frac{7}{16}$-in.-thick nut are used, what is the thickness of the metal if the nut is flush with the bolt after tightening?

114. Pins of $2\frac{3}{8}$ in. and $3\frac{7}{16}$ in. were cut from a drill rod 12 in. long. If $\frac{1}{16}$ in. of waste is allowed for each cut, how many inches of drill rod are left?

115. A piece of tapered stock has a diameter of $2\frac{5}{16}$ in. at one end and a diameter of $\frac{55}{64}$ in. at the other end. What is the difference in the diameters?

116. Four lengths measuring $6\frac{1}{4}$ in., $9\frac{3}{16}$ in., $7\frac{1}{8}$ in., and $5\frac{9}{32}$ in. are cut from 48 in. of copper tubing. How much copper tubing remains? Disregard waste.

3-7

Multiply and reduce answers to lowest terms. Convert improper fractions to whole or mixed numbers.

117. $\frac{1}{3} \times \frac{7}{8}$ **118.** $\frac{2}{5} \times \frac{7}{10}$ **119.** $\frac{7}{9} \times \frac{3}{8}$

120. $\frac{2}{3} \times \frac{5}{8} \times \frac{3}{16}$ **121.** $\frac{15}{16} \times \frac{4}{5} \times \frac{2}{3}$ **122.** $\frac{5}{7} \times \frac{3}{4}$

123. $\frac{7}{16} \times 18$ **124.** $\frac{3}{16} \times 184$ **125.** $1\frac{1}{2} \times \frac{4}{5}$

126. In a concrete mixture, $\frac{4}{7}$ of the total volume is sand. How much sand is needed for 135 cubic yards (yd^3) of concrete?

127. Concrete blocks are 8 in. high. If a $\frac{3}{8}$-in. mortar joint is used, how high will a wall of 12 courses of concrete blocks be? (*Hint:* There are 12 rows of mortar joints.)

128. An adjusting screw will move $\frac{3}{64}$ in. for each full turn. How far will it move in four turns?

129. A chef is making a dessert that is $\frac{3}{4}$ the original recipe. How much flour should be used if the original recipe calls for $3\frac{2}{3}$ cups of flour?

130. If an alloy is $\frac{3}{5}$ copper and $\frac{2}{5}$ zinc, how many pounds of each metal are in a casting weighing $112\frac{1}{2}$ lb?

3-8

Divide and reduce answers to lowest terms. Convert improper fractions to whole or mixed numbers.

131. $\frac{7}{8} \div \frac{3}{4}$ **132.** $\frac{4}{9} \div \frac{5}{16}$ **133.** $\frac{7}{8} \div \frac{3}{32}$

134. $8 \div \frac{2}{3}$ **135.** $18 \div \frac{3}{4}$ **136.** $35 \div \frac{5}{16}$

137. $5\frac{1}{10} \div 2\frac{11}{20}$ **138.** $27\frac{2}{3} \div \frac{2}{3}$ **139.** $7\frac{1}{5} \div 12$

140. On a house plan, $\frac{1}{4}$ in. represents 1 ft. Find the dimensions of a porch that measures $4\frac{1}{8}$ in. by $6\frac{1}{2}$ in. on the plan. (How many $\frac{1}{4}$'s are there in $4\frac{1}{8}$; how many $\frac{1}{4}$'s are there in $6\frac{1}{2}$?)

141. A pipe that is 12 in. long is to be cut into four equal parts. If $\frac{3}{16}$ in. is wasted per cut, what is the maximum length that each pipe can be? (It will take three cuts to divide the entire length into four equal parts.)

142. A stack of $\frac{5}{8}$-in. plywood is $21\frac{7}{8}$ in. high. How many sheets of plywood are in the stack?

143. A rod $1\frac{1}{8}$ yd long is cut into 6 equal pieces. What is the length of each piece?

144. If $7\frac{1}{2}$ gallons of liquid are distributed equally among five containers, what is the average number of gallons per container?

3-9

Divide and reduce answers to lowest terms. Convert improper fractions to whole or mixed numbers.

145. $\dfrac{\frac{1}{3}}{6}$ **146.** $\dfrac{4}{\frac{4}{5}}$ **147.** $\dfrac{8}{1\frac{1}{2}}$

148. $\dfrac{3\frac{1}{4}}{5}$ **149.** $\dfrac{2\frac{1}{5}}{8\frac{4}{5}}$ **150.** $\dfrac{16\frac{2}{3}}{3\frac{1}{3}}$

151. $\dfrac{12\frac{1}{2}}{100}$ **152.** $\dfrac{37\frac{1}{2}}{100}$

3-10

Change each fraction to three equivalent signed fractions.

153. $-\frac{3}{8}$ **154.** $\frac{-5}{9}$ **155.** $\frac{-7}{-8}$

Perform the indicated operations.

156. $\frac{-7}{8} + \frac{-3}{8}$

157. $\frac{5}{9} - \frac{-3}{7}$

158. $\frac{-5}{8} * \frac{-2}{3}$

159. $\frac{-4}{5} \div -\frac{7}{15}$

3-11

Use a calculator to perform the following operations.

160. $\frac{-11}{12} - \frac{^{-}7}{8}$

161. $1\frac{3}{5} \div \left(-7\frac{5}{8}\right)$

CHALLENGE PROBLEM

162. Len Smith has 180 ft of fencing and needs to build two square or rectangular holding yards for your two sheltie dogs. How should he design the two yards to get the largest area fenced in for each dog by using the 180 ft of wire?

CONCEPTS ANALYSIS

1. What two operations require a common denominator?
2. Explain how to find the reciprocal of a fraction.
3. What steps must be followed to find the reciprocal of a mixed number?
4. What number can be written as any fraction that has the same numerator and denominator?
5. What operation requires the use of the reciprocal of a fraction?
6. Name the operation that has each of the following for an answer: sum? difference? product? quotient?
7. What operation must be used to solve an applied problem if the total and one of the parts are given?
8. What does the denominator of a fraction indicate?
9. What does the numerator of a fraction indicate?
10. What kind of fraction has a value less than one?

Find, explain, and correct the mistakes in the following problems.

11. $\frac{5}{8} + \frac{1}{8} = \frac{6}{16} = \frac{3}{8}$

12.
$$\begin{array}{r} 12 \\ -\ 5\frac{3}{4} \\ \hline 7\frac{3}{4} \end{array}$$

13. $\frac{3}{5} \times 2\frac{1}{5} = 2\frac{3}{25}$

14. $\frac{5}{8} \div 4 = \frac{5}{8} \times \frac{4}{1} = \frac{5}{2} = 2\frac{1}{2}$

15.
$$\begin{array}{rcrcr} 12\frac{3}{4} & = & 12\frac{6}{8} & = & 11\frac{16}{8} \\ -\ 4\frac{7}{8} & = & 4\frac{7}{8} & = & 4\frac{7}{8} \\ \hline & & & & 7\frac{9}{8} = 7 + 1\frac{1}{8} = 8\frac{1}{8} \end{array}$$

CHAPTER SUMMARY

Objectives	What to Remember	Examples
Section 3-1		
1. Identify fraction terminology.	The numerator (top number) of a fraction represents the number of parts of the whole amount we are considering. The denominator (bottom number) of a fraction represents the number of parts a whole amount has been divided into. Proper fractions are less than 1. Improper fractions are equal to or larger than 1. A mixed number consists of a whole number and a common fraction written together and indicates addition of the whole number and fraction. A complex fraction has a fraction or mixed number in the numerator, or denominator, or both. The fraction line for complex fractions also indicates division. A decimal fraction is a number written with a decimal point to represent a fractional part whose denominator is 10 or a power of 10.	Identify the following as proper fractions, improper fractions, or mixed numbers: $\frac{5}{37}, \frac{7}{7}, \frac{12}{5}, 8\frac{5}{9}$. $\frac{5}{37}$, proper fraction; $\frac{7}{7}$, improper fraction; $\frac{12}{5}$, improper fraction; $8\frac{5}{9}$, mixed number. Write the following in decimal notation: $\frac{3}{10}, \frac{5}{100}, 4\frac{23}{100}$. $\frac{3}{10} = 0.3$, $\frac{5}{100} = 0.05$, $4\frac{23}{100} = 4.23$
Section 3-2		
1. Write equivalent fractions.	Equivalent fractions can be made by multiplying both the numerator and denominator by the same number.	Write three fractions equivalent to $\frac{7}{8}$. $\frac{7}{8} \times \frac{2}{2} = \frac{14}{16}$ $\frac{7}{8} \times \frac{3}{3} = \frac{21}{24}$ $\frac{7}{8} \times \frac{4}{4} = \frac{28}{32}$
2. Reduce fractions.	Fractions in which the numerator and denominator have a common factor can be reduced by dividing both numerator and denominator by the greatest common factor.	Reduce $\frac{12}{16}$. $\frac{12}{16} = \frac{12}{16} \div \frac{4}{4} = \frac{3}{4}$
Section 3-3		
1. Convert improper fractions to whole or mixed numbers.	To convert an improper fraction to a whole or mixed number, divide the numerator by the denominator. Write any remainder as a fraction having the original denominator as its denominator.	Convert the following to whole or mixed numbers: $\frac{18}{6}, \frac{15}{4}$. $\frac{18}{6} = 3$; $\frac{15}{4} = 3\frac{3}{4}$
2. Convert mixed numbers and whole numbers to improper fractions.	To convert a mixed number to an improper fraction: **1.** Multiply whole number by denominator. **2.** Add numerator to result of step 1. **3.** Place the sum from step 2 over the original denominator.	Change $4\frac{7}{8}$ to an improper fraction. $4\frac{7}{8} = [(8 \times 4) + 7]/8 = \frac{39}{8}$

Section 3-4		
1. Find common denominators.	To find the lowest common denominator, use the process shown in Chapter 1 for finding the least common multiple.	Find the lowest common denominator for $\frac{7}{18}$ and $\frac{5}{24}$. $18 = 2 \cdot 3 \cdot 3$ $24 = 2 \cdot 2 \cdot 2 \cdot 3$ $2 \cdot 2 \cdot 2 \cdot 3 \cdot 3 = 72$ The smallest number that can be divided evenly by both 18 and 24 is 72.
2. Compare fractions.	To compare fractions: **1.** Write the fractions as equivalent fractions with common denominators. **2.** Compare the numerators. The larger numerator indicates the larger fraction. To compare mixed numbers: **1.** Compare the whole number parts, if different. **2.** If the whole number parts are equal, write the fractions with common denominators. **3.** Compare the numerators.	Which fraction is smaller, $\frac{2}{5}$ or $\frac{5}{12}$? $\frac{2}{5} = \frac{24}{60}$ $\frac{5}{12} = \frac{25}{60}$ Since $\frac{24}{60}$ is smaller than $\frac{25}{60}$, $\frac{2}{5}$ is smaller than $\frac{5}{12}$.
Section 3-5		
1. Add fractions.	To add fractions, find the common denominator and convert each fraction to an equivalent fraction with the common denominator. Add the numerators and place the sum over the common denominator. Reduce the sum if possible.	Add: $\frac{1}{7} + \frac{3}{7} + \frac{2}{7} = \frac{6}{7}$ Add: $\frac{5}{8} + \frac{3}{4} = \frac{5}{8} + \frac{6}{8} = \frac{11}{8} = 1\frac{3}{8}$
2. Add mixed numbers.	To add mixed numbers: **1.** Convert each fractional part to an equivalent fraction with the LCD. **2.** Place the sum of the numerators over the LCD. **3.** Add the whole number parts. **4.** Write improper fraction from step 2 as a whole or mixed number; add the result to the whole number from step 3. **5.** Simplify if necessary.	Add $4\frac{3}{4} + 5\frac{2}{8} + 1\frac{1}{2}$. $4\frac{3}{4} = 4\frac{6}{8}$ $5\frac{2}{8} = 5\frac{2}{8}$ $1\frac{1}{2} = 1\frac{4}{8}$ $10\frac{12}{8}$ $\frac{12}{8} = 1\frac{4}{8}$ and $10 + 1\frac{4}{8} = 11\frac{1}{2}$

Section 3-6

1. Subtract fractions.

To subtract fractions:

1. Convert each fraction to an equivalent fraction that has the LCD as denominator.
2. Subtract the numerators.
3. Place the difference over the LCD.
4. Reduce if possible.

Subtract $\frac{5}{8} - \frac{7}{16}$.

$$\begin{array}{r} \frac{5}{8} = \frac{10}{16} \\ \frac{7}{16} = \frac{7}{16} \\ \hline \frac{3}{16} \end{array}$$

2. Subtract mixed numbers.

To subtract mixed numbers:

1. Write fractions as equivalent fractions with common denominators.
2. Borrow from the whole number and add to the fraction if the fraction in the minuend is smaller than the fraction in the subtrahend.
3. Subtract the fractions.
4. Subtract the whole numbers.
5. Simplify if necessary.

Subtract $5\frac{3}{8} - 3\frac{9}{16}$.

$$\begin{array}{r} 5\frac{3}{8} = 5\frac{6}{16} = 4\frac{22}{16} \\ 3\frac{9}{16} = 3\frac{9}{16} = 3\frac{9}{16} \\ \hline 1\frac{13}{16} \end{array}$$

Section 3-7

1. Multiply fractions.

To multiply fractions:

1. Reduce the numerator and denominator that have a common factor.
2. Multiply the numerators for the numerator of the product.
3. Multiply the denominators for the denominator of the product.
4. Be sure the product is reduced.

Multiply $\frac{4}{5} \times \frac{7}{10} \times \frac{15}{35}$.

$$\frac{\overset{2}{\cancel{4}}}{\underset{1}{\cancel{5}}} \times \frac{\overset{1}{\cancel{7}}}{\underset{5}{\cancel{10}}} \times \frac{\overset{3}{\cancel{15}}}{\underset{5}{\cancel{35}}} = \frac{6}{25}$$

2. Multiply mixed numbers.

To multiply fractions, whole numbers, and mixed numbers:

1. Write whole numbers as fractions with denominators of 1.
2. Write mixed numbers as improper fractions.
3. Reduce as much as possible.
4. Multiply the numerators for the numerator of the product.
5. Multiply the denominators for the denominator of the product.
6. Write product as whole number, mixed number, or fraction in lowest terms.

Multiply $4 \times 3\frac{1}{5} \times \frac{2}{7}$.

$$\frac{4}{1} \times \frac{16}{5} \times \frac{2}{7} = \frac{128}{35} = 3\frac{23}{35}$$

Section 3-8

1. Find reciprocals.

To write the reciprocal of a number, first express the number as a fraction; then interchange the numerator and denominator.

Find the reciprocal of $\frac{3}{5}$, 6, $2\frac{3}{4}$.

Reciprocal of $\frac{3}{5}$ is $\frac{5}{3}$; reciprocal of 6 is $\frac{1}{6}$; reciprocal of $2\frac{3}{4}$ or $\frac{11}{4}$ is $\frac{4}{11}$.

2. Divide fractions.	To divide fractions, replace the divisor with its reciprocal and change division to multiplication; then multiply.	Divide $\frac{4}{5} \div \frac{8}{9}$. $\frac{4}{5} \div \frac{8}{9} = \frac{4}{5} \times \frac{9}{8} =$ $\frac{1}{5} \times \frac{9}{2} = \frac{9}{10}$
3. Divide mixed numbers.	To divide mixed numbers: **1.** Write each mixed number as an improper fraction. **2.** Invert the divisor and change division to multiplication. **3.** Simplify if possible. **4.** Multiply.	Divide $4\frac{2}{3} \div 1\frac{1}{6}$. $4\frac{2}{3} \div 1\frac{1}{6} = \frac{14}{3} \div \frac{7}{6} = \frac{14}{3} \times \frac{6}{7} =$ $\frac{4}{1}$ or 4.
Section 3-9		
1. Simplify complex fractions.	To simplify a complex fraction: **1.** Write mixed numbers and whole numbers as improper fractions. **2.** Write division (as indicated by the fraction bar) as multiplication and invert the divisor. **3.** Simplify if possible. **4.** Multiply. **5.** Convert improper fraction to whole or mixed number if necessary.	Simplify the following complex fraction. $\frac{4\frac{1}{2}}{3\frac{3}{5}} = \frac{\frac{9}{2}}{\frac{18}{5}} = \frac{9}{2} \div \frac{18}{5} = \frac{9}{2} \times \frac{5}{18} =$ $\frac{5}{4} = 1\frac{1}{4}$
Section 3-10		
1. Change a signed fraction to an equivalent signed fraction.	If any two of the three signs of a signed fraction are changed, the value of the fraction is not changed.	Write three equivalent signed fractions for $-\frac{7}{8}$. $-\frac{7}{8} = -\frac{+7}{+8} =$ $\frac{+7}{-8}$ or $-\frac{-7}{-8}$ or $\frac{-7}{+8}$
2. Perform basic operations with signed fractions.	To add or subtract signed fractions, write equivalent fractions that have positive integers as denominators and have common denominators. Add or subtract the numerators using the rules for adding signed numbers. To multiply or divide signed fractions, multiply or divide the fractions, paying attention to the rules for multiplying or dividing signed numbers.	Add $\frac{-5}{8} + \frac{7}{8}$. $\frac{-5}{8} + \frac{7}{8} = \frac{2}{8} = \frac{1}{4}$
Section 3-11		
1. Use the fraction key on a calculator to perform operations with fractions.	Calculators that have a fraction key usually identifiy it as [a b/c]. To enter fractions or mixed numbers, enter each part, followed by the fraction key. Use the four operation keys in the usual way to perform operations with fractions on the calculator.	Multiply $\frac{3}{4} \times \frac{7}{8}$. 3 [a b/c] 4 [×] 7 [a b/c] 8 [=] ⇒ 21 ⌟ 32

WORDS TO KNOW

common fraction (p. 86)
numerator (p. 86)
denominator (p. 86)
fraction (p. 86)
improper fraction (p. 88)
proper fraction (p. 88)
mixed number (p. 88)
complex fraction (p. 88)
decimal fraction (p. 89)
mixed-decimal fraction (p. 89)
decimals (p. 90)
rational number (p. 90)
equivalent fractions (p. 92)
fundamental principle of fractions (p. 93)
reduce, lowest terms (p. 94)
common denominators (p. 99)
least common denominator (LCD) (p. 99)
tolerance (p. 106)
canceling (p. 115)
reciprocals (p. 119)
inverting (p. 120)
signed fractions (p. 126)
fraction key (p. 129)

CHAPTER TEST

Write a fraction to represent the following.

1. 3 out of 4 people in a survey

2. $7 \div 9$

Convert the following to mixed or whole numbers.

3. $\frac{9}{3}$

4. $\frac{14}{9}$

Convert the following to improper fractions.

5. $4\frac{6}{7}$

6. $3\frac{1}{10}$

Perform the following operations. When possible, simplify first. Make sure answers are in lowest terms.

7. $\frac{5}{6} \times \frac{3}{10}$

8. $\frac{3}{7} \times \frac{2}{9}$

9. $2\frac{2}{9} \times 1\frac{3}{4}$

10. $7 \times \frac{1}{3}$

11. $\frac{5}{12} \div \frac{5}{6}$

12. $\frac{3}{16} \div \frac{7}{8}$

13. $7\frac{1}{2} \div \frac{5}{9}$

14. $\dfrac{4\frac{2}{3}}{2\frac{1}{2}}$

15. $\frac{7}{12} + \frac{5}{6}$

16. $\frac{1}{3} + \frac{5}{9}$

17. $2\frac{3}{7} + 5 + \frac{1}{2}$

18. $\frac{3}{32} + 4 + 1\frac{3}{4}$

19. $\frac{7}{9} - \frac{2}{3}$

20. $6\frac{1}{4} - 2\frac{3}{4}$

21. $\frac{1}{2} - \frac{1}{3}$

22. $3\frac{1}{8} - 1\frac{7}{16}$

Determine which is larger. Show your work.

23. $\frac{7}{32}, \frac{5}{16}$

Arrange the fractions in order, beginning with the smallest. Show your work.

24. $\frac{5}{7}, \frac{10}{21}, \frac{3}{4}$

Solve the following problems.

25. Two of the seven security employees at the local community college received safety awards from the governor. Write a fraction to represent what part of employees received an award.

26. A candy-store owner mixed $1\frac{1}{2}$ pounds of caramels, $\frac{3}{4}$ pound of chocolates, and $\frac{1}{2}$ pound of candy corn. What was the total weight of the mixed candy?

27. A homemaker had $5\frac{1}{2}$ cups of sugar on hand to make a batch of cookies requiring $1\frac{2}{3}$ cup of sugar. How much sugar was left?

28. If $6\frac{1}{4}$ ft of wire is needed to make one electrical extension cord, how many extension cords can be made from $68\frac{3}{4}$ feet of wire?

29. A costume maker figures one costume requires $2\frac{2}{3}$ yards of red satin material. How many yardsof red satin material would be needed to make three costumes?

30. Will a $\frac{5}{8}$-in.-wide drill bit make a hole wide enough to allow a $\frac{1}{2}$-in. (outside diameter) copper tube to pass through?

CHAPTER

4 Decimal Fractions

Decimals or decimal numbers are really *decimal fractions*. The decimal point and the place values to the right represent a *fraction whose denominator is always 10 or some power of 10*, such as 100, 1000, and so on. *Mixed-decimal fractions*, like mixed numbers, contain a whole number as well as a fractional part.

For convenience, in this book we use the terms *decimal fraction, decimal number,* and *decimal* interchangeably. However, we must keep in mind that decimals are another way of writing fractions.

Since a fraction can be considered the division of the numerator by the denominator, the fraction $\frac{3}{8}$ is equivalent to $3 \div 8$. If we carry out the division, the quotient is 0.375, a decimal number. The mixed number $2\frac{1}{4}$ is equivalent to the whole number 2 plus the fractional part $1 \div 4$. If we carry out the division of the fractional part, the resulting mixed decimal number is 2.25.

The place-value structure of decimal fractions makes finding common denominators unnecessary. Since calculators often use only decimal fractions in their computation and are now widely available, decimal fractions are becoming more popular for both ordinary and technical uses.

4-1 DECIMAL NUMBERS AND THE PLACE-VALUE SYSTEM

Learning Objectives

L.O.1. Identify place values in decimal numbers.
L.O.2. Read decimal numbers.
L.O.3. Write fractions with power-of-10 denominators as decimal numbers.
L.O.4. Compare decimal numbers.

L.O.1. Identify place values in decimal numbers.

Identifying Place Values in Decimal Numbers

Earlier we used the place-value system to understand whole numbers. Examine the place-value chart shown in Fig. 4-1. We will use the place-value system to understand *decimal numbers*. In Fig. 4-1, to move from *left* to *right*, we divide by 10 to get the value of the next place. For example, to move from the hundreds place to the tens place, we have $100 \div 10 = 10$. To move from the tens place to the ones place, we have $10 \div 10 = 1$.

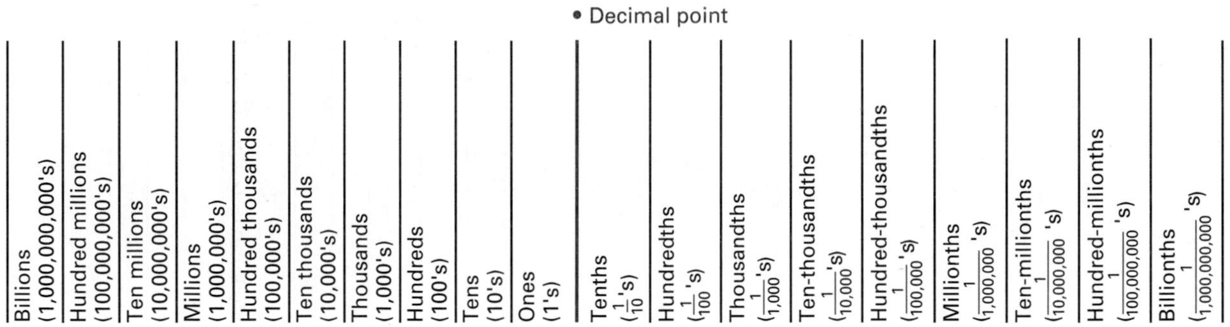

• Decimal point

Billions	Hundred millions	Ten millions	Millions	Hundred thousands	Ten thousands	Thousands	Hundreds	Tens	Ones	Tenths	Hundredths	Thousandths	Ten-thousandths	Hundred-thousandths	Millionths	Ten-millionths	Hundred-millionths	Billionths
(1,000,000,000's)	(100,000,000's)	(10,000,000's)	(1,000,000's)	(100,000's)	(10,000's)	(1,000's)	(100's)	(10's)	(1's)	($\frac{1}{10}$'s)	($\frac{1}{100}$'s)	($\frac{1}{1,000}$'s)	($\frac{1}{10,000}$'s)	($\frac{1}{100,000}$'s)	($\frac{1}{1,000,000}$'s)	($\frac{1}{10,000,000}$'s)	($\frac{1}{100,000,000}$'s)	($\frac{1}{1,000,000,000}$'s)

Figure 4-1 Place-value chart.

To extend the place-value chart on the right by moving from the ones place to the next place on the right, we have $1 \div 10 = \frac{1}{10}$. Thus, the place on the right of the ones place is called the *tenths place*. For the *hundredths* place we have $\frac{1}{10} \div 10$ or $\frac{1}{10} \times \frac{1}{10} = \frac{1}{100}$, and so on. A period (.) called a *decimal point* or *decimal* is placed between the ones place and tenths place so that we can identify the place value of each digit in a number.

Tips and Traps

The use of a period to separate the whole-number places from the decimal places is not a universally accepted notation. Some cultures use a comma instead.

Our notation for writing 32,495.8 may be written as 32.495,8 or 32 495,8.

The place values to the right of the ones place are fractions with denominators of 10 or some power of 10, such as 100, 1000, and so on. *In a decimal number, the digits to the right of the decimal point represent the numerator of the fraction.* The place value of the right-most digit indicates the denominator.

Rule 4-1.1 *To identify the place value of digits in decimal fractions:*

1. Mentally position the decimal number on the decimal place-value chart so that the decimal point of the number aligns with the decimal point on the chart.
2. Identify the place value of each digit according to its position on the chart.

EXAMPLE 4-1.1 Identify the place value of each digit in 32.4675 by applying the rule.

Solution: Comparing the number 32.4675 to the place-value chart in Fig. 4-2, we see that 3 is in the tens place, 2 is in the ones place, 4 is in the tenths place, 6 is in the hundredths place, 7 is in the thousandths place, and 5 is in the ten-thousandths place. ■

The place-value chart can be extended on the right side for smaller fractions in the same manner that it is extended on the left for larger numbers.

L.O.2. Read decimal numbers.

Reading Decimals

Examine the place-value chart in Fig. 4-2. Notice the *th* on the end of each place value on the right of the decimal point. When reading decimals, this *th* will indicate a decimal number. The word *and* is used to indicate the decimal point.

• Decimal point

Billions	Hundred millions	Ten millions	Millions	Hundred thousands	Ten thousands	Thousands	Hundreds	Tens	Ones	•	Tenths	Hundredths	Thousandths	Ten-thousandths	Hundred-thousandths	Millionths	Ten-millionths	Hundred-millionths	Billionths
								3	2	•	4	6	7	5					

Figure 4-2

To read a decimal number, we begin by reading the whole-number part; then we say *and* to indicate the decimal; then we read the decimal part. In reading the decimal part we use the same procedure as for the whole-number part, and end by reading the place value of the last digit in the decimal part. If the place value of the last digit in the decimal part is two words, these two words are hyphenated. For example, the ten-thousandths, hundred-thousandths, ten-millionths, and hundred-millionths places are all written with hyphens.

Rule 4-1.2 ***To read decimal numbers:***

1. Mentally align the number on the decimal place-value chart so that the decimal point of the number is directly under the decimal point on the chart.
2. Read the whole-number part.
3. Use *and* for the decimal point.
4. Read the decimal part like the whole-number part.
5. End by reading the *place value* of the last digit in the decimal part.

Tips and Traps

Informally, the decimal point is sometimes read as ''point.'' Thus 3.6 is read ''three point six.''

EXAMPLE 4-1.2 Read 52.386 by writing it in words.

Solution

1. Align the number on the chart (Fig. 4-3).

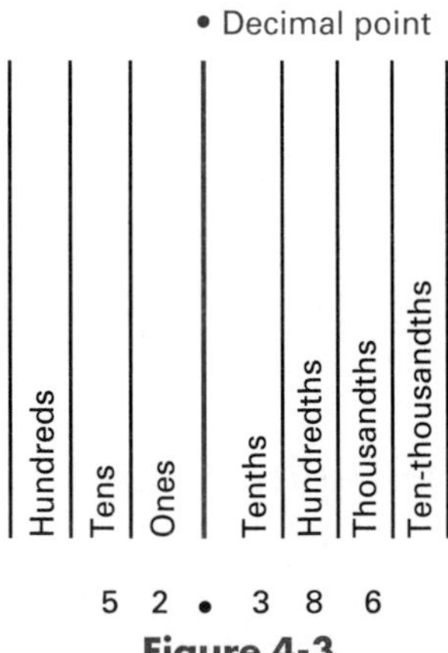

Figure 4-3

2. Read the whole-number part.

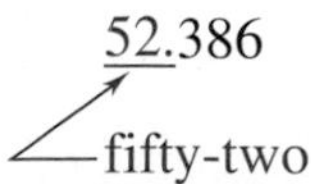

3. Use *and* for the decimal point.

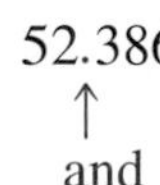

4. Read the decimal part like a whole-number part.

5. End by reading the *place value* of the last digit in the decimal part.

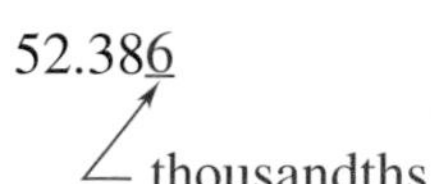

52.386 is "fifty-two and three hundred eighty-six thousandths." ■

EXAMPLE 4-1.3 Read 0.46 by writing it in words.

Solution

1. Align the number on the chart (Fig. 4-4).

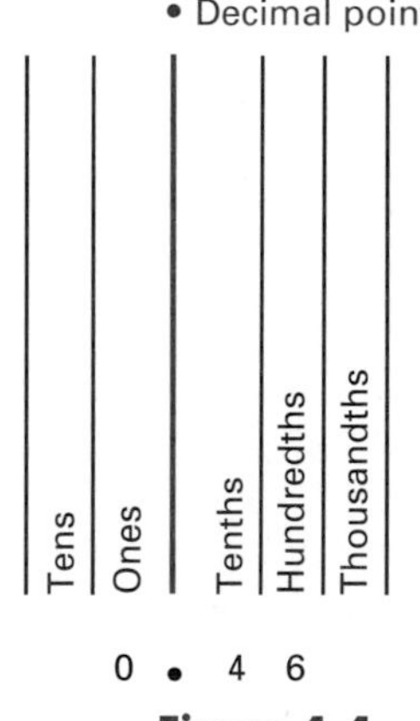

Figure 4-4

2. There is no whole-number part to read.

0.46
↑
no whole-number part

3. Do *not* read the decimal point.

0.46
↑
not read as *and*

4. Read the decimal part like a whole number.

0.46
↑
forty-six

5. End by reading the *place value* of the last digit in the decimal part.

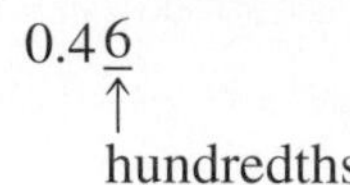

0.46 is ''forty-six hundredths.'' ■

EXAMPLE 4-1.4 Read 0.0162 by writing it in words.

Solution

1. Align the number on the chart (Fig. 4-5).

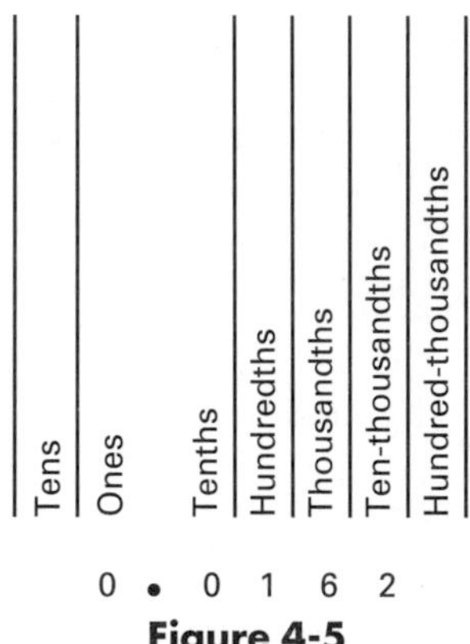

Figure 4-5

2. There is no whole-number part to read.

0.0162

3. Do *not* read the decimal point.

0.0162

4. Read the decimal part like a whole number.

0.0162
↑
one hundred sixty-two

5. End by reading the *place value* of the last digit in the decimal part.

0.0162
↑
ten-thousandths *(Be sure to use the hyphen.)*

0.0162 is ''one hundred sixty-two ten-thousandths.'' ■

When we read or write whole numbers in words, we do not use the word *and*. That is, 107 is "one hundred seven." When we write whole numbers using numerals, we usually omit the decimal point, even though the decimal point is understood to be at the end of the whole number. Therefore, any whole number, such as 32, can be written without a decimal (32) or with a decimal (32.).

L.O.3. Write fractions with power-of-10 denominators as decimal numbers.

Writing Fractions with Power-of-10 Denominators as Decimal Numbers

On the job we may encounter fractions with power-of-10 denominators like 10, 100, 1000, and so on. Examples of these are $\frac{1}{10}$, $\frac{75}{100}$, and so on. These fractions are so quickly changed into decimals that many workers prefer to write them as decimals and perhaps use a calculator to arrive at the answer or solution to the problem. If we encounter fractions like $\frac{1}{10}$ or $\frac{75}{100}$, we may want to use decimals instead. Here is a quick way to convert them.

Any fraction whose denominator is 10, 100, 1000, 10,000, and so on, can be written as a decimal number by writing the numerator and placing the decimal in the appropriate place to indicate the proper place value of the denominator. The decimal number will have the same number of digits in its decimal part as the fraction has zeros in its denominator.

EXAMPLE 4-1.5 Write $\frac{3}{10}$, $\frac{25}{100}$, $\frac{425}{100}$, and $\frac{3}{100}$ as decimal numbers.

Solution

$\frac{3}{10}$ is written 0.3.

$\frac{25}{100}$ is written 0.25.

$\frac{425}{100}$ is written 4.25.

$\frac{3}{100}$ is written 0.03. ■

Since $\frac{3}{10} = \frac{30}{100}$, we can see that $0.3 = 0.30$. Similarly, $0.7 = 0.70 = 0.700$.

Tips and Traps

If we affix zeros on the *right* end of a decimal number, we do not change the value of the number.

$$0.5 = 0.50 = 0.500$$

L.O.4. Compare decimal numbers.

Comparing Decimal Numbers

Our employment may sometimes require us to work with close tolerances. For instance, we may be working with wire and need to compare the diameters of several sizes of wire to arrange them in order from larger to smaller or smaller to larger. To do this, we need to be able to compare decimals, since wire size is often measured in decimals.

Rule 4-1.3 *To compare decimal numbers:*

1. Compare whole-number parts.
2. If whole-number parts are equal, compare digits place by place, starting at the tenths place and moving to the right.
3. Stop when two digits in the same place are different.
4. The digit that is larger determines the larger decimal number.

EXAMPLE 4-1.6 Compare the two numbers to see which is larger.

32.47 32.48

Solution: We begin by looking at the whole-number parts of the two numbers. In this case they are the same. Next, we look at the tenths place for each number. Both numbers have a 4 in the tenths place. Looking at the hundredths place, we see that 32.48 is the larger number because 8 is larger than 7. ■

EXAMPLE 4-1.7 Compare 0.4 and 0.07 to see which is larger.

Solution: Since the whole-number parts are the same (0), we compare the digits in the tenths place. 0.4 is larger because 4 is larger than 0. ■

If we write both numbers in the example above so that they have the same number of digits in the decimal part, they may be easier to compare.

$$0.4 = 0.40$$
$$0.07 = 0.07$$

Since 40 is larger than 7, 0.4 is larger than 0.07. This is equivalent to the process used in comparing common fractions. A common denominator is found and each fraction (or decimal fraction) is changed to an equivalent fraction having the common denominator.

Tips and Traps

Decimal fractions have a common denominator if they have the same number of digits to the right of the decimal point.

EXAMPLE 4-1.8 Compare 42.98 and 44.72 to see which is larger.

Solution: When comparing the whole-number parts, we see that 44 is larger than 42; therefore, 44.72 is larger than 42.98. ■

EXAMPLE 4-1.9 A measuring instrument called a micrometer is used to measure two objects. The measurements taken are 0.386 in. and 0.388 in. Which object is larger?

Solution: To compare the measurements above, we note that the digits in the tenths and hundredths places are the same. Comparing the digits in the thousandths places, we see that the object measuring 0.388 in. is larger. ■

SELF-STUDY EXERCISES 4-1

L.O.1.

1. What is the place value of the 7 in 32.407?
2. What is the place value of the 8 in 28.396?
3. What is the place value of the 4 in 3.00254?
4. What is the place value of the 3 in 457.2096532?
5. What is the place value of the 7 in 0.0387?

In the following problems, state what digit is in the place indicated.

6. Tens: 46.3079
7. Tenths: 2.0358
8. Thousandths: 520.0765
9. Ten-millionths: 3.002178356
10. Hundredths: 402.3786

L.O.2. Write the words for the following decimal numbers.

11. 21.387
12. 420.059
13. 0.89
14. 0.0568
15. 30.02379
16. 21.205085

Write the digits for the following numbers.

17. Three and forty-two hundredths
18. Seventy-eight and one hundred ninety-five thousandths.
19. Five hundred and five ten-thousandths
20. Seventy-five thousand, thirty-four hundred-thousandths

L.O.3. Write the following fractions as decimal numbers.

21. $\frac{5}{10}$
22. $\frac{23}{100}$
23. $\frac{7}{100}$
24. $\frac{683}{100}$
25. $\frac{79}{1000}$
26. $\frac{468}{1000}$
27. $\frac{587}{100}$
28. $\frac{108}{1000}$
29. $\frac{603}{100}$
30. $\frac{400}{100}$

L.O.4. Compare the following number pairs and identify the larger number.

31. 3.72, 3.68
32. 7.08, 7.06
33. 0.23, 0.3
34. 0.56, 0.5
35. 2.75, 2.65
36. 0.157, 0.2

Compare the following number pairs and identify the smaller number.

37. 8.9, 8.88
38. 0.25, 0.3
39. 0.913, 0.92
40. 0.761, 0.76
41. 5.983, 6.98
42. 1.972, 1.9735

Arrange the numbers in order from smaller to larger.

43. 0.23, 0.179, 0.314
44. 1.9, 1.87, 1.92
45. 72.1, 72.07, 73
46. Two micrometer readings are recorded as 0.837 in. and 0.81 in. Which is larger?
47. A micrometer reading for a part is 3.85 in. The specifications call for a dimension of 3.8 in. Which is larger, the micrometer reading or the specification callout?
48. A washer has an inside diameter of 0.33 in. Will it fit a bolt that has a diameter of 0.325 in.?
49. Aluminum sheeting can be purchased in 0.04-in. thickness or 0.035-in. thickness. Which thickness gives the thicker sheeting?
50. If No. 14 copper wire has a diameter of 0.064 in. and No. 10 wire has a diameter of 0.09 in., which has the larger diameter?

4-2 ROUNDING DECIMALS

Learning Objectives

L.O.1. Round decimal numbers by place value.
L.O.2. Round decimal numbers to one nonzero digit.

Whenever a decimal, or any number, is rounded, it becomes less accurate than the original decimal or number. Thus, if we round $3.99 to $4.00, the $4.00 is less accurate than the original $3.99. However, the difference between $3.99 and $4.00 is slight, only 1 cent. On the other hand, if we round $4.25 to $4.00, the difference here is greater and the rounded decimal is much more inaccurate, with a difference of 25 cents. Some job applications will require more accuracy in rounding than others. The rules for rounding numbers with decimals are similar to the rules for rounding whole numbers.

L.O.1. Round decimal numbers by place value.

Rounding Decimal Numbers by Place Value

When rounding a number to a certain place value, we must make sure that we are as accurate as our employer wants us to be. Generally, the size of the number and what it will be used for dictate to which decimal place it will be rounded.

Rule 4-2.1 ***To round a decimal number to a given place value:***

1. Locate the digit that occupies the rounding place. Then examine the digit that is to the immediate right.
2. If the digit to the right of the rounding place is 0, 1, 2, 3, or 4, do not change the digit in the rounding place. If the digit to the right of the rounding place is 5, 6, 7, 8, or 9, add 1 to the digit in the rounding place.
3. All digits on the *right* of the digit in the rounding place are replaced with zeros if they are to the left of the decimal point. Digits that are on the right of the digit in the rounding place *and* to the right of the decimal point are dropped.

Tips and Traps

When the digit in the rounding place is 9 and must be rounded up, it becomes 10. The 0 goes in place of the 9 and 1 is carried to the next place to the left.

EXAMPLE 4-2.1 Is 24.63 closer to 24.6 or 24.7?

Solution: We round to the tenths place. Applying Rule 4-2.1, we circle the 6, which is in the tenths place, to indicate the rounding place.

24.⑥3̲

Since the digit on the right of the 6 is 3, we do not change the 6.

24.⑥3̲
6

Next, we write all digits on the left of the 6 as they are.

24.⑥3̲
24.6

Since the 3, which is on the right of 6, is on the right of the decimal, it is dropped.
24.63 is closer to 24.6. ■

EXAMPLE 4-2.2 Round 46.879 to the hundredths place.

Solution

46.8⑦<u>9</u>	*(7 is in the hundredths place.)*
46.8⑦<u>9</u> 8	*(The next digit to the right is 9, so we add 1 to 7.)*
46.8⑦<u>9</u> 46.88	*(All digits on the left of 7 are written as they are.)*
46.88	*(The 9 is dropped since it is on the right of the rounding place and to the right of the decimal.)* ■

EXAMPLE 4-2.3 Round 32.6 to the tens place.

Solution

③<u>2</u>.6	*(3 is in the tens place.)*
③<u>2</u>.6 3	*(3 is unchanged since 2, the next digit to the right, is less than 5.)*
30	*(2 is replaced with a zero because it is before the decimal. 6 is dropped because it is to the right of the decimal.)* ■

EXAMPLE 4-2.4 Round 15.8 to the nearest whole number.

Solution: Rounding to the nearest whole number means rounding to the *ones* place.

1⑤.<u>8</u> 1 6	*(5 is in the ones place. 1 is added to 5 and 8 is dropped because it is after the decimal.)* ■

EXAMPLE 4-2.5 Round \$293.48 to the nearest dollar.

Solution: When we round to the nearest dollar, we are rounding to the *ones* place.

\$29③.<u>4</u>8

\$293 ■

EXAMPLE 4-2.6 Round \$71.8986 to the nearest cent.

Solution: One cent is 1 hundredth of a dollar, so to round to the nearest cent is to round to the *hundredths* place.

\$71.8⑨<u>8</u>6

\$71.90 ■

EXAMPLE 4-2.7 A motor has a horsepower rating of 30.497. Round this rating to the nearest whole number.

Solution

3⓪.<u>4</u>97

30 ■

L.O.2. Round decimal numbers to numbers with one nonzero digit.

Rounding Decimal Numbers to One Nonzero Digit

If we want to get a very quick estimate, we sometimes round to one nonzero digit. For example, if we have items priced at \$2.39, \$3.56, \$5.92, and \$2.13, we can get a quick estimate of the total by rounding to \$2.00, \$4.00, \$6.00, and \$2.00, respectively. Notice that all four rounded amounts have only one digit that is not zero. When we round numbers so that only one digit is not zero, we say that we are rounding to ''one nonzero digit'' or one significant digit.

Look at the numbers below. Notice that the *first* digit on the left is a *nonzero digit* (not a zero); the other digits are all zeros.

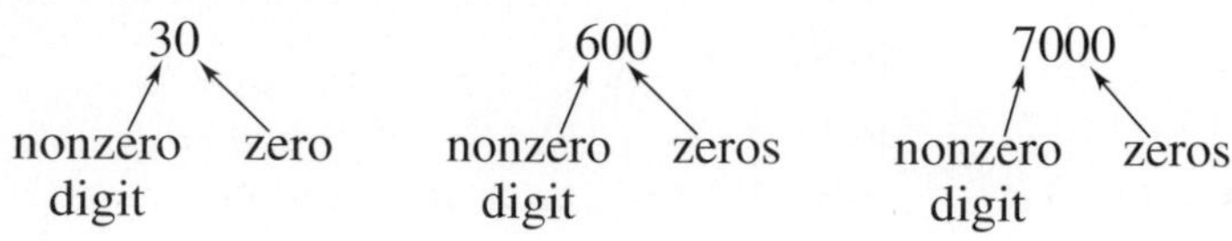

> **Rule 4-2.2** ***To round to one nonzero digit:***
>
> 1. Starting at the *left*, find the *first* digit that is not zero. This digit will be a 1, 2, 3, 4, 5, 6, 7, 8, or 9.
> 2. Round the number to the place value of the first nonzero digit.
> 3. All digits to the right of the rounding place are replaced with zeros up to the decimal point. Any digits that are on the right of the digit in the rounding place *and* follow the decimal point are dropped.

EXAMPLE 4-2.8 Round 4.23 to one nonzero digit.

Solution

1. Find the first digit from the left that is not zero.

   4.23
   ↑

   (*The 4 is the first nonzero digit.*)
2. The first nonzero digit, 4, is circled to mark the rounding place.

   ④.23

3. Round the number to the place value of the first nonzero digit.

   ④.2̲3 (*Nothing is added to 4 since the next digit to the right is less than 5.*)

4. Drop the digits on the right of the decimal point.

   4 ■

EXAMPLE 4-2.9 Round 78.4 to one nonzero digit.

Solution

1. Find the first nonzero digit from the left.

   78.4
   ↑

2. The first nonzero digit is 7.

   ⑦8.4

3. Round to the place value of the first nonzero digit.

   ⑧8.4 *(1 is added to 7 since the next digit on the right is 5 or more.)*

4. Replace each digit between the rounded digit and the decimal point with a zero. Drop all digits after the decimal point.

   80 *(0 replaces the digit between 8 and the decimal point. Drop the digits* after *the decimal point.)* ■

EXAMPLE 4-2.10 Round 0.83 to one nonzero digit.

Solution

0.⑧3 *(Remember to locate the first nonzero digit.)*
0.8 ■

EXAMPLE 4-2.11 If the price of a typewriter is $497.95, estimate this cost to one nonzero digit.

Solution

④97.95
$500 ■

Tips and Traps

When any amount has been rounded, it is no longer an exact amount. The rounded amount is now an *approximate* amount.

SELF-STUDY EXERCISES 4-2

L.O.1. Round to the place value indicated.

1. Nearest hundred: 468
2. Nearest hundred: 6248
3. Nearest thousand: 8263
4. Nearest ten thousand: 429,207
5. Nearest thousand: 39,748
6. Nearest ten thousand: 39,748
7. Nearest million: 285,487,412
8. Nearest ten: 468
9. Nearest billion: 82,629,426,021
10. Nearest ten million: 297,384,726

Round the decimals to the nearest whole number.

11. 42.7
12. 367.43
13. 7.983
14. 103.06
15. 2.9

Round the decimals to the nearest tenth.

16. 8.05
17. 12.936
18. 42.574
19. 83.23
20. 5.997

Round the decimals to the nearest hundredth.

21. 7.036
22. 42.065
23. 0.792
24. 3.198
25. 7.773

Round the decimals to the nearest thousandth.

26. 0.2173 **27.** 0.0196 **28.** 1.5085
29. 4.2378 **30.** 7.0039

Round to the nearest dollar.

31. $219.46 **32.** $82.93 **33.** $507.06
34. $2.83 **35.** $5.96

Round to the nearest cent.

36. $8.237 **37.** $0.291 **38.** $0.528
39. $5.796 **40.** $238.9238

41. A micrometer measure is listed as 0.7835 in. Round this measure to the nearest thousandth.

42. The diameter of an object is measured as 3.817 in. If specifications call for decimals to be expressed in hundredths, write this measure according to specifications.

43. To the nearest tenth, what is the current of a 2.836-A motor?

L.O.2. Round the numbers to numbers with one nonzero digit.

44. 483 **45.** 7.89 **46.** 62.5
47. 0.537 **48.** 0.0086
49. 0.095 **50.** 3.07 **51.** 52
52. 83.09 **53.** 52.8

54. If round steak costs $2.78 per pound, what is the cost per pound to the nearest dollar?

55. An estimate calls for converting 23.077 to an approximate number. What is the approximate number rounded to one nonzero digit?

56. The following amounts of monthly rainfall (in inches) were identified for these months: January, 0.355; March, 1.785; May, 0.45; July, 1.409; September, 0.07; and December, 2.018. Round each amount to one nonzero digit to identify the months with the most similar rainfall.

57. A researcher calculated the average response times (in seconds) as automobile drivers applied the brake when first seeing a road hazard. The following average times were calculated: driver A, 0.0275; driver B, 0.0264; driver C, 0.0234; driver D, 0.0284; and driver E, 0.0379. Round each time to one nonzero digit to help identify the subjects who were most similar in response time.

4-3 ADDING AND SUBTRACTING DECIMALS

Learning Objectives

L.O.1. Add decimals.
L.O.2. Subtract decimals.

L.O.1. Add decimals.

Adding Decimals

We add whole numbers by adding in columns all digits in the ones place, then adding all digits in the tens place, and so on. To add decimal numbers, we will follow the same procedure. When decimal numbers are aligned in this manner, the decimal points all fall in the same vertical line. Aligning decimal points has the same effect as using *like* denominators when adding (or subtracting) fractions.

Rule 4-3.1 *To add decimals:*

Arrange the numbers so that the decimal points are in one vertical line. Then add each column. Align the decimal for the sum in the same vertical line.

EXAMPLE 4-3.1 Find the sum of 4.3 + 13.6 + 0.3.

Solution: Align the numbers so that the decimals are in a vertical line. Then add each column and place the decimal in the sum.

$$\begin{array}{r} 4.3 \\ 13.6 \\ \underline{0.3} \\ 18.2 \end{array}$$

■

EXAMPLE 4-3.2 Add 42.3 + 17 + 0.36.

Solution

$$\begin{array}{l} 42.3 \\ 17 \\ \underline{\ \ 0.36} \end{array}$$

(Note that the decimal in 17 is understood to be at the right end.)

We should be very careful in aligning the digits. Careless writing can cause unnecessary errors. To avoid difficulty in adding the columns, we may write each number so that all have the same number of decimal places by adding zeros on the right.

$$\begin{array}{r} 42.30 \\ 17.00 \\ \underline{0.36} \\ 59.66 \end{array}$$

■

EXAMPLE 4-3.3 The total current that flows through an electrical circuit is found by adding the individual currents flowing through the circuit. If a circuit has 3.16 A, 4.06 A, 8.105 A, and 8 A of electricity flowing through it, what is the total current flowing through the circuit?

Solution

$$\begin{array}{r} 3.16 \\ 4.06 \\ 8.105 \\ \underline{8} \\ 23.325 \end{array}$$

The total current flowing through the circuit is 23.325 A. ■

L.O.2. Subtract decimals.

Subtracting Decimals

When subtracting decimals, we align the digits just as we do when adding decimals. Again, this is equivalent to finding common denominators for decimal fractions. Aligning the decimal points is sufficient.

> **Rule 4-3.2** ***To subtract decimals:***
>
> Arrange the numbers so that the decimal points are in the same vertical line. Then subtract each column and place the decimal in the difference.

EXAMPLE 4-3.4 Subtract 8.29 from 13.76.

Solution

$$\begin{array}{r} 13.76 \\ -\ \ 8.29 \\ \hline 5.47 \end{array}$$ *(Notice how the decimals are aligned.)*

■

EXAMPLE 4-3.5 Subtract 7.18 from 15.

Solution: In this problem we must be careful in aligning the decimals properly. Since 15 is a whole number, its decimal point will be placed after the 5.

$$\begin{array}{r} 15 \\ -\ \ 7.18 \\ \hline \end{array}$$

To subtract, we put zeros in the tenths and hundredths places of 15 and then borrow.

$$\begin{array}{r} 15.00 \\ -\ \ 7.18 \\ \hline 7.82 \end{array}$$

■

When a worker machines an object using a blueprint as a guide, a certain amount of variation from the blueprint specification is allowed for the machining process. This amount of variation is called *tolerance*. Thus, if a blueprint calls for a part to be 9.47 in. with a tolerance of ±0.05, this means that the actual part can be 0.05 in. *more* than the specification or 0.05 in. *less* than the specification. To find the *largest* possible size of the object, we add 9.47 + 0.05 = 9.52 in. To find the *smallest* possible size of the object, we subtract 9.47 − 0.05 = 9.42 in. The dimensions, 9.52 in. and 9.42 in., are called the *limit dimensions* of the object.

EXAMPLE 4-3.6 Find the limit dimensions of an object with a blueprint specification of 8.097 in. and a tolerance of ±0.005. (This is often written 8.097 ± 0.005.)

Solution

$$\begin{array}{r} 8.097 \\ -\ \ 0.005 \\ \hline 8.092 \text{ in.} \end{array} \qquad \begin{array}{r} 8.097 \\ +\ \ 0.005 \\ \hline 8.102 \text{ in.} \end{array}$$

The limit dimensions are 8.092 in. and 8.102 in.

■

General Tips for Using the Calculator

Adding and subtracting decimal numbers on a calculator is similar to whole numbers with the exception of entering the decimal point. The decimal point key [·] is pressed before the appropriate digit. It is not necessary to enter the zero *before* a decimal point in numbers of less than 1, such as 0.5 or 0.6. The two operations of Example 4-3.6 may be entered as follows.

Scientific calculator:

[AC] 8 [·] 097 [−] [·] 005 [=] ⇒ 8.092
[AC] 8 [·] 097 [+] [·] 005 [=] ⇒ 8.102

Graphing calculator:

[AC] 8 [·] 097 [−] [·] 005 [EXE] ⇒ 8.092
[AC] 8 [·] 097 [+] [·] 005 [EXE] ⇒ 8.102

SELF-STUDY EXERCISES 4-3

L.O.1. Add.

1. 4.2 + 3.6 + 7.9
2. 12.8 + 13.52 + 7.86
3. 3.9 + 4.02 + 0.21
4. 8.9 + 6.72 + 3.58 + 68.2
5. 83.37 + 42 + 1.6 + 3
6. 7 + 4.2 + 14.6 + 0.23
7. 23.9 + 54.3
8. 24.5 + 21.2
9. 205.03 + 56.305
10. 309.01 + 47.602
11. 0.784 + 5 + 1.2
12. 0.225 + 7 + 2.3
13. 900.75 + 225.85
14. 300.25 + 113.35
15. 3.7 + 0.6 + 4.8 + 9
16. 7.1 + 0.2 + 5.5 + 6

17. $\begin{array}{r} 0.70868 \\ +\ 0.10937 \\ \hline \end{array}$

18. $\begin{array}{r} 0.83967 \\ +\ 0.33675 \\ \hline \end{array}$

19. $\begin{array}{r} 51.006 \\ +\ \ 4.507 \\ \hline \end{array}$

20. $\begin{array}{r} 82.005 \\ +\ \ 3.406 \\ \hline \end{array}$

21. A pipe that is 0.103 in. thick has an inside diameter of 2.871 in. Find the outside diameter of the pipe.

22. The total current in a parallel circuit is found by adding the individual currents. If a circuit has individual currents of 3.98 A, 2.805 A, and 8.718 A, find the total current.

23. A part-time hourly worker earned $25.97 on Monday, $7.48 on Tuesday, $5.88 on Wednesday, $65.45 on Thursday, and $76.47 on Friday. Find the total week's wages.

24. A residential lot that measures 100.8 ft, 87.3 ft, 104.7 ft, and 98.6 ft on each of its four sides is to be fenced. How many feet of fencing are required?

25. A technician purchased an ac voltage sensor for $11.95, a grounded outlet analyzer for $5.99, and a neon circuit tester for $1.99. How much did she pay for all three items?

26. Julio purchased a $\frac{3}{32}$-in. submini stereo plug adapter for $2.99 and a $\frac{3}{32}$-in. submini mono plug adapter for $1.76. If the sales tax was $0.29, what was his total bill?

L.O.2. Subtract.

27. $\begin{array}{r} 14.86 \\ -\ \ 7.93 \\ \hline \end{array}$

28. $\begin{array}{r} 20.07 \\ -\ \ 4.236 \\ \hline \end{array}$

29. $\begin{array}{r} 813.673 \\ -\ \ \ \ 9.98 \\ \hline \end{array}$

30. $\begin{array}{r} 84 \\ -\ 27.86 \\ \hline \end{array}$

31. $\begin{array}{r} 5.079 \\ -\ 0.985 \\ \hline \end{array}$

32. Subtract 8.8 from 12.7.

33. Subtract 24.38 from 316.2.

34. Subtract 13.5 from 21.

35. Subtract 67.2 from 378.

36. Find the difference between 42 and 37.6.

37. One box of rivets weighs 52.6 lb and another box weighs 37.5 lb. The first box of rivets weighs how much more than the second box?

38. The length of an object is 12.09 in. according to a blueprint. If the tolerance is ± 0.01 in., what are the limit dimensions of the object?

39. Two lengths of copper tubing measure 63.6 cm and 3.77 cm. What is the difference in their lengths?

40. Find the limit dimensions of an object whose dimension is 4.195 in. $\pm$ 0.006 in.

41. Steel rods of 0.38 decimeter (dm) and 1.9 dm are cut from a steel rod 2.5 dm long. If the cutting waste is ignored, how long is the piece that is left?

42. If a current of 3.95 A is removed from a 15.5-A parallel circuit, find the remaining current in the circuit.

43. If a hardbound novel costs $27.50 and the softcover edition costs $18.75, how much can be saved by buying the softcover edition?

44. A student buys electronic supplies for an engineering technology class for $27.75 and pays with $30. How much is the change?

45. If the sheet music for a popular song costs $5.25, how much change would be returned if the purchaser pays with a $10 bill? Disregard sales tax.

4-4 MULTIPLYING DECIMALS

Learning Objectives

L.O.1. Multiply any decimal numbers.
L.O.2. Multiply decimal numbers by a power of 10.

L.O.1. Multiply any decimal numbers.

Multiplying Decimal Numbers

One way we can multiply decimal numbers is by writing them in their fractional form and then multiplying. To multiply 0.8×0.32, for example, we can write $\frac{8}{10} \times \frac{32}{100}$. Multiplying, we have $\frac{256}{1000}$, which is written as 0.256 in decimal form.

Before going on, let's use these three numbers to make some observations about decimal fractions and the place-value chart in Fig. 4-6. Recall that the number of zeros in the denominator is the same as the number of places to the right of the decimal point.

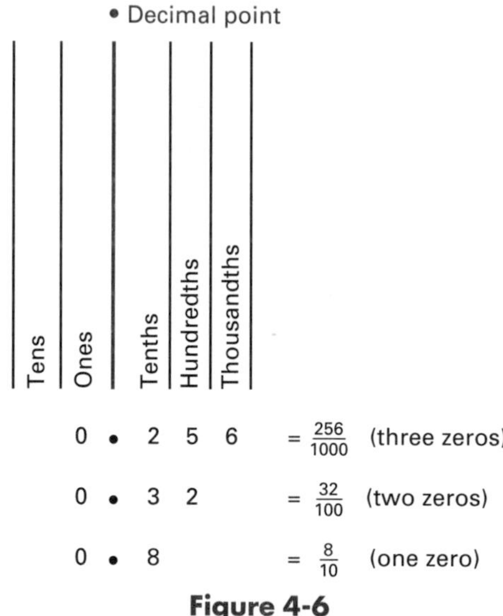

Figure 4-6

Examining this example, we see that 0.8 has one digit on the right of the decimal, while 0.32 has two digits on the right of the decimal. These two factors together have a total of three digits on the right of the decimal. If we compare this total with the number of digits on the right of the decimal in the answer, 0.256, we see that they are the same. This fact allows us to multiply decimal numbers without coverting them to fractions.

$$\begin{array}{rl} 0.32 & 2 \text{ places afer decimal} \\ \times \quad 0.8 & \underline{1} \text{ place after decimal} \\ \hline 0.256 & 3 \text{ places after decimal} \end{array}$$

$$0.256 = \frac{256}{1000}$$

Therefore, we may use the following rule to multiply decimal fractions instead of converting the decimals to common fractions. The results are the same.

> **Rule 4-4.1** *To multiply decimal numbers:*
>
> 1. Align the numbers as if they were whole numbers and multiply.
> 2. Then count the total number of digits on the right of the decimal in each factor.
> 3. Place the decimal in the product so that the number of decimal places is the sum of the number of decimal places in the factors.

EXAMPLE 4-4.1 Multiply 1.36×0.2.

Solution

$$\begin{array}{r} 1.36 \\ \times \quad 0.2 \\ \hline 0.272 \end{array}$$

(Note that in multiplication the decimals do not *have to be in a straight line. Place a zero in the ones place so that the decimal point will not be overlooked.)* ■

EXAMPLE 4-4.2 Multiply 0.309×0.17.

Solution

$$\begin{array}{r} 0.309 \\ \times \quad 0.17 \\ \hline 2163 \\ 309 \\ \hline 0.05253 \end{array}$$

(We did not have enough digits in the product, so a zero was inserted on the left *to give the appropriate number of decimal places.)* ■

EXAMPLE 4-4.3 Fifteen pieces of copper tubing are to be cut so that each is 12.7 in. long. How much copper tubing is needed if we disregard waste?

Solution

$$\begin{array}{r} 12.7 \\ \times \quad 15 \\ \hline 63\ 5 \\ 127 \\ \hline 190.5 \end{array}$$

(Note that the first factor has one digit to the right of the decimal point.)

190.5 in. of tubing are needed. ■

EXAMPLE 4-4.4 The outside diameter of a pipe is 7.82 cm (Fig. 4-7). If the pipe wall is 1.56 cm thick, find the inside diameter of the pipe.

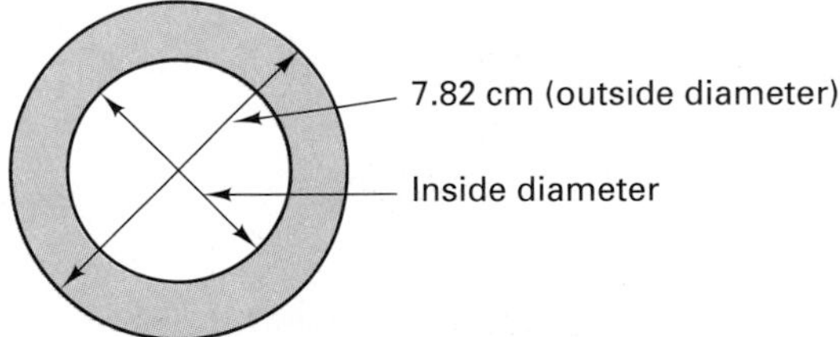

Figure 4-7

Solution: We double the thickness $(1.56 \times 2 = 3.12)$ to find the total thickness. This total is subtracted from the outside diameter to give the inside diameter.

$$\begin{array}{r} 7.82 \\ -\ 3.12 \\ \hline 4.70 \end{array}$$

The inside diameter is 4.70 cm. ■

General Tips for Using the Calculator

Using a calculator to multiply decimal numbers, as in addition and subtraction, requires insertion of the [·] key where it appears in any number being multiplied. Look at Example 4-4.1 worked on a calculator.

Scientific calculator:

[AC] 1 [·] 36 [×] [·] 2 [=] ⇒ 0.272

Graphing calculator:

[AC] 1 [·] 36 [×] [·] 2 [EXE] ⇒ 0.272

L.O.2. Multiply decimal numbers by a power of 10.

Multiplying Decimal Numbers by a Power of 10

If 100 sheets of plywood cost \$10.57 per sheet, we find the total cost by multiplying. \$10.57 × 100 = \$1057. When we solve problems involving multiplying by 10, 100, 1000, and so on, we should be able to perform the calculations mentally by just moving the decimal point. Calculations such as the one above can be done more quickly mentally than if a calculator were used.

Rule 4-4.2 ***To multiply a decimal number by a power of 10;***

Move the decimal point in the number to the *right* as many places as the 10, 100, 1000, or so on, has zeros.

EXAMPLE 4-4.5 Multiply 4.237 × 10.

Solution: Applying the shortcut rule, we move the decimal one place to the *right*.

■

EXAMPLE 4-4.6 Multiply 36.2 × 1000.

Solution: We have three zeros in 1000.

$$36.200 \times 1000 = 36{,}200$$

■

EXAMPLE 4-4.7 If one concrete block weighs 10.8 lb, find the weight of 100 concrete blocks.

Solution

$$10.80 \times 100 = 1080$$

The blocks weigh 1080 lb. ■

SELF-STUDY EXERCISES 4-4

L.O.1. Multiply.

1. 53.4 × 0.29

2. 37.7 × 1.5

3. 9.27 × 0.35

4. $\begin{array}{r} 0.215 \\ \times\ 0.27 \\ \hline \end{array}$

5. $\begin{array}{r} 0.271 \\ \times\ 0.32 \\ \hline \end{array}$

6. 3.7×1.93

7. 2.78×0.03
8. 73.806×2.305
9. 1.9067×0.2013
10. 8.2037×0.602
11. 42×0.73
12. 81×7.37
13. 81.7×0.621
14. 72.6×0.532
15. 326×1.04
16. 261×2.07
17. 6.35×40
18. 9.21×20
19. 93.07×0.01
20. 76.02×0.01
21. 586.35×0.001
22. 732.65×0.001
23. 2.0037×4.6
24. 3.0062×3.8
25. 7.04×0.0025
26. 6.01×0.0045
27. 0.457×2.003
28. 0.387×1.009
29. 0.05×0.003
30. 0.07×0.008

31. A pipe has an outside diameter of 4.327 in. Find the inside diameter if the pipe wall is 0.185 in. thick.
32. A plastic pipe has an inside diameter of 4.75 in. Find the outside diameter if the pipe wall is 0.25 in. thick.
33. Electrical switches cost $5.70 wholesale. If the retail price is $7.99 each, how much would an electrician save over retail by buying 12 switches at the wholesale price?
34. How much No. 24 electrical wire is needed to cut eight pieces each 18.9 in. long?
35. A retailer purchases 15 cases of potato chips at $8.67 per case. If the chips were sold at $12.95 per case, how much profit did the retailer make?
36. A contractor purchases 500 yd^3 of concrete at $21.00 per cubic yard, 24 yd^3 of sand at $5.00 per cubic yard, and 36 yd^3 of fill dirt at $3.00 per cubic yard. What was the total cost of the materials?
37. A micrometer is designed so that the screw thread makes the spindle advance 0.025 in. in one complete turn. How far will the spindle advance in 15 complete turns?
38. A 4 ft × 8 ft sheet of 1-in.-thick steel weighs 1307 lb and costs $0.30 a pound. Find the cost of five of these sheets.
39. If six 2.5-A devices are connected in parallel, what is the total amperage (A) of the circuit?
40. Sequoia ordered 24 BNC T-adapters at $3.25 each. How much was the total order?

L.O.2. Solve these problems mentally.

41. 42.3×10
42. 78.36×10
43. 48×10
44. 52.36×100
45. 4.867×100
46. 2.9×100
47. 42.385×1000
48. 0.6×1000
49. 29.73×1000
50. $4.23 \times 10{,}000$
51. $0.8346 \times 10{,}000$
52. $8.723 \times 100{,}000$
53. 5.93×1000
54. 68.725×100
55. 4.235×10

56. One hundred shrubs are needed to landscape the grounds of an office building. If each shrub costs $6.97, find the cost of 100 shrubs.
57. If trailer light bulbs cost $0.55 each, what is the cost of 1000 bulbs?
58. A dress manufacturer purchased 10,000 meters (m) of fabric at $1.27 per meter. How much did the fabric cost?
59. Wattage is the product of voltage and amperage. How many watts of electricity are used in a 100-volt circuit if the amperage is 3.5?
60. Can a 42-ft flatbed trailer whose maximum pay load is 48,000 lb haul 100 steel beams each weighing 478.6 lb?

4-5 DIVIDING DECIMALS

Learning Objectives

L.O.1. Divide decimal numbers with zero remainder.
L.O.2. Divide decimal numbers whose remainders are not zero.
L.O.3. Divide decimal numbers by a power of 10.

Dividing decimals is similar to dividing whole numbers. However, as in multiplication, we must be very careful about the placement of the decimal points in the quotient.

L.O.1. Divide decimal numbers with zero remainder.

Dividing Decimal Numbers with Zero Remainder

To understand the placement of the decimal, consider the example $3.6 \div 1.2$. This division can be written as the fraction $\frac{3.6}{1.2}$. If we multiply the numerator and denominator of the fraction by 10, we make each number a whole number. Since $\frac{10}{10} = 1$, we do not change the value of the fraction.

$$\frac{3.6}{1.2} \times \frac{10}{10} = \frac{3.6 \times 10}{1.2 \times 10} = \frac{36}{12} = 3$$

Now consider the same division problem in long-division form.

$$1.2\overline{)3.6}$$

The decimal point is shifted to the end of the divisor, 1.2, which in this case is one place. (Remember that multiplying by 10 results in moving the decimal one place to the right.) The dividend is also multiplied by 10 which shifts the decimal one place. The decimal in the quotient is written directly above the new decimal in the dividend. The decimal point should be placed in position *before* the division is started.

$$\begin{array}{r} 3. \\ 1.2.\overline{)3.6.} \end{array}$$

The reason for moving the decimal point in the divisor is to make the divisor a *whole* number to simplify the division process. This changes the value of the divisor, so the decimal point in the dividend must be moved the same number of places so as not to change the value of the division problem. This is similar to changing regular fractions to equivalent fractions—both numerator (dividend) and denominator (divisor) are multiplied by the same factor.

Rule 4-5.1 *To divide decimal numbers written in long-division form:*

1. Shift the decimal in the divisor so that it is on the right side of all digits. (By shifting the decimal, we are multiplying by 10, 100, 1000, and so on.)
2. Shift the decimal in the dividend to the right as many places as the decimal was shifted in the divisor. (Affix zeros if necessary.)
3. Write the decimal point in the answer directly above the shifted position of the decimal in the dividend. (Do this *before* dividing.)
4. Divide as you would in whole numbers.

EXAMPLE 4-5.1 Divide $4.8 \div 6$.

Solution: Insert a decimal point above the decimal point in the dividend.

$$\begin{array}{r} . \\ 6\overline{)4.8} \end{array}$$

$$\begin{array}{r} .8 \\ 6\overline{)4.8} \end{array} \quad \textit{(Divide.)}$$

When the divisor is a whole number, the decimal is understood to be on the right side of 6 and is not shifted. When the divisor is a whole number, the decimal point in the dividend is not shifted. The decimal is placed in the quotient directly above the decimal in the dividend. ■

EXAMPLE 4-5.2 Divide 3.12 ÷ 1.2.

Solution

```
1.2 |3.12        (Shift the decimal in the divisor and dividend.)

      .
 12 |31.2        (Place the decimal in the quotient. Then divide.)

     2.6
 12 |31.2
     24
      7 2
      7 2
```

■

EXAMPLE 4-5.3 Including a waste allowance, a room requires 472 ft^2 of wallpaper. How many single rolls will be needed for the job if a roll covers 29.5 ft^2 of surface?

Solution

```
29.5 |472.0      (Write a decimal point after the 2 and affix one zero.)

           .
29.5 |472.0      (Move the decimal and place it in the quotient.)

       16.
 295 |4720
      295
      1770
      1770
```

L.O.2. Divide decimal numbers whose remainders are not zero.

Dividing Decimal Numbers Whose Remainders Are Not Zero

Not all divisions have a remainder of zero. When divisions have a nonzero remainder, we have two alternatives. We can write the remainder as a fraction by placing it over the divisor and adding the fraction to the quotient, or we can round the quotient to a specified decimal place. We must choose one of these alternatives before dividing.

> **Rule 4-5.2** ***To express a remainder as a fraction:***
>
> 1. Divide to the specified place value.
> 2. Affix zeros to the dividend after the decimal if needed to carry out the division.
> 3. Write the remainder over the divisor as a fraction after the last place in the quotient. Reduce the fraction to lowest terms if appropriate.

EXAMPLE 4-5.4 Divide to the hundredths place and write the remainder as a fraction.

6.9 |23

Solution

$$\begin{array}{r} 3.33\,\tfrac{23}{69} \\ 6.9\overline{)\,23.0\,00} \\ \underline{20\,7} \\ 2\,3\,0 \\ \underline{2\,0\,7} \\ 2\,30 \\ \underline{2\,07} \\ 23 \end{array}$$

or $3.33\frac{1}{3}$

(Zeros are affixed so that the division can be carried to the decimal place specified, often hundredths.)

■

Rule 4-5.3 ***To round a quotient to a place value:***

1. Divide to one place past the place specified.
2. Affix zeros to the dividend after the decimal if needed to carry out the division.
3. Round the quotient to the place specified.

EXAMPLE 4-5.5 Divide and round the quotient to the nearest tenth.

$$3.2\overline{)\,15.27}$$

Solution

$$\begin{array}{r} 4.77 \\ 3.2\overline{)\,15.2\,70} \\ \underline{12\,8} \\ 2\,4\,7 \\ \underline{2\,2\,4} \\ 2\,30 \\ \underline{2\,24} \\ 6 \end{array}$$

(Since we are rounding to the nearest tenth, we divide to the hundredths place.)

4.77 rounds to 4.8. ■

General Tips for Using the Calculator

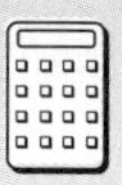

Division of decimal numbers on a calculator does *not* require moving any of the decimal points. Let's use a calculator for Example 4-5.5. Remember to enter the dividend *first*.

Scientific calculator:

[AC] 15 [·] 27 [÷] 3 [·] 2 [=] 4.771875

Graphing calculator:

[AC] 15 [·] 27 [÷] 3 [·] 2 [EXE] 4.771875

Rounding to the nearest tenth, we get 4.8.

L.O.3. Divide decimal numbers by a power of 10.

Dividing Decimal Numbers by a Power of Ten

Dividing by numbers such as 10, 100, 1000, and so on, can be done very quickly if we remember a shortcut rule. Look at the following example.

EXAMPLE 4-5.6 Divide 3.2 ÷ 10, 78.9 ÷ 100, and 52,900 ÷ 1000.

Solution

$$\begin{array}{r} 0.32 \\ 10\overline{)3.20} \\ \underline{3\,0} \\ 20 \\ \underline{20} \end{array} \qquad \begin{array}{r} 0.789 \\ 100\overline{)78.900} \\ \underline{70\,0} \\ 8\,90 \\ \underline{8\,00} \\ 900 \\ \underline{900} \end{array} \qquad \begin{array}{r} 52.9 \\ 1000\overline{)52900.0} \\ \underline{50000} \\ 2900 \\ \underline{2000} \\ 900\,0 \\ \underline{900\,0} \end{array}$$

$$3.2 \div 10 = 0.32$$

$$78.9 \div 100 = 0.789$$

$$52{,}900 \div 1000 = 52.9$$

■

If we compare the decimal point in the dividend with the decimal in the quotient, we notice that the decimal point in the quotient has been moved to the left as many places as the divisor (10, 100, 1000, and so on) has zeros.

> **Rule 4-5.4** ***To divide a decimal number by a power of 10:***
>
> Move the decimal to the *left* as many places as the divisor has zeros.

Compare this rule to Rule 4-4.2 for multiplying decimal numbers by 10, 100, 1000, and so on. For multiplication, the decimal moves to the *right*, whereas for division the decimal moves to the *left*.

EXAMPLE 4-5.7 If 100 lb of floor cleaner costs \$63, what is the cost of 1 lb?

Solution: Use the shortcut rule to divide by 100.

$$63 \div 100 = .63 = \$0.63$$

The cost for 1 lb is \$0.63. ■

SELF-STUDY EXERCISES 4-5

L.O.1. Divide.

1. $3\overline{)3.78}$ **2.** $26\overline{)80.34}$ **3.** $21\overline{)1.323}$

4. $1.2\overline{)342}$ **5.** $4.8\overline{)28.32}$ **6.** $7.6\overline{)342}$

7. $6.3\overline{)68.67}$ **8.** $0.23\overline{)0.0437}$ **9.** $0.09\overline{)0.0954}$

10. $0.41\overline{)0.1353}$ **11.** $0.32\overline{)8}$ **12.** $1449 \div 0.07$

13. $8118 \div 0.09$ **14.** $4066 \div 0.38$ **15.** $8.28 \div 4.6$

16. A pipe 7.3 ft long weighs 43.8 lb. What is the weight of 1 ft of pipe?

17. A room requires 770.5 ft^2 of wallpaper, including waste. How many whole single rolls are needed for the job if a roll covers 33.5 ft^2 of surface?

18. The feed per revolution of a drill is 0.012 in. A hole 7.2 in. deep will require how many revolutions of the drill?

19. A piece of channel iron 5.6 ft long is cut into eight pieces. Assuming that there is no waste, what is the length of each piece?

20. If voltage is wattage divided by amperage, find the voltage in a 300-watt circuit drawing 3.2 amps.

L.O.2. Divide to the hundredths place and write the remainder as a fraction.

21. $1.3\overline{)25.8}$
22. $2.4\overline{)4.37}$
23. $0.06\overline{)156.07}$
24. $41.7 \div 21$
25. $0.23 \div 2.9$

Divide and round the quotient to the place indicated.

26. Nearest tenth: $0.43\overline{)72.8}$
27. Nearest hundredth: $25\overline{)3.897}$
28. Nearest whole number: $4.1\overline{)34.86}$
29. Nearest cent: $5\overline{)\$4.823}$
30. Nearest dollar: $17\overline{)\$24.98}$
31. If 12 lathes cost \$6895, find the cost of each lathe to the nearest dollar.
32. If 12 electrolytic capacitors cost \$23.75, find the cost of one capacitor to the nearest cent.
33. To find the depth of an American Standard screw thread, the number of threads per inch is divided into 0.6495. Find the depth of thread of a screw that has 8 threads per inch. Round to the nearest ten-thousandth.
34. A stack of 40 sheets of Fiberglas is 6.9 in. high. What is the thickness of one sheet to the nearest tenth of an inch?

L.O.3. Use the shortcut rule for division to obtain the answers. Do the work mentally.

35. $10\overline{)3.19}$
36. $10\overline{)4.027}$
37. $10\overline{)5}$
38. $100\overline{)4.917}$
39. $100\overline{)2.081}$
40. $1000\overline{)0.921}$
41. $1000\overline{)8.35}$
42. $1000\overline{)0.2003}$
43. $10\overline{)0.0004}$
44. $1000\overline{)8}$
45. $4.592 \div 10$
46. $239.6 \div 100$
47. $84 \div 10{,}000$
48. $39.4 \div 10{,}000$
49. $0.00005 \div 100{,}000$
50. If 100 lb of potatoes wholesale for \$12.60, what is the wholesale price of the potatoes per pound?
51. A 1000-ft-long coil of wire weighs 21.75 lb. What is the per foot weight of the wire?
52. 10,000 linear feet (lin ft) of lumber is sold for \$2580. What is the cost per foot of the lumber?
53. If resistance (in ohms) is voltage divided by amperage, what is the resistance (in ohms) in a circuit carrying 114.58 volts and having 10 amperes of current?
54. If 100 sheets of plywood measure 75 in. thick, what is the thickness of one sheet of plywood?

4-6 FINDING AVERAGES AND ESTIMATING

Learning Objectives

L.O.1. Find the numerical average.
L.O.2. Estimate sums, differences, products, and quotients of decimals.

L.O.1. Find the numerical average.

Finding the Numerical Average

Sometimes a worker needs to use approximate numbers for one reason or another. Two commonly used approximations are averages and estimates. Averages are often used for comparisons, such as comparing the average mileage different cars get per gallon of gasoline. There are several different kinds of averages. Here we refer to the average that is known as the arithmetic average or mean. Estimates are used for getting a quick but rough idea of what the answer to a problem is *before* working the problem.

In most courses students take, their numerical grade is determined by a process called *averaging*. If the grades for a course are 92, 87, 76, 88, 95, and 96, we can find the average grade by adding the grades and dividing by the number of grades. Since we have six grades, the sum is divided by 6.

$$\frac{92 + 87 + 76 + 88 + 95 + 96}{6} = \frac{534}{6} = 89$$

Rule 4-6.1 ***To find the average of a group of numbers or measures:***

Add the numbers or measures and divide by the number of addends.

EXAMPLE 4-6.1 A car involved in an energy efficiency study had the following miles per gallon (mpg) listings for five tanks of gasoline: 21.7, 22.4, 26.9, 23.7, and 22.6 mpg. Find the average miles per gallon for the five tanks of gasoline.

Solution

$$\frac{21.7 + 22.4 + 26.9 + 23.7 + 22.6}{5} = \frac{117.3}{5} = 23.46$$

$$= 23.5 \text{ mpg} \qquad \textit{(rounded)}$$

The average is 23.5 mpg.

Tips and Traps

If the numbers being averaged are expressed to the same place value, the answer is usually rounded to that place value.

General Tips for Using the Calculator

An average can be found in a continuous sequence on a calculator. One important step is to use the [=] or [EXE] key to find the sum *before* dividing by the number of the measures. Let's apply the calculator to Example 4-6.1.

Scientific calculator:

[AC] 21 [.] 7 [+] 22 [.] 4 [+] 26 [.] 9 [+] 23 [.] 7 [+] 22 [.] 6 [=] [÷] 5 [=] ⇒ 23.46

Graphing calculator:

[AC] 21 [.] 7 [+] 22 [.] 4 [+] 26 [.] 9 [+] 23 [.] 7 [+] 22 [.] 6 [EXE] [÷] 5 [EXE] ⇒ 23.46

L.O.2. Estimate sums, differences, products, and quotients of decimals.

Estimating Sums, Differences, Products, and Quotients of Decimals

When estimating sums, differences, products, and quotients of decimals, we round the numbers in the problem and then perform the operation. *The only time we round first is in estimating.* We may round to any convenient place value or convert to numbers with one nonzero digit, whichever is more practical.

When estimating sums, the addends are rounded and *then* the rounded addends are added.

EXAMPLE 4-6.2 Estimate the sum by rounding to the nearest whole number.

$$42.3 + 16.86 + 21.3 + 50.215$$

Solution: Round each addend to the nearest whole number (ones place) and then add.

$$42 + 17 + 21 + 50 = 130$$

The estimated sum is 130. ■

When estimating differences, the minuend and subtrahend are rounded and *then* subtracted. Again, in estimating, we *round first*.

EXAMPLE 4-6.3 Estimate the difference by rounding to the nearest whole number.

$$4.789 - 2.9041$$

Solution: Round each number to the nearest whole number and then subtract.

$$5 - 3 = 2$$

The estimated difference is 2.

When estimating products, the multiplicand and multiplier (or factors) are rounded and *then* the values are multiplied. When estimating products, we often round the factors to one nonzero digit and then multiply so the estimate can be done mentally. In estimating, rounding is done *first*.

EXAMPLE 4-6.4 Estimate the product of 642.8 × 21.7 by converting each factor to a number with one nonzero digit.

Solution: 642.8 rounds to 600; 21.7 rounds to 20.

$$\begin{array}{r} 600 \\ \times \quad 20 \\ \hline 12{,}000 \end{array}$$

The estimated product is 12,000. ■

When estimating quotients, we round the divisor and the dividend. The divisor and dividend are usually rounded to one nonzero digit. We then divide to get the *first* digit of the quotient. Zeros are used to complete the whole-number part of the quotient. Here, too, the rounding is done *first*.

EXAMPLE 4-6.5 Estimate the quotient of 46.87 divided by 0.36 by rounding the divisor and dividend to numbers with one nonzero digit.

Solution

$0.36\overline{)46.87}$ *(Round to one nonzero digit and proceed as in Rule 4-5.1 for placement of the decimal.)*

$$\begin{array}{r} 1. \\ 0.4\overline{)50.0} \end{array}$$

$$\begin{array}{r} 100. \\ 4\overline{)500.} \end{array}$$ *(After getting the first digit, the remaining places are filled with zeros up to the decimal point.)* ■

EXAMPLE 4-6.6 Estimate the quotient of 9238.7 divided by 47 by converting to numbers with one nonzero digit.

Solution

$47\overline{)9238.7}$

$50\overline{)9000.}$ *(Round the divisor and dividend to one nonzero digits and proceed as in Rule 4-5.1 for placement of the decimal.)*

$$\begin{array}{r} 100. \\ 50\overline{)9000.} \end{array}$$

Proper placement of the first digit in the quotient is necessary to obtain the correct number of zeros. ■

Tips and Traps

In Example 4-6.6 the remainder of the first step in the division process would be 40 or $\frac{40}{50}$. A more reasonable estimate would be to say that the quotient is between 100 and 200, and is actually closer to 200 since $\frac{40}{50}$ is more than $\frac{1}{2}$.

SELF-STUDY EXERCISES 4-6

L.O.1. Find the average of the following. Round to the same place value as that used in the problems.

1. Temperature readings: 82.5°, 76.3°, 79.8°, 84.7°, 80.8°, 78.8°, 80.0°

2. Test scores: 86, 73, 95, 85

3. Weight of cotton bales: 515 lb, 468 lb, 435 lb, 396 lb

4. Monthly income: $873.46, $598.21, $293.85, $546.83, $695.83, $429.86, $955.34, $846.95, $1025.73, $1152.89, $957.64, $807.25

5. Average rainfall: 1.25 in., 0.54 in., 0.78 in., 2.35 in., 4.15 in., 1.09 in.

6. Amperes of current: 3.0 A, 2.5 A, 3.5 A, 4.0 A, 4.5 A

L.O.2. Estimate the following sums and differences by rounding to the place specified.

7. Nearest whole number: 42.87 + 39.073 + 82.9 + 21.6

8. Nearest dollar: $6.42 + $2.87 + $4.80

9. Nearest ten dollars: $342.87 + $29.95 + $127.82

10. Nearest whole number: 523.8 − 64.97

11. Nearest ten: 873.9 − 581.6

12. Nearest dollar: $21.93 − $16.42

13. Nearest hundred dollars: $928.97 − $297.52

Estimate the following products by converting each factor to a number with one nonzero digit.

14. 423 × 86 **15.** 27.9 × 13.4 **16.** 0.397 × 1.5

Estimate the following quotients by rounding the divisor and dividend to numbers with one nonzero digit.

17. $4.2\overline{)83.7}$ **18.** $93\overline{)479.8}$ **19.** $82\overline{)879.82}$

20. 523 ÷ 6.8 **21.** 21.7 ÷ 0.73

22. A contractor paid $32.87, $51.92, $86.97, and $52.98 for supplies. Estimate the total cost of the supplies to the nearest 10 dollars.

23. If a construction company needs 50,297 bricks, 28,957 bricks, 45,872 bricks, and 48,957 bricks for four jobs, estimate to the nearest thousand the number of bricks needed for all jobs.

24. A 42.8-ft piece of copper tubing is cut from a coil of copper tubing that is 192.5 ft long. Estimate to the nearest foot how much copper tubing remains in the coil.

25. If 4089 steel beams each weigh 85.8 lb, estimate the total weight by converting to numbers with one nonzero digit.

26. If light switches cost $1.85 each, estimate how many light switches can be purchased for $60.

4-7 CHANGING DECIMALS TO FRACTIONS AND FRACTIONS TO DECIMALS

Learning Objectives

L.O.1. Change decimals to fractions.
L.O.2. Change fractions and mixed numbers to decimals.

Depending on the circumstances of each mathematical problem encountered on the job, a worker may need to use fractions in some cases and decimals in other cases. It is necessary, then, to be able to convert decimals into fractions and fractions into decimals.

L.O.1. Change decimals to fractions.

Changing Decimals to Fractions

When a decimal is written as a fraction, the number of digits on the right of the decimal point determines the denominator of the fraction. In the decimal 0.23 there are two digits on the right of the decimal, which indicates that the denominator of the fraction will be 100. (One hundred has two zeros.) Decimal numbers with one digit on the right of the decimal have one zero in the denominator. Decimal numbers with three digits on the right of the decimal have three zeros in the denominator, and so on.

Rule 4-7.1 *To convert a decimal number to a fraction:*

1. Write the digits without the decimal point as the numerator.
2. Write the denominator as a power of 10 with as many zeros as places on the right of the decimal point.
3. Reduce or convert to a mixed number.

EXAMPLE 4-7.1 Write as a fraction 0.4, 0.73, 0.075, and 2.3.

Solution

$$0.4 = \frac{4}{10} = \frac{2}{5}$$ *(Tenths indicates a denominator of 10.)*

$$0.73 = \frac{73}{100}$$ *(Hundredths indicates a denominator of 100.)*

$$0.075 = \frac{75}{1000} = \frac{3}{40} \qquad \textit{(Thousandths indicates a denominator of 1000.)}$$

$$2.3 = \frac{23}{10} = 2\frac{3}{10} \qquad \textit{(A mixed decimal will be written as a mixed number.)}$$

■

Sometimes the decimals that need to be changed to fractions contain fractions. For example, $0.33\frac{1}{3}$ contains the fraction $\frac{1}{3}$. To convert these decimal numbers to fractions, we write them as complex fractions.

EXAMPLE 4-7.2 Write $0.33\frac{1}{3}$ as a fraction.

Solution

$$0.33\frac{1}{3} = \frac{33\frac{1}{3}}{100} = 33\frac{1}{3} \div 100$$

(Count places for digits only. That is, do not count the fraction $\frac{1}{3}$ as a place.)

Following our rule for dividing mixed numbers, write $33\frac{1}{3}$ as an improper fraction. Then invert the divisor 100 and multiply.

$$\frac{\overset{1}{\cancel{100}}}{3} \times \frac{1}{\underset{1}{\cancel{100}}} = \frac{1}{3}$$

■

L.O.2. Change fractions and mixed numbers to decimals.

Changing Fractions and Mixed Numbers to Decimals

With increased calculator usage, we may find it more convenient to work with decimals. Therefore, we will often change fractions to decimals when we make certain computations.

The bar separating the numerator and denominator of a fraction indicates division. $\frac{2}{5}$ also means $2 \div 5$ or $5\overline{)2}$.

If we write a decimal after the 2 and affix a zero, we can divide.

$$\begin{array}{r} 0.4 \\ 5\overline{)2.0} \end{array}$$

Therefore, $\frac{2}{5} = 0.4$.

> **Rule 4-7.2** ***To convert a fraction to a decimal number:***
>
> **1.** Place a decimal point after the numerator.
> **2.** Divide the numerator by the denominator using long division.
> **3.** Affix zeros to the numerator as needed for division.

Example 4-7.3 Change $\frac{7}{8}$ to a decimal.

Solution

$$7 \div 8 \quad \text{or} \quad \begin{array}{r} 0.875 \\ 8\overline{)7.000} \\ \underline{6\,4} \\ 60 \\ \underline{56} \\ 40 \\ \underline{40} \end{array}$$

(Affix zeros until the division terminates; that is, it has no remainder.) ■

> **Rule 4-7.3** *To convert a mixed number to a decimal number:*
>
> **1.** The whole-number part remains the same.
> **2.** Convert only the fraction part by using Rule 4-7.2.

Example 4-7.4 Change $3\frac{2}{5}$ to a mixed decimal.

Solution

$$\frac{2}{5} = \begin{array}{r} 0.4 \\ 5\overline{)2.0} \end{array} \quad \text{or} \quad 0.4$$

Then, $3\frac{2}{5} = 3.4$. ■

When we change some fractions to decimals, the quotient does not terminate.

Example 4-7.5 Change $\frac{1}{3}$ and $\frac{4}{11}$ to decimals.

Solution

$$1 \div 3 \quad \text{or} \quad \begin{array}{r} 0.3333 \\ 3\overline{)1.0000} \\ \underline{9} \\ 10 \\ \underline{9} \\ 10 \\ \underline{9} \\ 1 \end{array} \quad \text{or} \quad 0.33\frac{1}{3}$$

(No matter how many zeros are affixed, the remainder is still 1.)

When fractions are changed into decimals that do not terminate, the decimals are called *repeating decimals*.

$$4 \div 11 \quad \text{or} \quad 11\overline{)4.000000} \quad \text{with quotient } 0.363636 \quad \text{or} \quad 0.36\frac{4}{11}$$

(In this repeating decimal, two digits repeat.)

```
     0.363636
11 | 4.000000
     3 3
       70
       66
        40
        33
         70
         66
          40
          33
           70
           66
            4
```

■

Repeating decimals can be written with a line over the digit(s) that repeat or three dots at the end to indicate that they repeat.

$$\frac{1}{3} = 0.\overline{3} \quad \text{or} \quad 0.333333\ldots$$

$$\frac{4}{11} = 0.\overline{36} \quad \text{or} \quad 0.3636\ldots$$

General Tips for Using the Calculator

Fractions may be converted to decimal numbers by dividing the numerator by the denominator on a calculator. Remember to enter the dividend (numerator) *first*. The fractions in Example 4-7.5 can be converted to decimals on a calculator.

Scientific calculator:

[AC] 1 [÷] 3 [=] ⇒ 0.333333333
[AC] 4 [÷] 11 [=] ⇒ 0.363636363

Graphing calculator:

[AC] 1 [÷] 3 [EXE] ⇒ 0.333333333
[AC] 4 [÷] 11 [EXE] ⇒ 0.363636363

SELF-STUDY EXERCISES 4-7

L.O.1. Change the following decimals to their fraction or mixed-number equivalents, and reduce answers to lowest terms.

1. 0.5	**2.** 0.1	**3.** 0.2
4. 0.7	**5.** 0.25	**6.** 0.025
7. 3.9	**8.** 4.8	**9.** $0.66\frac{2}{3}$
10. $0.87\frac{1}{2}$	**11.** $0.37\frac{1}{2}$	**12.** $0.12\frac{1}{2}$

13. $0.16\frac{2}{3}$

14. 0.375

15. 0.625

16. A measure of 0.75 in. represents what fractional part of an inch?

17. What fraction represents a measure of 0.1875 ft?

18. The length of a screw is 2.375 in. Represent this length as a mixed number.

19. An instrument weighs $0.83\frac{1}{3}$ lb. Write this as a fraction of a pound.

20. Some sheet metal is 0.3125 in. thick. What is the thickness expressed as a fraction?

21. A predrilled PC board is 3.125 in. long. Write the length as a mixed number.

L.O.2. Change these fractions and mixed numbers to decimals.

22. $\frac{2}{5}$

23. $\frac{3}{10}$

24. $\frac{7}{8}$

25. $\frac{5}{8}$

26. $\frac{9}{20}$

27. $\frac{49}{50}$

28. $\frac{21}{100}$

29. $3\frac{7}{8}$

30. $1\frac{7}{16}$

31. $4\frac{9}{16}$

Write these fractions as repeating decimals.

32. $\frac{2}{3}$

33. $\frac{3}{11}$

34. $\frac{7}{9}$

Write these fractions as decimals to the hundredths place. Express the remainder as a fraction.

35. $\frac{5}{6}$

36. $\frac{7}{12}$

37. An aerial map shows a building measuring $2\frac{3}{64}$ in. on one side. If decimal measures were used, what would the side of the building measure on the map?

38. A blueprint shows a wall measuring $1\frac{1}{32}$ in. If decimal measures were used, what would the wall measure on the blueprint?

39. The property tax rate in one state is $4\frac{1}{2}\%$ per assessed value. What is the tax rate per assessed value expressed as a mixed-decimal percent?

40. A worker expects a $3\frac{3}{4}\%$ raise. Express the fractional part of the percent as a decimal percent.

41. A plan specifies allowing a gap of $\frac{1}{8}$ in. between vinyl flooring and the wall for expansion. What should the gap be in decimal notation?

42. A carpenter cuts a piece of oak trim $\frac{7}{16}$ in. too short. How much is the shortage in decimal notation?

4-8 SIGNED DECIMALS

Learning Objective

L.O.1. Perform basic operations with signed decimals.

Our understanding of decimal numbers can be extended to include signed decimal numbers if we review the rules of basic operations for signed integers.

L.O.1. Perform basic operations with signed decimals.

Perform Basic Operations with Signed Decimals

The same rules for performing the basic operations with integers can be used with signed decimals.

EXAMPLE 4-8.1 Add -5.32 and -3.24.

Solution

$$\begin{array}{r} -5.32 \\ \underline{-3.24} \\ -8.56 \end{array}$$

(Align the decimals and use the rule for adding numbers with like signs.) ■

EXAMPLE 4-8.2 Subtract -3.7 from 8.5.

Solution

$$8.5 - (-3.7) = 8.5 + 3.7 = 14.2$$ ■

EXAMPLE 4-8.3 Multiply 3.91 and -7.1.

Solution

$$\begin{array}{r} 3.91 \\ \underline{-\ 7.1} \\ 391 \\ \underline{2737\ } \\ 27.761 \end{array}$$

(Use the rule for multiplying numbers with unlike signs.)

Or use the calculator.

Scientific Calculator:

3 [.] 9 1 [×] 7 [.] 1 [+/−] [=] ⇒ 27.761

Graphing Calculator:

3 [.] 9 1 [×] [−] 7 [.] 1 [EXE] ⇒ 27.761 ■

EXAMPLE 4-8.4 Divide: $(-1.2) \div (-0.4)$

Solution

Determine the sign of the quotient according to the rule for dividing signed numbers. Division of two negative numbers results in a positive quotient.

Perform the division of decimals.

$$0.4\overline{)1.2}\quad = 3.$$

(Shift decimal points.) ■

SELF STUDY EXERCISES 4-8

L.O.1. Perform the indicated operations with signed decimals.

1. $5.823 - 32.12$
2. $-8.32 + 7.21$
3. $-84.23 - 7.21$
4. 34.6×-3.2
5. -7.2×8.2
6. -83.1×-4.1
7. $83.2 \div -3$
8. $-0.826 \div -2$
9. $-3.2 + 7.8(-3.2 + 0.2)$
10. $4.23 - 4.2/-1.2$

ASSIGNMENT EXERCISES

4-1

Write the following fractions as decimal numbers.

1. $\frac{3}{10}$
2. $\frac{15}{100}$
3. $\frac{4}{100}$
4. $\frac{75}{1000}$
5. $\frac{21}{10}$
6. $\frac{652}{100{,}000}$
7. What is the place value of 6 in 21.836?
8. What is the place value of 5 in 13.0586?
9. In 13.7213, what digit is in the tenths place?
10. In 15.02167, what digit is in the ten-thousandths place?
11. Write the words for 6.803.
12. Write the words for 0.0712.
13. The decimal equivalent of $\frac{5}{8}$ is six hundred twenty-five thousandths. Write this decimal using digits.
14. The thickness of a sheet of aluminum is forty-thousandths of an inch. Write this thickness as a decimal number.
15. Which of these decimal numbers is larger? 4.783 or 4.79?
16. Which of these decimal numbers is smaller? 0.83 or 0.825?
17. Write these decimal numbers in order of size from smaller to larger. 0.021, 0.0216, 0.02.
18. Two measurements of an object are recorded. If the measures are 4.831 in. and 4.820 in., which measure is larger?
19. The decimal equivalent of $\frac{7}{8}$ is 0.875. The decimal equivalent of $\frac{6}{7}$ is approximately 0.857. Which fraction is larger?
20. Two parts are machined from the same stock. They measure 1.023 in. and 1.03 in. after machining. Which part has been machined more; that is, which part is now smaller?

4-2

Round the numbers to the place indicated.

21. Nearest ten: 324
22. Nearest thousand: 28,714
23. Nearest thousand: 6882
24. Nearest ten million: 497,283,016
25. Nearest hundred: 468
26. Nearest hundred: 8236
27. Nearest ten thousand: 49,238
28. Nearest thousand: 49,238
29. Nearest billion: 26,500,000,129
30. Nearest ten thousand: 248,217
31. Nearest tenth: 41.378
32. Nearest hundredth: 7.0893
33. Nearest hundredth: 6.8957
34. Nearest thousandth: 1.078834
35. Nearest ten-thousandth: 23.46097
36. Nearest tenth: 0.09783
37. Round 24.237 to the nearest whole number.
38. Round $8.9378 to the nearest cent.
39. Round $42.98 to the nearest dollar.
40. Round $0.9986 to the nearest cent.
41. Round 83.052 to tens.
42. Round 0.097032 to hundred-thousandths.
43. Round 12.83 to one nonzero digit.
44. Round 7.93 to one nonzero digit.
45. Round 0.0736 to one nonzero digit.
46. Round the micrometer reading of 1.876 to hundredths.
47. What is the cost to the nearest dollar of paper that sells for $3.19 a ream?
48. An average is calculated to be 3.9765. What is this average rounded to the nearest tenth?
49. Round the micrometer reading of 1.438 to tenths.
50. A blueprint specification calls for all decimals to be expressed as hundredths. Write the measure, 21.0974 in., according to the specifications.

4-3

Add.

51. $7.43 + 2.19 + 42.38 + 16.21$
52. $3.47 + 42.32 + 3.82 + 4.09$

53. 6.2 + 32.7 + 46.82 + 0.29 + 4.237

54. 42 + 3.6 + 2.1 + 7.83

55. 86.3 + 9.2 + 70.02 + 3 + 2.7

Subtract.

56. 21.34 − 16.73

57. 15.934 − 12.807

58. 284.73 − 79.831

59. 423 − 72.687

60. 293.86 − 148

Add or subtract these measurements and round to the decimal place specified.

61. Nearest hundredth: 42.3 + 21.78 + 12.9205

62. Nearest tenth: 217.4 − 121.68

63. Nearest whole number: 21.6 + 7 + 18.92

64. Nearest whole number: 347 − 138.2

65. Nearest whole number: 473 + 84.3

66. One cubic foot (ft^3) of ice weighs 57.5 lb, while 1 ft^3 of seawater weighs 62.5 lb. The water weighs how much more than the ice?

67. A blueprint calls for the length of a part to be 8.296 in. with a tolerance of ±0.005 in. What are the limit dimensions of the part?

68. An air conditioner uses 10.4 kW, a stove uses 15.3 kW, a washer uses 2.9 kW, and a dryer uses 6.3 kW. What is the total number of kilowatts used by the electrical machines?

Use Fig. 4-8 for Exercises 69 to 75.

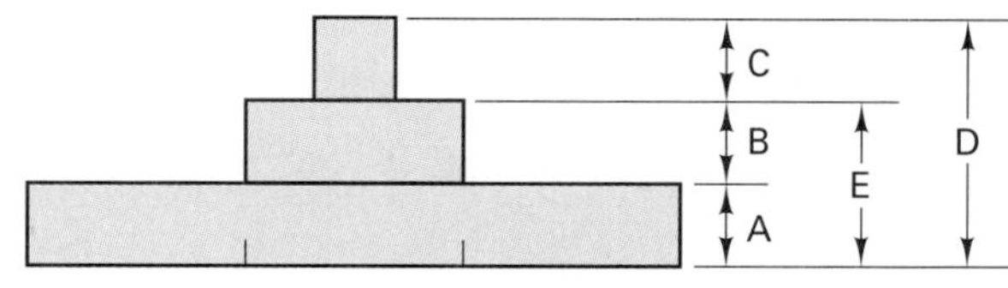

Figure 4-8

69. Find the length of *A* if *D* = 4.237 in., *B* = 1.861 in., and *C* = 1.946 in.

70. What is the dimension of *D* if *A* = 1.896 in., *B* = 2.315 in., and *C* = 1.732 in.?

71. What is the dimension of *A* if *E* = 4.86 in. and *B* = 1.972 in.?

72. Find the length of *D* if *C* = 4.23 in. and *E* = 4.17 in.

73. Give the limit dimensions of *D* if *D* = 8.935 in. with a ±0.005-in. tolerance.

74. What dimension should be listed for *B* if *A* measures 3.72 in. and *E* measures 5.98 in.?

75. What dimension should be listed for *C* if *D* measures 3.7 in. and *E* measures 1.6 in.?

4-4

Multiply.

76. 3.72 × 4.9

77. 8.36 × 0.25

78. 96.5 × 82

79. 0.237 × 0.09

80. 8.352 × 0.24

81. 21.3 × 17

82. 40.2 × 1.05

83. 0.029 × 0.076

84. 0.523 × 4.6

85. 23.86 × 0.98

Find the products mentally.

86. 3.95 × 10

87. 4.082 × 1000

88. 0.0283 × 10,000

89. 8.5 × 10,000

90. 0.9301 × 100,000

91. If plastic pipe costs $0.85 per foot, find the cost of 1000 ft of pipe.

92. Diesel fuel costs $1.65 per gallon. What is the cost of 1000 gal of fuel?

93. If a steel tape expands 0.00014 in. for each inch when heated, how much will a tape 864 in. long expand?

94. What is the total cost of 10,000 bricks at $0.155 per brick if the delivery charge is $50?

95. A piecework employee averages 178.6 pieces per day. If the employee earns $0.28 per item, how much is earned in 5 days?

96. If each strip of tongue-and-groove flooring covers a width of 2.5 in., how many inches can be covered by 35 strips of flooring?

97. The monthly payment for a car is \$159.32. What is the amount paid if 36 monthly payments are made?

98. The outside diameter of a piston ring is 3.475 in. If the ring wall is 0.845 in. thick, what is the inside diameter?

99. In a barometer, a height of 1 millimeter of mercury (mm Hg) represents an atmospheric pressure of 0.0193 pounds per square inch (lb/in.^{2}). A barometer reading of 760.104 mm Hg represents normal atmospheric pressure. What is the normal atmospheric pressure in pounds per square inch? Round to hundredths.

4-5

Divide.

100. $8\overline{)76.8}$

101. $39\overline{)2.34}$

102. $0.12\overline{)0.1236}$

103. $0.69 \div 0.03$

104. $735 \div 2.1$

Divide and round the quotient to the place indicated.

105. Nearest whole number: $7\overline{)82.9}$

106. Nearest tenth: $0.3\overline{)8.2}$

107. Nearest hundredth: $5.9\overline{)213.92}$

108. Nearest cent: \$9.87 $\div$ 16

109. Nearest dollar: \$21.83 $\div$ 12

Use the shortcut rule for dividing by 10, 100, 1000, and so on, to obtain the answers.

110. $100\overline{)34.6}$

111. $10\overline{)482.3}$

112. $1000\overline{)23.86}$

113. $593.8 \div 100$

114. $0.82 \div 10$

115. If 36 hammers cost \$322.92, find the cost in dollars and cents of each hammer.

116. Find the depth of thread of an American Standard screw that has 12 threads per inch. Round to the nearest ten-thousandth. (To find the depth, the number of threads per inch is divided into 0.6495.)

117. If it takes 5.75 gal of pesticide to treat a 10-acre field, what part of a gallon is needed per acre?

118. If 2.5 in. on a map represents 50 miles, how many miles are represented per inch?

4-6

Estimate by rounding to the place indicated.

119. Nearest ten: $43.8 + 21.9 + 15.7 + 72.8$

120. Nearest dollar: \$87.38 + \$21.96 + \$17.52

121. Nearest whole number: $59.83 - 21.7$

122. Nearest dollar: \$127.95 − \$89.68

123. One nonzero digit: 42.1×8.73

124. Nearest dollar: \$5.97 $\times$ 32

125. One nonzero digit: $5.2\overline{)83.76}$

126. One nonzero digit: \$29.78 $\div$ 12

127. Find the average of 42.7, 26.9, 83.4, and 51.9 to the nearest tenth.

128. Find the average whole-number score if grades of 97, 83, 76, 92, 83, and 95 were made.

129. Find the average measure if measures of 42.34 ft, 38.97 ft, 51.95 ft, and 61.88 ft were recorded. Round to the nearest hundredth.

130. Estimate the total cost of a shirt at \$15.97, a tie at \$9.85, and a pair of socks at \$2.35 in dollars.

131. Five light fixtures cost \$74.98, \$23.72, \$51.27, \$125.36, and \$85.93. Find the average cost of the fixtures to the nearest cent.

132. Two models of an electric stove sell for \$497.89 and \$685.50. To the nearest 10 dollars, estimate the difference in price.

133. A package of meat weighing 5.82 lb sells for \$2.19 per pound. Estimate in dollars the cost of the meat.

134. A room requires 942.35 ft^2 of wallpaper. Estimate to one nonzero digit the number of rolls required if one roll covers 33.5 ft^2 of surface.

4-7

Change the following decimals to their fraction or mixed-number equivalents and reduce answers to lowest terms.

135. 0.4 **136.** 0.9 **137.** 0.75

138. 2.7 **139.** $0.83\frac{1}{3}$ **140.** $0.62\frac{1}{2}$

141. 0.875 **142.** $3.33\frac{1}{3}$

Change the fractions and mixed numbers to decimals.

143. $\frac{7}{8}$ **144.** $\frac{1}{2}$ **145.** $5\frac{1}{4}$

146. $\frac{7}{20}$ **147.** $\frac{73}{100}$ **148.** $\frac{7}{10}$

Write the fractions as repeating decimals.

149. $\frac{1}{3}$ **150.** $\frac{5}{11}$ **151.** $\frac{5}{9}$

Write the fractions as decimals to the hundredths place. Express the remainders as fractions.

152. $\frac{1}{6}$ **153.** $\frac{5}{12}$ **154.** $\frac{2}{3}$

The following fractions are often used in measurements. Convert them to their terminating decimal equivalents.

155. $\frac{1}{64}$ **156.** $\frac{1}{32}$ **157.** $\frac{3}{64}$

158. $\frac{1}{16}$ **159.** $\frac{5}{64}$ **160.** $\frac{3}{32}$

161. $\frac{7}{64}$ **162.** $\frac{1}{8}$

The following decimals are often used in measurements. Convert them to their fraction equivalents and reduce.

163. 0.6875 **164.** 0.5625 **165.** 0.4375
166. 0.25 **167.** 0.375 **168.** 0.3125
169. 0.625

170. A current was recorded in a report as $3\frac{1}{5}$ amperes. By convention, amperes are expressed in decimals. What is the mixed-decimal equivalent of this current?

171. A worker must cut $\frac{7}{8}$ in. from a length of a galvanized gutter. What is this measure expressed as a decimal?

172. An electronic project plan showed a wire measurement as 7.0625 in. Express the wire measurement as a mixed fractional number in lowest terms.

173. A piece of sheet metal is 0.34375 in. thick. Convert this measurement to a fraction of an inch.

4-8

Perform the indicated operations with signed decimals. Apply the order of operations when appropriate.

174. $-8.2 + 7.3$ **175.** $-3.2 - 0.3$
176. $-3.5 \times 3.8 - 7.2 + 8.1 \times (-0.3)$ **177.** $-4.5 \div (-0.09) + (-4.3) - (-7.8)$

CHALLENGE PROBLEMS

178. Use a calculator to evaluate the following powers.
(a) 0.03^2 **(b)** 0.07^2 **(c)** 0.005^2
(d) 0.009^2 **(e)** 0.02^3 **(f)** 0.004^3

179. Use a calculator to find the square root of the following decimals.
(a) $\sqrt{0.64}$ **(b)** $\sqrt{0.09}$
(c) $\sqrt{0.0009}$ **(d)** $\sqrt{0.0625}$
(e) $\sqrt{0.000016}$ **(f)** $\sqrt{0.4}$

CONCEPTS ANALYSIS

1. Explain how the concept of like fractions can be used to compare decimals.
2. How are the procedures for adding fractions and adding decimals related?
3. How are the procedures for adding whole numbers and adding decimals related?
4. Examine the pattern developed by the decimal equivalents of proper fractions with a denominator of 11.
5. Without making any calculations, do you think 0.004 is a perfect square? Why or why not?
6. Without making any calculations, do you think 0.008 is a perfect cube? Why or why not?
7. Can you find at least one exception to the generalization that a perfect square decimal will have an even number of decimal places? Illustrate your answer.

CHAPTER SUMMARY

Objectives	What to Remember	Examples
Section 4-1		
1. Identify place values in decimal numbers.	Align decimal point in number under decimal point in place-value chart. Chart shows the value of each place in number.	In 3.75, 3 is in the ones place, 7 in the tenths place and 5 in the hundredths place.
2. Read and write decimal numbers.	Read whole-number part, *and* for decimal point, decimal part, and place value of right-most digit.	Write 32.075 in words. Thirty-two and seventy-five thousandths.
3. Write fractions with power-of-10 denominators as decimal numbers.	Write digits on right of decimal as numerator. Denominator will be a power of ten with as many zeros as decimal places.	Express $\frac{53}{1000}$ as a decimal: 0.053.
4. Compare decimal numbers.	Compare decimals by comparing the digits in the same place beginning from the left of each number.	Which decimal is larger, 0.23 or 0.225? Both numbers have the same digit, 2, in the tenths place. 0.23 is larger since it has a 3 in the hundredths place, while 0.225 has a 2 in that place.
Section 4-2		
1. Round decimal numbers by place value.	Round as in whole numbers; however, digits on the right side of the digit in the rounding place and after the decimal point are dropped rather than replaced with zeros.	Round 5.847 to the nearest tenth. 8 is in the tenths place. 4 is less than 5, so leave 8 as is and drop the 4 and 7: 5.8

2. Round to one nonzero digit.

To round decimals to one nonzero digit is to round to the first place from the left that is not zero.

Round 0.0372 to one nonzero digit. 3 is the first nonzero digit from the left. 7 is 5 or greater, so change 3 to 4 and drop the remaining digits on the right: 0.04

Section 4-3

1. Add decimals.

Add decimal numbers by arranging the addends so that the decimal points are in the same vertical line.

Add: $43.35 + 3.7 + 0.462$

$$\begin{array}{r} 43.35 \\ 3.7 \\ +\ \ 0.462 \\ \hline 47.512 \end{array}$$

2. Subtract decimals.

Subtract decimal numbers by arranging the minuend and subtrahend so that the decimal points are in the same vertical line.

Subtract: $53.824 - 4.0423$

$$\begin{array}{r} 53.824 \\ -\ \ 4.0423 \\ \hline 49.7817 \end{array}$$

Section 4-4

1. Multiply any decimal numbers.

To multiply decimal numbers, multiply the numbers as in whole numbers and determine the number of decimal digits in both numbers. Place the decimal in the product to the left of the same number of digits, counting from the right.

Multiply: 3.25×0.53

$$\begin{array}{r} 3.25 \\ \times\ \ 0.53 \\ \hline 9\,75 \\ 162\,5 \\ \hline 1.72\,25 \end{array}$$

2. Multiply decimal numbers by a power of 10.

To multiply a decimal number by a power of 10, shift the decimal to the *right* as many digits as the power of 10 has zeros. Affix zeros if necessary.

Multiply: 23.52×1000

23520 or 23,520

Section 4-5

1. Divide decimal numbers with zero remainder.

When dividing by a decimal number, shift the decimal in the divisor to the right end; shift the decimal in the dividend the same number of places to the right. Place the decimal in the quotient.

Divide:

$$\begin{array}{r} 3.8 \\ 2.1\overline{)7.98} \\ 6\,3 \\ \hline 1\,6\,8 \\ 1\,6\,8 \\ \hline \end{array}$$

2. Divide decimal numbers whose remainders are not zero.

Divide one place past the specified place value. Affix zeros in dividend if necessary. Round to specified place.

Divide and round to tenths:

$$\begin{array}{r} 1.70 \approx 1.7 \\ 15\overline{)25.60} \\ 15 \\ \hline 10\,6 \\ 10\,5 \\ \hline 10 \end{array}$$

Or divide to specified place and express the remainder as a fraction consisting of remainder over divisor.

Divide to tenths and express the remainder as a fraction:

$$\begin{array}{r} 1.7\frac{1}{15} \\ 15\overline{)25.6} \\ 15 \\ \hline 10\,6 \\ 10\,5 \\ \hline 1 \end{array}$$

3. Divide decimal numbers by a power of 10.

To divide a decimal by a power of 10, shift the decimal to the *left*. Note that this process is just the opposite as that for multiplication.

Divide: $3.52 \div 10 = 0.352$

Section 4-6

1. Find the numerical average.	Find the sum of the values and divide the sum by the number of values. Handle the remainder as directed.	Find the average of 74, 65, and 85. $74 + 65 + 85 = 224$ $224 \div 3 = 74\frac{2}{3}$ or 74.67
2. Estimate sums, differences, products, and quotients of decimals.	Round values to desired place. Then perform the operation.	$74.2 + 9.5$ (whole numbers), $74 + 10 = 84$ $67.9 - 35.7$ (tens), $70 - 40 = 30$ 14.6×2.6 (ones), $15 \times 3 = 45$ $16.45 \div 3.2$ (one nonzero digit), $20 \div 3 = 6$

Section 4-7

1. Change decimals to fractions.	To change a decimal to a fraction (or mixed number) in lowest terms, write the number without the decimal as the numerator and make the denominator have the same number of zeros as the decimal has decimal places.	Write 0.23 as a fraction: $\frac{23}{100}$
2. Change fractions and mixed numbers to decimals.	To change any fraction to a decimal, divide the denominator into the numerator by placing a decimal after the last digit of the numerator and affixing zeros as needed.	Change $\frac{5}{8}$ to a decimal: $5.000 \div 8 = .625$ (4 8, 20, 16, 40, 40)

Section 4-8

1. Perform basic operations with signed decimals	Use the same rules that were used to perform basic operations with integers.	Add $4.37 + (-2.91)$: $4.37 - 2.91 = 1.46$

WORDS TO KNOW

decimal (p. 144)
decimal fraction (p. 144)
decimal number (p. 144)
decimal point (p. 145)
decimal place (p. 145)
approximate amount (p. 155)
tolerance (p. 158)
limit dimensions (p. 158)
numerical average (p. 169)
estimating (p. 169)

CHAPTER TRIAL TEST

1. Write the digits for seven and twenty-seven thousandths.
2. Write the digits for two hundred four millionths.
3. Which number is larger, 7.13 or 7.2?
4. Which number is smaller, 5.09 or 5.1?
5. Round 2.875 to the nearest hundredth.
6. Round 4.7284 to the nearest whole number.
7. Round $4.834 to the nearest cent.
8. Round 48.3284 to the nearest tenth.
9. Round $21.89 to the nearest dollar.
10. Round $2.987 to the nearest cent.
11. Round 4.018 to the nearest hundredth.
12. Round $20.68 to the nearest dollar.
13. Find the sum: $4.1 + 3 + 7.86$.
14. Find the sum: $2.37 + 5 + 17.8$.
15. Find the difference: $21 - 17.96$.
16. Find the difference: $187.2 - 37.84$.

17. Add these measurements and round the sum to the nearest tenth: $4.8 + 3.72 + 0.973$.

18. Add these measurements and round the sum to the nearest hundredth: $28.375 + 2.3842 + 12 + 9.6$

19. Find the product: 7.97×0.23.

20. Find the product: 47.872×23.

21. Find the product: 4.08×0.05.

22. Find the product: 5.03×0.08.

23. Find the product: 42.73×1000.

24. Find the product: $0.23875 \times 10{,}000$.

25. $25\overline{)27.75}$

26. $37\overline{)126.54}$

27. $42.3 \div 0.03$

28. $179.97 \div 0.07$

29. Round answer to the nearest tenth: $7.2\overline{)83.41}$.

30. Round answer to the nearest hundredth: $3.6\overline{)14.25}$.

31. $52.38 \div 10{,}000$

32. $0.3489 \div 100$

33. Find the average of these test scores: 82, 95, 76, 84, 72, and 91. Round to the nearest whole number.

34. Find the average of these temperatures: 72, 78, 84, 75, 88. Round to the nearest whole number.

35. Estimate the sum by rounding to the nearest whole number: $3.85 + 7.46$.

36. Estimate the sum by converting each number to a rounded number having only one nonzero digit: $47.8 + 32.87 + 12.9$.

37. Estimate the difference by rounding each number to the nearest tenth: $0.87 - 0.328$.

38. Estimate the difference by converting each number to a rounded number having only one nonzero digit: $84.387 - 29.8$.

39. Estimate the product by rounding each decimal to a number having only one nonzero digit: 42.38×27.9.

40. Estimate the product by rounding each number to a number having only one nonzero digit: 47.5×0.043.

41. Estimate the quotient by rounding the divisor and dividend to numbers with one nonzero digit: $37.2\overline{)2987.5}$.

42. Estimate the quotient by rounding the divisor and dividend to numbers with one nonzero digit: $0.43\overline{)2.847}$.

43. Change $\frac{4}{5}$ to a decimal.

44. Change $\frac{3}{8}$ to a decimal.

45. Change $\frac{5}{7}$ to a decimal rounded to the nearest hundredth.

46. Change 0.9 to a fraction.

47. Change 0.175 to a fraction.

48. Change $0.62\frac{1}{2}$ to a fraction.

49. Change $0.33\frac{1}{3}$ to a fraction.

50. The blueprint specification for a machined part calls for its thickness to be 1.485 in. with a tolerance of ± 0.010 in. What are the limit dimensions of the part?

51. Heating oil costs \$1.75 per gallon. What is the cost of 10,000 gal?

52. A construction job requires 16 pieces of reinforced steel, each 7.96 ft in length. What length of steel is required for the job?

53. Add: $3.8 + (-2.7)$

54. Add: $-4.17 - 3.4$

55. Multiply: -4.7×2.3

CHAPTER

5 Percents

In the chapters on fractions and decimals we learned some very useful ways to express parts of quantities and to compare quantities. In this chapter we continue by standardizing our comparison of parts of quantities. We do this by expressing quantities in relation to a standard unit of 100. We can use this relationship, called a *percent*, to solve many different types of technical problems.

5-1 CHANGING NUMBERS TO PERCENT EQUIVALENTS

Learning Objective

L.O.1. Change any number to its percent equivalent.

The word *percent* means "per hundred" or "for every hundred." 35 percent means 35 per hundred or 35 out of every hundred or $\frac{35}{100}$. 100 percent represents one hundred out of 100 parts or $\frac{100}{100}$ or 1 whole quantity. The symbol % is used to represent "percent."

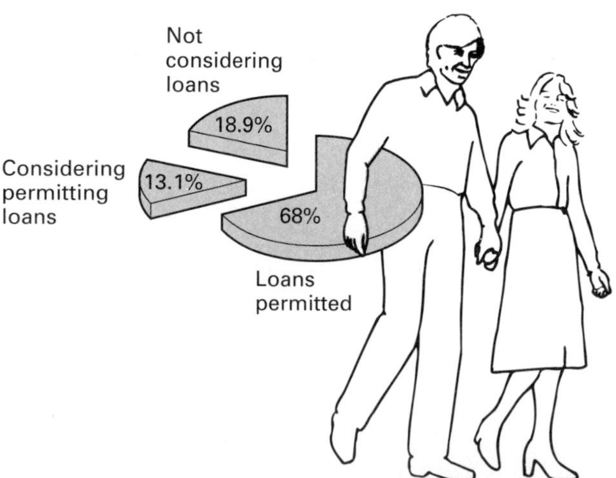

A look at percents that shape your finances. *Source:* Profit Sharing Council of America.

■ **Definition 5-1.1 Percent.** A **percent** is a fractional part of a hundred expressed with a percent sign (%).

L.O.1. Change any number to its percent equivalent.

Changing Any Number to Its Percent Equivalent

It often happens on the job that we need to use percents, but the numbers we have are not expressed as percents. For example, suppose that an electronic parts supply house wants to keep its inventory of general-purpose soldering irons always at 1200. After a large order was filled, the inventory dropped to 800 soldering irons and the parts manager ordered one-half of that amount to bring the inventory back up to 1200. We can express this $\frac{1}{2}$ as a percent by multiplying by 100.

$$\frac{1}{2} \times 100 = \frac{1}{\cancel{2}_1} \times \frac{\cancel{100}^{50}}{1} = 50\%$$

> **Rule 5-1.1** *To change any number to its percent equivalent:*
>
> Multiply by 100.

This rule can be applied to change any type of number—such as fractions, decimals, whole numbers, or mixed numbers—to a percent equivalent.

Example 5-1.1 Change the following fractions to percent equivalents: $\frac{1}{4}, \frac{1}{3}, \frac{3}{8}, \frac{4}{7}, \frac{1}{200}$, and $\frac{3}{1000}$.

Solution

$$\frac{1}{4} \times 100 = \frac{1}{\cancel{4}_1} \times \frac{\cancel{100}^{25}}{1} = 25\%$$

$$\frac{1}{3} \times 100 = \frac{1}{3} \times \frac{100}{1} = \frac{100}{3} = 33\frac{1}{3}\%$$

$$\frac{3}{8} \times 100 = \frac{3}{\cancel{8}_2} \times \frac{\cancel{100}^{25}}{1} = \frac{75}{2} = 37\frac{1}{2}\%$$

$$\frac{4}{7} \times 100 = \frac{4}{7} \times \frac{100}{1} = \frac{400}{7} = 57\frac{1}{7}\%$$

$$\frac{1}{200} \times 100 = \frac{1}{\cancel{200}_2} \times \frac{\cancel{100}^{1}}{1} = \frac{1}{2}\%$$

($\frac{1}{2}$% means $\frac{1}{2}$ of every hundredth or $\frac{1}{2}$ of 1%.)

$$\frac{3}{1000} \times 100 = \frac{3}{\cancel{1000}_{10}} \times \frac{\cancel{100}^{1}}{1} = \frac{3}{10}\%$$

($\frac{3}{10}$% means $\frac{3}{10}$ of every hundredth or $\frac{3}{10}$ of 1%.) ■

EXAMPLE 5-1.2 Change the following decimals to percent equivalents: 0.3, 0.25, 0.07, 0.006, and 0.0025.

Solution: We will use the shortcut procedure to multiply by 100: move the decimal point two places to the right.

$0.3 \times 100 = 030. = 30\%$

$0.25 \times 100 = 025. = 25\%$

$0.07 \times 100 = 007. = 7\%$

$0.006 \times 100 = 000.6 = 0.6\%$

(0.6% means 0.6 of every hundredth or 0.6 of 1%.)

$0.0025 \times 100 = 000.25 = 0.25\%$

(0.25% means 0.25 of every hundredth or 0.25 of 1%.) ■

EXAMPLE 5-1.3 Change the following whole numbers to their percent equivalents: 1, 3, 7.

Solution

$1 \times 100 = 100\%$ *(100 out of 100 or all of something)*

$3 \times 100 = 300\%$

When we talk of more than 100%, we are talking about more than one whole quantity. Thus 300% is three whole quantities or three times a quantity.

$7 \times 100 = 700\%$ *(7 whole quantities or 7 times a quantity)* ■

EXAMPLE 5-1.4 Change the following mixed numbers and decimals to their percent equivalents: $1\frac{1}{4}$, $3\frac{2}{3}$, 5.3, and 5.12.

Solution

$$1\frac{1}{4} \times 100 = \frac{5}{\cancel{4}_1} \times \frac{\cancel{100}^{25}}{1} = 125\%$$

$$3\frac{2}{3} \times 100 = \frac{11}{3} \times \frac{100}{1} = \frac{1100}{3} = 366\frac{2}{3}\%$$

$$5.3 \times 100 = 530. = 530\%$$

$$5.12 \times 100 = 512. = 512\%$$

■

SELF-STUDY EXERCISES 5-1

L.O.1.

1. If $\frac{2}{5}$ of the electricians in a city are self-employed, what percent are self-employed?

2. If $\frac{7}{10}$ of the bricklayers in a city are male, what percent are male?

Change the following numbers to their percent equivalents.

3. $\frac{5}{8}$

4. $\frac{7}{9}$

5. $\frac{7}{1000}$

6. $\frac{1}{350}$

7. 0.2

8. 0.14

9. 0.007

10. 0.0125

11. 5

12. 8

13. $1\frac{1}{3}$

14. $3\frac{1}{2}$

15. $4\frac{3}{10}$

16. $2\frac{1}{5}$

17. 3.05

18. 7.2

19. 15.1

20. 36.25

5-2 CHANGING PERCENTS TO NUMERICAL EQUIVALENTS

Learning Objectives

L.O.1. Change any percent less than 100% to its numerical equivalent.
L.O.2. Change any percent of 100% or more to its numerical equivalent.

Percents are used as a convenient way of expressing the relationship of any quantity to 100. They are excellent time-savers in making comparisons or stating problems in the technologies. However, we cannot use percents as such when solving a problem. Instead, we need to first convert the percents to fractional, decimal, whole, or mixed-number equivalents.

L.O.1. Change any percent less than 100% to its numerical equivalent.

Changing Quantities Less Than 100% to Numerical Equivalents

Rule 5-2.1 *To change a percent to a numerical equivalent:*

Divide by 100.

The numerical equivalent of a percent can be expressed in fractional or decimal form. In the following example, several percents are changed to both their fractional and decimal equivalents.

EXAMPLE 5-2.1 Change the following percents to their fractional and decimal equivalents: 75%, 38%, and 5%.

Solution

	Fractional equivalent	*Decimal equivalent*	
75%	$75 \div 100$	$75 \div 100$	
	$\frac{\overset{3}{\cancel{75}}}{1} \times \frac{1}{\underset{4}{\cancel{100}}} = \frac{3}{4}$	$.75 = 0.75$	*(Use shortcut procedure for dividing by 100: move decimal point two places to left.)*

38%	$38 \div 100$	$38 \div 100$
	$\frac{\overset{19}{\cancel{38}}}{1} \times \frac{1}{\underset{50}{\cancel{100}}} = \frac{19}{50}$	$.38 = 0.38$
5%	$5 \div 100$	$5 \div 100$
	$\frac{\overset{1}{\cancel{5}}}{1} \times \frac{1}{\underset{20}{\cancel{100}}} = \frac{1}{20}$	$.05 = 0.05$

■

Some percents will change more conveniently to one numerical equivalent than to another. In solving problems, we normally change the percent to the most convenient numerical equivalent for solving the problem. In the following examples, both the fractional and decimal equivalents of the percent will be given, and you can judge for yourself when fractional equivalents are more convenient than decimal equivalents, and vice versa.

EXAMPLE 5-2.2 Change the following percents to their fractional and decimal equivalents: $33\frac{1}{3}\%$, $37\frac{1}{2}\%$.

Solution

	Fractional equivalent	*Decimal equivalent*
$33\frac{1}{3}\%$	$33\frac{1}{3} \div 100$	$33\frac{1}{3} \div 100$
	$\frac{\overset{1}{\cancel{100}}}{3} \times \frac{1}{\underset{1}{\cancel{100}}} = \frac{1}{3}$	$.33\frac{1}{3} = 0.33\frac{1}{3}$

Remember, a decimal point separates whole quantities from fractional parts. Therefore, there is an *unwritten* decimal between 33 and $\frac{1}{3}$. Since $\frac{1}{3}$ will not change to a terminating decimal equivalent, using the decimal equivalent of $33\frac{1}{3}\%$ will create some difficult calculations.

	Fractional equivalent	*Decimal equivalent*
$37\frac{1}{2}\%$	$37\frac{1}{2} \div 100$	$37\frac{1}{2} \div 100 = 0.37\frac{1}{2}$
	$\frac{\overset{3}{\cancel{75}}}{2} \times \frac{1}{\underset{4}{\cancel{100}}} = \frac{3}{8}$	

Decimal equivalents are desirable when computations are made on a calculator. However, the decimal $0.37\frac{1}{2}$ would still not be adaptable to most calculators because of the fractional part that remains. For that reason, when a mixed-number percent is changed to decimal form, first change the fractional part of the mixed number to its decimal equivalent.

$$\frac{1}{2} = 0.5 \qquad 2\overline{)1.0}^{\,0.5}$$

Thus, $37\frac{1}{2}\% = 37.5\%$. Then, divide by 100.

$$37.5 \div 100 = 0.375$$

In this form, the decimal equivalent may be used on a calculator. ■

Example 5-2.3 Change 5.25% to its fractional and decimal equivalents.

Solution

	Fractional equivalent	*Decimal equivalent*
5.25%	$5.25\% = 5\frac{25}{100}\% = 5\frac{1}{4}\%$	$5.25\% = 5.25 \div 100 = 0.0525$

$$5\frac{1}{4}\% = 5\frac{1}{4} \div 100$$
$$= \frac{21}{4} \times \frac{1}{100}$$
$$= \frac{21}{400}$$

■

Example 5-2.4 Change the following percents to their fractional and decimal equivalents: $\frac{1}{2}\%$ and 0.25%.

Solution

	Fractional equivalent	*Decimal equivalent*
$\frac{1}{2}\%$	$\frac{1}{2}\% = \frac{1}{2} \div 100 = \frac{1}{2} \times \frac{1}{100} = \frac{1}{200}$	$\frac{1}{2}\% = 0.5\% = 0.5 \div 100 = 0.005$
0.25%	$0.25\% = \frac{25}{100}\% = \frac{1}{4}\%$	$0.25\% = 0.25 \div 100 = 0.0025$

$$\frac{1}{4}\% = \frac{1}{4} \div 100$$
$$= \frac{1}{4} \times \frac{1}{100}$$
$$= \frac{1}{400}$$

■

L.O.2. Change any percent of 100% or more to its numerical equivalent.

Changing Quantities of 100% or More to Numerical Equivalents

When a quantity is 100% or more, the fractional and decimal equivalents will be equal to or more than the whole number 1. The numerical equivalents will be whole numbers, mixed numbers, or mixed decimals. Again, in solving problems, use the more convenient numerical equivalent.

EXAMPLE 5-2.5 Change the following percents to their whole-number equivalents or to both their mixed-number and mixed-decimal equivalents: 700%, 375%, $233\frac{1}{3}\%$, and $462\frac{1}{2}\%$.

Solution

700%

$$700 \div 100 = \frac{\overset{7}{\cancel{700}}}{1} \times \frac{1}{\underset{1}{\cancel{100}}} = 7$$

or

$$700 \div 100 = 7$$

	Mixed-number equivalent	*Mixed-decimal equivalent*
375%	$375 \div 100$	$375 \div 100 = 3.75$

$$\frac{\overset{15}{\cancel{375}}}{1} \times \frac{1}{\underset{4}{\cancel{100}}} = \frac{15}{4} = 3\frac{3}{4}$$

	Mixed-number equivalent
$233\frac{1}{3}\%$	$233\frac{1}{3} \div 100$

$$\frac{\overset{7}{\cancel{700}}}{3} \times \frac{1}{\underset{1}{\cancel{100}}} = \frac{7}{3} = 2\frac{1}{3}$$

To save time and to allow more mental calculation, consider $233\frac{1}{3}\%$ to be $200\% + 33\frac{1}{3}\%$.

$$200\% = 2$$

$$33\frac{1}{3}\% = \frac{1}{3}$$

Thus, $200\% + 33\frac{1}{3}\% = 2 + \frac{1}{3} = 2\frac{1}{3}$.

Mixed-decimal equivalent

$$233\frac{1}{3} \div 100 = 2.33\frac{1}{3}$$

Since $\frac{1}{3}$ is equivalent to a nonterminating decimal, the mixed-decimal equivalent $2.33\frac{1}{3}$ would not be very practical in problem solving.

Mixed-number equivalent

$$462\frac{1}{2}\% \qquad 462\frac{1}{2} \div 100 = \frac{\overset{37}{\cancel{925}}}{2} \times \frac{1}{\underset{4}{\cancel{100}}} = \frac{37}{8} = 4\frac{5}{8}$$

Again, it is easier to consider $462\frac{1}{2}\%$ to be $400\% + 62\frac{1}{2}\%$.

$$400\% = 4$$

$$62\frac{1}{2}\% \qquad 62\frac{1}{2} \div 100 = \frac{\overset{5}{\cancel{125}}}{2} \times \frac{1}{\underset{4}{\cancel{100}}} = \frac{5}{8}$$

Then $462\frac{1}{2}\% = 4 + \frac{5}{8} = 4\frac{5}{8}$.

Mixed-decimal equivalent

$$462\frac{1}{2}\% = 462.5\%$$

$$462.5 \div 100 = 4.625$$

■

Example 5-2.6 Change 187.5% to its mixed-number and mixed-decimal equivalents.

Solution

Mixed-number equivalent

$$187.5\% = 187\frac{1}{2}\% \qquad \left(0.5\% = \frac{1}{2}\%\right)$$

$$187\frac{1}{2}\% = 100\% + 87\frac{1}{2}\%$$

$$= 1 + \frac{7}{8} \qquad \left(100\% = 1;\ 87\frac{1}{2}\% = 87\frac{1}{2} \div 100 = \frac{\overset{7}{\cancel{175}}}{2} \times \frac{1}{\underset{4}{\cancel{100}}} = \frac{7}{8}\right)$$

$$= 1\frac{7}{8}$$

Thus, $187.5\% = 1\frac{7}{8}$.

Mixed-decimal equivalent

$$187.5\% = 187.5 \div 100 = 1.875$$ ■

SELF-STUDY EXERCISES 5-2

L.O.1. Change the following to both fractional and decimal equivalents.

1. 36% **2.** 45% **3.** 20%

4. 75% **5.** $6\frac{1}{4}\%$ **6.** 62.5%

7. $66\frac{2}{3}\%$ **8.** 0.6% **9.** $\frac{1}{5}\%$

10. 0.05% **11.** $8\frac{1}{3}\%$ **12.** 18.75%

L.O.2. Change to equivalent whole numbers.

13. 800% **14.** 400%

Change to both mixed-number and mixed-decimal equivalents.

15. 250% **16.** 425% **17.** 176%

18. 380% **19.** $137\frac{1}{2}\%$ **20.** 387.5%

Change to mixed-number equivalents.

21. $166\frac{2}{3}\%$ **22.** $316\frac{2}{3}\%$

Change to mixed-decimal equivalents.

23. 115.3% **24.** $212\frac{1}{2}\%$ **25.** $106\frac{1}{4}\%$

5-3 COMMON EQUIVALENTS

Learning Objective

L.O.1. Substitute one common equivalent for another.

L.O.1. Substitute one common equivalent for another.

Substituting Common Equivalents of Fractions, Percents, and Decimals

Since many percents, fractions, and decimals are frequently used, it would be helpful to know these equivalents from memory. Knowing these will save much time on the job. Table 5-1 is a list of the most commonly used percents and equivalents. The equivalents were determined by the procedures used in Section 5-2.

To help us remember these equivalents, we will group them differently and suggest some mental calculations as memory aids.

$$50\% = \frac{1}{2} = 0.5$$

This is one of the easiest to learn because of its easy comparison to money. One dollar is 100 cents. One-half dollar is 50 cents. One-half dollar is $0.50 in dollar-and-cent notation.

$$25\% = \frac{1}{4} = 0.25$$

$$75\% = \frac{3}{4} = 0.75$$

Again, relating these to money, one-fourth of a dollar is 25 cents or $0.25, and three-fourths of a dollar is 75 cents or $0.75.

TABLE 5-1 Common Percent, Fraction, and Decimal Equivalents

Percent	Fraction	Decimal	Percent	Fraction	Decimal
10%	$\frac{1}{10}$	0.1	60%	$\frac{3}{5}$	0.6
20%	$\frac{1}{5}$	0.2	$66\frac{2}{3}\%$	$\frac{2}{3}$	$0.66\frac{2}{3}$ or 0.667[a]
25%	$\frac{1}{4}$	0.25	70%	$\frac{7}{10}$	0.7
30%	$\frac{3}{10}$	0.3	75%	$\frac{3}{4}$	0.75
$33\frac{1}{3}\%$	$\frac{1}{3}$	$0.33\frac{1}{3}$ or 0.333[a]	80%	$\frac{4}{5}$	0.8
40%	$\frac{2}{5}$	0.4	90%	$\frac{9}{10}$	0.9
50%	$\frac{1}{2}$	0.5	100%	$\frac{1}{1}$	1.0

[a]These decimals can be expressed with fractions or rounded decimals. Their fraction equivalents may be preferred because they are exact.

Ten percent and multiples of 10 percent are usually easy to remember.

$$10\% = \frac{1}{10} = 0.1 \qquad 60\% = \frac{6}{10} = \frac{3}{5} = 0.6$$

$$20\% = \frac{2}{10} = \frac{1}{5} = 0.2 \qquad 70\% = \frac{7}{10} = 0.7$$

$$30\% = \frac{3}{10} = 0.3 \qquad 80\% = \frac{8}{10} = \frac{4}{5} = 0.8$$

$$40\% = \frac{4}{10} = \frac{2}{5} = 0.4 \qquad 90\% = \frac{9}{10} = 0.9$$

$$50\% = \frac{5}{10} = \frac{1}{2} = 0.5 \qquad 100\% = \frac{1}{1} = 1$$

When 3 is divided into 100, the result is $33\frac{1}{3}$.

$$3\overline{)100} = 33\frac{1}{3}$$

$$33\frac{1}{3}\% = \frac{1}{3} = 0.33\frac{1}{3}$$

Two times $33\frac{1}{3}$ is $66\frac{2}{3}$.

$$66\frac{2}{3}\% = \frac{2}{3} = 0.66\frac{2}{3}$$

SELF-STUDY EXERCISES 5-3

L.O.1. Fill in the missing percents, fractions, or decimals *from memory*. Work this exercise only after memorizing the equivalents in Table 5-1.

	Percent	*Fraction*	*Decimal*		*Percent*	*Fraction*	*Decimal*
1.	10%			**2.**		$\frac{1}{4}$	
3.			0.2	**4.**		$\frac{1}{3}$	
5.	50%			**6.**		$\frac{4}{5}$	
7.			0.75	**8.**	$66\frac{2}{3}\%$		
9.			1	**10.**		$\frac{3}{10}$	
11.	40%			**12.**			0.7
13.		$\frac{9}{10}$		**14.**	60%		
15.		$\frac{1}{1}$		**16.**			0.25
17.		$\frac{2}{3}$		**18.**	70%		
19.			0.5	**20.**		$\frac{3}{5}$	
21.			0.1	**22.**	20%		
23.			$0.33\frac{1}{3}$	**24.**			$0.66\frac{2}{3}$
25.		$\frac{3}{4}$		**26.**	30%		
27.			0.9	**28.**		$\frac{1}{5}$	
29.			0.6	**30.**	100%		

5-4 FINDING RATE, BASE, AND PERCENTAGE

Learning Objectives

L.O.1. Solve percentage proportion for any missing element.
L.O.2. Use percentage proportion to find the percentage.
L.O.3. Use percentage proportion to find the rate.

L.O.4. Use percentage proportion to find the base.
L.O.5. Use percentage proportion to solve applied problems.

So far we have changed numbers into percents and percents into their fractional and decimal equivalents. We are now ready to start using percents to solve work-related problems common to the technologies. The key to solving problems with percents is to know the three basic elements of such problems.

L.O.1. Solve percentage proportion for any missing element.

Solving the Percentage Proportion

All problems involving percents will have three basic elements: the *rate*, the *base*, and the *percentage*. Knowing what each element is and how all three are related helps us solve problems with percents.

The *rate* is the percent; the *base* is the number that represents the original or total amount; the *percentage* represents part of the base.

It is possible to consolidate all three elements into one formula called the *percentage proportion.*

The relationship among the rate, percentage, base, and the standard unit of 100 can be represented by two fractions that are equal to each other.

Formula 5-4.1 *Percentage proportion:*

$$\frac{R}{100} = \frac{P}{B}$$

where P = percentage or part

R = rate or percent

B = base or total

Two fractions equal to each other form *a proportion*. Each fraction in a proportion is called *a ratio*.

■ **DEFINITION 5-4.1 Ratio.** A **ratio** is a fraction comparing a quantity or measure in the numerator to a quantity or measure in the denominator.

■ **DEFINITION 5-4.2 Proportion.** A **proportion** is a mathematical statement that shows two fractions or ratios are equal.

If we have a base of 75, a percentage of 15, and a rate of 20%, the two equal fractions or ratios would be

rate or percent → 20 ; percentage or part → 15

$$\frac{20}{100} = \frac{15}{75}$$

standard unit of 100 → 100 ; base or total → 75

Notice that if we multiply diagonally across the equal sign (cross multiplication), we find that the resulting *cross products* of the proportion are equal.

$$\frac{20}{100} \times \frac{15}{75}$$

$$20 \times 75 = 15 \times 100$$

$$1500 = 1500$$

In a percent problem, if we know any two of the elements, we can find the third element by using cross multiplication. (See Chapter 9 for more information on proportions.) This is called *solving* the proportion.

> **Rule 5-4.1** ***To solve the percentage proportion:***
>
> **1.** Cross multiply to find the cross products.
> **2.** Divide the product of the two numbers by the factor with the letter.

EXAMPLE 5-4.1 Find the percentage if the rate is 20% and the base is 75.

Solution

$$\frac{R}{100} = \frac{P}{B} \qquad \textit{(Set up proportion.)}$$

$$\frac{20}{100} = \frac{P}{75} \qquad \textit{(Substitute the known elements.)}$$

$$20 \times 75 = 100 \times P \qquad \textit{(Cross multiply.)}$$

$$1500 = 100 \times P$$

$$\frac{1500}{100} = P \qquad \textit{(Divide to find P.)}$$

$$15 = P$$

Therefore, the percentage is 15. ■

Tips and Traps

When using the calculator to solve percent problems, the percent key is *not* used if we work the problem using the percentage proportion. We can check by putting the 15 in the proportion in place of P and reducing both ratios. If the ratios are equal to each other, we have a true proportion and the solution is correct.

$$\frac{20}{100} = \frac{P}{75}$$

$$\frac{20}{100} = \frac{15}{75} \qquad \textit{(Reduce each ratio.)}$$

$$\frac{1}{5} = \frac{1}{5}$$

General Tips for Using the Calculator

Proportions can be solved using a calculator with a continuous series of steps.

$$\frac{20}{100} = \frac{P}{75}$$

20 [×] 75 [÷] 100 [=] ⇒ 15

EXAMPLE 5-4.2 Find the rate if the percentage is 14 and the base is 7.

Solution

$$\frac{R}{100} = \frac{P}{B} \qquad \textit{(Set up proportion.)}$$

$$\frac{R}{100} = \frac{14}{7} \qquad \textit{(Substitute known elements.)}$$

$$7 \times R = 14 \times 100 \qquad \textit{(Cross multiply.)}$$

$$7 \times R = 1400$$

$$R = \frac{1400}{7} \qquad \textit{(Divide to find R.)}$$

$$R = 200$$

Since R is the rate, 200 represents 200%. ■

EXAMPLE 5-4.3 Find the base if the rate is $\frac{1}{2}$% and the percentage is 40.

Solution

$$\frac{R}{100} = \frac{P}{B} \qquad \textit{(Set up proportion.)}$$

$$\frac{\frac{1}{2}}{100} = \frac{40}{B} \qquad \textit{(Substitute known elements.)}$$

$$\frac{1}{2} \times B = 40 \times 100 \qquad \textit{(Cross multiply.)}$$

$$\frac{1}{2} \times B = 4000$$

$$B = \frac{4000}{\frac{1}{2}} \qquad \textit{(Divide to find B.)}$$

$$B = \frac{4000}{1} \times \frac{2}{1} \qquad \textit{(Multiply by reciprocal.)}$$

$$B = 8000$$

The base is 8000. ■

The percentage proportion has several advantages. The obvious advantage is that we can use only one formula, the percentage proportion, to solve all three types of problems. Another advantage is that we can standardize our approach to percentage problems—have one basic way to solve them that works in all cases.

This general approach to percentage problems has other advantages, also. One is that the standard 100 takes care of having to convert decimal numbers to percents or convert percents to decimal equivalents, as is done in traditional approaches to these problems. This approach also simplifies the steps. For example, all problems involve one multiplication (cross multiplication) and one division step (division by whichever number the letter is multiplied by). This approach also includes either multiplication by 100 or division by 100, which can be done mentally.

Let's review several examples in which problems are solved using the percentage proportion. In these examples, pay particular attention to the clues that suggest whether the missing element is the rate, percentage, or base. Also notice the suggestions for using decimal versus fractional equivalents, as well as making calculations by hand versus by calculator.

L.O.2. Use percentage proportion to find the percentage.

Finding the Percentage

EXAMPLE 5-4.4 25% of 180 is what number?

Solution: In this example, 25% is the rate and the only other decision that has to be made is whether 180 is the base or the percentage. The missing number is 25% of, or part of, 180. Therefore, 180 is the base. The word *of* is a key word to look for in identifying the base. We are finding part *of* a quantity (the base).

Now, set up the proportion.

$$\frac{25}{100} = \frac{P}{180}$$

$$25 \times 180 = 100 \times P$$

$$4500 = 100 \times P$$

$$\frac{4500}{100} = P$$

$$45 = P$$

45 is 25% of 180. ■

Knowing fractional equivalents can save time when making calculations by hand. Let's substitute $\frac{1}{4}$ for 25% or 25 hundredths $\left(\frac{25}{100}\right)$ and rework this example.

$$\frac{1}{4} = \frac{P}{180}$$

$$1 \times 180 = 4 \times P$$

$$180 = 4 \times P$$

$$\frac{180}{4} = P$$

$$45 = P$$

(*The left ratio is the fractional equivalent of* $\frac{25}{100}$ *reduced to lowest terms,* $\frac{1}{4}$.)

If a calculator is available, the decimal equivalent for the percent can be used to save calculation time. 25% = 0.25. To employ our same technique of using proportions in problems involving percents, we can write 0.25 as $\frac{0.25}{1}$. (To divide by 1 does not change the value of a number.)

$$\frac{0.25}{1} = \frac{P}{180}$$ *(The left ratio employs the decimal equivalent of $\frac{25}{100}$, that is, 0.25.)*

$$0.25 \times 180 = P$$

$$45 = P$$

Thus, the percentage is 45 in all cases. ■

EXAMPLE 5-4.5 $33\frac{1}{3}\%$ of 282 is what number?

Solution: The rate is $33\frac{1}{3}\%$. The key word *of* tells us that 282 is the base. The percentage is missing.

$$\frac{33\frac{1}{3}}{100} = \frac{P}{282}$$

If the fractional equivalent for $33\frac{1}{3}\%$ or $\frac{33\frac{1}{3}}{100}$ is known from memory, we can simplify the calculations by substituting the fractional equivalent for $\frac{33\frac{1}{3}}{100}$ in the proportion. $\left(\frac{33\frac{1}{3}}{100} = \frac{1}{3}\text{; see Example 5-2.2.}\right)$

$$\frac{1}{3} = \frac{P}{282}$$

$$1 \times 282 = 3 \times P$$

$$\frac{282}{3} = P$$

$$94 = P$$

Thus, the percentage is 94. ■

Tips and Traps

Even with a calculator, it would still be better to use the fractional equivalent. This is because the decimal equivalent of $33\frac{1}{3}\%$ is a repeating decimal, and to use a repeating decimal equivalent on a calculator requires rounding at some point.

If we use 0.33 for the value of $\frac{1}{3}$, then $0.33 \times 282 = 93.06$, which is less than the exact answer of 94. If we round 93.06 to a whole number, we get 93, which is still less than 94. If we use 0.333 for the value of $\frac{1}{3}$, then $P = 93.906$, which can be rounded to the whole number 94 (the exact answer). If we round 93.06 and 93.906 to tenths, we get 93.1 and 93.9, still less than the exact 94.

Since the desired degree of accuracy may vary depending on the technical application, it is advisable to find the exact answer on the calculator by using the fractional equivalent.

EXAMPLE 5-4.6 27.5% of 152 is what number?

Solution: We are given the rate and the base, which follows *of*. We are seeking the percentage.

$$\frac{27.5}{100} = \frac{P}{152}$$

$$27.5 \times 152 = 100 \times P$$

$$4180 = 100 \times P$$

$$\frac{4180}{100} = P$$ *(Use the shortcut rule for dividing by 100.)*

$$41.8 = P$$ *(Ending 0 is not necessary.)*

$$41\frac{4}{5} = P$$ *(Answer can be expressed as a mixed decimal or mixed number.)*

$41\frac{4}{5}$ is 27.5% of 152. ■

Percents that are less than 1% or more than 100% follow the same procedures and cautions as any other percents. However, there are some time-savers we can use to make our work easier and faster when dealing with these percents. Again note the suggestions about using calculators versus solving problems by hand and using fractional versus decimal equivalents.

It is easy to find 1% of any number.

1% of 175 is

$$\frac{1}{100} = \frac{P}{175}$$

$$175 = 100 \times P$$

$$\frac{175}{100} = P$$

$$1.75 = P$$

To find 1% of a number, divide the number by 100 or move the decimal two places to the left.

Since 1% of a number can be found mentally, when working with a percent that is less than 1%, we will first find 1% of the number. The answer we are seeking must be less than 1% of the number. This estimating procedure will be very useful in checking the decimal placement. For instance, let's find $\frac{1}{4}$% of 875. $\frac{1}{4}$% means $\frac{1}{4}$ of 1%. 1% of 875 is 8.75 (move decimal two places to left). $\frac{1}{4}$% of 875 would be $\frac{1}{4}$ of 8.75, or a little more than the whole number 2. We can use this estimate to check the work in the following example.

EXAMPLE 5-4.7 $\frac{1}{4}\%$ of 875 is what number?

Solution: $\frac{1}{4}\%$ and 0.25% are equivalent. Either can be used in solving this problem. Since we are given the rate and the base, we need to find the percentage.

$$\frac{\frac{1}{4}}{100} = \frac{P}{875} \quad \text{or} \quad \frac{0.25}{100} = \frac{P}{875}$$

$$\frac{1}{4} \times 875 = 100 \times P \qquad 0.25 \times 875 = 100 \times P$$

$$\frac{875}{4} = 100 \times P \qquad 218.75 = 100 \times P$$

$$\frac{\frac{875}{4}}{100} = P \qquad \frac{218.75}{100} = P$$

$$2.1875 = P$$

$$\frac{875}{4} \div 100 = P$$

$$\frac{\overset{35}{\cancel{875}}}{4} \times \frac{1}{\underset{4}{\cancel{100}}} = P$$

$$\frac{35}{16} = P$$

$$2\frac{3}{16} = P$$

Therefore, the percentage is $2\frac{3}{16}$ or 2.1875. ■

100% of a number is 1 times the number or the number itself. When working with percents that are larger than 100%, we can also quickly make a rough estimate of the answer. For instance, 325% of 86 is what number?

325% of 86 is more than three times 86. 100% of 86 = 86. Thus, 325% of 86 must be more than 3 times 86 or more than 258.

EXAMPLE 5-4.8 325% of 86 is what number?

Solution: Again, we are finding the percentage, which is indicated by the key word *is*.

Option 1

$$\frac{325}{100} = \frac{P}{86}$$

$$325 \times 86 = 100 \times P$$

$$\frac{27950}{100} = P$$

Option 2

$$\frac{13}{4} = \frac{P}{86} \left(325\% = 3\frac{1}{4} = \frac{13}{4}\right)$$

$$13 \times 86 = 4 \times P$$

$$1118 = 4 \times P$$

Option 1	Option 2
$279.5 = P$	$\frac{1118}{4} = P$
$279\frac{1}{2} = P$	$279.5 = P$
	$279\frac{1}{2} = P$

Thus, the percentage is $279\frac{1}{2}$ or 279.5. ■

L.O.3. Use percentage proportion to find the rate.

Finding the Rate

EXAMPLE 5-4.9 What percent of 48 is 24?

Solution: The missing element in this problem is the rate. Is 48 the percentage (part) or the base (total)? 48 follows the key word *of* and is the base. 24 follows the key word *is* and is the percentage.

Set up the proportion.

$$\frac{R}{100} = \frac{24}{48}$$
$$R \times 48 = 100 \times 24$$
$$R \times 48 = 2400$$
$$R = \frac{2400}{48}$$
$$R = 50$$

Since R is the rate, 50 represents 50%. ■

EXAMPLE 5-4.10 What percent is 82.5 of 132?

Solution: The missing element is the rate. 82.5 is the percentage (part) and 132 is the base (total). Then,

$$\frac{R}{100} = \frac{82.5}{132}$$
$$8250 = 132 \times R \qquad \textit{(Use the shortcut for multiplying by 100.)}$$
$$\frac{8250}{132} = R$$
$$62.5 = R$$

Thus, the rate is 62.5% or $62\frac{1}{2}\%$. ■

EXAMPLE 5-4.11 What percent is 2 out of 600?

Solution: The rate is missing. The 2 represents the part or percentage and 600 is the base.

$$\frac{R}{100} = \frac{2}{600}$$
$$R \times 600 = 2 \times 100$$
$$R \times 600 = 200$$
$$R = \frac{200}{600}$$
$$R = \frac{1}{3}$$

Thus, the rate is $\frac{1}{3}$%. ■

EXAMPLE 5-4.12 261 is what percent of 87?

Solution: The rate is missing in this problem. Does 261 represent the percentage or the base? 261 is associated with the key word *is* and represents the percentage or part. 87 is associated with the key word *of* and represents the base (one total amount). Since the percentage is larger than the base, the rate will be more than 100%.

$$\frac{R}{100} = \frac{261}{87}$$
$$26{,}100 = 87 \times R$$
$$\frac{26{,}100}{87} = R$$
$$300 = R$$

Thus, the rate is 300%. ■

In actual practice, many people use the percentage proportion as a guide to finding percents (R) and percentages (P), but skip to the final calculations without writing out the whole proportion. Here are two commonly used shortcuts.

Shortcut 1: To find the percent of a number (base), such as 20% of 15, change the percent to a decimal and multiply it and the number.

$$20\% \text{ of } 15 \qquad (\textit{Rate} \times \textit{base})$$
$$0.20 \times 15 = 3$$

Shortcut 2: To find what percent one number (part) is of another (base), such as 3 is what percent of 15, divide the part by the base. Then move the decimal point two places right.

$$\frac{3 \text{ (part)}}{15 \text{ (base)}} = 0.2 = 20\% \qquad \textit{(Divide and change decimal to percent.)}$$

General Tips for Using the Calculator

Some scientific calculators and most graphing calculators do not have a percent key. Percent problems are still performed efficiently on these calculators by entering the percent in decimal notation or changing the calculator display to percent notation. Even when a calculator has a percent key the use of the [%] key may vary. The most common practice is to mentally convert percent notation to decimal notation or vice versa. The percent key is most often used in place of the equal key. In addition to performing the same purpose as the equal key, the % key also uses the decimal equivalent of the rate when the rate is used in the problem.

28% of 37 is what number?

Scientific calculator using [%] *key:*

[AC] 28 [×] 37 [%] ⇒ 10.36

Scientific calculator using [=]:

[AC] [.] 28 [×] 37 [=] ⇒ 10.36

Graphing calculator using [EXE] *key:*

[AC] [.] 28 [×] 37 [EXE] ⇒ 10.36

15 is what percent of 52?

Scientific calculator: The [%] key converts the decimal answer to a percent.

[AC] 15 [÷] 52 [%] ⇒ 28.84615385 *(a percent)*

15 is approximately 28.8% of 52.

Graphing calculator: The [EXE] key gives a decimal answer. You must change the decimal to a percent.

[AC] 15 [÷] 52 [EXE] ⇒ 0.2884615385 *(a decimal)*

0.2884615385 is approximately 28.8%.

L.O.4. Use percentage proportion to find the base.

Finding the Base

EXAMPLE 5-4.13 20% of what number is 45?

Solution: This time we know the rate, 20%, and the percentage, 45. We are looking for the base, as indicated by the key word *of*.

Option 1

$$\frac{20}{100} = \frac{45}{B}$$
$$20 \times B = 100 \times 45$$
$$20 \times B = 4500$$
$$B = \frac{4500}{20}$$
$$B = 225$$

Option 2

$$\frac{1}{5} = \frac{45}{B}$$
$$B = 5 \times 45$$
$$B = 225$$

Option 3

$$\frac{0.2}{1} = \frac{45}{B}$$
$$45 = 0.2 \times B$$
$$\frac{45}{0.2} = B$$
$$225 = B$$

Thus, the base is 225. ■

EXAMPLE 5-4.14 $\frac{3}{4}\%$ of what number is 11.25?

Solution: The rate is $\frac{3}{4}\%$ and the percentage is 11.25, as signaled by the key word *is*. The base is missing.

Option 1

$$\frac{\frac{3}{4}}{100} = \frac{11.25}{B}$$

$$\frac{3}{4} \times B = 1125$$

$$B = \frac{1125}{\frac{3}{4}}$$

$$B = \frac{\overset{375}{\cancel{1125}}}{1} \times \frac{4}{\underset{1}{\cancel{3}}}$$

$$B = 1500$$

Option 2

$$\frac{0.75}{100} = \frac{11.25}{B}$$

$$0.75 \times B = 1125$$

$$B = \frac{1125}{0.75}$$

$$B = 1500$$

Thus, the base is 1500. ■

EXAMPLE 5-4.15 398.18 is 215% of what number?

Solution: Here, we are looking for the base, as the key word *of* lets us know. We are given the percentage and the rate.

$$\frac{215}{100} = \frac{398.18}{B}$$

$$215 \times B = 39{,}818$$

$$B = \frac{39{,}818}{215}$$

$$B = 185.2 \quad \text{or} \quad 185\frac{1}{5}$$

Therefore, the base is 185.2 or $185\frac{1}{5}$. ■

L.O.5. Use percentage proportion to solve applied problems.

Solving Applied Problems Using the Percentage Proportion

In solving applied problems, the most difficult task is identifying the two parts that are given and determining which part is missing. Then the proportion can be set up and solved. Let's examine several examples of applied problems involving percents.

EXAMPLE 5-4.16 If a type of solder contains 55% tin, how many pounds of tin are needed to make 10 lb of solder?

Solution: In this problem, the *total amount* or the *base* is the 10 lb of solder. We know that 55% of this 10 lb of solder is tin. That is, the *rate* of tin in the solder is 55%. We want to find how much or what part of the 10 lb of solder is tin. This means that we are trying to find the *percentage* or *part*.

We write the proportion.

$$\frac{55}{100} = \frac{P}{10}$$

$$55 \times 10 = 100 \times P$$

$$550 = 100 \times P$$

$$\frac{550}{100} = P$$

$$5\frac{1}{2} = P$$

Thus, there are $5\frac{1}{2}$ lb of tin in 10 lb of solder. ■

EXAMPLE 5-4.17 If a 150-horsepower (hp) engine delivers only 105 hp to the driving wheels of a car, what is the efficiency of the engine?

Solution: *Efficiency* means the *percent* the output (105 hp) is of the total amount (150 hp) the engine is capable of delivering. So the base amount is 150 hp, the part or percentage delivered is 105 hp, and the percent of 150 represented by 105 is the rate or efficiency.

$$\frac{R}{100} = \frac{105}{150}$$

$$R \times 150 = 100 \times 105$$

$$R \times 150 = 10{,}500$$

$$R = \frac{10{,}500}{150}$$

$$R = 70$$

The engine is 70% efficient. ■

EXAMPLE 5-4.18 The effective value of current or voltage in an ac circuit is 71.3% of the maximum voltage. If a voltmeter shows a voltage of 110 volts (V) in a circuit, what is the maximum voltage?

Solution: 71.3% of the maximum voltage is 110 V. The maximum voltage is the *base*, and the amount of voltage shown in the voltmeter is 110 V or the *percentage*.

$$\frac{71.3}{100} = \frac{110}{B}$$

$$71.3 \times B = 100 \times 110$$

$$71.3 \times B = 11{,}000$$

$$B = \frac{11{,}000}{71.3}$$

$B = 154.277\ldots$

or

$B = 154$ V *(to the nearest volt)*

Thus, the maximum voltage is 154 V. ■

SELF-STUDY EXERCISES 5-4

L.O.1.

1. Find the percentage if the base is 75 and the rate is 5%.
2. Find the percentage if the base is 25 and the rate is 2.5%.
3. What is the rate if the base is 10.5 and the percentage is 7?
4. Find the rate when the base is 80 and the percentage is 30.
5. If the percentage is 4.75 and the rate is $33\frac{1}{3}\%$, find the base.
6. Find the base when the rate is 15% and the percentage is 52.5.
7. If the rate is $12\frac{1}{2}\%$ and the base is 75, find the percentage.
8. If the percentage is 11 and the rate is 5%, find the base.
9. Find the rate when the percentage is 15 and the base is 75.
10. What is the base if the percentage is 35 and the rate is 17.5%?

L.O.2.

11. 20% of 375 is what number?
12. 75% of 84 is what number?
13. $66\frac{2}{3}\%$ of 309 is what number?
14. 34.5% of 336 is what number?
15. $\frac{3}{4}\%$ of 90 is what number?
16. 0.2% of 470 is what number?
17. What number is 134% of 115?
18. Find 275% of 84.
19. 400% of 231 is what number?
20. $37\frac{1}{2}\%$ of 920 is what number?

L.O.3.

21. What percent of 348 is 87?
22. What percent of 350 is 105?
23. 72 is what percent of 216?
24. 28 is what percent of 85 (to the nearest tenth percent)?
25. 37.8 is what percent of 240?
26. What percent of 175 is 28?
27. 32 is what percent of 4000?
28. What percent is 2 out of 300?
29. What percent of 125 is 625?
30. 173.55 is what percent of 156?

L.O.4.

31. 50% of what number is 36?
32. 60% of what number is 30?
33. $12\frac{1}{2}\%$ of what number is 43?
34. 15.87 is 34.5% of what number?
35. $\frac{2}{3}\%$ of what number is $2\frac{2}{5}$?
36. 0.3% of what number is 0.825?
37. 150% of what number is $112\frac{1}{2}$?
38. 43% of what number is 107.5?
39. 92 is 500% of what number?

40. $133\frac{1}{3}\%$ of what number is 348?

L.O.5. Solve the following problems.

41. Cast iron contains 4.25% carbon. How much carbon is contained in a 25-lb bar of cast iron?

42. 3645 ball bearings are manufactured during one day. After being inspected, 121 of these ball bearings were rejected as imperfect. What percent of the ball bearings was rejected? (Round to the nearest whole percent.)

43. An engine operating at 82% efficiency transmits 164 hp. What is the engine's maximum capacity in horsepower?

44. If wrought iron contains 0.07% carbon, how much carbon is in a 30-lb bar of wrought iron?

45. The voltage of a generator is 120 V. If 6 V are lost in a supply line, what is the rate of voltage loss?

46. A contractor figures it costs $\frac{1}{2}\%$ of the total cost of a job to make a bid. What would be the cost of making a bid on a $115,000 job?

47. A certain ore yields an average of 67% iron. How much ore is needed to obtain 804 lb of iron?

48. A contractor makes a profit of $12,350 on a $115,750 job. What is the percent of profit? (Round to the nearest whole percent.)

49. 385 defective parts were produced during a day. If 4% of the parts produced were defective, how many parts were produced in all?

50. In a welding shop, 104,000 welds are made. If 97% of them are acceptable, how many are acceptable?

5-5 INCREASES AND DECREASES

Learning Objectives

L.O.1. Find the increase or decrease in percent problems.
L.O.2. Find the new amount directly in percent problems.
L.O.3. Find the rate or the base in increase or decrease problems.

Percents are often used in problems dealing with increases or decreases. For instance, if a TV repair shop is advised that its recent order for 250 power cords will be reduced by 14% because of a shortage of copper wire, the shop needs to find out how many power cords this decrease will amount to. The shop may have to order additional cords from another supplier.

L.O.1. Find the increase or decrease in percent problems.

Finding the Amount of Increase or Decrease

In working with increases or decreases, the *original amount* (250 power cords) will be the *base*. The *percentage* (power cords *not* received) will be the amount of *change* (*increase* or *decrease*). The *new amount* will be the original amount plus or minus the amount of change.

EXAMPLE 5-5.1 Pipefitters are to receive a 9% increase in wages per hour. If they were making $9.25 an hour, what will be the amount of increase per hour (to the nearest cent)? Also, what will be the new wage per hour? The original wage per hour is the base, and we want to find the amount of increase (percentage).

Solution

$$\frac{9}{100} = \frac{P}{9.25}$$ *(P represents amount of increase and 9% is the rate of increase.)*

$$9 \times 9.25 = 100 \times P$$

$$83.25 = 100 \times P$$

$$\frac{83.25}{100} = P$$

$$\$.8325 = P \quad \text{or} \quad \$0.83 \text{ to the nearest cent}$$

The pipefitters will receive an $0.83 per hour increase in wages.

$\$9.25 + \$0.83 = \$10.08$ *(New amount = original amount + amount of increase.)*

Their new hourly wage will be $10.08. ■

EXAMPLE 5-5.2 Molten iron shrinks 1.2% while cooling. What is the cooled length of a piece of iron if it is cast in a 24-cm pattern?

Solution: First, we will find the amount of shrinkage (amount of decrease, percentage). The original amount, 24 cm, is the base.

$$\frac{1.2}{100} = \frac{P}{24}$$ *(P represents amount of decrease and 1.2% is the rate of decrease.)*

$$1.2 \times 24 = 100 \times P$$

$$28.8 = 100 \times P$$

$$\frac{28.8}{100} = P$$

$$0.288 = P$$

The amount of shrinkage is 0.288 cm. Then the length of the cooled piece (new amount) is

$24 - 0.288 = 23.712$ cm *(New amount = original amount − amount of decrease.)* ■

EXAMPLE 5-5.3 Uncut earth is hard, packed soil. As it is dug, the volume increases or swells. A contractor figures that there will be a 20% earth swell when a mixture of uncut loam and clay soil is excavated. If 150 cubic yards (yd^3) of uncut earth are to be removed, taking into account the earth swell, how many cubic yards will have to be hauled away?

Solution: To find the amount of earth swell (amount of increase, percentage), find 20% of 150 yd^3. 150 yd^3 is the original amount or base.

(P represents amount of increase and 20% is the rate of increase.)

Option 1

$$\frac{20}{100} = \frac{P}{150}$$

$$20 \times 150 = 100 \times P$$

$$3000 = 100 \times P$$

$$\frac{3000}{100} = P$$

$$30 = P$$

Option 2

$$\frac{1}{5} = \frac{P}{150}$$

$$150 = 5 \times P$$

$$\frac{150}{5} = P$$

$$30 = P$$

Option 3

$$\frac{0.2}{1} = \frac{P}{150}$$

$$0.2 \times 150 = P$$

$$30 = P$$

Thus, the earth will swell 30 yd^3 when cut. To find the amount of earth to be hauled away, add 150 yd^3 and 30 yd^3.

$150 + 30 = 180$ yd^3 to be hauled away *(New amount = original amount + amount of increase.)* ■

L.O.2. Find the new amount directly in percent problems.

Finding the New Amount Directly

When knowing the amount of increase is not necessary, the new amount can be figured directly.

Remember, all of a quantity is 100%. If a quantity is to be increased by 20%, then the new amount will be 100% + 20% or 120% of the original amount.

If 150 yd^3 of uncut earth (base) increases by 20% when cut, the new amount (percentage) will be 120% of 150 yd^3.

Option 1

$$\frac{120}{100} = \frac{P}{150}$$

$$\frac{6}{5} = \frac{P}{150}$$

$$6 \times 150 = 5 \times P$$

$$900 = 5 \times P$$

$$\frac{900}{5} = P$$

$$180 = P$$

Option 2

$$\frac{1.2}{1} = \frac{P}{150}$$

(P represents new amount, since 120% is the new rate. That is, 120% of 150 is the new rate.)

$$1.2 \times 150 = P$$

$$180 = P$$

Thus, we can find the 180 yd^3 to be hauled away directly. ■

EXAMPLE 5-5.4 A drying process causes a 2% weight loss in a casting. If the wet casting weighs 130 kg, how much will the dried casting weigh?

Solution: The amount of weight loss (amount of decrease) is 2% of 130 kg.

Option 1

$$\frac{2}{100} = \frac{P}{130}$$

(P represents amount of decrease, since 2% is rate or percent of decrease.)

$$2 \times 130 = 100 \times P$$

$$260 = 100 \times P$$

$$\frac{260}{100} = P$$

$$2.6 = P$$

Thus, the casting will have a weight loss of 2.6 kg. The dried casting will weigh

$130 - 2.6 = 127.4$ kg *(New amount = original amount − amount of decrease.)* ■

Since knowing the amount of weight loss is not necessary, the dried casting weight could have been calculated directly.

The original weight equals 100%. There will be 2% weight loss, so the dried weight will be 100% − 2%, or 98% of the original weight.

Option 2

$$\frac{98}{100} = \frac{P}{130}$$

(P represents new amount, since 98% is the percent *that the new amount is of the original weight. That is, 98% of 130 is the new amount.)*

$$98 \times 130 = 100 \times P$$

$$\frac{12{,}740}{100} = P$$

$$127.4 = P$$

Thus, the weight of the dried casting is 127.4 kg. ■

EXAMPLE 5-5.5 A 3% error is acceptable for a machine part to be usable. If the part is intended to be 57 cm long, what is the range of measures that is acceptable for this part?

Solution: The machine part can be ±3% from the ideal length of 57 cm. The range of acceptable measures is found by finding the smallest acceptable measure and the largest acceptable measure. The symbol ± is read "plus or minus." It means that the item can be more (+) or less (−) than the designated amount. In this case, the part can be 3% longer than 57 cm or 3% shorter than 57 cm and still be usable. The smallest acceptable value is 97% of the ideal length (100% − 3%).

$$\frac{97}{100} = \frac{P}{57}$$

(P represents smallest acceptable amount, since 97% is the smallest acceptable percent.*)*

$$5529 = 100 \times P$$

$$\frac{5529}{100} = P$$

$$55.29 = P$$

Thus, the smallest acceptable measure is 55.29 cm. The largest acceptable value is 103% of the ideal length (100% + 3%).

$$\frac{103}{100} = \frac{P}{57}$$

(P represents largest acceptable amount, since 103% is the largest acceptable percent.*)*

$$5871 = 100 \times P$$

$$\frac{5871}{100} = P$$

$$58.71 = P$$

Thus, the largest acceptable value is 58.71 cm. The range of acceptable measures is from 55.29 cm to 58.71 cm.

We can also work this problem by first finding the amount of acceptable error in centimeters. Then we subtract this amount from the intended length of the machine part to get the smallest acceptable measure and add it to get the largest acceptable measure.

3% of 57 is the amount of acceptable error.

$\frac{3}{100} = \frac{P}{57}$ *(P represents acceptable error, since 3% is the acceptable error percent.)*

$171 = 100 \times P$

$1.71 = P$

57 ± 1.71 represents the range of acceptable measures.

$57 - 1.71 = 55.29$ *(Smallest acceptable amount = original amount − amount of acceptable error.)*

The smallest acceptable value is 55.29 cm.

$57 + 1.71 = 58.71$ *(Largest acceptable amount = original amount + amount of acceptable error.)*

The largest acceptable value is 58.71 cm. ■

L.O.3. Find the rate or the base in increase or decrease problems.

Finding the Rate or the Base

There are many kinds of increase or decrease problems that involve finding either the rate or the base.

EXAMPLE 5-5.6 A worn brake lining is measured to be $\frac{3}{32}$ in. thick. If the orignal thickness was $\frac{1}{4}$ in., what is the percent of wear?

Solution: First, the amount of wear is $\frac{1}{4} - \frac{3}{32}$.

$\frac{8}{32} - \frac{3}{32} = \frac{5}{32}$ *(Find common denominator. Subtract.)*

The amount of wear (decrease) is the percentage. The base is the original amount.

$$\frac{R}{100} = \frac{\frac{5}{32}}{\frac{1}{4}}$$ *(R represents percent of wear. $\frac{5}{32}$ is the amount of wear.)*

$$\frac{1}{4} \times R = \frac{\overset{25}{\cancel{100}}}{1} \times \frac{5}{\underset{8}{\cancel{32}}}$$

$$\frac{1}{4} \times R = \frac{125}{8}$$

$$R = \frac{125}{8} \div \frac{1}{4}$$

$$R = \frac{125}{\underset{2}{\cancel{8}}} \times \frac{\overset{1}{\cancel{4}}}{1}$$

$$R = 62\frac{1}{2}$$

The percent of wear is $62\frac{1}{2}\%$. ■

EXAMPLE 5-5.7 During the month of May, an electrician made a profit of $1525. In June, the same electrician made a profit of $1708. What is the percent of increase?

Solution: The amount of increase is $1708 − $1525 = $183. Then $183 is what percent of the *original* amount?

$$\frac{R}{100} = \frac{183}{1525}$$

(R represents percent of increase. 183 is the amount of increase.)

$$R \times 1525 = 18{,}300$$

$$R = \frac{18{,}300}{1525}$$

$$R = 12$$

Thus, the percent of increase is 12%. ■

EXAMPLE 5-5.8 An output of 141 hp is required for an engine. If there is a 6% loss of power, what amount of input horsepower (or base) is needed?

Solution: The input horsepower is the original amount, or 100%. If there is a 6% loss of power, the output horsepower is 94% of the input horsepower. (100% − 6% = 94%.)

$$\frac{94}{100} = \frac{141}{B}$$

(The percent must be expressed as the percent that the output is of the input (94%). 141 is the amount of output. B represents the base or input.)

$$94 \times B = 100 \times 141$$

$$94 \times B = 14{,}100$$

$$B = \frac{14{,}100}{94}$$

$$B = 150$$

Thus, an input of 150 hp is required. ■

In the applications used in this section on increase and decrease, we can see how important it is to understand what 100% of something really means. Many applications of percents use the fact that 100% of a quantity is the entire quantity. This knowledge lets us compute many quantities directly, without having to first figure the increase or decrease separately.

SELF-STUDY EXERCISES 5-5

L.O.1.

1. Find the amount of increase if 432 is increased by 25%.
2. If 78 is increased by 40%, find the new amount.
3. Find the amount of decrease if 68 is decreased by 15%.
4. If 135 is decreased by 75%, what is the new amount?

L.O.2.

5. Steel rods shrink 10% when cooled from furnace temperature to room temperature. If a tie rod is 30 in. long at furnance temperature, how long is the cooled tie rod?
6. 1650 board feet of 1-in. × 8-in. common boards are needed to subfloor a house. If 10% extra flooring is needed to allow for waste when the boards are laid square, how much flooring should be ordered?
7. If 17% extra flooring is needed to allow for waste when the boards are laid diagonally, how much flooring should be ordered to cover 2045 board feet of floor? Answer to the nearest whole board foot.
8. 25,400 bricks are required for a construction job. If 2% more bricks are needed to allow for breakage, how many bricks must be ordered?
9. When making an estimage on a job, the contractor wants to make a 10% profit. If all the estimated costs are $15,275, what is the total bid of cost and profit for the job?
10. Rock must be removed from a highway right of way. If 976 cubic yards (yd^3) of unblasted rock are to be removed, how many cubic yards is this after blasting? Blasting causes a 40% swell in volume.

L.O.3.

11. The cost of a pound of nails increased from $2.36 to $2.53. What was the percent of increase to the nearest whole-number percent?
12. A contractor estimated materials and labor for a job to cost $5385. An estimate one year later for the same job was $7808, due to inflation. Find the percent of increase due to inflation to the nearest whole number.
13. A chicken farmer bought 2575 baby chicks. Of this number 2060 lived to maturity. What percent of loss was experienced by the chicken farmer?
14. An electrician recorded costs of $1297 for a job. If she received $1232 for the job, what was the percent of money lost on the job? Round to the nearest whole number.
15. An engine that has a 4% loss of power has an output of 336 hp. What is the input (base) horsepower of the engine?
16. A contractor figures that 10 yd^3 of sand are needed for a job. If a 5% allowance for waste must be included, how much sand must be ordered?
17. A floor normally requiring 2580 board feet is to be laid diagonally. If a 17% waste allowance is necessary for flooring laid diagonally, how much flooring must be ordered? (Round to nearest whole number.)
18. A shop foreman records a 14% loss on rivets for waste. If the shop needs 25 lb of rivets, how many pounds must be ordered to compensate for loss due to waste?
19. Steel bars shrink 10% when cooled from furnance temperature to room temperature. If a cooled steel bar is 36 in. long, how long was it when it was formed?
20. The cost of No. 1 pine studs increased from $3.85 each to $4.62 each. Find the percent of increase.

5-6 BUSINESS APPLICATIONS

Learning Objectives

L.O.1. Calculate sales tax and payroll deductions.
L.O.2. Calculate discount and commission.
L.O.3. Calculate interest on loans and investments.

Employment in any profession will, at some time or other, require using mathematics for business purposes. This is because every company, plant, organization, corporation, partnership, or proprietorship is itself a business of some sort and is concerned with such matters as ordering and paying for supplies and equipment, figuring trade discounts on materials purchased, and processing the payroll accurately and on time.

L.O.1. Calculate sales tax and payroll deductions.

Finding Sales Tax and Payroll Deductions

One of the most common applications of percents in business settings is that of *sales tax*. Most states, counties, and cities add to the purchase price of many items a certain percent for sales tax. Then the total amount that a purchaser pays is the price of the item plus the sales tax.

Example 5-6.1 A 5% sales tax is levied on an order of building supplies costing $127.32. What is the amount of sales tax to be paid? What is the total bill?

Solution: To find the amount of sales tax to be paid, we need to find 5% of $127.32. The amount of sales tax is the percentage, and the cost of the supplies is the base.

$$\frac{5}{100} = \frac{P}{127.32}$$ *(P represents amount of sales tax. 5% is the percent or sales tax rate.)*

$$5 \times 127.32 = 100 \times P$$

$$636.60 = 100 \times P$$

$$\frac{636.60}{100} = P$$

$$\$6.366 = P$$

When dealing with money, the amounts are usually rounded to the nearest cent. Thus, the sales tax is $6.37.

To find the total amount to be paid, we add the cost of the supplies and the sales tax.

$$\$127.32 + \$6.37 = \$133.69$$ *(Total bill = cost of supplies + sales tax.)*

We could have found the total bill directly by considering the cost of the supplies as 100%. Since a 5% sales tax will be added, the total bill will be 105% of the cost of the supplies. Again, $127.32 is the base.

$$\frac{105}{100} = \frac{P}{127.32}$$ *(P represents total bill, since 105% is the total rate.)*

$$105 \times 127.32 = 100 \times P$$

$$13368.6 = 100 \times P$$ *(Zero in the hundredth place can be dropped.)*

$$\frac{13368.6}{100} = P$$

$$\$133.686 = P$$

$$\$133.69 = P$$ *(to the nearest cent)* ■

Other forms of taxes that involve most workers are withholding tax (income tax) and social security tax (FICA). These two taxes are normally deducted from a worker's paycheck.

Example 5-6.2 If the rate of social security tax (FICA) is 6.2% of the first $55,500 of earnings in a given year, how much social security tax is withheld on a weekly paycheck of $425?

Solution: The amount of pay before any deductions are made is called the *gross pay*, and it is the base. The amount of social security tax (FICA) will be the percentage.

$$\frac{6.2}{100} = \frac{P}{425}$$ *(P represents the amount of FICA deduction.)*

$$6.2 \times 425 = 100 \times P$$
$$2635 = 100 \times P$$
$$\frac{2635}{100} = P$$
$$\$26.35 = P$$

Therefore, $26.35 will be withheld for FICA. ■

Many employers take various *payroll deductions* from employees's paychecks. Some are required, such as withholding tax, FICA, and retirement contributions. Others are made as a convenience to the employee, such as insurance payments, union dues, and credit union or bank deposits.

The employee's total earnings before any deductions are made are called *gross pay*. The ''take-home'' pay or the amount after deductions is called *net pay*.

EXAMPLE 5-6.3 An employee's total weekly earnings are $475 and the take-home pay is $365.75. What percent of the gross pay are the total deductions?

Solution: There are two ways to approach this problem. One way is to first subtract the net pay from the gross pay to find the amount of the deductions.

$\$475 - 365.75 = 109.25$ *(total deductions)*

Then we find the percent that the deductions are of the gross pay. The amount of deductions is the percentage, and the gross pay is the base.

$$\frac{R}{100} = \frac{109.25}{475}$$ *(R represents percent of deductions. $109.25 is the amount of the deductions.)*

$$R \times 475 = 100 \times 109.25$$
$$R \times 475 = 10925$$
$$R = \frac{10925}{475}$$
$$R = 23$$

Thus, the deductions are 23% of the gross pay. Another way of approaching this problem is to find the percent the net pay (percentage) is of the gross pay (base).

$$\frac{R}{100} = \frac{365.75}{475}$$ *(R represents percent of net pay. $365.75 is the percentage or the amount of net pay.)*

$$R \times 475 = 100 \times 365.75$$
$$R \times 475 = 36575$$
$$R = \frac{36575}{475}$$
$$R = 77$$

The net pay is 77% of the gross pay. Since the gross pay is 100% and the net pay is 77%, the deductions are 100% − 77%, or 23% of the gross pay. ■

L.O.2. Calculate discount and commission.

Finding Discounts and Commissions

Many businesses give customers discounts for various reasons. These discounts may be given as incentives for the customers to pay cash or to pay within a certain time period. *Discounts* are also given to increase sales, to reduce inventories, or to move seasonal stock. In most cases the discounted price is determined by deducting a certain percent of the original price from the original price. The original price is the base.

EXAMPLE 5-6.4 A contractor is given a 3% discount for paying cash for the building supplies that are bought. If the total bill before the discount is \$143.38, what is the amount that the contractor will pay in cash?

Solution: Again, this problem can be approached two different ways. We can find the cash discount (percentage) by finding 3% of 143.38.

$$\frac{3}{100} = \frac{P}{143.38} \qquad \textit{(P represents the amount of discount and 3\% is the rate of discount.)}$$

$$3 \times 143.38 = 100 \times P$$

$$430.14 = 100 \times P$$

$$\frac{430.14}{100} = P$$

$$4.3014 = P$$

$$\$4.30 = P \qquad \textit{(to the nearest cent)}$$

Thus, the contractor pays \$143.38 − \$4.30 or \$139.08. Another approach in solving this problem is to again utilize our knowledge of 100%. The total bill is 100% and a 3% cash discount is given. Therefore, the discounted price is 97% of the original price (base).

$$\frac{97}{100} = \frac{P}{143.38} \qquad \textit{(P represents the discounted price and 97\% is the rate of the discounted price.)}$$

$$97 \times 143.38 = 100 \times P$$

$$13{,}907.86 = 100 \times P$$

$$\frac{13{,}907.86}{100} = P$$

$$139.0786 = P$$

$$\$139.08 = P \qquad \textit{(to the nearest cent)}$$

■

Persons in the sales profession are sometimes paid a salary based on the amount of sales that are made. This salary is usually a certain percent of the total sales and is called a *commisson*.

EXAMPLE 5-6.5 A salesperson receives a 6% commission on all sales that are made. If this salesperson sells \$15,575 in merchandise during a given pay period, what is the commission?

Solution: The amount of total sales during the pay period (\$15,575) is the base. The commision or part of the total sales that the salesperson receives in wages is the percentage.

$$\frac{6}{100} = \frac{P}{15{,}575} \qquad \textit{(P represents amount of commission and 6\% is the rate of commission.)}$$

$$6 \times 15{,}575 = 100 \times P$$

$$93{,}450 = 100 \times P$$

$$\frac{93{,}450}{100} = P$$

$$\$934.50 = P$$

The salesperson's commission for the pay period is $934.50. ■

EXAMPLE 5-6.6 An automotive parts salesperson earns a salary of $125 per week and 8% commission on all sales over $2500 per week. The sales during one week were $4875. What is the salesperson's salary for that week?

Solution: First, we need to determine the amount of sales on which the salesperson will earn a commission.

$$\$4875 - 2500 = \$2375$$

The salesperson receives an 8% commission on $2375 (base).

$$\frac{8}{100} = \frac{P}{2375} \qquad \textit{(P represents amount of commission earned. \$2375 is the base on which commission is earned.)}$$

$$8 \times 2375 = 100 \times P$$

$$19{,}000 = 100 \times P$$

$$\frac{19{,}000}{100} = P$$

$$\$190.00 = P \qquad \textit{(commission)}$$

The salesperson will receive a $125 base salary and $190 in commission.

$$\$125 + 190 = \$315$$

The salary for that week will be $315. ■

L.O.3. Calculate interest on loans and investments.

Finding Interest on Loans and Investments

Interest is the amount charged for borrowing or loaning money, or the amount of money earned when money is saved or invested. The *amount of interest* is the *percentage*, the total amount invested or borrowed is the *principal* or *base*, and the *percent of interest* is the *rate*.

The rate of interest is always expressed as a percent per time period. For example, the rate of interest on a loan may be 12% per year or per annum. The rate of interest or finance charge on a charge account may be $1\frac{1}{2}\%$ per month.

There are many different ways of figuring interest. Most banks or loan institutions use compound interest; some institutions figure interest using the exact time of a loan; some use an approximate time of a loan such as 30 days per month or 360 days per year. However, the Truth-in-Lending Law requires that all businesses equate their interest rate to an annual simple interest rate known as the *annual percentage rate* (APR). This allows consumers to compare rates of various institutions and to understand exactly what rate they are earning or are being charged.

Simple interest for one time period can be found by using our basic percentage proportion.

Example 5-6.7 Find the interest for 1 year on a loan of \$5000 if the interest rate is 15% per year.

Solution: \$5000 is the principal of the loan or the base. We are looking for the interest (percentage).

$$\frac{15}{100} = \frac{P}{5000}$$ *(P represents amount of interest. 15% is the interest rate.)*

$$15 \times 5000 = 100 \times P$$

$$75{,}000 = 100 \times P$$

$$\frac{75{,}000}{100} = P$$

$$\$750 = P$$

The interest on the loan is \$750. ■

Example 5-6.8 A credit-card service charges a finance charge (interest) of $1\frac{1}{2}\%$ per month on the average daily balance of the account. If the average daily balance on an account is \$157.48, what is the finance charge for the month?

Solution

$$\frac{1\frac{1}{2}}{100} = \frac{P}{157.48}$$ *(P represents finance charge, or interest, for one month. Remember, 1.5 can be substituted for $1\frac{1}{2}$ if desired.)*

$$1.5 \times 157.48 = 100 \times P$$

$$236.22 = 100 \times P$$

$$\frac{236.22}{100} = P$$

$$2.3622 = P$$

$$\$2.36 = P$$ *(to the nearest cent)*

The finance charge is \$2.36. ■

Example 5-6.9 \$1250 is invested for 6 months at an interest rate of $8\frac{1}{2}\%$ per year. Find the simple interest earned.

Solution: First, we will find the simple interest for one time period, which is one year.

$$\frac{8\frac{1}{2}}{100} = \frac{P}{1250}$$

$$8\frac{1}{2} \times 1250 = 100 \times P$$

(P represents interest for one year, since $8\frac{1}{2}\%$ is the rate for one year. Remember, 8.5 may be substituted for $8\frac{1}{2}$.)

$$10{,}625 = 100 \times P$$

$$\frac{10{,}625}{100} = P$$

$$\$106.25 = P$$

The interest for a full year is \$106.25. Now we will find what portion of a year is represented by 6 months. There are 12 months in a year, so 6 months is $\frac{6}{12}$ or $\frac{1}{2}$ year.

To find the interest for 6 months, we find $\frac{1}{2}$ of the interest for a full year, or $\frac{1}{2}$ of \$106.25.

$$\frac{1}{2} \times \frac{106.25}{1} = \frac{106.25}{2} = \$53.125 \text{ or } \$53.13$$

(to the nearest cent)
(Do not forget this step when amount of interest is figured for more or less time than the time period of the interest rate.)

The interest earned is \$53.13. ■

Example 5-6.10 For one month a business received \$3.07 in interest on an account balance of \$245.75. What is the monthly rate of interest?

Solution: \$3.07 is the interest or percentage, and \$245.75 is the principal or base.

$$\frac{R}{100} = \frac{3.07}{245.75}$$

(R represents monthly rate *of interest, since \$3.07 is the* monthly *interest.)*

$$R \times 245.75 = 100 \times 3.07$$

$$R \times 245.75 = 307$$

$$R = \frac{307}{245.75}$$

$$R = 1.249$$

$$R = 1.25\%$$

(to the nearest hundredth of a percent)

The monthly rate is 1.25%. ■

SELF-STUDY EXERCISES 5-6

L.O.1.

1. Find the sales tax and the total bill on an order of office supplies costing \$75.83 if the tax rate is 6%. Round to the nearest cent.
2. Materials to landscape a property total \$785.84. What is the total bill if the sales tax rate is $5\frac{1}{2}\%$? Round to the nearest cent.
3. If the rate of social security tax is 6.2%, find the tax on gross earnings of \$375.80. Round to the nearest cent.
4. An employee's gross earnings for a pay period are \$895.65. The net pay for this salary is \$675.23. What percent of the gross pay are the total deductions? Round to the nearest whole percent.

5. An employee has a net salary of $576.89 and a gross salary of $745.60. What percent of the gross salary is the total of the deductions? Round to the nearest whole percent.

L.O.2.

6. A manufacturer will give a 2% discount to customers paying cash. If a parts store paid cash for an order totaling $875.84, what amount was saved? Calculate to the nearest cent.

7. Find the cash price for an order totaling $3985.57 if a 3% discount is offered for cash orders. Calculate to the nearest cent.

8. What is the commission earned by a salesperson who sells $18,890 in merchandise if a 5% commission is paid on all sales?

9. A manufacturer's representative is paid a salary of $140 per week and 7% commission on all sales over $3200 per week. The sales for a given week were $7412. What is the representative's salary for that week?

10. A parts house manager is paid a salary of $2153 monthly plus a bonus of 1% of the net earnings of the business. Find the total salary for a month when the net earnings of the business were $105,275.

L.O.3. Solve the following problems.

11. An electrician purchases $650 worth of electrical materials. A finance charge of $1\frac{1}{2}\%$ per month is added to the bill. What is the finance charge for 1 month?

12. $10,000 is invested for 3 months at 12% per year. How much interest is earned?

13. A carpenter is charged $10.24 in interest on a credit-card account with a $584.87 average daily balance. What is the rate of interest? Round to the nearest hundredth of a percent.

14. Find the interest on a loan of $2450 at 15% per year for 1 year.

15. Find the interest on a loan of $5840 at 12% per year for 2 years, 6 months.

ELECTRONICS APPLICATION

Percentage Error

Technicians are often asked to find the percentage error in components and in parts of circuits.

When you calculate a value (called the *theoretical value* or the *predicted value*) and then measure that value (called the *measured* or *actual value*), the results are seldom exactly the same. It is not enough to say that measured values were close or not close to the predicted values. You need to give a percentage answer. This is called percentage error or percentage difference.

$$\% \text{ error} = \frac{\text{actual value} - \text{theoretical value}}{\text{theoretical value}} \times 100$$

or

$$\% \text{ error} = \frac{\text{measured value} - \text{coded value}}{\text{coded value}} \times 100$$

Suppose that you have a circuit in which you have calculated and then measured three voltages drops, or potential differences in voltage. These voltage drops are called V_1 and V_2 and V_3, and they are read as "V sub 1" and "V sub 2" and "V sub 3." They mean "voltage drop number one" and "voltage drop number two" and "voltage drop number three." The subscripts are numbers written "sub" or "under" the line and are simply there for indentification purposes. Subscripts are an easy way of making complicated sets of numbers easier to read. By using sequential numbers, we can arrange the numbers in an order that is easy to use.

As long as you understand what is meant, you can also write $V1$ and $V2$ and $V3$. The convention is that a letter coming first means the number following the letter is meant to be a type of subscript or identifier. This is particularly important when programming computers.

Suppose that your calculations show that you should get

$$V_1 = 20\text{ V}, \qquad V_2 = 30\text{ V}, \qquad V_3 = 40\text{ V}$$

When measurements are made using an voltmeter, the results are

$$V_1 = 21\text{ V}, \qquad V_2 = 29\text{ V}, \qquad V_3 = 43\text{ V}$$

First consider V_1. Since $V_{1\_meas} = 21$ V and $V_{1\_coded} = 20$ V,

$$\%\text{ error} = \frac{21\text{ V} - 20\text{ V}}{20\text{ V}} \times 100 = \frac{1}{20} \times 100 = \frac{100}{20} = +5\%$$

Next consider V_2. Since $V_{2\_meas} = 29$ V and $V_{2\_coded} = 30$ V,

$$\%\text{ error} = \frac{29\text{ V} - 30\text{ V}}{30\text{ V}} \times 100 = \frac{-1}{30} \times 100 = \frac{-100}{30} = -3.3\%$$

Next consider V_3. Since $V_{3\_meas} = 43$ V and $V_{3\_coded} = 40$ V,

$$\%\text{ error} = \frac{43\text{ V} - 40\text{ V}}{40\text{ V}} \times 100 = \frac{300}{40} = 7.5\%$$

Your writeup could say that all measurements were within 10% of what was predicted, which means they were very close indeed.

Now calculate the % error for three more voltage drops, each with an actual difference of 3 volts.

What is the % error if $V_{4\_meas} = 403$ V and $V_{4\_coded} = 400$ V?

$$\%\text{ error} = \frac{403\text{ V} - 400\text{ V}}{400\text{ V}} \times 100 = \frac{300}{400} = 0.75\%, \quad \text{or less than } 1\%$$

Very close indeed.

What is the % error if $V_{5\_meas} = 3997$ V and $V_{5\_coded} = 4000$ V? The actual measurement difference is only 3 volts. But

$$\%\text{ error} = \frac{3997\text{ V} - 4000\text{ V}}{4000\text{ V}} \times 100 = \frac{-300}{4000}$$

$$= -0.075\%, \quad \text{or less than } 1\%$$

which is extremely close.

What is the % error if $V_{6\_meas} = 1$ V and $V_{6\_coded} = 4$ V? The actual measurement difference is only 3 volts. But

$$\% \text{ error} = \frac{1\text{ V} - 4\text{ V}}{4\text{ V}} \times 100 = \frac{-300}{4} = -75\%, \quad \text{or extremely high error}$$

Notice that when the measured value was more than the theoretical value, as with V_1, V_3, V_4, and V_5, the % error ended up being positive. And when the measured value was less than the theoretical value, as with V_2 and V_6, the % error ended up being negative.

Exercises

Fill in the % errors in the following data table and then decide if each error is high, low, or in the middle. Low is closer to zero, middle is closer to −10 or +10, and high is less than −10 or greater than +10.

Sample Data Table for Recording Voltages

	Voltage Drop	Theoretical Value (V_c)	Measured Value (V_m)	% Error	High, Low, or Middle?
1.	V_1	50	55		
2.	V_2	50	46		
3.	V_3	500	505		
4.	V_4	500	480		
5.	V_5	250	200		
6.	V_6	250	260		
7.	V_7	25	26		
8.	V_8	25	23		
9.	V_9	2	1.5		
10.	V_{10}	2	2.2		

Answers For Exercises

	Voltage Drop	Theoretical Value (V_c)	Measured Value (V_m)	% Error	High, Low, or Middle?
1.	V_1	50	55	10	Middle
2.	V_2	50	46	−8	Middle
3.	V_3	500	505	1	Low
4.	V_4	500	480	−4	Low
5.	V_5	250	200	−20	High
6.	V_6	250	260	4	Low
7.	V_7	25	26	4	Low
8.	V_8	25	23	−8	Middle
9.	V_9	2	1.5	−25	High
10.	V_{10}	2	2.2	10	Middle

ASSIGNMENT EXERCISES

5-1

1. If $\frac{1}{3}$ of the light bulbs in a shipment are defective, what percent are defective?

2. If $\frac{87}{100}$ of a company's employees are present on a certain day, what percent are present?

Change the following numbers to their percent equivalents.

3. $\frac{3}{4}$

4. 0.7

5. 2

6. $4\frac{1}{3}$

7. 0.06

8. 125

9. 3.7

10. $\frac{5}{6}$

11. 0.0004

12. $6\frac{4}{5}$

13. $\frac{1}{750}$

14. 17.3

15. 10

16. $\frac{7}{500}$

17. $\frac{1}{8}$

18. 0.0315

19. $\frac{5}{11}$

20. 0.35

21. A recipe is $\frac{3}{5}$ water. What percent is water?

22. Pastry is $\frac{7}{8}$ water. What percent is water?

23. One brand of dog food is $\frac{1}{6}$ liver. What percent is liver?

24. A pie recipe calls for $\frac{2}{3}$ of the ingredients to be fresh fruit. What percent is fruit?

25. In a photographic mixture, 0.45 of the mixture is distilled water. What percent is water?

26. If 0.12 of a beverage is alcohol, what percent is alcohol?

27. In a baked pastry product, 0.6 of the ingredients is water. What percent is water?

28. Approximately $\frac{1}{5}$ of the big game hunters are successful. What percent are successful?

29. One survey revealed $\frac{3}{10}$ of the motor vehicles in a certain state used regular gasoline. What percent used regular gasoline?

30. If 0.3 of a salad dressing is vinegar, what percent is vinegar?

31. A coffee is packaged so that 0.005 of the contents is additives. What percent is additives?

32. If 0.001 of the ingredients of a certain tea is spice, what percent of the ingredients is spice?

33. A professor marked a test paper $\frac{24}{30}$. What percent is the test score?

5-2

Change the following to both fractional and decimal equivalents.

34. 72%

35. 65%

36. 40%

37. 25%

38. $12\frac{1}{2}\%$

39. 37.5%

40. $\frac{2}{3}\%$

41. 0.2%

42. $\frac{3}{5}\%$

43. 0.08%

44. $16\frac{2}{3}\%$

45. 31.25%

Change to equivalent whole numbers.

46. 300%

47. 200%

Change to both mixed-number and mixed-decimal equivalents.

48. 275%

49. 450%

50. 124%

51. 260%

52. $112\frac{1}{2}\%$

53. 462.5%

Change to mixed-number equivalents.

54. $183\frac{1}{3}\%$

55. $333\frac{1}{3}\%$

Change to mixed-decimal equivalents.

56. 227.2%

57. $318\frac{3}{4}\%$

58. $108\frac{1}{5}\%$

59. One estimate is that 21.5% of country music performers are female. Find the decimal equivalent of this percent.

60. A survey of automobile drivers indicated that 73.8% disliked vehicles whose computerized equipment "talked" to them in a simulated human voice. Find the decimal equivalent of this percent.

61. A worker pays $7\frac{3}{4}\%$ of wages earned into a retirement fund. Find the decimal equivalent of $7\frac{3}{4}\%$.

62. The sales tax rate in one city is $8\frac{1}{4}\%$. Find the decimal equivalent of $8\frac{1}{4}\%$.

63. A credit-card interest rate was 19.525%. Find the decimal equivalent of 19.525%.

64. Owner financing on a farm was offered at 9.275%. Find the decimal equivalent of 9.275%.

65. It is estimated that consumer prices increased by nearly 125% during a certain time period. Find the decimal equivalent of 125%.

66. Housing costs in one area over a 15-year period increased by 275%. Find the decimal equivalent of 275%.

67. It is estimated that motorhome owners camp out 140% more than owners of any other recreational vehicle. Express 140% as a fraction.

68. The cost of living increased 340% during a certain time period. Express 340% as a fraction.

5-3

Fill in the missing percents, fractions, or decimals *from memory.* Work this exercise only after memorizing the equivalents in Table 5-1.

	Percent	*Fraction*	*Decimal*		*Percent*	*Fraction*	*Decimal*
69.		$\frac{1}{5}$		**70.**			$0.66\frac{2}{3}$
71.			0.4	**72.**	25%		
73.		$\frac{3}{5}$		**74.**	70%		
75.		$\frac{1}{10}$		**76.**	$33\frac{1}{3}\%$		
77.		$\frac{1}{1}$		**78.**		$\frac{1}{2}$	
79.	30%			**80.**			0.8
81.		$\frac{9}{10}$		**82.**		$\frac{3}{4}$	

Find the missing percents, fractions, or decimals.

	Percent	*Fraction*	*Decimal*		*Percent*	*Fraction*	*Decimal*
83.	$12\frac{1}{2}\%$			**84.**		$\frac{3}{8}$	
85.	5%			**86.**			0.875
87.	$62\frac{1}{2}\%$			**88.**		$\frac{1}{16}$	

5-4

Solve the following problems involving percents.

89. 5% of 480 is what number?

90. $62\frac{1}{2}$% of 120 is what number?

91. $\frac{1}{4}$% of 175 is what number?

92. $233\frac{1}{3}$% of 576 is what number?

93. 39 is what percent of 65?

94. What percent of 118 is 42.48?

95. What percent of 65 is 162.5?

96. 80% of what number is 116?

97. 24% of what number is 19.92?

98. 7.56 is $6\frac{3}{4}$% of what number?

99. 260% of what number is 395.2?

100. 3 is 0.375% of what number?

101. 38.25 is what percent of 250?

102. 83% of 163 is what number?

103. What percent of 26 is 130?

104. 4.75% of 348.2 is what number?

105. $10\frac{1}{3}$% of what number is 8.68?

106. Specifications for bronze call for 80% copper. How much copper is needed to make 300 lb of bronze?

107. 84 lb of 224 lb of an alloy is zinc. What percent of the alloy is zinc?

108. When laying a subfloor using common 1-in. × 8-in. boards laid diagonally, 17% is allowed for waste. How many board feet will be wasted out of 1250 board feet?

109. 27 out of 2374 pieces produced by a particular machine were defective. What percent (to the nearest hundredth of a percent) were defective?

110. The voltage loss in a line is 2.5 V. If this is 2% of the generator voltage, what is the generator voltage?

111. If a family spends 28% of its income on food, how much of a $950 paycheck goes for food?

112. If a freshman class of 1125 college students is made up of 8% international students, how many international students are in the class?

113. A city prosecuted 1475 individuals with traffic citations. If 36,875 individuals received traffic citations, what percent were prosecuted?

114. A survey studied 600 people for their views on nuclear power plants near their towns. Of these, 75 people said that they approved of nuclear power plants near their towns. What percent approved?

115. In a certain college 67 students made the dean's list. If this was 33.5% of the student body, what was the total number of students in the college?

116. It is estimated that only 19% of the licensed big game hunters on a state wildlife management area are successful. If 95 big game hunters were successful, what was the total number of big game hunters?

5-5

117. A rough casting weighs 32.7 kg. After finishing on a lathe, it weighs 29.3 kg. Find the percent of weight loss to the nearest whole number percent.

118. A steel beam expands 0.01% of its length when exposed to the sun. If a beam measures 49.995 ft after being exposed to the sun, what is its cooled length?

119. A contractor ordered 800 board feet of lumber for a job that required 750 board feet. What percent of the required lumber was ordered for waste to the nearest whole number?

120. A brickmason received a 12% increase in wages, amounting to $35.40. Find the amount of wages received before the increase. Find the amount of wages received after the increase.

121. A lathe costing $600 was sold for $516. What was the percent of decrease in the price of the lathe?

122. According to specifications, a machined part may vary from its specified measure by ± 0.4% and still be usable. If the specified measure of the part is 75 in. long, what is the range of measures that is acceptable for the part?

123. A wet casting weighing 145 kg has a 2% weight loss in the drying process. How much will the dried casting weigh?

124. A mixture of uncut loam and clay soil will have a 20% earth swell when it is excavated. If 300 cubic yards of uncut earth are to be removed, how many cubic yards will have to be hauled away if the earth swell is taken into account?

125. A blueprint specification for a part lists its overall length as 62.5 cm. If a tolerance of $\pm 0.8\%$ is allowed, find the limit dimensions of the length of the part.

126. A book that used to sell for $18.50 now sells for 20% more. How much does the book now sell for?

127. A computer disk that once sold for $2.25 now sells for 25% less. How much does the computer disk now sell for?

128. A paperback dictionary originally sold for $4. It now sells for $1 more. What is the percent of the increase?

129. A laptop computer that was originally $2400 is now $300 more. What is the percent of the increase?

130. When earth is dug up, it usually increases in bulk or expands by 20%. How much earth will a contractor have to haul away if 150 cubic feet are dug up?

131. A shipping carton is rated to hold 50 pounds. A stronger carton that can hold 15% more weight will be used for added safety. How much weight will the new carton hold?

132. A 20-inch bar of iron measured 20.025 inches when it was heated. What was the percent of increase?

133. A pearl necklace measuring 18 inches was exchanged for one measuring 24 inches. What was the percent of increase?

134. A casting weighed 130 ounces when first made. After it dried, it weighed 127.4 ounces. What was the percent of weight loss caused by drying?

135. A dieter went from 168 pounds to 160 pounds in one week. To the nearest tenth of a percent, what was the percent of weight loss?

136. Workers took a 10% pay reduction to help their company stay open during economic hard times. What was the reduced annual salary of a worker who originally earned $35,000?

137. A motorist traded an older car with 350 horsepower for a new car with 17.4% less horsepower. What is the horsepower of the new car to the nearest whole number?

5-6

Solve the following problems.

138. Discontinued paneling is sale priced at 25% off the regular price. If the regular price is $12.50 per sheet, what is the sale price per sheet rounded to the nearest cent?

139. A parts distributor is paid a weekly salary of $250 plus an 8% commission on all sales over $2500 per week. What is the distributor's salary for a given week if the sales for that week totaled $4873?

140. An employee's gross earnings for a month are $1750. If the net pay is $1237, what percent of the gross pay is the total of the deductions? Round to the nearest whole percent.

141. An order of lumber totaled $348.25. If a 5% sales tax is added to the bill, what is the total bill? Round to the nearest cent.

142. A builder purchased a concrete mixer for $785. The builder did not know the sales tax rate, but the total bill was $828.18. Find the sales tax rate. Round to the nearest tenth of a percent.

143. An employer must match employees' social security contributions. If the weekly payroll is $27,542 and if the FICA rate is 6.2%, what are the employer's contributions? (All employees' year-to-date salaries were under maximum salary subject to the FICA tax.)

144. A contractor can receive a 2% discount on a monthly sand and gravel bill of $1655.75 if the bill is paid within 10 days of the statement date. How much is saved by paying the bill within the 10-day period? Figure to the nearest cent.

145. A businessperson is charged a $4.96 monthly finance charge on a bill of $283.15. What is the monthly interest rate on the account? Round to the nearest hundredth of a percent.

146. Find the interest on a loan of $3200 if the annual interest rate is 16% and the loan is for 9 months.

147. A property owner sold a house for $127,500 and paid off all outstanding mortgages. $52,475 cash was left from this transaction. If the money was invested at 14% interest for 18 months, how much interest was earned on this investment?

148. The sales tax in one city is 8.75% of the purchase price. How much is the sales tax on a purchase of $78.56?

149. A small town charges 3.25% of the purchase price for sales tax. What is the sales tax on a purchase of $27.45?

150. Interest for one year on a loan of $2400 was $396. What was the interest rate?

151. A $2000 certificate of deposit earned $170 interest in one year. What was the interest rate?

152. If a loan for $500 at 18.75% interest per annum (year) is paid out in 3 months, how much is the interest?

153. What is the interest on a business loan for $6500 at 9.75% interest per annum (year) paid in 7 months?

154. A salesperson earns a commission of 15% of the total monthly sales. If the salesperson earned $2145, how much were the total sales?

155. A sales clerk in a store is paid a salary plus 3% commission. If the sales clerk earned $10.65 commission for a weekend, how much were the sales?

CHALLENGE PROBLEMS

156. A homeowner with an annual family income of $35,500 spends in a year $6900 for a home mortgage, $950 for property taxes, $380 for homeowner's insurance, $2400 for utilities, and $200 for maintenance and repair. To the nearest tenth, what percent of the homeowner's annual income is spent for housing?

157. A motorist with an annual income of $18,250 spends each year $3420 on automobile financing, $652 on gasoline, $625 on insurance, and $150 on maintenance and repair. To the nearest tenth, what percent of the motorist's annual income is used for automotive transportation?

158. A 78-pound alloy of tin and silver contains 69.3 pounds of tin. Find the percent of silver in the alloy to the nearest tenth of a percent.

159. There are 25 women in a class of 35 students. Find the percent of men in the class to the nearest tenth of a percent.

160. A librarian checked out 25 books in a bookmobile. The books included 5 mysteries, 2 science fiction novels, 8 biographies, and 10 classics. Mysteries and biographies accounted for what percent of the books checked out?

CONCEPTS ANALYSIS

1. Under what conditions are two fractions proportional?

2. Solving a proportion with one missing term requires two computations. In the proportion $\frac{R}{100} = \frac{65}{26}$, what are the two computations that are to be performed to find the value of R?

3. Give some clues for determining if a value in a percent problem represents a rate.

4. Give some clues for determining if a value in a percent problem represents a percentage.

5. Give some clues for determining if a value in a percent problem represents a base.

6. Is the amount of sales tax required on a purchase in your state determined by a percent? What is the sales tax rate in your state?

7. If the total bill, including sales tax, on a purchase is 107% of the original amount, what is the sales tax rate?

8. If a dress is marked 25% off the original price, what percent of the original price does the buyer pay?

9. If a quantity increases 50%, is the new amount twice the original amount? Explain your answer.

10. If the cost of an item decreases 50%, is the new amount half the original amount? Explain your answer.

Find and explain any mistakes in the following. Rework the incorrect problems correctly.

11. $\frac{R}{100} = \frac{4}{5}$

$$\frac{R}{25} = \frac{1}{5}$$
$$5 \times R = 25 \times 1$$
$$5 \times R = 25$$
$$R = 25 \div 5$$
$$R = 5\%$$

12. $3\% = 0.3$

13. What percent of 25 is 75?

$$\frac{R}{100} = \frac{25}{75}$$
$$75 \times R = 100 \times 25$$
$$75 \times R = 2500$$
$$R = 2500 \div 75$$
$$R = 33\frac{1}{3}\%$$

14. 26 is 0.5% of what number?

$$\frac{0.5}{100} = \frac{26}{B}$$
$$0.5 \times 26 = 100 \times B$$
$$13 = 100 \times B$$
$$13 \div 100 = B$$
$$0.13 = B$$

15. If the cost of a \$15 shirt increases 10%, what is the new cost of the shirt?

$$\frac{10}{100} = \frac{P}{15}$$
$$100 \times P = 10 \times 15$$
$$100 \times P = 150$$
$$P = 150 \div 100$$
$$P = 1.5$$

The shirt increased \$1.50.

CHAPTER SUMMARY

Objectives	What to Remember	Example
Section 5-1		
1. Change any number to its percent equivalent.	Multiply the number by 100. For a shortcut, move decimal two places to the right.	Change $\frac{1}{2}$, 1.2, and 7 to percents. $\frac{1}{2} \times 100 = \frac{100}{2} = 50\%$ $1.2 \times 100 = 120\%$ $7 = 700\%$
Section 5-2		
1. Change any percent less than 100% to its numerical equivalent.	Divide by 100. Reduce if possible. For a shortcut, move decimal two places to the left.	Change 7% to a fraction. $7 \div 100 = \frac{7}{100}$ Change $1\frac{1}{4}\%$ to a fraction. $1\frac{1}{4} \div 100 = \frac{5}{4} \div 100 =$

Objectives	What to Remember	Example
		$\frac{5}{4} \times \frac{1}{100} =$ $\frac{5}{400} = \frac{1}{80}$ Convert 3.5% to a decimal. 3.5% = 0.035
2. Change any percent of 100% or more to its numerical equivalent.	Divide by 100. Reduce if possible. For a shortcut, move decimal two places to the left.	Convert 245% to a mixed number. $245 \div 100 = \frac{245}{100} = 2\frac{45}{100} = 2\frac{9}{20}$ Convert 124.5% to a decimal. 124.5% = 1.245 Convert $100\frac{1}{4}\%$ to a decimal. $100\frac{1}{4}\% = 100.25\%$ $100.25 \div 100 = 1.0025$
Section 5-3		
1. Substitute one common equivalent for another.	Memorize common equivalents, such as $\frac{1}{4} = 0.25$ or 25% and $\frac{1}{5} = 0.20$ or 20%, to make some calculations easier.	$25\% \times 60 =$ $\frac{1}{4} \times 60 =$ $\frac{60}{4} = 15$
Section 5-4		
1. Solve percentage proportion for any missing element.	Multiply diagonally across the equal sign to find the cross products. Divide the product of the two numbers by the factor with the R, P, or B.	$\frac{R}{100} = \frac{5}{25}$ $25 \times R = 5 \times 100$ $25 \times R = 500$ $R = \frac{500}{25}$ $R = 20\%$
2. Use percentage proportion to find the percentage.	Use key words like *of*, signaling base, and *is*, signaling percentage, to help identify missing element. Then set up the percentage proportion with P for percentage. The percentage is a part of the whole amount.	What amount is 5% of \$200? *Of* identifies \$200 as the base. The percent or rate is given. The missing element must be the percentage. $\frac{5}{100} = \frac{P}{200}$ $5 \times 200 = P \times 100$ $1000 = P \times 100$ $\frac{1000}{100} = P$ $\$10 = P$
	Use a shortcut. Multiply base by decimal equivalent of the percent.	$5\% \times 200 = P$ $0.05 \times 200 = 10$
3. Use percentage proportion to find the rate.	Use key words like *of*, signaling base, and *is*, signaling percentage, to help identify missing element. Then set up the percentage proportion with R for rate. The rate is a percent.	What percent of 6 is 2? *Is* suggests 2 is the percentage. *Of* identifies 6 as the base. $\frac{R}{100} = \frac{2}{6}$ $6 \times R = 2 \times 100$

$$6 \times R = 200$$
$$R = \frac{200}{6}$$
$$R = 33\frac{1}{3}\%$$

Use a shortcut. Divide percentage by the base.

$$\frac{2}{6} = 0.33\frac{1}{3} = 33\frac{1}{3}\%$$

4. Use percentage proportion to find the base.

Use key words like *of*, signaling base, and *is*, signaling percentage, to help identify missing element. Then set up the percentage proportion with B for base. The base is the total amount.

12 is 24% of what number? The percent is given. 12 is the part or percentage and is suggested by *is*. *Of* identifies ''what number'' as the missing base.

$$\frac{24}{100} = \frac{12}{B}$$
$$24 \times B = 12 \times 100$$
$$24 \times B = 1200$$
$$B = \frac{1200}{24}$$
$$B = 50$$

5. Use percentage proportion to solve applied problems.

Use key words like *of*, signaling base, and *is*, signaling percentage, to help identify missing element. Then set up the percentage proportion with R, P, or B as the missing element. The rate is a percent. The percentage is a part of the whole amount. The base is the total amount.

LaQuita sold 12 boxes of candy for a school project. If she started with 25 boxes, what percent of the candy did she sell? The percent is missing. 25 boxes is the total amount, or base. 12 is the part or percentage.

$$\frac{R}{100} = \frac{12}{25}$$
$$25 \times R = 12 \times 100$$
$$25 \times R = 1200$$
$$R = \frac{1200}{25}$$
$$R = 48\%$$

Section 5-5

1. Find the increase or decrease in percent problems.

The original amount is the base. The new amount is the original amount plus the increase or the original amount less the decrease. Subtract the new amount and the original amount to find the amount of increase or decrease.

Julio made \$15.25 an hour but took a pay cut to keep his job. If he now makes \$12.20 an hour, what was the percent of the pay cut? \$15.25 − \$12.20 = \$3.05 (amount of decrease). \$3.05 is the part (percentage) being considered. \$15.25 is the original amount, or base.

$$\frac{R}{100} = \frac{3.05}{15.25}$$
$$15.25 \times R = 3.05 \times 100$$
$$15.25 \times R = 305$$
$$R = \frac{305}{15.25}$$
$$R = 20\%$$

2. Find the new amount directly in percent problems.	Add percent of increase to 100% or subtract percent of decrease from 100%. Use this new percent in the percentage proportion.	A project requires 5 lb of galvanized nails. If 15% of the nails will be wasted, how many pounds must be purchased? 5 lb is the base, the original amount. The percent is 100% + 15%, or 115%. The amount of nails needed is the percentage, or part. $$\frac{115}{100} = \frac{P}{5}$$ $$100 \times P = 5 \times 115$$ $$100 \times P = 575$$ $$P = \frac{575}{100}$$ $$P = 5.75 \text{ lb}$$
3. Find the rate or the base in increase or decrease problems.	Subtract the original amount and new amount to find the amount of increase or decrease. Then set up percentage proportion, using R or B for the missing element.	A 5-in. power edger blade now measures $4\frac{3}{4}$ in. What is the percent of wear? $4\frac{4}{4} - 4\frac{3}{4} = \frac{1}{4}$. The part or percentage is $\frac{1}{4}$. The base or original amount is 5 in. $$\frac{R}{100} = \frac{\frac{1}{4}}{5}$$ $$5 \times R = \frac{1}{4} \times 100$$ $$5 \times R = 25$$ $$R = \frac{25}{5}$$ $$R = 5\%$$ A PC lost 20% of its memory when a set of RAM chips was removed. If the PC now has 512K of memory, what was the original memory? Current memory is 100% − 20% = 80% of original memory, or base. 512K is the part, or percentage. $$\frac{80}{100} = \frac{512}{B}$$ $$80 \times B = 512 \times 100$$ $$80 \times B = 51{,}200$$ $$B = \frac{51{,}200}{80}$$ $$B = 640\text{K}$$

Section 5-6

1. Calculate sales tax and payroll deductions.

The rate is the percent of tax or deduction. The base is the amount of the purchase or the gross pay. The percentage is amount of tax or deduction. Solve with a percentage proportion.

If sales tax is 7%, find the tax on a purchase of $30.

$$\frac{7}{100} = \frac{P}{30}$$
$$7 \times 30 = 100 \times P$$
$$210 = 100 \times P$$
$$\frac{210}{100} = P$$
$$\$2.10 = P$$

Total cost is $30 + $2.10 = $32.10.

If 8% of gross pay is deducted for retirement, how much is deducted from a monthly salary of $1650? What is the net pay?

$$\frac{8}{100} = \frac{P}{1650}$$
$$8 \times 1650 = 100 \times P$$
$$13{,}200 = 100 \times P$$
$$\frac{13{,}200}{100} = P$$
$$\$132 = P$$

Net pay is $1650 − $132 = $1518.

2. Calculate discount and commission.

The rate is the percent of discount or commission. The base is the price or amount of sales. The percentage is the amount of the discount or commission. Solve with a percentage proportion using *P* as the missing element.

A $45 dress is on sale at 20% off. How much is the discount? What is the sale price?

$$\frac{20}{100} = \frac{P}{45}$$
$$100 \times P = 20 \times 45$$
$$100 \times P = 900$$
$$P = \frac{900}{100}$$
$$P = \$9$$

Sale price is $45 − $9 = $36.

Chang sold $5000 of furniture and earned a 3% commission. How much is his commission?

$$\frac{3}{100} = \frac{P}{5000}$$
$$100 \times P = 3 \times 5000$$
$$100 \times P = 15{,}000$$
$$P = \frac{15{,}000}{100}$$
$$P = \$150$$

3. Calculate interest on loans and investments.	The rate is the percent of interest. The base is the amount of the loan or investment. The percentage is the amount of interest. Solve with a percentage proportion using P as the missing element.	Boris borrowed \$2500 at 8.5% interest for one year. How much did he have to repay? $\frac{8.5}{100} = \frac{P}{2500}$ $100 \times P = 8.5 \times 2500$ $100 \times P = 21{,}250$ $P = \frac{21{,}250}{100}$ $P = \$212.50$ He repaid \$2500 + \$212.50 = \$2712.50. Interest on an investment is found the same way.

WORDS TO KNOW

percent (p. 186)
percent equivalent (p. 187)
numerical equivalent (p. 189)
rate (p. 197)
base (p. 197)
percentage (p. 197)
ratio (p. 197)
proportion (p. 197)
percentage proportion (p. 197)
cross products (p. 197)
new amount (p. 210)
percent of increase or decrease (p. 214)
sales tax (p. 217)
FICA (p. 217)
gross pay (p. 217)
payroll deductions (p. 218)
discount (p. 219)
commission (p. 219)
interest (p. 220)

CHAPTER TRIAL TEST

Change the following numbers to their percent equivalents.

1. $\frac{7}{16}$ **2.** 0.003 **3.** $3\frac{1}{5}$

Change the following percents to their fractional or mixed-number equivalents.

4. $8\frac{1}{3}\%$ **5.** $\frac{1}{4}\%$ **6.** $366\frac{2}{3}\%$

Change the following percents to their decimal or mixed-decimal equivalents.

7. 0.5% **8.** 160% **9.** 80%

Solve the following problems involving percents.

10. 10% of 150 is what number?

11. What number is $6\frac{1}{4}\%$ of 144?

12. What percent of 275 is 33?

13. 45.75 is 15% of what number?

14. 55 is what percent of 11?

15. 250% of what number is 287.5?

16. What percent of 360 is 1.2?

17. 245% of what number is 164.4? Round to the nearest hundredth.

18. 5.4% of 57 is what number?

19. \$15,000 is invested at 14% per year for 3 months. How much interest is earned on the investment?

20. A casting measuring 48 cm when poured shrinks to 47.4 cm when cooled. What is the percent of decrease?

21. An electronic parts salesperson earned \$175 in commission. If commission is 7% of the sales, how much did the salesperson sell?

22. Electronic parts increased 15% in cost during a certain period, amounting to an increase of \$65.15 on one order. How much would the order have cost before the increase? Round to the nearest cent.

23. In 1993 an area vocational school had an enrollment of 325 men and 123 women. In 1994 there were 149 women. What was the percent of increase of women students? Round to the nearest hundredth.

24. The payroll for an electrical shop for one week was \$1500. If federal income and FICA taxes averaged 28%, how much was withheld from the \$1500?

25. During one period a bakery rejected 372 items as unfit for sale. In a following period the bakery rejected only 323 items, a decrease in unfit bakery items. What was the percent of decrease? Round to the nearest hundredth.

26. Materials to landscape a new home cost \$643.75. What is the amount of tax if the rate is 6%? Round to the nearest cent.

27. A casting weighed 36.6 kg. After milling, it weighs 34.7 kg. Find the percent of weight loss to the nearest whole percent.

28. After soil was excavated for a project, it swelled 15%. If 275 yd^3 were excavated, how many cubic yards of soil were there after excavation?

29. The total bill for machinist supplies was \$873.92 before a discount of 12%. How much was the discount? Round to the nearest cent.

30. A business paid a \$5.58 finance charge on a monthly balance of \$318.76. What was the monthly rate of interest? Round to the nearest hundredth.

CHAPTER

6 Direct Measurement

Measurements are used in business, in industry, and in our everyday lives. Increased international trade has caused a need for a worldwide standard of measurements. The International System of Units (SI), more commonly called the *metric system*, has been adopted by most nations of the world, including the United States, and a conversion to this system of measurement is gradually taking place.

However, the United States has used a nonmetric system of measurement for many years, and this system is still the customary system of measurement for many businesses and industries. This system is called the *English* or the *U.S. customary system of measurement*, and we begin our study of measurements with this system.

6-1 THE U.S. CUSTOMARY SYSTEM OF MEASUREMENT

Learning Objectives

L.O.1. Identify uses of U.S. customary system measures of length, weight, and capacity.
L.O.2. Convert from one U.S. customary unit of measure to another.
L.O.3. Write mixed U.S. customary measures in standard notation.

The *U.S. customary* or *English system* of measurement has always been used in the United States. Many of the units of measure within this system have become obsolete through the years. In our discussion of the U.S. customary system of measurement, we will include only those measures that are currently in use.

L.O.1. Identify uses of U.S. customary measures of length, weight, and capacity.

Introduction to the U.S. Customary System Measures of Length, Weight, and Capacity

There are three basic units in the U.S. customary system that are commonly used to measure length. They are the inch, the foot, and the mile. Table 6-1 gives the relationships among these measurements of length.

Inch: An inch is slightly less than the width of a quarter coin (Fig. 6-1). Historically, an inch was considered to be the width of a man's thumb. Inches are used to measure lengths that vary by increments the size of the width of a man's thumb or the width of a U.S. quarter. Some examples of lengths measured in inches are the sizes of men's trousers, belts, shirts, and jackets. Also, the diagonal

TABLE 6-1 U.S. Customary Units of Length or Distance

12 inches (in.)[a]	=	1 foot (ft)[b]
3 feet (ft)	=	1 yard (yd)
36 inches (in.)	=	1 yard (yd)
5280 feet (ft)	=	1 mile (mi)

[a]The symbol ″ means inches (8″ = 8 in.) or seconds (60″ = 60 seconds).

[b]The symbol ′ means feet (3′ = 3 ft) or minutes (60′ = 60 minutes).

distance in inches from corner to corner of a television screen determines its size, such as 13-in. or 19-in. television screens. Photographs and picture frames, such as 5 × 7 or 8 × 10, are measured in inches. Lumber sizes, like 2 × 4's or 2 × 12's, are indicated in inches.

Not actual size

Figure 6-1

Foot: A foot is 12 inches, or about the width of a placemat for a dining table (Fig. 6-2). The measure was originally based on the length of a human foot. Feet are used to measure larger lengths varying in size by increments of 12 inches. Lumber, for instance, is sold by the foot. A 2 × 4 is usually sold in lengths of 8 ft, 12 ft, 16 ft, or 20 ft. Human heights are also often measured in feet (and inches), such as 5 ft, 6 in.; 5 ft, 11 in.; or 6 ft, 2 in. Elevation, such as the height of mountains and the altitude of airplanes, is designated in feet.

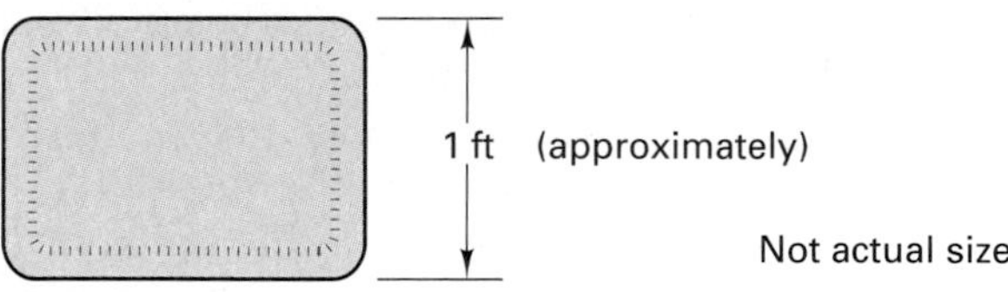

Not actual size

Figure 6-2

Mile: A mile is 5280 feet, or the length of approximately eight city blocks (Fig. 6-3). The mile was originally used by the Romans, who considered a mile to be 1000 paces of 5 feet each. Long distances are measured in miles, such as the distances between cities and the lengths traveled on a road, street, or highway. The mile is also used to designate the speed of a vehicle and speed limits, such as 55 miles per hour (mph) or 30 mph.

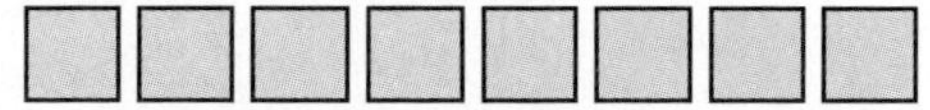

1 mile – about eight city blocks   Not actual size

Figure 6-3

Another length measure that is used is the *yard*. The yard is three feet or 36 inches. Yards are used in similar circumstances as feet, as well as measures for fabric and football fields.

TABLE 6-2 U.S. Customary Units of Weight or Mass

16 ounces (oz)	=	1 pound (lb)
2000 pounds (lb)	=	1 ton (T)

Many goods are exchanged or sold according to their amount of weight or mass. There are three commonly used measuring units for weight or mass in the U.S. customary system. They are the ounce, pound, and ton. Table 6-2 gives the relationships among these measures of weight or mass.

Ounce: The ounce is $\frac{1}{16}$ of a pound and is used to measure objects or products that vary in weight by increments of this weight. It is the approximate weight of three packaged individual servings of artificial sweetener (Fig. 6-4). (The troy ounce, which will not be used in this text, measures precious metals and is $\frac{1}{12}$ of a pound. *Ounce* is derived from a Latin word for $\frac{1}{12}$.) Ounces are used for measures of lightweight items like first-class letters, tubes of toothpaste or medicinal ointments, canned goods, and dry-packaged foods like pasta, candy, and gravy mixes.

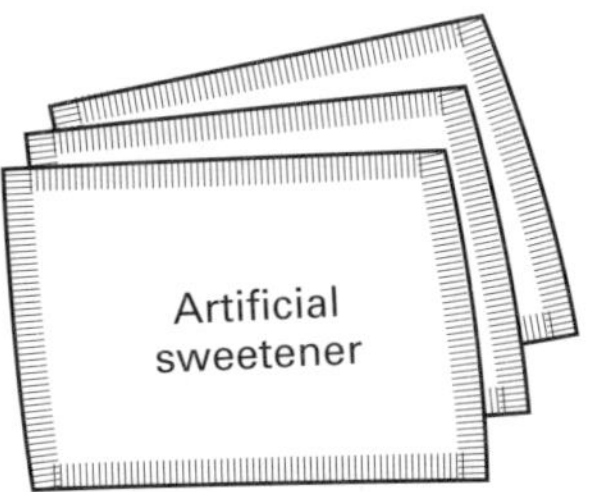

1 ounce – approximately three packets of artificial sweetener

Figure 6-4

Pound: A pound is equal to 16 ounces. It is used to measure heavier items that vary by increments of this weight. The term is derived from an old English word for weight. A pound is the weight of a 4.25-in. tall can of cut green beans or a paper container of coffee (Fig. 6-5). Many small, medium, and large items of merchandise are measured in pounds, like grocery and market items, people's weights, and air pressure in automotive tires. The pound is also used to calculate shipping charges for merchandise and parcel-post packages.

1 lb – about one can of cut green beans

Figure 6-5

Ton: A ton is 2000 pounds. The word originally meant a weight or measure. The typical American-made station wagon may help you to visualize this measure somewhat (Fig. 6-6), since an American-made station wagon may weigh 3 tons (although such vehicle weights are usually expressed in pounds). The ton is used for extremely heavy items, such as very large volumes of grain, the weights of huge animals like elephants (4 to 7 tons), and the weights by which coal and iron are sold.

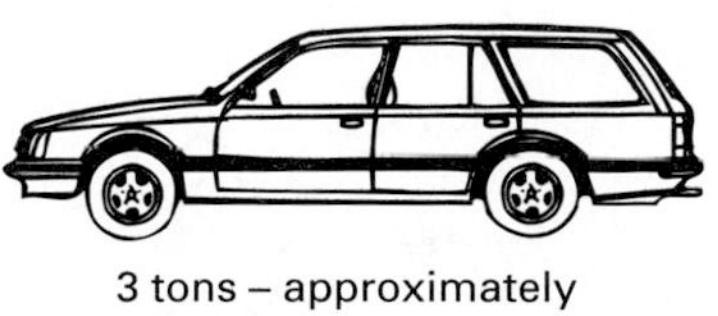

3 tons – approximately

Figure 6-6

The U.S. customary system of measurement has several units that are commonly used to measure capacity or volume. The system includes measuring units for both liquid and dry capacity measures; however, the dry capacity measures are seldom used. It is more common to express dry measures in terms of weight than in terms of capacity.

The U.S. customary units of measure for capacity or volume that are commonly used are the ounce, cup, pint, quart, and gallon. Table 6-3 gives the relationships among the liquid measures for capacity or volume. In the U.S. customary system the term *ounce* is used to represent both a weight measure and a liquid capacity measure. The measures are different and have no common relationship. The context of the problem will suggest whether the unit for weight or capacity is meant.

Ounce: A liquid ounce is a volume, not a weight. It is about the quantity of two small bottles of fingernail polish or perfume (Fig. 6-7). Liquid ounces are used for bottled medicine, canned or bottled carbonated beverages, baby bottles and formula, and similar-size quantities.

1 ounce – approximately two small bottles of fingernail polish

Figure 6-7

Cup: A cup is equal to 8 ounces. The term is derived from an old English word for tub, a kind of container. This measure is about the volume of a coffee

TABLE 6-3 U.S. Customary Units of Liquid Capacity or Volume

8 ounces (oz)	=	1 cup (c)
2 cups (c)	=	1 pint (pt)
2 pints (pt)	=	1 quart (qt)
4 cups (c)	=	1 quart (qt)
4 quarts (qt)	=	1 gallon (gal)

cup (Fig. 6-8). Cups are most often used for 8-oz quantities in cooking. Measuring cups used for cooking usually divide their contents into cups and portions of cups.

1 cup – approximately one large coffee cup

Figure 6-8

Pint: A pint is two cups. Its name comes from an Old English word meaning the spot that marks a certain level in a measuring device. It is the quantity of a *medium*-size paper container of milk (Fig. 6-9). Liquids like milk, automobile motor additives, produce like strawberries, ice cream, and similar quantities may be packaged in pint containers. In some localities, shucked oysters are sold by the pint.

1 pint – one medium size container of milk

Figure 6-9

Quart: A quart is two pints, or four cups, or 32 ounces. The term is derived from an Old English word for fourth. It is a fourth of a gallon. Milk, as well as various citrus juices, is often sold in quart containers (Fig. 6-10). Motor oil and ice cream are also packaged in quart containers. Quarts are also used in measuring liquid quantities in cooking and the capacities of cookware like mixing bowls, pots, and casserole dishes.

1 quart – one container of milk

Figure 6-10

Gallon: A gallon is four quarts. It is based on the ''wine gallon'' of British origin. Paint, varnish, and stain are often sold in gallon cans (Fig. 6-11). Motor fuel and heating oil are usually sold by the gallon. Large quantities of liquid propane and various chemicals are sold by the gallon. Often, statistics on liquid consumption, such as water or alcoholic beverages, are reported in gallons consumed per individual during a specified time period.

1 gallon – one large can of paint
Figure 6-11

EXAMPLE 6-1.1 Indicate the most reasonable measuring units for the following:

1. Package of rice
2. Height of Uncle Harry
3. A hippo
4. Sugar for a cake
5. Expensive French perfume
6. Distance between Memphis and New Orleans
7. Bottle of suntan lotion
8. Container of eggnog
9. Sherbet in grocery store freezer
10. Tank of gasoline
11. Man's dress shirt size
12. Material for draperies

Solution

1. Pounds or ounces
2. Feet, or feet and inches
3. Tons or pounds
4. Cups
5. Ounces
6. Miles
7. Ounces
8. Gallon, quart, or pint
9. Gallon or pint
10. Gallons
11. Inches
12. Yards

■

L.O.2. Convert from one U.S. customary unit of measure to another.

Converting from One U.S. Customary Unit to Another

Technical applications of units of measure often call for changing from one unit to another unit. Converting from one unit of measure to another unit of measure in the U.S. customary system can be done by deciding whether the beginning unit of measure is larger or smaller than the desired unit of measure. When changing to a larger unit of measure, there will be fewer units and a division step is required. When changing to a smaller unit of measure, there will be more units and a multiplication step is required. Since this process is often confusing, an alternative process can be used that employs a *unity ratio*.

■ **Definition 6-1.1 Unity Ratio.** A **unity ratio** is a ratio of measures whose value is 1.

A ratio is a fraction. A unity ratio, then, is a fraction with one unit of measure in the numerator and a different, but equivalent, unit of measure in the denominator. Some examples of unity ratios are

$$\frac{12 \text{ in.}}{1 \text{ ft}}, \quad \frac{1 \text{ ft}}{12 \text{ in.}}, \quad \frac{3 \text{ ft}}{1 \text{ yd}}, \quad \frac{1 \text{ mi}}{5280 \text{ ft}}$$

In each unity ratio the value of the numerator equals the value of the denominator. When we have a ratio with the numerator and denominator equal, the value of the ratio is 1. We call these ratios *unity ratios* because their value is 1. The word *unity* means 1.

When we convert from one unit of measure to another unit, we will use a unity ratio that contains the original unit and the new unit.

Rule 6-1.1 ***To change from one U.S. customary unit of measure to another:***

1. Set up the original amount as a fraction with the original unit of measure in the numerator.
2. Multiply this by a unity ratio with the original unit in the denominator and the new unit in the numerator.
3. Reduce like units of measure and all numbers wherever possible.

Example 6-1.2 Find the number of inches in 5 ft.

Solution: To make this conversion, we multiply 5 ft by a unity ratio that contains both inches and feet, since these are the two units involved.

Since 5 ft is a whole number, it is written with 1 as the denominator. The original unit is placed with the 5 in the *numerator* of the first fraction.

$$\frac{5 \text{ ft}}{1}\left(\frac{\quad}{\quad}\right)$$

The unit of measure we are changing *from* (feet) is placed in the *denominator* of the unity ratio. This placement will allow the original units to reduce later.

$$\frac{5 \text{ ft}}{1}\left(\frac{\quad}{\text{ft}}\right)$$

The unit we are changing *to* (inches) is placed in the *numerator* of the unity ratio.

$$\frac{5 \text{ ft}}{1}\left(\frac{\text{in.}}{\text{ft}}\right)$$

Now we place in the unity ratio the numerical values that make these two units of measure equivalent. (1 ft = 12 in.) Then we complete the calculation, reducing wherever possible.

$$\frac{5\ \cancel{\text{ft}}}{1}\left(\frac{12\ \text{in.}}{1\ \cancel{\text{ft}}}\right) = 60\ \text{in.}$$

Thus, 5 ft = 60 in. ■

EXAMPLE 6-1.3 How many inches are in 4.5 ft?

Solution

$$\frac{4.5\ \cancel{\text{ft}}}{1}\left(\frac{12\ \text{in.}}{1\ \cancel{\text{ft}}}\right) = 4.5\ (12\ \text{in.}) = 54\ \text{in.}$$

Thus, 4.5 ft = 54 in. ■

There are times when it is necessary to convert a U.S. customary unit to some U.S. unit other than the next larger or smaller unit of measure. For example, suppose that a dressmaker needs to know how many yards are in a certain number of inches of fabric. One way to solve this problem is to convert inches to feet with one unity ratio and then convert feet to yards with another unity ratio.

> **Rule 6-1.2**
>
> To change from one U.S. customary unit to one other than the next larger or smaller unit, proceed as before, but multiply the original amount by as many unity ratios as needed to attain the new U.S. customary unit.

EXAMPLE 6-1.4 How many inches are in $2\frac{1}{3}$ yd?

Solution: First, we write $2\frac{1}{3}$ as an improper fraction.

$$2\frac{1}{3}\ \text{yd} = \frac{7}{3}\ \text{yd}$$

Then we multiply the improper fraction by two unity ratios. One unity ratio is needed to convert yards to feet, and the other is needed to convert feet to inches. Note the original unit in the *numerator* of the improper fraction, $\frac{7}{3}$.

$$\frac{7\ \text{yd}}{3}\left(\frac{\text{ft}}{\text{yd}}\right)\left(\frac{\text{in.}}{\text{ft}}\right)$$

Next, we insert the numerical values that make the units equivalent in each unity ratio. (3 ft = 1 yd; 12 in. = 1 ft.) Then we complete the calculation, reducing wherever possible.

$$\frac{7\ \cancel{\text{yd}}}{\underset{1}{\cancel{3}}}\left(\frac{\overset{1}{\cancel{3}}\ \cancel{\text{ft}}}{1\ \cancel{\text{yd}}}\right)\left(\frac{12\ \text{in.}}{1\ \cancel{\text{ft}}}\right) = 84\ \text{in.}$$

Thus, $2\frac{1}{3}$ yd = 84 in.

However, if we used the conversion factor 36 in. = 1 yd, we could eliminate one unity ratio from the calculation. That is, we could convert $2\frac{1}{3}$ yd $\left(\frac{7}{3}\text{ yd}\right)$ to inches as follows:

$$\frac{7\text{ yd}}{3}\left(\frac{36\text{ in.}}{1\text{ yd}}\right)$$

$$\frac{7\,\cancel{\text{yd}}}{\underset{1}{\cancel{3}}}\left(\frac{\overset{12}{\cancel{36}}\text{ in.}}{1\,\cancel{\text{yd}}}\right) = 84\text{ in.}$$

Thus, $2\frac{1}{3}$ yd = 84 in. ■

EXAMPLE 6-1.5 How many pounds are in 84 oz?

Solution: We multiply 84 by a unity ratio with both ounces and pounds, since these are the two units involved. (1 lb = 16 oz.)

$$\frac{84\text{ oz}}{1}\left(\frac{\text{lb}}{\text{oz}}\right)$$

$$\frac{\overset{21}{\cancel{84}}\,\cancel{\text{oz}}}{1}\left(\frac{1\text{ lb}}{\underset{4}{\cancel{16}}\,\cancel{\text{oz}}}\right) = \frac{21}{4}\text{ lb} = 5\frac{1}{4}\text{ lb}$$

Thus, 84 oz = $5\frac{1}{4}$ lb. ■

EXAMPLE 6-1.6 Find the number of ounces in 2 tons.

Solution: We multiply 2 by one unity ratio to convert tons to pounds and by a second unity ratio to convert pounds to ounces. (1 T = 2000 lb; 1 lb = 16 oz.)

$$\frac{2\text{ T}}{1}\left(\frac{\text{lb}}{\text{T}}\right)\left(\frac{\text{oz}}{\text{lb}}\right)$$

$$\frac{2\,\cancel{\text{T}}}{1}\left(\frac{2000\,\cancel{\text{lb}}}{1\,\cancel{\text{T}}}\right)\left(\frac{16\text{ oz}}{1\,\cancel{\text{lb}}}\right) = 2(2000)(16)\text{ oz} = 64{,}000\text{ oz}$$

Thus, 2 T = 64,000 oz. ■

EXAMPLE 6-1.7 How many pints are in $4\frac{1}{2}$ quarts?

Solution: First, we write $4\frac{1}{2}$ qt as an improper fraction. Then we multiply it by a unity ratio with quarts and pints, since these are the two units involved. (2 pt = 1 qt.) Note the original unit is in the *numerator* of the improper fraction, $\frac{9}{2}$.

$$4\frac{1}{2}\text{ quarts} = \frac{9}{2}\text{ quarts}$$

$$\frac{9\text{ qt}}{2}\left(\frac{\text{pt}}{\text{qt}}\right)$$

$$\frac{9\ \cancel{\text{qt}}}{\underset{1}{\cancel{2}}}\left(\frac{\overset{1}{\cancel{2}}\text{ pt}}{1\ \cancel{\text{qt}}}\right) = 9\text{ pints}$$

Thus, $4\frac{1}{2}$ qt = 9 pt. ■

EXAMPLE 6-1.8 Find the number of ounces in 2 gallons.

Solution: Here, four unity ratios are employed to convert gallons to quarts, quarts to pints, pints to cups, and cups to ounces. The values are obtained from Table 6-3.

$$\frac{2\text{ gal}}{1}\left(\frac{\text{qt}}{\text{gal}}\right)\left(\frac{\text{pt}}{\text{qt}}\right)\left(\frac{\text{c}}{\text{pt}}\right)\left(\frac{\text{oz}}{\text{c}}\right)$$

$$\frac{2\ \cancel{\text{gal}}}{1}\left(\frac{4\ \cancel{\text{qt}}}{1\ \cancel{\text{gal}}}\right)\left(\frac{2\ \cancel{\text{pt}}}{1\ \cancel{\text{qt}}}\right)\frac{2\ \cancel{\text{c}}}{1\ \cancel{\text{pt}}}\left(\frac{8\text{ oz}}{1\ \cancel{\text{c}}}\right) = 2(4)(2)(2)(8)\text{ oz} = 256\text{ oz}$$

Thus, 2 gal = 256 oz.

We could work this same problem with fewer unity ratios if we made some preliminary calculations with the conversion factors from Table 6-3. For example, since 4 c = 1 qt and 8 oz = 1 c, we could multiply 4 (cups) by 8 (ounces per cup) to find the number of ounces in a quart. That is, $4 \times 8 = 32$ oz. Thus, 32 oz = 1 qt.

We can now find the number of ounces in 2 gallons with fewer unity ratios.

$$\frac{2\text{ gal}}{1}\left(\frac{4\text{ qt}}{1\text{ gal}}\right)\left(\frac{32\text{ oz}}{1\text{ qt}}\right)$$

$$\frac{2\ \cancel{\text{gal}}}{1}\left(\frac{4\ \cancel{\text{qt}}}{1\ \cancel{\text{gal}}}\right)\left(\frac{32\text{ oz}}{1\ \cancel{\text{qt}}}\right) = 256\text{ oz}$$

Thus, 2 gal = 256 oz. ■

Tips and Traps: Two Shortcuts

1. To change a U.S. customary unit to a desired *smaller* unit: Multiply the number of larger units by the number of desired smaller units that equals 1 larger unit.

2 yd = ______ ft *(The desired smaller unit is feet. 3 ft = 1 yd.)*

2×3 ft = 6 ft *(Multiply number of yards by 3 ft.)*

(Dimension analysis: $\frac{2\ \cancel{yd}}{1} \times \frac{3\ ft}{1\ \cancel{yd}} = 6\ ft$*)*

Thus, 2 yd = 6 ft.

2. To change a U.S. customary unit to a desired *larger* unit: Divide the original measure by the number of smaller units that equals 1 desired larger unit.

12 ft = ______ yd *(The desired larger unit is yards. 1 yd = 3 ft.)*

$$\frac{12 \text{ ft}}{3 \text{ ft}} = \frac{\overset{4}{\cancel{12}}\cancel{\text{ft}}}{\underset{1}{\cancel{3}}\cancel{\text{ft}}} = 4 \text{ yd}$$

(Divide original measure by measure having number of smaller units equal to 1 larger unit.)

(Dimension analysis: $\frac{12\cancel{ft}}{1} \times \frac{1\ yd}{3\cancel{ft}} = 4\ yd$*)*

Thus, 12 ft = 4 yd.

L.O.3. Write mixed U.S. customary measures in standard notation.

Standard Notation

Some measures are expressed by using two or more units of measure. For example, the weight of an object may be recorded as 5 lb 3 oz.

■ **Definition 6-1.2 Mixed Measures.** Measures expressed by using two or more units of measure are called **mixed measures**.

■ **Definition 6-1.3 Standard Notation. Standard notation** means that in a mixed measure each unit of measure is converted to the next larger unit of the mixed measure whenever possible.

That is, if an answer is 3 ft 15 in., the 15 in. is converted to 1 ft 3 in. and the 1 ft is added to the 3 ft.

$$15 \text{ in.} = \underbrace{12 \text{ in.}}_{1 \text{ ft}} + 3 \text{ in.}$$

Thus, in standard notation, 3 ft 15 in. is 4 ft 3 in.

Example 6-1.9 Express 9 ft 17 in. in standard notation.

Solution: Since 17 in. is more than 1 ft, we convert the mixed measure to *standard notation* by writing the 17 in. as 1 ft 5 in. and then combining the 1 ft with 9 ft to get the mixed measure of 10 ft 5 in.

Thus, 9 ft 17 in. = 10 ft 5 in. in standard notation. ■

Example 6-1.10 Express 8 lb 20 oz in standard notation.

Solution

8 lb 20 oz *(20 oz = 1 lb 4 oz)*

Thus, 8 lb 20 oz = 9 lb 4 oz in standard notation. ■

Example 6-1.11 Express 1 gal 5 qt in standard notation.

Solution

1 gal 5 qt *(5 qt = 1 gal 1 qt)*

Thus, 1 gal 5 qt = 2 gal 1 qt. ■

If the mixed measure contains three or more different measures, then conversion to standard notation may require two steps.

Example 6-1.12 Express 2 yd 4 ft 16 in. in standard notation.

Solution

2 yd 4 ft 16 in.

2 yd 5 ft 4 in. (16 in. = 1 ft 4 in.)

3 yd 2 ft 4 in. (5 ft = 1 yd 2 ft)

Thus, 2 yd 4 ft 16 in. = 3 yd 2 ft 4 in. ■

SELF-STUDY EXERCISES 6-1

L.O.1. Identify the appropriate U.S. customary measure for each of the following.

1. Shipping weight of a sofa
2. Liquid medicine in a bottle
3. Package of dried beans
4. Large home aquarium
5. Container of motor oil
6. Shipment of coal
7. Sack of potatoes
8. Size of a casserole dish
9. Height of a college professor
10. Milk for a cake recipe
11. Man's belt size
12. Cloth for a Mardi Gras costume
13. Hourly speed of a train
14. Weight of a first-class letter
15. Shaving lotion
16. Distance from home to school across town
17. Package of macaroni
18. Pork roast
19. Parcel-post package
20. Tube of ointment

L.O.2. Using unity ratios, convert the given measures to the new units of measure.

21. 4 ft = ______ in.

22. 7 yd = ______ ft

23. $2\frac{1}{2}$ mi = ______ yd

24. Find the number of yards in 28 ft.

25. If a car with 0 miles on the odometer was driven 10,560 ft, how many miles would register on the odometer?

Using any method, convert the given measures to the new units of measure.

26. How many ounces are in 3 lb?

27. Find the number of pounds in $36\frac{4}{5}$ oz.

28. How many ounces are in 45.8 lb?

29. A can of vegetables weighs 1.2 lb. How many ounces is this?

30. The net weight of a can of corn is 17 oz. If a case contains 24 cans, what is the net weight of a case in ounces? in pounds?

31. How many quarts are in 5 gal?

32. How many pints are in $6\frac{1}{2}$ qt?

33. Find the number of gallons in 6 qt.

34. How many pints are in $7\frac{1}{2}$ gal?

35. How many gallons are in 72 pt?

36. How many ounces are in 2 gal?

37. How many yards are in 54 in.?

38. How many inches are in 3 yd?

39. How many feet are in $1\frac{1}{3}$ mi?

40. How many cups are in $1\frac{1}{2}$ gal?

L.O.3. Express the following measures in standard notation.

41. 2 ft 20 in.

42. 1 mi 6375 ft

43. 2 lb $19\frac{1}{2}$ oz

44. 1 gal 5 qt

45. 1 gal 3 qt 2 pt
46. 1 T 2500 lb
47. 2 yd 1 ft 23 in.
48. 1 qt 5 c 10 oz
49. 2 ft 10 in.
50. 5 lb 25 oz
51. 3 gal 5 qt 48 oz
52. 6 qt 20 oz
53. 1 mi 265 yd 4500 ft 25 in.
54. 2 ft 40 in.
55. 2 T 2600 lb 15 oz
56. 3 lb 21 oz
57. 2 lb 5 oz
58. 1 yd 4 ft 16 in.
59. 2 yd 2 ft 11 in.
60. 1 mi 875 ft 6 in.

6-2 ADDING AND SUBTRACTING U.S. CUSTOMARY MEASURES

Learning Objectives

L.O.1. Add U.S. customary measures.
L.O.2. Subtract U.S. customary measures.

Once we can convert from one U.S. customary unit to another, we can perform addition, subtraction, multiplication, and division in the U.S. customary system.

L.O.1. Add U.S. customary measures.

Adding Measures

We can add U.S. customary measures *only* when their units are the same. Before we can add 3 ft and 2 in., we need to change one of the measures so that they are both expressed as inches or as feet.

Rule 6-2.1 ***To add unlike U.S. customary measures:***

1. Convert to a common U.S. customary unit.
2. Add.

EXAMPLE 6-2.1 Add 3 ft + 2 in.

Solution: Since 2 in. is a fraction of a foot, we can avoid working with fractions by converting the measures with the larger unit of measure (feet) to the smaller unit of measure (inches).

$$\frac{3\,\cancel{\text{ft}}}{1}\left(\frac{12\text{ in.}}{1\,\cancel{\text{ft}}}\right) = 36\text{ in.}$$

3 ft = 36 in., so we have

$$36\text{ in.} + 2\text{ in.} = 38\text{ in.}$$

■

Rule 6-2.2 ***To add mixed U.S. customary measures:***

1. Align the measures vertically so that the common units are written in the same vertical column.
2. Add.
3. Express the sum in standard notation.

EXAMPLE 6-2.2 Add 7 ft 9 in. + 3 ft 10 in.

Solution

$$\begin{array}{r} 7 \text{ ft } \ \ 9 \text{ in.} \\ + \ 2 \text{ ft } 10 \text{ in.} \\ \hline 9 \text{ ft } 19 \text{ in.} \end{array} \qquad (19 \text{ in.} = 1 \text{ ft } 7 \text{ in.})$$

Thus, 9 ft 19 in. = 10 ft 7 in. ■

EXAMPLE 6-2.3 Add and write the answer in standard form.

$$6 \text{ lb } 7 \text{ oz} + 3 \text{ lb } 13 \text{ oz}$$

Solution

$$\begin{array}{r} 6 \text{ lb } \ \ 7 \text{ oz} \\ + \ 3 \text{ lb } 13 \text{ oz} \\ \hline 9 \text{ lb } 20 \text{ oz} \end{array} \qquad (20 \text{ oz} = 1 \text{ lb } 4 \text{ oz})$$

Thus, 9 lb 20 oz = 10 lb 4 oz. ■

L.O.2. Subtract U.S. customary measures.

Subtracting Measures

To add measures, we must express different units of measure in a common unit. To subtract measures, we must also express different units in a common unit of measure.

Rule 6-2.3 ***To subtract unlike U.S. customary units:***

1. Convert to a common U.S. customary unit.
2. Subtract.

EXAMPLE 6-2.4 Subtract 15 in. from 2 ft.

Solution: Since changing 15 in. to feet will give us a mixed number, it is easier if we convert 2 ft to inches.

$$2 \text{ ft} = 24 \text{ in.}$$

$$24 \text{ in.} - 15 \text{ in.} = 9 \text{ in.}$$ ■

Rule 6-2.4 ***To subtract mixed U.S. customary measures:***

1. Align the measures vertically so that the common units are written in the same vertical column.
2. Subtract.

EXAMPLE 6-2.5 Subtract 5 ft 3 in. from 7 ft 4 in.

Solution: Align the measures that are alike in a vertical line and then subtract.

$$\begin{array}{r} 7 \text{ ft } 4 \text{ in.} \\ - \ 5 \text{ ft } 3 \text{ in.} \\ \hline 2 \text{ ft } 1 \text{ in.} \end{array}$$ ■

Sometimes, when subtracting mixed measures, we must subtract a larger unit from a smaller unit. In this case we must apply to the mixed measures our knowledge of borrowing.

> **Rule 6-2.5** ***To subtract a larger U.S. customary unit from a smaller unit in mixed measures:***
>
> 1. Align the common measures in vertical columns.
> 2. Borrow one unit from the next larger unit of measure in the minuend, convert to the equivalent smaller unit of measure, and add it to the smaller unit of measure.
> 3. Subtract the measures.

EXAMPLE 6-2.6 Subtract 3 lb 12 oz from 7 lb 8 oz.

Solution

$$\begin{array}{r} 7\text{ lb } \ 8\text{ oz} \\ -\ 3\text{ lb } 12\text{ oz} \\ \hline \end{array}$$

We always begin subtracting with the smallest unit of measure, which should be on the *right*. In this example, we notice that 12 oz cannot be subtraced from 8 oz. To perform this subtraction, we rewrite 7 lb as 6 lb 16 oz. Then we combine the 16 oz with 8 oz to increase the 8 oz to 24 oz.

$$\begin{array}{lr} 7\text{ lb } \ 8\text{ oz} = 6\text{ lb } 16\text{ oz} + 8\text{ oz} = & 6\text{ lb } 24\text{ oz} \\ 3\text{ lb } 12\text{ oz} = & -\ 3\text{ lb } 12\text{ oz} \\ \hline & 3\text{ lb } 12\text{ oz} \end{array}$$

■

SELF-STUDY EXERCISES 6-2

L.O.1. Add and write the answer in standard notation.

1. 5 oz + 2 lb

2. 4 ft + 7 in.

3. $\begin{array}{r} 8\text{ lb } 2\text{ oz} \\ +\ 7\text{ lb } 9\text{ oz} \\ \hline \end{array}$

4. $\begin{array}{r} 5\text{ ft } 45\text{ in.} \\ +\ 7\text{ ft } 30\text{ in.} \\ \hline \end{array}$

5. $\begin{array}{r} 5\text{ qt } 1\ \text{ pt} \\ +\ 2\text{ qt } 1\tfrac{1}{2}\text{ pt} \\ \hline \end{array}$

6. $\begin{array}{r} 8\text{ gal } 3\text{ qt} \\ +\ 5\text{ gal } 2\text{ qt} \\ \hline \end{array}$

7. $\begin{array}{r} 7\text{ yd } 2\text{ ft} \\ +\ 1\text{ yd } 2\text{ ft} \\ \hline \end{array}$

8. $\begin{array}{r} 3\text{ c } 6\text{ oz} \\ +\ 1\text{ c } 3\text{ oz} \\ \hline \end{array}$

9. $\begin{array}{r} 1\text{ ft } 13\text{ in.} \\ +\ 2\text{ ft } 15\text{ in.} \\ \hline \end{array}$

10. $\begin{array}{r} 4\text{ yd } 2\text{ ft } \ 7\text{ in.} \\ +\ 2\text{ yd } 1\text{ ft } 10\text{ in.} \\ \hline \end{array}$

Solve the following problems. When necessary, express the answers in standard notation.

11. A plumber has a 2-ft length of copper tubing and a 7-in. length of copper tubing. What is the total?

12. How long are two extension cords together if one is 6 ft and the other is 60 in.?

13. A mixture for hamburgers contains 2 lb 8 oz of ground round steak and 3 lb 7 oz of regular ground beef. How much does the hamburger mixture weigh?

14. One infant twin weighed 6 lb 1 oz and the other weighed 5 lb 15 oz at birth. What was their total weight?

L.O.2. Subtract and write the answer in standard form.

15. 2 ft − 18 in.

16. 2 qt − 3 pt

17. 3 yd − 7 ft

18. 6 lb − 18 oz

19. $\begin{array}{r} 3 \text{ lb } 12 \text{ oz} \\ - \ 2 \text{ lb } \ \ 6 \text{ oz} \\ \hline \end{array}$

20. $\begin{array}{r} 12 \text{ lb } \ \ 7 \text{ oz} \\ - \ \ 5 \text{ lb } 12 \text{ oz} \\ \hline \end{array}$

21. $\begin{array}{r} 3 \text{ ft } 8 \text{ in.} \\ - \ 2 \text{ ft } 9 \text{ in.} \\ \hline \end{array}$

22. $\begin{array}{r} 19 \text{ lb } 12 \text{ oz} \\ - \ \ 8 \text{ lb } 15 \text{ oz} \\ \hline \end{array}$

23. $\begin{array}{r} 2 \text{ ft } 30 \text{ in.} \\ - \ 1 \text{ ft } 40 \text{ in.} \\ \hline \end{array}$

24. $\begin{array}{r} 5 \text{ gal } 3 \text{ qt } \ 1 \text{ pt} \\ - \ 1 \text{ gal } 3 \text{ qt } 1\frac{1}{2} \text{ pt} \\ \hline \end{array}$

25. A seamstress had a length of cloth 5 ft long. What was its length after cutting off 10 in.?

26. A cabinet maker cut 9 in. from a 3-ft shelf. How long was the shelf after it was cut?

27. If a computer with an 80386 processor can sort a list of names in 1 min 30 sec and a computer with an 80486 processor can do the same job in 45 sec, how much time is saved by using the 80486 computer?

28. A car with a 2.0-L engine can accelerate a certain distance in 1 min 27 sec. A car with a 5.0-L engine can accelerate the same distance in 48 sec. How much faster is the car with the 5.0-L engine?

29. A package containing a laser printer weighed 74 lb 3 oz. The container and packing material weighed 4 lb 12 oz. How much did the laser printer weigh?

30. To weigh a pet poodle, Stacey held the dog while standing on a scale. If the scale showed 132 lb 6 oz and Stacey weighs 115 lb 8 oz, how much does the poodle weigh?

6-3 MULTIPLYING AND DIVIDING U.S. CUSTOMARY MEASURES

Learning Objectives

L.O.1. Multiply a U.S. customary measure by a number.
L.O.2. Multiply a U.S. customary measure by a measure.
L.O.3. Divide a U.S. customary measure by a number.
L.O.4. Divide a U.S. customary measure by a measure.
L.O.5. Change from one U.S. customary rate measure to another.

L.O.1. Multiply a U.S. customary measure by a number.

Multiplying Measures by a Number

In addition to adding and subtracting U.S. customary measures, we often need to multiply or divide those measures by some number.

Problems often arise in various fields that require us to multiply a measure by a given number. For example, a carpenter may buy cypress siding in lengths of 18 ft 6 in. To find the total length of six pieces of the siding, we multiply 18 ft 6 in. by 6. We do this by multiplying 6 by the numbers associated with each unit of measure.

Rule 6-3.1 ***To multiply a U.S. customary measure by a given number:***

1. Multiply the numbers associated with each unit of measure by the given number.
2. Write the resulting measure in standard notation.

EXAMPLE 6-3.1 A container has a capacity of 21 gal 3 qt. What is the total capacity of eight such containers?

Solution

$$\begin{array}{r} 21 \text{ gal } 3 \text{ qt} \\ \times \quad 8 \\ \hline 168 \text{ gal } 24 \text{ qt} \end{array} \quad (24 \text{ qt} = 6 \text{ gal})$$

$$168 \text{ gal } 24 \text{ qt} = 168 \text{ gal} + 6 \text{ gal} = 174 \text{ gal in standard notation}$$

The eight containers have a combined capacity of 174 gal. ■

L.O.2. Multiply a U.S. customary measure by a measure.

Multiplying Measures by a Measure

Length measures can be multiplied by like length measures to produce square measures. Square measures indicate areas.

in. × in. = in.^{2} ft × ft = ft^2 yd × yd = yd^2

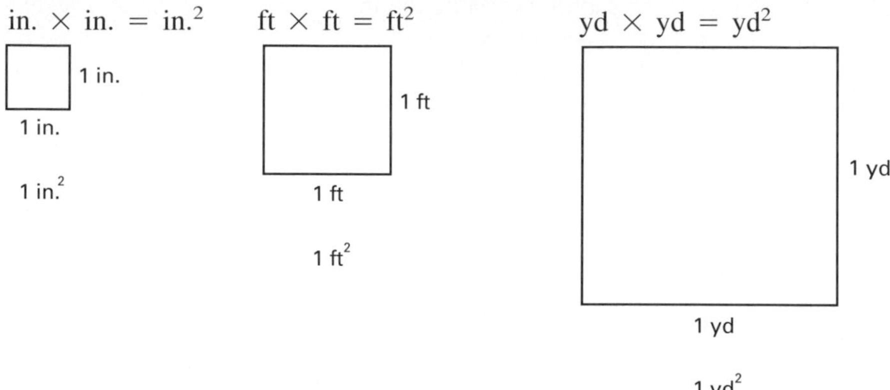

> **Rule 6-3.2** ***To multiply a length measure by a like length measure:***
>
> **1.** Multiply the numbers associated with each like unit of measure.
> **2.** The product will be a square unit of measure.

EXAMPLE 6-3.2 A desk top is 2 ft × 3 ft. What is the number of square feet in the surface?

Solution

$$2 \text{ ft} \times 3 \text{ ft} = 6 \text{ ft}^2$$ ■

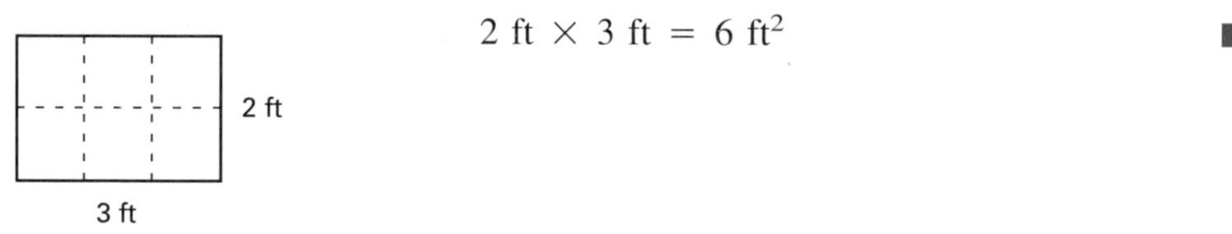

L.O.3. Divide a U.S. customary measure by a number.

Dividing Measures by a Number

We are frequently required to divide measures by a number. For example, if a recipe calls for 2 gal 2 qt of a liquid ingredient and the recipe is halved (divided by 2), what amount of this ingredient should be used? To solve this problem, we divide the measure by 2.

> **Rule 6-3.3** ***To divide a U.S. customary measure by a given number that divides evenly into each measure:***
>
> **1.** Divide the numbers associated with each unit of measure by the given number.
> **2.** Write the resulting measure in standard notation.

EXAMPLE 6-3.3 How much milk is needed for a half-recipe if the original recipe calls for 2 gal 2 qt?

Solution

$$2\overline{)2 \text{ gal } 2 \text{ qt}}$$ *(We divide each measure by 2.)*

$$\begin{array}{r} 1 \text{ gal } 1 \text{ qt} \\ 2\overline{)2 \text{ gal } 2 \text{ qt}} \end{array}$$

The half-recipe requires 1 gal 1 qt of the ingredient. ■

If a given number does not divide evenly into each measure, there will be a remainder. For instance, if we divided 5 gal by 2, the 2 would divide into the 5 gal 2 times with 1 gal left over as a remainder.

Rule 6-3.4 ***To divide U.S. customary measures by a given number that does not divide evenly into each measure:***

1. Set up the problem and proceed as in long division.
2. When a remainder occurs after subtraction, convert the remainder to the same unit used in the next smaller measure and add it to the next smaller measure.
3. Then continue by dividing the given number into this next smaller measure.
4. If a remainder occurs when the smallest unit is divided, express the remainder as a fractional part of the smallest unit.

EXAMPLE 6-3.4 Divide 5 gal 2 qt by 2.

Solution

$$\begin{array}{rlr} & 2 \text{ gal} & 3 \text{ qt} \\ 2 & \overline{)5 \text{ gal}} & \overline{2 \text{ qt}} \\ & \underline{4 \text{ gal}} & \\ & 1 \text{ gal} = & \underline{4 \text{ qt}} \\ & & 6 \text{ qt} \\ & & \underline{6 \text{ qt}} \end{array}$$

(4 qt + 2 qt = 6 qt
6 qt ÷ 2 = 3 qt)

■

EXAMPLE 6-3.5 Divide 5 gal 3 qt 1 pt by 3.

Solution

$$\begin{array}{rlrr} & 1 \text{ gal} & 3 \text{ qt} & 1\frac{2}{3} \text{ pt} \\ 3 & \overline{)5 \text{ gal}} & \overline{3 \text{ qt}} & \overline{1 \quad \text{pt}} \\ & \underline{3 \text{ gal}} & & \\ & 2 \text{ gal} = & \underline{8 \text{ qt}} & \\ & & 11 \text{ qt} & \\ & & \underline{9 \text{ qt}} & \\ & & 2 \text{ qt} = & \underline{4 \quad \text{pt}} \\ & & & 5 \quad \text{pt} \\ & & & \underline{3 \quad \text{pt}} \\ & & & 2 \end{array}$$

(3 qt + 8 qt = 11 qt
11 qt ÷ 3 = 3 qt, remainder 2 qt)

(1 pt + 4 pt = 5 pt
5 pt ÷ 3 = 1 pt, remainder 2)

(Write the remainder, 2, as a fraction, $\frac{2}{3}$, and add to 1 pint to get $1\frac{2}{3}$ pints.)

Thus, 1 gal 3 qt $1\frac{2}{3}$ pt is the solution. ■

EXAMPLE 6-3.6 Divide 38 yd by 6.

Solution: This problem can be worked two ways, depending on what units are desired. If yards is the only unit desired, we do this:

$$\begin{array}{r} 6\frac{2}{6}\text{ yd} \quad\text{or}\quad 6\frac{1}{3}\text{ yd} \\ 6\overline{)38\text{ yd}} \\ \underline{36\text{ yd}} \\ 2 \end{array}$$

If the result needs to be expressed in terms of feet and inches as well as yards, we do this:

$$\begin{array}{rl} 6\text{ yd} & \quad 1\text{ ft} \\ 6\overline{)38\text{ yd}} & \\ \underline{36\text{ yd}} & \\ 2\text{ yd} & = 6\text{ ft} \\ & \quad \underline{6\text{ ft}} \end{array}$$

■

L.O.4. Divide a U.S. customary measure by a measure.

Dividing Measures by a Measure

In addition to dividing measures by a number, we are often called on to divide measures by a measure. For instance, if tubing is manufactured in lengths of 8 ft 4 in. and a part is 10 in. long, how many parts can be cut from the length of tubing if we do not account for waste? To solve problems such as this, we must express both measures in the same unit, just as we did when we added and subtracted measures. We generally convert to the *smallest* unit used in the problem.

Rule 6-3.5 ***To divide a U.S. customary measure by a U.S. customary measure:***

1. Convert both measures to the same unit if they are different.
2. Write the division as a fraction, including the common unit in the numerator and the denominator.
3. Reduce the units and divide the numbers.

EXAMPLE 6-3.7 Divide 8 ft 4 in. by 10 in.

Solution: Converting 8 ft 4 in. to inches, we have

$$\frac{8\,\cancel{\text{ft}}}{1}\left(\frac{12\text{ in.}}{1\,\cancel{\text{ft}}}\right) = 96\text{ in.}$$

$$96\text{ in.} + 4\text{ in.} = 100\text{ in.}$$

Dividing, we have

$$100\text{ in.} \div 10\text{ in.}$$

If we write this division in fraction form, we can more easily see that the common units will reduce. In other words, the answer will be a number (not a measure) telling *how many times* one measure divides into another.

$$\frac{100 \cancel{\text{in.}}}{10 \cancel{\text{in.}}} = 10$$

Therefore, we see that 10 parts of equal length can be cut from the tubing. ■

L.O.5. Change from one U.S. customary rate measure to another.

Converting Rate Measures to Other Rate Measures

A rate measure is a ratio of two different kinds of measures. It is often referred to simply as a *rate*. Some examples of rates are 55 miles per hour, 20 cents per mile, and 3 gallons per minute. In each of these rates the word *per* means *divided by*.

The rate 55 miles per hour means 55 miles $\div$ 1 hour or $\frac{55 \text{ mi}}{1 \text{ hr}}$. The rate 20 cents per mile means 20 cents $\div$ 1 mile or $\frac{20 \text{ cents}}{1 \text{ mi}}$. The rate 3 gallons per minute means 3 gallons $\div$ 1 minute or $\frac{3 \text{ gal}}{1 \text{ min}}$.

Notice that every rate measure requires two units of measure, one in the numerator and one in the denominator. Sometimes it is necessary to express the numerator and/or denominator of a rate measure in other units of measure. We can do this by using unity ratios. Measures of time are universally accepted. All major countries use the same units of measure for time. The basic units of time are the year, month, week, day, hour, minute, and second. Table 6-4 gives the relationships among the units of measure for time. You will need these measures when working with rates.

Rule 6-3.6 ***To convert one U.S. customary rate measure to another U.S. customary rate measure:***

1. Set up the original rate measure equal to the new rate measure without its numerical values.
2. Compare the units of both numerators and both denominators to determine which original units will change.
3. Multiply each original measure that changes by a unity ratio containing the new unit so that the unit to be changed will reduce.

TABLE 6-4 Time Measures

1 year (yr)	=	12 months (mo)
1 week (wk)	=	7 days (da)
1 day (da)	=	24 hours (hr)
1 hour (hr)	=	60 minutes (min)[a]
1 minute (min)	=	60 seconds (sec)[b]

[a]The symbol ′ means feet (3′ = 3 ft) or minutes (60′ = 60 minutes).

[b]The symbol ″ means inches (8″ = 8 in.) or seconds (60″ = 60 seconds).

EXAMPLE 6-3.8 $\frac{40 \text{ mi}}{1 \text{ hr}} = \underline{\quad} \frac{\text{ft}}{\text{hr}}$?

Solution: First, determine which unit of measure in the rate is to be changed by comparing the first numerator with the second one and the first denominator with the second one. The miles are to be changed to feet here. Next, write a unity ratio containing feet and miles so that miles will reduce. Remember, a unity ratio is equivalent to 1, and when we multiply a quantity by 1, we do not change the value of the quantity.

$$\frac{40 \text{ mi}}{1 \text{ hr}}\left(\frac{\underline{\quad} \text{ ft}}{\underline{\quad} \text{ mi}}\right)$$ *[Set up the unity ratio to reduce miles (mi); then fill in the appropriate numbers.]*

$$\frac{40 \cancel{\text{mi}}}{1 \text{ hr}}\left(\frac{5280 \text{ ft}}{1 \cancel{\text{mi}}}\right) = \frac{40}{1 \text{ hr}}\left(\frac{5280 \text{ ft}}{1}\right) = 211{,}200 \frac{\text{ft}}{\text{hr}}$$ *(Reduce units of measure.)* ■

EXAMPLE 6-3.9 Convert $\frac{960 \text{ ft}}{\text{hr}}$ to $\frac{\text{ft}}{\text{min}}$.

Solution: In this example, the part of the rate measure that needs to be changed is hours in the denominator. Therefore, we set up our unity ratio with hours in the numerator so that this unit will reduce, and minutes is placed in the denominator to replace hours.

$$\frac{960 \text{ ft}}{\text{hr}}\left(\frac{\underline{\quad} \text{ hr}}{\underline{\quad} \text{ min}}\right)$$ *[Set up the unity ratio to reduce hours (hr); then fill in appropriate numbers.]*

$$\frac{\overset{16}{\cancel{960}} \text{ ft}}{1 \cancel{\text{hr}}}\left(\frac{1 \cancel{\text{hr}}}{\underset{1}{\cancel{60}} \text{ min}}\right) = 16 \text{ ft}\left(\frac{1}{\text{min}}\right) = 16 \frac{\text{ft}}{\text{min}}$$ *(Reduce units of measure and numbers.)*

Check the rate units in the answer to be sure that they are the units desired. ■

When both the numerator and denominator of a rate measure change, we need to multiply by at least two unity ratios to make the conversion. We first multiply by a unity ratio that will reduce the numerator of the original rate measure. Then we multiply by a unity ratio that will reduce the denominator of the original rate measure.

EXAMPLE 6-3.10 Change $\frac{3 \text{ pt}}{\text{sec}}$ to $\frac{\text{qt}}{\text{min}}$.

Solution: Since we are converting the numerator from pints to quarts, we use a unity ratio with quarts in the numerator and pints in the denominator to reduce pints in the original problem.

$$\frac{3 \text{ pt}}{\text{sec}}\left(\frac{1 \text{ qt}}{2 \text{ pt}}\right)$$

To complete the conversion, we multiply by the unity ratio that will change *seconds* in the denominator to *minutes*. This means that seconds must be reduced, so our unity ratio must have seconds in the *numerator*.

$$\frac{3 \text{ pt}}{\text{sec}}\left(\frac{1 \text{ qt}}{2 \text{ pt}}\right)\left(\frac{60 \text{ sec}}{1 \text{ min}}\right)$$

$$\frac{3 \cancel{\text{pt}}}{\cancel{\text{sec}}}\left(\frac{1 \text{ qt}}{\underset{1}{\cancel{2}} \cancel{\text{pt}}}\right)\left(\frac{\overset{30}{\cancel{60}} \cancel{\text{sec}}}{1 \text{ min}}\right) = 3(1 \text{ qt})\left(\frac{30}{\text{min}}\right) = 90 \frac{\text{qt}}{\text{min}}$$

Occasionally, a rate measure requires more than two unity ratios to convert it to another rate measure.

EXAMPLE 6-3.11 Convert $\frac{6 \text{ pt}}{\text{sec}}$ to $\frac{\text{gal}}{\text{min}}$.

Solution: Here we need two unity ratios to convert pints to quarts and then quarts to gallons. Then we need a third unity ratio to convert seconds to minutes.

$$\frac{6 \text{ pt}}{\text{sec}}\left(\frac{1 \text{ qt}}{2 \text{ pt}}\right)\left(\frac{1 \text{ gal}}{4 \text{ qt}}\right)\left(\frac{60 \text{ sec}}{1 \text{ min}}\right)$$

$$\frac{\overset{3}{\cancel{6}}\ \cancel{\text{pt}}}{\cancel{\text{sec}}}\left(\frac{1\ \cancel{\text{qt}}}{\underset{1}{\cancel{2}}\ \cancel{\text{pt}}}\right)\left(\frac{1 \text{ qal}}{\underset{1}{\cancel{4}}\ \cancel{\text{qt}}}\right)\left(\frac{\overset{15}{\cancel{60}}\ \cancel{\text{sec}}}{1 \text{ min}}\right) = 3(1)(1 \text{ gal})\left(\frac{15}{\text{min}}\right) = 45\frac{\text{gal}}{\text{min}}$$

■

SELF-STUDY EXERCISES 6-3

L.O.1. Multiply and write answers for mixed measures in standard notation.

1. 12 mi × 5

2. 18 gal × 6

3. 21 lb × 12

4. 3 qt 1 pt × 4

5. 7 lb 3 oz × 8

6. 5 gal 2 qt × 7

7. 8 gal 3 qt × 5

8. 7 ft 3 in. × 8

9. Tuna is packed in 1 lb 8 oz cans. If a case contains 24 cans, how much does a case weigh?

10. A car used 1 qt 1 pt of oil on each of 5 days. Find the total oil used.

L.O.2. Multiply.

11. 5 in. × 7 in.

12. 12 ft × 9 ft

13. 15 yd × 12 yd

14. 4 mi × 27 mi

15. A room is to be covered with square linoleum tiles that are 1 ft by 1 ft. If the room is 18 ft by 21 ft, how many tiles (square feet) are needed?

16. A customized van has a standard fuel tank and an auxiliary fuel tank each with a capacity of 23 gal 9 oz. How much gasoline is needed to fill both tanks?

17. Latonya has three containers, each containing 1 qt 3 pt of photographic solution. How much total photographic solution is in all three containers?

18. A package of nails weighs 1 lb 4 oz. How much would five packages weigh?

19. If a history textbook weighs 2 lb 8 oz, how much would three of the same books weigh?

L.O.3. Divide.

20. 12 gal ÷ 2

21. 3 days 6 hr ÷ 2

22. 20 yd 2 ft 6 in. ÷ 2

23. 4 yd 1 ft 9 in. ÷ 3

24. 4 gal 3 qt 1 pt ÷ 6

25. 60 gal ÷ 9 (express in gallons only)

26. 5 hr ÷ 3 (express in hours and minutes)

27. If eight trucks require 42 qt of oil for each to get a complete oil change, how many quarts are required for each truck?

28. Sixty feet of wire are required to complete eight jobs. If each job requires an equal amount of wire, find the amount of wire required for one job. (Express in feet and inches.)

29. A vat holding 10 gal 2 qt of chemical solution is emptied equally into three tanks. How many gallons and quarts are in each tank?

30. How many pieces of $\frac{1}{2}$-in. OD (outside diameter) plastic pipe 8 in. long can be cut from a piece 72 in. long?

31. A roll of No. 14 electrical cable 150 ft long is divided into 30 equal sections. How long is each section?

32. A brick mason has a container with 6 gal 2 qt 10 oz of muriatic acid that will be stored in two smaller containers for use in cleaning a brick wall. How much acid will be stored in each smaller container?

33. For a family picnic, Mr. Sonnier prepared 96 lb 12 oz of boiled crawfish. He brought the crawfish to the picnic site in four containers containing equal amounts. How much did the crawfish in each container weigh?

L.O.4. Divide.

34. 36 ft ÷ 12 ft

35. 51 in. ÷ 3 in.

36. 2 ft 8 in. ÷ 4 in.

37. 6 lb 12 oz ÷ 6 oz

38. 2 ft 6 in. ÷ 10 in.

39. How many 6-in. pieces can be cut from 48 in. of pipe?

40. How many 2-lb boxes can be filled from 18 lb of nails?

41. How many 15-oz. cans are in a case if the case weighs 22 lb 8 oz?

42. How many 8-dollar tickets can be purchased for 72 dollars?

43. How many pieces of wood 5 in. long can be cut from a piece 45 in. long?

L.O.5. Work the following problems.

44. $\frac{60 \text{ qt}}{\text{sec}} = \underline{\qquad} \frac{\text{gal}}{\text{sec}}$

45. $\frac{45 \text{ lb}}{\text{hr}} = \underline{\qquad} \frac{\text{lb}}{\text{min}}$

46. $\frac{3 \text{ mi}}{\text{hr}} = \underline{\qquad} \frac{\text{ft}}{\text{hr}}$

47. $\frac{144 \text{ lb}}{\text{min}} = \underline{\qquad} \frac{\text{oz}}{\text{min}}$

48. $\frac{30 \text{ gal}}{\text{min}} = \underline{\qquad} \frac{\text{qt}}{\text{sec}}$

49. $\frac{30 \text{ lb}}{\text{day}} = \underline{\qquad} \frac{\text{oz}}{\text{hr}}$

50. $\frac{8 \text{ in}}{\text{sec}} = \underline{\qquad} \frac{\text{ft}}{\text{min}}$

51. $\frac{5 \text{ mi}}{\text{min}} = \underline{\qquad} \frac{\text{ft}}{\text{sec}}$

52. A car traveling at the rate of 30 mph (miles per hour) is traveling how many feet per second?

53. A pump that can pump $45 \frac{\text{gal}}{\text{hr}}$ can pump how many quarts per minute?

54. A pump can dispose of sludge at the rate of $3200 \frac{\text{lb}}{\text{hr}}$. How many pounds can be disposed of per minute?

55. If water flows through a pipe at the rate of $50 \frac{\text{gal}}{\text{min}}$, how many gallons will flow per second?

6-4 INTRODUCTION TO THE METRIC SYSTEM

Learning Objectives

L.O.1. Identify uses of metric measures of length, weight, and capacity.
L.O.2. Convert from one metric unit of measure to another.
L.O.3. Make calculations with metric measures.

In the metric system there is a standard unit for each type of measurement. The *meter* is used for length or distance, the *gram* is used for weight or mass, and the *liter* is used for capacity or volume. There is also a series of prefixes that are added to the standard units to indicate measures greater than the standard units or less than the standard units.

Many consider the metric system to be easier to use than the U.S. customary system. One reason for this is that all measures greater than or less than the standard unit being used are in powers of 10. Changing from larger to smaller or from smaller to larger measures only requires multiplying or dividing by 10 or a power of 10. Another reason the metric system is easier to use than the U.S.

customary system is that the prefixes used with the standard unit represent the power of 10 by which the standard unit is divided or multiplied.

■ **DEFINITION 6-4.1 Metric System.** The **metric system** is an international system of measurement that uses standard units and prefixes to indicate their powers of 10.

L.O.1. Identify uses of metric measures of length, weight, and capacity.

Metric Measures and Their Uses

Before we start our study of metric measurements, look at some of the prefixes that are used in the metric system. Keep in mind that the prefix will have the same meaning no matter which unit (meter, gram, or liter) the prefix is attached to. The prefixes used in this chapter for *smaller* units than the standard unit are as follows:

deci- $\frac{1}{10}$ of

centi- $\frac{1}{100}$ of

milli- $\frac{1}{1000}$ of

The prefixes used in this chapter for *larger* units than the standard unit are as follows:

deka- 10 times (some books use *deca*)

hecto- 100 times

kilo- 1000 times

Other prefixes are used with very large and very small measures.

The prefixes can be related to our decimal system of numeration. Let's compare our decimal place-value chart with these prefixes (Fig. 6-12). The standard unit (whether meter, gram, or liter) corresponds to the *ones* place. All the places to the left are powers of the standard unit. That is, the value of *deka-* is 10 times the standard unit; the value of *hecto-* is 100 times this unit; the value of *kilo-* is 1000 times this unit; and so on. All the places to the right of the standard unit are subdivisions of the standard unit. That is, the value of *deci-* is $\frac{1}{10}$ of the standard unit; the value of *centi-* is $\frac{1}{100}$ of this unit; the value of *milli-* is $\frac{1}{1000}$ of this unit; and so on.

• Decimal point

Thousands (1000)	Hundreds (100)	Tens (10)	Units or ones (1)	Tenths ($\frac{1}{10}$)	Hundredths ($\frac{1}{100}$)	Thousandths ($\frac{1}{1000}$)
Kilo-	Hecto-	Deka-	STANDARD UNIT	Deci-	Centi-	Milli-

Figure 6-12

EXAMPLE 6-4.1 Give the value of the following metric units in terms of the prefix and the standard unit (gram, liter, or meter).

(a) Kilogram (kg) = 1000 times 1 gram or 1000 g

(b) Deciliter (dL) = $\frac{1}{10}$ of 1 liter or 0.1 L

(c) Hectometer (hm) = 100 times 1 meter or 100 m

(d) Dekaliter (dkL) = 10 times 1 liter or 10 L

(e) Milliliter (mL) = $\frac{1}{1000}$ of 1 liter or 0.001 L

(f) Centigram (cg) = $\frac{1}{100}$ of 1 gram or 0.01 g ■

Remember that all metric measurements of length, weight, and volume are expressed either as one of the standard units alone or with a suitable prefix.

Meter: The *meter* is the standard unit for measuring length. Both *meter* and *metre* are acceptable spellings for this unit of measure; however, we will use the spelling *meter* throughout this book. A meter is about 39.37 in. or 3.37 in. $\left(\text{approximately } 3\frac{1}{3} \text{ in.}\right)$ longer than a yard (Fig. 6-13). The meter is the appropriate unit to measure lengths and distances like room dimensions, land dimensions, lengths of poles, heights of mountains, and heights of buildings. The abbreviation for meter is *m*.

Not actual size

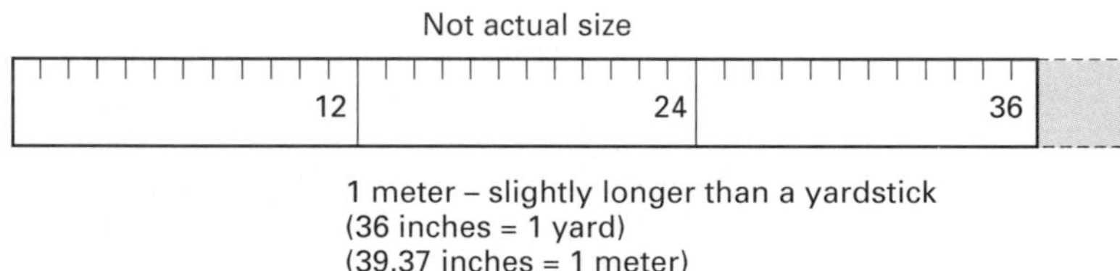

1 meter – slightly longer than a yardstick
(36 inches = 1 yard)
(39.37 inches = 1 meter)

Figure 6-13

Kilometer: To measure long distances, a larger measuring unit is needed. A *kilometer* is 1000 meters and is used for these longer distances (Fig. 6-14). The abbreviation for kilometer is *km*. The prefix *kilo* attached to the word *meter* means 1000. The distance from one city to another, one country to another, or one landmark to another would be measured in kilometers. Driving at a speed of 55 miles (or 90 km) per hour, we would travel 1 km in about 40 seconds (sec). An average walking speed is 1 km in about 10 minutes (min).

Not actual size

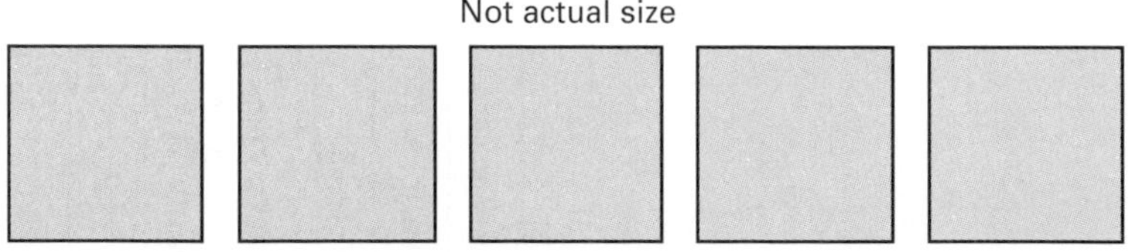

1 kilometer – about five city blocks

Figure 6-14

Centimeter: When measuring objects less than 1 m long, the most common unit used is the *centimeter (cm)*. Since the prefix *centi* means "$\frac{1}{100}$ of," a centimeter is one hundredth of a meter. A centimeter is about the width of a thumbtack head, somewhat less than $\frac{1}{2}$ in. (Fig. 6-15). Centimeters are used to measure medium-sized objects such as tires, clothing, textbooks, and television pictures.

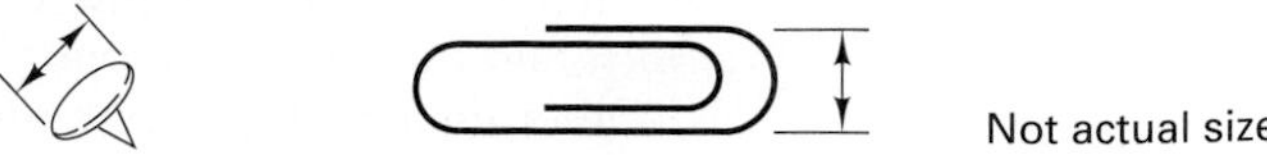

1 centimeter – about the width of a thumbtack or large paper clip

Figure 6-15

Millimeter: Many objects are too small to be measured in centimeters, so an even smaller unit of measure is needed. A *millimeter (mm)* is "$\frac{1}{1000}$ of" a meter. It is about the thickness of a plastic credit card or a dime coin (Fig. 6-16). Certain film sizes, bolt and nut sizes, the length of insects, and similar items are measured in millimeters.

Other units and their abbreviations are decimeter, dm; dekameter, dkm; and hectometer, hm.

Not actual size

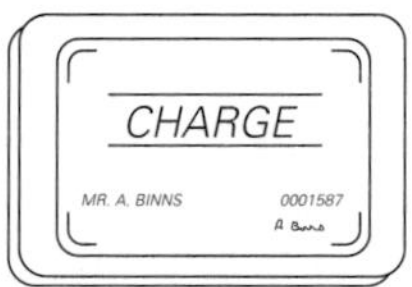

1 millimeter – about the thickness of a dime or a charge card

Figure 6-16

Gram: The standard unit for measuring weights in the metric system is the *gram* (Fig. 6-17). A gram is described as the weight of 1 cubic centimeter (cm^3) of water. A cubic centimeter is a cube that has each edge equal to 1 centimeter in length. It is a cube a little smaller than a sugar cube. The abbreviation *g* is used for gram. The gram is used for measuring small or light objects such as a paper clip, a cube of sugar, a coin, and a bar of soap.

1 gram – about the weight of two paper clips

Figure 6-17

Kilogram: A *kilogram (kg)* is 1000 grams (Fig. 6-18). Since a cube 10 cm on each edge can be divided into 1000 cm^3, the weight of the amount of water required to fill this cube would be 1000 grams or 1 kilogram. A kilogram is approximately 2.2 lb. The kilogram is probably the most often used unit for measuring weight. It is used for the weight of people, meat, sacks of flour, automobiles, and so on.

Not actual size

1 kilogram – about the weight of an average book with a one-inch spine

Figure 6-18

Milligram: Since the standard unit for measuring weight, the gram, is used only to measure small objects, the *milligram* (1/1000 of a gram) is used to measure *very* small objects. Milligrams are much too small for ordinary uses; however, pharmacists use milligrams (*mg*) when measuring small amounts of drugs, vitamins, and other medications (Fig. 6-19).

Other units and their abbreviations are decigram, dg; centigram, cg; dekagram, dkg; and hectogram, hg.

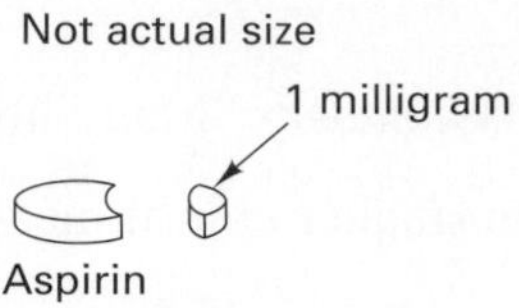

Figure 6-19

Measures of Capacity

Liter: A *liter (L)* is the volume of a cube 10 cm on each edge. It is the standard metric unit of capacity (Fig. 6-20). Like the meter, it may be spelled *liter* or *litre*, but we will use the spelling *liter*. Since a cube 10 cm on each edge filled with water weighs approximately 1 kg, then 1 L of water weighs about 1 kg. One liter is just a little larger than a liquid quart (a U.S. customary unit for measuring capacity). Soft drinks are often sold in 2-liter bottles, gasoline is sold by the liter at some service stations, and numerous other products are sold in liter containers.

1 liter – about the volume of a quart of milk or a soft drink in a plastic bottle

Figure 6-20

Milliliter: Since a liter is 1000 cm^3, $\frac{1}{1000}$ of a liter, or a *milliliter*, has the same capacity or volume as a cubic centimeter (Fig. 6-21). Most liquid medicine is labeled and sold in milliliters (*mL*) or cubic centimeters (cc or cm^3). Medicines, perfumes, and other very small quantities are measured in milliliters.

Other units and their abbreviations are deciliter, dL; centiliter, cL; dekaliter, dkL; hectoliter, hL; and kiloliter, kL.

There are other multiples and standard units in the metric system, but these are the ones frequently used.

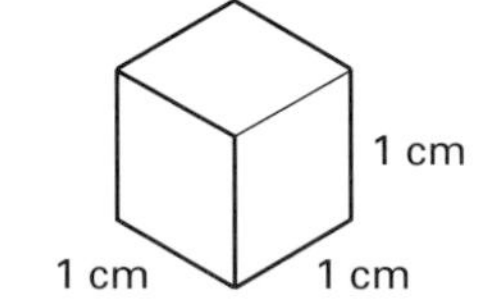

1 cubic centimeter = 1 milliliter

Figure 6-21

EXAMPLE 6-4.2 Choose the most reasonable metric measure for each of the following.

1. Distance from Jackson, Mississippi, to New Orleans
   (a) 322 m **(b)** 322 km **(c)** 322 cm **(d)** 322 mm
2. Weight of an adult woman
   (a) 56 g **(b)** 56 mg **(c)** 56 kg **(d)** 56 dkg
3. Width of color slide film
   (a) 35 cm **(b)** 35 m **(c)** 35 mm **(d)** 35 km
4. Bottle of eye drops
   (a) 30 dL **(b)** 30 dkL **(c)** 30 L **(d)** 30 mL
5. Weight of a pill
   (a) 352 mg **(b)** 352 dg **(c)** 352 g **(d)** 352 kg

Solution

1. **(b)** 322 km (about 200 mi)
2. **(c)** 56 kg (about 120 lb)
3. **(c)** 35 mm (roll film for automatic cameras)
4. **(d)** 30 mL (about 1 oz)
5. **(a)** 352 mg (one regular-strength aspirin) ■

L.O.2. Convert from one metric unit of measure to another.

Changing Metric Units

To understand the processes involved in changing one metric unit into another metric unit, let's arrange the units into a place-value chart like the one we used for decimals (Fig. 6-22). The units are arranged left to right from largest to smallest.

• Decimal point

Kilo-	Hecto-	Deka-	STANDARD UNIT	Deci-	Cent-	Milli-
1000 Standard units	100 Standard units	10 Standard units	Meter, gram, liter	$\frac{1}{10}$ of a Standard unit	$\frac{1}{100}$ of a Standard unit	$\frac{1}{1000}$ of a Standard unit

Figure 6-22

As we move from any place in the chart one place to the *right*, the metric unit changes to the next smaller unit. In effect, the larger unit is broken down into 10 smaller units. In other words, we are *multiplying* the larger unit by 10 when we move one place to the right.

For example, suppose that we wanted to change 2 m to decimeters. Since a decimeter is $\frac{1}{10}$ of a meter, there must be 10 dm in 1 meter. If it takes 10 dm to make 1 m, how many decimeters does it take to make 2 m? It takes twice as many as it does to make 1 m.

$$2 \times 10 = 20$$

Therefore, 2 m = 20 dm.

Instead of multiplying by 10, we can use the metric-value chart as directed in the following rule because, when changing to a smaller unit, we will always be moving to the right on the chart.

> **Rule 6-4.1** *To change from one metric unit to the* **next smaller** *unit:*
>
> **1.** Mentally position the measure on the metric-value chart so that the decimal point is immediately *after* the original measuring unit.
> **2.** To change to the next smaller unit, move the decimal one place to the right. (Affix a zero if necessary.) The decimal *follows* the new unit.

EXAMPLE 6-4.3 52 dkm = ____ m?

Solution: Align the number on the metric-value chart (Fig. 6-23). Since 52 is a whole number, the *understood* decimal is after the 52. Recall that a shortcut to multiplying by 10 is to move the decimal point *one* place to the right. Since 520 is a whole number, the decimal point is understood after the 520.

Thus, 52 dkm = 520 m. ■

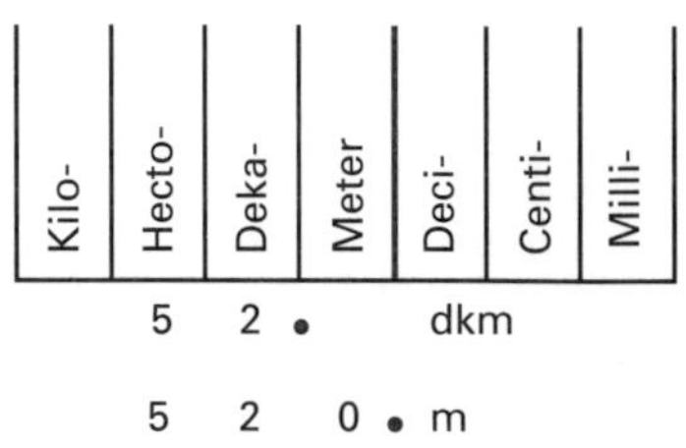

Figure 6-23

EXAMPLE 6-4.4 2.3 g = ____ dg?

Solution: Align the number on the metric-value chart (Fig. 6-24). The decimal is placed immediately *after* the measuring unit. Move the decimal one place to the right.

Thus, 2.3 g = 23 dg. ■

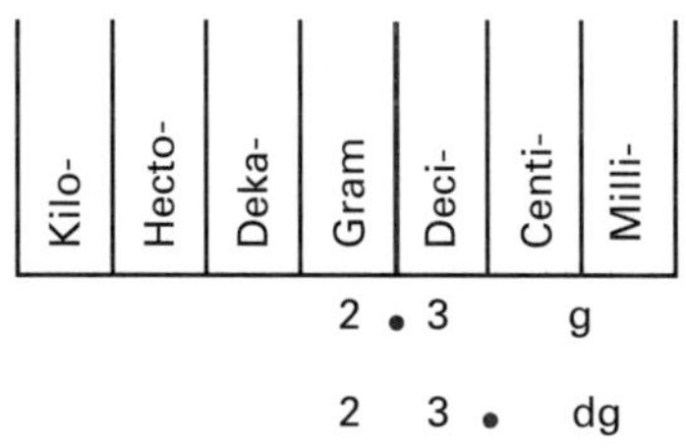

Figure 6-24

EXAMPLE 6-4.5 73 cL = ____ mL?

Solution: Align the number on the metric-value chart (Fig. 6-25). The decimal is placed immediately *after* the measuring unit. Don't forget to affix a zero if necessary after moving the decimal.

Thus, 73 cL = 730 mL. ■

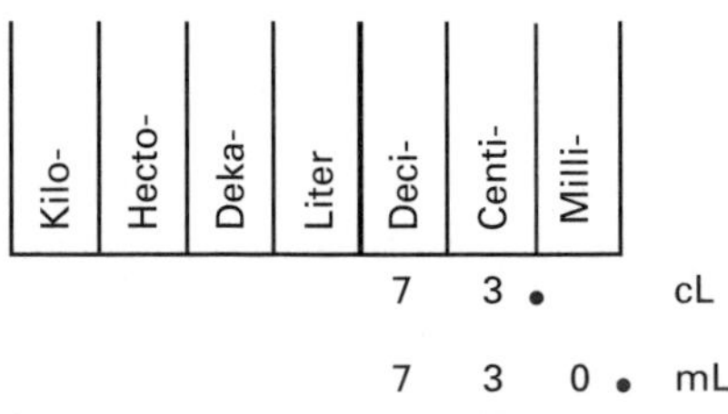

Figure 6-25

To expand on the last rule for changing to the next smaller metric unit, let's look at examples that do not change to the next smaller unit. Suppose that we needed to change 4 L to centiliters. The first step to the right would change 4 L to 40 dL. The next step to the right would change the 40 dL to 400 cL. Now let's extend our rule to cover this type of problem.

> **Rule 6-4.2** ***To change from one metric unit to* any smaller *metric unit:***
>
> 1. Mentally position the measure on the metric-value chart so that the decimal is immediately *after* the original measuring unit.
> 2. Move the decimal to the right so that it *follows* the new measuring unit. (Affix zeros if necessary.)

EXAMPLE 6-4.6 43 dkm = ____ cm?

Solution: Place 43 dkm on the chart so that the last digit is in the dekameter place. That is, the understood decimal that follows the 3 will be *after* the dekameter place (see Fig. 6-26). To change to centimeters, move the decimal point so that it falls *after* the centimeters place. Fill in the empty places with zeros (see Fig. 6-27). Recall that a shortcut to multiplication by powers of 10 is to move the decimal point one place to the right for each time 10 is used as a factor.

Thus, 43 dkm = 43,000 cm. ■

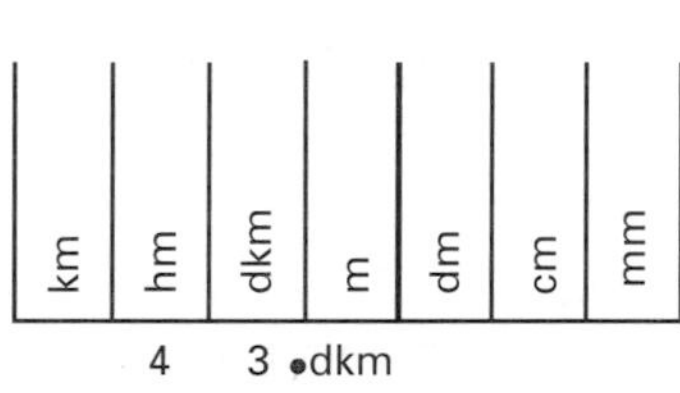

Figure 6-26

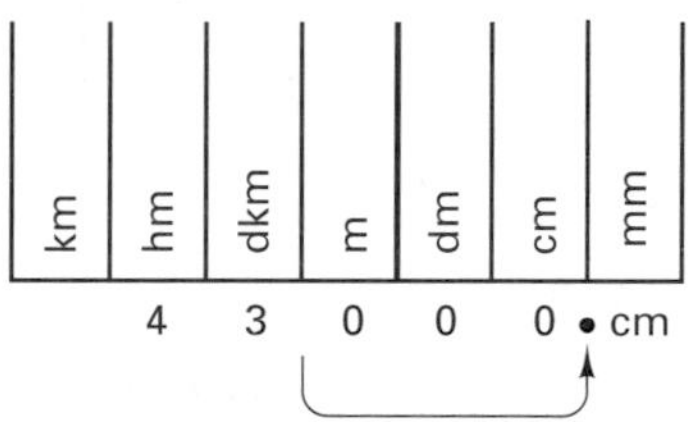

Figure 6-27

EXAMPLE 6-4.7 2.5 dg ____ mg? (Note the decimal location.)

Solution: Place the number 2.5 on the chart so that the decimal point falls *after* the decigrams place (see Fig. 6-28). (Note the decimal after the decigrams place.) To change to milligrams, shift the decimal two places to the right so that it falls *after* the milligrams place (see Fig. 6-29). (Note the decimal after the milligrams place.)

Thus, 2.5 dg = 250 mg. ■

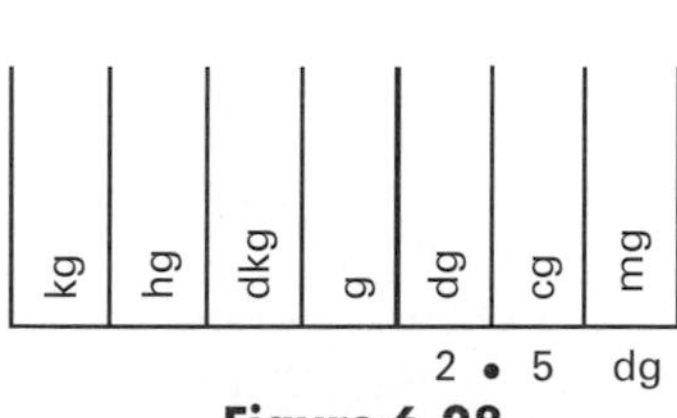

Figure 6-28

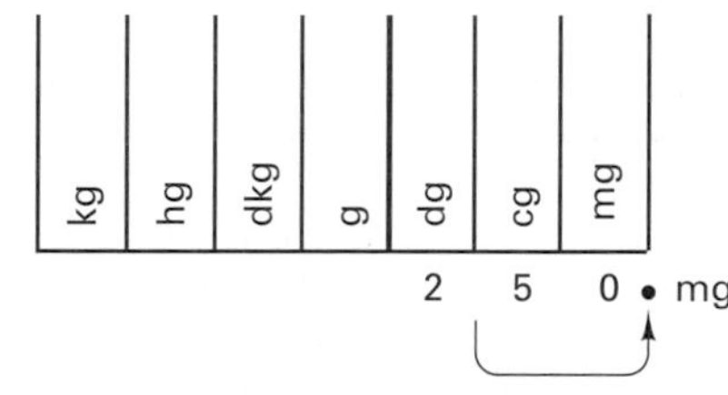

Figure 6-29

On the metric-value chart as we move one place to the *left*, the metric unit changes to the next larger unit. In effect, the smaller units are consolidated into one larger unit 10 times larger than each smaller unit. In other words, we are *dividing* the smaller unit by 10 when we move one place to the left.

For instance, suppose that we wanted to change 20 mm to centimeters. Since 1 millimeter is $\frac{1}{10}$ of a centimeter, there are 10 mm in 1 cm. If 1 cm equals 10 mm, how many centimeters are there in 20 mm? There are as many centimeters as there are 10's in 20 mm. ($20 \div 10 = 2$.) Therefore, 20 mm = 2 cm.

Because we are dividing by 10, we can use the metric-value chart as directed in the following rule since, when changing to a larger unit, we will always be moving to the left on the chart.

> **Rule 6-4.3** *To change from one metric unit to the* **next larger** *metric unit:*
>
> 1. Mentally position the measure on the metric-value chart so that the decimal point is immedatiely *after* the original measuring unit.
> 2. To change to the next larger unit, move the decimal one place to the left. (Affix a zero if necessary.) The decimal *follows* the new unit.

EXAMPLE 6-4.8 60 dkg = ____ hg?

Solution: Place the number on the chart so that the decimal is *after* the dekagrams place (Fig. 6-30). (Note the understood decimal after the dekagrams place.) Now move the decimal one place to the *left* so that it falls *after* the hectograms place (see Fig. 6-31).

Therefore, 60 dkg = 6 hg. ■

kg	hg	dkg	g	dg	cg	mg
	6	0 • dkg				

Figure 6-30

kg	hg	dkg	g	dg	cg	mg
	6 •	0	hg			

Figure 6-31

We can also expand this rule to cover converting a metric unit of measure to some unit other than the next larger unit.

> **Rule 6-4.4** *To change from one metric unit to* **any larger** *metric unit:*
>
> 1. Mentally position the measure on the metric-value chart so that the decimal is immediately *after* the original measuring unit.
> 2. Move the decimal to the left so that it *follows* the new measuring unit. (Affix zeros if necessary.)

EXAMPLE 6-4.9 3495 L = ____ kL?

Solution: The number is placed on the chart so that the digit 5 is in the liters place. That is, the decimal point falls after the liters place (see Fig. 6-32). (Note the understood decimal point after the liters place.) To change to kiloliters, move the decimal *three* places to the left so that the decimal falls *after* the kiloliters place (see Fig. 6-33). Recall that a shortcut to division by powers of 10 is to move the decimal point one place to the left for each time 10 is used as a divisor.

Thus, 3495 L = 3.495 kL. ■

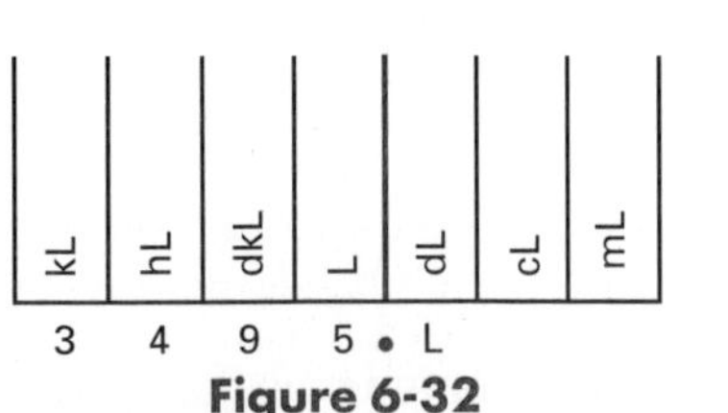

Figure 6-32

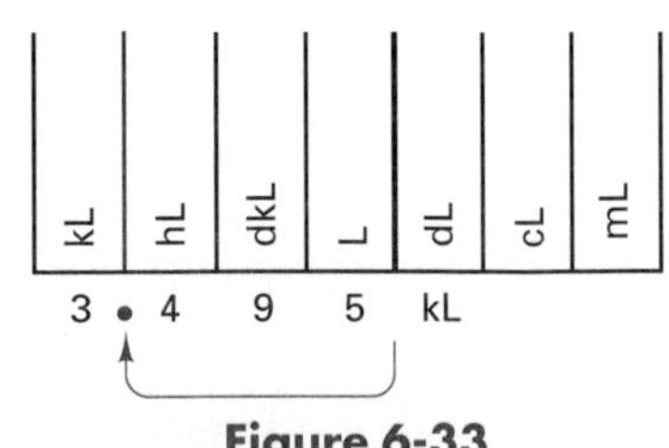

Figure 6-33

EXAMPLE 6-4.10 2.78 cm = ____ dkm?

Solution: Place the number on the chart so that the decimal is *after* the centimeters place (see Fig. 6-34). (Note the decimal position after the centimeters place.) Move the decimal three places to the left so that it falls *after* the dekameters place (see Fig. 6-35). (Note the decimal position after the dekameters place.)

Therefore, 2.78 cm = 0.00278 dkm. ■

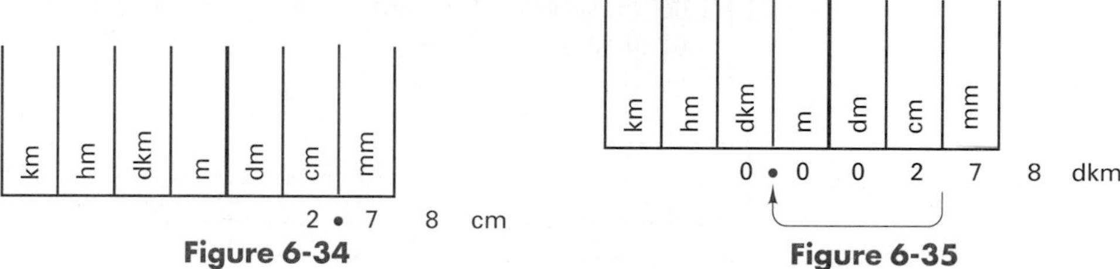

Figure 6-34

Figure 6-35

To summarize, we may change from one metric unit to another metric unit of measure of the same type by positioning the measure on the metric-value chart so that the decimal is immediately *after* the original measuring unit. Then we move the decimal to the right or left as necessary so that it *follows* the new measuring unit. Zeros may be affixed if necessary.

We may also change from one metric unit to another without thinking of the number positioned on the metric-value chart. Instead we mentally locate the original measuring *unit* on the metric-value chart and move to the desired new measuring *unit*. As we move, we carefully note the number of units we move past and the direction in which we move. Then we locate the decimal point in the original metric measure and shift the decimal in the same direction and the same number of units that we moved above.

Tips and Traps

To determine the movement of the decimal point when changing from one metric unit to another, answer the questions:

1. *How far* is it from the original unit to the new unit (how many places)?
2. *Which way* is the movement on the chart?

Change 28.392 cm to m.

How far is it from cm to m? Two places
Which way? Left

Move the decimal in the original measure *two places to the left.*

28.392 cm = 0.28392 m

EXAMPLE 6-4.11 4.29 hg = ____ dg?

Solution: Look at the metric-value chart and locate the hectogram place. Then move to the *right* to the decigram place. This move is *three places to the right*; thus we move the decimal in 4.29 *three places to the right.*

4.29 hg = 4290. dg

■

L.O.3. Make calculations with metric measures.

Basic Operations with Metric Measures

Up to this point, we have been comparing metric units and changing from one unit to a larger or smaller unit. Now we will use metric units in problems involving addition, subtraction, multiplication, and division. Our employment will often require us to perform these basic mathematical operations.

When adding or subtracting measures, it is important that we remember to add or subtract only *like* or *common* measures. By this we mean that the same measuring unit is used. For instance, 5 cm and 3 cm are *like* measures. In contrast, 17 kg and 4 hg are *unlike* measures.

Rule 6-4.5 *To add or subtract* **like** *or* **common** *metric measures:*

Add or subtract the numerical values. The answer will have the common unit of measure. *Examples:*

$$7 \text{ cm} + 4 \text{ cm} = 11 \text{ cm}$$

$$15 \text{ dkg} - 3 \text{ dkg} = 12 \text{ dkg}$$

Rule 6-4.6 *To add or subtract* **unlike** *metric measures:*

First change the measures to a common unit of measure. Then add or subtract the numerical values. The answer will have the common unit of measure.

EXAMPLE 6-4.12 Add 9 mL + 2 cL.

Solution: Change cL to mL; that is 2 cL = 20 mL.
Thus, 9 mL + 20 mL = 29 mL. ■

Look at this example again:

$$9 \text{ mL} + 2 \text{ cL}$$

We could have changed milliliters to centiliters:

$$9 \text{ mL} = 0.9 \text{ cL}$$

Then,

0.9 cL + 2 cL = *(Remember, in addition and subtraction, decimals must be in a straight line. There is an understood decimal after the addend 2.)*

```
  0.9 cL
  2   cL
  ------
  2.9 cL
```

Does 2.9 cL = 29 mL? Yes. Therefore, our answers are equivalent. ■

EXAMPLE 6-4.13 Subtract: 14 km − 34 hm.

Solution: Change hm to km; that is, 34 hm = 3.4 km.

```
   14.0 km
 −  3.4 km
 ---------
   10.6 km
```

(Caution: Watch alignment of decimals.)

The difference is 10.6 km or 106 hm. ■

Tips and Traps

Look at this problem:

$$2 \text{ cm} + 5 \text{ g}$$

Can 2 cm be added to 5 g? Can a common unit of measure be found for centimeters and grams? Centimeters measure length and grams measure weight; therefore, no common unit of measure can be found. Thus, we cannot add 2 cm + 5 g.

Since multiplication is a shortcut for repeated addition, if we wanted to know the total length of three pieces of wood that are each 15 cm long, we could multiply 15 cm by 3. That is, 15 cm $\times$ 3 = 45 cm.

> **Rule 6-4.7** ***To multiply a metric measure by a number:***
>
> Multiply the numerical values. The answer will have the same unit of measure as the measure.

Suppose we had a tube that was 30 cm long and we needed it cut into five equal parts. How long would each part be? To solve this problem, we need to divide 30 cm by 5.

$$\frac{30 \text{ cm}}{5} = 6 \text{ cm}$$

> **Rule 6-4.8** ***To divide a metric measure by a number:***
>
> Divide the numerical values. The answer will have the same unit of measure as the measure.

All types of measures can be divided by a number since all we are doing is dividing a quantity into parts.

There are situations where it becomes necessary to divide a measure by a measure. To give an example, suppose that we had a board 6 m long and we wanted to cut it into sections with each section 2 m long. How many sections would there be?

$$6 \text{ m} \div 2 \text{ m}$$

or

$$\frac{6 \not{\text{m}}}{2 \not{\text{m}}} = 3 \qquad \textit{(Units reduce.)}$$

Notice that the answer, 3, does not have a units label. That is because we are looking for *how many* 2-m sections there are.

> **Rule 6-4.9** ***To divide a metric measure by a like measure:***
>
> Divide the numerical values. The answer will be a number that tells how many parts there are. If unlike measures are used, we must first change to like measures.

EXAMPLE 6-4.14 Solve the following problems.

(a) Four micrometers each weighing 752 g will fit into a shipping carton. If the shipping carton and filler weigh 217 g, what is the total weight of the shipment?
(b) A 5-m-long board is to be cut into four equal parts. How long is each part?
(c) How many 25-g packages of seed can be made from 2 kg of seed?

Solution

(a) Total weight = carton and filler + four micrometers

$$= 217 \text{ g} + (4 \times 752 \text{ g})$$
$$= 217 \text{ g} + 3008 \text{ g}$$
$$= 3225 \text{ g or } 3.225 \text{ kg}$$

Thus, the total weight is 3225 g or 3.225 kg.

(b) $\frac{5 \text{ m}}{4} = 1.25 \text{ m}$ *(Divide length by number of parts.)*

Thus, each part is 1.25 m long.

(c) $\frac{2 \text{ kg}}{25 \text{ g}} = \frac{2000 \not{g}}{25 \not{g}}$ *(Convert kg to g; then divide. Reduce units.)*

$$= 80$$

Thus, 80 packages of seed can be made. ■

SELF-STUDY EXERCISES 6-4

L.O.1.

1. Give the value of the following metric units.
(a) kilometer (km) **(b)** dekaliter (dkL)
(c) decigram (dg) **(d)** millimeter (mm)
(e) hectogram (hg) **(f)** centiliter (cL)

Choose the most reasonable metric measure for each of the following.

2. Height of a 6-year-old child
(a) 1.5 km **(b)** 1.5 m **(c)** 1.5 cm **(d)** 1.5 mm

3. Diameter of a dime
(a) 1.5 m **(b)** 1.5 cm **(c)** 1.5 mm **(d)** 1.5 km

4. Distance from Los Angeles to San Francisco
(a) 800 m **(b)** 800 mm **(c)** 800 km **(d)** 800 cm

5. Length of a pencil
(a) 20 km **(b)** 20 cm **(c)** 20 m **(d)** 20 mm

6. Overnight accumulation of snowfall
(a) 8 cm **(b)** 8 m **(c)** 8 km

7. Width of home video tape
(a) 8 m **(b)** 8 km **(c)** 8 cm **(d)** 8 mm

8. Weight of the average male adult
(a) 70 mg **(b)** 70 kg **(c)** 70 g

9. Weight of a can of tuna
(a) 184 mg **(b)** 184 kg **(c)** 184 g

10. Weight of a teaspoonful of sugar
(a) 2 g **(b)** 2 kg **(c)** 2 mg

11. Weight of a vitamin C tablet
(a) 250 g **(b)** 250 mg **(c)** 250 kg

12. Weight of a dinner plate
(a) 350 kg **(b)** 350 mg **(c)** 350 g

13. Weight of a sack of dog food
(a) 25 g **(b)** 25 kg **(c)** 25 mg

14. Volume of a tank of gasoline
(a) 60 L **(b)** 60 mL

15. Dose of cough syrup
(a) 100 L **(b)** 100 mL

16. Glass of juice
(a) 0.12 mL (b) 0.12 L

L.O.2. Change to the measure indicated. In using the metric-value chart, the decimal is placed immediately *after* the measuring unit.

17. 4 m = ____ dm
18. 7 g = ____ dg
19. 58 km = ____ hm
20. 8 hL = ____ dkL
21. 0.25 km = ____ hm
22. 21 dkL = ____ L
23. 8.5 cm = ____ mm
24. 14.2 dg = ____ cg
25. How many milliliters are there in 15.3 cL?
26. How many meters are there in 46 dkm?
27. How many dekagrams are there in 7.5 hg?
28. How many millimeters are there in 16 cm?

Change each measure to the unit of measure indicated. In using the metric-value chart, the decimal is placed immediately *after* the measuring unit.

29. 4 L = ____ cL
30. 8 m = ____ mm
31. 58 km = ____ m
32. 8 hg = ____ g
33. 0.25 km = ____ m
34. 21 dkg = ____ dg
35. 10.25 hm = ____ cm
36. 8.33 L = ____ mL
37. 2 km = ____ mm
38. 0.7 g = ____ cg
39. Change 2.36 hL to liters.
40. Change 0.467 dkm to centimeters.
41. Change 3.8 kg to decigrams.
42. Change 13 dkm to centimeters.

Change each measure to the unit indicated. In using the metric-value chart, the decimal is placed immediately *after* the measuring unit.

43. 28 m = ____ dkm
44. 238 hL = ____ kL
45. 101 mg = ____ cg
46. 60 hm = ____ km
47. 29 dkL = ____ hL
48. 192.5 g = ____ dkg
49. 17 cm = ____ dm
50. How many decimeters are in 4389 cm?
51. How many grams are in 47 dg?
52. How many deciliters are in 2.25 cL?

Change each measure to the measure indicated. In using the metric-value chart, the decimal is placed immediately *after* the measuring unit.

53. 2743 mm = ____ m
54. 385 g = ____ kg
55. 15 dkm = ____ km
56. 8 cL = ____ L
57. 296,484 m = ____ hm
58. 29.83 dg = ____ dkg
59. 0.3 cm = ____ dkm
60. 40 dL = ____ kL
61. 2857 mg = ____ kg
62. 15,285 m = ____ km
63. Change 297 cm to hectometers.
64. Change 0.03 mL to liters.

L.O.3. Add or subtract these measures.

65. 3 m + 8 m
66. 7 hL + 5 hL
67. 15 cg − 9 cg
68. 2 dm + 4 cm
69. 5 cL + 9 mL
70. 4 m + 2 L
71. 14 kL − 39 hL
72. 1 g − 45 cg
73. 3 cg − 5 mL
74. 7 km + 2 m
75. A dress takes 3.5 m of fabric. If the bolt of fabric has 42 m left after the dress fabric is cut, how much was on the bolt to begin with?
76. 653 dkL of orange juice concentrate are removed from a vat containing 8 kL of the concentrate. How much concentrate remains in the vat?

Multiply.

77. 43 m × 12
78. 3.4 m × 12
79. 50.32 dm × 3
80. A plot of ground is divided into seven plots, each with road frontage of 138.5 m. What is the total road frontage of the plot of ground?

Divide.

81. 39 m ÷ 3

82. $\frac{54 \text{ cL}}{6}$

83. A block of silver weighing 978 g is cut into six equal pieces. How much does each piece weigh?

84. Two pieces of steel each 12 m long are cut into a total of 60 equal pieces. How long is each piece?

85. 2.5 cg ÷ 0.5 cg

86. 3 m ÷ 10 cm

87. How many 250-mL prescriptions can be made from a container of 4 L of decongestant?

88. How many 500-g containers are needed to contain 40 kg of grass seed?

Solve the following.

89. Add 4.6 cL + 5.28 dL of photographic developer.

90. Add 3 m + 2 dkm of fabric for draperies.

91. Subtract 19.8 km − 32.3 hm of paved highway.

92. Subtract 13 kL − 39 hL of stored liquid.

93. Multiply 0.25 cL of cologne by 5.

94. Multiply 35-mm film size by 2.

95. A length of satin fabric 30 dm long is cut into 4 equal pieces. How long is each peice?

96. A metal rod 4 m long is divided into 5 equal sections. How long is each section?

97. How many containers of jelly can be made from 8500 L of jelly if each container holds 4 dL of jelly?

98. How many 2-kg vials of hydrochloric acid (HCL) can be obtained from 38 kg of HCL?

6-5 METRIC–U.S. CUSTOMARY COMPARISONS

Learning Objective

L.O.1. Convert between U.S. customary measures and metric measures.

Learning to convert from the U.S. customary system to the metric system, or vice versa, is important because both systems are used in the United States. These conversions from U.S. customary to metric or metric to U.S. customary are more difficult than working entirely within one system. However, this can be expected, just as it is more difficult to convert American money to French money or French money to American money than to understand and use one kind of money only.

L.O.1. Convert between U.S. customary measures and metric measures.

Using Conversion Factors

To convert measures from the U.S. customary system to the metric system, and vice versa, we only need one *conversion factor* for each type of measurement (length, weight, and capacity).

For length:

1 m = 1.09 yd (to the nearest hundredth)

For weight:

1 kg = 2.20 lb (to the nearest hundredth)

For capacity:

(dry measure)
1 L = 0.91 dry qt (to the nearest hundredth)

(liquid measure)
1 L = 1.06 liquid qt (to the nearest hundredth)

Dry quarts are less common than liquid quarts; if the type of quart (dry or liquid) is not specified, use liquid quarts.

Other conversion factors, such as centimeters to inches and kilometers to miles, can be derived from these factors using unity ratios. Naturally, the more conversion factors we have before us, the better; but one per type of measurement is all that is necessary.

Other conversion factors that we may use are listed in Table 6-5. The numbers in parentheses are rounded to ten-thousandths and can be used if greater accuracy is desired.

Tips and Traps

We do *not* need to memorize these conversion factors. They will be given on tests or will be available on the job. The important thing is to remember to use the appropriate unity ratio.

Rule 6-5.1 ***To convert metric measures to U.S. customary measures or U.S. customary measures to metric measures:***

1. Select the appropriate conversion factor. (For convenience, select the conversion factor that allows the new measure to be in the *numerator* of the unity ratio and places a 1 in the denominator.)
2. Set up a unity ratio so that the measure we are changing from reduces out and the new measure remains.

TABLE 6-5 Metric–U.S. Customary Conversion Factors

Metric to U.S. customary units	U.S. customary to metric units
Measure of Length	
1 meter = 39.37 inches	1 inch = 25.4 millimeters
1 meter = 3.28 feet (3.2808)	1 inch = 2.54 centimeters
1 meter = 1.09 yards (1.0936)	1 inch = 0.0254 meter
1 centimeter = 0.39 or 0.4 inch (0.3937)	1 foot = 0.3 meter (0.3048)
1 millimeter = 0.04 inch (0.03937)	1 yard = 0.91 meter (0.9144)
1 kilometer = 0.62 mile (0.6214)	1 mile = 1.61 kilometers (1.6093)
Capacity: Liquid Measure	
1 liter = 1.06 liquid quarts (1.0567)	1 liquid quart = 0.95 liter (0.9463)
Capacity: Dry Measure	
1 liter = 0.91 dry quart (0.9081)	1 dry quart = 1.1 liters (1.1012)
Measure of Weight	
1 gram = 0.04 ounce (0.0353)	1 ounce = 28.35 grams (28.3495)
1 kilogram = 2.2 pounds (2.2046)	1 pound = 0.45 kilogram (0.4536)

EXAMPLE 6-5.1 Change 145 m to yards.

Solution

$$\frac{145\text{ m}}{1}\left(\frac{\quad\text{yd}}{\text{m}}\right) \quad \text{(use 1 m = 1.09 yd)}$$

$$\frac{145\not{\text{m}}}{1}\left(\frac{1.09\text{ yd}}{1\not{\text{m}}}\right) = 145 \times 1.09\text{ yd}$$

$$= 158.05\text{ yd}$$

(Using 1 yd = 0.91 m puts 0.91 m in the denominator and would involve division instead of multiplication. Multiplication is usually more convenient than division and results when the new measure is in the numerator of the unity ratio. Also, our answer would vary slightly because of rounding procedures.) ■

EXAMPLE 6-5.2 Change 3 in. to centimeters.

Solution

$$\frac{3\text{ in.}}{1}\left(\frac{\quad\text{cm}}{\text{in.}}\right) \quad \text{(use 1 in. = 2.54 cm)}$$

(Using 1 cm = 0.39 in. involves division. Answer would vary slightly.)

$$\frac{3\not{\text{in.}}}{1}\left(\frac{2.54\text{ cm}}{1\not{\text{in.}}}\right) = 3 \times 2.54\text{ cm}$$

$$= 7.62\text{ cm}$$ ■

EXAMPLE 6-5.3 A sack of ready-to-use concrete weighs 90 lb. How many kilograms does it weigh?

Solution

$$\frac{90\text{ lb}}{1}\left(\frac{0.45\text{ kg}}{1\text{ lb}}\right)$$

(Using 1 kg = 2.2 lb involves division. Answer would vary slightly.)

$$\frac{90\text{ lb}}{1}\left(\frac{0.45\text{ kg}}{1\text{ lb}}\right) = 90(0.45\text{ kg}) = 40.5\text{ kg}$$

Thus, the concrete weighs about 40.5 kg. ■

EXAMPLE 6-5.4 A 76-L aquarium is about to be filled with water. How many gallons is its capacity?

$$\frac{76\text{ L}}{1}\left(\frac{1.06\text{ qt}}{1\text{ L}}\right)\left(\frac{1\text{ gal}}{4\text{ qt}}\right)$$

(Since Table 6-5 does not give a conversion factor for gallons, use two unity ratios.)

$$\frac{\overset{19}{\not{76}}\not{\text{L}}}{1}\left(\frac{1.06\text{ qt}}{1\not{\text{L}}}\right)\left(\frac{1\text{ gal}}{\underset{1}{\not{4}}\text{ qt}}\right) = 20.14\text{ gal or 20 gal}$$

Thus, the aquarium is a 20-gal aquarium. ■

SELF-STUDY EXERCISES 6-5

L.O.1. Change to the units indicated.

1. 9 m to inches
2. 120 m to yards
3. 42 km to miles
4. 6 L to liquid quarts
5. 10 dry qt to liters
6. 27 kg to pounds

7. 50 lb to kilograms

8. 7 in. to centimeters

9. 18 ft to meters

10. 39 mL to ounces (*Hint:* 1 liquid qt = 32 oz.)

11. $5\frac{3}{4}$ gal to liters (*Hint:* 1 gal = 4 liquid qt.)

12. A spool of wire contains 100 ft of wire. How many meters of wire are on the spool?

13. A sheet of metal weighing 60 lb weighs how many kilograms?

14. Two cities that are 150 mi apart are how many kilometers apart?

15. A road 30 m wide is how many yards wide?

16. A tourist in Europe traveled 200 km, 60 km, and 120 km by car. How many total miles was this?

17. A Mexican resident jogged 5 km, 4 km, and 3 km. How many miles did the Mexican resident jog?

18. A container holds 12 dry quarts. How many liters will the container hold?

19. A spool of electrical wire contains 100 m of wire. How many feet of wire are on the spool?

6-6 READING INSTRUMENTS USED TO MEASURE LENGTH

Learning Objectives

L.O.1. Read the English rule.
L.O.2. Read the metric rule.

L.O.1. Read the English rule.

The English Rule

Most of us are familiar with the English rule. In elementary and high school we called it simply a *ruler*. Although we used it for drawing straight lines, it was intended primarily for measuring lengths.

The most common English rule uses an inch as the standard unit. Each inch is subdivided into fractional parts, usually 8, 16, 32, or 64. Let's examine carefully the portion of the English rule illustrated in Fig. 6-36.

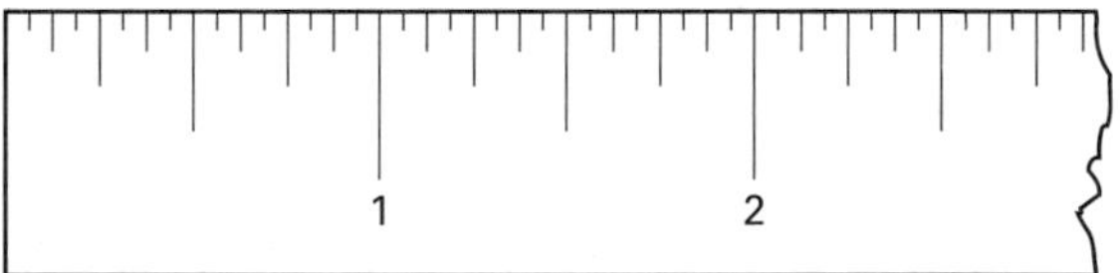

Figure 6-36

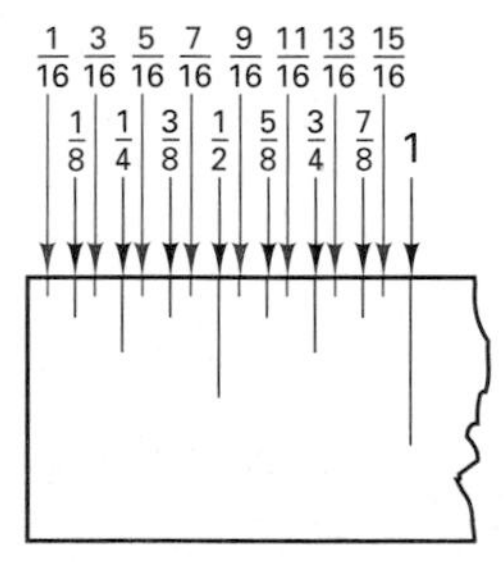

Figure 6-37

Each inch is divided into 16 equal parts; thus, each part is $\frac{1}{16}$ in.; that is, the first mark from the left edge represents $\frac{1}{16}$ in. The left end of the rule represents zero (0). (Some rules leave a small space between zero and the end of the rule.)

The second mark from the left edge of the rule represents $\frac{2}{16}$ or $\frac{1}{8}$ in. This mark is slightly longer than the first mark. Look at Fig. 6-37, which has each of the division marks labeled.

The fourth mark from the left is labeled $\frac{1}{4}$. That is, $\frac{4}{16} = \frac{1}{4}$. In each case, fractions are always reduced to lowest terms. Notice that the $\frac{1}{4}$ mark is slightly longer than the $\frac{1}{8}$ mark.

On the rule the division marks are made different lengths to make the rule easier to read. The shortest marks represent fractions that, in lowest terms, are sixteenths $\left(\frac{1}{16}, \frac{3}{16}, \frac{5}{16}, \frac{7}{16}, \frac{9}{16}, \frac{11}{16}, \frac{13}{16}, \frac{15}{16}\right)$. The marks representing fractions that reduce to eighths are slightly longer than the sixteenths marks $\left(\frac{1}{8}, \frac{3}{8}, \frac{5}{8}, \frac{7}{8}\right)$. Next, the marks representing fractions that reduce to fourths are slightly longer than the eighths marks $\left(\frac{1}{4}, \frac{3}{4}\right)$. The marks representing one-half $\left(\frac{1}{2}\right)$ are longer than the fourths marks, and the inch marks are longer than the one-half marks. Look at Fig. 6-37 again and pay particular attention to the lengths of the division marks.

In many of the examples and exercises that follow, we will measure *line segments* whose beginning and end are identified by capital letters, such as line segment AB in Fig. 6-38.

EXAMPLE 6-6.1 Measure line segment AB (Fig. 6-38a).

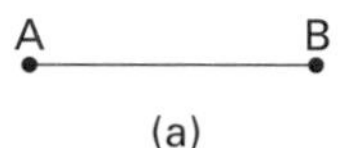

(a)

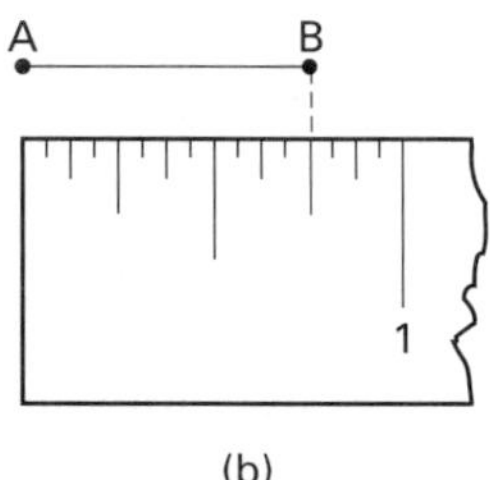

(b)

Figure 6-38

Solution: Align the left edge of the rule with point A (see Fig. 6-38b). Point B aligns with the mark $\frac{12}{16}$ or $\frac{3}{4}$. Remember, it is not necessary to count each mark if we notice from the length of the division mark that the fraction will reduce to fourths. ■

EXAMPLE 6-6.2 Measure line segment CD (Fig. 6-39).

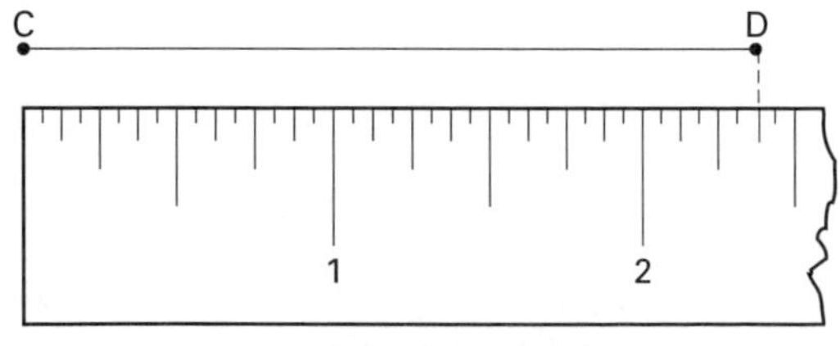

Figure 6-39

Solution: Align point C with zero. Line segment CD goes past the 2-in. mark, but not up to the 3-in. mark. Therefore, the measure of CD will be a mixed number between 2 and 3. Point D is $\frac{3}{8}$ in. past 2. Thus, CD is $2\frac{3}{8}$ in. ■

Tips and Traps

A line segment may not always line up exactly with a division mark. If this is the case, we must use eye judgment and decide which mark is closer to the end of the line segment.

EXAMPLE 6-6.3 Measure line segment EF (Fig. 6-40).

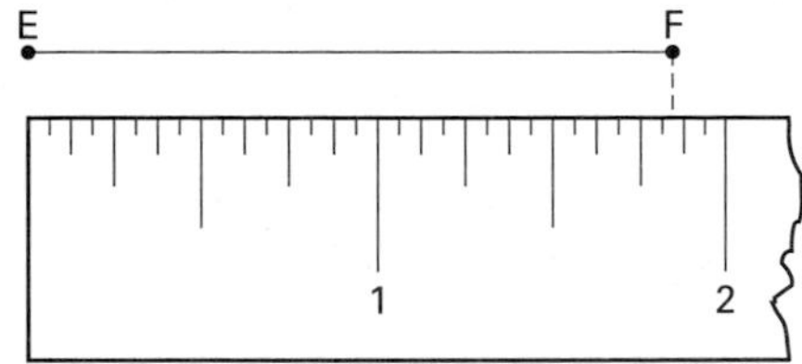

Figure 6-40

Solution: Point F lines up between $1\frac{13}{16}$ and $1\frac{7}{8}$. Remember, measurements are always approximations; but, using our best eye judgment, point F seems closer to $1\frac{13}{16}$ than $1\frac{7}{8}$. We will say EF is $1\frac{13}{16}''$ *to the nearest sixteenth of an inch.* ■

In practice, measurements are considered acceptable if they are within a desired *tolerance.* In the example above, since the smallest division is $\frac{1}{16}$, the desired tolerance would normally be plus or minus one-half of one-sixteenth or $\pm\frac{1}{32}\left(\frac{1}{2} \text{ of } \frac{1}{16} = \frac{1}{32}\right)$. That is, the acceptable measure can be $\frac{1}{32}$ more than or $\frac{1}{32}$ less than the ideal measure.

If the ideal measure is $1\frac{13}{16}$ in. and the tolerance is $\frac{1}{32}$ in., the *range* of acceptable values is from $1\frac{13}{16} - \frac{1}{32}$ to $1\frac{13}{16} + \frac{1}{32}$.

$$1\frac{13}{16} = 1\frac{26}{32}$$

$$1\frac{26}{32} - \frac{1}{32} = 1\frac{25}{32} \qquad 1\frac{26}{32} + \frac{1}{32} = 1\frac{27}{32}$$

The acceptable range is from $1\frac{25}{32}$ to $1\frac{27}{32}$.

EXAMPLE 6-6.4 Measure line segment GH (Fig. 6-41).

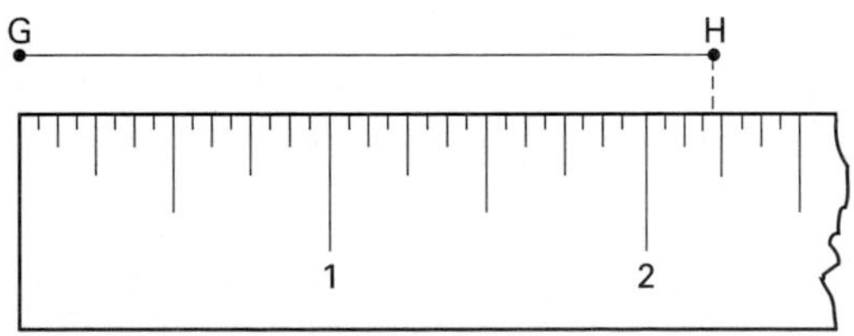

Figure 6-41

Solution: Point H lines up between $2\frac{3}{16}$ and $2\frac{1}{4}$. Using eye judgment, it seems that point H is halfway between $2\frac{3}{16}$ and $2\frac{1}{4}$. What mixed number is halfway between $2\frac{3}{16}$ and $2\frac{1}{4}$? Let's find out by comparing the two measures in thirty-seconds.

$$2\frac{3}{16} = 2\frac{6}{32}$$

$$2\frac{1}{4} = 2\frac{8}{32}$$

Therefore, $2\frac{7}{32}$ is halfway between $2\frac{3}{16}$ and $2\frac{1}{4}$. The measure of GH could be expressed as $2\frac{7}{32}$. However, both $2\frac{3}{16}$ and $2\frac{1}{4}$ would be within the desired tolerance of $\pm\frac{1}{32}$.

$$2\frac{7}{32} + \frac{1}{32} = 2\frac{8}{32} \quad \text{or} \quad 2\frac{1}{4}$$

$$2\frac{7}{32} - \frac{1}{32} = 2\frac{6}{32} \quad \text{or} \quad 2\frac{3}{16}$$

All three measures, $2\frac{3}{16}$, $2\frac{7}{32}$, and $2\frac{1}{4}$, would be correct approximations. ■

L.O.2. Read the metric rule.

The Metric Rule

The metric rule illustrated in Fig. 6-42 shows centimeters as the major divisions, represented by the longest lines. Each centimeter is divided into 10 millimeters, represented by the shortest lines. A line slightly longer than the millimeter line divides each centimeter into two equal parts of 5 mm each.

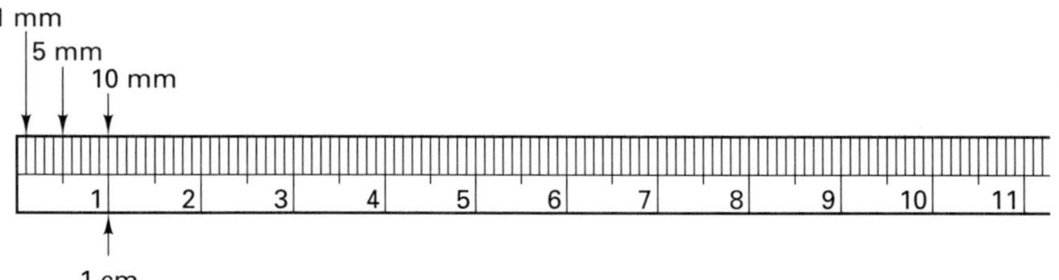

Figure 6-42

The metric rule is read like the English rule with the exception that only two measures are indicated, centimeters and millimeters. Other measures are calculated in relation to millimeters or centimeters. Thus, every 10 cm or 100 mm make 1 decimeter (dm).

EXAMPLE 6-6.5 Find the length of the line segment AB (Fig. 6-43).

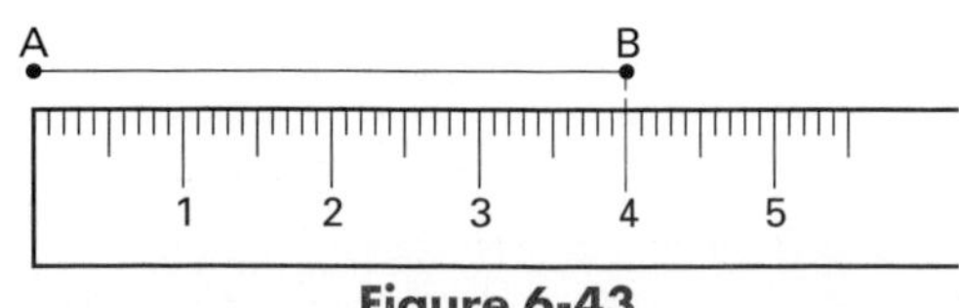

Figure 6-43

Solution: The line segment *AB* extends to the 4-cm mark. Thus, line segment *AB* is 4 cm (or 40 mm). ■

EXAMPLE 6-6.6 Find the length of line segment *CD* (Fig. 6-44) to the nearest millimeter.

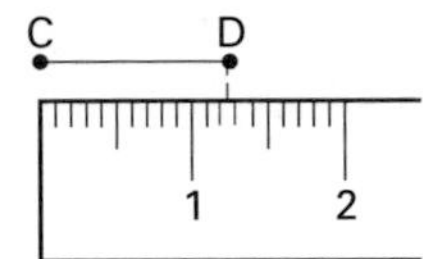

Figure 6-44

Solution: The line segment *CD* extends to between 12 mm and 13 mm. It appears to be closer the 12-mm mark by eye judgment. Thus, line segment *CD* is 12 mm to the nearest millimeter. The measure could also be written as 1.2 cm. ■

EXAMPLE 6-6.7 Find the length of the line segment *EF* (Fig. 6-45) if the acceptable tolerance is ± 0.5 mm.

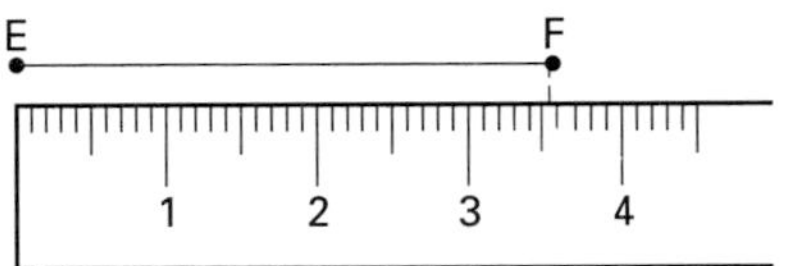

Figure 6-45

Solution: The end of line segment *EF* falls approximately halfway between the 35-mm mark and the 36-mm mark, by eye judgment, measuring about 35.5 mm. Since the tolerance is ± 0.5 mm, both 35 and 36 mm would be acceptable measures of the line segment *EF*. Thus, 35 mm, 35.5 mm, and 36 mm would be acceptable approximations for line segment *EF*. ■

SELF-STUDY EXERCISES 6-6

L.O.1. Measure the line segments 1 through 10 in Fig. 6-46 to the nearest sixteenth of an inch (tolerance $= \pm\frac{1}{32}$ in.).

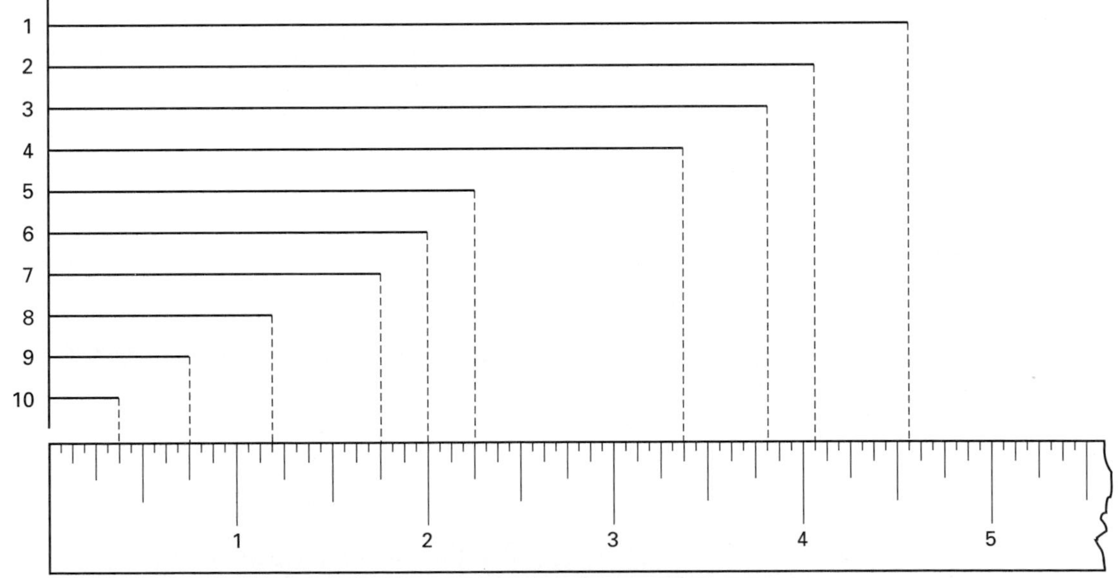

Figure 6-46

L.O.2. Measure the line segments 11 through 20 in Fig. 6-47 to the nearest millimeter. Measures can be expressed in millimeters or centimeters.

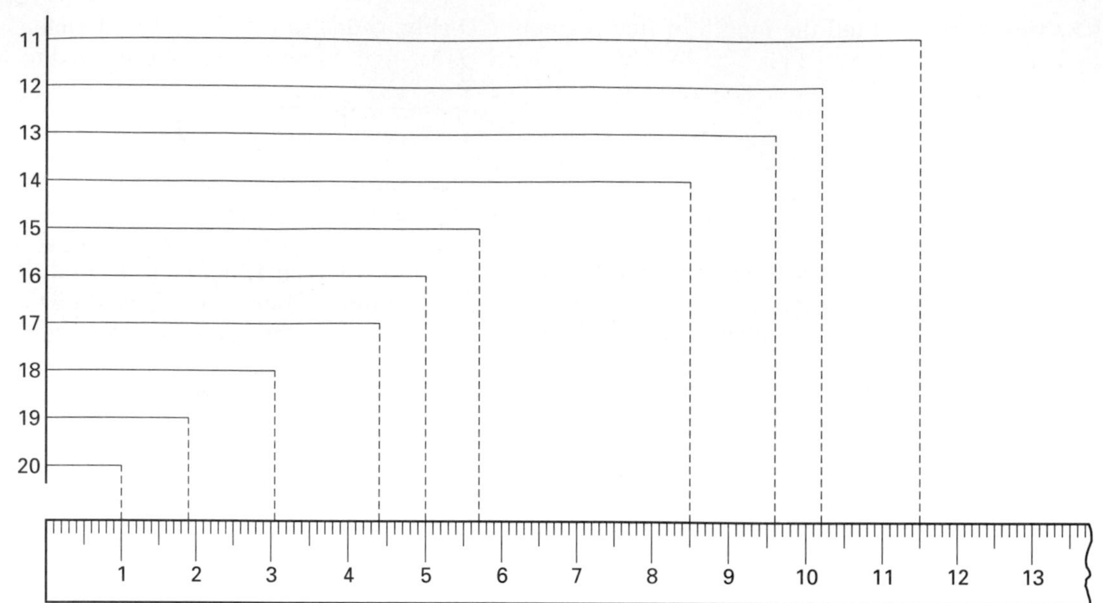

Figure 6-47

ASSIGNMENT EXERCISES

6-1

Identify the appropriate U.S. customary measure for each of the following.

1. Package of spaghetti
2. Tank of gasoline
3. Container of motor oil
4. Distance from work to the library
5. Package of taco shells
6. Porterhouse steak
7. Shipment of iron
8. Sack of flour
9. Size of an aluminum pot
10. Height of a tree
11. Sugar for a pie recipe
12. Man's shirt size
13. Cloth for a pair of kitchen curtains
14. Hourly speed of an aircraft
15. Shaving cream

Using unity ratios, convert the given measures to the new units of measure.

16. 8 ft = ____ in.

17. 12 ft = ____ yd

18. 11 yd = ____ ft

19. $1\frac{1}{5}$ mi = ____ ft

20. How many feet of wire are needed to put a fence along a property line $2\frac{1}{4}$ mi long?

21. How many ounces are in 5 lb?

22. An object weighing $57\frac{3}{5}$ lb weighs how many ounces?

23. Find the number of pounds in 680 oz.

24. A can of fruit weighs 22.4 oz. How many pounds is this?

25. The net weight of a can of peas is 19 oz. If a case contains 16 cans, what is the net weight of a case in ounces? in pounds?

26. How many quarts are in 8 pt?

27. How many pints are in $7\frac{1}{2}$ qt?

28. Find the number of gallons in 15 qt.

29. Find the number of pints in 3 gal.

30. How many gallons are in 36 pt?

31. A cereal company has 12 T of prepared grain and wants to market it as packages of several ounces each. How many ounces of prepared grain is this?

32. A sack of concrete mix weighs 70 lb. How many ounces does a sack of concrete mix weigh?

33. How many feet of wire are needed to fence a property line $1\frac{1}{4}$ mi long?

34. A cook has 1 qt of vegetable oil. The recipe used requires 2 c of oil. How many recipes can be made from the quart of oil?

Express the following measures in standard notation.

35. 6 ft 17 in.

36. 1 mi 5375 ft

37. 12 lb $17\frac{1}{2}$ oz

38. 2 gal 7 qt

39. 1 gal 2 qt 5 pt

40. 2 T 3100 lb

41. 3 yd 2 ft 16 in.

42. 1 qt 3 c 12 oz

43. $3\frac{1}{4}$ ft 10 in.

44. 2 lb 21 oz

45. 1 gal 3 qt 48 oz

6-2

Add or subtract as indicated. Write answers in standard form.

46. 12 oz + 2 lb

47. 8 ft − 49 in.

48. 4 gal + 3 qt

49.
$$\begin{array}{r} 2 \text{ ft } 9 \text{ in.} \\ + \ 8 \text{ ft } 2 \text{ in.} \\ \hline \end{array}$$

50.
$$\begin{array}{r} 7 \text{ lb } 8 \text{ oz} \\ + \ 5 \text{ lb } 9 \text{ oz} \\ \hline \end{array}$$

51.
$$\begin{array}{r} 5 \text{ gal } 3 \text{ qt} \\ + \ 2 \text{ gal } 3 \text{ qt} \\ \hline \end{array}$$

52.
$$\begin{array}{r} 7 \text{ ft } 9 \text{ in.} \\ - \ 4 \text{ ft } 6 \text{ in.} \\ \hline \end{array}$$

53.
$$\begin{array}{r} 4 \text{ lb } 9 \text{ oz} \\ - \ 3 \text{ lb } 11 \text{ oz} \\ \hline \end{array}$$

54.
$$\begin{array}{r} 4 \text{ yd } 1 \text{ ft } 8 \text{ in.} \\ - \ 2 \text{ yd } 2 \text{ ft } 11 \text{ in.} \\ \hline \end{array}$$

55. A rug 12 ft 6 in. long must fit in a room whose length is 10 ft 9 in. How much should be trimmed from the rug to make it fit the room?

56. Two packages to be sent air express each weigh 5 lb 4 oz. What is the shipping weight of the two packages?

57. A water hose purchased for an RV was 2 ft long. What was its length after cutting off 7 in.?

58. A vinyl flooring installer cut 19 in. from a piece of vinyl 13 ft long. How long was the vinyl piece after it was cut?

6-3

Multiply and write answers for mixed measures in standard form.

59.
$$\begin{array}{r} 42 \text{ ft} \\ 12 \text{ ft} \\ \hline \end{array}$$

60.
$$\begin{array}{r} 8 \text{ lb } 3 \text{ oz} \\ 9 \\ \hline \end{array}$$

61.
$$\begin{array}{r} 9 \text{ in.} \\ 7 \text{ in.} \\ \hline \end{array}$$

62.
$$\begin{array}{r} 10 \text{ gal } 3 \text{ qt} \\ 7 \\ \hline \end{array}$$

Divide.

63. 20 yd 2 ft 6 in. ÷ 2

64. 5 gal 3 qt 2 pt ÷ 6

65. 65 ft ÷ 12 (Write answer in feet.)

66. 21 ft ÷ 4 (Write answer in feet and inches.)

67. If 18 lb of candy are divided equally into four boxes, express the weight of the contents of each box in pounds and ounces.

68. If 32 equal lengths of pipe are needed for a job and each length is to be 2 ft 8 in., how many feet of pipe are needed for the job?

Work the following problems.

69. 14 ft ÷ 4 ft

70. 2 mi 120 ft ÷ 15 ft

71. 400 lb ÷ 90 lb

72. 6 yd 2 ft ÷ 5 ft

73. $5\frac{\text{mi}}{\text{min}} = \_\_\_\frac{\text{mi}}{\text{hr}}$

74. $2520\,\frac{\text{gal}}{\text{hr}} = \_\_\_\frac{\text{qt}}{\text{hr}}$

75. $88\frac{\text{ft}}{\text{sec}} = \_\_\_\frac{\text{mi}}{\text{hr}}$

76. $18\frac{\text{mi}}{\text{gal}} = \_\_\_\frac{\text{ft}}{\text{gal}}$

77. A pump that can move water at the rate of $75\frac{\text{gal}}{\text{hr}}$ can move how many gallons per minute?

78. A plane that travels at the rate of 240 mph is traveling how many feet per second?

79. How many quarts of milk are needed for a recipe that calls for 3 pt of milk?

80. How many $\frac{1}{2}$-oz servings of jelly can be made from a $1\frac{1}{2}$-lb container of jelly?

6-4

Give the prefix that relates each of the following numbers to the standard unit.

81. 1000 times
82. $\frac{1}{10}$ of
83. $\frac{1}{1000}$ of
84. 10 times
85. $\frac{1}{100}$ of
86. 100 times

Give the value of the prefixes of the following units of measure based on a standard measuring unit.

87. dekameter (dkm)
88. hectogram (hg)
89. milligram (mg)
90. centigram (cg)
91. kiloliter (kL)
92. deciliter (dL)

Choose the most reasonable answer.

93. Height of the Washington Monument
(a) 200 m (b) 200 cm (c) 200 mm (d) 200 km

94. Height of Mt. Rushmore
(a) 1.6 km (b) 1.6 m (c) 1.6 cm (d) 1.6 mm

95. Weight of an egg
(a) 50 g (b) 50 kg (c) 50 mg

96. Weight of a saccharin tablet
(a) 50 kg (b) 50 mg (c) 50 g

97. Weight of a man's shoe
(a) 0.25 g (b) 0.25 mg (c) 0.25 kg

98. Carton of milk
(a) 4 L (b) 4 mL

99. Bottle of perfume
(a) 50 L (b) 50 mL

Change to the unit indicated.

100. 0.4 dkm = _____ hm
101. 67.1 m = _____ dkm
102. 4 m = _____ dm
103. 2.3 m = _____ mm
104. 5 cm = _____ mm
105. 0.123 hm = _____ mm
106. How many millimeters are in 0.432 km?
107. 23 dkm = _____ mm
108. 42.7 cm = _____ dkm
109. 41,327 dkm = _____ km
110. A board is 1.82 m long. How many centimeters are in the board?
111. 394.5 g = _____ hg
112. 2.7 hg = _____ dg
113. 3,000,974 cg = _____ kg

Perform the operations indicated.

114. 25 mm − 14 mm
115. 12 g + 5 m
116. 17 mg − 8 mL
117. 8 g − 52 cg
118. 4.3 dkg × 7
119. 6.83 cg × 9
120. $\frac{18\text{ cm}}{9}$
121. 7.5 kg ÷ 0.5 kg
122. $\frac{8\text{ hL}}{20\text{ L}}$
123. 34 hL ÷ 4
124. 2.4 m ÷ 5 cm

125. Fabric must be purchased to make seven garments, each requiring 2.7 m of fabric. How much fabric must be purchased?

126. Candy weighing 526 g is mixed with candy weighing 342 g. What is the weight of the mixture?

127. A recipe calls for 5 mL of vanilla flavoring and 24 cL of milk. How much liquid is this?

128. Twenty boxes, each weighing 42 kg, are to be moved. How much weight must be moved?

129. A metal rod 42 m long is cut into seven equal pieces. How long is each piece?

130. Thirty-two kilograms of a chemical are distributed equally among 16 students. How many kilograms of chemical does each student receive?

131. A serving of punch is 25 cL. How many servings can be obtained from 25 L of punch?

132. How many containers of jelly can be made from 8548 L of jelly if each container holds 4 dL of jelly?

133. A bolt of fabric contains 6.8 dkm. If a shirt requires 1.7 m, how many shirts can be made from the bolt?

6-5

Make the following conversions.

134. 7 m = _____ inches

135. 215 m = _____ yards

136. 69 km = _____ miles

137. 15 L = _____ liquid quarts

138. 12 dry qt = _____ liters

139. 32 kg = _____ pounds

140. 10 lb = _____ kilograms

141. 9 in. = _____ centimeters

142. 21 ft = _____ meters

143. 14.8 dkL = _____ quarts

144. $3\frac{1}{2}$ gal = _____ liters

145. How many meters are in 200 ft of pipe?

146. Concrete weighing 90 lb weighs how many kilograms?

147. Two cities 175 mi apart are how many kilometers apart?

148. A room 10 m wide is how many feet wide?

6-6

Measure line segments 149 through 158 in Fig. 6-48 (tolerance = $\pm\frac{1}{32}$ in.).

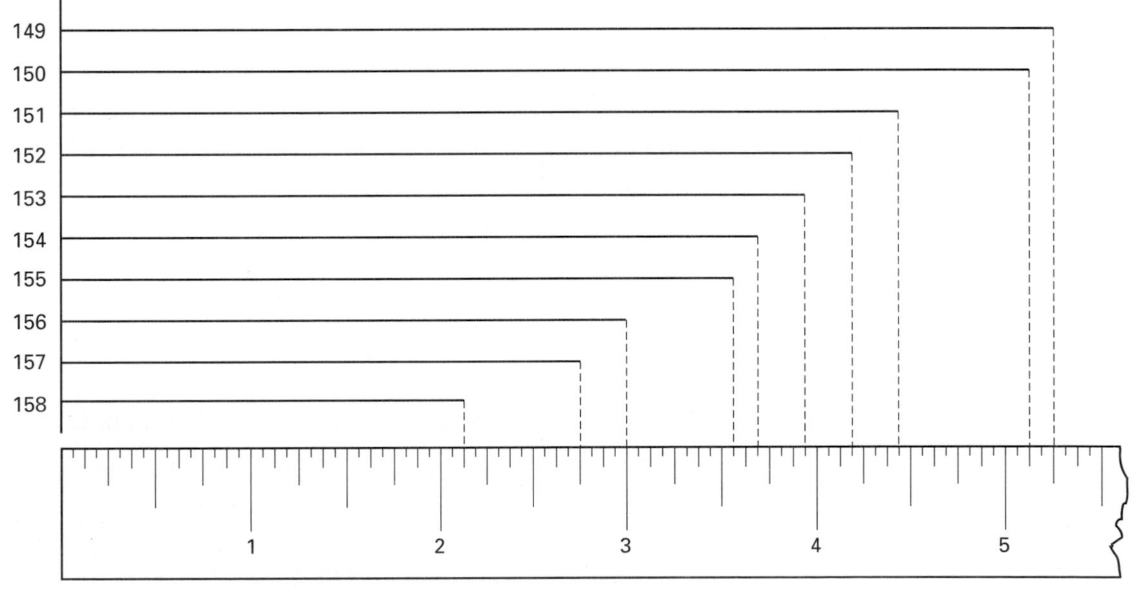

Figure 6-48

Measure line segments 159 through 168 in Fig. 6-49 to the nearest millimeter.

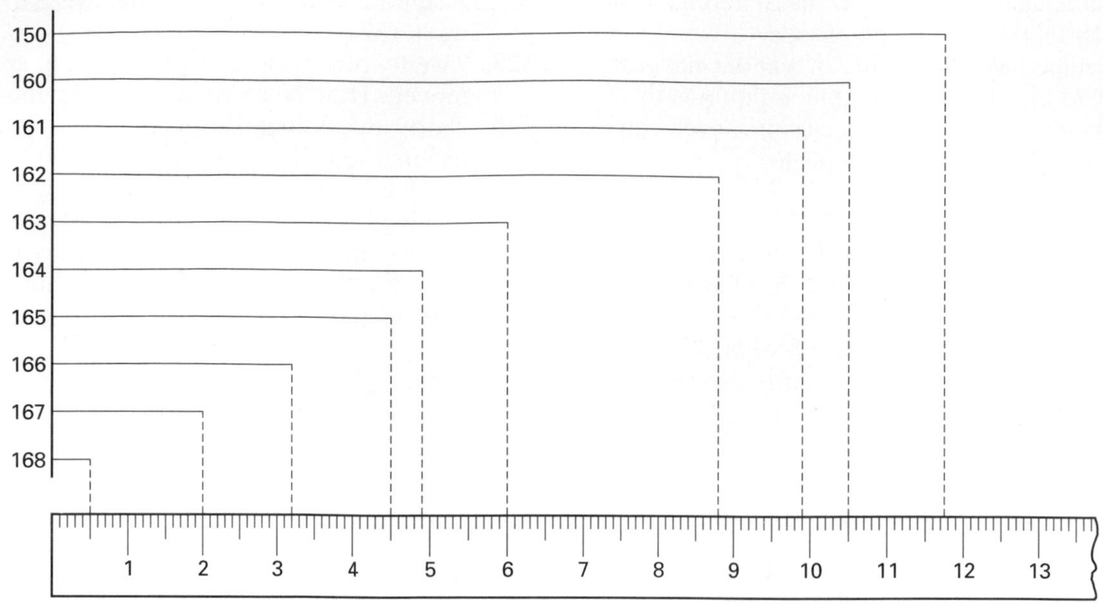

Figure 6-49

CHALLENGE PROBLEMS

169. Compare and contrast the U.S. customary system of measurement and the metric system of measurement.

170. Why do you think the metric system has not fully replaced the U.S. customary system in the United States?

CONCEPTS ANALYSIS

1. Give four items that would be measured with pounds.

2. Give four items that would be measured with feet.

3. If you were building a house, what units of linear measure would you be likely to use?

4. Explain how metric units of measure of length, weight, and capacity are similar.

5. Which measure is longer, a yard or a meter?

6. Which metric measure would you use if you were dispensing a liquid medicine?

7. If your medicine bottle reads 5 mg, is the measure more likely to be liquid or a capsule?

Identify the mistake, explain why it is wrong, and correct the mistake in each of the following.

8. 3.252 dkL = 325.2 kL

9. 418 yd = ____ ft?

$$418 \not{yd} \times \frac{3 \text{ ft}}{1 \not{yd}} = 1234 \text{ yd}$$

10. 24 pt = ____ gal?

$$24 \not{pt} \times \frac{1 \text{ gal}}{4 \not{pt}} = 6 \text{ gal}$$

11. A stack of plywood is 4 ft high. If each sheet of plywood is $\frac{3}{4}$ in. thick, how many pieces of plywood are in the stack?

4 ft = 48 in. sheets. The plywood is 48 in. tall. $48 \times \frac{3}{4} = 36$. There are 36 sheets of plywood in the stack.

CHAPTER SUMMARY

Objectives	What to Remember	Examples
Section 6-1		
1. Identify uses of U.S. customary measures of length, weight, and capacity.	Associate small, medium, and large objects with units designating small, medium, and large amounts.	Nose drops, oz; sofa, lb; coal, T; travel, mi; belt, in.; window, ft or in.; cloth, yd; oil, qt; fuel, gal; cooking, c
2. Convert from one U.S. customary unit of measure to another.	**1.** Write the original measure in the numerator of a fraction with 1 in the denominator. **2.** Multiply by a unity ratio with the original unit of measure in the denominator and the new unit in the numerator.	Change 5 ft to inches. $\frac{5\text{ ft}}{1} \times \frac{12\text{ in.}}{1\text{ ft}} = 60\text{ in.}$ Change 285 ft to yards. $\frac{285\text{ ft}}{1} \times \frac{1\text{ yd}}{3\text{ ft}} = 95\text{ yd}$
	Multiplication can be used to change to smaller units. Division can be used to change to larger units. A unity ratio (or ratios) can be used to change to any unit.	Change $3\frac{1}{2}$ quarts to cups. $3\frac{1}{2}\text{ quarts} = \frac{7}{2}$ $\frac{7\text{ qt}}{2} \times \frac{2\text{ pt}}{1\text{ qt}} \times \frac{2\text{ c}}{1\text{ pt}} = 14\text{ c}$
3. Write mixed U.S. customary measures in standard notation.	Each unit is converted to next larger unit when possible.	2 ft 16 in. = 3 ft 4 in. (16 in. = 1 ft 4 in.)
Section 6-2		
1. Add U.S. customary measures.	To add or subtract U.S. customary units: **1.** Convert the measures to the same unit of measure. **2.** Add or subtract.	Add 5 lb and 7 oz. $\frac{5\text{ lb}}{1} \times \frac{16\text{ oz}}{1\text{ lb}} = 80\text{ oz}$ 80 oz + 7 oz = 87 oz
2. Subtract U.S. customary measures.	To add or subtract mixed U.S. customary measures: **1.** Align like measures in columns. **2.** Add or subtract, carrying or borrowing as necessary.	Subtract 2 ft 8 in. from 7 ft. 6 ft 12 in. − 2 ft 8 in. = 4 ft 4 in. (Borrow.)
Section 6-3		
1. Multiply a U.S. customary measure by a number.	To multiply a measure by a number: **1.** Multiply each measure with a different unit by the number. **2.** Express the answer in standard notation.	Multiply 2 gal 3 qt by 5. 2 gal 3 qt × 5 = 10 gal 15 qt or 13 gal 3 qt
2. Multiply a U.S. customary measure by a measure.	To multiply a length measure by a like length measure: **1.** Multiply the numbers associated with each like unit of measure. **2.** The product will be a square unit of measure.	Multiply 5 yd by 12 yd. $5\text{ yd} \times 12\text{ yd} = 60\text{ yd}^2$

3. Divide a U.S. customary measure by a number.

To divide a measure by a number:
1. Divide the largest unit by the number.
2. Convert any remainder to the next smaller unit.
3. Repeat steps 1 and 2 until no other measures are left.
4. Express any remainder as a fraction of the smallest measure.

Divide 7 lb 5 oz by 5.

$$\begin{array}{r|ll} & 1 \text{ lb} & 7\frac{2}{5} \text{ oz} \\ \hline 5 & 7 \text{ lb} & 5 \text{ oz} \\ & \underline{5 \text{ lb}} & \\ & 2 \text{ lb} = & \underline{32 \text{ oz}} \\ & & 37 \text{ oz} \\ & & \underline{35 \text{ oz}} \\ & & 2 \end{array}$$

4. Divide a U.S. customary measure by a measure.

To divide a measure by a measure:
1. Convert the measures to the same unit
2. Divide.

How many 5-oz glasses of juice can be poured from 1 gallon of juice?

$$\frac{1 \text{ gal}}{1} \times \frac{4 \text{ qt}}{1 \text{ gal}} \times \frac{2 \text{ pt}}{1 \text{ qt}} \times \frac{2 \text{ c}}{1 \text{ pt}} \times \frac{8 \text{ oz}}{1 \text{ c}} = 128 \text{ oz}$$

$$\frac{128 \text{ oz}}{5 \text{ oz}} = 25\frac{3}{5} \text{ glasses}$$

5. Change from one U.S. customary rate measure to another.

1. Set up original rate equal to new rate without its numerical values.
2. Decide which units will change.
3. Multiply original rate by unity ratio or ratios with new unit to reduce old unit.

Change 10 mi/hr to ft/hr.

$$\frac{10 \text{ mi}}{1 \text{ hr}} \text{ to } \frac{\text{ft}}{\text{hr}}$$

$$\frac{10 \text{ mi}}{1 \text{ hr}}\left(\frac{5280 \text{ ft}}{1 \text{ mi}}\right)$$

52,800 ft/hr

Section 6-4

1. Identify uses of metric measures of length, weight, and capacity.

Associate small, medium, and large objects with units designating small, medium, and large amounts. Powers of 10 prefixes are used, such as *kilo* for 1000.

Perfume, mL; soda, L; travel, km; racetrack, m; pill, mg; potatoes, kg; eye drops, mL

2. Convert from one metric unit of measure to another.

Move the decimal point in the original measure to the left or right as many places as necessary to move from the original measuring unit to the new unit on the metric chart of prefixes.

Change 5.04 cl to liters. Move the decimal two places to the left.

$$5.04 \text{ cL} = 0.0504 \text{ L}$$

3. Make calculations with metric measures.

To add or subtract metric measures:
1. Change measures to measures with like units if necessary.
2. Add or subtract.

$7 \text{ km} + 34 \text{ m} =$
$7000 \text{ m} + 34 \text{ m} = 7034 \text{ m}$
or
$7 \text{ km} + 0.034 \text{ m} = 7.034 \text{ km}$

To multiply or divide a metric measure by a number, multiply or divide the numbers and keep the same measuring unit.

$7 \text{ mg} \times 3 = 21 \text{ mg}$

To divide a metric measure by a measure:
1. Change measures to measures with like units if necessary.
2. Divide the numbers, canceling the units of measure. The answer will be a number.

$5 \text{ m} \div 25 \text{ cm}$

$$\frac{500 \text{ cm}}{25 \text{ cm}} = 20$$

Section 6-5

1. Convert between U.S. customary measures and metric measures.

1. Set up the original amount as a fraction with the original unit of measure in the numerator and 1 in the denominator.
2. Multiply by a unity ratio with the original unit of measure in the denominator and the new unit in the numerator.

How many feet are in 14 meters?

$$\frac{14\ \not{m}}{1} \times \frac{3.28\text{ ft}}{1\ \not{m}} = 45.92\text{ ft}$$

Section 6-6

1. Read the English rule.

Align rule along object. Count the number of inches and fractional parts $\left(\frac{1}{32}\text{'s}, \frac{1}{16}\text{'s}, \frac{1}{8}\text{'s}, \frac{1}{4}\text{'s, and so on}\right)$ to determine approximate length. Use eye judgment to estimate closeness of object to a mark on rule.

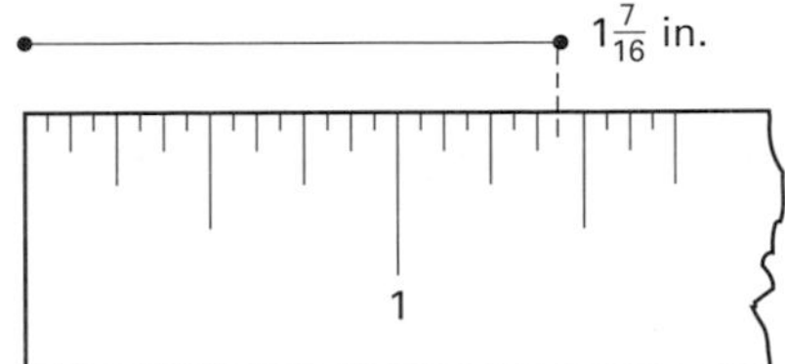

Figure 6-50

2. Read the metric rule.

Align rule along object. Count the number of millimeters and/or centimeters (10 mm = 1 cm) to determine approximate length. Use eye judgment to estimate closeness of object to a mark on rule.

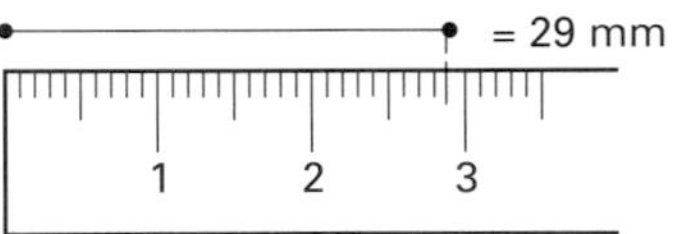

Figure 6-51

WORDS TO KNOW

English system (p. 238)
U.S. customary system (p. 238)
unity ratio (p. 244)
mixed measures (p. 248)
standard notation (p. 248)
unlike units (p. 250)
like units (p. 250)
rate measure (p. 257)
metric system (p. 260)
meter (p. 260)
gram (p. 260)
liter (p. 260)
deci- (p. 261)
centi- (p. 261)
milli- (p. 261)
deka- (p. 261)
hecto- (p. 261)
kilo- (p. 261)
kilometer (p. 262)
centimeter (p. 262)
millimeter (p. 263)
kilogram (p. 263)
milligram (p. 263)
milliliter (p. 264)
conversion factors (p. 274)
English rule (p. 277)
metric rule (p. 280)

CHAPTER TRIAL TEST

Change to the measure indicated (See Tables 6-1 to 6-4 for conversion factors).

1. 3 ft = _____ in.
2. 36 oz = _____ lb
3. 32 qt = _____ gal
4. Add: 2 yd 6 ft 10 in. + 3 ft 7 in. Write your answer in standard notation.
5. $60\frac{\text{gal}}{\text{min}} = \_\_\_\_\_ \frac{\text{qt}}{\text{sec}}$
6. 21 in. ÷ 3 in.
7. A 495-ft section of highway is to be resurfaced. How many yards is this?
8. How many quarts are contained in a 55-gal drum?
9. If an automobile travels at 55 mph, how many feet is it traveling per second?
10. A spark plug wire kit contains $18\frac{1}{2}$ ft of wire in a coil. The directions call for cutting off individual lengths of 28 in. each. How many 28-in. wires can be cut from the coil?

11. Find the measure of the line segment AB in Fig. 6-52 (tolerance $= \pm\frac{1}{32}$ in.).

12. Select the appropriate U.S. customary measure for the contents of a swimming pool.

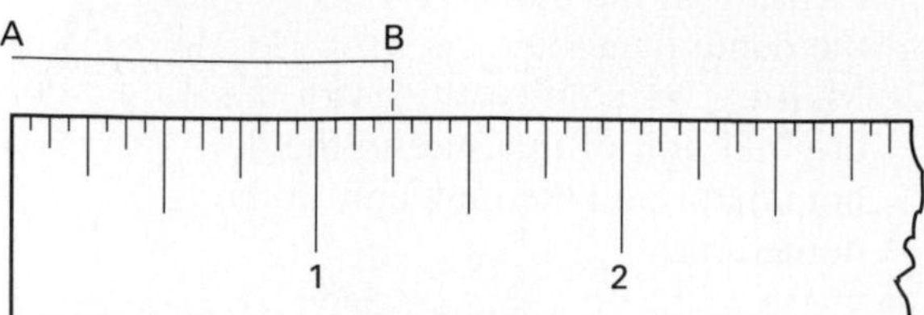

Figure 6-52

13. Write 1 gal 3 qt 6 c 20 oz in standard notation.

14. Which is the appropriate metric measure for eye drops: 28 mL, 28 dL, 28 L, or 28 kL?

Give the prefix for each of the following.

15. $\frac{1}{10}$ of standard unit

16. $10 \times$ standard unit

Change to the metric unit indicated.

17. 298 m = _____ km

18. 8 dm = _____ mm

19. 5.2 dL of liquid are poured from a container holding 10 L. How many liters of liquid remain in the container?

20. A bar of soap weighs 175 g. How much do 15 bars of soap weigh (in kilograms)?

21. 75 mi = _____ km

22. 25 kg = _____ lb

23. 4 L = _____ pt

24. How many liters are contained in a 55-gal drum?

25. A 0.243-caliber bullet travels $2450 \frac{\text{ft}}{\text{sec}}$ for the first 200 yd. Using conversion factors, change $2450 \frac{\text{ft}}{\text{sec}}$ to $\frac{\text{meters}}{\text{sec}}$.

CHAPTER

7 Perimeter and Area

There are occasions in almost every technical field when we need to figure how much surface space a straight-sided, flat figure, called a polygon, occupies or what is the combined length of its sides. Most of us will have to work with such geometric figures and their measurements at one time or another. It is to our advantage to be familiar with them and the formulas that can save both time and money on the job and at home.

7-1 SQUARES, RECTANGLES, AND PARALLELOGRAMS

Learning Objectives

L.O.1. Find the perimeter and area of a square.
L.O.2. Find the perimeter and area of a rectangle.
L.O.3. Find the perimeter and area of a parallelogram.

A *polygon* is a plane or flat closed figure described by straight line segments and angles. Perhaps among the most familiar and easy-to-work-with polygons are the *parallelogram*, the *rectangle*, and the *square*. We will begin with a definition of terms.

■ **DEFINITION 7-1.1 Parallelogram.** A **parallelogram** is a four-sided polygon whose opposite sides are parallel (Fig. 7-1).

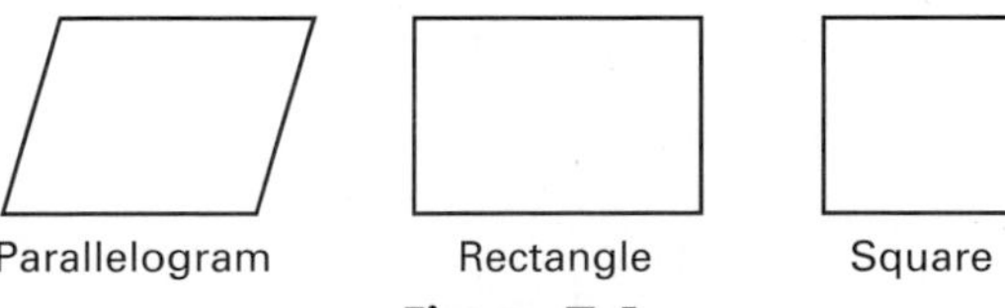

Figure 7-1

■ **DEFINITION 7-1.2 Rectangle.** A **rectangle** is a parallelogram whose angles are all right angles (Fig. 7-1).

■ **DEFINITION 7-1.3 Square.** A **square** is a parallelogram whose sides are all of equal length and whose angles are all right angles (Fig. 7-1).

■ **Definition 7-1.4 Perimeter.** The **perimeter** is the total length of the sides of a plane figure (Fig. 7-2).

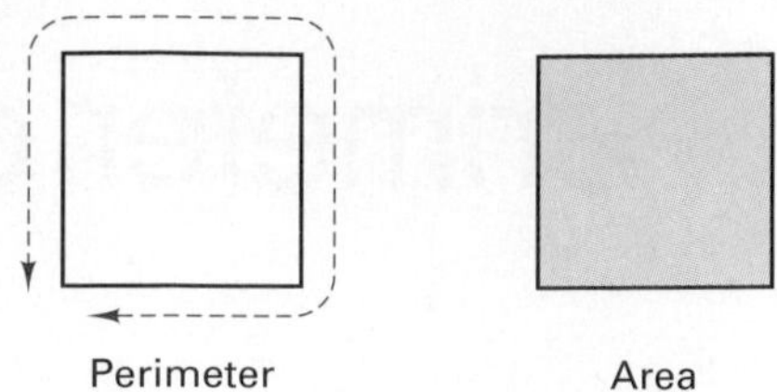

Figure 7-2

■ **Definition 7-1.5 Area.** The **area** is the amount of surface of a plane figure (Fig. 7-2).

L.O.1. Find the perimeter and area of a square.

Perimeter and Area of a Square

If a carpenter had to install baseboard molding in a newly constructed den or family room, he or she would need to know how many feet of baseboard molding were needed. The carpenter needs to know the distance around the room, that is, the perimeter of the room. According to the plans, the room is square and measures 18 ft on each side (Fig. 7-3).

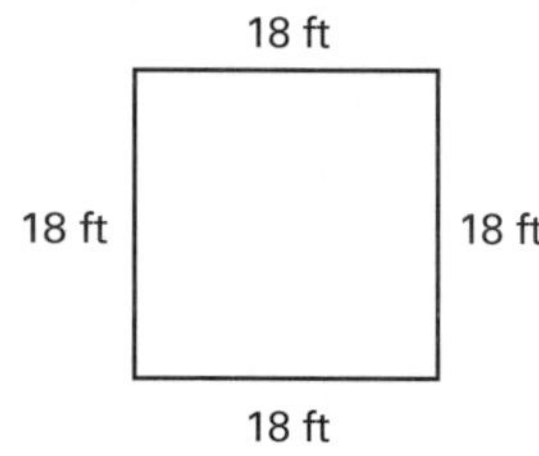

Figure 7-3

The perimeter, or distance around the room, would be figured by adding together the length of each wall. 18 ft + 18 ft + 18 ft + 18 ft = 72 ft. Since all sides are equal and there are four sides, we can multiply 4×18.

> **Formula 7-1.1** *Perimeter of a square:*
>
> $$P = 4s$$

An expression like $4s$ means 4 times the value represented by s.

Example 7-1.1 How much aluminum edge molding is needed to surround a stainless steel kitchen sink that measures 40 cm on each side?

Solution

$P = 4s$

$P = 4(40 \text{ cm})$ *(Substitute measurement of one side. Multiply.)*

$P = 160 \text{ cm}$

The sink requires 160 cm of molding. ■

On the other hand, the area of a square is the entire amount of surface that the square occupies. If we had a square that measured 6 in. on a side, it would be represented as in Fig. 7-4.

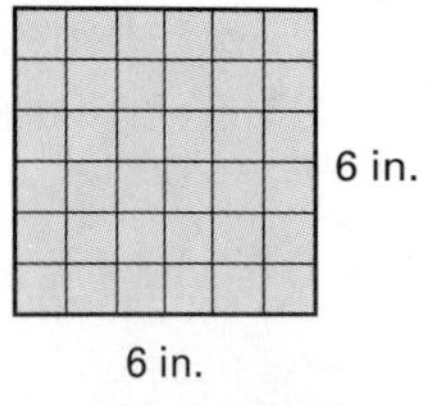

Figure 7-4

Each small square in the larger figure is a *square inch*, that is, a square measuring 1 in. on each side. If we count the number of square inches in the larger figure, we find 36. This is the area of the square in question. The area can be expressed as 36 sq in. or 36 in.^{2}. Instead of counting each square inch in the larger square, we could have arrived at 36 sq in. or 36 in.^{2} by multiplying one side by the other (6 in. $\times$ 6 in. = 36 in.^{2}). From this we get the formula for the area of a square.

Tips and Traps

The superscript 2 (small raised 2) following a unit of measure is a number that indicates *square measure* or area. Thus, 5 in.^{2} = 5 square inches, 23 ft^2 = 23 square feet, 120 cm^2 = 120 square centimeters, and so on. The number 2 following a unit measure is merely a shortcut way of expressing a measure that has *already been "squared."*

Formula 7-1.2 *Area of square:*

$$A = s^2$$

The expression s^2 means $s \times s$, just as 5^2 means 5×5.

EXAMPLE 7-1.2 Lynn Fly, who installs vinyl flooring, discovered that one of the squares in the flooring pattern was damaged and decided to cut out the damaged square and replace it. The damaged square measured 20 cm on each side. What is the area of the square to be replaced?

Solution

$A = s^2$

$A = (20 \text{ cm})^2$ *(Substitute measurement of one side. Square the measurement.)*

$A = 400 \text{ cm}^2$

The vinyl square measures 400 cm^2. ■

Tips and Traps

In evaluating a formula the measuring units are often omitted from the written steps. However, it is very important to analyze the measuring units of the problem and to determine the measuring unit of the solution.

Example 7-1.1

$P = 4s$

$P = 4(40)$ *(The measuring unit is centimeters and a measure is multiplied by a number.)*

$P = 160 \text{ cm}$ *(The measuring unit of the solution is centimeters.)*

Example 7-1.2

$A = s^2$

$A = (20)^2$ *(The measuring unit is centimeters and a measure is squared or a measure is multiplied by a measure.)*

$A = 400 \text{ cm}^2$ *(The measuring unit of the solution is square centimeters.)*

L.O.2. Find the perimeter and area of a rectangle.

Perimeter and Area of a Rectangle

Finding the perimeter of a rectangle is similar to finding the perimeter of a square. That is, we add the measures of the four sides, the longer sides being called the *length* (l) and the shorter sides being called the *width* (w), as shown in Fig. 7-5.

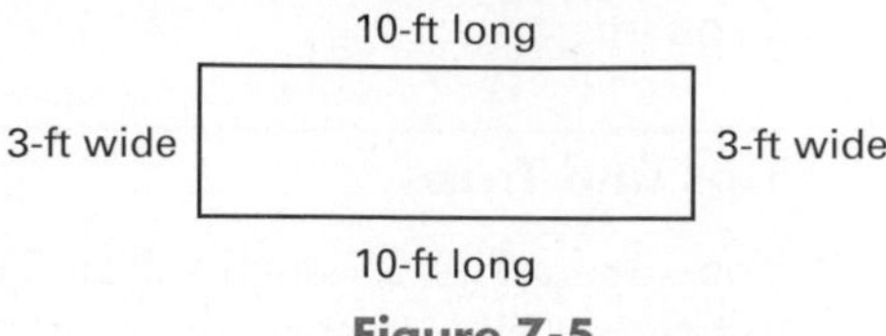

Figure 7-5

$$P = 10 + 3 + 10 + 3 = 26 \text{ ft}$$

We notice that the perimeter is composed of two lengths and two widths:

$$P = 2l + 2w$$

The distributive property shows that $2(l + w)$ and $2l + 2w$ are equivalent.

> **Formula 7-1.3** *Perimeter of a rectangle:*
> $$P = 2(l + w)$$

The formula means that *both* the length *and* the width are multiplied by 2.

EXAMPLE 7-1.3 A shop that makes custom picture frames has an order for a frame whose outside measurements will be 42 in. by 30 in. How many inches of picture frame molding will be needed for the job?

Solution

$P = 2(l + w)$

$P = 2(42 + 30)$ *(Substitute. Perform operations.)*

$P = 2(72)$

$P = 144$

In analyzing the dimensions, inches are added to inches giving the results in inches. Then, the inches are multiplied by a number and the final result is inches.

Thus, 144 in. are needed for the project. ■

In some applications, we may need to decrease the total perimeter to account for doorways and other openings in the perimeter.

EXAMPLE 7-1.4 A chain-link fence is to be installed around a yard measuring 25 m by 30 m. A gate 3 m wide will be installed to allow entrance of an automobile. The gate comes preassembled from the manufacturer. How much fencing does the installer need to put up the fence?

Solution

$P = 2(l + w) - 3$ *(Subtract length of gate.)*

$P = 2(30 + 25) - 3$ *(Substitute. Perform operations.)*

$P = 2(55) - 3$

$P = 110 - 3$

$P = 107 \text{ m}$

The job requires 107 m of fencing. ■

The area of a rectangle can be found by dividing it into smaller squares and counting the number of squares (Fig. 7-6). By count, there are 15 square meters or 15 m^2 in the rectangle. Multiplying length by width also gives us 15 m^2.

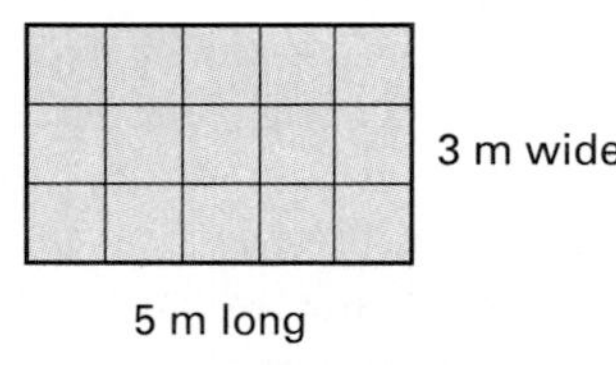

Figure 7-6

> **Formula 7-1.4** *Area of a rectangle:*
>
> $$A = lw$$

The expression lw represents length times width.

EXAMPLE 7-1.5 A carpet installer needs to carpet a room measuring 16 ft by 20 ft. Projecting out from one wall is a fireplace whose hearth measures 3 ft by 6 ft (Fig. 7-7). How many square yards of carpet does the installer require for the job? How much is wasted?

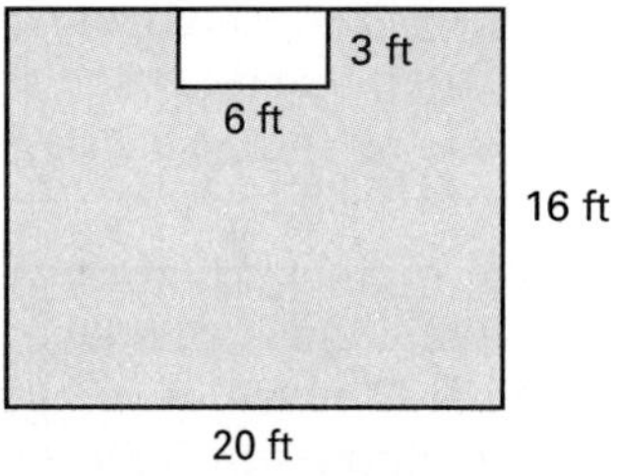

Figure 7-7

Solution: This problem has several steps. First, compute the area of the room without considering the area of the hearth. This is the amount of carpet needed for the job. Second, compute the area of the fireplace hearth. This is the amount wasted. Finally, analyze the dimensions to see if the result is expressed using the desired measuring unit.

Let A_1 = area of room and A_2 = area of hearth.

Tips and Traps

The subscript (small lowered 1) following a letter is a descriptive notation. Thus, A_1 and A_2 are two different areas. The number of the subscript does not affect the calculations of the problem.

Letters or words can also be used as subscripts.

$$A_{\text{room}} = \text{area of room}, \qquad A_{\text{hearth}} = \text{area of hearth}$$

$A_1 = lw$

$A_1 = 20 \times 16$ *(Substitute and multiply.)*

$A_1 = 320 \text{ ft}^2$ *(area of room)*

$A_2 = lw$

$A_2 = 3 \times 6$ *(Substitute and multiply.)*

$A_2 = 18 \text{ ft}^2$ *(area of hearth)*

However, since carpet is sold in square yards, we must convert the square footage to square yards. Since $9 \text{ ft}^2 = 1 \text{ yd}^2$, we can use a *unity ratio* to convert square feet to square yards.

A unity ratio, as we recall from Chapter 6, is a fraction with one unit of measure in the numerator and a different, but equivalent, unit of measure in the denominator. A unity ratio contains the original unit and the new unit. Now we can complete Example 7-1.5 using the unity ratio $\frac{1 \text{ yd}^2}{9 \text{ ft}^2}$.

$$\frac{320 \cancel{\text{ft}^2}}{1} \times \frac{1 \text{ yd}^2}{9 \cancel{\text{ft}^2}} = \frac{320 \text{ yd}^2}{9} = 35\frac{5}{9} \text{ yd}^2, \qquad \frac{\overset{2}{\cancel{18}} \cancel{\text{ft}^2}}{1} \times \frac{1 \text{ yd}^2}{\underset{1}{\cancel{9}} \cancel{\text{ft}^2}} = 2 \text{ yd}^2$$

This job requires $35\frac{5}{9}$ yd² of carpet. Of this, 2 yd² is waste that can be used for carpeting smaller areas such as closets. ■

In other applications, we may find that additional calculations are necessary in addition to finding one or more areas. For instance, in estimating the amount of paint, wallcovering, tongue-and-groove flooring, roofing, shingles, siding, floor tiles, bricks, and similar building materials, consideration must be given to the decrease or increase in area coverage of these materials upon installation. Generally, such decreases and increases are standardized by the industry and made known to the user. For example, a paint manufacturer will indicate on the product label the number of square feet of coverage on bare wood and a different number of square feet of coverage on previously painted surfaces, or a roofing shingle manufacturer will indicate how many bundles of shingles are needed per 100 ft² of roof depending on the amount of shingle overlap desired.

EXAMPLE 7-1.6 If a $\frac{1}{2}$-in. mortar joint is used to build a wall of 2-in. × 4-in. × 8-in. bricks, six bricks will cover one square foot. If a $\frac{1}{4}$-in. mortar joint is used, seven bricks will be needed to cover 1 square foot. How many bricks would be needed to build a 50-ft × 14-ft wall with $\frac{1}{2}$-in. joints? With $\frac{1}{4}$-in. joints?

Solution

$$A = 50 \times 14 \qquad \textit{(Find area of wall.)}$$

$$A = 700 \text{ ft}^2$$

$$\frac{700 \cancel{\text{ft}^2}}{1} \times \frac{6 \text{ bricks}}{1 \cancel{\text{ft}^2}} = 4200 \text{ bricks} \qquad \textit{(Find number of bricks at 6 per ft}^2\textit{, using a unity ratio.)}$$

$$\frac{700 \cancel{\text{ft}^2}}{1} \times \frac{7 \text{ bricks}}{1 \cancel{\text{ft}^2}} = 4900 \text{ bricks} \qquad \textit{(Find number of bricks at 7 per ft}^2\textit{, using a unity ratio.)}$$

Using $\frac{1}{2}$-in. mortar joints, 4200 bricks are needed. Using $\frac{1}{4}$-in. mortar joints, 4900 bricks are needed. ■

EXAMPLE 7-1.7 Asphalt roofing shingles are sold in bundles. The number of bundles needed to cover a square (100 ft^2) depends on the amount of overlap when the shingles are installed. If the overlap allows 4 in. of each shingle to be exposed, then four bundles are needed per square. With a 5-in. exposure, 3.2 bundles are needed per square. Figure the number of bundles for a 4-in. exposure and for a 5-in. exposure for a roof measuring 30 ft × 20 ft.

Solution

$$A = 30 \times 20 \qquad \textit{(Find area of roof.)}$$

$$= 600 \text{ ft}^2$$

$$\frac{\overset{6}{\cancel{600}} \cancel{\text{ft}^2}}{1} \times \frac{1 \text{ square}}{\underset{1}{\cancel{100}} \cancel{\text{ft}^2}} = 6 \text{ squares} \qquad \textit{[Find number of squares (100 ft}^2\textit{ areas), using a unity ratio.]}$$

$$6 \times 4 = 24 \text{ bundles} \qquad \textit{(Find number of bundles for 4-in. exposure.)}$$

$$6 \times 3.2 = 19.2 \textit{ or } 20 \text{ bundles} \qquad \textit{(Find number of bundles for 5-in. exposure.)}$$

Therefore, 24 bundles of shingles are required for a 4-in. exposure, and 20 bundles (from 19.2 bundles) are required for a 5-in. exposure. ■

L.O.3. Find the perimeter and area of a parallelogram.

Perimeter and Area of a Parallelogram

The perimeter of a parallelogram, like the perimeter of a rectangle, is the sum of its four sides. Instead of naming the sides of the parallelogram length and width, we will use the terms *base* and *adjacent side*. Notice the locations of the base and adjacent side in Fig. 7-8.

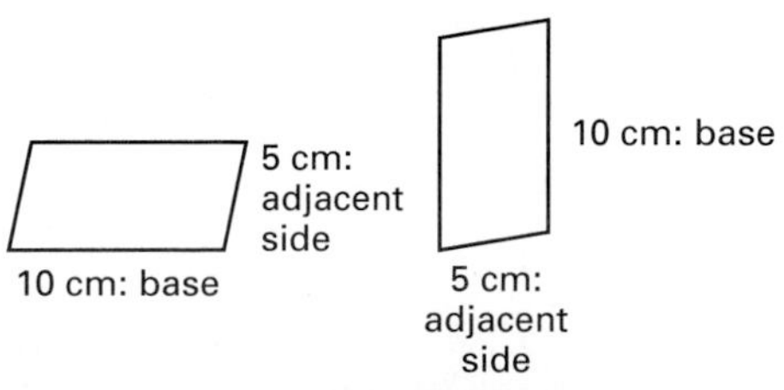

Figure 7-8

■ **Definition 7-1.6 Base.** The **base** of any polygon is a horizontal side or a side that would be horizontal if the polygon's orientation were modified.

■ **Definition 7-1.7 Adjacent Side.** The **adjacent side** of any polygon is a side that has an end point in common with the base.

The formula for the perimeter of a parallelogram is twice the sum of the base and adjacent side.

> **Formula 7-1.5** *Perimeter of a parallelogram:*
>
> $$P = 2(b + s)$$

This formula is similar to the formula for the perimeter of a rectangle. However, for the area of a parallelogram, we must use still other terminology and a somewhat different formula than we used for the area of a rectangle. Study the parallelogram in Fig. 7-9.

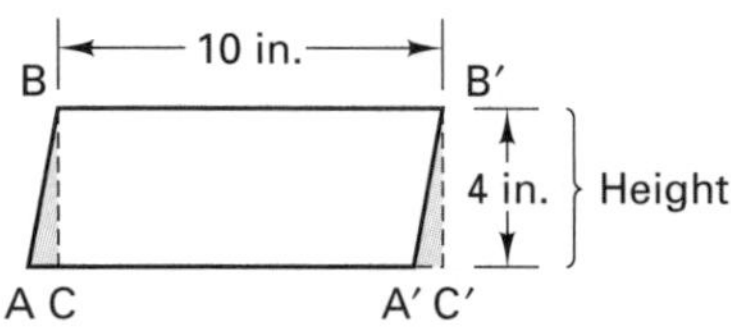

Figure 7-9

The *height* of a parallelogram or of any other geometric figure must be clearly understood and defined. Sometimes height is called *altitude*.

■ **Definition 7-1.8 Height.** The **height** of a polygon is the perpendicular distance from the base to the highest point of the polygon above the base.

If triangle ABC ($\triangle ABC$) in Fig. 7-9 were transposed to the right side of the parallelogram, we would have a rectangle. The area, therefore, would be the product of the *base* times the *height*. That is, $10 \times 4 = 40$ in.^{2}.

> **Formula 7-1.6** *Area of a parallelogram:*
>
> $$A = bh$$

Example 7-1.8 Find the perimeter and the area of a parallelogram with a base of 16 in., an adjacent side of 8 in., and a height of 7 in.

Solution: Visualize the parallelogram.

8 in.
7 in.
16 in.

$P = 2(b + s)$	$A = bh$	
$P = 2(16 + 8)$	$A = 16 \times 7$	*(Substitute.)*
$P = 2(24)$	$A = 112$ in.^{2}	
$P = 48$ in.		

■

General Tips for Using the Calculator

Many calculations for area and perimeter require only multiplication. However, the area of a square and the perimeters of rectangles and parallelograms require other operations.

Area of a Square

Assume a square with 5-in. sides.

Scientific calculator:

[AC] 5 [x^2] ⇒ 25

Graphing calculator:

[AC] 5 [x^2] [EXE] ⇒ 25

Note that [EXE] or [ENTER] is required on a graphing calculator. Also, some calculators may require the use of another key to activate the [x^2] key, particularly if the symbol x^2 is written above the key, $\overset{x^2}{[\sqrt{x}]}$. The key that activates the operations above a key may be labeled shift, 2nd, or INV.

Perimeter of a Rectangle or Parallelogram

Assume a figure 10 cm long and 5 cm on the short side.

Scientific calculator:

[AC] 5 [+] 10 [=] [×] 2 [=] ⇒ 30

The sum is entered first.

Graphing calculator:

[AC] 2 [(] 5 [+] 10 [)] [EXE] ⇒ 30

SELF-STUDY EXERCISES 7-1

L.O.1. Find the perimeter and the area of the following figure.

1. (Square figure with sides labeled 3 cm and 3 cm)

2. Name the figure in Exercise 1.

Solve the following problems involving perimeter and area.

3. Marjorie Young will wallpaper a laundry room 8 ft by 8 ft by 8-ft high. How many square feet of paper will be needed if there are 63 ft^2 of openings in the room?

4. A square parking lot is to have curbs built on all four sides. If the lot is 150 ft on each side, how many feet of curb are needed? Allow 10 ft for a driveway into the parking lot.

5. Making no allowances for bases, the pitcher's mound, or the home plate area, how many square yards of artificial turf are needed to resurface an infield at an indoor baseball stadium? The infield is 90 ft on each side. (9 ft^2 = 1 yd^2.)

6. Ted Davis is a farmer who wants to apply fertilizer to a 40-acre field with dimensions $\frac{1}{4}$ mi × $\frac{1}{4}$ mi. Find the area in square miles.

7. If $\frac{1}{4}$ mi is 1320 ft, how many feet of fencing would be needed to enclose the field in Exercise 6, assuming that a 12-ft steel gate will be installed?

8. A 36-in. × 36-in. ceramic tile shower stall will be installed. How many 4-in. × 4-in. tiles will be needed to cover the floor? Disregard the drain opening.

9. A border of 4-in. × 4-in. wall tiles surrounds the floor of the shower in Exercise 8. How many tiles are needed for this border?

10. A 20-in. × 20-in. central heating and air conditioning return air vent will be installed in a wall. Find the perimeter and area of the wall opening.

L.O.2. Find the perimeter and the area of the following figure.

11.

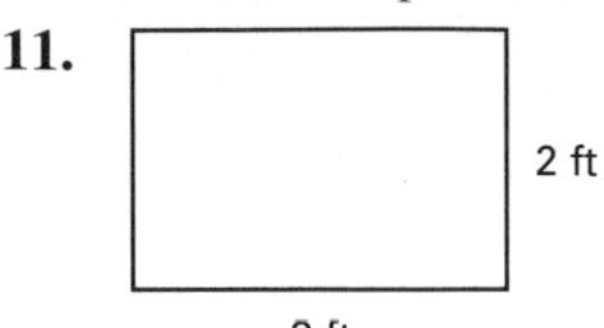

12. Name the figure in Exercise 11.

Solve the following problems involving perimeter and area.

13. A rectangular parking lot is 340 ft by 125 ft. Find the number of square feet in the parking lot.

14. A room is 15 ft by 12 ft. How many square feet of flooring are needed for the room?

15. Wallpaper will be used to paper a small kitchen 9 ft by 10 ft by 8-ft high. How many square feet of paper will be needed if there are 63 ft^2 of openings in the kitchen?

16. How many feet of quarter-round molding are needed to finish around the baseboard after sheet vinyl flooring is installed if the room is 16 ft by 18 ft and there are three 3-ft-wide doorways?

17. The dimensions of a sun porch are 9 ft 6 in. by 15 ft. How many board feet of 1-in.-thick tongue-and-groove flooring are needed for the porch? (Board feet for 1-in.-thick lumber is same number as ft^2.) Disregard waste.

18. The swimming pool in Fig.7-10 measures 32 ft by 18 ft. How much fencing is needed, including material for a gate, if the fence is to be built 7 ft from each side of the pool?

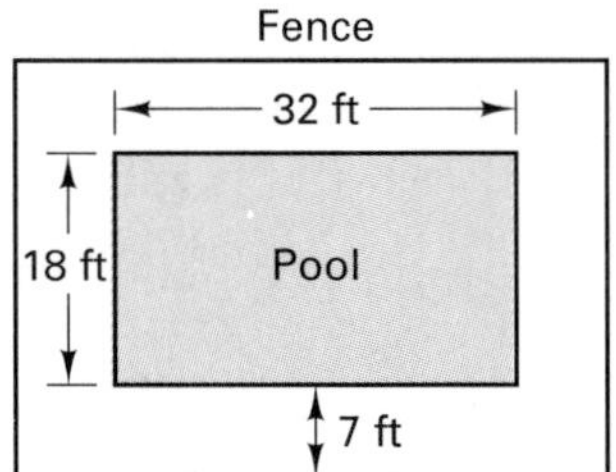

Figure 7-10

19. A Formica tabletop measures 40 in. by 62 in. How many feet of edge trim would be needed? (12 in. = 1 ft)

20. A countertop requires rolled edging to be installed on all four sides. How much rolled edging material is needed if the countertop measures 25 in. × 40 in.?

L.O.3. Find the perimeter and area of the following figure.

21.

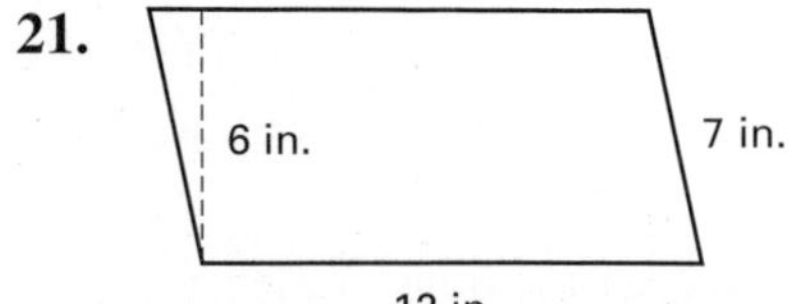

22. Name the figure in Exercise 21.

Solve the following problems involving perimeter and area.

23. An illuminated sign advertising a place of business is shaped like a parallelogram with a base of 48 in. and an adjacent side of 30 in. How many feet of aluminum molding are needed to frame the sign?

24. Find the area of a parking lot shaped like a parallelogram 200 ft long if the perpendicular distance between the sides is 45 ft (Fig. 7-11).

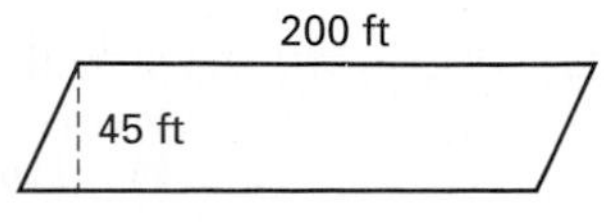

Figure 7-11

25. A customized van will have a window cut in each side in the shape of a parallelogram with a base of 20 in. and an adjacent side of 11 in. How many inches of trim are needed to surround the two windows?

26. A design in a marble floor will feature parallelograms with a base of 48 in. and a height of 24 in. If the figure will be made of 8-in. × 8-in. marble tiles, how many tiles will be needed for each figure, assuming no waste?

27. A machine cuts sheet metal into parallelograms measuring 6 cm along the base and 3 cm in height. How many can be cut from a piece of sheet metal 100 cm × 200 cm assuming no waste?

28. A contemporary building will have a window in the shape of a parallelogram with a base of 50 in. and an adjacent side of 30 in. How many inches of trim are needed to surround the window?

29. A table for a reading lab will have a top in the shape of a parallelogram with a base of 36 in. and an adjacent side of 18 in. How many inches of edge trim are needed to surround the table top?

30. Signs in the shape of parallelograms with a base of 24 in. and a height of 10 in. will be cut from sheet metal. How many whole signs can be cut from a piece of sheet metal 48 in. × 48 in.? Illustrate your answer with a drawing.

7-2 AREA AND CIRCUMFERENCE OF A CIRCLE

Learning Objectives

L.O.1. Find the circumference of a circle.
L.O.2. Find the area of a circle.
L.O.3. Solve applied problems involving circles.

Tanks, rings, pipes, pillars, arches, gears, pulleys, wheels, and similar items are all based on the *circle* in one way or another. In many cases the circle or some part of the circle is found in composite figures in combination with one or more polygons. In this section we will study circles, their circumferences, and their areas.

Most of us are familiar with circles and some of their properties. However, for technical applications, it is necessary to clarify what a circle is and what some of its properties are. Refer to Fig. 7-12.

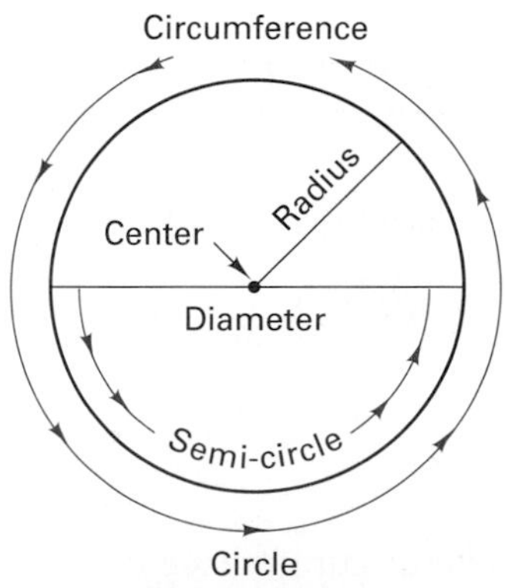

Figure 7-12

■ **DEFINITION 7-2.1 Circle.** A **circle** is a closed curved line whose points lie in a plane and are the same distance from the *center* of the figure.

■ **DEFINITION 7-2.2 Radius.** A **radius** (plural: *radii*, pronounced ''ray · dē · ī'') is a line segment from the center of a circle to a point on the circle. It is half the diameter.

■ **DEFINITION 7-2.3 Diameter.** The **diameter** of a circle is the line segment from a point on the circle through the center to another point on the circle.

■ **DEFINITION 7-2.4 Circumference.** The **circumference** of a circle is the perimeter or length of the closed curved line that forms the circle.

■ **DEFINITION 7-2.5 Semicircle.** A **semicircle** is one-half a circle and is created by drawing a diameter.

L.O.1. Find the circumference of a circle.

Circumference of a Circle

The circle is a geometric form that has a special relationship between its circumference and its diameter. If we divide the circumference of any circle by the diameter, the quotient is always the same number. This number is a nonrepeating, nonterminating decimal approximately equal to 3.1415927 to seven decimal places. The Greek letter π (pronounced ''pie'') is used to represent this value. Convenient approximations often used in calculations involving π are $3\frac{1}{7}$ and 3.14.

Since the quotient of the circumference and the diameter is π and since division and multiplication are inverse operations, the circumference equals π times the diameter. Because the diameter is twice the radius, the circumference can also equal π times twice the radius.

Formula 7-2.1 *Circumference of a circle:*

$$C = \pi d \quad \text{or} \quad C = 2\pi r$$

Tips and Traps

All calculations involving π will be approximations. Many calculators include a π function where π to seven or more decimal places can be used by pressing a single key. Other calculators may require use of another key to activate the $\boxed{\pi}$ key. However, in making calculations by hand or with a calculator not having the $\boxed{\pi}$ function, 3.14 is sometimes adequate. **This text uses the calculator value 3.141592654 for π.**

EXAMPLE 7-2.1 What is the circumference of a circle whose diameter is 1.3 m? Round to tenths.

Solution

$C = \pi d = \pi(1.3) = 4.08407045$ *(The $\boxed{\pi}$ key on scientific or graphing calculators is used for value of π.)*

The circumference is 4.1 m (rounded). ■

Tips and Traps

For ease in using a calculator, convert mixed U.S. customary linear measurements to their decimal equivalents if they terminate conveniently. For instance, if a diameter is 7 ft 6 in., convert it to 7.5 ft (from $7\frac{6}{12}$ ft, in which $\frac{6}{12} = 0.5$).

EXAMPLE 7-2.2 Find to the nearest hundredth the circumference of a circle whose radius is 1 ft 9 in.

Solution

$C = 2\pi r$

$C = 2\pi(1.75)$ *(Substitute for π and r. $1\frac{9}{12}$ ft = 1.75 ft)*

$C = 10.99557429$

$C = 11.00$ in. *(rounded)*

or

$C = \pi d$

$C = \pi(3.5)$ *(Diameter is 2 × radius; 1.75 × 2 = 3.5.)*

$C = 10.99557429$

$C = 11.00$ *(rounded)* ■

L.O.2. Find the area of a circle.

Area of a Circle

The area of a circle, like the circumference, is obtained from the relationships within the circle itself. If we divide a circle into two semicircles and then subdivide each semicircle into a number of pie-shaped pieces, we have something like *A* in Fig. 7-13. If we then spread out the upper pie-shaped pieces and lower pie-shaped pieces, we would have a figure like *B* in Fig. 7-13. Now if we pushed the upper pieces and lower pieces together, the result would approximate a rectangle whose length is $\frac{1}{2}$ the circumference and whose width is the radius. Thus, the area of the circle would be approximately the area of the rectangle, that is, length times width. Since the length of the rectangle is one-half the circumference and the width is the radius, the area of a circle equals one-half the circumference times the radius.

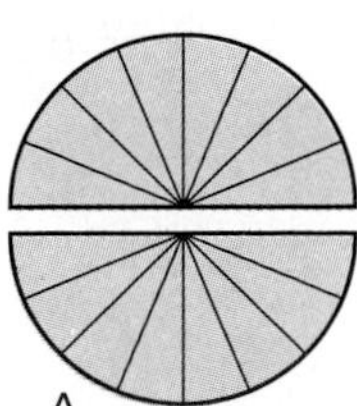

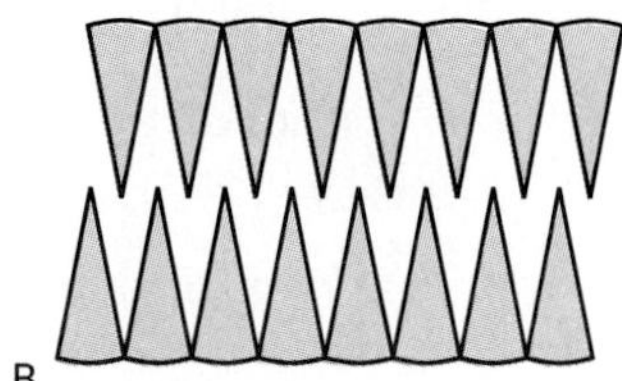

Figure 7-13

$$A = \frac{1}{2}C \times r$$ *(Since the formula for circumference is $C = 2\pi r$, we can substitute $2\pi r$ for C.)*

$$A = \frac{1}{2}(2\pi r)(r)$$

$$A = \frac{1}{\cancel{2}_1}(\cancel{2}^1\pi r)(r)$$ *(Multiply. Reduce where possible.)*

$$A = \pi r^2$$

Formula 7-2.2 *Area of a circle:*

$$A = \pi r^2$$

Recall that r^2 means $r \times r$.

EXAMPLE 7-2.3 Find the area of a circle whose radius is 8.5 m. Round to tenths.

Solution

$$A = \pi r^2 = \pi(8.5)^2 = \pi(72.25) = 226.9800692 \text{ m}^2$$

The area of the circle is 227.0 m^2. ■

Recall in the order of operations that powers precede multiplication.

Tips and Traps

When mixed U.S. customary measures do not convert to convenient, terminating decimal equivalents, a continuous calculator sequence may be preferred.

EXAMPLE 7-2.4 Find the area to the nearest hundredth of the top of a circular tank with a diameter of 12 ft 8 in.

Solution: If we are using a calculator, we can calculate the area using continuous steps as shown.

$$A = \pi r^2$$

$$A = (\pi)\left(\frac{12 + \frac{8}{12}}{2}\right)^2$$ *(Follow order of operations. 8 in. $= \frac{8}{12}$ ft. The diameter, $12 + \frac{8}{12}$, is divided by 2 to equal the radius.)*

$$A = 126.012772 \text{ ft}^2$$ *(scientific calculator result)*

$$A = 126.01 \text{ ft}^2$$ *(rounded)* ■

The area of the top of the tank is 126.01 ft^2 or 125.95 ft^2 (basic calculator using $\pi = 3.14$). ■

General Tips for Using the Calculator

The calculator sequence for Example 7-2.4 follows:

Scientific calculator:

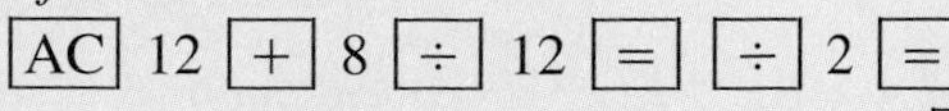

[x^2] [×] [π] [=] ⇒ 126.012772

Graphing calculator:

[AC] 12 [+] 8 [÷] 12 [EXE] [÷] 2 [EXE] [x^2]

[×] [π] [EXE] ⇒ 126.0127721

An alternate sequence of keystrokes will result if the parentheses keys are used. Experiment with your calculator to find other calculator sequences.

L.O.3. Solve applied problems involving circles.

Selected Applications of the Circle

EXAMPLE 7-2.5 A 15-in.-diameter wheel has a 3-in. hole in the center. Find the area of a side of the wheel to the nearest tenth (Fig. 7-14).

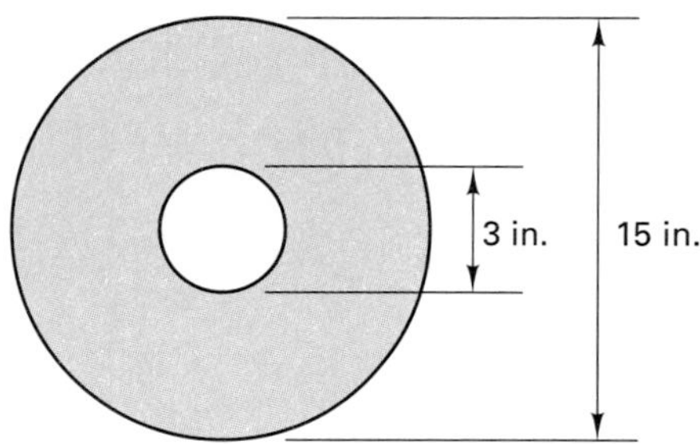

Figure 7-14

Solution: We are asked to find the area of the shaded portion of the wheel in Fig. 7-14. To do so, we find the area of the larger circle (diameter 15 in.) and *subtract* from it the area of the smaller circle (diameter 3 in.). The shaded portion is called a *ring*.

$$A_1 = \pi r^2 \quad \left(r = \frac{15}{2} = 7.5\right) \qquad A_2 = \pi r^2 \quad \left(r = \frac{3}{2} = 1.5\right)$$

$$A_1 = \pi(7.5)^2 \qquad A_2 = \pi(1.5)^2$$

$$A_1 = \pi(56.25) \qquad A_2 = \pi(2.25)$$

$$A_1 = 176.7145868 \text{ in.}^2 \qquad A_2 = 7.068583471 \text{ in.}^2$$

Area of wheel (ring) $= A_1 - A_2$

$$A = 176.7145868 - 7.068583471 = 169.6460033 \quad \text{or} \quad 169.6 \text{ in.}^2$$

The area of the wheel (ring) is 169.6 in.^{2} ■

EXAMPLE 7-2.6 A bandsaw has two 25-cm wheels spaced 90 cm between centers (Fig. 7-15). Find the length of the continuous saw blade.

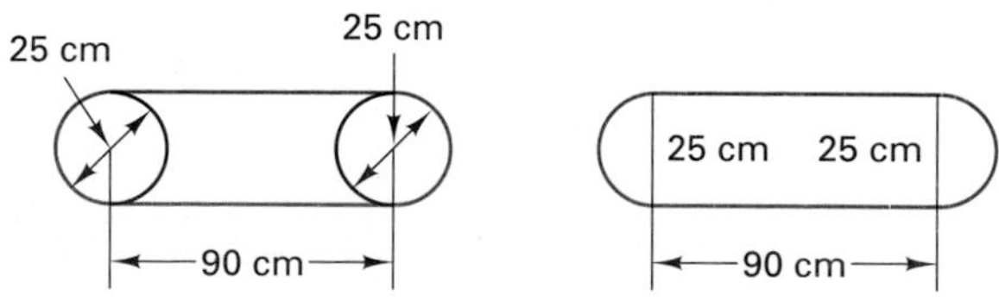

Figure 7-15

Solution: This layout is a composite figure. The figure consists of a semicircle at each end and a rectangle in the middle. It is called a *semicircular-sided* figure. Since the two semicircles equal one whole circle, we need to find the circumference of one circle (wheel) and add it to the lengths of the two sides of the rectangle.

$C = \pi d$	Total length of blade $= C + 2l$
$C = \pi(25)$	Total length of blade $= 78.53981634 + 2(90)$
$C = 78.53981634$ cm	Total length of blade $= 258.5$ cm *(rounded)*

The bandsaw blade is 258.5 cm in length. ■

EXAMPLE 7-2.7 Find the cutting speed of a lathe if a piece of work 7 in. in diameter turns on a lathe at 75 rpm.

Solution: The *cutting speed* is the speed of a tool that passes over the work, such as the speed of a sander as it sands (passes over) a piece of wood. If the cutting speed is too fast or too slow, safety and quality are impaired. The formula for cutting speed is CS $= C$ (in feet) $\times$ rpm

CS = cutting speed

C = circumference or one revolution (in feet)

rpm = revolutions per minute

Cutting speed is measured in *feet per minute* (ft/min).

$C = \pi d$

$C = \pi(7)$

$C = 21.99114858$ in.

$C = 1.832595715$ ft

(Convert 21.99114858 in. into feet using a unity ratio.

$$\frac{21.99114858 \text{ in.}}{1} \times \frac{1 \text{ ft}}{12 \text{ in.}} = 1.832595715 \text{ ft.})$$

CS $= C \times$ rpm $= 1.832595715 \times 75 = 137$ ft/min. *(rounded)*

The cutting speed of the lathe is approximately 137 ft/min. ■

EXAMPLE 7-2.8 A 4-in.-inside-diameter pipe and a 6-in.-inside-diameter pipe flow into a third pipe that carries off the flow of the two pipes. What should be the interior diameter (to the nearest tenth) of the third pipe so that it accepts the *combined flow*?

Solution: The cross-sectional area of the third pipe must be at least the same as the sum of the cross-sectional areas of the two smaller pipes.

Combined area $= A_1 + A_2$

$$\left(r_1 = \frac{4}{2} = 2;\ r_2 = \frac{6}{2} = 3\right)$$

$A_1 = \pi r_1^2$	$A_2 = \pi r_2^2$
$A_1 = \pi(2)^2$	$A_2 = \pi(3)^2$
$A_1 = \pi(4)$	$A_2 = \pi(9)$
$A_1 = 12.56637061 \text{ in.}^2$	$A_2 = 28.27433388 \text{ in.}^2$

Combined area $= A_1 + A_2 = 40.8407045 \text{ in.}^2$

Now, use the combined area to find the diameter ($2 \times$ radius) of the third pipe.

A variation of the formula for the area of a circle is $r = \sqrt{\frac{A}{\pi}}$. This variation allows you to find the radius of a circle when the area is known. An in-depth study of the techniques for finding variations of formulas will be given in Chapter 13.

$r = \sqrt{\frac{40.8407045}{\pi}}$ *(Substitute for A and π.)*

$r = \sqrt{13}$ in.

$r = 3.605551275$ in. *(from calculator)*

$d = 2r$ *(Since diameter is twice the radius, we find diameter by using $d = 2r$.)*

$d = 2(3.605551275) = 7.211102551$ in.

The diameter of the third pipe needs to be slightly more than 7.2 in. for the combined flow. ■

SELF-STUDY EXERCISES 7-2

L.O.1. Find the circumference of circles with the following dimension. Round to tenths.

1. diameter = 8 cm

2. radius = 3 in.

3. radius = 1.5 ft

4. diameter = 5.5 m

L.O.2. Find the shaded area of the following figures to the nearest tenth.

5.

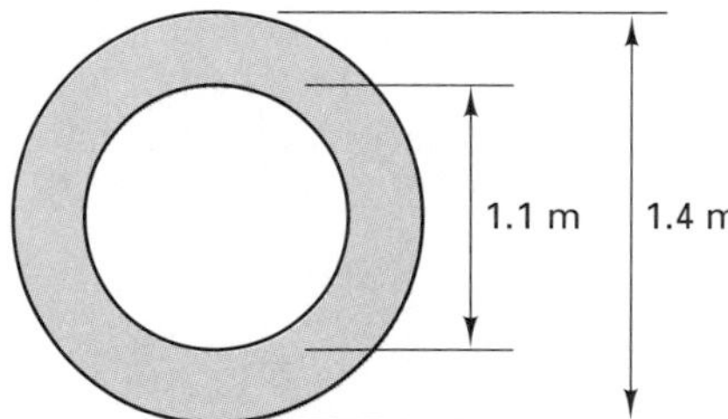

6.

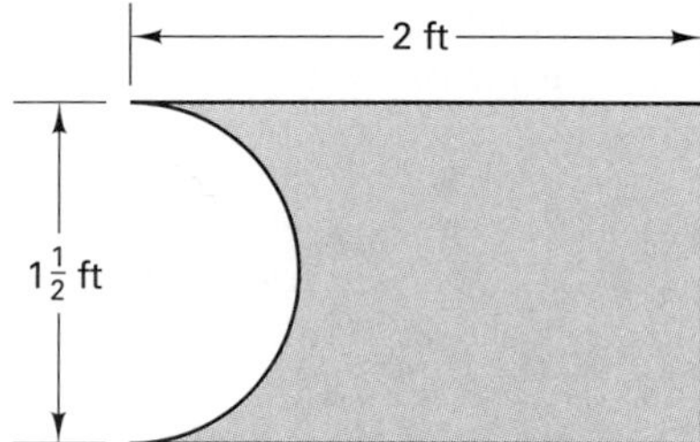

7.

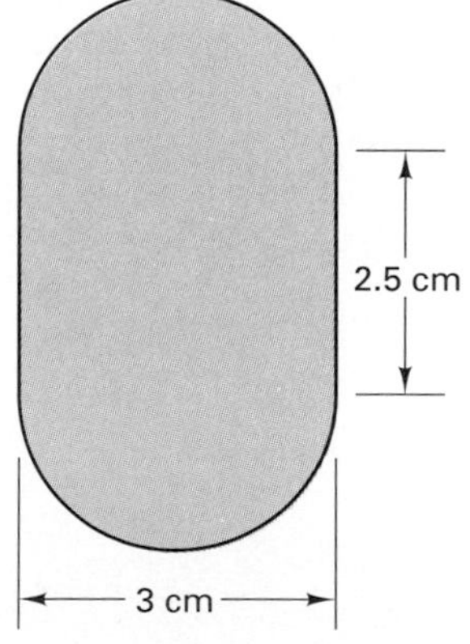

8.

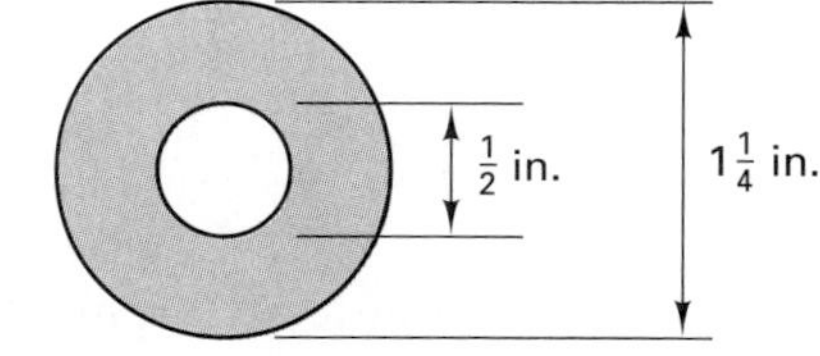

L.O.3. Solve the following applied problems.

9. A swimming pool is in the form of a semicircular-sided figure. Its width is 20 ft and the parallel portions of the sides are each 20 ft (Fig. 7-16). What is the area of a 5-ft-wide walk surrounding the pool?

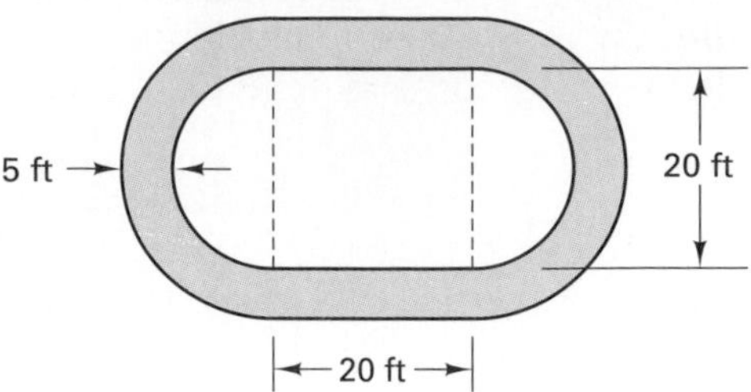

Figure 7-16

10. A belt connecting two 9-in.-diameter drums on a conveyor system needs replacing. How many inches must the new belt be if the centers of the drums are 10 ft apart?

11. Find the area of the ring formed by the cross-cut section of Fig. 7-17.

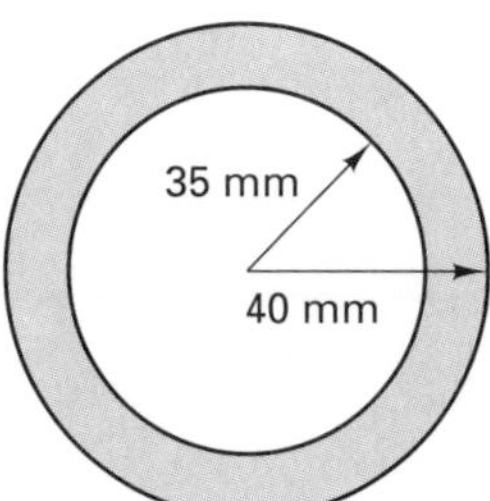

Figure 7-17

12. A 2-in.-inside-diameter pipe and a 4-in.-inside-diameter pipe empty into a third pipe whose inside diameter is 5 in. Is the third pipe large enough for the combined flow? (Justify your answer.)

13. The wall of a galvanized water pipe is 3.5 mm thick. If the outside circumference is 68 mm, what is the area (to nearest tenth) of the shaded inside cross-cut section (Fig. 7-18)?

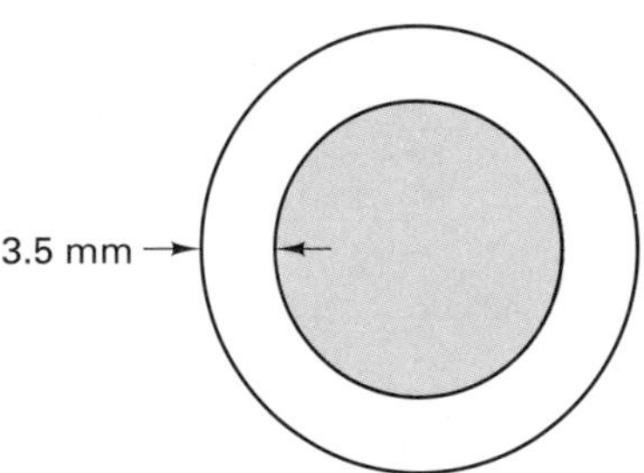

Figure 7-18

14. A large pipe whose interior cross-sectional area is 20 in.^{2} empties into two smaller pipes that each have an interior diameter of 4 in. Are the smaller pipes together large enough to carry off the flow from the larger pipe? (Justify your answer.)

15. Cutting speed, when applied to a grinding wheel, is called *surface speed*. What is the surface speed in ft/min of a 9-in.-diameter grinding wheel revolving at 1200 rpm? (Surface speed = circumference in feet × rpm.)

16. A polishing wheel 12 in. in diameter revolves at 500 rpm. What is the surface speed? (See Exercise 15 for formula.)

17. A lamp lights up effectively an area 10′ in diameter (Fig. 7-19). How many square feet does the lamp light effectively?

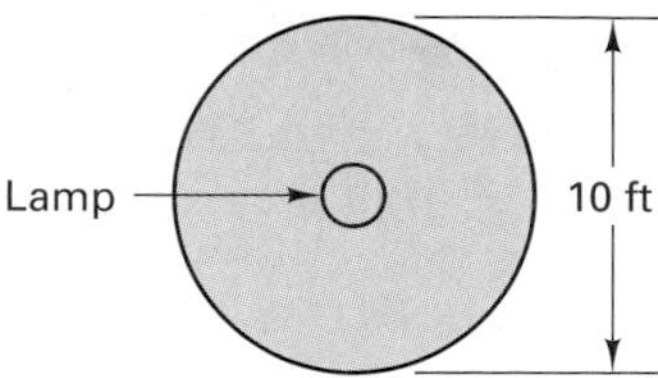

Figure 7-19

18. A steel sleeve with an outside diameter of 3 in. measures $2\frac{1}{2}$ in. across the inside. The sleeve will be babbitted to fit a 1.604-in.-diameter motor shaft. What must the thickness of the babbitt be (in hundredths)? The babbitt is shaded in Fig. 7-20. A babbitt is a lining of babbitt metal, a soft antifriction alloy, that allows two metal parts to fit snugly together.

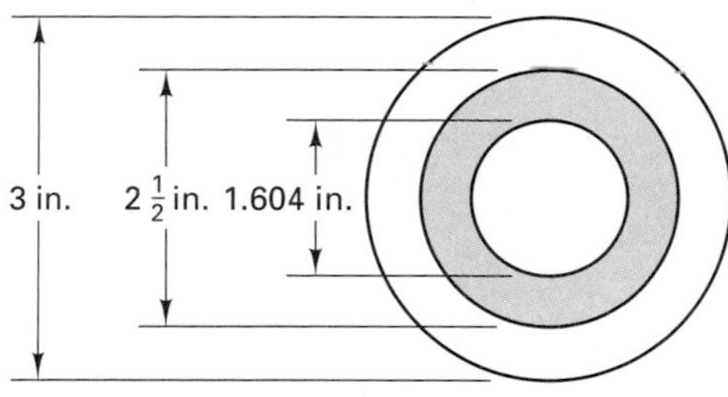

Figure 7-20

19. Two 12-cm-diameter drums are connected by a belt to form a conveyor system. The centers of the drums are 2 m apart. How long must a replacement belt be (to the nearest tenth of a meter)?

ASSIGNMENT EXERCISES

7-1

Find the perimeter and the area of the following figures.

1. 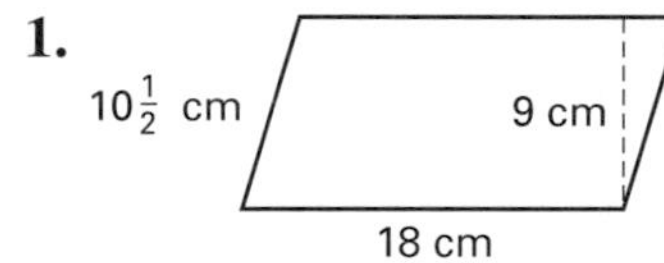

2. 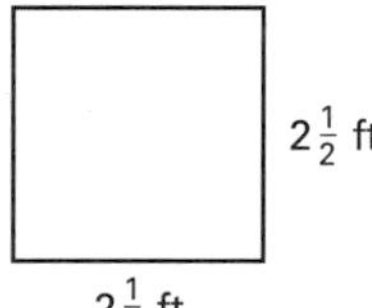

3. 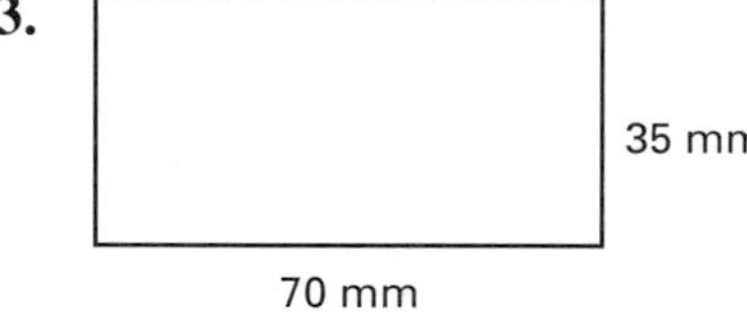

4. If a parking lot for a new hospital measures 275 ft by 150 ft, how many square feet need to be blacktopped?

5. In Exercise 4, find the amount of curbing needed to surround the parking lot if 14 feet are allowed for a driveway.

6. A hall wall with no windows or doors measures 25 ft by 8-ft high. Find the number of square feet to be covered if paneling is installed on the two walls.

7. A den 18 ft by $16\frac{1}{2}$ ft is to be carpeted. How many square yards of carpeting are needed?

8. Vincent Ores, a contractor, is to brick the front of a television repair shop that has a doorway measuring 7 ft by 6 ft. How many bricks are needed if the store front is 20 ft by 12 ft and the bricks cover at the rate of 6 per square foot using $\frac{1}{2}$-in. mortar joints?

9. A roof measuring 16 ft by 20 ft is to be covered with asphalt roofing cement. How much would the project cost if the asphalt roofing cement spreads at the rate of 150 square feet per gallon and costs $4.75 per gallon? The cement is purchased by the gallon only.

10. Estimate the number of feet of roll fencing needed to fence a square storage area measuring $15\frac{1}{2}$ ft on a side. A preassembled gate 4-ft wide will be installed.

11. A dining room wall 12 ft long by 8 ft high will be wallpapered. How many single rolls (24 in. by 20 ft) of wallpaper will be needed? Assume there is no waste.

12. A square area of land is 10,000 m^2. If a baseball field must be at least 99. 1 m along each foul line to the park fence, is the square adequate for regulation baseball?

13. Debbie Murphy is building a contemporary home with four front windows, each in the form of a parallelogram. If each window has a 5-ft base and a height of 2 ft, how many square feet of the 25-ft × 11-ft wall will require stain?

14. If the stain used on the wall in Exercise 13 is applied at a cost of $2.75 per square yard, find the cost to the nearest dollar of staining the front wall.

7-2

Find the perimeter (or circumference) and the area of the following figures. Round to hundredths.

15.

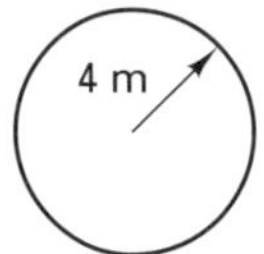

16.

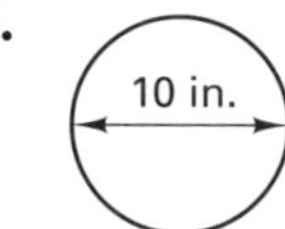

17.

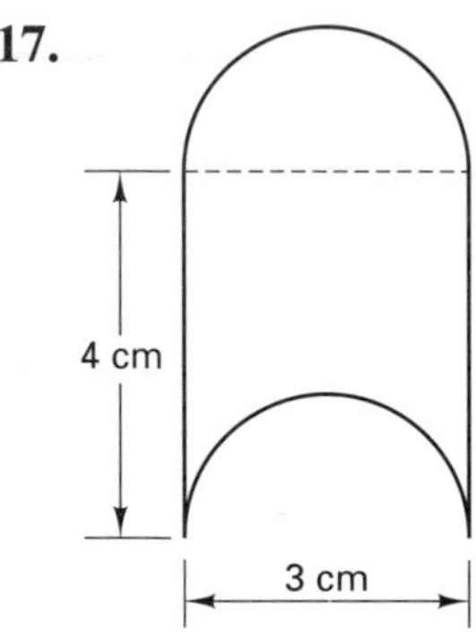

18.

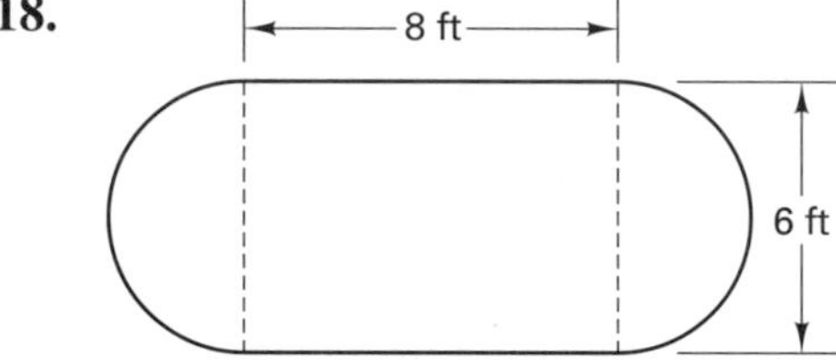

19. A circle has a diameter of 2.5 m. If the diameter were increased 1.2 m, what would be the difference in their areas (to the nearest tenth)?

20. If a circular flower bed has a radius of 24 in., how many $5\frac{3}{4}$-in.-long flat bricks would be needed to enclose it at the rate of two bricks per 12 in.?

21. Three water hoses whose interior cross-sectional areas are 1 in.^{2}, $1\frac{1}{2}$ in.^{2}, and 2 in.^{2} empty into a larger hose. What inside diameter must the larger hose be to carry away the combined flow of the three smaller hoses? Round to tenths.

22. Two 35-mm-diameter drums connect a conveyor belt. If the centers of the drums are 70 cm apart, how many cm long is the conveyor belt? Round to tenths.

23. A $\frac{1}{4}$-in. electric drill with variable speed control turns as slowly as 25 rpm. If an abrasive disk with a 5-in. diameter is attached to the drill drive shaft, what is the disk's slowest cutting speed in ft/min? Round to the nearest whole number. (Cutting speed = circumference in feet × rpm.)

24. What is the cross-sectional area of the opening in a round flue tile whose inside diameter is 8 in.? Round any part of an inch to the next tenth of an inch.

25. Triple-wall galvanized chimney pipe for prefabricated sheet-metal fireplaces has a circumference of 47 in. What is the inside diameter of the fire-stop spacer through which the chimney pipe passes as it goes through the ceiling? Round to the nearest tenth.

26. Find the area of the cross section of a wire whose diameter is $\frac{1}{16}$ in. (Express the answer to the nearest thousandth.)

27. Find the area of the shaded portion of the sidewalk in Fig. 7-21. The shaded portion is one-fourth of a circle. Express the answer in square yards rounded to tenths. ($144 \text{ in.}^2 = 1 \text{ ft}^2$, $9 \text{ ft}^2 = 1 \text{ yd}^2$.)

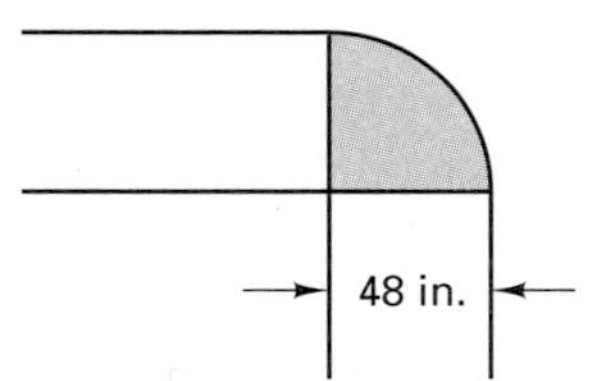

Figure 7-21

28. Find the area of the composite layout shown in Fig. 7-22. Round the answer to the nearest tenth.

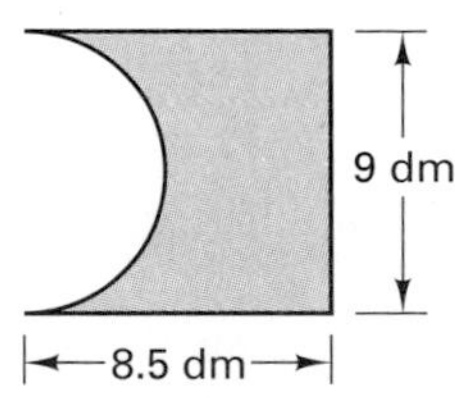

Figure 7-22

29. If a cogwheel makes one complete revolution and its radius is 9.4 in., how long is the path traveled by any one point on the cogwheel? Round to tenths.

CHALLENGE PROBLEMS

30. Lou Ferrante plans to build a dog pen. He has 360 feet of fencing and would like to enclose as much area as possible for his dog. What length and width should he make the dog pen?

31. Tosha Riddle, an interior designer needs to order a table cloth to cover a circular table that is 24 inches from the floor and 18 inches across the top. The cloth must drape the table so that it just touches the floor. Determine the shape needed for the cloth. Find the size of the fabric needed for the cloth.

32. Given that the formula for finding the area of a triangle is $A = \frac{1}{2}bh$, find the area of the shaded portion in the figure for Problem 6 of the Concepts Analysis.

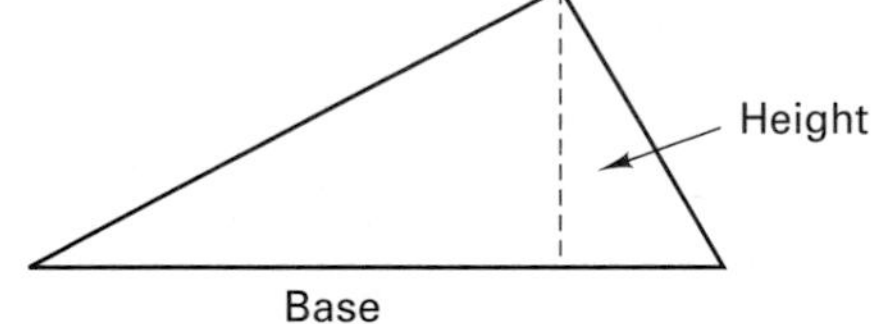

CONCEPTS ANALYSIS

1. Describe five activities or jobs that would require you to find the perimeter of a shape.

2. Describe five activities or jobs that would require you to find the area of a shape.

3. Draw five parallelograms of different sizes and cut out each parallelogram. Cut each parallelogram into 2 pieces by cutting through two corners or vertices. Describe similarities and/or differences between the two pieces of each parallelogram.

4. Each piece of the parallelogram in Problem 3 forms a triangle. Use the comparisons from Problem 3 to write a formula to find the area of a triangle as it relates to a parallelogram.

5. Drawing freely on the formulas for the circumference of a circle, devise your own formulas to find the radius and the diameter of a circle.

6. Explain how you could find the area of the following composite figure.

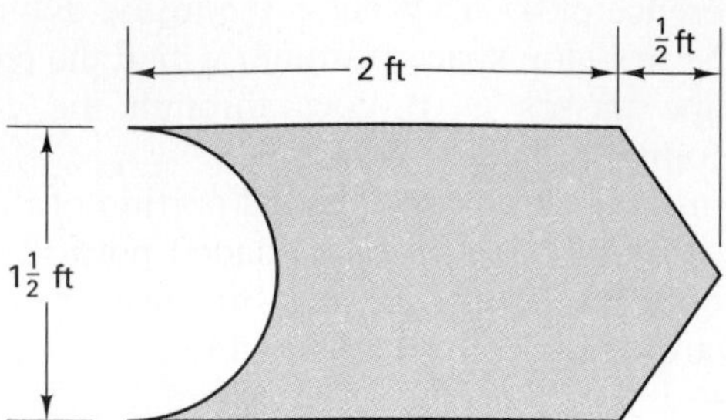

7. Discuss the similarities and differences of a square and a rectangle.

8. Discuss the similarities and differences of a rectangle and a parallelogram.

9. Draw a shape that has the same relationship to a parallelogram as a square has to a rectangle.

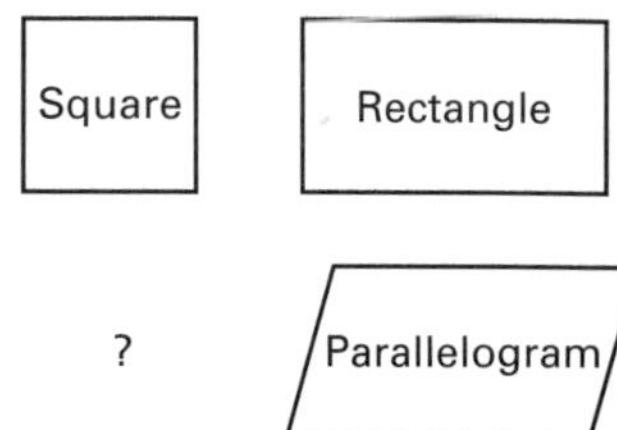

10. The new shape in Problem 9 is called a rhombus. List the properties of a rhombus. Write a formula for finding the area and perimeter of a rhombus.

CHAPTER SUMMARY

Objectives	What to Remember	Examples
Section 7-1		
1. Find the perimeter and area of a square.	Formula for perimeter: $P = 4s$. s is length of a side. Formula for area: $A = s^2$.	Find the perimeter and area of a square 6.5 cm on a side. $P = 4s$ $P = 4(6.5)$ $P = 26$ cm Find the area of a square 1.5 in. on a side. $A = s^2$ $A = 1.5^2$ $A = 2.25$ in.^{2}
2. Find the perimeter and area of a rectangle.	Formula for perimeter: $P = 2(l + w)$. l is length of long side. w is width, or length of short side. Formula for area: $A = lw$.	A rectangular flower bed is 8 ft $\times$ 3 ft. How many feet of edging are needed to surround the bed? $P = 2(l + w)$ $P = 2(8 + 3)$ $P = 2(11)$ $P = 22$ ft How much landscaping fabric is needed to cover the bed? $A = lw$ $A = 8 \times 3$ $A = 24$ ft^2

3. Find the perimeter and area of a parallelogram.

Formula for perimeter: $P = 2(b + s)$. b is the base, or length of long side. s is length of a short side.

Formula for area: $A = bh$. b is the base. H is the height, or length of a perpendicular distance between the bases.

Find the perimeter of a parallelogram with a base of 1 m and a side of 0.5 m.

$$P = 2(b + s)$$
$$P = 2(1 + 0.5)$$
$$P = 2(1.5)$$
$$P = 3 \text{ m}$$

Find the area of a sign shaped like a parallelogram with a base of 36 in. and a height of 12 in.

$$A = bh$$
$$A = 36(12)$$
$$A = 432 \text{ in.}^2$$

Section 7-2

1. Find the circumference of a circle.

Formula for circumference: $C = \pi d$ or $C = 2\pi r$. d is the diameter or distance across the center of a circle. r is the radius or half the diameter.

Find the circumference of a circle with a 3-cm diameter.

$$C = \pi d$$
$$C = \pi(3)$$
$$C = 9.4 \text{ cm (rounded)}$$

What is the distance around a circle with an 18-in. radius?

$$C = 2\pi r$$
$$C = 2\pi (18)$$
$$C = 113.1 \text{ (rounded)}$$

2. Find the area of a circle.

Formula for area: $A = \pi r^2$.

Find the area of a circle whose diameter is 3 m. First, find the radius: $3 \div 2 = 1.5$.

$$A = \pi r^2$$
$$A = \pi(1.5)^2$$
$$A = 7.07 \text{ m}^2 \text{ (rounded)}$$

3. Solve applied problems involving circles.

Some problems may involve composite figures and require several areas or circumferences or perimeters to be used.

A belt moving two pulleys 5 in. in diameter must be replaced. How long must the belt be if the centers of the pulleys are 18 in. apart? This setup forms a semicircular-sided figure. Both semicircles form one full circle. Find its circumference and add the upper and lower distances ($2 \times 18 = 36$).

$$P = \pi d$$
$$P = \pi(5)$$
$$P = 15.70796327$$

$15.70796327 + 36 = 51.7$ in. (rounded). The belt must be 51.7 in.

WORDS TO KNOW

square (p. 291)
rectangle (p. 291)
parallelogram (p. 291)
area (p. 292)
perimeter (p. 292)
base (p. 297)
adjacent side (p. 297)
height (p. 298)
circle (p. 301)
diameter (p. 301)
radius (p. 301)
center (p. 301)
circumference (p. 301)
semicircle (p. 301)

CHAPTER TRIAL TEST

Find the perimeters and areas of the following figures.

1.

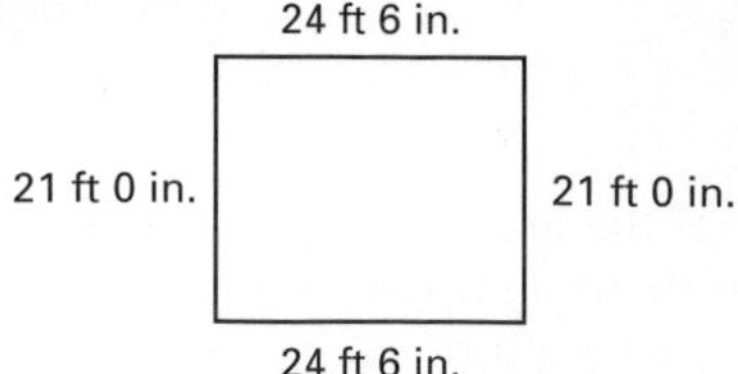

2.

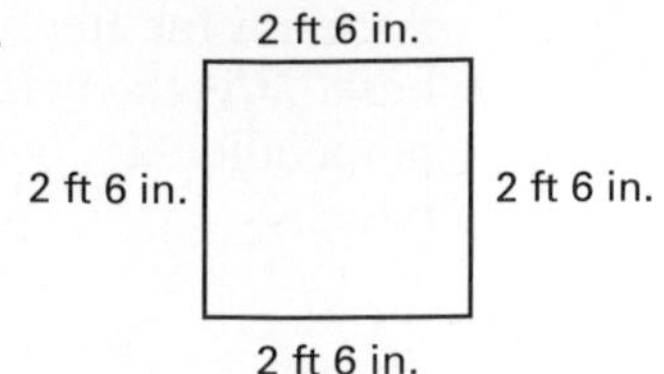

3.

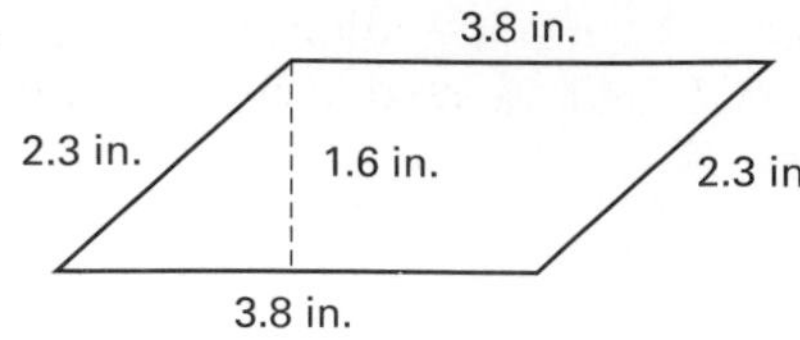

4.

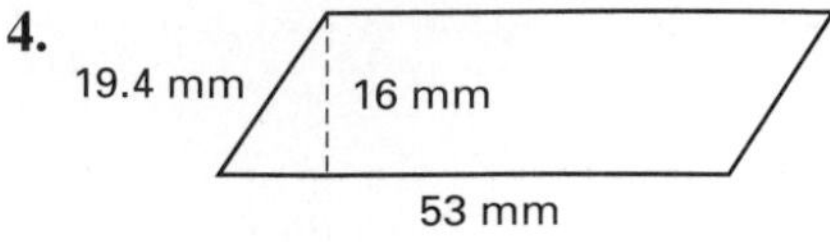

5. A company charges \$3.00 labor per square yard to install carpet and padding. If a dining room is 12 ft $\times$ 13 ft, what would the labor cost to install the padding and carpet? (9 $\text{ft}^2 = 1\ \text{yd}^2$.)

6. How many squares (100 ft^2) of siding would be needed for four sides of a garage 18 ft $\times$ 15 ft $\times$ 9 ft high if there are 125 ft^2 of openings? Allow one-fourth extra siding for overlap and waste.

7. How many square feet of floor space are there in the plan shown in Fig. 7-24?

8. A machine stamps 1-in. squares from sheet metal. How many can it cut from a 3-ft $\times$ 4-ft sheet of metal?

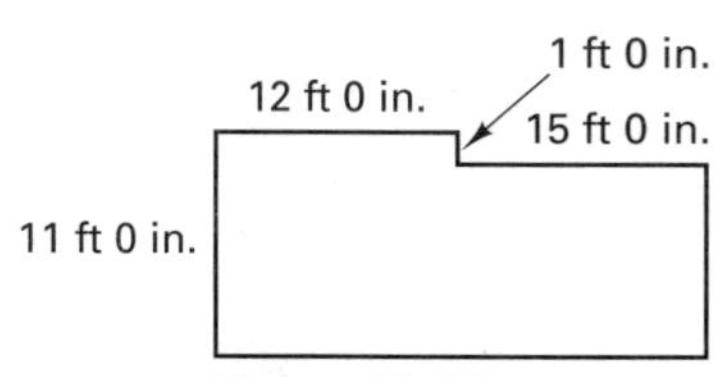

Figure 7-24

Use Fig. 7-25 to identify the numbered parts of the circle.

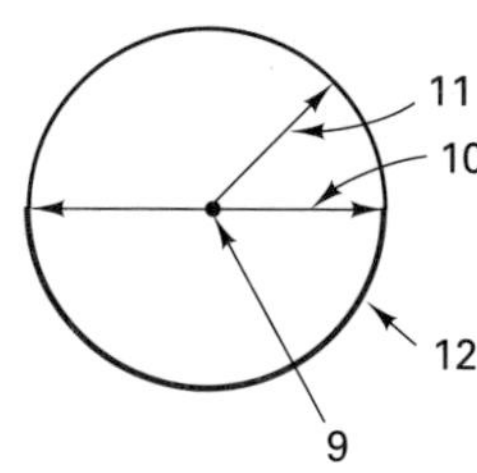

Figure 7-25

9. ____ **10.** ____ **11.** ____

12. ____

Solve these problems and round, when necessary, to hundredths.

13. Find the circumference of a circle whose radius is 6 in.

14. Find the area of the circle of Exercise 13.

15. What is the perimeter of a roller-skating rink with the dimensions shown in Fig. 7-26?

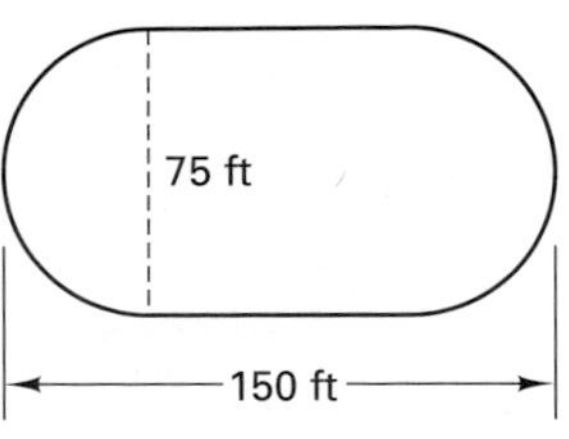

Figure 7-26

16. Find the area to the nearest square foot of the skating rink in Exercise 15.

17. A water pipe with an inside diameter of $\frac{5}{8}$ in. and an outside diameter of $\frac{7}{8}$ in. is cut off at one end. What is the cross-sectional area of the ring formed by the walls of the pipe?

18. Find the area of the shaded portion of the tiled walk that surrounds a rectangular swimming pool (Fig. 7-27).

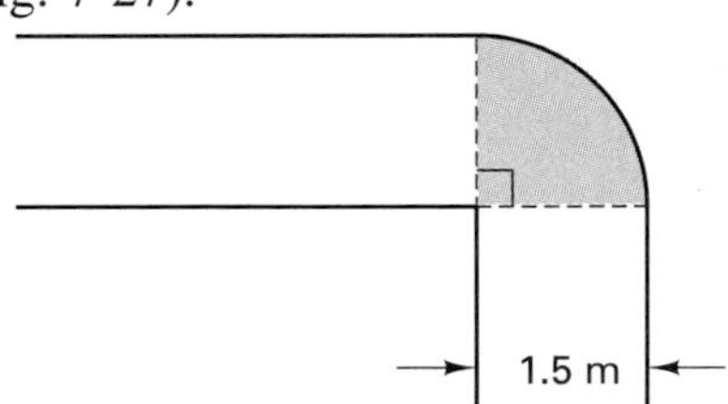

Figure 7-27

19. Find the area of a washer with an inside diameter of $\frac{1}{4}$ in. and an outside diameter of $\frac{1}{2}$ in.

20. If it costs \$1.25 to sod each square foot of a circular lawn (Fig. 7-28), how much would it cost, to the nearest cent, if the lawn is $6\frac{1}{2}$ ft wide and surrounds a round goldfish pond having a diameter of 12 ft?

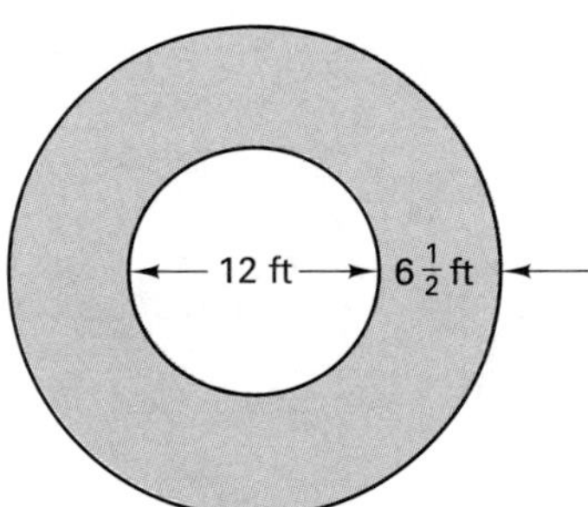

Figure 7-28

CHAPTER

8 Interpreting and Analyzing Data

A graph is a means of showing information visually. Some graphs show how our tax dollars arc divided among various government services or how the population of an area has grown over the years. Other graphs are used to show equations, inequalities, and their solutions.

8-1 READING CIRCLE, BAR, AND LINE GRAPHS

Learning Objectives

L.O.1. Read circle graphs.
L.O.2. Read bar graphs.
L.O.3. Read line graphs.

Many of the graphs we encounter in our employment give us useful information at a glance. However, we must be able to read, or interpret, the graphs properly in order to use them to our benefit. Three common graphs used to represent data are the circle graph, the bar graph, and the line graph.

L.O.1. Read circle graphs.

Circle Graphs

■ **DEFINITION 8-1.1 Circle Graph.** A **circle graph** uses a divided circle to show pictorially how a total amount is divided into parts.

The complete circle represents the total amount of one whole quantity. Then the circle is divided into parts so that the sum of all the parts equals the whole quantity. These parts can be expressed as fractions, decimals, or percents. Figure 8-1 illustrates a circle graph.

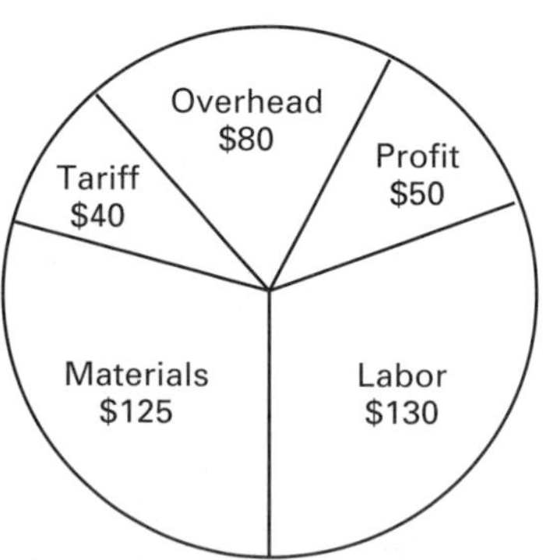

Figure 8-1 Distribution of wholesale price of $425 color television receiver.

To "read" a graph is to examine the information on the graph.

Rule 8-1.1 ***To read a circle, bar, or line graph:***

1. Examine the title of the graph to find out what information is shown.
2. Examine the parts to see how they relate to one another and to the whole amount.
3. Examine the labels for each part of the graph and any explanatory remarks that may be given.
4. Use the given parts to calculate additional numerical amounts.

EXAMPLE 8-1.1 Answer the following questions using Fig. 8-1.

(a) What percent of the wholesale price is the cost of labor?
(b) What percent of the wholesale price is the cost of materials?
(c) What would the wholesale price be if no tariff (tax) was paid on imported parts?

Solution

(a) $$\frac{R}{100} = \frac{130}{425}$$ *(R represents percent of wholesale price that is labor cost.)*

$$R \times 425 = 13{,}000$$

$$R = \frac{13{,}000}{425}$$

$$R = 30.58823529$$

$$R = 30.6\% \text{ (labor)}$$

(b) $$\frac{R}{100} = \frac{125}{425}$$ *(R represents percent of wholesale price that is materials cost.)*

$$R \times 425 = 12{,}500$$

$$R = \frac{12{,}500}{425}$$

$$R = 29.41176471$$

$$R = 29.4\% \text{ (materials)}$$

(c) Price − tariff = 425 − 40 = \$385 (cost without tariff) ■

L.O.2. Read bar graphs.

Bar Graphs

Each type of graph developed historically because of a need for the kind of information it made accessible to the user. The bar graph is no exception.

■ **DEFINITION 8-1.2 Bar Graph.** A **bar graph** uses two or more bars to show pictorially how two or more amounts compare to each other rather than to a total.

Numerical amounts are represented by the lengths of the bars. The bars can be drawn either horizontally or vertically.

The axis (*reference line*) along the length of the bars is scaled with numerical amounts. Then the other axis (reference line) along the base of the bars will be used to label the bars. Figure 8-2 illustrates a bar graph with horizontal bars.

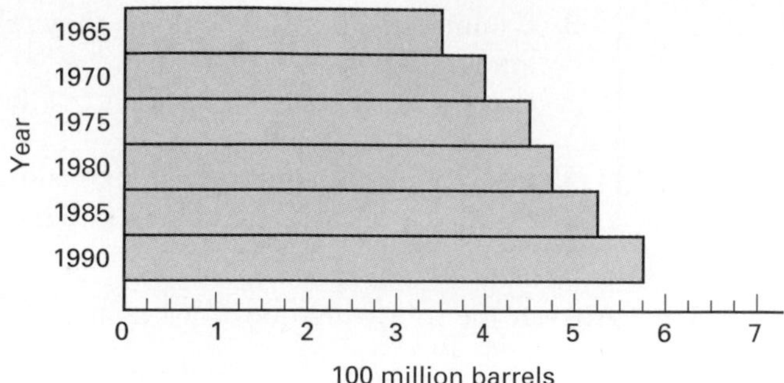

Figure 8-2 Company oil production.

EXAMPLE 8-1.2 Answer the following questions using Fig. 8-2.

(a) How many more 100 million barrels of oil are indicated for the Company in 1990 than in 1970?
(b) Judging from the graph, should Company oil production in 1995 be more or less than in 1990?
(c) How many 100 million barrels of oil did the Company produce in 1970?

Solution

(a) 1990 production − 1970 production = 5.75 − 4.00 = 1.75 hundred million barrels.
(b) More, since the trend has been toward greater production.
(c) Four hundred million barrels in 1970. ■

L.O.3. Read line graphs.

Line Graphs

A third kind of graph we encounter in our reading of industrial reports, handbooks, and the like is the line graph.

■ **DEFINITION 8-1.3 Line Graph.** A **line graph** uses one or more lines to show changes in data.

The horizontal axis or reference line in a line graph usually represents periods of time or specific times. The vertical axis or reference line is usually scaled to represent numerical amounts. Line graphs are used to pictorially represent trends in data to quickly show high values and low values. Figure 8-3 illustrates a line graph.

EXAMPLE 8-1.3 Answer the following questions using Fig. 8-3.

(a) If the film is to be developed to a contrast of 0.5, how long must it be developed?
(b) If the film is developed for 13 min, what is its degree of contrast?
(c) If normal conditions require a 0.58 degree of contrast, how long should film be developed?

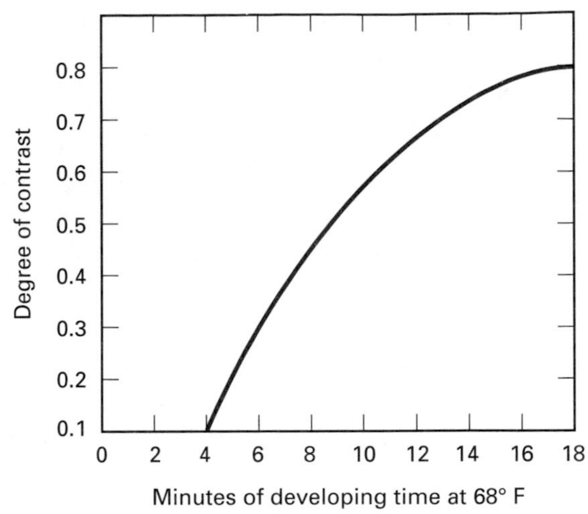

Figure 8-3 Film development table.

Solution

(a) 9 min (midpoint between 8 and 10, where the curved line intersects the axis for 0.5).
(b) 0.7 degrees of contrast (where the curved line intersects the axis for 0.7 and 13 min).
(c) 10 min (where the curved line intersects the approximate 0.58-degree axis and 10-min axis). ■

SELF-STUDY EXERCISES 8-1

L.O.1. Use Fig. 8-4 to answer Exercises 1 to 3.

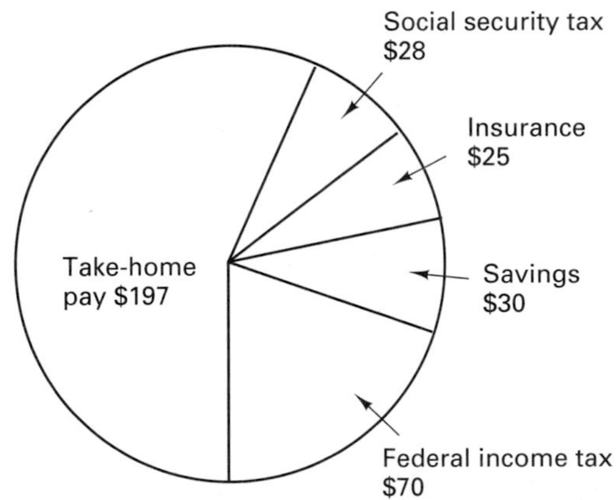

Figure 8-4 Distribution of weekly salary of $350.

1. What percent of the gross salary goes into savings? Round to tenths.

2. What percent of the take-home pay is federal income tax? Round to tenths.

3. What percent of the gross pay is the take-home pay? Round to tenths.

L.O.2. Use Fig. 8-5 to answer Exercises 4 to 6.

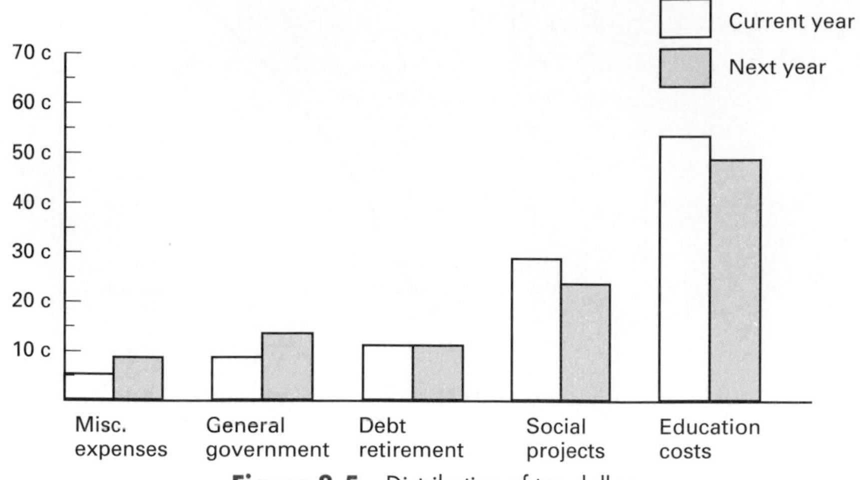

Figure 8-5 Distribution of tax dollars.

4. What expenditure is expected to be the same next year as this year?
5. What two expenditures are expected to increase next year?
6. What two expenditures are expected to decrease next year?

L.O.3. Use Fig. 8-6 to answer Exercises 7 to 10.

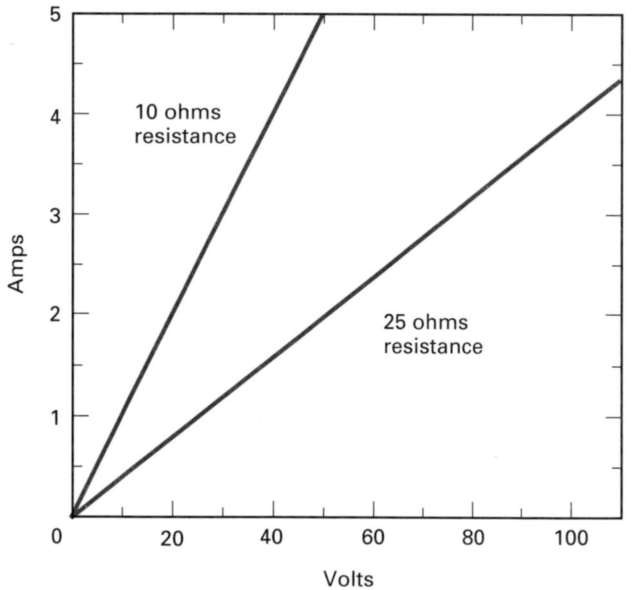

Figure 8-6 Amperage produced by voltage across two resistances.

7. How many amperes of current are produced by 50 V when the resistance is 10 ohms?
8. How many volts are needed to produce 2 A of current when the resistance is 25 ohms?
9. Approximately how many volts are required to produce a current of 3.5 A when the resistance is 10 ohms?
10. Find the resistance when 100 V are needed to produce a current of 4 A.

8-2 FREQUENCY DISTRIBUTIONS, HISTOGRAMS, AND FREQUENCY POLYGONS

Learning Objectives

L.O.1. Interpret and make frequency distributions.
L.O.2. Interpret and make histograms.
L.O.3. Interpret and make frequency polygons.

L.O.1. Interpret and make frequency distributions.

Interpreting and Making Frequency Distributions

Suppose for a class of 25 students the instructor recorded the following grades:

76	91	71	83	97
87	77	88	93	77
93	81	63	79	74
77	76	97	87	89
68	90	84	88	91

It would be difficult for the instructor to make sense of all these numbers as they appear above. But the instructor can arrange the scores into several smaller groups, called *class intervals*. The word *class* means a special group, just as a group of students in a course is a class.

These scores can be grouped into class intervals such as 60–64, 65–69, 70–74, 75–79, 80–84, 85–89, 90–94, and 95–99. Each class interval has an odd number of scores. The *middle score* of each interval is a *class midpoint*.

The instructor can now *tally* the number of scores that fall into each class interval to get a *class frequency*, the number of scores in each class interval.

A compilation of class intervals, midpoints, tallies, and class frequencies is called a *frequency distribution*.

EXAMPLE 8-2.1 Answer the following questions about the frequency distribution in Table 8-1.

(a) How many students scored 70 or above?

$$2 + 6 + 3 + 5 + 5 + 2 = 23$$

23 students scored 70 or above.

TABLE 8-1 Frequency Distribution of 25 Scores

Class Interval	Midpoint	Tally	Class Frequency
60–64	62	/	1
65–69	67	/	1
70–74	72	//	2
75–79	77	~~////~~ /	6
80–84	82	///	3
85–89	87	~~////~~	5
90–94	92	~~////~~	5
95–99	97	//	2

(b) How many students made A's (90 or higher)?

$$5 + 2 = 7$$

7 made A's (90 or higher).

(c) What percent of the total grades were A's (90's)?

$$\frac{7 \text{ A's}}{25 \text{ total}} = \frac{7}{25} = 0.28 = 28\% \text{ A's}$$

(d) Was the test too difficult?
The relatively high number of 90's (7) compared to the relatively low number of 60's (2) suggests that the test was not too difficult for the class.

(e) What is the ratio of F's (60's) to A's (90's)?

$$\frac{2 \text{ F's}}{7 \text{ A's}} = \frac{2}{7}$$

The ratio is $\frac{2}{7}$. ■

EXAMPLE 8-2.2 Students in a history class reported their credit-hour loads as shown. Make a frequency distribution of their credit hours. Credit hours carried: 3, 12, 15, 3, 6, 6, 12, 9, 12, 9, 6, 3, 12, 18, 6, 9.

Solution: To have a class interval with an easy-to-find midpoint, use an odd-numbered interval. Here an interval of 5 is used. That is, 0–4 contains five possibilities: 0, 1, 2, 3, and 4. The middle number is the midpoint, 2. Make a tally mark for each time the credit hours of each student falls in the interval. Then count the tally marks to get the frequency (Table 8-2). ■

TABLE 8-2

Class Interval	Midpoint	Tally	Class Frequency
0–4	2	///	3
5–9	7	~~////~~ //	7
10–14	12	////	4
15–19	17	//	2

L.O.2. Interpret and make histograms.

Interpreting and Making Histograms

A *histogram* is a bar graph in which the two scales are class intervals and class frequencies. The frequency distribution from Example 8-2.2 can be made into a histogram. The frequencies in this histogram form the vertical scale. The intervals form the horizontal scale.

EXAMPLE 8-2.3 Answer the following questions about the histogram in Fig. 8-7.

(a) How many students carried 5 to 9 hours?
7 carried 5 to 9 hours.

(b) How many students carried less than 15 hours?
3 + 7 + 4 = 14 carried less than 15 hours.

(c) What percent of the total carried 10 to 14 hours?

$$\frac{4}{16} = \frac{1}{4} = 0.25 = 25\%$$

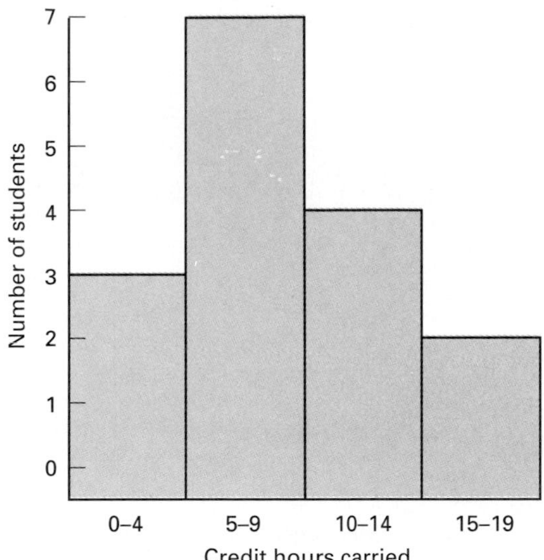

Figure 8-7 Credit hours carried by 16 students in history class.

(d) What is the ratio of students carrying 0 to 4 hours to those carrying 5 to 9 hours?

$$\frac{3 \text{ with } 0 \text{ to } 4}{7 \text{ with } 5 \text{ to } 9} = \frac{3}{7}$$

The ratio is $\frac{3}{7}$. ■

EXAMPLE 8-2.4 A company gives vacation leave based on employee years of service. The employees fall into four categories: 0–2 years, 8 employees; 3–5 years, 6 employees; 6–8 years, 4 employees; 9–11 years, 2 employees. Make a histogram showing this information (see Fig. 8-8).

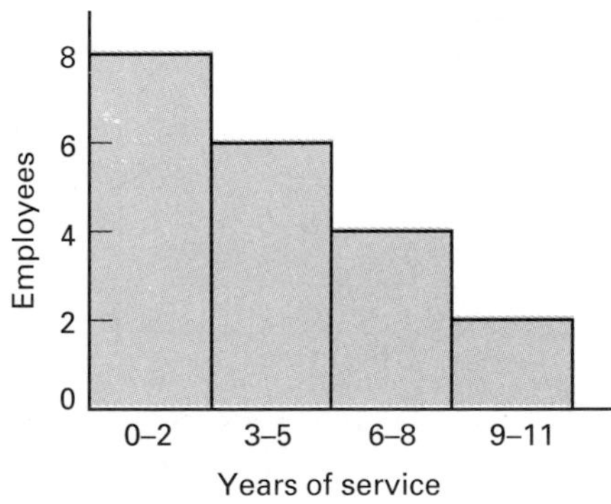

Figure 8-8 Years of service of 20 employees.

Solution: The years of service are already arranged in class intervals and so may be used as given. The class frequency or number of employees in each interval is also provided and so may be used as given. Therefore, it is *not* necessary to make a frequency distribution. The class intervals form the horizontal scale and the frequencies form the vertical scale. ■

L.O.3. Interpret and make frequency polygons.

Interpreting and Making Frequency Polygons

A *frequency polygon* is a line graph made by identifying each class midpoint, marking it with a dot, and connecting the dots with a line. The histogram from Example 8-2.4 may be used to make a frequency polygon. The class midpoints may be considered an average for each class interval.

EXAMPLE 8-2.5 Make a frequency polygon for the histogram in Example 8-2.4.

Solution: First, identify the class midpoints with a dot, as in Fig. 8-9.

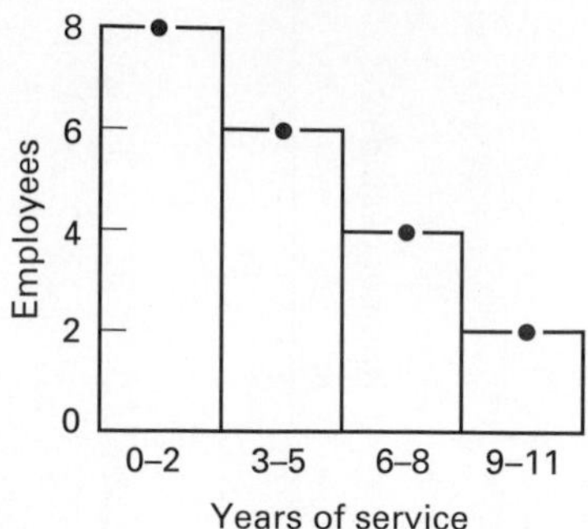

Figure 8-9 Years of service of 20 employees.

Connect the dots with a line, as in Fig. 8-10.

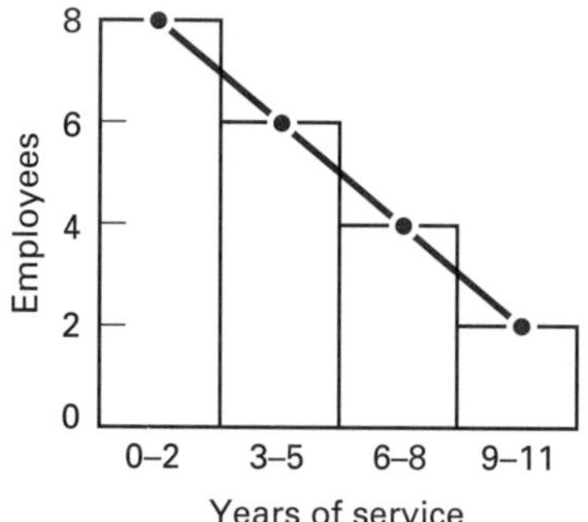

Figure 8-10 Years of service of 20 employees.

Write the line graph without the bars, as in Fig. 8-11. ■

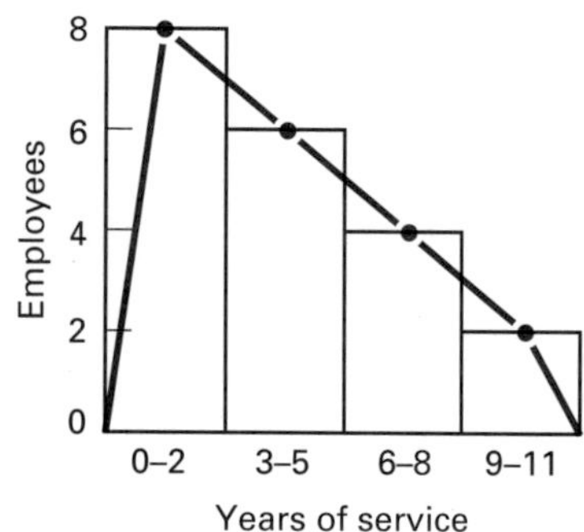

Figure 8-11 Years of service of 20 employees.

EXAMPLE 8-2.6 Answer the following questions on the frequency polygon in Example 8-2.5.

(a) How many employees have 5 or fewer years of service?
$8 + 6 = 14$ with 5 or fewer years of service.

(b) What is the ratio of employees with 2 or fewer years of service to employees with 9 or more years of service?

$$\frac{8 \text{ with } 0 \text{ to } 2 \text{ years}}{2 \text{ with } 9 \text{ to } 11 \text{ years}} = \frac{8}{2} = \frac{4}{1}$$

The ratio is $\frac{4}{1}$.

(c) What percent of the number of employees with 2 or fewer years of service is the number of employees with 9 or more years of service?

$$\frac{2}{8} = \frac{1}{4} = 0.25 = 25\%$$

(d) What is the average number of years of service for employees in the 6 to 8 years of service bracket?
7 is the midpoint or average for the interval. ■

SELF-STUDY EXERCISES 8-2

L.O.1. Use the frequency distribution in Table 8-3 to answer the questions that follow. The distribution shows the ages of 25 college students in an algebra class.

TABLE 8-3 Frequency Distribution of 25 Ages

Class Interval	Midpoint	Tally	Class Frequency
38–40	39	/	1
35–37	36	/	1
32–34	33	//	2
29–31	30	///	3
26–28	27	//	2
23–25	24	~~////~~ /	6
20–22	21	~~////~~ //	7
17–19	18	///	3

1. How many students are 22 or younger?
2. How many students are older than 34?
3. What is the ratio of the number of students 38 to 40 to the number 17 to 19?
4. What is the ratio of the smallest class frequency to the largest class frequency?
5. What percent of the total class are students age 17 to 19?
6. What percent of the total class are students age 20 to 22?
7. What two age groups make up the smallest number of students in the class?
8. What two age groups make up the largest number of students in the class?
9. How many students are over age 28?
10. How many students are under age 26?

Use the following hourly pay rates of 33 support employees in a private college to complete a frequency distribution using the format shown in Table 8-4.

\$5	\$5	\$9	\$6	\$5	\$5	\$5	\$5	\$6	\$6	\$7	\$7	\$5	\$5
\$10	\$9	\$6	\$10	\$7	\$15	\$5	\$5	\$8	\$5	\$6	\$8	\$5	\$5
\$11	\$12	\$6	\$14	\$4									

TABLE 8-4 Pay Rates of 33 Support Employees

	Class Interval	Midpoint	Tally	Class Frequency
11.	\$13–15	______	______	______
12.	\$10–12	______	______	______
13.	\$7–9	______	______	______
14.	\$4–6	______	______	______

Use the following 40 test scores of two physics classes to complete a frequency distribution using the format in Table 8-5.

57 91 76 89 82 59 72 88 76 84 67 59 77 66 56 76
77 84 85 79 69 88 75 58 85 65 67 66 93 83 69 81
80 64 78 76 72 90 79 90

TABLE 8-5 Test Scores of 40 Physics Students

	Class Interval	Midpoint	Tally	Class Frequency
15.	91–95	______	______	______
16.	86–90	______	______	______
17.	81–85	______	______	______
18.	76–80	______	______	______
19.	71–75	______	______	______
20.	66–70	______	______	______
21.	61–65	______	______	______
22.	56–60	______	______	______

L.O.2. Complete the following.

23. Use the information from Exercises 11 through 14 to make a histogram. Use the reference lines in Fig. 8-12 as guides.

24. Use the information from Exercises 15 through 22 to make a histogram.

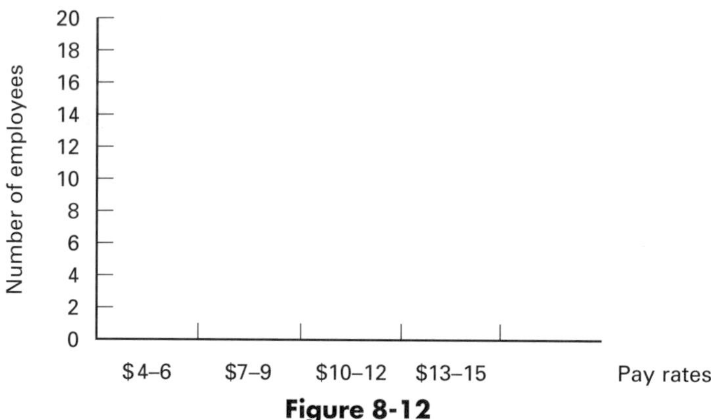

Figure 8-12

L.O.3.

25. Use the information from Exercises 11 through 14 to make a frequency polygon. Use the reference lines in Fig. 8-12 as guides.

26. Use the information from Exercises 15 through 22 to make a frequency polygon.

8-3 FINDING STATISTICAL MEASURES

Learning Objectives

L.O.1. Find the arithmetic mean or arithmetic average.
L.O.2. Find the median, the mode, and the range.

L.O.1. Find the arithmetic mean or arithmetic average.

Finding the Arithmetic Mean or Average

An *average* is an approximate number that represents a quantity considered typical or representative of several related quantities.

> **Rule 8-3.1** *To find the* **arithmetic mean** *or average:*
>
> **1.** Add the quantities.
> **2.** Divide their sum by the number of quantities.

EXAMPLE 8-3.1 Find the average of each group of quantities.

(a) Test scores: 67, 86, 77, 72, 81
Since there are 5 scores, find their sum and divide by 5:

$$\frac{67 + 86 + 77 + 72 + 81}{5} = \frac{383}{5} = 76.6$$

(b) Pounds: 21, 33, 12.5, 35.2 (to nearest hundredth)
Since there are 4 weights, find their sum and divide by 4.

$$\frac{21 + 33 + 12.5 + 35.2}{4} = \frac{101.7}{4} = 25.425$$

Thus, the average weight is 25.43 (to nearest hundredth). ■

EXAMPLE 8-3.2 An automobile used 41 gallons of regular gasoline on a trip of 876 miles. What was the average miles per gallon to the nearest tenth?

Solution: The number of miles for each of 41 gallons of gasoline has already been added (876). So we divide by 41.

$$\frac{876}{41} = 21.4 \quad (\textit{to nearest tenth})$$

Thus, the car averaged 21.4 miles per gallon on the trip. ■

EXAMPLE 8-3.3 A college student made the following final grades for the past term: B, A, A, C. Find the student's QPA (quality point average) to the nearest hundredth.

Solution: To find the QPA for a term, the quality points for the letter grade of each course are multiplied by the credit hours of each course to obtain the total quality points for each course. This total is divided by the total credit hours earned. The quality points awarded are A = 4, B = 3, and C = 2.

Subject	Grade	Hours		Points per Hour		Total Points
Algebra 101	B	3	×	3	=	9
History 102	A	3	×	4	=	12
Education 301	A	3	×	4	=	12
English 101	C	4	×	2	=	8
		13 hours				41 points

$$\frac{41}{13} = 3.153846154 \quad \text{or} \quad 3.15 \quad (\textit{to nearest hundredth})$$

Thus, the student's QPA for the term is 3.15. ■

EXAMPLE 8-3.4 A community college student has the following grades in Physics 101: 73, 84, 80, 62, and 70. What grade is needed on the last test for a C or 75 average?

Solution: One way to find out the needed grade is to assume that each grade is 75 for the 6 tests:

$$\frac{75 + 75 + 75 + 75 + 75 + 75}{6} = \frac{450}{6} = 75$$

Then find the sum of the first five tests actually taken:

$$73 + 84 + 80 + 62 + 70 = 369$$

Subtract to find the difference between this sum and 450. The difference is the needed score.

$$450 - 369 = 81$$

The student must make 81. To check, assume 81 on the last test and find the average.

$$\frac{73 + 84 + 80 + 62 + 70 + 81}{6} = \frac{450}{6} = 75$$

Thus, 81 is the score needed on the last test. ■

L.O.2. Find the median, the mode, and the range.

Finding the Median, Mode, and Range

Besides the mean or arithmetic average, we sometimes use the *median* and the *mode* to describe data.

The *median* is the middle quantity when the given quantities are arranged in order of size.

> **Rule 8-3.2** *To find the median:*
>
> 1. Arrange the quantities in order of size.
> 2. If the number of quantities is odd, the median is the middle quantity.
> 3. If the number of quantities is even, the median is the average of the two middle quantities.

EXAMPLE 8-3.5 The following scores were made on a test: 67, 89, 78, 65, 74. What is the median score?

Solution: In order of size:

89
78
74 ← median or middle score
67 in odd number of scores
65 ■

EXAMPLE 8-3.6 The following temperatures were recorded: 56°, 48°, 66°, and 62°. What is the median temperature?

Solution: Since the number of temperatures is even, find the average of the two middle scores.

In order of size: 48

$$\left.\begin{matrix}56\\62\end{matrix}\right\}\frac{56+62}{2}=\frac{118}{2}=59$$

66

Thus, the median temperature is 59°. ■

The *mode* is the most frequently occurring quantity among the quantities considered.

Rule 8-3.3 ***To find the mode:***

1. Identify the quantity that occurs most frequently as the mode.
2. If no quantity occurs more than another quantity, there is no mode for those quantities.
3. If more than one quantity occurs with the same frequency which is the greatest frequency, the set of quantities or scores will have more than one mode.

EXAMPLE 8-3.7 The following hourly pay rates are used at fast-food restaurants: cooks, \$4.25; servers, \$3.50; bussers, \$3.50; dishwashers. \$3.75; managers \$6.50. Find the mode.

Solution: The hourly pay rate of \$3.50 occurs more than any other rate. It is the mode. ■

EXAMPLE 8-3.8 The following daily shifts are worked by employees at a mall clothing store: 4 hours, 6 hours, 8 hours. Find the mode.

Solution: Since no shift occurs more than another, there is no mode. ■

The mean, the median, and the mode are also refered to as *measures of central tendency*. Another group of statistical measures are *measures of variation or dispersion*. One of these measures is the *range*. The range is the difference between the lowest quantity and the highest quantity in a set of data.

EXAMPLE 8-3.9 Find the range for the data described in Example 8-3.7.

Solution: The high value is \$6.50. The low value is \$3.50.

$$\text{range} = \$6.50 - \$3.50 = \$3.00$$ ■

Another measure of variation or dispersion is *standard deviation*. Standard deviation is introduced in most statistics courses.

Tips and Traps

A mistake often made in making conclusions or inferences from statistical measures is to examine only one statistic, such as the mean. To get a complete picture of the data requires looking at more than one statistic.

SELF-STUDY EXERCISES 8-3

L.O.1. Find the mean of the given quantities. Round to hundredths if necessary.

1. 12, 14, 16, 18, 20
2. 13, 15, 17, 19, 21
3. 68, 54, 73, 69
4. 85, 68, 77, 65
5. 37.6, 29.8
6. 65.3, 67.9
7. 32°F, 41°F, 54°F
8. 10°C, 13°C, 15°C
9. \$27, \$32, \$65, \$29, \$21
10. \$32, \$43, \$22, \$63, \$36
11. 11 in., 17 in., 16 in.
12. 9 in., 7 in., 8 in.
13. Scores: 56, 67, 79
14. Scores: 63, 59, 76

Solve the following problems.

15. A baseball player made 276 home runs over a 16-year period. What was the average number of home runs per year to the nearest tenth?

16. An automobile salesperson sold 163 new cars over a 6-month period. What was the average number of cars sold per month to the nearest tenth?

17. An automobile used 32 gallons of regular gasoline on a trip of 786 miles. What was the average miles per gallon to the nearest tenth?

18. A pickup truck used 25 gallons of regular gasoline on a trip of 256 miles. What was the average miles per gallon to the nearest tenth?

L.O.2. Find the median for each group of numbers.

19. 32, 56, 21, 44, 87
20. 78, 23, 56, 43, 38
21. 12, 21, 14, 18, 15, 16
22. 21, 33, 18, 32, 19, 44
23. \$22, \$35, \$45, \$30, \$29
24. \$66, \$54, \$76, \$55, \$69

25. The following hourly pay rates are used at fast-food restaurants: cooks, \$4.75; servers, \$4.85; bussers, \$4.75; dishwashers, \$4.85; managers, \$8.25. Find the median pay rate.

26. The following hourly pay rates are used at a locally owned store: clerks, \$5.45; bookkeepers, \$6.25; operators, \$4.95; assistant managers, \$7.95. Find the median pay rate.

Find the mode for each group of numbers.

27. 2, 4, 6, 2, 8, 2
28. 5, 12, 5, 5, 20
29. 21, 32, 67, 34, 23, 22
30. 32, 45, 41, 23, 56, 77
31. \$56, \$67, \$32, \$78, \$67, \$20, \$67, \$56
32. \$32, \$87, \$67, \$32, \$32, \$87, \$77, \$22

33. The following weekend work shifts are in effect at a mall clothing store: 4 hours in AM, 6 hours in PM, 4 hours in PM. Find the mode for the number of hours.

34. The following prices are in effect at a fast-food restaurant: \$1.75, hamburgers; \$1.97, hot ham sandwiches; \$2.38, chicken fillet sandwiches; \$1.97, roast beef sandwiches. Find the mode.

Find the range for each group of numbers.

35. 22, 36, 41, 41, 17
36. 28, 33, 36, 13, 28
37. 10, 23, 12, 17, 13, 16
38. 23, 23, 18, 32, 29, 14
39. \$25, \$15, \$25, \$40, \$19
40. \$36, \$44, \$26, \$52, \$19
41. 23°F, 37°F, 29°F, 54°F, 46°F, 71°F, 67°F

8-4 COUNTING TECHNIQUES AND SIMPLE PROBABILITIES

Learning Objectives

L.O.1. Count the number of ways objects in a set can be arranged.
L.O.2. Determine the chance of an event occurring if an activity is repeated over and over.

L.O.1. Count the number of ways objects in a set can be arranged.

Counting Techniques

A *set* is a well-defined group of objects or *elements*. The numbers 2, 4, 6, 8, and 10 can be a set of even numbers. Women, men, and children can be a set of people. A, B, and C can be a set of capital letters.

Counting, in this section, refers to determining all the possible ways the elements in a set can be arranged.

One way to find out is to *list* all possible arrangements and then count the number of arrangements.

EXAMPLE 8-4.1 List and count the ways the elements in the set *A*, *B*, and *C* can be arranged?

Solution

When A is first, there are only two possibilities, *BC* and *CB*:
ABC
ACB

When *B* is first, there are only two possibilities. *AC* and *CA*:
BAC
BCA

When *C* is first, there are only two possibilities, *AB* and *BA*:
CAB
CBA

Therefore, *A*, *B*, and *C* can be arranged in six ways. Each of these ways can also be called a set. ■

If more than three elements are in the set, the procedure becomes more challenging. It may be helpful to use a *tree diagram*, which allows each new set of possibilities to branch out from a previous possibility.

EXAMPLE 8-4.2 List and count the ways the elements in the set *W*, *X*, *Y*, and *Z* can be arranged?

Solution

4 choices	3 choices	2 choices	1 choice	
W	*X*	*Y*	*Z*	= *WXYZ*
		Z	*Y*	= *WXZY*
	Y	*Z*	*X*	= *WYZX*
		X	*Z*	= *WYXZ*
	Z	*X*	*Y*	= *WZXY*
		Y	*X*	= *WZYX*
X	*W*	*Y*	*Z*	= *XWYZ*
		Z	*Y*	= *XWZY*
	Y	*W*	*Z*	= *XYWZ*
		Z	*W*	= *XYZW*
	Z	*W*	*Y*	= *XZWY*
		Y	*W*	= *XZYW*
Y	*W*	*X*	*Z*	= *YWXZ*
		Z	*X*	= *YWZX*
	X	*W*	*Z*	= *YXWZ*
		Z	*W*	= *YXZW*
	Z	*X*	*W*	= *YZXW*
		W	*X*	= *YZWX*
Z	*W*	*X*	*Y*	= *ZWXY*
		Y	*X*	= *ZWYX*
	X	*W*	*Y*	= *ZXWY*
		Y	*W*	= *ZXYW*
	Y	*W*	*X*	= *ZYWX*
		X	*W*	= *ZYXW*
				24 possibilities or sets

■

As evident in Example 8-4.2, the greater the number of elements in a set, the greater the complexity and time required to list all possible arrangements.

A faster way to obtain a count of the possible ways the elements in any set can be arranged is by using logic and common sense.

In Example 8-4.2 we have four possibilities for the first letter, W, X, Y, or Z. For each of the four possible first letters, we have three choices left. For each of these three letters, we have two choices left. Finally, for each of these two letters, we have one choice left.

By multiplying the number of possible letter choices for each position, we can determine the total number of possibilities without listing them: $4 \cdot 3 \cdot 2 \cdot 1 = 24$.

EXAMPLE 8-4.3 A coin is tossed three times. With each toss, the coin falls heads up or tails up. How many combinations of heads and tails are there with three tosses of the coin?

Solution: There are three tosses. Each toss has only two possibilities, heads or tails. That is,

1st toss	2nd toss	3rd toss
2 possibilities	2 possibilities	2 possibilities

By multiplying the number of possible outcomes for each toss, we get $2 \cdot 2 \cdot 2 = 8$. So 8 combinations are possible.

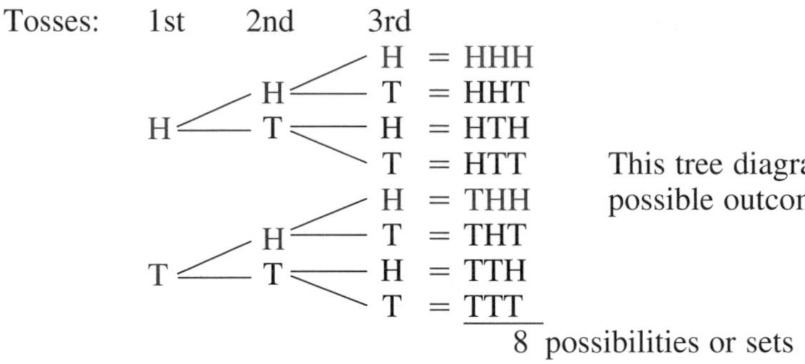

Tosses:	1st	2nd	3rd	
	H	H	H	= HHH
			T	= HHT
		T	H	= HTH
			T	= HTT
	T	H	H	= THH
			T	= THT
		T	H	= TTH
			T	= TTT
				8 possibilities or sets

This tree diagram illustrates the possible outcomes of three tosses. ■

EXAMPLE 8-4.4 Henry has three ties: a red tie, a blue tie, and a green tie. He also has three shirts: a white shirt, a pink shirt, and a yellow shirt. How many combinations of shirts and ties are possible?

Solution: If we start with the shirts, there are three possibilities (white, pink, and yellow). For each shirt, there are three possible ties (red, blue, and green). So we have $3 \cdot 3 = 9$ possible combinations. ■

EXAMPLE 8-4.5 Given the digits 0, 1, 2, 3, 4, and 5, how many three-digit numbers can be formed without repeating a digit?

Solution: The numbers to be formed contain three digits, so there are three positions to fill. We have six digits to work with. The first digit can be one of six. For each of these six digits, there are five possible second digits. For each of the five second digits, there are four possible third digits.

Positions:	1st	2nd	3rd
Possibilities:	6	5	4

By multiplying the possibilities for each position, we get $6 \cdot 5 \cdot 4 = 120$ possible combinations or sets. ■

L.O.2. Determine the chance of an event occurring if an activity is repeated over and over.

Simple Probabilities

Probability means the chance of an event occurring if an activity is repeated over and over. The probability of an event occurring is expressed as a ratio or a percent.

Weather forecasters generally use percents, such as when they forecast a 60% chance of rain or a 20% chance of snow. This text will use ratios like $\frac{3}{5}$ for 3 chances out of 5 or $\frac{2}{3}$ for 2 chances out of 3.

When a coin is tossed, two outcomes are possible, heads or tails. But only one side will be on top. The probability of tossing heads is one out of two, or $\frac{1}{2}$.

Rule 8-4.2

To express the probability of event A occurring, make a ratio with the number of A's in the set divided by the total number of elements in the set. Express the ratio in lowest terms.

EXAMPLE 8-4.6 When a die is rolled, what is the probability that a three will appear?

Solution: A die has six sides numbered by dots to represent 1 through 6. Each side is an element in the set of six sides. So there is a total of six elements in the set. Only one element is a three. The probability of rolling a three is $\frac{1}{6}$. ■

EXAMPLE 8-4.7 A holiday gift shopper wrapped eight men's ties in separate boxes. There were two solid-color ties and six striped ties. If the gift boxes were given at random to eight men, what is the probability of a man receiving a solid-color tie?

Solution: The total elements in the set are eight. There are two solid-color elements. The probability of receiving a solid-color tie is $\frac{2}{8}$, which reduces to $\frac{1}{4}$. ■

EXAMPLE 8-4.8 A grade-school teacher has a box of 144 wooden pencils. Normally, three of the pencils will have broken leads. What is the probability that the first pencil picked will have broken lead? If the lead is broken on the first pencil selected, what is the probability of picking a second pencil with broken lead?

Solution: On the first pick, the probability of picking one pencil with broken lead is $\frac{3}{144}$. We assume that the first pick was a pencil with broken lead, and so this leaves a total of 143 pencils and now only two have broken leads. On the second pick, the probability of picking a pencil with broken lead is $\frac{2}{143}$. ■

SELF-STUDY EXERCISES 8-4

L.O.1. Complete the following exercises on counting techniques.

1. List and count all the combinations possible for Keaton, Brienne, and Renee to be seated in three adjacent seats at a basketball game.

2. List and count all the combinations possible for arranging books A, B, C, and D on a shelf.

3. Count all the combinations possible for Jim Riddle to arrange a T-shirt, a sport shirt, a dress shirt, and a sweater on a shelf for display.

4. How many combinations are possible for arranging containers of lettuce, carrots, tomatoes, bell peppers, and celery in a row on a salad bar.

5. Part of a manufacturing process involves five steps. The steps can be arranged in any order. The company efficiency officer wants to determine the most efficient order for the five steps. How many combinations of steps are possible?

L.O.2. Complete the following exercises on simple probability.

6. A drawing will be held to award door prizes. If the names of 24 people are in the pool for the drawing, what is the probability that David's name will be pulled at random for the first prize? If his name is pulled, what is the probability that Gaynell's name will be pulled next?

7. A box of greeting cards contains 20 friendship cards, 10 get-well cards, and 10 congratulations cards. What is the probability of picking a get-well card at random?

8. Mimi tosses 21 pet food coupons, 16 dishwashing detergent coupons, and 11 cereal coupons into a container. What is the probability of reaching in and picking a cereal coupon?

9. A jar holds 10 lock washers and 15 flat washers. What is the probability of drawing a lock washer at random?

10. A TV quiz program puts all questions in a box. If the box contains five hard questions, five average questions, and five easy questions, what is the probability of being asked an easy question?

ASSIGNMENT EXERCISES

8-1

Use Fig. 8-13 to answer Exercises 1 to 4.

1. In what year(s) did women use more sick days than men?

2. In what year(s) did men use about five sick days?

3. In what year(s) did men use more sick days than women?

4. What was the greatest number of sick days for men?

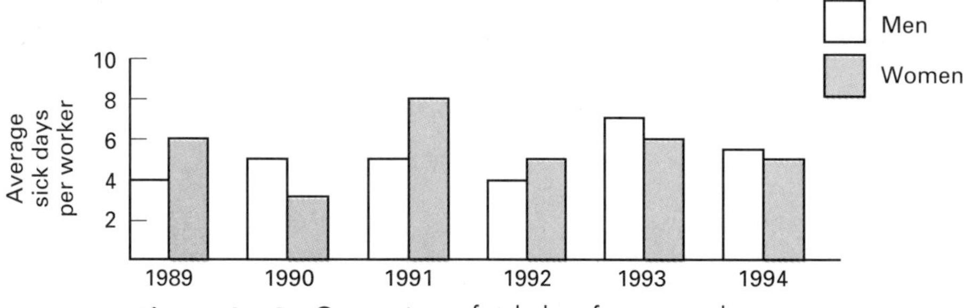

Figure 8-13 Comparison of sick days for men and women.

Use Fig. 8-14 to answer Exercises 5 to 7.

5. What percent of the total cost is the cost of the lot? Round to the nearest tenth.

6. What percent of the total cost is the cost of the house? Round to the nearest tenth.

7. The cost of the lot and landscaping represents what percent of the total cost? Round to the nearest tenth.

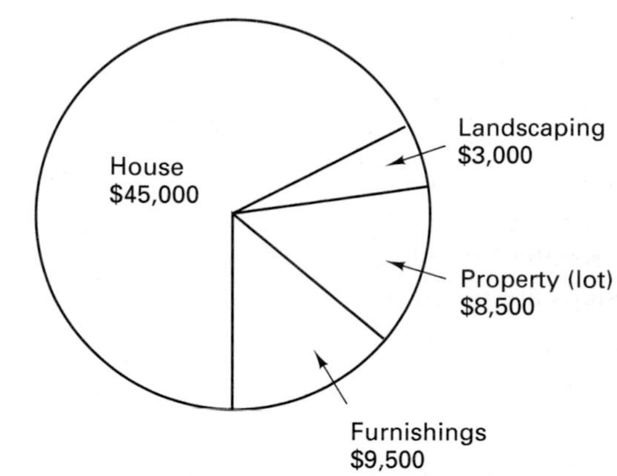

Figure 8-14 Distribution of costs for a $66,000 home.

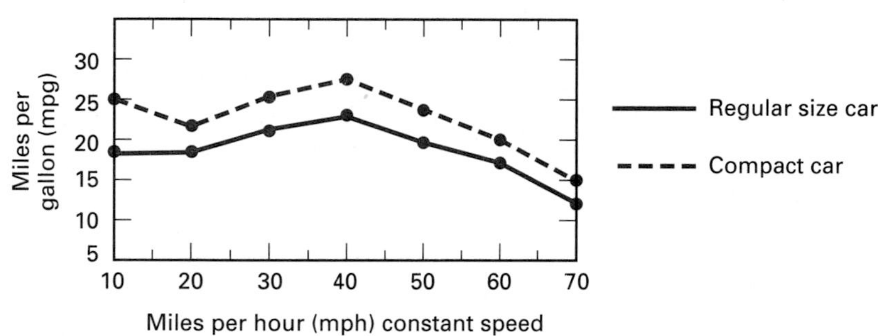

Figure 8-15 Automobile gasoline mileage comparisons.

Use Fig. 8-15 to answer Exercises 8 to 10.

8. What speed gave the highest gasoline mileage for both types of automobiles?

9. What speed gave the lowest gasoline mileage for both types of automobiles?

10. At what speed did the first noticeable decrease in gasoline mileage occur? Which car showed this decrease?

8-2

Use the histogram in Fig. 8-16 to answer the questions that follow about a public school system's classroom computers and their memory expressed in megabytes (M).

11. How many computers have 64M to 128M of memory?

12. How many computers have 512M to 640M of memory?

13. How many computers does the system have?

14. How many computers have more than 128M of memory?

15. What is the ratio of the number of 512M–640M computers to the number of 256M–384M computers?

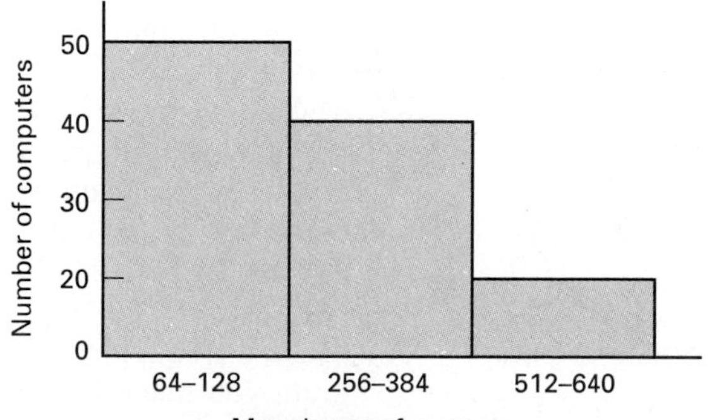

Figure 8-16 Public school system's classroom computers and memory.

A regional photograph association is composed of 54 members of several clubs. Make a frequency distribution of the members' ages: 17, 18, 20, 21, 21, 24, 24, 29, 29, 29, 31, 31, 33, 33, 34, 35, 35, 38, 38, 38, 39, 41, 42, 43, 43, 43, 43, 45, 45, 47, 47, 48, 48, 48, 49, 50, 51, 51, 52, 56, 56, 58, 58, 60, 60, 62, 64, 64, 65, 66, 68, 70, 71, 71. Use the format shown in Table 8-6.

TABLE 8-6 Ages of Club Members of Regional Photography Association

	Class Interval	Midpoint	Tally	Class Frequency
16.	66–75	________	________	________
17.	56–65	________	________	________
18.	46–55	________	________	________
19.	36–45	________	________	________
20.	26–35	________	________	________
21.	16–25	________	________	________

22. Make a histogram of the members' ages from the previous frequency distribution. Use the reference lines in Fig. 8-17 as guides.

23. Make a frequency polygon from the histogram in Fig. 8-17.

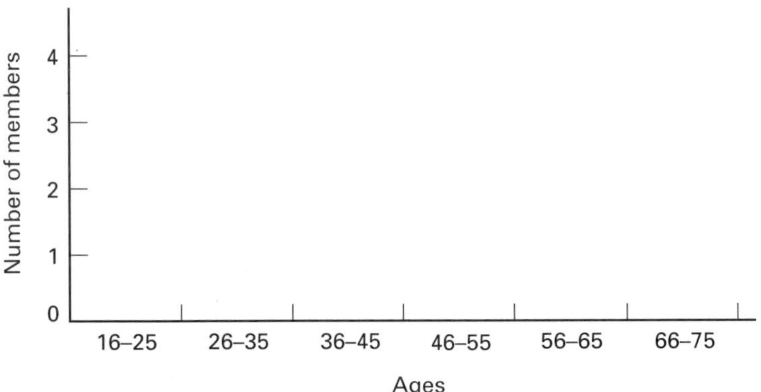

Figure 8-17 Ages of members of photography association.

Answer the following based on the information in Exercises 22 and 23.

24. In what age group is the least number of members?

25. In what age group is the greatest number of members?

26. How many members are under age 36?

27. How many members are over age 55?

28. What is the ratio of the number of members age 66 to 75 to members age 16 to 25?

29. What is the ratio of the number of members age 66 to 75 to the number of members age 46 to 55?

30. What percent (to the nearest tenth of a percent) of the members age 46 to 55 are the members age 16 to 25?

31. What percent (to the nearest tenth of a percent) of the total number of members are the members age 46 to 55?

32. Make a frequency distribution of the following scores on an English usage test: 68, 70, 74, 77, 78, 82, 82, 84, 86, 86, 86, 88, 89, 90, 90, 93.

33. Make a frequency distribution of the following mileages (miles per gallon, or mpg) reported one week by customers to an automotive rental company for six-cylinder cars: 22, 22, 23, 23, 23, 24, 24, 25, 26, 26, 27, 29, 29, 30, 30, 31, 31.

34. Make a histogram with the information from Exercise 33.

35. Make a frequency polygon based on the histogram you made for Exercise 34.

8-3

Solve the following problems.

36. Jimmy Smith made 176 baskets over a 16-game period. What was the average number of baskets per game to the nearest tenth?

37. Ben Duke sold 63 new cars over a 5-month period. What was the average number of cars sold per month to the nearest tenth?

38. An automobile used 22 gallons of regular gasoline on a trip of 358 miles. What was the average miles per gallon to the nearest tenth?

39. A delivery truck used 21 gallons of regular gasoline on a trip of 289 miles. What was the average miles per gallon to the nearest tenth?

40. The following hourly pay rates are used at fast-food restaurants: cooks, \$6.25; servers, \$6.95; bussers, \$5.20; dishwashers, \$5.20; managers, \$8.25. Find the median pay rate.

41. The following hourly pay rates are used at a locally owned store: office assistants, \$5.85; bookkeepers, \$6.20; cashiers, \$5.45; assistant managers, \$7.90. Find the median pay rate.

42. The following weekend work shifts are in effect at a mall clothing store: 3 hours in AM, 6 hours in PM, 3 hours in PM. Find the mode.

43. The following prices are in effect at a fast-food restaurant: \$1.85, hamburgers; \$1.98, hot ham sandwiches; \$2.28, chicken filet sandwiches; \$1.85, roast beef sandwiches. Find the mode.

44. Kim Collier, a student at a technical college, made the following final grades for the past term. Find the student's QPA (quality point average) to the nearest hundredth.

Subject	Grade	Hours	Points per Hour
Electronics 101	A	4	4
Circuits 201	A	4	4
Algebra 101	B	4	3

45. Tami Murphy made the following final grades for the past term. Find Tami's QPA (quality point average) to the nearest hundredth.

Subject	Grade	Hours	Points per Hour
English 201	A	3	4
History 202	B	3	3
Philosophy 101	A	3	4

46. Janice Van Dyke has the following grades in Algebra 102: 98, 82, 87, 72, and 82. What grade is needed on the last test for a B or 85 average?

47. Sarah Smith has the following grades in American History: 79, 73, 71, 78, and 86. What grade is needed on the last test for a C or 75 average?

48. Find the mean, median, and mode for these test scores: 67, 87, 76, 89, 70, 69, 82.

8-4

49. Susan Duke's new clothes consist of three blazers, four skirts, and two sweaters. How many three-piece outfits can she make from the new clothes?

50. For her first professional job, Martha Deskin, a recent college graduate purchased four pairs of shoes, three business suits, and five shirts. How many clothing combinations are possible?

51. There are three magazines on horses and ten magazines on fashion in Dr. Nelson Campany's waiting room. If Shirley Riddle sends her toddler to get two magazines, what is the probability that the child will randomly pick a fashion magazine on the first draw? If the child is successful, what is the probability of the child picking a fashion magazine on the second draw?

52. Two coins are tossed three times. Count the number of combinations of heads and tails with three tosses of the two coins.

53. If three coins are tossed, what is the number of possible combinations of heads and tails?

54. Cy Pipkin is puzzled over a true–false question on a test and does not know the answer. What is the probability that he will pick the right answer by chance?

55. Suppose a computer operator has five 3.5-inch floppy disks and knows one is filled, but cannot remember which one. If he picks one and puts it in the computer, what is the probability of his picking the filled disk?

56. A box of wrapped candies is said to contain 5 cherry cordials, 10 cremes, and 15 chocolate-covered caramels. What is the probability of picking a cherry cordial?

57. A box of stationery contains 12 sheets imprinted with an animal scene, 24 sheets imprinted with a floral design, and 12 plain sheets. What is the probability of picking a sheet with an animal scene at random?

ELECTRONICS APPLICATION

Use of Calculator to Find Percentage Error

If your calculator has more than one memory, you may have to use two or more keystrokes since you must tell the calculator which memory to use. For instance, [STO] 1 and [RCL] 1 are used to store in and recall from memory 1. [STO] 2 and [RCL] 2 are used to store in and recall from memory 2.

The symbol [/] or [÷] means "divided by," so use the division key on the calculator. That symbol, when used in a fraction such as $\frac{3}{4}$, means that 3 is divided by 4. The 3 is the numerator or dividend, and the 4 is the denominator or divisor. It makes sense that

$$\frac{3}{4} = 0.75 = \frac{75}{100} = 75\%$$

The equation to be used for percentage error is

$$\%\ \text{error} = \frac{\text{measured value} - \text{coded value}}{\text{coded value}} \times 100$$

Use a pair of parenthesis to group the difference in the numerator in the equation so it can be used as a fraction. Notice the difference between the left parenthesis, [(], and the right parentheses, [)]. The left parenthesis is also called an open parenthesis, and the right parenthesis is also called a closed parenthesis.

Since the coded value appears twice in the equation, it is entered only once and then stored in memory. Many calculator sequences will work. The following calculator sequence is easy to use since each value is entered only once:

coded value [STO] [(] measured value [−] [RCL] [)] [/] [RCL] [×] 100 [=]

Suppose $R_{1\_meas} = 21.5\ \text{k}\Omega$ and $R_{1\_coded} = 20\ \text{k}\Omega$. Upon analyzing the dimensions of the measures, the units of measure cancel. Multiplying by 100 converts the value to a percent.

20 [STO] [(] 21.5 [−] [RCL] [)] [/] [RCL] [×] 100 [=] ⇒ 7.5

The percentage error is 7.5%.

Exercises

Use your calculator to fill in the % error in the following data table of resistance measurements using two or more significant digits. Resistance values are often given with a related tolerance, which means a maximum acceptable percentage of error. Decide for each of the following if the value of the resistance is within or outside the tolerance for the given resistor.

Data Table for Resistance Values

	Resistor	Theoretical (coded) Resistance	Tolerance %	Measured Resistance	% Error	Within Tolerance?
1.	R_1	250 Ω	10	260 Ω		
2.	R_2	250 Ω	10	240 Ω		
3.	R_3	47 kΩ	10	48 kΩ		
4.	R_4	47 kΩ	10	46 kΩ		
5.	R_5	51 kΩ	10	51.5 kΩ		
6.	R_6	51 kΩ	10	59 kΩ		
7.	R_7	35 kΩ	10	26 kΩ		
8.	R_8	35 kΩ	10	29 kΩ		
9.	R_9	2500 Ω	10	2560 Ω		
10.	R_{10}	2500 Ω	10	2490 Ω		
11.	R_{11}	21 kΩ	10	26.3 kΩ		
12.	R_{12}	21 kΩ	10	20.2 kΩ		

Answers for Exercises

Data Table for Resistance Values

	Resistor	Theoretical (coded) Resistance	Tolerance %	Measured Resistance	% Error	Within Tolerance?
1.	R_1	250 Ω	10	260 Ω	4.0	Yes
2.	R_2	250 Ω	10	240 Ω	−4.0	Yes
3.	R_3	47 kΩ	10	48 kΩ	2.1	Yes
4.	R_4	47 kΩ	10	46 kΩ	−2.1	Yes
5.	R_5	51 kΩ	10	51.5 kΩ	0.98	Yes
6.	R_6	51 kΩ	10	59 kΩ	1.6	Yes
7.	R_7	35 kΩ	10	26 kΩ	−26	No
8.	R_8	35 kΩ	10	29 kΩ	−17	No
9.	R_9	2500 Ω	10	2560 Ω	2.4	Yes
10.	R_{10}	2500 Ω	10	2490 Ω	−0.40	Yes
11.	R_{11}	21 kΩ	10	26.3 kΩ	25	No
12.	R_{12}	21 kΩ	10	20.2 kΩ	−3.8	Yes

CHALLENGE PROBLEM

One complete circle is made up of 360 degrees. A measuring instrument called a *protractor* is used to draw angles of various sizes.

58. Use a protractor to make a circle graph of the sources of a construction company's yearly income of $150,000 showing 50% from small business construction, 25% from home remodeling, 15% from local government jobs, and 10% from miscellaneous work.

CONCEPTS ANALYSIS

1. What type of information does a circle graph show?

2. Give a situation where it would be appropriate to organize the data in a circle graph.

3. What type of information does a bar graph show?

4. Give a situation where it would be appropriate to organize the data in a bar graph.

5. What type of information does a line graph show?

6. Give a situation where it would be appropriate to organize the data in a line graph.

7. Explain the similarities and differences between a histogram and a frequency polygon.

8. Explain the differences among the three types of averages: the mean, the median, and the mode.

9. Show by example how using a tree diagram for counting gives the same result as multiplying the number of choices for each position in the set.

10. If there are 5 red marbles and 7 blue marbles in a bag, is the probability of drawing a red marble $\frac{5}{7}$? Explain your answer.

CHAPTER SUMMARY

Objective	What to Remember	Examples
Section 8-1		
1. Read circle graphs.	A circle graph compares parts to a whole.	If the total monthly revenue at a used car dealership is $75,000, what is the revenue from trucks? 35% of 75,000 = 0.35×75000 = $26,250

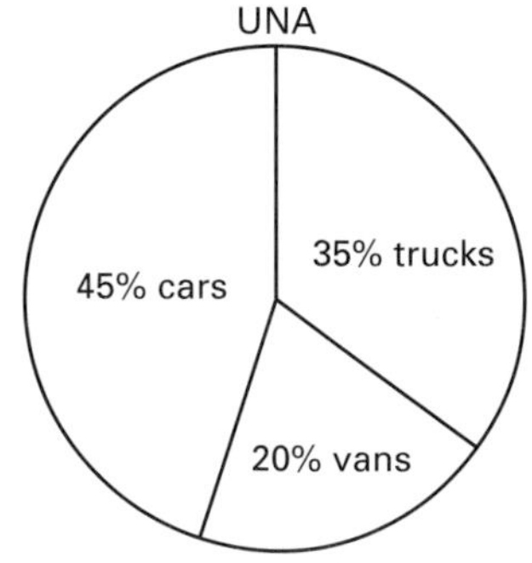

2. Read bar graphs.

Bar graphs compare items to each other.

What is the ratio of men's salaries to women's salaries in the pants department?

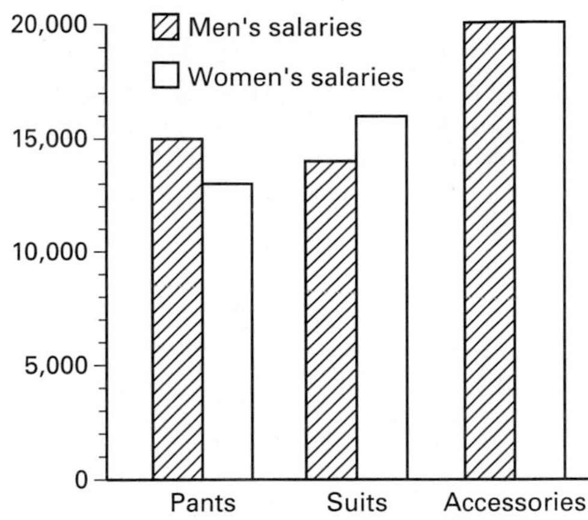

Men's salary in pants department = $15,000

Women's salary in pants department = $13,000

Ratio of men's salary to women's salary $= \dfrac{15}{13}$

3. Read line graphs.

A line graph shows how an item changes with time.

Make a line graph using the following average prices for textbooks.

Fall: 1990, $34; 1991, $37; 1992, $41; 1993, $42

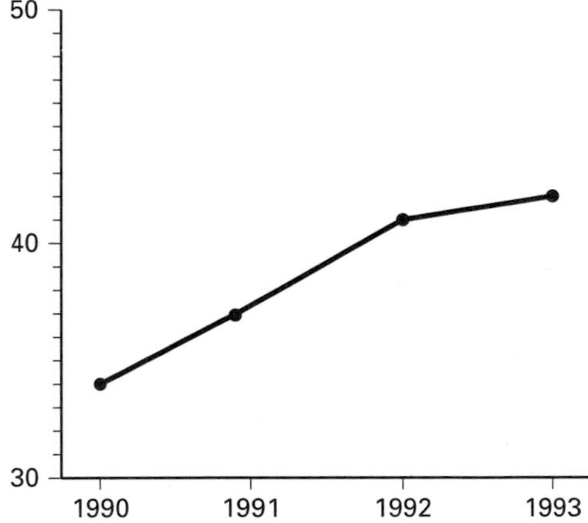

Section 8-2

1. Interpret and make frequency distributions.

To make a frequency distribution, determine the appropriate interval for classifying the data. Tally the data.

Make a frequency distribution with the following data indicating leave days for State College employees.

2 2 4 4 4 5 5 6 6 8
8 8 9 12 12 12 14 15
20 20

Annual Leave Days of 20 State College Employees

Class Interval	Tally	Class Frequency
16–20	//	2
11–15	~~////~~	5
6–10	~~////~~ /	6
1–5	~~////~~ //	7

2. Interpret and make histograms.

A histogram is a bar graph representing data in a frequency distribution.

Make a histogram using the data for State College; then draw a frequency polygon for the same data.

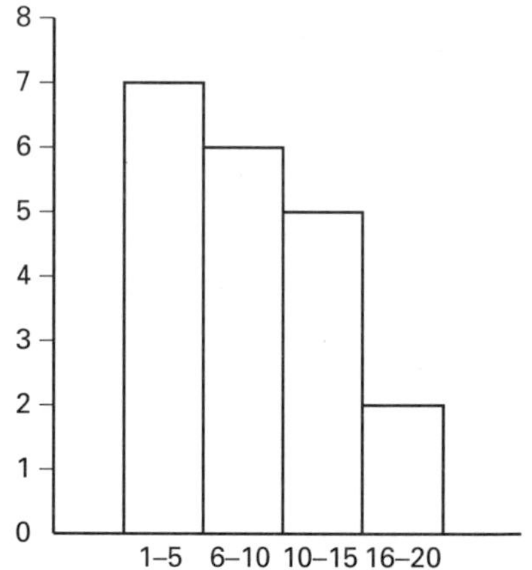

3. Interpret and make frequency polygons.

A frequency polygon is a line graph connecting the midpoints of each class interval in a frequency distribution.

Draw a frequency polygon using the data for the State College.

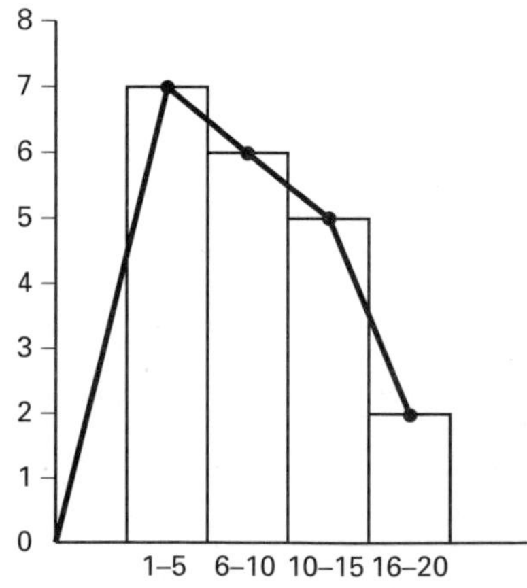

Section 8-3

1. Find the arithmetic mean or arithmetic average.	Add the quantities; divide by the number of quantities.	Find the arithmetic mean of the following test scores: 76, 86, 93, 87, 68, 76, 88 $76 + 86 + 93 + 87 + 68 + 76 + 88 = 574$ $574 \div 7 = 82$ The mean is 82.
2. Find the median, the mode, and the range.	The median of an odd number of quantities is the middle quantity when the quantities are arranged in order of size. For an even number of quantities, average the two middle quantities.	Find the median of the previous set of test scores. 68, 76, 76, <u>86</u>, 87, 88, 93 The median is 86.
	The mode is the quantity that occurs most frequently. A set of quantities may have no mode or more than one mode.	The mode is 76.
	The range is the difference between the largest and smallest quantity.	The range is $93 - 68 = 25$.

Section 8-4

1. Count the number of ways objects in a set can be arranged.	Multiply the number of positions possible for each object in the set.	Renee Smith's closet has two new blazers (navy and red) and four new skirts (gray, black, tan, and brown). How many outfits can she make from the new clothes? 1 blazer + 1 skirt = 1 outfit $\begin{pmatrix}\text{blazer}\\\text{choices}\end{pmatrix} \cdot \begin{pmatrix}\text{skirt}\\\text{choices}\end{pmatrix} = \begin{pmatrix}\text{possible}\\\text{outfits}\end{pmatrix}$ $3 \cdot 4 = 12$
2. Determine the probability of an event occurring if an activity is repeated over and over.	The probability of an event occurring is the ratio of the number of possible successful outcomes to the number of possible outcomes.	A box contains 30 chocolate caramel candies and 70 regular caramel candies. What is the probability of picking a regular caramel? $\frac{70}{100} = \frac{7}{10}$ The probability of picking a regular caramel is $\frac{7}{10}$.

WORDS TO KNOW

graph, (p. 316)
circle graph, (p. 316)
bar graph, (p. 317)
line graph, (p. 318)
class intervals, (p. 321)
class midpoint, (p. 321)
tally, (p. 321)
class frequency, (p. 321)
frequency distribution, (p. 321)
histogram, (p. 322)
frequency polygon, (p. 323)
statistical measurements, (p. 326)
average, (p. 326)
arithmetic mean or arithmetic average, (p. 326)
median, (p. 328)
mode, (p. 329)
measures of central tendency, (p. 329)
range, (p. 329)
measures of variation or dispersion, (p. 329)
set, (p. 330)
elements, (p. 330)
counting, (p. 331)
tree diagram, (p. 331)
probability, (p. 333)

CHAPTER TRIAL TEST

1. A ______ graph is used to show how different data items are related to each other.
2. A ______ graph is used to show how a whole quantity is related to its parts.
3. A ______ graph is used to show how an item or items change.

Answer Problems 4 and 5 from the line graph of Fig. 8-18.

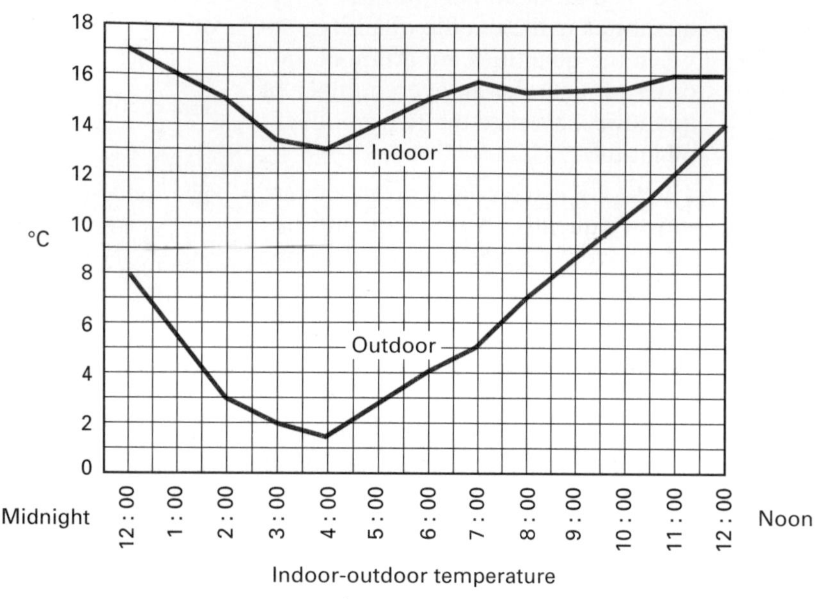

Figure 8-18

4. How many degrees warmer was it indoors at midnight than outdoors?
5. What was the change in outdoor temperature between 11:00 A.M. and noon?

Answer the following questions on the manufacturing costs for the electronic game as shown in the circle graph in Figure 8-19.

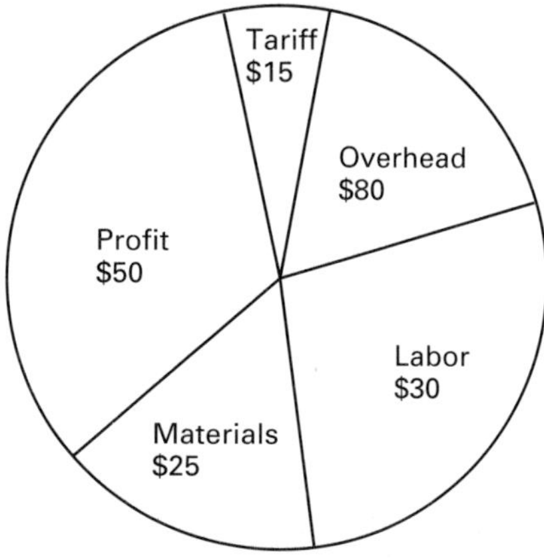

Figure 8-19 Manufacturing costs of a $200 electronic game.

6. What percent of the total cost is materials?
7. What would the profit be if there were no tariff or tax on imported parts?
8. How much are overhead and materials together?
9. What percent of the total cost is the profit?
10. What is the ratio of labor to total cost?
11. What is the ratio of the tariff to total cost?

Use the bar graph in Fig. 8-20 to answer the questions about the academic-year starting salaries of women and men college professors in various academic departments of a certain college.

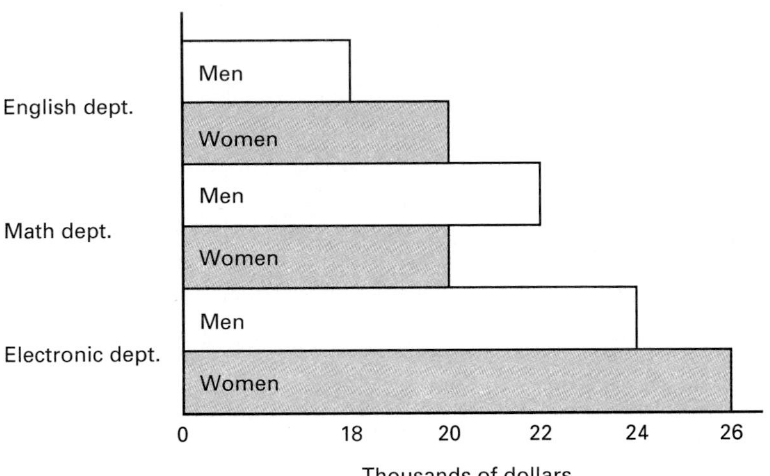

Figure 8-20 Salaries of women and men college professors.

12. In what departments do men make more than women?

13. In what departments do women make more than men?

14. What percent of women's salaries is men's salaries in the English Department (to the nearest tenth of a percent)?

15. What percent of men's salaries is women's salaries in the Electronics Department (to the nearest tenth of a percent)?

16. What is the ratio of men's salaries to women's salaries in the Mathematics Department?

17. What is the ratio of men's salaries to women's salaries in the English Department?

18. Make a line graph to illustrate the following numerical information about the average prices of two-, three-, and four-bedroom homes in a certain subdivision. Use the reference lines in Fig. 8-21 as guides.

2-bedroom homes, average \$40,000
3-bedroom homes, average \$50,000
4-bedroom homes, average \$60,000

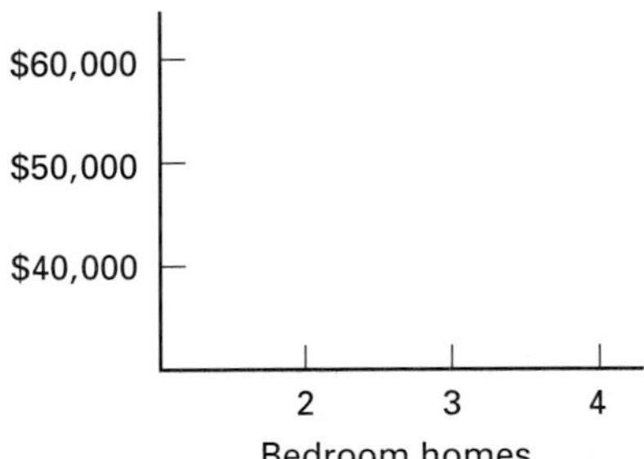

Figure 8-21 Prices of homes by number of bedrooms.

Use the frequency distribution shown in Table 8-7 to answer the questions that follow. The distribution shows the number of correct answers on a 30-question test in a science class.

TABLE 8-7 Frequency Distribution of Correct Answers

Class Interval	Midpoint	Tally	Class Frequency
26–30	28	///	3
21–25	23	////	4
16–20	18	~~////~~ //	7
11–15	13	///	3
6–10	8	//	2
1–5	3	/	1

19. How many students scored less than 16 correct?

20. How many students scored more than 25 correct?

21. What is the ratio of the number of students scoring 6 to 10 correct to those scoring 21 to 25 correct?

22. What percent of the total scored 21 to 30 correct?

Use the following ages of 24 children in a day-care center to complete a frequency distribution, using the format shown in Table 8-8.

1 1 1 1 1 1 1 2 2 2 3 3 3 3 3
4 4 4 4 5 5 5 5 6

TABLE 8-8 Ages of 24 Day-care Children

	Class Interval	Midpoint	Tally	Class Frequency
23.	4–6	________	________	________
24.	1–3	________	________	________

Use the histogram in Fig. 8-22 to answer the questions that follow about a company's software programs and the number of employees trained to use them.

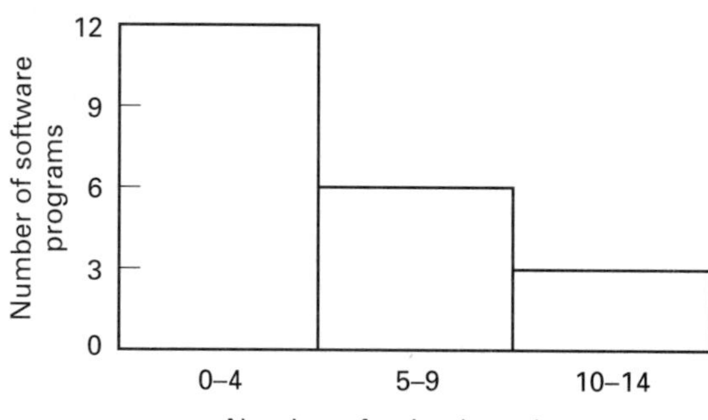

Figure 8-22 Company employees trained to use software.

25. How many employees at the most can use only three programs?

26. How many programs does the company have?

27. How many employees does the company have?

28. Find the mean, median, and mode for these test scores: 77, 87, 77, 89, 70, 69, 82.

29. Find the mean, median, and mode for these test scores: 81, 78, 69, 75, 81, 93, 68.

30. List and count the ways the elements in the set *L*, *M*, *N*, and *O* can be arranged.

31. Rayford has two ties: a red tie and a blue tie. He also has three shirts: a white shirt, a green shirt, and a yellow shirt. How many combinations of shirts and ties are possible?

32. Millie has in her purse three green eye shadows, four white eye shadows, and two black eye shadows. What is the probability of her pulling out a black eye shadow?

33. An envelope contains the names of two men and three women to be interviewed for a promotion. The interviewer wants to pull names to determine the order of the interviews. What is the probably of pulling a woman's name first?

34. If a small boy has one red marble and three yellow marbles in his pocket, what is the probability of his pulling out the red one on the first try?

CHAPTER

9 Signed Numbers and Linear Equations

Signed numbers are a useful tool for many technical applications. When a quantity or number is unknown, we can often use other known numbers to find the missing number or quantity. We use *equations* to solve these problems.

9-1 SIGNED NUMBERS

Learning Objectives

L.O.1. Perform basic operations with signed numbers.
L.O.2. Simplify expressions using the order of operations.

L.O.1. Perform basic operations with signed numbers.

Signed Numbers

The basic operations of signed numbers are applied in solving equations. Before studying the techniques for solving various types of equations, we will briefly review the rules for the basic operations of signed numbers. For a more thorough coverage of these rules, refer to Chapter 2, Section 3-10, and Section 4-8.

Rule 9-1.1 ***Basic operations for signed numbers:***

To *add* signed number having *like* signs:

1. Add their absolute values.
2. Give the answer the common sign.

To *add* signed numbers having *unlike* signs:

1. Subtract the smaller absolute value from the larger absolute value.
2. Give the answer the sign of the number with the larger absolute value.

To *subtract* signed numbers:

1. Change the subtraction to an equivalent addition by changing the sign of the subtrahend to its opposite.
2. Use the appropriate rule for adding signed numbers.

To *multiply* two signed numbers having *like* signs:

1. Multiply their absolute values.
2. Make the sign of the product positive.

To *multiply* two signed numbers having *unlike* signs:

1. Multiply their absolute values.
2. Make the sign of the product negative.

To *divide* signed numbers having *like* signs:

1. Divide their absolute values.
2. Make the quotient positive.

To *divide* signed numbers having *unlike* signs:

1. Divide their absolute values.
2. Make the quotient negative.

EXAMPLE 9-1.1 Perform the indicated operations.

(a) $-5 - 7$ **(b)** $-6 + 9$ **(c)** $4 - (-2)$ **(d)** $-3(7)$ **(e)** $\frac{-18}{-3}$

Solution

(a) $-5 - 7 = -12$ *(Interpret as -5 plus -7. Apply rule for adding numbers with like signs.)*

(b) $-6 + 9 = 3$ *(Apply rule for adding numbers with unlike signs.)*

(c) $4 - (-2) =$ *(Change to an equivalent addition.)*
$4 + 2 = 6$ *(Apply rule for adding numbers with like signs.)*

(d) $-3(7) = -21$ *(Apply rule for multiplying numbers with unlike signs.)*

(e) $\frac{-18}{-3} = 6$ *(Apply rule for dividing numbers with like signs.)* ■

L.O.2. Simplify expressions using the order of operations.

Order of Operations

When more than one operation is included in a problem, we must perform the operations in a specified order. We will briefly review the order of operations. For a more thorough coverage, refer to Section 2-6.

Rule 9-1.2 ***Order of operations for signed numbers:***

Perform operations in the following order as they appear from left to right:

1. Parentheses used as groupings and other grouping symbols
2. Exponents (powers and roots)
3. Multiplications and divisions
4. Additions and subtractions

EXAMPLE 9-1.2 Evaluate the following expressions.

(a) $3 + 2(5 - 8)$ **(b)** $7 - (4 - 9)$ **(c)** $4(-5) - \frac{8}{-2} + 2(5 + 2)$

Solution

(a) $3 + 2(5 - 8) =$ *(Perform operation inside parentheses.)*
$3 + 2(-3) =$ *(Multiply numbers with unlike signs.)*
$3 + -6 = -3$ *(Add numbers with unlike signs.)*

(b) $7 - (4 - 9) =$ *(Perform operation inside parentheses.)*
$7 - (-5) =$ *(Change to equivalent addition.)*
$7 + 5 = 12$ *(Add numbers with like signs.)*

(c) $4(-5) - \frac{8}{-2} + 2(5 + 2) =$ *(Perform operation inside parentheses.)*

$4(-5) - \frac{8}{-2} + 2(7) =$ *(Perform multiplications and divisions as they appear from left to right.)*
$-20 - (-4) + 14 =$ *(Change to equivalent addition.)*
$-20 + 4 + 14 = -4$ *(Perform additions and subtractions as they appear from left to right.)* ■

SELF-STUDY EXERCISES 9-1

L.O.1. Perform the indicated operations.

1. $-3 - 12$
2. $-8 + 7$
3. $-4 - (-2)$
4. $5(-2)$
5. $-3(11)$
6. $-7(-6)$
7. $\frac{35}{-7}$
8. $\frac{-48}{-6}$
9. $-\frac{4}{5}\left(-\frac{2}{7}\right)$
10. $\frac{-12.88}{5.6}$

L.O.2. Perform the indicated operations.

11. $3 + 5 - 8 + 7 - 1$
12. $-2 - 4 + 7 - 1$
13. $4 + 7(-3)$
14. $4(-5) + 3 - 2$
15. $3 + \frac{8}{-2} + 5$
16. $4 - (3 + 7)$
17. $3(2 - 5) + 1$
18. $6(-2) - 2(8 + 3)$
19. $\frac{-16}{-4} - 7(3)$
20. $7 + 2(3 - 5) + 7 - \frac{30}{-5}$

9-2 SIMPLIFYING ALGEBRAIC EXPRESSIONS

Learning Objectives

L.O.1. Identify equations, terms, factors, constants, and variables.
L.O.2. Add and subtract like terms.
L.O.3. Apply the distributive principle to simplify expressions.

Algebraic Terminology

Before we begin to work with equations to solve problems, we need to understand some of the basic concepts and terminology of equations.

In mathematics we often use the symbol "=" to show quantities are "equal to" each other. We can write the statement "five is equal to two plus three" in symbols as $5 = 2 + 3$. This symbolic statement is called an *equation.* "Symbolic" means we are using symbols like "=" and "+."

■ **DEFINITION 9-2.1 Equation.** An **equation** is a symbolic statement that two expressions or quantities are equal in value. This statement may be true or false.

EXAMPLE 9-2.1 Verify that the following statements are true equations.

(a) $3(8) = 24$ **(b)** $12 - 3 = 9$

Solution: To verify that these statements are true equations, we must *evaluate* or find the value of the expression, or quantity, on each side of the equal sign. To be a true equation, the value of the left side of the equation must equal the value of the right side of the equation.

(a) $3(8) = 24$ *(3 times 8 is 24.)*
$24 = 24$ *(The equation is true.)*

(b) $12 - 3 = 9$ *(12 minus 3 is 9.)*
$9 = 9$ *(The equation is true.)* ■

Some types of equations contain a *missing* number or *unknown* number. We *solve* an equation by finding the missing number that makes the equation *true.* The missing number is represented in the equation by a letter such as x, a, or z. In each equation, the letter has a certain (unknown) value. The same letter may be used in several equations, and its value in each equation depends on the other numbers and relationships in the equation. Because the value of the letter varies with each equation, the letter is often called a *variable.* Because the value of the letter is unknown until the equation is solved, the letter is also called the *unknown* in the equation. The *solution,* also called the *root,* of an equation is the value of the variable that makes the equation true.

■ **DEFINITION 9-2.2 Variable or Unknown.** A letter that represents an unknown or missing number is called a **variable** or **unknown**.

■ **DEFINITION 9-2.3 Root or Solution.** The **root** or **solution** of an equation is the value of the variable that makes the equation true.

EXAMPLE 9-2.2 Solve the following equations.

(a) $x = 3 + 8$ **(b)** $\frac{18}{3} = n$ **(c)** $y = 3(4) - 5$

Solution: Since the value of one side of an equation must equal the value of the other side of the equation, we solve the equation, that is, find the missing number, or variable, by evaluating the opposite side of the equation.

(a) $x = 3 + 8$ *(8 added to 3 is 11.)*
$x = 11$

(b) $\frac{18}{3} = n$ *(18 divided by 3 is 6.)*
$6 = n$

(c) $y = 3(4) - 5$	*(Remember, multiplication is worked first in the order of operations. 3 times 4 is 12.)*
$y = 12 - 5$	*(Add 12 and −5.)*
$y = 7$	*(12 plus − 5 = 7.)*

Since both sides of an equation are equivalent, it does not matter which side comes first in the equation. In these examples, for instance, $x = 3 + 8$ means the same as $3 + 8 = x$; $\frac{18}{3} = n$ means the same as $n = \frac{18}{3}$; and $y = 3(4) - 5$ means the same as $3(4) - 5 = y$. In other words, $n = 5$ is equivalent to $5 = n$, and so on. The unknown, or variable, may appear on either the left or right side of an equation. ■

Rule 9-2.1 ***Symmetric property of equality:***

If the sides of an equation are interchanged, equality is maintained.

If $a = b$, then $b = a$.

Numbers that are multiplied together are called *factors*. Since division is the inverse operation of multiplication, it can be rewritten as an equivalent product. $\frac{3}{y}$ can be rewritten as $3 \cdot \frac{1}{y}$. Thus, division can be expressed as multiplication of factors. An algebraic expression, or quantity, can be written showing a known factor multiplied times a variable, or unknown factor, such as $5x$. The number 5 is the known factor multiplied times some unknown number that we represent with the letter x.

If an algebraic expression contains only factors, then we refer to this expression as a *term*.

$2xy$ is a term containing the factors 2, x, and y.

$\frac{2x}{y}$ is a term containing factors 2, x, and $\frac{1}{y}$.

$2(x + 3)$ is a term containing the factors 2 and $(x + 3)$. Since $(x + 3)$ is grouped, it is one quantity and is considered as a factor. However, if the distributive principle is applied, and each term within parentheses is multiplied by 2, then $2(x + 3)$ becomes $2x + 6$. In this expression there are *two* terms, $2x$ and 6.

Tips and Traps

When a term is the product of letters alone (such as ab or xyz) or the product of a number and one or more letters (such as $3x$ or $2ab$), parentheses or other symbols of multiplication are usually omitted. Thus, ab means a times b, $3x$ means 3 times x, and so on.

In algebra it is important to be able to distinguish between terms and factors.

■ **Definition 9-2.4 Factors.** Algebraic expressions indicating multiplication of quantities are called **factors**.

■ **Definition 9-2.5 Terms.** Algebraic expressions indicating single quantities or quantities that are added or subtracted are called **terms**.

A term can be a single letter; a single number; the product of numbers and letters; the product of numbers, letters, and/or groupings; or the quotient of numbers, letters, and/or groupings. The fraction line implies the numerator and/or the denominator is grouped.

Tips and Traps

Plus and minus signs that are *not* within a grouping separate terms.

$3a + b$	$3(a + b)$	$\dfrac{3}{a + b}$
two terms	one term	one term

EXAMPLE 9-2.3 Identify the terms in the following expressions by drawing a box around each term.

(a) $a - 2$
(b) $3x + 5$
(c) $2x - 4(x + 7)$
(d) $2ab + 4a - b + 2(a + b)$
(e) $\dfrac{x}{2}$
(f) $\dfrac{2a + 1}{3}$
(g) $\dfrac{2a}{3} + 1$

Solution

(a) $\boxed{a} - \boxed{2}$ *(Terms are separated by "+" and "−" signs that are not within a grouping.)*

(b) $\boxed{3x} + \boxed{5}$

(c) $\boxed{2x} - \boxed{4(x + 7)}$

(d) $\boxed{2ab} + \boxed{4a} - \boxed{b} + \boxed{2(a + b)}$

(e) $\boxed{\dfrac{x}{2}}$

(f) $\boxed{\dfrac{2a + 1}{3}}$ *(The numerator or denominator of a fraction is a grouping, such as 2a + 1).*

(g) $\boxed{\dfrac{2a}{3}} + \boxed{1}$ ■

Terms that contain only numbers are called *number terms* or *constants*, and terms that contain only letters or that contain both numbers and letters are called *letter terms* or *variable terms.*

■ **DEFINITION 9-2.6 Number Terms or Constants.** A term that contains only numbers is called a **number** term or **constant**.

■ **DEFINITION 9-2.7 Letter Term or Variable Term.** A term that contains only one letter, several letters used as factors, or a combination of letters and numbers used as factors is called a **letter term** or **variable term**.

If a term contains more than one factor, each factor is the *coefficient* of the other factors. Any factor(s) in a term containing more than one factor may be grouped as the coefficient of the remaining factor(s). In the term $2ab$:

- ☐ 2 is the *coefficient* of ab.
- ☐ a is the *coefficient* of $2b$.
- ☐ b is the *coefficient* of $2a$.
- ☐ $2b$ is the *coefficient* of a.
- ☐ $2a$ is the *coefficient* of b.
- ☐ ab is the *coefficient* of 2.

■ **DEFINITION 9-2.8 Coefficient.** Each factor of a term containing more than one factor is the **coefficient** of the remaining factor(s).

In algebraic expressions, we are usually interested in the *numerical coefficient* of a term. In the term $2ab$, 2 is the *numerical coefficient* of ab.

Recall that mathematics makes use of signs, both positive and negative. If a term is positive, such as $2a$, there is no need to write the "+" sign before the term. But if the term is negative, such as $-5b$, then the "−" sign *must* be expressed. In the term $2a$, the numerical coefficient is 2, which is positive. In the term $-5b$, the numerical coefficient is negative (-5).

If the term is of the type $-\frac{a}{3}$ or $\frac{2b}{7}$, the term is the product of a fractional coefficient and a letter. That is, the term $-\frac{a}{3} = -\frac{1}{3}\left(\frac{a}{1}\right)$ or $-\frac{1}{3}a$ and the term $\frac{2b}{7} = \frac{2}{7}\left(\frac{b}{1}\right)$ or $\frac{2}{7}b$. The *numerical coefficient* of a is $-\frac{1}{3}$. The *numerical coefficient* of b is $\frac{2}{7}$. Similarly, in the term $\frac{4n}{5}$ the *numerical coefficient* of n is $\frac{4}{5}$.

■ **DEFINITION 9-2.9 Numerical Coefficient.** The numerical factor of a term is the **numerical coefficient**. Unless otherwise specified, the word *coefficient* will be used to mean the *numerical coefficient*. The coefficient is a signed number. The coefficient should be written *in front* of the variable.

EXAMPLE 9-2.4 Identify the numerical coefficient in each term.

(a) $2x$
(b) $-3ab$
(c) $4(x + 3)$
(d) $-\frac{n}{3}$
(e) $\frac{2b}{5}$

Solution

(a) The coefficient of x is 2.
(b) The coefficient of ab is -3.
(c) The coefficient of $(x + 3)$ is 4.
(d) The term $-\frac{n}{3}$ is the same as $-\frac{1}{3}n$. Therefore, the coefficient of n is $-\frac{1}{3}$.
(e) $\frac{2b}{5}$ is the same as $\frac{2}{5}\left(\frac{b}{1}\right)$ or $\frac{2}{5}b$. Thus, the coefficient of b is $\frac{2}{5}$. ■

When a letter term has no written numerical coefficient, the coefficient is understood to be 1. That is, the coefficient of x is 1. $x = 1x$. Similarly, the coefficient of ab is 1. $ab = 1ab$. Likewise, the numerical coefficient of $-x$ is -1. $-x = -1x$. The coefficient of $-bc$ is -1. $-bc = -1bc$.

L.O.2. Add and subtract like terms.

Combining Like Terms

We saw in adding and subtracting fractions that we added or subtracted fractions with like or common denominators. Similarly, in adding and subtracting measures we needed like units of measure. With variable or letter terms, we can only add or subtract like terms.

■ **DEFINITION 9-2.10 Like Terms.** Terms are **like terms** if they are number terms or if they are letter terms with exactly the same letter factors.

$4y$ and $2y$ are like terms because both contain the same letter (y). Similarly, 3 and 1 are like terms because both are numbers. We *cannot* combine 3 and $4y$ because they are unlike terms (number term and letter term). We *cannot* combine $4y$ and $2x$ because they are unlike terms (different letters).

Rule 9-2.1 *To combine numerical terms:*

1. Change subtraction to addition if appropriate.
2. Add the numerical terms using the appropriate rule for adding signed numbers.
3. The sum will be a number term.

Rule 9-2.2 *To combine like letter terms:*

1. Change subtraction to addition if appropriate.
2. Add the numerical coefficients of the like letter terms using the appropriate rule for adding signed numbers.
3. The sum will have the same letter factors as the like terms being added.

EXAMPLE 9-2.5 Simplify the following expressions by combining like terms.

(a) $5a + 2a - a$
(b) $3x + 5y + 8 - 2x + y - 12$

Solution

(a) $5a + 2a - a = 6a$ *(Add coefficients. $5 + 2 - 1 = 6$.)*

(b) $3x + 5y + 8 - 2x + y - 12$
$= x + 6y - 4$

(Add like terms.
$3x - 2x = x$
$5y + y = 6y$
$8 - 12 = -4$) ■

L.O.3. Apply the distributive principle to simplify expressions.

Applying the Distribution Principle

The distributive principle can be extended to include multiplying number terms and letter terms. For now, we will only examine situations in which we are multiplying number and letter terms by number terms.

EXAMPLE 9-2.6 Simplify by applying the distributive principle.

(a) $3(5x - 2)$ **(b)** $-7(x + y - 3z)$ **(c)** $-(-3x + 4)$

Solution

(a) $3(5x - 2) = 3(5x) - 3(2)$
$= 15x - 6$ *(Apply distributive principle.)*

(b) $-7(x + y - 3z) = -7(x) - 7(y) - 7(-3z)$
$= -7x - 7y + 21z$ *(Apply distributive principle.)*

(c) $-(-3x + 4) = -1(-3x) - 1(+4)$
$= 3x - 4$ *(Apply distributive principle.)* ■

SELF-STUDY EXERCISES 9-2

L.O.1. Solve the following equations.

1. $n = 4 + 7$ **2.** $m = 8 - 2$ **3.** $5 - 9 = y$

4. $\frac{12}{2} = x$ **5.** $p = 2(5) - 1$ **6.** $b = 6 - 3(5)$

Identify the terms in the following expressions by drawing a box around each term.

7. $7 + c$ **8.** $4a - 7$ **9.** $3x - 2(x + 3)$

10. $\frac{a}{3}$ **11.** $7xy + 3x - 4 + 2(x + y)$ **12.** $14x + 3$

13. $\frac{7}{a + 5}$ **14.** $\frac{4x}{7} + 5$

Identify the numerical coefficient in each of the following terms.

15. $5x$ **16.** $-4xy$ **17.** $\frac{n}{5}$

18. $\frac{2a}{7}$ **19.** $6(x + y)$ **20.** $-\frac{4}{5}x$

L.O.2. Simplify the following expressions by combining like terms.

21. $3a - 7a + a$ **22.** $-8x + y - 3y$

23. $5x - 3y + 2x + y$ **24.** $-4a + b + 9 + a - b - 3$

25. $2x + 8 - x + 4 - x - 2$ **26.** $3a + 5b + 8c + 1$

L.O.3. Simplify the following expressions by applying the distributive principle and combining like terms if appropriate.

27. $5(2x - 4)$ **28.** $4 + 3(2y + 3)$

29. $3 - (4x - 2)$ **30.** $5 - 2(6a + 1)$

9-3 SOLVING BASIC EQUATIONS

Learning Objectives

L.O.1. Solve basic equations.
L.O.2. Check the solutions of equations.

In Section 9-2 we saw how an equation like $x = 3 + 8$ could be solved by performing the calculations on the side opposite the variable. In doing so, we got $x = 11$ and the equation was solved because we found what value the variable x was equal to in this equation. Equations on the job will not all be this simple. We will examine other types of equations, beginning with the basic equation.

Solving Basic Equations

L.O.1. Solve basic equations.

■ **DEFINITION 9-3.1 Basic Equation.** An equation that consists of only a variable and its coefficient on one side of the equal sign and only a number term on the other side of the equal sign is called a **basic equation**.

Some examples of basic equations are:

$$5x = 15 \qquad -2 = 4b$$

$$2x = 11 \qquad -3y = 6$$

$$8 = 3a$$

A basic equation states that a number times a variable equals a product. In the equation $5x = 15$, we have the number 5 times the variable x equal to the product 15. We can determine mentally that the missing number must be 3; however, we want to learn a procedure for solving even the simplest of equations so that we can apply the same procedure to more complicated equations.

An equation is *solved* when the letter is alone on one side of the equation. That is, the coefficient of the letter term is 1. The number that is on the side opposite the letter is called the *root* or *solution* of the equation.

In the equation $5x = 15$, the numerical coefficient of the letter term is 5. Since we would like the numerical coefficient of x to be 1, *we multiply both sides of the equation by the reciprocal of 5*, which is $\frac{1}{5}$. Recall that multiplication of a number by the reciprocal, or multiplicative inverse, of that number gives us 1.

$$\left(\frac{1}{5}\right)5x = 15\left(\frac{1}{5}\right)$$

$$\left(\frac{1}{\cancel{5}_1}\right)\overset{1}{\cancel{5}}x = \overset{3}{\cancel{15}}\left(\frac{1}{\cancel{5}_1}\right) \qquad \textit{(Reduce where possible.)}$$

$$1x = 3 \qquad \textit{(Work multiplication on each side.)}$$

$$x = 3$$

The basic principle that underlies this procedure and all other procedures for solving equations is to preserve the equality of each side of the equation.

Basic Principle of Equality

Whenever we perform an operation on one side of the equation, we must perform the same operation on the other side to preserve the equality of the two sides.

For example, if we had a scale with 1 oz on one side and 1 oz on the other side, the scale would be balanced. If we increased or decreased the weight on one side, we would need to do the same on the other side or one side would be heavier than the other and the equality of both sides would be lost (see Fig. 9-1).

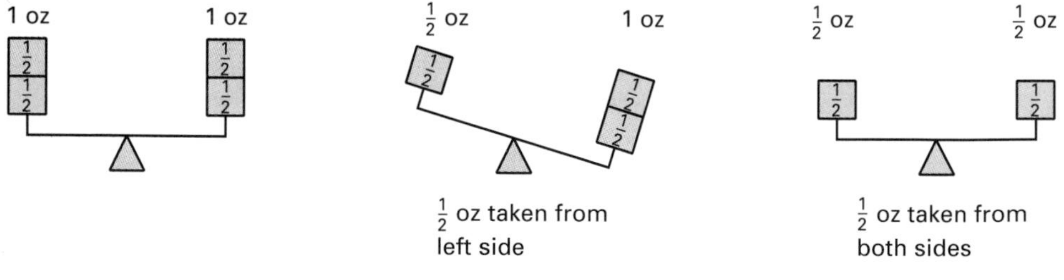

Figure 9-1

Let's look at the equation $5x = 15$ again. Notice that the numerical coefficient 5 is "attached" to the unknown, x, by multiplication. Since division is the inverse operation of multiplication, we can solve the equation by *dividing both sides of the equation by 5*. This procedure also gives us a numerical coefficient of 1 for x.

$$\frac{5x}{5} = \frac{15}{5}$$

$$\frac{\overset{1}{\cancel{5}}x}{\underset{1}{\cancel{5}}} = \frac{\overset{3}{\cancel{15}}}{\underset{1}{\cancel{5}}} \quad \textit{(Reduce within each fraction.)}$$

$$1x = 3$$

$$x = 3$$

In short, we can solve a basic equation by multiplication or by division, as indicated in Rule 9-3.1.

> **Rule 9-3.1** ***To solve a* basic equation*:***
>
> Divide both sides of the equation by the coefficient of the letter term;
>
> or
>
> Multiply both sides of the equation by the reciprocal of the coefficient of the letter term.

This rule is based on the *multiplication axiom.* The multiplication axiom states that both sides of an equation may be multiplied by the same *nonzero* quantity without changing the equality of the two sides. And, since multiplication and division are inverse operations, to divide by a number is the same as to multiply by its multiplicative inverse—its reciprocal.

Let us now solve several basic equations using multiplication or division.

EXAMPLE 9-3.1 Solve the equation $4x = 48$.

Solution

$$4x = 48$$

$$\left(\frac{1}{4}\right)4x = 48\left(\frac{1}{4}\right)$$ *(Multiply both sides of the equation by the reciprocal of the coefficient of the letter term.)*

$$x = 12$$

The same equation can be solved by dividing both sides of the equation by the coefficient of the letter term.

$$4x = 48$$

$$\frac{4x}{4} = \frac{48}{4}$$ *(Divide both sides by 4.)*

$$1x = 12$$ *(This step is usually omitted.)*

$$x = 12$$

■

When the equation contains only whole numbers or decimals, it is more convenient to divide by the coefficient of the letter than to multiply by the reciprocal of the coefficient of that letter.

EXAMPLE 9-3.2 Solve the equation $45 = 0.5x$.

Solution

$$\frac{45}{0.5} = \frac{0.5x}{0.5} \qquad \textit{(Divide both sides by 0.5.)}$$

$$90 = x$$

■

Multiplication and division are effective no matter which side of the equation contains the letter term. However, once the equation is solved, many people prefer to put the letter on the left ($x = 90$ instead of $90 = x$). Either way is correct.

EXAMPLE 9-3.3 Solve the equation $-6x = 42$.

Solution

$$\frac{-6x}{-6} = \frac{42}{-6} \qquad \textit{(Divide both sides by } -6.\textit{)}$$

$$x = -7 \qquad \textit{(Apply the rules for dividing signed numbers.)}$$

■

EXAMPLE 9-3.4 Solve the equation $4x = -22$.

Solution

$$4x = -22$$

$$\frac{4x}{4} = \frac{-22}{4} \qquad \textit{(Divide both sides by 4.)}$$

$$x = -\frac{11}{2} \qquad \left(-5\frac{1}{2} \textit{ or } -5.5\right)$$

■

Tips and Traps

The solution of many equations will not be a whole number. When this is the case, the solution can be left as a *proper* or *improper fraction* in lowest terms. The solution can also be changed to its decimal or mixed-number equivalent. When applied problems are solved, we will be given an appropriate form for expressing our answers. In the meantime, we will leave the solutions as fractions in lowest terms.

EXAMPLE 9-3.5 Solve the equation $\frac{1}{4}n = 7$.

Solution: We will solve this equation two ways, using multiplication and division.

Using multiplication: $\frac{1}{4}n = 7$

$$\left(\frac{4}{1}\right)\frac{1}{4}n = 7\left(\frac{4}{1}\right) \qquad \left(\textit{The reciprocal of } \frac{1}{4} \textit{ is } \frac{4}{1}.\right)$$

$$n = 28$$

Using division: $\frac{1}{4}n = 7$

$$\frac{\frac{1}{4}n}{\frac{1}{4}} = \frac{7}{\frac{1}{4}} \qquad \left(7 \div \frac{1}{4} = \frac{7}{1} \times \frac{4}{1} = 28\right)$$

$$n = 28$$

■

Tips and Traps

When the equation contains a fraction, it is usually preferable to multiply by the reciprocal of the coefficient of the variable.

EXAMPLE 9-3.6 Solve the equation $-n = 25$.

Solution: The numerical coefficient is -1, so the equation is not solved. The numerical coefficient must be $+1$ for the equation to be solved.

$$\frac{-1n}{-1} = \frac{25}{-1}$$

$$n = -25$$

■

L.O.2. Check the solutions of equations.

Checking Solutions of Equations

When an equation has been solved, the solution, or *root*, can be checked to avoid careless mistakes.

> **Rule 9-3.2**
>
> To check the solution of an equation, substitute the root in place of the letter in the original equation and perform all indicated operations. If the solution is correct, the value of the left side should equal the value of the right side.

EXAMPLE 9-3.7 The solution for the equation $4x = 48$ (Example 9-3.1) was found to be 12. Check or verify this root.

Solution

$4x = 48$ *(Substitute 12 for x.)*

$4(12) = 48$ *(Perform the multiplication.)*

$48 = 48$ *(The root is verified if the final equality is true.)* ■

EXAMPLE 9-3.8 The solution for the equation $-6x = 42$ (Example 9-3.3) was found to be -7. Check or verify this root.

Solution

$-6x = 42$

$-6(-7) = 42$ *(Substitute.)*

$42 = 42$

■

EXAMPLE 9-3.9 The solution for the equation $4x = -22$ (Example 9-3.4) was found to be $-\frac{11}{2}$. Check or verify this root.

Solution

$$4x = -22$$

$$4\left(-\frac{11}{2}\right) = -22 \qquad \left[\textit{Substitute.}\ \left(\frac{\overset{2}{\cancel{4}}}{1}\right)\left(-\frac{11}{\underset{1}{\cancel{2}}}\right) = \frac{-22}{1}.\right]$$

$$-22 = -22$$

■

EXAMPLE 9-3.10 The solution of the equation $\frac{1}{4}n = 7$ (Example 9-3.5) was found to be 28. Check or verify this solution.

Solution

$$\frac{1}{4}n = 7$$

$$\frac{1}{4}(28) = 7 \qquad \left[\left(\frac{1}{\underset{1}{\cancel{4}}}\right)\left(\frac{\overset{7}{\cancel{28}}}{1}\right) = \frac{7}{1}\right]$$

$$7 = 7$$

■

SELF-STUDY EXERCISES 9-3

L.O.1. Solve the following equations. Use multiplication or division. Show your work.

1. $7x = 56$
2. $2x = 18$
3. $15 = 3x$
4. $36 = 1.2n$
5. $-3a = 27$
6. $-18 = 9b$
7. $-7c = -49$
8. $-72 = 8x$
9. $5x = 32$
10. $36 = 8y$
11. $-6n = 15$
12. $-28 = -3b$
13. $-x = 7$
14. $-2 = -y$
15. $\frac{1}{3}x = 5$
16. $4 = \frac{1}{2}y$
17. $\frac{3}{5}x = 2$
18. $-\frac{3}{4}y = 5$
19. $-\frac{1}{2}x = -6$
20. $21 = \frac{3}{8}n$

L.O.2. **21.–40.** Check or verify each solution or root for the even-numbered equations in Self-Study Exercises 9-3.

9-4 ISOLATING THE VARIABLE IN SOLVING EQUATIONS

Learning Objectives

L.O.1. Solve equations with like terms on the same side of the equation.
L.O.2. Solve equations with like terms on opposite sides of the equation.

Equations are not always basic equations. Sometimes they are more involved and must be *simplified* into the form of the basic equation so that they can be solved. This section will show how to simplify or change into basic equations certain types of equations that are not basic equations.

L.O.1. Solve equations with like terms on the same side of the equation.

Equations with Like Terms on the Same Side

Recall that a basic equation has a single letter term for one side and a single number term for the opposite side. If an equation has more than one letter term on one side or more than one number term on the opposite side, we *combine* the terms on each side to form a basic equation.

We can combine only *like* terms. By "combine," we mean to add them according to the signed number rules.

EXAMPLE 9-4.1 Solve the equation $4x - x = 7 + 8$.

Solution

$4x - x = 7 + 8$ *(The coefficient of x is −1.)*

$\underbrace{4x - x} = \underbrace{7 + 8}$ *(Think of 4x plus −1x = 3x. 7 + 8 = 15.)*

$3x = 15$ *(We now have a basic equation.)*

The basic equation can be solved as before with multiplication or division on both sides. Division is more convenient with whole numbers.

$\frac{3x}{3} = \frac{15}{3}$ *(We divide both sides by the numerical coefficient of x to "undo" multiplication.)*

$x = 5$ ■

L.O.2. Solve equations with like terms on opposite sides of the equation.

Equations with Like Terms on Opposite Sides

In many equations, like terms do not appear on the same side of the equal sign. In such cases the like terms must be manipulated so that they appear on the same side of the equal sign *before* they can be combined. The basic purpose is to *isolate* the variable or letter terms on one side of the equation and thereby obtain a basic equation to be solved.

Terms can be moved from one side of an equation to another by applying the *addition axiom.* The addition axiom states that the same amount can be added to both sides of an equation without changing the equality of the two sides.

EXAMPLE 9-4.2 Solve the equation $2x + 4 = 8$.

Solution: The like terms 4 and 8 are on opposite sides of the equal sign. To isolate the $2x$, we need to eliminate 4. Therefore, if we had some way to move the number term 4 to the right side of the equation, all the number terms would be on one side and the letter term would be isolated on the opposite side.

We can move the 4 from the letter term side to the opposite side by adding the *opposite* of 4 (that is, −4) to *both* sides. Remember, whatever we do to one side, we must do to the other side.

$2x + 4 = 8$

$2x \underbrace{+ 4 - 4} = \underbrace{8 - 4}$ *(Add −4, the opposite of 4, to both sides.)*

$2x + 0 = 4$ *(4 − 4 = 0, 8 − 4 = 4)*

$2x = 4$ *(2x + 0 = 2x)*

Now we have a *basic equation* that we can solve with multiplication or division on both sides. Here, division is used.

$$\frac{2x}{2} = \frac{4}{2} \quad \textit{(We divide both sides by the numerical coefficient.)}$$

$$x = 2$$

■

EXAMPLE 9-4.3 Solve the equation $9 - 4x = 8x$.

Solution: In this equation the like terms $4x$ and $8x$ are on opposite sides of the equation. We want to manipulate the terms in this equation so that both letter terms are isolated on one side of the equation and the number term is on the other side. The simplest way of accomplishing this is to eliminate $-4x$ from the left side of the equation by adding its opposite to both sides.

$$9 - 4x = 8x$$

$$9 \underbrace{- 4x + 4x} = \underbrace{8x + 4x} \quad \textit{(Add 4x, the opposite of } -4x\textit{, to both sides.)}$$

$$9 + 0 = 12x \quad (-4x + 4x = 0,\ 8x + 4x = 12x)$$

$$9 = 12x \quad (9 + 0 = 9)$$

$$\frac{9}{12} = \frac{12x}{12} \quad \textit{(Divide both sides by coefficient of x.)}$$

$$\frac{9}{12} = x \quad \textit{(Reduce to lowest terms.)}$$

$$\frac{3}{4} = x \quad \text{or} \quad x = \frac{3}{4}$$

■

EXAMPLE 9-4.4 Solve the equation $7 - 5x = 12$.

Solution: The like terms are 7 and 12. Add -7 to both sides of the equation to isolate the variable.

$$7 - 5x = 12$$

$$7 - 7 - 5x = 12 - 7$$

$$0 - 5x = 5$$

$$-5x = 5$$

$$\frac{-5x}{-5} = \frac{5}{-5} \quad \textit{(Note that both sides are divided by } -5\textit{, since the coefficient is negative.)}$$

$$x = -1$$

■

Notice in the preceding examples that each time we add to both sides the opposite of a term it is eliminated on one side of the equation and appears as the *opposite* on the other side of the equation. Knowing this, we can *transpose* as a shortcut for adding opposites to both sides. "Transpose" means "move across as the opposite of."

Tips and Traps

A shortcut to applying the addition property of equations is called *transposing*. When the opposite of a term is to be added to both sides of an equation, the term may be omitted from one side of the equation and written as its opposite on the other side. That is, the term is *transposed* to the other side of the equation as its opposite.

In other words, we can *mentally* add the opposite term to both sides but show only part of the procedure.

Let's rework Examples 9-4.2, 9-4.3, and 9-4.4 by transposing like terms to isolate the variable.

$$2x + 4 = 8$$

$$2x = 8 - 4 \qquad \textit{(4 is transposed to the right as } -4.\textit{)}$$

$$2x = 4$$

$$\frac{2x}{2} = \frac{4}{2}$$

$$x = 2$$

$$9 - 4x = 8x$$

$$9 = 8x + 4x \qquad \textit{(} -4x \textit{ is transposed to the right as } +4x.\textit{)}$$

$$9 = 12x$$

$$\frac{9}{12} = \frac{12x}{12}$$

$$\frac{3}{4} = x$$

$$7 - 5x = 12$$

$$-5x = 12 - 7 \qquad \textit{(7 is transposed to the right as } -7.\textit{)}$$

$$-5x = 5$$

$$\frac{-5x}{-5} = \frac{5}{-5}$$

$$x = -1$$

No matter how many terms are contained in an equation, our main strategy in solving the equation is to (1) combine like terms on each side of the equation, (2) transpose terms to isolate the variable or letter terms on one side and the number terms on the other, (3) combine like terms on each side to get a basic equation, and then (4) solve the basic equation by multiplication or division.

EXAMPLE 9-4.5 Solve the equation $9x + 2 = 6x - 10$.

Solution: We must first decide on which side of the equation to isolate the letter terms. Once that decision has been made, the number terms must be collected on the opposite side. We will solve this equation two ways, by isolating the letter terms on the left and by isolating the letter terms on the right. The main point is that, no matter which side is chosen for the letter terms, the number terms must be on the opposite side.

Letter Terms Isolated on Left

$$9x + 2 = 6x - 10$$

$$9x - 6x = -10 - 2 \qquad \textit{(2 and 6x are transposed.)}$$

$$3x = -12$$

$$\frac{3x}{3} = \frac{-12}{3}$$

$$x = -4$$

Letter Terms Isolated on Right

$$9x + 2 = 6x - 10$$

$$2 + 10 = 6x - 9x \qquad \textit{(9x and } -10 \textit{ are transposed.)}$$

$$12 = -3x$$

$$\frac{12}{-3} = \frac{-3x}{-3}$$

$$-4 = x$$

■

EXAMPLE 9-4.6 Solve the equation $2x - 3 + 5x = 8 - 6x$.

Solution

$$2x - 3 + 5x = 8 - 6x$$

$$7x - 3 = 8 - 6x \qquad \textit{(Combine like terms 2x and 5x on left side.)}$$

$$7x + 6x = 8 + 3 \qquad \textit{(Transpose to collect like terms.)}$$

$$13x = 11 \qquad \textit{(Combine like terms to get a basic equation.)}$$

$$\frac{13x}{13} = \frac{11}{13} \qquad \textit{(Divide by the coefficient of x.)}$$

$$x = \frac{11}{13}$$

■

Number terms and coefficients in equations can also be fractions and decimals.

EXAMPLE 9-4.7 Solve $\frac{5}{6} = -3k$.

Solution

$$\frac{5}{6} = -3k$$

$$\left(-\frac{1}{3}\right)\frac{5}{6} = \left(-\frac{1}{3}\right)\left(-\frac{3k}{1}\right)$$

$$-\frac{5}{18} = k$$

$\left(\textit{Recall that a 1 is understood as the denominator of any whole number written as a fraction, so the reciprocal of } -3 \textit{ is } -\frac{1}{3}.\right)$ ■

EXAMPLE 9-4.8 Solve $\frac{1}{7}H + \frac{2}{7}H = \frac{1}{3}$.

Solution

$$\frac{1}{7}H + \frac{2}{7}H = \frac{1}{3}$$

(Check that the numerical coefficients have the same denominator. In this case they do.)

$$\frac{3}{7}H = \frac{1}{3}$$

(Combine the numerators of the numerical coefficients $\left(\frac{1}{7} + \frac{2}{7} = \frac{3}{7}\right)$*, retaining the variable H.)*

$$\left(\frac{\overset{1}{\cancel{7}}}{\underset{1}{\cancel{3}}}\right)\frac{\overset{1}{\cancel{3}}}{\underset{1}{\cancel{7}}}H = \left(\frac{7}{3}\right)\frac{1}{3}$$

$\left(\textit{Multiply both sides by the reciprocal of } \frac{3}{7}.\right)$

$$H = \frac{7}{9}$$

■

Example 9-4.9 Solve $9 + \frac{1}{4}b = \frac{3}{4}b.$

Solution

$$9 + \frac{1}{4}b = \frac{3}{4}b$$ *(Note that the letter terms are on opposite sides.)*

$$9 = \frac{3}{4}b - \frac{1}{4}b$$ *(Transpose. The denominators of the letter terms are the same.)*

$$9 = \frac{2}{4}b$$ *(Combine the letter terms.)*

$$9 = \frac{1}{2}b$$ *(Reduce.)*

$$\left(\frac{2}{1}\right)9 = \left(\frac{2}{1}\right)\frac{1}{2}b$$ $\left(\textit{Multiply both sides by the reciprocal of } \frac{1}{2}.\right)$

$$18 = b$$

■

Example 9-4.10 Solve $1.1 = 3.4 + R.$

Solution

$$1.1 = 3.4 + R$$ *(Transpose.)*

$$1.1 - 3.4 = R$$ *(Combine terms.)*

$$-2.3 = R$$

■

Example 9-4.11 Solve $0.09x = 0.54.$

Solution

$$0.09x = 0.54$$

$$\frac{0.09x}{0.09} = \frac{0.54}{0.09}$$ *(Divide both sides by coefficient.)*

$$x = 6$$

$$0.09.\overline{)0.54.}\quad 6.$$

■

SELF-STUDY EXERCISES 9-4

L.O.1. Solve the following equations.

1. $3x + 7x = 60$ **2.** $42 = 8m - 2m$ **3.** $5a - 6a = 3$
4. $3m - 9m = 3$ **5.** $y + 3y = 32$ **6.** $0 = 2x - x$

L.O.2. Solve the following equations.

7. $b + 6 = 5$ **8.** $1 = x - 7$ **9.** $5t - 18 = 12$
10. $10 - 2x = 4$ **11.** $2y + 7 = 17$ **12.** $3x + 7 = x$
13. $3a - 8 = 7a$ **14.** $10x + 18 = 8x$ **15.** $4x + 7 = 8$
16. $4x = 5x + 8$ **17.** $2t + 6 = t + 13$ **18.** $12 + 5x = 6 - x$
19. $4y - 8 = 2y + 14$ **20.** $8 - 7y = y + 24$ **21.** $\frac{2x}{3} = 18$

22. $\dfrac{y + 1}{2} = 7$

23. $\dfrac{8 - R}{76} = 1$

24. $P = \dfrac{1}{2} + \dfrac{1}{3}$

25. $x + \dfrac{1}{7}x = 16$

26. $\dfrac{2}{5} - x = \dfrac{1}{2}x + \dfrac{4}{5}$

27. $\dfrac{3}{7}m - \dfrac{1}{2} = \dfrac{2}{3}$

28. $\dfrac{1}{4}s = \dfrac{1}{4} + \dfrac{1}{10} + \dfrac{1}{20}$

29. $m = 2 + \dfrac{1}{4}m$

30. $\dfrac{7}{9} + 3 = \dfrac{1}{2}T$

31. $2.3x = 4.6$

32. $0.8R = 0.6$ (round to nearest tenth)

33. $0.33x + 0.25x = 3.5$ (round to nearest hundredth)

34. $0.04x = 0.08 - x$ (round to nearest hundredth)

35. $0.47 = R + 0.4R$ (round to nearest hundredth)

9-5 APPLYING THE DISTRIBUTIVE PRINCIPLE IN SOLVING EQUATIONS

Learning Objective

L.O.1. Solve equations that contain parentheses.

When an equation contains an addition or subtraction in parentheses and that quantity in parentheses is multiplied by another factor, we have an example of the *distributive principle*. (See Chapter 1.)

L.O.1. Solve equations that contain parentheses.

Applying the Distributive Principle in Solving Equations

If an equation contains a distributive multiplication, such as $2(y - 3)$ or $4(z + 1)$, the first step in solving the equation is to apply the distributive principle.

EXAMPLE 9-5.1 Solve the equation $3(x + 2) = 18$.

Solution

$$3(x + 2) = 18 \qquad \textit{(Each term in parentheses must be multiplied by 3.)}$$

$$3x + 6 = 18 \qquad \left[\begin{aligned} 3(x + 2) &= 3(x) + 3(2) \\ &= 3x + 6 \end{aligned}\right]$$

$$3x = 18 - 6$$

$$3x = 12$$

$$\frac{3x}{3} = \frac{12}{3}$$

$$x = 4$$

■

EXAMPLE 9-5.2 Solve the equation $5x + 7 = 2(3 - x) - 13$.

Solution

$$5x + 7 = 2(3 - x) - 13 \qquad \textit{(Each term in parentheses must be multiplied by 2.)}$$

$$5x + 7 = 6 - 2x - 13 \qquad \left[\begin{aligned} 2(3 - x) &= 2(3) - 2(x) \\ &= 6 - 2x \end{aligned}\right]$$

$$5x + 7 = -7 - 2x$$

$$5x + 2x = -7 - 7$$

$$7x = -14$$

$$\frac{7x}{7} = \frac{-14}{7}$$

$$x = -2$$

■

Recall from Section 9-3 that the root of an equation can be checked by substituting the root for the letter in each place it appears in the equation and then evaluating both sides of the equation. If the root (solution) is correct, both sides of the equation will equal the same number.

Let's check the root $x = -2$ for Example 9-5.2.

$$5x + 7 = 2(3 - x) - 13$$

$$5(-2) + 7 = 2(3 - (-2)) - 13$$ *[Substitute -2 for x. Because the grouping $2(3 - (-2))$ has two negatives, be careful with the signs.*

$$5(-2) + 7 = 2(3 + (+2)) - 13$$ *(Evaluate grouping first.)*

$$5(-2) + 7 = 2(5) - 13$$ *(Perform all multiplications next.)*

$$-10 + 7 = 10 - 13$$ *(Add last.)*

$$-3 = -3$$

Since $-3 = -3$, the solution checks.

In solving any equation, it is very important to be careful with the sign of each term. The algebraic sign of any term is the sign that comes before the term. The following examples show that much care should be taken in dealing with the signs of terms.

EXAMPLE 9-5.3 Solve the equation $6 - (x + 3) = 2x$.

Solution: Since $(x + 3)$ is in parentheses, it is a grouping that is handled as one term. The sign of the term is negative and the numerical coefficient is understood to be -1.

The first step in solving this equation is to apply the distributive principle by multiplying $(x + 3)$ by its understood coefficient, -1.

$$6 - (x + 3) = 2x$$

$$6 - 1(x + 3) = 2x$$ *[-1 is understood before $(x + 3)$.]*

$$6 - x - 3 = 2x$$ $$\left[\begin{aligned} -(x + 3) &= -1(x) - 1(3) \\ &= -x - 3 \end{aligned}\right]$$

$$6 - 3 = 2x + x$$

$$3 = 3x$$

$$\frac{3}{3} = \frac{3x}{3}$$

$$1 = x$$

■

Rule 9-5.1 ***To multiply a sum or difference in parentheses by a coefficient using a shortcut:***

1. Consider the sign between terms as the sign of the second term.
2. Multiply each term in the parentheses by the coefficient using the rules for multiplying signed numbers.

Example 9-5.4 Solve and check the equation $28 = 7x - 3(x - 4)$.

Solution

$$28 = 7x - 3(x - 4)$$ *(Each term in parentheses is multiplied by −3.)*

$$28 = 7x - 3x + 12$$

$$28 - 12 = 4x$$ *(Combine like terms on right and then transpose.)*

$$16 = 4x$$

$$\frac{16}{4} = \frac{4x}{4}$$

$$4 = x$$

As equations get more involved, the importance of checking becomes more apparent. To check the root 4, we will substitute 4 for x in the equation.

$$28 = 7x - 3(x - 4)$$

$$28 = 7(4) - 3(4 - 4)$$ *(Substitute 4 for x. Then simplify the grouping 4 − 4 = 0.)*

$$28 = 7(4) - 3(0)$$ *(Multiply first; then subtract.)*

$$28 = 28 - 0$$

$$28 = 28$$ ■

The following rule summarizes some strategies for solving equations.

Rule 9-5.2 ***To solve equations use the following strategies as needed:***

1. Apply the distributive principle.
2. Combine like terms on each side of the equation.
3. Transpose terms to collect the variable or letter terms on one side and number terms on the other.
4. Combine like terms on each side to get a basic equation.
5. Solve the basic equation by multiplication or division.

SELF-STUDY EXERCISES 9-5

L.O.1. Solve the following equations.

1. $3(7 + x) = 30$
2. $15 = 5(2 - y)$
3. $6(3x - 1) = 12$
4. $4t = 2(7 + 3t)$
5. $2x = 7 - (x + 6)$
6. $6x - 2(x - 3) = 30$
7. $4a + 5 = 3(2 + a) - 4$
8. $8x - (3x - 2) = 12$
9. $15 - 3(2x + 2) = 6$
10. $(2b + 1) = 5(b - 4) + 1$

9-6 STRATEGIES FOR PROBLEM SOLVING

Learning Objectives

L.O.1. Translate phrases into symbolic expressions.
L.O.2. Translate statements into equations.

The use of letters to represent numbers and manipulating symbols to find unknown values is called *symbolic manipulation* or algebra. Algebra can be used to solve many different types of problems. Our first task is to translate a given English statement or phrase into an algebraic equation or expression.

Our study will begin by translating phrases into symbols. One strategy for mastering this task is to read carefully. Skimming or careless reading can cause us to overlook key words and, as a result, write incorrect algebraic expressions or equations.

L.O.1. Translate phrases into symbolic expressions.

Writing Algebraic Expressions

Writing algebraic expressions or solving problems using algebra requires us to *translate* verbal statements or phrases into *symbolic expressions*. To do this, an understanding of the meaning of a word in a mathematical context is essential. An important step in the task of translating verbal statements or phrases into symbols is to understand the meaning of *each* word in the statement or phrase.

Let's examine some key words or phrases that distinguish between addition and subtraction.

Addition: the sum of, plus, increased by, more than, added to, exceeds, longer, total, heavier, older, wider, taller, gain, greater than, more, expands

Subtraction: less than, decreased by, subtracted from, the difference between, diminished by, take away, reduced by, less, minus, shrinks, younger, lower, shorter, narrower, slower, loss

EXAMPLE 9-6.1 Write the following expressions using algebraic symbols.

(a) The sum of 5 and x.
(b) 7 more than y.
(c) b added to 3.
(d) A bar x inches long expands 0.03 in. after heating.
(e) The large diameter of a taper exceeds the small diameter (x) by 0.75 in.

Solution: Since addition is commutative, the order of terms does not matter. Therefore, there can be more than one acceptable expression. However, the exact translation is preferred and is listed first below.

(a) $5 + x$ or $x + 5$.
(b) $y + 7$ or $7 + y$.
(c) $3 + b$ or $b + 3$.
(d) If x represents the length of a bar before heating, then $x + 0.03$ in., or 0.03 in. $+ x$ represents the length of the bar after heating.
(e) If x represents the small diameter of a taper, then $x + 0.75$ in., or 0.75 in. $+ x$ represents the large diameter of the taper. ■

EXAMPLE 9-6.2 Write the following expressions using algebraic symbols.

(a) The difference between x and 2.
(b) y decreased by 9.
(c) 4 less than b.
(d) k subtracted from 15.
(e) A wet casting that weighs x kg will lose 0.7 kg during the drying process. Express the weight of the dry casting.

Solution: Subtraction is *not* commutative. Therefore, the order of terms is important in writing expressions involving subtractions.

(a) $x - 2$. When an expression is stated as the difference between two numbers, in arithmetic we take the smaller number from the larger. However, in algebra, we interpret this as the *first term minus the second.*
(b) $y - 9$. Nine is taken from y.
(c) $b - 4$. Notice the order of these. Four is taken from b.
(d) $15 - k$. The variable, k, is subtracted from 15.
(e) $x - 0.7$. The amount of weight loss is taken from the initial weight of the casting. ■

In algebra, multiplication is ordinarily expressed without the times sign ($\times$) that denotes multiplication. When two numbers are multiplied, parentheses or a raised dot are used to indicate multiplication.

When a number and a letter or letters are multiplied, no multiplication sign or parentheses are necessary. Also, since multiplication is commutative, the order of the factors does not matter. However, we write the numerical factor first, if there is one, and usually arrange letter factors alphabetically.

In algebra, division is usually written in fractional form. The division signs, $\overline{)\quad}$ or $\div$, are not often used in an algebraic expression or equation. Division is *not* commutative, so the placement of the divisor and dividend is very important.

Words and phrases that imply multiplication and division are:

Multiplication:	times, multiply, of, the product of, multiplied by
Division:	divide, divided by, divided into

Some words that imply multiplication or division by specific numbers are twice (2 times), double (2 times), triple (3 times), and half of ($\frac{1}{2}$ times or divided by 2).

EXAMPLE 9-6.3 Write the following expressions using algebraic symbols.

(a) $4x$ times y.
(b) The product of 7 and b.
(c) Twice d.
(d) Half of a.
(e) 5 divided by n.
(f) 4 divided into m.
(g) One chain saw weighs p pounds. Express the weight of 7 chain saws.
(h) A pipe x cm long is divided into four equal parts. Express the length of each part.

Solution

(a) $4xy$
(b) $7b$
(c) $2d$
(d) $\frac{1}{2}a$ or $\frac{a}{2}$
(e) $\frac{5}{n}$
(f) $\frac{m}{4}$
(g) $7p$
(h) $\frac{x}{4}$ ■

Parentheses are often necessary in writing verbal expressions in symbols. Parentheses are used to show the operational order. Remember, operations that are grouped in parentheses precede operations of multiplication and division.

EXAMPLE 9-6.4 Express "twice the sum of x and 5" in symbols.

Solution: In this expression the sum of x and 5 should be found before the result is multiplied by 2. Parentheses are necessary to show this.

$$2(x + 5) \quad \text{or} \quad 2(5 + x)$$

Without the parentheses, the expression $2a + 5$ has an entirely different meaning. In this case, only a is multiplied by 2 and then 5 is added to that result. ■

EXAMPLE 9-6.5 A machine part weighs 2.7 kg and is to be shipped in a carton weighing x kg. Find the total shipping weight of three cartons and their contents.

Solution: Each part will have a total shipping weight of $(2.7 + x)$ kg. Three parts would total $3(2.7 + x)$ kg. ■

L.O.2. Translate statements into equations.

Writing Algebraic Equations

When a statement indicates that two expressions are equal, we can write the statement as an equation. There are several key words or phrases that imply equality.

Equality: equals, is, is equal to, was, are, were, the result is, what is left is, the same as, gives, giving, makes, leaves, leaving

Let's extend Example 9-6.4 into an equation.

EXAMPLE 9-6.6 Twice the sum of a number and 5 is 22. Find the number.

Solution

READ THE PROBLEM CAREFULLY.
ANALYZE THE PROBLEM.

1. What are you asked to find in the problem?

A missing number that meets certain conditions

Then let some symbol (usually, a letter of the alphabet) *temporarily* hold the position in the problem that the number would hold.

$$x = \text{missing number}$$

UNDERSTAND ALL THE WORDS IN THE PROBLEM.

2. What does the word "twice" mean?

Two times

DESCRIBE "SYMBOLICALLY" THE CONDITIONS OF THE PROBLEM.

3. What are the conditions or restrictions on x?
Look at the wording carefully! Is twice the number increased by five *or* is the sum of the number and five doubled?

$2x + 5$	or	$2(x + 5)$
Five more than twice a number		Twice the *sum* of a number and five

This problem states "twice the sum of a number and 5."
Thus, the conditions are $2(x + 5)$.

4. What are the results of the calculation of the conditions?

"is 22"

NOW WRITE THE EQUATION.

5. The equation is

$$2(x + 5) = 22$$

SOLVE THE EQUATION.

6. Work each step carefully.

$$2(x + 5) = 22$$
$$2x + 10 = 22 \quad \textit{(Distribute.)}$$
$$2x = 22 - 10 \quad \textit{(Transpose.)}$$
$$2x = 12 \quad \textit{(Solve for x.)}$$
$$x = 6$$

VERIFY THE ANSWER.

7. Does the answer, 6, meet the conditions of the problem?
Twice the sum of the number and 5 is 22.

$$2(6 + 5) = 22$$
$$2(11) = 22$$
$$22 = 22$$

Yes. ■

EXAMPLE 9-6.7 Write the following statements as equations.

(a) Three more than a number equals 10.
(b) If 2 is subtracted from 4 times a number, the result is 18.
(c) A 100-lb package was divided into three parts. Two of the parts were equal and the third part was 10 lb more than either of the equal parts.

Solution

(a) Three more than a number equals 10.

$$x + 3 = 10$$

(b) $\underbrace{\text{If 2 is subtracted from 4 times a number,}}_{4x - 2}$ $\underbrace{\text{the result is}}_{=}$ $\underbrace{\text{18.}}_{18}$

$$4x - 2 = 18$$

(c) First, let's list the facts of the problem.

$$100 \text{ lb} = \text{total weight of package}$$
$$x = \text{weight of one equal part}$$
$$x = \text{weight of other equal part}$$
$$x + 10 = \text{weight of third part}$$

The sum of the three parts equals the total:

$$x + x + x + 10 = 100$$

or, if we combine letter terms, $3x + 10 = 100$. ■

Now let's extend Example 9-6.5 into an applied problem.

EXAMPLE 9-6.8 A machine part weighs 2.7 kg and is to be shipped in a carton weighing x kg. If the total weight of three packaged machine parts is 9.3 kg, how much does each carton weigh? (See Fig. 9-2.)

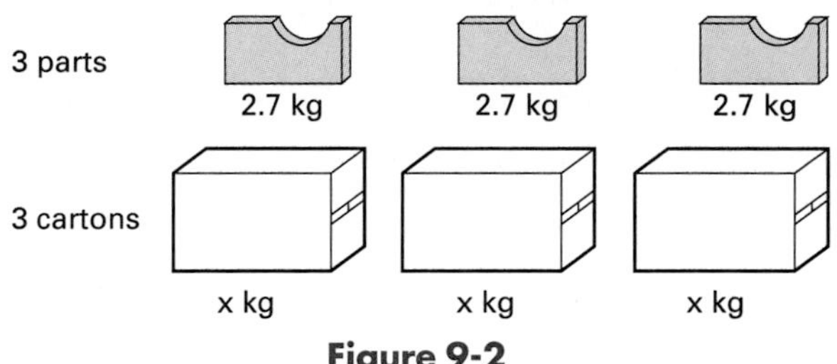

Figure 9-2

Solution Apply problem-solving tips of Example 9-6.6.

READ THE PROBLEM CAREFULLY.

VISUALIZE THE PROBLEM. (See Fig. 9-2.)

ANALYZE THE PROBLEM.

1. What are you asked to find?

The weight of one shipping carton

x = weight of one carton

UNDERSTAND ALL THE WORDS IN THE PROBLEM.

2. What does "each carton" mean?

Carton without the part

What does "packaged part" mean?

Part plus carton

DESCRIBE "SYMBOLICALLY" THE CONDITIONS OF THE PROBLEM.

3. What are the conditions or restrictions on x?

$$2.7 \text{ kg} = \text{weight of one part}$$
$$3 = \text{number of parts to be shipped}$$
$$2.7 + x = \text{weight of one packaged part (part + carton)}$$

EXAMINE THE OPTIONS.

$$(2.7 + x) + (2.7 + x) + (2.7 + x) = \text{total shipping weight}$$

or *preferably*

$$3(2.7 + x) = \text{total shipping weight}$$

4. What are the results of the calculation of the conditions?

is 9.3

NOW WRITE THE EQUATION.

5. The equation is $3(2.7 + x) = 9.3$.

SOLVE THE EQUATION.

6. Work each step carefully.

$$3(2.7 + x) = 9.3$$

$$3(2.7) + 3x = 9.3 \qquad \textit{(Distribute.)}$$

$$8.1 + 3x = 9.3 \qquad \textit{(Make calculation.)}$$

$$3x = 9.3 - 8.1 \qquad \textit{(Rearrange terms.)}$$

$$\frac{3x}{3} = \frac{1.2}{3} \qquad \textit{(Solve for x.)}$$

$$x = 0.4$$

VERIFY THE ANSWER.

7. Does the weight of 3 parts and 3 shipping cartons equal a total weight of 9.3 kg?

$$3(2.7 + 0.4) = 9.3$$

$$3(3.1) = 9.3$$

$$9.3 = 9.3$$

Yes, the answer meets the conditions of the problem. ■

Now, we will summarize some important tips to use in problem solving.

Tips and Traps

Steps in Problem Solving

1. Read the problem carefully.
2. Understand all the words in the problem.
3. Analyze the problem:
   What are you asked to find?
   What facts are given?
   What facts are implied?
4. Visualize the problem.
5. Describe the conditions of the problem "symbolically."
6. Examine the options.
7. Develop a *plan* for solving the problem.
8. Write your *plan* "symbolically." That is, write an equation.
9. Solve the equation.
10. Verify your answer with the conditions of the problem.

SELF-STUDY EXERCISES 9-6

L.O.1. Write the following in algebraic symbols.

1. The sum of x and 3.

2. The difference between x and y.

3. Four more than x.

4. Seven less than x.

5. Y decreased by 10.

6. Twice a.

7. The product of 7 and m is doubled.

8. The difference between a and 15.

9. A bar x dm long expands 0.07 dm after heating.

10. The small diameter of a taper is 0.25 dm less than the large diameter, x.

11. The number of employees at one plant exceeds the number at another plant by 50. The second plant has x employees.

12. 0.025 cm is machined from a part x cm thick.

13. Thirty implants each weighing x oz are packed in a box weighing 0.8 oz. What is the weight of four packaged boxes?

14. What is the weight of x lengths of wire if each length is 5 oz less than y?

L.O.2. Write the following statements as equations and solve.

15. The sum of x and 4 equals 12. Find x.

16. 4 less than 2 times a number is 6. Find the number.

17. Three parts totaling 27 lb are packaged for shipping. Two parts weigh the same. The third part weighs 3 lb less than each of the two equal parts. Find the weight of each part.

18. How many gallons of water must be added to 24 gal of pure cleaner to make 60 gal of diluted cleaner?

19. A wet casting weighing 4.03 kg weighs 3.97 kg after drying. Write and solve an algebraic equation to find the weight loss due to drying.

20. A plumber needs three times as much perforated pipe as solid pipe to lay a drain line 400 ft long. How much of each type of pipe is needed?

ASSIGNMENT EXERCISES

9-1

Perform the indicated operations.

1. $-8 + 17$ **2.** $-3 - 12$ **3.** $6 - (-7)$

4. $-3(-18)$ **5.** $11(-7)$ **6.** $\frac{-52}{2}$

7. $5 - 7(3) + 4 - 7$ **8.** $3 + 5(8 - 15) - 5$ **9.** $6(-5) - (7 - 3)$

10. $3(2 - 8) - 5(6 + 1)$

9-2

Solve the following equations.

11. $y = 5 + 3$ **12.** $k = 7 - 5$ **13.** $\frac{14}{2} = x$

14. $3 - 10(2) = b$ **15.** $x = 6(3) - 5$ **16.** $c = 3 - 8$

Identify the terms in the following expressions by drawing a box around each term.

17. $5 + k$ **18.** $5x - 3$ **19.** $6a - 4(a - 7)$

20. $\frac{5}{m}$ **21.** $9mx - 2m + 8 + 6(m + x)$ **22.** $12x - 32$

23. $\frac{x + 5}{a + 3}$ **24.** $\frac{3x}{11} + 8$

Identify the numerical coefficients in each of the following terms.

25. $7M$　**26.** $-8ab$　**27.** $\frac{y}{8}$

28. $\frac{3m}{11}$　**29.** $12(2x - 3y)$　**30.** $-\frac{7}{8}d$

Combine like terms.

31. $5x - 3x + 7x$　**32.** $-3a + 2a - 8a$　**33.** $4x - 3y + 2z - 7x + y$

34. $-6a + 3b - 2c + 3b - c$　**35.** $7 + 3x - 2 + 8y + 6 + 7x - y$

Simplify and combine like terms.

36. $5(3a + 7)$　**37.** $-7(-3x - 2)$　**38.** $-3(5g - 7)$

39. $-(7x - 3)$　**40.** $5 - 3(7b + 6) - 12b$

9-3

Solve the following equations using multiplication or division. Check the answers

41. $3x = 21$　**42.** $4x = -28$　**43.** $-15 = 2b$

44. $-5y = 30$　**45.** $-7y = -49$　**46.** $-5 = -m$

47. $8 = -x$　**48.** $0.6 = -a$　**49.** $3 = \frac{1}{5}x$

50. $-\frac{2}{7}x = 8$　**51.** $-\frac{3}{8}x = -24$　**52.** $\frac{1}{2}x = -5$

53. $42 = -\frac{6}{7}x$　**54.** $-\frac{5}{8}x = -10$　**55.** $\frac{1}{7}x = 12$

9-4

Solve the following equations.

56. $5y - 7y = 14$　**57.** $4x + x = 25$　**58.** $36 = 9a - 5a$

59. $2b - 7b = 10$　**60.** $0 = 4t - t$　**61.** $21 = x + 2x$

62. $8x - 3x = 6 + 9$　**63.** $20 - 4 = 2x - 6x$　**64.** $13 - 27 = 3x - 10x$

65. $x + 7 = 10$　**66.** $x - 5 = 3$　**67.** $x - 3 = -4$

68. $3 + y = -5$　**69.** $1 = a - 4$　**70.** $t + 7 = 12$

71. $3x + 4 = 19$　**72.** $4x - 3 = 9$　**73.** $15 - 3x = -6$

74. $5 = 3x - 7$　**75.** $-7 = 6x - 31$　**76.** $4 = 7 - 4x$

77. $-12 = -8 - 2x$　**78.** $2x + 6 = x$　**79.** $5x - 12 = 9x$

80. $12x + 27 = 3x$　**81.** $3x + 9 = 10$　**82.** $7x = 8x + 4$

83. $4y + 8 = 3y - 4$　**84.** $10 + 4x = 5 - x$　**85.** $7 - 4y = y + 22$

86. $7x - 1 = 4x + 17$　**87.** $4x - 3 = 2x + 6$　**88.** $y - 5 = 6y + 30$

89. $8 - 2y = 15 - 3y$　**90.** $6 - 7x = 15 - x$　**91.** $5x - 12 = 2x + 15$

92. $18x - 21 = 15x + 33$　**93.** $7x - 5 + 2x = 3 - 4x + 12$

94. $2x - 3 + 15 = 7x - 8 - 6x$　**95.** $3x - 5x + 2 = 6x - 5 + 12x$

96. $\frac{R}{7} - 6 = -R$　**97.** $0 = \frac{8}{9}c + \frac{1}{4}$　**98.** $\frac{2}{15}P - P = 4$

99. $6.7y - y = 8.4$ *(round to tenths)*　**100.** $0.9R = 0.3$ *(round to tenths)*

101. $0.86 = R + 0.4R$ *(round to hundredths)*　**102.** $0.04y = 0.02 - y$ *(round to hundredths)*

9-5

Solve the following equations.

103. $18 = 6(2 - y)$　**104.** $4(6 + x) = 36$　**105.** $3x = 3(9 + 2x)$

106. $4(5x - 1) = 16$　**107.** $7x - 3(x - 8) = 28$　**108.** $4a = 8 - (a + 7)$

109. $5x = 7 + (x + 5)$　**110.** $3(x + 2) - 5 = 2x + 7$　**111.** $4(3 - x) = 2x$

112. $3(2x - 4) = 4x - 6$　**113.** $-2(4 - 2x) = -16$　**114.** $3(2x + 1) = -3$

115. $5(3 - 2x) = -5$　**116.** $-16 = -2(-2x + 4)$　**117.** $8 = 6 - 2(3x - 1)$

118. $4x - (x + 3) = 3$　**119.** $3(x - 1) = 18 - 2(x + 3)$　**120.** $-(x - 1) = 2(x + 7)$

121. $-(2x + 1) = -7$
122. $2 + 3(x - 4) = 2x - 5$
123. $7 = 3 + 4(x + 2)$
124. $7(x + 2) = -6 + 2x$
125. $3(4x + 3) = 3 - 4(x - 1)$
126. $3(2 - x) - 1 = 4(3 - x)$

9-6

Write the following expressions in algebraic symbols.

127. 5 increased by x.
128. 21 added to y.
129. A number, y, diminished by 7.
130. The difference between 5 and x.
131. A number, x, exceeded by 5.
132. The difference between a and 7.
133. The product of m and xy.
134. A chain is 8 in. longer than x.
135. A number, y, divided into 8.
136. A container holds 8 gal more than x.
137. A number, x, is tripled.
138. The length of a rectangular plate is 5 in. longer than the width, w.
139. 8 divided by x.
140. Twice the sum of x and 7.

Write the following statements as equations and solve.

141. The difference between x and 6 equals 8. Find x.
142. Twice a number increased by 5 is 17. Find the number.
143. Five times the sum of x and 6 is 42 more than x. Find x.
144. How many gallons of water must be added to 46 gal of pure alcohol to make 100 gal of alcohol solution?
145. If one technician works 3 hours less than another and the total of their hours is 51, how many hours has each technician worked?
146. The shorter side of a carpenter's square in the shape of an L is 6 in. shorter than the longer side. If the total length of the L-shaped tool is 24 in., what is the measure of each side of the square?

ELECTRONICS APPLICATION

Kirchhoff's Laws

Kirchhoff's current law (KCL) and Kirchhoff's voltage law (KVL) form the basis of all electronics. The only difficult thing about the two laws is spelling Kirchhoff—notice that there are two h's and two f's. The rest is easy.

Kirchhoff's current law (KCL) says that *the sum of all currents at a node equals zero.* A node is like an intersection near your house. At 3 A.M. there are no cars in the intersection. If each entering car counts +1, and each exiting car counts −1, then the sum of all the cars will equal zero. KCL works the same way, only with current, which is measured in amperes (A). Since an ampere is rather large, measurements are often in milliamperes (mA), or thousandths of an ampere.

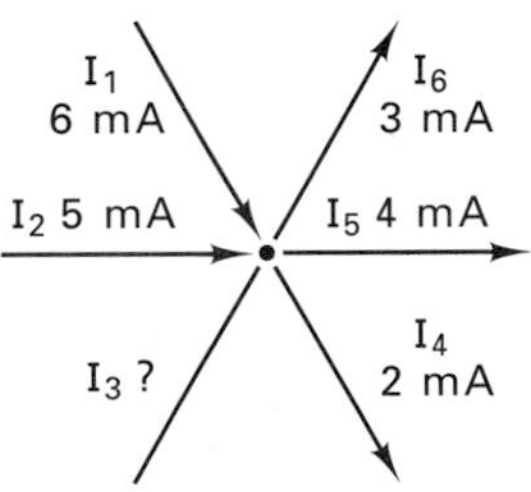

Circuit 1

Arrows are often used to indicate current. *Let each entering arrow have a + sign and each exiting arrow have a − sign.* Assume that there is a node with two currents entering the node ($I_1 = 6$ mA, $I_2 = 5$ mA), an unknown current called

I_3, and three currents exiting the node ($I_4 = 2$ mA, $I_5 = 4$ mA, and $I_6 = 3$ mA). As an equation,

$$\begin{aligned} I_1 \quad + \quad I_2 \quad + I_3 + \quad I_4 \quad + \quad I_5 \quad + \quad I_6 \quad &= 0 \\ 6 \text{ mA} + 5 \text{ mA} + I_3 - 2 \text{ mA} - 4 \text{ mA} - 3 \text{ mA} &= 0 \end{aligned}$$

To find I_3,

$$I_3 = 0 - I_1 - I_2 - I_4 - I_5 - I_6$$

Inserting the numbers,

$$\begin{aligned} I_3 &= 0 - 6 \text{ mA} - 5 \text{ mA} + 2 \text{ mA} + 4 \text{ mA} + 3 \text{ mA} \\ I_3 &= -2 \text{ mA} \end{aligned}$$

This means that I_3 *is exiting the node and equals 2 mA*. To prove this, the numbers need to be reinserted into the original equation.

$$\begin{aligned} I_1 \quad + \quad I_2 \quad + \quad I_3 \quad + \quad I_4 \quad + \quad I_5 \quad + \quad I_6 \quad &= 0 \\ 6 \text{ mA} + 5 \text{ mA} - 2 \text{ mA} - 2 \text{ mA} - 4 \text{ mA} - 3 \text{ mA} &= 0 \qquad \text{It works!} \end{aligned}$$

Kirchhoff's voltage law (KVL) says that *the sum of all voltages in any loop equals zero*. Picture a bug walking around any loop in a circuit. Voltage is always measured as a drop in potential across something, so there is a plus sign at one end of the something and a minus at the other end. As the bug walks around the circuit, *the first sign encountered is the one used*. The equations for KVL look like those for KCL. The only difference is that a closed loop is used, rather than a node.

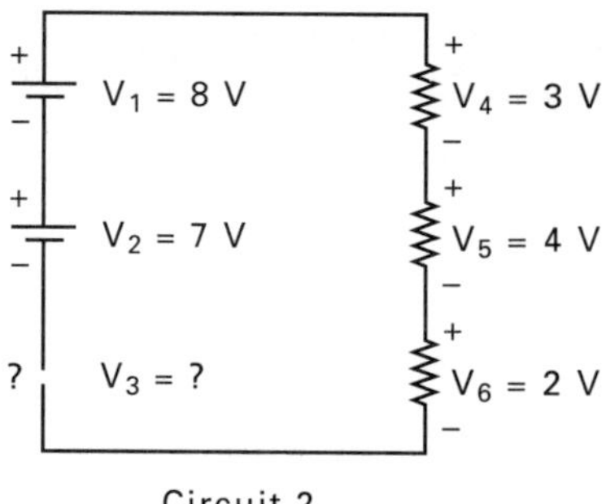

Circuit 2

Look at circuit 2, which has two voltage sources ($V_1 = 8$ V, $V_2 = 7$ V), an unknown voltage called V_3, and three voltage drops ($V_4 = 3$ V, $V_5 = 4$ V, $V_6 = 2$ V). Assume you start at the top of the circuit and go down the left side to the bottom and then up the right side. As an equation, then,

$$\begin{aligned} V_1 \quad + \quad V_2 \quad + V_3 + \quad V_6 \quad + \quad V_5 \quad + \quad V_4 \quad &= 0 \\ 8 \text{ V} + 7 \text{ V} + V_3 - 2 \text{ V} - 4 \text{ V} - 3 \text{ V} &= 0 \end{aligned}$$

Notice that, if you started at the top and first went down the right side, across the bottom, and then up the left side, the sum still equals zero.

$$\begin{aligned} V_4 + V_5 + V_6 - V_3 - V_2 - V_1 &= 0 \\ +3 \text{ V} + 4 \text{ V} + 2 \text{ V} - V_3 - 7 \text{ V} - 8 \text{ V} &= 0 \end{aligned}$$

Solving the first equation for V_3, we have

$$V_3 = 0 - V_1 - V_2 - V_4 - V_5 - V_6$$

Inserting the numbers,

$$V_3 = 0 - 8\text{ V} - 7\text{ V} + 3\text{ V} + 4\text{ V} + 2\text{ V}$$
$$V_3 = -6\text{ V}$$

This means that V_3 *is a voltage drop of 6 V*. To prove this, the numbers need to be reinserted into the original equation.

$$V_1 + V_2 + V_3 + V_4 + V_5 + V_6 = 0$$
$$8\text{ V} + 7\text{ V} - 6\text{ V} - 3\text{ V} - 4\text{ V} - 2\text{ V} = 0 \qquad \text{It works!}$$

Exercises

Redraw each of the eight circuits (including circuits 1 and 2). Find the missing currents and voltages in all the circuits and draw the correct component on the circuit.

For each circuit, write a complete equation using all the values to show that the algebraic sum of the currents at a node or of the voltages in a loop equals zero.

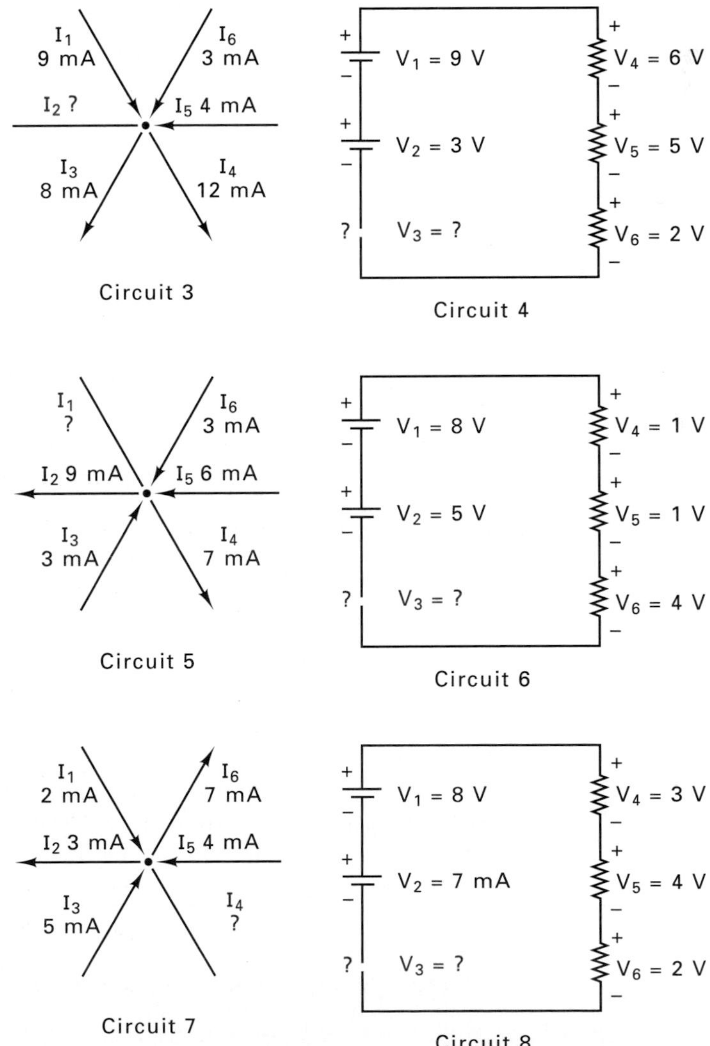

Answers for Exercises

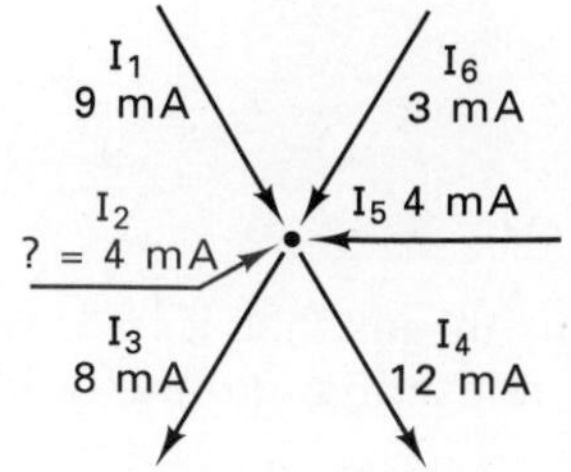

Circuit 3

$$I_1 + I_2 + I_3 + I_4 + I_5 + I_6 = 0$$
$$9\text{ mA} \underline{+ 4\text{ mA}} - 8\text{ mA} - 12\text{ mA} + 4\text{ mA} + 3\text{ mA} = 0$$

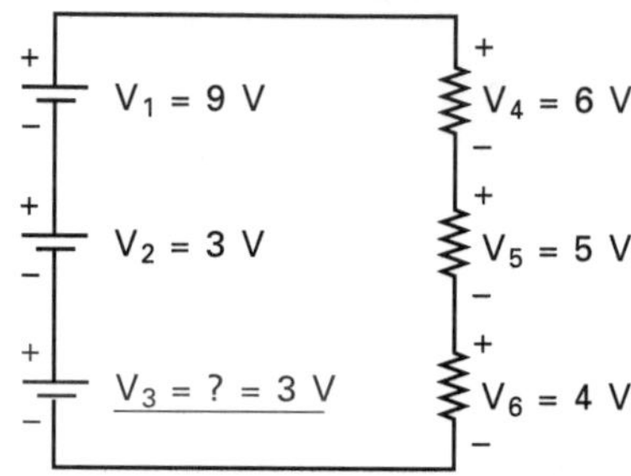

Circuit 4

$$V_1 + V_2 + V_3 + V_4 + V_5 + V_6 = 0$$
$$9\text{ V} + 3\text{ V} \underline{+ 3\text{ V}} - 4\text{ V} - 5\text{ V} - 6\text{ V} = 0$$

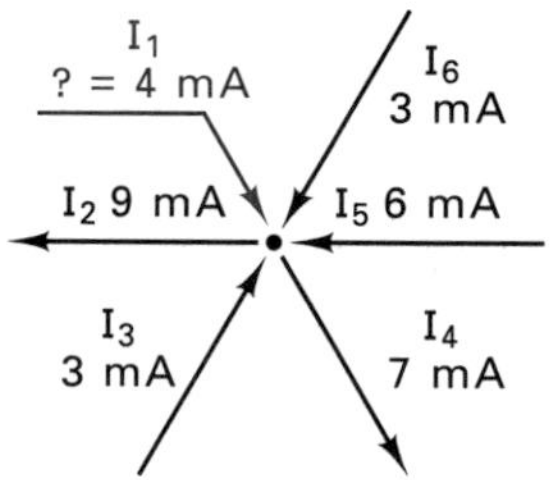

Circuit 5

$$I_1 + I_2 + I_3 + I_4 + I_5 + I_6 = 0$$
$$\underline{4\text{ mA}} - 9\text{ mA} + 3\text{ mA} - 7\text{ mA} + 6\text{ mA} + 3\text{ mA} = 0$$

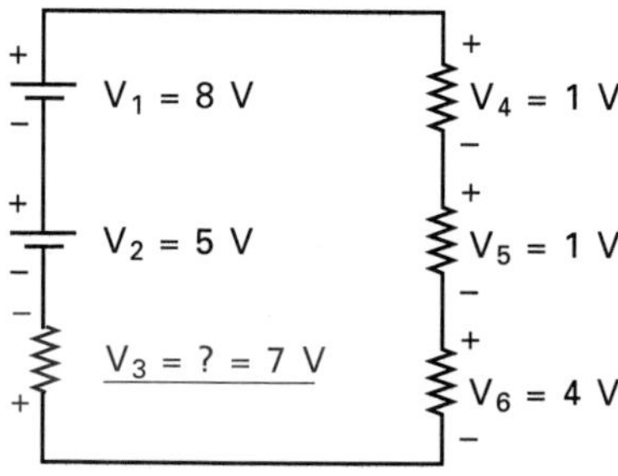

Circuit 6

$$V_1 + V_2 + V_3 + V_4 + V_5 + V_6 = 0$$
$$8\text{ V} + 5\text{ V} \underline{- 7\text{ V}} - 4\text{ V} - 1\text{ V} - 1\text{ V} = 0$$

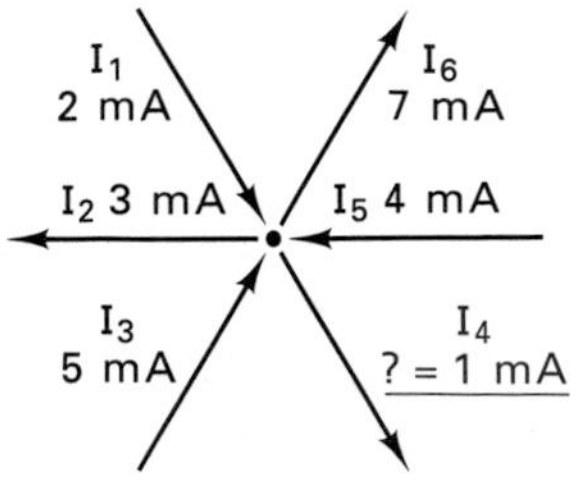

Circuit 7

$$I_1 \quad + \quad I_2 \quad + \quad I_3 \quad + \quad I_4 \quad + \quad I_5 \quad + \quad I_6 \quad = 0$$
$$2\text{ mA} - 3\text{ mA} + 5\text{ mA} \underline{- 1\text{ mA}} + 4\text{ mA} - 7\text{ mA} = 0$$

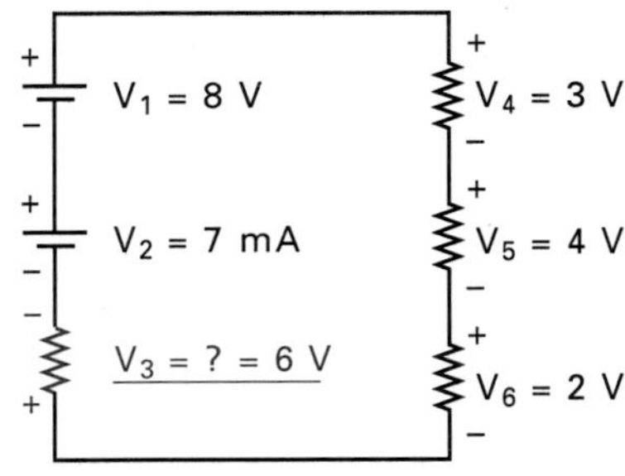

Circuit 8

$$V_1 \ + \ V_2 \ + \ V_3 \ + \ V_4 \ + \ V_5 \ + \ V_6 \ = 0$$
$$8\text{ V} + 7\text{ V} \underline{- 6\text{ V}} - 2\text{ V} - 4\text{ V} - 3\text{ V} = 0$$

CHALLENGE PROBLEMS

Make up a word problem or situation that would use the following equations in the solution.

147. $2x + 7 = 47$

148. $3(x + 5) + 2 = 80$

CONCEPTS ANALYSIS

1. What is a basic equation and how do you solve it?

2. Write the following in the proper sequence for solving equations. An item can be used more than once.
   - **(a)** Apply the addition axiom to arrange the equation so that number terms are on one side and letter terms are on the other.
   - **(b)** Apply the multiplication axiom by dividing both sides by the coefficient of the letter term or by multiplying both sides by the reciprocal of the coefficient of the letter term.
   - **(c)** Apply the distributive principle to remove any parentheses.
   - **(d)** Combine like terms that are on the same side of the equal sign.

3. Write an example of an equation and solve it to illustrate the use of the multiplication axiom. If several steps are in the solution, identify the step that applies the multiplication axiom.

4. Write an example of an equation and solve it to illustrate the use of the addition axiom. If several steps are in the solution, identify the step that applies the addition axiom.

5. Give some examples of words that generally imply addition.

6. Give some examples of words that generally imply subtraction.

7. Give some examples of words that generally imply multiplication.

8. Give some examples of words that generally imply division.

9. Why is it so important to arrange subtraction and division problems in the proper order?

10. In a basic equation like $\frac{3}{4}x = \frac{5}{8}$, solve the equation in two different ways. Why are the two ways equivalent? What property of equality was used?

CHAPTER SUMMARY

Objectives	What to Remember	Examples
Section 9-1		
1. Perform basic operations with signed numbers.	Review rules for addition, subtraction, multiplication, and division of integers (Chapter 2).	Perform the operations: $-4 + 7 = 3$ $-6 - (-2) =$ $-6 + 2 = -4$ $-3(-8) = 24$ $-8(2) = -16$ $\frac{-12}{4} = -3$ $\frac{-28}{-7} = 4$

2. Simplify expressions using the order of operations.	Perform operations as they appear from left to right: **1.** Parentheses (groupings) **2.** Exponents (powers and roots) **3.** Multiplications and divisions **4.** Additions and subtractions	$4(-2) - 2 + 2(-8 + 1)$ $= 4(-2) - 2 + 2(-7)$ $= -8 - 2 - 14$ $= -24$
Section 9-2		
1. Identify equations, terms, factors, constants, and variables.	Review definitions.	Identify the terms, factors, constants, and variables in the following expression: $2x + \frac{x}{7} - 3$ Terms: $2x, \frac{x}{7}, -3$ Factors: 2 and x, x and $\frac{1}{7}$ Constant: -3 Variable : x
2. Add and subtract like terms.	Add and subtract only like terms. Combine by adding the coefficients.	Simplify: $3x + 4y - 5y + 7x = 10x - y$
3. Apply the distributive property to simplify expressions.	Multiply each term in the grouping by the factor in front of the grouping.	Simplify: $3(5x - 3) - 2(x + 4)$ $= 15x - 9 - 2x - 8$ $= 13x - 17$
Section 9-3		
1. Solve basic equations.	Multiply both sides of the equation by the reciprocal of the coefficient of the letter term *or* divide both sides of the equation by the coefficient of the letter term. This process applies the multiplication axiom.	Solve: $\frac{x}{5} = 8$ $\frac{5}{1}\left(\frac{x}{5}\right) = 8\left(\frac{5}{1}\right)$ $x = 40$ $-8x = 24$ $\frac{-8x}{-8} = \frac{24}{-8}$ $x = -3$ $\frac{1}{3}x = \frac{4}{9}$ $\left(\frac{3}{1}\right)\left(\frac{1}{3}x\right) = \left(\frac{4}{9}\right)\left(\frac{3}{1}\right)$ $x = \frac{4}{3}$ $7.5 = 2.5x$ $\frac{7.5}{2.5} = \frac{2.5x}{2.5}$ $3 = x$ $\frac{x}{0.6} = 2.9$ $0.6\left(\frac{x}{0.6}\right) = 0.6(2.9)$ $x = 1.74$

2. Check the solutions of equations.	Substitute the value of the variable for the variable in each place that it appears in the equation. Perform operations on both sides of the equation. The two sides of the equation should be equal.	Verify that $x = 3$ is the solution for the equation $2x - 1 = 5$. $2(3) - 1 = 5$ $6 - 1 = 5$ $5 = 5$
Section 9-4		
1. Solve equations with like terms on the same side of the equation.	Combine the like terms that are on the same side of the equation. Solve the remaining basic equation.	Solve. $3x - 5x = 12$ $-2x = 12$ $\frac{-2x}{-2} = \frac{12}{-2}$ $x = -6$
2. Solve equations with like terms on opposite sides of the equation.	Use the addition axiom to move letter terms to one side of the equation and number terms to the other. Combine like terms. Solve the remaining basic equation.	Solve. $x - 5 = 7$ $x - 5 + 5 = 7 + 5$ $x = 12$ Solve. $x - \frac{3}{8} = \frac{5}{8}$ $x = \frac{5}{8} + \frac{3}{8}$ $x = \frac{8}{8}$ $x = 1$ Solve. $7.2 = x - 3.5$ $7.2 + 3.5 = x$ $10.7 = x$ Solve. $3x - 5 = 5x + 7$ $3x - 5x = 7 + 5$ $-2x = 12$ $\frac{-2x}{-2} = \frac{12}{-2}$ $x = -6$
Section 9-5		
1. Solve equations that contain parentheses.	Remove parentheses using the distributive principle. Then, continue solving the remaining equation.	Solve. $3(2x - 5) = 8x + 7$ $6x - 15 = 8x + 7$ $6x - 8x = 7 + 15$ $-2x = 22$ $\frac{-2x}{-2} = \frac{22}{-2}$ $x = -11$
Section 9-6		
1. Translate phrases into symbolic expressions.	Use key words to translate phrases into mathematical symbols.	Write an algebraic expression for five more than two times a number. $2x + 5$

2. Translate statements into equations.	Use key words to translate statements into symbolic equations.	A shipment of textbooks to a college bookstore was sent in two boxes weighing 37 pounds total. One box weighed 9 pounds more than the other. What was the weight of each box? $$\begin{aligned} x &= \text{weight of one box} \\ x + 9 &= \text{weight of other box} \\ x + x + 9 &= 37 \\ 2x + 9 &= 37 \\ 2x &= 37 - 9 \\ 2x &= 28 \\ \frac{2x}{2} &= \frac{28}{2} \\ x &= 14 \text{ pounds} \\ x + 9 &= 23 \text{ pounds} \end{aligned}$$

WORDS TO KNOW

like signs (p. 347)
unlike signs (p. 347)
order of operations (p. 348)
equation (p. 350)
variable (p. 350)
solution (p. 350)
root (p. 350)
symmetric property of equality (p. 351)
factor (p. 351)
term (p. 351)
number terms (p. 352)
constants (p. 352)
coefficient (p. 352)
numerical coefficient (p. 353)
like terms (p. 353)
distributive principle (p. 354)
basic equation (p. 355)
basic principle of equality (p. 356)
multiplication axiom (p. 357)
addition axiom (p. 361)
transposing (p. 362)
distributive principle (p. 366)
symbolic expressions (p. 369)

CHAPTER TRIAL TEST

Perform the indicated operations.

1. $-8 - 5$

2. $-3(12)$

3. $\frac{-28}{-7}$

4. $-3 + 4(2 + 8)$

5. $7 - 5(3 - 6) + \frac{-8}{2}$

Draw a box around each term.

6. $5x - 3$

7. $4ab - \frac{2}{x} + 3(x + 7)$

Identify the numerical coefficient in each of the following terms.

8. $8x$

9. $\frac{x}{7}$

10. $5(x + y)$

Combine like terms.

11. $7m - 3m + m$

12. $4x - 3y + z - x - 2y + 2z$

13. $3x - 5(2x + 4)$

Write the following expressions in algebraic notation.

14. 5 added to x

15. $2y$ decreased by 9

16. Seven times the difference between a and 2

17. The sum of k and 8

Solve the following equations.

18. $5x = 30$

19. $-8y = 72$

20. $\frac{x}{2} = 5$

21. $-\frac{3}{5}m = -6$

22. $x = 5 - 8$

23. $5x + 2x = 49$

24. $y - 2y = 15$

25. $2x = 3x - 7$

26. $5 - 2x = 3x - 10$

27. $5x + 3 - 7x = 2x + 4x - 11$

28. $3(x + 4) = 18$

29. $2(x - 7) = -10$

30. $7 - (x - 3) = 4x$

31. $3x + 2 = 4(x - 1) - 1$

32. $4.5x - 3.4 = 2.1x - 0.4$

33. 5 added to x equals 16. Find x.

34. Twice the sum of a number and 3 is 10. Find the number.

35. One tank holds four times as many gallons as a smaller tank. If the tanks hold 2580 gal together, how much does each tank hold?

CHAPTER

10 Equations with Fractions and Decimals

Thus far, we have used mostly whole numbers in equations. However, technical mathematics will require us to deal with a variety of fractional and decimal quantities in solving certain kinds of equations, as this chapter will illustrate.

10-1 SOLVING FRACTIONAL EQUATIONS BY CLEARING THE DENOMINATORS

Learning Objectives

L.O.1. Solve fractional equations by clearing the denominators.
L.O.2. Solve work problems with fractional equations.

L.O.1. Solve fractional equations by clearing the denominators.

Clearing Equations of Fractions

A technique for solving equations that contain fractions is to *clear* the equation of all denominators in the first step. Then the resulting equation, which contains no fractions, can be solved using previously learned techniques.

This process, also called clearing the fractions, is based on basic principles of arithmetic and algebra. First, we will look at an example that contains only one fraction.

EXAMPLE 10-1.1 Solve the following equation by clearing the denominator or fraction first.

$$2x + \frac{3}{4} = 1$$

Solution: We can *clear* an equation of a single fraction or denominator by multiplying the *entire* equation by the denominator of the fraction. *This means multiplying each term by the denominator.*

$$2x + \frac{3}{4} = 1$$

$$4(2x) + 4\left(\frac{3}{4}\right) = 4(1) \qquad \textit{(Multiply each term by denominator 4.)}$$

$$4(2x) + \overset{1}{\cancel{4}}\left(\frac{3}{\underset{1}{\cancel{4}}}\right) = 4(1) \qquad \textit{(Reduce and multiply.)}$$

$$8x + 3 = 4$$

The new equation contains no fraction and can be solved using methods learned previously.

$$8x + 3 = 4$$
$$8x = 4 - 3$$
$$8x = 1$$
$$\frac{8x}{8} = \frac{1}{8}$$
$$x = \frac{1}{8}$$

To check, substitute $\frac{1}{8}$ for x in the original equation.

$$2x + \frac{3}{4} = 1$$
$$2\left(\frac{1}{8}\right) + \frac{3}{4} = 1$$
$$\frac{2}{8} + \frac{3}{4} = 1 \qquad \left(\textit{Reduce, } \frac{2}{8} = \frac{1}{4}.\right)$$
$$\frac{1}{4} + \frac{3}{4} = 1 \qquad \left(\frac{1}{4} + \frac{3}{4} = \frac{4}{4} = 1\right)$$
$$1 = 1$$

■

> **Rule 10-1.1** ***To clear an equation of a single fraction or denominator:***
>
> Multiply both sides of the equation by the denominator.

EXAMPLE 10-1.2 Solve $\frac{4d}{3} = 12$.

Solution

$$(\overset{1}{\not{3}})\frac{4d}{\underset{1}{\not{3}}} = (3)(12) \qquad \textit{(Multiply both sides by the denominator 3. Reduce where possible.)}$$
$$4d = 36 \qquad \textit{(a basic equation)}$$
$$\frac{4d}{4} = \frac{36}{4}$$
$$d = 9$$

Checking the root in the original equation, we get

$$\frac{4(\overset{3}{\not{9}})}{\underset{1}{\not{3}}} = 12$$
$$12 = 12$$

■

An advantage of this rule is that it can be applied to equations with a fraction on only one side when the unknown is in the *denominator*. For example, we can use this procedure to solve $\frac{3}{7x} = 4$. Note that the term $\frac{3}{7x}$ is *not* the product of $\frac{3}{7}$ and x, since $\frac{3}{7}\left(\frac{x}{1}\right) = \frac{3x}{7}$. Therefore, the numerical coefficient of x is *not* $\frac{3}{7}$, and so we solve the equation by clearing the denominator.

EXAMPLE 10-1.3 Solve $\frac{10}{x} = 2$.

Solution

$$\frac{10}{x} = 2$$

$$(\overset{1}{\cancel{x}})\frac{10}{\underset{1}{\cancel{x}}} = (x)2 \qquad \textit{(Multiply both sides by denominator x. Reduce where possible.)}$$

$$10 = 2x \qquad \textit{(a basic equation)}$$

$$\frac{10}{2} = \frac{2x}{2}$$

$$5 = x$$

Checking the root in the original equation, we get

$$\frac{\overset{2}{\cancel{10}}}{\underset{1}{\cancel{5}}} = 2$$

$$2 = 2$$

■

Again, the solution checks because the equality of the two sides of the original equation was maintained since both sides were multiplied by the same factor, x, when we cleared the equation of the fraction.

Tips and Traps

When we solve equations with a variable in the denominator, we may get a "root" that will not make a true statement when substituted in the original equation. Solutions or roots that do not make a true statement in the original equation are called **extraneous roots. When solving an equation with the variable in the denominator, we *must* check the root to see if it makes a true statement in the original equation**.

A more complex variation of this kind of equation is one in which the fraction side of the equation contains more than one term in its numerator or denominator. Rule 10-1.1 applies in this case also.

EXAMPLE 10-1.4 Solve $\frac{12}{Q+6} = 1$.

Solution

$$\frac{12}{Q+6} = 1$$

$$(\overset{1}{\cancel{Q+6}})\frac{12}{\underset{1}{\cancel{Q+6}}} = (Q+6)1$$ *(Multiply both sides of the equation by the denominator, the quantity Q + 6. Reduce where possible.)*

$$12 = Q + 6$$

$$12 - 6 = Q$$ *(Transpose.)*

$$6 = Q$$ *(Since the numerical coefficient of Q is already 1, the equation is solved.)*

Checking the root, we get

$$\frac{12}{6+6} = 1$$

$$\frac{12}{12} = 1$$

$$1 = 1$$ ■

EXAMPLE 10-1.5 Solve $3 = \frac{4n-7}{n}$.

Solution

$$3 = \frac{4n-7}{n}$$

$$(n)3 = (\overset{1}{\cancel{n}})\frac{4n-7}{\underset{1}{\cancel{n}}}$$ *(Multiply both sides by denominator n. Reduce where possible.)*

$$3n = 4n - 7$$

$$3n - 4n = -7$$ *(Transpose.)*

$$-n = -7$$

$$\frac{-1n}{-1} = \frac{-7}{-1}$$ *(Since the numerical coefficient of n is −1, we divide both sides by −1 to solve. Be careful with signs.)*

$$n = 7$$

Check to see if the root makes a true statement in the original equation.

$$3 = \frac{4(7)-7}{7}$$

$$3 = \frac{28-7}{7}$$

$$3 = \frac{21}{7}$$

$$3 = 3$$

The root makes a true statement; therefore, it is a true root of the equation. ■

In the next example we will look at an equation that has more than one fraction.

Example 10-1.6 Solve the following equation by clearing fractions or denominators first.

$$-\frac{1}{4}x = 9 - \frac{2}{3}x$$

Solution: The two fractions have denominators of 4 and 3. To clear *both* fractions or denominators at the same time, we multiply each term of the *entire* equation by both denominators or by the product of 4 and 3. This product is called the ***least common* multiple** (LCM) of 4 and 3 and is found the same way as the least common denominator.

$$-\frac{1}{4}x = 9 - \frac{2}{3}x$$

Multiply each term by both denominators.

$$(4)(3)\left(-\frac{1}{4}x\right) = (4)(3)(9) - (4)(3)\left(\frac{2}{3}x\right)$$

Next, reduce as you would in multiplying fractions in arithmetic.

$$(\overset{1}{\cancel{4}})(3)\left(-\frac{1}{\underset{1}{\cancel{4}}}x\right) = (4)(3)(9) - (4)(\overset{1}{\cancel{3}})\left(\frac{2}{\underset{1}{\cancel{3}}}x\right)$$

Multiply the remaining factors

$$-3x = 108 - 8x$$

The new equation contains no fractions. Solve the new equation.

$$-3x = 108 - 8x$$

$$-3x + 8x = 108$$

$$5x = 108$$

$$\frac{5x}{5} = \frac{108}{5}$$

$$x = \frac{108}{5}$$

To check, substitute $\frac{108}{5}$ for x in the original equation.

$$-\frac{1}{4}x = 9 - \frac{2}{3}x$$

$$-\frac{1}{\underset{1}{\cancel{4}}}\left(\frac{\overset{27}{\cancel{108}}}{5}\right) = 9 - \frac{2}{\underset{1}{\cancel{3}}}\left(\frac{\overset{36}{\cancel{108}}}{5}\right)$$

$$-\frac{27}{5} = 9 - \frac{72}{5}$$

At this point we can choose to work with the fractions or we can *clear* the fractions. Let's clear the fractions.

$$(5)\left(-\frac{27}{5}\right) = 5(9) - (5)\left(\frac{72}{5}\right)$$

$$(\overset{1}{\cancel{5}})\left(-\frac{27}{\underset{1}{\cancel{5}}}\right) = 5(9) - (\overset{1}{\cancel{5}})\left(\frac{72}{\underset{1}{\cancel{5}}}\right)$$

$$-27 = 45 - 72$$

$$-27 = -27$$ ■

> **Rule 10-1.2** *To clear an equation of fractions:*
>
> Multiply each term of the *entire* equation by the least common multiple (LCM) of the denominators of the equation.

Clearing fractions creates larger numbers in the new equation, but all numbers will be integers instead of fractions.

Example 10-1.7 Solve the following equation by clearing fractions or denominators first.

$$\frac{1}{R} = \frac{1}{4} + \frac{1}{10} + \frac{1}{20}$$

Solution: The LCM is $20R$, since R, 4, 10, and 20 will all divide evenly into $20R$. Multiply each term in the *entire* equation by $20R$ and reduce. Using the LCM is simpler than multiplying by all four denominators.

$$\frac{1}{R} = \frac{1}{4} + \frac{1}{10} + \frac{1}{20}$$

$$20R\left(\frac{1}{R}\right) = 20R\left(\frac{1}{4}\right) + 20R\left(\frac{1}{10}\right) + 20R\left(\frac{1}{20}\right)$$

$$20 = 5R + 2R + R$$

$$\frac{20}{8} = \frac{8R}{8}$$ *(Combine terms and divide by coefficient 8.)*

$$\frac{20}{8} = R$$

$$\frac{5}{2} = R$$ *(in lowest terms)*

When an equation is multiplied by a variable, we must check for extraneous roots. We substitute $\frac{5}{2}$ for R in the original equation.

$$\frac{1}{R} = \frac{1}{4} + \frac{1}{10} + \frac{1}{20}$$

$$\frac{1}{\frac{5}{2}} = \frac{1}{4} + \frac{1}{10} + \frac{1}{20} \qquad \left[\frac{1}{\frac{5}{2}} = (1)\left(\frac{2}{5}\right) = \frac{2}{5}\right]$$

$$\frac{2}{5} = \frac{5}{20} + \frac{2}{20} + \frac{1}{20}$$

$$\frac{2}{5} = \frac{8}{20}$$

$$\frac{2}{5} = \frac{2}{5}$$

■

Tips and Traps

Sometimes applied problems using fractional equations have their solutions expressed as decimal numbers rather than as fractions. In these cases we need to perform the indicated division represented by the fractional solution.

The equation in Example 10-1.7 is derived from the formula for finding total resistance in a parallel dc circuit with three branches rated at 4, 10, and 20 ohms, respectively. Since ohms are expressed in decimal numbers, our fractional solution should be converted to a decimal equivalent.

$$R = \frac{5}{2}$$

$$R = 2.5 \text{ ohms}$$

L.O.2. Solve work problems with fractional equations.

Solving Work Problems

Our knowledge of fractional equations now enables us to solve problems. The basic idea used to solve work problems is:

> **Formula 10-1.1** *Formula for amount of work:*
>
> Rate × time = amount of work

If car A travels 50 miles in 1 hour, then car A's *rate* of work (travel) is 50 miles per 1 hour, or $\frac{50 \text{ mi}}{1 \text{ hr}}$ expressed as a fraction. If car A travels for 3 hours, then the rate times the time $\left(\frac{50 \text{ mi}}{1 \cancel{\text{hr}}} \times 3 \cancel{\text{hr}}\right)$ equals the total amount of work (travel) completed. Car A traveled (worked) 150 miles:

$$\text{rate} \times \text{time} = \text{amount of work}$$

$$\frac{50 \text{ mi}}{1 \cancel{\text{hr}}} \times 3 \cancel{\text{hr}} = 150 \text{ mi}$$

Example 10-1.8 A carpenter can install 1 door in 3 hours. How many doors can the carpenter install in 30 hours?

Solution

$$\text{rate} = 1 \text{ door per 3 hours or } \frac{1}{3} \text{ door per hour}$$
$$\text{time} = 30 \text{ hours}$$
$$\text{rate} \times \text{time} = \text{amount of work}$$
$$\frac{1 \text{ door}}{\overset{}{\underset{1}{\cancel{3}}}\ \cancel{\text{hr}}} \times \overset{10}{\cancel{30}}\ \cancel{\text{hr}} = 10 \text{ doors}$$

Thus, 10 doors can be installed in 30 hours. ■

If two workers or machines do a job together, we find the fractional part of the job each does. Combined, the fractional parts equal 1 total job. The basic formula for solving this kind of work problem is:

Formula 10-1.2 *Formula for completing one job when working together:*

$$\begin{pmatrix} A\text{'s} \\ \text{portion} \\ \text{of job} \end{pmatrix} + \begin{pmatrix} B\text{'s} \\ \text{portion} \\ \text{of job} \end{pmatrix} = 1 \text{ completed job}$$

or

$$\begin{pmatrix} A\text{'s} \\ \text{rate} \times \text{time} \end{pmatrix} + \begin{pmatrix} B\text{'s} \\ \text{rate} \times \text{time} \end{pmatrix} = 1 \text{ completed job}$$

Example 10-1.9 Pipe 1 can fill a tank in 6 min and pipe 2 can fill the same tank in 8 min. How long would it take for both pipes together to fill the tank?

Solution: Pipe 1 fills the tank at a rate of 1 tank per 6 min or $\frac{1}{6}$ tank per minute. Pipe 2 fills the tank at a rate of 1 tank per 8 min or $\frac{1}{8}$ tank per minute. Since the amount of work = rate × time, if we let T = time (in minutes) for both pipes together to fill the tank, then $\frac{1}{6}T$ = part of tank filled by pipe 1 and $\frac{1}{8}T$ = part of tank filled by pipe 2. Both amounts together equal 1 completed job (a full tank). Thus,

$$\underset{\text{pipe 1's rate} \times \text{time}}{\frac{1}{6}T} + \underset{\text{pipe 2's rate} \times \text{time}}{\frac{1}{8}T} = \underset{\text{completed job}}{1}$$

Then we clear the fractions. The LCM is 24.

$$(24)\left(\frac{1}{6}T\right) + (24)\left(\frac{1}{8}T\right) = (24)(1)$$

$$\overset{4}{(\cancel{24})}\left(\frac{1}{\underset{1}{\cancel{6}}}T\right) + \overset{3}{(\cancel{24})}\left(\frac{1}{\underset{1}{\cancel{8}}}T\right) = (24)(1)$$

$$4T + 3T = 24$$

$$7T = 24$$

$$\frac{7T}{7} = \frac{24}{7}$$

$$T = \frac{24}{7}\left(\text{or } 3\frac{3}{7}\right) \text{ min}$$

Both pipes working together can fill the tank in $3\frac{3}{7}$ min. ■

Tips and Traps

In solving ordinary equations, a root that is an improper fraction like $\frac{24}{7}$ is left as an improper fraction. However, when solving applied problems such as work problems, it is better to change improper fractions like $\frac{24}{7}$ into mixed numbers like $3\frac{3}{7}$ because $3\frac{3}{7}$ min is easier to understand than $\frac{24}{7}$ min.

Now, continuing with work problems, suppose the pipes described in the problem above had opposite functions; that is, suppose as pipe 1 is filling the tank, pipe 2 is emptying the tank. In this case, we would subtract the fractional part pipe 2 removes from the fractional part pipe 1 puts into the tank. This combined action, if pipe 1 fills at a faster rate than pipe 2 empties, will result in a full tank. The formula for solving this work problem is:

Formula 10-1.3 ***Formula for completing one job when working in opposition:***

$$\begin{pmatrix} A\text{'s} \\ \text{portion} \\ \text{of job} \end{pmatrix} - \begin{pmatrix} B\text{'s} \\ \text{portion} \\ \text{of job} \end{pmatrix} = 1 \text{ completed job}$$

$$\begin{pmatrix} A\text{'s} \\ \text{rate} \times \text{time} \end{pmatrix} - \begin{pmatrix} B\text{'s} \\ \text{rate} \times \text{time} \end{pmatrix} = 1 \text{ completed job}$$

Here, we *subtract* the second fractional part of work from the first.

EXAMPLE 10-1.10 Apply Formula 10-1.3 to the pipes described in Example 10-1.9 if they are working in opposition.

Solution: Pipe 1's rate in Example 10-1.9 is $\frac{1}{6}$ tank per minute and pipe 2's rate is $\frac{1}{8}$ tank per minute. Remember, rate × time = amount of work. Therefore, if we let T = the time to fill the tank with both pipes in operation, and pipe 1 fills as pipe 2 empties, our equation becomes:

$$\underset{\text{Pipe 1's rate} \times \text{time}}{\frac{1}{6}T} - \underset{\text{Pipe 2's rate} \times \text{time}}{\frac{1}{8}T} = \underset{\text{completed job}}{1}$$

$$(24)\left(\frac{1}{6}T\right) - (24)\left(\frac{1}{8}T\right) = (24)(1)$$

$$\overset{4}{\cancel{(24)}}\left(\frac{1}{\underset{1}{\cancel{6}}}T\right) - \overset{3}{\cancel{(24)}}\left(\frac{1}{\underset{1}{\cancel{8}}}T\right) = (24)(1)$$

$$4T - 3T = 24$$

$$T = 24 \text{ min}$$

With both pipes working, one filling and the other emptying, the tank can be filled in 24 min. ■

SELF-STUDY EXERCISES 10-1

L.O.1. Solve the following fractional equations.

1. $\frac{2}{7}x = 8$

2. $-7 = \frac{21}{33}p$

3. $\frac{1}{3}r = \frac{6}{7}$

4. $0 = -\frac{2}{5}c$

5. $-\frac{5}{3}m = 9$

6. $-9m = \frac{5}{3}$

7. $\frac{5}{8}t = 1$

8. $-\frac{5}{7}p = -\frac{11}{21}$

9. $10 = -\frac{1}{35}t$

10. $\frac{5}{12}z = 20$

11. $\frac{2x}{3} = 18$

12. $\frac{7}{Q} = 21$

13. $\frac{y+1}{2} = 7$

14. $\frac{7}{p-4} = -8$

15. $0 = \frac{x}{4}$

16. $-\frac{8}{P} = -72$

17. $\frac{P}{-8} = -72$

18. $-8 = \frac{4B}{B-6}$

19. $\frac{3P}{7} = 12$

20. $\frac{8-R}{76} = 1$

21. $\frac{2}{9}c + \frac{1}{3}c = \frac{3}{7}$

22. $-\frac{1}{4}x = 9 - \frac{2}{3}x$

23. $\frac{2}{7}y + \frac{3}{8} = \frac{1}{7}y + \frac{5}{3}$

24. $\frac{1}{3}x + \frac{1}{2}x = \frac{20}{3}$

25. $\frac{7}{R} - \frac{2}{R} = -1$

26. $S = \frac{1}{15} + \frac{1}{5} + \frac{1}{30}$

27. $18 - \frac{1}{4}x = \frac{1}{2}$

28. $\frac{1}{7}H - \frac{1}{3}H = 0$

29. $\frac{7}{16}h + \frac{1}{9} = \frac{1}{3}$

30. $x + \frac{1}{4}x = 8$

31. $3y + 9 = \frac{1}{4}y$

32. $18 = \frac{4}{3x} - \frac{3}{2x}$

33. $\frac{2}{7}p + 1 = \frac{1}{3}p$

34. $S = \frac{1}{10} + \frac{1}{25} + \frac{1}{50}$

35. $-x + \frac{1}{7} = \frac{1}{2}x$

36. $0 = 1 + \frac{2}{9}c - c$

37–42. Check or verify the roots of Exercises 12, 14, 16, 18, 25, and 32.

L.O.2. Set up a fractional equation for each of the following. Solve each equation.

43. A licensed electrician can install 8 light fixtures in 2 hr. How many light fixtures can be installed in 20 hr?

44. A printing press can produce 1 day's newspaper in 4 hr. A higher-speed press can do 1 day's newspaper in 2 hr. How much time would it take both presses to do the 1 job?

45. One machine can pack 1 day's salmon catch in 8 hr. A second machine can pack 1 day's catch in only 5 hr. How much time would it require for 1 day's catch to be packed if both machines were used together?

46. A painter can paint a house in 6 days. Another painter takes 8 days to paint the same house. If they work together, how much time would it take them to paint the house?

47. A tank has two pipes entering it and one leaving it. Pipe 1 fills the tank in 3 min. Pipe 2 takes 7 min to fill the same tank. Pipe 3, however, empties the tank in 21 min. How much time does it take to fill the tank with all three pipes operating at the same time?

10-2 SOLVING DECIMAL EQUATIONS BY CLEARING THE DECIMALS

Learning Objectives

L.O.1. Solve decimal equations by clearing decimals.
L.O.2. Solve applied problems with decimal equations.

Solving Decimal Equations

A technique similar to clearing fractions or denominators is to clear the equation of all decimals before starting to solve the equation. Like the equation cleared of all fractions, the equation cleared of all decimals contains only whole numbers. It can then be solved using techniques previously studied.

> **Rule 10-2.1** *To clear an equation of decimals:*
>
> Multiply each term of the *entire* equation by the least common multiple (LCM) of the fractional amounts following the decimal points.

EXAMPLE 10-2.1 Solve the following equation by first clearing the equation of decimals.

$$1.1 = 3.4 + R$$

Solution: In $1.1 = 3.4 + R$, the LCM is 10, since both decimals are in tenths (have denominators of 10).

Next, we clear the equation of decimals by multiplying the *entire* equation by 10. Using the shortcut rule for multiplying by 10, move the decimal one place to the right. Again, *each term* is multiplied by 10.

$$1.1 = 3.4 + R$$
$$10(1.1) = 10(3.4) + 10R$$
$$11 = 34 + 10R$$

The resulting equation contains no decimals. Solve the resulting equation.

$$11 = 34 + 10R$$
$$11 - 34 = 10R$$
$$-\frac{23}{10} = \frac{10R}{10}$$
$$-2.3 = R$$

To check, -2.3 is substituted for R in the original equation.

$$1.1 = 3.4 + R$$
$$1.1 = 3.4 + (-2.3)$$
$$1.1 = 1.1$$

■

Tips and Traps

Since digits to the right of the decimal point represent fractional amounts whose denominators are determined by place value position, the LCM for all the decimal numbers in an equation will be the denominator of the place value indicated by the fractional amount with the most digits to the right of the decimal.

This procedure of clearing decimals allows us to avoid having to divide an amount by a decimal, which is typically a common source of error.

EXAMPLE 10-2.2 Solve the following equation by first clearing the equation of decimals.

$$0.38 + 1.1y = 0.6$$

Solution: The largest indicated denominator and thus the LCM for the equation is 100. Then we clear the equation of decimals by multiplying the *entire* equation by 100. Using the shortcut rule for multiplying by 100, move the decimal two places to the right. Do this for each term.

$$0.38 + 1.1y = 0.6$$
$$100(0.38) + 100(1.1y) = 100(0.6)$$
$$38 + 110y = 60$$
$$110y = 60 - 38$$
$$\frac{110y}{110} = \frac{22}{110}$$
$$y = 0.2$$

■

L.O.2. Solve applied problems with decimal equations.

Solving Applied Problems

A number of applied problems are solved with decimal equations. One common problem involves interest on a loan or on an investment.

A basic tool for interest problems is the interest formula. It resembles the percentage proportion after cross multiplication (Section 5-4), but it includes the time over which the money was borrowed or invested. If we know three of the elements, we can find the fourth element.

Formula 10-2.1 *Formula for Interest*

$$I = PRT$$

where I = **interest**, P = **principal**, R = **rate** or percent, and T = **time.**

For comparison to the percentage proportion, interest is the part or percentage and the principal is the base.

Example 10-2.3 A $1000 investment is made for $2\frac{1}{2}$ years at 8.25%. Find the amount of interest.

Solution: In using the interest formula, change the percent to a decimal. Then substitute the values in the formula.

$$I = PRT$$

$$= \$1000 \times 0.0825 \times 2.5 \text{ years}$$ *(8.25% = 0.0825; $2\frac{1}{2}$ years = 2.5 years LCM = 10,000)*

$$(10{,}000)I = 1000 \times (10{,}000)(0.0825) \times 2.5$$ *(Since the right side of the equation has only one term and multiplication is commutative and associative, multiply any factor on the right by the LCM.)*

$$10{,}000I = 1000 \times 825 \times 2.5$$ *(Note that we still have a decimal, 2.5. So not all decimals have been cleared by this procedure.)*

$$10{,}000I = 2{,}062{,}500$$

$$\frac{10{,}000I}{10{,}000} = \frac{2{,}062{,}500}{10{,}000}$$

$$I = \$206.25$$

The interest for $2\frac{1}{2}$ years is $206.25. ■

Tips and Traps

This example illustrates that in some cases clearing the decimals is not the most efficient means to solve an equation with decimals. The procedure may produce extremely large, cumbersome numbers and may not clear all the decimals. A calculator will give the solution more quickly and more efficiently if we proceed as if the equation contained whole numbers. The calculator is another tool for helping us solve mathematical problems.

Let's use the interest formula to solve some problems without clearing the decimals. In this case, the rate is always expressed as a decimal.

EXAMPLE 10-2.4 Gomez earned \$360 on a two-year investment at 12% interest. How much did he invest?

Solution: Substitute the interest (\$360), the time (2 years), and the rate (12%) in the interest formula.

$$I = \text{PRT}$$

$$360 = P \times 0.12 \times 2 \qquad \textit{(The base, P, is missing; 12\% = 0.12)}$$

$$360 = 0.24P$$

$$\frac{360}{0.24} = \frac{0.24P}{0.24}$$

$$\$1500 = P$$

Thus, Gomez invested \$1500. ■

EXAMPLE 10-2.5 Aetna Photo Studio borrowed \$3500 for some darkroom equipment and had to pay \$1890 in interest over a three-year period. What was the interest rate?

Solution: Substitute the principal (\$3500), interest (\$1890), and time (3 years) in the interest formula.

$$I = PRT$$

$$1890 = 3500 \times R \times 3 \qquad \textit{(The rate, R, is missing.)}$$

$$1890 = 10{,}500R$$

$$\frac{1890}{10{,}500} = \frac{10{,}500R}{10{,}500}$$

$$0.18 = R \qquad \textit{(0.18 = 18\%)}$$

Thus, the interest rate was 18%. ■

EXAMPLE 10-2.6 Ms. Boudreaux wants to earn \$500. If she invests \$3000 at 5%, how long will she have to wait?

Solution: Substitute the interest (\$500), rate (5%), and principal (\$3000) in the interest formula.

$$I = PRT$$

$$500 = 3000 \times 0.05 \times T \qquad \textit{(5\% = 0.05; the time, T, is missing.)}$$

$$500 = 150T$$

$$\frac{500}{150} = \frac{150T}{150}$$

$$3.\overline{33} = T \qquad \textit{(}3.\overline{33}\textit{ years} = 3\frac{1}{3}\textit{ years)}$$

Thus, Ms. Boudreaux must wait $3\frac{1}{3}$ years (or 3 years 4 months). ■

General Tips for Using the Calculator

If the interest is missing, the calculator sequence is to multiply the three known factors (*PRT*). If any other element is missing, the following calculator sequences may be used. Let's use Example 10-2.6 to illustrate.

Scientific calculator:

[AC] 3000 [×] .05 [=] [m+] [CE/C] 500 [÷] [mR] [=] ⇒ 3.33333333

The [m+] key stores an amount in memory and [mR] recalls the amount.

Graphing calculator:

3000 [×] .05 [EXE]
500 [÷] [ANS] [EXE] ⇒ 3.33333333

The [ANS] key inserts the answer to the *last* calculation. A similar sequence is used for finding principal or rate.

Not all problems deal with interest. Many other problems may be solved with equations containing decimal numbers. Let's examine a few.

EXAMPLE 10-2.7 The distance formula is distance = rate × time. If a car was driven 82.5 mi at 55 miles per hour (mph), how long did the trip take?

Solution

$$\text{distance} = \text{rate} \times \text{time}$$

$$82.5 = 55 \times T \qquad \textit{(Time, T, is unknown.)}$$

$$82.5 = 55T$$

$$\frac{82.5}{55} = \frac{55T}{55}$$

$$1.5 = T$$

Thus, the trip took 1.5 or $1\frac{1}{2}$ hours.

This and other formulas may be solved in the manner of the interest formula, $I = PRT$. If all the elements are given but one, the missing one can be found. ■

EXAMPLE 10-2.8 The formula for electrical power is $V = W/A$: voltage (V) equals wattage (W) divided by amperage (A). Find the voltage to the nearest hundredth needed for a circuit of 1280 W with a current of 12.23 A.

Solution

$$V = \frac{W}{A}$$

$$V = \frac{1280}{12.23}$$

$$V = 104.6606705 \quad \text{or} \quad 104.66 \text{ V}$$ ■

EXAMPLE 10-2.9 Juan earned $30.38 working 6.75 hr at a fast-food restaurant. What was his hourly wage?

Solution: We know that hourly wage × hours worked = pay, so we can set up an equation using W to represent hourly wage.

hourly wage × hours worked = amount of pay *(The hourly wage, W, is missing.)*

$$W \times 6.75 = 30.38$$

$$6.75W = 30.38$$

$$\frac{6.75W}{6.75} = \frac{30.38}{6.75}$$

$$W = 4.500740741 \quad \text{or} \quad \$4.50 \text{ per hour}$$ ■

SELF-STUDY EXERCISES 10-2

L.O.1. Solve the following equations.

1. $2.3x = 4.6$

2. $0.8R = 0.6$ *(round to nearest tenth)*

3. $0.33x + 0.25x = 3.5$ *(round to nearest hundredth)*

4. $0.3a = 4.8$

5. $1.5p = 7$ *(round to nearest tenth)*

6. $0.04x = 0.08 - x$ *(round to nearest hundredth)*

7. $0.4p = 0.014$

8. $0.47 = R + 0.4R$ *(round to nearest hundredth)*

9. $2.3 = 5.6 + y$

10. $4.3 = 0.3x - 7.34$

11. $2x + 3.7 = 10.3$

12. $0.16 + 2.3x = -0.3$

13. $1.5x + 2.1 = 3$

14. $3.82 - 2.5y = 1$

15. $0.15p = 2.4$

L.O.2. Solve the following problems using decimal equations.

16. If the formula for force is force = pressure × area, how many pounds of force are produced by a pressure of 35 psi on a piston whose surface area is 2.5 in.^{2}? Express the pounds of force as a decimal number.

17. The circumference of a circle equals the product of π times the diameter. If a steel rod has a diameter of 1.5 in., what is the circumference of the rod to the nearest hundredth? (Circumference is the distance around a circle.)

18. The distance formula is distance = rate × time. If a trucker drove 682.5 mi at 55 miles per hour (mph), how long did she drive? (Answer to nearest whole number.)

19. The distance formula is distance = rate × time. If a tractor-trailer rig was driven 422.5 mi at 65 miles per hour (mph) on expressways, how long did the trip take?

20. Electrical resistance in ohms (Ω) is voltage (V) divided by amperes (A). Find the resistance to the nearest tenth for a motor with a voltage of 12.4 volts requiring 1.5 amps.

21. Rita earned $39.75 working 7.5 hr at a college bookstore. What was her hourly wage?

22. The formula for electrical power is $V = W/A$: voltage (V) equals wattage (W) divided by amperage (A). Find the voltage to the nearest hundredth needed for a circuit of 500 W with a current of 3.2 A.

23. Find the interest paid on a loan of $2400 for 1 year at an interest rate of 11%.

24. Find the interest paid on a loan of $800 at $8\frac{1}{2}\%$ interest for 2 years.

25. Find the total amount of money (maturity value) that the borrower will pay back on a loan of $1400 at $12\frac{1}{2}\%$ interest for 3 years.

26. Find the rate of interest on an investment of $2500 made by Connie Honda for a period of 2 years if she received $612.50 in interest.

27. Maddy Brown needed start-up money for her bakery. She borrowed $12,000 for 30 months and paid $360 interest on the loan. What interest rate did she pay?

28. Raul Fletes needed money to buy lawn equipment. He borrowed \$500 for 7 months and paid \$53.96 in interest. What was the rate of interest?

29. Linda Davis agreed to lend money to Alex Luciano at a special interest rate of 9%, on the condition that he borrow enough that he would pay her \$500 in interest over a 2-year period. What was the minimum amount Alex could borrow?

30. Rob Thweatt needed money for college. He borrowed \$6000 at 12% interest. If he paid \$360 interest, what was the duration of the loan?

10-3 USING PROPORTIONS TO SOLVE PROBLEMS

Learning Objectives

L.O.1. Solve fractional equations that are proportions.
L.O.2. Solve problems with direct proportions.
L.O.3. Solve problems with inverse proportions.
L.O.4. Solve problems that involve similar triangles.

L.O.1. Solve fractional equations that are proportions.

Solving Fractional Equations that are Proportions

Another common type of equation that contains fractions is a *proportion.* An example of a proportion is

$$\frac{N}{5} = \frac{1}{12}$$

In a proportion, *each side* of the equation is a fraction or ratio. (See Section 5-4 for an introduction to ratio and proportion.)

The term *ratio* is often used when the fraction represents quantities or measures in solving (word) problems. We will examine this use of ratios and proportions in objective 2. But for now, our concern is with solving an equation expressed as a proportion.

The relationship of the terms of the ratios or fractions in a proportion such as

$$\frac{6}{9} = \frac{2}{3}$$

can also be stated as

"6 is to 9 as 2 is to 3"

Still another way of representing a proportion is

$$6:9 = 2:3, \quad \text{which is read} \quad \text{"6 is to 9 as 2 is to 3"}$$

As noted in Section 5-4, in a proportion the *cross products* are equal. A cross product is the product of the numerator of one ratio and the denominator of the other. We state this relationship symbolically in the following property:

Property 10-3.1 ***Property of Proportions:***

If $\frac{a}{b} = \frac{c}{d}$, then $ad = bc$, $\quad b, d \neq 0$

That is,

$$\text{if } \frac{6}{9} = \frac{2}{3}$$
$$\text{then } 6(3) = 9(2)$$
$$\text{or } 18 = 18$$

Another way of stating this is that the product of the *extremes* (end factors a and d) equals the product of the *means* (middle factors b and c).

We can use this property in solving equations that are in the form of a proportion.

EXAMPLE 10-3.1 Solve the following proportions.

(a) $\frac{x}{4} = \frac{9}{6}$ **(b)** $\frac{4x}{5} = \frac{17}{20}$

(c) $\frac{x - 2}{x + 8} = \frac{3}{5}$ **(d)** $\frac{3}{x} = 7$

Solution

(a) $\frac{x}{4} = \frac{9}{6}$

$6x = 36$ *(Cross multiply. 6(x) = 6(x); 9(4) = 36*

$\frac{6x}{6} = \frac{36}{6}$ *(Solve for x.)*

$x = 6$

(b) $\frac{4x}{5} = \frac{17}{20}$

$80x = 85$ *(Cross multiply. 4x(20) = 80x; 17(5) = 85*

$\frac{80x}{80} = \frac{85}{80}$ *(Solve for x.)*

$x = \frac{85}{80}$ *(Reduce.)*

$= \frac{17}{16}$

(c) $\frac{x - 2}{x + 8} = \frac{3}{5}$

$5(x - 2) = 3(x + 8)$ *(Cross multiply.)*

$5x - 10 = 3x + 24$ *(Distribute.)*

$5x - 3x = 24 + 10$ *(Transpose.)*

$2x = 34$ *(Combine terms.)*

$\frac{2x}{2} = \frac{34}{2}$ *(Solve for x.)*

$x = 17$

(d) $\frac{3}{x} = 7$

The whole number 7 can be written as the improper fraction $\frac{7}{1}$. Then the equation is in the form of a proportion.

$$\frac{3}{x} = \frac{7}{1}$$

$$3 = 7x \qquad \textit{(Cross multiply.)}$$

$$\frac{3}{7} = \frac{7x}{7} \qquad \textit{(Solve for x.)}$$

$$\frac{3}{7} = x$$

■

We saw above how to solve equations in the form of proportions. We solved them by cross multiplication. Now we study how to set up proportions in order to solve problems. We will look at two kinds of proportions, *direct proportions* and *inverse proportions*. Let's look at direct proportions first.

L.O.2. Solve problems with direct proportions.

Direct Proportions or Direct Variation

■ **DEFINITION 10-3.1 Direct Proportion.** A **direct proportion** is one in which the quantities being compared are directly related so that as one quantity increases (or decreases) the other quantity also increases (or decreases). This relationship is also called a *direct variation*.

For example, the miles a vehicle travels are *directly proportional* to the amount of gasoline used. That is, as the number of miles traveled *increases*, the amount of gasoline used by the vehicle also *increases*. Or, as the number of miles traveled by the vehicle *decreases*, the amount of gasoline needed also *decreases*.

The ratios used in setting up proportions to solve problems may either compare *like* measures, such as comparing miles traveled on one trip to miles traveled on a second trip (352 mi/276 mi), or *unlike* measures, such as comparing the cost of a roast to its weight ($24/8 lb). Either may be used in a direct proportion. The unknown quantity is designated by a letter, such as x.

Rule 10-3.1 ***To set up a direct proportion:***

1. The numerator and the denominator of each ratio or fraction are written in the *same* order.
2. The two ratios are made equal to each other.

Let's look at how we set up direct proportions in order to solve problems.

EXAMPLE 10-3.2 A truck will travel 102 mi on 6 gal of gasoline. How far will it travel on 30 gal of gasoline?

Solution I: In setting up a direct proportion, the relationship in each fraction is expressed in the same order; for example, miles per gallon on the left is compared to miles per gallon on the right. Each ratio compares *unlike* measures, miles to gallons.

$$\frac{102 \text{ mi}}{6 \text{ gal}} = \frac{x \text{ mi}}{30 \text{ gal}} \qquad \frac{distance_1}{gasoline_1} = \frac{distance_2}{gasoline_2}$$

$$\frac{102}{6} = \frac{x}{30} \qquad \left(\frac{mi}{gal} = \frac{mi}{gal}\right)$$

$$102(30) = 6x \qquad (Cross\ multiply.\ mi(gal) = gal(mi))$$

$$3060 = 6x$$

$$\frac{3060}{6} = \frac{6x}{6} \qquad \left(\frac{mi(gal)}{gal} = \frac{gal(mi)}{gal}\right)$$

$$510 = x \qquad (mi = mi)$$

Thus, 510 mi can be traveled on 30 gal of gasoline.

Solution II: Another way of solving this problem is to use two ratios comparing *like* measures. That is, the relationship of miles to miles on the left is compared to the relationship of gallons to gallons on the right.

$$\frac{102 \text{ mi}}{x \text{ mi}} = \frac{6 \text{ gal}}{30 \text{ gal}} \qquad \left(\frac{distance_1}{distance_2} = \frac{gasoline_1}{gasoline_2}\right)$$

$$\frac{102}{x} = \frac{6}{30} \qquad \left(\frac{mi}{mi} = \frac{gal}{gal}\right)$$

$$102(30) = 6x \qquad (Cross\ multiply.\ mi(gal) = gal(mi))$$

$$3060 = 6x \qquad \left(\frac{mi(gal)}{gal} = \frac{gal(mi)}{gal}\right)$$

$$510 = x \qquad (x \text{ is expressed in miles.})$$

When using ratios comparing *like* measures, the numerators of each ratio are related (102 mi and 6 gal) and the denominators are related (x mi and 30 gal) ■

EXAMPLE 10-3.3 If a piece of stock tapers 1 in. for every 24 in. of length, what is the amount of taper of a 30-in. piece of stock? (See Fig. 10-1).

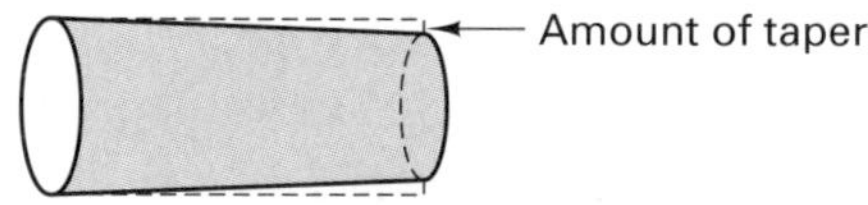

Figure 10-1

Solution: The amount of taper and the length of the stock are directly proportional. As the amount of taper increases, the length increases. Therefore, we set up our proportion so that the numerators and denominators of both fractions are in the same order.

$$\frac{1\text{-in. taper}}{24\text{-in. length}} = \frac{x\text{-in. taper}}{30\text{-in. length}} \quad \text{or} \quad \frac{1\text{-in. taper}}{x\text{-in. taper}} = \frac{24\text{-in. length}}{30\text{-in. length}}$$

$$\frac{1}{24} = \frac{x}{30} \qquad\qquad \frac{1}{x} = \frac{24}{30}$$

$$1(30) = 24x \quad (Cross\ multiply.) \quad 1(30) = 24x$$

$$30 = 24x \qquad\qquad 30 = 24x$$

$$\frac{30}{24} = \frac{24x}{24} \qquad\qquad \frac{30}{24} = \frac{24x}{24}$$

$$\frac{5}{4} = x \text{ or } 1\frac{1}{4} = x \qquad\qquad \frac{5}{4} = x \text{ or } 1\frac{1}{4} = x$$

The amount of taper for a 30-in. length of stock is $1\frac{1}{4}$ in. ■

EXAMPLE 10-3.4 The ratio of chicory to coffee in a New Orleans coffee mixture is 1 : 8. If the coffee company uses 75 lb of chicory for a batch of the coffee mixture, how many pounds of coffee are needed?

Solution: As with the previous proportions, this is a direct proportion because the amount of coffee will increase as the amount of chicory increases. We know one ratio already, $\frac{1}{8}$, that is, 1 lb of chicory per 8 lb of coffee. We start with this known ratio.

$$\frac{1 \text{ lb chicory}}{8 \text{ lb coffee}} = \frac{75 \text{ lb chicory}}{x \text{ lb coffee}}$$ *(The ratio on the right compares amount of chicory to amount of coffee needed.)*

$$\frac{1}{8} = \frac{75}{x}$$

$$1x = 75(8)$$ *(Cross multiply.)*

$$x = 600$$

Thus, the coffee mixture requires 600 lb of coffee. ■

EXAMPLE 10-3.5 A car traveled 385.6 mi on 15.5 gal of gasoline. How far can it travel on 10 gal of gasoline? Round to the nearest tenth of a mile.

Solution: This is a direct proportion, since gasoline use increases as the number of miles driven increases.

$$\frac{385.6 \text{ mi}}{15.5 \text{ gal}} = \frac{x \text{ mi}}{10 \text{ gal}}$$

$$\frac{385.6}{15.5} = \frac{x}{10}$$

$$(385.6)(10) = (15.5)(x)$$ *(Cross multiply.)*

$$3856 = 15.5x$$

$$10(3856) = 10(15.5x)$$ *(Clear the decimal.)*

$$38560 = 155x$$

$$\frac{38560}{155} = \frac{155x}{155}$$ *(Divide both sides by coefficient.)*

$$248.7741935 = x$$

$$x = 248.8$$ *(rounded)*

Thus, the car can travel 248.8 mi. ■

L.O.3. Solve problems with inverse proportions.

Inverse Proportion or Inverse Variation

Now that we have seen how to set up a direct proportion, let's turn our attention to inverse proportions, where two quantities are *inversely proportional.*

■ **DEFINITION 10-3.2 Inverse Proportion.** An **inverse proportion** is one in which the quantities being compared are inversely related so that as one quantity increases the other quantity decreases, or as one decreases the other increases. This relationship is also called *inverse variation.*

For example, as we *increase* pressure on materials such as foam rubber, gases, loose dirt, and so on, the materials become compressed and so *decrease* in size or volume. Or, as we *decrease* pressure on these same materials, they *increase* in size or volume as they expand. This is the opposite, or inverse, of direct proportion.

Another example is that of having 3 workers frame a house in 2 weeks. If the contractor *increases* the number of workers to 6, the time it takes would *decrease* to 1 week, assuming that the workers work at the same rate of speed. In other words, the time required is *inversely proportional* to the number of workers on the job.

Unlike the ratios in direct proportion, the ratios in an inverse proportion *must compare like measures* and be written *in inverse order*, such that the numerator of one corresponds to the denominator of the other.

Rule 10-3.2 *To set up an inverse proportion:*

1. Each ratio or fraction is expressed in *like* measures.
2. The numerator and denominator of one ratio or fraction is written in *inverse* order.
3. The two ratios or fractions are equal to each other.

Let's look at how we set up inverse proportions in order to solve problems. The problems will be ones of *inverse variation*, meaning that one quantity increases (decreases) as the other decreases (increases).

EXAMPLE 10-3.6 If the intensity of a light illuminating a wall is 30 candelas when the light source is 10 ft from the wall, to what level does the intensity in candelas decrease if the light source is 15 ft from the wall?

Solution: The relationship here is one of *inverse variation*. As the distance between the wall and light source *increases*, the intensity *decreases*. *Like* measures are compared.

$$\frac{30 \text{ candelas at } 10 \text{ ft}}{x \text{ candelas at } 15 \text{ ft}} = \frac{15 \text{ ft}}{10 \text{ ft}} \qquad \frac{\text{intensity}_1}{\text{intensity}_2} = \frac{\text{distance}_2}{\text{distance}_1}$$

Notice that the left ratio contains the *same units* of measure (candelas) and the right ratio also contains the *same units* of measure (feet). Also, observe that the measures in the right ratio are in *inverse* order from the order of the measures in the left ratio. That is, the numerator of the left ratio relates to the denominator of the right ratio, and the denominator of the left ratio relates to the numerator of the right ratio.

Solving, we get

$$\frac{30}{x} = \frac{15}{10} \qquad \left(\frac{\textit{candelas}}{\textit{candelas}} = \frac{\textit{ft}}{\textit{ft}}\right)$$

$$30(10) = 15x \qquad (\textit{Cross multiply. candelas(ft)} = \textit{ft(candelas)})$$

$$300 = 15x$$

$$\frac{300}{15} = x \qquad \left(\frac{\textit{candelas(ft)}}{\textit{ft}} = \textit{candelas}\right)$$

$$x = 20 \text{ candelas at } 15 \text{ ft}$$

■

EXAMPLE 10-3.7 If 5 machines take 12 days to complete a job, how long will it take for 8 machines to do the job?

Solution: As the number of machines *increases*, the amount of time required to do the job *decreases*. Thus, the quantities are *inversely proportional*.

$$\frac{5 \text{ machines}}{8 \text{ machines}} = \frac{x \text{ days for 8 machines}}{12 \text{ days for 5 machines}}$$ *(Each ratio uses like measures. The ratios are in inverse order.)*

$$\frac{5}{8} = \frac{x}{12}$$ $$\left(\frac{\textit{machines}}{\textit{machines}} = \frac{\textit{days}}{\textit{days}}\right)$$

$$5(12) = 8x$$ *(Cross multiply. machines(days) = machines(days))*

$$60 = 8x$$

$$\frac{60}{8} = x$$ $$\left(\frac{\textit{machines(days)}}{\textit{machines}} = \textit{days}\right)$$

$$x = \frac{15}{2} \quad \text{or} \quad 7\frac{1}{2} \text{ days}$$

It will take $7\frac{1}{2}$ days for 8 machines to do the job. ■

Gears and pulleys involve *inverse* relationships. Suppose that a large gear and a small gear are in mesh or a large pulley is connected by a belt to a smaller pulley. The larger gear or pulley has the *slower* speed, and the smaller gear or pulley has the *faster* speed. A bicycle is an example. The larger gear or pulley is turned by the pedals and this motion is transmitted to the smaller gear or pulley turning the rear wheel.

If the diameter (greatest distance across a circular object) of the larger gear or pulley is increased, the larger size would mean that the larger gear or pulley would take even longer to make a complete revolution. However, it would turn the smaller gear or pulley even faster. Thus, the larger gear or pulley would revolve *slower* while the smaller one would revolve *faster*.

EXAMPLE 10-3.8 A gear measuring 10 in. across is in mesh with a gear measuring 5 in. across (Fig. 10-2). If the larger gear has a speed of 25 revolutions per minute (rpm), at how many rpm does the smaller gear turn?

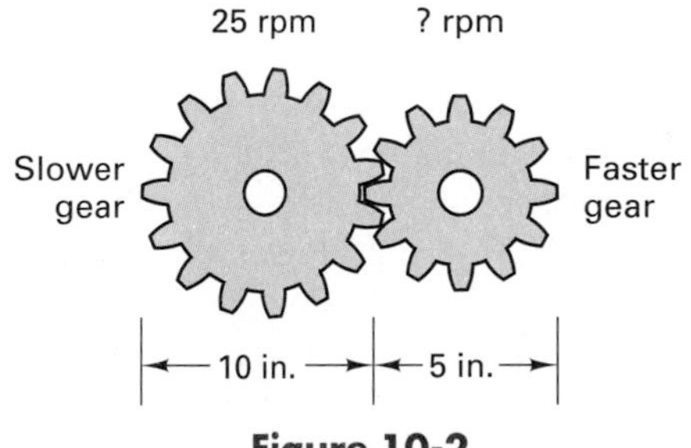

Figure 10-2

Solution: Since gears in mesh are *inversely* related, we set up an inverse proportion. Each ratio uses like measures and the ratios are in inverse order.

$$\frac{25 \text{ rpm of larger gear}}{x \text{ rpm of smaller gear}} = \frac{5 \text{ in. of smaller gear}}{10 \text{ in. of larger gear}}$$

$$\frac{25}{x} = \frac{5}{10}$$

$$25(10) = 5x$$ *(Cross multiply.)*

$$250 = 5x$$

$$\frac{250}{5} = x$$

$$50 = x$$

Thus, the smaller gear turns at 50 rpm. ■

Example 10-3.9 A small pulley that measures 6 in. in diameter (across) turns 500 rpm and drives a larger pulley at 300 rpm. What is the diameter of the larger pulley?

Solution: The pulleys are *inversely* related. The *faster* (smaller) pulley drives the *slower* (larger) pulley. Again, both ratios must be expressed in like measures and be in inverse order.

$$\frac{\text{500 rpm of smaller pulley}}{\text{300 rpm of larger pulley}} = \frac{x\text{-in. diameter of larger pulley}}{\text{6-in. diameter of smaller pulley}}$$

$$\frac{500}{300} = \frac{x}{6}$$

$$500(6) = 300x \qquad \textit{(Cross multiply.)}$$

$$3000 = 300x$$

$$\frac{3000}{300} = x$$

$$10 = x$$

Thus, the diameter of the larger pulley is 10 in. ■

Tips and Traps

When we encounter proportional relationships among quantities, we should first check to see whether the relationship is direct or inverse. If it is *direct*, we make the ratios equal to one another and keep the same order in both ratios. If it is *inverse*, we make the ratios equal to one another *but* use the same units in each ratio and express one ratio in inverse order to the other.

L.O.4. Solve problems that involve similar triangles.

Similar and Congruent Triangles

Many applications involving proportions are based on the properties of similar triangles. Before looking at applied problems using similar triangles, we will examine the definitions of similar and congruent triangles. *Congruent triangles* have the same size and shape. *Similar triangles* have the same shape but not the same size.

Every triangle has six parts, three angles and three sides. Each angle or side of one similar or congruent triangle has a corresponding angle or side in the other similar or congruent triangle. The symbol for showing congruency is ≅ and is read "is congruent to." The symbol for showing similarity is ~ and is read "is similar to."

Examine the following figures.

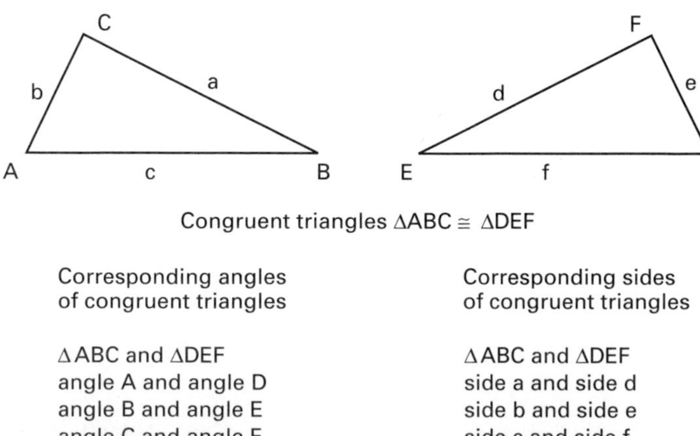

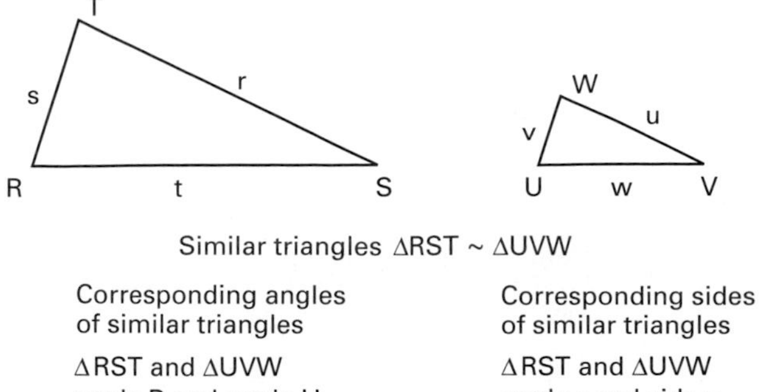

Similar triangles ΔRST ~ ΔUVW

Corresponding angles of similar triangles	Corresponding sides of similar triangles
ΔRST and ΔUVW	ΔRST and ΔUVW
angle R and angle U	angle r and side u
angle S and angle V	angle s and side v
angle T and angle W	angle t and side w

■ **Definition 10-3.1 Congruent Triangles.** Two triangles are **congruent** if they have the same size and shape. Each angle of one triangle is equal to its corresponding angle in the other triangle. Each side of one triangle is equal to its corresponding side in the other triangle.

■ **Definition 10-3.2 Similar Triangles.** Two triangles are **similar** if they have the same shape, but different size. Each angle of one triangle is equal to its corresponding angle in the other triangle. Each side of one triangle is directly proportional to its corresponding side in the other triangle.

In Fig. 10-3, $\triangle ABC$ is similar to $\triangle MNQ$. The angles that correspond are $\angle A$ and $\angle M$, $\angle B$ and $\angle N$, $\angle C$ and $\angle Q$. They correspond because each pair is equal. Since the triangles are similar, their corresponding sides are proportional. Corresponding sides are opposite equal angles. The corresponding sides in Fig. 10-3 are $\overline{BC}$ and $\overline{NQ}$, $\overline{AC}$ and $\overline{MQ}$, $\overline{AB}$ and $\overline{MN}$. Because the corresponding sides are proportional, their ratios are equal. For example, we can write the proportion

$$\frac{BC}{NQ} = \frac{AC}{MQ} = \frac{AB}{MN}$$

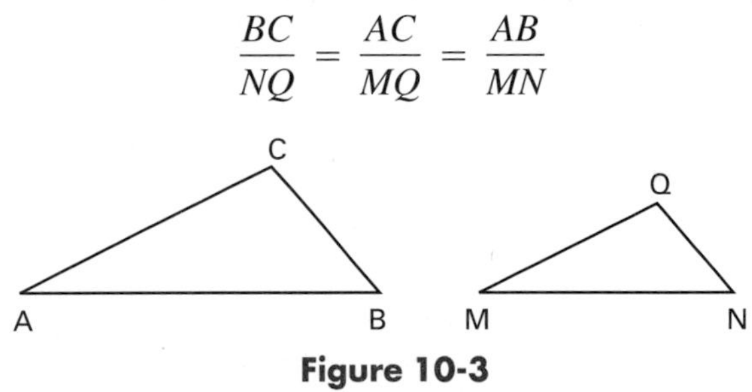

Figure 10-3

> **Rule 10-3.3**
>
> If two triangles are *similar*, the ratios of their corresponding sides are equal.

We can use this property of similar triangles to solve many problems.

Example 10-3.10 Find the missing side in Fig. 10-4 if $\triangle ABC \sim \triangle DEF$.

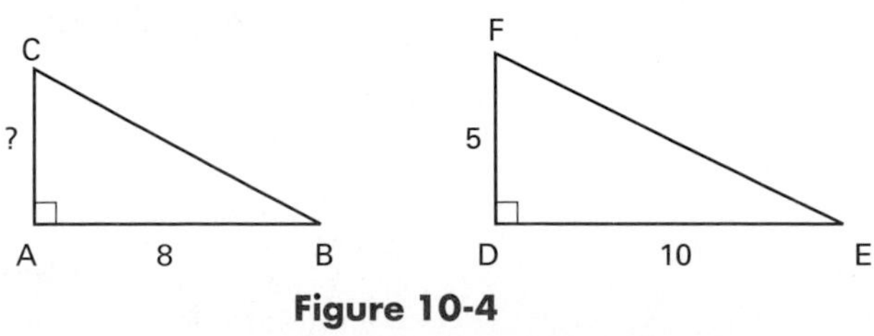

Figure 10-4

Solution

$\dfrac{b}{e} = \dfrac{c}{f}$ *(Write ratios of corresponding sides in a proportion. For convenience, single letters may be used to identify sides in this and similar proportions. Side a is opposite ∠A, side b is opposite ∠B, and so on.)*

$\dfrac{b}{5} = \dfrac{8}{10}$ *(Substitute values and find missing side b.)*

$10b = 8(5)$ *(Cross multiply.)*

$10b = 40$

$b = 4$ ■

Example 10-3.11 A tree surgeon must know the height of a tree to determine which way it should fall so that it would not endanger lives, traffic, or property. A 6-ft pole makes a 4-ft shadow on the ground when the tree makes a 20-ft shadow (Fig. 10-5). What is the height of the tree?

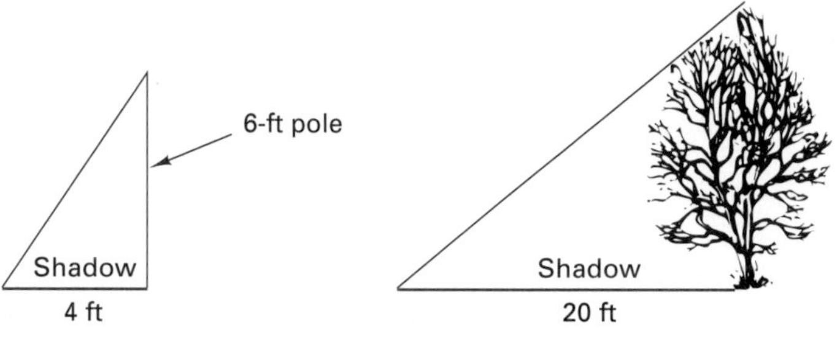

Figure 10-5

Solution: Since the triangles formed are similar, we can use a proportion to solve the problem. Let $x =$ height of tree.

$$\frac{x \text{ (height of tree)}}{6 \text{ (height of pole)}} = \frac{20 \text{ (shadow of tree)}}{4 \text{ (shadow of pole)}}$$

$4x = 6(20)$ *(Cross multiply.)*

$4x = 120$

$x = 30$

The tree is 30 ft tall. ■

Example 10-3.12 A building lies between points *A* and *B*, so the distance between these points cannot be measured directly by a surveyor. Find the distance using the similar triangles shown in Fig. 10-6.

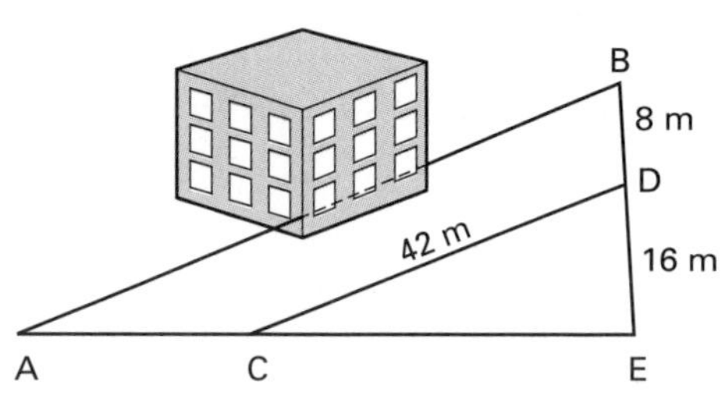

Figure 10-6

Solution: $\triangle ABE \sim \triangle CDE$. Note that CD must be made parallel to AB for the triangles to be similar.

Visualize the triangles as separate triangles.

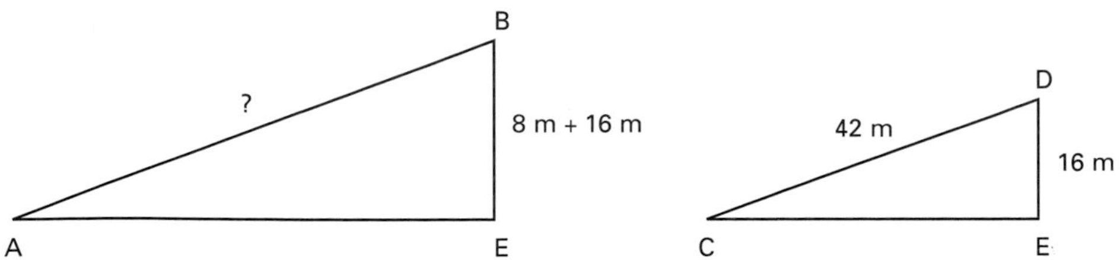

$$\frac{CD}{AB} = \frac{DE}{BE}$$

$$\frac{42 \text{ m}}{AB} = \frac{16 \text{ m}}{24 \text{ m}} \qquad (BE = 16 + 8 = 24)$$

$$16AB = 42(24)$$

$$16AB = 1008$$

$$AB = 63 \text{ m}$$

■

SELF-STUDY EXERCISES 10-3

L.O.1. Solve the following proportions.

1. $\frac{x}{5} = \frac{9}{15}$

2. $\frac{3x}{16} = \frac{3}{8}$

3. $\frac{x - 1}{x + 6} = \frac{4}{5}$

4. $\frac{5}{x} = 8$

5. $\frac{2}{7} = \frac{x - 4}{x + 3}$

6. $\frac{2x + 1}{8} = \frac{3}{7}$

7. $\frac{3x - 2}{3} = \frac{2x + 1}{3}$

8. $\frac{5}{2x - 2} = \frac{1}{8}$

9. $\frac{8}{3x + 2} = \frac{8}{14}$

10. $\frac{2x}{8} = \frac{3x + 1}{7}$

L.O.2. Solve the following problems using proportions.

11. If 7 cans of dog food sell for \$4.13, how much will 10 cans sell for?

12. If 6 cans of coffee sell for \$22.24, how much will 20 cans sell for?

13. A mechanic took 7 hours to tune up 9 fuel-injected engines. At this rate, how many fuel-injected engines can be tuned up in 37.5 hours? Round to the nearest whole number.

14. A costume maker took 9 hours to make 4 headpieces for a Mardi Gras ball. At this rate, how many complete headpieces can be made in 35 hours?

15. How far can a family travel in 5 days if it travels at the rate of 855 miles in 3 days?

16. How far can a tractor-trailer rig travel in 8 days if it travels at the rate of 1680 miles in 4 days?

17. How much crystallized insecticide does 275 acres of farmland need if the insecticide treats 50 acres per 100 pounds?

18. How much fertilizer does 2625 square feet of lawn need if the fertilizer treats 1575 square feet per gallon? Express answer to the nearest tenth of a gallon.

L.O.3. Solve the following problems using proportions.

19. The fan pulley and alternator pulley are connected by a fan belt on an automobile engine. The fan pulley is 225 cm in diameter and the alternator pulley is 125 cm in diameter. If the fan pulley turns at 500 rpm, how many revolutions per minute does the faster alternator pulley turn?

20. The volume of a certain gas is inversely proportional to the pressure on it. If the gas has a value of 160 in.^{3} under a pressure of 20 pounds per square inch (psi), what will the volume be if the pressure is decreased to 16 psi?

21. A pulley that measures 15 in. across (diameter) turns at 1600 rpm and drives a larger pulley at the rate of 1200 rpm. What is the diameter of the larger pulley in this inverse relationship?

22. Six painters can trim the exterior of all the new brick homes in a subdivision in 9 weeks. The contractor wants to have the homes ready in just 3 weeks and so needs more painters. How many painters are needed for the job? Assume that the painters all work at the same rate.

23. Two groundskeepers take 25 hr to prepare a golf course for a tournament. How long would it take 5 groundskeepers to prepare the golf course?

24. A gear measures 5 in. across. It turns another gear 2.5 in. across. If the larger gear has a speed of 25 revolutions per minute (rpm), what is the rpm of the smaller gear?

25. A small pulley 3 in. in diameter turns 250 revolutions per minute (rpm) and drives a larger pulley at 150 rpm. What is the diameter (distance across) of the larger pulley?

26. Two painters working at the same speed can paint 800 ft^2 of wall space in 6 hr. If a third painter paints at the same speed, how long would it take the three of them to paint the same wall space?

27. Three machines can complete a printing project in 5 hr. How many machines would be needed to finish the same project in 3 hr?

28. A gear measures 6 in. across. It is in mesh with another gear with a diameter of 3 in. If the larger gear has a speed of 60 revolutions per minute (rpm), what is the rpm of the smaller gear?

29. Harold can assemble parts in 30 min. If he gets help from assistants who work at his rate and together they complete the assembly in 6 min, how many helpers did he get?

30. A 4.5-in. pulley turning at 1000 revolutions per minute (rpm) is belted to a larger pulley turning at 500 rpm. What is the size of the larger pulley?

L.O.4. Write the corresponding parts not given for the congruent triangles in problems 31–33.

31.

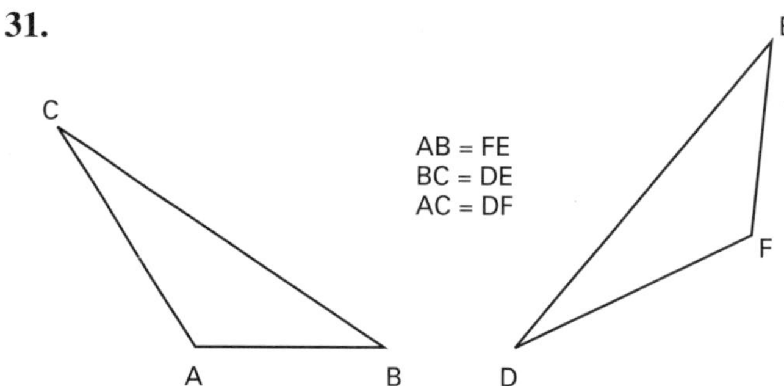

32.

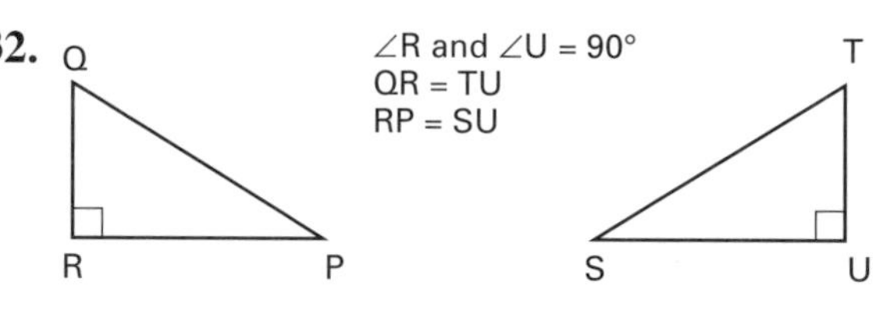

33.

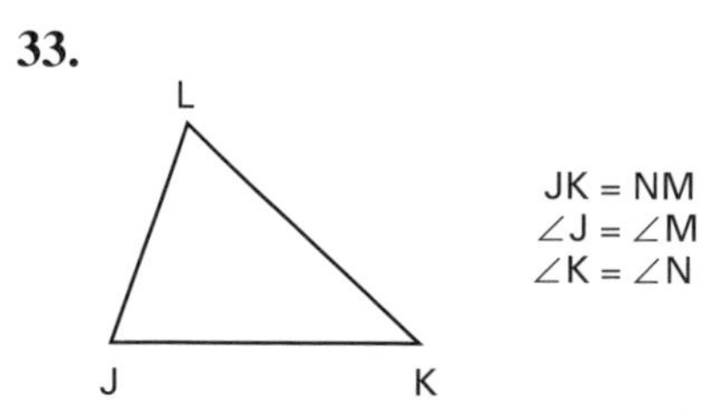

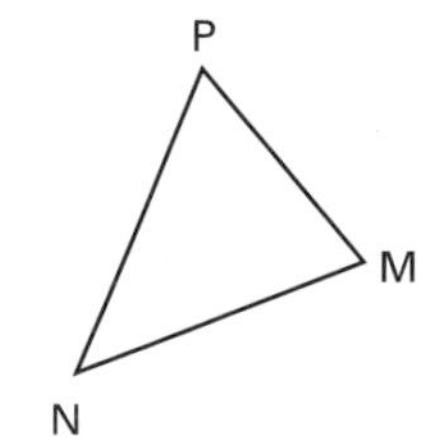

34. $\triangle ABC \sim \triangle DEF$. $\angle A = \angle E$, $\angle C = \angle F$. Find a and d. Use Fig. 10-7.

C, F, 15, 12, A, 10, B, D, 6, E

Figure 10-7

35. $\triangle ABC \sim \triangle DEC$. $\angle 1$ and $\angle 2$ have the same measure. Find DC and DE. (*Hint:* Let $DC = x$ and $AC = x + 3$. Use Fig. 10-8.)

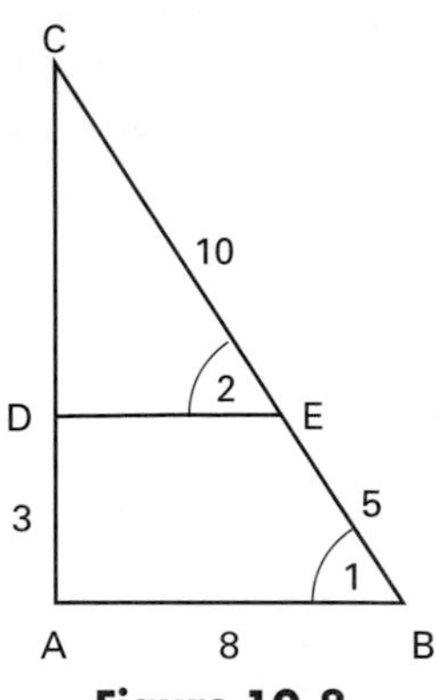

Figure 10-8

36. Find the height of a tree that makes a 30-ft shadow when a 6-ft 6-in. pole makes a shadow of 3 ft.

ELECTRONICS APPLICATION

Voltage-divider Equation

One handy equation often used in electronics is the voltage-divider equation. This enables you to directly calculate the voltage at a particular place in the circuit without first finding the current. With a simple circuit with one source, V_t, and two voltage drops, V_1 (across R_1) and V_2 (across R_2), the equation for V_2 is

$$V_2 = \frac{R_2 V_t}{R_1 + R_2}$$

To solve this equation for R_2, you must follow through several steps. First, eliminate denominators. Multiply both sides by the full denominator and then cancel. That gets everything on one line.

$$(R_1 + R_2)V_2 = \frac{(R_1 + R_2)R_2 V_t}{R_1 + R_2}$$

$$(R_1 + R_2)V_2 = R_2 V_t$$

Distribute and gather all terms that include R_2 on the right side of the equal sign and all terms that do not include R_2 on the left side of the equal sign. Then, since all the terms on the right of the equal sign contain R_2, the R_2 can be factored out.

$$R_1 V_2 + R_2 V_2 \qquad = R_2 V_t$$

$$R_1 V_2 = R_2(V_t - V_2) = R_2 V_t - R_2 V_2$$

Since the right side of the equal sign contains a factor times R_2, both sides of the equation can be divided by that factor.

$$\frac{R_1 V_2}{(V_t - V_2)} = \frac{R_2(V_t - V_2)}{(V_t - V_2)}; \frac{R_1 V_2}{(V_t - V_2)} = R_2$$

Leave the parentheses since they help us remember to use parentheses when using the calculator.

Does this equation work? Let's say that $R_1 = 51\ \Omega$, $V_2 = 7$ V, and $V_t = 10$ V. Then

$$R_2 = \frac{(51\ \Omega)(7\ \text{V})}{10\ \text{V} - 7\ \text{V}}$$

$$R_2 = 119\ \Omega$$

Volts cancel leaving only ohms. Now check in the original formula.

$$V_2 = \frac{R_2 V_t}{R_1 + R_2};\ 7\ \text{V} = \frac{(119\ \Omega)(10\ \text{V})}{51\ \Omega + 119\ \Omega} \qquad \textit{(correct)}$$

Exercises

For each of the following, first solve using letters to get the general equation. Then prove that the new equation is correct by using the given numbers to find a numerical answer. Prove the numerical answer obtained by putting the values back in the original equation.

1. Solve for R_1 in the equation $V_1 = \dfrac{R_1 V_t}{R_1 + R_2}$. What is R_1 if $V_1 = 7$ V, $V_t = 9$ V, and $R_2 = 8$ kΩ?
2. Solve for R_2 in the equation $V_2 = \dfrac{R_2 V_t}{R_1 + R_2}$. What is R_2 if $V_2 = 3$ V, $V_t = 8$ V, and $R_1 = 120\ \Omega$?

   Note: the next problems use the current-divider formula, which uses G (measured in siemens) and I (measured in amperes).

3. Solve for G_1 in the equation $I_1 = \dfrac{G_1 I_t}{G_1 + G_2}$. What is G_1 if $I_1 = 5$ mA, $I_t = 11$ mA, and $G_2 = 8$ mS?
4. Solve for G_2 in the equation $I_2 = \dfrac{G_2 I_t}{G_1 + G_2}$. What is G_2 if $I_2 = 15$ mA, $I_t = 35$ mA, and $G_1 = 4$ mS?

Answers for Exercises

1.

$$R_1 = \frac{R_2 V_1}{(V_t - V_1)};\ 28\ \text{k}\Omega = \frac{8\ \text{k}\Omega\ (7\ \text{V})}{(9\ \text{V} - 7\ \text{V})}$$

Proof: $$V_1 = \frac{R_1 V_t}{R_1 + R_2};\ 7\ \text{V} = \frac{(28\ \text{k}\Omega)(9\ \text{V})}{28\ \text{k}\Omega + 8\ \text{k}\Omega};$$

2.

$$R_2 = \frac{R_1 V_2}{(V_t - V_2)};\ 72\ \Omega = \frac{120\ \Omega\ (3\ \text{V})}{(8\ \text{V} - 3\ \text{V})}$$

Proof: $$V_2 = \frac{R_2 V_t}{R_1 + R_2};\ 3\ \text{V} = \frac{(72\ \text{Ohms})(8\ \text{V})}{120\ \text{Ohms} + 72\ \text{Ohms}};$$

3. $G_1 = \frac{G_2 I_1}{(I_t - I_1)}$; $6.67\text{ mS} = \frac{8\text{ mS }(5\text{ mA})}{(11\text{ mA} - 5\text{ mA})}$

Proof: $I_1 = \frac{G_1 I_t}{G_1 + G_2}$; $5\text{ mA} = \frac{(6.67\text{ mS})(11\text{ mA})}{6.67\text{ mS} + 8\text{ mS}}$;

4. $G_2 = \frac{G_1 I_2}{(V_t - I_2)}$; $3\text{ mS} = \frac{4\text{ mS}(15\text{ mA})}{(35\text{ mA} - 15\text{ mA})}$

Proof: $I_2 = \frac{G_2 I_t}{G_1 + G_2}$; $15\text{ mA} = \frac{(3\text{ mS})(35\text{ mA})}{3\text{ mS} + 4\text{ mS}}$;

ASSIGNMENT EXERCISES

10-1

Solve the following fractional equations.

1. $P = \frac{1}{2} + \frac{1}{3}$

2. $x + \frac{1}{7}x = 16$

3. $\frac{2}{5} - x = \frac{1}{2}x + \frac{4}{5}$

4. $\frac{R}{7} - 6 = -R$

5. $\frac{3}{7}m - \frac{1}{2} = \frac{2}{3}$

6. $0 = \frac{8}{9}c + \frac{1}{4}$

7. $\frac{2}{15}P - P = 4$

8. $\frac{1}{4}S = \frac{1}{4} + \frac{1}{10} + \frac{1}{20}$

9. $m = 2 + \frac{1}{4}m$

10. $\frac{7}{9} + 3 = \frac{1}{2}T$

11–15. Check or verify Exercises 2, 4, 6, 8, and 10.

Solve the following equations.

16. $6x + \frac{1}{4} = 5$

17. $2x + 4 = \frac{1}{2}$

18. $\frac{2}{5}x + 6 = \frac{2}{3}x$

19. $\frac{3}{10}x = \frac{1}{8}x + \frac{2}{5}$

20. $\frac{2}{5} + 3x = \frac{1}{10} - x$

21. $\frac{1}{x} = \frac{2}{5} + \frac{3}{10}$

22. $\frac{1}{R} = \frac{1}{10} + \frac{1}{3} + \frac{1}{6}$

23. $\frac{2}{P} = \frac{1}{2} + \frac{1}{4} - \frac{5}{12}$

24. $\frac{3}{x} + 4 = \frac{1}{5} - 7$

25. $\frac{5}{12}x - \frac{3}{4} = \frac{1}{9} - \frac{2}{3}x$

Set up a fractional equation for each of the following. Solve each equation.

26. A plastic tube fills a container in 10 min. A drain in the container, however, empties the container in 30 min. If the drain is open, how long would it take to fill the container?

27. Melissa can complete a project in 3 hr. Henry can complete the same project in 7 hr. How long would it take Melissa and Henry working together to complete the project?

28. One optical scanner can read a stack of sheets in 20 min. A second scanner can read the same stack in 12 min. How much time would it take for both scanners together to process the 1 stack of sheets?

29. An apprentice electrician can install 5 light fixtures in 2 hr. How many light fixtures can be installed in 10 hr?

30. A brick mason can erect a retaining wall in 6 hr. The brick mason's apprentice can do the same job in 10 hr. How much time would it take both of them working together to erect a retaining wall?

Solve the following equations.

31. $3.4 = 1.5 + T$

32. $2y + 2.9 = 11.7$

33. $2.3x - 4.1 = 0.5$

34. $0.22 + 1.6x = -0.9$

35. $6.8 = 0.2y - 8.64$

36. $1.4x - 7.2 = 3.5x - 4.3$

37. $0.3x - 2.15 = 0.8x + 3.75$

38. $1.3x + 2 = 8.6x - 3.24$

39. $2.7 - x = 5 + 2x$

40. $4x - 3.2 + x = 3.3 - 2.4x$

41. $6.7y - y = 8.4$ *(round to tenths)*

42. $0.9R = 0.3$ *(round to tenths)*

43. $\frac{4x}{0.7} = \frac{3}{1.2}$ *(round to hundredths)*

44. $0.86 = R + 0.4R$ *(round to hundredths)*

45. $0.04y = 0.02 - y$ *(round to hundredths)*

46. $\frac{x}{6} = \frac{1.8}{3}$

47. $\frac{2.1}{x} = \frac{4.3}{7}$ *(round to tenths)*

48. The distance formula is distance = rate × time. If a motorhome traveled 350.8 mi at 50 miles per hour (mph), how long to the nearest hour did the trip take?

49. Electrical resistance in ohms (Ω) is voltage (V) divided by amperes (A). Find the resistance of a small motor with a voltage of 8.5 volts requiring 0.5 amps.

50. Lester earned $29.69 working $4\frac{3}{4}$ hr at a department store. What was his hourly wage?

51. The formula for electrical power is $V = W/A$: voltage (V) equals wattage (W) divided by amperage (A) or $V = \frac{W}{A}$. Find the voltage to the nearest hundredth needed for a circuit of 385 W with a current of 3.5 A.

52. If the formula for simple interest is interest = principal × rate × time, find the interest to the nearest cent on a principal of $1000 at a 19.5% rate for 1.5 years.

53. If the formula for force is force = pressure × area, how many pounds of force are produced by a pressure of 30 psi on a piston whose surface area is 12.5 in.^{2}?

10-2

Solve the following proportions.

54. $\frac{x}{6} = \frac{5}{3}$

55. $\frac{3x}{8} = \frac{3}{4}$

56. $\frac{x - 3}{x + 6} = \frac{2}{5}$

57. $\frac{7}{x} = 6$

58. $\frac{2}{3} = \frac{x + 3}{x - 7}$

59. $\frac{4x + 3}{15} = \frac{1}{3}$

60. $\frac{3x - 2}{3} = \frac{2x + 1}{4}$

61. $\frac{5}{4x - 3} = \frac{3}{8}$

62. $\frac{8}{3x - 2} = \frac{2}{3}$

63. $\frac{4x}{7} = \frac{2x + 3}{3}$

64. $\frac{2x}{3x - 2} = \frac{5}{8}$

65. $\frac{5x}{3} = \frac{2x + 1}{4}$

66. $\frac{5}{9} = \frac{x}{2x - 1}$

67. $\frac{7}{x} = \frac{5}{4x + 3}$

68. $\frac{3}{5} = \frac{2x - 3}{7x + 4}$

10-3

Solve the following problems using proportions.

69. If 15 machines can complete a job in 6 weeks, how many machines are needed to complete the job in 4 weeks?

70. In preparing for a banquet for 30 people, Cedric Henderson uses 9 lb of potatoes. How many pounds of potatoes will be needed for a banquet for 175 people?

71. A car with a speed control device travels 100 mi at 50 mi per hour (mph). The trip takes 2 hr. If the car traveled at 40 mph, how much time would the driver need to reach the same destination?

72. A 6-ft person casts a 5-ft shadow on the ground. How tall is a nearby tree that casts a 30-ft shadow?

73. There are 25 women in a class of 35 students. If this is typical of all classes in the college, how many women are enrolled in the college if it has 6300 students altogether?

74. It takes five people 7 days to clear an acre of land of debris left by a tornado; inversely, more people can do the job in less time. How long would it take seven people all working at the same rate?

75. A gear whose diameter is 45 cm is in mesh with a gear whose diameter is 30 cm. If the larger gear turns at 1000 rpm, how many revolutions does the smaller, faster gear turn per minute?

76. A certain cloth sells at a rate of 3 yd for $7.00. How many yards can Clemetee Whaley buy for $35.00?

77. Three workers take 5 days to assemble a shipment of microwave ovens; inversely, more workers can do the job in less time. How long will it take five workers to do the same job?

78. An architect's drawing is scaled at $\frac{3}{4}$ in. = 6 ft. What is the actual height of a door that measures $\frac{7}{8}$ in. on the drawing?

79. A CAD program scales a blueprint so that $\frac{5}{8}$ inch = 2 feet. What is the actual measure of a wall that is shown as $1\frac{5}{16}$ inches on the blueprint?

80. A 10-in. pulley makes 900 revolutions every minute. It drives a larger pulley at 500 rpm. What is the diameter of the larger pulley in this inverse relationship?

81. On a recent trip, a mid-size automobile used 51.2 gal of unleaded gasoline. If the trip involved a total of 845 mi, how many gallons of gasoline would be used for a longer trip of 1350 mi? Round to tenths.

82. A pulley whose diameter is 3.5 in. is belted to a pulley whose diameter is 8.5 in. In this inverse relationship, if the smaller, faster pulley turns at the rate of 1200 rpm, what is the rpm of the slower pulley? Round to nearest whole number.

83. A contractor estimates that a painter can paint 300 ft^2 of wall space in 3.5 hr. How many hours should the contractor estimate for the painter to paint 425 ft^2? Express answer to nearest tenth.

84. A coffee company mixes 1.6 lb of chicory with every 3.5 lb of coffee. At this ratio, how many pounds of chicory are needed to mix with 2500 lb of coffee? Round to the nearest whole number.

85. A wire 825 ft long has a resistance of 1.983 ohms (Ω). How long is a wire of the same diameter if the resistance is 3.247 ohms? Round to the nearest whole foot.

86. The ceramic tile ordered for a job weighs 510 lb. If the average weight of the ceramic tile is 4.25 lb per square foot, how many square feet are estimated for this job?

87. A gear is 4 in. across. It turns a smaller gear 2 in. across. If the larger gear has a speed of 30 revolutions per minute (rpm), what is the rpm of the smaller gear?

88. A small pulley gear with a 6 in. diameter turns 350 revolutions per minute (rpm) and drives a larger pulley at 150 rpm. What is the diameter (distance across) of the larger pulley?

89. Two machines can complete a printing project in 6 hr. How many machines would be needed to finish the same project in 4 hr?

90. Waylon can install a hard drive in 10 PCs in 6 hr. He gets help from assistants who work at his rate and together they complete the installation in 2 hr. How many helpers did he get?

91. Write proportions for the similar triangles in Fig. 10-9.

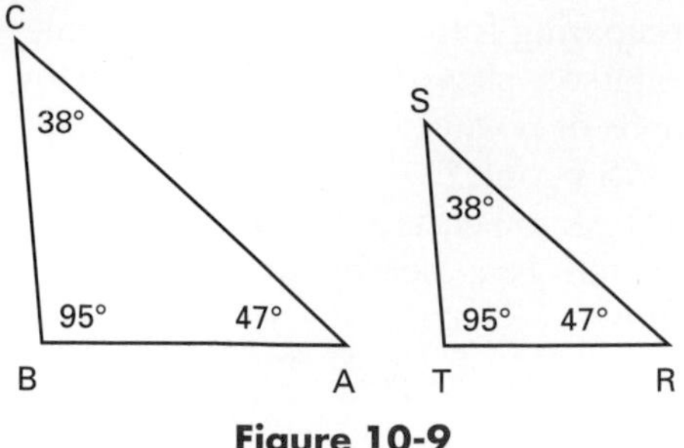

Figure 10-9

92. $\triangle LMN \sim \triangle XYZ$, $\angle L = \angle X$, $\angle M = \angle Y$. Find m and x (Fig. 10-10).

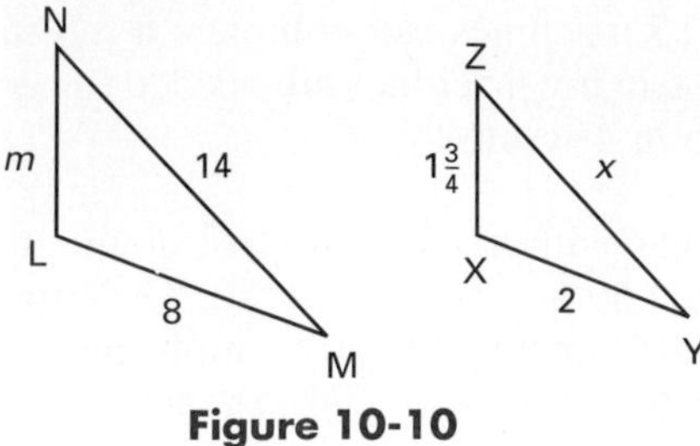

Figure 10-10

93. Find AB if $\triangle DEC \sim \triangle AEB$ (Fig. 10-11).

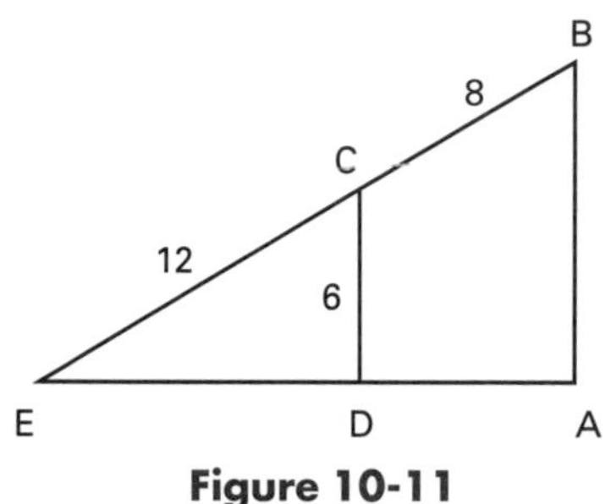

Figure 10-11

CHALLENGE PROBLEMS

94. Make up a word problem that can be solved with a direct proportion. Include a lawn mower, tanks of gasoline, and acres to be moved.

95. A gear turning at 130 rpm has 50 teeth. It is in mesh with another gear that turns at 65 rpm. How many teeth does the other gear have?

96. Make up a word problem that can be solved with an inverse proportion. Include a belt, pulleys, rpm's, and diameters of the pulleys.

97. The ratio of water to antifreeze in a mixture of radiator solution is 2 to 5. If the radiator is filled with 10 gallons of liquid, how much is water and how much is antifreeze?

CONCEPTS ANALYSIS

1. What is the rule for clearing an equation of fractions or denominators?
2. When the variable is in the denominator of a fraction, why should the root always be checked?
3. What is an extraneous root?
4. How can the LCM for a decimal equation be easily identified?
5. What is the rule for clearing an equation of decimals?
6. Is it always wise to clear an equation of decimals? Explain.
7. State the simple interest formula in words and symbols.
8. Express in symbols the property of a proportion.
9. Explain the difference between a direct and an inverse proportion.
10. Explain how to set up a direct proportion and an inverse proportion.
11. Give some examples of situations that are directly proportional.
12. Give some examples of situations that are inversely proportional.

Find a mistake in the following examples. Explain the mistake and correct it.

13.
$$\frac{2}{3}x = \frac{4}{5}$$
$$\left(\frac{3}{2}\right)\frac{2}{3}x = \left(\frac{4}{5}\right)\left(\frac{2}{3}\right)$$
$$x = \frac{8}{15}$$

14.
$$0.5x + x = 3.7$$
$$0.6x = 3.7$$
$$\frac{0.6x}{0.6} = \frac{3.7}{0.6}$$
$$x = 6.167 \quad \text{(rounded)}$$

15.
$$3(0.2x - 3.1) = 4.7$$
$$10[3(0.2x - 3.1)] = 10[4.7]$$
$$30(2x - 31) = 47$$
$$60x - 930 = 47$$
$$60x = 930 + 47$$
$$60x = 977$$
$$\frac{60x}{60} = \frac{977}{60}$$
$$x = 16.283 \quad \text{(rounded)}$$

CHAPTER SUMMARY

Objectives	What to Remember	Examples
Section 10-1 **1.** Solve fractional equations by clearing the denominators.	To clear the equation of denominators: multiply the *entire* equation by the least common multiple (LCM) of the denominators of the equation. If a variable is in a denominator, check the root to see if it makes a true statement and is not an *extraneous root.*	$\frac{3}{a} + \frac{1}{3} = \frac{2}{3a}$ LCM $= 3a$ $(3a)\left(\frac{3}{a}\right) + (3a)\left(\frac{1}{3}\right) = (3a)\left(\frac{2}{3a}\right)$ $\overset{3}{(\cancel{3a})}\left(\frac{3}{\underset{1}{\cancel{a}}}\right) + \overset{a}{(\cancel{3a})}\left(\frac{1}{\underset{1}{\cancel{3}}}\right) = \overset{1}{(\cancel{3a})}\left(\frac{2}{\underset{1}{\cancel{3a}}}\right)$ $9 + a = 2$ $a = 2 - 9$ $a = -7$ Check: $\frac{3}{-7} + \frac{1}{3} = \frac{2}{3(-7)} =$ $\frac{3}{-7} + \frac{1}{3} = \frac{2}{-21}$ LCM $= 21$ $(21)\left(\frac{3}{-7}\right) + (21)\left(\frac{1}{3}\right) = (21)\left(\frac{2}{-21}\right)$ $\overset{-3}{(\cancel{21})}\left(\frac{3}{\underset{1}{\cancel{-7}}}\right) + \overset{7}{(\cancel{21})}\left(\frac{1}{\underset{1}{\cancel{3}}}\right) = \overset{-1}{(\cancel{21})}\left(\frac{2}{\underset{1}{\cancel{-21}}}\right)$ $-9 + 7 = -2$
	The root makes a true statement.	$-2 = -2$

2. Solve work problems with fractional equations.	Amount of work: rate $\times$ time = amount of work A *rate* is a ratio such as 1 card per 15 min or $\frac{1 \text{ card}}{15 \text{ min}}$. Cancel units when solving.	Lashonda can install a PC video card in 15 min. How many cards can she install in $1\frac{1}{2}$ hours? $\text{rate} \times \text{time} = \begin{matrix}\text{amount}\\ \text{of work}\end{matrix}$ $\left(1\frac{1}{2}\text{ hr} = 90\text{ min}\right)$ $\frac{1\text{ card}}{15\text{ min}} \times 90\text{ min} = 6\text{ cards}$
	Completing one job when working together: (A's rate $\times$ time) + (B's rate $\times$ time) = 1 job Let T = time. Let 1 = 1 completed job.	Galenda assembles a product in 10 min and Marcus assembles the product in 15 min. How long will it take them together to assemble one product? Galenda's time $\times$ rate: $\frac{1}{10}T$ Marcus's time $\times$ rate: $\frac{1}{15}T$ $\frac{1}{10}T + \frac{1}{15}T = 1$ LCM = 30 $(30)\left(\frac{1}{10}T\right) + (30)\left(\frac{1}{15}T\right) = (30)(1)$ $(\overset{3}{\cancel{30}})\left(\frac{1}{\underset{1}{\cancel{10}}}T\right) + (\overset{2}{\cancel{30}})\left(\frac{1}{\underset{1}{\cancel{15}}}T\right) = (30)(1)$ $3T + 2T = 30$ $5T = 30$ $\frac{5T}{5} = \frac{30}{5}$ $T = 6\text{ min}$
	Completing one job when working in opposition: (A's rate $\times$ time) $-$ (B's rate $\times$ time) = 1 job Solve as above but *subtract* the two amounts of work.	An inlet valve can fill a vat in 2 hr. A drain valve can empty the vat in 5 hr. With both valves open, how long will it take for the vat to fill? Inlet valve's rate $\times$ time: $\frac{1}{2}T$ Drain. Valve's rate $\times$ time: $\frac{1}{5}T$

$$\frac{1}{2}T - \frac{1}{5}T = 1$$

$$\text{LCM} = 10$$

$$(10)\left(\frac{1}{2}T\right) - (10)\left(\frac{1}{5}T\right) = (10)1$$

$$\overset{5}{\cancel{(10)}}\left(\frac{1}{\underset{1}{\cancel{2}}}T\right) - \overset{2}{\cancel{(10)}}\left(\frac{1}{\underset{1}{\cancel{5}}}T\right) = (10)1$$

$$5T - 2T = 10$$

$$3T = 10$$

$$\frac{3T}{3} = \frac{10}{3}$$

$$T = 3\frac{1}{3}\text{ hr}$$

Section 10-2

1. Solve decimal equations by clearing decimals.

To clear the equation of decimals: multiply the *entire* equation by the least common multiple (LCM) of the fractional amounts following the decimal point. (The fractional amount with the most places after the decimal point will be the LCM.)

$$3.5x + 2.75 = 10$$

$$\text{LCM} = 100$$

$$(100)(3.5x) + (100)(2.75) = (100)(10)$$

$$350x + 275 = 1000$$

$$350x = 1000 - 275$$

$$350x = 725$$

$$\frac{350x}{350} = \frac{725}{350}$$

$$x = 2.071428571$$

$$x = 2.07 \quad \text{(rounded)}$$

2. Solve applied problems with decimal equations.

In many cases, decimal equations set up to solve problems may be treated as if the decimals were whole numbers.

John worked 2.5 hr at \$8.50 per hour. What was his pay? Let P = pay.

$$P = 2.5(8.50)$$

$$P = \$21.25$$

Interest formula: $I = PRT$, where I is interest, R is rate, P is principal, and T is time. If any three elements are given, the fourth one can be found. A calculator is useful in making the calculations.

Sequoia earned \$300 on an investment of \$1500 for 2 years. What interest rate was she paid?
Given: principal (\$1500), interest (\$300), and time (2 years). The rate (R) is missing.

$$I = PRT$$

$$300 = 1500 \times R \times 2$$

$$300 = 3000R$$

$$\frac{300}{3000} = \frac{3000R}{3000}$$

$$0.1 = R$$

Change decimal to a percent:
0.1 = 10% (Move decimal 2 places to right.)

Section 10-3

1. Solve fractional equations that are proportions	Property of proportions: If $\frac{a}{b} = \frac{c}{d}$, then $ad = bc$ $(b, d \neq 0)$ To solve: **1.** Find cross products. **2.** Divide both sides by coefficient of letter term.	$\frac{4}{2} = \frac{x}{6}$ $(2)(x) = (4)(6)$ $2x = 24$ $\frac{2x}{2} = \frac{24}{2}$ $x = 12$
2. Solve problems with direct proportions.	Verify problem involves direct proportion: as one amount increases (decreases), other amount increases (decreases). Set up and solve direct proportion: **1.** Write numerator and denominator of each ratio or fraction in *same* order. **2.** Make the two ratios or fractions equal to each other.	Kinta makes \$72 for 8 hr of work. How many hours must he work to make \$150? Proportion is direct: as hours increase, pay increases. $\frac{\$72}{8 \text{ hr}} = \frac{\$150}{x \text{ hr}}$ $\frac{72}{8} = \frac{150}{x}$ $(72)(x) = (8)(150)$ $72x = 1200$ $\frac{72x}{72} = \frac{1200}{72}$ $x = 16\frac{2}{3}$ hr
3. Solve problems with inverse proportions.	Verify problem involves inverse proportion: as one amount increases (decreases), other amount decreases (increases). Set up and solve inverse proportion: **1.** Write each ratio or fraction in *like* measures. **2.** Write each ratio or fraction in *inverse* order **3.** Make the two ratios or fractions equal to each other.	Jane can paint a room in 4 hr. If she has two helpers who also can paint a room in 4 hr. how long would it take all three to paint the same room? Proportion is inverse: as number of painters increase, time decreases. $\frac{1 \text{ painter}}{3 \text{ painters}} = \frac{X \text{ hr}}{4 \text{ hr}}$ (inverse ratio) $\frac{1}{3} = \frac{X}{4}$ $(3)(X) = (1)(4)$ $3X = 4$ $\frac{3X}{3} = \frac{4}{3}$ $X = 1\frac{1}{3}$ hr
4. Solve problems that involve similar triangles.	Each angle of one triangle is equal to its corresponding angle in the other triangle. Each side of one triangle is proportional to its corresponding side in the other triangle.	Find the height of a building that makes a shadow of 25 m when a meter stick makes a shadow of 1.5 m.

$$\frac{\text{x-m building}}{\text{25-m building shadow}} = \frac{\text{1-m stick}}{\text{1.5-m stick shadow}}$$

$$\frac{x}{25} = \frac{1}{1.5}$$

$$1.5\,x = 25$$

$$x = \frac{25}{1.5}$$

$$x = 16.67 \text{ m}$$

WORDS TO KNOW

extraneous root (p. 389)
least common multiple (LCM) (p. 391)
work rate (p. 393)
interest (p. 399)
principal (p. 399)
time (p. 399)
ratio (p. 403)
proportions (p. 403)
property of proportions (p. 403)
direct proportion (p. 405)
inverse proportion (p. 407)
similar triangles (p. 410)
congruent triangles (p. 410)

CHAPTER TRIAL TEST

Solve the following equations containing fractions.

1. $\frac{3}{8}y = 6$

2. $4 = \frac{1}{3}x + 2$

3. $\frac{3a}{7} = 9$

4. $\frac{R}{7} = \frac{2}{5}$

5. $\frac{3 + Q}{1} = \frac{4}{5}$

6. $\frac{3}{y + 2} = \frac{2}{3}$

7. $\frac{8}{y + 2} = -7$

8. $\frac{4}{5}z + z = 8$

9. $\frac{2}{7}x = \frac{1}{2}x + 4$

10. $-\frac{2}{3}x = \frac{1}{4}x - 11$

11. $5x + \frac{3}{5} = 2$

12. $3x + 2 = \frac{2}{3}$

13. $\frac{3}{5}x + \frac{1}{10}x = \frac{1}{3}$

14. $\frac{1}{x} = \frac{1}{3} + \frac{5}{6}$

Solve the following equations containing decimal numbers. Round to hundredths when necessary.

15. $1.3x = 8.02$

16. $4.5y + 1.1 = 3.6$

17. $\frac{1.2}{x} = 4.05$

18. $\frac{3.8}{6} = \frac{0.05}{R}$

19. $0.18x = 300 - x$

20. $4.3 = 7.6 + x$

21. $3x + 1.4 = 8.9$

22. $7.9 = 0.5x - 8.35$

23. $0.23 + 7.1x = -0.8$

Solve the following problems involving fractions, decimal numbers, and proportions.

24. A pipe can fill 1 tank in 4 hr. If a second pipe can empty 1 tank in 6 hr, how long will it take for the tank to fill with both pipes operating?

25. A 9-in. gear is in mesh with a 4-in. gear. If the larger gear makes 75 rpm, how many revolutions per minute does the smaller gear make in this inverse relationship?

26. One employee can wallpaper a room in 2 hr. Another employee can wallpaper the same room in 3 hr. How long will it take both employees to wallpaper the room when working together?

27. Using the formula, pressure $= \frac{\text{force}}{\text{area}}$, how much pressure does a force of 32.75 lb exert on a surface of 24.65 in.^{2} in a hydraulic system? Answer in pounds per square inch (psi) rounded to hundredths.

28. A one-year loan for the purchase of an electronically controlled assembly machine in a factory cost the management \$2758 in simple interest. If the interest rate was 19% how much did the machine cost (principal)? Round to the nearest dollar. Interest = principal × rate × time.

29. If a compact car used 62.5 L of unleaded gasoline to travel 400 mi, how many liters of gasoline would the driver use to travel 350 mi? Round to tenths.

30. If three workers take 8 days to complete a job, how many workers would be needed to finish the same job in only 6 days if each worked at the same rate? (More workers take fewer days.)

31. Resistance in a parallel dc circuit equals voltage divided by amperes. What is the resistance (in ohms) if the voltage is 40 V and amperage is 3.5 A? Express answer as a decimal rounded to thousandths.

32. The ratio of men to women in technical and trade occupations is estimated to be 3 to 1, that is, $\frac{3}{1}$. If 56,250 men are employed in such occupations in a certain city, how many employees are women?

33. If an ice maker produces 75 lb of ice in $3\frac{1}{2}$ hr, how many pounds of ice would it produce in 5 hr? Round to the nearest whole number.

34. Find HI if $\triangle ABC \sim \triangle GHI$ (Fig. 10-12).

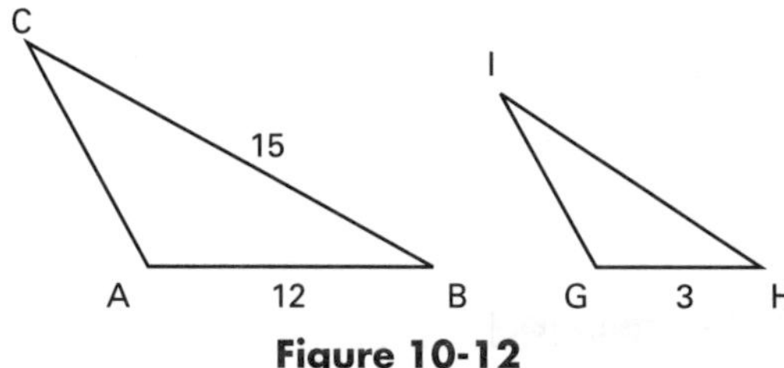

Figure 10-12

35. Find DB if $\triangle ABE \sim \triangle CDE$, $CD = 9$, $AB = 12$, $DE = 15$ (Fig. 10-13).

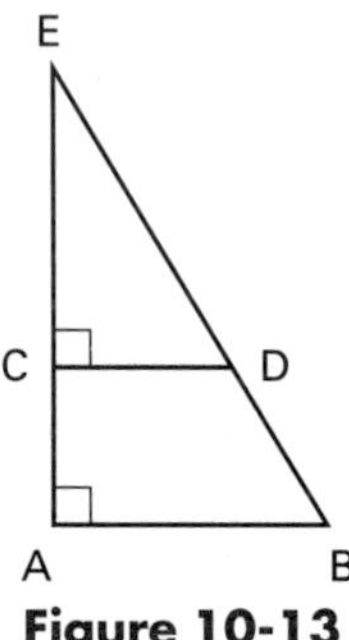

Figure 10-13

36. A model of a triangular part is shown in Fig. 10-14. Find side x if the part is to be similar to the model.

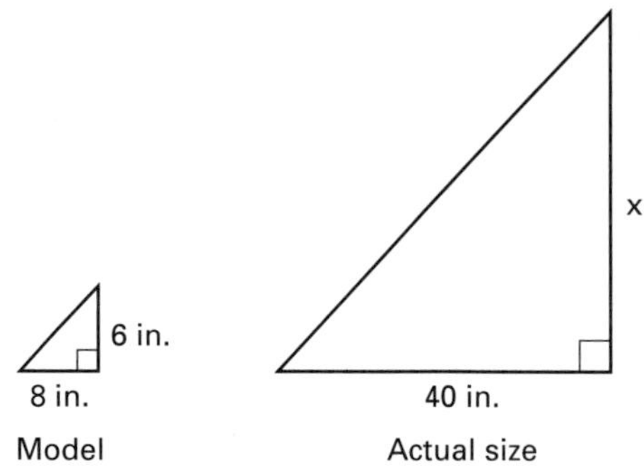

Figure 10-14

CHAPTER

11 Powers and Logarithms

In Chapter 1, we saw that repeated multiplication formed *powers*. In this chapter, we will extend our study of powers to include variable powers and logarithms. Powers are found in numerous technical applications that use scientific notation and occur frequently in formulas used to solve technical problems.

11-1 LAWS OF EXPONENTS

Learning Objectives

L.O.1. Find a specified power of natural numbers, whole numbers, integers, fractions, and decimals.
L.O.2. Multiply powers with like bases.
L.O.3. Divide powers with like bases.
L.O.4. Find a power of a power.

L.O.1. Find a specified power of natural numbers, whole numbers, integers, fractions, and decimals.

Powers

We have seen in various equations and formulas that a variable can be all types of numbers: natural numbers, whole numbers, integers, fractions, decimals, and signed numbers. When raising a variable to a power, we must consider all the types of numbers that the variable can be. Also, as we study other types of numbers, we will look at their powers.

EXAMPLE 11-1.1 Complete the chart by evaluating x^2, x^3, and x^4 for the following values of x.

Solution

x	x^2	x^3	x^4
5	$(5)(5) = 25$	$(5)(5)(5) = 125$	$(5)(5)(5)(5) = 625$
-3	$(-3)(-3) = 9$	$(-3)(-3)(-3) = -27$	$(-3)(-3)(-3)(-3) = 81$
$\frac{1}{3}$	$\left(\frac{1}{3}\right)\left(\frac{1}{3}\right) = \frac{1}{9}$	$\left(\frac{1}{3}\right)\left(\frac{1}{3}\right)\left(\frac{1}{3}\right) = \frac{1}{27}$	$\left(\frac{1}{3}\right)\left(\frac{1}{3}\right)\left(\frac{1}{3}\right)\left(\frac{1}{3}\right) = \frac{1}{81}$
0.7	$(0.7)(0.7) = 0.49$	$(0.7)(0.7)(0.7) = 0.343$	$(0.7)(0.7)(0.7)(0.7) = 0.2401$
-2.1	$(-2.1)(-2.1) = 4.41$	$(-2.1)(-2.1)(-2.1) = -9.261$	$(-2.1)(-2.1)(-2.1)(-2.1) = 19.4481$

■

General Tips for Using the Calculator

The general power key on a scientific or graphing calculator can be used to find a power of a number, for example, $(-2.1)^4$.

Scientific: 2.1 [+/−] [y^x] 4 [=] ⇒ 19.4481

Graphing: [(] [−] 2.1 [)] [y^x] 4 [EXE] ⇒ 19.4481 *(Parentheses are necessary.)*

As we begin our study of powers, we will first examine expressions with natural-number exponents. Then we will broaden our study to include exponents that are whole numbers, integers, fractions, decimals, and other types of numbers.

There are several laws of exponents that apply to terms that have *like* bases. The *laws of exponents* are contained in the *rules* and *definitions* presented in this section.

L.O.2. Multiply powers with like bases.

Multiplication of Powers with Like Bases

To multiply 2^3 by 2^2 using repeated multiplication, we have 2(2)(2) times 2(2) or 2(2)(2)(2)(2). Another way of writing this is $2^3 \times 2^2 = 2^5$.

We can multiply the terms x^4 and x^2. Even though the value of x is not known, $x^4(x^2)$ is $x(x)(x)(x)$ times $x(x)$ or $x(x)(x)(x)(x)(x)$.

$$x^4(x^2) = x^6$$

The following rule is a shortcut for using repeated multiplication.

Rule 11-1.1 ***To multiply powers that have like bases:***

The exponent of the product is the *sum* of the exponents of the powers and the base of the product remains unchanged. This rule can be stated symbolically as

$$a^m(a^n) = a^{m+n}$$

where a, m, and n are any types of numbers.

Example 11-1.2 Write the products of the following expressions.

(a) $y^4(y^3)$ **(b)** $a(a^2)$ **(c)** $b(b)$ **(d)** $x(x^3)(x^2)$ **(e)** $m(m)(m^2)$

Solution

(a) $y^4(y^3) = y^{4+3} = y^7$

(b) $a(a^2) = a^{1+2} = a^3$ *(Remember, when there is no written exponent, the exponent is 1.)*

(c) $b(b) = b^{1+1} = b^2$

(d) $x(x^3)(x^2) = x^{1+3+2} = x^6$

(e) $m(m)(m^2) = m^{1+1+2} = m^4$ ■

Tips and Traps

The previous rule for exponents applies *only* to expressions with *like* bases. Thus, $x^2(y^3)$ can only be written as x^2y^3. No other simplification can be made.

L.O.3. Divide powers with like bases.

Division of Powers with Like Bases

In reducing fractions, any factors that are common to both the numerator and denominator will cancel or reduce to a factor of 1. This concept is used when powers that have like bases are divided.

EXAMPLE 11-1.3 Reduce or simplify the following fractions. That is, perform the division indicated by each fraction.

(a) $\frac{x^5}{x^2}$ **(b)** $\frac{a^2}{a}$ **(c)** $\frac{b^7}{b^5}$ **(d)** $\frac{m^3}{m^3}$ **(e)** $\frac{y^3}{y^4}$

Solution

(a) $\frac{x^5}{x^2} = \frac{x(x)(x)(\not{x})(\not{x})}{\not{x}(\not{x})} = x^3$

(b) $\frac{a^2}{a} = \frac{a(\not{a})}{\not{a}} = a \quad (\text{or } a^1)$

(c) $\frac{b^7}{b^5} = \frac{b(b)(\not{b})(\not{b})(\not{b})(\not{b})(\not{b})}{(\not{b})(\not{b})(\not{b})(\not{b})(\not{b})} = b^2$

(d) $\frac{m^3}{m^3} = \frac{(\not{m})(\not{m})(\not{m})}{(\not{m})(\not{m})(\not{m})} = 1$ *(All factors of m reduce to 1.)*

(e) $\frac{y^3}{y^4} = \frac{(\not{y})(\not{y})(\not{y})}{(y)(\not{y})(\not{y})(\not{y})} = \frac{1}{y}$ *(The denominator has more factors of y than the numerator.)* ■

Rule 11-1.2 ***To divide powers that have like bases:***

The exponent of the quotient is the *difference* between the exponents of the dividend and divisor and the base of the quotient remains unchanged. This rule can be stated symbolically as

$$\frac{a^m}{a^n} = a^{m-n}, \qquad a \neq 0$$

Now let's apply this rule to the problems in Example 11-1.3.

(a) $\frac{x^5}{x^2} = x^{5-2} = x^3$

(b) $\frac{a^2}{a} = a^{2-1} = a^1 \quad \text{or} \quad a$

(c) $\frac{b^7}{b^5} = b^{7-5} = b^2$

(d) $\frac{m^3}{m^3} = m^{3-3} = m^0$

(e) $\frac{y^3}{y^4} = y^{3-4} = y^{-1}$

When Rule 11-1.2 is applied to Example 11-1.3, problem (d) results in a zero exponent. The *zero exponent* indicates a *zero power.*

■ **DEFINITION 11-1.1 Zero Exponent.** Any nonzero number with an exponent of zero is equal to 1.

$$a^0 = 1, \qquad a \neq 0$$

Tips and Traps

Looking at problem (d) of Example 11-1.3 we see that our definition of a zero exponent is consistent with the following rule, which we learned in arithmetic.

$$\frac{n}{n} = 1$$

Or, in words: Any nonzero number divided by itself is 1.

$$\frac{m^3}{m^3} = m^{3-3} = m^0 = 1$$

EXAMPLE 11-1.4 Reduce or simplify the following.

(a) $\frac{a^5}{a^5}$ **(b)** $\frac{x^2}{x^2}$ **(c)** $\frac{y}{y}$

Solution

(a) $\frac{a^5}{a^5} = a^{5-5} = a^0 = 1$

(b) $\frac{x^2}{x^2} = x^{2-2} = x^0 = 1$

(c) $\frac{y}{y} = y^{1-1} = y^0 = 1$ ■

When Rule 11-1.2 is applied to Example 11-1.3, problem (e) results in a negative exponent. The *negative exponent* indicates the reciprocal of the power.

■ **DEFINITION 11-1.2 Negative Exponent.** Any nonzero number with a **negative exponent** is equal to the reciprocal of that number with the positive exponent.

$$a^{-n} = \frac{1}{a^n}, \qquad a \neq 0$$

or

$$\frac{1}{a^{-n}} = a^n, \qquad a \neq 0$$

Tips and Traps

In a division of like bases, when the larger number of factors appears in the denominator, there will be factors still remaining in the denominator after reducing. The factors can be left in the denominator with a positive exponent $\left(\frac{1}{a^n}\right)$ or they can be written with a negative exponent (a^{-n}). That is,

$$\frac{y^3}{y^4} = y^{3-4} = y^{-1} \quad \text{or} \quad \frac{1}{y^1} \quad \text{or} \quad \frac{1}{y}$$

We generally prefer answers to be written with positive exponents.

Example 11-1.5 Write the following quotients using positive exponents.

(a) $\frac{x^5}{x^8}$ **(b)** $\frac{a}{a^4}$ **(c)** $\frac{b}{b^2}$ **(d)** $\frac{y^{-3}}{y^2}$ **(e)** $\frac{x^3}{x^{-5}}$

Solution

(a) $\frac{x^5}{x^8} = x^{5-8} = x^{-3} = \frac{1}{x^3}$

(b) $\frac{a}{a^4} = a^{1-4} = a^{-3} = \frac{1}{a^3}$

(c) $\frac{b}{b^2} = b^{1-2} = b^{-1} = \frac{1}{b}$

(d) $\frac{y^{-3}}{y^2} = y^{-3-2} = y^{-5} = \frac{1}{y^5}$

(e) $\frac{x^3}{x^{-5}} = x^{3-(-5)} = x^{3+5} = x^8$ ■

Tips and Traps

Again, the inverse relationship of multiplication and division will allow us flexibility in applying the laws of exponents.

Look again at Example 11-1.5a.

$$\frac{x^5}{x^8} \text{ is } x^5 \div x^8 \text{ or } x^5 \cdot \frac{1}{x^8} \text{ or } x^5 \cdot x^{-8}$$

Then, the multiplication law of exponents can be used.

$$x^5 \cdot x^{-8} = x^{5+(-8)} = x^{-3}$$

A factor can be moved from a numerator to a denominator (or vice versa) by changing the sign of the exponent.

$$x^{-2} = \frac{1}{x^2}, \quad \frac{1}{x^{-3}} = x^3$$

This property DOES NOT APPLY to a term that is part of a numerator or denominator.

$$\frac{x^{-2}+1}{3} \text{ DOES NOT EQUAL } \frac{1}{3x^2}.$$

$$\frac{x^{-2}+1}{3} = \frac{x^{-2}}{3} + \frac{1}{3} = \frac{1}{3x^2} + \frac{1}{3}$$

L.O.4. Find a power of a power.

Powers of Powers

In the term $(2^3)^2$, we have a power raised to a power. 2^3 is the base and 2 is the exponent.

$$(2^3)^2 \text{ is } (2^3)(2^3) = 2^{3+3} = 2^6 \text{ or } 64$$

Let's look at some other examples of raising powers to a power.

Example 11-1.6 Find the following powers of powers by first expressing the term as repeated multiplication and then applying the multiplication law of exponents.

(a) $(3^2)^3$ **(b)** $(x^3)^4$ **(c)** $(a^2)^2$ **(d)** $(y^3)^2$ **(e)** $(n^3)^5$

Solution

(a) $(3^2)^3 = (3^2)(3^2)(3^2) = 3^{2+2+2} = 3^6 = 729$
(b) $(x^3)^4 = (x^3)(x^3)(x^3)(x^3) = x^{3+3+3+3} = x^{12}$
(c) $(a^2)^2 = (a^2)(a^2) = a^{2+2} = a^4$
(d) $(y^3)^2 = (y^3)(y^3) = y^{3+3} = y^6$
(e) $(n^3)^5 = (n^3)(n^3)(n^3)(n^3)(n^3) = n^{3+3+3+3+3} = n^{15}$ ■

From Example 11-1.6 we can see a pattern developing. This leads us to the following rule.

Rule 11-1.3 *To raise a power to a power:*

Multiply exponents and keep the same base.

$$(a^m)^n = a^{mn}$$

Applying this rule to the problems in Example 11-1.6, we have

(a) $(3^2)^3 = 3^{2(3)} = 3^6 = 729$
(b) $(x^3)^4 = x^{3(4)} = x^{12}$
(c) $(a^2)^2 = a^{2(2)} = a^4$
(d) $(y^3)^2 = y^{3(2)} = y^6$
(e) $(n^3)^5 = n^{3(5)} = n^{15}$

There are other laws of exponents that are extensions or applications of the three laws we have already examined. While these additional laws are not necessary, they are useful tools for simplifying expressions containing exponents. These laws are expressed in the rules that follow.

Rule 11-1.4 *To raise a fraction or quotient to a power:*

Raise both the numerator and denominator to that power.

$$\left(\frac{a}{b}\right)^n = \frac{a^n}{b^n}$$

EXAMPLE 11-1.7 Raise the following fractions to the indicated power.

(a) $\left(\frac{2}{3}\right)^2$ **(b)** $\left(\frac{x}{y^3}\right)^3$ **(c)** $\left(\frac{-3}{4}\right)^2$ **(d)** $\left(\frac{-1}{3}\right)^3$

Solution

(a) $\left(\frac{2}{3}\right)^2 = \frac{2^2}{3^2} = \frac{4}{9}$

(b) $\left(\frac{x}{y^3}\right)^3 = \frac{x^3}{(y^3)^3} = \frac{x^3}{y^9}$

(c) $\left(\frac{-3}{4}\right)^2 = \frac{(-3)^2}{4^2} = \frac{9}{16}$ $[(-3)(-3) = +9]$

(d) $\left(\frac{-1}{3}\right)^3 = \frac{(-1)^3}{3^3} = \frac{-1}{27}$ $\quad [(-1)(-1)(-1) = -1]$ ■

Rule 11-1.5 *To raise a product to a power:*

Raise each factor to the indicated power.

$$(ab)^n = a^n b^n$$

EXAMPLE 11-1.8 Raise the following products to the indicated powers.

(a) $(ab)^2$ **(b)** $(a^2b)^3$ **(c)** $(xy^2)^2$ **(d)** $(3x)^2$ **(e)** $(2x^2y)^3$ **(f)** $(-5xy)^2$

Solution

(a) $(ab)^2 = a^{1(2)}b^{1(2)} = a^2b^2$
(b) $(a^2b)^3 = a^{2(3)}b^{1(3)} = a^6b^3$
(c) $(xy^2)^2 = x^{1(2)}y^{2(2)} = x^2y^4$
(d) $(3x)^2 = 3^{1(2)}x^{1(2)} = 3^2x^2 = 9x^2$
(e) $(2x^2y)^3 = 2^{1(3)}x^{2(3)}y^{1(3)} = 2^3x^6y^3 = 8x^6y^3$
(f) $(-5xy)^2 = (-5)^{1(2)}x^{1(2)}y^{1(2)} = (-5)^2x^2y^2 = 25x^2y^2$ ■

In applying the laws of exponents, it is very important to recognize important differences.

Tips and Traps

It is very important to understand what the laws of exponents *do not* include.

The laws apply to factors.

$$(a + b)^3 \text{ DOES NOT EQUAL } a^3 + b^3$$

The laws apply to like bases.

$$a^2(b^3) \text{ DOES NOT EQUAL } ab^5 \text{ or } (ab)^5$$

An exponent only affects the one factor or grouping immediately to the left.

$$3x^2 \text{ and } (3x)^2 \text{ ARE NOT EQUAL.}$$

In the term $3x^2$, the numerical coefficient 3 is multiplied times the square of x. In the term $(3x)^2$, $3x$ is squared. Thus, $(3x)^2 = 3^2x^2 = 9x^2$.

A negative coefficient of a base is not affected by the exponent.

$$-x^3 \text{ means } -(x)(x)(x)$$

If $x = 4$, $-x^3 = -(4)^3$ or $-(64) = -64$.
If $x = -4$, $-x^3 = -(-4)^3$ or $-(-64) = 64$.

SELF-STUDY EXERCISES 11-1

L.O.1. Evaluate the following powers:

1. 5^3
2. $\left(\frac{2}{5}\right)^2$
3. $(-3)^4$
4. $(0.3)^2$
5. $(-1.2)^3$
6. $(-8)^3$

L.O.2. Write the products of the following.

7. $x^3(x^4)$ **8.** $m(m^3)$ **9.** $a(a)$
10. $x^2(x^3)(x^5)$ **11.** $y(y^2)(y^3)$ **12.** $a^2(b)$

L.O.3. Write the quotients of the following. Write answers with positive exponents.

13. $\frac{y^7}{y^2}$ **14.** $\frac{x^5}{x}$ **15.** $\frac{a^3}{a^4}$
16. $\frac{b^6}{b^5}$ **17.** $\frac{m^2}{m^2}$ **18.** $\frac{x}{x^3}$
19. $\frac{y^5}{y}$ **20.** $\frac{n^2}{n^{-5}}$ **21.** $\frac{x^{-4}}{x^6}$
22. $\frac{n^2}{n^3}$ **23.** $\frac{x^7}{x^0}$ **24.** $\frac{x^{-3}}{x^2}$
25. $\frac{x^{-4}}{x^{-2}}$

L.O.4. Raise the following terms to the indicated power.

26. $(x^3)^2$ **27.** $(y^7)^0$ **28.** $\left(-\frac{1}{2}\right)^3$
29. $\left(-\frac{2}{7}\right)^2$ **30.** $\left(\frac{a}{b}\right)^4$ **31.** $(2m^2n)^3$
32. $\left(\frac{x^2}{y}\right)^3$ **33.** $\left(\frac{3}{5}\right)^2$ **34.** $(-2a)^2$
35. $(x^2y)^3$

11-2 BASIC OPERATIONS WITH ALGEBRAIC EXPRESSIONS CONTAINING POWERS

Learning Objectives

L.O.1. Add or subtract like terms containing powers.
L.O.2. Multiply or divide algebraic expressions containing powers.

Adding and Subtracting Like Terms

L.O.1. Add or subtract like terms containing powers.

Now that variables with exponents have been introduced, we must broaden our concept of *like terms.* For letter terms to be like terms, the letters as well as the exponents of the letters must be exactly the same. That is, $2x^2$ and $-4x^2$ are like terms because the x's are both squared. But $2x^2$ and $4x$ are not like terms because the x's do not have the same exponents.

EXAMPLE 11-2.1 Tell whether the following pairs of terms are like terms.

(a) $3a^2b$ and $-\frac{2}{3}a^2b$ **(b)** $-8xy^2$ and $7x^2y$
(c) $10ab^2$ and $(-2ab)^2$ **(d)** $5x^4y^3$ and $-2y^3x^4$
(e) $2x^2$ and $3x^3$

Solution

(a) $3a^2b$ and $-\frac{2}{3}a^2b$ are like terms. All letters and their exponents are the same.
(b) $-8xy^2$ and $7x^2y$ are *not* like terms. In $-8xy^2$ the exponent of x is 1 and in $7x^2y$ the exponent of x is 2. Also, the exponents of y are not the same.
(c) $10ab^2$ and $(-2ab)^2$ are *not* like terms. In the term $10ab^2$ only the b is squared. In $(-2ab)^2$ the entire term is squared and the result of the squaring is $4a^2b^2$.
(d) $5x^4y^3$ and $-2y^3x^4$ are like terms. Both x factors have an exponent of 4 and both y factors have an exponent of 3. Since multiplication is commutative, the order of factors does not matter.
(e) $2x^2$ and $3x^3$ are *not* like terms. The exponents of x are not the same. ■

When an algebraic expression contains several terms, we simplify the expression as much as possible by combining any like terms.

Rule 11-2.1

To combine like terms, combine the coefficients. The letter factors and exponents do not change.

EXAMPLE 11-2.2 Simplify the following algebraic expressions by combining like terms.

(a) $5x^3 + 2x^3$ **(b)** $x^5 - 4x^5$ **(c)** $a^3 + 4a^2 + 3a^3 - 6a^2$
(d) $3m + 5n - m$ **(e)** $y + y - 5y^2 + y^3$

Solution

(a) $5x^3 + 2x^3 = 7x^3$ *($5x^3$ and $2x^3$ are like terms; therefore, add the coefficients 5 and 2. The answer will have the same letter factor and exponent as the like terms.)*

(b) $x^5 - 4x^5 = -3x^5$ *(x^5 and $-4x^5$ are like terms. $1 - 4 = -3$. The answer will have the same letter factor and exponent as the like terms.)*

(c) $a^3 + 4a^2 + 3a^3 - 6a^2 = 4a^3 - 2a^2$ *(a^3 and $3a^3$ are like terms.) ($4a^2$ and $-6a^2$ are like terms.)*

(d) $3m + 5n - m = 2m + 5n$ *($3m$ and $-m$ are like terms.)*

(e) $y + y - 5y^2 + y^3 = 2y - 5y^2 + y^3$ *(y and y are like terms.)* ■

When an algebraic expression contains a grouping that is immediately preceded by a minus sign, this means we are subtracting the entire grouping. Recall that the grouping is multiplied by -1, the implied coefficient of the grouping. This causes *each* sign within the grouping to be changed to its opposite and at the same time removes the parentheses. If the grouping is preceded by a plus sign or an unexpressed positive sign, parentheses are removed without changing signs, as if each term in the grouping were multiplied by $+1$, the implied coefficient.

EXAMPLE 11-2.3 Simplify the following expressions.

(a) $y^2 + 2y - (3y^2 + 5y)$
(b) $(4ab - 5) - (2ab - 6)$
(c) $(2x^2 + x) + (3x^2 - 2x) - (x^2 + 3x)$
(d) $(m^3 - 3m^2 - 5m + 4) - (4m^3 - 2m^2 - 5m + 2)$

Solution

(a) $y^2 + 2y - (3y^2 + 5y) =$ *(Distribute implied coefficient of −1.)*

$y^2 + 2y - 1(3y^2 + 5y) =$

$y^2 + 2y - 3y^2 - 5y =$ *(Combine like terms.)*

$-2y^2 - 3y$

(b) $(4ab - 5) - (2ab - 6) =$ *(Distribute implied coefficient of −1.)*

$(4ab - 5) - 1(2ab - 6) =$

$4ab - 5 - 2ab + 6 =$ *(Combine like terms.)*

$2ab + 1$

(c) $(2x^2 + x) + (3x^2 - 2x) - (x^2 + 3x) =$ *(Distribute implied coefficient −1.)*

$(2x^2 + x) + (3x^2 - 2x) - 1(x^2 + 3x) =$

$2x^2 + x + 3x^2 - 2x - x^2 - 3x =$ *(Combine like terms.)*

$4x^2 - 4x$

(d) $(m^3 - 3m^2 - 5m + 4) - (4m^3 - 2m^2 - 5m + 2) =$ *(Distribute implied coefficient of −1.)*

$(m^3 - 3m^2 - 5m + 4) - 1(4m^3 - 2m^2 - 5m + 2) =$

$m^3 - 3m^2 - 5m + 4 - 4m^3 + 2m^2 + 5m - 2 =$ *(Combine like terms.)*

$-3m^3 - m^2 + 2$ ■

L.O.2. Multiply or divide algebraic expressions containing powers.

Multiplying and Dividing Algebraic Expressions Containing Powers

Rule 11-2.1 *To multiply or divide algebraic expressions:*

1. The coefficients are multiplied or divided using the rules for signed numbers.
2. The letter factors are multiplied or divided using the laws of exponents.

Look at the following examples.

Example 11-2.4 Multiply or divide as indicated. Express answers with positive exponents.

(a) $(4x)(3x^2)$ **(b)** $(-6y^2)(2y^3)$ **(c)** $(-a)(-3a)$ **(d)** $\frac{2x^4}{x}$

(e) $\frac{-6y^5}{2y^3}$ **(f)** $\frac{-4x}{4x^2}$ **(g)** $\frac{-5x^4}{15x^2}$ **(h)** $\frac{3x}{12x^3}$

Solution

(a) $(4x)(3x^2) = 4(3)(x^{1+2}) = 12x^3$
(b) $(-6y^2)(2y^3) = -6(2)(y^{2+3}) = -12y^5$
(c) $(-a)(-3a) = -1(-3)(a^{1+1}) = 3a^2$
The coefficient of $-a$ is -1. The exponent of $-a$ is 1.

(d) $\frac{2x^4}{x} = \frac{2}{1}(x^{4-1}) = 2x^3$

The coefficient of x is 1. The coefficients of the letter factors are divided using the rules for dividing signed numbers. The exponent of x is 1. The letter factors are divided using the laws of exponents.

(e) $\frac{-6y^5}{2y^3} = \left(\frac{-6}{2}\right)(y^{5-3}) = -3y^2$

(f) $\frac{-4x}{4x^2} = \left(\frac{-4}{4}\right)(x^{1-2}) = -1x^{-1} \quad \text{or} \quad -x^{-1} = -\frac{1}{x}$

This should be written as $-\frac{1}{x}$ or $\frac{-1}{x}$ so as to have a positive exponent.

(g) $\frac{-5x^4}{15x^2} = \left(\frac{-5}{15}\right)(x^{4-2}) = -\frac{1}{3}x^2 \quad \text{or} \quad -\frac{x^2}{3}$

Remember that $\frac{-1}{3}x^2$ is the same as $-\frac{1}{3}\left(\frac{x^2}{1}\right) = -\frac{x^2}{3}$.

(h) $\frac{3x}{12x^3} = \left(\frac{3}{12}\right)(x^{1-3}) = \frac{1}{4}x^{-2} = \frac{1}{4x^2}$

Since $\frac{1}{4}x^{-2} = \frac{1}{4}\left(\frac{1}{x^2}\right)$, this should be written as $\frac{1}{4x^2}$. ■

If factors are to be multiplied by more than one term, the distributive principle is applied.

EXAMPLE 11-2.5 Perform the multiplications.

(a) $2x(x^2 - 4x)$ (b) $-2y^3(2y^2 + 5y - 6)$ (c) $4a(3a^3 - 2a^2 - a)$

Solution

(a) $2x(x^2 - 4x) = 2x^3 - 8x^2$

(b) $-2y^3(2y^2 + 5y - 6) = -4y^5 - 10y^4 + 12y^3$

(c) $4a(3a^3 - 2a^2 - a) = 12a^4 - 8a^3 - 4a^2$ ■

When more than one term is divided by a single term, we must divide *each* term in the dividend (numerator) by the divisor (denominator). Actually, each term in the numerator represents the numerator of a separate fraction with the given denominator.

EXAMPLE 11-2.6 Perform the divisions.

(a) $\frac{4x^2 - 2x + 8}{2}$ (b) $\frac{18a^4 + 15a^3 - 9a^2 - 12a}{3a}$

(c) $\frac{3x^3 - x^2}{x^3}$ (d) $\frac{6x^3 + 2x^2}{2x^2}$

Solution

(a) $\frac{4x^2 - 2x + 8}{2} = \frac{\overset{2}{\cancel{4}}x^2}{\underset{1}{\cancel{2}}} - \frac{\overset{1}{\cancel{2}}x}{\underset{1}{\cancel{2}}} + \frac{\overset{4}{\cancel{8}}}{\underset{1}{\cancel{2}}} = 2x^2 - x + 4$

(b) $\dfrac{18a^4 + 15a^3 - 9a^2 - 12a}{3a} = \dfrac{18a^4}{3a} + \dfrac{15a^3}{3a} - \dfrac{9a^2}{3a} - \dfrac{12a}{3a}$
$= 6a^3 + 5a^2 - 3a - 4$

(c) $\dfrac{3x^3 - x^2}{x^3} = \dfrac{3x^3}{x^3} - \dfrac{x^2}{x^3} = 3 - x^{-1}$ or $3 - \dfrac{1}{x}$

(d) $\dfrac{6x^3 + 2x^2}{2x^2} = \dfrac{6x^3}{2x^2} + \dfrac{2x^2}{2x^2} = 3x + 1$ ■

SELF-STUDY EXERCISES 11-2

L.O.1. Simplify the following expressions.

1. $3a^2 + 4a^2$
2. $5x^3 - 2x^3$
3. $b^2 + 3a^2 + 2b^2 - 5a^2$
4. $3a - 2b - a$
5. $x - 3x - 2x^2 - 3x^2$
6. $3a^2 - 2a^2 + 4a^2$
7. $x^2 + 3y - (2x^2 + 5y)$
8. $4m^2 - 2n^2 - (2m^2 - 3n^2)$
9. $7a + 3b + 8c + 2a - (b + 2c)$
10. $5x + 3y + (7x - 2z)$

L.O.2. Perform the multiplication or division as indicated.

11. $7x(2x^2)$
12. $(-2m)(-m^2)$
13. $(-3m)(7m)$
14. $(-y^3)(2y^3)$
15. $\dfrac{6x^4}{3x^2}$
16. $\dfrac{-5a^2}{10a}$
17. $\dfrac{-7x}{-14x^3}$
18. $\dfrac{-9x^5}{12x^2}$
19. $3x(x - 6)$
20. $4x(3x^2 - 7x + 8)$
21. $-4x(2x - 3)$
22. $2x^2(5 + 2x)$
23. $\dfrac{6x^2 - 4x}{2x}$
24. $\dfrac{12x^5 - 6x^3 - 3x^2}{3x^2}$
25. $\dfrac{7x^4 - x^2}{x^3}$
26. $\dfrac{8x^4 + 6x^3}{2x^2}$

11-3 POWERS OF 10 AND SCIENTIFIC NOTATION

Learning Objectives

L.O.1. Multiply and divide by powers of 10.
L.O.2. Change a number from scientific notation to ordinary notation.
L.O.3. Change a number from ordinary notation to scientific notation.
L.O.4. Multiply and divide numbers in scientific notation.

L.O.1. Multiply and divide by powers of 10.

Powers of 10

We have learned that our number system is based on the number 10. That is, each place value in our number system is a *power of 10*. Look at the place-value chart (Fig. 11-1) to see how our number system relates to powers of 10.

The ones place is 10^0. This is consistent with the definition of zero exponents (Definition 11-1.1). The tenths place is $\dfrac{1}{10}$ or 10^{-1}, which is consistent with the definition of negative exponents (Definition 11-1.2).

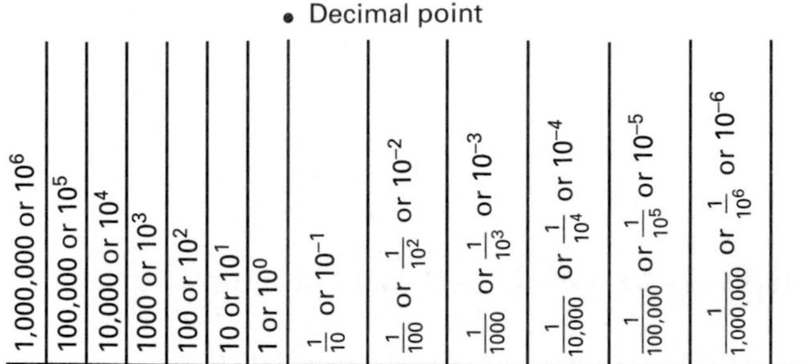

Figure 11-1

Notice the relationship between the exponent of 10 and the number of zeros in the ordinary number.

$$1{,}000{,}000 = 10^6$$
$$100{,}000 = 10^5$$
$$10{,}000 = 10^4$$
$$1000 = 10^3$$
$$100 = 10^2$$
$$10 = 10^1$$
$$1 = 10^0$$
$$\frac{1}{10} = 10^{-1}$$
$$\frac{1}{100} = 10^{-2}$$
$$\frac{1}{1000} = 10^{-3}$$
$$\frac{1}{10{,}000} = 10^{-4}$$
$$\frac{1}{100{,}000} = 10^{-5}$$
$$\frac{1}{1{,}000{,}000} = 10^{-6}$$

That is, the absolute value of the exponent equals the number of zeros in the ordinary number.

We learned in arithmetic that we could quickly multiply or divide by 10, 100, 1000, and so on, by shifting the decimal point. When multiplying or dividing by a power of 10, the exponent of 10 tells how many places the decimal is to be shifted and in what direction.

Rule 11-3.1 *To multiply by a power of 10:*

Shift the decimal point to the *right* the number of places indicated by a *positive* exponent, or to the *left* the number of places indicated by a *negative* exponent.

Rule 11-3.2 *To divide by a power of 10:*

Change the division to an equivalent multiplication and use the rule for multiplying by a power of 10.

EXAMPLE 11-3.1 Perform the multiplication or division by using powers of 10.

(a) 275×10 **(b)** 0.18×100 **(c)** 2.4×1000 **(d)** 43×0.1
(e) $3.14 \div 10$ **(f)** $0.48 \div 100$ **(g)** $20.1 \div 1000$

Solution

(a) $275 \times 10 = 275 \times 10^1 = 2750$
Since the exponent is $+1$, move the decimal one place to the right.
(b) $0.18 \times 100 = 0.18 \times 10^2 = 18$
Since the exponent is $+2$, move the decimal two places to the right.
(c) $2.4 \times 1000 = 2.4 \times 10^3 = 2400$
Since the exponent is $+3$, move the decimal three places to the right.
(d) $43 \times 0.1 = 43 \times 10^{-1} = 4.3$
Since the exponent is -1, move the decimal one place to the left.
(e) $3.14 \div 10$
First, change the division to an equivalent multiplication.

$$3.14 \div 10 = 3.14 \times \frac{1}{10}$$

Next, express the fraction $\frac{1}{10}$ as a power of 10.

$$3.14 \times 10^{-1}$$

Since the exponent of 10 is -1, move the decimal one place to the *left*.

$$= 0.314$$

(f) $0.48 \div 100 = 0.48 \times \frac{1}{100} = 0.48 \times 10^{-2} = 0.0048$

(g) $20.1 \div 1000 = 20.1 \times \frac{1}{1000} = 20.1 \times 10^{-3} = 0.0201$ ■

A power of 10 can be multiplied or divided by another power of 10 by following the laws of exponents.

EXAMPLE 11-3.2 Multiply or divide by using the laws of exponents.

(a) $10^5(10^2)$ **(b)** $10^{-1}(10^2)$ **(c)** $10^0(10^3)$ **(d)** $\frac{10}{10^3}$

(e) $\frac{10^5}{10^4}$ **(f)** $\frac{10^2}{10^2}$ **(g)** $\frac{10^{-2}}{10^3}$

Solution

(a) $10^5(10^2) = 10^{5+2} = 10^7$
(b) $10^{-1}(10^2) = 10^{(-1+2)} = 10^1$ or 10

(c) $10^0(10^3) = 10^{0+3} = 10^3$

(d) $\frac{10}{10^3} = 10^{1-3} = 10^{-2}$ or $\frac{1}{10^2}$

(e) $\frac{10^5}{10^4} = 10^{5-4} = 10^1$ or 10

(f) $\frac{10^2}{10^2} = 10^{2-2} = 10^0$ or 1

(g) $\frac{10^{-2}}{10^3} = 10^{-2-3} = 10^{-5}$ or $\frac{1}{10^5}$ ■

L.O.2. Change a number from scientific notation to ordinary notation.

Scientific Notation and Ordinary Notation

Powers of 10 are used in technological applications in a special form that saves time in performing operations with very large or very small numbers. This special form for writing such numbers is called *scientific notation.*

To write a number in scientific notation, the decimal is shifted so that the number has a numerical value of at least 1 but less than 10. That is, the whole-number part of the value is a single nonzero digit. Then that value is multiplied by the appropriate power of 10.

■ **DEFINITION 11-3.1 Scientific Notation.** A number is expressed in **scientific notation** if it is the product of two factors and the first factor is a number greater than or equal to 1 but less than 10 and the second factor is a power of 10.

Tips and Traps

When numbers between 0 and 1 are written in scientific notation, negative exponents are used. The first factor in scientific notation will always have only one nonzero digit on the left of the decimal.

EXAMPLE 11-3.3 Which of the following terms is expressed in scientific notation?

(a) 4.7×10^2 **(b)** 0.2×10^{-1} **(c)** 3.4×5^2 **(d)** 2.7×10^0
(e) 34×10^4 **(f)** 8×10^{-6}

Solution

(a) 4.7 is more than 1 but less than 10. 10^2 is a power of 10. Thus, the term is in scientific notation.
(b) Even though 10^{-1} is a power of 10, this term is not in scientific notation because the first factor (0.2) is less than 1 and therefore is *not* more than or equal to 1.
(c) 3.4 is more than 1 and less than 10, but 5^2 is not a power of 10. Thus, this term is not in scientific notation.
(d) 2.7 is more than 1 and less than 10. 10^0 is a power of 10. Thus, this term is in scientific notation.
(e) 34 is greater than 10. Even though 10^4 is a power of 10, this term is not in scientific notation because the first factor (34) is 10 or more.
(f) 8 is more than 1 and less than 10. 10^{-6} is a power of 10. Thus, this term is in scientific notation. ■

When a number is written strictly according to place value, it is called an *ordinary number*.

Rule 11-3.3 ***To change from a number written in scientific notation to an ordinary number:***

1. Perform the indicated multiplication by moving the decimal point in the first factor the appropriate number of places. Insert zeros as necessary.
2. Omit the power-of-10 factor.

Remember, when multiplying by a power of 10, the exponent of 10 tells us how many places and in which direction to move the decimal.

EXAMPLE 11-3.4 Change the following to ordinary numbers.

(a) 3.6×10^4 **(b)** 2.8×10^{-2} **(c)** 1.1×10^0 **(d)** 6.9×10^{-5}
(e) 9.7×10^6

Solution

(a) $3.6 \times 10^4 = 36000. = 36{,}000$
The decimal is moved four places to the right.
(b) $2.8 \times 10^{-2} = .028 = 0.028$
The decimal is moved two places to the left.
(c) $1.1 \times 10^0 = 1.1$
The decimal is moved no (zero) places.
(d) $6.9 \times 10^{-5} = .000069 = 0.000069$
The decimal is moved five places to the left.
(e) $9.7 \times 10^6 = 9700000. = 9{,}700{,}000$
The decimal is moved six places to the right. ■

L.O.3. Change a number from ordinary notation to scientific notation.

Writing Ordinary Numbers in Scientific Notation

If we want to express an ordinary number in scientific notation, we are basically reversing the procedures we used before. Shifting the decimal point changes the value of a number. The power-of-ten factor is used to offset or balance this change. It is important to maintain the original value of the number.

Rule 11-3.4 ***To change from a number written in ordinary notation to scientific notation:***

1. Indicate where the decimal should be positioned in the ordinary number so that the number is valued at 1 or between 1 and 10 by inserting a caret (∧) in the proper place.
2. Determine how many places and in which direction the decimal shifts *from* the new position (caret) *to* the old position (decimal point). This number will be the exponent of the power of 10.

Tips and Traps

Moving the decimal in the ordinary number changes the value of the number unless you balance the effect of the move in the power-of-10 factor.

From the new to the old position indicates the proper number of places and the direction (positive or negative) for balancing with the power-of-10 factor.

Remember the word "NO." Count from "New" to "Old."

$3{,}800 = 3.8 \times 10^3$ *($3_\wedge 800. \times 10^3$ $N \to O = +3$)*

$0.0045 = 4.5 \times 10^{-3}$ *($0.004_\wedge 5 \times 10^{-3}$ $N \to O = -3$)*

Example 11-3.5 Express the following numbers in scientific notation.

(a) 285 **(b)** 0.007 **(c)** 9.1 **(d)** 85,000 **(e)** 0.00074

Solution

(a) $285 \to 2_\wedge 85.$

The unwritten decimal is after the 5. Place the caret between 2 and 8 so that the number 2.85 is between 1 and 10.

$$2.85 \times 10^?$$

Count *from* the caret *to* the decimal to determine the exponent of 10. A move two places to the right represents the exponent +2.

$$2.85 \times 10^2$$

(b) $0.007 \to 0.007_\wedge$

7 is between 1 and 10. Count *from* the caret *to* the decimal. A move three places to the left represents the exponent −3.

$$7 \times 10^{-3}$$

(c) 9.1

9.1 is already between 1 and 10, so the decimal does not move. That is, the decimal moves zero places.

$$9.1 \times 10^0$$

(d) $85{,}000 \to 8_\wedge 5000.$

$$8.5 \times 10^4$$

From the caret *to* the decimal is four places to the right.

(e) $0.00074 \to 0.0007_\wedge 4$

$$7.4 \times 10^{-4}$$

From the caret *to* the decimal is four places to the left.

Occasionally a number will be in power-of-10 notation, but it will not be in scientific notation because the first factor is not equal to 1 or between 1 and 10. When this is the case, shift the decimal to the proper place and adjust the original power-of-10 factor appropriately.

Tips and Traps

The expression "between 1 and 10" means any number that is more than 1 but less than 10. The first factor in scientific notation is equal to 1 OR between 1 and 10.

EXAMPLE 11-3.6 Express the following in scientific notation.

(a) 37×10^5 **(b)** 0.03×10^3

Solution

(a) $3_\wedge 7 \times 10^5 = 37. \times 10^1 \times 10^5$ $(N \rightarrow O = +1)$
$= 3.7 \times 10^6$
(b) $0.03_\wedge \times 10^3 = 0.03 \times 10^{-2} \times 10^3$ $(N \rightarrow O = -2)$
$= 3 \times 10^1$ ■

L.O.4. Multiply and divide numbers in scientific notation.

Multiplication and Division Involving Numbers in Scientific Notation

Multiplication and division can be performed with numbers expressed in scientific notation without having to convert the numbers in scientific notation to ordinary numbers.

Rule 11-3.5 *To multiply numbers in scientific notation:*

1. Multiply the first factors.
2. Multiply the power-of-10 factors by using the laws of exponents.
3. Examine the first factor of the product (step 1) to see if its value is equal to 1 or between 1 and 10.
   (a) If so, write the results of steps 1 and 2.
   (b) If not, shift the decimal so that the first factor is equal to 1 or between 1 and 10 and adjust the exponent of the power of 10 accordingly.

EXAMPLE 11-3.7 Perform the multiplications.

(a) $(4 \times 10^2)(2 \times 10^3)$ **(b)** $(3.7 \times 10^3)(2.5 \times 10^{-1})$
(c) $(8.4 \times 10^{-2})(5.2 \times 10^{-3})$

Solution

(a) $(4 \times 10^2)(2 \times 10^3) = 8 \times 10^5$
Since 8 is between 1 and 10, we do not make any adjustments.
(b) $(3.7 \times 10^3)(2.5 \times 10^{-1}) = 9.25 \times 10^2$
Since 9.25 is between 1 and 10, we do not make any adjustments.
(c) $(8.4 \times 10^{-2})(5.2 \times 10^{-3}) = 43.68 \times 10^{-5}$
43.68 is not between 1 and 10. Therefore we must make adjustments.

$$43.68 \rightarrow 4_\wedge 3.68 \quad \text{or} \quad 4.368 \times 10^1$$

Now we multiply 4.368×10^1 times 10^{-5}.

$$4.368 \times 10^1 \times 10^{-5} = 4.368 \times 10^{1-5}$$
$$= 4.368 \times 10^{-4}$$ ■

Division involving numbers in scientific notation follows a similar procedure.

> **Rule 11-3.6** *To divide numbers in scientific notation:*
>
> 1. Divide the first factors.
> 2. Divide the powers-of-10 factors by using the laws of exponents.
> 3. Examine the first factor of the quotient (step 1) to see if its value is equal to 1 or between 1 and 10.
>   **(a)** If so, write the results of steps 1 and 2.
>   **(b)** If not, shift the decimal so that the first factor is equal to 1 or between 1 and 10 and adjust the exponent of the power of 10 accordingly.

EXAMPLE 11-3.8 Perform the divisions.

(a) $\dfrac{3 \times 10^5}{2 \times 10^2}$ **(b)** $\dfrac{1.44 \times 10^{-3}}{6 \times 10^{-5}}$ **(c)** $\dfrac{9.6 \times 10^{29}}{3.2 \times 10^{111}}$ **(d)** $\dfrac{1.25 \times 10^3}{5}$

Solution

(a) $\dfrac{3 \times 10^5}{2 \times 10^2} = \dfrac{3}{2} \times 10^{5-2} = 1.5 \times 10^3$

Since 1.5 is between 1 and 10, no adjustments are necessary.

(b) $\dfrac{1.44 \times 10^{-3}}{6 \times 10^{-5}} = \dfrac{1.44}{6} \times 10^{-3-(-5)} = 0.24 \times 10^{-3+5} = 0.24 \times 10^2$

0.24 is less than 1, so adjustments are necessary.

$$0.24 \rightarrow 0.2_{\wedge}4 = 2.4 \times 10^{-1}$$

Then $2.4 \times 10^{-1} \times 10^2 = 2.4 \times 10^1$.

(c) $\dfrac{9.6 \times 10^{29}}{3.2 \times 10^{111}} = \dfrac{9.6}{3.2} \times 10^{29-111} = 3 \times 10^{-82}$

(d) $\dfrac{1.25 \times 10^3}{5}$

Remember, 5 is the same as 5×10^0.

$$\frac{1.25 \times 10^3}{5 \times 10^0} = \frac{1.25}{5} \times 10^{3-0} = 0.25 \times 10^3$$

0.25 is less than 1.

$$0.25 \rightarrow 0.2_{\wedge}5 = 2.5 \times 10^{-1}$$

Then, $2.5 \times 10^{-1} \times 10^3 =$

$$2.5 \times 10^{-1+3} = 2.5 \times 10^2$$

■

General Tips for Using the Calculator

The power-of-ten key, labeled [EXP] or [EE] on most calculators, is a shortcut key for entering the following keys:

[×] 10 [x^y]

This key is only used for powers of ten and only the *exponent* of 10 is entered. If you enter [×] 10 then [EXP], your answer will have one extra factor of 10.

Parts (a), (b), and (d) of Example 11-3.8 can be worked using the power-of-10 key on most calculators.

(a) 3 [EXP] 5 [÷] 2 [EXP] 2 [=] ⇒ 1500

This result will then need to be expressed in scientific notation if desired: $1500 = 1.5 \times 10^3$.

Part (c) cannot be done with all calculators because some calculators will accept no more than two digits for an exponent.

The internal program of a calculator has predetermined how the output of a calculator will be displayed. For example, even if you would like an answer displayed in scientific notation, it may fall within the guidelines to be displayed as an ordinary number. You must make the conversion to scientific notation. The reverse may also be true.

We will now apply our knowledge of scientific notation to solving applied problems.

EXAMPLE 11-3.9 A star is 5.7 light-years from Earth. If 1 light-year is 5.87×10^{12} miles, how many miles from Earth is the star?

Solution: To solve this problem, we express the given information in a proportion of two fractions relating light-years to miles.

$$\frac{1 \text{ light-year}}{5.87 \times 10^{12} \text{ mi}} = \frac{5.7 \text{ light-years}}{x \text{ mi}}$$

$$\frac{1}{5.87 \times 10^{12}} = \frac{5.7}{x} \qquad \textit{(Cross multiply.)}$$

$$x = 5.7(5.87 \times 10^{12})$$

$$x = 33.459 \times 10^{12} \qquad \textit{(Perform scientific notation adjustment.)}$$

$$x = 3.3459 \times 10^1 \times 10^{12}$$

$$x = 3.3459 \times 10^{13}$$

$$\text{or } 3.3 \times 10^{13} \qquad \textit{(First factor is rounded to tenths.)}$$

■

EXAMPLE 11-3.10 An angstrom unit is 1×10^{-7} mm. How many millimeters are in 14.82 angstrom units?

Solution: We will set up a proportion of fractions relating angstrom units to millimeters.

$$\frac{1 \text{ angstrom}}{1 \times 10^{-7} \text{ mm}} = \frac{14.82 \text{ angstroms}}{x \text{ mm}}$$

$$\frac{1}{1 \times 10^{-7}} = \frac{14.82}{x} \quad \textit{(Cross multiply.)}$$

$$x = 14.82 \times 10^{-7}$$

$$x = 1.482 \times 10^{1} \times 10^{-7} \quad \textit{(Adjust.)}$$

$$x = 1.482 \times 10^{-6} \text{ mm}$$

■

EXAMPLE 11-3.11 One coulomb (C) is approximately 6.28×10^{18} electrons. How many coulombs are in 2.512×10^{21} electrons?

Solution: Setting up the proportion, we have

$$\frac{1 \text{ C}}{6.28 \times 10^{18} \text{ electrons}} = \frac{x \text{ C}}{2.512 \times 10^{21} \text{ electrons}}$$

$$\frac{1}{6.28 \times 10^{18}} = \frac{x}{2.512 \times 10^{21}} \quad \textit{(Cross multiply.)}$$

$$(6.28 \times 10^{18})(x) = 2.512 \times 10^{21} \quad \textit{(Divide by coefficient of x.)}$$

$$x = \frac{2.512 \times 10^{21}}{6.28 \times 10^{18}}$$

$$x = 0.4 \times 10^{3} \quad \textit{(Adjust.)}$$

$$x = 4 \times 10^{-1} \times 10^{3}$$

$$x = 4 \times 10^{2} \quad \textit{(Write as ordinary number.)}$$

$$x = 400 \text{ C}$$

■

SELF-STUDY EXERCISES 11-3

L.O.1. Multiply or divide as indicated.

1. 0.37×10^{2} **2.** 1.82×10^{3} **3.** 5.6×10^{-1}
4. 142×10^{-2} **5.** 78×10^{4} **6.** 62×10^{0}
7. $4.6 \div 10^{4}$ **8.** $6.1 \div 10$ **9.** $7.2 \div 10^{1}$
10. $42 \div 10^{0}$ **11.** $10^{4}(10^{6})$ **12.** $10^{-3}(10^{-4})$
13. $10^{0}(10^{-3})$ **14.** $10^{-3}(10^{4})$ **15.** $10(10^{2})$
16. $\frac{10^{4}}{10^{2}}$ **17.** $\frac{10}{10^{4}}$ **18.** $\frac{10^{4}}{10^{4}}$
19. $\frac{10^{-2}}{10^{3}}$ **20.** $\frac{10^{0}}{10^{1}}$

L.O.2. Write the following as ordinary numbers.

21. 4.3×10^{2} **22.** 6.5×10^{-3} **23.** 2.2×10^{0}
24. 7.3×10 **25.** 9.3×10^{-2} **26.** 8.3×10^{4}
27. 5.8×10^{-3} **28.** 8×10^{4} **29.** 6.732×10^{0}
30. 5.89×10^{-3}

L.O.3. Express the following in scientific notation.

31. 392 **32.** 0.02 **33.** 7.03
34. 42,000 **35.** 0.081 **36.** 0.0021

37. 23.92 **38.** 0.101 **39.** 1.002
40. 721

Write in scientific notation.

41. 42×10^4 **42.** 32.6×10^3 **43.** 0.213×10^2
44. 0.0062×10^{-3} **45.** $56{,}000 \times 10^{-3}$

L.O.4. Perform the indicated operations. Express answers in scientific notation.

46. $(6.7 \times 10^4)(3.2 \times 10^2)$ **47.** $(1.6 \times 10^{-1})(3.5 \times 10^4)$ **48.** $(5.0 \times 10^{-3})(4.72 \times 10^0)$

49. $(8.6 \times 10^{-3})(5.5 \times 10^{-1})$ **50.** $\dfrac{3.15 \times 10^5}{4.5 \times 10^2}$ **51.** $\dfrac{4.68 \times 10^3}{7.2 \times 10^7}$

52. $\dfrac{4.55 \times 10^{-1}}{6.5 \times 10^{-4}}$ **53.** $\dfrac{7.84 \times 10^{-2}}{9.8 \times 10^0}$

54. A star is 5.5 light years from Earth. If one light year is 5.87×10^{12} miles, how many miles from Earth is the star?

55. An angstrom (Å) unit is 1×10^{-7} mm. How many angstrom units are in 4.2×10^{-5} mm?

11-4 POLYNOMIALS

Learning Objectives

L.O.1. Identify polynomials, monomials, binomials, and trinomials.
L.O.2. Identify the degree of terms and polynomials.
L.O.3. Arrange polynomials in descending order.

L.O.1 Identify polynomials, monomials, binomials, and trinomials.

Identifying Polynomials

A special type of algebraic expression that contains variables and exponents is a *polynomial.*

■ **Definition 11-4.1 Polynomial.** A **polynomial** is an algebraic expression in which the exponents of the variables are nonnegative integers.

Example 11-4.1 Identify which expressions are polynomials. If an expression is not a polynomial, explain why.

(a) $5x^2 + 3x + 2$ **(b)** $5x - \dfrac{3}{x}$

(c) 9 **(d)** $-\dfrac{1}{2}x + 3x^{-2}$

Solution

(a) $5x^2 + 3x + 2$ is a polynomial.
(b) $5x - \dfrac{3}{x}$ is not a polynomial because the term $-\dfrac{3}{x}$ is equivalent to $-3x^{-1}$. A polynomial cannot have a variable with a negative exponent.
(c) 9 is a polynomial since it is equivalent to $9x^0$ and the exponent, zero, is a nonnegative integer.
(d) $-\dfrac{1}{2}x + 3x^{-2}$ is not a polynomial because $3x^{-2}$ has a negative exponent. ■

Some polynomials have special names depending on the number of terms contained in the polynomial.

■ **DEFINITION 11-4.2 Monomial.** A **monomial** is a polynomial containing one term.

$$3, \quad -2x, \quad 5ab, \quad 7xy^2, \quad \frac{x}{y}, \quad \frac{3a^2}{4}$$

are monomials.

■ **DEFINITION 11-4.3 Binomial.** A **binomial** is a polynomial containing two terms.

$$x + 3, \quad 2x^2 - 5x, \quad x + \frac{y}{4}, \quad 3(x - 1) + 2$$

are binomials.

■ **DEFINITION 11-4.4 Trinomial.** A **trinomial** is a polynomial containing three terms.

$$a + b + c, \quad x^2 - 3x + 4, \quad x + \frac{2a}{b} - 5$$

are trinomials.

EXAMPLE 11-4.2 Identify each of the following expressions as a monomial, binomial, or trinomial.

(a) $x^2y - 1$ **(b)** $4(x - 2)$

(c) $\dfrac{2x + 5}{2y}$ **(d)** $3x^2 - x + 1$

(e) $3x^2 - (x + 1)$

Solution

(a) $x^2y - 1$ *Binomial*

(b) $4(x - 2)$ *Monomial. If the distributive property is applied, the expression will become a binomial.* $4(x - 2) = 4x - 8$

(c) $\dfrac{2x + 5}{2y}$ *This one-term expression is not a monomial, since it is not a polynomial (the exponent of y is −1 if in the numerator).*

(d) $3x^2 - x + 1$ *Trinomial*

(e) $3x^2 - (x + 1)$ *Binomial* ■

L.O.2. Identify the degree of terms and polynomials.

Degree of Terms and Polynomials

The *degree of a term* that has only one variable with a nonnegative exponent is the same as the exponent of the variable.

$$3x^4, \text{ fourth degree} \qquad -5x, \text{ first degree}$$

Number terms other than zero have a degree of 0. (A variable to the zero power is implied.)

$$5 = 5x^0$$

If a term has more than one variable, the degree of the term is the sum of the exponents of the variable factors.

$2xy$, second degree $\qquad -2ab^2$, third degree

EXAMPLE 11-4.3 Identify the degree of each term in the polynomial $5x^3 + 2x^2 - 3x + 3$.

Solution

$5x^3$, degree 3 (or third degree)
$2x^2$, degree 2 (or second degree)
$-3x$, degree 1 (or first degree)
3, degree 0 ■

Special names are associated with terms of degree 0, 1, 2, and 3. A *constant* term has degree 0. A *linear* term has degree 1. A *quadratic* term has degree 2. A *cubic* term has degree 3.

The *degree of a polynomial* that has only one variable and only positive integral exponents is the degree of the term with the largest exponent. A *linear polynomial* has degree 1. A *quadratic polynomial* has degree 2. A *cubic polynomial* has degree 3.

$5x^3 - 2$ has a degree of 3 and is a cubic polynomial.

$x + 7$ has a degree of 1 and is a linear polynomial.

$7x^2 - 4x + 5$ has a degree of 2 and is a quadratic polynomial.

EXAMPLE 11-4.4 Identify the degree of the following polynomials.

(a) $5x^4 + 2x - 1$ **(b)** $3x^3 - 4x^2 + x - 5$

(c) 7 **(d)** $x - \frac{1}{2}$

Solution

(a) $5x^4 + 2x - 1$ has a degree of 4.
(b) $3x^3 - 4x^2 + x - 5$ has a degree of 3 and is a cubic polynomial.
(c) 7 has a degree of 0 and is a constant.
(d) $x - \frac{1}{2}$ has a degree of 1 and is a linear polynomial. ■

L.O.3. Arrange polynomials in descending order.

Descending Order of Polynomials

The terms of a polynomial are customarily arranged in order based on the degree of each term of the polynomial. The terms can be arranged beginning with the term with the highest degree (*descending order*) or beginning with the term with the lowest degree (*ascending order*).

Tips and Traps

Polynomials are most often arranged in descending order so that the degree of the polynomial is the degree of the first term.

The first term of a polynomial arranged in descending order is called the *leading term* of the polynomial. The coefficient of the leading term of a polynomial is called the *leading coefficient*.

EXAMPLE 11-4.5 Arrange each polynomial in descending order and identify the degree, the leading term, and the leading coefficient of the polynomial.

(a) $5x + 3x^3 - 7 + 6x^2$ **(b)** $x^4 - 2x + 3$
(c) $4 + x$ **(d)** $x^2 + 5$

Solution

(a) $5x + 3x^3 - 7 + 6x^2 = 3x^3 + 6x^2 + 5x - 7.$ *(Third degree, leading term is $3x^3$, leading coefficient is 3).*

(b) $x^4 - 2x + 3$ is already in descending order. *(Fourth degree, leading term is x^4, leading coefficient is 1.)*

Tips and Traps

Missing terms have a coefficient of 0.
In Example 11-4.5(b), $x^4 - 2x + 3$ is the same as $x^4 + 0x^3 + 0x^2 - 2x^1 + 3x^0$.

(c) $4 + x = x + 4.$ *(First degree or linear polynomial, leading term is x, leading coefficient is 1.)*

(d) $x^2 + 5$ is already in descending order. *(Second degree or quadratic polynomial, leading term is x^2, leading coefficient is 1.)*

or

$x^2 + 0x^2 + 5x^0$ ■

SELF-STUDY EXERCISES 11-4

L.O.1. Identify each of the following expressions as a monomial, binomial, or trinomial.

1. $2x^3y - 7x$ **2.** $5xy^2 + 8y$ **3.** $3xy$
4. $7ab$ **5.** $5(3x - y)$ **6.** $(x - 5)(2x + 4)$
7. $5x^2 - 8x + 3$ **8.** $7y^2 + 5y - 1$ **9.** $\frac{4x - 1}{5}$
10. $\frac{x}{6} - 5$ **11.** $4(x^2 - 2) + x^3$ **12.** $3x^2 - 7(x - 8)$

L.O.2. Identify the degree of each term in each polynomial.

13. $6x$ **14.** $8x^2$ **15.** $6x^2 - 8x + 12$
16. $7x^3 - 8x + 12$ **17.** $x - 12$ **18.** $3x^2 - 8$
19. 15 **20.** 21 **21.** $2x - \frac{1}{4}$
22. $8x^2 + \frac{5}{6}$

Identify the degree of the following polynomials.

23. $5x^2 + 8x - 14$ **24.** $x^3 - 8x^2 + 5$ **25.** $9 - x^3 + x^6$
26. $12 - 15x^2 - 7x^5$ **27.** $2x - \frac{4}{5}x^2$ **28.** $\frac{7}{8} - x$

L.O.3. Arrange each polynomial in descending order and identify the degree, leading term, and leading coefficient of the polynomial.

29. $5x - 3x^2$ **30.** $7 - x^3$ **31.** $4x - 8 + 9x^2$
32. $5x^2 + 8 - 3x$ **33.** $7x^3 - x + 8x^2 - 12$ **34.** $7 - 15x^4 + 12x$
35. $-7x + 8x^6 - 7x^3$ **36.** $15 - 14x^8 + x$

11-5 EXPONENTIAL EXPRESSIONS

Learning Objectives

L.O.1. Evaluate formulas with at least one exponential term.
L.O.2. Evaluate formulas that contain a power of the natural exponential, e.
L.O.3. Solve exponential equations in the form $b^x = b^y$, where $b > 0$ and $b \neq 1$.

Many scientific, technical, and business phenomena have the property of exponential growth. That is, the growth rate does not remain constant as certain physical properties increase. Instead, the growth rate increases exponentially. For example, notice the difference between $2x$ and 2^x when x increases.

$2x$	2^x
$2(1) = 2$	$2^1 = 2$
$2(2) = 4$	$2^2 = 4$
$2(3) = 6$	$2^3 = 8$
$2(4) = 8$	$2^4 = 16$
$2(5) = 10$	$2^5 = 32$
$2(6) = 12$	$2^6 = 64$

2^x is said to increase *exponentially.*

Before the easy availability of scientific calculators, logarithms were used extensively to make numerical calculations. Exponential and logarithmic expressions are found in many formulas. Even with the use of a scientific calculator, an understanding of exponential and logarithmic expressions is necessary. In this section, we will evaluate expressions containing exponential expressions with the scientific or graphing calculator.

L.O.1. Evaluate formulas with at least one exponential term.

Formulas with Exponential Expressions

Many formulas have terms that contain exponents. An *exponential expression* is an expression that contains at least one term that has a *variable exponent.* A variable exponent is an exponent that has at least one letter factor. Exponential expressions can be evaluated on a scientific calculator by using the key $\boxed{y^x}$ or $\boxed{x^y}$. The base of the term is entered first, then $\boxed{x^y}$, followed by the value of the exponent.

A commonly used formula that contains a variable exponent is the formula for calculating the compound amount for compound interest. *Compound interest* is the interest that is calculated at the end of each period and then added to the principal for the next period. The *compound amount* is the principal and interest accumulated over a period of time. The accumulated amount A that a principal P will be worth at the end of t years, when invested at an interest rate r and compounded n times a year, is given by the following formula.

Formula 11-5.1 *Compound Amount*

$$A = P\left(1 + \frac{r}{n}\right)^{nt}$$

where A = accumulated amount
P = original principal
t = time in years
r = rate per year
n = compounding periods per year

EXAMPLE 11-5.1 Using the formula $A = P\left(1 + \frac{r}{n}\right)^{nt}$ and a scientific calculator, find the accumulated amount on an investment of \$1500, invested at an interest rate of 9% for 3 years, if the interest is compounded quarterly.

Solution: Using the compound amount formula,

$$A = P\left(1 + \frac{r}{n}\right)^{nt}$$

$$A = 1500\left(1 + \frac{0.09}{4}\right)^{(4)(3)} \quad \textit{(9\% = 0.09 \quad Quarterly means 4 times a year.)}$$

$$A = 1500(1 + 0.0225)^{12}$$

$$A = 1500(1.0225)^{12} \quad (1.0225 \ \boxed{x^y}\ 12 = 1.30604999)$$

$$A = 1500(1.30604999)$$

$$A = 1959.07 \quad \textit{(rounded)}$$

The accumulated amount of the \$1500 invested after 3 years is \$1959.07 to the nearest cent. ■

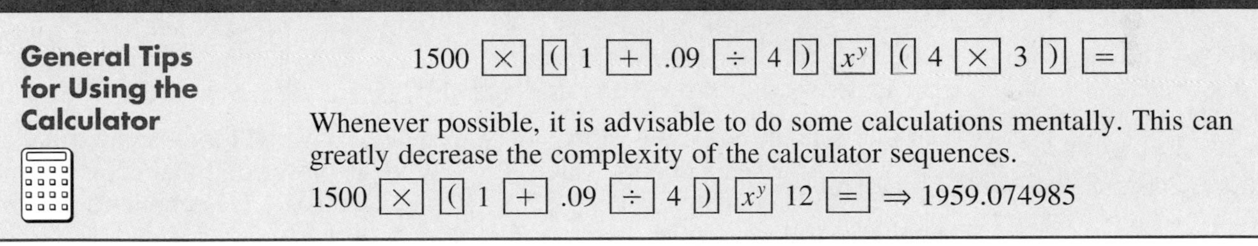

General Tips for Using the Calculator

1500 $\boxed{\times}$ $\boxed{(}$ 1 $\boxed{+}$ .09 $\boxed{\div}$ 4 $\boxed{)}$ $\boxed{x^y}$ $\boxed{(}$ 4 $\boxed{\times}$ 3 $\boxed{)}$ $\boxed{=}$

Whenever possible, it is advisable to do some calculations mentally. This can greatly decrease the complexity of the calculator sequences.

1500 $\boxed{\times}$ $\boxed{(}$ 1 $\boxed{+}$ .09 $\boxed{\div}$ 4 $\boxed{)}$ $\boxed{x^y}$ 12 $\boxed{=}$ ⇒ 1959.074985

L.O.2. Evaluate formulas that contain a power of the natural exponential, e.

The Natural Exponential

In many applications involving circles, the irrational number π (approximately equal to 3.14159) is used. Another irrational number, e, arises in the discussion of many physical phenomena. Many formulas will contain a power of the natural exponential e.

Exponential change is an interesting phenomenon. Let's look at the value of the expression $(1 + 1/n)^n$ as n gets larger and larger.

n	$(1 + 1/n)^n$	result
1	$(1 + 1/n)^n$	2
2	$(1 + 1/n)^n$	2.25
3	$(1 + 1/n)^n$	2.37037037
10	$(1 + 1/n)^n$	2.59374246
100	$(1 + 1/n)^n$	2.704813829
1000	$(1 + 1/n)^n$	2.716923932
10,000	$(1 + 1/n)^n$	2.718145927

The value of the expression changes very little as the value of n gets larger. We can say that the value approaches a given number. We will call the number e, the *natural exponential*. The natural exponential, e, like π, is an irrational number and will never terminate or repeat as more decimal places are examined.

■ **DEFINITION 11-5.1** The **natural exponential, e.** The **natural exponential**, e, is the limit that the value of the expression $(1 + 1/n)^n$ approaches as n gets larger and larger without bound.

To evaluate formulas containing the natural exponential e, we will use a scientific or graphing calculator. The key is generally labeled $\boxed{e^x}$, and the exponent is entered first on the scientific calculator. On some graphing calculators, the exponent is entered after pressing the exponential key.

EXAMPLE 11-5.2 The formula for the atmospheric pressure (in millimeters of mercury) is $P = 760e^{-0.00013h}$, where h is the height in meters. Find the atmospheric pressure at 100 meters above sea level ($h = 100$).

Solution: Substitute 100 for h in the formula.

$$\begin{aligned} P &= 760e^{-0.00013h} \\ &= 760e^{-0.00013(100)} \\ &= 760e^{-0.013} \\ &= 760(0.987084135) && (e^{-0.013} = 0.987084135) \\ &= 750.1839427 && \text{(one calculator sequence } 760\ \boxed{\times}\ .013\ \boxed{+/-}\ \boxed{e^x}\ \boxed{=}\ ) \end{aligned}$$

Thus, the atmospheric pressure at 100 meters above sea level is 750.18 mm. ■

L.O.3. Solve exponential equations in the form $b^x = b^y$, where $b > 0$ and $b \neq 1$.

Solving Exponential Equations by Equating Exponents of Like Bases

To illustrate the properties of exponential equations, we will look at the equation $2^x = 32$. In the equation the value of x will be the power of 2 that gives a result of 32. We can rewrite 32 as 2^5. Thus, $2^x = 2^5$. To solve the equation for x, we apply the following rule:

Rule 11-5.1

To solve an exponential equation in the form of $b^x = b^y$, apply the following property and solve for x. If $b^x = b^y$, and $b > 0$ and $b \neq 1$, then $x = y$.

Tips and Traps

It is important to notice that this property is appropriate only when the bases are alike, positive, and not equal to one.
Why must we exclude $b = 1$?

If $1^5 = 1^8$, then does $5 = 8$?

EXAMPLE 11-5.3 Solve the equation $2^x = 32$.

Solution: Rewrite 32 as a power of 2.

$$32 = 2^5$$

Then apply Rule 11-5.1.

$$\text{If } 2^x = 2^5, \quad \text{then} \quad x = 5.$$

Check to see if the solution is appropriate.

$$\text{Does } 2^5 = 32?$$

Yes, so 5 is the correct solution. ■

EXAMPLE 11-5.4 Solve the equation, $3^{x+1} = 27$.

Solution: Rewrite 27 as a power of 3.

$$27 = 3^3$$

Then, if $3^{x+1} = 3^3$, $x + 1 = 3$.
Solve for x.

$$x + 1 = 3$$
$$x = 3 - 1$$
$$x = 2$$

Check to see if the solution, 2, is correct.

$$\text{Does } 3^{2+1} = 27?$$
$$3^3 = 27$$

Thus, the solution is correct. ■

It will not always be possible to rewrite an exponential equation as an equation with like bases. In such cases, other methods for solving the exponential equation are used.

SELF-STUDY EXERCISES 11-5

L.O.1. Using the compound amount formula, $A = P\left(1 + \frac{r}{n}\right)^{nt}$, find the accumulated amount for the following.

1. Principal = \$1500, rate = 10%, compounded annually, time = 5 years.
2. Principal = \$1750, rate = 8%, compounded quarterly, time = 2 years.
3. The number of grams of a chemical that will dissolve in a solution is given by the formula $C = 100e^{0.02t}$, where t = temperature in degrees Celsius. Evaluate when:
   (a) $t = 10$ **(c)** $t = 25$
   (b) $t = 20$ **(d)** $t = 30$
4. The compound amount when an investment is compounded continually (every instant) is expressed by the formula $A = Pe^{ni}$, where A = compounded amount, P = principal, n = number of years, and i = interest rate per year. Find the compound amount when:
   (a) Principal = \$1000, interest = 9%, for 2 years
   (b) Principal = \$1500, interest = 10%, for 6 months.

Evaluate using a scientific calculator.

5. 4^3 **6.** 3^{-5} **7.** 5^{10}

8. 8^{-3} **9.** $9^{2.5}$ **10.** $10^{-\frac{5}{2}}$

L.O.2. A formula for electric current is $i = 1.50e^{-200t}$, where t is time in seconds. Calculate the current for the following times. Express answers in scientific notation.

11. 1 second **12.** 1.1 seconds **13.** 0.5 second

Evaluate using a scientific calculator. Round to hundredths.

14. e^2 **15.** e^{-3} **16.** $e^{0.21}$
17. $e^{-3.5}$

L.O.3. Solve for x.

18. $3^x = 3^7$ **19.** $5^x = 5^{-3}$ **20.** $3^{x+4} = 3^6$
21. $2^{x-3} = 2^7$ **22.** $6^x = 6^7$ **23.** $3^x = 27$
24. $2^x = 64$ **25.** $3^x = \frac{1}{81}$ **26.** $2^x = \frac{1}{64}$
27. $4^{3x} = 128$ **28.** $3^{2x} = 243$ **29.** $4^{3-x} = \frac{1}{16}$
30. $2^{4-x} = \frac{1}{16}$

11-6 LOGARITHMIC EXPRESSIONS

Learning Objectives

L.O.1. Write exponential expressions as equivalent logarithmic expressions.
L.O.2. Write logarithmic expressions as exponential expressions.
L.O.3. Evaluate common and natural logarithmic expressions using a calculator.
L.O.4. Evaluate logarithms with a base other than 10 or e.
L.O.5. Evaluate formulas containing at least one logarithmic term.
L.O.6. Simplify logarithmic expressions by using the properties of logarithms.

Logarithms were first introduced as a relatively fast way of carrying out lengthy calculations. The scientific calculator has diminished the importance of logarithms as a computational device; however, the importance of logarithms in advanced mathematics, electronics, and theoretical work is more evident than ever. Many formulas use logarithms to show the relationships of physical properties. Also, logarithms are used to solve many exponential equations.

L.O.1. Write exponential expressions as equivalent logarithmic expressions.

Writing Exponential Expressions as Equivalent Logarithmic Expressions

In Chapter 1, we looked at powers of whole numbers and roots of perfect powers. To expand on that information, we will extend our study to logarithms. The three basic components of a power are the base, exponent, and result of exponentiation or power. Since the word ''*power*'' is often used to mean more than one concept, we will use the **result of exponentiation** terminology.

exponent

$3^4 = 81 \leftarrow$ result of exponentiation (power)

base

When finding a power or the result of exponentiation, you are given the base and exponent. When finding a root, you know the base (radicand) and the index of the root (exponent). A third type of calculation will be to find the exponent when you are given the base and the result of exponentiation. The exponent in this process is called the *logarithm*.

In the logarithmic form $\log_b x = y$, b is the *base*, y is the *exponent* or *logarithm*, and x is the *result of exponentiation*.

In the exponential form $x = b^y$, b is also the base, y is the exponent, and x is the result of the exponentiation.

Rule 11-6.1 ***To convert an exponential expression to a logarithmic equation:***

If $x = b^y$, then $\log_b x = y$.

1. The exponent in the exponential form is the variable solved for in the logarithmic form.
2. The base in the exponential form is the base in the logarithmic form.
3. The variable solved for in the exponential form is the result of exponentiation in logarithmic form. This result is sometimes referred to as the argument.

EXAMPLE 11-6.1 Rewrite the following in logarithmic form.

(a) $2^4 = 16$ **(b)** $3^2 = 9$ **(c)** $2^{-2} = \frac{1}{4}$

Solution

(a) Base $= 2$, exponent $= 4$, result of exponentiation $= 16$

$$\log_2 16 = 4$$

(b) Base $= 3$, exponent $= 2$, result of exponentiation $= 9$

$$\log_3 9 = 2$$

(c) Base $= 2$, exponent $= -2$, result of exponentiation $= \frac{1}{4}$

$$\log_2 \frac{1}{4} = -2$$

■

L.O.2. Write logarithmic expressions as exponential expressions.

Writing Logarithmic Expressions as Exponential Expressions

The inverse of the process in the preceding objective will change a logarithmic expression to an exponential expression.

Rule 11-6.2 ***To convert a logarithmic equation to an exponential equation:***

$\log_b x = y$ converts to $x = b^y$, provided that $b > 0$ and $b \neq 1$.

1. The dependent variable (solved for variable) in the logarithmic form is the exponent in the exponential form.
2. The base in the logarithmic form is the base in the exponential form.
3. The result of the exponentiation in logarithmic form is the dependent variable (solved for variable) in exponential form.

EXAMPLE 11-6.2 Rewrite each of the following in exponential form.

(a) $\log_2 32 = 5$ **(b)** $\log_3 81 = 4$ **(c)** $\log_5 \frac{1}{25} = -2$

(d) $\log_{10} 0.001 = -3$

Solution

(a) Base = 2, exponent = 5, result of exponentiation = 32

$$2^5 = 32$$

(b) Base = 3, exponent = 4, result of exponentiation = 81

$$3^4 = 81$$

(c) Base = 5, exponent = −2, result of exponentiation = $\frac{1}{25}$

$$5^{-2} = \frac{1}{25}$$

(d) Base = 10, exponent = −3, result of exponentiation = 0.001

$$10^{-3} = 0.001$$

■

When the base of a logarithm is 10, the logarithm is referred to as a *common logarithm*. If the base is omitted in a logarithmic expression, the base is assumed to be 10. Thus, $\log_{10} 1000 = 3$ is normally written as $\log 1000 = 3$.

A logarithm with a base of *e* is a *natural logarithm* and is abbreviated as *ln*. Scientific or graphing calculators normally have an [ln] key.

On a scientific calculator, expressions containing common logarithms can be evaluated using the [log] key. Expressions containing natural logarithms can be evaluated using the [ln] key. Expressions containing logarithms with a base different from 10 can sometimes be evaluated by first converting the logarithm to an equivalent exponential equation and then applying Rule 11-5.1. For example, $\log 100 = x$ can be written as $10^x = 100$. Then, by inspection, $10^x = 10^2$ and $x = 2$.

L.O.3. Evaluate common and natural logarithmic expressions using a calculator.

Finding Logarithms with a Calculator

Calculators are used almost exclusively when using logarithms.

EXAMPLE 11-6.3 Evaluate the following.

(a) log 10,000 **(b)** log 0.000001 **(c)** $\log_4 256$ **(d)** $\log_3 \frac{1}{27}$

Solution

(a) Using the [log] key,

log 10,000 = 4 *(Scientific: Enter 10000 [log].)*

General Tips for Using the Calculator

Scientific and graphing calculators generally have two logarithm keys, [log] for *common logarithms* and [ln] for *natural logarithms.*

Evaluate log 2 and ln 2.

Scientific:
To find the common log of 2, first enter the result of exponentiation and then the [log] key.

$$2 \boxed{\log} \Rightarrow 0.301029996$$

To find the natural log of 2, enter the result of exponentiation, 2, and then the [ln] key.

$$2 \boxed{\ln} \Rightarrow 0.693147181$$

Graphing:
To find the common log of 2, press the [log] key, followed by the result of exponentiation, 2.

$$\boxed{\log}\ 2\ \boxed{\text{EXE}} \Rightarrow 0.301299957$$

To find the natural log of 2, press the [ln] key followed by the result of exponentiation, 2.

$$\boxed{\ln}\ 2\ \boxed{\text{EXE}} \Rightarrow 0.6931471806$$

(b) Using the [log] key,

$$\log 0.000001 = -6 \qquad \textit{(Scientific: Enter .000001 } \boxed{\log}\textit{.)}$$

(c) Rewriting as an exponential equation and solving the exponential equation,

$$\log_4 256 = x \text{ is } 4^x = 256 \qquad (256 = 4^4)$$
$$4^x = 4^4$$
$$x = 4$$

(d) Rewriting as an exponential equation and solving the exponential equation,

$$\log_3 \frac{1}{27} = x \text{ is } 3^x = \frac{1}{27} \qquad \left(27 = 3^3;\ \frac{1}{27} = 3^{-3}\right)$$
$$3^x = 3^{-3}$$
$$x = -3$$

■

EXAMPLE 11-6.4 Evaluate the following using a scientific calculator. Express answers to the nearest ten-thousandth.

(a) *ln* 5 **(b)** *ln* 4.5 **(c)** *ln* 948

Solution: Enter the result of exponentiation or the argument of the natural logarithm and then press the [ln] key.

(a) *ln* 5 = 1.6094 **(b)** *ln* 4.5 = 1.5041 **(c)** *ln* 948 = 6.8544 ■

L.O.4. Evaluate logarithms with a base other than 10 or e.

Logarithms with Other Bases

To find the logarithm for a base other than 10 or e using a calculator, a conversion formula is necessary.

Rule 11-6.3 ***To evaluate a logarithm with a base b other than 10 or e:***

$$\log_b a = \frac{\log a}{\log b}$$

A similar process can be used with natural logarithms.

$$\log_b a = \frac{\ln a}{\ln b}$$

EXAMPLE 11-6.5 Find $\log_7 343$

Solution

$$\log_7 343 = \frac{\log 343}{\log 7}$$

Scientific: 343 [log] [÷] 7 [log] [=] ⇒ 3
Graphing: [log] 343 [÷] [log] 7 [EXE] ⇒ 3 ■

L.O.5. Evaluate formulas containing at least one logarithmic term.

Formulas Using Logarithms

Many formulas in science make use of both common and natural logarithms.

EXAMPLE 11-6.6 The loudness of sound is measured by a unit called a *decibel*. A very faint sound, called the *threshold sound*, is assigned an intensity of I_o. Other sounds have an intensity of I, which is a specified number times the threshold sound ($I = nI_o$). Then the decibel rate is given by the formula bel $= 10 \log \frac{I}{I_o}$. Find the decibel rating for sounds having the following intensities (I).

(a) a whisper, $110I_o$
(b) a voice, $230I_o$
(c) a busy street, $9{,}000{,}000I_o$
(d) loud music, $875{,}000{,}000{,}000I_o$
(e) a jet plane at takeoff, $109{,}000{,}000{,}000{,}000I_o$

Solution

(a) $I = 110I_o$

$$\begin{aligned} bel &= 10 \log \frac{110I_o}{I_o} \\ &= 10 \log 110 \qquad (\log 110 = 2.041392685) \\ &= 20 \text{ (rounded)} \end{aligned}$$

(b) $I = 230I_o$

$$\begin{aligned} bel &= 10 \log \frac{230I_o}{I_o} \\ &= 10 \log 230 \qquad (\log 230 = 2.361727836) \\ &= 24 \text{ (rounded)} \end{aligned}$$

(c) $I = 9{,}000{,}000I_o$

$$bel = 10 \log \frac{9{,}000{,}000I_o}{I_o}$$

$$bel = 10 \log 9{,}000{,}000 \qquad (\log 9{,}000{,}000 = 6.954242509)$$

$$bel = 70 \text{ (rounded)}$$

(d) $I = 875{,}000{,}000{,}000I_o$

$$bel = 10 \log \frac{875{,}000{,}000{,}000I_o}{I_o}$$

$$bel = 10 \log 875{,}000{,}000{,}000$$

$$bel = 10 \log (8.75 \times 10^{11})$$

Scientific: 8.75 [EXP] 11 [log] [×] 10 [=] ⇒ 119.4200805
Graphing: 10 [log] 8.75 [EXP] 11 [EXE] ⇒ 119.4200805

$$bel = 119 \quad \text{(rounded)}$$

(e) $I = 109{,}000{,}000{,}000{,}000I_o = (1.09 \times 10^{14})I_o$

$$bel = 10 \log \frac{(1.09 \times 10^{14})I_o}{I_o}$$

$$bel = 10 \log (1.09 \times 10^{14})$$

$$bel = 10(14.0374265)$$

$$bel = 140 \text{ (rounded)}$$

■

L.O.6. Simplify logarithmic expressions by using the properties of logarithms.

Properties of Logarithms

Many applications of logarithms and exponential expressions require the understanding of the properties of logarithms.

Since a logarithm is an exponent, the laws of exponents are appropriate in simplifying logarithmic expressions. A few manual manipulations can be made before using the calculator. Similar laws are appropriate for both common and natural logarithms and logarithms with bases other than 10 and *e*.

Rule 11-6.4 ***Properties of Logarithms:***

$$\log_b mn = \log_b m + \log_b n$$

$$\log_b m/n = \log_b m - \log_b n$$

$$\log_b m^n = n \log_b m$$

$$\log_b b = 1$$

We can illustrate these laws with examples using a calculator.

Example 11-6.7 Show that the following statements are true by using a calculator:

(a) $\log 6 = \log 2 + \log 3$
(b) $\log 2 = \log 6 - \log 3$

(c) $\log 2^3 = 3 \log 2$
(d) $\log 20 = 1 + \log 2$

Solution

(a) $\log 6 = \log 2(3) = \log 2 + \log 3$
6 [log] [=] $\Rightarrow$ 0.77815125, 2 [log] [+] 3 [log] = $\Rightarrow$ 0.77815125

(b) $\log 2 = \log \frac{6}{3} = \log 6 - \log 3$
2 [log] [=] $\Rightarrow$ 0.301029995, 6 [log] [−] 3 [log] = $\Rightarrow$ 0.301029995

(c) $\log 2^3 = 3 \log 2$
2 [x^y] 3 [=] [log] $\Rightarrow$ 0.903089987, 3 [×] 2 [log] = $\Rightarrow$ 0.903089987

(d) $\log 20 = \log 10 + \log 2 = 1 + \log 2$
20 [log] [=] $\Rightarrow$ 1.301029996, 1 + 2 [log] [=] $\Rightarrow$ 1.301029996

SELF-STUDY EXERCISES 11-6

L.O.1. Rewrite the following as logarithmic equations.

1. $3^2 = 9$ **2.** $2^5 = 32$ **3.** $9^{\frac{1}{2}} = 3$

4. $16^{\frac{1}{4}} = 2$ **5.** $4^{-2} = \frac{1}{16}$ **6.** $3^{-4} = \frac{1}{81}$

L.O.2. Rewrite the following as exponential equations.

7. $\log_3 81 = 4$ **8.** $\log_{12} 144 = 2$ **9.** $\log_2 \frac{1}{8} = -3$

10. $\log_5 \frac{1}{25} = -2$ **11.** $\log_{25} \frac{1}{5} = -0.5$ **12.** $\log_4 \frac{1}{2} = -0.5$

Solve for x by using an equivalent exponential expression.

13. $\log_4 64 = x$ **14.** $\log_3 x = -4$ **15.** $\log_6 36 = x$

16. $\log_7 \frac{1}{49} = x$ **17.** $\log_5 x = 4$ **18.** $\log_4 \frac{1}{256} = x$

L.O.3. Evaluate the following with a scientific calculator. Express the answer to the nearest ten-thousandth.

19. log 3 **20.** log 6 **21.** log 2.4
22. log 4.2 **23.** log 150 **24.** log 0.0012
25. *ln* 4 **26.** *ln* 2.5 **27.** *ln* 0.15
28. *ln* 275 **29.** *ln* 100

L.O.4. Evaluate the following logarithms with a calculator.

30. $\log_5 125$ **31.** $\log_3 729$ **32.** $\log_{\frac{1}{2}} 0.03125$

33. $\log_7 49$ **34.** $\log_8 56$

L.O.5. The intensity of an earthquake is measured on the *Richter scale* by the formula

$$\text{Richter scale rating} = \log \frac{I}{I_o}$$

where I_o is the measure of the intensity of a very small (faint) earthquake.

Find the Richter scale ratings of earthquakes having the following intensities:

35. $1000I_o$ **36. (b)** $100{,}000I_o$ **37.** $100{,}000{,}000I_o$

L.O.6. Solve for x by using an equivalent logarithmic expression and the facts $\log_3 2 = 0.631$ and $\log_3 5 = 1.465$.

38. $\log_3 10 = x$ **39.** $\log_3 8 = x$ **40.** $\log_3 6 = x$

ELECTRONICS APPLICATION

Prefixes Used in Technical Fields

Technical people must often deal with numbers that are very large or very small. It is difficult to read and to write numbers that have many decimal places in them, both to the left and to the right of the decimal point. To make things easier, these numbers are usually changed to the power-of-10 form, that is, the number times 10 to some power, which shows how many places you are away from the decimal point. Then the power-of-10 factor can be replaced with a letter that stands for that value.

$$9.876 \times 10^3 = 9.876 \times 1000 = 9876$$

The 3 is the exponent and can be read as +3, which means the decimal point moves three places to the right from where it is in the original 9.876; so you end up with 9876. The 10^3 can be replaced with k for kilo which means thousand. If the exponent is +6, then the 10^6 can be replaced with M for mega (million), and the decimal point moves 6 places to the right. If the exponent is +9, then the 10^9 can be replaced with G for giga (billion), and the decimal point moves 9 places to the right.

$$\begin{aligned}
9.876 \times 10^3 &= 9.876\text{ k} = 9.876 \times 1000 = 9876\\
9.876 \times 10^6 &= 9.876\text{ M} = 9.876 \times 1{,}000{,}000 = 9{,}876{,}000\\
9.876 \times 10^9 &= 9.876\text{ G} = 9.876 \times 1{,}000{,}000{,}000 = 9{,}876{,}000{,}000\\
9.876 \times 10^{12} &= 9.876\text{ T} = 9.876 \times 1{,}000{,}000{,}000{,}000\\
&= 9{,}876{,}000{,}000{,}000\\
9.876 \times 10^{15} &= 9.876\text{ P} = 9.876 \times 1{,}000{,}000{,}000{,}000{,}000\\
&= 9{,}876{,}000{,}000{,}000{,}000
\end{aligned}$$

If the exponent is negative, then the decimal point must move to the left. The prefix that replaces 10^{-3} is m for milli, and the prefix that replaces 10^{-6} is μ for micro. (The Greek letter mu, μ, was used to get an "m" sound for the prefix micro.)

$$\begin{aligned}
9.876 \times 10^{-3} &= 9.876\text{ m} = 9.876 \times 0.001 = 0.009876\\
9.876 \times 10^{-6} &= 9.876\ \mu = 9.87 \times 0.000001 = 0.000009876\\
9.876 \times 10^{-9} &= 9.876\text{ n} = 9.876 \times 0.000000001 = 0.000000009876\\
9.876 \times 10^{-12} &= 9.876\text{ p} = 9.876 \times 0.000000000001\\
&= 0.000000000009876\\
9.876 \times 10^{-15} &= 9.876\text{ f} = 9.876 \times 0.000000000000001\\
&= 0.000000000000009876
\end{aligned}$$

Our technology is moving so fast that it is important to learn the words to keep up with developments. Measurements are being made today in many fields in femto seconds; 1 femto second (1 f sec) means that the 1 is in the fifteenth place to the right of the decimal point by 15 zeros. This means that if it takes 1 femto second to do a calculation, then 1 peta can be done in 1 sec.

Don't try to memorize the prefixes. But be able to use them.

$$10^{15} = \text{P} = \text{peta} = \text{quadrillion}$$
$$10^{12} = \text{T} = \text{tera} = \text{trillion}$$
$$10^{9} = \text{G} = \text{giga} = \text{billion}$$
$$10^{6} = \text{M} = \text{mega} = \text{million}$$
$$10^{3} = \text{k} = \text{kilo} = \text{thousand}$$
$$10^{0} = (\text{unit} = \text{no prefix})$$
$$10^{-3} = \text{m} = \text{milli} = \text{thousandth}$$
$$10^{-6} = \mu = \text{micro} = \text{millionth}$$
$$10^{-9} = \text{n} = \text{nano} = \text{billionth}$$
$$10^{-12} = \text{p} = \text{pico} = \text{trillionth}$$
$$10^{-15} = \text{f} = \text{femto} = \text{quadrillionth}$$

Exercises on Prefixes

$$\text{P} = 10^{15} \quad \text{T} = 10^{12} \quad \text{G} = 10^{9} \quad \text{M} = 10^{6} \quad \text{k} = 10^{3}$$
$$\text{m} = 10^{-3} \quad \mu = 10^{-6} \quad \text{n} = 10^{-9} \quad \text{p} = 10^{-12} \quad \text{f} = 10^{-15}$$

Convert each of the following numbers into a basic unit with no prefix. Follow the examples. The letter that follows the prefix is a unit that you keep. These are common units used in electronics.

1. 9753 kW = 9,753,000 W
2. 15.68 μA = 0.00001568 A
3. 4.567 mA = ________ A
4. 3.471 MW = ________ W
5. 65.42 μS = ________ S
6. 4892 pF = ________ F
7. 15.89 km = ________ m
8. 3.51 GV = ________ V
9. 53.2 kΩ = ________ Ω
10. 1.30 fS = ________ S
11. 5.69 nH = ________ H
12. 75.6 pF = ________ F
13. 24.56 mm = ________ m
14. 876.54 μS = ________ S

Answers for Exercises on Prefixes

$P = 10^{15}$ $T = 10^{12}$ $G = 10^{9}$ $M = 10^{6}$ $k = 10^{3}$
$m = 10^{-3}$ $\mu = 10^{-6}$ $n = 10^{-9}$ $p = 10^{-12}$ $f = 10^{-15}$

3. 4.567 mA	=	0.004567	A
4. 3.471 MW	=	3,471,000	W
5. 65.42 μS	=	0.00006542	S
6. 4892 pF	=	0.000000004892	F
7. 15.89 km	=	15,890	m
8. 3.51 GV	=	3,510,000,000	V
9. 53.2 kΩ	=	53,200	Ω
10. 1.30 fS	=	0.00000000000000130	S
11. 5.69 nH	=	0.00000000569	H
12. 75.6 pF	=	0.0000000000756	F
13. 24.56 mm	=	0.02456	m
14. 876.54 μS	=	0.00087654	S

ASSIGNMENT EXERCISES

11-1

Evaluate using a calculator.

1. 3^4 **2.** 2^{-10} **3.** 5^{12}
4. 16^{-2} **5.** 10^{-4} **6.** 15^4
7. 12^8 **8.** 143^{-4} **9.** $\left(\frac{3}{5}\right)^4$
10. $(-7)^5$

Perform the indicated operations. Write answers with positive exponents.

11. $x^5 \cdot x^5$ **12.** $x^2(x^4)$ **13.** $x^{-1}(x^7)$
14. $\frac{x^7}{x^4}$ **15.** $\frac{x^8}{x^5}$ **16.** $\frac{x^3}{x^5}$
17. $(x^3)^4$ **18.** $(-x^3)^3$ **19.** $(x^{-3})^{-5}$

11-2

Simplify the following.

20. $4x^3 + 7x - 3x^3 - 5x$ **21.** $8x - 2x^4 - 3x^3 + 5x - x^3$ **22.** $5x - 3x + 7x^2 - 8x$
23. $4x^2 - 3y^2 + 7x^2 - 8y^2$ **24.** $4x^3(-3x^4)$ **25.** $-7x^8(-3x^{-2})$
26. $\frac{12x^5}{6x^3}$ **27.** $\frac{12x^7}{-18x^4}$ **28.** $\frac{11x^4}{22x^7}$
29. $5x(2x^2 + 3x - 4)$ **30.** $8x^3(2x - 6)$ **31.** $\frac{6x^3 - 12x^2 + 21x}{3x}$
32. $\frac{25y^5 - 85y^3 + 70y^2}{-5y}$

11-3

Perform the indicated operations. Express as ordinary numbers.

33. $10^5 \cdot 10^7$ **34.** $10^{-2} \cdot 10^8$ **35.** $10^7 \cdot 10^{-10}$
36. 4.2×10^5 **37.** $8.73 \div 10^{-3}$ **38.** $5.6 \div 10^{-2}$

Change to scientific notation.

39. 52,000 **40.** 160 **41.** 0.00017

Perform the indicated operations and write the result in scientific notation.

42. $(4.2 \times 10^5)(3.9 \times 10^{-2})$

43. $(7.8 \times 10^{53})(5.6 \times 10^{72})$

44. $\dfrac{5.2 \times 10^8}{6.1 \times 10^5}$

45. $\dfrac{1.25 \times 10^3}{3.7 \times 10^{-8}}$

11-4

46. Describe a polynomial and give an example.

47. Is the expression $5x^3 - 3x^{-2}$ a polynomial? Why or why not?

48. Arrange the following polynomials in descending order and identify the degree of each polynomial, the leading term, and the leading coefficient.
(a) $5x + 3x^3 - 8 + x^2$
(b) $3y^5 - 7y - 8y^4 + 12$

11-5

A formula for electric current is $i = 1.50e^{-200t}$, where t is time in seconds. Calculate the current for the following times. Express answers in scientific notation.

49. 0.07 second **50.** 0.2 second **51.** 0.4 second

52. The number of grams of a chemical that will disolve in a solution is given by the formula $C = 100e^{0.05t}$, where t = temperature in degrees Celsius. Evaluate when:
(a) $t = 10$ **(b)** $t = 20$
(c) $t = 45$ **(d)** $t = 50$.

53. The compound amount when an investment is compounded continually (every instant) is expressed by the formula $A = Pe^{ni}$, where A = compounded amount, P = principal, n = number of years, and i = interest rate per year. Find the compound amount when:
(a) Principal = \$2000, interest = 8%, for 3 years.
(b) Principal = \$500, interest = 12%, for 9 months.

Solve for x.

54. $2^x = 2^6$ **55.** $3^x = 3^{-2}$ **56.** $2^{x+3} = 2^7$

57. $4^{x-2} = 4^2$ **58.** $5^{2x-1} = 5^2$ **59.** $6^{3x+2} = 6^{-3}$

60. $2^x = 16$ **61.** $3^x = 81$ **62.** $3^x = \dfrac{1}{9}$

63. $2^x = \dfrac{1}{32}$ **64.** $4^{2x} = 64$ **65.** $5^{3x} = 125$

66. $3^{4-x} = \dfrac{1}{27}$ **67.** $6^{2-x} = \dfrac{1}{36}$

Evaluate.

68. e^3 **69.** e^{-4} **70.** $e^{-0.12}$

71. e^{-10}

11-6

Rewrite the following as logarithmic equations.

72. $5^2 = 25$ **73.** $3^4 = 81$ **74.** $81^{\frac{1}{2}} = 9$

75. $27^{\frac{1}{3}} = 3$ **76.** $5^{-3} = \dfrac{1}{125}$ **77.** $4^{-3} = \dfrac{1}{64}$

78. $8^{-\frac{1}{3}} = \frac{1}{2}$

79. $9^{-\frac{1}{2}} = \frac{1}{3}$

80. $121^{\frac{1}{2}} = 11$

81. $12^{-2} = \frac{1}{144}$

82. Write a true exponential expression; convert it to logarithmic form.

Rewrite the following as exponential equations. Verify if the equation is true.

83. $\log_{11} 121 = 2$

84. $\log_3 81 = 4$

85. $\log_{15} 1 = 0$

86. $\log_{25} 5 = \frac{1}{2}$

87. $\log_7 7 = 1$

88. $\log_3 3 = 1$

89. $\log_4 \frac{1}{16} = -2$

90. $\log_2 \frac{1}{16} = -4$

91. $\log_9 \frac{1}{3} = -0.5$

92. $\log_{16} \frac{1}{4} = -0.5$

Evaluate the following with a scientific calculator. Express answer to the nearest ten-thousandth.

93. log 5

94. log 3.8

95. log 180

96. log 0.0015

97. log 0.4

98. *ln* 12

99. *ln* 270

100. *ln* 0.134

101. *ln* 0.8

102. *ln* 80

103. $\log_5 30$

104. $\log_7 120$

Solve for x by using an equivalent exponential expression.

105. $\log_4 16 = x$

106. $\log_7 49 = x$

107. $\log_7 x = 3$

108. $\log_5 x = -2$

109. $\log_6 \frac{1}{36} = x$

110. $\log_4 \frac{1}{64} = x$

111. The intensity of an earthquake is measured on the Richter scale by the formula

$$\text{Richter scale rating} = \log \frac{I}{I_o}$$

where I_o is the measure of the intensity of a very small (faint) earthquake. Find the Richter scale ratings of the earthquakes having the following intensities;
(a) $100I_o$ **(b)** $10{,}000I_o$
(c) $150{,}000{,}000I_o$

Solve for x by using an equivalent logarithmic expression and the facts $\log_2 3 = 1.585$ and $\log_2 7 = 2.807$.

112. $\log_2 21$

113. $\log_2 9$

114. $\log_2 6$

CHALLENGE PROBLEM

115. The Environmental Protection Agency (EPA) monitors atmosphere and soil contamination by dangerous chemicals. When possible, the chemical contamination is decomposed by using microorganisms which change the chemicals so they are no longer harmful. A particular microorganism can reduce the contamination level to about 65% of the existing level every 30 days. A soil test for a contaminated site shows 72,000,000 units per cubic meter of soil.

(a) Write a formula for determining the contamination level after x 30-day periods: the initial contamination times the percent reduction (65% or 0.65) raised to the xth power.
(b) What is the level of contamination after 60 days? after 150 days?
(c) A ''safe level'' is 60,000 units of contamination per cubic meter of soil. Estimate how long it would take for the soil to reach this safe level. Discuss your method of arriving at the estimate.

CONCEPTS ANALYSIS

Write the following laws of exponents and properties in words. Also, give an example illustrating each.

1. $a^m \cdot a^n = a^{m+n}$

2. $\dfrac{a^m}{a^n} = a^{m-n}, \quad a \neq 0$

3. $a^0 = 1, \quad a \neq 0$

4. $a^{-n} = \dfrac{1}{a^n}, \quad a \neq 0$

5. $(a^m)^n = a^{mn}$

6. $\left(\dfrac{a}{b}\right)^n = \dfrac{a^n}{b^n}, \quad b \neq 0$

7. $(ab)^n = a^n b^n$

8. What two conditions must be satisfied before a number is in scientific notation?

9. Explain in words what is meant by the following mathematical statement: If $b^x = b^y$, $b > 0$ and $b \neq 1$, then $x = y$. Write an equation that meets these conditions and solve it.

10. Show symbolically the relationship between an exponential equation and a logarithm. Give an example illustrating this relationship.

11. Explain the difference between a common logarithm and a natural logarithm.

Find the mistake in the following examples. Explain the mistake and correct it.

12. $\dfrac{9x^3 - 12x^2 + 3x}{3x} = 3x^2 - 4x$

13. $x^5(x^3) = x^{15}$

14. $\dfrac{3.4 \times 10^5}{2 \times 10^{-2}} = 1.7 \times 10^3$

15.
$$\begin{aligned} 2^{x-3} &= 4^2 \\ x - 3 &= 2 \\ x &= 2 + 3 \\ x &= 5 \end{aligned}$$

SUMMARY

Objectives	What to Remember	Examples
Section 11-1		
1. Find a specified power of natural numbers, whole numbers, integers, fractions, and decimals.	Use the general power key to find a power of a number.	Use the general power key to find the following powers: 7^3, $(-8)^3$, 0.12^4 7 [x^y] 3 [=] ⇒ 343. 8 [+/−] [x^y] 3 [=] ⇒ −512. 0.12 [x^y] 4 [=] ⇒ 0.0002073

2. Multiply powers with like bases.	To multiply powers with like bases, add the exponents and keep the common base as the base of the product.	Multiply. $x^5(x^{-7}) = x^{5+(-7)} = x^{-2}$
3. Divide powers with like bases.	To divide powers with like bases, subtract the exponents and keep the common base as the base of the quotient.	Divide. $\frac{a^7}{a^4} = a^{7-4} = a^3$
4. Find a power of a power.	To find a power of a power, multiply the exponents and keep the same base.	Simplify. $(y^5)^4 = y^{5(4)} = y^{20}$
Section 11-2		
1. Add or subtract like terms containing powers.	Like terms are terms that not only have the same letter factors, but that also have the same exponent. Like terms are added or subtracted by adding or subtracting the coefficients of the like terms and keeping the variable factors and their exponents exactly the same.	Simplify. $3x^4 + 8x^2 - 7x^2 + 2x^4 = 5x^4 + x^2$
2. Multiply or divide algebraic expressions containing powers.	To multiply or divide algebraic expressions containing powers, multiply or divide the coefficients; then add or subtract the exponents of like bases.	Simplify. $(3x^4y^5)(7x^2yz) = 21x^6y^6z$
Section 11-3		
1. Multiply and divide by powers of 10.	To multiply by powers of 10, add the exponents and keep the base of 10. To divide by powers of 10, subtract the exponents and keep the base of 10. To use the calculator for these operations, use the EXP or EE key.	Multiply. $10^6(10)^7 = 10^{13}$
2. Change a number from scientific notation to ordinary notation.	To change a number from scientific notation to ordinary notation, perform the indicated multiplication by moving the decimal point in the first factor the appropriate number of places. Insert zeros as necessary. Omit the power of 10 factor.	Write 3.27×10^{-4} as an ordinary number. Shift the decimal 4 places to the *left.* 0.000327
3. Change a number from ordinary notation to scientific notation.	To change a number written in ordinary notation to scientific notation, insert a caret in the proper place to indicate where the decimal should be positioned so that the number is valued at 1 or between 1 and 10. Then determine how many places and in which direction the decimal shifts from the new position (caret) to the old position (decimal point). This number will be the exponent of the power of 10.	Write 54,000 in scientific notation. $5_\wedge4000. = 5.4 \times 10^4$

4. Multiply and divide numbers in scientific notation.

To multiply numbers in scientific notation, multiply the first factors; then multiply the powers of 10 by using the laws of exponents. Next, examine the first factor of the product to see if its value is equal to 1 or between 1 and 10. If the factor is 1 or between 1 and 10, the process is complete. If the factor is not 1 or between 1 and 10, shift the decimal so that the factor is equal to 1 or between 1 and 10 and adjust the exponent of the power of 10 accordingly.

Multiply.

$(4.5 \times 10^{89})(7.5 \times 10^{36}) =$
$33.75 \times 10^{125} =$
3.375×10^{126}
$3_\wedge375 \times 10^{1} \times 10^{125}$

To divide numbers in scientific notation, use steps similar to multiplication, but apply the rule for the division of signed numbers and the division law of exponents for division.

Divide.

$$(3 \times 10^{-3}) \div (4 \times 10^{2}) =$$
$$0.75 \times 10^{-5} =$$
$$07_\wedge5 \times 10^{-1} \times 10^{-5} =$$
$$7.5 \times 10^{-6}$$

Section 11-4

1. Identify polynomials, monomials, binomials, and trinomials.

Polynomials are algebraic expressions in which the exponents of the variable are nonnegative integers. A monomial is a polynomial with a single term. A binomial is a polynomial containing two terms. A trinomial is a polynomial containing three terms.

Give an example of a polynomial, a monomial, a binomial, and a trinomial:

polynomial: $4x^3 + 6x^2 - x + 3$
monomial: $8x$
binomial: $6x + 3$
trinomial: $8x^2 - 4x - 3$

2. Identify the degree of terms and polynomials.

The degree of a term containing one variable is the exponent of the variable. The degree of a polynomial that has only one variable is the degree of the term that has the largest exponent.

What is the degree of the polynomial listed in the previous example?
The highest exponent is 3; thus the degree of the polynomial is 3.

3. Arrange polynomials in descending order.

To arrange polynomials in descending order, list the term that has the highest degree first, the term that has the next highest degree second, and so on, until all terms have been listed.

Arrange the following polynomial in descending order.
$4 - 2x + 7x^5 - 3x^2 + x^3$
Descending order.
$7x^5 + x^3 - 3x^2 - 2x + 4$

Section 11-5

1. Evaluate formulas with at least one exponential term.

A scientific or graphing calculator can be used along with the order of operations to evaluate formulas with at least one exponential term.

Use Formula 11-5.1 for compound interest to find the compound amount for a loan of \$5000 for 3 years at an annual interest rate of 6% if the principal is compounded semiannually.

$$A = P\left(1 + \frac{r}{n}\right)^{nt}$$
$$A = 5000\left(1 + \frac{0.06}{2}\right)^{2(3)}$$
$$A = 5000(1.03)^6$$
$$A = 5000(1.194052297)$$
$$A = \$5970.26 \text{ (rounded)}$$

2. Evaluate formulas that contain a power of the natural exponential e.	Use a scientific or graphing calculator to evaluate formulas containing the natural exponential e.	Use the formula $P = 760e^{-0.00013h}$ for atmospheric pressure to find P at 50 meters above sea level (h). $P = 760e^{-0.00013(50)}$ $P = 760(0.9935210793)$ $P = 755.08$ (rounded)
3. Solve exponential equations in the form $b^x = b^y$, where $b > 0$ and $b \neq 1$.	To solve an exponential equation in the form $b^x = b^y$, apply the property that states, when the bases are equal, the exponents are equal. If the bases are not equal, rewrite the bases as powers so that the bases are equal, simplify, and then the exponents will be equal.	Solve the equation $4^{3x+1} = 8$. Rewrite the bases: $4 = 2^2$ and $8 = 2^3$ $2^{2(3x+1)} = 2^3$ So, $2(3x + 1) = 3$ $6x + 2 = 3$ $6x = 1$ $x = \frac{1}{6}$

Section 11-6

1. Write exponential expressions as equivalent logarithmic expressions.	To write exponential expressions as equivalent logarithmic expressions, use the following format: $x = b^y$ converts to $\log_b x = y$.	Write $16 = 2^4$ in logarithmic form. $\log_2 16 = 4$
2. Write logarithmic expressions as exponential expressions.	To write logarithmic expressions as exponential expressions, use the following format: $\log_b x = y$ converts to $x = b^y$.	Write $\log_5 125 = 3$ in exponential form. $5^3 = 125$
3. Evaluate common and natural logarithmic expressions using a calculator.	Use the [log] key to find common logarithms and the [ln] key to find natural logarithms. Most scientific calculators require the number to be entered, followed by the [log] or [ln] key. Most graphing calculators require the [log] or [ln] key to be entered followed by the number.	Use a calculator to find log 25. Scientific calculator steps: 25 [log] ⇒ 1.397940009
4. Evaluate logarithms with a base other than 10 or e.	To evaluate logarithms with bases other than 10 or e, divide the common or natural log of the number by the common or natural log of the base.	Evaluate $\log_4 64$: $\log_4 64 = \frac{\log 64}{\log 4}$ $= 3$
5. Evaluate formulas containing at least one logarithmic term.	To evaluate formulas containing at least one logarithmic term, use a calculator and follow the order of operations.	Use the formula bel $= 10 \log \left(\frac{I}{I_0}\right)$ to find the decibel rating for a sound that is 350 times the threshold sound ($350I_o$). $\text{bel} = 10 \log \left(\frac{350I_o}{I_o}\right)$ $bel = 10 \log 350$ $bel = 25.44$ (rounded)

6. Simplify logarithmic expressions by using the properties of logarithms.	The laws of exponents also apply to expressions with logarithms.	Write log (3)(8) in another way. Use a calculator to verify the equation is true.
		log (3)(8) = log 3 + log 8

$\log_b mn = \log_b m + \log_b n$

$\log_b \frac{m}{n} = \log_b m - \log_b n$

$\log_b m^n = n \log_b m$

WORDS TO KNOW

laws of exponents, (p. 428)
zero exponent, (p. 429)
negative exponent, (p. 430)
like terms, (p. 434)
powers of 10, (p. 438)
scientific notation, (p. 441)
ordinary number, (p. 442)
polynomials, (p. 448)
monomial, (p. 449)
binomial, (p. 449)
trinomial, (p. 449)
degree of a term, (p. 449)
constant term, (p. 450)
linear term, (p. 450)
quadratic term, (p. 450)
cubic term, (p. 450)
degree of a polynomial, (p. 450)
quadratic polynomial, (p. 450)
descending order, (p. 450)
ascending order, (p. 450)
leading term, (p. 450)
leading coefficient, (p. 450)
exponential expression, (p. 452)
compound interest, (p. 452)
compound amount, (p. 452)
natural exponential, (p. 454)
exponential equation, (p. 456)
result of exponentiation, (p. 456)
logarithmic expression, (p. 457)
common logarithm, (p. 458)
natural logarithm, (p. 458)
decibel, (p. 460)
threshold sound, (p. 460)

CHAPTER TRIAL TEST

Perform the indicated operations. Write the answers with positive exponents.

1. $(x^4)(x)$

2. $x^7(x^5)$

3. $\frac{x^6}{x^3}$

4. $\frac{x^0}{x^2}$

5. $\left(\frac{4}{7}\right)^2$

6. $(6a^2b)^2$

7. $\left(\frac{x^2}{y}\right)^2$

8. $3x(4x^3)$

9. $(-2a^4)(3a^2)$

10. $\frac{12x^2}{4x^3}$

11. $4a(3a^2 - 2a + 5)$

12. $\frac{60x^3 - 45x^2 - 5x}{5x}$

13. $(10^3)^2$

14. $\frac{10^{-5}}{10^3}$

Write as ordinary numbers.

15. 42×10^3

16. 0.83×10^2

17. 420×10^{-2}

18. 21×10^{-3}

19. $42 \div 10^3$

20. $8.4 \div 10^{-2}$

Write in scientific notation.

21. 240

22. 5.2301

23. 0.00086

24. 39×10^5

25. 783×10^{-5}

Perform the indicated operations. Express the answers in scientific notation.

26. $(5.9 \times 10^5)(3.1 \times 10^4)$

27. $(7.2 \times 10^{-3})(4.1 \times 10^2)$

28. $\dfrac{2.87 \times 10^5}{3.5 \times 10^7}$

29. $\dfrac{5.25 \times 10^4}{1.5 \times 10^2}$

30. A star is 5.9 light-years from Earth. If 1 light-year is 9.45×10^{12} km, how many kilometers from Earth is the star?

31. The total resistance (in ohms) of a dc series circuit equals the total voltage divided by the total amperage. If the total voltage is 3×10^3 V and the total amperage is 2×10^{-3} A, find the total resistance (in ohms) expressed as an ordinary number.

Evaluate using a scientific calculator.

32. 5^{-8}

33. $15^{\frac{3}{2}}$

34. $e^{-0.25}$

35. 12^5

Solve for x.

36. $4^x = 4^{-3}$

37. $2^{x-4} = 2^5$

38. $3^x = \dfrac{1}{9}$

39. $2^{2x-1} = 8$

Rewrite the following as logarithmic equations.

40. $2^8 = 256$

41. $4^{-\frac{1}{2}} = \dfrac{1}{2}$

Rewrite the following as exponential equations.

42. $\log_5 625 = 4$

43. $\log_3 \dfrac{1}{27} = -3$

Evaluate the following with a scientific calculator. Express the answer to the nearest ten-thousandth.

44. $\log 4.8$

45. $ln\ 32$

Solve for x by using an equivalent exponential expression and a scientific calculator.

46. $\log_4 x = -2$

47. $\log_6 216 = x$

Use a calculator to evaluate the following. Round to ten thousandths.

48. $\log_3 5$

49. $\log_2 6$

50. $\log_7 2$

51. $\log_8 21$

52. The formula for population growth of a certain species of insects in a controlled research environment is

$$P = 1{,}000{,}000e^{0.05t}$$

where t = time in weeks. Find the projected population after (a) 2 weeks and (b) 3 weeks.

53. The revenue in thousands of dollars from sales of a product is approximated by the formula

$$S = 125 + 83 \log (5t + 1)$$

where t is the number of years after a product was marketed. Find the projected revenue from sales of the product after 3 years.

54. The height in meters of the male members of a certain group is approximated by the formula

$$h = 0.4 + \log t$$

where t represents age in years for the first 20 years of life ($1 \leq t \leq 20$). Find the projected height of a 10-year-old male.

CHAPTER

12 Roots and Radicals

Frequently, squares and square roots of quantities are used in solving equations. In this chapter we discuss both *square roots* and *square root radicals*, as well as other roots and radicals.

12-1 SQUARE ROOTS

Learning Objectives

L.O.1. Find the square root of nonnegative real numbers.
L.O.2. Find the square root of variables.

Recall from Chapter 1 that when a number has an exponent of 2 we say it is ''squared'' or that the number is used as a factor 2 times. That is, we square a number by multiplying the number times itself, such as $5^2 = 5(5) = 25$. The 5 is called the *square root* of 25.

■ **DEFINITION 12-1.1 Square Root.** The **square root** of a number is that number which, when multplied times itself, equals the original number.

L.O.1. Find the square root of nonnegative real numbers.

Finding Square Roots of Nonnegative Real Numbers

Squaring a number and finding the square root of a number are inverse operations. $3^2 = 9$, and the square root of 9 is 3 (since 3 times 3 is 9).

Every positive number actually has *two* square roots. We have just seen that $3(3) = 9$, so 3 is a square root of 9. Also, $(-3)(-3) = 9$, so -3 is also a square root of 9.

Similarly, $7^2 = 7(7) = 49$ and $(-7)^2 = (-7)(-7) = 49$. Thus, the square roots of 49 are $+7$ and -7.

Since every positive number has two square roots, we can express both roots by using the symbol $\pm$. This symbol is read *plus or minus*. Thus, the square root of 49 is ± 7. One symbol used to indicate square roots is the radical sign, $\sqrt{}$. The expression that is placed under the radical sign is called the *radicand*.

The square root radical sign has an understood *index* or root of 2. The *square root radical symbol* $\sqrt{}$ can also be shown as $\sqrt[2]{}$. Similarly, the cube root radical symbol has an index of 3, $\sqrt[3]{}$; the fourth root radical symbol is $\sqrt[4]{}$; and so on.

Square roots and other roots can also be expressed as powers with fractional exponents. The radicand is the base of the power, the numerator of the exponent is 1, and the denominator of the exponent is the index of the root.

$$\sqrt{x} = x^{\frac{1}{2}}, \qquad \sqrt[3]{x} = x^{\frac{1}{3}}, \qquad \sqrt[4]{x} = x^{\frac{1}{4}}$$

We will extend our discussion of fractional exponents later in this chapter.

The square roots of nonnegative numbers are *real numbers*. The square roots of negative numbers are called *imaginary numbers*. Imaginary numbers are discussed in greater detail later in this chapter.

The positive square root of 64 is indicated by $\sqrt{64}$ or $64^{\frac{1}{2}}$. The negative square root of 64 is indicated by $-\sqrt{64}$ or $-(64^{\frac{1}{2}})$. If we wish to indicate the positive and negative square roots of 64, we write $\pm\sqrt{64}$ or $\pm(64^{\frac{1}{2}})$. The positive square root of a number is called the *principal square root*.

The procedure for finding square roots is more complicated than the other basic operations of adding, subtracting, multiplying dividing, and raising to powers. Because of the availability of inexpensive calculators that will give square roots, square roots on the job are rarely calculated by hand. In our discussion we will rely on calculators to find square roots and other roots.

In working with squares and square roots, we often mention the term *perfect square.*

■ **Definition 12-1.2 Perfect Square.** A **perfect square** is a nonnegative number or expression whose square root is an exact real number or expression.

Exact numbers can be whole numbers, fractions, decimals, letter terms, or quantities expressed in scientific notation.

Example 12-1.1 Generate a list of whole numbers from 0 through 400 that are perfect squares.

Solution

n	**Perfect Square (n^2)**	n	**Perfect Square (n^2)**
0	0	11	121
1	1	12	144
2	4	13	169
3	9	14	196
4	16	15	225
5	25	16	256
6	36	17	289
7	49	18	324
8	64	19	361
9	81	20	400
10	100		

Although repeated use of whole-number perfect squares will naturally result in remembering this list, the list can be generated at will if the numbers are forgotten.

■

A fraction is a *perfect square* when both the numerator and denominator are perfect squares. The square root of a perfect-square fraction can be found by taking the square roots of both the numerator and denominator.

General Tips for Using the Calculator

There is more than one way to find the square root of a number using a calculator. The most direct and practical method is to use the square-root key, $\boxed{\sqrt{\ }}$.

Scientific: 1. Enter the radicand.
2. Press the square-root key, $\boxed{\sqrt{\ }}$.

Graphing: 1. Press the square-root key, $\boxed{\sqrt{\ }}$.
2. Enter the radicand.
3. Execute, $\boxed{\text{EXE}}$ or $\boxed{\text{Enter}}$.

Find the square root of 169.

Scientific: 169 $\boxed{\sqrt{\ }} \Rightarrow 13.$

Graphing: $\boxed{\sqrt{\ }}$ 169 $\boxed{\text{EXE}} \Rightarrow 13.$

Now we will look at other methods for finding roots using a calculator. Although these methods are not practical for square roots, they will help us to learn the capabilities of our calculator. We will examine the scientific calculator. Using the general-root key, $\boxed{\sqrt[y]{x}}$:

1. Enter radicand.
2. Press general-root key, $\boxed{\sqrt[y]{x}}$.
3. Enter index of root.
4. Press equal.

$$169 \ \boxed{\sqrt[y]{x}} \ 2 \ \boxed{=} \Rightarrow 13$$

Using the fractional-exponent key $\boxed{x^{\frac{1}{y}}}$:

1. Enter radicand.
2. Press fractional-exponent key, $\boxed{x^{\frac{1}{y}}}$.
3. Enter denominator of fractional exponent.
4. Press equal.

$$169 \ \boxed{x^{\frac{1}{y}}} \ 2 \ \boxed{=} \Rightarrow 13$$

Using the general-power key $\boxed{x^y}$.

1. Enter radicand.
2. Press general-power key, $\boxed{x^y}$.
3. Enter fractional exponent or decimal equivalent.
4. Press equal.

$$169 \ \boxed{x^y} \ 1 \ \boxed{a\frac{b}{c}} \ 2 \ \boxed{=} \Rightarrow 13$$

or

$$169 \ \boxed{x^y} \ \boxed{(} \ 1 \ \boxed{\div} \ 2 \ \boxed{)} \ \boxed{=} \Rightarrow 13$$

or

$$169 \ \boxed{x^y} \ .5 \ \boxed{=} \Rightarrow 13$$

EXAMPLE 12-1.2 Find the principal square roots of the following fractions.

(a) $\frac{4}{9}$ **(b)** $\frac{1}{16}$ **(c)** $\frac{36}{49}$

Solution

(a) $\sqrt{\frac{4}{9}} = \frac{\sqrt{4}}{\sqrt{9}} = \frac{2}{3}$ **(b)** $\sqrt{\frac{1}{16}} = \frac{\sqrt{1}}{\sqrt{16}} = \frac{1}{4}$

(c) $\sqrt{\frac{36}{49}} = \frac{\sqrt{36}}{\sqrt{49}} = \frac{6}{7}$ ■

Perfect-square decimals are harder to recognize. Since $0.1 \times 0.1 = 0.01$, $\sqrt{0.01} = 0.1$. Simlarly, $\sqrt{0.16} = 0.4$ and $\sqrt{0.0081} = 0.09$. In practice, a calculator is generally used even with perfect-square decimals and numbers written in scientific notation.

The square roots of perfect squares are rational numbers and exact amounts. The square roots of numbers that are not perfect squares are *irrational numbers.* The notation used to represent the exact amount of an irrational number is called radical or exact notation. To express the exact amount of irrational numbers, we use the radical sign. For example, $\sqrt{2}$ is an exact amount while 1.414213562 is an approximate value for $\sqrt{2}$.

EXAMPLE 12-1.3 Find the square root of the following to the nearest thousandth.

(a) $\sqrt{15}$ **(b)** $-\sqrt{3.47}$ **(c)** $\pm\sqrt{0.175}$

Solution

	Scientific Calculator Steps
(a) $\sqrt{15} = 3.873$	15 [$\sqrt{\ }$]
(b) $-\sqrt{3.47} = -1.863$	3.47 [$\sqrt{\ }$] [+/−]
(c) $\pm\sqrt{0.175} = \pm 0.418$	.175 [$\sqrt{\ }$] *(± sign must be affixed manually.)*

■

L.O.2. Find the square root of variables.

Finding Square Roots of Variables

The square root of a variable is found by using the fractional exponent notation and the laws of exponents. Since at this point we are only considering real numbers, we will *assume* that the variables represent positive values.

$$\sqrt{x^2} = (x^2)^{\frac{1}{2}} = x^1 = x$$
$$\sqrt{x^4} = (x^4)^{\frac{1}{2}} = x^2$$
$$\sqrt{x^6} = (x^6)^{\frac{1}{2}} = x^3$$

EXAMPLE 12-1.4 Find the square root of the following positive variables.

(a) $\sqrt{x^8}$ **(b)** $-\sqrt{x^{12}}$ **(c)** $\pm\sqrt{x^{18}}$ **(d)** $\sqrt{\frac{a^2}{b^4}}$

Solution

(a) $\sqrt{x^8} = (x^8)^{\frac{1}{2}} = x^4$

(b) $-\sqrt{x^{12}} = -(x^{12})^{\frac{1}{2}} = -x^6$

(c) $\pm\sqrt{x^{18}} = \pm(x^{18})^{\frac{1}{2}} = \pm x^9$

(d) $\sqrt{\frac{a^2}{b^4}} = \frac{(a^2)^{\frac{1}{2}}}{(b^4)^{\frac{1}{2}}} = \frac{a}{b^2}$ ■

Tips and Traps

A positive variable with an even exponent is a perfect square.

To find the square root of a variable factor, multiply the exponent by $\frac{1}{2}$.

SELF-STUDY EXERCISES 12-1

L.O.1. Find the principal square root of each of the following.

1. $\frac{1}{25}$ **2.** $\frac{49}{81}$ **3.** $\frac{121}{144}$

4. 0.04 **5.** 0.25 **6.** 0.0016

Find the square root of each of the following. Round to the nearest thousandth.

7. $\sqrt{19}$ **8.** $-\sqrt{5.268}$ **9.** $\pm\sqrt{3.472}$

10. $-\sqrt{0.87}$

L.O.2. Find the square root of each of the following positive variables.

11. $\sqrt{x^{16}}$ **12.** $\sqrt{x^6}$ **13.** $-\sqrt{x^{30}}$

14. $\pm\sqrt{x^8}$ **15.** $\sqrt{\frac{a^{12}}{b^{22}}}$

12-2 SIMPLIFYING SQUARE-ROOT RADICALS WITH PERFECT-SQUARE FACTORS IN THE RADICAND

Learning Objectives

L.O.1. Apply the inverse property of squaring and square roots.
L.O.2. Simplify square-root radicals containing perfect-square factors.

L.O.1. Apply the inverse property of squaring and square roots.

Inverse Property of Squares and Square Roots

When simplifying fractions, we reduced factors that were common to both the numerator and denominator. We will simplify radicals when the radicand has perfect-square factors. To simplify radicals, we will apply the *inverse property of squares and square roots:* that is, squaring and finding the square root are inverse operations.

Rule 12-2.1 *Inverse property of squares and square roots:*

If the square root of a positive number is squared, the result is the original number.

$$(\sqrt{x})^2 = \left(x^{\frac{1}{2}}\right)^2 = x, \text{ for positive values of } x$$

If the square root of a squared positive number is taken, the result is the original number.

$$\sqrt{x^2} = (x^2)^{\frac{1}{2}} = x, \quad \text{for positive values of } x.$$

Example 12-2.1 Simplify the following expressions.

(a) $(\sqrt{6})^2$ **(b)** $\sqrt{5^2}$ **(c)** $(\sqrt{a})^2$ **(d)** $\left(\sqrt{\frac{1}{2}}\right)^2$ **(e)** $\sqrt{R^2}$

Solution

(a) $(\sqrt{6})^2 = 6$ **(b)** $\sqrt{5^2} = 5$ **(c)** $(\sqrt{a})^2 = a$

(d) $\left(\sqrt{\frac{1}{2}}\right)^2 = \frac{1}{2}$ **(e)** $\sqrt{R^2} = R$ ■

L.O.2. Simplify square-root radicals containing perfect-square factors.

Simplifying Square-Root Radicals

Some radicands will be made up of factors that are perfect squares, while others will not be perfect squares. In the latter case, it may be useful *to factor* the radicand.

■ **Definition 12-2.1 To factor.** To **factor an algebraic expression** is to write it as the indicated product of two or more factors, that is, as a multiplication.

Some radicands that are not perfect squares can be *factored* into a perfect square times another factor. We simplify radicals by taking the square root of as many perfect-square factors as possible. The factors that are not perfect squares are left under the radical sign, and the square root of the perfect-square factors will be outside the radical: $\sqrt{4x} = 2\sqrt{x}$; $\sqrt{3x^2} = x\sqrt{3}$.

A number such as 8 is not a perfect square; however, it does have a perfect square factor. 4 is a factor of 8 and 4 is a perfect square. Thus, $\sqrt{8}$ can be written as $\sqrt{4 \cdot 2} = 2\sqrt{2}$.

Similarly, x^3 is not a perfect square because the exponent is not even. However, any power having an odd exponent larger than 1 has a perfect square factor. Remember, $a^m a^n = a^{m+n}$. Thus, the perfect-square factor of any factor with an odd number larger than 1 for an exponent can be written as a^{n-1} (where n is odd). The remaining factor will be written as a^1. Therefore, $a^{n-1} \cdot a^1 = a^n$

Thus, $a^3 = a^2a^1$. Then $\sqrt{a^3}$ can be simplified as $\sqrt{a^2 \cdot a^1} = a\sqrt{a}$. To find a^2 and a^1 as factors of the power a^3, we subtract the exponent 1 from the odd exponent 3.

We will simplify any radical expression by taking the square root of ALL perfect-square factors. This procedure can be summarized as a rule.

Rule 12-2.2 *To simplify square-root radicals containing perfect-square factors:*

1. If the radicand is a perfect square, express it as a square root without the radical sign.
2. If the radicand is *not* a perfect square, factor the radicand into as many perfect-square factors as possible. The square root of the perfect-square factors will appear *outside* the radical and the other factors will stay *inside* (under) the radical sign.

If the radicand is *not* a perfect square and *cannot* be factored into one or more perfect-square factors, it is already in simplified form.

EXAMPLE 12-2.2 Simplify the following radicals.

(a) $\sqrt{21ab^2}$ **(b)** $\sqrt{32}$ **(c)** $\sqrt{y^7}$ **(d)** $\sqrt{18x^5}$
(e) $\sqrt{75xy^3z^5}$ **(f)** $\sqrt{7}$

Solution: Some perfect squares that can be used in simplifying radicals are 4, 9, 16, 25, 36, 49, 64, 81.

(a) $\sqrt{21ab^2} = \sqrt{21}(\sqrt{a})(\sqrt{b^2}) = \sqrt{21}(\sqrt{a})(b) = b\sqrt{21a}$
Factors removed from under the radical sign are written before the radical. Since the factors 21 and a are not perfect squares and have no perfect-square factors, they are written under the same radical sign.

(b) $\sqrt{32}$
What is the *largest* perfect-square factor of 32? 4 is a factor of 32, but 16 is also a factor of 32. Use the *largest* perfect-square factor.

$$\sqrt{32} = \sqrt{16 \cdot 2} = \sqrt{16}(\sqrt{2}) = 4\sqrt{2}$$

(c) $\sqrt{y^7}$
The largest perfect square factor of y^7 is y^{7-1} or y^6.

$$\sqrt{y^7} = \sqrt{y^6 \cdot y^1} = \sqrt{y^6}(\sqrt{y})$$
$$= y^3\sqrt{y}$$

(d) $\sqrt{18x^5} = \sqrt{9 \cdot 2 \cdot x^4 \cdot x^1}$
$= \sqrt{9}(\sqrt{2})(\sqrt{x^4})(\sqrt{x})$
$= 3(\sqrt{2})(x^2)\sqrt{x}$
$= 3x^2\sqrt{2x}$

In this example we showed each step in our simplifying process. However, we customarily do most of these steps mentally.

$$\sqrt{18x^5} = \sqrt{9 \cdot 2 \cdot x^4 \cdot x} = 3x^2\sqrt{2x}$$

Write the square roots of the perfect-square factors outside the radical sign. The other factors are written under the radical sign.

(e) $\sqrt{75xy^3z^5} = \sqrt{25 \cdot 3 \cdot x \cdot y^2 \cdot y \cdot z^4 \cdot z}$
$= 5yz^2\sqrt{3xyz}$

(f) $\sqrt{7} = \sqrt{7}$ *(7 contains no perfect-square factors.)*

Tips and Traps

- Whole-number perfect squares: 1, 4, 9, 16, 25, 36, 49, 64, 81, 100, 121, 144, ...
- One is a factor of any number. $8 = 8 \cdot 1$. To factor using the perfect square 1 does not simplify a radicand.
- Any variable with an exponent higher than 1 is a perfect square or has a perfect-square factor.
- Perfect-square variables:

$$x^2, x^4, x^6, x^8, x^{10}, \ldots$$

- Variables with a perfect-square factor:

$$x^3 = x^2 \cdot x^1, \qquad x^5 = x^4 \cdot x^1$$
$$x^7 = x^6 \cdot x^1, \qquad x^9 = x^8 \cdot x^1$$

- A convenient way to keep up with perfect-square factors is to circle them. The square roots of circled factors are written outside the radical sign. The uncircled factors stay in the radicand as is.

$$\sqrt{75ab^4c^3} = \sqrt{\textcircled{25} \cdot 3 \cdot a \cdot \textcircled{b^4} \cdot \textcircled{c^2} \cdot c^1}$$
$$= 5b^2c\sqrt{3ac}$$

Example 12-2.3 Simplify the following.

(a) $\sqrt{7x^2}$ (b) $\sqrt{9a}$ (c) $\sqrt{32m^5n^6}$

Solution

(a) $\sqrt{7\textcircled{x^2}} = x\sqrt{7}$

(b) $\sqrt{\textcircled{9}a} = 3\sqrt{a}$

(c) $\sqrt{32m^5n^6} = \sqrt{\textcircled{16} \cdot 2 \cdot \textcircled{m^4} \cdot m \cdot \textcircled{n^6}} = 4m^2n^3\sqrt{2m}$ ■

SELF-STUDY EXERCISES 12-2

L.O.1. Simplify the following expressions. Variables are positive.

1. $(\sqrt{7})^2$ **2.** $\sqrt{19^2}$ **3.** $\sqrt{x^2}$

4. $\left(\sqrt{\dfrac{1}{5}}\right)^2$ **5.** $(\sqrt{j})^2$ **6.** $(\sqrt{5.3})^2$

L.O.2. Simplify.

7. $\sqrt{24}$ **8.** $\sqrt{98}$ **9.** $\sqrt{48}$

10. $\sqrt{x^9}$ **11.** $\sqrt{y^{15}}$ **12.** $\sqrt{12x^3}$

13. $\sqrt{56a^5}$ **14.** $\sqrt{72a^3x^4}$ **15.** $\sqrt{44x^5y^2z^7}$

16. Design a square-root radical with the product of a constant factor that is not a perfect square, but has a perfect-square factor, and a variable factor with an odd exponent greater than 1. Then simplify.

12-3 BASIC OPERATIONS WITH SQUARE-ROOT RADICALS

Learning Objectives

L.O.1. Add or subtract square-root radicals.
L.O.2. Multiply square-root radicals
L.O.3. Divide square-root radicals.
L.O.4. Rationalize a numerator or a denominator.

L.O.1. Add or subtract square-root radicals.

Adding and Subtracting Radicals

As you recall, when adding or subtracting measures or algebraic terms, we can only add or subtract like quantities. Similarly, only *like* radicals can be added or subtracted. Square-root radicals are *like* radicals whenever the radicands are identical.

■ **DEFINITION 12-3.1 Like Radicals.** When the radicands are identical and the radicals have the same order or index, the radicals are **like radicals**.

Rule 12-3.1 ***To add or subtract like square-root radicals:***

Add or subtract the coefficients of the radicals. The common radical will be used in the solution.

$$a\sqrt{b} + c\sqrt{b} = (a + c)\sqrt{b}$$

EXAMPLE 12-3.1 Add or subtract the following radicals when possible.

(a) $3\sqrt{7} + 2\sqrt{7}$ **(b)** $4\sqrt{2} + \sqrt{2}$ **(c)** $3 + \sqrt{3}$
(d) $\frac{7}{8}\sqrt{5} - \frac{3}{8}\sqrt{5}$ **(e)** $3\sqrt{11} - 3\sqrt{11}$ **(f)** $2\sqrt{2} - \sqrt{3}$

Solution

(a) $3\sqrt{7} + 2\sqrt{7} = 5\sqrt{7}$
(b) $4\sqrt{2} + \sqrt{2} = 5\sqrt{2}$
When no coefficient is written in front of a radical, the coefficient is 1.
(c) $3 + \sqrt{3}$
No addition can be performed. These are not like radicals.
(d) $\frac{7}{8}\sqrt{5} - \frac{3}{8}\sqrt{5} = \frac{4}{8}\sqrt{5}$ or $\frac{1}{2}\left(\frac{\sqrt{5}}{1}\right) = \frac{\sqrt{5}}{2}$
(e) $3\sqrt{11} - 3\sqrt{11} = 0\sqrt{11} = 0$
Zero times any number is zero.
(f) $2\sqrt{2} - \sqrt{3}$
The terms cannot be combined. These are not like radicals. ■

We can add or subtract square-root radical expressions only if they have *like* radicands. Sometimes we can simplify the radical expressions and obtain like radicands. If we obtain like radicands, then we can add or subtract.

EXAMPLE 12-3.2 Add or subtract the following radical expressions.

(a) $12\sqrt{5} + 3\sqrt{20}$ **(b)** $\sqrt{3} - \sqrt{27}$ **(c)** $6\sqrt{3} + 2\sqrt{8}$

Solution

(a) $12\sqrt{5} + 3\sqrt{20}$ *($\sqrt{20}$ can be simplified.)*
$12\sqrt{5} + 3\sqrt{4 \cdot 5}$
$12\sqrt{5} + 3 \cdot 2\sqrt{5}$
$12\sqrt{5} + 6\sqrt{5}$
$18\sqrt{5}$

(b) $\sqrt{3} - \sqrt{27}$ *($\sqrt{27}$ can be simplified.)*
$\sqrt{3} - \sqrt{9 \cdot 3}$
$\sqrt{3} - 3\sqrt{3}$
$-2\sqrt{3}$

(c) $6\sqrt{3} + 2\sqrt{8}$ *($\sqrt{8}$ can be simplified.)*
$6\sqrt{3} + 2\sqrt{4 \cdot 2}$
$6\sqrt{3} + 2 \cdot 2\sqrt{2}$
$6\sqrt{3} + 4\sqrt{2}$ *(Terms cannot be combined. Radicals are still unlike radicals.)* ■

L.O.2. Multiply square-root radicals.

Multiplying Radicals

When multiplying two square-root radicals, the expressions under the radical signs (radicands) are multiplied together. Numbers in front of the radical signs are coefficients of the radicals and are multiplied separately.

Rule 12-3.2 ***To multiply square-root radicals:***

1. The coefficients are multiplied to give the coefficient of the product.
2. The radicands are multiplied to give the radicand of the product.

$$a\sqrt{b} \cdot c\sqrt{d} = ac\sqrt{bd}$$

EXAMPLE 12-3.3 Multiply the following radicals.

(a) $\sqrt{3} \cdot \sqrt{5}$ (b) $\sqrt{\frac{7}{8}} \cdot \sqrt{\frac{2}{3}}$ (c) $3\sqrt{2} \cdot 4\sqrt{3}$
(d) $\sqrt{3} \cdot \sqrt{12}$ (e) $2\sqrt{2} \cdot 3\sqrt{5}$

Solution

(a) $\sqrt{3} \cdot \sqrt{5} = \sqrt{15}$

(b) $\sqrt{\frac{7}{8}} \cdot \sqrt{\frac{2}{3}} = \sqrt{\frac{14}{24}} = \sqrt{\frac{7}{12}}$ *(Reduce.)*

(c) $3\sqrt{2} \cdot 4\sqrt{3} = 12\sqrt{6}$ *(Multiply coefficients 3 and 4. Then multiply radicands 2 and 3.)*

(d) $\sqrt{3} \cdot \sqrt{12} = \sqrt{36} = 6$ *(36 is a perfect square.)*

(e) $2\sqrt{2} \cdot 3\sqrt{5} = 6\sqrt{10}$ ■

L.O.3. Divide square-root radicals.

Dividing Radicals

When dividing square-root radicals, we follow a similar procedure. Coefficients and radicands are divided (or reduced) separately.

> **Rule 12-3.3** *To divide square-root radicals:*
>
> 1. The coefficients are divided to give the coefficient of the quotient.
> 2. The radicands are divided to give the radicand of the quotient.
>
> $$\frac{a\sqrt{b}}{c\sqrt{d}} = \frac{a}{c}\sqrt{\frac{b}{d}}$$

EXAMPLE 12-3.4 Divide the following radicals.

(a) $\dfrac{\sqrt{12}}{\sqrt{4}}$ **(b)** $\dfrac{\sqrt{\frac{2}{3}}}{\sqrt{\frac{7}{4}}}$ **(c)** $\dfrac{3\sqrt{6}}{6}$ **(d)** $\dfrac{5\sqrt{20}}{\sqrt{10}}$ **(e)** $\dfrac{\sqrt{50}}{\sqrt{2}}$

Solution

(a) $\dfrac{\sqrt{12}}{\sqrt{4}} = \sqrt{\dfrac{12}{4}} = \sqrt{3}$

(b) $\dfrac{\sqrt{\frac{2}{3}}}{\sqrt{\frac{7}{4}}} = \sqrt{\dfrac{\frac{2}{3}}{\frac{7}{4}}} = \sqrt{\dfrac{2}{3}\left(\dfrac{4}{7}\right)}$ $\quad \left(\dfrac{2}{3} \div \dfrac{7}{4} = \dfrac{2}{3} \cdot \dfrac{4}{7}\right)$

$\sqrt{\dfrac{8}{21}} = \sqrt{\dfrac{4 \cdot 2}{21}} = \dfrac{2\sqrt{2}}{\sqrt{21}}$

(c) $\dfrac{3\sqrt{6}}{6}$

The 6 in the denominator *will not* divide into the 6 in the numerator. The 6 in the numerator is a radicand and the 6 in the denominator is not. However, the 3 and the denominator 6 can be divided (or reduced) because they are both outside the radical. Remember, $\sqrt{6}$ is not 6, but 2.449 (the square root of 6).

$\dfrac{\overset{1}{\cancel{3}}\sqrt{6}}{\underset{2}{\cancel{6}}} = \dfrac{\sqrt{6}}{2}$ *(A coefficient of 1 does not have to be written in front of the radical.)*

(d) $\dfrac{5\sqrt{20}}{\sqrt{10}}$

The 5 and 10 are not divided or reduced because the 10 is a radicand and the 5 is not. However, the 20 and 10 can be divided since they are both square-root radicands.

$$\frac{5\sqrt{20}}{\sqrt{10}} = 5\sqrt{2}$$

(e) $\frac{\sqrt{50}}{\sqrt{2}} = \sqrt{25} = 5$ ■

L.O.4. Rationalize a numerator or a denominator.

Rationalizing a Numerator or a Denominator

Radicals are often simplified whenever they appear in the *denominator* of a fraction. Thus, $\frac{1}{\sqrt{3}}$ is not a simplified radical. Also, $\sqrt{\frac{2}{5}}$ is not simplified since it is equivalent to $\frac{\sqrt{2}}{\sqrt{5}}$ in which the denominator contains a radical. Whenever this condition occurs, we can change the denominator to a rational number by *rationalizing* the denominator.

Rule 12-3.4 ***To rationalize a denominator:***

1. Remove perfect-square factors from all radicands.
2. Multiply the denominator by another radical so that the resulting radicand is a perfect square.
3. To preserve the value of the fraction, multiply the numerator by the same radical. Thus, we multiply by an equivalent of 1.
4. Reduce the resulting fraction, if possible.

Example 12-3.5 Rationalize all denominators. Simplify the answers if possible.

(a) $\frac{1}{\sqrt{3}}$ (b) $\frac{5}{\sqrt{7}}$ (c) $\frac{2}{\sqrt{x}}$ (d) $\frac{4}{\sqrt{8}}$ (e) $\frac{5\sqrt{2}}{x^2\sqrt{3x}}$ (f) $\sqrt{\frac{2}{3}}$

Solution

(a) $\frac{1}{\sqrt{3}} = \frac{1}{\sqrt{3}} \cdot \frac{\sqrt{3}}{\sqrt{3}} = \frac{\sqrt{3}}{3}$ $\quad (\sqrt{3} \cdot \sqrt{3} = (\sqrt{3})^2 = 3)$

Remember, a radical times itself always equals the radicand of the original radical.

(b) $\frac{5}{\sqrt{7}} = \frac{5}{\sqrt{7}} \cdot \frac{\sqrt{7}}{\sqrt{7}} = \frac{5\sqrt{7}}{7}$ $\quad (\sqrt{7} \cdot \sqrt{7} = (\sqrt{7})^2 = 7)$

(c) $\frac{2}{\sqrt{x}} = \frac{2}{\sqrt{x}} \cdot \frac{\sqrt{x}}{\sqrt{x}} = \frac{2\sqrt{x}}{x}$ $\quad (\sqrt{x} \cdot \sqrt{x} = (\sqrt{x})^2 = x)$

(d) $\frac{4}{\sqrt{8}} = \frac{4}{\sqrt{4 \cdot 2}} = \frac{4}{2\sqrt{2}} = \frac{2}{\sqrt{2}}$ *(Simplify perfect-square factors then rationalize the denominator.)*

$$\frac{2}{\sqrt{2}} \cdot \frac{\sqrt{2}}{\sqrt{2}} = \frac{2\sqrt{2}}{2} = \sqrt{2}$$

(e) $\frac{5\sqrt{2}}{x^2\sqrt{3x}} = \frac{5\sqrt{2}}{x^2\sqrt{3x}} \cdot \frac{\sqrt{3x}}{\sqrt{3x}} = \frac{5\sqrt{6x}}{x^2 \cdot 3x} = \frac{5\sqrt{6x}}{3x^3}$

(f) $\sqrt{\frac{2}{3}} = \frac{\sqrt{2}}{\sqrt{3}} \cdot \frac{\sqrt{3}}{\sqrt{3}} = \frac{\sqrt{6}}{3}$ ■

When is rationalizing a denominator practical? Before calculators and computers, a radical in the denominator meant dividing by a decimal approximation of an irrational number. Calculations were more difficult and answers were generally less accurate. Are calculator values rounded to ten-thousandths less accurate if the denominator is not rationalized? Examine the next example.

EXAMPLE 12-3.6 Compare the approximate value to the nearest ten-thousandth of the following expressions:

$$\frac{1}{\sqrt{2}} \text{ and } \frac{\sqrt{2}}{2}$$

Solution

Scientific: 1 ÷ 2 [√] [=] ⇒ 0.70710678
2 [√] ÷ 2 [=] ⇒ 0.70710678

Graphing: 1 ÷ [√] 2 [EXE] ⇒ 0.7071067812
[√] 2 ÷ 2 [EXE] ⇒ 0.7071067812

The radical expressions are equivalent. ■

Tips and Traps

In reality, the importance of rationalizing radical expressions in technical applications is minimal. Approximate values and calculators and computers are most often used.

The most important use of rationalizing is to find alternate, but equivalent, representations of expressions.

Occasionally, it is desirable to *rationalize a numerator.* A procedure similar to rationalizing a denominator is used.

EXAMPLE 12-3.7 Change the following expression to an equivalent expression with a rational numerator.

$$\frac{\sqrt{3}}{5}$$

Solution

$\frac{\sqrt{3}}{5} \cdot \frac{\sqrt{3}}{\sqrt{3}} = \frac{3}{\sqrt{15}}$ *(Rationalize the numerator.)* ■

Traditionally, radical expressions are in simplest form if:

1. There are no perfect-square factors in any radicand.
2. There are no fractional radicands.
3. The denominator of a radical expression is rational or contains no radicals.

A logical sequence for simplifying radical expressions follows:

Rule 12-3.5 *To simplify a radical expression:*

1. Reduce radicands and coefficients whenever possible.
2. Simplify expressions with perfect-square factors in the radicand.
3. Reduce radicands and coefficients whenever possible.
4. Rationalize denominators that contain radicals.
5. Reduce radicands and coefficients whenever possible.

EXAMPLE 12-3.8 Perform the operation and simplify if possible.

(a) $\sqrt{y^3} \cdot \sqrt{8y^2}$ **(b)** $\sqrt{\frac{1}{3}} \cdot \sqrt{\frac{8}{3x}}$

Solution

(a) $\sqrt{y^3} \cdot \sqrt{8y^2} = \sqrt{8y^5} = \sqrt{4 \cdot 2 \cdot y^4 \cdot y} = 2y^2\sqrt{2y}$

(b) $\sqrt{\frac{1}{3}} \cdot \sqrt{\frac{8}{3x}} = \sqrt{\frac{8}{9x}} = \frac{\sqrt{8}}{\sqrt{9x}} = \frac{\sqrt{4 \cdot 2}}{\sqrt{9 \cdot x}} = \frac{2\sqrt{2}}{3\sqrt{x}} \cdot \frac{\sqrt{x}}{\sqrt{x}} = \frac{2\sqrt{2x}}{3x}$ ■

SELF-STUDY EXERCISES 12-3

L.O.1. Add or subtract. Simplify radicals where necessary.

1. $5\sqrt{3} + 7\sqrt{3}$
2. $8\sqrt{5} - 12\sqrt{5}$
3. $4\sqrt{7} + 3\sqrt{7} - 5\sqrt{7}$
4. $2\sqrt{3} - 8\sqrt{5} + 7\sqrt{3}$
5. $9\sqrt{11} - 3\sqrt{6} + 4\sqrt{6} - 12\sqrt{11}$
6. $44\sqrt{2} + \sqrt{3} - \sqrt{2} + 5\sqrt{3}$
7. $2\sqrt{3} + 5\sqrt{12}$
8. $7\sqrt{5} + 2\sqrt{45}$
9. $4\sqrt{63} - \sqrt{7}$
10. $3\sqrt{6} - 2\sqrt{54}$
11. $8\sqrt{2} - 3\sqrt{28}$
12. $2\sqrt{3} + \sqrt{48}$
13. $3\sqrt{5} + 4\sqrt{180}$
14. $7\sqrt{98} - 2\sqrt{2}$
15. $6\sqrt{40} - 2\sqrt{90}$
16. $\sqrt{12} - \sqrt{27}$

L.O.2. Multiply and simplify if possible.

17. $5\sqrt{2} \cdot 3\sqrt{5}$
18. $8\sqrt{3} \cdot 5\sqrt{12}$
19. $5\sqrt{3x} \cdot 4\sqrt{5x^2}$
20. $2x\sqrt{3x^4} \cdot 7x^2\sqrt{8x}$

Perform the indicated operations and simplify if possible.

21. $\sqrt{x^2} \cdot \sqrt{3x}$
22. $\sqrt{\frac{2}{3}} \cdot \sqrt{\frac{4}{5y}}$
23. $\sqrt{\frac{1}{x}} \cdot \sqrt{\frac{8}{7x}}$
24. $\sqrt{\frac{4}{9y^2}} \cdot \sqrt{\frac{1}{2y}}$
25. $\sqrt{\frac{8x}{3}} \cdot \sqrt{\frac{2x^2}{3}}$
26. $\sqrt{7} \cdot \sqrt{x^2}$
27. $\sqrt{\frac{4x^2}{x^3}} \cdot \sqrt{12x}$
28. $\sqrt{\frac{1}{2}} \cdot \sqrt{\frac{x^2}{3}}$
29. $\sqrt{\frac{y^3}{2}} \cdot \sqrt{\frac{2}{7y}}$
30. $\sqrt{8x} \cdot \sqrt{x^2}$

L.O.3. Divide and simplify.

31. $\frac{4\sqrt{12}}{2\sqrt{6}}$ **32.** $\frac{15\sqrt{24}}{9\sqrt{2}}$ **33.** $\frac{12x^2\sqrt{8x}}{15x\sqrt{6x^3}}$

34. $\frac{6x^4\sqrt{25x^3}}{2x\sqrt{16x^2}}$

L.O.4. Rationalize the denominator in the following.

35. $\frac{5}{\sqrt{3}}$ **36.** $\frac{6}{\sqrt{5}}$ **37.** $\frac{1}{\sqrt{8}}$

38. $\frac{\sqrt{5}}{\sqrt{11}}$ **39.** $\frac{\sqrt{8}}{\sqrt{12}}$ **40.** $\frac{5x}{\sqrt{3x}}$

41. $\frac{2x^2}{\sqrt{7x^2}}$ **42.** $\frac{4x^5}{\sqrt{12x^3}}$

Rationalize the numerator in the following.

43. $\frac{\sqrt{3}}{4}$ **44.** $\frac{\sqrt{5}}{\sqrt{7}}$ **45.** $\frac{\sqrt{11}}{\sqrt{7}}$

46. $\frac{\sqrt{21}}{\sqrt{7}}$ **47.** $\frac{5\sqrt{3}}{10}$

12-4 ROOTS AND RATIONAL EXPONENTS

Learning Objectives

L.O.1. Write radicals in fractional exponent form.
L.O.2. Perform operations with expressions with rational exponents.

L.O.1. Write radicals in fractional exponent form.

Writing Radicals in Fractional Exponent Form

The same properties and laws of exponents also apply to radical terms; however, it is difficult to see the comparison while the terms are in radical form. Also, many manipulations and simplifications of radical expressions can be accomplished more easily when the radical expressions are written in an alternative form. Expressions containing *fractional exponents*, often called *rational exponents*, are an alternative way of expressing radical expressions.

With fractional exponents, the laws of exponents as well as all arithmetic properties of fractions can be applied. For example, when raising a quantity to a power, multiply exponents. The exponent is then reduced to lowest terms.

Suppose that $\sqrt{x}$ is raised to the fourth power.

$$(x^{\frac{1}{2}})^4 = x^{\frac{4}{2}} = x^2 \qquad \left(\frac{1}{2} \times 4 = \frac{4}{2}\right)$$

Again, we apply the laws of exponents and arithmetic properties of fractions.

What happens if $\sqrt{x}$ is raised to the third power (cubed)?

$$(x^{\frac{1}{2}})^3 = x^{\frac{3}{2}}$$

This exponent is left in improper fraction form, and it brings up the need for further interpretation of the meaning of a rational or fractional exponent.

Rule 12-4.1 ***To convert between fractional exponent notation and radical notation:***

The numerator of a rational or fractional exponent represents the power of the radicand, and the denominator of the rational or fractional exponent represents the index or order of the root.

$$x^{\frac{\text{power}}{\text{root}}} = \sqrt[\text{root}]{x^{\text{power}}}$$

Then, is $(\sqrt{x})^2$ the same as $\sqrt{x^2}$? *Yes*, for $x \geq 0$.

$$(x^{\frac{1}{2}})^2 = x^{\frac{2}{2}} = x, \qquad (x^2)^{\frac{1}{2}} = x^{\frac{2}{2}} = x$$

Rule 12-4.2 ***To simplify radicals using fractional exponents:***

1. Convert the radicals to equivalent expressions using fractional exponents.
2. Apply the laws of exponents and the arithmetic of fractions.
3. Convert simplified expressions back to radical notation if desired.

EXAMPLE 12-4.1 Convert the following radical expressions to equivalent expressions using fractional exponents and simplify.

(a) $\sqrt[3]{x}$ **(b)** $\sqrt[5]{2y}$ **(c)** $(\sqrt{ab})^3$ **(d)** $\sqrt[4]{16b^8}$
(e) $(\sqrt[3]{27xy^5})^4$

Solution

(a) $\sqrt[3]{x} = x^{\frac{1}{3}}$ **(b)** $\sqrt[5]{2y} = (2y)^{\frac{1}{5}}$ or $2^{\frac{1}{5}}y^{\frac{1}{5}}$

(c) $(\sqrt{ab})^3 = (ab)^{\frac{3}{2}}$ or $a^{\frac{3}{2}}b^{\frac{3}{2}}$

(d) $\sqrt[4]{16b^8} = (2^4b^8)^{\frac{1}{4}} = (2^4)^{\frac{1}{4}}(b^8)^{\frac{1}{4}} = 2b^2$

The fourth root of 16 is the number that is used as a factor 4 times to equal 16. By inspection (or with a scientific calculator), $\sqrt[4]{16} = 2$.

(e) $(\sqrt[3]{27xy^5})^4 = (3^3x^1y^5)^{\frac{4}{3}} = (3^3)^{\frac{4}{3}}x^{\frac{4}{3}}(y^5)^{\frac{4}{3}} = 3^4x^{\frac{4}{3}}y^{\frac{20}{3}} = 81x^{\frac{4}{3}}y^{\frac{20}{3}}$

A numerical coefficient that is a perfect power, in this case a perfect cube, can be evaluated. However, if the coefficient is not a perfect power, we usually do not evaluate it unless specifically instructed to do so with a scientific calculator. In this case, 27 is a perfect cube; $3^3 = 27$. Therefore, $27^{\frac{4}{3}} = (3^3)^{\frac{4}{3}} = (3)^4 = 81$. Thus, we could write the expression as $81x^{\frac{4}{3}}y^{\frac{20}{3}}$. ■

L.O.2. Perform operations with expressions with rational exponents.

Performing Operations with Expressions with Rational Exponents

Other laws of exponents are applied to fractional exponents. Remember, the laws of exponents apply to factors having *like bases*. The following example illustrates other laws of exponents applied to fractional exponents.

EXAMPLE 12-4.2 Perform the following operations. Express answers with positive exponents in lowest terms.

(a) $(x^{\frac{3}{2}})(x^{\frac{1}{2}})$ **(b)** $(3a^{\frac{1}{2}}b^3)^2$ **(c)** $\dfrac{x^{\frac{1}{2}}}{x^{\frac{1}{3}}}$ **(d)** $\dfrac{10a^3}{2a^{\frac{1}{2}}}$

Solution

(a) $(x^{\frac{3}{2}})(x^{\frac{1}{2}}) = x^{\frac{3}{2}+\frac{1}{2}}x^{\frac{4}{2}} = x^2$ **(b)** $(3a^{\frac{1}{2}}b^3)^2 = 3^2ab^6 = 9ab^6$

(c) $\dfrac{x^{\frac{1}{2}}}{x^{\frac{1}{3}}} = x^{\frac{1}{2}-\frac{1}{3}} = x^{\frac{1}{6}}$ $\left(\dfrac{1}{2} - \dfrac{1}{3} = \dfrac{3}{6} - \dfrac{2}{6} = \dfrac{1}{6}\right)$

(d) $\dfrac{10a^3}{2a^{\frac{1}{2}}} = 5a^{3-\frac{1}{2}} = 5a^{\frac{5}{2}}$ $\left(3 - \dfrac{1}{2} = \dfrac{3}{1} - \dfrac{1}{2} = \dfrac{6}{2} - \dfrac{1}{2} = \dfrac{5}{2}\right)$ ■

Exponents can also be expressed as decimals: $\sqrt{x} = x^{\frac{1}{2}} = x^{0.5}$

EXAMPLE 12-4.3 Perform the following operations. Express answers with positive exponents.

(a) $a^{2.3}(a^3)$ **(b)** $(5a^{3.5}b^{0.5})^2$

Solution

(a) $a^{2.3}(a^3) = a^{2.3+3} = a^{5.3}$

What does $a^{5.3}$ mean? If written as an improper fraction, $a^{5.3} = a^{5\frac{3}{10}} = a^{\frac{53}{10}}$.

This means we take the tenth root of a to the 53rd power or $\sqrt[10]{a^{53}}$.

(b) $(5a^{3.5}b^{0.5})^2 = 5^2a^{3.5(2)}b^{0.5(2)}$
$= 25a^7b$ ■

SELF-STUDY EXERCISES 12-4

L.O.1. Convert the following radical expressions to equivalent expressions using fractional exponents and simplify.

1. $\sqrt[4]{a}$ **2.** $\sqrt[3]{m}$ **3.** $\sqrt[5]{3x}$
4. $\sqrt[4]{7a}$ **5.** $(\sqrt{xy})^4$ **6.** $\sqrt[3]{27a^6}$
7. $\sqrt[3]{32x^7}$ **8.** $(\sqrt{36x^3y^4})^4$ **9.** $(\sqrt[3]{16a^4b^7})^2$

L.O.2. Perform the following operations. Express answers with positive exponents in lowest terms.

10. $(a^{\frac{5}{2}})(a^{\frac{1}{2}})$ **11.** $(b^{\frac{3}{5}})(b^{\frac{1}{5}})$ **12.** $(4x^{\frac{1}{3}}y^4)^3$

13. $(2m^{\frac{1}{5}}n^2)^5$ **14.** $\dfrac{a^{\frac{3}{5}}}{a^{\frac{1}{5}}}$ **15.** $\dfrac{b^{\frac{1}{5}}}{b^{\frac{3}{5}}}$

16. $\dfrac{m^{\frac{7}{8}}}{m^{\frac{3}{4}}}$ **17.** $\dfrac{x^{\frac{1}{3}}}{x^{-\frac{5}{6}}}$ **18.** $\dfrac{a^2}{a^{\frac{1}{3}}}$

19. $\dfrac{14a^4}{7a^{\frac{3}{5}}}$ **20.** $\dfrac{8a^{\frac{2}{3}}}{24a^5}$ **21.** $\dfrac{6a^{\frac{7}{8}}}{27a^2}$

12-5 COMPLEX AND IMAGINARY NUMBERS

Learning Objectives

L.O.1. Write imaginary numbers using the letter i.
L.O.2. Raise imaginary numbers to powers.
L.O.3. Write real and imaginary numbers in complex form, $a + bi$.
L.O.4. Combine complex numbers.

L.O.1. Write imaginary numbers using the letter i.

Imaginary Numbers

Taking the square root of a negative number introduces a new type of number. This new type of number is called an imaginary number. Before we expand our number system to include imaginary numbers, we will look at the types of *real numbers*.

Real numbers include both rational and irrational numbers. Rational numbers include fractions, decimals, integers, whole numbers, and natural numbers. Real numbers and imaginary numbers make up a new set of numbers called *complex numbers*. Figure 12-1 illustrates the types of numbers that are included in complex numbers.

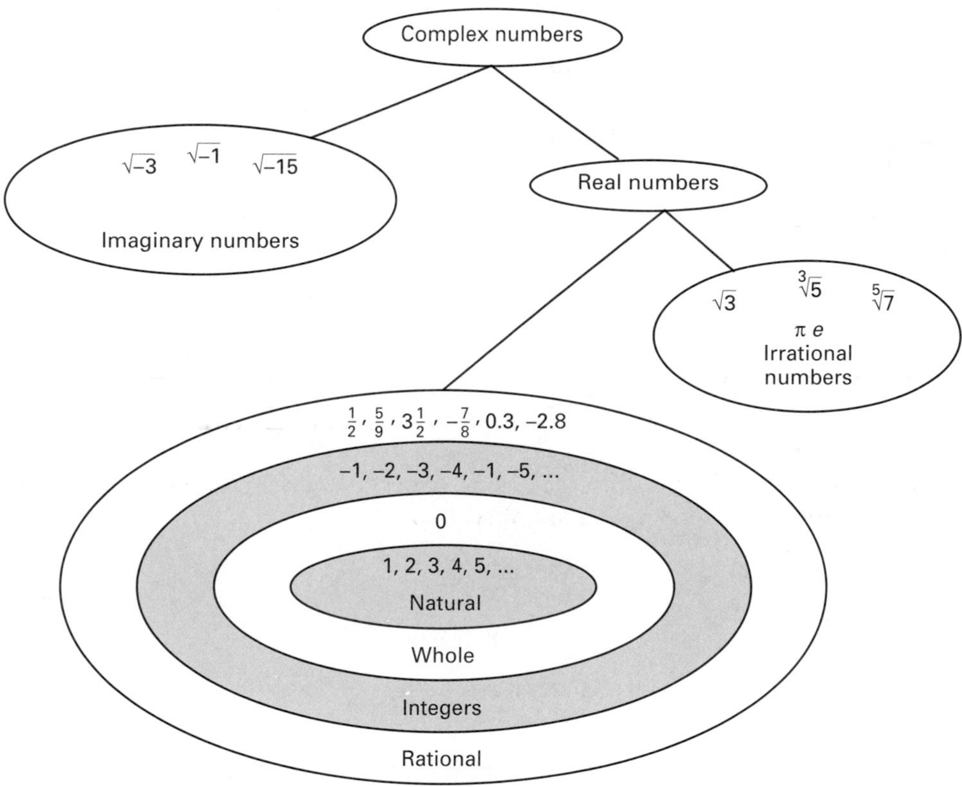

Figure 12-1

The square root of a negative number, say $\sqrt{-16}$, can be simplified as $\sqrt{-1 \cdot 16}$ or $4\sqrt{-1}$. Since the square root of negative one is an *imaginary* factor and 4 is a real factor, we will use the letter i to represent $\sqrt{-1}$ and rewrite $4\sqrt{-1}$ as $4i$. Similarly, $\sqrt{-4} = \sqrt{-1 \cdot 4} = 2\sqrt{-1} = 2i$.

■ **Definition 12-5.1 Imaginary Number.** An **imaginary number** has a factor of $\sqrt{-1}$, which is represented by i. ($i = \sqrt{-1}$)

Tips and Traps

Electronics and other applications of imaginary numbers use j to represent $\sqrt{-1}$ rather than i.

EXAMPLE 12-5.1 Rewrite the following imaginary numbers using the letter i for $\sqrt{-1}$.

(a) $\sqrt{-9}$ **(b)** $\sqrt{-25}$ **(c)** $\sqrt{-7}$

Solution

(a) $\sqrt{-9} = \sqrt{-1 \cdot 9} = 3\sqrt{-1} = 3i$

(b) $\sqrt{-25} = \sqrt{-1 \cdot 25} = 5\sqrt{-1} = 5i$

(c) $\sqrt{-7} = \sqrt{-1 \cdot 7} = \sqrt{7} \cdot \sqrt{-1} = \sqrt{7}i$ or $i\sqrt{7}$ · ■

Tips and Traps

$\sqrt{7i}$ and $\sqrt{7}i$ do not represent the same amount. Since it is easy to confuse the two terms, when the coefficient of an imaginary number is an irrational number, it is appropriate to write the i factor first: $\sqrt{7}i = i\sqrt{7}$

L.O.2. Raise imaginary numbers to powers.

Powers of Imaginary Numbers

Some powers of imaginary numbers are real numbers. Examine the pattern that develops with powers of i.

$$i = \sqrt{-1}$$
$$i^2 = (\sqrt{-1})^2 = -1$$
$$i^3 = i^2 \cdot i^1 = -1i = -i$$
$$i^4 = i^2 \cdot i^2 = -1(-1) = 1$$
$$i^5 = i^4 \cdot i^1 = 1(i) = i$$
$$i^6 = i^4 \cdot i^2 = 1(-1) = -1$$
$$i^7 = i^4 \cdot i^3 = 1(-i) = -i$$
$$i^8 = i^4 \cdot i^4 = 1(1) = 1$$
$$i^9 = i^4 \cdot i^4 \cdot i^1 = 1(1)(i) = i$$
$$i^{10} = i^4 \cdot i^4 \cdot i^2 = 1(1)(-1) = -1$$
$$i^{11} = i^4 \cdot i^4 \cdot i^3 = 1(1)(-i) = -i$$
$$i^{12} = i^4 \cdot i^4 \cdot i^4 = 1(1)(1) = 1$$

Rule 12-5.1 *To simplify a power of* **i:**

1. Divide the exponent by 4 and examine the **remainder**.
2. The power of i will simplify as follows, based on the remainder in step 1.

remainder of 0 $\Rightarrow$ 1

remainder of 1 $\Rightarrow$ i

remainder of 2 $\Rightarrow$ -1

remainder of 3 $\Rightarrow$ $-i$

EXAMPLE 12-5.2 Simplify the following powers of i.

(a) i^{15} **(b)** i^{20} **(c)** i^{33} **(d)** i^{18}

Solution

(a) $i^{15} = -i$ *(15 ÷ 4 = 3 R3, remainder of 3* $\Rightarrow -i$*)*
(b) $i^{20} = 1$ *(20 ÷ 4 = 5, remainder 0* $\Rightarrow 1$*)*
(c) $i^{33} = i$ *(33 ÷ 4 = 8 R1, remainder 1* $\Rightarrow i$*)*
(d) $i^{18} = -1$ *(18 ÷ 4 = 4 R2, remainder 2* $\Rightarrow -1$*)* ■

L.O.3. Write real and imaginary numbers in complex form, $a + bi$.

Complex Numbers

A complex number has two parts, a real part and an imaginary part.

■ **DEFINITION 12-5.2 Complex Number.** A **complex number** is a number that can be written in the form of $\boldsymbol{a + bi}$, where a and b are real numbers and i is $\sqrt{-1}$.

If $a = 0$, then $a + bi$ is the same as $0 + bi$ or bi, which is an imaginary number. If $b = 0$, then $a + bi$ is the same as $a + 0 \cdot i$ or a, which is a real number. Thus, real numbers and imaginary numbers are also complex numbers.

EXAMPLE 12-5.3 Rewrite the following in the form $a + bi$.

(a) 5 **(b)** $\sqrt{-36}$ **(c)** $-3i^2$

Solution

(a) $5 = 5 + 0i$ *(*$a = 5, b = 0$*)*
(b) $\sqrt{-36} = 6i = 0 + 6i$ *(*$a = 0, b = 6$*)*
(c) $-3i^2 = -3(-1) = 3 = 3 + 0i$ *(*$a = 3, b = 0$*)* ■

L.O.4. Combine complex numbers.

Combining Complex Numbers

Complex numbers are combined by adding *like* parts. Real parts are added together and imaginary parts are added together.

Rule 12-5.2 ***To combine complex numbers:***

1. Add real parts for the real part of the answer.
2. Add imaginary parts for the imaginary part of the answer.

$$(a + bi) + (c + di) = (a + c) + (b + d)i$$

EXAMPLE 12-5.4 Combine the following:

(a) $(3 + 5i) + (8 - 2i)$ **(b)** $(-3 + i) - (7 - 4i)$
(c) $-2 + (5 + 6i)$

Solution

(a) $(3 + 5i) + (8 - 2i) = 11 + 3i$ *(*$3 + 8 = 11; 5i - 2i = 3i$*)*
(b) $(-3 + i) - (7 - 4i) =$ *(Distribute −1.)*
$(-3 + i) + (-7 + 4i) = -10 + 5i$ *(*$-3 - 7 = -10; i + 4i = 5i$*)*
(c) $-2 + (5 + 6i) = 3 + 6i$ *(*$-2 + 5 = 3$*)* ■

SELF-STUDY EXERCISES 12-5

L.O.1. Write the following imaginary numbers using the letter i. Simplify if possible.

1. $\sqrt{-25}$ **2.** $\sqrt{-36}$ **3.** $\sqrt{-64x^2}$
4. $\sqrt{-32y^5}$

L.O.2. Simplify the following powers of i.

5. i^{17} **6.** i^{20} **7.** i^{24}
8. i^{32}

L.O.3. Write the following real and imaginary numbers in complex form.

9. 15 **10.** $33i$ **11.** $5 + \sqrt{-4}$
12. $8 + \sqrt{-32}$ **13.** $7 - \sqrt{-3}$ **14.** $-7i^3$
15. $4i^6$

L.O.4. Combine the following complex numbers.

16. $(4 + 3i) + (7 + 2i)$ **17.** $\sqrt{12} - 5\sqrt{-3} + (\sqrt{8} - \sqrt{-27})$
18. $12 - 3i - (8 - 4i)$ **19.** $15 + 8i - (3 - 12i)$
20. $(4 + 7i) - (3 - 2i)$

12-6 EQUATIONS WITH SQUARES AND SQUARE ROOTS

Learning Objectives

L.O.1. Solve equations with squared letter terms.
L.O.2. Solve equations with radical terms.

L.O.1. Solve equations with squared letter terms.

Equations with Squared Letter Terms

In solving equations with squared letter terms, we will use the same procedures for solving equations that we used in Chapter 9.

When squared letter terms are involved, we need to take one more step. After the equation is simplified to a basic equation, we will take the square root of both sides of the equation to find the roots. In doing so, the solutions will be square roots expressed with $\pm$ to show both the negative and the positive square-root values.

Rule 12-6.1 ***To solve an equation containing only squared variable terms and number terms:***

1. Perform all normal steps to isolate the squared variable and to obtain a numerical coefficient of 1.
2. Take the square root of both sides.

EXAMPLE 12-6.1 Solve the following equations:

(a) $x^2 = 4$ **(b)** $x^2 + 9 = 25$ **(c)** $5x^2 + 10 = 30$

(d) $\frac{3}{25} = \frac{x^2}{48}$ **(e)** $18 = \frac{x^2}{5}$

Solution

(a) $x^2 = 4$ *(Take square root of both sides.)*

$x = \pm 2$

(b) $x^2 + 9 = 25$ *(Isolate squared variable term.)*

$x^2 = 25 - 9$ *(Combine like terms.)*

$x^2 = 16$ *(Take square root of both sides.)*

$x = \pm 4$

(c) $5x^2 + 10 = 30$ *(Isolate squared variable term.)*

$5x^2 = 30 - 10$ *(Combine like terms.)*

$5x^2 = 20$ *(Divide by coefficient of 5.)*

$$\frac{5x^2}{5} = \frac{20}{5}$$

$x^2 = 4$

$x = \pm 2$

(d) $\frac{3}{25} = \frac{x^2}{48}$ *(To solve a proportion, cross multiply.)*

$3(48) = 25x^2$

$144 = 25x^2$

$$\frac{144}{25} = \frac{25x^2}{25}$$

$$\frac{144}{25} = x^2$$

mc2$\pm\frac{12}{5} = x$

(e) $18 = \frac{x^2}{5}$

$\frac{18}{1} = \frac{x^2}{5}$ *(Write as a proportion.)*

$18(5) = x^2$

$90 = x^2$ *($\sqrt{90} = \sqrt{9 \cdot 10} = 3\sqrt{10}$)*

$\pm\sqrt{90} = x$ *(exact answer)*

$\pm 3\sqrt{10} = x$

or

$\pm 9.487 = x$ *(approximate answer)* ■

L.O.2. Solve equations with radical terms.

Equations with Radicals

In this discussion we consider only equations containing a radical isolated on either or both sides of the equation. Whenever this is the case, we solve the equation by first squaring both sides of the equation. This will rid the equation of radicals. Then we can use previously learned procedures to finish solving the equation.

Rule 12-6.2 ***To solve an equation containing square-root radicals that are isolated on either or both sides of the equation:***

1. Square both sides to eliminate any radical.
2. Perform all normal steps to isolate the variable and solve the equation.

EXAMPLE 12-6.2 Solve the following equations.

(a) $\sqrt{x} = \dfrac{3}{4}$ **(b)** $\sqrt{x - 2} = 5$ **(c)** $\sqrt{\dfrac{x}{3}} = 7$ **(d)** $\sqrt{\dfrac{5}{x + 1}} = 10$

(e) $\sqrt{5x^2 - 36} = 8$ **(f)** $\sqrt{1.7x^2} = \sqrt{6.8}$

Solution

(a) *(Square both sides of equation.)*

$$\begin{aligned} \sqrt{x} &= \frac{3}{4} \\ (\sqrt{x})^2 &= \left(\frac{3}{4}\right)^2 \\ x &= \frac{9}{16} \end{aligned}$$

(b) *(Square both sides of equation.)* *(Any square-root radical squared equals the radicand.)*

$$\begin{aligned} \sqrt{x - 2} &= 5 \\ (\sqrt{x - 2})^2\text{mc-1} &= 5^2 \\ x - 2 &= 25 \\ x &= 25 + 2 \\ x &= 27 \end{aligned}$$

(c) *(Square both sides of equation.)*

$$\begin{aligned} \sqrt{\frac{x}{3}} &= 7 \\ \left(\sqrt{\frac{x}{3}}\right)^2 &= 7^2 \\ \frac{x}{3} &= 49 \\ \frac{x}{3} &= \frac{49}{1} \\ x &= 3(49) \\ x &= 147 \end{aligned}$$

(d) *(Square both sides of equation.)*

$$\begin{aligned} \sqrt{\frac{5}{x + 1}} &= 10 \\ \left(\sqrt{\frac{5}{x + 1}}\right)^2 &= 10^2 \\ \frac{5}{x + 1} &= 100 \\ \frac{5}{x + 1} &= \frac{100}{1} && \textit{(Cross multiply.)} \\ 5 &= 100(x + 1) && \textit{(Distribute.)} \\ 5 &= 100x + 100 && \textit{(Transpose.)} \\ 5 - 100 &= 100x \\ -95 &= 100x \\ \frac{-95}{100} &= \frac{100x}{100} \\ -\frac{19}{20} &= x \end{aligned}$$

In the previous problem each step was written out to help you recall the procedures in solving the equation. However, you should always do as many steps as possible mentally.

(e)
$$\begin{aligned}
\sqrt{5x^2 - 36} &= 8 \\
(\sqrt{5x^2 - 36})^2 &= 8^2\text{mc-1} && \textit{(Do this step mentally.)} \\
5x^2 - 36 &= 64 \\
5x^2 &= 64 + 36 \\
5x^2 &= 100 \\
\frac{5x^2}{5} &= \frac{100}{5} && \textit{(Do this step mentally.)} \\
x^2 &= 20 \\
x &= \pm\sqrt{20} && \textit{(Exact answer simplified} = \pm 2\sqrt{5}.) \\
x &= \pm 4.472 && \textit{(approximate answer)}
\end{aligned}$$

(f)
$$\begin{aligned}
\sqrt{1.7x^2} &= \sqrt{6.8} \\
1.7x^2 &= 6.8 \\
x^2 &= \frac{6.8}{1.7} \\
x^2 &= 4 \\
x &= \pm\text{mc-12}
\end{aligned}$$

■

EXAMPLE 12-6.3 Solve $\sqrt{2.4Y^2} = 5.4$.

Solution

$$\begin{aligned}
\sqrt{2.4Y^2} &= 5.4 \\
(\sqrt{2.4Y^2})^2 &= (5.4)^2 \\
2.4Y^2 &= 29.16 \\
Y^2 &= \frac{29.16}{2.4} \\
Y^2 &= 12.15 \\
Y &= \pm 3.486 && \textit{(approximate answer)}
\end{aligned}$$

■

EXAMPLE 12-6.4 Shirley Riddle needs to prepare a flower bed for planting. She needs a square bed with an area of 128 ft^2. How long should each side of the bed be?

Solution

$$\begin{aligned}
A &= s^2 && \textit{(formula for area of square)} \\
128 &= s^2 \\
\sqrt{128} &= s && \textit{(Exact answer simplified} = 8\sqrt{2}.) \\
11.3 &= s && \textit{(approximate answer)}
\end{aligned}$$

■

Tips and Traps

Exact Answers versus Approximate Answers

An answer is an exact amount when no rounding has been done. If an exact answer is desired, give the answer in simplest form.

If the answer has been rounded, it becomes an approximate answer. Approximate answers are often more meaningful than exact answers, especially in applied problems.

One Answer versus Two Answers

Look again at the six problems in Example 12-6.2. Parts (a) to (d) have variables to the first power after the radical is removed. Therefore, there will be *at most* one answer. Parts (e) and (f) have variables that are squared when the radical is removed. These equations will have *at most* two roots or solutions.

Extraneous Roots

The techniques of squaring both sides of an equation and taking the square root of both sides of an equation are equivalent to multiplying or dividing by a variable, thus introducing the possibility of an extraneous root. Be sure to check your roots.

SELF-STUDY EXERCISES 12-6

L.O.1. Solve the following equations. Use a calculator to evaluate to the nearest thousandth when necessary.

1. $y^2 = 25$ **2.** $-9 + 3x^2 = 18$ **3.** $3 = \frac{36}{x^2}$

4. $\frac{3}{81} = \frac{x^2}{12}$ **5.** $3 + R^2 = 12$ **6.** $\frac{1}{2}P^2 = \frac{3}{7}$

7. $\frac{4}{9} = \frac{x^2}{3}$ **8.** $16 - T^2 = 0$ **9.** $5x^2 - 10 = 20$

10. $5x^2 - 28 = 8$

L.O.2. Solve the following equations. Use a calculator if necessary. Evaluate to the nearest thousandth.

11. $\sqrt{x} = 9$ **12.** $\sqrt{\frac{y}{3}} = 8$ **13.** $8 = \sqrt{y - 2}$

14. $\sqrt{3x^2 - 3} = 9$ **15.** $\sqrt{\frac{3}{x + 2}} = 12$ **16.** $\sqrt{x^2} = 2$

17. $10 = \sqrt{2x}$ **18.** $\sqrt{\frac{2}{3}x^2} = 4$ **19.** $\sqrt{1.3y^2} = 2.4$

20. $\sqrt{y^2} = 2.9$

ASSIGNMENT EXERCISES

12-1

Find the principal square root.

1. $\frac{16}{25}$ **2.** $\frac{4}{9}$ **3.** $\frac{1}{49}$

4. 0.09 **5.** 0.49 **6.** 0.0081

Find the square root and round to the nearest tenth.

7. $\sqrt{35}$ **8.** $-\sqrt{4.3}$ **9.** $\pm\sqrt{7.32}$

10. $-\sqrt{0.37}$

Find the square root of the following if all variables represent positive numbers.

11. $\sqrt{y^{12}}$ **12.** $\sqrt{a^{10}}$ **13.** $-\sqrt{b^{18}}$

14. $\pm\sqrt{\frac{x^4}{y^6}}$

12-2

Simplify the following expressions.

15. $(\sqrt{5})^2$
16. $(\sqrt{x^5})^2$
17. $\sqrt{x^2}$
18. $(\sqrt{8})^2$
19. $\sqrt{9P^3}$
20. $\sqrt{8^2}$
21. $\sqrt{18a^2b}$
22. $\sqrt{12x^2y^3}$
23. $\sqrt{32x^5y^2}$
24. $\sqrt{63x^4y^7}$
25. $\sqrt{75x^{10}y^9}$
26. $\sqrt{125xy^3}$

12-3

Add or subtract the following radicals.

27. $3\sqrt{7} - 2\sqrt{28}$
28. $\sqrt{2} - \sqrt{8}$
29. $2\sqrt{6} + 3\sqrt{54}$
30. $3\sqrt{5} - 2\sqrt{45}$
31. $4\sqrt{3} - 8\sqrt{48}$
32. $\sqrt{40} + \sqrt{90}$
33. $5\sqrt{8} - 3\sqrt{50}$
34. $5\sqrt{7} - 4\sqrt{63}$
35. $3\sqrt{2} - 5\sqrt{32}$

Multiply the following radicals and simplify.

36. $\sqrt{6} \cdot \sqrt{3}$
37. $2\sqrt{8} \cdot 3\sqrt{6}$
38. $2\sqrt{a} \cdot \sqrt{b}$
39. $5\sqrt{3} \cdot 8\sqrt{7}$
40. $2\sqrt{3} \cdot 5\sqrt{18}$
41. $-8\sqrt{5} \cdot 4\sqrt{30}$

Divide the following radicals and simplify.

42. $\dfrac{4\sqrt{8}}{2}$
43. $\dfrac{3\sqrt{5}}{2\sqrt{20}}$
44. $\dfrac{2\sqrt{90}}{\sqrt{5}}$
45. $\dfrac{6\sqrt{18}}{8\sqrt{12}}$
46. $\dfrac{14\sqrt{56}}{7\sqrt{7}}$
47. $\dfrac{5\sqrt{48}}{20\sqrt{20}}$
48. $\dfrac{\sqrt{9x}}{\sqrt{3x}}$
49. $\dfrac{\sqrt{3y^3}}{\sqrt{y^3}}$
50. $\dfrac{81\sqrt{2}}{9\sqrt{4}}$
51. $\left(\sqrt{\dfrac{9}{16}}\right)^2$
52. $\sqrt{\dfrac{9c^4}{25y^6}}$

Rationalize the denominator and simplify.

53. $\dfrac{5}{\sqrt{17}}$
54. $\dfrac{1}{\sqrt{8}}$
55. $\dfrac{\sqrt{7}}{\sqrt{12}}$
56. $\dfrac{\sqrt{3}}{\sqrt{7x}}$

Rationalize the numerator and simplify.

57. $\dfrac{\sqrt{3}}{8}$
58. $\sqrt{7}$
59. $\dfrac{\sqrt{24}}{5}$
60. $\dfrac{\sqrt{15}}{5\sqrt{7}}$

12-4

Write the following using fractional exponents.

61. $\sqrt{x}$
62. $\sqrt[3]{x^5}$
63. $\sqrt[5]{x^4}$

Write the following in radical format.

64. $x^{\frac{5}{8}}$
65. $y^{\frac{3}{5}}$
66. $a^{\frac{1}{4}}$

Convert the following radicals to equivalent expressions using fractional exponents.

67. $\sqrt[3]{x}$
68. $\sqrt[4]{p}$
69. $\sqrt[5]{4y}$
70. $(\sqrt{ab})^6$
71. $\sqrt[3]{8b^{12}}$
72. $(\sqrt{49x^2y^3})^4$

Perform the following operations. Express answers with positive exponents in lowest terms.

73. $(a^{\frac{1}{2}})(a^{\frac{3}{2}})$ **74.** $(x^{\frac{4}{3}})(x^{\frac{2}{3}})$ **75.** $y^{\frac{3}{4}} \cdot y^{\frac{1}{4}}$

76. $y^{\frac{5}{8}} \cdot y^{\frac{1}{8}}$ **77.** $(3x^{\frac{1}{4}}y^2)^3$ **78.** $(2x^{\frac{3}{4}}y)^2$

79. $(4ax^{\frac{1}{2}})^3$ **80.** $(x^{\frac{1}{2}})^{\frac{1}{3}}$ **81.** $\dfrac{x^{\frac{3}{4}}}{x^{\frac{1}{4}}}$

82. $\dfrac{x^{\frac{1}{6}}}{x^{\frac{5}{6}}}$ **83.** $\dfrac{a^{\frac{5}{6}}}{a^{-\frac{1}{3}}}$ **84.** $\dfrac{a^{\frac{7}{10}}}{a^{\frac{2}{5}}}$

85. $\dfrac{x^{\frac{5}{8}}}{x^{\frac{3}{4}}}$ **86.** $\dfrac{y^{\frac{1}{3}}}{y^{\frac{5}{6}}}$ **87.** $\dfrac{a^3}{a^{\frac{1}{3}}}$

88. $\dfrac{a^2}{a^{\frac{3}{5}}}$ **89.** $\dfrac{12a^4}{6a^{\frac{1}{2}}}$ **90.** $\dfrac{27x^3}{9x^{\frac{2}{3}}}$

91. $\dfrac{15a^{\frac{3}{5}}}{10a^5}$ **92.** $\dfrac{14a^{\frac{5}{6}}}{24a^2}$

12-5

Write the following numbers using the letter i. Simplify if possible.

93. $\sqrt{-100}$ **94.** $-\sqrt{-16x^2}$ **95.** $\pm\sqrt{-24y^7}$

Simplify the powers of i.

96. i^5 **97.** i^{14} **98.** i^{98}

99. i^{77}

Write the following as complex numbers in simplified form.

100. 5 **101.** $15i$ **102.** $3 + \sqrt{-9}$

103. $-12i^5$ **104.** $-6i^{11}$

Simplify.

105. $(5 + 3i) + (2 - 7i)$ **106.** $(4 - i) - (3 - 2i)$

107. $(7 - \sqrt{-9}) + (4 + \sqrt{-16})$

12-6

Solve the following equations containing squares and radicals. Use a calculator if needed. Evaluate to the nearest thousandth.

108. $q^2 = 81$ **109.** $x^2 - 36 = 0$ **110.** $3x^2 - 2 = 7$

111. $x^2 - 4 = 0$ **112.** $7 + P^2 = 107$ **113.** $x^2 + 4 = 29$

114. $\sqrt{x - 1} = 7$ **115.** $18 = 2x^2$ **116.** $\sqrt{P + 2} = 12$

117. $\sqrt{\dfrac{27}{2}} = x$ **118.** $\sqrt{3 + x} = 14$ **119.** $\sqrt{q + 3} = 7$

120. $\sqrt{\dfrac{1}{4x}} = 2$ **121.** $\sqrt{1.3x^2} = 11.7$ **122.** $\sqrt{x^2 + 1} = 5$

123. $\sqrt{x^2 + 2} = 9$ **124.** $\sqrt{3 + y^2} = 10$ **125.** $\sqrt{Q^2 - 1} = 0$

126. $0 = \sqrt{z^2 - 4}$ **127.** $\sqrt{2 + y^2} = 8$

CHALLENGE PROBLEMS

128. List all whole numbers less than 100 that are not perfect squares, but have perfect square factors. Then write the square root of each number and simplify the square root.

Perform the indicated operation and express in simplest form.

129. $5i(3 - 4i)$

130. $12(7 + 2i)$

CONCEPTS ANALYSIS

Write the following rules in words. Assume that all radicands represent positive values.

1. $a\sqrt{b} \cdot c\sqrt{d} = ac\sqrt{bd}$

2. $\dfrac{a\sqrt{b}}{c\sqrt{d}} = \dfrac{a}{c}\sqrt{\dfrac{b}{d}};\quad c, d \neq 0$

3. $a\sqrt{b} + c\sqrt{b} = (a + c)\sqrt{b}$

4. $(\sqrt{x})^2 = x$ or $\sqrt{x^2} = x$, for positive values of x.

5. List the conditions for a radical expression to be in simplest form.

6. What does it mean to *rationalize* a denominator? What calculations or manipulations with fractions are easier if the denominator is a rational number?

7. What property of equality can be used to solve an equation that contains a squared variable term?

8. When should the procedure of taking the square root of both sides of an equation be used?

9. What property of equality can be used to solve an equation that contains a square root radical?

10. When should the procedure of squaring both sides of an equation be used?

11. Write the following property in words.

$$x^{\frac{1}{n}} = \sqrt[n]{x}$$

12. Write the following property in words.

$$x^{\frac{m}{n}} = \sqrt[n]{x^m} \text{ or } (\sqrt[n]{x})^m$$

Illustrate with numerical examples the following properties of radicals by using the laws of exponents and fractional exponents. Assume that all radicands are positive values.

13. $\sqrt[n]{xy} = \sqrt[n]{x} \cdot \sqrt[n]{y}$

14. $\sqrt[m]{\sqrt[n]{x}} = \sqrt[n]{\sqrt[m]{x}} = \sqrt[mn]{x}$

15. $\sqrt[n]{\dfrac{x}{y}} = \dfrac{\sqrt[n]{x}}{\sqrt[n]{y}}$

16. $x^{-\frac{m}{n}} = \dfrac{1}{x^{\frac{m}{n}}}$

CHAPTER SUMMARY

Objectives	What to Remember	Examples
Section 12-1		
1. Find the square root of nonnegative real numbers.	Nonnegative real numbers have a positive and a negative square root. The positive square root is called the principal square root. The sign in front of the radical symbol indicates which square root is to be considered.	Give the square root of the following: $\sqrt{25}$, $-\sqrt{81}$, $\pm\sqrt{64}$ $\sqrt{25} = 5$, $-\sqrt{81} = -9$, $\pm\sqrt{64} = \pm 8$

2. Find the square root of variables.	Variable factors are perfect squares if the exponent is divisible by 2. To find the square root of a perfect square variable with an exponent, take half of the exponent and keep the same base.	Give the square root of the following. x^6, x^{10}, x^{24} $\sqrt{x^6} = x^3$, $\sqrt{x^{10}} = x^5$, $\sqrt{x^{24}} = x^{12}$
Section 12-2		
1. Apply the inverse property of squaring and square roots.	If the square root of a positive number is squared, the result is the original number. If the square root is taken of a squared positive number, the result is the original number.	Simplify the following expressions. $(\sqrt{7})^2$, $\sqrt{13^2}$, $\left(\sqrt{\frac{3}{5}}\right)^2$, $\sqrt{m^2}$ $(\sqrt{7})^2 = 7$, $\sqrt{13^2} = 13$, $\left(\sqrt{\frac{3}{5}}\right)^2 = \frac{3}{5}$, $\sqrt{m^2} = m$
2. Simplify square root radicals containing perfect square factors.	To simplify square root radicals containing perfect square factors, factor constants and variables using the largest possible perfect square factor of the constant and of each variable. Note that the largest perfect square of a variable will be written with the largest possible even-numbered exponent.	Simplify the following radical expressions. $\sqrt{98}$, $\sqrt{x^{13}}$, $\sqrt{72y^9}$, $5\sqrt{12}$ $\sqrt{98} = \sqrt{49(2)} = 7\sqrt{2}$; $\sqrt{x^{13}} = \sqrt{x^{12}(x)} = x^6\sqrt{x}$, $\sqrt{72y^9} = \sqrt{36 \cdot 2 \cdot y^8 \cdot y} = 6y^4\sqrt{2y}$; $5\sqrt{12} = 5\sqrt{4 \cdot 3} = 5 \cdot 2\sqrt{3} = 10\sqrt{3}$
Section 12-3		
1. Add or subtract square root radicals.	To add or subtract square root radicals, first simplify all radical expressions; then add or subtract the coefficients of like radical terms. Like radical terms are terms that have exactly the same factors under the radical and have the same index.	Add or subtract. $3\sqrt{5x} + 7\sqrt{5x}$; $2\sqrt{18} - 5\sqrt{8}$ $3\sqrt{5x} + 7\sqrt{5x} = 10\sqrt{5x}$; $2\sqrt{18} - 5\sqrt{8} = 2\sqrt{9 \cdot 2} - 5\sqrt{4 \cdot 2} = 2 \cdot 3\sqrt{2} - 5 \cdot 2\sqrt{2} = 6\sqrt{2} - 10\sqrt{2} = -4\sqrt{2}$
2. Multiply square root radicals.	To multiply radicals that have the same index, multiply the coefficients and write as the coefficient of the product, and multiply the radicands and write as the radicand of the product; then simplify the radical.	Multiply. $5\sqrt{7} \cdot 8\sqrt{14} = 40\sqrt{98} = 40\sqrt{49 \cdot 2} = 40 \cdot 7\sqrt{2} = 280\sqrt{2}$
3. Divide square root radicals.	To divide radicals that have the same index, divide the coefficients for the coefficient of the quotient; then divide the radicands for the radicand of the quotient. Simplify any remaining radical expressions that can be simplified; then simplify any resulting coefficients that can be simplified.	Divide. $\frac{12\sqrt{75}}{8\sqrt{6}}$ $\frac{12\sqrt{75}}{8\sqrt{6}} = \frac{3\sqrt{3 \cdot 25}}{2\sqrt{3 \cdot 2}}$ $\frac{3\sqrt{25}}{2\sqrt{2}} = \frac{3 \cdot 5}{2\sqrt{2}} = \frac{15}{2\sqrt{2}}$
4. Rationalize a numerator or a denominator.	To rationalize a denominator of a radical expression, remove perfect-square factors from all radicands, multiply the denominator by another radical so that the resulting radicand is a perfect square, and multiply the numerator by the same radical the denominator was multiplied by; simplify the resulting radicals, and reduce the resulting fraction, if possible.	Rationalize the denominator. $\sqrt{\frac{3}{5}} = \frac{\sqrt{3}}{\sqrt{5}} \cdot \frac{\sqrt{5}}{\sqrt{5}} = \frac{\sqrt{15}}{5}$

	To rationalize a numerator, remove perfect-square factors from all radicands, multiply the numerator by another radical so that the resulting radicand is a perfect square, and multiply the denominator by the same radical the numerator was multiplied by; simplify the radicals and reduce the resulting fraction, if possible.	Rationalize the numerator. $\sqrt{\frac{3}{5}} = \frac{\sqrt{3}}{\sqrt{5}} \cdot \frac{\sqrt{3}}{\sqrt{3}} = \frac{\sqrt{9}}{\sqrt{15}} = \frac{3}{\sqrt{15}}$
Section 12-4		
1. Write radicals in fractional exponent form.	To write a radical expression as an expression with fractional exponents, the radicand is the base of the expression. The exponent of the radicand and the index of the radical are used to make a fractional exponent for the base. The exponent of the radicand is the numerator, and the index is the denominator of the fractional exponent.	Write in fractional exponent form. $\sqrt[5]{3^2} = 3^{\frac{2}{5}}$ Write in radical form: $5^{\frac{1}{2}} = \sqrt{5}$; $7^{\frac{3}{5}} = \sqrt[5]{7^3}$ or $\sqrt[5]{343}$
2. Perform operations with expressions with rational exponents.	All the rules governing exponents that were used for integral exponents apply to expressions with rational exponents.	Simplify. $x^{\frac{1}{3}} \cdot x^{\frac{2}{3}} = x^{\frac{1}{3}+\frac{2}{3}} = x^{\frac{3}{3}} = x$; $\frac{x^{\frac{5}{8}}}{x^{\frac{3}{8}}} = x^{\frac{5}{8}-\frac{3}{8}} = x^{\frac{2}{8}} = x^{\frac{1}{4}}$
Section 12-5		
1. Write imaginary numbers using the letter i.	The letter i is used to represent $\sqrt{-1}$. Thus, the square root of negative numbers can be expressed as imaginary numbers and simplified using the letter i. Be careful to distinguish when the negative is *outside* the radical and when it is *under* the radical.	Simplify. $\sqrt{-48} = \sqrt{-1 \cdot 16 \cdot 3} = 4i\sqrt{3}$
2. Raise imaginary numbers to powers.	Imaginary numbers can be raised to powers by examining the remainder when the exponent is divided by 4. Review Rule 12-5.1.	Simplify: $i^{17} = i^{16} \cdot i = i$; $i^{42} = i^{40} \cdot i^2 = i^2 = -1$
3. Write real and imaginary numbers in complex form, $a + bi$.	A complex number is a number that can be written in the form $a + bi$, where a and b are real numbers and i is $\sqrt{-1}$. Either a or b can be zero. If a is zero, the number is an imaginary number; if b is zero, the number is a real number.	Rewrite the following as complex numbers. $\sqrt{-81}$, 38, $\sqrt{9}$, $\sqrt{-19}$ $\sqrt{-81} = 9i$ or $0 + 9i$; $38 = 38 + 0i$; $\sqrt{9} = 3$ or $3 + 0i$; $\sqrt{-19} = i\sqrt{19}$ or $0 + i\sqrt{19}$
4. Combine complex numbers.	To combine complex numbers, add the real parts for the real part of the result; then add the imaginary parts for the imaginary part of the result. Be careful with signs when subtracting complex numbers.	Simplify: $(5 + 3i) + (7 - 8i) = (5 + 7) + (3 - 8)i = 12 - 5i$; $(3 - 8i) - (4 + 6i) = (3 - 4) + (-8 - 6)i = -1 - 14i$

Section 12-6

1. Solve equations with squared letter terms.	To solve an equation containing only squared variable terms and number terms, perform all normal steps to isolate the squared variable and to obtain a numerical coefficient of 1 for the squared variable. Then take the square root of both sides. Note, when taking the square root of both sides, be sure to use both the positive and the negative square roots of the constant. Thus, two solutions are possible. Simplify radicals for exact solutions; use a calculator for approximate solutions.	Solve: $3x^2 + 5 = 29$ $3x^2 = 24$ $x^2 = 8$ $x = \pm\sqrt{8}$ Exact: $x = \pm 2\sqrt{2}$ Approximate: $x = \pm 2.828$
2. Solve equations with radical terms.	To solve a basic equation containing square root radicals, square both sides to eliminate any radical; then perform all normal steps to isolate the variable and solve the equation.	Solve: $\sqrt{2x + 1} = 5$ $(\sqrt{2x + 1})^2 = 5^2$ $2x + 1 = 25$ $2x = 24$ $x = 12$

WORDS TO KNOW

square root, (p. 474)
radicand, (p. 474)
plus or minus symbol, (p. 474)
index, (p. 474)
square root radical symbol, (p. 474)
radical symbol, (p. 474)
principal square root, (p. 475)
real numbers, (p. 475)
imaginary numbers, (p. 475)
perfect square, (p. 475)
fractional exponents, (p. 475)
inverse property of squares and square roots, (p. 478)
factor an algebraic expression, (p. 479)
like radicals, (p. 482)
rationalize a denominator, (p. 485)
rationalize a numerator, (p. 486)
rational exponents, (p. 488)
decimal exponents, (p. 490)
imaginary numbers, (p. 491)
real numbers, (p. 491)
complex numbers, (p. 491)
the letter *i*, (p. 491)
remainder, (p. 492)
$a + bi$, (p. 493)
real part of a complex number, (p. 493)
imaginary part of a complex number, (p. 493)

CHAPTER TRIAL TEST

Perform the indicated operations. Simplify if possible. Rationalize the denominators if needed.

1. $2\sqrt{7} \cdot 3\sqrt{2}$

2. $\dfrac{\sqrt{8}}{\sqrt{2}}$

3. $4\sqrt{3} + 2\sqrt{3}$

4. $3\sqrt{8} - 4\sqrt{8}$

5. $10\sqrt{7} - 10\sqrt{7}$

6. $\dfrac{4\sqrt{2}}{\sqrt{3}}$

7. $\sqrt{6x} \cdot \sqrt{8x^3}$

8. $\sqrt{3y^3} \cdot \sqrt{15y^2}$

9. $\dfrac{6\sqrt{8}}{2\sqrt{3}}$

10. $\dfrac{3\sqrt{a}}{\sqrt{b}}$

11. $\dfrac{3\sqrt{5}}{2} \cdot \dfrac{7}{\sqrt{3x}}$

12. $\sqrt{x^3} \cdot \sqrt{7x^2}$

13. $\dfrac{5\sqrt{6}}{\sqrt{5x}}$

Solve the following equations containing radicals or squared terms. Evaluate to the nearest thousandth.

14. $x^2 = 144$

15. $\sqrt{5 - x^2} = 2$

16. $8 = y^2 - 1$

17. $7 = \sqrt{Q}$

18. $\sqrt{\frac{x^2}{2}} = 6$

19. $\sqrt{\frac{6}{y}} = \sqrt{\frac{2}{3}}$

20. $\sqrt{4x} = 20$

21. $7x^2 = 343$

22. $\sqrt{144p^2 - 1} = 0$

23. $x^2 - 4 = 32$

Use the calculator to find the square roots. Round to the nearest thousandth.

24. $\sqrt{16.349}$

25. $\sqrt{99.601}$

Convert the following radical expressions to equivalent expressions using fractional exponents and simplify.

26. $\sqrt[6]{x}$

27. $\sqrt[3]{27x^{15}}$

28. $(\sqrt{5x^4})^6$

29. $\sqrt[5]{x^{10}y^{15}z^{30}}$

30. $\sqrt[4]{x^{12}}$

Perform the following operations. Express answers with positive exponents in lowest terms.

31. $a^{\frac{4}{5}} \cdot a^{\frac{1}{5}}$

32. $(2x^{\frac{1}{4}}y)^4$

33. $(125x^{\frac{1}{2}}y^6)^{\frac{1}{3}}$

34. $\dfrac{b^{\frac{3}{4}}}{b^{\frac{1}{4}}}$

35. $\dfrac{k^{\frac{1}{5}}}{k^{\frac{3}{5}}}$

36. $\dfrac{c^4}{c^{\frac{1}{3}}}$

37. $\dfrac{12x^{\frac{3}{5}}}{6x^{-\frac{2}{5}}}$

38. $\dfrac{(x^{\frac{2}{3}}y^{\frac{1}{4}})(x^{\frac{1}{3}}y^{\frac{3}{4}})}{xy}$

39. $\dfrac{r^{-\frac{1}{5}}s^{\frac{1}{3}}}{r^{\frac{3}{5}}s^{-\frac{5}{3}}}$

Simplify.

40. i^{23}

41. i^{39}

42. i^{88}

Add or subtract.

43. $(5 + 3i) - (8 - 2i)$

44. $(7 - i) + (4 - 3i)$

45. $(15 - 3i) + (7i)$

CHAPTER

13 Formulas and Applications

In Chapters 5, 7, and 10, we solved a number of applied problems with percent, area, perimeter, and interest *formulas*, although we did not always call them formulas. The equation for work problems, rate $\times$ time $=$ amount of work, was a formula. Another formula was the one used to find force: force $= \frac{\text{pressure}}{\text{area}}$.

Formulas are really equations that have been used so frequently to solve certain types of problems that they have become the accepted means of solving these problems. Most formulas are expressed with one or more letter terms rather than words, such as $C = \frac{5}{9}(F - 32)$ and $F = \frac{9}{5}C + 32$.

13-1 FORMULA EVALUATION

Learning Objective

L.O.1. Evaluate formulas for a given variable.

L.O.1. Evaluate formulas for a given variable.

Basic Formula Evaluation

To *evaluate* a formula is to substitute known numerical values for some variables and perform the indicated operations to find the value of the variable in question. In so doing, we may need to use any and all of the steps and procedures we learned previously for solving equations.

Many formulas involve more than one operation. When evaluating such formulas, we follow the rules for the order of operations.

Rule 13-1.1 *Order of operations:*

Perform operations in the following order as they appear from left to right.

1. Parentheses used as groupings and other grouping symbols
2. Exponents (powers and roots)
3. Multiplications and divisions
4. Additions and subtractions

This section will illustrate several formula evaluations to help develop a concept of formula evaluation. Subsequent sections will focus on more specific formulas and their evaluations.

The percentage proportion studied previously (Chapter 5) is actually a formula and shows how any formula written as a proportion may be evaluated. This may be a good place to begin formula evaluation.

EXAMPLE 13-1.1 Evaluate the formula $\frac{R}{100} = \frac{P}{B}$ if $R = 20$ and $P = 10$.

Solution

$$\frac{R}{100} = \frac{P}{B}$$

$$\frac{20}{100} = \frac{10}{B} \qquad \text{(Substitute values.)}$$

$$20(B) = 10(100) \qquad \text{(Cross multiply.)}$$

$$20B = 1000$$

$$\frac{20B}{20} = \frac{1000}{20}$$

$$B = 50$$

Thus, the base is 50. ■

The formulas in Examples 13-1.2 to 13-1.4 are variations of the percentage proportion. In the first formula, the rate is isolated.

EXAMPLE 13-1.2 Evaluate the formula $R = \frac{100P}{B}$ if $P = \$75$ and $B = \$125$.

Solution

$$R = \frac{100P}{B} \qquad \textit{(Rate = 100 times the percentage divided by the base.)}$$

$$R = \frac{100(75)}{125} \qquad \text{(Substitute values.)}$$

$$R = \frac{7500}{125}$$

$$R = 60$$

Thus, the rate is 60%. ■

Sometimes the letter we are asked to solve for is not isolated. In this case, we substitute the given values and isolate the variable as in any equation. The next example shows the variable in the numerator of a fraction.

EXAMPLE 13-1.3 Solve the formula $P = \frac{RB}{100}$ for B if $P = 45$ and $R = 30$.

Solution

$$P = \frac{RB}{100} \qquad \textit{(Percentage = rate times base divided by 100.)}$$

$$45 = \frac{30B}{100} \quad \textit{(Substitute values.)}$$

$$45(100) = \frac{30B}{100}(100) \quad \textit{(Multiply to eliminate denominator.)}$$

$$4500 = \frac{30B}{\cancelto{1}{100}}(\cancelto{1}{100}) \quad \textit{(Reduce whenever possible.)}$$

$$\frac{4500}{30} = \frac{30B}{30}$$

$$150 = B$$

Therefore, the base is 150. ■

The variable may also be in the denominator of a fraction.

EXAMPLE 13-1.4 Evaluate the formula $B = \frac{100P}{R}$ if $P = 60$ m and $B = 200$ m.

Solution

$$B = \frac{100P}{R} \quad \textit{(Base = 100 times percentage divided by the rate.)}$$

$$200 = \frac{100(60)}{R} \quad \textit{(Substitute values.)}$$

$$200 = \frac{6000}{R}$$

$$200(R) = \frac{6000}{R}(R) \quad \textit{(Multiply to eliminate denominator.)}$$

$$200R = \frac{6000}{\cancelto{1}{R}}(\cancelto{1}{R})$$

$$200R = 6000$$

$$\frac{200R}{200} = \frac{6000}{200}$$

$$R = 30$$

Therefore, the rate is 30%. ■

Let's look at the formula for the perimeter of a rectangle (Chapter 7). Again, we will solve for a letter that is not isolated. In this example, the variable is within parentheses and must be removed from the parentheses.

EXAMPLE 13-1.5 Solve the formula $P = 2(l + w)$ for w if $P = 12$ ft and $l = 4$ ft.

Solution

$$P = 2(l + w) \quad \textit{(Perimeter of a rectangle = two times the sum of the length and the width.)}$$

$$12 = 2(4 + w) \quad \textit{(Substitute values.)}$$

$$12 = 8 + 2w \qquad \textit{(Apply distributive principle.)}$$
$$12 - 8 = 2w \qquad \textit{(Isolate the term with the variable.)}$$
$$4 = 2w$$
$$\frac{4}{2} = \frac{2w}{2}$$
$$2 = w$$

Therefore, the width is 2 ft. ■

The formula for area of a circle (Chapter 7) contains a squared variable. In the next example, we will solve for the squared variable.

EXAMPLE 13-1.6 Evaluate the formula $A = \pi r^2$ for r if $A = 144$ in.^{2}.

Solution

$$A = \pi r^2 \qquad \textit{(Area of a circle = } \pi \textit{ times the radius squared.)}$$
$$144 = \pi r^2 \qquad \textit{(Substitute values.)}$$
$$\frac{144}{\pi} = \frac{\pi r^2}{\pi} \qquad \textit{(Isolate the variable.)}$$
$$\frac{144}{\pi} = r^2$$
$$r = \sqrt{\frac{144}{\pi}} \qquad \textit{(Take square root of both sides. For convenience, rewrite with r on left side.)}$$
$$r = \sqrt{45.83662361}$$
$$r = 6.770275003$$

Thus, the radius is 6.77 in. (rounded). ■

A variation of the same formula for the area of a circle allows us to solve for the variable under a square root radical.

EXAMPLE 13-1.7 Evaluate the formula $r = \sqrt{\frac{A}{\pi}}$ for A if $r = 6.25$ cm.

Solution

$$r = \sqrt{\frac{A}{\pi}} \qquad \textit{(Radius = square root of the area of the circle divided by } \pi.)$$
$$6.25 = \sqrt{\frac{A}{\pi}} \qquad \textit{(Substitute values.)}$$
$$(6.25)^2 = \left(\sqrt{\frac{A}{\pi}}\right)^2 \qquad \textit{(Square both sides.)}$$
$$39.0625 = \frac{A}{\pi}$$
$$39.0625(\pi) = \frac{A}{\cancel{\pi}_1}\,(\cancel{\pi}^1) \qquad \textit{(Eliminate denominator.)}$$

$$39.0625\pi = A$$
$$122.718463 = A$$

Thus, the area is 122.72 cm^2 (rounded). ■

The interest formula (Chapter 5) may be used to illustrate a variable that is one of several factors.

EXAMPLE 13-1.8 Evaluate the formula $I = PRT$ for P if $I = \$94.50$, $R = 21\%$, and $T = \frac{1}{2}$ year.

Solution: For convenience in using a calculator, convert $\frac{1}{2}$ year to 0.5 year. In this formula, the rate is expressed as a decimal equivalent, 0.21.

$$I = PRT$$
$$94.50 = P(0.21)(0.5) \quad \textit{(Substitute values.)}$$
$$94.50 = 0.105P$$
$$\frac{94.50}{0.105} = \frac{0.105P}{0.105}$$
$$900 = P$$

Therefore, the principal is \$900. ■

SELF-STUDY EXERCISES 13-1

L.O.1. Evaluate the following formulas. Use $A = 5$, $B = 6$, and $C = 8$.

1. $X = B + (3C - A)$

2. $X = A^2 - (2C - B)$

3. $X = C(B^2 + 4AC) - 3$

4. $X = \frac{3(C - A)}{3} + B$

5. $X = \frac{2(C - A)^2}{A - 1}$

Evaluate the interest formula $I = PRT$ using the following values.

6. Find the interest if $P = \$800$, $R = 15.5\%$, and $T = 2\frac{1}{2}$ years.

7. Find the rate if $I = \$427.50$, $P = \$1500$, and $T = 2$ years.

8. Find the time if $I = \$236.25$, $P = \$750$, and $R = 10.5\%$.

9. Find the principal if $I = \$838.50$, $R = 21\frac{1}{2}\%$, and $T = 1\frac{1}{2}$ years.

Evaluate the percentage formula $P = \frac{RB}{100}$ using the following values.

10. Find the percentage if $R = 15$ and $B = 600$ lb.

11. Find the rate if $P = 24$ kg and $B = 300$ kg.

12. Find the base if $P = \$250$ and $R = 7.4$. Round to hundredths.

13. Evaluate the rate formula $R = \frac{100P}{B}$ for P if $R = 16$ and $B = 85$.

14. Evaluate the base formula $B = \frac{100P}{R}$ for R if $B = \$2200$ and $P = \$374$.

15. Find the length of a rectangular work area if the perimeter is 180 in. and the width is 24 in. Evaluate the formula $P = 2(l + w)$.

16. Find the radius of a circle whose area is 254.5 cm^2 using the formula $r = \sqrt{\frac{A}{\pi}}$. Round to the nearest tenth.

17. Find the cost (C) if the markup (M) on an item is \$5.25 and the selling price (S) is \$15.75. Use the formula $M = S - C$.

18. Using the formula for the side of a square, $s = \sqrt{A}$, find the length of a side of a square field whose area is $\frac{1}{16}$ mi^2.

19. Evaluate the formula for the area of a circle, $A = \pi r^2$, if $r = 7$ in. Round to the nearest tenth.

20. Evaluate the formula for the area of a square, $A = s^2$, if $s = 2.5$ km.

21. Using the markup formula $M = S - C$, find the selling price (S) if the markup (M) is \$12.75 and the cost (C) is \$36.

13-2 FORMULA REARRANGEMENT

Learning Objective

L.O.1. Rearrange formulas to solve for a given variable.

L.O.1. Rearrange formulas to solve for a given variable.

Basic Formula Rearrangement

The evaluation of formulas often requires solving an equation to find the desired letter term.

Formula rearrangement generally refers to isolating a letter term other than the one already isolated (if any) in the formula. Solving formulas in this manner shortens our work when doing repeated formula evaluations. Since we will not evaluate the formulas after we rearrange them, we will not go into what each formula represents. Here we are interested only in the techniques used in the rearrangement of the formulas. After the formulas are solved for the desired variable, the formula is rewritten with the variable on the left side for convenience.

The following formulas require transposition and careful handling of signs.

Tips and Traps

It may help to think of the letter term we are solving for as the unknown or variable, and to think of the other letter terms *as if* they were the number terms in an ordinary equation.

EXAMPLE 13-2.1 Solve the formula $M = S - C$ for S.

Solution

$$M = S - C \quad \textit{(Transpose to isolate S.)}$$

$$M + C = S \quad \textit{(The coefficient of S is positive, so formula is solved.)}$$

$$S = M + C \quad \textit{(S is rewritten on left for convenience.)}$$

■

EXAMPLE 13-2.2 Solve the formula $M = S - C$ for C.

Solution

$$M = S - C \quad \textit{(Transpose to isolate C.)}$$

$$M - S = -C$$

$$\frac{M - S}{-1} = \frac{-C}{-1} \quad \textit{(The coefficient of C is negative, so we need to divide both sides by } -1.\textit{)}$$

$$-M + S = C \quad \textit{(Note effect on signs of division by } -1.\textit{)}$$

$$S - M = C \quad \textit{(Write the positive term first.)}$$

$$C = S - M \quad \textit{(C is rewritten on left for convenience.)}$$

■

Sometimes a formula contains a term of several factors and we need to solve for one of those factors. In this case, treat the factor being solved for as the variable and the other factors as its coefficient.

EXAMPLE 13-2.3 Solve for R in the formula $I = PRT$.

Solution

$$I = PRT \quad \textit{(Since we are solving for R, PT is its coefficient.)}$$

$$\frac{I}{PT} = \frac{PRT}{PT} \quad \textit{(Divide both sides by coefficient of variable.)}$$

$$\frac{I}{PT} = R \quad \textit{(Rewrite with R on the left.)}$$

$$R = \frac{I}{PT}$$

■

Sometimes formulas contain addition (or subtraction) and multiplication in which we must use the distributive principle to solve for a particular letter. In the formula $A = P(1 + ni)$, we must use the distributive principle to solve for either n or i because each appears inside the grouping.

EXAMPLE 13-2.4 Solve for n in $A = P(1 + ni)$.

Solution

$$A = P(1 + ni)$$

$$A = P + Pni \quad \textit{(Use the distributive principle to remove n from parentheses.)}$$

$$A - P = Pni \quad \textit{(Transpose P to isolate the term with n.)}$$

$$\frac{A - P}{Pi} = \frac{Pni}{Pi} \quad \textit{(Divide both sides by coefficient of n.)}$$

$$\frac{A - P}{Pi} = n \quad \textit{(Rewrite.)}$$

$$n = \frac{A - P}{Pi}$$

■

Sometimes we will use the distributive principle *in reverse* to rearrange formulas. If we wish to solve the formula $S = FC + VC$ for C, we must first rewrite the formula by using the distributive principle because the C appears in two terms. This is the same as expressing $FC + VC$ as the product of C and $(F + V)$.

EXAMPLE 13-2.5 Solve for C in $S = FC + VC$.

Solution

$S = FC + VC$ [*Since C appears in two terms, use the distributive principle to express the terms as the product of C and (F + V).*]

$$S = C(F + V)$$

$$\frac{S}{F + V} = \frac{C(F + V)}{F + V}$$ *[Since we are solving for C, divide both sides by coefficient of C, which is (F + V).]*

$$\frac{S}{F + V} = C$$

$$C = \frac{S}{F + V}$$ ■

When the formula contains division, the denominator is eliminated before further steps are taken.

EXAMPLE 13-2.6 Solve the formula $R = \frac{V}{I}$ for V.

Solution

$$R = \frac{V}{I}$$

$$(I)R = \frac{V}{\not{I}}(\not{I})$$ *(Multiply both sides by denominator I to eliminate it.)*

$$IR = V$$

$$V = IR$$ ■

Tips and Traps

Many formulas express relationships between two similar measurements by noting the different measurements with *subscripts*.

The average temperature for a 3-hour period is found by adding the temperatures for each of the periods and dividing by 3.

$$\text{Average temp.} = \frac{\text{temp. 1st hour} + \text{temp. 2nd hour} + \text{temp. 3rd hour}}{3}$$

Using an abbreviated form of the formula, we have

$$t_{\text{av}} = \frac{t_1 + t_2 + t_3}{3}$$

Since we are referring to temperatures throughout the formula, we use subscripts to distinguish the various temperatures.

To read this formula with subscripts we say "t sub av equals the sum of t sub one plus t sub two plus t sub three, all divided by three."

When we wish to rearrange formulas that are written in the form of a proportion, we can use the property of proportions that allows cross multiplication. Once we have cross multiplied, we use methods previously discussed to solve for the appropriate letter.

EXAMPLE 13-2.7 Solve for T_1 in the formula $\frac{T_1}{T_2} = \frac{V_1}{V_2}$.

Solution

$$\frac{T_1}{T_2} = \frac{V_1}{V_2}$$ *(Watch the subscripts.)*

$$T_1V_2 = T_2V_1 \quad \textit{(Cross multiply.)}$$

$$\frac{T_1V_2}{V_2} = \frac{T_2V_1}{V_2} \quad \textit{(Divide both sides by coefficient of } T_1 \textit{ to solve for } T_1.\textit{)}$$

$$T_1 = \frac{T_2V_1}{V_2}$$

■

To solve a formula for a letter that is squared, we first solve for the squared letter and then take the square root of both sides of the formula.

Example 13-2.8 Solve for a in the formula $c^2 = a^2 + b^2$.

Solution

$$c^2 = a^2 + b^2$$

$$c^2 - b^2 = a^2 \quad \textit{(Transpose } b^2 \textit{ to isolate letter being solved for.)}$$

$$\sqrt{c^2 - b^2} = \sqrt{a^2} \quad \textit{(Take square root of both sides.)}$$

$$\sqrt{c^2 - b^2} = a$$

$$a = \sqrt{c^2 - b^2}$$

■

To rearrange a formula with a single term on each side, when one side is a square root, we *begin* by squaring both sides of the formula. Then we solve for the desired letter by using the same techniques used earlier to rearrange formulas.

Example 13-2.9 Solve for A in the formula $s = \sqrt{A}$.

Solution

$$s = \sqrt{A}$$

$$s^2 = (\sqrt{A})^2$$

[Square both sides to eliminate the $\sqrt{\ }$. Notice that $(\sqrt{A})^2 = \sqrt{A} \cdot \sqrt{A} = A$.]

$$s^2 = A$$

$$A = s^2$$

■

SELF-STUDY EXERCISES 13-2

L.O.1. Rearrange the following formulas.

1. Solve $E = X + I$ for X.
2. Solve $A = \frac{M}{A} - K$ for K.
3. Solve $S = 2\pi rh$ for r.
4. Solve $m = bx + by$ for b.
5. Solve $A = m(x + 2y)$ for y.
6. Solve $\frac{T_1}{T_2} = \frac{V_1}{V_2}$ for T_2.
7. Solve $V = \pi r^2 h$ for r.
8. Solve $m = \sqrt{XY}$ for X.
9. Solve $\frac{R}{100} = \frac{P}{B}$ for R.
10. Solve $P = 2(b + s)$ for b.
11. Solve $C = 2\pi r$ for r.
12. Solve $A = lw$ for l.
13. Solve $R = AC - BC$ for C.
14. Solve $A = s^2$ for s.
15. Solve $D = RT$ for R.
16. Solve $S = P - D$ for D.
17. Solve $A = \pi r^2$ for r.
18. Solve $a^2 + b^2 = c^2$ for c.

19. The formula for finding the amount of a repayment on a loan is $A = I + P$, where A is the amount of the repayment, I is the interest, and P is the principal. Solve the formula for interest.

20. The formula for finding interest is $I = PRT$, where I represents interest, P represents principal, R represents rate, and T represents time. Rearrange the formula to find the time.

13-3 TEMPERATURE FORMULAS

Learning Objectives

L.O.1. Make Celsius/Kelvin temperature conversions.
L.O.2. Make Rankine/Fahrenheit temperature conversions.
L.O.3. Convert Fahrenheit to Celsius temperatures.
L.O.4. Convert Celsius to Fahrenheit temperatures.

Many technical applications require conversions between different temperature scales. We will be concerned with temperature conversions involving the Celsius, Kelvin, Rankine, and Fahrenheit scales.

L.O.1. Make Celsius/Kelvin temperature conversions.

Celsius/Kelvin Conversions

One scale used to measure temperature in the metric system of measurement is called the *Kelvin* scale. Units on this scale are abbreviated with a capital K (without the symbol ° because these units are called *Kelvins*) and are measured from absolute zero, the temperature at which *all* heat is said to be removed from matter. Another metric temperature scale is the *Celsius* scale (abbreviated °C), which has as its zero the freezing point of water. The Kelvin and Celsius scales are related such that absolute zero on the Kelvin scale is the same as −273°C on the Celsius scale. Each unit of change on the Kelvin scale is equal to 1 degree of change on the Celsius scale. That is, the size of a Kelvin and a Celsius degree are the same on both scales (see Fig. 13-1).

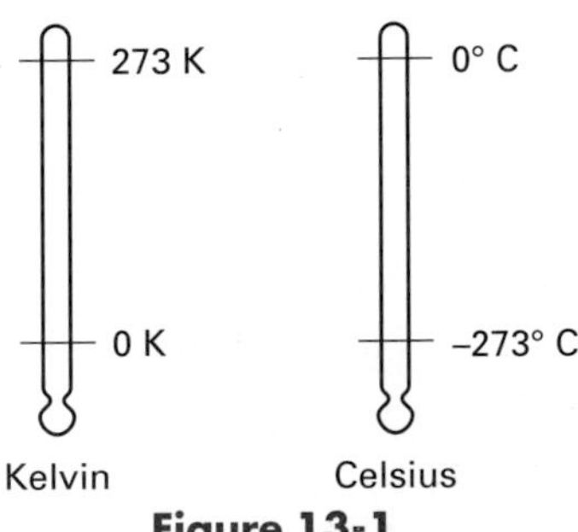

Figure 13-1

Formula 13-3.1 ***To convert Celsius to Kelvin:***

$$K = °C + 273$$

Although temperature conversions can be made using unity ratios, we generally use formulas to make temperature conversions rather than unity ratios.

EXAMPLE 13-3.1 A temperature of 21°C on the Celsius scale is the same as what temperature on the Kelvin scale?

Solution: According to the formula K = °C + 273, we can find the Kelvin temperature by *adding* 273 to the Celsius temperature.

$$K = 21 + 273$$
$$K = 294$$

Thus, 21°C = 294 K. ■

Formula 13-3.2 *To convert Kelvin to Celsius:*

$$°C = K - 273$$

EXAMPLE 13-3.2 A temperature of 300 K on the Kelvin scale corresponds to what temperature on the Celsius scale?

Solution: To find the Celsius temperature, we *subtract* 273 from the Kelvin reading.

$$°C = K - 273$$
$$°C = 300 - 273$$
$$°C = 27$$

Thus, 300 K = 27°C. ■

L.O.2. Make Rankine/Fahrenheit temperature conversions.

Rankine/Fahrenheit Conversions

The U.S. customary system temperature scale that starts at absolute zero is called the *Rankine* scale. It is related to the more familiar *Fahrenheit* scale, which places the freezing point of water at 32°. The Rankine (°R) and Fahrenheit (°F) scales have the same relationship as the Kelvin and Celsius scales in the metric system. That is, 1 degree of change on the Rankine scale is equal to 1 degree of change on the Fahrenheit scale. Absolute zero (the zero for the Rankine scale) corresponds to 460 degrees *below* zero on the Fahrenheit scale.

Formula 13-3.3 *To convert Fahrenheit to Rankine:*

$$°R = °F + 460$$

EXAMPLE 13-3.3 What temperature on the Rankine scale corresponds to 40°F on the Fahrenheit scale?

Solution: To find the Rankine temperature, we *add* 460 to the Fahrenheit temperature reading.

$$°R = °F + 460$$
$$°R = 40 + 460$$
$$°R = 500$$

Thus, 40°F = 500°R. ■

Formula 13-3.4 *To convert Rankine to Fahrenheit:*

$$°F = °R - 460$$

EXAMPLE 13-3.4 What temperature on the Fahrenheit scale corresponds to 650°R on the Rankine scale?

Solution: To find the Fahrenheit temperature, we *subtract* 460 from the Rankine temperature.

$$°F = °R - 460$$
$$°F = 650 - 460$$
$$°F = 190$$

Thus, 650°R = 190°F. ■

L.O.3. Convert Fahrenheit to Celsius temperatures.

Fahrenheit to Celsius Conversions

The Celsius and Fahrenheit scales are the most common temperature scales used for reporting air and body temperatures. Since we still use both scales, we need to be able to convert Fahrenheit temperatures to Celsius and Celsius temperatures to Fahrenheit. The formulas used for converting temperatures using these two scales are more complicated than the previous ones because 1 degree of change on the Celsius scale does *not* equal 1 degree of change on the Fahrenheit scale.

Formula 13-3.5 *To convert Fahrenheit to Celsius:*

$$°C = \frac{5}{9}(°F - 32)$$

If we are given a temperature reading in the Fahrenheit scale and need to change it to the Celsius scale, we use the formula $°C = \frac{5}{9}(°F - 32)$.

EXAMPLE 13-3.5 Change 212°F (the boiling point of water) to degrees Celsius.

Solution: We begin by writing 212 in place of the °F in the formula.

$$°C = \frac{5}{9}(°F - 32)$$

$$°C = \frac{5}{9}(212 - 32)$$

The first step in the order of operations is to work groupings, so we need to subtract 212 − 32 which makes the value of the grouping 180.

Recall that a number, $\frac{5}{9}$, written in front of a grouping (°F − 32) means that we *multiply* the number $\frac{5}{9}$ by the value of the grouping.

$$°C = \frac{5}{9}(180) \qquad \left[\frac{5}{9}(180) \text{ means } \frac{5}{9} \times 180\right]$$

The next step in the order of operations is to multiply or divide.

$$°C = \frac{5}{9}\left(\frac{180}{1}\right) \quad \text{or} \quad \frac{5}{\cancel{9}_1} \times \frac{\cancel{180}^{20}}{1} = 100$$

$$°C = 100$$

Thus, 212°F = 100°C.

The order of these operations is reviewed in Chapter 1 and Section 13-1. ■

EXAMPLE 13-3.6 Change 77°F to degrees Celsius.

Solution

$$°C = \frac{5}{9}(°F - 32)$$

$$°C = \frac{5}{9}(77 - 32) \qquad \textit{(Write 77 in place of °F.)}$$

$$°C = \frac{5}{9}(45) \qquad \textit{(Work grouping.)}$$

$$°C = \frac{5}{\cancel{9}_1} \times \frac{\cancel{45}^{5}}{1} \qquad \textit{(Multiply. Reduce if possible.)}$$

$$°C = 25$$

Thus, 77°F = 25°C. ■

L.O.4. Convert Celsius to Fahrenheit temperatures.

Celsius to Fahrenheit Conversions

When a temperature is expressed in Celsius and needs to be expressed in Fahrenheit, we use the formula $°F = \frac{9}{5}°C + 32$ and follow the rules of the order of operations as indicated in the examples. When no groupings are present, multiplication is done before addition.

> **Formula 13-3.6** *To convert Celsius to Fahrenheit:*
>
> $$°F = \frac{9}{5}°C + 32$$

EXAMPLE 13-3.7 Change 100°C to degrees Fahrenheit.

Solution

$$°F = \frac{9}{5}°C + 32$$ *(A number, $\frac{9}{5}$, written in front of a letter, C, indicates multiplication.)*

$$°F = \frac{9}{5}(100) + 32 \qquad \textit{(Write 100 in place of C.)}$$

$$°F = \frac{\overset{1}{\not{9}}}{\underset{1}{\not{5}}} \times \frac{\overset{20}{\not{100}}}{1} + 32 \quad \textit{(Multiply first, reducing if possible.)}$$

$$°F = 180 + 32 \quad \textit{(Add.)}$$

$$°F = 212$$

Thus, 100°C = 212°F. ■

EXAMPLE 13-3.8 Change 25°C to degrees Fahrenheit.

Solution

$$°F = \frac{9}{5}°C + 32$$

$$°F = \frac{9}{5}(25) + 32 \quad \textit{(Write 25 in place of C.)}$$

$$°F = \frac{9}{\underset{1}{\not{5}}} \times \frac{\overset{5}{\not{25}}}{1} + 32 \quad \textit{(Multiply first, reducing if possible.)}$$

$$°F = 45 + 32 \quad \textit{(Add.)}$$

$$°F = 77$$

Thus, 25°C = 77°F. ■

SELF-STUDY EXERCISES 13-3

L.O.1. Make the following temperature conversions.

1. 82°C = ____ K
2. 438 K = ____ °C
3. 17°C = ____ K
4. 273 K = ____ °C
5. 98 K ____ °C
6. 71°C = ____ K
7. 60°C ____ K
8. 192 K = ____ °C

L.O.2. Make the following temperature conversions.

9. 460°F = ____ °R
10. 98°F = ____ °R
11. 710°R = ____ °F
12. 920°R = ____ °F
13. 180°F = ____ °R
14. 212°F = ____ °R
15. 350°R = ____ °F
16. 600°R = ____ °F
17. 0°R = ____ °F
18. 32°F = ____ °R

L.O.3. Change the following Fahrenheit temperatures to Celsius.

19. 95°F
20. 32°F
21. 113°F
22. 41°F
23. 59°F
24. 50°F
25. 149°F
26. 122°F
27. 176°F
28. 248°F

L.O.4. Change the following Celsius temperatures to Fahrenheit.

29. 70°C
30. 15°C
31. 45°C
32. 50°C
33. 20°C
34. 215°C
35. 310°C
36. 410°C
37. 185°C
38. 0°C

13-4 GEOMETRIC FORMULAS

Learning Objectives

L.O.1. Find the perimeter and area of a trapezoid.
L.O.2. Find the perimeter and area of a triangle.
L.O.3. Use the Pythagorean theorem to find the missing side of a right triangle.
L.O.4. Find the area of prisms and cylinders.
L.O.5. Find the volume of prisms and cylinders.
L.O.6. Find the area and volume of a sphere.
L.O.7. Find the area and volume of a cone.

Geometry involves the study and measurement of shapes. We have already seen in Chapter 7 four geometric figures: the square, the rectangle, the parallelogram, and the circle. We used formulas to find their perimeters and areas. These formulas are listed next.

Formulas

Square
Perimeter: $P = 4s$ Area: $A = s^2$

Rectangle
Perimeter: $P = 2(l + w)$ Area: $A = lw$

Parallelogram
Perimeter: $P = 2(b + s)$ Area: $A = bh$

Circle
Circumference: $C = \pi d$ or $2\pi r$ Area: $A = \pi r^2$

Geometric layouts and figures used in technical occupations are not limited to squares, rectangles, and parallelograms. Other common figures are *trapezoids* and *triangles*.

L.O.1. Find the perimeter and area of a trapezoid.

Figure 13-2

Perimeter and Area of a Trapezoid

Trapezoids are used in the construction of many objects such as picture frames (illustrated in Fig. 13-2), table tops, windows, and roofs.

■ **DEFINITION 13-4.1 Trapezoid.** A **trapezoid** is a four-sided polygon having only two parallel sides.

Unlike a parallelogram, the trapezoid's four sides may all be of unequal size, as illustrated in Fig. 13-3.

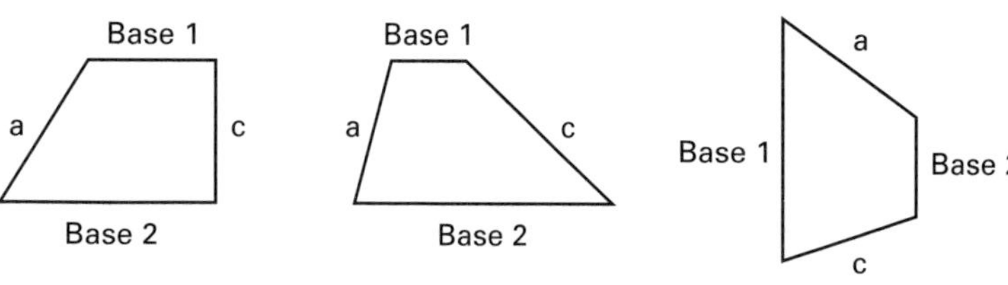

Figure 13-3

Note that the parallel sides are called bases (b_1 and b_2) with the subscripts 1 and 2 used to distinguish them. The nonparallel sides are designated side a and side c. The perimeter is the sum of the lengths of all four sides, or $b_1 + b_2 + a + c$.

Since all four sides may be unequal, neither the bases nor the nonparallel sides may be combined to simplify the formula.

> **Formula 13-4.1** *Perimeter of a trapezoid:*
>
> $$P = b_1 + b_2 + a + c$$

EXAMPLE 13-4.1 The rear window of a pickup truck is a trapezoid whose shorter base is 40 in., whose longer base is 48 in., and whose nonparallel sides are each 16 in. How many inches of rubber gasket are needed to surround the window?

Solution

$P = b_1 + b_2 + a + c$

$P = 40 + 48 + 16 + 16$ *(Substitute in formula.)*

$P = 120$ in.

The rubber gasket to surround the rear window must be 120 in. long. ■

The area of a trapezoid can be represented by dividing it into squares (Fig. 13-4). However, if we recognize that by placing two congruent trapezoids end to end we form a parallelogram, then the area of one trapezoid is half the area of the larger parallelogram. Since the area of a parallelogram is the height times the base, the area of a trapezoid is *one-half* the height times the ''base'' of the parallelogram formed by joining two equal trapezoids end to end.

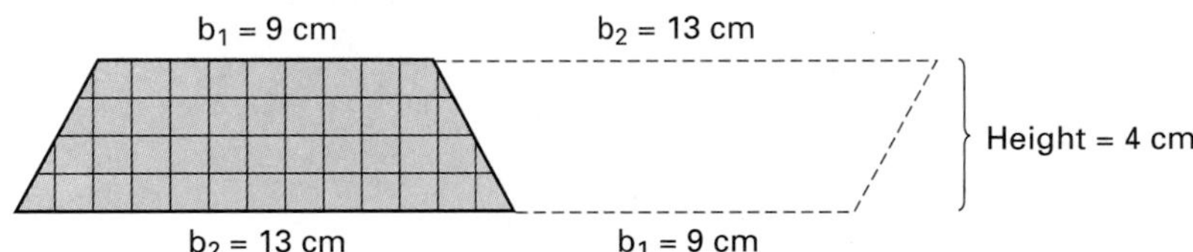

Figure 13-4

> **Formula 13-4.2** *Area of a trapezoid:*
>
> $$A = \frac{1}{2}h(b_1 + b_2)$$

EXAMPLE 13-4.2 A trapezoidal table top for special use in a reading classroom measures 60 cm on the shorter base and 90 cm on the longer base. If the height (perpendicular distance between the bases) is 50 cm, how much Formica is needed to re-cover the top?

Solution

$A = \frac{1}{2}h(b_1 + b_2)$

$A = \frac{1}{2}(50)(60 + 90)$ *(Substitute in formula.)*

$A = \frac{1}{\not{2}_1}(\not{50}^{25})(150)$ *(Reduce where possible.)*

$A = 3750 \text{ cm}^2$

Thus, 3750 cm^2 of Formica are needed. ■

L.O.2. Find the perimeter and area of a triangle.

Perimeter and Area of a Triangle

■ **Definition 13-4.2 Triangle.** A **triangle** is a polygon that has three sides.

It also has three vertices. The sides of a triangle form three angles.

The *degree* is a unit used to measure angles. A more thorough discussion of the degree used as an angle measure is found in Chapter 21.

Tips and Traps

In any triangle, no matter what its shape or size, the sum of the measures of these three angles is 180°.

Classifying Triangles by Angles

There are two primary separations of triangles according to angles. These are triangles that have one right (90°) angle and triangles that do not have a right angle. For this discussion, see Fig. 13-5.

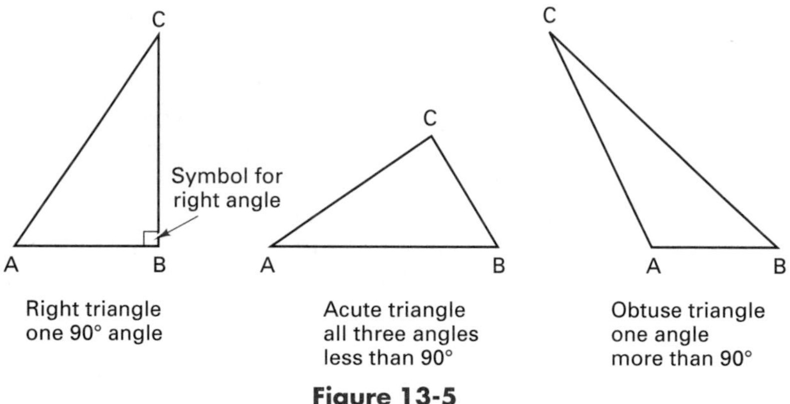

Figure 13-5

■ **Definition 13-4.3 Right Triangle.** A **right triangle** is a triangle that has one right (90°) angle. The sum of the other two angles is 90°.

One side of a right triangle is used quite frequently in mathematics and deserves special attention. This side is the *hypotenuse.*

■ **Definition 13-4.4 Hypotenuse.** The side of a right triangle that is opposite the right angle is called the **hypotenuse**.

The other sides are called *legs*, or the vertical side may be called the *altitude* and the horizontal side the *base*. The right angle is usually marked ⌉ or ⌈ .

Right triangles are used in many technologies. Surveyors, carpenters, and navigators are but a few who frequently use right triangles to solve problems involving relationships among lines and angles.

■ **Definition 13-4.5 Oblique Triangle.** An **oblique triangle** is any triangle that does not contain a right angle.

There are two kinds of oblique triangles, *acute* and *obtuse*. Acute triangles have each angle less than 90°. Obtuse triangles have one angle more than 90°. For a more thorough discussion of acute and obtuse angles and triangles, see Chapter 21.

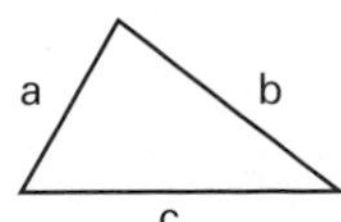

Figure 13-6

The perimeter of a triangle is the sum of the lengths of the three sides (Fig. 13-6).

> **Formula 13-4.3** *Perimeter of a triangle:*
>
> $$P = a + b + c$$

EXAMPLE 13-4.3 The triangular gable ends of an apartment unit in a complex under construction will be outlined in contrasting trim. If each gable is 32 ft wide and 18 ft on each side, how many feet of contrasting wood trim are needed for each gable? Disregard overlap at corners (Fig. 13-7).

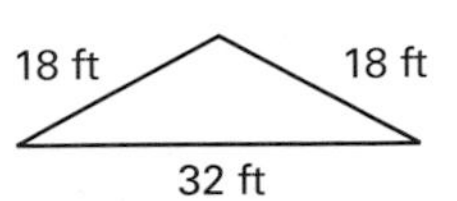

Figure 13-7

Solution

$$P = a + b + c$$

$$P = 18 + 32 + 18 \qquad \textit{(Substitute in formula.)}$$

$$P = 68 \text{ ft}$$

Each gable will require 68 linear feet of trim. ■

The area of a triangle, like the area of a trapezoid, can be figured by dividing it into square units (Fig. 13-8). Notice that one side is designated as the base and that the height is the perpendicular distance from the base to the top. Like the trapezoid, the triangle forms a parallelogram if an identical triangle is placed beside the original triangle. A triangle, then, is half a parallelogram, and its area can be expressed as half the area of a parallelogram, that is, one-half the base times the height.

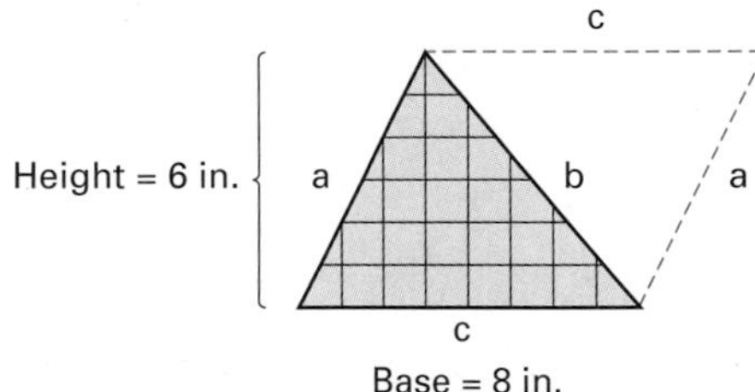

Figure 13-8

> **Formula 13-4.4** *Area of a triangle:*
>
> $$A = \frac{1}{2}bh$$

EXAMPLE 13-4.4 A gable end of a house has a rise of 7 ft 6 in. and a span of 30 ft 6 in. What is its area? The rise is the height of the gable and the span is the base of the gable (Fig. 13-9).

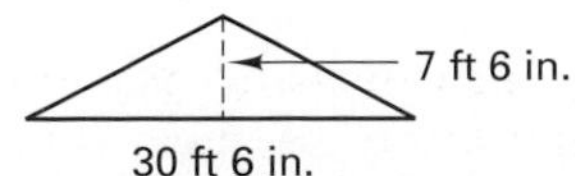

Figure 13-9

Solution

$A = \frac{1}{2}bh$ *(Substitute in formula after converting to one common unit of measure. 7 ft 6 in. = 7.5 ft, 30 ft 6 in. = 30.5 ft since 1 ft = 12 in.)*

$A = \frac{1}{2}(30.5)(7.5)$

$A = 0.5(30.5)(7.5)$

$A = 114.375 \text{ ft}^2$

The area of a gable is 114.375 ft^2. ■

In a right triangle, the sides adjacent to the hypotenuse or the sides that form the right angle are perpendicular to each other. Thus, one is the height and the other is the base.

If a triangle is known to be a right triangle, the height and the base can be identified even if not labeled as such.

EXAMPLE 13-4.5 Find the area of a piece of sheet metal cut as a right triangle. One side is 12 in., one is 16 in., and the third is 20 in. (Fig. 13-10).

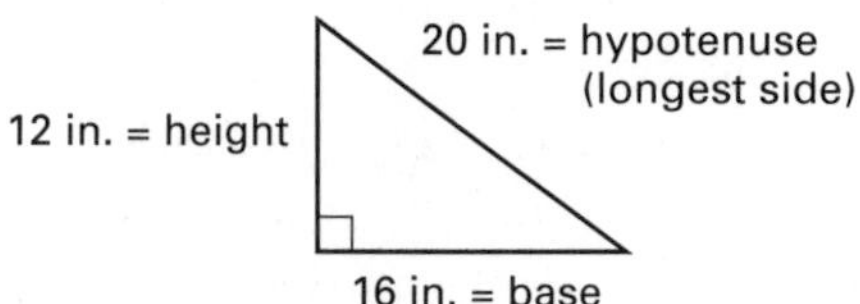

(Designate one side as the base and another as the height. The longest side is the hypotenuse.)

Figure 13-10

Solution

$A = \frac{1}{2}bh$

$A = \frac{1}{\cancel{2}_1}(\cancel{16}^{8})(12)$ *(Substitute in formula.)*

$A = 96 \text{ in.}^2$

The area of the sheet metal is 96 in.^{2}. ■

The height of a triangle need not be measured along one side such as in a right triangle or inside the triangle such as in the gable of Example 13-4.4. In some triangles, the height is measured along an imaginary line outside the triangle from the base to the highest point of the triangle. The triangles illustrated in Fig. 13-11 are examples of triangles with the height measured outside the triangle itself. The area is calculated using the standard formula. Let's figure the area of $\triangle DEF$. ($\triangle$ means triangle.)

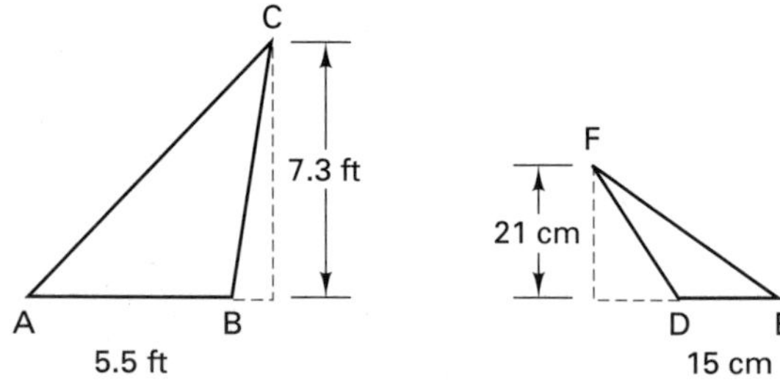

Figure 13-11

EXAMPLE 13-4.6 The base of $\triangle DEF$ in Fig. 13-11 is 15 cm and the height is 21 cm. Find the area.

Solution

$$A = \frac{1}{2}bh$$

$$A = \frac{1}{2}(15)(21) \qquad \text{(Substitute in formula.)}$$

$$A = \frac{1}{2}(315)$$

$$A = 157.5 \text{ cm}^2$$

The area of $\triangle DEF$ is 157.5 cm^2. ■

L.O.3. Use the Pythagorean theorem to find the missing side of a right triangle.

The Pythagorean Theorem

One of the most famous and useful theorems in mathematics is the Pythagorean theorem. It is named for the Greek mathematician, Pythagoras.

■ **DEFINITION 13-4.6 Pythagorean Theorem.** The **Pythagorean theorem** states that the square of the hypotenuse of a right triangle is equal to the sum of the squares of the two legs of the triangle.

In symbols, the Pythagorean theorem may be expressed as the formula

$$c^2 = a^2 + b^2$$

where c is the hypotenuse and a and b are legs.

This theorem, illustrated in Fig. 13-12, has numerous applications in many technical fields. When working with the Pythagorean theorem, we will know two sides of a right triangle and be expected to find the remaining side.

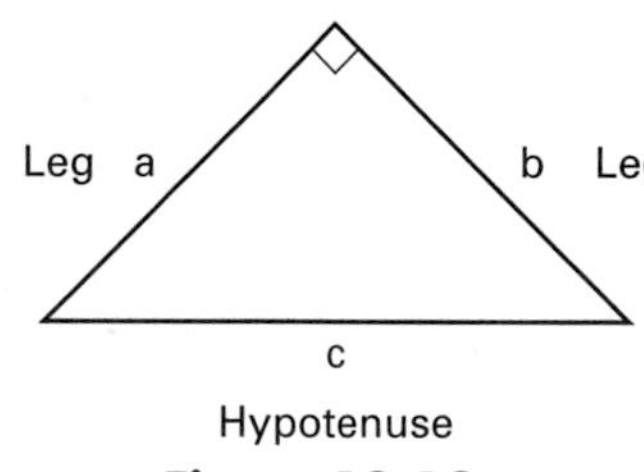

Figure 13-12

EXAMPLE 13-4.7 In $\triangle ABC$ (Fig. 13-13), if $AB = 4$ ft and $AC = 3$ ft, find BC.

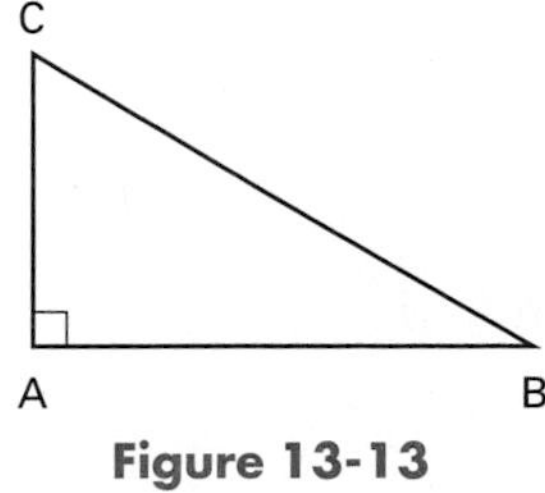

Figure 13-13

Solution

$$(BC)^2 = (AC)^2 + (AB)^2 \qquad \text{(State theorem symbolically.)}$$

$$(BC)^2 = 3^2 + 4^2 \qquad \text{(Substitute values.)}$$

$$(BC)^2 = 9 + 16$$

$$(BC)^2 = 25$$

$$BC = \sqrt{25} \qquad \text{(Take square root of both sides.)}$$

$$BC = 5 \text{ ft} \qquad \text{(Use only positive square root.)}$$

■

EXAMPLE 13-4.8 If $a = 8$ mm and $c = 17$ mm, find b (Fig. 13-14).

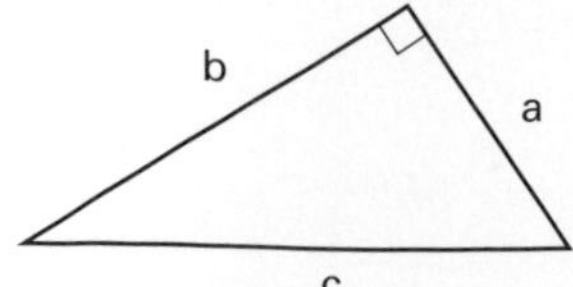

Figure 13-14

Solution

$$c^2 = a^2 + b^2 \quad \textit{(State theorem symbolically.)}$$
$$17^2 = 8^2 + b^2 \quad \textit{(Substitute values.)}$$
$$289 = 64 + b^2$$
$$289 - 64 = b^2 \quad \textit{(Transpose to isolate.)}$$
$$225 = b^2$$
$$\sqrt{225} = b \quad \textit{(Take square root of both sides.)}$$
$$15 \text{ mm} = b$$

■

Many times we are confronted with a problem that contains ''hidden'' triangles. In these cases, we need to visualize the triangle or triangles in the problem. Often, we have to draw one or more of the sides of the ''hidden'' triangle as an aid in solving the problem.

EXAMPLE 13-4.9 Find the center-to-center distance between pulleys A and C (Fig. 13-15).

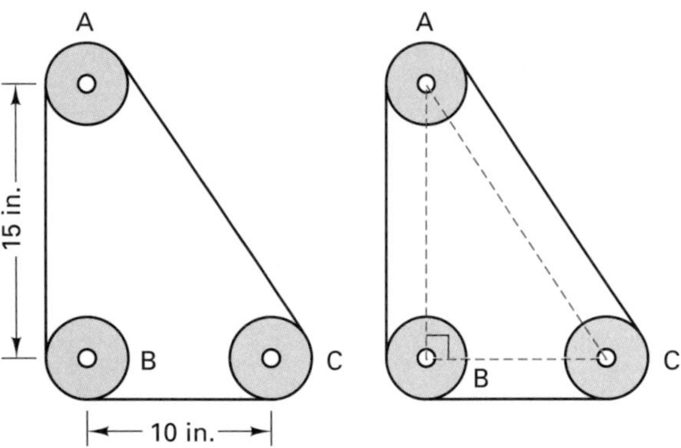

Figure 13-15

Solution: Connect the center points of the three pulleys to form a right triangle. The hypotenuse is the distance between pulleys A and C. We use the Pythagorean theorem and substitute the given values for the two known sides.

$$(AC)^2 = (AB)^2 + (BC)^2 \quad \textit{(State theorem symbolically.)}$$
$$(AC)^2 = 15^2 + 10^2 \quad \textit{(Substitute values.)}$$
$$(AC)^2 = 225 + 100$$
$$(AC)^2 = 325 \quad \textit{(Take square root of both sides.)}$$
$$AC = \sqrt{325}$$
$$AC = 18.0277564$$

Thus, the distance from pulley A to pulley C is 18.0 in. (to the nearest tenth). ■

EXAMPLE 13-4.10 The head of a bolt is a square 0.5 in. on a side (distance across flats). What is the distance from corner to corner (distance across corners)? (See Fig. 13-16.)

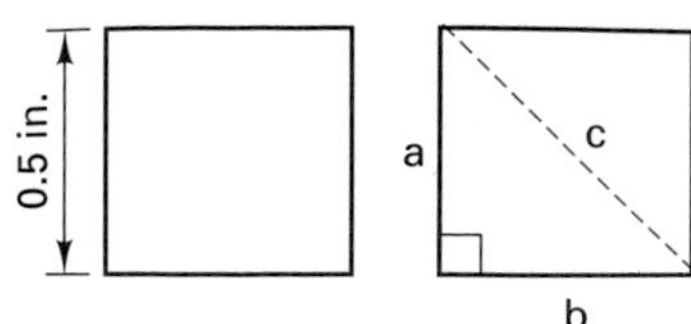

Figure 13-16

Solution: The sides of the bolt head form the legs of a triangle if a diagonal line is drawn from one corner to the opposite corner. This diagonal forms the hypotenuse of the right triangle. Since the legs of the triangle are known to be 0.5 in. each, we can substitute in the Pythagorean theorem to find the diagonal line, the hypotenuse, which is the distance across corners.

$c^2 = a^2 + b^2$ *(State theorem symbolically.)*

$c^2 = 0.5^2 + 0.5^2$ *(Substitute values.)*

$c^2 = 0.25 + 0.25$

$c^2 = 0.5$ *(Take square root of both sides.)*

$c = \sqrt{0.5}$

$c = 0.707106781$

Thus, the distance across corners is 0.71 in. (to the nearest hundredth). ■

Sometimes right triangles are used in technical fields to represent certain relationships, such as forces acting on an object at right angles or electrical and electronic phenomena related in the way the three sides of a right triangle are related. Let's look at an example.

EXAMPLE 13-4.11 Forces A and B come together at a right angle to produce force C (Fig. 13-17). If force A is 74.8 lb and the resulting force C is 91.5 lb, what is force B?

Figure 13-17

Solution: Since the forces are related in the way the sides of a right triangle are related, we may use the Pythagorean theorem to find the missing force B, a leg of the triangle.

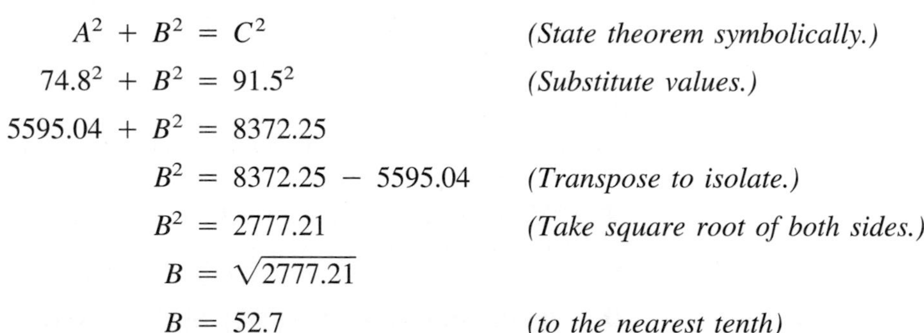

$A^2 + B^2 = C^2$ *(State theorem symbolically.)*

$74.8^2 + B^2 = 91.5^2$ *(Substitute values.)*

$5595.04 + B^2 = 8372.25$

$B^2 = 8372.25 - 5595.04$ *(Transpose to isolate.)*

$B^2 = 2777.21$ *(Take square root of both sides.)*

$B = \sqrt{2777.21}$

$B = 52.7$ *(to the nearest tenth)*

Thus, force B is 52.7 lb. ■

L.O.4. Find the area of prisms and cylinders.

Surface Area of Prisms and Cylinders

Common household items like ice cubes, cardboard storage boxes, and toy building blocks are examples of solid geometric figures classified generally as *prisms*. Cans and pipes are examples of *cylinders*.

■ **DEFINITION 13-4.7 Prism.** A **prism** is a solid whose bases (ends) are parallel, congruent polygons and whose faces (sides) are parallelograms, rectangles, or squares. In a *right prism* the faces are perpendicular to the bases.

■ **DEFINITION 13-4.8 Cylinder.** A (right circular) **cylinder** is a solid with a curved surface and two circular bases such that the height is perpendicular to the bases.

■ **DEFINITION 13-4.9 Height.** The **height** of a solid with two bases is the shortest distance between the two bases.

In right circular cylinders and in right prisms the height is the same as the length of a side or face. However, in oblique prisms the height is the perpendicular distance between the bases and is different from the length of a side or face (see Fig. 13-18).

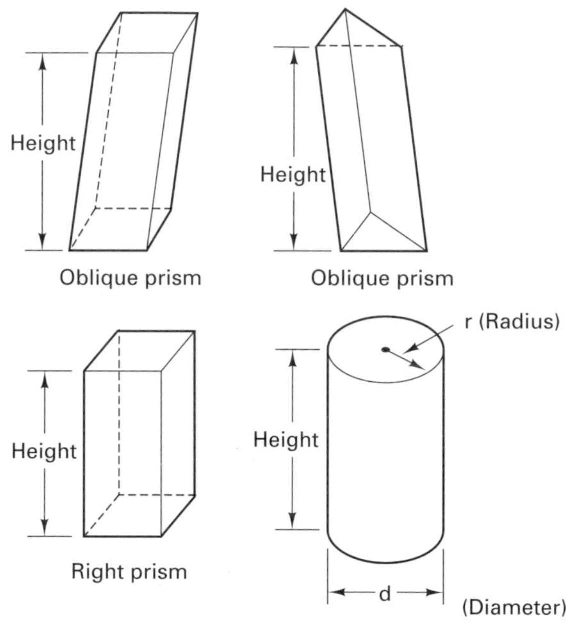

Figure 13-18

Some jobs require us to find the area of a solid figure. The area of a solid figure can refer to just the area of the *sides* of the figure. Or area can refer to the overall area, including the bases along with the sides.

■ **DEFINITION 13-4.10 Lateral Surface Area.** The **lateral surface area** (LSA) of a solid figure is the area of its sides only.

■ **DEFINITION 13-4.11 Total Surface Area.** The **total surface area** (TSA) of a solid figure is the area of the sides plus the area of its base or bases.

To find the lateral surface area, all we need to do is find the sum of the areas of each side using the formulas from Chapter 7. However, we can also find the lateral surface area (LSA) of a prism or cylinder by multiplying the perimeter of the base times the height of the solid.

In the formulas and examples that follow, we will consider only *right* prisms and *right* circular cylinders.

> **Formula 13-4.5** *Lateral surface area of a right prism or cylinder:*
>
> $\text{LSA} = ph, \qquad p = \text{perimeter of base}, \; h = \text{height of the solid figure}$

To get the total surface area (TSA), we add to the lateral surface area the areas of the two bases.

> **Formula 13-4.6** *Total surface area of a right prism or cylinder:*
>
> $\text{TSA} = ph + 2B, \qquad h = \text{height of solid figure},$
>
> $p = \text{perimeter of base}, \qquad B = \text{area of base}$

EXAMPLE 13-4.12 Find the lateral surface area of a rectangular shipping carton measuring 24 in. in length, 12 in. in width, and 20 in. in height (Fig. 13-19).

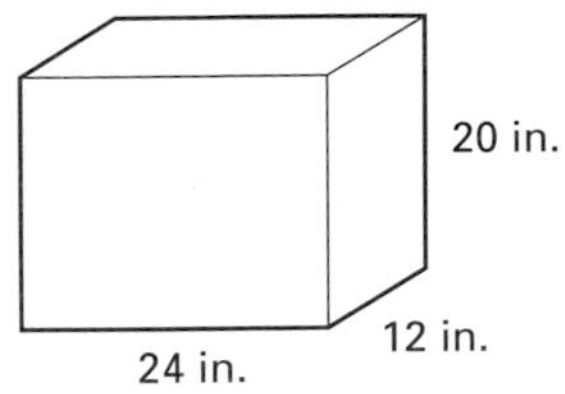

Figure 13-19

Solution

$\text{LSA} = ph$

$\text{LSA} = (2l + 2w)h$ *(Since the base is a rectangle, its perimeter is $2l + 2w$.)*

$\text{LSA} = [2(24) + 2(12)]20$ *(Substitute numerical values.)*

$\text{LSA} = [72]20$

$\text{LSA} = 1440 \text{ in.}^2$

The lateral surface area of the carton is 1440 in.^{2}. ■

EXAMPLE 13-4.13 How many square centimeters of sheet metal are required to manufacture a can whose radius is 4.5 cm and whose height is 9 cm? Assume no waste or overlap.

Solution

$\text{TSA} = ph + 2B$ *(Total surface area is needed.)*

$\text{TSA} = 2\pi rh + 2\pi r^2$ *($p = 2\pi r$, $B = \pi r^2$.)*

$\text{TSA} = 2(\pi)(4.5)(9) + 2(\pi)(4.5)^2$ *(Substitute values.)*

$\text{TSA} = 254.4690049 + 127.2345025$

$\text{TSA} = 381.70 \text{ cm}^2$ *(rounded)*

The can requires 381.70 cm^2 of sheet metal. ■

EXAMPLE 13-4.14 Find the total surface area of the triangular prism shown in Fig. 13-20.

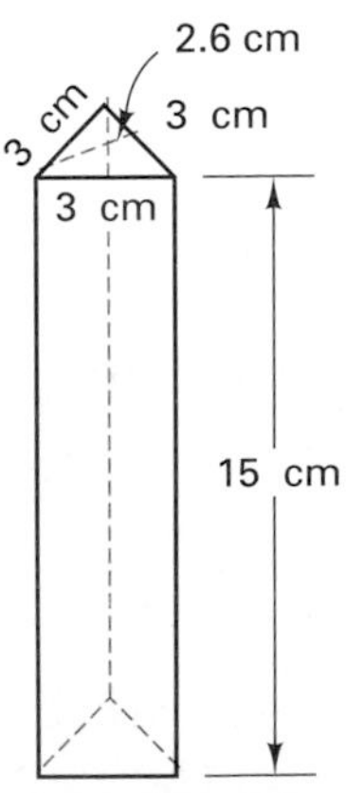

Figure 13-20

Solution

$\text{TSA} = ph + 2B$

$\text{TSA} = 9(15) + (2)\left(\frac{1}{2}\right)(3)(2.6)$ *[Perimeter of triangular base is $3 + 3 + 3 = 9$ cm. Area of triangular base is $\frac{1}{2}bh$, or $\frac{1}{2}(3)(2.6)$.]*

$\text{TSA} = 135 + 7.8$

$\text{TSA} = 142.8 \text{ cm}^2$

The total surface area of the triangular prism is 142.8 cm^2. ■

L.O.5. Find the volume of prisms and cylinders.

Volume of Prisms and Cylinders

At times we need to know the *volume* of an object, such as a container, to estimate, for example, how many of these containers can be loaded into a given size storage area or shipped in a tractor-trailer rig of certain dimensions.

■ **DEFINITION 13-4.12 Volume.** The **volume** of a solid geometric figure is the amount of space it occupies, measured in terms of three dimensions (length, width, and height).

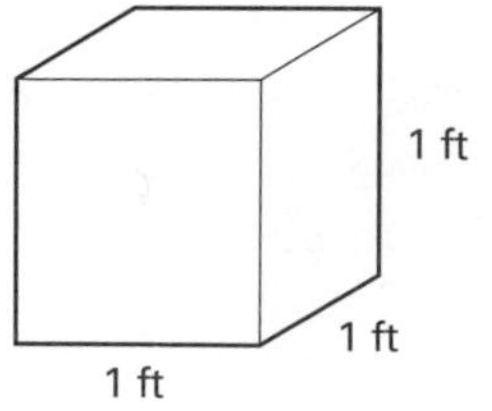

One cubic foot or 1 ft^3

Figure 13-21

If we had a rectangular box measuring 1 ft long, 1 ft wide, and 1 ft high (Fig. 13-21), it would be a cube representing 1 cubic foot (ft^3). Its volume is calculated by multiplying length $\times$ width $\times$ height, or $1 \times 1 \times 1 = 1$ ft^3. This cubic measure is often indicated by the exponent 3 following the unit of measure, meaning that the measure has *already been "cubed."* It is merely a shortcut for indicating such cubic measures as ft^3 = cubic feet, in.^{3} = cubic inches, cm^3 = cubic centimeters, and so on.

From this concept of 1 ft^3 or 1 cubic foot comes the formula for the volume of a rectangular box as length $\times$ width $\times$ height or $V = lwh$. Note that $l \times w$ is actually the formula for the area of the rectangle (or square) that forms the base of the rectangular box. If the base of the prism were a triangle, pentagon, hexagon, or other polygon, we would have to use the appropriate formula for its area. In the case of a cylinder, we would use the formula for the area of a circle since its base is a circle. Thus, the general formula for the volume of *any* right prism or cylinder is as follows.

Formula 13-4.7 *Volume of right prism or cylinder:*

$V = Bh$, B = area of base, h = height of prism or cylinder

EXAMPLE 13-4.15 Find the volume of the triangular prism in Example 13-4.14, whose height is 15 cm and whose bases are triangles 3 cm on a side and 2.6 cm in height.

Solution

$V = Bh_2$ [*h_1 = 2.6 cm (height of prism base), h_2 = 15 cm (height of prism)*]

$V = \left(\frac{1}{2}bh_1\right)h_2$ *(Substitute the formula for the area of the triangular base.)*

$V = \left[\frac{1}{2}(3)(2.6)\right]15$ *(Substitute values.)*

$V = 58.5 \text{ cm}^3$

The volume of the prism is 58.5 cm^3. ■

EXAMPLE 13-4.16 What is the cubic-inch displacement (space occupied) of a cylinder whose diameter is 5 in. and whose height is 4 in.?

Solution

$V = Bh$

$V = \pi r^2 h$ *(Substitute the formula for the area of the circular base.)*

$$V = \pi(2.5)^2(4) \quad \textit{(Substitute values; } r = \frac{1}{2} \textit{ diameter, or 2.5.)}$$

$$V = 78.5 \text{ in.}^3 \quad \textit{(rounded)}$$

The cylinder displacement is 78.5 in.^{3}. ■

L.O.6. Find the area and volume of a sphere.

Surface Area and Volume of a Sphere

Soccer balls, golf balls, tennis balls, baseballs, and ball bearings are *spheres.* Spheres are also used as tanks to store gas and water because spheres hold the greatest volume for a specified amount of surface area.

■ **DEFINITION 13-4.13 Sphere.** A **sphere** is a solid formed by a curved surface whose points are all equidistant from a point inside called the *center* (Fig. 13-22).

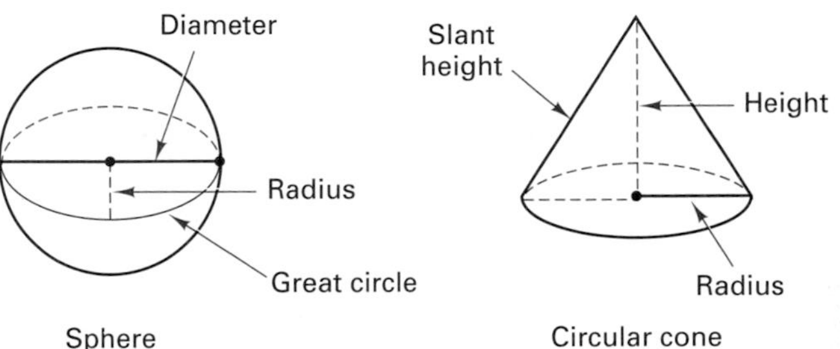

Figure 13-22

■ **DEFINITION 13-4.14 Great Circle.** The **great circle** divides the sphere in half at its greatest diameter and is formed by a plane through the center of the sphere.

A sphere does not have bases like prisms and cylinders. The surface area of a sphere includes *all* the surface and so there is only one formula. Because of the relationship of the sphere to the circle, the formula includes elements of the formula for the area of a circle. The total surface area of the sphere is four times the area of a circle with the same radius.

> **Formula 13-4.8** *Total surface area of a sphere:*
>
> $$\text{TSA} = 4\pi r^2$$

The formula for the volume of a sphere also contains elements found in formulas for a circle. The volume of a sphere is calculated with the following formula.

> **Formula 13-4.9** *Volume of a sphere:*
>
> $$V = \frac{4\pi r^3}{3}$$

Note that the radius is *cubed*, or raised to the power of 3, which indicates volume.

EXAMPLE 13-4.17 Find the surface area and volume of a sphere whose diameter is 90 cm.

Solution

$$\text{TSA} = 4\pi r^2$$

$$\text{TSA} = 4(\pi)(45)^2 \quad \textit{(Substitute values. } \tfrac{1}{2}d = r, \textit{ so } r = 45.)$$

$$\text{TSA} = 25{,}447 \text{ cm}^2 \quad \textit{(rounded)}$$

$$V = \frac{4\pi r^3}{3}$$

$$V = \frac{4(\pi)(45)^3}{3} \quad \textit{(Substitute values.)}$$

$$V = 381{,}704 \text{ cm}^3 \quad \textit{(rounded)}$$

■

L.O.7. Find the area and volume of a cone.

Surface Areas and Volume of a Cone

One example of an object based on the *cone* is the funnel or the circular rain cap placed on top of stove vent pipes extending through the roofs of many homes.

■ **DEFINITION 13-4.15 Cone.** A (right) **cone** is a solid whose base is a circle and whose side surface tapers to a point, called the *vertex* or *apex*, and whose height is a perpendicular line between the base and apex (Fig. 13-22).

In Fig. 13-22, note that in addition to the perpendicular height of the cone, there is also the *slant height* of a cone.

■ **DEFINITION 13-4.16 Slant Height.** The **slant height** of a cone is the distance along the side from the base to the apex.

The lateral surface area of a cone equals the circumference of the base times $\frac{1}{2}$ the slant height, or LSA $= 2\pi r\frac{s}{2}$, which is simplified as follows.

Formula 13-4.10 *Lateral surface area of a cone:*

$$\text{LSA} = \pi rs, \qquad r = \text{radius}, \qquad s = \text{slant height}$$

The total surface area, then, is the lateral surface area plus the area of the base.

Formula 13-4.11 *Total surface area of a cone:*

$$\text{TSA} = \pi rs + \pi r^2$$

where r = radius of circular base
s = slant height
πr^2 = area of base

The volume of a cone is equal to one-third the area of the base times the *height of the cone* (*not* the slant height).

Formula 13-4.12 *Volume of a cone:*

$$V = \frac{\pi r^2 h}{3}$$

where πr^2 = area of circular base
h = height of cone

Example 13-4.18 Find the lateral surface area, total surface area, and volume of a cone whose diameter is 8 cm, height 6 cm, and slant height 7 cm. Round to hundredths.

Solution

$$\text{LSA} = \pi r s$$

$$\text{LSA} = (\pi)(4)(7) \qquad \textit{(Substitute values. } r = \tfrac{1}{2}d\textit{, or 4.)}$$

$$\text{LSA} = 87.9645943 \text{ cm}^2$$

$$\text{TSA} = \pi r s + \pi r^2$$

$$\text{TSA} = 87.9645943 + (\pi)(4)^2 \qquad \textit{(Substitute values.)}$$

$$\text{TSA} = 138.23 \text{ cm}^2 \qquad \textit{(rounded)}$$

$$V = \frac{\pi r^2 h}{3}$$

$$V = \frac{(\pi)(4)^2(6)}{3} \qquad \textit{(Substitute values.)}$$

$$V = 100.53 \text{ cm}^3 \qquad \textit{(rounded)}$$

■

Example 13-4.19 Find the weight of the cast-iron solid shown in Fig. 13-23 if cast iron weighs 0.26 lb per cubic inch. Round to the nearest whole pound.

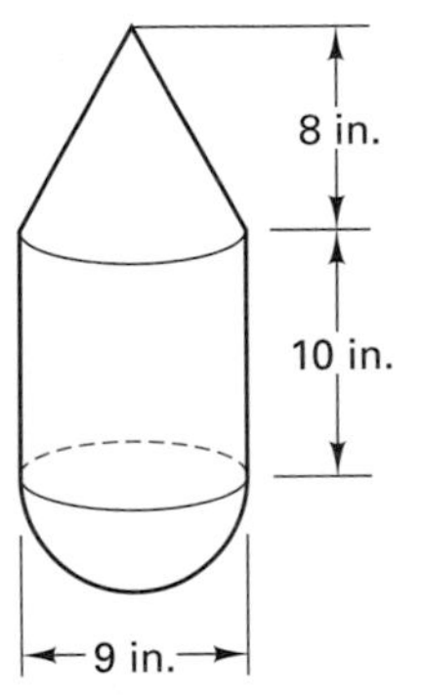

Figure 13-23

Solution: The solution requires finding the volume of the cone that forms the top of the solid, the volume of the cylinder that forms the middle portion of the solid, and the volume of the hemisphere (half sphere) that forms the bottom of the solid.

$$V_{\text{cone}} = \frac{\pi r^2 h}{3} \qquad V_{\text{cylinder}} = \pi r^2 h \qquad V_{\text{hemisphere}} = \frac{1}{2}\left(\frac{4\pi r^3}{3}\right)$$

$$V_{\text{cone}} = \frac{(\pi)(4.5)^2(8)}{3} \qquad V_{\text{cylinder}} = (\pi)(4.5)^2(10) \qquad V_{\text{hemisphere}} = \frac{1}{2}\left(\frac{4(\pi)(4.5)^3}{3}\right)$$

$$V_{\text{cone}} = 169.6460033 \text{ in.}^3 \quad V_{\text{cylinder}} = 636.1725124 \text{ in.}^3 \quad V_{\text{hemisphere}} = 190.8517537 \text{ in.}^3$$

$$\text{Total volume} = V_{\text{cone}} + V_{\text{cylinder}} + V_{\text{hemisphere}}$$

$$\text{Total volume} = 996.6702694 \text{ in.}^3$$

Convert to pounds:

$$\frac{996.6702694 \text{ in.}^3}{1} \times \frac{0.26 \text{ lb}}{1 \text{ in.}^3} = 259 \text{ lb.} \qquad \textit{(rounded)}$$

To the nearest whole pound, the solid cast-iron figure weighs 259 lb. ■

SELF-STUDY EXERCISES 13-4

L.O.1. Find the perimeter and the area of the following figures. Round to hundredths.

1.

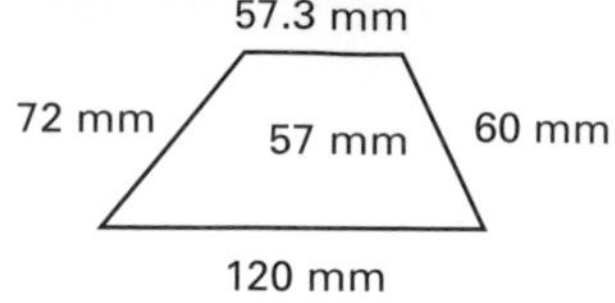

2.

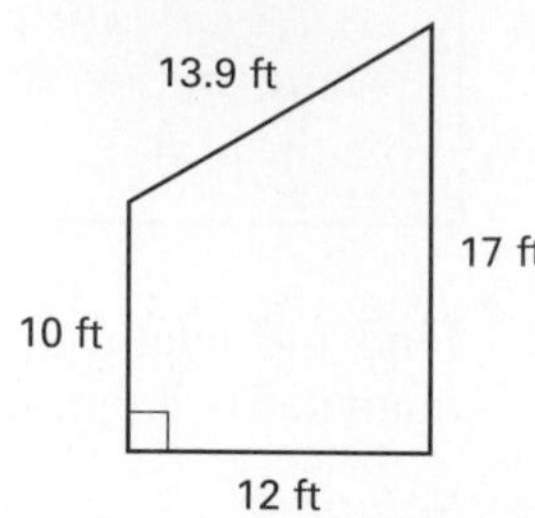

Solve the following problems involving perimeter and area.

3. The six glass panes in a kitchen light fixture each measure $4\frac{1}{2}$ in. along the top and 10 in. along the bottom. The top and bottom are parallel. The height of each pane is 8 in. What is the combined area of the six trapezoidal panes?

4. A section of a hip roof is a trapezoid measuring 38 ft at the bottom, 14 ft at the top, and 10 ft high. Find the area of this section of the roof in square feet.

5. A lot in an urban area is 60 ft wide. The sides are 120 ft and 154 ft, and they are perpendicular to the width of the lot. Find the area of the trapezoidal property.

6. A swimming pool is fashioned in a trapezoidal design. The parallel sides are $18\frac{1}{2}$ ft and 31 ft. The other sides are $24\frac{1}{2}$ ft and 24 ft. What is the perimeter of the pool?

L.O.2. Find the perimeter and area of the triangle in Figure 13-24. Round to hundredths.

7.

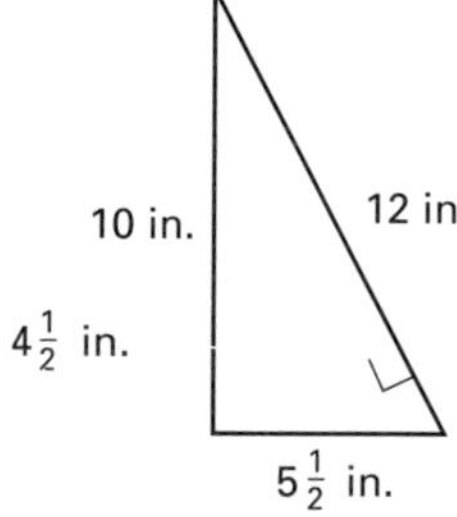

Figure 13-24

8.

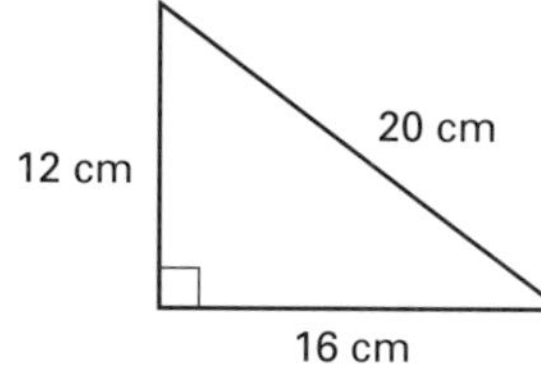

9. Find the area of the triangle in Fig. 13-25.

Figure 13-25

Solve the following problems involving perimeter and area.

10. Max Cisneros is planning a patio which will adjoin the sides of his L-shaped home. One side of the home is 24 feet and the other is 18 feet. The shape of the patio is triangular. Draw a representation of the patio and find the perimeter of the triangle if its hypotenuse is 30 feet. Find the number of square feet of surface area that need to be covered with concrete.

11. A louver, or triangular vent, for a gable roof measures 6 ft 6 in. wide and stands 3 ft high. What is the area in square feet of the vented portion of the gable?

12. A mason lays tile in the form of a triangle. The height of the triangle is 12 ft and the base is 5 ft. What is the area?

13. A metal worker cuts a triangular plate with a base of 11 in. and a height of $4\frac{1}{2}$ in. from a piece of metal. What is the area of the plate?

14. If aluminum siding costs $6.75 a square yard installed, how much would it cost to put the siding on the two triangular gable ends of a roof under construction? Each gable has a span (base) of 30 ft 6 in. and a rise (height) of 7 ft 6 in. Any portion of a square yard is rounded to the next highest square yard.

15. Find the height of a triangle if its base is 15 ft 6 in. and its area is 62 ft^2.

16. Find the perimeter of a triangle whose sides are 2 ft 6 in., 1 yd 8 in., and 4 ft 6 in.

L.O.3. Find the missing side of the right triangle. Round the final answers to the nearest thousandth. Use Fig. 13-26 for Exercises 17 through 19.

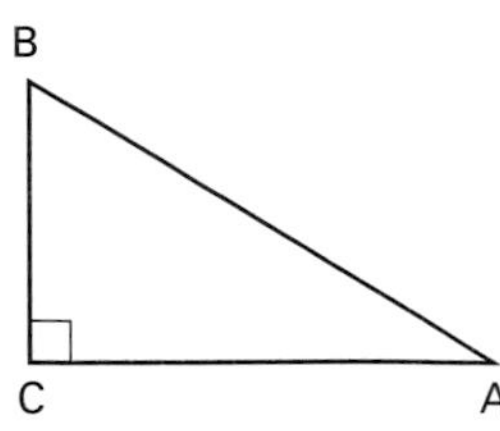

Figure 13-26

17. $AC = ?$
$BC = 7$ cm
$AB = 25$ cm

18. $AC = 24$ mm
$BC = ?$
$AB = 26$ mm

19. $AC = 15$ yd
$BC = 8$ yd
$AB = ?$

Use Fig. 13-27 for Exercises 20 through 22. Find the missing dimensions. Round to thousandths where appropriate.

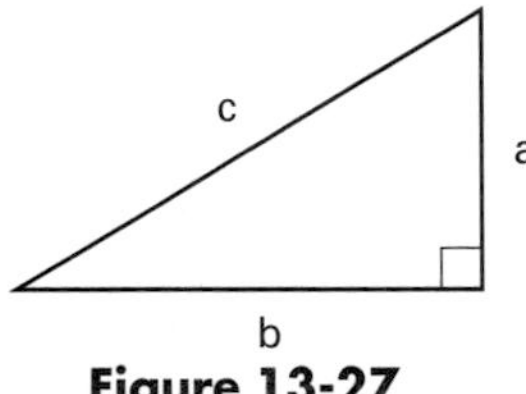

Figure 13-27

20. $a = 5$ cm
$b = 4$ cm
$c = ?$

21. $a = ?$
$b = 9$ m
$c = 11$ m

22. $a = 9$ ft
$b = ?$
$c = 15$ ft

Solve the following problems. Round the final answers to the nearest thousandth if necessary.

23. A light pole will be braced with a wire that is to be tied to a stake in the ground 18 ft from the base of the pole, which extends 26 ft above the ground. If the wire is attached to the pole 2 ft from the top, how much wire must be used to brace the pole? (See Fig. 13-28.)

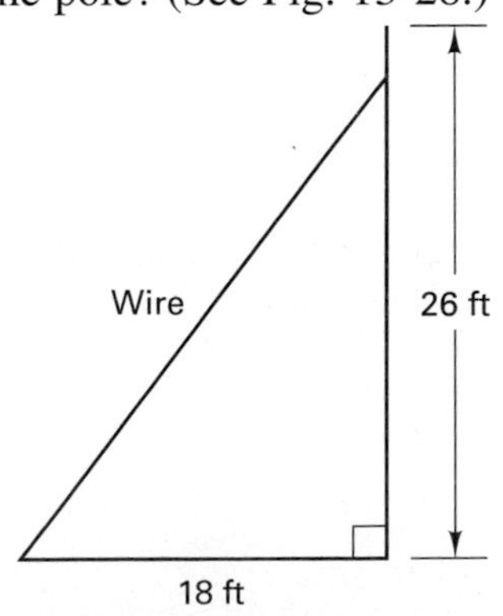

Figure 13-28

24. Find the center-to-center distance between holes A and C in a sheet metal plate if the distance between the centers of A and B is 16.5 cm and the distance between the centers of B and C is 36.2 cm (Fig. 13-29).

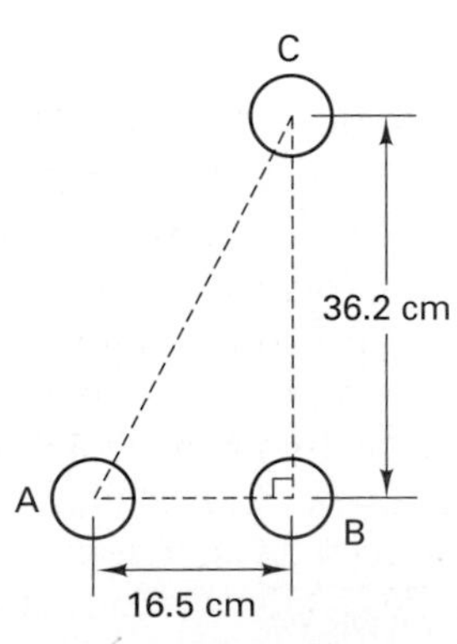

Figure 13-29

25. Find the length of a rafter that has a 10-in. overhang if the rise of the roof is 10 ft and the joists are 48 ft long (Fig. 13-30).

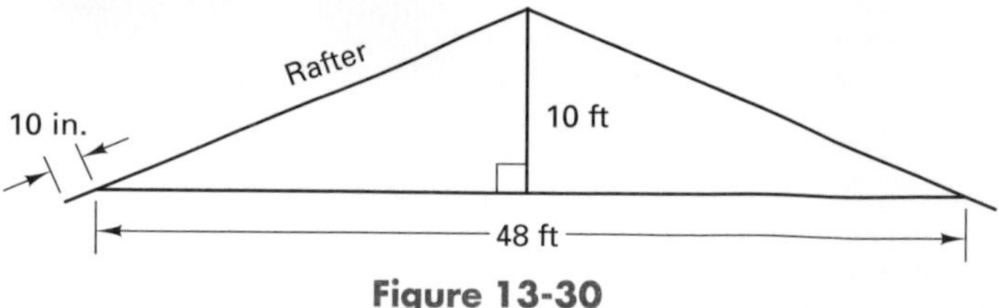

Figure 13-30

26. A stair stringer is 8 ft high and extends 10 ft from the wall (Fig. 13-31). How long will the stair stringer be?

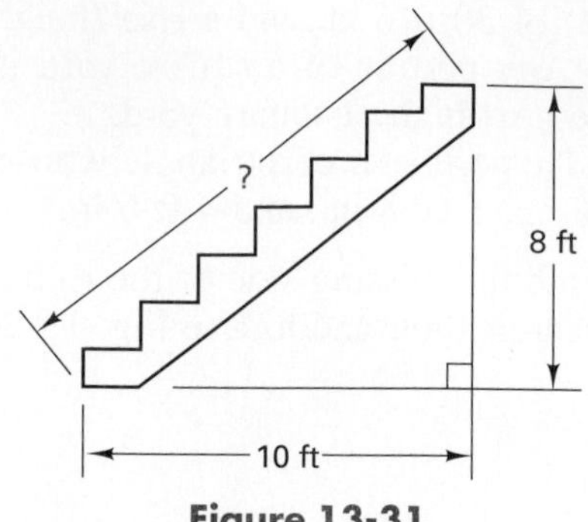

Figure 13-31

27. A machinist wishes to strengthen an L bracket that is 5 cm by 12 cm by welding a brace to each end of the bracket (Fig. 13-32). How much metal rod is needed for the brace?

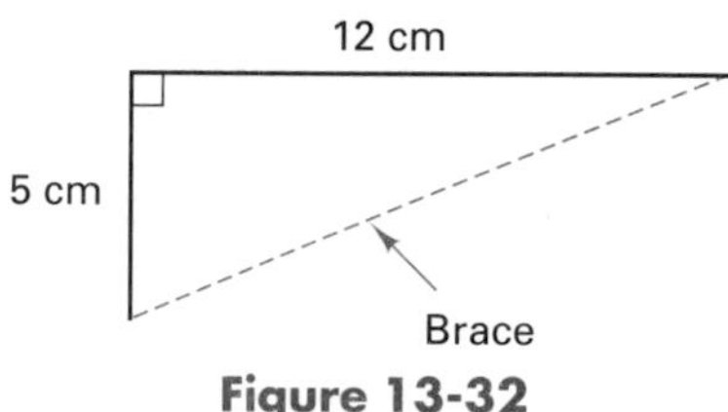

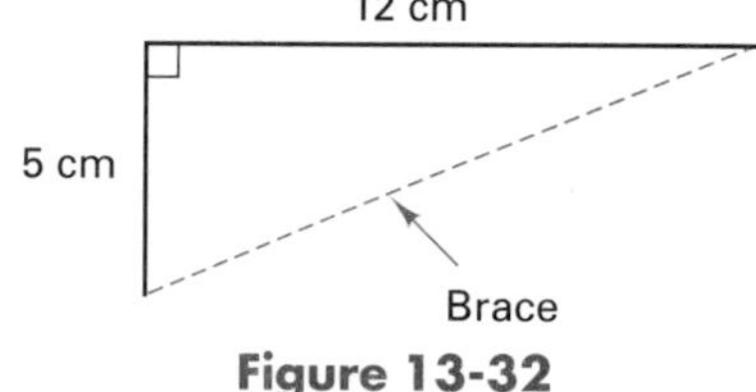

Figure 13-32

28. To make a rectangular table more stable, a diagonal brace is attached to the underside of the table surface. If the table is 27 dm by 36 dm, how long is the brace?

29. Find the length of the side of the largest square nut that can be milled from a piece of round stock whose diameter is 15 mm (Fig. 13-33).

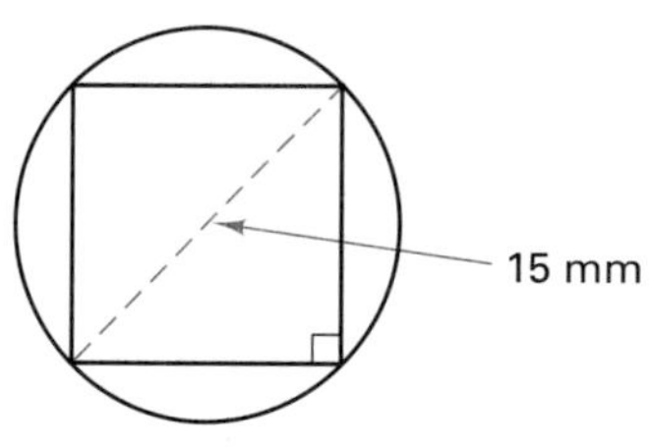

Figure 13-33

30. A rigid length of electrical conduit must be shaped as shown in Fig. 13-34 to clear an obstruction. What is the total length of the conduit needed? (*Hint:* Do not forget to include *AB* and *CD* in the total length.)

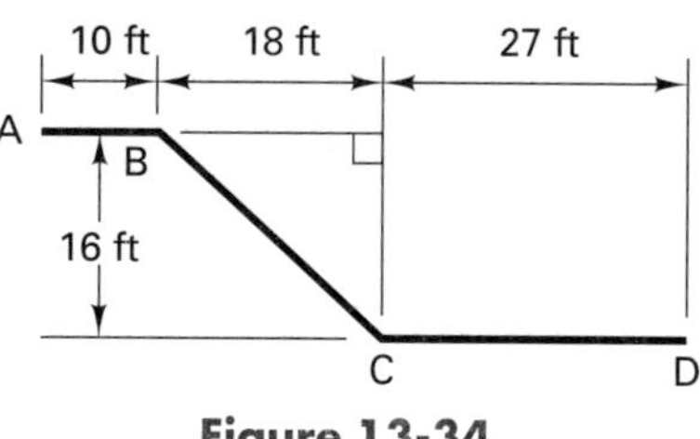

Figure 13-34

31. The vector diagram in Fig. 13-35 is used in electrical applications. Find the voltage of E_a.

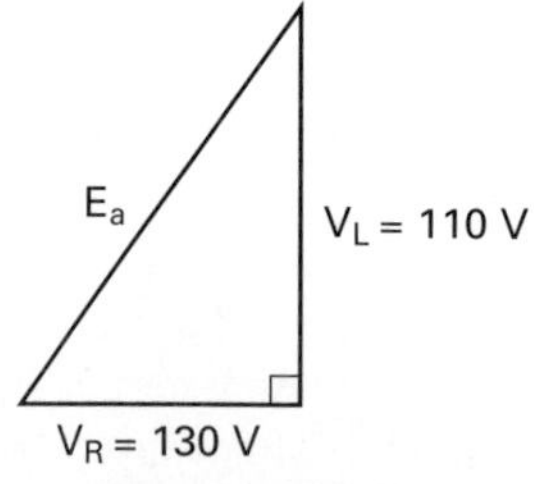

Figure 13-35

L.O.4. Find the lateral surface area and total surface area of the solids in the following figures. Round to the nearest hundredth if necessary.

32.

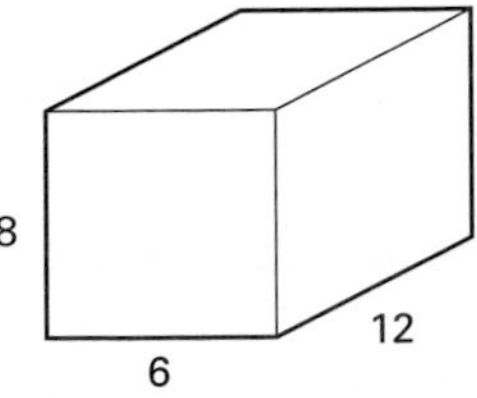

33.

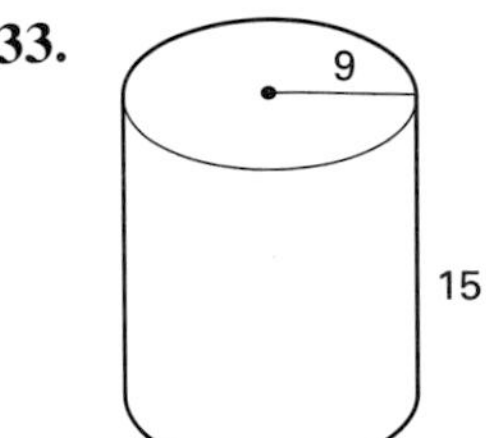

Solve the following problems.

34. What is the total surface area of an oil storage tank 35 ft in diameter and 12 ft tall? Round to the nearest whole number.

35. What is the total surface area of a triangular prism whose height is 6 in. and whose triangular base measures 2 in. on each side with a height of 1.7 in?

36. A solid chocolate candy bar in the form of a prism is enclosed in a clear wrapper. The manufacturer wants to wrap the sides of the bar with paper indicating the company name, ingredients, and weight of the bar. What is the lateral area to be wrapped if the bar measures as shown in Fig. 13-36?

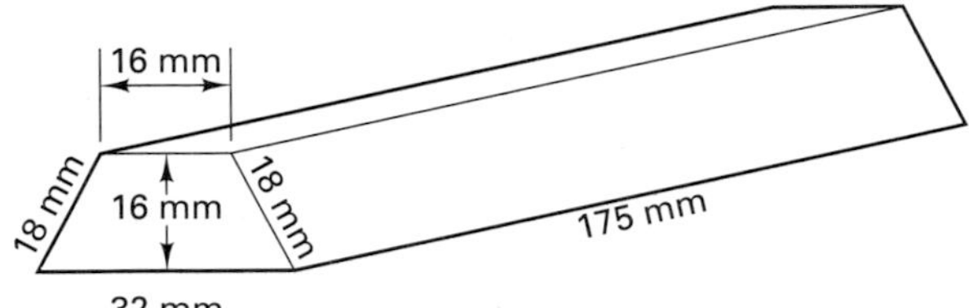

Figure 13-36

37. Find the number of square inches needed to make a paper label for an aluminum can $2\frac{1}{2}$ in. in diameter and $4\frac{3}{4}$ in. tall. Assume no waste or overlap. Round to tenths.

38. A wall-mounted three-way speaker is to be covered with wood-grained vinyl. What is the lateral area to be covered if the speaker is 19 in. high, 11 in. wide, and 10 in. deep? (*Hint:* Assume that the 11×19 rectangle is the base.)

39. If the side of a 1-lb coffee can will be imprinted to show manufacturer and contents, how many square inches of lateral surface may be imprinted if the can is 4 in. across and $5\frac{1}{2}$ in. high? Round to tenths.

L.O.5. Find the volume of the solids in the following figures. Round to the nearest hundredth if necessary.

40.

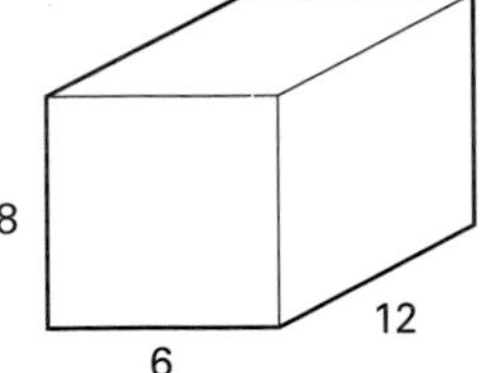

41.

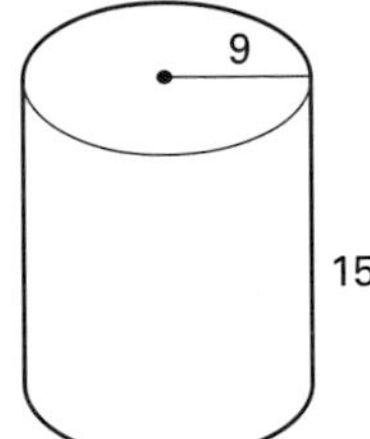

Solve the following problems.

42. How many cubic inches are in an aluminum can $2\frac{1}{2}$ in. in diameter and $4\frac{3}{4}$ in. tall? Round to tenths.

43. What is the volume of an oil storage tank 40 ft in diameter and 15 ft tall? Round to the nearest whole number.

44. A right pentagonal prism is 10 cm high. If the area of each pentagonal base is 32 cm^2, what is the volume of the prism?

45. What is the volume of a triangular prism whose height is 8 in. and whose triangular base measures 4 in. on each side with a height of 3.46 in.? Round to hundredths.

46. A cylindrical water well is 1200 ft deep and 6 in. across. How much soil and other material were removed? Round to the nearest cubic foot. (*Hint:* Convert measures to a common unit.)

L.O.6. Solve the following problems. Round to tenths.

47. Find the surface area of a sphere with radius of 5 cm.

48. Find the volume of a sphere with radius of 6 in.

49. How many square feet of steel are needed to manufacture a spherical water tank with a diameter of 45 ft?

50. If 1 ft^3 = 7.48 gal, how many gallons can the water tank in Exercise 49 hold?

51. A spherical propane tank has a diameter of 4 ft. How many square feet of surface area needs to be painted?

52. If a propane tank is filled to 90% of its capacity, how many gallons of propane will the tank in Exercise 51 hold? (1 ft^3 = 7.48 gal.)

L.O.7. Solve the following problems. Round to tenths.

53. Find the lateral surface area, total surface area, and volume of the cone in Fig. 13-37.

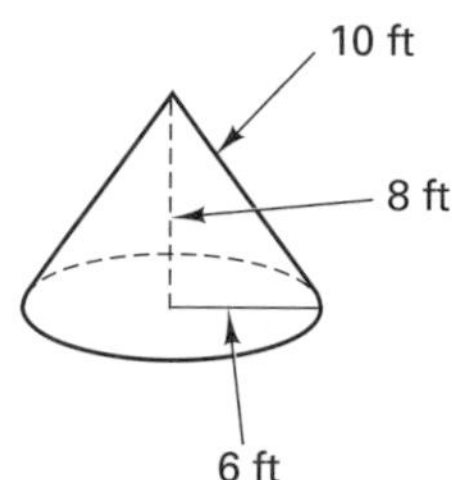

Figure 13-37

54. How many cubic feet are in a conical pile of sand 30 ft in diameter and 20 ft high?

55. How many square centimeters of sheet metal are needed to form a conical rain cap 25 cm in diameter if the slant height will be 15 cm?

56. A cone-shaped storage container holds a photographic chemical. If the container is 80 cm wide and 30 cm high, how many liters of the chemical does it hold if 1 L = 1000 cm^3?

57. Find the total surface area of a conical tank whose radius is 15 ft and whose slant height is 20 ft.

58. Find the height of a conical tank whose volume is 261.67 ft^3 and whose radius is 5 ft. (*Hint:* Rearrange the volume formula to find the height.)

59. A cylindrical water tower with a conical top and hemispheric bottom needs to be painted (Fig. 13-38). If the cost is \$2.19 per square foot, how much would it cost (to the nearest dollar) to paint the tank?

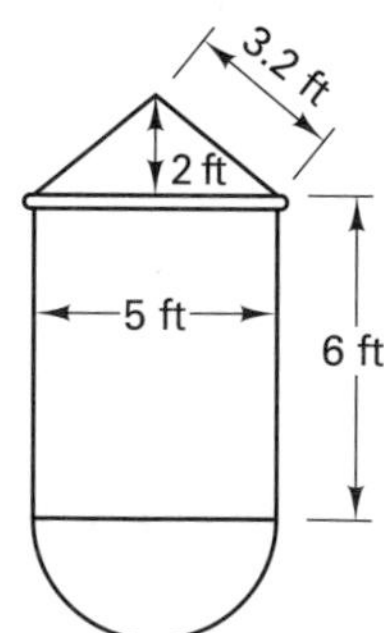

Figure 13-38

13-5 MISCELLANEOUS TECHNICAL FORMULAS

Learning Objective

L.O.1. Evaluate miscellaneous technical formulas.

In addition to temperature conversion formulas and geometric formulas, a variety of other formulas is used in technical applications. In this section, we will present a miscellaneous assortment of technical formulas and show how they are evaluated and, if necessary, rearranged.

L.O.1. Evaluate miscellaneous technical formulas.

Miscellaneous Technical Formulas

EXAMPLE 13-5.1 Evaluate the formula $E = \frac{I - P}{I}$ if $I = 24{,}000$ calories (cal) and $P = 8600$ cal.

Solution

$E = \frac{I - P}{I}$ *(Engine efficiency = difference between heat input and output divided by heat input.)*

$E = \frac{24{,}000 - 8600}{24{,}000}$ *(Substitute given values.)*

$E = \frac{15{,}400}{24{,}000}$ *(Perform numerator grouping.)*

$E = 0.642$ *(nearest thousandth)*

Thus, the engine efficiency is 0.642 (64.2% efficient). ■

EXAMPLE 13-5.2 Evaluate the formula $R_T = \frac{R_1R_2}{R_1 + R_2}$ if $R_1 = 10$ ohms and $R_2 = 6$ ohms.

Solution

$R_T = \frac{R_1R_2}{R_1 + R_2}$ *(Total resistance = product of 1st resistance and 2nd resistance divided by sum of 1st and 2nd resistances.)*

$R_T = \frac{10(6)}{10 + 6}$ *(Substitute given values.)*

$R_T = \frac{60}{16}$ *(Perform numerator and denominator groupings.)*

$R_T = 3.75$

Thus, the total resistance in the circuit is 3.75 ohms. ■

In the next formula, do not forget to square the D.

EXAMPLE 13-5.3 Evaluate the formula $H = \frac{D^2N}{2.5}$ if $D = 4$ and $N = 8$.

Solution

$H = \frac{D^2N}{2.5}$ *(Horsepower = diameter of cylinder squared times number of cylinders divided by 2.5.)*

$$H = \frac{4^2(8)}{2.5} \quad \textit{(Perform power operation.)}$$

$$H = \frac{16(8)}{2.5} \quad \textit{(Perform multiplication.)}$$

$$H = \frac{128}{2.5}$$

$$H = 51.2$$

The engine is rated at 51.2 hp. ■

Sometimes we are asked to solve for a letter term that is not isolated. Let's look again at the horsepower formula and solve for something other than horsepower.

EXAMPLE 13-5.4 Solve the formula $H = \frac{D^2N}{2.5}$ for D (in inches) if $H = 45$ and $N = 6$.

Solution

$$H = \frac{D^2N}{2.5}$$

$$45 = \frac{D^2(6)}{2.5}$$

$$(2.5)45 = \frac{D^2(6)}{2.5}(2.5) \quad \textit{(Multiply to eliminate denominator.)}$$

$$112.5 = 6D^2$$

$$\frac{112.5}{6} = D^2$$

$$18.75 = D^2$$

$$D = 4.330 \quad \textit{(rounded)}$$

Thus, the diameter of the piston is 4.330 in. ■

Tips and Traps

In practical problems where a negative square root is unrealistic, we use only the positive square root. In the preceding example, for instance, we cannot have a negative diameter of a piston.

Caution: The next formula includes both powers and roots and the solution should be checked for extraneous roots.

EXAMPLE 13-5.5 In the formula $Z = \sqrt{R^2 + X^2}$, solve for R (in ohms) if $Z = 12.4$ ohms and $X = 12$ ohms.

Solution

$$Z = \sqrt{R^2 + X^2} \quad \textit{(Impedance = square root of sum of resistance squared plus reactance squared.)}$$

$$12.4 = \sqrt{R^2 + (12)^2}$$

$$(12.4)^2 = (\sqrt{R^2 + 144})^2 \qquad \textit{(Eliminate radical.)}$$
$$153.76 = R^2 + 144 \qquad \textit{(Isolate } R^2.\textit{)}$$
$$153.76 - 144 = R^2$$
$$9.76 = R^2$$
$$R = 3.124 \qquad \textit{(rounded)}$$

Thus, the resistance is 3.1 ohms (rounded to tenths). ■

Note the many steps in the following formula and the importance of observing the order of operations. The formula is for finding the length of a belt connecting two pulleys.

EXAMPLE 13-5.6 Evaluate

$$L = 2C + 1.57(D + d) + \frac{D + d}{4C}$$

if $C = 24$ in., $D = 16$ in., and $d = 4$ in.

Solution

$$L = 2C + 1.57(D + d) + \frac{D + d}{4C}$$

(L = length of belt joining two pulleys,
C = distance between centers of pulleys,
D = diameter of large pulley,
d = diameter of small pulley; Fig. 13-39.)

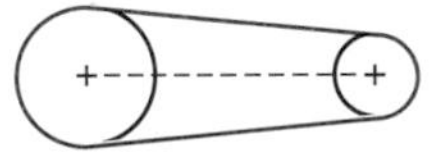

Figure 13-39

$$L = 2(24) + 1.57(16 + 4) + \frac{16 + 4}{4(24)}$$

$$L = 2(24) + 1.57(20) + \frac{20}{96} \qquad \textit{(Work groupings in parentheses, numerator, and denominator.)}$$

$$L = 48 + 31.4 + 0.20833333 \qquad \textit{(Work multiplication and division.)}$$

$$L = 79.61 \qquad \textit{(rounded)}$$

The pulley belt is 79.61 in. long. ■

SELF-STUDY EXERCISES 13-5

L.O.1. Complete the following.

1. Use the formula $R_t = \dfrac{R_1R_2}{R_1 + R_2}$ to find the total resistance (R_t) if one resistance (R_1) is 12 ohms and the second resistance (R_2) is 8 ohms.

2. What is the percent efficiency (E) of an engine if the input (I) is 25,000 calories and the output (P) is 9600 calories? Use the formula $E = \dfrac{I - P}{I}$.

3. Distance is rate times time, or $D = RT$. Find the rate if the distance traveled is 140 mi and the time traveled is 4 hr.

4. The formula for voltage (Ohm's law) is $E = IR$. Find the amperes of current (I) if the voltage (E) is 120 V and the resistance (R) is 80 ohms.

5. According to Boyle's law, if temperature is constant, the volume of a gas is inversely proportional to the pressure on it. Find the final volume (V_2) of a gas using the formula $\frac{V_1}{V_2} = \frac{P_2}{P_1}$ if the original volume (V_1) is 15 ft^3, the original pressure (P_1) is 60 lb per square inch (psi), and the final pressure (P_2) is 150 psi.

6. The formula for power (P) in watts ([illegible] $P = I^2R$. Find the current (I) in amperes [illegible] evice draws 63 W and the resistance (R) is [illegible] ms.

7. Use the formula $H = \frac{D^2N}{2.5}$ to find the number of cylinders (N) required in an engine of 3.2 hp (H) if the cylinder diameter (D) is 2 in.

8. The formula for the speed (s) of a driven pulley in revolutions per minute (rpm) is $s = \frac{DS}{d}$. Find the speed of a driven pulley with diameter (d) of 5 in. if the diameter (D) of the driving pulley is 10 in. and its speed (S) is 800 rpm.

9. If the distance (C) between the centers of the pulleys in Exercise 8 is 24 in., find the length (L) of the belt connecting them using the formula

$$L = 2C + 1.57(D + d) + \frac{D + d}{4C}$$

Round to hundredths.

10. Find the reactance (X) in ohms using the formula $Z = \sqrt{R^2 + X^2}$ if the impedance (Z) is 10 ohms and the resistance (R) is [illegible] ohms. Round to tenths.

ELECTRONICS APPLICATION

Ohm's Law

Ohm's law is derived from Kirchhoff's laws to form the basis for much of the work done in electronics. In its simplest forms, Ohm's law is represented by three equations:

$$E = IR, \qquad P = IE, \qquad G = \frac{1}{R}$$

E represents potential difference, which is measured in volts (V).

I represents current, which is measured in amperes (A).

P represents power, which is measured in watts (W).

R represents resistance, which is measured in ohms (Ω).

G represents conductance, which is measured in siemens (S).

First consider $E = IR$. The R can be replaced by $1/G$, so $E = I/G$. This can be rewritten as $I = GE$, which can be rearranged to get $G = I/E$.

The equation $P = IE$ can be rearranged to get $I = P/E$ and $E = P/I$. Also, IR can be substituted for E in the equation $P = IE$ to get $P = IIR = I^2R$, and $1/G$ can be substituted for R to get $P = I^2/G$.

Assume that you start with $P = I^2/G$ and wish to solve for I or G. Eliminate denominators (multiply through by G) to get $PG = I^2$. This gives both $G = I^2/P$ and also $I = \sqrt{PG}$.

Suppose that $P = 4$ mW and $G = 7\ \mu$S. Then $I = 167\ \mu$A.

$$\text{Proof:}\quad P = \frac{I^2}{G} = \frac{(167\ \mu\text{A})^2}{7\ \mu\text{S}} = 3984\text{ W or 4 mW} \qquad \textit{(rounded)}$$

Exercises

Solve for each of the following first in letters (for the general equation) and then prove your equation by finding the numerical answer. Prove the numerical answer obtained by substituting back into the original equation. Be sure to use the given prefixes.

1. Solve for I in $P = I^2R$. What is I if $P = 8$ W and $R = 2\ \Omega$?
2. Solve for G in $E = \frac{I}{G}$. What is G if $E = 12$ V and $I = 4$ mA?
3. Solve for P in $G = \frac{I^2}{P}$. What is P if $G = 8$ mS and $I = 6$ mA?
4. Solve for R in $P = I^2R$. What is R if $I = 4$ mA and $P = 8$ mW?
5. Solve for E in $P = \frac{E^2}{R}$. What is E if $P = 9$ W and $R = 63\ \Omega$?
6. Solve for E in $P = E^2G$. What is E if $P = 8$ W and $G = 2$ mS?
7. Solve for R_3 in $R_t = R_1 + R_2 + R_3$. What is R_3 if $R_t = 1410\ \Omega$, $R_1 = 420\ \Omega$, and $R_2 = 440\ \Omega$?
8. Solve for G_3 in $G_t = G_1 + G_2 + G_3$. What is G_3 if $G_t = 92$ mS, $G_1 = 22$ mS, and $G_2 = 24$ mS?
9. Solve for R_1 in $V_1 = \frac{R_1V_t}{R_1 + R_2}$. What is R_1 if $V_1 = 8$ V, $V_t = 12$ V, and $R_2 = 41\ \Omega$?
10. Solve for G_2 in $I_2 = \frac{G_2I_t}{G_1 + G_2}$. What is G_2 if $I_2 = 5$ mA, $I_t = 8$ mA, and $G_1 = 6$ mS?

Answers for Exercises

1. Solve for I in $P = I^2R$. What is I if $P = 8$ W and $R = 2\ \Omega$?

$$I = \sqrt{\frac{P}{R}} = \sqrt{\frac{8}{2}} = 2\text{ A} \qquad \text{Proof: } 2^2 \times 2 = 8$$

2. Solve for G in $E = \frac{I}{G}$. What is G if $E = 12$ V and $I = 4$ mA?

$$G = \frac{I}{E} = \frac{4\text{ mA}}{12\text{ V}} = 0.333\text{ mS} = 333\ \mu\text{S} \qquad \text{Proof: } \frac{4\text{ mA}}{0.333\text{ mS}} = 12\text{ V}$$

3. Solve for P in $G = \frac{I^2}{P}$. What is P if $G = 8$ mS and $I = 6$ mA?

$$P = \frac{I^2}{G} = \frac{(6\text{ mA})^2}{8\text{ mS}} = 4.5\text{ mW} \qquad \text{Proof: } \frac{(6\text{ mA})^2}{4.5\text{ mW}} = 8\text{ mS}$$

4. Solve for R in $P = I^2R$. What is R if $I = 4$ mA and $P = 8$ mW?

$$R = \frac{P}{I^2} = \frac{8\text{ mW}}{(4\text{ mA})^2} = 500\ \Omega \qquad \text{Proof: } (4\text{ mA})^2(500\ \Omega) = 8\text{ mW}$$

5. Solve for E in $P = \frac{E^2}{R}$. What is E if $P = 9$ W and $R = 63\ \Omega$?

$$E = \sqrt{PR} = \sqrt{9 \times 63} = 23.8\text{ V} \qquad \text{Proof: } \frac{(23.8\text{V})^2}{63\ \Omega} = 9\text{ W}$$

6. Solve for E in $P = E^2G$. What is E if $P = 8$ W and $G = 2$ mS?

$$E = \sqrt{\frac{P}{G}} = \sqrt{\frac{8 \text{ W}}{2 \text{ mS}}} = 63.2 \text{ V} \qquad \text{Proof:} \quad (63.2 \text{ V})^2(2 \text{ mS}) = 8 \text{ W}$$

7. Solve for R_3 in $R_t = R_1 + R_2 + R_3$. What is R_3 if $R_t = 1410\ \Omega$, $R_1 = 420\ \Omega$, and $R_2 = 440\ \Omega$?

$$R_3 = R_t - R_1 - R_2 = 1410\ \Omega - 420\ \Omega - 440\ \Omega = 550\ \Omega$$
Proof: $420\ \Omega + 440\ \Omega + 550\ \Omega = 1410\ \Omega$

8. Solve for G_3 in $G_t = G_1 + G_2 + G_3$. What is G_3 if $G_t = 92$ mS, $G_1 = 22$ mS, and $G_2 = 24$ mS?

$$G_3 = G_t - G_1 - G_2 = 92 \text{ mS} - 22 \text{ mS} - 24 \text{ mS} = 46 \text{ mS}$$
Proof: $22 \text{ mS} + 24 \text{ mS} + 46 \text{ mS} = 92 \text{ mS}$

9. Solve for R_1 in $V_1 = \dfrac{R_1V_t}{R_1 + R_2}$. What is R_1 if $V_1 = 8$ V, $V_t = 12$ V, and $R_2 = 41\ \Omega$?

$$R_1 = \frac{V_1R_2}{V_t - V_1} = \frac{8 \text{ V } (41\ \Omega)}{12 \text{ V} - 8 \text{ V}} = 82\ \Omega$$

$$\text{Proof:} \quad \frac{(82\ \Omega)(12 \text{ V})}{82\ \Omega + 41\ \Omega} = 8 \text{ V}$$

10. Solve for G_2 in $I_2 = \dfrac{G_2I_t}{G_1 + G_2}$. What is G_2 if $I_2 = 5$ mA, $I_t = 8$ mA, and $G_1 = 6$ mS?

$$G_2 = \frac{I_2G_1}{I_t - I_2} = \frac{5 \text{ mA } 6 \text{ mS}}{8 \text{ mA} - 5 \text{ mA}} = 10 \text{ mS}$$

$$\text{Proof:} \quad \frac{(10 \text{ mS})(8 \text{ mA})}{6 \text{ mS} + 10 \text{ mS}} = 5 \text{ mA}$$

ASSIGNMENT EXERCISES

13-1

Evaluate the following formulas. Use $D = 2.5$, $E = 3$, and $F = 4$.

1. $Y = E + (3F - D)$
2. $Y = F(E^2 + 4DF) - 3$
3. $Y = \dfrac{2(F - D)^2}{D - 1}$
4. $Y = \dfrac{3(F - D)}{3} + E$
5. $Y = D^2 - (2F - E)$

Evaluate the percentage formula $P = \dfrac{RB}{100}$ using the following values.

6. Find the percentage if $R = 10$ and $B = 300$ lb.
7. Find the rate if $P = 12$ kg and $B = 125$ kg.

8. Find the base if $P = \$28.05$ and $R = 8.5$.

9. Evaluate the rate formula $R = \frac{100P}{B}$ for P if $R = 12$ and $B = 90$.

10. Evaluate the base formula $B = \frac{100P}{R}$ for R if $B = \$5000$ and $P = \$450$.

Evaluate the interest formula $I = PRT$ using the following values.

11. Find the interest if $P = \$440$, $R = 16\%$, and $T = 2\frac{3}{4}$ years.

12. Find the rate if $I = \$2484$, $P = \$4600$, and $T = 3$ years.

13. Find the time if $I = \$387.50$, $P = \$1550$, and $R = 12.5\%$.

14. Find the principal if $I = \$1665$, $R = 18\frac{1}{2}\%$, and $T = 1\frac{1}{2}$ years.

15. Find the length of a rectangular work area if the perimeter is 160 in. and the width is 30 in. Use the formula $P = 2(l + w)$.

16. Find the radius of a circle whose area is 132.7 mm^2 using the formula $r = \sqrt{\frac{A}{\pi}}$. Round to the nearest tenth.

17. Find the cost (C) if the markup (M) on an item is \$25.75 and the selling price (S) is \$115.25. Use the formula $M = S - C$.

18. Using the formula for the side of a square, $s^2 = A$, find the length of a side of a square field whose area is $\frac{1}{4}$ mi^2.

19. Evaluate the formula for the area of a circle, $A = \pi r^2$, if $r = 5.5$ in. Round to the nearest tenth.

20. Evaluate the formula for the area of a square, $A = s^2$, if $s = 3.25$ km.

13-2

Solve the following formulas as indicated.

21. $D = F - (m + n)$ for F.
22. $H = C - S$ for S.
23. $V = lwh$ for w.
24. $R = h(p + 3q)$ for q.
25. $c = ah + ab$ for a.
26. $K = \frac{m + n}{p}$ for m.
27. $B = cr^2x$ for r.
28. $r = \sqrt{s^2 - t^2}$ for t.
29. $PB = A$ for B.
30. $V = lwh$ for h.
31. $I = Prt$ for r.
32. $s = c + m$ for c.
33. $s = r - d$ for r.
34. $s = r - d$ for d.
35. $v = v_0 - 32t$ for t.
36. $P = 2(l + w)$ for w.
37. $A = P(1 + rt)$ for t.
38. $V = \frac{1}{3}Bh$ for h.

39. The formula for finding the sale price on an item is $S = P - D$, where S is the sale price, P is the original price, and D is the discount. Solve the formula for the original price.

40. The formula for finding tax is $T = RM$, where T represents tax, R represents the tax rate, and M represents the marked price. Rearrange the formula to find the marked price.

13-3

Make the following temperature conversions.

41. 78°C = _____ K
42. 410 K = _____ °C
43. 12°F = _____ °R
44. 720°R = _____ °F
45. 95°C = _____ °F
46. 86°F = _____ °C

47. The boiling point of pure water is 100°C. What is the boiling point of water on the Fahrenheit scale?

48. The freezing point of benzene is 5°C. What is this temperature on the Fahrenheit scale?

49. Candy that should reach a temperature of 365°F should reach what temperature on the Celsius scale?

50. Summer road surface temperatures of 122°F would show as what reading on the Celsius scale?

13-4

Find the perimeter and the area of the following figures.

51.

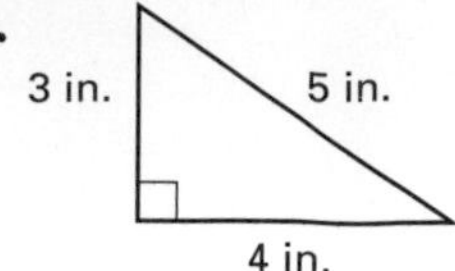

52.

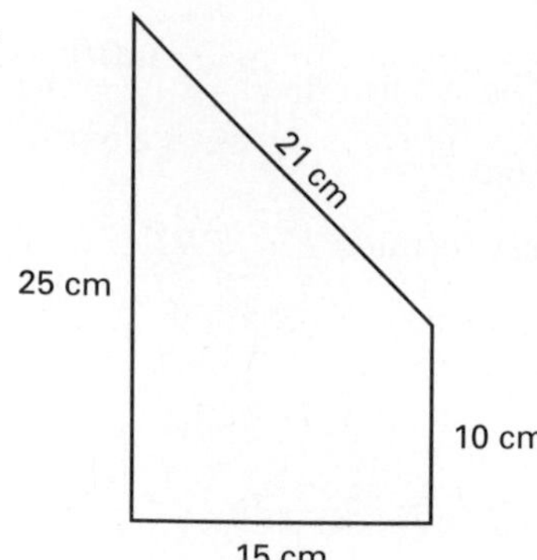

53.

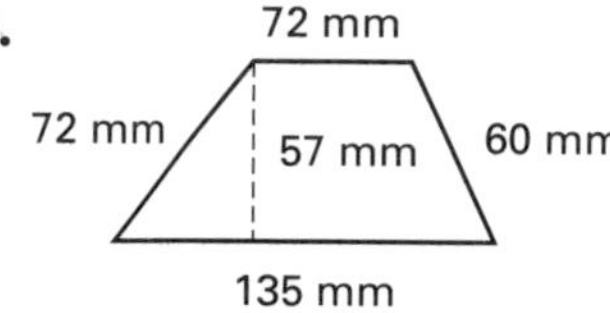

54.

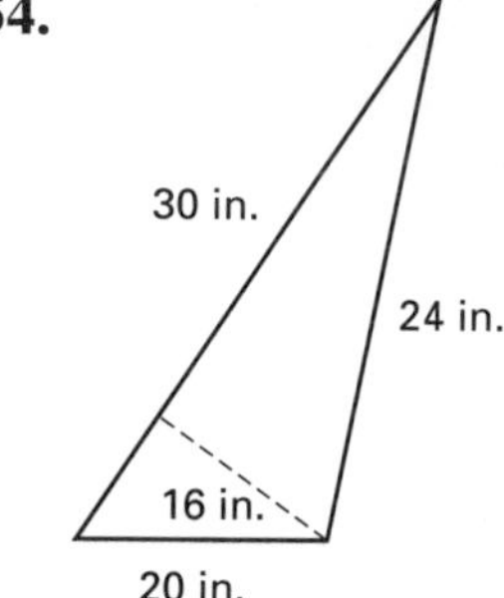

55. Find the area of the triangle in Fig. 13-40.

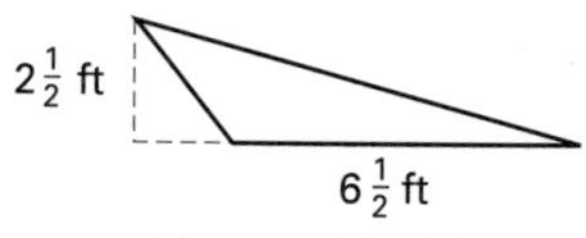

Figure 13-40

Solve the following problems involving perimeter and area.

56. A trapezoidal wall section of a contemporary home has a height of 9 ft, a top length of 7 ft, and a bottom length of 14 ft. Find the area of the wall section.

57. The sides of a triangle are $2\frac{1}{2}$ ft, $3\frac{3}{4}$ ft, and 5 ft. If the triangle is to be made into an advertising sign for a restaurant, how many feet of contrasting trim would be needed to outline the sign?

58. Using your knowledge of formula rearrangement and triangles, find the base of a triangle whose height is 26.25 ft and whose area is 236.25 ft^2.

59. If a triangular louver whose height is 3 ft and whose base is 6 ft 6 in. is installed in each gable end of a house, how many square feet of ventilation would the two louvers provide?

60. One trapezoidal side of a hip roof measures $21\frac{1}{2}$ ft along the ridge line (upper base) and $45\frac{1}{2}$ ft at the bottom (lower base). The height of the side is 16 ft. How many bundles of roofing shingles are necessary to cover the side at a 5-in. exposure if 3.2 bundles cover each square (100 ft^2)?

61. A stairway to the upstairs portion of a house has a run (base) of 10 ft and a rise (height) of $7\frac{1}{2}$ ft. How many sheets of 4-ft × 8-ft paneling are needed to finish the exposed side?

62. A gable end of a house has a span (base) of 30 ft and a rise (height) of $5\frac{1}{2}$ ft. If the gable is to be covered with 8-in. wide bevel siding with a $6\frac{1}{2}$-in. exposure to the weather, how many square feet of siding are needed if we assume a 23% loss for lap and waste? (Round any portion of a square foot to the next highest square foot.)

63. If the three sides of the gable end of a roof are each 16 ft, how many feet of trim would be needed to surround the gable end?

64. A trapezoidal flower bed has its longer base along the entire length of a 45-ft outside wall of a building. If the shorter base is 25 ft and the adjacent sides are each 11.5 ft, how many feet of garden edging material are needed to surround the flower bed if no edging is used against the building wall?

65. If the flower bed in Exercise 64 projects 5.7 ft from the building wall, how many square feet of surface are available for flowering plants?

Use Fig. 13-41 to solve the following exercises. Round the final answers to the nearest thousandth if necessary.

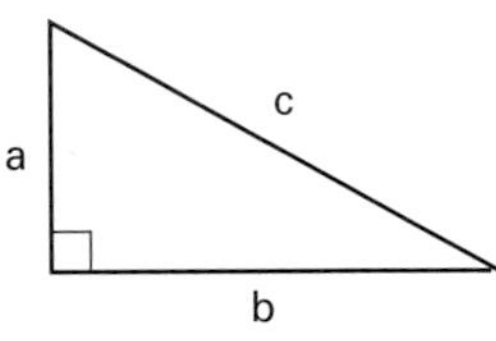

Figure 13-41

66. $a = 3$ m
$b = ?$
$c = 5$ m

67. $a = 9$ in.
$b = 12$ in.
$c = ?$

68. $a = 8$ cm
$b = 15$ cm
$c = ?$

69. $a = 7$ ft
$b = ?$
$c = 10$ ft

70. $a = 8$ mm
$b = ?$
$c = 17$ mm

71. $a = ?$
$b = 15$ yd
$c = 17$ yd

72. $a = ?$
$b = 12$ km
$c = 15$ km

73. $a = 11$ mi
$b = 17$ mi
$c = ?$

74. $a = 10$ in.
$b = 24$ in.
$c = ?$

75. $a = ?$
$b = 40$ cm
$c = 50$ cm

Solve the following problems. Round the final answers to the nearest thousandth if necessary.

76. If the base of a ladder is placed on the ground 4 ft from a house, how tall must the ladder be (to the nearest foot) to reach the chimney top that extends $18\frac{1}{2}$ ft from the ground?

77. In an automobile, there are three pulleys connected by one belt. The center-to-center distance between the pulleys farthest apart cannot be measured conveniently. The other center-to-center distances are 12 in. and 18 in. (Fig. 13-42). Find the distance between the pulleys farthest apart.

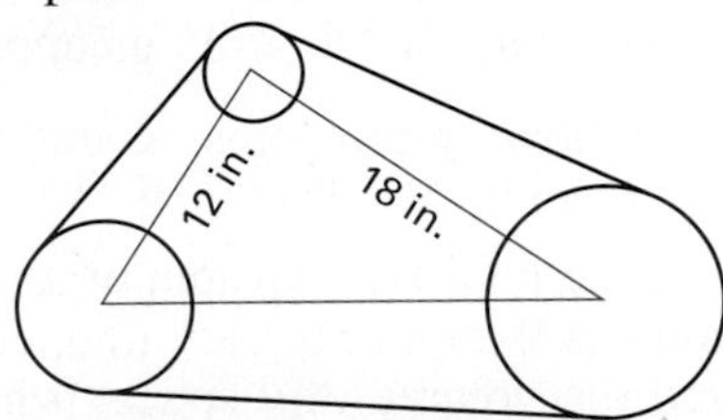

Figure 13-42

78. A central vacuum outlet is installed in one corner of a rectangular room that measures $9' \times 12'$. How long must the nonelastic hose be to reach all parts of the room?

79. Find the diameter of a piece of round steel from which a 3-in. square nut can be milled.

80. Find the distance across the corners of a square nut that is 7.9 mm on a side (Fig. 13-43).

81. Find the volume of the triangular prism of Fig. 13-44.

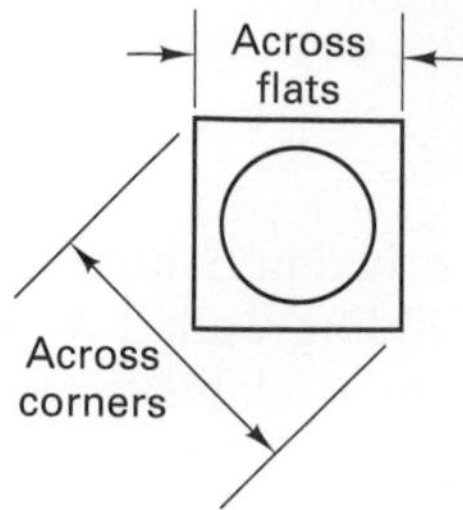

Figure 13-43

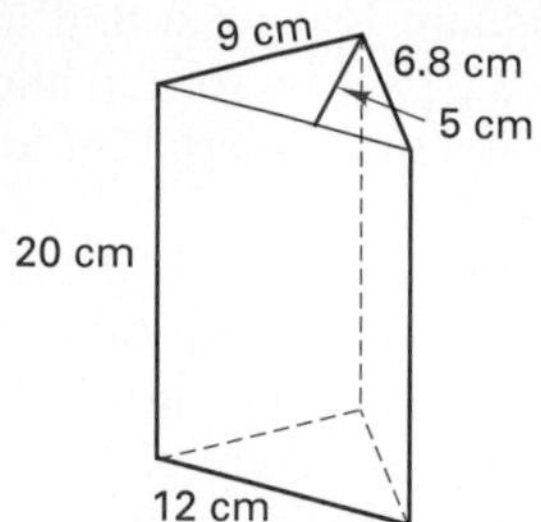

Figure 13-44

82. Find the lateral surface area of the triangular prism of Fig. 13-44.

83. Find the total surface area of the triangular prism of Fig. 13-44.

84. Find the lateral surface area of the cylinder of Fig. 13-45.

85. Find the total surface area of the cylinder of Fig. 13-45.

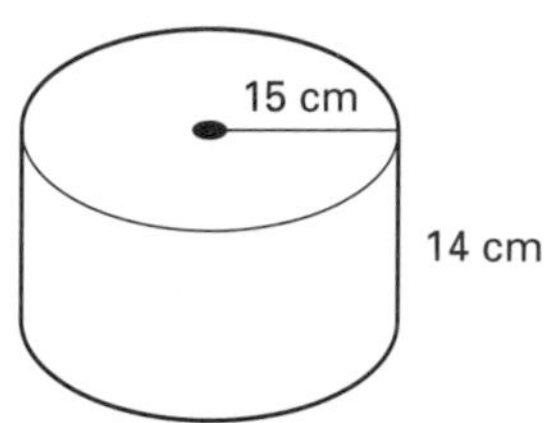

Figure 13-45

86. Find the volume of the cylinder of Fig. 13-45.

87. How many cubic yards of top soil are needed to cover an 85-ft by 65-ft area for landscaping if the topsoil will be 6 in. deep? Round to the nearest whole number. ($27 \text{ ft}^3 = 1 \text{ yd}^3$.)

88. If concrete weighs 160 lb per cubic foot, what is the weight of a concrete circular slab 4 in. thick and 15 ft across?

89. What is the lateral surface area of a hexagonal column 20 ft high that measures 6 in. on a side?

90. An interstate highway was repaired in one section 48 ft across, 25 ft long, and 8 in. deep. If concrete costs $25.50 per cubic yard, what is the cost of the concrete needed to repair the highway rounded to the nearest dollar? ($27 \text{ ft}^3 = 1 \text{ yd}^3$.)

91. A pipeline to carry oil between two towns 5 mi apart has an inside diameter of 18 in. If $1 \text{ mi} = 5280 \text{ ft}$ and $1 \text{ ft}^3 = 7.48 \text{ gal}$, how many gallons of oil will the pipeline hold (to the nearest gallon)?

92. How many barrels of oil will a tank hold if its height is $65\frac{1}{2}$ ft and its radius is 20 ft? Round to the nearest whole barrel. ($31.5 \text{ gal} = 1 \text{ barrel}$ and $1 \text{ ft}^3 = 7.48 \text{ gal}$.)

Solve the following problems. Round the final answer to the nearest tenth unless otherwise specified.

93. Find the total surface area of a sphere whose radius is 9 m.

94. Find the total surface area of a sphere whose diameter is 20 cm.

95. Find the volume of a sphere whose radius is 12 ft.

96. Find the volume of a sphere whose diameter is 30 cm.

97. Find the lateral surface area of a cone whose radius is 6 cm and whose slant height is 9 cm.

98. Find the total surface area of a cone whose radius is 4 m and whose slant height is 8 m.

99. Find the volume of a cone whose radius is 6 in. and whose height is 10 in.

100. If 1 gal = 231 in.^{3}, find the number of gallons that a conical oil container 18 in. high and 23 in. across can hold.

101. The entire exterior surface of a conical tank whose slant height is 12 ft and whose diameter is 18 ft will be painted. If the paint covers at a rate of 350 ft^2 per gallon, how many gallons of paint are needed for the job? Round any fraction of a gallon to the next whole gallon.

102. A hopper deposited sand in a cone-shaped pile with a diameter of 9′ 6″ and a height of 8′ 3″. To the nearest cubic foot, how much sand was deposited?

103. How many square feet of plastic material are needed to devise a wind-tunnel cone whose base is 6 ft across and whose height is 12 ft (Fig. 13-46). (*Hint:* To find the slant height, consider it to be the hypotenuse of a right triangle.)

104. To the nearest square foot, how much nylon material is needed to construct a conical tent-like pavillion 30 ft in diameter and 5 ft from the apex to the base? (*Hint:* See Exercise 103).

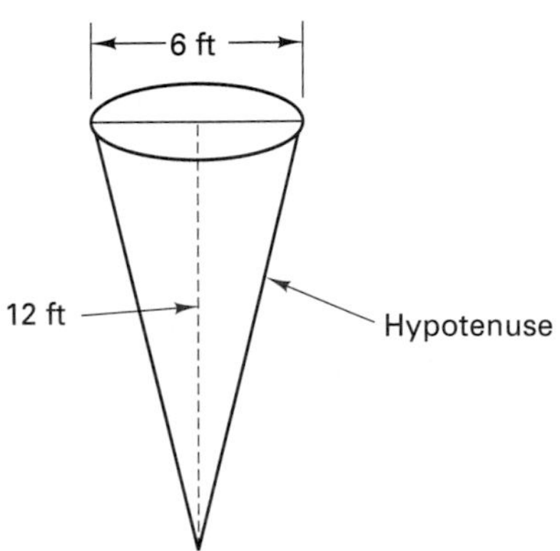

Figure 13-46

105. A No. 5 soccer ball has a diameter of 8.8 in. How much leather is needed to cover the surface?

106. How much does a 4-in. lead ball weigh if lead weighs 1 lb per 2.4 in.^{3}?

107. A spherical tank is anchored halfway in the ground. How many cubic feet of earth had to be excavated to install the tank? The tank measures 20 ft across. Round to the nearest whole cubic foot.

13-5

108. Use the formula $R_t = \dfrac{R_1R_2}{R_1 + R_2}$ to find the total resistance (R_t) if one resistance (R_1) is 10 ohms and the second resistance (R_2) is 9 ohms. Round to tenths.

109. The formula for voltage (Ohm's law) is $E = IR$. Find the amperes of current (I) if the voltage (E) is 220 V and the resistance (R) is 80 ohms.

110. Distance is rate times time, or $D = RT$. Find the rate if the distance traveled is 260 mi and the time traveled is 4 hr.

111. According to Boyle's law, if temperature is constant, the volume of a gas is inversely proportional to the pressure on it. Find the final volume (V_2) of a gas using the formula $\dfrac{V_1}{V_2} = \dfrac{P_2}{P_1}$ if the original volume (V_1) is 30 ft^3, the original pressure (P_1) is 75 lb per square inch (psi), and the final pressure (P_2) is 225 psi.

112. What is the percent efficiency (E) of an engine if the input (I) is 22,600 calories and the output (P) is 5600 calories? Use the formula $E = \frac{I - P}{I}$. Round to the nearest tenth of a percent.

113. The formula for power (P) in watts (W) is $P = I^2R$. Find the current (I) in amperes if a device draws 80 W and the resistance (R) is 8 ohms. Round to tenths.

114. Find the impedance (Z) in ohms using the formula $Z = \sqrt{R^2 + X^2}$ if the resistance (R) is 4 ohms and the reactance (X) is 7 ohms. Round to tenths.

115. The formula for the speed (s) of a driven pulley in revolutions per minute (rpm) is $s = \frac{DS}{d}$. Find the speed of a driven pulley with diameter (d) of 3 in. if the diameter (D) of the driving pulley is 7 in. and its speed (s) is 600 rpm.

116. If the distance (C) between the centers of the pulleys in Exercise 115 is 30 in., find the length (L) of the belt connecting them using the formula

$$L = 2C + 1.57(D + d) + \frac{D + d}{4C}$$

Round to hundredths.

117. Use the formula $H = \frac{D^2N}{2.5}$ to find the number of cylinders (N) required in an engine of 5 hp (H) if the cylinder diameter (D) is 2.5 in.

CHALLENGE PROBLEMS

118. Devise your own formulas for the following relationships.
   (a) An electrical power company computes the monthly charges by multiplying the kilowatts of power used times the cost per kilowatt and adds to that a fixed monthly fee.
   (b) A store figures monthly payments on a charge account by multiplying the interest rate times the unpaid balance and then adding purchases and subtracting payments.
   (c) Profit on the sale of a certain type of item is the product of the number of items sold and the difference between the selling price of the item and its cost to the seller.

119. Explain the usefulness of formula rearrangement in solving applied problems. Use at least one formula to illustrate.

CONCEPTS ANALYSIS

1. Briefly describe the procedure for evaluating a formula when numerical values are given for every variable in the formula except one.
2. Give some instances when it would be desirable to rearrange a formula.
3. What is the difference between two triangles that are congruent and two triangles that are similar? Draw a sketch illustrating two congruent triangles and two similar triangles.
4. How are proportions used to find missing parts of a triangle? Illustrate your answer by making up a problem involving similar triangles. Solve your problem.
5. State the Pythagorean theorem in words. Illustrate the Pythagorean theorem by drawing a right triangle, labeling its parts, and writing the theorem symbolically.
6. Make up a practical application that can be solved using the Pythagorean theorem.

Find the mistakes in the following problems. Explain the mistakes and rework the problem correctly.

7. In the formula $a = 3b - c$, find b if $a = 5$ and $c = 1$.

$$\begin{aligned} 5 &= 3b - 1 \\ 5 + 1 &= 3b \\ 6 &= 3b \\ 3 &= b \end{aligned}$$

8. Solve the formula for y. $3x + 2y = 6$

$$\begin{aligned} 3x &= -2y + 6 \\ x &= \frac{-2y + 6}{3} \\ x &= -\frac{2}{3}y + 2 \end{aligned}$$

9. In the right triangle ABC, find b if $a = 8$ mm and $c = 17$ mm.

$$\begin{aligned} c &= a + b \\ 17 &= 8 + b \\ 289 &= 64 + b \\ 289 - 64 &= b \\ 225 &= b \end{aligned}$$

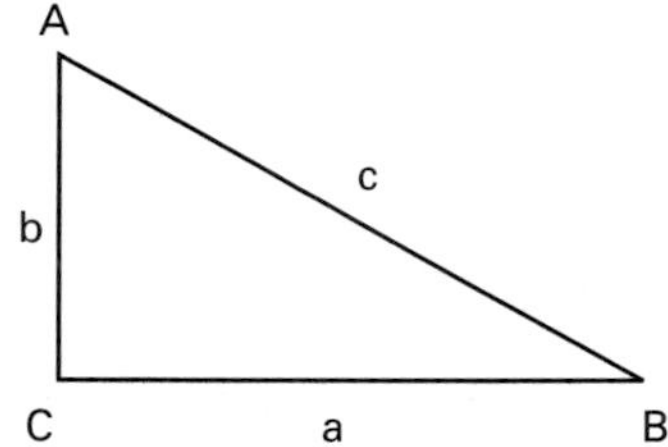

10. Solve the formula for force (F) if the diameter of the piston (D) is 3.5 in. and the pressure (P) is 125 lb per in.^{2} (psi).

$$F = \frac{PD^2}{1.27}$$

$$F = \frac{(125)(3.5)^2}{1.27}$$

$$F = \frac{437.5^2}{1.27}$$

$$F = \frac{191{,}406.25}{1.27}$$

$$F = 150{,}713.5827 \text{ lb}$$

CHAPTER SUMMARY

Objectives	What to Remember	Examples
Section 13-1		
1. Evaluate formulas for a given variable.	Substitute values. Solve using rules for solving equations and/or the order of operations.	Find the area (A) if the length (l) is 4 ft and the width (w) is 2 ft. $A = lw$ $A = (4)(2)$ $A = 8 \text{ ft}^2$
	Sometimes the letter to solve for must be isolated.	If the perimeter (P) of a square is 12 in., find the length of a side. $P = 4s$ $12 = 4s$ $\frac{12}{4} = \frac{4s}{4}$ $s = 3$ in.
Section 13-2		
1. Rearrange formulas to solve for a given variable.	Isolate the desired variable so that it appears on the left. This often makes evaluation simpler. Apply all rules for solving equations.	Solve the formula $S = \frac{R + P}{2}$ for R. $(2)S = \frac{R + P}{2}(2)$ $2S = R + P$ $2S - P = R$ $R = 2S - P$

Section 13-3

1. Make Celsius/Kelvin temperature conversions.

Use formulas:
$K = °C + 273$

Change 20°C to K.
$$K = °C + 273$$
$$K = 20 + 273$$
$$K = 293$$

$°C = K - 273$

Change 281 K to °C.
$$°C = K - 273$$
$$°C = 281 - 273$$
$$°C = 8$$

2. Make Rankine/Fahrenheit temperature conversions.

Use formulas:
$°R = °F + 460$

Change 50°F to °R
$$°R = °F + 460$$
$$°R = 50 + 460$$
$$°R = 510$$

$°F = °R - 460$

Change 750°R to °F.
$$°F = °R - 460$$
$$°F = 750 - 460$$
$$°F = 290$$

3. Convert Fahrenheit to Celsius temperatures.

Use formula:
$°C = \frac{5}{9}(°F - 32)$

Change 14°F to °C.
$$°C = \frac{5}{9}(°F - 32)$$
$$°C = \frac{5}{9}(14 - 32)$$
$$°C = \frac{5}{9}(-18)$$
$$°C = -10$$

4. Convert Celsius to Fahrenheit temperatures.

Use formula:
$°F = \frac{9}{5}°C + 32$

Change 28°C to °F.
$$°F = \frac{9}{5}°C + 32$$
$$°F = \frac{9}{5}(28) + 32$$
$$°F = \frac{252}{5} + 32$$
$$°F = 50.4 + 32$$
$$°F = 82.4$$

Section 13-4

1. Find the perimeter and area of a trapezoid.

Use formulas:
$P = b_1 + b_2 + a + c$

Find the perimeter (P) of a trapezoid whose bases are 10 cm and 8 cm and whose other sides (a, c) are each 7 cm.
$$P = b_1 + b_2 + a + c$$
$$P = 10 + 8 + 7 + 7$$
$$P = 32 \text{ cm}$$

$A = \frac{1}{2}h(b_1 + b_2)$

Find the area of a trapezoid whose bases are 11 and 18 in. and whose height is 12 in.
$$A = \frac{1}{2}h(b_1 + b_2)$$
$$A = \frac{1}{2}(12)(11 + 18)$$
$$A = \frac{1}{2}(12)(29)$$
$$A = 174 \text{ in.}^2$$

2. Find the perimeter and area of a triangle.	Use formulas: $P = a + b + c$	What is the perimeter (P) of a triangular roof vent whose sides a, b, and c are 6 ft, 4 ft, and 4 ft? $P = a + b + c$ $P = 6 + 4 + 4$ $P = 14$ ft
	$A = \frac{1}{2}(bh)$	A triangle has a base of 3 m and a height of 2 m. Find its area. $A = \frac{1}{2}(bh)$ $A = \frac{1}{2}(3)(2)$ $A = 3\text{ m}^2$
3. Use the Pythagorean theorem to find the missing side of a right triangle.	The square of the hypotenuse (of a right triangle) equals the sum of the squares of the other two sides. $c^2 = a^2 + b^2$	The hypotenuse (c) of a right triangle is 10 in. If side b is 6 in., find side a. $c^2 = a^2 + b^2$ $10^2 = a^2 + 6^2$ $100 = a^2 + 36$ $100 - 36 = a^2$ $64 = a^2$ $\sqrt{64} = a$ $a = 8$ in.
4. Find the area of prisms and cylinders.	Use formula for lateral surface area (area of sides): LSA $= ph$, where p is the perimeter of the base and h is the height.	Find the lateral surface area of a triangular prism whose base is 4 in. on each side and whose height is 10 in. LSA $= ph$ LSA $= (4 + 4 + 4)(10)$ LSA $= (12)(10)$ LSA $= 120\text{ in.}^2$
	Use formula for total surface area (sides plus bases): TSA $= ph + 2B$, where p is perimeter of the base, h is the height, and B is the area of a base.	Find the total surface area of the preceding triangle if the height of each base is 3.464 in. TSA $= ph + 2B$ TSA $= 120 + 2\left(\frac{1}{2}\right)(4)(3.464)$ TSA $= 120 + 13.856$ TSA $= 133.856\text{ in.}^2$
5. Find the volume of prisms and cylinders.	Use formula: $V = Bh$, where B is area of the base and h is the height.	Find the volume of a cylinder whose diameter is 20 mm and whose height is 80 mm. $V = Bh$ $V = \pi r^2 h$ $V = \pi(10)^2(80)$ $V = \pi(100)(80)$ $V = 25{,}132.74123\text{ mm}^3$
6. Find the area and volume of a sphere.	Use formula for total surface area: TSA $= 4\pi r^2$, where r is the radius.	What is the surface area of a sphere 6 in. in diameter? TSA $= 4\pi r^2$ TSA $= 4\pi(3)^2$ TSA $= 4\pi(9)$ TSA $= 113.10\text{ in.}^2$ (rounded)

Use formula for volume:

$V = \frac{4\pi r^3}{3}$, where r is the radius.

A spherical gas tank is 10 ft in diameter. Find its volume.

$$V = \frac{4\pi r^3}{3}$$
$$V = \frac{4\pi(5)^3}{3}$$
$$V = \frac{4\pi(125)}{3}$$
$$V = 523.60 \text{ ft}^3 \text{ (rounded)}$$

7. Find the area and volume of a cone.

Use formula for lateral surface area: LSA $= \pi rs$, where r is the radius and s is the slant height.

A conical pile of gravel has a diameter of 30 ft and a slant height of 40 ft. What is the lateral surface area?

$$\text{LSA} = \pi rs$$
$$\text{LSA} = \pi(15)(40)$$
$$\text{LSA} = 1884.96 \text{ ft}^2 \text{ (rounded)}$$

Use formula for total surface area: TSA $= \pi rs + \pi r^2$, where r is the radius and s is the slant height. πr^2 is the area of the base.

Find the total surface area of the preceding cone.

$$\text{TSA} = \pi rs + \pi r^2$$
$$\text{TSA} = 1884.955592 + \pi(15)^2$$
$$\text{TSA} = 1884.955592 + \pi(225)$$
$$\text{TSA} = 1884.955592 + 706.8583471$$
$$\text{TSA} = 2591.81 \text{ ft}^3 \text{ (rounded)}$$

Use formula for volume:

$V = \frac{\pi r^2 h}{3}$, where r is the radius and h is the height.

Find the volume of a cone whose radius is 6 in. and whose height is 12 in.

$$V = \frac{\pi r^2 h}{3}$$
$$V = \frac{\pi(6)^2(12)}{3}$$
$$V = \frac{\pi(36)(12)}{3}$$
$$V = 452.39 \text{ in.}^3 \text{ (rounded)}$$

Section 13-5

1. Evaluate miscellaneous technical formulas.

Apply all techniques for solving equations.

The formula for the distance (d) an object falls is $d = \frac{1}{2}gt^2$. Find the distance if gravity (g) is 32 ft per second squared and seconds (t) is 3.

$$d = \frac{1}{2}gt^2$$
$$d = \frac{1}{2}(32)(3)^2$$
$$d = \frac{1}{2}(32)(9)$$
$$d = 144 \text{ ft}$$

WORDS TO KNOW

formula (p. 506)
evaluation (p. 506)
rearrangement (p. 511)
subscripts (p. 513)
Celsius (p. 515)
Kelvin (p. 515)
Rankine (p. 516)
Fahrenheit (p. 516)
geometry (p. 520)
trapezoid (p. 520)
parallel (p. 520)
base (p. 520)
height (p. 521)
altitude (p. 522)
triangle (p. 522)
right triangle (p. 522)
hypotenuse (p. 522)
oblique triangle (p. 522)
degrees (p. 522)
leg (p. 522)
right (p. 522)
perimeter (p. 523)
area (p. 523)
Pythagorean theorem (p. 525)
pulleys (p. 526)
prism (p. 527)
cylinder (p. 527)
lateral surface area (p. 528)
total surface area (p. 528)
volume (p.530)
great circle (p. 531)
cone (p. 531)
sphere (p. 531)
slant height (p. 532)
horsepower (p. 539)

CHAPTER TRIAL TEST

1. The electrical resistance of a wire is found from the formula $R = \frac{PL}{A}$. Rearrange the formula to find the length L of the wire.

2. The formula for volume (V) of a solid rectangular figure is $V = lwh$ (length × width × height). If the volume of a mailing container must be 7.5 cm^3 and its length 1.5 cm and width 0.5 cm, what must its height be?

3. Engine displacement d is found from the formula $d = \pi r^2 sn$. Solve to find r (radius of the bore).

4. Using $d = 351$ in.^{3}, $s = 3.5$ in. stroke, and $n = 8$ cylinders, calculate to the nearest tenth the radius of the bore with the rearranged formula in Problem 3.

Use the following formulas for Problems 5 to 8:

$$K = °C + 273 \qquad °C = K - 273$$
$$°R = °F + 460 \qquad °F = °R - 460$$
$$°C = \tfrac{5}{9}(°F - 32) \qquad °F = \tfrac{9}{5}°C + 32$$

5. 88°C = ____ K

6. 104°F = ____ °C

7. 195°C = ____ °F

8. 65°F = ____ °R

9. A section of a hip roof is a trapezoid measuring 35 ft at the bottom, 15 ft at the top, and 10 ft high. Find the area of this section of the roof in square feet.

10. The gable end of a roof is a triangle whose rise (height) is 7 ft and whose span (base) is 24 ft. The gable end contains a window 24 in. × 36 in. How much of the gable area will need to be painted?

11. A metal rod is welded to a metal support (Fig. 13-47). If the two sides of the metal support are 8 dm and 6 dm, find the length of the metal rod needed to make the brace.

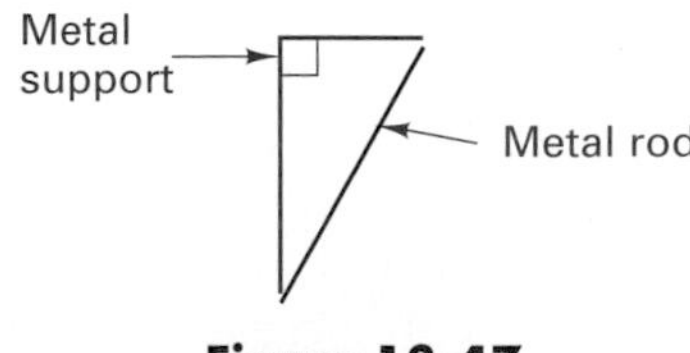

Figure 13-47

12. Find the perimeter of the polygons shown in the following figures (Fig. 13-48).

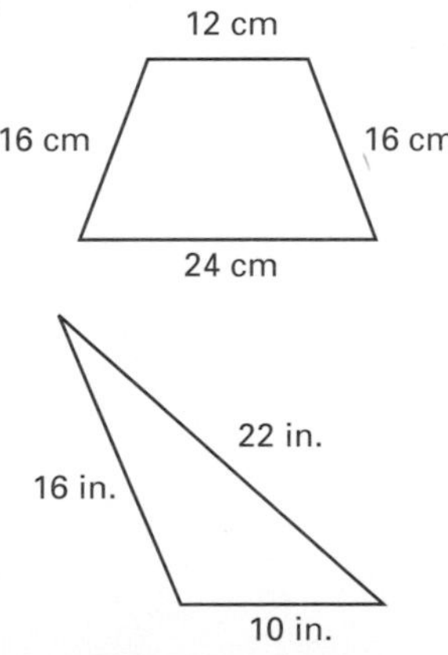

Figure 13-48

Solve the following problems.

13. A right pentagonal prism measures 1 in. on each side of its base and has a height of 10 in. What is its lateral surface area?

14. How much space is required to store 30 microcomputers crated in boxes that measure 62 cm by 70 cm by 50 cm?

15. A spherical tank 12 ft in diameter can hold how many gallons of fluid if 1 ft^3 = 7.48 gal? Answer to the nearest whole gallon.

16. The bases of a brass prism are equilateral triangles whose bases are 3 in. and whose heights are 2.6 in. If the prism's height is 8 in., what is the volume of the brass prism?

17. Hard coal broken into small pieces is dumped into a pile shaped like a cone. The base is 35 ft across and the pile stands 12 ft tall. How many cubic feet of coal are in the pile? Round to tenths.

18. A spherical gas storage tank 2.5 m wide needs to be sandblasted, primed, and refinished. The owner received an estimate of $8.50 per square meter. How much should the job cost the owner, to the nearest dollar?

19. A sheet metal worker wants to make a tin cone that will later be modified for use. If the base of the cone is to be 30 cm across and its height is 25 cm, what is the total surface area required to the nearest tenth? (*Hint:* Use your knowledge of a right triangle to find the slant height.)

20. A cold water pipe whose outside diameter is $\frac{7}{8}$ in. runs 23 ft in an attic. How many whole rolls of insulating wrap are needed if one roll covers $5\frac{1}{2}$ ft^2 and no allowance is made for overlap. (*Hint:* $\frac{7}{8}$ in. is $\frac{7}{8} \times \frac{1}{12} = 0.0729$ ft.)

21. A steel rod has a diameter of 20 in. Find the volume of steel in a 5-ft length of the rod in cubic feet to the nearest tenth.

22. Find the total surface area of a cylindrical storage tank if it is 30 ft tall and has a diameter of 12 ft. Round to hundredths.

23. Using the formula $P = \frac{1.27F}{D^2}$, calculate force F (in pounds) if the pressure P is 180 psi and the piston diameter is 3.25 in. Round to nearest hundredth.

24. Use the formula $R_t = \frac{R_1 R_2}{R_1 + R_2}$ to find the total resistance (R_t) if one resistance (R_1) is 9 ohms and the second resistance (R_2) is 8 ohms. Round to tenths.

25. If the efficiency (E) of an engine is 70% and the input (I) is 40,000 calories, find the output (P) in calories. Use the formula $E = \frac{I - P}{I}$.

CHAPTER

14 Products and Factoring

As we know, products are the results of multiplication and factors are the quantities that, when multiplied, give us a product. Knowledge of products and factoring is important in solving certain kinds of equations or performing certain kinds of tasks at work.

14-1 THE DISTRIBUTIVE PRINCIPLE AND COMMON FACTORS

Learning Objective

L.O.1. Factor an expression containing a common factor.

L.O.1. Factor an expression containing a common factor.

Factoring an Expression Containing a Common Factor

First, let's review some of the algebraic terminology from Chapter 11.

A *monomial* is a polynomial containing only one term, such as $5x^2$.

A *binomial* is a polynomial containing only two terms, such as $3a + 4$.

A *trinomial* is a polynomial containing only three terms, such as $4x^2 + x - 2$.

A *polynomial* is an expression with constants and variables with natural-number exponents and containing one or more terms with at least one variable term.

The multiplication problem $7a(3a + 2)$ is the indicated product of $7a$ and the binomial $3a + 2$. This form of the expression is called *factored* form. After the expression is multiplied, we have two terms written as the indicated sum $21a^2 + 14a$. This is called *expanded* form. To rewrite the binomial $21a^2 + 14a$ as the indicated product $7a(3a + 2)$ is to *factor* it.

Let's look at a general example of the distributive principle:

$$a(x + y) = ax + ay$$

Notice that a appears as a factor in both terms on the right side of the equal sign. When a factor appears in each of several terms, it is called a *common factor* of the terms. The distributive principle in reverse can be used to write the addition as a multiplication. In other words, we can *factor* the expression.

EXAMPLE 14-1.1 Write $3a + 3b$ in factored form.

Solution: Since 3 is the common factor in both terms, we can use the distributive principle to factor the binomial.

$$3a + 3b = 3(a + b)$$ ■

The distributive principle also applies if we have more than two terms.

EXAMPLE 14-1.2 Write $3ab + 9a + 12b$ in factored form.

Solution: Since 3 is the only factor that appears in all three terms of the trinomial, it is the common factor.

$$3ab + \underbrace{9a}_{3 \cdot 3} + \underbrace{12b}_{3 \cdot 4} = 3(ab + 3a + 4b)$$ ■

> **Rule 14-1.1** *To factor an expression containing a common factor:*
>
> 1. Find the *greatest* factor common to *each* term of the expression.
> 2. Divide each term by the common factor.
> 3. Rewrite the expression as the indicated product of the greatest common factor and the quotients in step 2.

When looking for a common factor, we always look for *all* common factors.

EXAMPLE 14-1.3 Factor $4a^2 + 6a$ completely.

Solution: Since 2 and a are common factors, the greatest common factor is $2a$.

$$4a^2 + 6a = 2a(2a + 3) \qquad \left(\frac{4a^2}{2a} = 2a, \frac{6a}{2a} = 3\right)$$

We should always check our factoring by multiplying.

$$2a(2a + 3) = 4a^2 + 6a$$

This process of checking can be done mentally. ■

EXAMPLE 14-1.4 Factor $2x^2 + 2x^3$ completely.

Solution: Since 2 and x^2 are common factors, the greatest common factor is $2x^2$.

$$2x^2 + 2x^3 = 2x^2(1 + x) \qquad \left(\frac{2x^2}{2x^2} = 1, \frac{2x^3}{2x^2} = x\right)$$

Note that a 1 remains in the grouping since $2x^2$ is the product of $2x^2$ and 1. ■

SELF-STUDY EXERCISES 14-1

L.O.1. Factor completely. Check.

1. $7a + 7b$	**2.** $m^2 + 2m$	**3.** $6x^2 + 3x$
4. $5ab + 10a + 20b$	**5.** $4ax^2 + 6a^2x$	**6.** $5a - 7ab$
7. $12a^2 - 15a + 6$	**8.** $3x^3 - 9x^2 - 6x$	**9.** $8a^2b + 14ab^3$
10. $3m^2 - 6m^3$		

14-2 MULTIPLYING POLYNOMIALS

Learning Objectives

L.O.1. Use the FOIL method to multiply two binomials.
L.O.2. Multiply polynomials to obtain three special products.

L.O.1. Use the FOIL method to multiply two binomials.

Multiplying Two Binomials

We can multiply two binomials by using the distributive principle. According to the distributive principle, each term of the first factor is multiplied by each term of the second factor. In the case where the first factor is a binomial, we use the distributive principle more than once.

EXAMPLE 14-2.1 Multiply $(x + 4)(x + 2)$.

Solution

$$\begin{aligned}(x + 4)(x + 2) &= x(x + 2) + 4(x + 2) && \text{(distributive principle)}\\ &= x^2 + 2x + 4x + 8 && \text{(distributive principle)}\\ &= x^2 + 6x + 8 && \text{(Combine like terms.)}\end{aligned}$$

■

FOIL Method

Rule 14-2.1 ***To multiply two binomials by the FOIL method:***

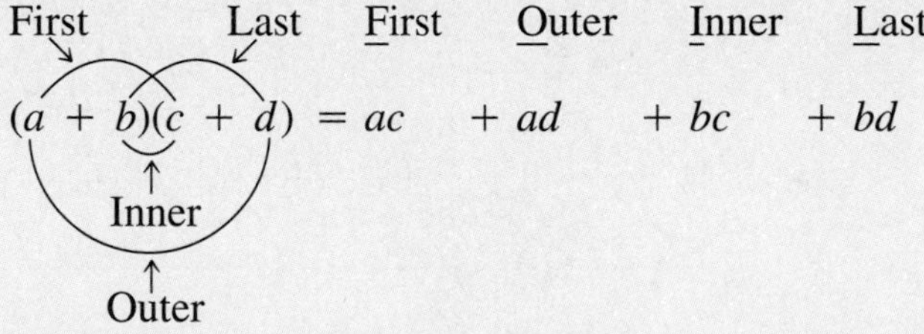

To help us remember a systematic way of accomplishing this multiplication, we can use the word **FOIL**, where F refers to the product of the *first* terms in each factor, O refers to the product of the two *outer* terms, I refers to the product of the *inner* terms, and L refers to the product of the *last* terms in each factor.

If the inner and outer products are like terms, they should be combined.

EXAMPLE 14-2.2 Use the FOIL method to multiply $(x + 5)(x + 3)$.

Solution

$$\begin{aligned}&\quad\ \ \underline{\text{F}}\quad\ \underline{\text{O}}\quad\ \ \underline{\text{I}}\quad\ \ \underline{\text{L}}\\(x + 5)(x + 3) &= x^2 + 3x + 5x + 15\\ &= x^2 + 8x + 15\end{aligned}$$

■

In Example 14-2.2, all signs were positive. But take special care with the signs in the next example.

EXAMPLE 14-2.3 Use the FOIL method to multiply $(2x - 3)(x + 1)$.

Solution

$$\begin{aligned}(2x - 3)(x + 1) &= \overset{\text{F}}{2x^2} + \overset{\text{O}}{2x} - \overset{\text{I}}{3x} - \overset{\text{L}}{3} \\ &= 2x^2 - x - 3\end{aligned}$$

■

L.O.2. Multiply polynomials to obtain three special products.

Special Products

When we multiply the sum of two terms by the difference of the same two terms, we can make some special observations about the product.

$$(x + 3)(x - 3) = \overset{\text{F}}{x^2} \underbrace{- \overset{\text{O}}{3x} + \overset{\text{I}}{3x}}_{0} - \overset{\text{L}}{9} = x^2 - 9$$

Examine the final product. Notice that the product has only *two* terms and is a *difference*. The sum of the outer and inner products is zero. Also notice that the terms of the product are *perfect squares*. Now compare the product with its factors. The first term of the product is the square of the first term in either of its factors. The second term in the product is the square of the second term in either of its factors.

We call pairs of factors like $(a + b)(a - b)$ the sum and difference of the same two terms or *conjugate pairs*. The product is called "the difference of two perfect squares."

Rule 14-2.2 ***To mentally multiply the sum and difference of the same two terms:***

1. Square the first term.
2. Insert a minus sign.
3. Square the second term.

EXAMPLE 14-2.4 Find the products mentally.

(a) $(x + 2)(x - 2)$ **(b)** $(a - 4)(a + 4)$ **(c)** $(b + 9)(b - 9)$
(d) $(2m + 3)(2m - 3)$ **(e)** $(2a + y)(2a - y)$

Solution: Since each of these is the sum and difference of the same two terms, the product will be the difference of two perfect squares.

(a) $(x + 2)(x - 2) = x^2 - 4$ **(b)** $(a - 4)(a + 4) = a^2 - 16$
(c) $(b + 9)(b - 9) = b^2 - 81$ **(d)** $(2m + 3)(2m - 3) = 4m^2 - 9$
(e) $(2a + y)(2a - y) = 4a^2 - y^2$ ■

When we multiply the same two binomials, we can also make some special observations. Since $x \cdot x = x^2$, $(x + 3)(x + 3) = (x + 3)^2$. The quantity $(x + 3)^2$ is called a *binomial square*.

EXAMPLE 14-2.5 Find the products of the binomial squares.

(a) $(x + 2)^2$ **(b)** $(x + 5)^2$ **(c)** $(x - 3)^2$ **(d)** $(2x - 5)^2$
(e) $(3a - 2b)^2$

Solution: To find the products, write each binomial square as two factors; then use the FOIL method to multiply.

(a) $(x + 2)^2 = (x + 2)(x + 2) = x^2 + 4x + 4$ $\quad$ *$(2x + 2x = 4x)$*
(b) $(x + 5)^2 = (x + 5)(x + 5) = x^2 + 10x + 25$ $\quad$ *$(5x + 5x = 10x)$*
(c) $(x - 3)^2 = (x - 3)(x - 3) = x^2 - 6x + 9$ $\quad$ *$(-3x - 3x = -6x)$*
(d) $(2x - 5)^2 = (2x - 5)(2x - 5)$
$= 4x^2 - 20x + 25$ $\quad$ *$(-10x - 10x = -20x)$*
(e) $(3a - 2b)^2 = (3a - 2b)(3a - 2b)$
$= 9a^2 - 12ab + 4b^2$ $\quad$ *$(-6ab - 6ab = -12ab)$* ■

Notice that each product is a trinomial (three terms). These are special trinomials called *perfect square trinomials.* In each, the first and last terms are positive and perfect squares and result from squaring the terms of the binomial. To get the middle term of the trinomial, we combine the outer and inner products. Since the outer and inner products are the same, the middle term can more easily be found by doubling the product of the two terms of the binomial. Let's apply these observations to an example.

EXAMPLE 14-2.6 Find the product of $(2x + 5)^2$.

Solution

$(2x + 5)^2 = 4x^2 + \_\_\_ + 25$ $\quad$ *(Square binomial terms.)*

$(2x + 5)^2 = 4x^2 + 20x + 25$ $\quad$ *(Double product of binomial terms. $2x(5) = 10x$; $2(10x) = 20x$.)* ■

Rule 14-2.3 ***To mentally square a binomial:***

1. Square the first term.
2. Double the product of the two terms.
3. Square the second term.

EXAMPLE 14-2.7 Square the following binomials using the steps given in Rule 14-2.3.

(a) $(x + 2)^2$ $\quad$ **(b)** $(2x - 3)^2$ $\quad$ **(c)** $(5x + 1)^2$
(d) $(3x - 4)^2$ $\quad$ **(e)** $(7x + 2)^2$

Solution

	Square first term		Double product of terms		Square second term
	$(x)^2$		$2 \cdot (2x)$		$(2)^2$
(a) $(x + 2)^2 =$	x^2	$+$	$4x$	$+$	4
	$(2x)^2$		$2 \cdot (-6x)$		$(-3)^2$
(b) $(2x - 3)^2 =$	$4x^2$	$-$	$12x$	$+$	9
	$(5x)^2$		$2 \cdot (5x)$		$(1)^2$
(c) $(5x + 1)^2 =$	$25x^2$	$+$	$10x$	$+$	1
	$(3x)^2$		$2 \cdot (-12x)$		$(-4)^2$
(d) $(3x - 4)^2 =$	$9x^2$	$-$	$24x$	$+$	16
	$(7x)^2$		$2 \cdot (14x)$		$(2)^2$
(e) $(7x + 2)^2 =$	$49x^2$	$+$	$28x$	$+$	4

■

Now let's look at the special products of the following two polynomials:

$$(a + b)(a^2 - ab + b^2)$$

The binomial is the sum of two terms and the trinomial's first and last terms are the squares of the two terms of the binomial, and its middle term is the product of the two terms of the binomial preceded by a negative sign.

$$(a - b)(a^2 + ab + b^2)$$

The binomial is the difference of two terms and the trinomial's first and last terms are the squares of the two terms of the binomial, and its middle term is the product of the two terms of the binomial preceded by a positive sign.

Tips and Traps

Special products are problems that fit a specific pattern. If the problem fits the pattern, then the product is predictable and can be done mentally. Let's look carefully at the pattern developed by the special products:

$$(a + b)(a^2 - ab + b^2) \quad \text{and} \quad (a - b)(a^2 + ab + b^2)$$

In addition to the first and last terms of the trinomial being the squares of the two terms of the binomial and the middle term being the product of the two terms, a key element in these two special cases is that, when the binomial is a *sum* $(+)$, the middle term of the trinomial is *negative* $(-)$, and when the binomial is a *difference* $(-)$, the middle term of the trinomial is *positive* $(+)$.

Observe the products when the binomial and trinomial are multiplied by an extension of the FOIL method or repeated applications of the distributive principle.

EXAMPLE 14-2.8 Multiply $(a + b)(a^2 - ab + b^2)$.

Solution: Multiply each term in the trinomial by each term in the binomial.

First Middle Last

$$\begin{aligned}(a + b)(a^2 - ab + b^2) = a^3 &- a^2b + ab^2 \\ &+ a^2b - ab^2 + b^3 \\ \hline a^3 &\qquad\qquad\quad + b^3\end{aligned}$$

(The four middle terms add to zero.)

Thus, $(a + b)(a^2 - ab + b^2) = a^3 + b^3$, the *sum of two perfect cubes.* ■

EXAMPLE 14-2.9 Multiply $(a - b)(a^2 + ab + b^2)$.

Solution: Multiply each term in the trinomial by each term in the binomial.

First Middle Last

$$\begin{aligned}(a - b)(a^2 + ab + b^2) = a^3 &+ a^2b + ab^2 \\ &- a^2b - ab^2 - b^3 \\ \hline a^3 &\qquad\qquad\quad - b^3\end{aligned}$$

(The four middle terms add to zero.)

Thus, $(a - b)(a^2 + ab + b^2) = a^3 - b^3$, the *difference of two perfect cubes.* ■

Rule 14-2.4 *To mentally multiply a binomial and a trinomial of the types $(a + b)(a^2 - ab + b^2)$ and $(a - b)(a^2 + ab + b^2)$:*

1. Cube the first term of the binomial.
2. If the binomial is a sum, insert a plus sign. If it is a difference, insert a minus sign.
3. Cube the second term of the binomial.

EXAMPLE 14-2.10 Mentally multiply $(x - y)(x^2 + xy + y^2)$.

Solution: Cube both terms of the binomial. Since the binomial is a difference, insert a minus sign between the cubes.

$$(x - y)(x^2 + xy + y^2) = x^3 - y^3$$ ■

EXAMPLE 14-2.11 Mentally multiply $(m + p)(m^2 - mp + p^2)$.

Solution: Cube both terms of the binomial. Since the binomial is a sum, insert a plus sign between the cubes.

$$(m + p)(m^2 - mp + p^2) = m^3 + p^3$$ ■

EXAMPLE 14-2.12 Mentally multiply $(4 - d)(16 + 4d + d^2)$.

Solution: 16 is the same as 4^2, so this problem is worked like Example 14-2.10: Cube the 4 and the d; insert a minus sign between the cubes since the binomial is a difference.

$(4 - d)(16 + 4d + d^2) = 64 - d^3 \quad (4^3 = 64)$ ■

EXAMPLE 14-2.13 Mentally multiply $(a + 3)(a^2 - 3a + 9)$.

Solution: 9 is the same as 3^2, so this problem is worked like Example 14-2.11: Cube the a and the 3; insert a plus sign between the cubes since the binomial is a sum.

$(a + 3)(a^2 - 3a + 9) = a^3 + 27 \quad (3^3 = 27)$ ■

SELF-STUDY EXERCISES 14-2

L.O.1. Use the FOIL method to find the products. Practice combining the outer and inner products mentally.

1. $(a + 3)(a + 8)$
2. $(x - 4)(x + 5)$
3. $(y - 7)(y - 3)$
4. $(2a - 3)(a + 4)$
5. $(3a - 2b)(a - 2b)$
6. $(5x - y)(c - 5y)$
7. $(3x - 4)(2x - 3)$
8. $(a - b)(2a - 5b)$
9. $(7 - m)(3 - 7m)$
10. $(5 - 2x)(8 - x)$
11. $(x + 7)(x + 4)$
12. $(y - 7)(y - 5)$
13. $(m + 3)(m - 7)$
14. $(3b - 2)(x + 6)$
15. $(4r - 5)(3r + 2)$
16. $(5 - x)(7 - 3x)$
17. $(4 - 2m)(1 - 3m)$
18. $(2 + 3x)(3 + 2x)$
19. $(x + 3)(2x - 5)$
20. $(5x - 7y)(4x + 3y)$
21. $(2a + 3b)(7a - b)$
22. $(5a + 2b)(6a - 5b)$
23. $(9x - 2y)(3x + 4y)$
24. $(5x - 8y)(4x - 3y)$
25. $(7m - 2n)(3m + 5n)$

L.O.2. Find the special products mentally.

26. $(a + 3)(a - 3)$
27. $(2x + 3)(2x - 3)$
28. $(a - y)(a + y)$

29. $(4r + 5)(4r - 5)$
30. $(5x + 2)(5x - 2)$
31. $(7 + m)(7 - m)$
32. $(2 - 3x)^2$
33. $(3x + 4)^2$
34. $(Q + L)^2$
35. $(a^2 + 1)^2$
36. $(2d - 5)^2$
37. $(3a + 2x)^2$
38. $(3x - 7)(3x + 7)$
39. $(6 + Q)^2$
40. $(y - 5x)(y + 5x)$
41. $(4 - 3j)^2$
42. $(3m - 2p)(3m + 2p)$
43. $(m^2 + p^2)^2$
44. $(2a - 7c)(2a + 7c)$
45. $(9 - 13a)^2$

Use the extension of the FOIL method to find the products. Perform the work mentally.

46. $(x + p)(x^2 - xp + p^2)$
47. $(Q - L)(Q^2 + QL + L^2)$
48. $(3 + a)(9 - 3a + a^2)$
49. $(2x - 4p)(4x^2 + 8xp + 16p^2)$
50. $(3m - 2)(9m^2 + 6m + 4)$
51. $(6 + p)(36 - 6p + p^2)$
52. $(5y - p)(25y^2 + 5yp + p^2)$
53. $(x + 2y)(x^2 - 2xy + 4y^2)$
54. $(Q - 6)(Q^2 + 6Q + 36)$
55. $(3T + 2)(9T^2 - 6T + 4)$

14-3 FACTORING SPECIAL PRODUCTS

Learning Objectives

L.O.1. Recognize and factor the difference of two perfect squares.
L.O.2. Recognize and factor a perfect square trinomial.
L.O.3. Recognize and factor the sum or difference of two perfect cubes.

To factor any of the special products of the previous section requires us to apply the inverse of the process. Instead of multiplying a pair of factors to obtain the difference of two perfect squares or a perfect square trinomial or the sum or difference of two perfect cubes, we start with the difference of two perfect squares or a perfect square trinomial or the sum or difference of two perfect cubes and "work back" to the factors that produce these special products.

To rewrite a special product in factored form requires us to recognize the product as a pattern. Once we identify the special product, then we must know the pattern of the product in factored form.

L.O.1. Recognize and factor the difference of two perfect squares.

Recognizing and Factoring the Difference of Two Perfect Squares

The difference of two perfect squares, such as $x^2 - a^2$, factors into the product of two binomials, $(x + a)(x - a)$. These factors are the sum and difference of the same two terms. The terms of each binomial are the square roots of the terms of the expression to be factored. Before we can factor such a special product, we must be able to recognize an expression as the difference of two perfect squares.

EXAMPLE 14-3.1 Identify the special products that are the difference of two perfect squares in the following examples.

(a) $x^2 - 9$ **(b)** $a^2 + 49$ **(c)** $m^2 - 27$ **(d)** $3y^2 - 25$
(e) $9x^2 - 4$ **(f)** $4x^2 - 4x + 1$

Solution

(a) Difference of two perfect squares.
(b) Not a special product; this is a *sum*, not a difference.
(c) Not a special product; 27 is not a perfect square.
(d) Not a special product; in $3y^2$, 3 is not a perfect square.
(e) Difference of two perfect squares.
(f) Not the difference of two perfect squares; this is not a binomial. ■

Rule 14-3.1 *To factor the difference of two perfect squares:*

1. Take the square root of the first term.
2. Take the square root of the second term.
3. Write one factor as the *sum* of the square roots found in steps 1 and 2, and write the other as the *difference* of the square roots found in steps 1 and 2.

EXAMPLE 14-3.2 Factor the following special products which are the differences of two perfect squares.

(a) $a^2 - 9$ **(b)** $x^2 - 36$ **(c)** $4x^2 - 1$
(d) $16m^2 - 49$ **(e)** $9a^2 - 25b^2$

Solution

(a) $(a + 3)(a - 3)$ **(b)** $(x + 6)(x - 6)$ **(c)** $(2x + 1)(2x - 1)$
(d) $(4m + 7)(4m - 7)$ **(e)** $(3a + 5b)(3a - 5b)$ ■

Tips and Traps

Because multiplication is commutative, the factors given as answers for Example 14-3.2 may be expressed in any order, such as $(a + 3)(a - 3)$ or $(a - 3)(a + 3)$, $(x + 6)(x - 6)$ or $(x - 6)(x + 6)$, and so on.

L.O.2. Recognize and factor a perfect-square trinomial.

Recognizing and Factoring Perfect-Square Trinomials

A trinomial is a *perfect-square trinomial* if the first and last terms are positive perfect squares and the absolute value of the middle term is *twice* the product of the square roots of the first and last terms. Again, we need to be able to distinguish these special products from other expressions before we can factor them.

EXAMPLE 14-3.3 Tell why the following trinomials are *not* perfect square trinomials.

(a) $x^2 + 2x - 1$ **(b)** $4x^2 + 6x + 9$
(c) $x^2 - 5x + 4$ **(d)** $-4x^2 - 4x + 1$

Solution

(a) The last term, -1, is negative. This term must be positive in a perfect square trinomial.
(b) The middle term, $6x$, is not twice the product of $2x$ and 3.
(c) The middle term, $-5x$, is not twice the product of x and 2.
(d) The first term, $-4x^2$, is negative. It should be positive. ■

EXAMPLE 14-3.4 Verify that the following trinomials are perfect square trinomials.

(a) $x^2 + 14x + 49$ **(b)** $4m^2 - 12m + 9$ **(c)** $9x^2 + 24xy + 16y^2$

Solution

(a) First and last terms, x^2 and 49, are positive perfect squares. Middle term, $14x$, has an absolute value of twice the product of the square roots of x^2 and 49. $2(7x) = 14x$.

(b) First and last terms, $4m^2$ and 9, are positive perfect squares. Middle term, $-12m$, has an absolute value of twice the product of the square roots of $4m^2$ and 9. $2(2m \cdot 3) = 12m$.

(c) First and last terms, $9x^2$ and $16y^2$, are positive perfect squares. Middle term, $24xy$, has an absolute value of twice the product of the square roots of $9x^2$ and $16y^2$. $2(3x \cdot 4y) = 24xy$. ■

> **Rule 14-3.2** *To factor a perfect-square trinomial:*
>
> **1.** Write the square root of the first term.
> **2.** Write the sign of the middle term.
> **3.** Write the square root of the last term.
> **4.** Indicate the square of this binomial quantity.

EXAMPLE 14-3.5 Factor the perfect-square trinomials in Example 14-3.4.

Solution

(a) $x^2 + 14x + 49$

Square root of *first* term	Sign of *middle* term	Square root of *last* term	Indicate *square* of this quantity
x	$+$	7	$(x + 7)^2$

(b) $4m^2 - 12m + 9$

$\downarrow\ \downarrow\ \downarrow$

$(2m - 3)^2$

(c) $9x^2 + 24xy + 16y^2$

$\downarrow\ \downarrow\ \downarrow$

$(3x + 4y)^2$

L.O.3. Recognize and factor the sum or difference of two perfect cubes.

Recognizing and Factoring the Sum or Difference of Two Perfect Cubes

The sum or difference of two perfect cubes, such as $x^3 + y^3$ or $a^3 - b^3$, factors into the product of a binomial and a trinomial, $(x + y)(x^2 - xy + y^2)$ or $(a - b)(a^2 + ab + b^2)$. Before we can factor these special products, we must be able to recognize an expression as the sum or difference of two perfect cubes.

EXAMPLE 14-3.6 Identify the special products that are the sum or difference of two perfect cubes in the following examples.

(a) $a^3 - 27$ **(b)** $8x^3 + 9$ **(c)** $8y^3 - 125$
(d) $64 - 8a^3$ **(e)** $a^3 + 9b^3$ **(f)** $c^3 + y^3 - 27$
(g) $27x^3 + 8$ **(h)** $3a^3 - 64$

Solution

(a) Difference of two perfect cubes.
(b) Not a special product; 9 is not a perfect cube.
(c) Difference of two perfect cubes.
(d) Difference of two perfect cubes.
(e) Not a special product; $9b^3$ is not a perfect cube.
(f) Not a special product; this is a trinomial.
(g) Sum of two perfect cubes.
(h) Not a special product; $3a^3$ is not a perfect cube. ■

Rule 14-3.3 *To factor the sum of two perfect cubes:*

1. Write the binomial factor as the *sum* of the cube roots of the two terms.
2. Write the trinomial factor as the square of the first term from step 1, *minus* the product of the two terms from step 1, plus the square of the second term from step 1.

$$a^3 + b^3 = (a + b)(a^2 - ab + b^2)$$

Rule 14-3.4 *To factor the difference of two perfect cubes:*

1. Write the binomial factor as the *difference* of the cube roots of the two terms.
2. Write the trinomial factor as the square of the first term from step 1, *plus* the product of the two terms from step 1, plus the square of the second term from step 1.

$$a^3 - b^3 = (a - b)(a^2 + ab + b^2)$$

EXAMPLE 14-3.7 Factor the following special products, which are the sum or difference of two perfect cubes.

(a) $x^3 + y^3$ **(b)** $27a^3 - 8$
(c) $64p^3 - q^3$ **(d)** $m^3 + 125n^3$

Solution

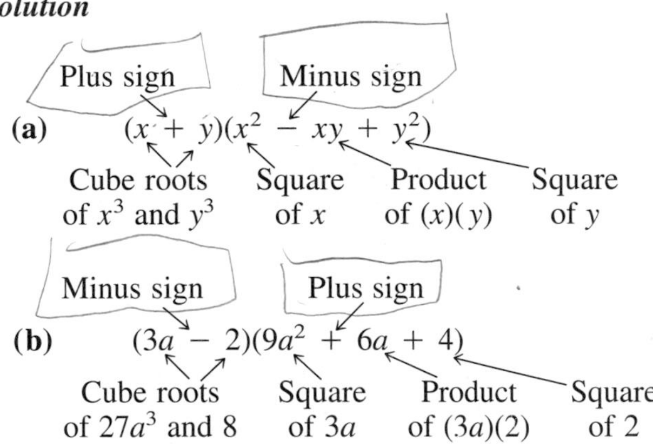

(c) $(4p - q)(16p^2 + 4pq + q^2)$
(d) $(m + 5n)(m^2 - 5mn + 25n^2)$

SELF-STUDY EXERCISES 14-3

L.O.1. Identify the special products that are the difference of two perfect squares.

1. $r^2 - s^2$ **2.** $d^2 - 4d + 10$ **3.** $4y^2 - 16$
4. $36m^2 - 9n^2$ **5.** $9 + 4a^2$ **6.** $64p^2 - q^2$

Factor the following special products.

7. $y^2 - 49$ **8.** $16x^2 - 1$ **9.** $9a^2 - 100$
10. $4m^2 - 81n^2$ **11.** $9x^2 - 64y^2$ **12.** $25x^2 - 64$
13. $100 - 49x^2$ **14.** $4x^2 - 49y^2$ **15.** $121m^2 - 49n^2$
16. $81x^2 - 169$

L.O.2. Verify which of the following trinomials are or are not perfect-square trinomials.

17. $4y^2 + 2y + 16$
18. $9m^2 - 24mn + 16n^2$
19. $16a^2 + 8a - 1$
20. $-9r^2 + 12r + 4$
21. $y^2 - 14y + 49$
22. $p^2 + 10p + 25$

Factor the following special products.

23. $x^2 + 6x + 9$
24. $x^2 + 14x + 49$
25. $x^2 - 12x + 36$
26. $x^2 - 16x + 64$
27. $4a^2 + 4a + 1$
28. $25x^2 - 10x + 1$
29. $9m^2 - 48m + 64$
30. $4x^2 - 36x + 81$
31. $x^2 - 12xy + 36y^2$
32. $4a^2 - 20ab + 25b^2$

L.O.3. Identify the special products that are the sum or difference of two perfect cubes.

33. $8b^3 - 125$
34. $y^2 - 14y + 49$
35. $T^3 + 27$
36. $c^3 + 16$
37. $125x^3 - 8y^3$
38. $64 + a^3$

Factor the following special products, which are the sum or difference of two perfect cubes.

39. $8m^3 - 343$
40. $y^3 - 125$
41. $Q^3 + 27$
42. $c^3 + 1$
43. $125d^3 - 8p^3$
44. $a^3 + 64$
45. $216a^3 - b^3$
46. $x^3 + Q^3$
47. $8p^3 - 125$
48. $27 - 8y^3$

14-4 FACTORING GENERAL TRINOMIALS

Learning Objectives

L.O.1. Factor general trinomials whose squared term has a coefficient of 1.
L.O.2. Remove common factors after grouping an expression.
L.O.3. Factor a general trinomial by grouping.
L.O.4. Factor a general trinomial by trial and error.
L.O.5. Factor any binomial or trinomial that is not prime.

We often need to factor expressions that do not fit the special cases described in Section 14-3. We will use factoring to solve quadratic equations (in Chapter 15) and to simplify algebraic fractions. Trinomials that are not perfect square trinomials are called *general trinomials.*

L.O.1 Factor general trinomials whose squared term has a coefficient of 1.

Strategies for Factoring General Trinomials Whose Squared Terms Have a Coefficient of 1

Recall the FOIL method of multiplying two binomials, $(x + 3)(x + 2)$. x^2 is the product of the First terms of each binomial. $2x + 3x = 5x$ is the sum of the products of the Outer and Inner terms. $+6$ is the product of the Last terms of each binomial.

The trinomial is $x^2 + 5x + 6$. If we wish to factor this trinomial, we begin by indicating that it factors into two binomials.

$$(\quad)(\quad)$$

If the sign of the *third* term is positive, the signs of the two binomial factors will be the same. The sign of the middle term will tell us which sign to use. In the trinomial $x^2 + 5x + 6$, the sign of the 6 is positive, so the signs will be alike.

Since the sign of the $5x$ is positive, the signs of both binomial factors will be positive.

$$(\; + \;)(\; + \;)$$

Next, we write the factors of the *first* term in the *first* position of each binomial.

$$(x + \quad)(x + \quad)$$

The factors of x^2 are x and x.

Then we write the factors of the *last* term in the *last* position of each binomial. Although we have several pairs whose product is $+6(+3, +2; -3, -2; +6, +1; -6, -1)$, *we choose the pair that will give us the correct middle term including its sign when we multiply the two binomials together*. The positive sign on the third term indicates that we need two factors whose *sum* is $+5$. So we choose $+3$ and $+2$.

$(x + 3)(x + 2) = x^2 + 5x + 6$ *(Only this choice gives correct middle term, $+5x$.)*

$$\left.\begin{aligned}(x - 3)(x - 2) &= x^2 - 5x + 6\\(x + 6)(x + 1) &= x^2 + 7x + 6\\(x - 6)(x - 1) &= x^2 - 7x + 6\end{aligned}\right\}$$ *(These choices give wrong sign and/ or wrong numerical value for middle term.)*

Therefore, the factors of the trinomial $x^2 + 5x + 6$ are $(x + 3)(x + 2)$. Since multiplication is commutative, the two factors may be written in the opposite order, $(x + 2)(x + 3)$.

Tips and Traps

In the examples that follow, pay particular attention to the signs of the middle and last terms of the trinomials and their effect on the signs of the binomial factors.

EXAMPLE 14-4.1 Factor $x^2 + 6x + 5$.

Solution

1. $(\quad)(\quad)$ *(Write parentheses for two binomial factors.)*
2. $(x + \quad)(x + \quad)$ *(The positive sign on $+5$ indicates like signs, and the positive sign on the $+6x$ indicates both signs are positive. Write factors of x^2 as first term of each binomial.)*
3. $(x + 5)(x + 1)$ [*Write factors of $+5$ ($+5$ and $+1$) as last term of each binomial. Note, $+5$ of the trinomial has two sets of factors: $+5$, $+1$ and -5, -1. We choose $+5$ and $+1$ because the sign of the middle term is positive.*]
4. $(x + 5)(x + 1) = x^2 + 6x + 5$ *(Multiply the binomials to check factoring.)*

The correct factoring of $x^2 + 6x + 5$ is $(x + 5)(x + 1)$. ■

EXAMPLE 14-4.2 Factor $x^2 - 4x + 3$.

Solution

1. $(\quad)(\quad)$	*(Indicate binomial factors.)*
2. $(x - \quad)(x - \quad)$	*(The positive sign on +3 indicates both signs are the same; the negative sign on −4x indicates we should use negative signs. Write factors of x^2 as first term of each binomial.)*
3. $(x - 1)(x - 3)$	*[Write factors of +3 (−1 and −3) as last terms. Note, +3 of trinomial has two sets of factors: −1, −3 and +1, +3. We chose −1 and −3 because the sign of the middle term is negative.]*
4. $(x - 1)(x - 3) = x^2 - 4x + 3$	*(Check by multiplying.)*

The correct factoring of $x^2 - 4x + 3$ is $(x - 1)(x - 3)$. ■

EXAMPLE 14-4.3 Factor $x^2 - 2x - 8$.

Solution

1. $(\quad)(\quad)$	*(Indicate binomial factors.)*
2. $(x - \quad)(x + \quad)$	*(Write factors of x^2. The negative sign of −8 indicates the signs will be different.)*
3. $(x - 2)(x + 4)$ $(x - 4)(x + 2)$ $(x - 8)(x + 1)$ $(x - 1)(x + 8)$	*(To find factors of −8 that will give a middle term of −2x, find the two factors of 8 that have a difference of 2. Then use the negative sign with the larger factor.)*
4. $(x - 4)(x + 2) = x^2 - 2x - 8$	

The correct factoring of $x^2 - 2x - 8$ is $(x - 4)(x + 2)$. ■

When trinomials have terms whose coefficients have several pairs of factors, such as in the trinomial $12a^2 + 28a + 15$, more cases have to be considered. Two methods of factoring general trinomials are common. The one method is systematic, while the other requires trial and error. Before we use the systematic method, we need to learn to factor by grouping.

L.O.2. Remove common factors after grouping an expression.

Removing Common Factors after Grouping

Some algebraic expressions that have no factors common to every term in the expression may have common factors for some group of terms. Some of these cases will allow the expression to be written in factored form. In the expression $2x^2 - 2xb + ax - ab$, there are four terms and there is no factor common to each term. To factor, write the expression as two terms by *grouping* the first two terms and the last two terms; then look for common factors in each grouping.

$$(2x^2 - 2xb) + (ax - ab)$$

The first grouping $(2x^2 - 2xb)$ has a common factor of $2x$ and can be written in factored form as $2x(x - b)$. The second grouping $(ax - ab)$ has a common factor of a and can be written in factored form as $a(x - b)$. If we look at the entire expression $2x(x - b) + a(x - b)$, we see the common binomial factor $(x - b)$. When we factor this common factor from both terms, we have $(x - b)(2x + a)$. The expression is now written as a product of two factors. We can check the factoring by using the distributive principle or the FOIL method of multiplying.

$$(x - b)(2x + a) = 2x^2 - 2xb + ax - ab$$

This result is the same as the original expression.

EXAMPLE 14-4.4 Write the following expressions in factored form by using grouping.

(a) $mx + 2m - 4x - 8$ **(b)** $y^2 + 2xy + 3y + 6x$
(c) $3x^2 - 9x - 7x + 21$ **(d)** $2x^2 + 8x + 5y - 15$

Solution

(a) $mx + 2m - 4x - 8$

Group the four-termed expression into two terms.

$$(mx + 2m) + (-4x - 8)$$

Factor out any common factor in each of the two terms.

$$m(x + 2) + -4(x + 2)$$

or

$$m(x + 2) - 4(x + 2)$$

Write the expression as a single term in factored form by factoring out the common factor $(x + 2)$.

$$(x + 2)(m - 4)$$

Check the result by using the distributive principle.

$$(x + 2)(m) + (x + 2)(-4)$$

$$mx + 2m + -4x - 8$$

or

$$mx + 2m - 4x - 8$$

Tips and Traps

When factoring by grouping, manipulate the signs of the common factor so that the lead coefficient of the binomial is positive.

$$-3x + 9$$

Factor as $-3(x - 3)$, NOT $3(-x + 3)$.

(b) $y^2 + 2xy + 3y + 6x$

$(y^2 + 2xy) + (3y + 6x)$	*(Group into two terms.)*
$y(y + 2x) + 3(y + 2x)$	*(Factor out common factor in each term.)*
$(y + 2x)(y + 3)$	*(Factor into one term by factoring out common binomial factor.)*

(c) $3x^2 - 9x - 7x + 21$

$(3x^2 - 9x) + (-7x + 21)$ *(Group into two terms.)*

$3x(x - 3) + -7(x - 3)$ *(Factor each term.)*

$3x(x - 3) - 7(x - 3)$

Notice that -7 was factored to give the term $-7(x - 3)$. Factor $(x - 3)$ from both terms.

$(x - 3)(3x - 7)$ *(Factor into one term by factoring out common binomial factor.)*

(d) $2x^2 + 8x + 5y - 15$

$(2x^2 + 8x) + (5y - 15)$ *(Group into two terms.)*

$2x(x + 4) + 5(y - 3)$ *(Factor each term.)*

In this example the two terms do not have a common factor. Thus, the expression cannot be factored into one term. Even if we rearrange the terms, we will not be able to write the expression as a single term. ■

L.O.3. Factor a general trinomial by grouping.

Factoring General Trinomials by Grouping

We can now use this systematic method of factoring by grouping to factor general trinomials. First, we will factor trinomials of the type $ax^2 + bx + c$.

Rule 14-4.1 *To factor a general trinomial by grouping:*

1. Multiply the coefficient of the first term by the coefficient of the last term.
2. Factor the product from step 1 into two factors:
   (a) whose *sum* is the coefficient of the middle term if the sign of the last term is positive, or
   (b) whose *difference* is the coefficient of the middle term if the sign of the last term is negative.
3. Rewrite the trinomial so that the middle term is the sum or difference from step 2.
4. Divide the trinomial from step 3 into two groups with a common factor in each.
5. Factor out the common factor.

We learned earlier that trinomials that have a positive third term will factor into two binomials whose signs are alike.

EXAMPLE 14-4.5

Factor $6x^2 + 19x + 10$.

Solution: To factor this trinomial by grouping, we first multiply the coefficients of the first and third terms.

$$6(10) = 60$$

Next, we factor 60 so that the *sum* of the factors is 19, the coefficient of the middle term. The 60 should be factored systematically to find the desired pair of factors.

$$
\begin{aligned}
60 = {} & (1)60 \\
& (2)30 \\
& (3)20 \\
& (4)15 \\
& (5)12 \\
& (6)10
\end{aligned}
$$

Of the pairs of factors listed, only one pair, 4(15), has a sum of 19. Now we rewrite the trinomial $6x^2 + 19x + 10$ so that the middle term, $19x$, is written as $4x + 15x$. Then we factor by grouping.

$$
\begin{gathered}
6x^2 + 19x + 10 \\
6x^2 + 4x + 15x + 10 \\
(6x^2 + 4x) + (15x + 10) \\
2x(3x + 2) + 5(3x + 2) \\
(3x + 2)(2x + 5)
\end{gathered}
$$

The factoring can be checked by using the distributive principle or the FOIL method of multiplying. ■

EXAMPLE 14-4.6 Factor $20x^2 - 23x + 6$.

Solution: First, find the product of 20 and 6.

$$20(6) = 120$$

List the factor pairs of 120 and select the pair that has a sum of 23.

$$
\begin{aligned}
120 = {} & (1)\ 120 \\
& (2)\ \ 60 \\
& (3)\ \ 40 \\
& (4)\ \ 30 \\
& (5)\ \ 24 \\
& (6)\ \ 20 \\
& (8)\ \ 15^* \\
& (10)\ 12
\end{aligned}
$$

From the list of factors, we choose 8(15), whose sum, 8 + 15, is 23, the coefficient of the middle term. Next, we rewrite the trinomial.

$$
\begin{aligned}
& 20x^2 - 23x + 6 \\
& 20x^2 - 8x - 15x + 6 \qquad (-23x = -8x - 15x) \\
& (20x^2 - 8x) + (-15x + 6) \\
& 4x(5x - 2) - 3(5x - 2) \\
& (5x - 2)(4x - 3)
\end{aligned}
$$

The factoring should be checked by distributing.

If the third term of the trinomial has a negative sign, we use the same procedure but look for two factors whose *difference* is the coefficient of the middle term. ■

Example 14-4.7 Factor $10x^2 + 19x - 15$.

Solution: Find the product of 10 and 15.

$$10(15) = 150$$

Systematically factor 150 to find two factors whose difference is 19.

$$\begin{aligned} 150 = {} & (1)\ 150 \\ & (2)\ 75 \\ & (3)\ 50 \\ & (5)\ 30 \\ & (6)\ 25^* \qquad (25 - 6 = 19) \\ & (10)\ 15 \end{aligned}$$

The factors 25 and 6 have a difference of 19. When we rewrite the trinomial, we rewrite $19x$ as $25x - 6x$.

$$10x^2 + 19x - 15$$
$$10x^2 + 25x - 6x - 15$$
$$(10x^2 + 25x) + (-6x - 15)$$
$$5x(2x + 5) - 3(2x + 5)$$
$$(2x + 5)(5x - 3)$$

■

Now, let's factor by grouping a trinomial whose middle term and end term are negative. Since the last term is negative, we will look for two factors whose *difference* is the coefficient of the middle term.

Example 14-4.8 Factor $x^2 - 2x - 8$.

Solution: First, find the product of 1 and 8.

$$1(8) = 8$$

Factor 8 so that the difference of the absolute values of the factors is 2.

$$(1)8$$
$$(2)4 \qquad (|4| - |2| = 2)$$

From the list of factors, we choose (2)4. We rewrite the middle term, $-2x$, as $-4x + 2x$. Next we rewrite the trinomial.

$$x^2 - 4x + 2x - 8 \qquad (-2x = -4x + 2x)$$
$$(x^2 - 4x) + (2x - 8)$$
$$x(x - 4) + 2(x - 4)$$
$$(x - 4)(x + 2)$$

The factoring should be checked by distributing. ■

Tips and Traps

The following is a variation of the factoring by grouping method.

Factor $6x^2 + 7x - 20$.

$6x^2 + 7x - 20$	*[Multiply the coefficients of the first and third terms. 6(−20) = −120.]*
$\dfrac{(6x \quad)(6x \quad)}{6}$	*(Use the coefficient of the first term as the coefficient of the first term in each binomial. This gives us an extra factor of 6, and we will compensate by dividing the expression by 6.)*
	(Look for factors of −120 whose algebraic sum is +7.)

1	*120*	
2	*60*	
3	*40*	
4	*30*	
5	*24*	
6	*20*	
8	*15**	$-8 + 15 = 7$
10	*12*	

$\dfrac{(6x - 8)(6x + 15)}{6}$	*(Use the factors of −120 whose algebraic sum is 7 as the coefficients of the second term of each binomial.)*
$\dfrac{2(3x - 4)(3)(2x + 5)}{6}$	*(Factor common factors from each binomial.)*
$\dfrac{\not{6}(3x - 4)(2x + 5)}{\not{6}}$	*(The factors of the trinomial will be the two binomial factors. The factor, 6, reduces.)*
$(3x - 4)(2x + 5)$	($6x^2 + 7x - 20$ *in factored form*)
FOIL to check factoring:	$6x^2 + 15x - 8x - 20 = 6x^2 + 7x - 20$

Why does this work? The coefficient of the first term in the original trinomial, 6, was used in each binomial. Thus, the resulting product would have an extra factor of 6. This extra factor of 6 must be removed to obtain the correct factors. *This method is only effective if common factors are factored from the trinomial first*. Common factors of the original trinomial are part of the final answer.

Factor $30x^2 + 8 - 32x$.

$30x^2 - 32x + 8$	*(Arrange in descending powers of x.)*
$2(15x^2 - 16x + 4)$	*(Factor any common factors.)*
$\dfrac{2(15x \quad)(15x \quad)}{15}$	*(Multiply the coefficients of the first and third terms of the trinomial that results after the common factor has been removed, 15(4) = 60. Be sure to keep the common factor, 2, as a factor in the answer. An extra factor of 15 is produced, and we will compensate by dividing by 15.)*
	(Find factors of 60 whose algebraic sum is −16.)

1	*60*	
2	*30*	
3	*20*	
4	*15*	
5	*12*	
6	*10**	$-6 + (-10) = -16$

$$\frac{2(15x - 6)(15x - 10)}{15}$$

(Use the factors of 60 whose algebraic sum is −16 as the second term in each binomial. Remember, at this point an extra factor of 15 is still present.)

$$\frac{2(\not{3})(5x - 2)(\not{5})(3x - 2)}{\not{15}}$$

(Factor common factors from each binomial.)

$$2(5x - 2)(3x - 2)$$

(Reduce extra factor of 15: 3(5). Common factor of 2 remains in answer.)

L.O.4. Factor a general trinomial by trial and error.

Factoring Trinomials by Trial and Error

Rule 14-4.2 ***To factor any general trinomial by trial and error:***

1. Indicate binomial factors with double parentheses ( )( ).
2. List all possible factors of the first and last terms of the trinomial.
3. List all possible binomial factors using factors from step 2.
4. Multiply binomial factors from step 3 to find the pair that gives the correct middle term of the trinomial.
5. Check all work to be sure that the signs are correct.

To factor the trinomial $2a^2 + 5a + 3$ by *trial and error*, we place the factors of the first term, $2a^2$, in the first position of each binomial factor.

$$(2a \quad)(a \quad)$$

We place the factors of the last term, 3, in the second position of each binomial factor.

$$(2a \quad 3)(a \quad 1)$$

or

$$(2a \quad 1)(a \quad 3)$$

We have to consider *both* possible arrangements of the factors. The signs of 3 and 1 are both positive, since all signs in the trinomial are positive.

$$(2a + 3)(a + 1)$$

or

$$(2a + 1)(a + 3)$$

To determine the correct factoring, we multiply each product to see which yields $2a^2 + 5a + 3$.

$$\begin{aligned}(2a + 3)(a + 1) &= 2a^2 + 2a + 3a + 3\\ &= 2a^2 + 5a + 3\\ (2a + 1)(a + 3) &= 2a^2 + 6a + a + 3\\ &= 2a^2 + 7a + 3\end{aligned}$$

The first factoring, $(2a + 3)(a + 1)$, is correct.

Tips and Traps

To make our work easier and less time consuming, we should learn to obtain the outer and inner products and add them *mentally*.

EXAMPLE 14-4.9 Factor $6x^2 + 7x - 5$ by listing all possible factors and then adding the outer and inner products to find the correct factoring.

Solution

	(middle term)	
$(6x + 5)(x - 1)$	$-x$	*(Since the sign of the last term, -5, is negative, we must use unlike signs for the factors.)*
$(6x - 5)(x + 1)$	$+x$	
$(6x + 1)(x - 5)$	$-29x$	
$(6x - 1)(x + 5)$	$+29x$	
$(3x + 5)(2x - 1)$	$+7x$	
$(3x - 5)(2x + 1)$	$-7x$	
$(3x + 1)(2x - 5)$	$-13x$	
$(3x - 1)(2x + 5)$	$+13x$	

Figure 14-1

The fifth pair of factors, $(3x + 5)(2x - 1)$, gives the correct middle term, $+7x$. Thus, $(3x + 5)(2x - 1)$ is the correct factoring of $6x^2 + 7x - 5$. ■

L.O.5. Factor any binomial or trinomial that is not prime.

Factoring Any Binomial or Trinomial That Is Not Prime

Rule 14-4.3 *To factor any binomial or trinomial:*

1. First, factor out the greatest common factor (if any) using Rule 14-1.1.
2. Check the binomial or trinomial to see if it is a special product.
   (a) If it is the difference of two perfect squares, use Rule 14-3.1.
   (b) If it is a perfect-square trinomial, use Rule 14-3.2.
   (c) If it is the sum or difference of two perfect cubes, use Rule 14-3.3 or Rule 14-3.4.
3. If there is no special product, use Rule 14-4.1 to factor by grouping or use Rule 14-4.2 for ''trial-and-error'' factoring of a general trinomial.
4. Check to see that each factor cannot be factored further.

We used the word *prime* to identify positive numbers that have no factors except themselves and 1. When the word *prime* refers to algebraic expressions, it describes algebraic expressions that have only themselves and 1 as factors.

EXAMPLE 14-4.10 Completely factor $4x^3 - 2x^2 - 6x$.

Solution: We first look for a common factor. $2x$ is the common factor.

$$2x(2x^2 - x - 3)$$

Now we factor the trinomial but *keep* the $2x$ factor also.

$$2x(2x - 3)(x + 1)$$ ■

EXAMPLE 14.4.11 Factor completely $12x^2 - 27$.

Solution: Look for a common factor. 3 is the common factor.

$$3(4x^2 - 9)$$

Now factor the difference of two perfect squares.

$$3(2x - 3)(2x + 3)$$ ■

EXAMPLE 14-4.12 Factor completely $-18x^2 + 24x^2 - 8x$.

Solution: Look for a common factor: $-2x$. We often factor a negative when there are more negative terms than positive terms or when the term with the highest degree is negative.

$$-2x(9x^2 - 12x + 4)$$

Now factor the perfect-square trinomial.

$$-2x(3x - 2)^2$$ ■

Tips and Traps

Why do we factor? There is general disagreement on the importance of factoring polynomials.

For factoring:

- Many problem-solving techniques that use factoring can be done more quickly than with the formula-based technique.

- Properties and patterns are generally easier to recognize from factored form.
- Computers (and even humans) can often evaluate factored expressions more quickly.

Evaluate the following when $x = 4$ and $y = 3$.

Factoring technique:	Formula-based technique:
$4x(2x + y)(x - 2y)$	$8x^3 - 12x^2y - 8xy^2$
$4 \cdot 4(2 \cdot 4 + 3)(4 - 2 \cdot 3)$	$8 \cdot 4^3 - 12 \cdot 4^2 \cdot 3 - 8 \cdot 4 \cdot 3^2$
$16(11)(-2)$	$8 \cdot 64 - 12 \cdot 16 \cdot 3 - 8 \cdot 4 \cdot 9$
-352	$512 - 576 - 288$
	-352

Against factoring:

- Real-world applications rarely have numbers and expressions that will factor.
- Factoring techniques apply to special cases. Formula-based techniques apply in general.
- With the accessibility of programmable calculators and inexpensive computer software, formula-based techniques are more practical.

SELF-STUDY EXERCISE 14-4

L.O.1. Factor.

1. $x^2 + 5x + 6$
2. $x^2 - 11x + 28$
3. $x^2 + 8x + 12$
4. $x^2 - 4x + 3$
5. $x^2 + 8x + 7$
6. $x^2 + 7x + 10$
7. $x^2 - x - 12$
8. $y^2 - 3y - 10$
9. $y^2 - y - 6$
10. $a^2 + a - 12$
11. $b^2 + 2b - 3$
12. $14 - 5b - b^2$
13. $x^2 - 7x + 12$
14. $x^2 - x - 30$
15. $x^2 + 11x + 18$
16. $x^2 - 9x + 18$
17. $x^2 - 7x - 18$
18. $x^2 + 17x - 18$
19. $x^2 + 9x + 20$
20. $x^2 - 12x + 20$
21. $x^2 - 10x + 16$
22. $x^2 - 17x + 16$
23. $x^2 - 13x - 14$
24. $x^2 - 5x - 14$

L.O.2. Factor the following polynomials by removing common factors after grouping.

25. $x^2 + xy + 4x + 4y$
26. $6x^2 + 4x - 3xy - 2y$
27. $3mx + 5m - 6nx - 10n$
28. $30xy - 35y - 36x + 42$
29. $x^2 - 2x + 8x - 16$
30. $6x^2 - 2x - 21x + 7$
31. $x^2 - 4x + x - 4$
32. $8x^2 - 4x + 6x - 3$
33. $x^2 - 5x + 4x - 20$
34. $3x^2 - 6x + 5x - 10$

L.O.3. Factor the following trinomials by grouping.

35. $3x^2 + 7x + 2$
36. $3x^2 + 14x + 8$
37. $6x^2 + 13x + 6$
38. $x^2 - 18x + 32$
39. $6x^2 - 17x + 12$
40. $2x^2 - 9x + 10$
41. $6x^2 - 13x + 5$
42. $p^2 + 5p - 36$
43. $6x^2 - 11x + 5$
44. $8x^2 + 26x + 15$
45. $15x^2 - 22x - 5$
46. $y^2 - 15y + 36$
47. $2x^2 - 5x - 7$
48. $12x^2 + 8x - 15$
49. $10x^2 + x - 3$
50. $Q^2 - 7Q - 44$

L.O.4. Factor the following trinomials by trial and error.

51. $6x^2 + 7xy - 10y^2$
52. $6a^2 - 17ab - 14b^2$
53. $18x^2 - 3x - 10$
54. $20x^2 - xy - 12y^2$
55. $x^2 + x - 56$
56. $a^2 + 21a + 38$

L.O.5. Factor the following polynomials completely. Be sure to look for common factors first and then look for special cases.

57. $4x - 4$
58. $x^2 + x - 6$
59. $2x^2 + x - 3$
60. $x^2 - 9$
61. $4x^2 - 16$
62. $m^2 + 2m - 15$

63. $2a^2 + 6a + 4$
64. $b^2 + 6b + 9$
65. $16m^2 - 8m + 1$
66. $x^2 + 8x + 7$
67. $2m^2 + 5m + 2$
68. $2m^2 - 5m - 3$
69. $2a^2 - 3a - 5$
70. $3x^2 + 10x - 8$
71. $6x^2 + x - 15$
72. $8x^2 + 10x - 3$

ASSIGNMENT EXERCISES

14-1

Factor completely. Check.

1. $5x + 5y$
2. $2x + 5x^2$
3. $12m^2 - 8n^2$
4. $25x^2y - 10xy^3 + 5xy$
5. $2a^3 - 14a^2 - 2a$
6. $30a^3 - 18a^2 - 12a$
7. $15x^3 - 5x^2 - 20x$
8. $9a^2b^3 - 6a^3b^2$
9. $18a^3 + 12a^2$
10. $a^2b + ab^2$

14-2

Use the FOIL method to find the products. Combine the outer and inner products mentally.

11. $(x + 7)(x + 4)$
12. $(y - 7)(y - 5)$
13. $(m + 3)(m - 7)$
14. $(3b - 2)(x + 6)$
15. $(4r - 5)(3r + 2)$
16. $(5 - x)(7 - 3x)$
17. $(4 - 2m)(1 - 3m)$
18. $(2 + 3x)(3 + 2x)$
19. $(x + 3)(2x - 5)$
20. $(5x - 7y)(4x + 3y)$
21. $(2a + 3b)(7a - b)$
22. $(5a + 2b)(6a - 5b)$
23. $(9x - 2y)(3x + 4y)$
24. $(5x - 8y)(4x - 3y)$
25. $(7m - 2n)(3m + 5n)$

Find the following special products mentally.

26. $(y - 4)(y + 4)$
27. $(6x - 5)(6x + 5)$
28. $(3m + 4)(3m - 4)$
29. $(7y + 11)(7y - 11)$
30. $(5x - 2y)(5x + 2y)$
31. $(8a - 5b)(8a + 5b)$
32. $(12r - 7s)(12r + 7s)$
33. $(8x + y)(8x - y)$
34. $(5m + 11n)(5m - 11n)$
35. $(3y - 4z)(3y + 4z)$
36. $(x + 8)^2$
37. $(x + 9)^2$
38. $(x - 7)^2$
39. $(x - 3)^2$
40. $(2x - 3)^2$
41. $(4x - 15)^2$
42. $(5 + 3m)^2$
43. $(8 + 7m)^2$
44. $(5x - 13)^2$
45. $(4x - 11)^2$

Use the extension of the FOIL method to find the products.

46. $(K + L)(K^2 - KL + L^2)$
47. $(g - h)(g^2 + gh + h^2)$
48. $(4 + a)(16 - 4a + a^2)$
49. $(2H - 3T)(4H^2 + 6HT + 9T^2)$
50. $(3a - 5)(9a^2 + 15a + 25)$
51. $(6 + i)(36 - 6i + i^2)$
52. $(9y - p)(81y^2 + 9yp + p^2)$
53. $(z + 2t)(z^2 - 2zt + 4t^2)$
54. $(g - 2)(g^2 + 2g + 4)$
55. $(7T + 2)(49T^2 - 14T + 4)$

14-3

Identify the special products that are the difference of two perfect squares.

56. $25m^2 - 4n^2$
57. $64 + 4a^2$
58. $4f^2 - 9g^2$
59. $H^2 - G^2$
60. $s^2 - 4s + 10$
61. $64b^2 - 49$

Verify which of the following trinomials are or are not perfect square trinomials.

62. $16c^2 + 8c - 1$
63. $-9x^2 + 12x + 4$
64. $4d^2 + 2d + 16$
65. $9t^2 - 24tp + 16p^2$
66. $a^2 - 14a + 49$
67. $j^2 + 10j + 25$

Identify the special products that are the sum or difference of two perfect cubes.

68. $R^3 + 81$
69. $125a^3 - 8b^3$
70. $64 + m^3$
71. $8z^3 - 125$
72. $d^3 - 12d + 36$
73. $64W^3 + 27$

Factor the expressions that are special products. Explain why the other expressions are not special products.

74. $x^2 - 81$
75. $25y^2 - 4$
76. $100a^2 - 8ab^2$
77. $a^2b^2 + 49$
78. $121 - 9m^2$
79. $a^2 + 2a + 1$

80. $4x^2 + 12x + 9$
81. $16c^2 - 24bc + 9b^2$
82. $9y^2 - 12y - 4$
83. $n^2 + 169 - 26n$ (*Hint:* Rearrange.)
84. $16d^2 - 20d + 25$
85. $36a^2 + 84ab + 49b^2$
86. $4x^2 - 25y^2$
87. $49 - 14x + x^2$
88. $9x^2y^2 - 49z^2$
89. $64 + 25x^2$
90. $4x^2 + 12xy + 9y^2$
91. $16x^2 + 24x + 9y^2$
92. $36 - x^2$
93. $49 - 81y^2$
94. $64x^2 - 25y^2$
95. $9x^2 - 100y^2$
96. $a^2 - 10a + 25$
97. $9x^2 - 6xy + y^2$
98. $25 - 16a^2b^2$
99. $9x^2y^2 - z^2$
100. $x^2 - y^2$
101. $x^2 + 4x + 4$
102. $\frac{1}{4}x^2 - \frac{1}{9}y^2$
103. $\frac{4}{25}x^2 - \frac{1}{16}y^2$
104. $27v^3 - 8$
105. $T^3 - 8$
106. $8r^3 + 27$
107. $d^3 + 729$
108. $125c^3 - 216d^3$
109. $27K^3 + 64$

14-4

Factor the following trinomials.

110. $x^2 + 10x + 21$
111. $x^2 + 11x + 24$
112. $x^2 + 29x + 28$
113. $x^2 + 13x + 30$
114. $x^2 - 13x + 40$
115. $x^2 - 9x + 8$
116. $x^2 - 17x - 18$
117. $x^2 - 11x - 26$
118. $x^2 + x - 30$
119. $x^2 + 5x - 24$
120. $6x^2 + 25x + 14$
121. $6x^2 + 25x + 4$
122. $4x^2 - 23x + 15$
123. $5x^2 - 34x + 24$
124. $3x^2 - x - 14$
125. $6x^2 - x - 35$
126. $3x^2 + 11x - 4$
127. $7x^2 - 13x - 24$

Factor the following polynomials. Look for common factors and special cases.

128. $5mn - 25m$
129. $9a^2 - 100$
130. $a^2 - b^2$
131. $2x^2 - 3x - 2$
132. $b^3 - 8b^2 - b$
133. $a^2 - 81$
134. $x^2 - 14x + 13$
135. $y^2 - 14y + 49$
136. $m^2 - 3m + 2$
137. $b^2 + 8b + 15$
138. $x^2 - 13x + 30$
139. $169 - m^2$
140. $5x^2 + 13x + 6$
141. $x^2 - 4x - 32$
142. $x^2 + 9x + 14$
143. $x^2 + 19x - 20$
144. $x^2 + 8x - 20$
145. $2x^2 - 4x - 16$
146. $5x^2 - 20$
147. $2x^3 - 10x^2 - 12x$

CHALLENGE PROBLEMS

148. Explain the value of factoring a general trinomial by the grouping method as opposed to the trial-and-error method.

149. What is the value of being able to recognize the special products when we factor algebraic expressions?

150. A standard-sized rectangular swimming pool is 25 ft long and 15 ft wide. This gives a water-surface area of 375 ft^2. A customer wants to examine some options for varying the size of the pool. If x is the amount of adjustment to the length and y is the amount of adjustment to the width, find the following.

(a) Write a formula in both factored and expanded form for finding the water-surface area of a pool that is adjusted x feet in length and y feet in width.

(b) Write a formula using one variable in both factored and expanded form for the water-surface area of a pool when the length and width are increased the same amount.

(c) Will the formulas in parts (a) and (b) work for both increasing and decreasing the size of the pool?

(d) Illustrate your answer in part (c) with numerical examples.

CONCEPTS ANALYSIS

1. Explain how the FOIL process for multiplying two binomials is an application of the distributive property.
2. (a) How is the distributive property used to multiply a binomial and a trinomial?
   (b) How is the distributive property used to multiply two trinomials?
3. Give an example of the product of a binomial and a trinomial to illustrate the procedure given in Question 2a.
4. Give an example of the product of two trinomials to illustrate the procedure given in Question 2b.
5. List the properties of a binomial that is the difference of two perfect squares.
6. List the properties of a perfect-square trinomial.
7. Explain in sentence form how the sign of the third term of a general trinomial affects the signs between the terms of its binomial factors.
8. What do we mean if we say a polynomial is prime?
9. Write a brief comment explaining each step of the following example.

   Factor $10x^2 - 19x - 12$ using grouping.

   (1) $\underline{120}$
   1, 120
   2, 60
   3, 40
   4, 30
   5, 24* *(Factors whose*
   6, 20 *difference = 19)*
   8, 15
   10, 12
   $\underline{10x^2 - 19x - 12}$

   (2) $10x^2 + 5x - 24x - 12$
   (3) $5x(2x + 1) - 12(2x + 1)$
   (4) $(2x + 1)(5x - 12)$
10. Find a mistake in each of the following. Briefly explain the mistake, then work the problem correctly.
    (a) $(3x - 2y)^2 = 9x^2 + 4y^2$
    (b) $5xy^2 - 45x = 5x(y^2 - 9)$
    (c) $2x^2 - 28x + 96 = (2x - 12)(x - 8)$
    (d) $x^2 - 5x - 6 = (x - 2)(x - 3)$

CHAPTER SUMMARY

Objectives	What to Remember	Examples
Section 14-1		
1. Factor an expression containing a common factor.	To factor an expression containing a common factor: **1.** Find the largest factor common to each term of the expression. **2.** Rewrite the expression as the indicated product of the largest common factor and the remaining quantity.	Factor $3ab + 9b^2$ $3b(a + 3b)$

Section 14-2

1. Use the FOIL method to multiply two binomials.	To multiply two binomials by the FOIL method: Multiply First terms, Outer terms, Inner terms, and Last terms. Combine like terms.	Multiply $(3a - 2)(a + 3)$. F O I L $3a^2 + 9a - 2a - 6$ $3a^2 + 7a - 6$
2. Multiply polynomials to obtain three special products.	To multiply the sum and difference of the same two terms: **1.** Square first term. **2.** Insert minus sign. **3.** Square second term.	Multiply $(3b - 2)(3b + 2)$. $9b^2 - 4$
	To square a binomial: **1.** Square first term. **2.** Double product of the two terms. **3.** Square second term.	Multiply $(2a - 3)^2$. $4a^2 - 12a + 9$
	To multiply a binomial and a trinomial of the types $(a + b)(a^2 - ab + b^2)$ and $(a - b)(a^2 + ab + b^2)$: **1.** Cube first term of binomial. **2.** If the binomial is a sum, insert a plus sign; if the binomial is a difference, insert a minus sign. **3.** Cube second term of binomial.	Multiply $(8b - 2)(64b^2 + 16b + 4)$. $512b^3 + 8$ Multiply $(2x + 5)(4x^2 - 10x + 25)$. $8x^3 - 125$

Section 14-3

1. Recognize and factor the difference of two perfect squares.	The difference of two perfect squares, $4z^2 - 9$, factors into two binomials, $(2z - 3)(2z + 3)$. To factor the difference of two perfect squares: **1.** Take square root of first term. **2.** Take square root of second term. **3.** Write one factor as the sum and one factor as the difference of the square roots from steps 1 and 2.	Identify which expression is the special product, the difference of two squares. (a) $36 - 27A^2$: No, $27A^2$ is not a perfect square. (b) $9c^2 - 4$: Yes, both terms are perfect squares. Factor $9c^2 - 4$. $(3c + 2)(3c - 2)$
2. Recognize and factor a perfect-square trinomial.	In a perfect square trinomial, the first and last terms are positive perfect squares and the absolute value of the middle term is twice the product of the square roots of the first and last terms, such as $a^2 + 2ab + b^2$.	Identify which expression is a perfect square trinomial. (a) $4x^2 + 18x + 9$: Not a perfect square trinomial, the middle term is not twice the product of the square roots of the first and last terms. (b) $9a^2 + 6a + 4$: First and last terms are perfect squares. Middle term is twice the product of the square roots of the first and last terms, so we factor.
	To factor a perfect square trinomial: **1.** Write square root of first term. **2.** Write sign of middle term. **3.** Write square root of last term. **4.** Indicate square of the quantity.	Factor $9a^2 + 6a + 4$. $(3a + 2)^2$

3. Recognize and factor the sum or difference of two perfect cubes.	The sum of two perfect cubes is two perfect cubes with a plus sign between them. The difference of two perfect cubes is two perfect cubes with a minus sign between them. Each factors into a binomial and a trinomial. To factor the sum (difference) of two perfect cubes: **1.** Write binomial factor as the sum (difference) of the cube roots of the two terms. **2.** Write the trinomial factor as the square of first term from step 1, insert minus sign if factoring a sum (plus sign if factoring a difference). Write the product of the two terms from step 1, plus the square of the second term from step 1.	Identify the sum or difference of two perfect cubes. (a) $b^3 - 27$: Yes, both terms are perfect cubes. (b) $X^3 - 6$: No, 6 is not a perfect cube. (c) $8 + y^3$: Yes, both terms are perfect cubes. Factor $8 + y^3$. $(2 + y)(4 - 2y + y^2)$ Factor $b^3 - 27$. $(b - 3)(b^2 + 3b + 9)$
Section 14-4		
1. Factor general trinomials whose squared term has a coefficient of 1.	To factor a general trinomial whose squared term has coefficient of 1: Factor using FOIL method in reverse. Use factors of third term that will give desired middle term and sign.	Factor $a^2 - 5a + 6$. $(\quad)(\quad)$ $(a \quad 3)(a \quad 2)$ $(a - 3)(a - 2)$ These factors of 6, -3 and -2, give $-5a$ as middle term.
2. Remove common factors after grouping an expression.	Arrange terms of expression into groups whose terms each have a common factor. Rewrite the expression as an indicated product of the common factor and the remaining quantity. Factor common binomial factor.	Factor $2x^2 - 4x + xy - 2y$ completely: $(2x^2 - 4x) + (xy - 2y)$ $2x(x - 2) + y(x - 2)$ $(x - 2)(2x + y)$
3. Factor a general trinomial by grouping.	To factor a general trinomial by grouping: **1.** Multiply coefficients of first term and last term. **2.** Factor product from step 1 into two factors (a) whose sum is the coefficient of middle term if last term is positive, or (b) whose difference is the coefficient of middle term if the last term is negative. **3.** Rewrite trinomial so middle term is sum or difference from step 2. **4.** Write trinomial from step 3 into two groups with a common factor in each. **5.** Factor out the common factor. **6.** Factor common binomial factor.	Factor $6x^2 - 17x + 12$. $6 \cdot 12 = 72$ $1 \cdot 72$ $2 \cdot 36$ $3 \cdot 21$ $4 \cdot 18$ $6 \cdot 12$ $8 \cdot 9$* $6x^2 - 8x - 9x + 12$ $(6x^2 - 8x) + (-9x + 12)$ $2x(3x - 4) - 3(3x - 4)$ $(3x - 4)(2x - 3)$

4. Factor a general trinomial by trial and error.	To factor a general trinomial by trial and error: **1.** Indicate binomial factors by ( )( ). **2.** List all possible factors of first and last terms. **3.** List all possible binomial factors. **4.** Multiply binomial factors to find product which is original trinomial. **5.** Check to be sure signs are correct.	Factor $3x^2 - 13x - 10$. ( )( ) Factors of $3x^2$: $3x(x)$ Factors of 10: 1(10) 2(5) $(3x - 1)(x + 10)$ | $(3x - 10)(x + 1)$ $(3x + 1)(x - 10)$ | $(3x + 10)(x - 1)$ $(3x - 5)(x + 2)$ | $(3x - 2)(x + 5)$ $(3x + 5)(x - 2)$ | $(3x + 2)(x - 5)$
5. Factor any binomial or trinomial that is not prime.	To factor any polynomial: **1.** Factor out the greatest common factor (if any). **2.** Check for any special products. **3.** If no special product, factor by grouping or trial and error.	Factor $12x^3 + 6x^2 - 18x$ completely. $6x(2x^2 + x - 3)$ $6x(2x + 3)(x - 1)$ (After removing common factor, grouping or trial and error can be used to factor trinomial.)

WORDS TO KNOW

monomial (p. 557)
binomial (p. 557)
trinomial (p. 557)
polynomial (p. 557)
factor (p. 557)
common factor (p. 557)
FOIL (p. 559)
special products (p. 560)
perfect squares (p. 560)
conjugate pairs (p. 560)
binomial square (p. 560)
perfect square trinomial (p. 561)
perfect cubes (p. 562)
square root (p. 565)
cube root (p. 566)
general trinomial (p. 568)
grouping (p. 570)
trial and error (p. 576)
prime (p. 578)

CHAPTER TRIAL TEST

Find the following products.

1. $3(x + 2y)$
2. $7x(x - 5)$
3. $7x(2x^2 + 3x - 5)$
4. $2x^2(2x + 3)$
5. $(m - 7)(m + 7)$
6. $(3x - 2)(3x + 2)$
7. $(a + 3)^2$
8. $(2x - 7)^2$
9. $(x - 3)(2x - 5)$
10. $(7x - 3)(2x + 1)$
11. $(x - 2)(x^2 + 2x + 4)$
12. $(3y + 4)(9y^2 - 12y + 16)$
13. $(5a - 3)(25a^2 + 15a + 9)$
14. $(a + 6)(a^2 - 6a + 36)$

Factor completely.

15. $7x^2 + 8x$
16. $6ax + 15bx$
17. $7a^2b - 14ab$
18. $9x^2 - 42x + 49$
19. $x^2 + 9x + 8$
20. $x^3 + 8x$
21. $x^2 - 5x - 36$
22. $x^2 + 5x - 24$
23. $6x^2 - 5x - 6$
24. $x^3 - 27$
25. $3x^2 + 23x + 30$
26. $12y^2 - 27$
27. $a^2 + 16ab + 64b^2$
28. $y^2 - y - 6$
29. $b^2 - 3b - 10$
30. $x^2 - 7x + 10$
31. $3x^2 + x - 4$
32. $a^2 - 4a + 3$
33. $3m^2 - 5m + 2$
34. $4c^2 - 28c + 49$
35. $27a^3 - 8$
36. $3x^2 + 11xy + 3y^2$

CHAPTER

15 Solving Quadratic and Higher-Degree Equations

Most of the equations we have been solving so far are called linear equations. In such equations, the variable appears only to the first power, such as x or y. However, when we solve problems on the job, we sometimes have to solve equations that are not linear equations. The equations we will study in this chapter include quadratic and higher-degree equations. In these equations, the variable appears to a power greater than the first power, such as x^2 or y^2.

15-1 QUADRATIC EQUATIONS

Learning Objectives

L.O.1. Write quadratic equations in standard form.
L.O.2. Identify the coefficients of the quadratic, linear, and constant terms of a quadratic equation.

L.O.1. Write quadratic equations in standard form.

Standard Form of a Quadratic Equation

The quadratic equations in this chapter will have only one variable. Before we look at the kinds of quadratic equations we will study in this chapter, let's first look at a definition of this type of equation.

■ **DEFINITION 15-1.1 Quadratic Equation.** A **quadratic equation** is an equation in which at least one letter term is raised to the second power and no letter terms have a power higher than 2. The *standard form* for a quadratic equation is $ax^2 + bx + c = 0$, where a, b, and c are real numbers and $a > 0$.

EXAMPLE 15-1.1 Arrange the following quadratic equations in standard form.

(a) $3x^2 - 3 + 5x = 0$ **(b)** $7x - 2x^2 - 8 = 0$
(c) $12 = 7x^2 - 4x$ **(d)** $-3x^2 - 3x + 2 = 0$
(e) $7x^2 = 3$ **(f)** $x^2 = 4x$

Solution

(a) $3x^2 - 3 + 5x = 0$

$3x^2 + 5x - 3 = 0$ *(Arrange the terms in descending powers of x.)*

(b) $7x - 2x^2 - 8 = 0$

$-2x^2 + 7x - 8 = 0$ *(Arrange the terms in descending powers of x.)*

$-1(-2x^2 + 7x - 8) = -1(0)$ *(Multiply both sides by −1 to make coefficient of x^2 term positive.)*

$2x^2 - 7x + 8 = 0$ *(Standard form)*

(c) $12 = 7x^2 - 4x$

$0 = 7x^2 - 4x - 12$ *(Rearrange the terms so that one side of the equation equals zero.)*

$7x^2 - 4x - 12 = 0$ *(Interchange sides of equation if desired.)*

(d) $-3x^2 - 3x + 2 = 0$

$-1(-3x^2 - 3x + 2) = -1(0)$ *(Multiply both sides by −1.)*

$3x^2 + 3x - 2 = 0$

(e) $7x^2 = 3$

$7x^2 - 3 = 0$ *(Rearrange so that one side equals zero.)*

(f) $x^2 = 4x$

$x^2 - 4x = 0$ *(Rearrange so that one side equals zero.)* ■

L.O.2. Identify the coefficients of the quadratic, linear, and constant terms of a quadratic equation.

Types of Quadratic Equations

There are several methods for solving quadratic equations. Some methods can be used for any type of quadratic equation, but are very time consuming. Other methods are quicker, but only apply to certain types of quadratic equations. In choosing an appropriate method, it is helpful to be able to recognize similar and different characteristics of quadratic equations.

In the standard form of quadratic equations, $ax^2 + bx + c = 0$, we have three types of terms.

1. ax^2, is a *quadratic* term. That is, the degree of the term is 2. In standard form this term is the leading term and a is the leading coefficient.
2. bx is a *linear* term. That is, the degree of the term is 1, the coefficient b.
3. c is a *number* term or *constant*. That is, the degree of the term is 0. The coefficient is c ($c = cx^0$).

A quadratic equation that has all three types of terms is sometimes referred to as a *complete quadratic equation.*

$ax^2 + bx + c = 0$ *(complete quadratic equation)*

A quadratic equation that has a quadratic term (ax^2) and a linear term (bx), but no constant (c) is sometimes referred to as an *incomplete quadratic equation.*

$ax^2 + bx = 0$ *(incomplete quadratic equation)*

A quadratic equation that has a quadratic term (ax^2) and a constant (c), but no linear term (bx) is sometimes referred to as a *pure quadratic equation.*

$ax^2 + c = 0$ *(pure quadratic equation)*

It will be helpful in planning our strategy for solving a quadratic equation to be able to identify the types of terms in a quadratic equation and to identify a, b, and c.

EXAMPLE 15.1-2 Write each equation in standard form and identify a, b, and c.

(a) $3x^2 + 5x - 2 = 0$ **(b)** $5x^2 = 2x - 3$
(c) $3x = 5x^2$ **(d)** $-6x^2 + 2x = 0$
(e) $5x^2 - 4 = 0$ **(f)** $x^2 = 9$
(g) $x^2 + 3x = 5x$ **(h)** $7 - x^2 = 3 + x$

Solution

(a) $3x^2 + 5x - 2 = 0$ *(In standard form)*
$a = 3, \quad b = 5, \quad c = -2$

(b) $5x^2 = 2x - 3$
$5x^2 - 2x + 3 = 0$ *(Write in standard form.)*
$a = 5, \quad b = -2, \quad c = 3$

(c) $3x = 5x^2$
$0 = 5x^2 - 3x$ *(Write in standard form.)*
$5x^2 - 3x = 0$ *(Interchange sides of equation.)*
$a = 5, \quad b = -3, \quad c = 0$

(d) $-6x^2 + 2x = 0$
$-1(-6x^2 + 2x = 0)$ *(Multiply by -1.)*
$6x^2 - 2x = 0$ *(Standard form)*
$a = 6, \quad b = -2, \quad c = 0$

(e) $5x^2 - 4 = 0$ *(Standard form)*
$a = 5, \quad b = 0, \quad c = -4$

(f) $x^2 = 9$
$x^2 - 9 = 0$ *(Write in standard form.)*
$a = 1, \quad b = 0, \quad c = -9$

(g) $x^2 + 3x = 5x$
$x^2 + 3x - 5x = 0$ *(Rearrange.)*
$x^2 - 2x = 0$ *(Combine like terms.)*
$a = 1, \quad b = -2, \quad c = 0$

(h) $7 - x^2 = 3 + x$
$-x^2 - x + 7 - 3 = 0$ *(Rearrange.)*
$-x^2 - x + 4 = 0$ *(Combine like terms.)*
$-1(-x^2 - x + 4 = 0)$ *(Multiply by -1.)*
$x^2 + x - 4 = 0$ *(Standard form)*
$a = 1, \quad b = 1, \quad c = -4$ ■

SELF-STUDY EXERCISES 15-1

L.O.1. Arrange the following quadratic equations in standard form.

1. $5 - 4x + 7x^2 = 0$ **2.** $8x^2 - 3 = 6x$ **3.** $7x^2 = 5$
4. $x^2 = 6x - 8$ **5.** $x^2 - 3x = 6x - 8$ **6.** $8 - x^2 + 6x = 2x$
7. $3x^2 + 5 - 6x = 0$ **8.** $5 = x^2 - 6x$ **9.** $x^2 - 16 = 0$
10. $8x^2 - 7x = 8$ **11.** $8x^2 + 8x - 2 = 8$ **12.** $0.3x^2 - 0.4x = 3$

L.O.2. Write each equation in standard form and identify a, b, and c.

13. $5x = x^2$ **14.** $7x - 3x^2 = 5$ **15.** $7x^2 - 4x = 0$
16. $8 + 3x^2 = 5x$ **17.** $5x - 6 = x^2$ **18.** $11x^2 = 8x$
19. $x = x^2$ **20.** $9x^2 - 7x = 12$ **21.** $5 = x^2$

22. $-x^2 - 6x + 3 = 0$

23. $0.2x - 5x^2 = 1.4$

24. $\frac{2}{3}x^2 - \frac{5}{6}x = \frac{1}{2}$

25. $1.3x^2 - 8 = 0$

26. $\sqrt{3}x^2 + \sqrt{5}x - 2 = 0$

Write equations in standard form for the following.

27. Coefficient of the quadratic term is 8; coefficient of the linear term is -2; the constant term is -3.

28. $a = 5, \quad b = 2, \quad c = -7$.

29. The constant is zero, the coefficient of the linear term is 3, and the coefficient of the quadratic term is 1.

15-2 SOLVING PURE QUADRATIC EQUATIONS: $ax^2 + c = 0$

Learning Objective

L.O.1. Solve pure quadratic equations ($ax^2 + c = 0$).

L.O.1. Solve pure quadratic equations ($ax^2 + c = 0$).

Using the Square-Root Method

To solve pure quadratic equations, we solve for the squared letter and then take the square root of both sides of the equation. Rule 12-4.1 applies since all the equations in Section 12-4 are pure quadratic equations, but not necessarily in standard form.

In Section 12-4 and in each of the following examples, we obtain *two* answers. When solving any quadratic equation in one variable, we may have as many as two answers. In every example we will check all answers.

EXAMPLE 15-2.1 Solve $3y^2 = 27$.

Solution

$$3y^2 = 27$$

$$y^2 = 9 \qquad \textit{(Solve for squared letter.)}$$

$$y = \pm 3 \qquad \textit{(Take square root of both sides.)}$$

Check $y = +3$	Check $y = -3$
$3y^2 = 27$	$3y^2 = 27$
$3(3)^2 = 27$	$3(-3)^2 = 27$
$3(9) = 27$	$3(9) = 27$
$27 = 27$	$27 = 27$

■

Rule 15-2.1 ***To solve a pure quadratic equation ($ax^2 + c = 0$):***

1. Rearrange the equation, if necessary, so that quadratic or squared-letter terms are on one side and constants or number terms are on the other side of the equation.
2. Combine like terms, if appropriate.
3. Solve for the squared letter so that the coefficient is $+1$.
4. Apply the square-root principle of equality by taking the square root of both sides.
5. Check both answers.

Some quadratic equations may involve fractions or decimal numbers, such as the next two examples. Note how each is handled.

EXAMPLE 15-2.2 Solve $4x^2 - 9 = 0$.

Solution

$4x^2 - 9 = 0$

$4x^2 = 9$ *(Transpose.)*

$x^2 = \frac{9}{4}$ *(Solve for squared letter.)*

$x = \pm\frac{3}{2}$ *(Take square root of numerator and denominator if they are perfect square whole numbers.)*

$x = \pm 1.5$ *(Then divide if decimal answer is desired.)*

Check $x = +1.5$	Check $x = -1.5$
$4x^2 - 9 = 0$	$4x^2 - 9 = 0$
$4(1.5)^2 - 9 = 0$	$4(-1.5)^2 - 9 = 0$
$4(2.25) - 9 = 0$	$4(2.25) - 9 = 0$
$9 - 9 = 0$	$9 - 9 = 0$
$0 = 0$	$0 = 0$

In some applications the fractional answer may be more convenient. In other applications the decimal answer may be more convenient. ■

EXAMPLE 15-2.3 Solve $0.04y^2 = 0.25$.

Solution

$0.04y^2 = 0.25$

$y^2 = \frac{0.25}{0.04}$ *(Solve for squared letter.)*

$y = \pm\frac{0.5}{0.2}$ *(Take square root of numerator and denominator if they are perfect squares.)*

$y = \pm 2.5$ *(Divide.)*

The square roots of decimal numbers like 0.25 and 0.04 are perfect square decimal numbers.

Here is another way to work the problem. We may divide *first* and then take the square root of this quotient.

$y^2 = \frac{0.25}{0.04}$

$y^2 = 6.25$ *(Divide.)*

$y = \pm 2.5$

Calculator check $y = +2.5$	Calculator check $y = -2.5$
$0.04y^2 = 0.25$	$0.04y^2 = 0.25$
$0.04(2.5)^2 = 0.25$	$0.04(-2.5)^2 = 0.25$
$0.04(6.25) = 0.25$	$0.04(6.25) = 0.25$
$0.25 = 0.25$	$0.25 = 0.25$

■

Not all pure quadratic equations have *real solutions*. Our study will only include equations with real solutions.

EXAMPLE 15-2.4 Solve $3m^2 + 48 = 0$

Solution

$$3m^2 + 48 = 0$$

$$3m^2 = -48 \quad \textit{(Isolate variable.)}$$

$$m^2 = \frac{-48}{3} \quad \textit{(Solve for } m^2\textit{.)}$$

$$m^2 = -16$$

$$m = \pm\sqrt{-16} \quad \textit{(Apply square root principle.)}$$

There are *no real solutions* to this equation. The solutions would be imaginary numbers. ■

SELF-STUDY EXERCISES 15-2

L.O.1. Solve the following equations. Round to thousandths when necessary.

1. $x^2 = 9$
2. $x^2 - 49 = 0$
3. $9x^2 = 64$
4. $16x^2 - 49 = 0$
5. $0.09y^2 = 0.81$
6. $0.04x^2 = 0.81$
7. $2x^2 = 10$
8. $3x^2 + 4 = 7$
9. $5x^2 - 8 = 12$
10. $7x^2 - 2 = 19$

11. Describe the process for solving a pure quadratic equation.

12. What is the relationship between the roots of a pure quadratic equation?

15-3 SOLVING INCOMPLETE QUADRATIC EQUATIONS: $ax^2 + bx = 0$

Learning Objective

L.O.1. Solve incomplete quadratic equations ($ax^2 + bx = 0$).

L.O.1. Solve incomplete quadratic equations ($ax^2 + bx = 0$).

Using the Zero-Product Property with Incomplete Quadratic Equations

Incomplete quadratic equations like $x^2 + 2x = 0$ can be solved by factoring $x^2 + 2x$. Since $x^2 + 2x$ has a common factor of x, we can factor the expression to

$$x(x + 2) = 0$$

We now have two factors whose product is 0. If we multiply two factors and get a product of 0, one or both factors *must* be 0, since $0 \cdot 0 = 0$ as well as $0 \cdot a = 0$, where a is any number. This is called the *zero-product property.*

Rule 15-3.1 *Zero-product property.*

If $ab = 0$, then a or b or both equal zero.

If $x = 0$, we can check by substituting 0 for x wherever x occurs in the original equation.

$$\begin{aligned} x^2 + 2x &= 0 \\ 0^2 + 2(0) &= 0 \\ 0 + 0 &= 0 \\ 0 &= 0 \end{aligned}$$

If $x + 2 = 0$, we first solve for x.

$$\begin{aligned} x + 2 &= 0 \\ x &= -2 \end{aligned}$$

Since $x = -2$, we can check by substituting -2 for x wherever x occurs in the original equation.

$$\begin{aligned} x^2 + 2x &= 0 \\ (-2)^2 + 2(-2) &= 0 \qquad [(-2)^2 = +4] \\ 4 + (-4) &= 0 \\ 0 &= 0 \end{aligned}$$

Therefore, we find that the original equation checks with either value of x. This means that x may be 0 or -2 in this incomplete quadratic equation.

Rule 15-3.2 ***To solve an incomplete quadratic equation ($ax^2 + bx = 0$):***

1. If necessary, transpose so that all terms are on one side of the equation in standard form with zero on the other side.
2. Factor the expression that always has a common factor and set each factor equal to zero.
3. Solve for the variable in each equation formed in step 2.

EXAMPLE 15-3.1 Solve $2x^2 = 5x$ for x.

Solution

$$\begin{aligned} 2x^2 &= 5x \\ 2x^2 - 5x &= 0 && \textit{(Transpose into standard form.)} \\ x(2x - 5) &= 0 && \textit{(Factor common factor.)} \\ x = 0 \quad 2x - 5 &= 0 && \textit{(Set each factor equal to zero.)} \\ x = 0 \qquad 2x &= 5 && \textit{(Solve each equation.)} \\ x &= \frac{5}{2} \end{aligned}$$

Check $x = 0$ Check $x = \frac{5}{2}$

$2x^2 = 5x$ $2x^2 = 5x$

$$2(0)^2 = 5(0) \qquad 2\left(\frac{5}{2}\right)^2 = 5\left(\frac{5}{2}\right)$$

$$2(0) = 5(0) \qquad \overset{1}{\cancel{2}}\left(\frac{25}{\underset{2}{\cancel{4}}}\right) = \frac{25}{2}$$

$$0 = 0 \qquad \frac{25}{2} = \frac{25}{2}$$

■

EXAMPLE 15-3.2 Solve $4x^2 - 8x = 0$ for x.

Solution

$$4x^2 - 8x = 0$$

$$4x(x - 2) = 0 \qquad \textit{(Factor common factors.)}$$

$$4x = 0 \quad x - 2 = 0 \qquad \textit{(Set each factor equal to zero.)}$$

$$x = \frac{0}{4} \qquad x = 2 \qquad \textit{(Solve each equation.)}$$

$$x = 0$$

Check $x = 0$	Check $x = 2$
$4(0)^2 - 8(0) = 0$	$4(2)^2 - 8(2) = 0$
$4(0) - 8(0) = 0$	$4(4) - 8(2) = 0$
$0 - 0 = 0$	$16 - 16 = 0$
$0 = 0$	$0 = 0$

■

SELF-STUDY EXERCISES 15-3

L.O.1. Solve the following equations.

1. $x^2 - 3x = 0$
2. $3x^2 = 7x$
3. $5x^2 - 10x = 0$
4. $2x^2 + x = 0$
5. $8x^2 - 4x = 0$
6. $5x^2 = 15x$
7. $x^2 - 6x = 0$
8. $y^2 = -4y$
9. $3x^2 + 2x = 0$
10. $9x^2 - 12x = 0$

11. How does an incomplete quadratic equation differ from a pure quadratic equation?

12. Will one of the two roots of an incomplete quadratic equation always be zero? Justify your answer.

15-4 SOLVING COMPLETE QUADRATIC EQUATIONS BY FACTORING

Learning Objective

L.O.1 Solve complete quadratic equations by factoring ($ax^2 + bx + c = 0$).

L.O.1. Solve complete quadratic equations by factoring ($ax^2 + bx + c = 0$)

Using the Zero-Product Property with Complete Quadratic Equations

In this section we use knowledge of factoring trinomials, which we learned in Chapter 14.

> **Rule 15-4.1** *To solve a complete quadratic equation by factoring:*
>
> 1. If necessary, transpose terms so that all terms are on one side of the equation in standard form with zero on the other side.
> 2. Factor the trinomial, if possible, into the product of two binomials and set each factor equal to zero.
> 3. Solve for the variable in each equation formed in step 2.

EXAMPLE 15-4.1 Solve $x^2 + 6x + 5 = 0$ for x.

Solution

$$x^2 + 6x + 5 = 0$$

$$(x + 5)(x + 1) = 0 \qquad \textit{(Factor into the product of two binomials.)}$$

$$x + 5 = 0 \quad x + 1 = 0 \qquad \textit{(Set each factor equal to 0.)}$$

$$x = -5 \quad x = -1 \qquad \textit{(Solve for x in each equation.)}$$

$$\begin{array}{rr} \text{Check } x = -5 & \text{Check } x = -1 \\ x^2 + 6x + 5 = 0 & x^2 + 6x + 5 = 0 \\ (-5)^2 + 6(-5) + 5 = 0 & (-1)^2 + 6(-1) + 5 = 0 \\ 25 + (-30) + 5 = 0 & 1 + (-6) + 5 = 0 \\ -5 + 5 = 0 & -5 + 5 = 0 \\ 0 = 0 & 0 = 0 \end{array}$$

■

EXAMPLE 15-4.2 Solve $6x^2 + 4 = 11x$ for x.

Solution

$$6x^2 + 4 = 11x$$

$$6x^2 - 11x + 4 = 0 \qquad \textit{(Transpose into standard form.)}$$

$$(3x - 4)(2x - 1) = 0 \qquad \textit{(Factor into the product of two binomials.)}$$

$$3x - 4 = 0 \quad 2x - 1 = 0 \qquad \textit{(Set each factor equal to 0.)}$$

$$3x = 4 \quad 2x = 1 \qquad \textit{(Solve each equation.)}$$

$$x = \frac{4}{3} \quad x = \frac{1}{2}$$

$$\begin{array}{rr} \text{Check } x = \dfrac{4}{3} & \text{Check } x = \dfrac{1}{2} \\ 6x^2 + 4 = 11x & 6x^2 + 4 = 11x \\ 6\left(\dfrac{4}{3}\right)^2 + 4 = 11\left(\dfrac{4}{3}\right) & 6\left(\dfrac{1}{2}\right)^2 + 4 = 11\left(\dfrac{1}{2}\right) \\ \overset{2}{\cancel{6}}\left(\dfrac{16}{\underset{3}{\cancel{9}}}\right) + 4 = \dfrac{44}{3} & \overset{3}{\cancel{6}}\left(\dfrac{1}{\underset{2}{\cancel{4}}}\right) + 4 = \dfrac{11}{2} \end{array}$$

$$\left(4 = \frac{12}{3}\right) \quad \frac{32}{3} + \frac{12}{3} = \frac{44}{3} \qquad\qquad \frac{3}{2} + \frac{8}{2} = \frac{11}{2} \quad \left(4 = \frac{8}{2}\right)$$

$$\frac{44}{3} = \frac{44}{3} \qquad\qquad \frac{11}{2} = \frac{11}{2}$$

■

SELF-STUDY EXERCISES 15-4

L.O.1. Solve the following equations by factoring.

1. $x^2 + 5x + 6 = 0$
2. $x^2 - 6x + 9 = 0$
3. $x^2 - 5x - 14 = 0$
4. $x^2 + 3x - 18 = 0$
5. $x^2 + 7x + 12 = 0$
6. $y^2 - 8y = -15$
7. $a^2 - 13a - 14 = 0$
8. $b^2 - 9b = -18$
9. $2x^2 - 7x + 3 = 0$
10. $3x^2 + 13x + 4 = 0$
11. $10x^2 - x = 3$
12. $6x^2 + 11x + 3 = 0$
13. $2x^2 + 13x + 15 = 0$
14. $3x^2 - 10x + 8 = 0$
15. $6x^2 + 17x = 3$

15-5 SOLVING QUADRATIC EQUATIONS USING THE QUADRATIC FORMULA

Learning Objectives

L.O.1. Solve quadratic equations using the quadratic formula.
L.O.2. Solve applied problems that use quadratic equations.

L.O.1. Solve quadratic equations using the quadratic formula.

Using the Quadratic Formula

Quadratic equations may consist of trinomials that are difficult or impossible to factor. To solve such quadratic equations, we can use the *quadratic formula*. This formula results from solving the standard quadratic equation, $ax^2 + bx + c = 0$, by a method called *completing the square*.

Formula 15-5.1

Quadratic Formula: $x = \dfrac{-b \pm \sqrt{b^2 - 4ac}}{2a}$

Since the formula is derived from the standard quadratic equation $ax^2 + bx + c = 0$, the a, b, and c in the formula are the same a, b, and c in the standard quadratic equation. To evaluate the formula, we first identify a, b, and c.

Then we evaluate the formula to solve equations. Even though we have one formula, we still will have *two* values for x because the formula requires us to *add* or *subtract* the radical.

$$x = \frac{-b \pm \sqrt{b^2 - 4ac}}{2a}$$

Tips and Traps

When using the formula to solve problems, we should begin by writing the formula. This will help us remember the formula. When writing the formula, be sure to extend the fraction bar beneath the *entire* numerator.

Example 15-5.1 Use the quadratic formula to solve for x.

$$x^2 + 5x + 6 = 0$$

Solution

1. Identify a, b, and c:

$$a = 1, \quad b = 5, \quad c = 6$$

2. Write the formula:

$$x = \frac{-b \pm \sqrt{b^2 - 4ac}}{2a}$$

3. Substitute for a, b, and c:

$$x = \frac{-5 \pm \sqrt{5^2 - 4 \cdot 1 \cdot 6}}{2 \cdot 1}$$

4. Evalute the formula:

$$x = \frac{-5 \pm \sqrt{25 - 24}}{2}$$

$$x = \frac{-5 \pm \sqrt{1}}{2}$$

$$x = \frac{-5 \pm 1}{2}$$

At this point we separate the formula into two parts, one using the $+1$ and the other using the -1.

$$x = \frac{-5 + 1}{2} \qquad\qquad x = \frac{-5 - 1}{2}$$

$$x = \frac{-4}{2} \qquad\qquad x = \frac{-6}{2}$$

$$x = -2 \qquad\qquad x = -3$$

Check $x = -2$ or Check $x = -3$

$$x^2 + 5x + 6 = 0 \qquad\qquad x^2 + 5x + 6 = 0$$

$$(-2)^2 + 5(-2) + 6 = 0 \qquad\qquad (-3)^2 + 5(-3) + 6 = 0$$

$$4 - 10 + 6 = 0 \qquad\qquad 9 - 15 + 6 = 0$$

$$-6 + 6 = 0 \qquad\qquad -6 + 6 = 0$$

$$0 = 0 \qquad\qquad 0 = 0$$

■

Tips and Traps

This problem also could have been solved by factoring.

$$x^2 + 5x + 6 = 0$$

$$(x + 2)(x + 3) = 0 \qquad \textit{(Factor.)}$$

$$x + 2 = 0 \qquad x + 3 = 0 \qquad \textit{(Set each factor equal to zero.)}$$

$$x = -2 \qquad x = -3 \qquad \textit{(Solve each equation.)}$$

With good factoring skills, some quadratic equations are solved more quickly by factoring and using the zero-product property.

EXAMPLE 15-5.2 Use the quadratic formula to solve for x. Round answers to the nearest hundredth.

$$3x^2 - 3x - 7 = 0$$

Solution

1. Identify a, b, and c:

$$a = 3, \quad b = -3, \quad c = -7$$

2. Write the formula:

$$x = \frac{-b \pm \sqrt{b^2 - 4ac}}{2a}$$

3. Substitute in the formula:

$$x = \frac{+3 \pm \sqrt{(-3)^2 - 4(3)(-7)}}{2 \cdot 3} \qquad \textit{(}-(b) = -(-3) = +3\textit{)}$$

4. Evaluate the formula:

$$x = \frac{+3 \pm \sqrt{9 + 84}}{6} \qquad (-4(3)(-7) = +84)$$

$$x = \frac{+3 \pm \sqrt{93}}{6} \qquad \textit{(Use a calculator to find } \sqrt{93}.\textit{)}$$

$$x = \frac{+3 \pm 9.643650761}{6} \qquad (\pm 9.643650761 \textit{ is calculator value for } \sqrt{93}.)$$

$$x = \frac{3 + 9.643650761}{6} \qquad x = \frac{3 - 9.643650761}{6} \qquad \textit{(Separate into two parts.)}$$

$$x = \frac{12.6436590761}{6} \qquad x = \frac{-6.643650761}{6}$$

$$x = 2.11 \qquad x = -1.11$$

(rounded to nearest hundredth) *(rounded to nearest hundredth)*

Notice that rounding to hundredths takes place *after* the final calculation has been made.

Check $x = 2.11$

$$3x^2 - 3x - 7 = 0$$
$$3(2.11)^2 - 3(2.11) - 7 = 0$$
$$3(4.4521) - 6.33 - 7 = 0$$
$$13.3563 - 6.33 - 7 = 0$$
$$0.0263 \doteq 0$$

Check $x = -1.11$

$$3x^2 - 3x - 7 = 0$$
$$3(-1.11)^2 - 3(-1.11) - 7 = 0$$
$$3(1.2321) + 3.33 - 7 = 0$$
$$3.6963 + 3.33 - 7 = 0$$
$$0.0263 \doteq 0$$

■

Tips and Traps

The symbol $\doteq$ means "is approximately equal to."
If we had checked *before* rounding, the check would have been "closer."

General Tips for Using the Calculator

Use the full calculator value to check.

This can be done efficiently using a graphing calculator. Calculate the first root.

3 [+] [√] 93 [EXE] [÷] 6 [EXE] ⇒ 2.107275127

Check:

3 [ANS] [x^2] [−] 3 [ANS] [−] 7 [EXE] ⇒ −1.6E − 09 or −0.0000000016

Calculate the second root.

3 [−] [√] 93 [EXE] [÷] 6 [EXE] ⇒ −1.107275127

Check:

3 [ANS] [x^2] [−] 3 [ANS] [−] 7 [EXE] ⇒ 8.E − 12 or 0.000000000008

In the next example, the equation contains no squared term at first, but the squared term appears *after* the multiplication to eliminate the denominator.

EXAMPLE 15-5.3 Solve for x. Round answers to the nearest hundredth.

$$\frac{3}{5 - x} = 2x$$

Solution: We must first eliminate the denominator and then write the equation in standard form.

$$\frac{3}{5 - x} = 2x$$

$$\cancel{(5 - x)}\frac{3}{\cancel{5 - x}} = 2x(5 - x)$$ *(Multiply both sides by denominator to eliminate it.)*

$$3 = 2x(5 - x)$$

$$3 = 10x - 2x^2$$ *(After multiplication, we have a squared term.)*

$$2x^2 - 10x + 3 = 0$$ *(Write equation in standard form. Then proceed as usual.)*

$$a = 2, \quad b = -10, \quad c = 3$$

$$x = \frac{-b \pm \sqrt{b^2 - 4ac}}{2a}$$ *(Write formula.)*

$$x = \frac{+10 \pm \sqrt{(-10)^2 - 4(2)(3)}}{2(2)}$$ *(Substitute.)*

$$x = \frac{10 \pm \sqrt{100 - 24}}{4}$$

$$x = \frac{10 \pm \sqrt{76}}{4}$$

$$x = \frac{10 \pm 8.717797887}{4}$$ *(8.717797887 is calculator value for $\sqrt{76}$.)*

$$x = \frac{10 + 8.717797887}{4} \qquad x = \frac{10 - 8.717797887}{4}$$

$$x = \frac{18.717797887}{4} \qquad x = \frac{1.282202113}{4}$$

$$x = 4.68 \qquad x = 0.32$$ *(Round after final calculation.)*

Check $x = 4.68$ — Check $x = 0.32$

$$\frac{3}{5 - x} = 2x \qquad \frac{3}{5 - x} = 2x$$

$$\frac{3}{5 - (4.68)} = 2(4.68) \qquad \frac{3}{5 - (0.32)} = 2(0.32)$$

$$\frac{3}{0.32} = 9.36 \qquad \frac{3}{4.68} = 0.64$$

$$9.375 \doteq 9.36 \qquad 0.6410256 \doteq 0.64$$ ■

Tips and Traps

When an equation is multiplied by a factor that contains a variable, as in the previous example, an *extraneous root* can be introduced. An extraneous root is a root that will *not* check when substituted into the original equation. Thus, it is very important to check each root.

L.O.2. Solve applied problems that use quadratic equations.

Applications

Many applications result in quadratic equations.

EXAMPLE 15-5.4 Find the length and width of a rectangular table if the length is 8 inches more than the width and the area is 260 in.^{2}.

Solution

Let $x =$ the number of inches in the width

$x + 8 =$ the number of inches in the length

Area = length times width, or $A = lw$

$260 = (x + 8)(x)$ *(Substitute into area formula.)*

$260 = x^2 + 8x$

$x^2 + 8x - 260 = 0$

$a = 1, \quad b = 8, \quad c = -260$

$$x = \frac{-b \pm \sqrt{b^2 - 4ac}}{2a}$$

$$x = \frac{-8 \pm \sqrt{(8)^2 - 4(1)(-260)}}{2(1)}$$ *(Substitute into quadratic formula.)*

$$x = \frac{-8 \pm \sqrt{64 + 1040}}{2}$$

$$x = \frac{-8 \pm \sqrt{1104}}{2}$$

$$x = \frac{-8 \pm 33.22649545}{2}$$

$$x = \frac{-8 + 33.22649545}{2}$$

$$x = \frac{25.22649545}{2}$$ *(Measurements are positive, so use only the positive value.)*

$$x = 12.6$$ *(width rounded to nearest tenth inch)*

$$x + 8 = 20.6$$ *(length)*

Thus, the width is 12.6 in. and length is 20.6 in. ■

Tips and Traps

If a quadratic equation is used to solve certain applied problems in which a negative answer is not possible (such as in measures of length and width), the negative answer is discarded and only the positive answer is used.

SELF-STUDY EXERCISES 15-5

L.O.1. Solve the following quadratic equations by using the quadratic formula.

1. $3x^2 - 7x - 6 = 0$
2. $x^2 + x - 12 = 0$
3. $5x^2 - 6x = 11$
4. $x^2 - 6x + 9 = 0$
5. $x^2 - x - 6 = 0$
6. $8x^2 - 2x = 3$

Solve the following quadratic equations by using the quadratic formula. Round each final answer to the nearest hundredth.

7. $x^2 = -9x + 20$
8. $3x^2 + 6x + 1 = 0$
9. $\frac{2}{x} + \frac{5}{2} = x$
10. $\frac{2}{x} + 3x = 8$

L.O.2. Solve the following applied problems.

11. A rectangular table top is 3 cm longer than it is wide. Find the length and width to the nearest hundredth if the area is 47.5 cm^2. (Area = length × width, or $A = lw$.)

12. A rectangular instrument case has an area of 40 in.^{2}. If the length is 6 in. more than the width, find the dimensions (length and width) of the instrument case.

13. A bricklayer plans to build an arch with a span (s) of 8 m and a radius (r) of 4 m. How high (h) is the arch? (Use the formula $h^2 - 2hr + \frac{s^2}{4} = 0$.)

14. A rectangular patio slab contains 180 ft^2. If the length is 1.5 times the width, find the width and length of the slab to the nearest whole number.

15. A farmer normally plants a field 80 feet by 120 feet. This year the government requires the planting area to be decreased by 20% and the farmer chooses to decrease the width and length of the field by an equal amount. Show an illustration of the original field and the reduced-size field. If x is the amount by which the length and width is decreased, find the length and width of the new field.

16. The recommended dosage of a certain type of medicine is determined by the patient's weight. The formula to determine the dosage is given by $D = 0.1w^2 + 5w$, where D is the dosage in milligrams and w is the patient's body weight in kilograms. A doctor ordered a dosage of 1800 milligrams. This dosage is to be administered to what weight patient?

15-6 SELECTING AN APPROPRIATE METHOD FOR SOLVING QUADRATIC EQUATIONS

Learning Objective

L.O.1. Determine the nature of the roots of a quadratic equation by examining the discriminant.

L.O.1. Determine the nature of the roots of a quadratic equation by examining the discriminant.

Examining the Discriminant

In the previous section, several problems were solved using the quadratic formula when, in fact, another method for solving may have been quicker or easier. In general, the following suggestion is made for solving quadratic equations.

To solve a quadratic equation:

1. Write the equation in the form $ax^2 + bx + c = 0$.
2. Factor if possible.
3. If it is not possible to factor or if factoring is difficult, use the quadratic formula.

All types of quadratic equations can be solved using the completing-the-square method (not presented in this text) or the quadratic formula. Only certain types of quadratic equations can be solved by factoring or applying the square root principle. The radicand of the radical portion of the quadratic formula, $b^2 - 4ac$, is called the *discriminant*. The general characteristics of the quadratic equation can be determined by examining the discriminant.

Rule 15-6.1 *In the quadratic equation $ax^2 + bx + c = 0$:*

1. If $b^2 - 4ac \geq 0$, the equation has real-number roots.
   - **a.** If $b^2 - 4ac$ is a perfect square, the two roots are rational.
   - **b.** If $b^2 - 4ac = 0$, there is one rational root.
   - **c.** If $b^2 - 4ac$ is not a perfect square, the two roots are irrational.
2. If $b^2 - 4ac < 0$, the equation has no real-number roots.

EXAMPLE 15-6.1 Examine the discriminant of each of the following equations. If the equation has real-number solutions, find the solutions.

(a) $5x^2 + 3x - 1 = 0$ **(b)** $3x^2 + 5x = 2$
(c) $4x^2 + 2x + 3 = 0$

Solution

(a) $5x^2 + 3x - 1 = 0$ *($a = 5, b = 3, c = -1$)*

$$b^2 - 4ac = 3^2 - 4(5)(-1)$$ *(Examine discriminant.)*

$$= 9 + 20$$

$$= 29$$ *(irrational roots)*

$$x = \frac{-b \pm \sqrt{b^2 - 4ac}}{2a}$$ *(quadratic formula)*

$$x = \frac{-3 \pm \sqrt{29}}{2(5)}$$ *(Substitute.)*

$$x = \frac{-3 + \sqrt{29}}{10} \quad \text{or} \quad x = \frac{-3 - \sqrt{29}}{10}$$ *(exact roots)*

(b) $3x^2 + 5x = 2$

$3x^2 + 5x - 2 = 0$ *(Standard form: $a = 3$, $b = 5$, $c = -2$)*

$b^2 - 4ac = 5^2 - 4(3)(-2)$ *(Examine discriminant.)*

$= 25 + 24$

$= 49$ *(rational roots)*

$$x = \frac{-b \pm \sqrt{b^2 - 4ac}}{2a}$$ *(quadratic formula)*

$$x = \frac{-5 + \sqrt{49}}{2(3)} \qquad x = \frac{-5 - \sqrt{49}}{2(3)}$$

$$x = \frac{-5 + 7}{6} \qquad x = \frac{-5 - 7}{6}$$

$$x = \frac{2}{6} \qquad x = \frac{-12}{6}$$

$$x = \frac{1}{3} \qquad x = -2$$ *(exact roots)*

(c) $4x^2 + 2x + 3 = 0$ *($a = 4$, $b = 2$, $c = 3$)*

$b^2 - 4ac = 2^2 - 4(4)(3)$ *(Examine discriminant.)*

$= 4 - 48$

$= -44$ *(no real roots)* ■

SELF-STUDY EXERCISES 15-6

L.O.1. Examine the discriminant of each of the following equations. If the equation has real-number roots, find the roots. Round to hundredths if necessary.

1. $3x^2 + x - 2 = 0$
2. $x^2 - 3x = -1$
3. $2x^2 + x = 2$
4. $3x^2 - 2x + 1 = 0$
5. $x^2 - 3x - 7 = 0$
6. $3x^2 + 5x - 6 = 0$

7. Use the discriminant to write a quadratic equation that has real roots.

8. Use the discriminant to write a quadratic equation that has roots that are real, rational, and unequal.

15-7 SOLVING HIGHER-DEGREE EQUATIONS BY FACTORING

Learning Objective

L.O.1. Identify the degree of an equation.
L.O.2. Solve higher-degree equations by factoring.

L.O.1. Identify the degree of an equation.

The Degree of an Equation

In Section 15-1 we defined a *quadratic* equation as an equation that has at least one letter term raised to the second power and no letter terms have a power higher than 2. Similarly, a *cubic* equation is an equation that has at least one letter term raised to the third power and no letter terms have a power higher than 3.

A quadratic equation can also be referred to as a *second-degree* equation and a cubic equation as a *third-degree* equation.

■ **DEFINITION 15-7.1 Degree of an Equation.** The **degree of an equation** is the highest power of any letter term that appears in the equation.

When we solved linear or first-degree equations, we obtained at most one solution or root for the equation. When we solved quadratic or second-degree equations, we obtained at most two solutions or roots. A cubic or third-degree equation will have at most three solutions or roots.

EXAMPLE 15-7.1 State the degree of each of the following equations.

(a) $x^2 + 3x + 4 = 0$ **(b)** $x^3 = 27$ **(c)** $x^4 + 3x^3 + 2x^2 + x + 4 = 0$
(d) $x^8 = 256$ **(e)** $3x + 7 = 5x - 3$

Solution

(a) $x^2 + 3x + 4 = 0$
Quadratic or second-degree equation
(b) $x^3 = 27$
Cubic or third-degree equation
(c) $x^4 + 3x^3 + 2x^2 + x + 4 = 0$
Fourth-degree equation
(d) $x^8 = 256$
Eighth-degree equation
(e) $3x + 7 = 5x - 3$
Linear or first-degree equation ■

L.O.2. Solve higher-degree equations by factoring.

Solving Higher-Degree Equations

In our discussion, we will limit our study to a few basic types of *higher-degree equations*. A more thorough study is reserved for advanced mathematics courses. We will look only at higher-degree equations that are written in factored form or can easily be written in factored form.

We will solve these equations by using the zero-product property. That is, if the product of the factors is equal to 0, then all factors containing a variable may be equal to 0. (See Section 15-3.)

EXAMPLE 15-7.2 Solve the following equations by factoring.

(a) $x(x + 4)(x - 3) = 0$ **(b)** $5x(x - 7)(2x - 1) = 0$
(c) $2x^3 - 14x^2 + 20x = 0$ **(d)** $3x^3 - 27x = 0$

Solution

(a) $x(x + 4)(x - 3) = 0$ *(Already factored)*

$x = 0 \quad x + 4 = 0 \quad x - 3 = 0$ *(Set each factor equal to zero.)*

$x = 0 \quad x = -4 \quad x = 3$ *(Solve each equation.)*

(b) $5x(x - 7)(2x - 1) = 0$ *(Already factored)*

$5x = 0 \quad x - 7 = 0 \quad 2x - 1 = 0$ *(Set each factor to zero.)*

$\frac{5x}{5} = \frac{0}{5} \quad x = 7 \quad 2x = 1$ *(Solve each equation.)*

$x = 0 \quad \frac{2x}{2} = \frac{1}{2}$

$x = \frac{1}{2}$

(c)
$$\begin{aligned} 2x^3 - 14x^2 + 20x &= 0 \\ 2x(x^2 - 7x + 10) &= 0 \quad \textit{(Factor out common factors.)} \\ 2x(x - 5)(x - 2) &= 0 \quad \textit{(Factor trinomial.)} \end{aligned}$$

$2x = 0 \qquad x - 5 = 0 \qquad x - 2 = 0$ *(Set each factor equal to zero.)*

$\frac{2x}{2} = \frac{0}{2} \qquad x = 5 \qquad x = 2$ *(Solve each equation.)*

$x = 0$

(d)
$$\begin{aligned} 3x^3 - 27x &= 0 \quad \textit{(Factor out common factors.)} \\ 3x(x^2 - 9) &= 0 \quad \textit{(Factor difference of two squares.)} \\ 3x(x + 3)(x - 3) &= 0 \end{aligned}$$

$3x = 0 \qquad x + 3 = 0 \qquad x - 3 = 0$ *(Set each factor equal to zero.)*

$\frac{3x}{3} = \frac{0}{3} \qquad x = -3 \qquad x = 3$ *(Solve each equation.)*

$x = 0$

■

SELF-STUDY EXERCISES 15-7

L.O.1. State the degree of each of the following equations.

1. $x^2 + 2x - 3 = 0$
2. $3x + 2x = 5$
3. $x^4 = 42$
4. $2x - 7 = 4x = 3$
5. $3x^3 + 2x^4 + 3 = x^2$
6. $x^7 = 128$

L.O.2. Find the roots of the following equations. Factor if necessary.

7. $x(x - 2)(x + 3) = 0$
8. $2x(2x - 1)(x + 3) = 0$
9. $3x(2x - 5)(3x - 2) = 0$
10. $x^3 - 7x^2 + 10x = 0$
11. $3x^3 - 3x^2 = 18x$
12. $4x^3 + 10x^2 + 4x = 0$
13. $2x^3 - 18x = 0$
14. $12x^3 = 3x$
15. $16x^3 = 9x$

ASSIGNMENT EXERCISES

15-1

Identify the following quadratic equations as pure, incomplete, or complete.

1. $x^2 = 49$
2. $x^2 - 5x = 0$
3. $5x^2 - 45 = 0$
4. $3x^2 + 2x - 1 = 0$
5. $8x^2 + 6x = 0$
6. $5x^2 + 2x + 1 = 0$
7. $x^2 - 32 = 0$
8. $x^2 + x = 0$
9. $3x^2 + 6x + 1 = 0$
10. $x^2 + 7x + 9 = 0$

15-2

Solve the following equations. Round to thousandths when necessary.

11. $x^2 = 100$
12. $x^2 - 64 = 0$
13. $4x^2 = 9$
14. $64x^2 - 49 = 0$
15. $0.36y^2 = 1.09$
16. $0.16x^2 = 0.64$
17. $5x^2 = 40$
18. $2x^2 - 5 = 3$
19. $6x^2 + 4 = 34$
20. $5x^2 - 6 = 19$
21. $3x^2 = 12$
22. $3x^2 - 4 = 8$
23. $2x^2 = 34$
24. $5x^2 - 9 = 30$
25. $3y^2 - 36 = -8$
26. $2x^2 + 3 = 51$
27. $\frac{1}{2}x^2 = 8$
28. $\frac{2}{3}x^2 = 24$
29. $\frac{1}{4}x^2 - 1 = 15$
30. $\frac{2}{5}x^2 + 2 = 8$

15-3

Solve the following equations.

31. $x^2 - 5x = 0$
32. $4x^2 = 8x$
33. $6x^2 - 12x = 0$
34. $3x^2 + x = 0$
35. $10x^2 + 5x = 0$
36. $3y^2 = 12y$
37. $y^2 - 5y = 0$
38. $x^2 = 16x$
39. $12x^2 + 8x = 0$
40. $8x^2 - 12x = 0$
41. $x^2 + 3x = 0$
42. $4x^2 - 28x = 0$
43. $5x^2 = 45x$
44. $7x^2 = 28x$
45. $y^2 + 8y = 0$
46. $z^2 - 6z = 0$
47. $3m^2 - 5m = 0$
48. $4n^2 - 3n = 0$
49. $2x^2 = x$
50. $5y^2 = y$

15-4

Solve the following equations by factoring.

51. $x^2 - 4x + 3 = 0$
52. $x^2 + 7x + 12 = 0$
53. $x^2 + 3x = 10$
54. $x^2 - 7x + 12 = 0$
55. $x^2 + 7x = -6$
56. $x^2 + 3 = -4x$
57. $x^2 - 6x + 8 = 0$
58. $6y + 7 = y^2$
59. $6y^2 - 5y - 6 = 0$
60. $5y^2 + 23y = 10$
61. $10y^2 - 21y - 10 = 0$
62. $6x^2 - 16x + 8 = 0$
63. $4x^2 + 7x + 3 = 0$
64. $3x^2 = -7x + 6$
65. $12y^2 - 5y - 3 = 0$
66. $x^2 - 3x = 18$
67. $x^2 + 19x = 42$
68. $3x^2 + x - 2 = 0$
69. $3y^2 + y - 2 = 0$
70. $2x^2 - 4x - 6 = 0$
71. $2x^2 - 10x + 12 = 0$
72. $y^2 + 18y + 45 = 0$
73. $x^2 - 3x - 18 = 0$
74. $3x^2 - 9x - 30 = 0$
75. $2y^2 + 22y + 60 = 0$

15-5

Indicate the values for a, b, and c in the following quadratic equations.

76. $5x^2 + x + 6 = 0$
77. $x^2 - 2x = 8$
78. $x^2 - 7x + 12 = 0$
79. $x^2 + 3x = 4$
80. $3x^2 = 2x + 7$
81. $x^2 - 3x = -2$

Solve the following quadratic equations by using the quadratic formula.

82. $x^2 - 9x + 20 = 0$
83. $x^2 - 8x - 9 = 0$
84. $x^2 - 5x = -6$
85. $x^2 + 2x = 8$
86. $x^2 - x - 12 = 0$
87. $2x^2 - 3x - 2 = 0$

Solve the following quadratic equations by using the quadratic formula. Round each final answer to the nearest hundredth.

88. $3x^2 + 6x + 2 = 0$
89. $2x^2 - 3x - 1 = 0$
90. $5x^2 + 4x - 8 = 0$
91. $3x^2 + 5x + 1 = 0$

92. A bricklayer plans to build an arch with a span (s) of 10 m and a radius (r) of 5 m. How high (h) is the arch? (Use the formula $h^2 - 2hr + \frac{s^2}{4} = 0$.)

93. A rectangular kitchen contains 240 ft^2. If the length is two times the width, find the length and width of the room to the nearest whole number. (Area = length × width, or $A = lw$.)

94. What are the dimensions of a rectangular tool storage room if the area is 45.5 m^2 and the room is 0.5 m longer than it is wide? Round to nearest tenth.

95. Find the length and width of a piece of fiberglass if its length is three times the width and the area is 591 $in.^2$. Round to the nearest inch.

15-6

Use the discriminant of the quadratic formula to describe the roots of the following equations.

96. $x^2 - 3x + 2 = 0$
97. $x^2 + 8x + 16 = 0$
98. $2x^2 - 3x - 5 = 0$
99. $5x^2 - 100 = 0$
100. $3x^2 - 2x + 4 = 0$
101. $2x = 5x^2 - 3$

15-7

State the degree of each of the following equations.

102. $2x + 5x = 15$
103. $3x - 2x^3 + 8 = 0$
104. $16 = x^4$
105. $6 - 3x - 3 = 2x + 4$
106. $y^6 = 729$
107. $5y^8 + 2y^3 - 6 = y^2$

Find the roots of the following equations. Factor if necessary.

108. $x(x+2)(x-3)=0$

109. $2x(3x-2)(x-2)=0$

110. $3x(2x+1)(x+4)=0$

111. $2x^3+10x^2+12x=0$

112. $x^3=2x^2$

113. $2x^3+9x^2=5x$

114. $6x^3+3x^2-18x=0$

115. $3x^3+7x^2-6x=0$

116. $3x^3-x^2-2x=0$

117. $x^3+6x^2+8x=0$

118. $x^3-8x^2+15x=0$

119. $x^3-x^2-20x=0$

120. $x^3+x^2-20x=0$

121. $y^3-6y^2-27y=0$

122. $y^3-12y^2-45y=0$

123. $x^3-3x^2=40x$

124. $x^3-2x^2=15x$

125. $2x^3+12x^2+16x=0$

126. $4x^3+32x^2+60x=0$

CHALLENGE PROBLEM

127. Many objects are designed with dimensions according to the ratio of the *Golden Ratio*. Objects that have measurements according to this ratio are said to be most pleasing to the eye. The *Golden Rectangle* has dimensions of length (l) and height (h) that satisfy the following formula.

$$\frac{l+h}{l}=\frac{l}{h}$$

You can cross-multiply to obtain a quadratic equation for the Golden Rectangle.

You have been commissioned to construct a wall hanging for the lobby of a new office building lobby. The wall is 32 feet long and the ceiling is 20 feet high. The owners want the wall hanging to be at least 2 feet from the ceiling, the floor, and each of the side corners. They also want the wall hanging to have dimensions according to the Golden Ratio.

Using the given information, determine the largest-size wall hanging that can be placed in the lobby that also has the dimensions of the Golden Rectangle.

CONCEPTS ANALYSIS

1. State the zero-product property and explain how it applies to solving quadratic equations?

2. What is an extraneous root?

3. What technique for solving equations can produce an extraneous root?

4. When will a quadratic equation have no real solutions?

5. When will a quadratic equation have irrational roots?

6. When will a quadratic equation have rational roots?

Find a mistake in each of the following. Correct and briefly explain the mistake.

7.
$$2x^2-2x-12=0$$
$$2(x^2-x-6)=0$$
$$2(x+2)(x-3)=0$$
$$2=0,\quad x+2=0,\quad x-3=0$$
$$x=-2,\quad x=3$$
(The roots are 0, −2, and 3.)

8. $2x^3 + 5x^2 + 2x = 0$

$$\frac{x(x^2 + 5x + 2)}{x} = \frac{0}{x}$$

$$2x^2 + 5x + 2 = 0$$

$$(2x + 1)(x + 2) = 0$$

$$2x + 1 = 0 \qquad x + 2 = 0$$

$$2x = -1 \qquad x = -2$$

$$x = -\frac{1}{2}$$

(The roots are $-\frac{1}{2}$ *and* -2.*)*

9. Why are you not allowed to divide both sides of an equation by a variable?

10. What is the maximum number of roots that the equation $x^4 = 16$ *could* have? How many real roots does the equation have?

CHAPTER SUMMARY

Objectives	What to Remember	Examples
Section 15-1		
1. Write quadratic equations in standard form.	Write the equation with all terms on the left side of the equation and zero on the right arranged with the terms in descending powers arranged. The leading coefficient should be positive.	Write the equation $5x + 3 = 4x^2$ in standard form. $-4x^2 + 5x + 3 = 0$ $-1(-4x^2 + 5x + 3) = -1(0)$ $4x^2 - 5x - 3 = 0$
2. Identify the coefficients of the quadratic, linear, and constant terms of a quadratic equation.	The coefficient of the quadratic term is the coefficient of the squared variable; the coefficient of the linear term is the coefficient of the first-power variable and the constant term is its own coefficient.	Identify the coefficient of the quadratic and linear terms and identify the constant term in the following: $5x = 4x^2 - 3$ First, write the equation in standard form: $4x^2 - 5x - 3 = 0.$ The coefficient of the quadratic term is 4. The coefficient of the linear term is -5 and the constant term is -3.
Section 15-2		
1. Solve pure quadratic equations ($ax^2 + c = 0$).	To solve pure quadratic equations, solve for the squared variable; then take the square root of both sides of the equation. The two roots have the same absolute value.	Solve $2x^2 - 72 = 0.$ $2x^2 = 72$ $x^2 = 36$ $x = \pm 6$
Section 15-3		
1. Solve incomplete quadratic equations ($ax^2 + bx = 0$).	Arrange the equation in standard form. Factor the common factor. Then set each of the two factors equal to zero and solve for the variable. Both roots are rational and one root is always zero.	Solve $5x^2 - 15x = 0.$ $5x(x - 3) = 0$ $5x = 0 \quad x - 3 = 0$ $x = 0 \quad x = 3$

Section 15-4

1. Solve complete quadratic equations by factoring ($ax^2 + bx + c = 0$).

Arrange the quadratic equation in standard form and factor the trinomial. Then set each factor equal to zero and solve for the variable.

If the expression factors, the roots are rational.

Solve $2x^2 - 5x - 3 = 0$.

$$(x - 3)(2x + 1) = 0$$

$$x - 3 = 0 \qquad 2x + 1 = 0$$

$$x = 3 \qquad 2x = -1$$

$$x = -\frac{1}{2}$$

Section 15-5

1. Solve quadratic equations using the quadratic formula.

In the quadratic formula

$$x = \frac{-b \pm \sqrt{b^2 - 4ac}}{2a},$$

a is the coefficient of the quadratic term, b is the coefficient of the linear term, and c is the constant when the equation is written in standard form. The variable is x.

The characteristics of the roots are determined by examining the discriminant (Section 15-6).

Solve using the quadratic formula:

$$5x^2 + 7x - 6 = 0$$

$$a = 5, b = 7, c = -6$$

$$x = \frac{-7 \pm \sqrt{7^2 - 4(5)(-6)}}{2(5)}$$

$$x = \frac{-7 \pm \sqrt{49 + 120}}{10}$$

$$x = \frac{-7 \pm \sqrt{169}}{10}$$

$$x = \frac{-7 \pm 13}{10}$$

$$x = \frac{6}{10} \text{ or } \frac{3}{5}$$

$$x = \frac{-20}{10} \text{ or } -2$$

2. Solve applied problems that use quadratic equations.

Solving applied problems often requires knowledge of other mathematical formulas, for example, $A = lw$ (area of a rectangle).

Find the length and width of a rectangular parking lot if the length is to be 12 meters longer than the width and the area is to be 6,205 square meters.

Let x = number of meters in width
$x + 12$ = number of meters in length

$$x(x + 12) = 6205$$

$$x^2 + 12x = 6205$$

$$x^2 + 12x - 6205 = 0$$

Use formula with calculator.

$$x = \frac{-12 \pm \sqrt{12^2 - 4(1)(-6205)}}{2(1)}$$

$$x = \frac{-12 \pm \sqrt{144 + 24820}}{2}$$

$$x = \frac{-12 \pm \sqrt{24964}}{2}$$

$$x = \frac{-12 \pm 158}{2}$$

$$x = \frac{-12 + 158}{2}$$

$$x = \frac{146}{2}$$

$x = 73$ width *(Disregard negative root.)*

$x + 12 = 85$ length

Section 15-6

1. Determine the nature of the roots of a quadratic equation by examining the discriminant.	The radicand of the quadratic formula, $b^2 - 4ac$, is the discriminant of the quadratic equation. **1.** If $b^2 - 4ac \geq 0$, the equation has real-number solutions. **a.** If $b^2 - 4ac$ is a perfect square, the solutions are rational. **b.** If $b^2 - 4ac = 0$, there is one rational solution. **c.** If $b^2 - 4ac$ is not a perfect square, the roots are irrational. **2.** If $b^2 - 4ac < 0$, the equation has no real-number solutions.	Use the discriminant to determine the characteristics of the roots of the equation $3x^2 - 5x + 7 = 0$. $(-5)^2 - 4(3)(7) = 25 - 84$ $= -59$ The roots are not real.

Section 15-7

1. Solve higher-degree equations by factoring.	The higher degree equations discussed in this section will have a common variable factor and can be solved by factoring.	Solve. $x^3 + 2x^2 - 3x = 0$. $x(x^2 + 2x - 3) = 0$ $x(x + 3)(x - 1) = 0$ $x = 0 \quad x + 3 = 0 \quad x - 1 = 0$ $x = -3 \quad x = 1$

WORDS TO KNOW

quadratic equation (p. 586)
quadratic term (p. 587)
linear term (p. 587)
constant term (p. 587)
pure quadratic equation (p. 587)
incomplete quadratic equation (p. 587)
complete quadratic equation (p. 587)
real solutions (p. 591)
zero-product property (p. 591)
quadratic formula (p. 595)
discriminant (p. 601)
cubic equation (p. 602)
degree of an equation (p. 603)
higher-degree equations (p. 603)

CHAPTER TRIAL TEST

Identify the following quadratic equations as pure, incomplete, or complete.

1. $3x^2 = 42$
2. $7x^2 - 3x + 2 = 0$
3. $5x^2 = 7x$
4. $4x^2 - 1 = 0$

Solve the following quadratic equations. When necessary, round final answer to the nearest hundredth.

5. $x^2 = 81$
6. $x^2 - 36 = 0$
7. $9x^2 = 16$
8. $81x^2 - 64 = 0$
9. $0.09x^2 = 0.49$
10. $x^2 - 2x = 0$
11. $3x^2 - 6x = 0$
12. $12y^2 + 18y = 0$
13. $a^2 - 5a + 6 = 0$
14. $x^2 - 3x - 4 = 0$
15. $2x^2 + 12 = 11x$
16. $3x^2 - 11x - 4 = 0$

17. Find the diameter (d) in mils to the nearest hundredth of a copper wire conductor whose resistance (R) is 1.314 ohms and whose length (L) is 3642.5 ft. (Formula: $R = \frac{KL}{d^2}$, where K is 10.4 for copper wire.)

18. Find the radius (r) of a circle whose area (A) is 35.15 cm^2. Round answer to the nearest hundredth centimeter. (Formula: $A = \pi r^2$.)

19. What is the current in amps (I) to the nearest hundredth if the resistance (R) of the circuit is 52.29 ohms and the watts (W) used are 205? (Formula: $R = \frac{W}{I^2}$.)

Use the quadratic formula to solve the following equations. When necessary, round final answer to the nearest hundredth.

20. $2x^2 + 3x + 1 = 0$

21. $x^2 + 5x - 2 = 0$

22. $3x^2 - 6x - 1 = 0$

23. Find the length and width of a piece of sheet metal if the area is 8 ft^2 and the length is 3 ft longer than the width. Round to the nearest tenth. (Area = length × width, or $A = lw$.)

24. A square parcel of land is 156.25 m^2 in area. What is the length of a side to the nearest hundredth? (Use the formula $A = s^2$, where A is the area and s is the length of a side.)

25. In the formula $E = 0.5\,mv^2$, solve for v if $E = 180$ and $m = 10$.

Find the roots of the following equations.

26. $x(2x - 5)(x - 3) = 0$

27. $2x(5x - 6)(3x - 2) = 0$

28. $x(x - 1)(x + 2) = 0$

29. $6x^3 + 21x^2 = 45x$

30. $2x^3 - x^2 - 6x = 0$

CHAPTER

16 Rational Expressions

Once we become proficient in factoring algebraic expressions, we can use this skill to simplify *rational expressions*. Factoring was used in solving quadratic and higher-degree equations. Now we will see how factoring helps us in simplifying rational expressions, also called algebraic fractions.

16-1 SIMPLIFYING RATIONAL EXPRESSIONS

Learning Objective

L.O.1. Simplify, or reduce, rational expressions.

L.O.1. Simplify, or reduce, rational expressions.

Simplifying Rational Expressions

In arithmetic we learned to reduce fractions by removing any *factors* that were common to both the numerator and denominator. This process of removing common factors is also referred to as *reducing* or *dividing out common factors*. In our study of the laws of exponents, we also *reduced* or *simplified* algebraic fractions by reducing common factors. This process is accomplished whenever the terms of the fraction (the numerator and denominator) are written in factored form. Look at the following example.

EXAMPLE 16-1.1 Simplify or reduce the following fractions.

(a) $\frac{12}{15}$ **(b)** $\frac{24}{36}$ **(c)** $\frac{2x^2yz^3}{4xy^2z}$ **(d)** $\frac{9a^2b^3}{3ab}$

Solution

(a) $\frac{12}{15} = \frac{(2)(2)(3)}{(3)(5)}$ *(Write in factored form.)*

$\frac{(2)(2)(\cancel{3})}{(\cancel{3})(5)} = \frac{4}{5}$ *(Reduce common factors and multiply remaining factors.)*

(b) $\frac{24}{36} = \frac{(2)(2)(2)(3)}{(2)(2)(3)(3)}$ *(Write in factored form.)*

$\frac{(\cancel{2})(\cancel{2})(2)(\cancel{3})}{(\cancel{2})(\cancel{2})(\cancel{3})(3)} = \frac{2}{3}$ *(Reduce common factors and multiply remaining factors.)*

(c) $\dfrac{2x^2yz^3}{4xy^2z} = \dfrac{\not{2}x^{2-1}y^{1-2}z^{3-1}}{(2)(\not{2})}$ *(Write numbers in factored form and apply the laws of exponents to literal factors with like bases.)*

$\dfrac{xy^{-1}z^2}{2} = \dfrac{xz^2}{2y}$ *(Express all literal factors with positive exponents.)*

(d) $\dfrac{9a^2b^3}{3ab} = \dfrac{(3)(\not{3})a^{2-1}b^{3-1}}{\not{3}}$ *(Write numbers in factored form and apply the laws of exponents to literal factors with like bases.)*

$= 3ab^2$ ■

Before continuing, we must clearly understand that we are reducing common *factors*. This procedure *will not* apply to *addends* or *terms*. Look at the next example.

EXAMPLE 16-1.2 Does $\dfrac{5}{10}$ reduce to $\dfrac{3}{8}$ or $\dfrac{1}{2}$?

Solution: $\dfrac{5}{10}$ can be rewritten as addends or terms:

$$\frac{\not{2} + 3}{\not{2} + 8} = \frac{3}{8} \qquad \textit{(INCORRECT)}$$

If common addends are reduced,

$$\frac{5}{10} \neq \frac{3}{8}$$

$\dfrac{5}{10}$ can be rewritten as factors:

$$\frac{(1)(\not{5})}{(2)(\not{5})} = \frac{1}{2}$$

If common factors are reduced,

$$\frac{5}{10} = \frac{1}{2}$$

The correct answer is $\dfrac{1}{2}$. Common addends or terms *cannot* be reduced. ■

Now we will use our knowledge of factoring *polynomials* to simplify rational expressions. A *rational expression* is an algebraic fraction in which the numerator or denominator or both are polynomials.

First, we will look at some examples that are written in factored form.

EXAMPLE 16-1.3 Simplify the following rational expressions.

(a) $\dfrac{x(x + 2)}{(x + 2)(x + 3)}$ **(b)** $\dfrac{3ab(a + 4)}{6ab(a + 3)}$

(c) $\dfrac{2x^2(2x - 1)}{x^3(x - 1)}$ **(d)** $\dfrac{(x + 3)(x - 4)}{(x - 3)(x + 4)}$

Solution

(a) $$\frac{x(x+2)}{(x+2)(x+3)} = \frac{x\cancel{(x+2)}}{\cancel{(x+2)}(x+3)}$$ *(Reduce common binomial factors.)*

$$= \frac{x}{x+3}$$

The remaining x's cannot be reduced. The x in the numerator is a factor ($x = 1 \cdot x$); however, the x in the denominator is an *addend* or *term.*

(b) $$\frac{3ab(a+4)}{6ab(a+3)} = \frac{\overset{1}{\cancel{3}}\cancel{a}\cancel{b}(a+4)}{\underset{2}{\cancel{6}}\cancel{a}\cancel{b}(a+3)}$$ $[1(a + 4) = a + 4]$

$$= \frac{a+4}{2(a+3)} \text{ or } \frac{a+4}{2a+6}$$

(c) $$\frac{2x^2(2x-1)}{x^3(x-1)} \qquad \left(\frac{x^2}{x^3} = x^{2-3} = x^{-1} = \frac{1}{x}\right)$$

$$= \frac{2(2x-1)}{x(x-1)} \text{ or } \frac{4x-2}{x^2-x}$$

(d) $$\frac{(x+3)(x-4)}{(x-3)(x+4)}$$

There are no common factors; therefore, the fraction is in lowest terms. However, the factors in the numerator and denominator can be multiplied to give the following algebraic fraction:

$$\frac{x^2 - x - 12}{x^2 + x - 12}$$ ■

If an algebraic fraction, or rational expression, is not in factored form, our first task is to factor *completely* both the numerator and denominator. After the terms of an algebraic fraction are expressed as factors, we will reduce factors that are common to both the numerator and denominator.

EXAMPLE 16-1.4 Reduce each rational expression to its simplest form.

(a) $\dfrac{x+y}{4x^2+4xy}$ (b) $\dfrac{m-n}{m^2-n^2}$

(c) $\dfrac{a^2+b^2}{a^2-b^2}$ (d) $\dfrac{x^2-6x+9}{x^2-9}$

(e) $\dfrac{2a^2+5a-3}{a^2+11a+24}$ (f) $\dfrac{3x^2-12}{6x+12}$

Solution

(a) $$\frac{x+y}{4x^2+4xy} = \frac{(x+y)}{4x(x+y)}$$ *(Factor out common factor in denominator and group numerator.)*

$$\frac{\cancel{(x+y)}}{4x\cancel{(x+y)}} = \frac{1}{4x}$$ *[Reduce.* $x + y = 1(x + y)$*]*

(b) $\dfrac{m - n}{m^2 - n^2} = \dfrac{m - n}{(m + n)(m - n)}$ *(Factor the difference of squares.)*

$\dfrac{(\cancel{m - n})}{(m + n)(\cancel{m - n})} = \dfrac{1}{(m + n)}$ *(Reduce. m − n = 1(m − n.)]*

(c) $\dfrac{a^2 + b^2}{a^2 - b^2} = \dfrac{a^2 + b^2}{(a + b)(a - b)}$

The numerator, which is the *sum* of squares, will not factor. There are no factors common to both the numerator and denominator. Thus, the fraction is in simplest form:

$$\frac{a^2 + b^2}{a^2 - b^2}$$

(d) $\dfrac{x^2 - 6x + 9}{x^2 - 9} = \dfrac{(x - 3)(x - 3)}{(x + 3)(x - 3)}$ *(Write in factored form.)*

$\dfrac{(x - 3)(\cancel{x - 3})}{(x + 3)(\cancel{x - 3})} = \dfrac{x - 3}{x + 3}$ *(Reduce.)*

(e) $\dfrac{2a^2 + 5a - 3}{a^2 + 11a + 24} = \dfrac{(2a - 1)(a + 3)}{(a + 8)(a + 3)}$ *(Write in factored form.)*

$\dfrac{(2a - 1)(\cancel{a + 3})}{(a + 8)(\cancel{a + 3})} = \dfrac{2a - 1}{a + 8}$ *(Reduce.)*

(f) $\dfrac{3x^2 - 12}{6x + 12} = \dfrac{3(x^2 - 4)}{6(x + 2)} = \dfrac{3(x + 2)(x - 2)}{(3)(2)(x + 2)}$ *(Factor completely.)*

$= \dfrac{\cancel{3}(\cancel{x + 2})(x - 2)}{(\cancel{3})(2)(\cancel{x + 2})} = \dfrac{x - 2}{2}$ *(Reduce.)* ■

SELF-STUDY EXERCISES 16-1

L.O.1. Simplify the following rational expressions.

1. $\dfrac{8}{18}$ **2.** $\dfrac{9}{24}$ **3.** $\dfrac{4a^2b^3}{2ab}$

4. $\dfrac{x(x + 3)}{(x + 3)(x + 2)}$ **5.** $\dfrac{x + y}{2x + 2y}$ **6.** $\dfrac{3x^2(3x + 2)}{x^3(2x - 1}$

7. $\dfrac{a + b}{a^2 - b^2}$ **8.** $\dfrac{(x + 2)(x - 5)}{(x + 5)(x - 2)}$ **9.** $\dfrac{x^2 - 4x + 4}{x^2 - 4}$

10. $\dfrac{4x^2 - 16}{6x + 12}$

16-2 MULTIPLYING AND DIVIDING RATIONAL EXPRESSIONS

Learning Objective

L.O.1. Multiply and divide rational expressions.

L.O.1. Multiply and divide rational expressions.

Multiplying and Dividing Rational Expressions

Let us review the procedure that is used in multiplying or dividing fractions in arithmetic.

EXAMPLE 16-2.1 Multiply or divide as indicated. Express answers in simplest form.

(a) $\frac{5}{8} \cdot \frac{16}{25}$ **(b)** $7 \cdot \frac{3}{8}$ **(c)** $\frac{7}{10} \div \frac{4}{5}$ **(d)** $\frac{\frac{7}{16}}{2}$

Solution

(a) $\frac{5}{8} \cdot \frac{16}{25} = \frac{\overset{1}{\cancel{5}}}{\underset{1}{\cancel{8}}} \cdot \frac{\overset{2}{\cancel{16}}}{\underset{5}{\cancel{25}}}$ *(Reduce factors common to a numerator and denominator.)*

$= \frac{2}{5}$

If factors that are common to a numerator and denominator are reduced or canceled *before* multiplying, the product will be in lowest terms.

(b) $7 \cdot \frac{3}{8} = \frac{7}{1} \cdot \frac{3}{8}$ *(Nothing will reduce.)*

$\frac{7}{1} \cdot \frac{3}{8} = \frac{21}{8}$ or $2\frac{5}{8}$

(c) $\frac{7}{10} \div \frac{4}{5}$ *(Convert division to multiplication by using the reciprocal of the divisor.)*

$\frac{7}{10} \cdot \frac{5}{4} = \frac{7}{\underset{2}{\cancel{10}}} \cdot \frac{\overset{1}{\cancel{5}}}{4} = \frac{7}{8}$ *(Reduce factors common to a numerator and denominator.)*

(d) $\frac{\frac{7}{16}}{2}$ *(Rewrite complex fraction as a division.)*

$\frac{7}{16} \div 2 = \frac{7}{16} \div \frac{2}{1}$ *(Rewrite division as multiplication.)*

$= \frac{7}{16} \cdot \frac{1}{2} = \frac{7}{32}$ ■

When applying the process of multiplying or dividing fractions to rational expressions, the key fact to remember is that the numerators and denominators should be written in *factored* form whenever possible.

EXAMPLE 16-2.2 Multiply or divide as indicated.

(a) $\frac{15xy^2}{xy^3} \cdot \frac{(xy)^3}{5x}$

(b) $\frac{x^2 - 4x - 12}{2x - 12} \cdot \frac{x - 4}{x^2 + 4x + 4}$

(c) $\frac{4y^2 - 9}{2y^2} \div \frac{y^2 - 2y - 15}{4y^2 + 12y}$

(d) $\frac{2x - 6}{1 - x} \div \frac{x^2 - 2x - 3}{x^2 - 1}$

Solution

(a) $\dfrac{15xy^2}{xy^3} \cdot \dfrac{(xy)^3}{5x}$

$= \dfrac{15xy^2}{xy^3} \cdot \dfrac{x^3y^3}{5x} = \dfrac{15x^4y^5}{5x^2y^3}$ *(Apply laws of exponents.)*

$= \dfrac{\overset{3}{\cancel{15}}\,\overset{x^2}{\cancel{x^4}}\,\overset{y^2}{\cancel{y^5}}}{\underset{1}{\cancel{5}}\,\underset{1}{\cancel{x^2}}\,\underset{1}{\cancel{y^3}}}$ *(Reduce.)*

$= 3x^2y^2$

(b) $\dfrac{x^2 - 4x - 12}{2x - 12} \cdot \dfrac{x - 4}{x^2 + 4x + 4}$

$= \dfrac{(x + 2)(x - 6)}{2(x - 6)} \cdot \dfrac{x - 4}{(x + 2)(x + 2)}$ *(Factor.)*

$= \dfrac{\cancel{(x + 2)}\cancel{(x - 6)}}{2\cancel{(x - 6)}} \cdot \dfrac{x - 4}{\cancel{(x + 2)}(x + 2)}$ *(Reduce.)*

$= \dfrac{x - 4}{2(x + 2)}$ or $\dfrac{x - 4}{2x + 4}$

(c) $\dfrac{4y^2 - 9}{2y^2} \div \dfrac{y^2 - 2y - 15}{4y^2 + 12y}$

$= \dfrac{4y^2 - 9}{2y^2} \cdot \dfrac{4y^2 + 12y}{y^2 - 2y - 15}$ *(Convert to multiplication.)*

$= \dfrac{(2y + 3)(2y - 3)}{2y^2} \cdot \dfrac{4y(y + 3)}{(y + 3)(y - 5)}$ *(Factor.)*

$= \dfrac{(2y + 3)(2y - 3)}{\underset{y}{\cancel{2y^2}}} \cdot \dfrac{\overset{2}{\cancel{4y}}\cancel{(y + 3)}}{\cancel{(y + 3)}(y - 5)}$ *(Reduce.)*

$= \dfrac{2(2y + 3)(2y - 3)}{y(y - 5)}$ or $\dfrac{2(4y^2 - 9)}{y^2 - 5y}$

or $\dfrac{8y^2 - 18}{y^2 - 5y}$

(d) $\dfrac{2x - 6}{1 - x} \div \dfrac{x^2 - 2x - 3}{x^2 - 1}$

$= \dfrac{2x - 6}{1 - x} \cdot \dfrac{x^2 - 1}{x^2 - 2x - 3}$ *(Convert to multiplication.)*

$= \dfrac{2(x - 3)}{-1(x - 1)} \cdot \dfrac{(x + 1)(x - 1)}{(x + 1)(x - 3)}$ *(Factor. Note effect of factoring out -1 in denominator of first fraction so the $x - 1$ will reduce with $x - 1$ in numerator of second fraction.)*

$= \dfrac{2\cancel{(x - 3)}}{-1\cancel{(x - 1)}} \cdot \dfrac{\cancel{(x + 1)}\cancel{(x - 1)}}{\cancel{(x + 1)}\cancel{(x - 3)}}$ *(Reduce.)*

$= \dfrac{2}{-1} = -2$

■

SELF-STUDY EXERCISES 16-2

L.O.1. Multiply or divide the following fractions. Reduce to simplest terms.

1. $\dfrac{5x^2}{3y} \cdot \dfrac{2x}{3y}$

2. $\dfrac{a^2 - b^2}{4} \cdot \dfrac{12}{a + b}$

3. $\dfrac{2u + 2v}{5} \cdot \dfrac{10}{u + v}$

4. $\dfrac{a^2 - 49}{b^2 - 25} \cdot \dfrac{b - 5}{a + 7}$

5. $\dfrac{x^2 + 2x + 1}{5x - 5} \cdot \dfrac{15}{x + 1}$

6. $\dfrac{13r^2}{20a^2} \div \dfrac{39r^2}{5a}$

7. $\dfrac{a - b}{4} \div \dfrac{a - b}{2}$

8. $\dfrac{x}{x^2 - 4x + 4} \div \dfrac{1}{x - 2}$

9. $\dfrac{5a^2 - 5b^2}{a^2b^2} \div \dfrac{a + b}{10ab}$

10. $\dfrac{x^2 + 4x + 3}{x^2 - 4x - 5} \div \dfrac{x + 3}{x - 5}$

16-3 ADDING AND SUBTRACTING RATIONAL EXPRESSIONS

Learning Objective

L.O.1. Add and subtract rational expressions.

L.O.1. Add and subtract rational expressions.

Adding and Subtracting Rational Expressions

In adding and subtracting fractions in arithmetic, the key thing to remember is that we can only add or subtract fractions with like denominators. Whenever we have unlike fractions, we must first find a common denominator for the fractions. We then convert each fraction to an equivalent fraction having the common denominator, and add or subtract the numerators.

We will refresh our memory of the procedure for adding and subtracting fractions by working some arithmetic examples.

EXAMPLE 16-3.1 Add or subtract as indicated.

(a) $\dfrac{3}{8} + \dfrac{1}{8}$ **(b)** $\dfrac{7}{12} - \dfrac{1}{3}$ **(c)** $\dfrac{1}{4} + \dfrac{2}{5}$ **(d)** $5 - \dfrac{2}{3}$

Solution

(a) $\dfrac{3}{8} + \dfrac{1}{8}$

Since the denominators are the same, add the numerators and *keep* the *like* denominator.

$$\frac{3}{8} + \frac{1}{8} = \frac{4}{8} = \frac{1}{2}$$

(b) $\dfrac{7}{12} - \dfrac{1}{3}$

First, select a common denominator. Since 3 divides evenly into 12, 12 is the *least* common denominator.

$$\frac{1}{3} = \frac{?}{12} \qquad \begin{aligned} 1(4) &= 4 \\ 3(4) &= 12 \end{aligned}$$ *(Multiply numerator and denominator by same number, in this case, 4.)*

$$\frac{1}{3} = \frac{4}{12}$$

Then

$$\frac{7}{12} - \frac{4}{12} = \frac{3}{12} = \frac{1}{4}$$

(c) $\frac{1}{4} + \frac{2}{5}$

Since 4 and 5 have no common factors, the least common denominator will be their *product* or 20.

$$\frac{1}{4} = \frac{?}{20} = \frac{5}{20} \qquad \begin{aligned} 1(5) &= 5 \\ 4(5) &= 20 \end{aligned}$$ *(Multiply numerator and denominator by 5.)*

$$\frac{2}{5} = \frac{?}{20} = \frac{8}{20} \qquad \begin{aligned} 2(4) &= 8 \\ 5(4) &= 20 \end{aligned}$$ *(Multiply numerator and denominator by 4.)*

$$\frac{5}{20} + \frac{8}{20} = \frac{13}{20}$$

(d) $5 - \frac{2}{3}$

The whole number 5 is $\frac{5}{1}$ as a fraction. The least common denominator for 1 and 3 is 3.

$$\frac{5}{1} = \frac{?}{3} = \frac{15}{3} \qquad \begin{aligned} 5(3) &= 15 \\ 1(3) &= 3 \end{aligned}$$ *(Multiply numerator and denominator by 3.)*

$$\frac{15}{3} - \frac{2}{3} = \frac{13}{3} \quad \text{or} \quad 4\frac{1}{3}$$

■

Tips and Traps

In arithmetic, we often write an improper fraction as a mixed number. In algebra, we normally leave the answer as an improper fraction.

The procedures for adding and subtracting rational expressions are similar to the arithmetic procedures. We can only add or subtract terms with *like* denominators. If the denominators are *unlike*, change each term to an equivalent term with the least common denominator.

The least common denominator (LCD) is the smallest expression that each of the denominators will divide into evenly. To find the least common denominator, first factor each denominator into prime factors. Then list each prime factor the greatest number of times it appears in any one denominator. The product of these prime factors is the least common denominator.

EXAMPLE 16-3.2 Add $\frac{5}{2x} + \frac{3}{x} + \frac{9}{4}$.

Solution: The factors included in each denominator are 2 and x, x, and 2 and 2. The least common denominator is $(2)(2)(x)$, or $4x$.

First, change each fraction to an equivalent fraction with the common denominator.

$$\frac{5}{2x} = \frac{5(2)}{2x(2)} = \frac{10}{4x}$$ *(Multiply numerator and denominator by 2.)*

$$\frac{3}{x} = \frac{3(4)}{x(4)} = \frac{12}{4x}$$ *(Multiply numerator and denominator by 4.)*

$$\frac{9}{4} = \frac{9(x)}{4(x)} = \frac{9x}{4x}$$ *(Multiply numerator and denominator by x.)*

Then, add numerators and *keep* the like (common) denominator.

$$\frac{10}{4x} + \frac{12}{4x} + \frac{9x}{4x} = \frac{10 + 12 + 9x}{4x}$$

$$= \frac{22 + 9x}{4x}$$ *(Combine like terms in the numerator.)* ■

Note: The basic steps for adding rational expressions with unlike denominators are the same as the steps for adding fractions in arithmetic.

> **Rule 16-3.1** ***To add rational expressions with unlike denominators:***
>
> **1.** Find the *least common denominator (LCD).*
> **2.** Change *each* fraction to an equivalent fraction having the common denominator.
> **3.** Add numerators.
> **4.** Keep the same denominator.
> **5.** Reduce (or simplify) if possible.

Example 16-3.3 Add $\frac{4}{x + 3} + \frac{3}{x - 3}$.

Solution: The LCD is the product $(x + 3)(x - 3)$.

$$\frac{4}{x + 3} = \frac{4(x - 3)}{(x + 3)(x - 3)} = \frac{4x - 12}{(x + 3)(x - 3)}$$ *[Multiply numerator and denominator by $(x - 3)$.]*

$$\frac{3}{(x - 3)} = \frac{3(x + 3)}{(x - 3)(x + 3)} = \frac{3x + 9}{(x + 3)(x - 3)}$$ *[Multiply numerator and denominator by $(x + 3)$. Order does not matter in multiplication.]*

$$\frac{4x - 12}{(x + 3)(x - 3)} + \frac{3x + 9}{(x + 3)(x - 3)}$$ *(Add numerators.)*

$$= \frac{4x - 12 + 3x + 9}{(x + 3)(x - 3)} = \frac{7x - 3}{(x + 3)(x - 3)}$$ *(Combine like terms in numerator.)*

or $\frac{7x - 3}{x^2 - 9}$ ■

Some of the steps in Example 16-3.3 might have been combined into one step, thus requiring fewer *written* steps, but the example illustrates each step that must be performed, either mentally or on paper, to add the rational expressions. Look at the next example.

EXAMPLE 16-3.4 Subtract $\frac{x}{x-2} - \frac{5}{x+4}$.

Solution: The LCD is the product $(x - 2)(x + 4)$.

$$\frac{x}{x-2} = \frac{x(x+4)}{(x-2)(x+4)} = \frac{x^2+4x}{(x-2)(x+4)}$$ *[Multiply numerator and denominator by (x + 4).]*

$$\frac{5}{(x+4)} = \frac{5(x-2)}{(x+4)(x-2)} = \frac{5x-10}{(x-2)(x+4)}$$ *[Multiply numerator and denominator by (x − 2).]*

$$\frac{x^2+4x}{(x-2)(x+4)} - \frac{5x-10}{(x-2)(x+4)}$$

$$\frac{x^2+4x-(5x-10)}{(x-2)(x+4)}$$ *(Note, the ENTIRE numerator following the subtraction sign is subtracted.)*

$$\frac{x^2+4x-5x+10}{(x-2)(x+4)}$$ *(Be careful with the signs when parentheses are removed.)*

$$\frac{x^2-x+10}{(x-2)(x+4)} \text{ or } \frac{x^2-x+10}{x^2+2x-8}$$ *(Since $x^2 - x + 10$ will not factor, the fraction cannot be reduced.)* ■

SELF-STUDY EXERCISES 16-3

L.O.1. Add or subtract the following rational expressions. Reduce to simplest terms.

1. $\frac{3}{7} + \frac{2}{7}$ **2.** $\frac{3}{8} + \frac{7}{16}$ **3.** $\frac{2x}{3} + \frac{5x}{6}$

4. $\frac{3x}{8} + \frac{7x}{12}$ **5.** $\frac{2}{x} + \frac{3}{2}$ **6.** $\frac{5}{2x} + \frac{6}{x} + \frac{2}{3x}$

7. $\frac{5}{x+1} + \frac{4}{x-1}$ **8.** $\frac{6}{x+3} + \frac{4}{x-1}$ **9.** $\frac{5}{2x+1} - \frac{2}{x-1}$

10. $\frac{7}{x-6} - \frac{6}{x-5}$

16-4 COMPLEX RATIONAL EXPRESSIONS

Learning Objective

L.O.1. Simplify complex rational expressions.

L.O.1. Simplify complex rational expressions.

Simplifying Complex Rational Expressions

A *complex rational expression* is a rational expression that has a rational expression in its numerator or its denominator or both. Some examples of complex rational expressions are:

$$\frac{\frac{3+x}{5}}{x}, \quad \frac{5x^2}{\frac{2x}{x-1}}, \quad \frac{\frac{2}{x}+x}{5-\frac{3}{x}}$$

Fractions and rational expressions indicate division.

Rule 16-4.1 ***To simplify a complex rational expression:***

1. Add or subtract any rational expressions in the numerator and/or denominator.
2. Rewrite the complex rational expression as a division of rational expressions.
3. Convert the division of rational expressions to an equivalent multiplication of rational expressions.
4. Perform the multiplication.
5. Simplify the result.

EXAMPLE 16-4.1 Simplify.

(a) $\dfrac{\frac{3}{5}}{\frac{2}{3}}$ **(b)** $\dfrac{\frac{x+y}{5}}{y}$

(c) $\dfrac{4xy}{\frac{2x}{5}}$ **(d)** $\dfrac{\frac{x^2-y^2}{2x}}{\frac{x-y}{3x^2}}$

(e) $\dfrac{1+\frac{1}{x}}{\frac{2}{3}}$ **(f)** $\dfrac{\frac{2}{x}-\frac{5}{2x}}{\frac{3}{4x}+\frac{3}{x}}$

Solution

(a) $\dfrac{\frac{3}{5}}{\frac{2}{3}} =$

$\frac{3}{5} \div \frac{2}{3}$ *(Rewrite as division.)*

$\frac{3}{5} \cdot \frac{3}{2}$ *(Convert to equivalent multiplication.)*

$\frac{9}{10}$ *(Multiply.)*

Tips and Traps

Steps 1 and 2 can be combined by expressing the complex fraction as the product of the numerator and the reciprocal of the denominator.

$$\frac{\frac{a}{b}}{\frac{c}{d}} = \frac{a}{b} \cdot \frac{d}{c} \qquad (b, c, \text{ and } d \neq 0)$$

(b) $\dfrac{\frac{x + y}{5}}{y}$

$= \dfrac{x + y}{5} \cdot \dfrac{1}{y}$ *(Multiply numerator times reciprocal of denominator.)*

$= \dfrac{x + y}{5y}$

(c) $\dfrac{4xy}{\frac{2x}{5}}$

$= \dfrac{4xy}{1} \cdot \dfrac{5}{2x}$ *(Multiply numerator times reciprocal of denominator.)*

$= \dfrac{\overset{2}{\cancel{4x}}y}{1} \cdot \dfrac{5}{\cancel{2x}}$ *(Reduce.)*

$= \dfrac{10y}{1} = 10y$ *(Simplify.)*

(d) $\dfrac{\frac{x^2 - y^2}{2x}}{\frac{x - y}{3x^2}}$

$= \dfrac{x^2 - y^2}{2x} \cdot \dfrac{3x^2}{x - y}$ *(Multiply numerator times reciprocal of denominator.)*

$= \dfrac{(x + y)(x - y)}{2x} \cdot \dfrac{3x^2}{x - y}$ *(Factor.)*

$= \dfrac{(x + y)\cancel{(x - y)}}{2\cancel{x}} \cdot \dfrac{3\overset{x}{\cancel{x^2}}}{\cancel{x - y}}$ *(Reduce.)*

$= \dfrac{3x(x + y)}{2}$ or $\dfrac{3x^2 + 3xy}{2}$ *(Multiply.)*

(e) $\dfrac{1 + \frac{1}{x}}{\frac{2}{3}}$

$$= \frac{\frac{x}{x} + \frac{1}{x}}{\frac{2}{3}}$$

(Find common denominator for numerator $1 = \frac{\mathbf{x}}{\mathbf{x}}$.)

$$= \frac{\frac{x+1}{x}}{\frac{2}{3}}$$

(Add terms in numerator.)

$$= \frac{x+1}{x} \cdot \frac{3}{2}$$

(Multiply numerator times reciprocal of denominator.)

$$= \frac{3(x+1)}{2x} \quad \text{or} \quad \frac{3x+3}{2x}$$

(f) $$\frac{\frac{2}{x} - \frac{5}{2x}}{\frac{3}{4x} + \frac{3}{x}}$$

$$= \frac{\frac{4}{2x} - \frac{5}{2x}}{\frac{3}{4x} + \frac{12}{4x}}$$

(Find common denominator for numerator, 2x, and denominator, 4x.)

$$= \frac{\frac{4-5}{2x}}{\frac{3+12}{4x}} = \frac{\frac{-1}{2x}}{\frac{15}{4x}}$$

(Add or subtract fractions in numerator and in denominator.)

$$= \frac{-1}{2x} \cdot \frac{4x}{15}$$

(Multiply numerator times reciprocal of denominator.)

$$= \frac{-1}{\cancel{2x}_{1}} \cdot \frac{\overset{2}{\cancel{4x}}}{15}$$

(Reduce.)

$$= \frac{-2}{15} = -\frac{2}{15}$$

(Multiply.) ■

SELF-STUDY EXERCISES 16-4

L.O.1. Simplify.

1. $\dfrac{\frac{5}{9}}{\frac{3}{5}}$

2. $\dfrac{\frac{7}{8}}{\frac{5}{6}}$

3. $\dfrac{\frac{x-5}{4}}{3x}$

4. $\dfrac{\frac{6}{x-2}}{9x}$

5. $\dfrac{4x}{\frac{8x}{x+3}}$

6. $\dfrac{7x}{\frac{21x}{x-3}}$

7. $\dfrac{\dfrac{x^2 - 36}{5x}}{\dfrac{x + 6}{15x}}$

8. $\dfrac{\dfrac{x^2 - 5x}{8}}{\dfrac{2x - 10}{12x}}$

9. $\dfrac{2 + \dfrac{1}{x}}{\dfrac{3}{5}}$

10. $\dfrac{3 + \dfrac{2}{x}}{\dfrac{5}{8}}$

11. $\dfrac{\dfrac{3}{x} + \dfrac{5}{3x}}{\dfrac{3}{6x} - \dfrac{2}{x}}$

12. $\dfrac{\dfrac{x}{6} - \dfrac{3x}{9}}{\dfrac{5x}{12} + \dfrac{3x}{4}}$

16-5 SOLVING EQUATIONS WITH RATIONAL EXPRESSIONS

Learning Objectives

L.O.1. Exclude certain values as solutions of rational equations.
L.O.2. Solve rational equations with variable denominators.

L.O.1. Exclude certain values as solutions of rational equations.

Finding Excluded Values

A rational equation is an equation that contains one or more rational expressions. Some examples of rational equations are

$$\frac{1}{2} + \frac{1}{x} = \frac{1}{3}, \qquad x - \frac{3}{x} = 4, \qquad \frac{x}{x - 2} = \frac{1}{x + 3}$$

We solved such equations in Chapter 10 by first clearing the equation of all fractions. We cleared the fractions by multiplying both sides of the equation by the *least common multiple (LCM)* of all the denominators in the equation. Then we solved the equation as any other equation.

However, when the variable is in the denominator of the equation, an important caution must be observed.

Tips and Traps

Since division by zero is impossible, any value of the variable that makes the value of any denominator zero is excluded as a possible solution of an equation. Therefore, it is important (1) to determine which values are *excluded values* and cannot be used as possible solutions, and (2) to check each possible solution.

EXAMPLE 16-5.1 Determine the value or values that must be excluded as possible solutions of the following.

(a) $\dfrac{1}{x} + \dfrac{1}{2} = 5$ **(b)** $\dfrac{1}{x} = \dfrac{1}{x - 3}$

(c) $\dfrac{1}{x + 2} + \dfrac{1}{x - 2} = \dfrac{1}{x^2 - 4}$ **(d)** $\dfrac{3x}{x + 2} = 3 - \dfrac{5}{2x}$

Solution

(a) $\frac{1}{x} + \frac{1}{2} = 5$ *(Set each denominator containing a variable equal to 0.)*

$x = 0$ *(excluded value)*

(b) $\frac{1}{x} = \frac{1}{x - 3}$ *(Set each denominator containing a variable equal to 0.)*

$x = 0, \quad x - 3 = 0$ *(Solve each equation.)*

$x = 3$

Excluded values are 0 and 3.

(c) $\frac{1}{x + 2} + \frac{1}{x - 2} = \frac{1}{x^2 - 4}$ *(Set each denominator containing a variable equal to 0.)*

$x + 2 = 0 \quad x - 2 = 0 \quad x^2 - 4 = 0$ *(Solve each equation.)*

$x = -2 \quad x = 2 \quad (x + 2)(x - 2) = 0$

$x + 2 = 0 \quad x - 2 = 0$

$x = -2 \quad x = 2$

Excluded values are -2 and 2.

(d) $\frac{3x}{x + 2} = 3 - \frac{5}{2x}$ *(Set each denominator containing a variable equal to 0.)*

$x + 2 = 0 \quad 2x = 0$ *(Solve each equation.)*

$x = -2 \quad x = 0$

Excluded values are -2 and 0. ■

L.O.2. Solve rational equations with variable denominators.

Solving Rational Equations

When solving rational equations with variable denominators, find any values that must be excluded as possible solutions by setting each variable denominator equal to zero and solving the resulting equations. Finding excluded values is another way to check whether roots are *extraneous roots* (Chapter 10).

Even though we have found the excluded values, it is a good idea to check the solution to make sure the root makes a true statement in the original equation.

Example 16-5.2 Solve the following rational equations. Check.

(a) $\frac{2}{y} + \frac{1}{3} = 1$ **(b)** $\frac{2}{y - 2} = \frac{4}{y + 1}$

(c) $\frac{5}{x - 2} - \frac{3x}{x - 2} = -\frac{1}{x - 2}$

Solution

(a) $\frac{2}{y} + \frac{1}{3} = 1$

Excluded value: $y = 0$.

$$3y\left(\frac{2}{y}\right) + 3y\left(\frac{1}{3}\right) = 3y(1) \qquad \textit{(Multiply by LCM, 3y.)}$$

$$(3\cancel{y})\left(\frac{2}{\cancel{y}}\right) + (\cancel{3}y)\left(\frac{1}{\cancel{3}}\right) = 3y(1) \qquad \textit{(Cancel.)}$$

$$6 + y = 3y$$
$$6 = 3y - y$$
$$6 = 2y$$
$$\frac{6}{2} = \frac{2y}{2}$$
$$3 = y \qquad \textit{(Check, since 3 is not an excluded value.)}$$

Check: $\frac{2}{y} + \frac{1}{3} = 1$

$$\frac{2}{3} + \frac{1}{3} = 1 \qquad \textit{(Substitute 3 for y.)}$$
$$\frac{3}{3} = 1$$
$$1 = 1 \qquad \textit{(Solution checks.)}$$

(b) $\frac{2}{y - 2} = \frac{4}{y + 1}$

Excluded values: $y - 2 = 0 \qquad y + 1 = 0$

$y = 2 \qquad y = -1$

$$(y - 2)(y + 1)\frac{2}{y - 2} = (y - 2)(y + 1)\frac{4}{y + 1} \qquad \textit{[Multiply by LCM, (y - 2)(y + 1).]}$$

$$(\cancel{y - 2})(y + 1)\frac{2}{\cancel{y - 2}} = (y - 2)(\cancel{y + 1})\frac{4}{\cancel{y + 1}} \qquad \textit{(Cancel.)}$$

$$2(y + 1) = 4(y - 2)$$
$$2y + 2 = 4y - 8$$
$$2y - 4y = -8 - 2$$
$$-2y = -10$$
$$\frac{-2y}{-2} = \frac{-10}{-2}$$
$$y = 5 \qquad \textit{(Check, since 5 is not an excluded value.)}$$

Check: $\dfrac{2}{y - 2} = \dfrac{4}{y + 1}$

$\dfrac{2}{5 - 2} = \dfrac{4}{5 + 1}$ *(Substitute 5 for y.)*

$\dfrac{2}{3} = \dfrac{4}{6}$ *(Reduce $\frac{4}{6}$.)*

$\dfrac{2}{3} = \dfrac{2}{3}$ *(Solution checks.)*

(c) $\dfrac{5}{x - 2} - \dfrac{3x}{x - 2} = -\dfrac{1}{x - 2}$

Excluded value: $x - 2 = 0$

$x = 2$

$$(x - 2)\frac{5}{x - 2} - (x - 2)\frac{3x}{x - 2} = (x - 2)\left(-\frac{1}{(x - 2)}\right)$$ *(Multiply by LCM, x − 2, to clear the common denominator.)*

$$\cancel{(x-2)}\frac{5}{\cancel{x-2}} - \cancel{(x-2)}\frac{3x}{\cancel{x-2}} = \cancel{(x-2)}\left(-\frac{1}{\cancel{(x-2)}}\right)$$ *(Cancel.)*

$$5 - 3x = -1$$
$$-3x = -1 - 5$$
$$-3x = -6$$
$$\frac{-3x}{-3} = \frac{-6}{-3}$$
$$x = 2$$ *(This is an excluded value. The root will not check.)*

Check: $\dfrac{5}{x - 2} - \dfrac{3x}{x - 2} = -\dfrac{1}{x - 2}$

$\dfrac{5}{2 - 2} - \dfrac{3x}{2 - 2} = -\dfrac{1}{2 - 2}$ *(Substitute 2 for x.)*

$\dfrac{5}{0} - \dfrac{3(2)}{0} = \dfrac{1}{0}$ *(Division by zero is impossible. The equation has no solution.)* ■

A number of application problems can be solved by using rational equations. Let's look at an example.

EXAMPLE 16-5.3 Sally Lee purchased a package of 3.5-in. DSDD computer disks for \$16 and a smaller package of 3.5-in. HD computer disks for \$6. She paid 40 cents more per HD disk than she paid for each DSDD disk, and the DSDD package contained 15 more disks than the HD package. How many of each type of disk did she purchase? What was the price of each type of disk?

Solution: Let d = the number of HD disks. Let $d + 15$ = the number of DSDD disks. Cost ÷ number of disks = cost per disk. Cost per HD disk $\left(\dfrac{6}{d}\right)$ = cost per DSDD disk $\left(\frac{16}{d + 5}\right)$ plus 40 cents. Using these relationships, set up a rational equation.

$$\frac{6}{d} = \frac{16}{d + 15} + \frac{40}{100} \qquad \left(\textit{40 cents} = \frac{40}{100}\right)$$

$$\frac{6}{d} = \frac{16}{d + 15} + \frac{2}{5} \qquad \left(\frac{40}{100} = \frac{2}{5}\right)$$

$$(d)(d + 15)(5)\,\frac{6}{d} = (d)(d + 15)(5)\,\frac{16}{d + 15} + (d)(d + 15)(5)\,\frac{2}{5} \qquad \textit{[Multiply by LCM, (d)(d + 15)(5).]}$$

$$(\cancel{d})(d + 15)(5)\,\frac{6}{\cancel{d}} = (d)\cancel{(d + 15)}(5)\frac{16}{\cancel{d + 15}} + (d)(d + 15)(\cancel{5})\,\frac{2}{\cancel{5}} \qquad \textit{(Cancel.)}$$

$$\begin{aligned}
(d + 15)(5)(6) &= (d)(5)(16) + (d)(d + 15)(2) \\
30(d + 15) &= 80d + 2d(d + 15) && \textit{(Distribute.)} \\
30d + 450 &= 80d + 2d^2 + 30d \\
0 &= 80d + 2d^2 + 30d - 30d - 450 \\
0 &= 2d^2 + 80d - 450 \\
0 &= 2(d^2 + 40d - 225) \\
0 &= 2(d - 5)(d + 45)
\end{aligned}$$

$$d - 5 = 0 \qquad\qquad d + 45 = 0$$

$$d = 5 \text{ HD disks} \qquad\qquad d = -45 \qquad\qquad \textit{(Disregard negative value.)}$$

Thus,

$$\begin{aligned}
d + 15 = 5 + 15 &= 20 \text{ DSDD disks} \\
\$6 \div 5 &= \$1.20 \text{ per HD disk} \\
\$16 \div 20 &= \$0.80 \text{ per DSDD disk}
\end{aligned}$$

Check:

$$\frac{6}{d} = \frac{16}{d + 15} + \frac{2}{5}$$

$$\frac{6}{5} = \frac{16}{5 + 15} + \frac{2}{5} \qquad \textit{(Substitute 5 for d.)}$$

$$\frac{6}{5} = \frac{16}{20} + \frac{2}{5}$$

$$\frac{6}{5} = \frac{16}{20} + \frac{8}{20}$$

$$\frac{6}{5} = \frac{24}{20} \qquad \left(\frac{24}{20} = \frac{6}{5}\right)$$

$$\frac{6}{5} = \frac{6}{5} \qquad \textit{(The solution checks.)}$$

■

SELF-STUDY EXERCISES 16-5

L.O.1. Determine the value or values that must be excluded as possible solutions of the following.

1. $\frac{5}{x} + \frac{3}{5} = 7$

2. $\frac{6}{x} + 5 = \frac{5}{12}$

3. $\frac{9}{x - 5} = \frac{7}{x}$

4. $\frac{15}{2x} = \frac{6}{x + 9}$

5. $\frac{7}{x + 8} + \frac{2}{x - 8} = \frac{5}{x^2 - 64}$

6. $\frac{9}{x - 2} + \frac{5}{x + 2} = \frac{1}{x^2 - 4}$

7. $\frac{2x}{4x - 3} - 4 = \frac{5}{6x}$

8. $\frac{1}{9x} + \frac{8x}{5x + 15} = 11$

L.O.2. Solve the following equations. Check for extraneous roots.

9. $\frac{6}{7} + \frac{5}{x} = 1$

10. $\frac{4}{x} - \frac{3}{4} = \frac{1}{20}$

11. $\frac{2}{x} + \frac{5}{2x} - \frac{1}{2} = 4$

12. $\frac{5}{2x} - \frac{1}{x} = 6$

13. $\frac{2}{x - 3} + \frac{7}{x + 3} = \frac{21}{x^2 - 9}$

14. $\frac{1}{x^2 - 25} = \frac{3}{x - 5} + \frac{2}{x + 5}$

Use rational equations to solve the following problems. Check for extraneous roots.

15. Shawna can complete an electronic lab project in 3 days. Tom can complete the same project in 4 days. How many days will it take to complete the project if Shawna and Tom work together? (*Hint:* Rate × time = amount of work.)

16. A group of investors purchases land in Maine for $12,000. When four more investors join the group, the cost dropped $500 per person in the original group. How many investors were in the original group?

17. Gomez commutes 50 miles on a motorbike to a technical college, but on the return trip he travels 15 miles per hour faster and makes the trip in $\frac{3}{4}$ hour less time. Find Gomez's speed to and from school. (*Hint:* Time = distance ÷ rate.)

18. Pipe 1 can fill a tank in 2 minutes and pipe 2 can empty the tank in 6 minutes. How long would it take to fill the tank if both pipes were working? (*Hint:* Rate × time = amount of work.)

19. Kim bought $150 worth of medium-roast coffee and at the same cost per pound, $100 worth of dark-roast coffee. If she bought 25 pounds more of the medium-roast coffee than the dark-roast coffee, how many pounds of each did she buy and what was the cost per pound?

20. Find resistance$_2$ if resistance$_1$ is 10 ohms and resistance$_t$ is 3.75 ohms, using the formula $R_t = \frac{R_1R_2}{R_1 + R_2}$.

ASSIGNMENT EXERCISES

16-1

Simplify the following fractions.

1. $\frac{18}{24}$

2. $\frac{24}{42}$

3. $\frac{5a^2b^3c}{10a^3bc^2}$

4. $\dfrac{3x^2y}{9x^3y^3}$

5. $\dfrac{4xy(x-3)}{8xy(x+3)}$

6. $\dfrac{(x+7)(x-3)}{(x-3)(x-7)}$

7. $\dfrac{(x-4)(x+2)}{(x+2)(x-4)}$

8. $\dfrac{x+1}{3x+3}$

9. $\dfrac{m^2-n^2}{m^2+n^2}$

10. $\dfrac{3x^2+8x-3}{2x^2+5x-3}$

11. $\dfrac{x}{x+xy}$

12. $\dfrac{2x}{4x+6}$

13. $\dfrac{5x+15}{x+3}$

14. $\dfrac{x^3+2x^2-3x}{3x}$

15. $\dfrac{y^2+2y+1}{y+1}$

16. $\dfrac{y-1}{y^2-2y+1}$

17. $\dfrac{2x-6}{x^2+3x-18}$

18. $\dfrac{x^2-4x+4}{x^2-2x}$

19. $\dfrac{3x-9}{x-3}$

20. $\dfrac{4x-12}{x-3}$

16-2

Multiply or divide the following fractions. Reduce to simplest terms.

21. $\dfrac{3x^2}{2y}\cdot\dfrac{5x}{6y}$

22. $\dfrac{x^2-y^2}{6}\cdot\dfrac{18}{x-y}$

23. $\dfrac{9}{x+b}\cdot\dfrac{5x+5b}{3}$

24. $\dfrac{81-x^2}{16-f^2}\cdot\dfrac{4-f}{9+x}$

25. $\dfrac{4y^2-4y+1}{6y-6}\cdot\dfrac{24}{2y-1}$

26. $\dfrac{x-3}{x+5}\cdot\dfrac{2x^2+10x}{2x-6}$

27. $\dfrac{5-x}{x-5}\cdot\dfrac{x-1}{1-x}$

28. $\dfrac{3-x}{x-2}\cdot\dfrac{2x-4}{x-3}$

29. $\dfrac{x^2+6x+9}{x^2-4}\cdot\dfrac{x-2}{x+3}$

30. $\dfrac{x^2}{x^2-9}\cdot\dfrac{x^2-5x+6}{x^2-2x}$

31. $\dfrac{2a+b}{8}\div\dfrac{2a+b}{2}$

32. $\dfrac{17a^2}{21y^2}\div\dfrac{34a^2}{68}$

33. $\dfrac{y^2-2y+1}{y}\div\dfrac{1}{y-1}$

34. $\dfrac{x^2y^2}{3x^2-3y^2}\div\dfrac{8xy}{x-y}$

35. $\dfrac{y^2+6y+9}{y^2+4y+4}\div\dfrac{y+3}{y+2}$

36. $\dfrac{2x+2y}{3}\div\dfrac{x^2-y^2}{x-y}$

37. $\dfrac{3x^2+6x}{x}\div\dfrac{2x+4}{x^2}$

38. $\dfrac{x^2-7x}{x^2-3x-28}\div\dfrac{1}{-x-4}$

39. $\dfrac{y^2-16}{y+3}\div\dfrac{y-4}{y^2-9}$

40. $\dfrac{12x+24}{36x-36}\div\dfrac{6x+12}{8x-8}$

16-3

Add or subtract the following fractions. Reduce to simplest terms.

41. $\dfrac{2}{9}+\dfrac{4}{9}$

42. $\dfrac{2}{3}+\dfrac{5}{12}$

43. $\dfrac{3x}{7}+\dfrac{2x}{14}$

44. $\dfrac{5x}{3}-\dfrac{2x}{4}$

45. $\dfrac{3x}{4}+\dfrac{5x}{6}$

46. $\dfrac{3}{4}+\dfrac{7}{x}$

47. $\dfrac{5}{x}-\dfrac{7}{3}$

48. $\dfrac{3}{x}+\dfrac{5}{7}$

49. $\dfrac{3}{4x}+\dfrac{2}{x}+\dfrac{3}{6x}$

50. $\dfrac{6}{2x+3}+\dfrac{2}{2x-3}$

51. $\dfrac{7}{x-3}+\dfrac{3}{x+2}$

52. $\dfrac{4}{3x+2}-\dfrac{3}{x-2}$

53. $\dfrac{8}{x+3} - \dfrac{2}{x-4}$

54. $\dfrac{3}{x-2} + \dfrac{5}{2-x}$

55. $\dfrac{x}{x-5} - \dfrac{3}{5-x}$

16-4

Simplify.

56. $\dfrac{\frac{2}{7}}{\frac{3}{4}}$

57. $\dfrac{\frac{5}{x-3}}{4}$

58. $\dfrac{6ab}{\frac{3a}{4}}$

59. $\dfrac{\frac{x^2-4x}{6x}}{\frac{x-4}{8x^2}}$

60. $\dfrac{2+\frac{3}{x}}{\frac{2}{x}}$

61. $\dfrac{\frac{5}{x}-\frac{3}{4x}}{\frac{1}{3x}+\frac{2}{x}}$

62. $\dfrac{5-\frac{x-2}{x}}{\frac{x-4}{2x}-2}$

63. $\dfrac{\frac{3x}{6}-\frac{5}{x}}{\frac{x}{3}+\frac{4}{2x}}$

16-5

Determine the value or values that must be excluded as possible solutions of the following.

64. $\dfrac{3}{x} - \dfrac{4}{5} = 2$

65. $\dfrac{4}{x} = \dfrac{3}{x-2}$

66. $\dfrac{3}{x-5} - \dfrac{4}{x+5} = \dfrac{1}{x^2-25}$

67. $\dfrac{5x}{2x-1} - 6 = \dfrac{4}{3x}$

Solve the following equations. Check for extraneous roots.

68. $\dfrac{3}{x} + \dfrac{2}{3} = 1$

69. $\dfrac{4}{x} = \dfrac{1}{x+5}$

70. $\dfrac{3}{x-4} + \dfrac{1}{x+4} = \dfrac{1}{x^2-16}$

71. $-\dfrac{4x}{x+1} = 3 - \dfrac{4}{x+1}$

Use rational equations to solve the following problems. Check for extraneous roots.

72. Fugita can complete the assembly of 50 widgets in 6 days. Ohn can complete the same job in 8 days. How many days will it take to complete the project if Fugita and Ohn work together? (*Hint:* Rate × time = amount of work.)

73. Several students chipped in a total of $120 to buy a small refrigerator to use in the dorm. Later, another student joined the group, and the cost to the original group dropped $10 per person. How many students were in the original group?

CHALLENGE PROBLEMS

74. Pipe 1 can fill a tank in 4 minutes and pipe 2 can fill the same tank in 3 minutes. What proportion of the tank will be filled if both pipes were working for 1 minute? (*Hint:* Rate × time = amount of work.)

75. One machine can do a job in 5 hours alone. How many hours would it take a second machine to complete the job alone if both machines together can do the job in 3 hours?

CONCEPTS ANALYSIS

1. How are the properties $\frac{n}{n} = 1$ and $1 \times n = n$ applied in the following example: $\frac{4}{6} = \frac{2}{3}$?

2. Why is the following problem incorrect?

$$\frac{5}{10} = \frac{\cancel{2} + 3}{\cancel{2} + 8} = \frac{3}{8}$$

3. Write a brief comment for each step of the following example.

$$\frac{x^2 + 3x + 2}{x^2 - 9} \div \frac{2x^2 + 3x - 2}{2x^2 - 7x + 3} =$$

$$\frac{x^2 + 3x + 2}{x^2 - 9} \cdot \frac{2x^2 - 7x + 3}{2x^2 + 3x - 2} =$$

$$\frac{(x + 1)\cancel{(x + 2)}}{(x + 3)\cancel{(x - 3)}} \cdot \frac{\cancel{(2x - 1)}\cancel{(x - 3)}}{\cancel{(2x - 1)}\cancel{(x + 2)}} = \frac{x + 1}{x + 3}$$

Find the mistake in each of the following and briefly explain the mistake. Then rework the problem correctly.

4. $\frac{x}{x + 3} = \frac{1}{3}$

5. $\frac{x^2 - 4}{x^2 + 4x + 4} \div x + 2 = \frac{\cancel{(x + 2)}(x - 2)}{\cancel{(x + 2)}\cancel{(x + 2)}} \cdot \frac{\cancel{x + 2}}{1} = x - 2$

6. $\frac{5}{x + 2} + \frac{3}{x - 3} = \frac{8}{(x + 2)(x - 3)}$

7. $\frac{3}{x - 1} - \frac{5}{x + 1} =$

$$\frac{3(x + 1)}{(x - 1)(x + 1)} - \frac{5(x - 1)}{(x - 1)(x + 1)} =$$

$$\frac{3x + 3 - 5x - 5}{(x + 1)(x - 1)} =$$

$$\frac{-2x - 2}{(x + 1)(x - 1)} =$$

$$\frac{-2(x + 1)}{(x + 1)(x - 1)} = \frac{-2}{x - 1}$$

8. $\frac{3x + 2}{x - 5} = \frac{x - 4}{3x + 2}$

$$\frac{\cancel{3x + 2}}{x - 5} = \frac{x - 4}{\cancel{3x + 2}}$$

$$\frac{1}{x - 5} = \frac{x - 4}{1}$$

$$\frac{x - 4}{x - 5}$$

9. How is a division of rational expressions related to a multiplication of rational expressions?

10. In your own words, write a rule for multiplying rational expressions.

CHAPTER SUMMARY

Objectives	What to Remember	Examples
Section 16-1 **1.** Simplify, or reduce, rational expressions.	If the rational expression is not in factored form, factor completely the numerator and the denominator. Then reduce factors common to both numerator and denominator.	Simplify $\frac{a + b}{16a^2 + 16ab}$. $\frac{a + b}{16a(a + b)}$ (factored form) $\frac{\cancel{a + b}}{16a(\cancel{a + b})} = \frac{1}{16a}$
Section 16-2 **1.** Multiply and divide rational expressions.	Write the numerator and denominator in factored form whenever possible. For multiplication, reduce factors common to both numerator and denominator.	Multiply $\frac{x}{2x + 8} \cdot \frac{2}{x + 1}$. $\frac{x}{2(x + 4)} \cdot \frac{2}{x + 1} =$ (factored form) $= \frac{(\cancel{2})(x)}{\cancel{2}(x + 4)(x + 1)}$ (Multiply and cancel.) $= \frac{x}{(x + 4)(x + 1)}$ or $\frac{x}{x^2 + 5x + 4}$
	For division, multiply by reciprocal of second factor; then reduce factors common to both numerator and denominator.	Divide $\frac{x}{x + 2} \div \frac{2}{x + 2}$. $\frac{x}{\cancel{x + 2}} \cdot \frac{\cancel{x + 2}}{2} = \frac{x}{2}$ (Multiply by reciprocal. Cancel.)
Section 16-3 **1.** Add and subtract rational expressions.	Add (or subtract) only rational expressions with the same denominator. For rational expressions with unlike denominators: **1.** Find the LCD. **2.** Change each fraction to an equivalent one having the LCD. **3.** Add (or subtract) the numerators.	Perform the following operations: (a) $\frac{x}{x + 4} + \frac{2x}{x + 4}$ $\frac{x + 2x}{x + 4}$ (Add numerators.) $= \frac{3x}{x + 4}$

4. Keep same LCD.
5. Reduce if needed.

(b) $\dfrac{3}{x+2} - \dfrac{2}{x+1}$

[LCD $= (x+2)(x+1)$]

$$\frac{3}{x+2} = \frac{3(x+1)}{(x+2)(x+1)} = \frac{3x+3}{(x+2)(x+1)}$$

$$\frac{2}{x+1} = \frac{2(x+2)}{(x+1)(x+2)} = \frac{2x+4}{(x+1)(x+2)}$$

$$\frac{3x+3-(2x+4)}{(x+1)(x+2)} = \frac{3x+3-2x-4}{(x+1)(x+2)} = \frac{x-1}{(x+1)(x+2)}$$

Section 16-4

1. Simplify complex rational expressions.

To simplify complex rational expressions, add or subtract any rational expressions in the numerator and/or denominator. Then rewrite the expression as the division of two rational expressions. Next, rewrite the division as a multiplication of the numerator by the reciprocal of the denominator. Factor and perform the indicated multiplication.

Simplify.

$$\frac{\frac{x-3}{5x}}{\frac{4x-12}{15x^2}}$$

$$\frac{x-3}{5x} \div \frac{4x-12}{15x^2}$$

$$\frac{\cancel{x-3}}{\cancel{5x}} \cdot \frac{\overset{3x}{\cancel{15x^2}}}{4(\cancel{x-3})}$$

$$\frac{3x}{4}$$

Section 16-5

1. Exclude certain values as solutions of rational equations.

To find the values that make a denominator with a variable equal to zero, set each denominator with a variable equal to zero. Then solve the resulting equations. The excluded values obtained cannot be solutions of the rational equation.

Find the excluded values for the following:

(a) $4 = \dfrac{2}{3+y}$

$3 + y = 0$

$y = -3$ (excluded value)

(b) $\dfrac{2x}{x^2 - x - 6} = 8$

$x^2 - x - 6 = 0$

$(x+2)(x-3) = 0$

$x + 2 = 0 \qquad x - 3 = 0$

$x = -2 \qquad x = 3$

(-2, 3 excluded values)

2. Solve rational equations with variable denominators.	To solve rational equations with variable denominators, multiply each term of the equation by the LCM of all denominators in the equation. Solve the resulting equation. Check all solutions to determine which are true roots and which are extraneous roots.	Solve $\frac{2}{x-3} - \frac{1}{x+3} = \frac{1}{x^2-9}$. LCM is $x^2 - 9$ or $(x-3)(x+3)$. Excluded values are 3 and -3. $2(x+3) - 1(x-3) = 1$; $2x + 6 - x + 3 = 1$; $x + 9 = 1$; $x = -8$. Check: $\frac{2}{-8-3} - \frac{1}{-8+3} = \frac{1}{(-8)^2-9}$; $\frac{2}{-11} - \frac{1}{-5} = \frac{1}{64-9}$; $\frac{-2}{11} + \frac{1}{5} = \frac{1}{55}$; $\frac{-10}{55} + \frac{11}{55} = \frac{1}{55}$; $\frac{1}{55} = \frac{1}{55}$

WORDS TO KNOW

simplify (p. 611)
reduce (p. 611)
rational expression (p. 612)
polynomial (p. 612)
common factors (p. 612)
exponents (p. 612)
addends (p. 612)
terms (p. 612)
numerator (p. 612)
denominator (p. 612)
like fractions (p. 617)
LCD (p. 618)
complex rational expression (p. 620)
excluded values (p. 624)
extraneous roots (p. 625)

CHAPTER TRIAL TEST

Simplify the following expressions.

1. $\frac{x-3}{2x-6}$

2. $\frac{x^2-16}{x-4}$

3. $\frac{6x^2-11x+4}{2x^2+5x-3}$

4. $\frac{x^2-6x+8}{x-2}$

5. $\frac{(x-2)(x+4)}{(x+4)(x+2)}$

6. $\frac{y^2+x^2}{y^2-x^2}$

7. $\frac{6xy}{ab} \cdot \frac{a^2b}{2xy^2}$

8. $\frac{x^2-y^2}{2x+y} \cdot \frac{4x^2+2xy}{x+y}$

9. $\frac{x-2y}{x^3-3x^2y} \div \frac{x^2-4y^2}{x-3y}$

10. $\frac{2a^2-ab-b^2}{6x^2+x-1} \div \frac{a^2-b^2}{8x+4}$

11. $\frac{2x^2+3x+1}{x} \div \frac{x+1}{1}$

12. $\frac{4y^2}{2x} \cdot \frac{x}{8x}$

13. $\frac{1}{x+2} - \frac{1}{x-3}$

14. $\frac{2}{x+2} + \frac{3}{x-1}$

15. $\frac{3}{x} + \frac{1}{4}$

16. $\frac{2}{3y} - \frac{7}{y}$

17. $\frac{5}{3x - 2} + \frac{7}{x + 2}$

18. $\frac{2x}{3} - \frac{5x}{2}$

19. $\frac{x - 2y}{x^2 - 4y^2}$

20. $\frac{3}{2y} + \frac{2}{y} + \frac{1}{5y}$

21. $\frac{2x}{1 - \frac{3}{x}}$

22. $\frac{\frac{1}{a} + \frac{1}{b}}{ab}$

Determine the excluded values of x.

23. $\frac{5}{x} = \frac{2}{x + 3}$

24. $\frac{3}{x - 4} + \frac{5}{x + 4} = 6$

Solve the following rational equations. Check for extraneous roots.

25. $\frac{3x}{x - 2} + 4 = \frac{3}{x - 2}$

26. $\frac{x}{x + 3} = 5$

27. $\frac{x}{y - 2} = \frac{-4}{y + 1}$

Use rational equations to solve the following problems. Check for extraneous roots.

28. Henry can assemble four computers in 3 hours. Lester can assemble four computers in 4 hours. How long will it take to assemble four computers if both work together? (*Hint:* Rate $\times$ time $=$ amount of work.)

29. A medical group purchases land in Colorado for \$100,000. When five more persons joined the group, the cost dropped \$10,000 per person in the original group. How many persons were in the original group?

30. Cedric Partee drove 300 miles in one day while on a vacation to the mountains, but on the return trip by the same route he drove 10 miles per hour less and the return trip took 1 hour more time. Find Mr. Partee's speed to and from the mountains. (*Hint:* Time $=$ distance $\div$ rate.)

CHAPTER

17 Inequalities and Absolute Values

After a basic understanding of signed numbers, solving equations, and the laws of exponents, we will apply these concepts to another type of mathematical statement, inequalities. In this chapter we will examine inequalities while reviewing the basic procedures for solving equations.

17-1 INEQUALITIES AND SETS

Learning Objectives

L.O.1. Use set terminology.
L.O.2. Illustrate sets of numbers that are inequalities on a number line and write inequalities in interval notation.

In many mathematical applications there will be a need to deal with statements of inequality. An *inequality* is a mathematical statement that shows quantities that *are not equal*. The symbol $\neq$ is read ''is not equal to.''

$5 \neq 7$ Five *is not equal to* seven.

In most cases, to show that quantities are not equal to each other does not give us enough information. We may need to know which quantity is larger (or smaller). When we want specific information, such as 5 *is less than* 7, or 7 *is greater than* 5, we use specific inequality symbols, $<$ (is less than) and $>$ (is greater than).

$5 < 7$ Five *is less than* seven.

$7 > 5$ Seven *is greater than* five.

To help distinguish the symbols, we can think of them as arrowheads. The arrowhead *points* toward the smaller quantity. The wide opening *opens* toward the larger quantity.

As in equations, inequalities can have missing amounts that are represented by letters. To *solve* an inequality is to find the value or set of values of the unknown quantity that makes the statement true.

L.O.1. Use set terminology.

Sets

Before solving inequalities, we will introduce some terminology and notations used with sets of numbers. A *set* is a group or collection of items. For example, a set of days of the week that begin with the letter T would include Tuesday and Thursday. However, in this chapter we will examine sets of numbers. The items or numbers that belong to a set are called *members* or *elements of a set.* The description of a set should clearly distinguish between the elements that belong to the set and those that do not belong. This description can be given in words or by using various types of *set notation.* To illustrate the various types of set notation, we will examine the set of whole numbers between 1 and 8. One notation is to make a list or *roster* of the elements of a set. These elements are enclosed in braces and separated with commas.

$$\text{Set of whole numbers between 1 and 8} = \{2, 3, 4, 5, 6, 7\}$$

Another method for illustrating a set is with *set-builder notation*. The elements of the set are written in the form of an inequality using a variable to represent all the elements of the set. The types of numbers that are to be included in a set will be denoted by the following guide.

The following capital letters are used to denote the indicated set of numbers:

N = natural numbers	W = whole numbers
Z = integers	Q = rational numbers
I = irrational numbers	R = real numbers
M = imaginary numbers	C = complex numbers

Symbols that substitute for phrases and words that are often used in describing sets are:

$|$ is read "such that."

$\in$ is read " is an element of."

Now, we can "build" the set of whole numbers between 1 and 8 with the following symbolic statement.

$$\text{Set of whole numbers between 1 and 8} = \{x|x \in W \text{ and } 1 < x < 8\}$$

This statement is read "the set of values of x such that x is an element of the set of whole numbers and x is between 1 and 8."

A special set is the empty set. The *empty set* is a set containing no elements. Symbolically, the empty set is identified as { } or ϕ. An example of an empty set would be the set of whole numbers between 1 and 2. The set of rational numbers between 1 and 2 would include numbers like $1\frac{1}{2}$ and 1.3, but there are no whole numbers between 1 and 2. Thus the set of whole numbers between 1 and 2 is the empty set.

EXAMPLE 17-1.1 Answer the following statements as true or false.

(a) 5 is an element of the set of whole numbers.

(b) $\frac{3}{4}$ is an element of the set of Z.

(c) $-8 \in$ the set of real numbers with the property $\{x|x < -5\}$.

(d) 3.7 is an element of the set of rational numbers with the property $\{x|x > 3.7\}$.
(e) The set of prime numbers that are evenly divisible by 2 is the empty set.

Solution

(a) 5 is an element of the set of whole numbers. True.
(b) $\frac{3}{4}$ is an element of the set of integers. False. Integers include only whole numbers and their opposites.
(c) -8 is an element of the set of real numbers with the property $\{x|x < -5\}$. True. -8 is a real number and it is less than -5. Both conditions are satisfied.
(d) 3.7 is an element of the set of rational numbers with the property $\{x|x > 3.7\}$. False. 3.7 is a rational number. A number can equal itself, but a number cannot be greater than itself.
(e) The set of prime numbers that are evenly divisible by 2 is the empty set. False. 2 is a prime number, and 2 is divisible by 2. Therefore, the set contains the element 2. ■

L.O.2. Illustrate sets of numbers that are inequalities on a number line and write inequalities in interval notation.

Inequalities and Interval Notation

Another type of notation used to represent inequalities is *interval notation*. The two *boundaries* are separated by a comma and enclosed with a symbol that indicates whether the boundary is included or not. A parenthesis indicates that the boundary is not included, and a bracket indicates that a boundary is included. If there is no boundary in one or both directions, an infinity symbol, ∞, is used.

EXAMPLE 17-1.2 Represent the following sets of numbers on the number line and by using interval notation.

(a) $1 < x < 8$ **(b)** $1 \leq x \leq 8$ **(c)** $x < 3$
(d) $x \geq 3$ **(e)** all real numbers

Solution

(a) (b)

(a) $1 < x < 8$ is represented by $(1, 8)$. x is greater than 1, so 1 is not included. x is also less than 8, so 8 is not included. This can also be described as the values of x that are between 1 and 8.
(b) $1 \leq x \leq 8$ is represented by $[1, 8]$. x is greater than or equal to 1, so 1 is included. x is also less than or equal to 8, so 8 is included. This can also be described as values between 1 and 8, inclusive.

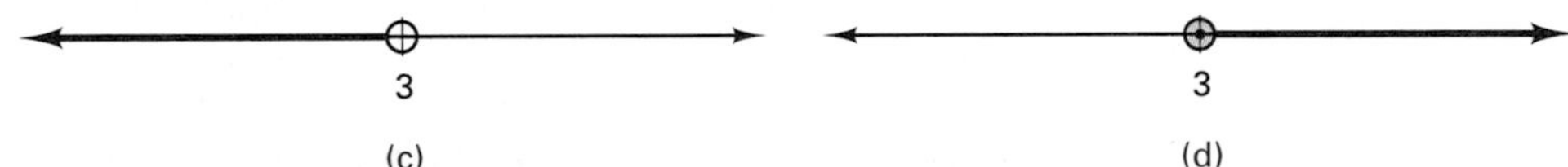

(c) (d)

(c) $x < 3$ is represented by $(-\infty, 3)$. x is less than 3, so 3 is not included. The values in the negative direction beyond 3 have no boundary.
(d) $x \geq 3$ is represented by $[3, +\infty)$ or $(3, \infty)$. x is greater than or equal to 3, so 3 is included. In the positive direction beyond 3 there is no boundary.

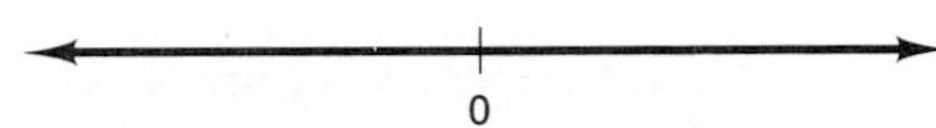

(e) All real numbers are represented by $(-\infty, +\infty)$ or $(-\infty, \infty)$. Real numbers extend in both directions without boundaries. ■

SELF-STUDY EXERCISES 17-1

L.O.1. Answer the following statements as true or false.

1. 0 is an element of the natural numbers.

2. $\frac{5}{8} \in N$.

3. $-6 \in Z$.

4. 5.3 is an element of the set of rational numbers with the property $\{x | x \leq 5.3\}$.

L.O.2. Represent the following sets on the number line and by using interval notation.

5. $8 \leq x \leq 12$

6. $x > 5$

7. $x \leq -2$

8. All real numbers greater than -6.

9. University Trailer Company had sales of $843,000 for the previous year. The projected sales for the current year are more than the previous year, but less than $1,000,000, which is projected for the next year. Express sales for the current year using interval notation.

17-2 SOLVING SIMPLE LINEAR INEQUALITIES

Learning Objective

L.O.1. Solve a simple linear inequality.

L.O.1. Solve a simple linear inequality.

Solving Simple Linear Inequalities

The procedures for solving simple inequalities are similar to the procedures for solving equations.

EXAMPLE 17-2.1 Find the set of numbers that makes the statement $x < 5 + 4$ true.

Solution

$x < 5 + 4$ *(Combine number terms.)*

$x < 9$ *(solution set)* ■

This means that the missing value, x, can be any amount that *is less than* 9 (8, 4, 8.75, 3, -6, and so on). The solution to an inequality is a *set* of numbers that satisfies the conditions of the statement. The number 9 in Example 17-2.1 represents the upper limit or boundary. In this case the *solution set* includes all numbers less than *but not equal to* 9.

On a number line the solution set is as shown in Fig. 17-1. The darkened solution set does not include the number 9. In interval notation the solution can be written as $(-\infty, 9)$.

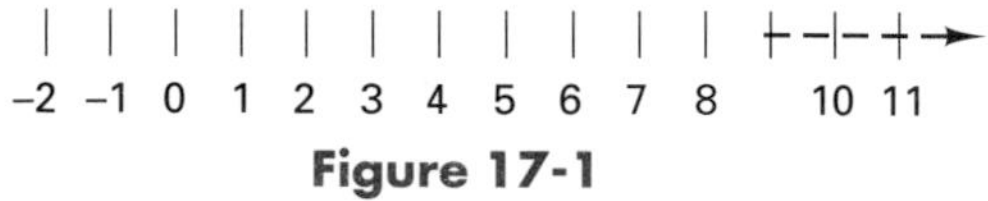

Figure 17-1

Another pair of symbols used to show inequalities is $\leq$ and $\geq$. The symbol $\leq$ is read "*is less than or equal to.*" The symbol $\geq$ is read "*is greater than or equal to.*" The statement $x \leq 9$ means that the solution set is 9 or any number less than 9.

Compare Figs. 17-1 and 17-2. In Fig. 17-1, nine *is not* part of the solution set; thus nine is represented on the number line with an open circle. In Fig. 17-2, nine *is* part of the solution set; thus nine is represented on the number line with a dot or darkened circle. The statement $x \leq 9$ is written in interval notation as $(-\infty, 9]$.

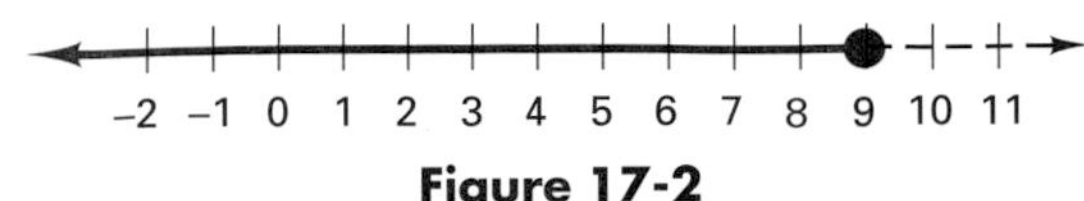

Figure 17-2

In the next several examples we illustrate the similarities between solving equations and solving inequalities. To understand the interpretation of the solution of an inequality, we will represent the solution set on the number line and write the solution in interval notation for several examples.

Example 17-2.2 Solve the following equation and inequality.

$$4x + 6 = 2x - 2 \qquad\qquad 4x + 6 > 2x - 2$$

Solution

$$4x + 6 = 2x - 2 \qquad\qquad 4x + 6 > 2x - 2$$

$$4x - 2x = -6 - 2 \qquad\qquad 4x - 2x > -6 - 2 \qquad \textit{(Transpose.)}$$

$$2x = -8 \qquad\qquad 2x > -8 \qquad \textit{(Combine.)}$$

$$\frac{2x}{2} = \frac{-8}{2} \qquad\qquad \frac{2x}{2} > \frac{-8}{2} \qquad \textit{(Divide.)}$$

$$x = -4 \qquad\qquad x > -4 \quad \text{or} \quad (-4, \infty)$$

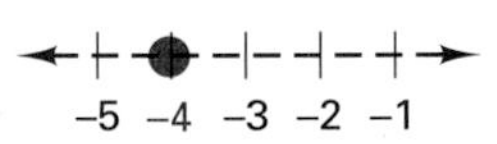

Figure 17-3

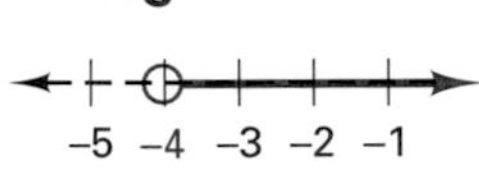

Figure 17-4

The solution for the equation is -4 (Fig. 17-3).

The solution set for the inequality is any number greater than, but not including, -4 (Fig. 17-4). ■

Example 17-2.3 Solve the following equation and inequality.

$$2(x - 3) = 12 \qquad\qquad 2(x - 3) \leq 12$$

Solution

$$2(x - 3) = 12 \qquad\qquad 2(x - 3) \leq 12$$

$$2x - 6 = 12 \qquad\qquad 2x - 6 \leq 12 \qquad \textit{(Distribute.)}$$

$$2x = 12 + 6 \qquad\qquad 2x \leq 12 + 6 \qquad \textit{(Transpose.)}$$

$$2x = 18 \qquad\qquad 2x \leq 18 \qquad \textit{(Combine.)}$$

$$\frac{2x}{2} = \frac{18}{2} \qquad\qquad \frac{2x}{2} \leq \frac{18}{2} \qquad \textit{(Divide.)}$$

$$x = 9 \qquad\qquad x \leq 9 \quad \text{or} \quad (-\infty, 9]$$

7 8 9 10

Figure 17-5

7 8 9 10

Figure 17-6

The solution for the equation is 9 (Fig. 17-5).

The solution set for the inequality is 9 *or* any number less than 9 (Fig. 17-6). ■

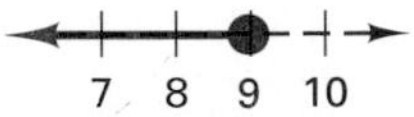

EXAMPLE 17-2.4 Solve the following equation and inequality.

$3x + 5 = 7x + 13 \qquad 3x + 5 < 7x + 13$

Solution

$3x + 5 = 7x + 13$	$3x + 5 < 7x + 13$	
$5 - 13 = 7x - 3x$	$5 - 13 < 7x - 3x$	*(Transpose.)*
$-8 = 4x$	$-8 < 4x$	*(Combine.)*
$\frac{-8}{4} = \frac{4x}{4}$	$\frac{-8}{4} < \frac{4x}{4}$	*(Divide.)*
$-2 = x$	$-2 < x$	
Recall, in equations we can rewrite $-2 = x$ as $x = -2$.	In equalities, to say -2 *is less than* x is the same as saying x *is greater than* -2.	*(We prefer writing the variable on the left.)*
$x = -2$	$x > -2$ or $(-2, \infty)$	■

Example 17-2.4 illustrates a very important difference in solving equations and solving inequalities. The sides of an equation can be interchanged without making any changes in the equal sign. This is due to the *symmetric* property of equality. That is, if $a = b$, then $b = a$.

The symmetric property *does not* apply to inequalities. If a *is less than* b, then b *is not less than* a. In fact, the opposite is true. If a *is less than* b, then b *is greater than* a. In symbols: if $a < b$, then $b > a$.

Rule 17-2.1 ***Interchanging the sides of an inequality:***

When the sides of an inequality are interchanged, the sense of the inequality is reversed.

If $a < b$, then $b > a$. If $a > b$, then $b < a$.

If $a \leq b$, then $b \geq a$. If $a \geq b$, then $b \leq a$.

The *sense of an inequality* is the appropriate comparison symbol: less than, greater than, less than or equal, and greater than or equal.

Let's now look at Example 17-2.4 again. This time we collect letter terms on the left in solving both the equation and the inequality.

EXAMPLE 17-2.5 Solve the following equation and inequality.

$3x + 5 = 7x + 13 \qquad 3x + 5 < 7x + 13$

Solution

$3x + 5 = 7x + 13$	$3x + 5 < 7x + 13$	
$3x - 7x = 13 - 5$	$3x - 7x < 13 - 5$	*(Transpose.)*
$-4x = 8$	$-4x < 8$	*(Combine.)*
$\frac{-4x}{-4} = \frac{8}{-4}$	$\frac{-4x}{-4} \; ? \; \frac{8}{-4}$	*(Divide.)*
$x = -2$	$x \; ? \; -2$	

Is the solution set for the inequality $x < -2$ or $x > -2$? To illustrate which solution set is appropriate, we will substitute a value that is less than -2 to see if the statement is true. Then we will substitute a value that is greater than -2 to see if the statement is true.

First, we try $x = -3$, which is less than -2.

$$3x + 5 < 7x + 13$$

$$3(-3) + 5 \overset{?}{<} 7(-3) + 13$$

$$-9 + 5 \overset{?}{<} -21 + 13$$

$$-4 \overset{?}{<} -8 \qquad \text{False}$$

This statement is false. -4 *is not less than* -8. Thus $x < -2$ is *not* the solution set.

Next, we try $x = -1$, which is greater than -2.

$$3x + 5 < 7x + 13$$

$$3(-1) + 5 \overset{?}{<} 7(-1) + 13$$

$$-3 + 5 \overset{?}{<} -7 + 13$$

$$2 \overset{?}{<} 6 \qquad \text{True}$$

This statement is true. 2 *is less than* 6. Thus $x > -2$ is the solution set. ■

In Example 17-2.5, we logically tested possible solutions to decide which relationship, $<$ or $>$, was appropriate. Another way to illustrate this property of inequalities is to look at specific numbers and the number line. We will start with a true statement and its representation on the number line.

$2 < 5$ 2 is to the left of 5.

Then, multiply each side of the inequality by the negative number -1.

$-1(2) < -1(5)$

$-2 < -5$ This is a false statement because -2 is to the right of -5.

To make the statement true, the sense of the inequality must be reversed.

$-2 > -5$

Rule 17-2.2 ***Multiplying or dividing an inequality by a negative number:***

If both sides of an inequality are multiplied or divided by a negative number, the sense of the inequality is reversed.

If $a < b$, then $-a > -b$. If $a > b$, then $-a < -b$.

If $a \leq b$, then $-a \geq -b$. If $a \geq b$, then $-a \leq -b$.

Inequalities are solved by performing similar procedures for solving equations except when the sides of the inequality are interchanged or when both sides of the inequality are multiplied or divided by a negative number.

> **Rule 17-2.3** *To solve inequalities:*
>
> **1.** Follow the same sequence of steps that would normally be used to solve a similar equation.
> **2.** The sense of the inequality remains the same unless the following situations occur:
> **a.** The sides of the inequality are interchanged.
> **b.** The steps used in solving the inequality require that the entire inequality (both sides) be multiplied or divided by a negative number.
> **3.** If either of the situations a or b in step 2 occurs in solving an inequality, *reverse* the sense of the inequality. That is, less than (<) becomes greater than (>), and vice versa.

EXAMPLE 17-2.6 Solve the following inequality.

$$4x - 2 \le 3(25 - x)$$

Solution

$4x - 2 \le 3(25 - x)$

$4x - 2 \le 75 - 3x$ *(Remove parentheses.)*

$4x + 3x \le 75 + 2$ *(Collect letter terms on left and number terms on right.)*

$7x \le 77$ *(Combine like terms.)*

$\frac{7x}{7} \le \frac{77}{7}$ *(Divide by coefficient of letter.)*

$x \le 11$ or $(-\infty, 11]$ *(solution set)* ■

EXAMPLE 17-2.7 Solve the following inequality.

$$2x - 3(x + 2) > 5x - (x - 5)$$

Solution

$2x - 3(x + 2) > 5x - (x - 5)$

$2x - 3x - 6 > 5x - x + 5$ *(Remove parentheses.)*

$-x - 6 > 4x + 5$ *(Combine like terms.)*

$-x - 4x > 5 + 6$ *(Transpose.)*

$-5x > 11$ *(Combine like terms.)*

$\frac{-5x}{-5} < \frac{11}{-5}$ *(Divide by coefficient of letter.)*

$x < -\frac{11}{5}$ or $\left(-\infty, -\frac{11}{5}\right)$ *(Since both sides have been divided by a negative number,* reverse sense *of inequality.)* ■

EXAMPLE 17-2.8 An electrically controlled thermostat is set so that the heating unit automatically comes on and continues to run when the temperature is equal to or below 72°F. At what Celsius temperatures will the heating unit come on? A formula relating Celsius and Fahrenheit temperatures is $°F = \frac{9}{5}°C + 32$.

Solution: Using the formula $°F = \frac{9}{5}°C + 32$, the heating unit will operate when the expression $\frac{9}{5}°C + 32$ is less than or equal to 72.

$$\frac{9}{5}C + 32 \leq 72$$

$$\frac{9}{5}C \leq 72 - 32 \qquad \textit{(Transpose.)}$$

$$\frac{9}{5}C \leq 40 \qquad \textit{(Combine like terms.)}$$

$$\frac{5}{9}\left(\frac{9}{5}\right)C \leq \left(\frac{5}{9}\right)(40) \qquad \textit{(Multiply by reciprocal of coefficient of letter.)}$$

$$C \leq \frac{200}{9}$$

$$C \leq 22.22° \quad \text{or} \quad (-\infty, 22.22] \qquad \textit{(to nearest hundredth)}$$

The heating unit will come on and continue to run if the temperature is less than or equal to 22.22°C. ■

SELF-STUDY EXERCISES 17-2

L.O.1. Solve the following inequalities. Show the solution set on a number line and by using interval notation.

1. $3x + 7x < 60$
2. $5a - 6a \leq 3$
3. $y + 3y \leq 32$
4. $b + 6 > 5$
5. $5t - 18 < 12$
6. $2y + 7 < 17$
7. $3a - 8 \geq 7a$
8. $4x + 7 \leq 8$
9. $2t + 6 \leq t + 13$
10. $4y - 8 > 2y + 14$
11. $3(7 + x) \geq 30$
12. $6(3x - 1) < 12$
13. $2x > 7 - (x + 6)$
14. $4a + 5 \leq 3(2 + a) - 4$
15. $15 - 3(2x + 2) > 6$

16. Duke Jones sold \$196 more than twice as much merchandise as Karen Goodwin. If Duke sold at least \$52,800, how much did Karen sell? Write an inequality to represent the facts and solve.

17-3 COMPOUND INEQUALITIES

Learning Objectives

L.O.1. Identify subsets of sets and perform set operations.
L.O.2. Solve compound inequalities with the conjunction condition.
L.O.3. Solve compound inequalities with the disjunction condition.

In many applications of inequalities, more than one condition is placed on the solution. As an example, when measurements are made, a range of acceptable values is specified. All measurements are approximations, and the range of acceptable values is generally stated as a tolerance. If an acceptable measurement is specified to be within ± 0.005 of an inch, then the range of acceptable values can be from 0.005 of an inch *less than* the ideal measurement to 0.005 of an inch *more than* the ideal measurement. This range can be stated symbolically as a *compound inequality*. A *compound inequality* is a mathematical statement that places more than one condition on the solution. The conditions placed on a compound inequality may use the connective *and* to indicate that both conditions must be met simultaneously. Such compound inequalities may be written as a continuous statement. The conditions placed on a compound inequality may use the connective *or* to indicate that either condition may be met. Such compound inequalities must be written as two separate statements using the connective *or*. If an ideal measurement is 3 in. with a tolerance of ± 0.005 in., the range of acceptable values would be

$$3 - 0.005 \leq x \leq 3 + 0.005$$
$$2.995 \leq x \leq 3.005$$

L.O.1. Identify subsets of sets and perform set operations.

Subsets and Set Operations

Before solving compound inequalities, we will look at some additional properties of sets. Another concept that is common when working with sets is the concept of subsets. The symbol $\subset$ is read "is a subset of." One set of items can be a subset of another set. A set is a *subset* of a second set if every element of the first set is also an element of the second set. If $A = \{1, 2, 3\}$ and $B = \{2\}$, then B is a subset of A. This is written in symbols as $B \subset A$ and is illustrated in Fig. 17-7.

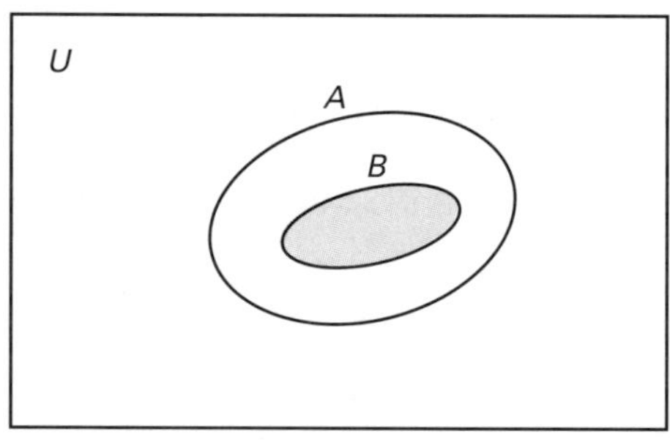

$B \subset A$

Figure 17-7

Two special sets are the universal set and the empty set. The *universal set* is the most inclusive of a group of sets. The universal set is sometimes identified as U. The *empty set* is a set containing no elements. The empty set is identified as $\{\ \}$ or ϕ.

> **Rule 17-3.1** ***Special set relationships:***
>
> Every set is a subset of itself: $A \subset A$
>
> The empty set is a subset of every set: $\phi \subset A$

EXAMPLE 17-3.1. Using the given set definitions answer the following statements as true or false.

$$U = \{0, 1, 2, 3, 4, 5, 6, 7, 8, 9\};\ A = \{1, 3, 5, 7, 9\};$$
$$B = \{0, 2, 4, 6, 8\};\ C = \{1, 2, 3\};\ D = \{1\};\ E = \{\ \ \}$$

(a) $A \subset U$ **(b)** $B \subset U$ **(c)** $A \subset B$ **(d)** $C \subset A$
(e) $C \subset U$ **(f)** $D \subset A$ **(g)** $E \subset U$ **(h)** $E \subset B$

Solution

(a) $A \subset U$ True; every element of A is also an element of U.
(b) $B \subset U$ True; every element of B is also an element of U.
(c) $A \subset B$ False; 1, 3, 5, 7, and 9 are not elements of B.
(d) $C \subset A$ False; 2 is not an element of A.
(e) $C \subset U$ True; every element of C is also an element of U.
(f) $D \subset A$ True; 1 is an element of A.
(g) $E \subset U$ True; the empty set is a subset of every set.
(h) $E \subset B$ True; the empty set is a subset of every set. ■

Two common set operations are union and intersection. The *union* of two sets is a set that includes all elements that appear in *either* of the two sets. Union is generally associated with the condition "or." The symbol for union is ∪. The shaded portion in Fig. 17-8 represents $A \cup B$.

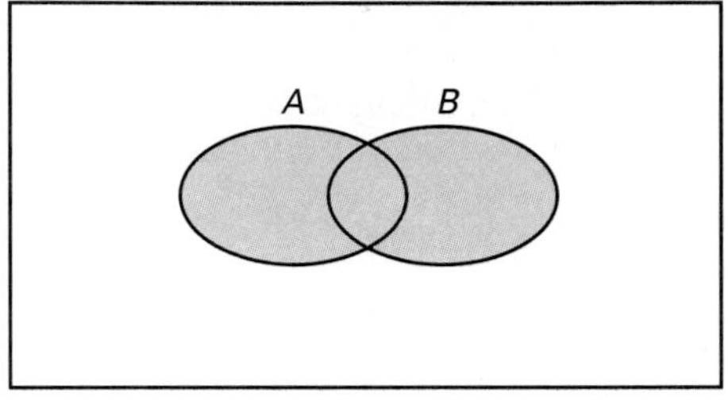

$A \cup B$
Figure 17-8

The *intersection* of two sets is a set that includes all elements that appear in *both* of the two sets. Intersection is generally associated with the condition "and." The symbol for intersection is ∩. The shaded portion of Fig. 17-9 represents $A \cap B$.

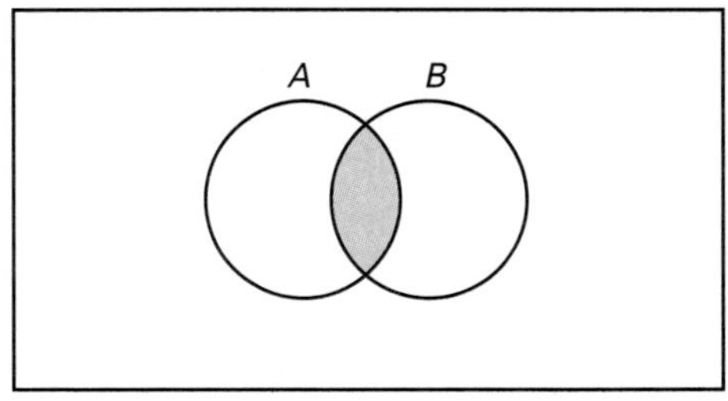

$A \cap B$
Figure 17-9

Example 17-3.2 If $A = \{1, 2, 3, 4, 5\}$, $B = \{1, 3, 5, 7, 9\}$, and $C = \{2, 4, 6, 8\ 10\}$, list the elements in the following sets.

(a) $A \cup B$ **(b)** $A \cup C$ **(c)** $B \cup C$
(d) $A \cap B$ **(e)** $A \cap C$ **(f)** $B \cap C$

Solution

(a) $A \cup B = \{1, 2, 3, 4, 5, 7, 9\}$
(b) $A \cup C = \{1, 2, 3, 4, 5, 6, 8, 10\}$

(c) $B \cup C = \{1, 2, 3, 4, 5, 6, 7, 8, 9, 10\}$
(d) $A \cap B = \{1, 3, 5\}$
(e) $A \cap C = \{2, 4\}$
(f) $B \cap C = \{\ \ \}$ or ϕ ■

Another set operation is complement. The *complement* of a set is a set that includes every element of the universal set that is *not* an element of the given set. The symbol for the complement of a set is ′ and is read "prime." A′ is represented by the shaded portion of Fig. 17-10.

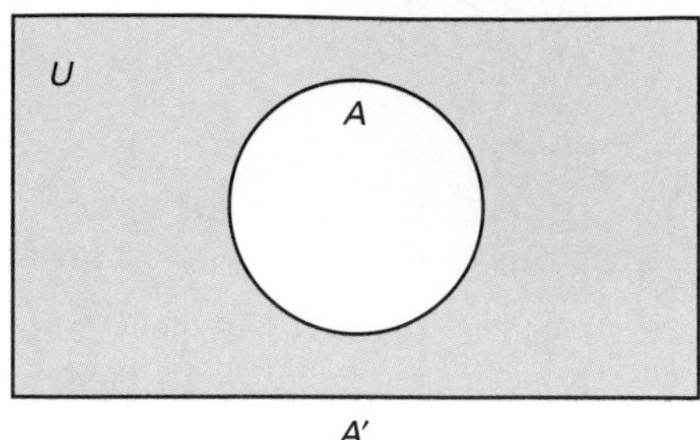

Figure 17-10

EXAMPLE 17-3.3 If $U = \{0, 1, 2, 3, 4, 5\}$, $A = \{1, 3, 5\}$, $B = \{0, 2, 4\}$, and $C = \{1, 2, 3\}$, list the elements in the following sets.

(a) A' (b) B' (c) C'
(d) $(A \cup B)'$ (e) $(A \cap C)'$

Solution

(a) $A' = \{0, 2, 4\}$
(b) $B' = \{1, 3, 5\}$
(c) $C' = \{0, 4, 5\}$
(d) $(A \cup B)'$
$A \cup B = \{0, 1, 2, 3, 4, 5\}$
$(A \cup B)' = \{\ \ \}$ or ϕ
(e) $(A \cap C)'$
$A \cap C = \{1, 3\}$
$(A \cap C)' = \{0, 2, 4, 5\}$ ■

L.O.2. Solve compound inequalities with the conjunction condition.

Conjunction

A compound inequality is a statement that places more than one condition on the variable of the inequality. The solution set is the set of values for the variable that meets all conditions of the problem.

One type of compound inequality is *conjunction.* A conjuction is an intersection or "and" relationship. That is, both conditions must be met simultaneously. A conjunction can be written as a continuous statement.

If $a < x$ and $x < b$, then $a < x < b$.
If $a > x$ and $x > b$, then $a > x > b$.

Similar compound inequalities may also use $\leq$ and $\geq$.

If $a \leq x$ and $x \leq b$, then $a \leq x \leq b$.
If $a \geq x$ and $x \geq b$, then $a \geq x \geq b$.

Rule 17-3.2 *To solve a compound inequality that is a conjunction:*

1. Separate the compound inequality into two simple inequalities using the conditions of the conjunction.
2. Solve each simple inequality.
3. Determine the solution set that includes the *intersection* of the solution sets of the two simple inequalities.

Example 17-3.4 Find the solution set for each of the following compound inequalities. Graph the solution set on a number line.

(a) $5 < x + 3 < 12$ **(b)** $-7 \leq x - 7 \leq 1$
(c) $-3 > 3x > -6$ **(d)** $32 \geq 5x + 7 \geq 17$
(e) $\frac{3}{4} \geq 2x - \frac{1}{2} \geq \frac{1}{4}$ **(f)** $0.5 < 4x + 1 < 3.8$

Solution

(a) $5 < x + 3 < 12$ *(Separate into two simple inequalities.)*

$5 < x + 3$ $x + 3 < 12$

$5 - 3 < x$ $x < 12 - 3$

$2 < x$ or $x > 2$ $x < 9$

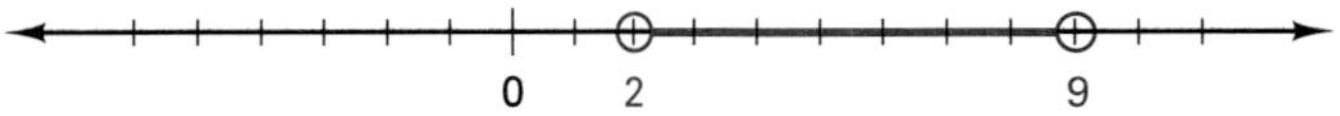

Figure 17-11

Write the solution set as a continuous statement.

$$2 < x < 9 \quad \text{or} \quad (2, 9)$$

(b) $-7 \leq x - 7 \leq 1$ *(Separate into two simple inequalities.)*

$-7 \leq x - 7$ $x - 7 \leq 1$

$-7 + 7 \leq x$ $x \leq 1 + 7$

$0 \leq x$ or $x \geq 0$ $x \leq 8$

Figure 17-12

Write the solution set as a continuous statement.

$$0 \leq x \leq 8 \quad \text{or} \quad [0, 8]$$

(c) $-3 > 3x > -6$ *(Separate into two simple inequalities.)*

$-3 > 3x$ $3x > -6$

$\frac{-3}{3} > \frac{3x}{3}$ $\frac{3x}{3} > \frac{-6}{3}$

$-1 > x$ or $x < -1$ $x > -2$

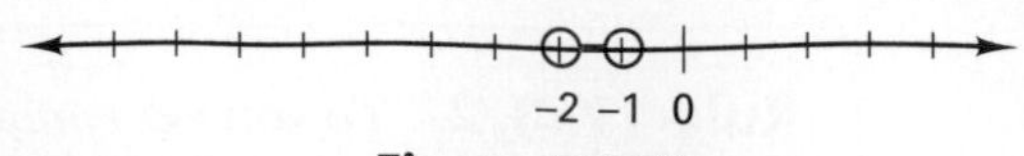

Figure 17-13

Write the solution set as a continuous statement.

$$-1 > x > -2 \quad \text{or} \quad -2 < x < -1 \quad \text{or} \quad (-2, -1)$$

Tips and Traps

Even though continuous compound inequalities can be written using greater than or less than, using less than follows the natural positions of the boundaries on the number line. If $a > b > c$, then $c < b < a$. For instance, if $3 > 0 > -2$, then $-2 < 0 < 3$.

(d) $32 \geq 5x + 7 \geq 17$ *(Separate into two simple inequalities.)*

$$32 \geq 5x + 7 \qquad 5x + 7 \geq 17$$

$$32 - 7 \geq 5x \qquad 5x \geq 17 - 7$$

$$25 \geq 5x \qquad 5x \geq 10$$

$$5 \geq x \text{ or } x \leq 5 \qquad x \geq 2$$

0 2 5

Figure 17-14

Write as a continuous statement.

$$5 \geq x \geq 2 \quad \text{or} \quad 2 \leq x \leq 5 \quad \text{or} \quad [2, 5]$$

(e) $\frac{3}{4} \geq 2x - \frac{1}{2} \geq \frac{1}{4}$ *(Clear fractions.)*

$$4\left(\frac{3}{4} \geq 2x - \frac{1}{2} \geq \frac{1}{4}\right)$$

$$3 \geq 8x - 2 \geq 1 \qquad \textit{(Separate into two simple inequalities.)}$$

$$3 \geq 8x - 2 \qquad 8x - 2 \geq 1$$

$$3 + 2 \geq 8x \qquad 8x \geq 1 + 2$$

$$5 \geq 8x \qquad 8x \geq 3$$

$$\frac{5}{8} \geq x \text{ or } x \leq \frac{5}{8} \qquad x \geq \frac{3}{8}$$

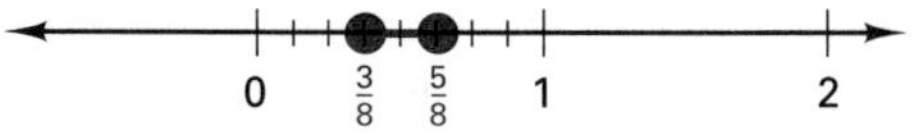

0 $\frac{3}{8}$ $\frac{5}{8}$ 1 2

Figure 17-15

Write as a continuous statement.

$$\frac{5}{8} \geq x \geq \frac{3}{8} \quad \text{or} \quad \frac{3}{8} \leq x \leq \frac{5}{8} \quad \text{or} \quad \left[\frac{3}{8}, \frac{5}{8}\right]$$

(f) $0.5 < 4x + 1 < 3.8$ *(Clear decimals.)*

$$10(0.5 < 4x + 1 < 3.8)$$

$$5 < 40x + 10 < 38$$ *(Separate into two simple inequalities.)*

$5 < 40x + 10$	$40x + 10 < 38$
$5 - 10 < 40x$	$40x < 38 - 10$
$-5 < 40x$	$40x < 28$
$\frac{-5}{40} < x$	$x < \frac{28}{40}$
$-0.125 < x$ or $x > -0.125$	$x < 0.7$

Tips and Traps

When a problem has decimals, you generally express the answer in decimals.

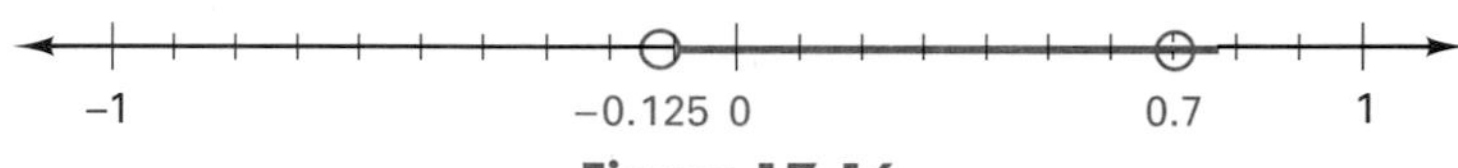

Figure 17-16

Write as a continuous statement.

$$-0.125 < x < 0.7 \quad \text{or} \quad (-0.125, 0.7)$$ ■

As with equations and simple inequalities, a compound inequality may not have a solution.

EXAMPLE 17-3.5 Find the solution set for the following compound inequality.

$$x + 1 < 1 < 3x - 2$$

Solution: Separate into two simple inequalities.

$x + 1 < 1$	$1 < 3x - 2$
$x < 1 - 1$	$2 + 1 < 3x$
$x < 0$	$3 < 3x$
	$1 < x$ or $x > 1$

Figure 17-17

It is impossible for both conditions to be met at the same time. Thus the compound inequality has no solution. ■

L.O.3. Solve compound inequalities with the disjunction condition.

Disjunction

Another type of compound inequality is *disjunction.* A disjunction is a union or "or" relationship. That is, either condition can be met to solve the disjunction. A disjunction *cannot* be written as a continuous statement.

Rule 17-3.3 *To solve a compound inequality that is a disjunction:*

1. Solve each simple inequality.
2. Determine the solution set that includes the union of the solution sets of the two simple inequalities.

EXAMPLE 17-3.6 Find the solution set for each of the following compound inequalities. Graph the solution set on a number line.

(a) $x + 3 < -2$ or $x + 3 > 2$ **(b)** $x - 5 \leq -3$ or $x - 5 \geq 3$

Solution

(a) $x + 3 < -2$ or $x + 3 > 2$ *(Solve each simple inequality.)*

$x < -2 - 3$ $x > 2 - 3$

$x < -5$ or $x > -1$ *(solution set)*

$(-\infty, -5)$ or $(-1, \infty)$

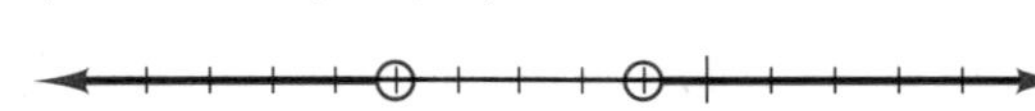

Figure 17-18

(b) $x - 5 \leq -3$ or $x - 5 \geq 3$ *(Solve each simple inequality.)*

$x \leq -3 + 5$ $x \geq 3 + 5$

$x \leq 2$ or $x \geq 8$ *(solution set)*

$(-\infty, 2]$ or $[8, \infty)$ ■

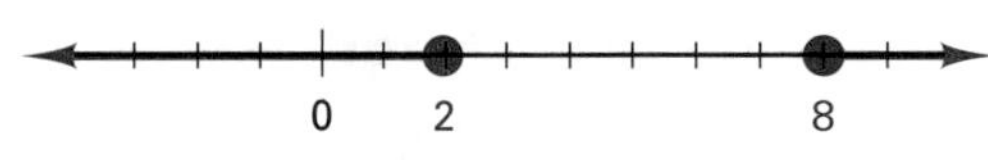

Figure 17-19

SELF-STUDY EXERCISES 17-3

Use the following sets for problems 1–10.

$U = \{-5, -4, -3, -2, -1, 0, 1, 2, 3, 4, 5\}$

$A = \{-4, -2, 0\}$, $B = \{1, 2, 3, 4, 5\}$

$C = \{2, -2\}$, $D = \{0\}$, $E = \{\ \}$

L.O.1. Answer the following statements as true or false and justify your answers.

1. $B \subset U$ **2.** $C \subset A$ **3.** $E \subset B$
4. $A \subset B$

List the elements in the following sets.

5. $A \cap B$ **6.** $C \cup D$ **7.** $A \cup B$
8. A' **9.** $(A \cap D)'$ **10.** $(B \cup C)'$

L.O.2. Solve the following compound inequalities. Show the solution set on a number line and by using interval notation.

11. $x - 1 < 5 < 2x + 1$ **12.** $2x - 7 < x < 3x - 4$ **13.** $x + 3 < 8 < 2x - 12$
14. $x + 2 \leq 7 \leq 2x - 3$ **15.** $2x + 3 < 15 < 3x - 9$ **16.** $1 < x - 2 < 5$

17. $0 < 5x < 15$
18. $4 \leq 6 - x \leq 8$
19. $6 < -2x < 12$
20. $1 < 6 - x < 3$

L.O.3.

21. $x + 3 < 5$ or $x - 7 > 2$
22. $x - 1 \leq 2$ or $x \geq 7$
23. $2x - 1 \leq 7$ or $3x \geq 15$
24. $5 - x \leq 2$ or $x + 1 \leq 2$
25. $5x - 2 \leq 3x + 1$ or $2x \geq 7$
26. The blueprint specifications for a part show it should have a measure of 5.27 cm with a tolerance of ± 0.05 cm. Express the limit dimensions of the part with a compound inequality. If x represents the measure of a different part on the blueprint with the same tolerance, express the limit dimensions with inequalities.

17-4 SOLVING QUADRATIC AND RATIONAL INEQUALITIES

Learning Objectives

L.O.1. Solve quadratic inequalities.
L.O.2. Solve rational inequalities.

Solving *quadratic* and *rational inequalities* has some basic similarities to solving quadratic and rational equations. The solution to these inequalities, like the solution to linear inequalities, is a *set* of numbers with a specific boundary.

L.O.1. Solve quadratic inequalities.

Quadratic Inequalities

To solve quadratic inequalities by factoring, we will begin as we did in solving quadratic equations. Rearrange the terms of the inequality so that the right side of the inequality is zero. Then write the terms of the left side in factored form. For example, in the inequality $x^2 < 5x$, we have $x^2 - 5x < 0$, which is $x(x - 5) < 0$.

We determine the boundaries or *critical values* of the solution set by determining the values of x that make each factor *equal* zero.

$$x = 0 \qquad x - 5 = 0$$
$$x = 5$$

The *critical values* are 0 and 5.

Now, plot these critical values on a number line.

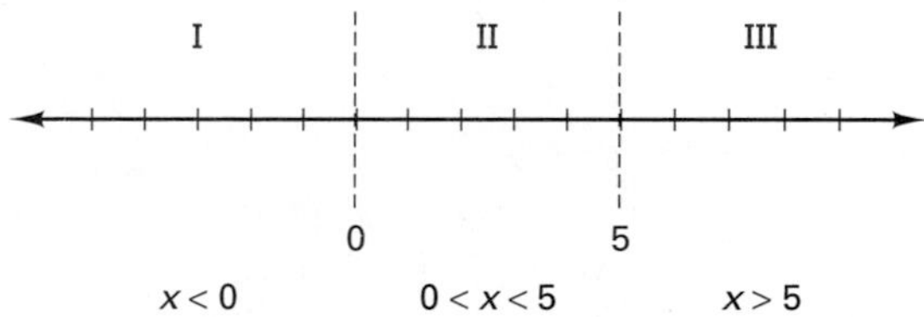

These values divide the number line into three regions. Region I indicates all values *less than* 0 ($x < 0$). Region II indicates values between 0 and 5 ($0 < x < 5$). Region III indicates values *greater than* 5 ($x > 5$).

The solution set for the inequality will include one or more of these regions. To determine which regions are in the solution set, select *any* point in each region, substitute that value into the inequality, and determine if the resulting inequality is a true or false statement.

Region I: $x < 0$

Suppose $x = -1$. (*Region I test point*)

$$x(x - 5) < 0$$
$$-1(-1 - 5) < 0$$
$$-1(-6) < 0$$
$$+6 < 0$$

The inequality is false.

Therefore, region I is not included in the solution set.

Region II: $0 < x < 5$

Suppose $x = +1$. (*Region II test point*)

$$x(x - 5) < 0$$
$$1(1 - 5) < 0$$
$$1(-4) < 0$$
$$-4 < 0$$

The inequality is true.

Therefore, region II, $0 < x < 5$, is in the solution set.

Region III: $x > 5$

Suppose $x = 6$. (*Region III test point*)

$$x(x - 5) < 0$$
$$6(6 - 5) < 0$$
$$6(1) < 0$$
$$6 < 0$$

The inequality is false.

Therefore, region III is not included in the solution set. The solution set includes only region II, $0 < x < 5$.

I II III

0 5

$0 < x < 5$

We summarize this procedure in the following rule:

Rule 17-4.1 ***To solve quadratic inequalities by factoring:***

1. Rearrange the inequality so that the right side of the inequality is zero (0).
2. Write the left side of the inequality in factored form.
3. Determine the critical values by finding the values that make each factor equal to zero (0).
4. Plot the critical values on a number line.
5. Test each region of values by selecting any point within a region, substituting that value into the inequality, solving the inequality, and deciding if the resulting inequality is a *true* statement.
6. The solution set for the quadratic inequality will be the region or regions that produce a *true* statement in step 5.

Look at the next example.

EXAMPLE 17-4.1 Solve the inequality $x^2 + 5x + 6 \leq 0$.

Solution: Write in factored form.

$$(x + 3)(x + 2) \leq 0$$

Determine the critical values.

$$x + 3 = 0 \qquad x + 2 = 0$$
$$x = -3 \qquad x = -2$$

Plot the critical values and label the corresponding regions.

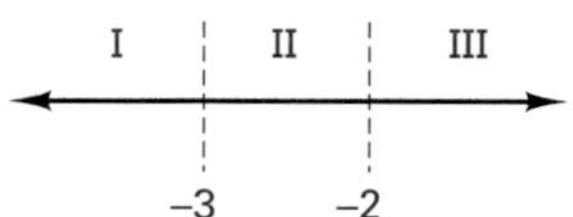

Region I: $x \leq -3$

Region II: $-3 \leq x \leq -2$

Region III: $x \geq -2$

Test each region.

Region I: $x \leq -3$

Suppose $x = -4$. (*Region I test point*)

$$(x + 3)(x + 2) \leq 0$$
$$(-4 + 3)(-4 + 2) \leq 0$$
$$(-1)(-2) \leq 0$$
$$2 \leq 0$$

The inequality is false.

Region II: $-3 \leq x \leq -2$

Suppose $x = -2.5$. *(Region II test point)*

$$(-2.5 + 3)(-2.5 + 2) \leq 0$$
$$(0.5)(-0.5) \leq 0$$
$$-0.25 \leq 0$$

The inequality is true.

Region III: $x \geq -2$

Suppose $x = -1$. *(Region III test point)*

$$(-1 + 3)(-1 + 2) \leq 0$$
$$(2)(1) \leq 0$$
$$2 \leq 0$$

The inequality is false.

The solution set is $-3 \leq x \leq -2$ or $[-3, -2]$. ■

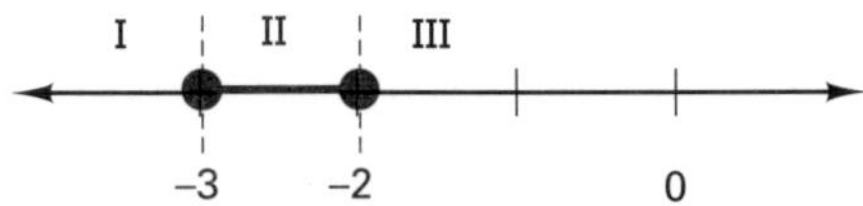

EXAMPLE 17-4.2 Solve the inequality $2x^2 + x - 6 > 0$.

Solution: Write in factored form.

$$(2x - 3)(x + 2) > 0$$

Determine the critical values.

$$2x - 3 = 0 \qquad x + 2 = 0$$
$$2x = 3 \qquad x = -2$$
$$x = \frac{3}{2}$$

Plot the critical values and label the corresponding regions.

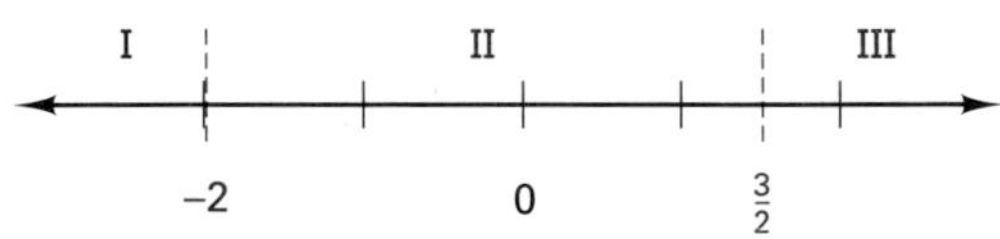

Region I: $x < -2$

Region II: $-2 < x < \frac{3}{2}$

Region III: $x > \frac{3}{2}$

Test each region.

Region I: $x < -2$

Suppose $x = -3$. *(Region I test point)*

$$(2x - 3)(x + 2) > 0$$
$$[2(-3) - 3][-3 + 2] > 0$$
$$[-6 - 3][-1] > 0$$
$$[-9][-1] > 0$$
$$9 > 0$$

The inequality is true.

Region II: $-2 < x < \frac{3}{2}$

Suppose $x = 0$. *(Region II test point)*

$$(2x - 3)(x + 2) > 0$$
$$[2(0) - 3][0 + 2] > 0$$
$$(0 - 3)(2) > 0$$
$$(-3)(2) > 0$$
$$-6 > 0$$

The inequality is false.

Region III: $x > \frac{3}{2}$

Suppose $x = 2$. *(Region III test point)*

$$(2x - 3)(x + 2) > 0$$
$$[2(2) - 3][2 + 2] > 0$$
$$(4 - 3)(4) > 0$$
$$(1)(4) > 0$$
$$4 > 0$$

The inequality is true.

The solution set is $x < -2$ or $x > \frac{3}{2}$ or $(-\infty, -2)$ or $\left(\frac{3}{2}, \infty\right)$. ■

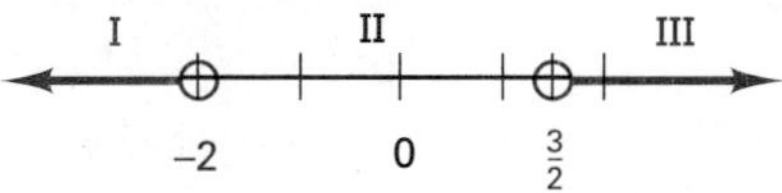

There are other ways of determining the solution set of a quadratic inequality. One way that is similar to finding a test point in each region is to determine the sign of each factor in each region. Let's examine the inequalities in Example 17-4.1 and 17-4.2 again.

EXAMPLE 17-4.3 Determine the solution set of the following by examining the signs of each factor in each region.

(a) $x^2 + 5x + 6 \leq 0$ **(b)** $2x^2 + x - 6 > 0$

Solution

(a) $x^2 + 5x + 6 \leq 0$

$(x + 3)(x + 2) \leq 0$

Critical values: $x = -3$

$x = -2$

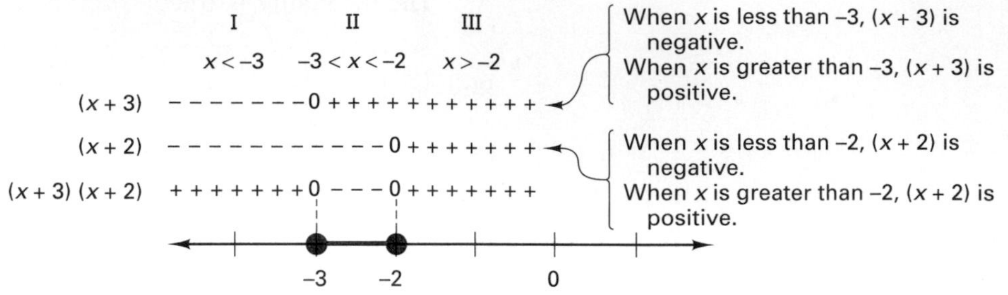

I
II
III
x < –3
–3 < x < –2
x > –2
(x + 3)
– – – – – – – 0 + + + + + + + + + + +
(x + 2)
– – – – – – – – – – – 0 + + + + + + + +
(x + 3) (x + 2)
+ + + + + + + 0 – – – 0 + + + + + + + +
–3
–2
0
When x is less than –3, (x + 3) is negative.
When x is greater than –3, (x + 3) is positive.
When x is less than –2, (x + 2) is negative.
When x is greater than –2, (x + 2) is positive.

The product $(x + 3)(x + 2)$ is positive in regions I and III.

The product $(x + 3)(x + 2)$ is negative in region II.

The solution set includes regions where the product is *negative*.

The solution set is $-3 \leq x \leq -2$.

(b) $2x^2 + x - 6 > 0$

$(2x - 3)(x + 2) > 0$

Critical values: $x = \frac{3}{2}$

$x = -2$

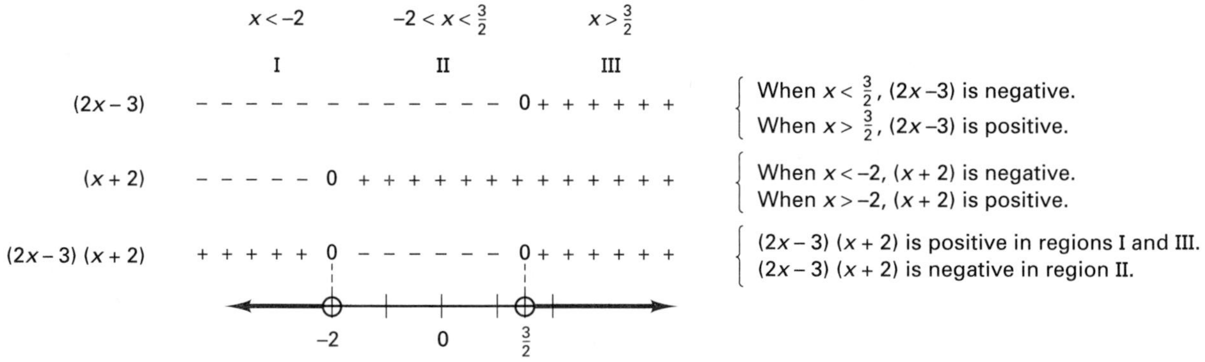

x < –2
–2 < x < 3/2
x > 3/2
I
II
III
(2x – 3)
– – – – – – – – – – – – – 0 + + + + + +
(x + 2)
– – – – – 0 + + + + + + + + + + + + + +
(2x – 3) (x + 2)
+ + + + + 0 – – – – – – – 0 + + + + + +
–2
0
3/2
When x < 3/2, (2x –3) is negative.
When x > 3/2, (2x –3) is positive.
When x < –2, (x + 2) is negative.
When x > –2, (x + 2) is positive.
(2x – 3) (x + 2) is positive in regions I and III.
(2x – 3) (x + 2) is negative in region II.

The solution set is $x < -2$ or $x > \frac{3}{2}$. ■

After examining several problems, you may notice a pattern for determining the solution set by inspection once the critical values are known. This pattern or generalization can then be applied to quadratic inequalities that do not factor.

Tips and Traps

When a quadratic inequality is in standard form, $ax^2 + bx + c < 0$, where $a > 0$ (positive), and the solutions to the quadratic equation $ax^2 + bx + c = 0$ are s_1 and s_2 and $s_1 < s_2$, then the solution set is $s_1 < x < s_2$.

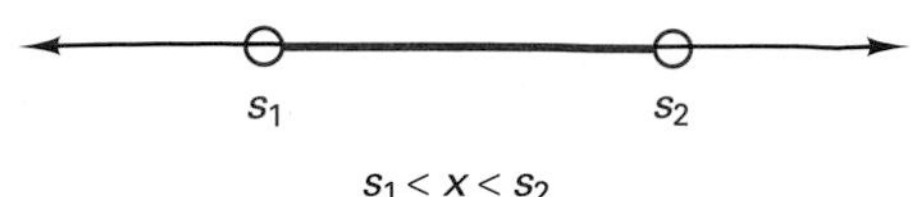

$s_1 < x < s_2$

When a quadratic inequality is in standard form, $ax^2 + bx + c > 0$, where $a > 0$ (positive), and the solutions to the quadratic equation $ax^2 + bx + c = 0$ are s_1 and s_2, and $s_1 < s_2$, then the solution set is $x < s_1$ or $x > s_2$.

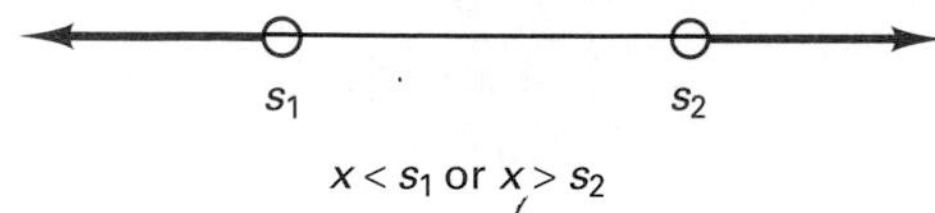

$x < s_1$ or $x > s_2$

Similar generalizations can be made for $ax^2 + bx + c \leq 0$ and $ax^2 + bx + c \geq 0$.

L.O.2. Solve rational inequalities.

Rational Inequalities

To solve a *rational inequality* like $\frac{x + 2}{x - 5} > 0$, we will find the critical values by setting each factor equal to zero.

Rule 17-4.2 ***To solve a rational inequality like*** $\frac{x + a}{x + b} < 0$ ***or*** $\frac{x + a}{x + b} > 0$***:***

1. Find critical values by setting the numerator and denominator equal to zero.
2. Solve each equation and use the solutions (critical values) to divide the number line into three regions.
3. Regions that have like signs for both the numerator and denominator are in the solution set of $\frac{x + a}{x + b} > 0$.
4. Regions that have unlike signs for the numerator and denominator are in the solution set of $\frac{x + a}{x + b} < 0$.

EXAMPLE 17-4.4 Solve the following inequalities.

(a) $\frac{x + 2}{x - 5} < 0$ **(b)** $\frac{x + 2}{x - 5} > 0$

Solution: For both inequalities:

$$x + 2 = 0 \qquad x - 5 = 0$$

$$x = -2 \qquad x = 5$$

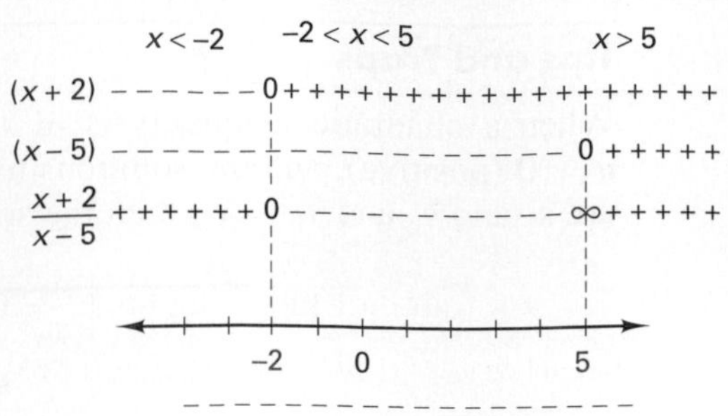

(a) Solution set:

For $\frac{x+2}{x-5} < 0$,

$-2 < x < 5$

or

$(-2, 5)$

(b) Solution set:

For $\frac{x+2}{x-5} > 0$,

$x < -2$ or $x > 5$

or

$(-\infty, -2)$ or $(5, \infty)$ ■

SELF-STUDY EXERCISES 17-4

L.O.1 Solve the following quadratic inequalities and graph the solution on a number line.

1. $(x - 2)(x + 3) < 0$
2. $(2x - 1)(x + 3) > 0$
3. $(2x - 5)(3x - 2) < 0$
4. $(y + 6)(y - 2) \geq 0$
5. $(a + 4)(a - 6) < 0$
6. $x^2 - 7x + 10 \leq 0$
7. $x^2 - x - 6 \geq 0$
8. $x^2 - 4x + 3 \leq 0$
9. $4x^2 - 21x \leq -5$
10. $6x^2 + x > 1$

L.O.2. Solve the following rational inequalities and graph the solution on a number line.

11. $\frac{x-3}{x+7} < 0$
12. $\frac{x+8}{x-3} > 0$
13. $\frac{2x-6}{3x-1} > 0$
14. $\frac{x-7}{x+7} < 0$
15. $\frac{x}{x-1} < 0$

17-5 EQUATIONS CONTAINING ONE ABSOLUTE-VALUE TERM

Learning Objective

L.O.1. Solve equations containing one absolute-value term.

L.O.1. Solve equations containing one absolute-value term.

Solving Equations Using One Absolute-Value Term

Earlier, absolute value was defined as the number of units from zero. That is, $|+7|$ or $|-7|$ is 7. If the *absolute value* of a term is indicated in an equation, then both possible values of the term must be considered in solving the equation. Look at the following example.

EXAMPLE 17-5.1 Solve the equation $|x| = 5$.

Solution: By definition of absolute value, the solutions are $x = +5$ or $x = -5$, which can be written symbolically as $x = \pm 5$. ■

Notice that the equation $|x| = 5$ has two roots, $+5$ and -5. To solve an equation that contains an absolute value, we must examine each of two cases. One is that the expression within the absolute-value symbols is positive, and the other is that the expression within the absolute-value symbols is negative. Look at the next example.

EXAMPLE 17-5.2 Find the roots for the equation $|x - 4| = 7$.

Solution: By definition of absolute value we have two cases to be examined.

Case 1: $x - 4 = 7$

Case 2: $-(x - 4) = 7$

We solve the equation given in each case to find the *two* roots.

$$\begin{aligned} x - 4 &= 7 \\ x &= 7 + 4 \\ x &= 11 \end{aligned} \qquad \begin{aligned} -(x - 4) &= 7 \\ -x + 4 &= 7 \\ -x &= 7 - 4 \\ -x &= 3 \\ x &= -3 \end{aligned}$$

Thus, the roots of the equation are 11 and -3. Each root will check to be a correct root.

To check each root, substitute each root separately into the original equation.

$$\begin{aligned} &\text{If } x = 11, \\ |x - 4| &= 7 \\ |11 - 4| &= 7 \\ |7| &= 7 \\ 7 &= 7 \end{aligned} \qquad \begin{aligned} &\text{If } x = -3, \\ |x - 4| &= 7 \\ |-3 - 4| &= 7 \\ |-7| &= 7 \\ 7 &= 7 \end{aligned}$$

■

EXAMPLE 17-5.3 Find the roots of the equation $|y + 3| - 5 = 6$.

Solution: Before we can separate the equation into two cases, we should *isolate* the absolute-value term. That is, we rearrange the terms of the equation so that the absolute-value term is alone on one side of the equation.

$$\begin{aligned} |y + 3| - 5 &= 6 \\ |y + 3| &= 6 + 5 \\ |y + 3| &= 11 \end{aligned}$$

Now, separate into two cases.

$$\begin{aligned} y + 3 &= 11 \\ y &= 11 - 3 \\ y &= 8 \end{aligned} \qquad \begin{aligned} -(y + 3) &= 11 \\ -y - 3 &= 11 \\ -y &= 11 + 3 \\ -y &= 14 \\ y &= -14 \end{aligned}$$

Thus the roots of the equation are 8 and -14. Each root will check to be a correct root. ■

The following rule summarizes the procedures used in Examples 17-5.1, 17-5.2, and 17-5.3

Rule 17-5.1 ***To solve an equation containing one absolute-value term:***

1. Isolate the absolute-value term on one side of the equation.
2. Separate the equation into two cases. One case considers the expression within the absolute-value symbols to be positive. The other case considers it to be negative.
3. Solve each case to obtain the *two* roots of the equation.
4. $|x| = b$ has no solution if b is negative ($b < 0$).

EXAMPLE 17-5.4 Solve the equation $|2x - 5| + 4 = 7$.

Solution

$$|2x - 5| + 4 = 7$$

$$|2x - 5| = 7 - 4 \quad \textit{(Isolate the absolute-value term.)}$$

$$|2x - 5| = 3$$

Separate into two cases. Solve each equation.

$$2x - 5 = 3 \qquad -(2x - 5) = 3$$

$$2x = 3 + 5 \qquad -2x + 5 = 3$$

$$2x = 8 \qquad -2x = 3 - 5$$

$$x = 4 \qquad -2x = -2$$

$$x = 1$$

Thus, the roots of the equation are 4 and 1. Each root will check to be a correct root. ■

Tips and Traps

If the absolute-value term is isolated before separating into two cases, the following property applies:

$$\text{If } |x| = a \text{ and } a \geq 0, \text{ then } x = \pm a$$

Applying this property to Example 17-5.4, we have

$$|2x - 5| + 4 = 7$$

$$|2x - 5| = 7 - 4$$

$$|2x - 5| = 3$$

Now, separate into two cases.

Case 1	*Case 2*
$2x - 5 = 3$	$2x - 5 = -3$
$2x = 3 + 5$	$2x = -3 + 5$
$2x = 8$	$2x = 2$
$x = 4$	$x = 1$

SELF-STUDY EXERCISES 17-5

L.O.1. Solve the following equations containing absolute values.

1. $|x| = 8$ **2.** $|x - 3| = 5$ **3.** $|x - 7| = 2$
4. $|2x - 3| = 9$ **5.** $|3x - 7| = 2$ **6.** $|x| - 4 = 7$
7. $|x - 2| - 5 = 3$ **8.** $-5 + |x - 4| = -2$ **9.** $|2x - 1| - 3 = 4$
10. $|3x - 2| + 1 = 6$

17-6 ABSOLUTE-VALUE INEQUALITIES

Learning Objectives

L.O.1. Solve absolute-value inequalities using the less than relationship.
L.O.2. Solve absolute-value inequalities using the greater than relationship.

L.O.1. Solve absolute-value inequalities using the less than relationship.

Absolute-Value Inequalities Using < or ≤

An inequality that contains an absolute-value term may have a set of values that are solutions to the inequalities. If the inequality is a "less than" or "less than or equal to" relationship, then the following property is used.

If $|x| < b$ and $b > 0$, then $-b < x < b$.

or

If $|x| \leq b$ and $b > 0$, then $-b \leq x \leq b$.

On a number line (Fig. 17-20), the solution set is a continuous set of values.

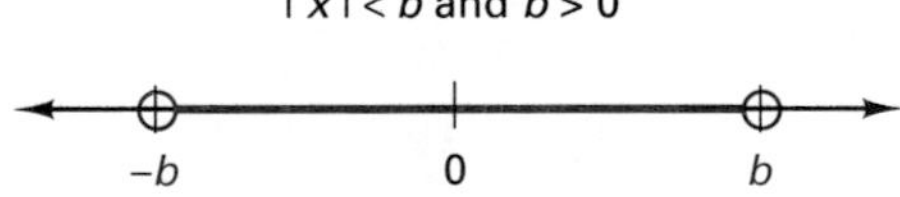

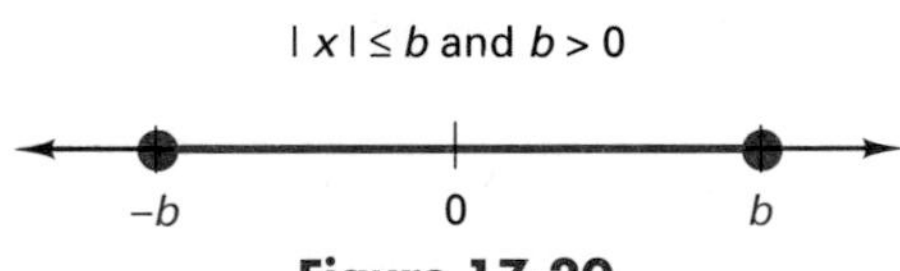

Figure 17-20

EXAMPLE 17-6.1 Find the solution set for each of the following inequalities.

(a) $|x| < 3$ **(b)** $|x + 5| < 8$ **(c)** $|x - 1| \leq 1$
(d) $|3x + 2| \leq 14$

Solution

(a) $|x| < 3$ *(Apply property of inequalities having < relationship.)*

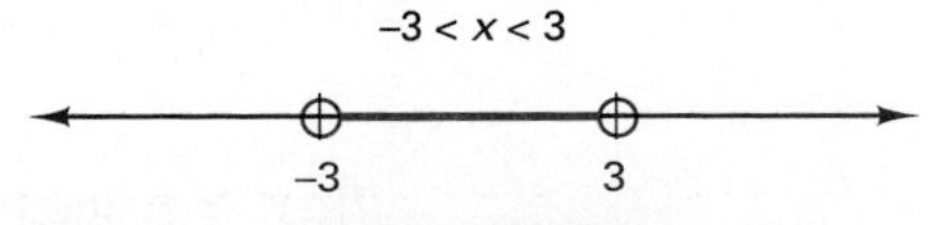

Figure 17-21

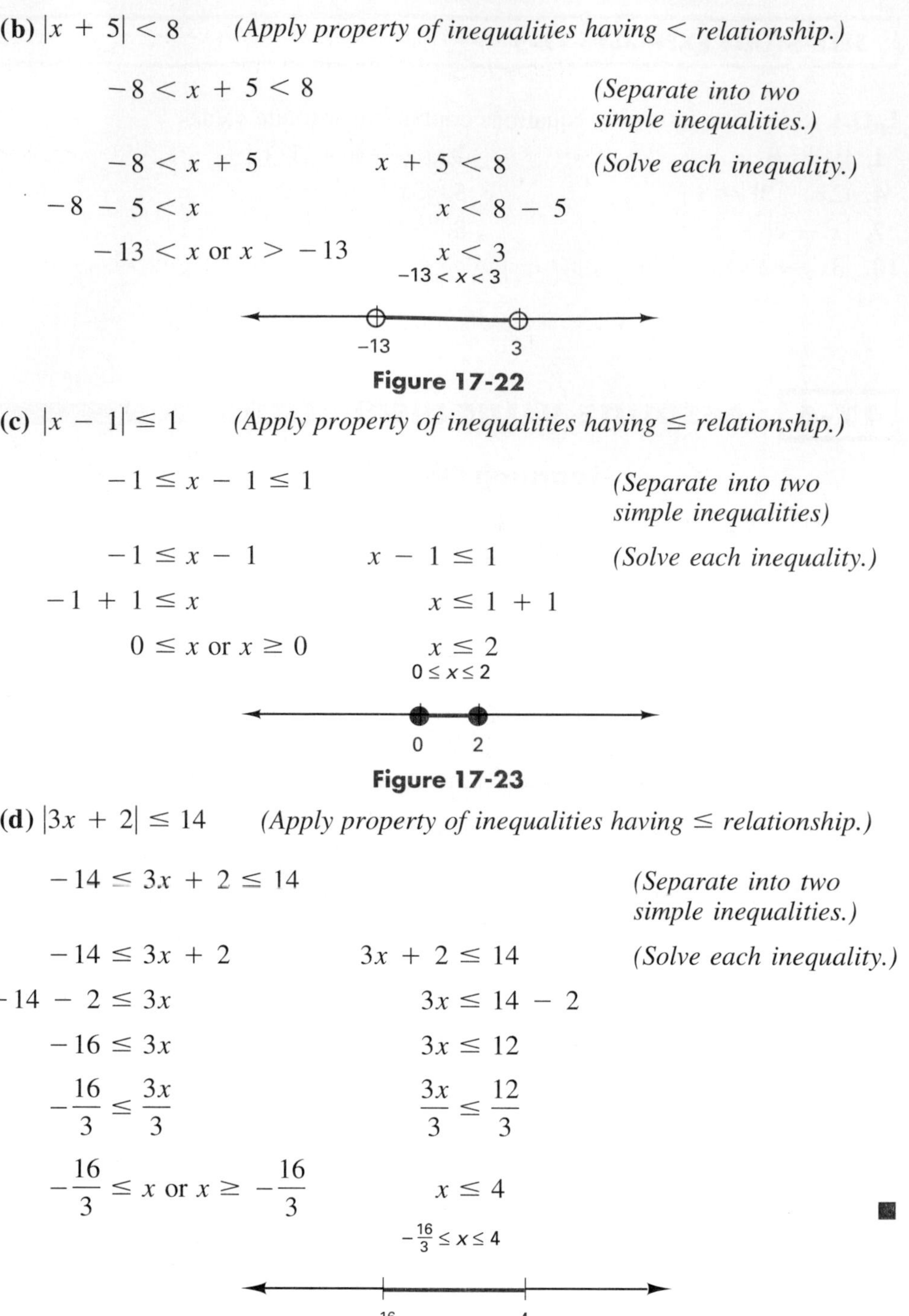

(b) $|x + 5| < 8$ *(Apply property of inequalities having < relationship.)*

$-8 < x + 5 < 8$ *(Separate into two simple inequalities.)*

$-8 < x + 5 \qquad x + 5 < 8$ *(Solve each inequality.)*

$-8 - 5 < x \qquad x < 8 - 5$

$-13 < x$ or $x > -13 \qquad x < 3$

$-13 < x < 3$

Figure 17-22

(c) $|x - 1| \leq 1$ *(Apply property of inequalities having ≤ relationship.)*

$-1 \leq x - 1 \leq 1$ *(Separate into two simple inequalities)*

$-1 \leq x - 1 \qquad x - 1 \leq 1$ *(Solve each inequality.)*

$-1 + 1 \leq x \qquad x \leq 1 + 1$

$0 \leq x$ or $x \geq 0 \qquad x \leq 2$

$0 \leq x \leq 2$

Figure 17-23

(d) $|3x + 2| \leq 14$ *(Apply property of inequalities having ≤ relationship.)*

$-14 \leq 3x + 2 \leq 14$ *(Separate into two simple inequalities.)*

$-14 \leq 3x + 2 \qquad 3x + 2 \leq 14$ *(Solve each inequality.)*

$-14 - 2 \leq 3x \qquad 3x \leq 14 - 2$

$-16 \leq 3x \qquad 3x \leq 12$

$-\frac{16}{3} \leq \frac{3x}{3} \qquad \frac{3x}{3} \leq \frac{12}{3}$

$-\frac{16}{3} \leq x$ or $x \geq -\frac{16}{3} \qquad x \leq 4$ ■

$-\frac{16}{3} \leq x \leq 4$

Figure 17-24

L.O.2. Solve absolute-value inequalities using the greater than relationship.

Absolute-Value Inequalities Using > or ≥

An absolute-value inequality having a "greater than" or "greater than or equal to" relationship will have a solution set that is represented by the extreme values on the number line.

If $|x| > b$ and $b > 0$, then $x < -b$ or $x > b$.

or

If $|x| \geq b$ and $b > 0$, then $x \leq -b$ or $x \geq b$.

The solution set for an absolute-value inequality having a $>$ or $\geq$ relationship is represented on the number lines in Figs. 17-25 and 17-26.

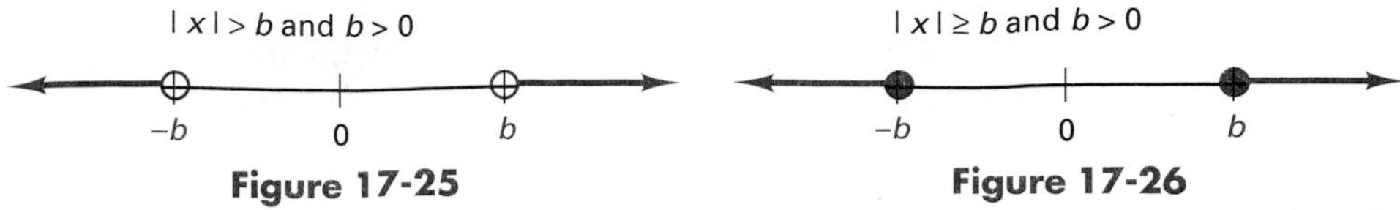

Figure 17-25

Figure 17-26

EXAMPLE 17-6.2 Find the solution set for each of the following inequalities.

(a) $|x| > 5$ **(b)** $|x - 2| > 3$ **(c)** $|x + 1| \geq 4$
(d) $|2x - 3| \geq 7$

Solution

(a) $|x| > 5$ *(Apply the property of inequalities having $>$ relationship.)*

$x < -5$ or $x > 5$

Figure 17-27

(b) $|x - 2| > 3$ *(Apply the property of inequalities having $>$ relationship.)*

$x - 2 < -3$ or $x - 2 > 3$ *(Solve each inequality.)*

$x < -3 + 2$ $\qquad x > 3 + 2$

$x < -1$ $\qquad x > 5$

$x < -1$ or $x > 5$

Figure 17-28

(c) $|x + 1| \geq 4$ *(Apply the property of inequalities having $\geq$ relationship.)*

$x + 1 \leq -4$ or $x + 1 \geq 4$ *(Solve each inequality.)*

$x \leq -4 - 1$ $\qquad x \geq 4 - 1$

$x \leq -5$ $\qquad x \geq 3$

$x \leq -5$ or $x \geq 3$

Figure 17-29

(d) $|2x - 2| \geq 7$ *(Apply the property of inequalities having $\geq$ relationship.)*

$2x - 2 \leq -7$ or $2x - 2 \geq 7$

$2x \leq -7 + 2$ $\qquad 2x \geq 7 + 2$

$2x \leq -5$ $\qquad 2x \geq 9$

$x \leq -\frac{5}{2}$ $\qquad x \geq \frac{9}{2}$ ■

$x \leq -\frac{5}{2}$ or $x \geq \frac{9}{2}$

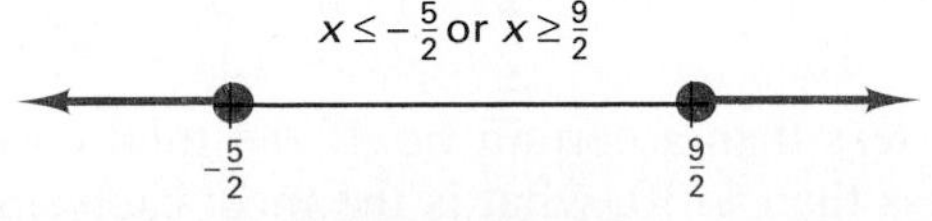

Figure 17-30

SELF-STUDY EXERCISES 17-6

L.O.1. Find the solution set for each of the following inequalities. Graph the solution set on a number line.

1. $|x| < 5$
2. $|x| < 1$
3. $|x| < 2$
4. $|x| \leq 7$
5. $|x + 3| < 2$
6. $|x + 4| < 3$
7. $|x - 3| < 4$
8. $|x - 2| < 3$
9. $|x - 5| \leq 6$
10. $|x - 1| \leq 7$
11. $|3x - 5| < 7$
12. $|2x - 4| < 3$
13. $|5x + 2| \leq 3$
14. $|4x + 7| \leq 5$

L.O.2. Find the solution set for each of the following inequalities. Graph the solution set on a number line.

15. $|x| > 2$
16. $|x| > 3$
17. $|x| \geq 5$
18. $|x| \geq 6$
19. $|x - 5| > 2$
20. $|x - 4| > 6$
21. $|x + 3| \geq 5$
22. $|x + 4| \geq 6$
23. $|2x - 5| \geq 0$
24. $|3x - 2| \geq 4$
25. $|5x + 8| \geq 2$
26. $|4x - 3| \geq 9$

Write the solution set for each of the following inequalities in set-builder notation and graph the solution set.

27. $|3x - 2| + 4 \leq 5$
28. $|2x - 5| - 3 \leq 2$
29. $|5x - 3| - 6 \geq 1$
30. $|4x - 6| + 4 \geq 5$
31. $|3x - 8| - 2 \leq -6$
32. $|5x + 7| + 3 \geq -6$

Find the solution set for each of the following inequalities.

33. $|x - 3| \geq 3$
34. $|4x + 1| > 1$
35. $|x - 4| \leq 3$
36. $|2x - 1| < 0$
37. $|x| < 7$
38. $|x + 8| < 3$
39. $|x| > 12$
40. $|x + 8| > 2$

ASSIGNMENT EXERCISES

17-1

1. Describe the empty set.
2. Write the symbols used to indicate the empty set.
3. Use symbols to write the following: Five is an element of the whole numbers.

Represent the following sets of numbers on the number line and using interval notation.

4. $x \leq 3$
5. $x > -7$
6. $-3 < x < 5$
7. $-4 \leq x < 2$
8. All real numbers
9. $-2 < x$
10. $-5 > x$

17-2

Solve the following inequalities. Show the solution set on a number line.

11. $42 > 8m - 2m$
12. $3m - 2m < 3$
13. $0 < 2x - x$
14. $1 \leq x - 7$
15. $10 - 2x \geq 4$
16. $3x + 4 > x$
17. $10x + 18 > 8x$
18. $4x \leq 5x + 8$
19. $12 + 5x > 6 - x$
20. $8 - 7y < y + 24$
21. $15 \geq 5(2 - y)$
22. $4t > 2(7 + 3t)$
23. $6x - 2(x - 3) \leq 30$
24. $8x - (3x - 2) > 12$
25. $2(b + 1) < 5(b - 4) + 1$
26. A shirt costs \$3 less than a certain tie. If the total cost for six ties and two shirts must be less than \$130, what is the most each shirt and tie can cost?

Use sets U, A, B, C, and D to give the elements in the following sets.

$$U = \{-2, -1, 0, 1, 2, 3, 4, 5\}$$
$$A = \{-2, -1\}, \quad B = \{0\}, \quad C = \{0, 1, 2, 3, 4\},$$
$$D = \{-1, 0, 1\}$$

27. $A \cap B$
28. $B \cup D$
29. $A \cap D$
30. A'
31. $(A \cup C)'$

Solve. Show the solution on the number line and by using interval notation.

32. $x - 7 < 2$
33. $x + 4 > 2$
34. $3x - 1 \leq 8$
35. $3x - 2 \leq 4x + 1$
36. $5 < x - 3 < 8$
37. $-3 < 2x - 4 < 5$
38. $-7 < x - 5 < 7$

17-3

Solve the following compound inequalities. Show the solution set on a number line and by using interval notation.

39. $x + 1 < 5 < 2x + 1$
40. $3x - 8 < x < 5x + 24$
41. $x + 2 < 7 < 2x - 15$
42. $x + 3 \leq 7 \leq 2x - 1$
43. $2x + 3 < 15 < 3x + 9$
44. $2 < 3x - 4 < 8$
45. $-5 \leq -3x + 1 < 10$
46. $2x + 3 \leq 5x + 6 < -3x - 7$
47. $-3 \leq 4x + 5 \leq 2$
48. $4x - 2 < x + 8 < 9x + 1$
49. $x + 3 < 5$ or $x > 8$
50. $x - 7 < 4$ or $x + 3 > 8$
51. $x - 3 < -12$ or $x + 1 > 9$
52. $3x < -4$ or $2x - 1 > 9$

17-4

Solve the following inequalities and graph the solution on a number line.

53. $(x - 5)(x - 2) > 0$
54. $(3x + 2)(x - 2) < 0$
55. $(3x + 1)(2x - 3) < 0$
56. $(5x - 6)(x + 1) \geq 0$
57. $(x + 1)(x - 2) \leq 0$
58. $x^2 + x - 12 < 0$
59. $2x^2 \leq 5x + 3$
60. $2x^2 - 3x \geq -1$
61. $2x^2 + 7x - 15 < 0$
62. $x^2 - 2x - 8 \geq 0$
63. $\dfrac{x - 7}{x + 1} < 0$
64. $\dfrac{x + 1}{x - 3} > 0$
65. $\dfrac{x}{x + 8} > 0$
66. $\dfrac{x - 8}{x + 1} < 0$

17-5

Solve the following equations containing absolute values.

67. $|x| = 12$
68. $|x - 9| = 2$
69. $|x + 3| = 7$
70. $|x + 4| = 11$
71. $|x - 8| = 12$
72. $|x + 7| = 3$
73. $|4x - 7| = 17$
74. $|2x + 3| = 5$
75. $|7x + 8| = 15$
76. $|4x + 1| = 9$
77. $|7x - 4| = 17$
78. $|6x - 2| = 3$
79. $|3x - 9| = 2$
80. $|x| + 8 = 10$
81. $|x| + 12 = 19$
82. $|2x| - 1 = 9$
83. $|x| - 9 = 7$
84. $|x - 4| - 10 = 6$
85. $-5 + |x - 3| = 2$
86. $|4x - 2| + 1 = 5$
87. $|4x - 3| - 12 = -7$
88. $|3x + 4| - 5 = 17$
89. $|x + 2| + 8 = 9$
90. $|5x + 2| + 4 = 21$

17-6

Find and graph the solution set for each of the following inequalities.

91. $|x - 3| < 4$
92. $|x + 7| > 3$
93. $|x - 4| - 3 < 5$
94. $|x - 1| < 9$
95. $|x - 3| < -4$

Write inequality statements for the following.

96. You earn more than \$35,000 annually.

97. The gross income for Riddle's market exceeds that of Smith's market and falls short of the income of Duke's market. Smith's gross income is \$108,000 and Duke's gross income is \$250,000. Write an inequality expressing these relationships.

CHALLENGE PROBLEM

98. You are a travel agent and have been asked to plan a trip for the local math group. The math group has raised \$8,700 to spend on the trip. Your agency charges a one-time setup fee of \$300. The cost per person will be \$620.

(a) Write an inequality that shows the relationship of the maximum cost of the trip and the costs charged by the travel agency for x persons.

(b) If the total cost of the trip is \$8,700, solve the inequality to determine how many math students can go on the trip.

CONCEPTS ANALYSIS

1. The symmetric property of equations states that if $a = b$ then $b = a$. Is an inequality symmetric? Illustrate your answer with an example.

2. In equations if $a = b$, then $ac = bc$, if c equals any real number. Is the statement, if $a < b$ then $ac < bc$, true if c is any real number? For what values of c is the statement true? For what values of c is the statement false?

3. Write in your own words two differences in the procedures for solving equations and for solving simple inequalities.

4. Explain the difference between the statements $x < 2$ and $x \leq 2$.

5. If $x < 2$ or $x > 10$, is it correct to write the statement as $2 > x > 10$? Why or why not? Explain your answer.

6. Explain the difference between an *and* relationship and an *or* relationshp in inequalities.

7. In your own words, write a procedure for solving a compound inequality such as $5 < x + 2 < 9$.

8. Find the mistake in the following. Correct and briefly explain the mistake.

$$3x + 5 < 5x - 7$$
$$3x - 5x < -7 - 5$$
$$-2x < -12$$
$$x < 6$$

9. In the inequality $x^2 + 6x + 5 < 0$, what roles do the numbers -5 and -1 play? Explain the solution for the inequality in words.

10. If p_1 and p_2 are real numbers that solve the equation $ax^2 + bx + c = 0$ and $p_1 < p_2$, then when will the solution of $ax^2 + bx + c < 0$ be $p_1 < x < p_2$? When will the solution be $x < p_1$ or $x > p_2$? Give an example to illustrate each answer.

CHAPTER SUMMARY

Objectives	What to Remember	Examples
Section 17-1		
1. Use set terminology.	Set-builder notation is sometimes used to show sets.	Use set-builder notation to show the following set: the set of integers between -2 and 3, including 3. $\{x \mid x \in Z \text{ and } -2 < x \leq 3\}$
2. Illustrate sets of numbers that are inequalities on a number line and write inequalities in interval notation.	Solutions to inequalities can be shown on the number line or by using interval notation.	Show the solution of $-2 < x \leq 3$ on the number line and by using interval notation. 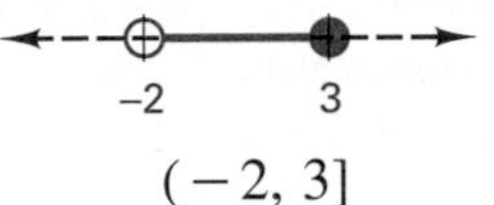 $(-2, 3]$
Section 17-2		
1. Solve a simple linear inequality.	To solve a simple linear inequality, isolate the letter as in solving equations with two exceptions. When reversing the sides of an inequality, reverse the sense of the inequality, and when multiplying or dividing by a negative number, reverse the sense of the inequality. Show both the algebra and graphic solutions of the inequality.	Solve. $4x - 1 < 7$ $4x < 8$ $x < 2$ 2 $-2x + 3 > 9$ $-2x > 6$ $x < -3$ -3
Section 17-3		
1. Identify subsets of sets and perform set operations.	Every set has the set itself and the empty set as subsets. Set operations include union (∪), intersection (∩), and complements (A′).	List the subsets of the set {3, 5, 8}: ∅, {3, 5, 8} (the set itself), {3}, {5}, {8}, {3, 5}, {3, 8), {5, 8}. If U = {1, 2, 3, 4, 5}, A = {1, 3}, and B = {2, 4}, find A ∪ B, A ∩ B, and A′. A ∪ B = {1, 2, 3, 4} A ∩ B = ∅ A′ = {2, 4, 5}
2. Solve compound inequalities with the conjunction condition.	To solve a compound inequality that is a conjunction, separate the compound inequality into two simple inequalities using the conditions of the conjunction. Solve each simple inequality. Determine the solution set that includes the *intersection* of the solution sets of the two simple inequalities. *Note:* If there is no overlap in the sets, the solution is the empty set.	Indicate the solution using a number line and by using interval notation. $-3 < x + 2 \leq 2$ $-3 < x + 2$ and $x + 2 \leq 2$ $-3 -2 < x, \quad x \leq 0$ $-5 < x$ or $x > -5$ $-5 < x \leq 0$ or $(-5, 0]$ -5 0

3. Solve compound inequalities with the disjunction condition.	To solve a compound inequality that is a disjunction: Solve each simple inequality. Determine the solution set that includes the union of the solution sets of the two simple inequalities.	Solve and indicate the solution set on the number line and by using interval notation. $x - 1 < 3$ or $x + 2 > 8$ $x < 4$ or $x > 6$ $(-\infty, 4)$ or $(6, \infty)$

Section 17-4

1. Solve quadratic inequalities.	To solve quadratic inequalities by factoring, rearrange the inequality so that the right side of the inequality is zero. Write the left side of the inequality in factored form. Determine the critical values that make each factor equal to zero. Test each region of values. The solution set for the quadratic inequality will be the region or regions that produce a true statement.	Solve $x^2 + 4x - 21 < 0$. $(x + 7)(x - 3) = 0$ $x + 7 = 0$ $x - 3 = 0$ $x = -7$ $x = 3$ I II III –7 3 Test region I. Let $x = -8$. $(-8)^2 + 4(-8) - 21 < 0?$ $64 - 32 - 21 < 0$ $11 < 0$. False. Test region II. Let $x = 0$. $0^2 + 4(0) - 21 < 0?$ $0 + 0 - 21 < 0$ $-21 < 0$. True. Test region III. Let $x = 4$. $4^2 + 4(4) - 21 < 0?$ $16 + 16 - 21 < 0$ $11 < 0$. False. –7 3 $(-7, 3)$
2. Solve rational inequalities.	To solve rational inequalities set both the numerator and denominator equal to zero and solve for the variable. These solutions are the boundaries for the solution region of the inequality solution. The region or regions that have like signs for the numerator and denominator are solutions for the $>$ and $\geq$ inequalities. The region or regions that have unlike signs for the numerator and denominator are solutions for the $<$ or $\leq$ inequalities.	Solve the rational inequality $\frac{x + 3}{x - 2} \leq 0$. $x = 2$ is an excluded value. $x + 3 = 0$ $x - 2 = 0$ $x = -3$ $x = 2$ –3 2 Region I has like signs $(\frac{-}{-})$ Region II has unlike signs $(\frac{+}{-})$ Region III has like signs $(\frac{+}{+})$ Unlike signs indicate < 0.

Section 17-5

1. Solve equations containing one absolute-value term.

To solve absolute-value equations, isolate the absolute-value expression then form two equations. One equation considers the expression within the absolute-value symbol to be positive; the other equation considers the expression to be negative. Solve each case to obtain the two roots of the equation.

Solve $|x - 3| + 1 = 5$.

$$|x - 3| + 1 = 5$$
$$|x - 3| = 4$$
$$x - 3 = 4 \quad \text{or} \quad -(x - 3) = 4$$
$$x = 7 \quad \text{or} \quad -x + 3 = 4$$
$$-x = 1$$
$$x = -1$$

Section 17-6

1. Solve absolute-value inequalities using the *less than* relationship.

To solve absolute-value inequalities using the less than relationship, use the property, if $|x| < b$ and $b > 0$; then $-b < x < b$.

Solve $|x - 3| < 4$.

$$-4 < x - 3 < 4$$
$$-4 < x - 3 \qquad x - 3 < 4$$
$$-1 < x \text{ or } x > -1 \qquad x < 7$$
$$-1 < x < 7$$

2. Solve absolute-value inequalities using the *greater than* relationship.

To solve absolute-value inequalities using the greater than relationship; use the property, if $|x| > b$ and $b > 0$, then $x < -b$ or $x > b$.

Solve $|x + 7| > 1$.

$$x + 7 < -1 \quad \text{or} \quad x + 7 > 1$$
$$x < -8 \quad \text{or} \quad x > -6$$

WORDS TO KNOW

inequality, (p. 637)
set, (p. 638)
member or element of a set, (p. 638)
set notation, (p. 638)
set-builder notation, (p. 638)
empty set, (p. 638)
interval notation, (p. 639)
boundary, (p. 639)
sense of an inequality, (p. 642)
compound inequality, (p. 646)
subset, (p. 646)
union of sets, (p. 647)
intersection of sets, (p. 647)
complement of a set, (p. 648)
conjunction, (p. 648)
disjunction, (p. 651)
quadratic inequality, (p. 653)
critical values, (p. 653)
rational inequality, (p. 659)
absolute value terms in equations, (p. 660)
property of absolute-value inequalities using $<$ or $\le$, (p. 663)
property of absolute-value inequalities using $>$ or $\ge$, (p. 664)

CHAPTER TRIAL TEST

Represent the following sets of numbers on the number line and by using interval notation.

1. $x \ge -12$
2. $-3 < x \le -2$

Solve the following inequalities. Graph the solution on a number line and write the solution using interval notation.

3. $3x - 1 > 8$
4. $2 - 3x \ge 14$
5. $10 < 2 + 4x$
6. $-5b > -30$
7. $\frac{1}{3}x + 5 \le 3$
8. $2(1 - y) + 3(2y - 2) \ge 12$
9. $5 - 3x < 3 - (2x - 4)$
10. $7 + 4x \le 2x - 1$
11. $-5 < x + 3 < 7$
12. $\frac{1}{4}x + 2 < 3 < \frac{1}{3}x + 9$
13. $3x - 1 \le 5 \le x - 5$
14. $(x + 4)(x - 2) < 0$
15. $(2x + 3)(x - 1) > 0$
16. $2y^2 + y < 15$
17. $2x - 3 < 1$ or $x + 1 > 7$
18. $2(x - 1) < 3$ or $x + 7 > 15$

Use the following sets to give the elements in the following sets.

$$U = \{1, 2, 3, 4, 5, 6, 7, 8, 9\} \quad A = \{5, 8, 9\}, \quad B = \{1, 2, 3, 4, 5, 6, 7, 8\}$$

19. $A \cup B$
20. $A \cap B$
21. $A \cap B'$
22. List all the subsets of set A.

Solve. Graph the solution set on a number line and write the solution set in interval notation.

23. $\dfrac{x-2}{x+5} < 0$

24. $\dfrac{x-3}{x} > 0$

Solve.

25. $|x| = 15$

26. $|x+8| = 7$

27. $|x| + 8 = 10$

Solve and graph the solution set on a number line and write the solution set in interval notation.

28. $|4x - 7| < 17$

29. $|x| + 8 > 10$

30. $|x+1| - 3 < 2$

CHAPTER

18 Graphing

A graph is a means of showing information visually. Some graphs show how our tax dollars are divided among various government services or how the population of an area has grown over the years. Other graphs are used to show equations, inequalities, and their solutions.

18-1 THE RECTANGULAR COORDINATE SYSTEM

Learning Objectives

L.O.1. Plot the points indicated by ordered pairs of numbers.
L.O.2. Identify the quadrant that contains a given point.
L.O.3. Find the coordinates of a given point.

We often use graphs to solve problems in technical applications. A common mathematical graph uses the *rectangular coordinate system*, which is similar to the lines of latitude and longitude used in locating places on a map and is the basis for line graphs in industrial reports, handbooks, and the like.

L.O.1 Plot the points indicated by ordered pairs of numbers.

Plotting Points

With the number line introduced in Chapter 1, we were able to graphically represent any numerical value on the line. By placing two number lines at 90° angles to each other and having them intersect at 0 on each line, we can locate any point on a plane or flat surface.

■ DEFINITION 18-1.1 **Rectangular Coordinate System.** The **rectangular coordinate system** is a plane (or flat) surface in which two number lines intersect at 90° angles and on which mathematical values may be plotted from the point of intersection that is 0 on both scales.

The point where both number lines intersect is called the *origin*. The *horizontal* line (the line that extends to the left and right from the origin) is called the *axis of abscissas*, or the *x-axis*. The *vertical* line (the line that goes up and down from the origin) is called the *axis of ordinates*, or the *y-axis*. The two axes divide the plane surface into four sections. Each section is called a *quadrant*. The number of units of horizontal movement from the origin is the *abscissa* or *x-coordinate*, and the number of units of vertical movement from the origin is the *ordinate* or *y-coordinate*. These two numbers are called the *coordinates* of the point.

Look at point A on Fig. 18-1. To locate point A from the origin, we have moved 5 units to the right and 3 units up. To locate point B from the origin, we have moved 4 units to the left and 1 unit up. To locate point C from the origin, we have moved 2 units to the left and 4 units down; and to locate point D from the origin, we have moved 3 units to the right and 2 units down.

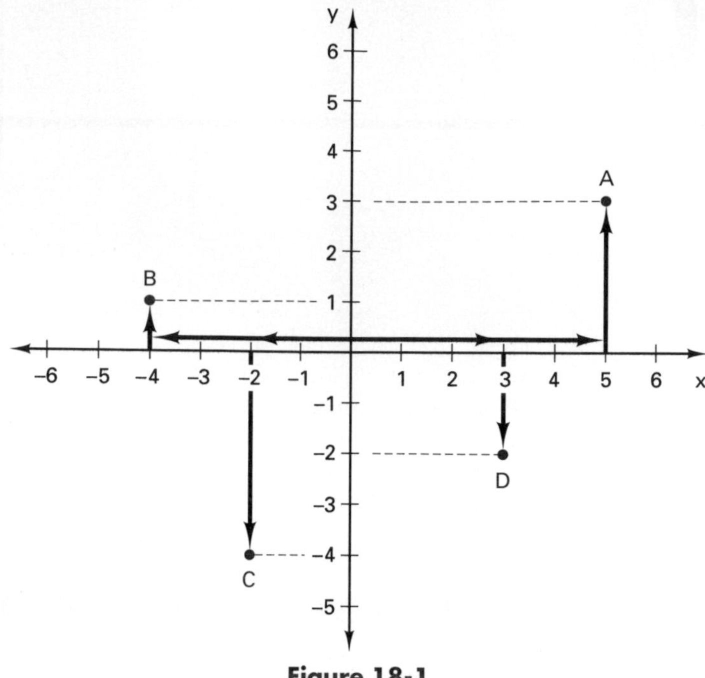

Figure 18-1

When locating any point, we determine the horizontal (x) movement first. Then we determine the vertical (y) movement. The point is written symbolically as (x, y). The coordinate pair (5, 3) represents point A in Fig. 18-1.

> **Rule 18-1.1** ***To plot a point,*** *that is, to position a given point in relation to the origin:*
>
> 1. Start at the origin and move horizontally (left or right) the number of units indicated by the x-coordinate.
> 2. From that point, move vertically (up or down) the number of units indicated by the y-coordinate.

In many cases the coordinates of a point may contain fractions, mixed numbers, or decimals. When plotting coordinates that are not whole numbers, we must make a close approximation of the actual value. The purpose of graphs is to give a pictorial view of data, and much of the information on a graph merely approximates the actual data.

EXAMPLE 18-1.1 Plot the following points.

Point A	(5, 2)	Point C	$(-1, -2)$	Point E	$\left(-3\frac{1}{2}, 2\frac{1}{2}\right)$
Point B	$(-3, 4)$	Point D	$(3, -6)$	Point F	$(3.5, -6.5)$

Solution: The solutions are shown in Fig. 18-2. ■

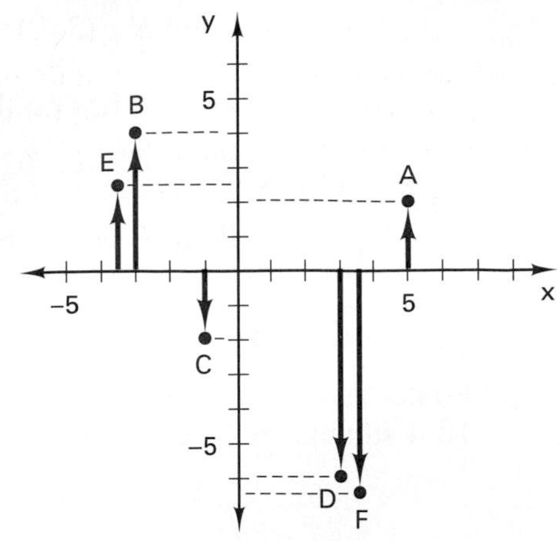

Figure 18-2

So far, all the points we have dealt with have fallen within one of the four quadrants. A point can, however, fall on one or both of the axes. This occurs when one or both of the coordinates are 0. A coordinate of 0 indicates no movement in a particular direction. Look at some examples of points with one or both coordinates equal to 0.

EXAMPLE 18-1.2 Plot the following points.

Point A (0, 5)	Point C $(-1, 0)$	Point E (3, 0)
Point B (5, 0)	Point D $(0, -1)$	Point F (0, 0)

Solution: Answers are shown in Fig. 18-3.

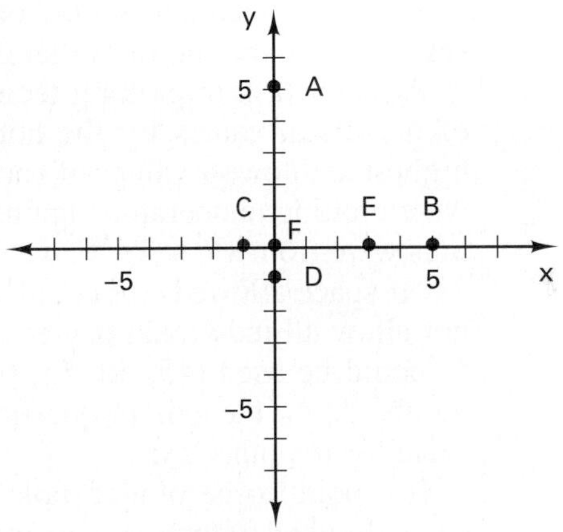

Figure 18-3

Point A (0, 5) no horizontal movement
$+5$ units vertically

Point A lies on the positive y-axis.

Point B (5, 0) $+5$ units horizontally
no vertical movement

Point B lies on the positive x-axis.

Point C $(-1, 0)$ -1 unit horizontally
no vertical movement

Point C lies on the negative x-axis.

Point D $(0, -1)$ no horizontal movement
-1 unit vertically

Point D lies on the negative y-axis.

Point E (3, 0) +3 units horizontally
no vertical movement
Point E lies on the positive x-axis.

Point F (0, 0) no horizontal movement
no vertical movement
Point F lies on both axes at the origin.

Tips and Traps

Points that have one or more coordinates of zero fall on one or both axes. Figure 18-4 summarizes these points.

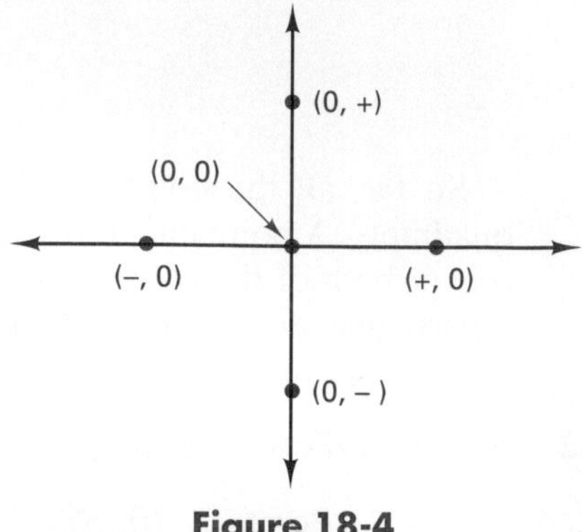

Figure 18-4

Often technical information from a data table is plotted on a line graph formed by a rectangular coordinate system to make the data easier to interpret. Each axis should be labeled and *scaled* or numbered in such increments that the complete set of points and the lines that connect the points cover most of the graph.

A major task in plotting technical data is selecting the x-scale and the y-scale of numerical values for the horizontal and vertical axes. In actual practice, the highest and lowest values of each variable to be plotted determine the scale. If we were plotting temperature against the days of the week, and the temperature readings were from 45° to 75°F, we would have a range of 31°, counting 45 up through 75. If space allowed, we could indicate all 31° on one axis. However, space may not allow all individual degrees to be indicated. Perhaps a scale in increments of 5° could be used (45, 50, 55, 60, 65, 70, 75), requiring just seven values to be numbered on the axis in question. A similar procedure can be used to select the scale for the other axis.

If a point to be plotted falls between marked intervals, the technician would use individual judgment to estimate the approximate location of the point and plot it accordingly.

For convenience, we will be told what data to put on the x-axis and y-axis, but on the job we should follow employer's guidelines.

EXAMPLE 18-1.3 From the data in the following table, plot a graph to show how horsepower (hp) increased as the revolutions per minute (rpm) of a certain test engine increased.

x-axis (rpm)	y-axis (hp)
250	10
500	22
750	33
1000	45
1250	55

Solution: The values for the rpm (x-) axis are already in 250-rpm increments and are used for the rpm scale. The hp values in the data table are not in even increments, but are at approximately 10-unit intervals; for this reason, the hp (y-) axis is scaled in multiples of 10. Values not falling on a y-axis interval are located in their approximate positions (see Fig. 18-5). ■

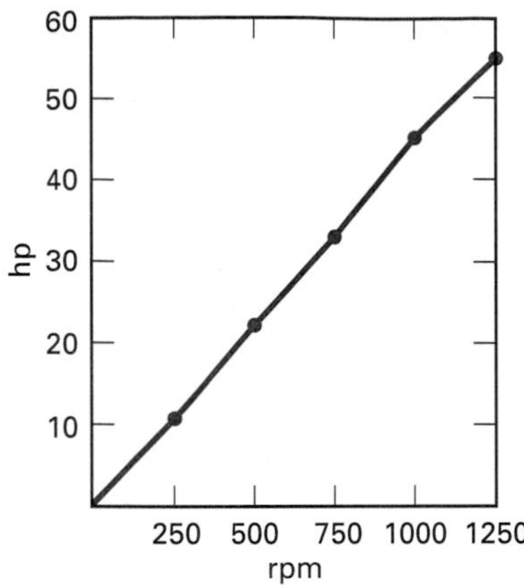

Figure 18-5

L.O.2. Identify the quadrant that contains a given point.

Quadrants

The rectangular coordinate system divides a plane into four regions called *quadrants*. We number the quadrants using Roman numerals starting in the upper-right region and proceeding *counterclockwise*. The coordinates of all points within a given quadrant will have the same algebraic signs.

Any point located in quadrant I represents horizontal movement to the right and vertical movement is up. Therefore, in quadrant I all x-coordinates and all y-coordinates are positive. In quadrant II, horizontal movement is to the left and vertical movement is up. Thus, in quadrant II all x-coordinates are negative and all y-coordinates are positive. In quadrant III (left and down), all x-coordinates are negative and all y-coordinates are negative. And in quadrant IV (right and down), all x-coordinates are positive and all y-coordinates are negative. These facts are summarized in Fig. 18-6.

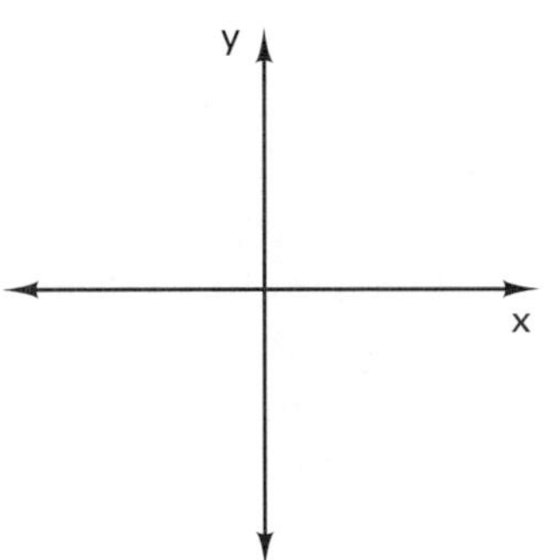

Figure 18-6

EXAMPLE 18-1.4 Determine the quadrant in which each of the following points would lie.

Point A	$(-2, 4)$	Point C	$(3, 2)$
Point B	$(7, -3)$	Point D	$(-1, -4)$

Solution

Point A	quadrant II	Point C	quadrant I
Point B	quadrant IV	Point D	quadrant III

■

L.O.3. Find the coordinates of a given point.

Coordinates of Points

When a point is positioned on the coordinate system, the x-coordinate is determined by the amount of horizontal movement from the origin. The y-coordinate is determined by the amount of vertical movement from the origin.

EXAMPLE 18-1.5 Give the coordinates of points A through L in Fig. 18-7 in (x, y) notation.

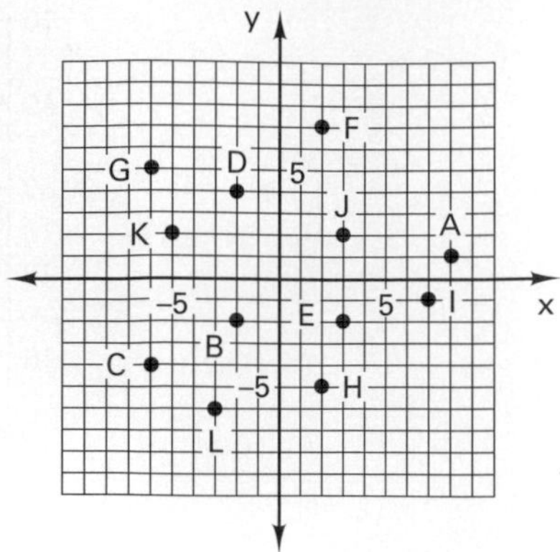

Figure 18-7

Solution

Point A	$(8, 1)$	Point E	$(3, -2)$	Point I	$(7, -1)$
Point B	$(-2, -2)$	Point F	$(2, 7)$	Point J	$(3, 2)$
Point C	$(-6, -4)$	Point G	$(-6, 5)$	Point K	$(-5, 2)$
Point D	$(-2, 4)$	Point H	$(2, -5)$	Point L	$(-3, -6)$ ■

SELF-STUDY EXERCISES 18-1

L.O.1. Using a set of coordinate axes, plot the following points and label them with the appropriate letter.

1. Point A $(-3, 2)$
2. Point B $(4, -7)$
3. Point C $(2, 1)$
4. Point D $(1, 6)$
5. Point E $(-2, -4)$

Plot the graph of the technical information contained in the data table below. Use increments of 0.5 for amps (1.5, 2.0, 2.5, . . .) and 20 for joules of heat (30, 50, 70, . . .). Number each point from 6 to 10. The data show the relationship between amperage and heat (joules) in an electrical circuit.

	x-axis (amps)	y-axis (joules)
6.	1.5	31.5
7.	2.0	56.
8.	2.3	74.06
9.	3.0	126.
10.	3.4	161.84

L.O.2. Give the quadrant for the following coordinate pairs.

11. $(-3, 4)$
12. $(7, -8)$
13. $(-5, -2)$
14. $(6, 3)$
15. $(-32, 7)$

L.O.3. Give the coordinates of the following points shown in Fig. 18-8.

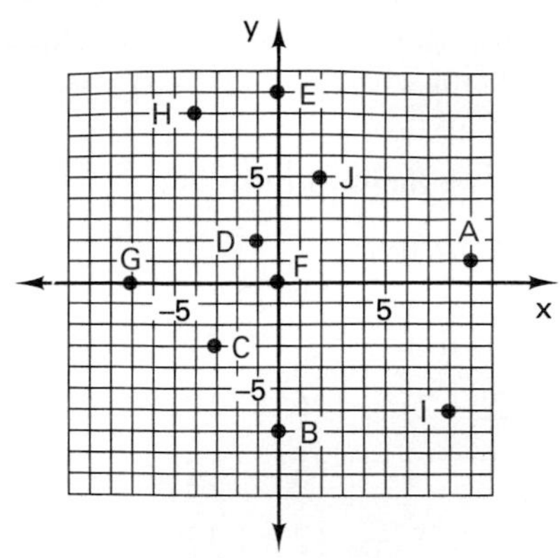

Figure 18-8

16. *A*
17. *B*
18. *C*
19. *D*
20. *E*
21. *F*
22. *G*
23. *H*
24. *I*
25. *J*

18-2 EQUATIONS IN TWO VARIABLES AND FUNCTION NOTATION

Learning Objectives

L.O.1. Check the solution of an equation with two variables.
L.O.2. Evaluate equations with two variables when given the value of one variable.
L.O.3. Interpret function notation and evaluate functions.

Most of the equations we have studied in this text have had only one variable, such as x or y, and took the form $3x - 2 = 7$ or $5y + 3 = 9$. However, equations may have two variables, both x *and* y, such as $2x + 3y = 12$. Both types of equations are *linear equations*.

■ **DEFINITION 18-2.1 Linear Equation.** A **linear equation** is one whose graph is a straight line.

A *linear equation in two variables* is an equation that can be written in the standard form

$$ax + by = c$$

where a, b, and c are real numbers and a and b are not both zero.

L.O.1. Check the solution of an equation with two variables.

Checking Solutions to Equations in Two Variables

The solutions to a linear equation in two variables are *ordered pairs* of numbers. There is one number for each variable. The numbers are ordered in the alphabetical order of the two variables rather than the order the letters appear in the equation. Thus, in the equation $y + x = 6$, a solution of (4, 2) indicates that $x = 4$ and $y = 2$. An ordered pair is enclosed in parentheses and the numbers are separated by a comma.

Solutions to equations in two variables are ordered pairs of numbers that make a true statement. An equation in two variables will have many solutions. We can check each solution to an equation in two variables by substituting the numerical values for the respective variables and performing the operations indicated.

EXAMPLE 18-2.1 Check the solution $(-2, 16)$ for the equation $2x + y = 12$.

Solution

$2x + y = 12$

$2(-2) + 16 = 12$ *(Substitute -2 for x and 16 for y.)*

$-4 + 16 = 12$

$12 = 12$ *(The ordered pair makes the equation true. So $(-2, 16)$ is a solution.)* ■

Now let's check several solutions for the same equation.

EXAMPLE 18-2.2 Check the ordered pairs $(3, -2)$, $(3, 1)$, and $(-2, -9)$ for the equation $y = 2x - 5$.

Solution: Check for $(3, -2)$.

$y = 2x - 5$

$-2 = 2(3) - 5$ *(Substitute 3 for x and -2 for y.)*

$-2 = 6 - 5$

$-2 = 1$ *(The ordered pair does not make the equation true. So $(3, -2)$ is not a solution.)*

Check for $(3, 1)$.

$y = 2x - 5$

$1 = 2(3) - 5$ *(Substitute 3 for x and 1 for y.)*

$1 = 6 - 5$

$1 = 1$ *(The ordered pair makes the equation true. So $(3, 1)$ is a solution.)*

Check for $(-2, -9)$.

$y = 2x - 5$

$-9 = 2(-2) - 5$ *(Substitute -2 for x and -9 for y.)*

$-9 = -4 - 5$

$-9 = -9$ *(The ordered pair makes the equation true. So $(-2, -9)$ is a solution.)* ■

L.O.2. Evaluate equations with two variables when given the value of one variable.

Solving an Equation in Two Variables When Given One Variable

If we are given the value of one variable for an equation with two variables, we can substitute the given value in the equation and solve the equation for the other variable.

EXAMPLE 18-2.3 Solve the equation $2x - 4y = 14$, if $x = 3$.

$$2x - 4y = 14$$

$$2(3) - 4y = 14 \qquad \textit{(Substitute 3 for x.)}$$

$$6 - 4y = 14$$

$$-4y = 14 - 6 \qquad \textit{(Subtract 6 from both sides.)}$$

$$-4y = 8$$

$$\frac{-4y}{-4} = \frac{8}{-4} \qquad \textit{(Divide both sides by } -4.)$$

$$y = -2$$

The solution is $(3, -2)$; $x = 3$ and $y = -2$. ■

The standard form of an equation in two variables, $ax + bx = c$, is such that both a and b are not zero. However, either a or b may be zero.

When a or b is zero, the variable for which zero is the coefficient is eliminated from the equation. Let's take the equation $5x + 3y = 15$ and replace the coefficient 3 with zero:

$$5x + 3y = 15$$

$$5x + 0(y) = 15 \qquad \textit{(Replace 3 with 0.)}$$

$$5x + 0 = 15 \qquad \textit{(0(y) = 0. 0 times any number is 0.)}$$

$$5x = 15$$

$$\frac{5x}{5} = \frac{15}{5} \qquad \textit{(Divide both sides by 5.)}$$

$$x = 3$$

Thus, an equation like $x = 3$ may be considered along with equations in two variables if we think of the coefficient of the missing variable term as zero:

$$x = 3 \quad \text{is considered as} \quad x + 0y = 3$$

Written with 0 as the coefficient of y, this form of the equation shows that, for any value of y, x is 3.

EXAMPLE 18-2.4 For the equation $x = 6$, complete the given ordered pairs.

Solution

Equation	Ordered Pairs			
$x = 6$	(, 2)	(, −3)	(, 1)	
	(6, 2)	(6, −3)	(6, 1)	*(x is 6 for any value of y.)* ■

A number of application problems can be solved with equations in two variables.

Example 18-2.5 A small business photocopied a report that included 25 black-and-white pages and 12 color pages. The cost was \$2.45. Letting b = the cost for a black-and-white page and c = the cost for a color page, set up an equation with two variables.

Solution

Cost of black-and-white copies: $25b$ (25 times cost per page)

Cost of color copies: $12c$ (12 times cost per page)

Total cost: black-and-white plus color = \$2.45

Equation: $25b + 12c = 2.45$ ■

An equation containing two variables shows the relationships between the two unknown amounts. If a value is known for either amount, the other amount can be found by solving the equation for the missing amount.

Example 18-2.6 Solve the equation from Example 18-2.5 to find the cost for each black-and-white copy if color copies cost \$0.10 each.

Solution

$$25b + 12c = 2.45$$

$$25b + 12(0.10) = 2.45 \qquad \textit{(Substitute 0.10 for c.)}$$

$$25b + 1.20 = 2.45$$

$$25b = 2.45 - 1.20 \qquad \textit{(Subtract 1.20 from both sides.)}$$

$$25b = 1.25$$

$$\frac{25b}{25} = \frac{1.25}{25} \qquad \textit{(Divide both sides by 25.)}$$

$$b = 0.05$$

Each black-and-white copy cost \$0.05. ■

L.O.3. Interpret function notation and evaluate functions.

Function Notation

Another way of writing an equation in two variables is in *function notation*. In the equation $f(x) = x + 6$ (read "a function of x is $x + 6$" or "f of x is $x + 6$"), one variable is x and the other is $f(x)$. The notation emphasizes that the value of $f(x)$ depends on the value of x. Thus, we call $f(x)$ the *dependent variable* and x the *independent variable*. Using the function $f(x) = x + 6$, $f(2) = x + 6$ is interpreted to mean that we find the value of the dependent variable $f(x)$ when $x = 2$.

$$f(x) = x + 6$$

$$f(2) = 2 + 6 \qquad \textit{(Substitute 2 for x.)}$$

$$= 8 \qquad \textit{(Combine.)}$$

Example 18-2.7 Find (a) $f(1)$, (b) $f(-2)$, and (c) $f(0)$ for the function $f(x) = 3x - 1$.

Solution

(a) $f(x) = 3x - 1$
$f(1) = 3(1) - 1$ *(Substitute 1 for x.)*
$= 3 - 1$
$= 2$

(b) $f(x) = 3x - 1$
$f(-2) = 3(-2) - 1$ *(Substitute −2 for x.)*
$= -6 - 1$
$= -7$

(c) $f(x) = 3x - 1$
$f(0) = 3(0) - 1$ *(Substitute 0 for x.)*
$= 0 - 1$
$= -1$

■

When using function notation our calculations are restricted to the right side of the equation. It is permissible to omit the left side of the equation in the steps of the solution.

We will study the definition of a function, some uses of functions, and graphing functions later. For now, we will concentrate on using the function notation and evaluating functions for given values.

SELF-STUDY EXERCISES

L.O.1. Determine which of the following ordered pairs are solutions for the equation $2x - 5y = 9$.

1. (4, 5) **2.** (17, 5) **3.** (12, 3)
4. (6, 1) **5.** (7, 1) **6.** (4.5, 0)

Determine which of the following ordered pairs are solutions for the equation $3y = 2 - 4x$.

7. $(2, -2)$ **8.** $(-2, 2)$ **9.** $(5, -6)$
10. $(-6, 5)$ **11.** $\left(0, \frac{2}{3}\right)$ **12.** $\left(6, -7\frac{1}{3}\right)$

Determine which of the following ordered pairs are solutions for the equation $4y - x = 7$.

13. (1, 2) **14.** $(-2, -12)$ **15.** $(-2, -15)$
16. (3, 5) **17.** (5, 3) **18.** (7, 0)

L.O.2. Solve the following equations.

19. $x - y = 3$, if $x = 8$ **20.** $x + y = 7$, if $x = 2$
21. $x + 2y = -4$, if $y = 1$ **22.** $x - 3y = 12$, if $y = 3$
23. $5x - y = 10$, if $x = 5$ **24.** $4x + y = 8$, if $x = 2$
25. $3x + 4y = 16$, if $y = -2$ **26.** $5x + 2y = 9$, if $y = -3$
27. $3x - y = 16$, if $x = 6$ **28.** $x + 3y = 12$, if $x = 9$ **29.** $\frac{2}{3}x - y = 6$, if $y = 2$
30. $x - \frac{3}{4}y = 9$, if $x = 0$ **31.** $6 = x + 3y$, if $y = -4$ **32.** $8 = 2x + y$, if $x = -2$

Complete the ordered pairs for each equation.

33. $x = 1$ (, 3); (, −2); (, 0); (, 1) **34.** $y = 4$ (−1,); (3,); (0,); (2,)
35. $y = -7$ (2,); (−3,); (0,); (5,) **36.** $x = 9$ (, −2); (, 3); (, 0); (, 2)

Solve the following using equations with two variables.

37. Diane purchased three shirts at one price and five belts at another price. Let s = the price of each shirt and b = the price of each belt. Each shirt cost $22. How much did each belt cost if the total purchase price was $126?

38. Paulette and Terry Fink paid $13 for four spark plugs and five quarts of motor oil. Let p = the cost of each plug and q = the cost of each quart of oil. If the oil cost $1 per quart, how much did each spark plug cost?

39. Paul and Donna were charged \$160 for four pairs of pants and two shirts. If the shirts cost \$12 each, how much did all four pairs of pants cost? Let p = the cost of each pair of pants and s = the cost of each shirt.

40. David and Deanna paid \$15,000 for an automobile. The purchase price included three extended warranty payments of \$400 each. How much did the car cost without the extended warranty? Let c = the cost of the car and w = the cost of each warranty payment.

L.O.3. Evaluate the following for $f(3)$, $f(-2)$, and $f(0)$.

41. $f(x) = 4x - 3$ **42.** $f(x) = 2x + 7$ **43.** $f(x) = -4x - 1$

44. $f(x) = -6x + 2$ **45.** $f(x) = \frac{2}{3}x + \frac{1}{2}$ **46.** $f(x) = \frac{3}{8}x + \frac{1}{4}$

47. $f(x) = -\frac{1}{5}x + 2$ **48.** $f(x) = -\frac{4}{5}x + 1$ **49.** $f(x) = 0.7x - 0.23$

50. $f(x) = 2.1x + 0.5$ **51.** $f(x) = -2.4x - 5$ **52.** $f(x) = -0.3x - 2.4$

53. If $f(x) = 5x + 7$, find $f(-2)$.

54. If $f(x) = -15x - 3$, find $f(-1)$.

55. If $f(x) = 7x - 4$, find $f(0)$.

56. If $f(x) = -6x + 1$, find $f(-5)$.

18-3 GRAPHING LINEAR EQUATIONS WITH TWO VARIABLES

Learning Objectives

L.O.1. Graph linear equations using the table of solutions method.
L.O.2. Graph linear equations using intercepts.
L.O.3. Graph linear equations using the slope–intercept method.
L.O.4. Graph linear equations using a graphing calculator.

Sometimes it is useful to graph equations, that is, plot the values of the variable or variables on the rectangular coordinate system. In this section we will graph *linear equations* with two variables.

L.O.1. Graph linear equations using the table-of-solutions method.

Graphing Linear Equations Using a Table of Solutions

Look at the equation $x + y = 7$. The ordered pair $x = 3$ and $y = 4$ is a solution of this equation. In an equation of this type, the ordered pair $x = 3$ and $y = 4$ is not the only solution for this equation. The ordered pair $x = 2$ and $y = 5$ also satisfies this equation $(2 + 5 = 7)$ and is a solution of the equation. Similarly, we can see that there are many solutions to this equation. Here is a list of some other solutions of this equation.

$x = 1, y = 6$	$(1 + 6 = 7)$	$x = 8, y = -1$	$(8 + (-1) = 7)$
$x = 0, y = 7$	$(0 + 7 = 7)$	$x = \frac{1}{2}, y = 6\frac{1}{2}$	$\left(\frac{1}{2} + 6\frac{1}{2} = 7\right)$
$x = -3, y = 10$	$(-3 + 10 = 7)$	$x = 2.4, y = 4.6$	$(2.4 + 4.6 = 7)$

Since there is an infinite number of solutions for an equation of this type, we can represent the solutions graphically. To do this, we consider each solution as a point on our rectangular coordinate system. Then we write some of the solutions of the equation in a *table of solutions*, as follows.

x	y
3	4
2	5
1	6
0	7
-3	10
8	-1
$\frac{1}{2}$	$6\frac{1}{2}$
2.4	4.6

Each pair of values in the table represents a point on the graph of the equation. These solutions can be written in coordinate notation as

$$
\begin{array}{ll}
(3,\ 4) & (-3,\ 10) \\
(2,\ 5) & (8,\ -1) \\
(1,\ 6) & \left(\frac{1}{2},\ 6\frac{1}{2}\right) \\
(0,\ 7) & (2.4,\ 4.6)
\end{array}
$$

Let's graph the points in relation to the coordinate axes of Fig. 18-9.

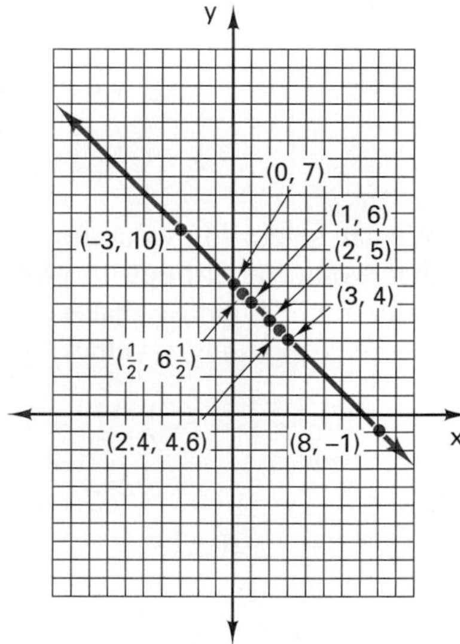

Figure 18-9

Notice that these eight solutions to the linear equation $x + y = 7$ can be connected by a straight line. This straight line can be extended in both directions to represent all the solutions of the equation. Any solution to this equation will fall on this straight line. If the line crosses the x-axis, the point of intersection is called the *x-intercept.* If the line crosses the y-axis, the point of intersection is called the *y-intercept.*.

The following two rules help us find solutions of linear equations and plot them on a rectangular coordinate system.

Rule 18-3.1 *To find a pair of values that are a solution of a linear equation containing two variables:*

1. Solve the equation for one of the variables. This will be the dependent variable.
2. Select a value for the other variable (the independent variable).
3. Calculate the corresponding value of the variable solved for in step 1.

Rule 18-3.2 *To graph a linear equation using a table of solutions:*

1. Determine at least three solutions of the equation and make a table of solutions.
2. Plot each solution included in the table of solutions as a point in relation to a pair of coordinate axes.
3. Connect these points with a straight line. (If the points do not all fall on the same line, check for calculation errors or plotting errors.)
4. Represent all the solutions by placing an arrow on each end of the line. This indicates that the line extends indefinitely in both directions.

Example 18-3.1 Graph the equation $2x + y = 12$.

Solution: First, we need to determine at least three solutions for this equation. We do this by selecting values for x and calculating the corresponding y-value. To simplify our calculations, we solve the equation for y, since y already has a coefficient of 1.

$$2x + y = 12$$

$$y = 12 - 2x \quad \text{or} \quad y = -2x + 12 \qquad \textit{(preferred)}$$

We can select any value for x, but if our selection creates a corresponding y-value that would be difficult to plot, we can disregard that solution and select another value for x. Suppose that we select the values -2, 0, and $+2$ for x. We then find the corresponding y-values by substituting in the equation $y = -2x + 12$.

When $x = -2$

$$\begin{aligned} y &= -2x + 12 \\ &= -2(-2) + 12 \\ &= +4 + 12 \\ &= 16 \end{aligned}$$

When $x = 0$

$$\begin{aligned} y &= -2x + 12 \\ &= -2(0) + 12 \\ &= -0 + 12 \\ &= 12 \end{aligned}$$

When $x = +2$

$$\begin{aligned} y &= -2x + 12 \\ &= -2(2) + 12 \\ &= -4 + 12 \\ &= 8 \end{aligned}$$

The first two solutions for x have relatively large corresponding y-values. We can avoid having to plot such large numbers by selecting different values for x, say 4 and 6.

When $x = 4$

$$\begin{aligned} y &= -2x + 12 \\ &= -2(4) + 12 \\ &= -8 + 12 \\ &= 4 \end{aligned}$$

When $x = 6$

$$\begin{aligned} y &= -2x + 12 \\ &= -2(6) + 12 \\ &= -12 + 12 \\ &= 0 \end{aligned}$$

Next, we make a table of the solutions that we are graphing:

x	y
2	8
4	4
6	0

Finally, we plot these solutions, connect them with a straight line, and place arrows on each end of the line as shown in Fig. 18-10.

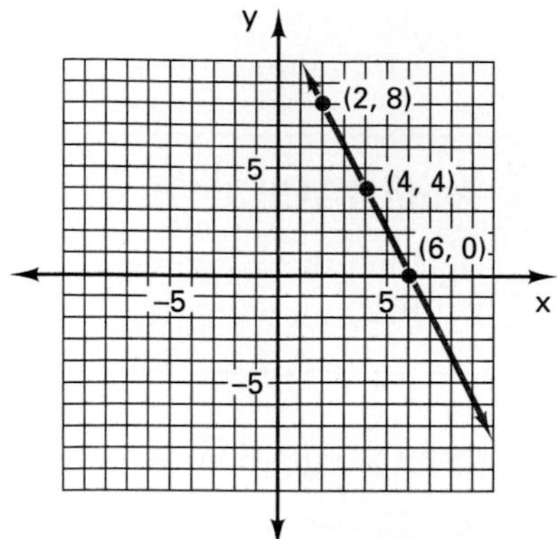

Figure 18-10

We can check our graph by finding another solution of the equation and seeing if it falls on the graph of the equation. When $x = 1$,

$$\begin{aligned} y &= -2(1) + 12 \\ &= -2 + 12 \\ &= 10 \end{aligned}$$

Since the solution (1, 10) falls on the graph of $2x + y = 12$, the solution checks. ■

When an equation is solved for y, it can also be written in function notation: $y = -2x + 12$ or $f(x) = -2x + 12$. A table of solutions for the function would be

x	$f(x)$
2	8
4	4
6	0

The graph of $f(x) = -2x + 12$ would be the same as Figure 18-10. The values for $f(x)$ are plotted on the y-axis.

Tips and Traps

The following general tips will be useful in using the table-of-solutions method of graphing linear equations in two variables.

1. Solve for y and select values for x.
2. Select values for x that are easy to work with.
3. Two points determine a straight line, but the third point is important to catch mistakes. The third point is a check point.

L.O.2. Graph linear equations using intercepts.

Graphing Using Intercepts

We can also graph a linear equation with two variables by finding only the x-intercept and the y-intercept, plotting these points, and drawing a line through them.

■ **Definition 18-3.1 x-intercept.** The **x-intercept** is the point on the x-axis through which the line of the equation passes $(x, 0)$.

■ **Definition 18-3.2 y-intercept.** The **y-intercept** is the point on the y-axis through which the line of the equation passes $(0, y)$.

The following rule may help.

Rule 18-3.3 *To find the intercepts of the axes:*

1. Find the x-intercept by letting $y = 0$ and solving for x.
2. Find the y-intercept by letting $x = 0$ and solving for y.

Example 18-3.2 Graph the equation $3x - y = 5$ by using the intercepts of each axis.

Solution

$3x - y = 5$ *(For x-intercept, let y = 0.)*

$3x - 0 = 5$

$3x = 5$

$x = \frac{5}{3}$ *(A mixed number or decimal is easier to plot than an improper fraction.)*

$x = 1\frac{2}{3}$

Thus, the x-intercept is $\left(1\frac{2}{3}, 0\right)$.

$3x - y = 5$

$3(0) - y = 5$ *(For y-intercept, let x = 0.)*

$-y = 5$

$y = -5$

Thus, the y-intercept is $(0, -5)$

Plot the two points, $\left(1\frac{2}{3}, 0\right)$ and $(0, -5)$ and draw the graph. See Fig. 18-11.

We can check by finding one other point in the usual way and plotting it. If it is on the line, the graph is correct. When $x = 2$,

$3x - y = 5$

$3(2) - y = 5$ *(Substitute 2 for x.)*

$6 - 5 = y$ *(Solve for y.)*

$1 = y$ *(The point (2, 1) checks.)* ■

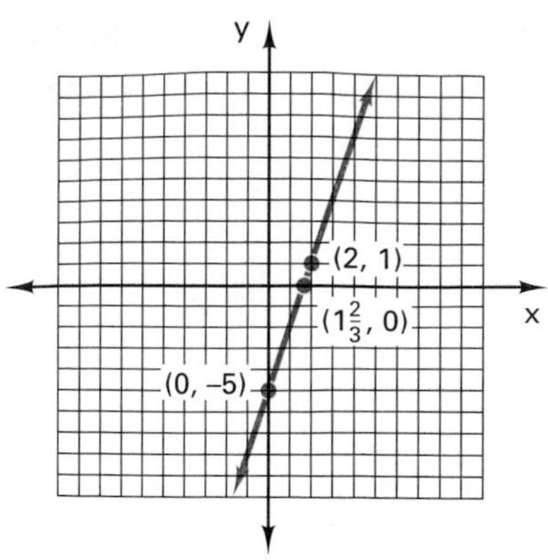

Figure 18-11

Tips and Traps

If both intercepts are (0, 0), they coincide on the origin, forming only *one point*. An additional point must be found in the usual way in order to have two distinct points to draw the line through. A third point is useful to check your work.

$$y = 7x$$

If $x = 0$, then $y = 7(0)$ or $y = 0$. Both the x- and y-intercepts are (0, 0). So let $x = 1$, then $y = 7(1)$ or $y = 7$. Plot the point (1, 7) and (0, 0) and draw the graph.

L.O.3. Graph linear equations using the slope–intercept method.

Graphing Using the Slope and *y*-Intercept

An equation in the form $y = mx + b$, where m and b are real numbers, is called the *slope–intercept form of the equation*. This form allows us to identify characteristics of the graph of this equation by inspection.

First, let's examine the characteristics identified by b. We will let $m = 1$ and look at the graphs of the equations in Figure 18-12.

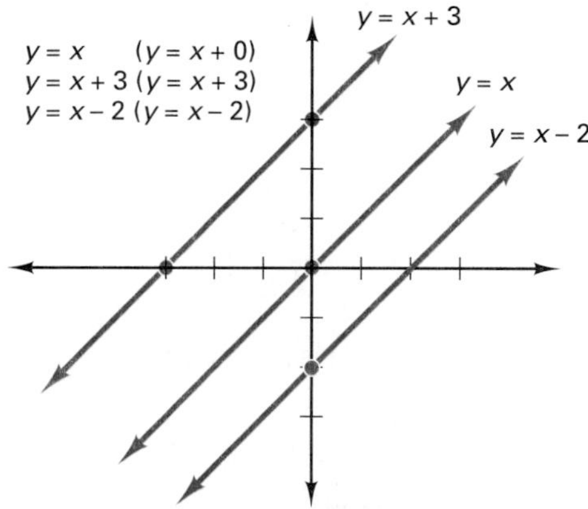

Figure 18-12

The three graphs have the same slope (steepness or slant), but they have different y-intercepts.

$y = x$	crosses y at	0;	y-intercept = (0, 0)
$y = x + 3$	crosses y at	$+3$;	y-intercept = (0, 3)
$y = x - 2$	crosses y at	-2;	y-intercept = (0, -2)

The constant (b) in an equation in the form $y = mx + b$ is the y-coordinate of the y-intercept.

Now, let's examine the characteristics identified by m. We will let $b = 0$ and look at the graphs of the equations in Figure 18-13.

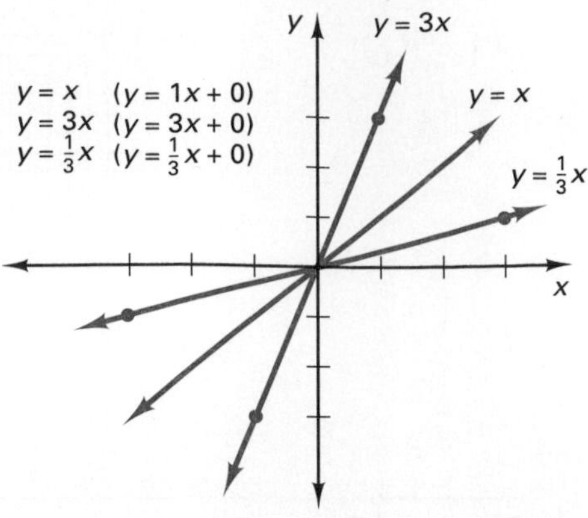

Figure 18-13

The three graphs have the same y-intercepts, but they have different slopes (steepness or slant). A formal definition for the slope of a line will be given in Chapter 19, but for now we will define the *slope* of a line as the ratio of the vertical change from one point to another to the horizontal change.

A slope of 3 (written as the ratio $\frac{3}{1}$) means a change of 3 vertical units for every 1 horizontal unit. A slope of $\frac{1}{3}$ means a change of 1 vertical unit for every 3 horizontal units (see Fig. 18-14).

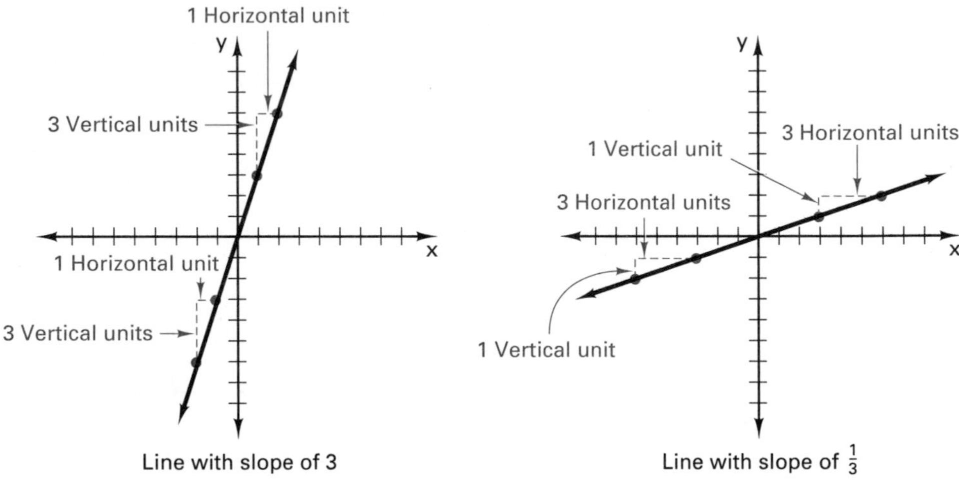

Figure 18-14

An equation in the slope–intercept form can be graphed by using just the slope and y-intercept.

Rule 18-3.4 ***To graph a linear equation in the form $y = mx + b$ by using the slope and y-intercept:***

1. Locate the y-intercept on the y-axis.
2. Using the slope, determine the amount of vertical and horizontal movement indicated.
3. From the y-intercept, locate additional points on the graph of the equation by counting the indicated vertical and horizontal movement.
4. Draw the line graph connecting the points.

EXAMPLE 18-3.3 Graph the following equations using the slope and y-intercept.

(a) $y = 5x + 2$ **(b)** $y = -\frac{3}{4}x - 3$ **(c)** $y = -2x + 1$

(d) $2x + y = 4$

Solution

(a) $y = 5x + 2$

y-intercept $= (0, 2)$.

Slope $= \frac{5}{1}$ or $\frac{-5}{-1}$.

$\frac{5}{1}$ indicates vertical movement of $+5$ and horizontal movement of $+1$.

$\frac{-5}{-1}$ indicates vertical movement of -5 and horizontal movement of -1.

See Figure 18-15.

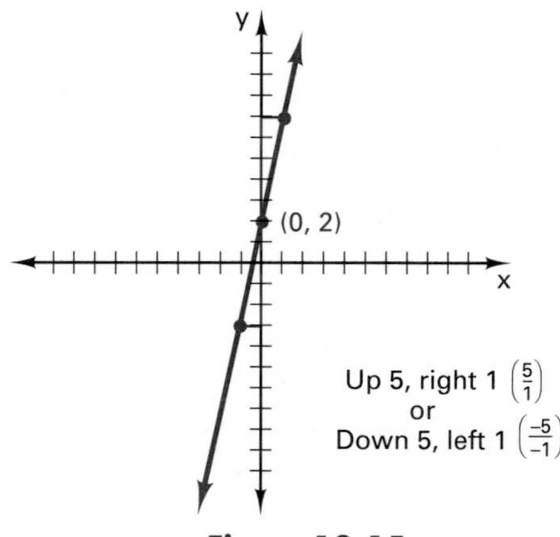

Figure 18-15

(b) $y = -\frac{3}{4}x - 3$

y-intercept $= (0, -3)$.

Slope $= \frac{-3}{4}$ or $\frac{3}{-4}$.

$\frac{-3}{4}$ indicates vertical movement of -3 and horizontal movement of $+4$.

$\frac{3}{-4}$ indicates vertical movement of $+3$ and horizontal movement of -4.

See Figure 18-16.

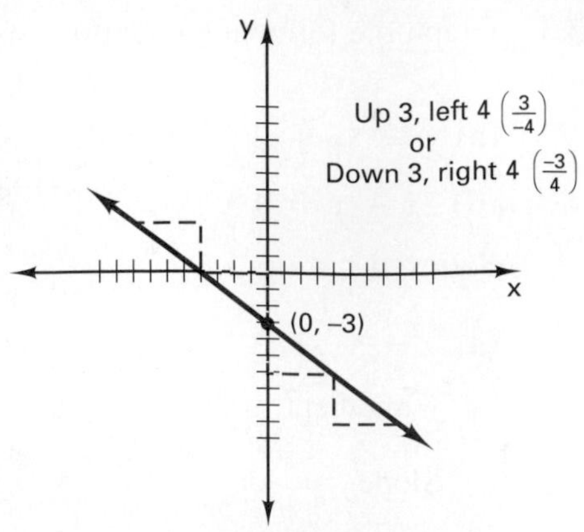

Figure 18-16

(c) $y = -2x + 1$

y-intercept $= (0, 1)$.

Slope $= \dfrac{-2}{1}$ or $\dfrac{2}{-1}$.

$\dfrac{-2}{1}$ indicates vertical movement of -2 and horizontal movement of $+1$.

$\dfrac{2}{-1}$ indicates vertical movement of $+2$ and horizontal movement of -1.

See Fig. 18-17.

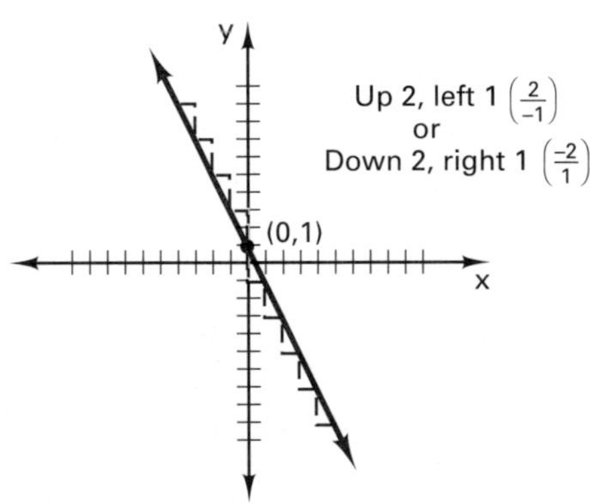

Figure 18-17

(d) $2x + y = 4$
$y = -2x + 4$ *(Solve for y.)*

y-intercept $= (0, 4)$

Slope $= \dfrac{-2}{1}$ or $\dfrac{2}{-1}$

$\dfrac{-2}{1}$ indicates vertical movement of -2 and horizontal movement of $+1$.

$\dfrac{2}{-1}$ indicates vertical movement of $+2$ and horizontal movement of -1.

See Figure 18-18.

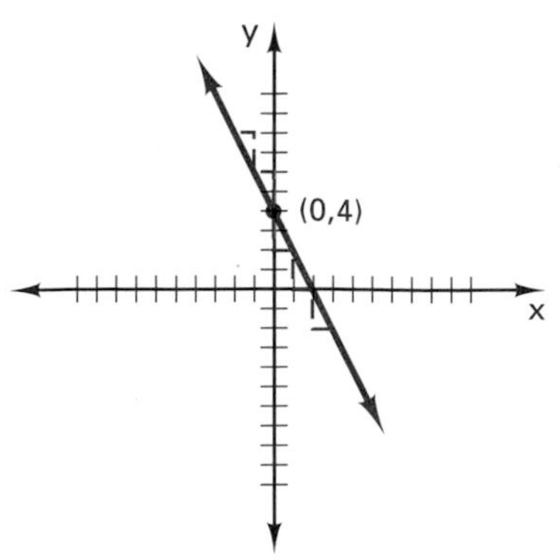

Figure 18-18

L.O.4. Graph linear equations using a graphing calculator.

Using a Graphing Calculator to Graph Linear Equations

A graphing calculator, such as the Casio 7700 or 7000 or the Texas Instruments 81 or 82, can be used to show the graphs of linear equations that can be written in the form $y = mx + b$. Also, computer software, such as Converge, Derive, Mathcad, and Mathematica, can also be used to show the graphs of linear equations. These calculators and computer software are tools that free us from the tedious task of graphing equations and allow us to focus on the patterns and properties of graphs.

Every model of graphing calculator and every brand of computer algebra software will have a different set of key strokes for graphing equations. You will need to refer to the owner's manual to adapt to a particular model or brand. However, we will illustrate the usefulness of these tools by showing a few features on the Casio 7700 and the TI-81.

EXAMPLE 18-3.4 Graph $y = 5x + 2$ using a graphing calculator.

Solution

Casio fx-7700G:

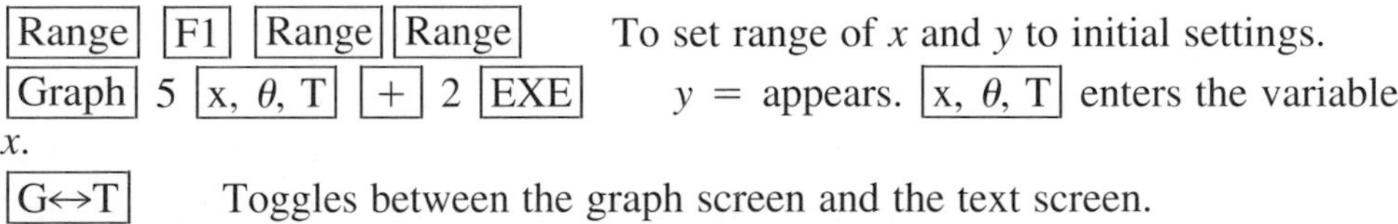

[Range] [F1] [Range] [Range] To set range of x and y to initial settings.
[Graph] 5 [x, θ, T] [+] 2 [EXE] $y =$ appears. [x, θ, T] enters the variable x.
[G↔T] Toggles between the graph screen and the text screen.

To see a closer view we will define a box and view the box.

Select the Zoom feature with function key [F2]. Select the box feature by selecting the [F1] option. The cursor (blinking dot) is at the origin. Use the left arrow key to move the cursor to a corner of the desired viewing box ($x = -1$, $y = -0.5$). Notice that the coordinates of x and y change as you move the cursor. When the cursor is in the desired place, press [EXE] to "nail down" the corner.

Next, move the up arrow key to an adjacent corner of the desired viewing box ($x = -1$, $y = 3$). A line connects the first corner and the second. Press [EXE], then use the right arrow key to move the cursor to the next adjacent corner ($x = 1$, $y = 3$). A box is forming when you move the arrow keys. Once the box is defined, press [EXE] to view the graph in the box. Be sure to include critical values like the x- and y-intercepts in the box. To see the approximate values of the x- and y-intercepts, use the [Trace] function on the calculator. Press [F1] (Trace). The coordinates of the leftmost point on the graph appear at the bottom of the screen. As you move the right arrow key, the cursor moves along the graph, and the coordinates change accordingly. Move the cursor to the value of y that is closest to 0 to determine the coordinates of the x-intercept.

$$y = -0.021276 \qquad \text{Closer to 0.}$$
$$y = 0.0851063$$

The approximate value of x when $y = 0$ is -0.4.

Next, find the value of x that is closest to 0 to determine the coordinates of the y-intercept.

$$x = -4.\text{E-}14 \quad (-0.00000000000004) \qquad \text{Closer to 0.}$$
$$x = 0.0212765$$

The approximate value of y is 2 when $x = 0$. If the desired accuracy is not achieved, you can define a viewing box just around the x-intercept or y-intercept for closer views. There are other ways to "zoom" critical values. Refer to your calculator owner's manual for more details.

To return to the initial range settings, press [Range] [F1] [Range] [Range].

To clear the graph from the calculator, press [F5] [EXE] [AC].

TI-81:

The keyboard layout of the TI-81 is different from the Casio fx-7700G, but the procedures for graphing are similar. To enter the equation to be graphed, use the [y=] key. The variable key is labeled [X|T].

[y=] 5 [X|T] [+] 2 [Graph]

For a closer view, use the [Zoom] key. Select option 1, box, by pressing 1. The cursor begins at the origin rather than on the graph. Use the right or left arrow to move the cursor to the right or left and the up or down arrow key to move the cursor up or down. Move the cursor to any corner of the box and press [ENTER]. Move the cursor to the diagonal corner of your desired box and press [ENTER]. The portion of the graph inside the box is displayed. To find the approximate values of the x- and y-intercepts, use the [Trace] key and the left or right arrow keys.

For more details, refer to your calculator owner's manual.

To clear the graph, return to the text screen using the [y=] key. Use the delete key, [Del], to delete the equation. The range can be changed using the [Range] key or the [Zoom] key.

To return to the default range, press [Zoom] and select the standard, option 6. To display a graph with equal intervals on x and y, select option 5, square, from the zoom menu. ■

SELF-STUDY EXERCISES 18-3

L.O.1. Graph the following equations. Make a table of at least three solutions before plotting.

1. $x + y = 9$ **2.** $x - y = 3$ **3.** $y = 2x - 4$
4. $2x - y = 5$ **5.** $x + 2y = 7$

L.O.2. Graph the following equations using the intercepts procedure.

6. $x + 3y = 5$ **7.** $\frac{1}{2}y = 4 + x$ **8.** $y = 3x - 1$

9. $5x = y + 2$ **10.** $x = -2y + 3$

L.O.3. Graph the following equations using the slope–intercept procedure.

11. $y = 2x - 3$ **12.** $y = -\frac{1}{2}x - 2$ **13.** $y = -\frac{3}{5}x$

14. $x - 2y = 3$ **15.** $2x + y = 1$

L.O.4. Verify the graphs for Exercises 1 through 15 using a graphing calculator.

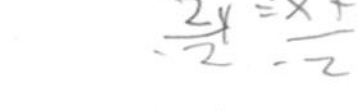

18-4 GRAPHING LINEAR INEQUALITIES WITH TWO VARIABLES

Learning Objectives

L.O.1. Graph linear inequalities with two variables using test points.
L.O.2. Graph linear inequalities with two variables using a graphing calculator.

The techniques used for graphing linear equations can be extended to graphing inequalities. Let's look at graphing linear inequalities.

L.O.1. Graph linear inequalities with two variables using test points.

Graphing Linear Inequalities Using Test Points

To graph a linear inequality, first graph a similar mathematical statement that substitutes an equal sign for the inequality symbol.

The graphical solution for a linear inequality includes all points on one side of the line representing the graph of the similar equation. If either the "less than or equal" (≤) or "greater than or equal" (≥) symbol is used, the points on the line are also included.

EXAMPLE 18-4.1 Graph the inequality $2x + y < 3$.

Solution: First graph the equation

$$2x + y = 3$$

Solve the equation for y.

$$y = -2x + 3$$

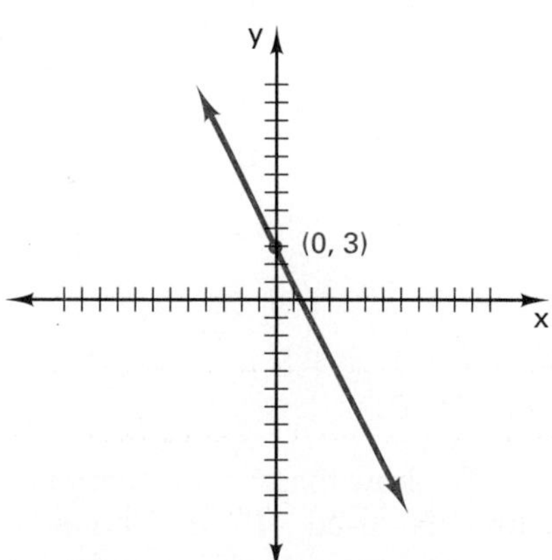

Slope = −2, y-intercept = (0, 3)

Figure 18-19

The graph of the equation $y = -2x + 3$ divides the surface into two parts and it is the *boundary* separating the parts. In graphing inequalities, one of the parts is included in the solution. To represent this solution, we will *shade* the appropriate part.

Now, select two points, one on each side of the boundary to determine which side of the line contains the set of points that solves the inequality. Suppose that we choose (2, 2) as the point above the boundary.

$$\text{Is } 2x + y < 3 \quad \text{when} \quad x = 2 \text{ and } y = 2?$$
$$2(2) + 2 < 3$$
$$4 + 2 < 3$$
$$6 < 3 \qquad \text{False statement}$$

Thus, the side of the line including the point (2, 2) represents the opposite inequality

$$2x + y > 3$$

Suppose that we choose (0, 0) as the point below the boundary.

$$\text{Is } 2x + y < 3 \quad \text{when} \quad x = 0 \text{ and } y = 0?$$
$$2(0) + 0 < 3$$
$$0 + 0 < 3$$
$$0 < 3 \qquad \text{True statement}$$

Thus, the side of the line including the point (0, 0) represents

$$2x + y < 3$$

To graph the solution set of the inequality, shade the portion of the graph that represents $2x + y < 3$, the given inequality (see Fig. 18-20). This is the side of the boundary where (0, 0) is located.

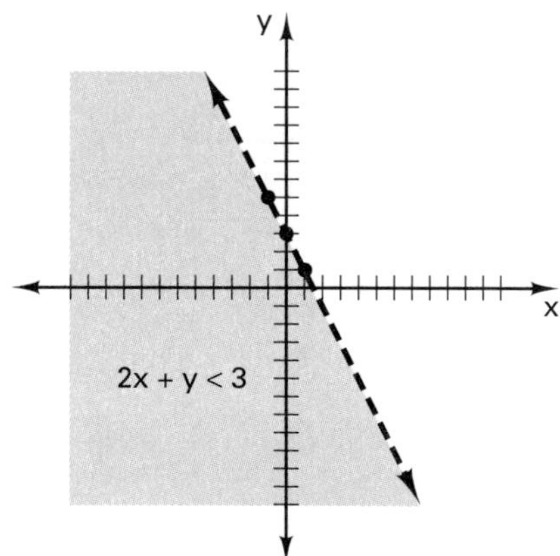

Figure 18-20

To show that the boundary is not included in the solution set, a dashed or broken line is used. If the boundary is to be included, as in the graph of $2x + y \leq 3$, a solid or unbroken line is used. Figure 18-21 illustrates the graph of $2x + y \leq 3$. ■

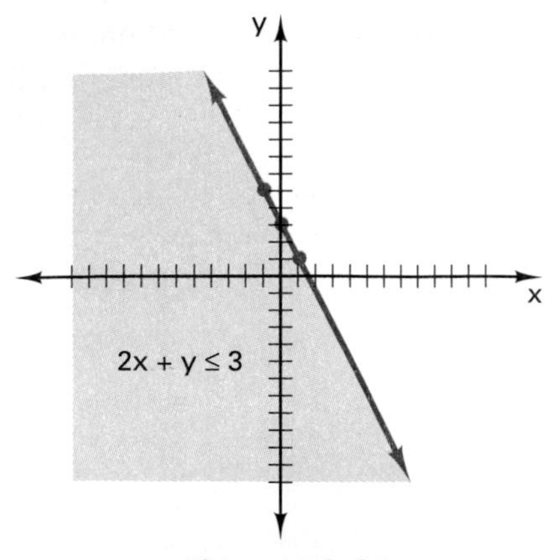

Figure 18-21

Tips and Traps

Choose numbers that are easy to work with when selecting the coordinates to determine the side of the boundary that represents the solution of the inequality.

EXAMPLE 18-4.2 Graph the inequality $y \geq 3x + 1$.

Solution: Graph the line $y = 3x + 1$.

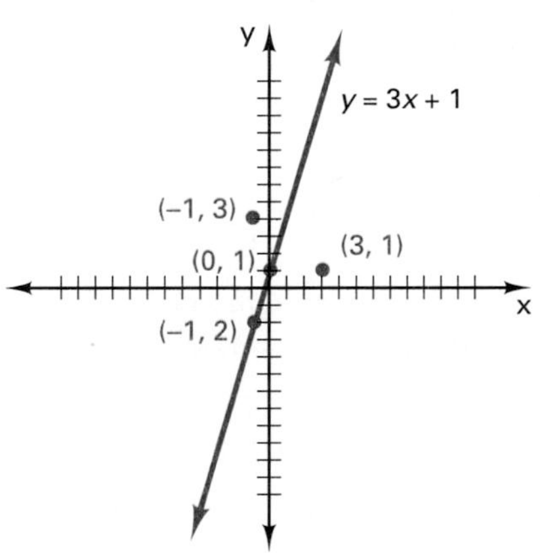

Figure 18-22

Since the inequality $y \geq 3x + 1$ *does* include the line $y = 3x + 1$, we will represent the line with an unbroken line (see Fig. 18-23).

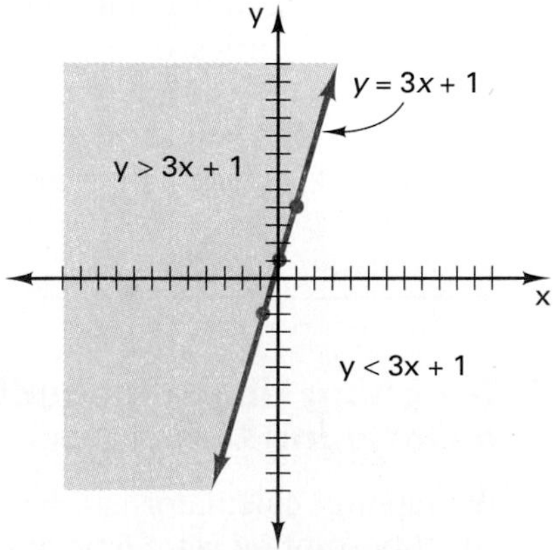

Figure 18-23

Now we select two points, one on each side of the boundary to determine the solution set of the inequality. Suppose that we choose (3, 1) as one of our points.

$$\text{Is } y \geq 3x + 1 \quad \text{when} \quad x = 3 \text{ and } y = 1?$$

$$1 \geq 3(3) + 1$$

$$1 \geq 9 + 1$$

$$1 \geq 10 \qquad \text{False statement}$$

Thus, the boundary and the side of the line including the point (3, 1) represents the opposite inequality

$$y \leq 3x + 1$$

Suppose that we choose $(-1, 3)$ as the other point.

$$\text{Is } y \geq 3x + 1 \quad \text{when} \quad x = -1 \text{ and } y = 3?$$

$$3 \geq 3(-1) + 1$$

$$3 \geq -3 + 1$$

$$3 \geq -2 \qquad \text{True statement}$$

Thus, the boundary and the side of the boundary including the point $(-1, 3)$ represents

$$y \geq 3x + 1$$

To show the solution set, we shade the portion of the graph that represents $y \geq 3x + 1$, the given inequality (see Fig. 18-23). The shaded portion along with the boundary line is the graph of $y \geq 3x + 1$. Notice, this is the same side of the line that contains the point $(-1, 3)$. ■

Tips and Traps

Do two points have to be tested to determine which region to shade?

If the first test point makes a true statement, the side of the boundary that includes the test point is shaded.

If the first test point makes a false statement, the side of the boundary that does not include the test point is shaded.

Thus, only one test point is necessary.

Other properties of inequalities can be applied when the inequality is in the form $y < mx + b$, $y \leq mx + b$, $y > mx + b$, or $y \geq mx + b$.

In general, when an inequality is in the form $y < mx + b$ or $y \leq mx + b$, the solution set includes all points *below* the boundary line. When an inequality is in the form $y > mx + b$ or $y \geq mx + b$, the solution set includes all points *above* the boundary line.

L.O.2. Graph linear inequalities with two variables using a graphing calculator.

Graphing Linear Inequalities Using a Graphing Calculator

A graphing calculator can be used to show the graph of the boundary of an inequality solution set. Some graphing calculators will also show the shaded region. Again, the owner's manual should be consulted to determine the capabilities of a given model or brand.

Using the Casio 7700, inequalities can be selected by accessing the menu system.

[Mode] [Shift] [÷] accesses inequalities mode.

The appropriate inequality is selected by the function keys.

[F1] $y >$

[F2] $y <$

[F3] $y \geq$

[F4] $y \leq$

The right side of the inequality is entered as before.

EXAMPLE 18-4.3 Use a graphing calculator to graph $2x + y < 3$.

Solution

Solve for y and write in the form $y < mx + b$.

$$y < -2x + 3$$

Clear previous graphs from the calculator.

Cls
[F5] [EXE]

Put the calculator in the inequality mode.

[Mode] [Shift] [÷]

Enter the inequality.

[Graph] [F2] [−] 2 [X,θ,T] [+] 3 [EXE]

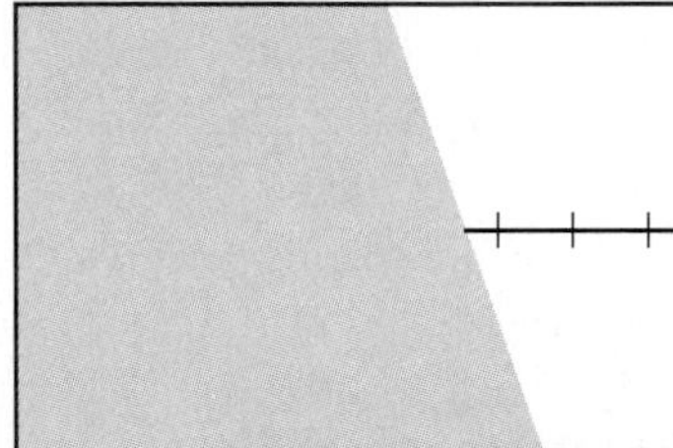

Figure 18-24

The calculator graph does not distinguish between a dashed or solid boundary line.

To put the calculator back in the equation mode, press

[Mode] [Shift] [+]. ■

SELF-STUDY EXERCISES 18-4

L.O.1. Graph the following linear inequalities using test points.

1. $x - y < 8$
2. $x + y > 4$
3. $3x + y < 2$
4. $2x + y \leq 1$
5. $x + 2y < 3$
6. $2x + 2y \geq 3$
7. $y \geq 2x - 3$
8. $y \geq -3x + 1$
9. $y > \frac{1}{2}x - 3$
10. $y > -\frac{3}{2}x + \frac{1}{2}$

L.O.2. Verify the graphs in Problems 1 through 10 using a graphing calculator.

18-5 GRAPHING QUADRATIC EQUATIONS AND INEQUALITIES

Learning Objectives

L.O.1. Identify nonlinear equations.
L.O.2. Graph quadratic equations using the table-of-solutions method.
L.O.3. Graph quadratic equations and inequalities using a graphing calculator.

L.O.1. Identify nonlinear equations.

Identifying Nonlinear Equations

In the previous sections, we discussed graphing of linear equations and inequalities. Linear equations have two basic characteristics.

1. No letter term will have a letter with an exponent greater than 1.
2. No letter term will have two or more different letter factors.

Quadratic equations also have two basic characteristics.

1. No letter term will have a degree greater than 2. Quadratic equations may have terms with two different letter factors as long as the degree of the term does not exceed 2.
2. At least one term of the equation must have degree 2.

EXAMPLE 18-5.1 Identify the following equations as linear, quadratic, or other nonlinear equations?

(a) $3x + 4y = 12$
(b) $3x + 4 = 8$
(c) $3x + 4xy + 2y = 0$
(d) $y = x^2 + 4$
(e) $y = 3x^3 - 2x^2 + x$

Solution

(a) $3x + 4y = 12$
This is a linear equation. The letter terms have neither an exponent greater than 1 nor two different letter factors in the same term.

(b) $3x + 4 = 8$
This is a linear equation. The only letter term does not have an exponent greater than 1 nor two different letter factors.

(c) $3x + 4xy + 2y = 0$
This is *not* a linear equation. The term $4xy$ has two different letter factors and a degree of 2. This is a quadratic equation.

(**d**) $y = x^2 + 4$

This is *not* a linear equation. The term x^2 has a degree of 2. The equation is a quadratic equation.

(**e**) $y = 3x^3 - 2x^2 + x$

This equation is neither linear nor quadratic. It has a term that has degree 3. The equation is called a *cubic equation.* ■

L.O.2. Graph quadratic equations using the table-of-solutions method.

Graphing Quadratic Equations Using the Table-of-Solutions Method

Linear equations graph into *straight* lines. We will examine the graphs of some equations that have a degree higher than 1.

The graph of an equation of a degree higher than 1 will be a *curved* line. The curved line can be a parabola, hyperbola, circle, ellipse, or an irregular curved line. We will not define these terms at this time, but Figure 18-25 illustrates them. The equation for a parabola that opens up or down is distinguished from other quadratic equations. The y variable has degree 1. The x variable must have one term with degree 2. Such a quadratic equation can be written in function notation, $f(x) = ax^2 + bx + c$.

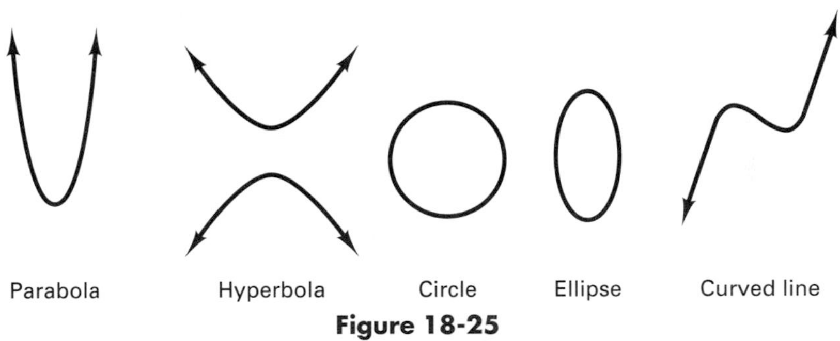

Figure 18-25

One method of graphing nonlinear equations is to form a table of solutions and plot the points. This method will require *more* than the three points that we customarily use in graphing linear equations or inequalities.

A parabola is symmetrical. That is, it can be folded in half and the two halves match. The fold line is called the *axis of symmetry.* For a parabola in the form $y = ax^2 + bx + c$, the equation of the axis of symmetry is $x = -\frac{b}{2a}$.

The point of the graph that crosses the axis of symmetry is the *vertex* of the parabola. Thus, the x-coordinate of the vertex of the parabola is $-\frac{b}{2a}$.

EXAMPLE 18-5.2 Graph the following equation using the table-of-solutions method.

$$y = x^2$$

Solution: Find the x-coordinate of the vertex, $-\frac{b}{2a}$. The equation $y = x^2$ can be written, $y = x^2 + 0x + 0$. Hence, $-\frac{b}{2a} = -\frac{0}{2(1)} = 0$. Make a table of values using at least *three* values of x on either side of the vertex and zero for the values of the x variable.

	x	y	x^2
	-3		
	-2		
	-1		
Vertex	0		
	$+1$		
	$+2$		
	$+3$		

Calculate the corresponding y values. The completed table is

x	y	x^2
-3	9	$(-3)^2 = 9$
-2	4	$(-2)^2 = 4$
-1	1	$(-1)^2 = 1$
Vertex 0	0	$0^2 = 0$
$+1$	1	$1^2 = 1$
$+2$	4	$2^2 = 4$
$+3$	9	$3^2 = 9$

Plot the points and trace the shape (see Fig. 18-26). ■

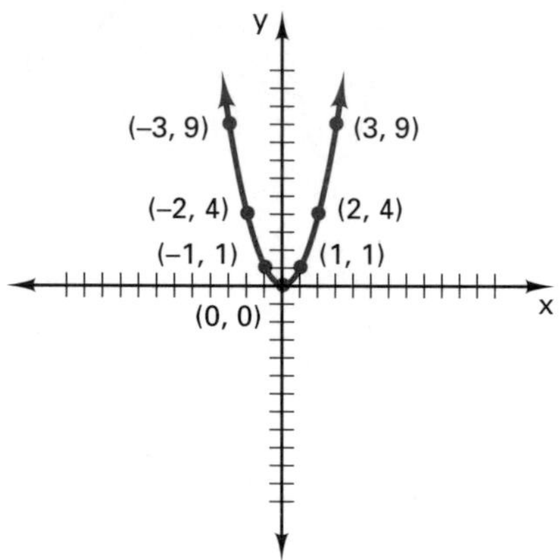

Figure 18-26

Example 18-5.3 Graph the following equation using the table-of-solutions method.

$$y = x^2 + x - 6$$

Solution: To examine some of the properties of quadratic equations, factor and solve the equation.

$$y = x^2 + x - 6$$
$$y = (x + 3)(x - 2)$$

To solve, let $y = 0$. Recall, when $y = 0$, the corresponding values of x are x-coordinates of x-intercepts.

$$0 = (x + 3)(x - 2)$$
$$x + 3 = 0 \qquad x - 2 = 0$$
$$x = -3 \qquad x = 2$$

Thus, the graph crosses the x-axis at -3 and 2.

For the parabola $y = x^2 + x - 6$, the equation of the axis of symmetry is $x = -\frac{b}{2a}$ or $x = -\frac{1}{2(1)} = -\frac{1}{2}$.

Now, let's form a table of solutions for $y = x^2 + x - 6$ that are centered around $x = -\frac{1}{2}$.

	x	y	$x^2 + x - 6$
	-3	0	$(-3)^2 + (-3) - 6 = 9 - 3 - 6 = 0$
	-2	-4	$(-2)^2 + (-2) - 6 = 4 - 2 - 6 = -4$
	-1	-6	$(-1)^2 + (-1) - 6 = 1 - 1 - 6 = -6$
Vertex	$-\frac{1}{2}$	$-6\frac{1}{4}$	$\left(-\frac{1}{2}\right)^2 + \left(-\frac{1}{2}\right) - 6 = \frac{1}{4} - \frac{1}{2} - 6 = -6\frac{1}{4}$
	0	-6	$0^2 + 0 - 6 = 0 + 0 - 6 = -6$
	1	-4	$1^2 + 1 - 6 = 1 + 1 - 6 = -4$
	2	0	$2^2 + 2 - 6 = 4 + 2 - 6 = 0$

By using the vertex and pairs of points equally spaced on each side of the vertex, we can shorten the number of calculations we need to make in the table-of-solutions method. See Fig. 18-27. ■

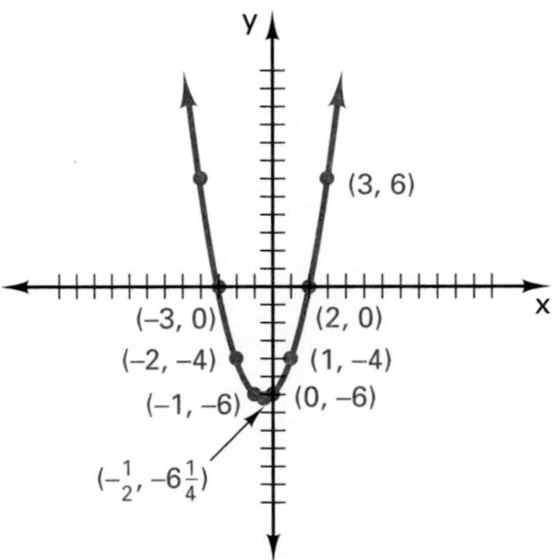

Figure 18-27

EXAMPLE 18-5.4 Graph the following equation using the table-of-values technique.

$$y = -x^2 + 3$$

Solution

Axis of symmetry: $x = -\frac{b}{2a}$ $\quad (y = -x^2 + 0x + 3)$

$$x = -\frac{0}{2(-1)}$$

$$x = 0$$

Make a table of values using at least *three* values less than and three values more than zero for the values of the x variable.

	x	y	$-x^2 + 3$
	-3	-6	$-(-3)^2 + 3 = -9 + 3 = -6$
	-2	-1	$-(-2)^2 + 3 = -4 + 3 = -1$
	-1	2	$-(-1)^2 + 3 = -1 + 3 = 2$
Vertex	0	3	$-(0)^2 + 3 = -0 + 3 = 3$
	$+1$	2	$-(1)^2 + 3 = -1 + 3 = 2$
	$+2$	-1	$-(2)^2 + 3 = -4 + 3 = -1$
	$+3$	-6	$-(3)^2 + 3 = -9 + 3 = -6$

x	y
-3	-6
-2	-1
-1	2
0	3
$+1$	2
$+2$	-1
$+3$	-6

apply concept of symmetry

Plot the points and trace the shape. See Fig. 18-28. ■

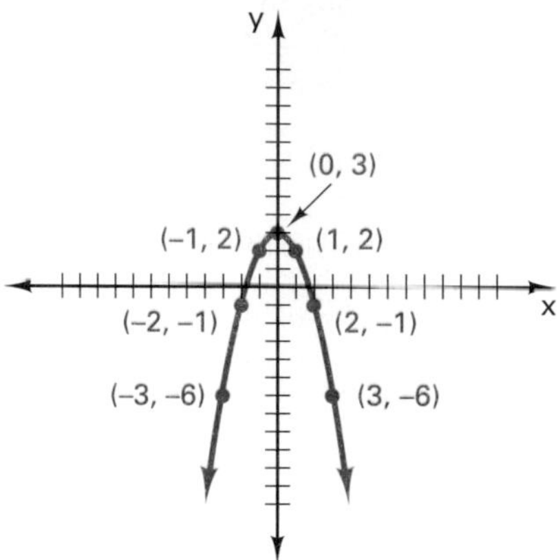

Figure 18-28

Tips and Traps

When the coefficient of the squared letter term is negative as in Example 18-5.4, the graph of the parabola opens downward. When the coefficient of the squared letter term is positive, the graph of the parabola opens upward as in Example 18-5.3.

L.O.3. Graph quadratic equations and inequalities using a graphing calculator.

Graphing Quadratic Equations and Inequalities Using a Graphing Calculator

Quadratic equations and inequalities are graphed on a graphing calculator using the same procedures as for linear equations and inequalities. The equation or inequality must first be solved for y.

EXAMPLE 18-5.5 Graph the following using a graphing calculator.

(a) $y = 2x^2 - 5x - 3$ **(b)** $y < 2x^2 - 5x - 3$
(c) $y \geq 2x^2 - 5x - 3$

Solution

(a) $y = 2x^2 - 5x - 3$

Casio 7700:

Keystrokes	
Cls [F5] [EXE]	Clear previous graphs.
INIT [Range] [F1] [Range] [Range]	Initiate range to preset range.
[Graph] 2 [x, θ, T] [x^2] [−] 5 [x, θ, T] [−] 3 [EXE]	

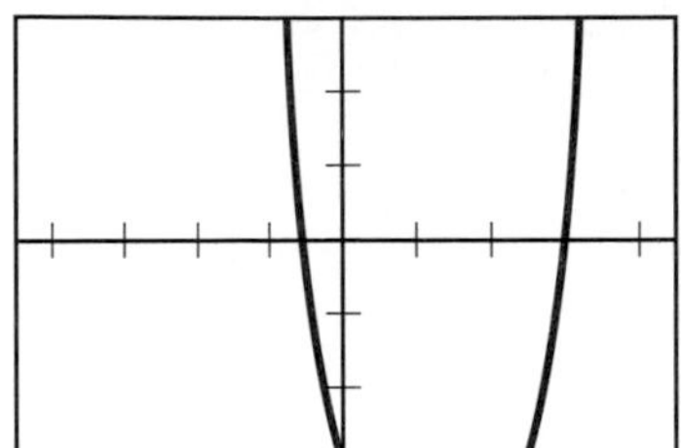

Figure 18-29

To view the portion of the parabola containing the vertex, change the range to show the x-values from -2 to $+4$ and the y-values from -10 to 5. Leave the scale at 1.

[Range] [−] 2 [EXE] 4 [EXE] [EXE] [−] 10 [EXE] 5 [EXE] [EXE] [Range]

Leave all settings on the second screen the same by repeatedly pressing the EXE key.

Graph again by pressing the left arrow key to remove the ''done'' indication. Then press [EXE].

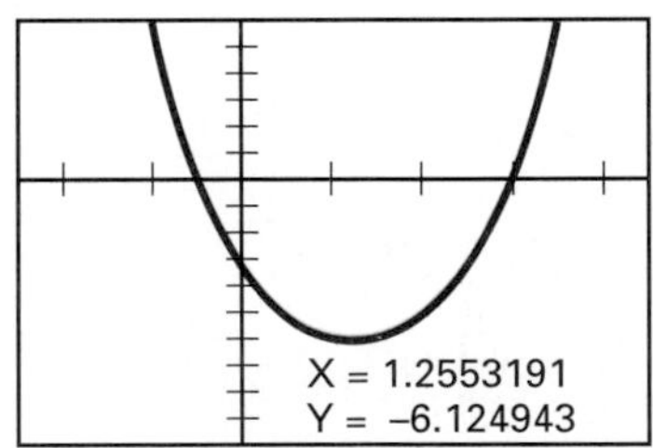

Figure 18-30

Different views can be seen by changing the range. More accurate values for x and y at critical points such as x- and y-intercepts and the vertex can be displayed by using the trace function with closer views of particular points.

(b) $y < 2x^2 - 5x - 3$

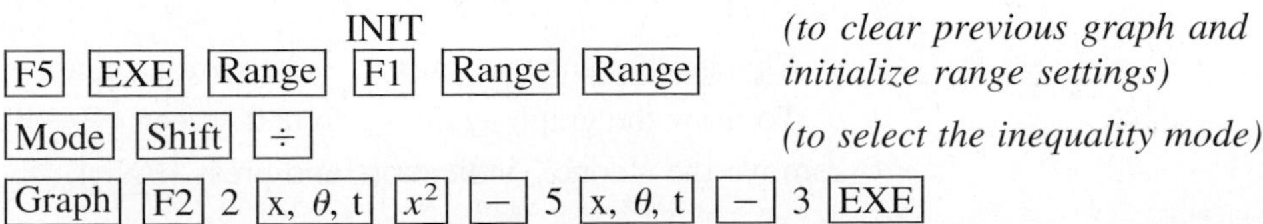

Keystrokes	
INIT [F5] [EXE] [Range] [F1] [Range] [Range]	*(to clear previous graph and initialize range settings)*
[Mode] [Shift] [÷]	*(to select the inequality mode)*
[Graph] [F2] 2 [x, θ, t] [x^2] [−] 5 [x, θ, t] [−] 3 [EXE]	

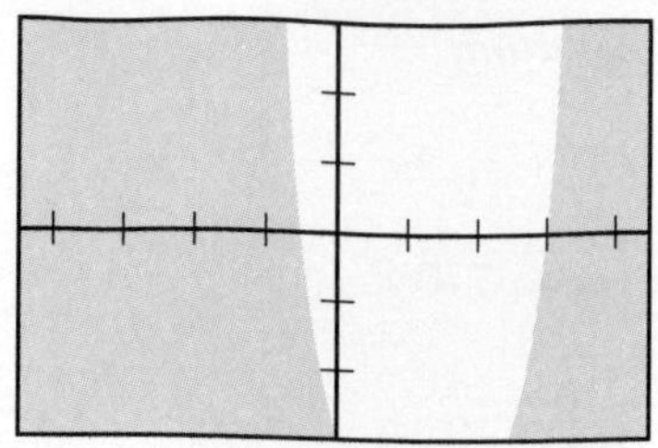

Figure 18-31

To see a different view, change the range settings to the settings used in part a. To draw the graph again, go to the text screen [G↔T], press the left arrow key to remove the "done" indication, and press [EXE].

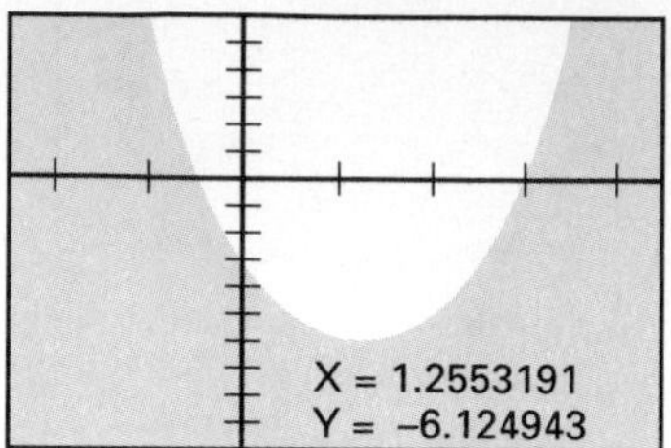

Figure 18-32

Tips and Traps

Other features of the calculator, such as zoom and box, allow you to view different portions of the graph. A smaller range allows a closer examination of critical points. A larger range allows a more complete overview of the graph.

On the Casio fx-7700G, the initial settings for the range that are programmed into the calculator give a view with x and y intervals equally spaced, thus having a minimum of distortion. Multiplying (or dividing) the x and y values by the same amount will allow different views with a minimum of distortion.

(c) $y \geq 2x^2 - 5x - 3$

[CLS] [EXE] [Range] [F1] [Range] [Range] *(to clear previous graph and initialize range settings)*

[Mode] [Shift] [÷] *(to select the inequality mode)*

[Graph] [F3] 2 [x, θ, t] [x^2] [−] 5 [x, θ, t] [−] 3 [EXE]

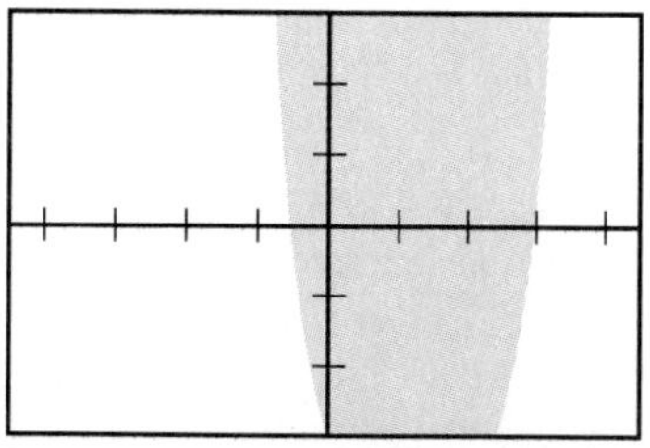

Figure 18-33

Change the range settings to −2 to 4 for x and −10 to 5 for y.

To draw the graph again, go to text screen [G↔T], press the left arrow key to remove the "done" indication, and press [EXE]. ■

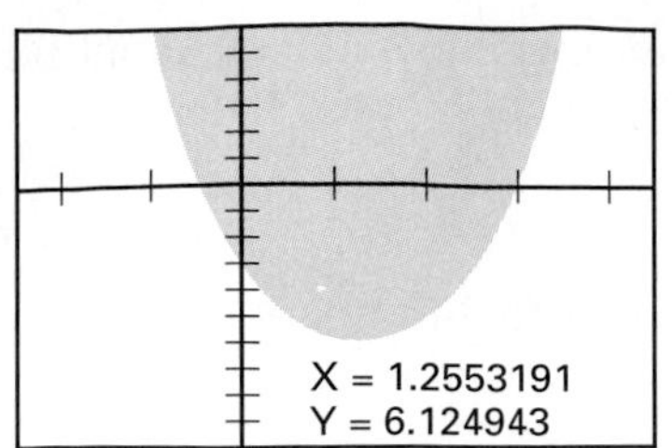

Figure 18-34

SELF-STUDY EXERCISES 18-5

L.O.1. Identify the following equations as linear, quadratic, or other.

1. $y = 3x - 7$
2. $y = x^2 - 4x + 2$
3. $x^2 + y = 5x$
4. $x^2 + y^2 - 3 = 2x$
5. $2x + 3 = 8$
6. $\frac{x}{y} = 3$

L.O.2. Use a table of solutions to graph the following quadratic equations.

7. $y = x^2$
8. $y = 3x^2$
9. $y = \frac{1}{3}x^2$
10. $y = -4x^2$
11. $y = -\frac{1}{4}x^2$
12. $y = x^2 - 4$
13. $y = x^2 + 4$
14. $y = x^2 - 6x + 9$
15. $y = -x^2 + 6x - 9$

L.O.3. Graph the following using a graphing calculator. Reset the range if necessary to show the vertex of the parabola.

16. $y = 3x^2 + 5x - 2$
17. $y = (2x - 3)(x - 1)$
18. $y > 2x^2 - 9x - 5$
19. $y > -2x^2 + 9x + 5$

ASSIGNMENT EXERCISES

18-1

Give the coordinates of the following points in Fig. 18-35.

1. A
2. B
3. C
4. D
5. E
6. F
7. G
8. H
9. I
10. J

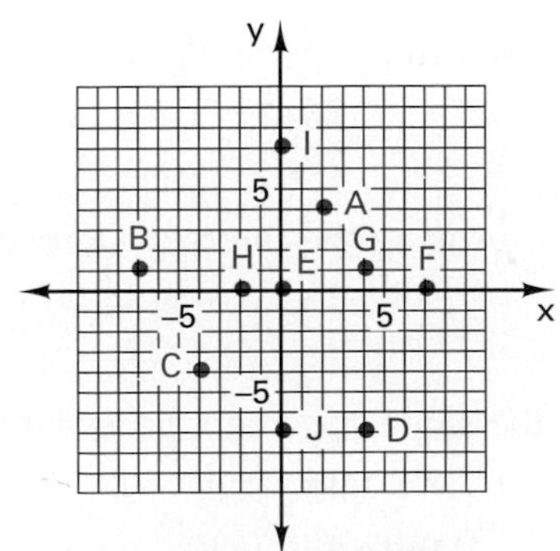

Figure 18-35

Using a set of coordinate axes, identify the quadrant for each point; then plot the points and label them with the appropriate letters.

11. Point A $(-3, 1)$ **12.** Point B $(5, -3)$ **13.** Point C $(-1, -7)$
14. Point D $(0, -6)$ **15.** Point E $(0, 0)$

16. Plot the graph of the technical information contained in the following data table. Use increments of 0.1 for time in seconds (0.1, 0.2, 0.3, . . .) and 1 for amps (4, 5, 6, . . .). The data show the time needed for amperage to build up in an ac circuit with a coil.

x-axis (seconds)	*y*-axis (amps)
0.1	4.2
0.2	6.6
0.3	8.0
0.4	8.9
0.5	9.3

17. A production manager has contracted to produce 7000 parts in 20 weeks. The table shows the cumulative weekly production totals for the first 8 weeks.

(a) Graph these data using the week number as the x-value and the production total as the y-value.

(b) Draw a straight line that best fits the relationship.

(c) By extending the line, predict whether the manager will reach her goal if she continues or if she needs to modify the production schedule or add another production line to achieve the goal.

Week number	Cumulative production
1	423
2	675
3	1055
4	1374
5	1511
6	1893
7	2305
8	2728

18-2

Which of the following coordinate pairs are solutions for the equation $2x - 3y = 12$?

18. $(1, -3)$ **19.** $(3, -2)$ **20.** $(0, 4)$
21. $(6, 0)$

22. In the equation $2x + y = 8$, find x when $y = 10$.

23. Find y in the equation $x - 3y = 5$ when $x = 8$.

24. For the function $f(x) = \frac{2}{3}x - 1$, find $f(6)$, $f(0)$, and $f\left(\frac{1}{4}\right)$.

18-3

Graph the following equations using a table of solutions.

25. $x + y = 3$ **26.** $x - y = 1$ **27.** $2x + y = 1$
28. $x - 2y = 4$ **29.** $y = 3x - 4$ **30.** $x = 5y$

Graph the following equations using the intercepts procedure.

31. $x = -4y - 1$ **32.** $x + y = -4$ **33.** $3x - y = 1$
34. $x = -4y$

Graph the following using the slope–intercept procedure.

35. $y = 5x - 2$ **36.** $y = -x$ **37.** $y = -3x - 1$
38. $y = \frac{1}{2}x + 3$ **39.** $x - y = 4$ **40.** $2y + 4 = -3$

41. $x - 2y = -1$

42. Verify Exercises 25 through 41 with a graphing calculator.

18-4

Graph the following linear inequalities using test points and verify with a graphing calculator.

43. $4x + y < 2$
44. $x + y > 6$
45. $3x + y \leq 2$
46. $x + 3y > 4$
47. $x - 2y < 8$
48. $3x + 2y \geq 4$
49. $y \geq 3x - 2$
50. $y \leq -2x + 1$
51. $y > \frac{2}{3}x - 2$

18-5

Use a table of solutions to graph the following quadratic equations and verify with a graphing calculator.

52. $y = x^2 - 1$
53. $y = x^2 + 1$
54. $y = -x^2 + 1$
55. $y = -x^2 - 1$
56. $y = 2x^2 + 3$
57. $y = 3x^2 - 5$
58. $y = x^2 - 2x + 1$
59. $y = -x^2 + 2x - 1$
60. $y = x^2 - 4x + 4$
61. $y = -x^2 + 4x - 4$

Use a calculator to graph the following. Set the range to display the vertex.

62. $y < x^2 - 2x + 1$
63. $y \geq -2x^2$
64. $y \leq \frac{1}{2}x^2 - 3$

CHALLENGE PROBLEMS

65. A 1-gallon can of indoor house paint is advertised to cover 400 square feet of wall surface.
    (a) Make a table to show the amount of wall surface area that can be covered by 1, 2, 3, . . . , or 10 gallons of paint.
    (b) Draw a graph of these data.
    (c) Use the graph to decide how many 1-gallon cans of paint it would take to cover 5500 square feet of wall surface.
    (d) The paint being used for this job can be purchased for $19.95 a gallon. Find the cost of the paint for the 5500 square feet of wall surface.
    (e) The sales tax rate is 8.25%. Calculate the total cost of the paint.

66. Using column 1 of the table below for the horizontal scale and one of the other columns for the vertical scale, select the column illustrated in graph A. . . . in graph B . . . in graph C.

Payment number	Payment amount	Applied to interest	Applied to principal	Balanced owed	Total interest
1	611.09	580.00	31.09	69,568.91	580.00
2	611.09	579.74	31.35	69,537.56	1,159.74
3	611.09	579.48	31.61	69,505.95	1,739.22
4	611.09	579.22	31.87	69,474.08	2,318.44
5	611.09	578.95	32.14	69,441.94	2,897.39
6	611.09	578.68	32.41	69,409.53	3,476.07
7	611.09	578.41	32.68	69,376.85	4,054.48
8	611.09	578.14	32.95	69,343.90	4,632.62
9	611.09	577.87	33.22	69,310.68	5,210.49
10	611.09	577.59	33.50	69,277.18	5,788.08

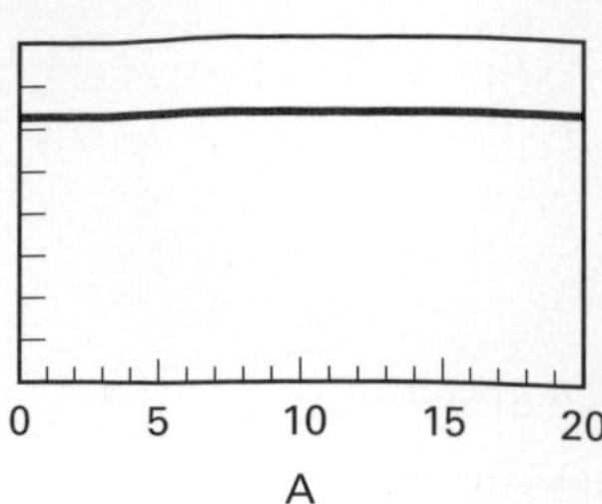

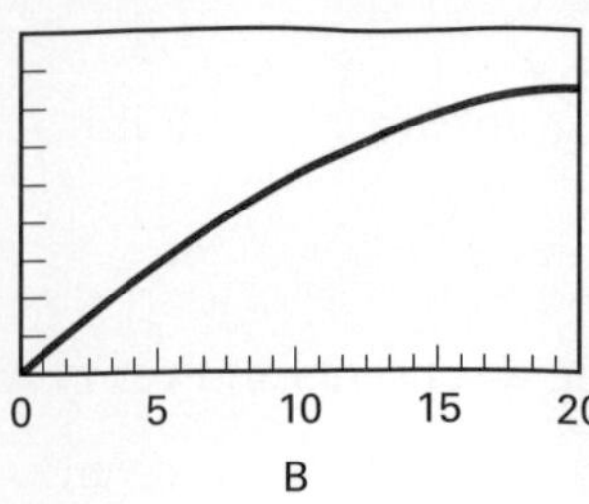

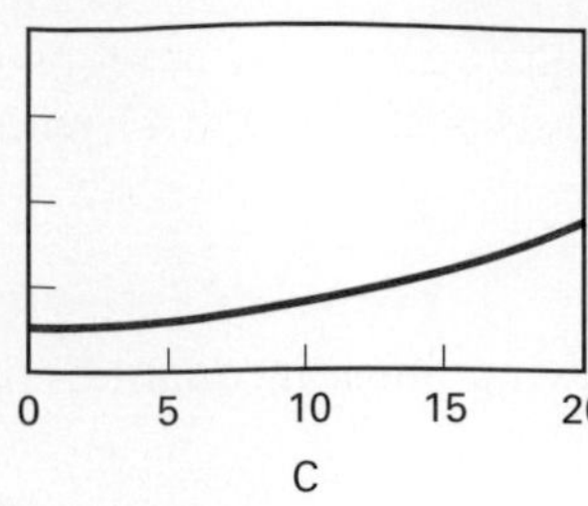

67. Sketch graphs to describe the pattern shown by the numbers in the two columns not illustrated by graphs A, B, and C.

CONCEPTS ANALYSIS

1. What do we mean when we say a point is represented by an ordered pair of numbers?
2. What does the graph of an equation represent?
3. Explain the procedure for plotting the point $(5, -2)$ on a grid representing the rectangular coordinate system.
4. How can you tell from an equation if the graph will be a straight line?
5. Discuss the similarities and differences in the table-of-solutions method, the intercepts method, and the slope–intercept method for graphing a linear equation.
6. Why is it generally helpful to solve an equation for y before using the table-of-solutions method?
7. How is the graph of a linear inequality different from the graph of a linear equation?
8. When graphing linear inequalities, how do you determine which portion of the graph to shade?
9. How is the graph of a quadratic equation different from the graph of a linear equation?
10. What do *axis of symmetry* and *vertex* refer to on the graph of a quadratic equation that represents a parabola?

CHAPTER SUMMARY

Objectives	What to Remember	Examples
Section 18-1		
1. Plot the points indicated by ordered pairs of numbers.	Begin at the origin, move the first amount or x-coordinate on the horizontal scale and the second amount or y-coordinate on the vertical scale.	Plot $(-3, 2)$ and $(2, -1)$ on the coordinate system. 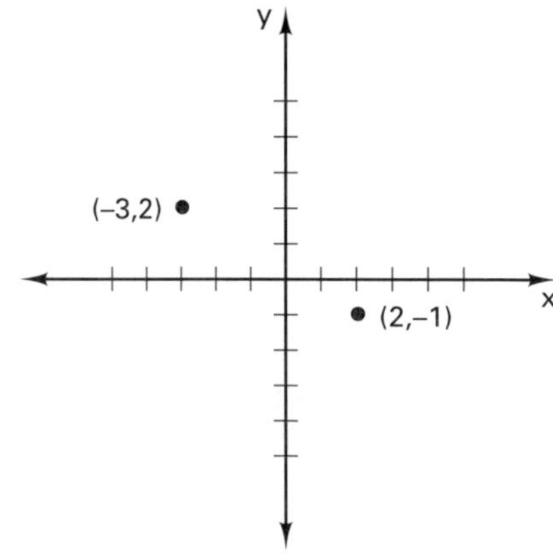

2. Identify the quadrant that contains a given point.

The signs of coordinates in each quadrant can be identified as follows: quadrant I $(+, +)$; quadrant II $(-, +)$; quadrant III $(-, -)$; quadrant IV $(+, -)$.

Indicate the quadrant that contains the points for the following coordinates: $(-3, 2)$, $(5, -2)$, $(3, 2)$, $(-3, -2)$.

$(-3, 2)$, II; $(5, -2)$, IV; $(3, 2)$, I; $(-3, -2)$, III

3. Find the coordinates of a given point.

To write the coordinates of a given point, begin at the origin and count the horizontal distance and record it as the first coordinate; then count the vertical distance and record it as the second coordinate in the ordered pair.

Write the coordinates for points *A* and *B* from the graph.

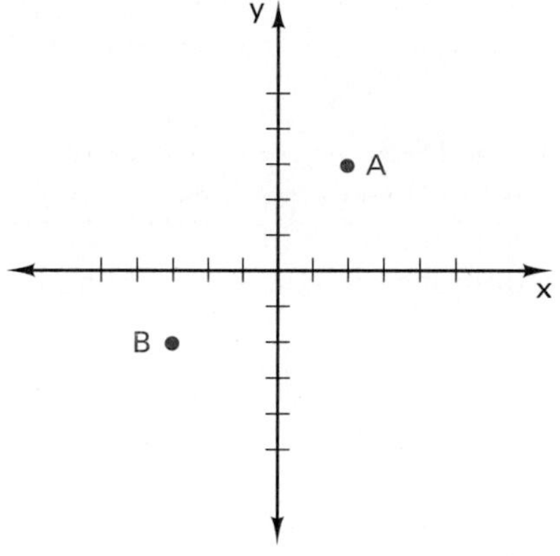

A (2, 3); *B* (−3, −2)

Section 18-2

1. Check the solution of an equation with two variables.

To check the solution of an equation in two variables, substitute the numerical value for *x* in each place it occurs and substitute the numerical value for *y* in each place it occurs. Simplify each side of the equation using the order of operations. The solution will make a true statement; that is, both sides of the equation will be equal to the same number.

A solution for the equation $3x - y = 1$ is (1, 2). Check to verify the solution.

$$3(1) - 2 = 1$$
$$3 - 2 = 1$$
$$1 = 1 \quad \text{True}$$

2. Evaluate equations with two variables when given the value of one variable.

To evaluate an equation when the value of one variable is given, substitute the numerical value of the given variable in place of that variable each place it occurs in the equation; then solve the equation for the other variable.

Evaluate the equation $4x - y = 6$ if $y = 14$.

$$4x - 14 = 6$$
$$4x = 20$$
$$x = 5$$

3. Interpret function notation and evaluate functions.

The notation $f(x)$ indicates a function and does not mean f times x. $f(x)$ is read "f of x." $f(x)$ has the same meaning as y.

Write the equation $x - 2y = 8$ in function notation. First, solve for y.

$$-2y = -x + 8$$
$$y = \frac{1}{2}x - 4$$
$$f(x) = \frac{1}{2}x - 4$$

To evaluate a function, replace the x with a numerical value every place it occurs in the function.

Evaluate the function

$$f(x) = \frac{1}{2}x - 4 \text{ for } x = 6.$$

$$f(6) = \frac{1}{2}(6) - 4$$
$$= 3 - 4$$
$$= -1$$

The function notation permits one to see both coordinates when evaluated. The x-coordinate is 6 when the y-coordinate is -1.

Section 18-3

1. Graph linear equations using the table-of-solutions method.

To graph a linear equation, solve the equation for y; then assign x values and determine the corresponding y-values.

Make a table of values to graph the equation $y = 2x - 1$.

x	y	
-1	-3	$y = 2(-1) - 1 = -2 - 1 = -3$
0	-1	$y = 2(0) - 1 = 0 - 1 = -1$
1	1	$y = 2(1) - 1 = 2 - 1 = 1$

2. Graph linear equations using intercepts.

Find the x-intercept by letting $y = 0$ and solving for x.

Graph the equation $y = 2x - 1$ by using the intercepts method.

x-intercept: $0 = 2x - 1$
$$1 = 2x$$
$$\frac{1}{2} = x$$
$$\left(\frac{1}{2}, 0\right)$$

Find the y-intercept by letting $x = 0$ and solving for y.

y-intercept: $y = 2(0) - 1$
$$y = 0 - 1$$
$$y = -1$$
$$(0, -1)$$

Plot the two points and draw the graph of the equation.

Plot the two points; then draw the graph.

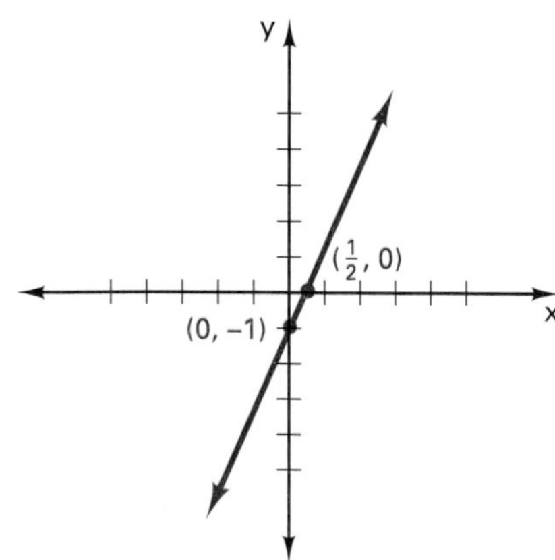

3. Graph linear equations using the slope–intercept method.

Write the equation in the form $y = mx + b$. The slope is m. In fraction form the numerator is the rise and the denominator is the run. The y-coordinate of the y-intercept is b.

To graph using the slope–intercept method, first locate the y-intercept or b by counting vertically along the y-axis. From this point, count the slope (rise, then run) and locate the second point. Draw the graph through the two points.

Use the slope–intercept method to graph $y = 2x - 1$. The y-intercept is -1, so count down 1 from the origin. The coordinates of this point are $(0, -1)$. From this point, rise 2 and run 1. The coordinates of the second point are $(1, 1)$. Connect the two points.

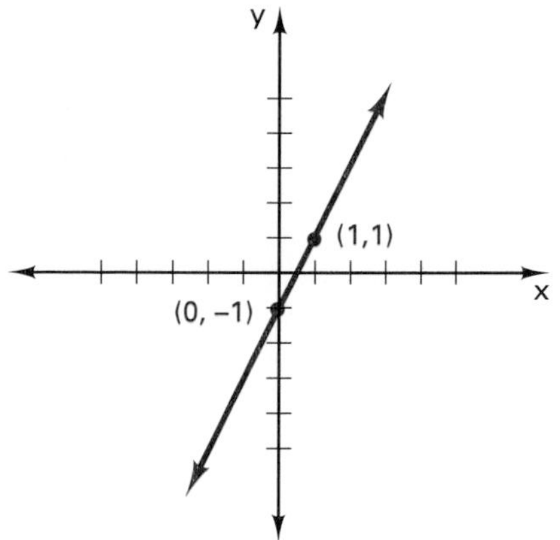

4. Graph linear equations using a graphing calculator.

First, solve the equation for y. Press the GRAPH key; then enter the right side of the equation. Next, press EXE to show the graph on the screen.

Graph the equation $y = 2x - 1$. Steps for the Casio fx-7700G are:

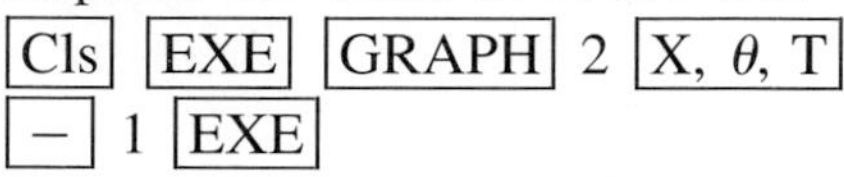
Cls EXE GRAPH 2 X, θ, T − 1 EXE

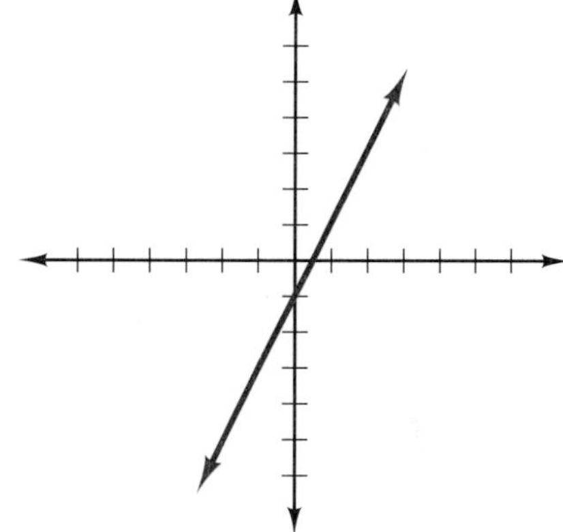

Section 18-4

1. Graph linear inequalities with two variables using test points.

Graph a linear inequality by first graphing the corresponding equation. Select a point on one side of the boundary line and substitute each coordinate in place of the appropriate letter. If the statement is true, shade the side of the boundary that contains the point. If false, shade the other side of the boundary. For $>$ or $<$ inequalities, make the boundary line dashed. For $\geq$ or $\leq$ inequalities, make the boundary line solid.

Graph $y \geq 2x - 1$.
Graph $y = 2x - 1$.
Make boundary solid.
Test (1, 3):

$$3 \geq 2(1) - 1$$
$$3 \geq 2 - 1$$
$$3 \geq 1 \qquad \text{True}$$

Shade the side of the boundary line that contains the point if the coordinates made a true statement.

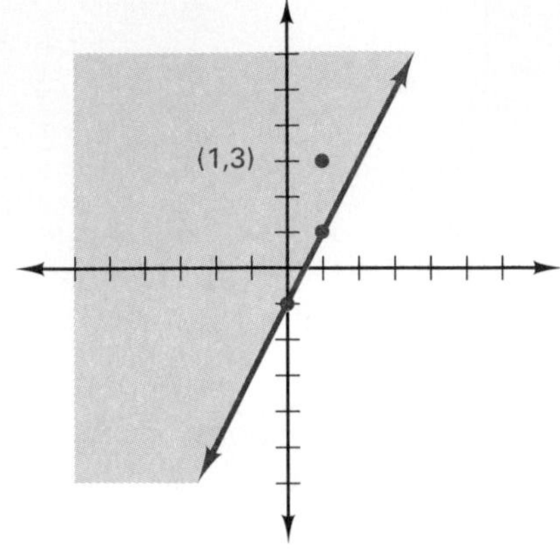

2. Graph linear inequalities with two variables using a graphing calculator.

Set the initial settings on the graphing calculator to graph inequalities.

Use the calculator to graph $y \leq 2x - 1$.

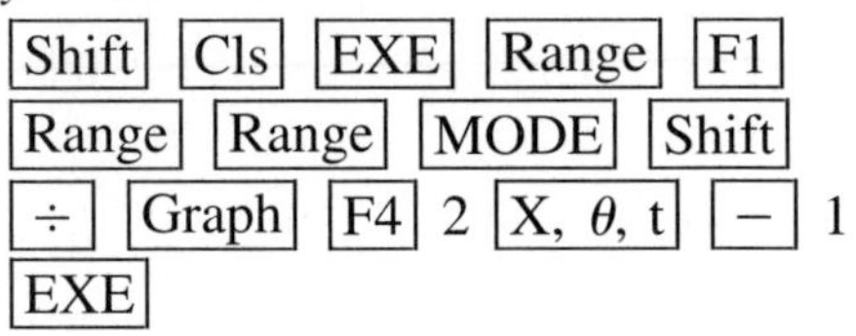

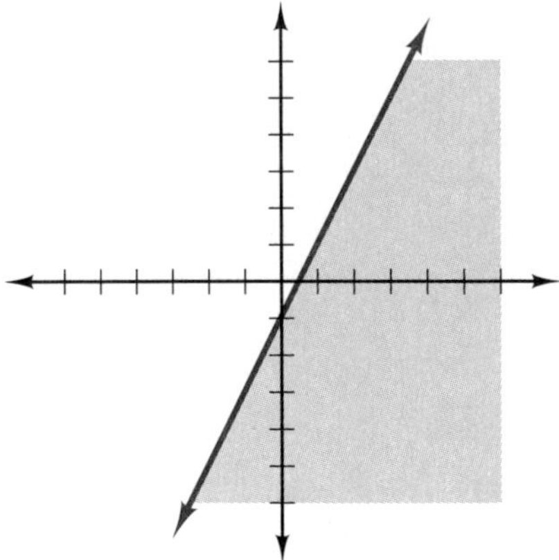

Section 18-5

1. Identify nonlinear equations.

Linear equations can have letter terms with an exponent of *one* only. And no term can have two letter factors. Quadratic equations must have at least one term with degree 2 and no terms with degree greater than 2.

Identify the linear and quadratic equations from the following.

(a) $2x - 3 = y$
(b) $x^2 - y = 8$
(c) $xy - 5x = 2$
(d) $y = 3x - 7$

(a) and (d) are linear equations.
(b) and (c) are quadratic equations.

2. Graph quadratic equations using the table-of-solutions method.	Before graphing quadratic equations, determine the x-coordinate of the vertex of the parabola by finding the value of $-\frac{b}{2a}$. Substitute the x-coordinate into the equation to find the corresponding y-value. Select two x-coordinates that are less than the vertex coordinate and two that are more than the vertex coordinate. Substitute these into the equation and find the corresponding y-value. Use the five ordered pairs to plot points, and join the points with a smooth curve.	Graph $y = x^2 - 6x + 8$. $-\frac{b}{2a} = -\frac{-6}{2(1)} = \frac{6}{2} = 3.$

x	y	
5	3	$5^2 - 6(5) + 8$
4	0	$4^2 - 6(4) + 8$
3	−1	$3^2 - 6(3) + 8$ (vertex)
2	0	$2^2 - 6(2) + 8$
1	3	$1^2 - 6(1) + 8$

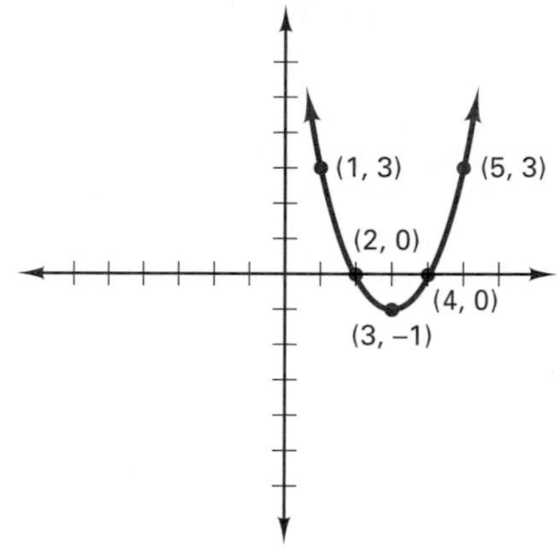

3. Graph quadratic equations and inequalities using a graphing calculator.	To graph quadratic equations using a graphing calculator, clear and initialize the calculator. [Cls] [EXE] [Range] [F1] [Range] [Range] Be sure calculator is in equation mode. [Mode] [Shift] [+] To graph quadratic inequalities, clear and initialize the calculator. Put the calculator in inequality mode.	Graph $y = x^2 - 6x + 8$. Clear and initialize the calculator. [Graph] [x, θ, t] [x²] [−] 6 [x, θ, t] [+] 8 [EXE] (See previous graph.) Graph $y > x^2 - 6x + 8$. Clear and initialize the calculator. [Cls] [EXE] [Range] [F1] [Range] [Range] [Mode] [Shift] [÷] [Graph] [F1] [x, θ, t] [x²] [−] 6 [x, θ, t] [+] 8 [EXE]

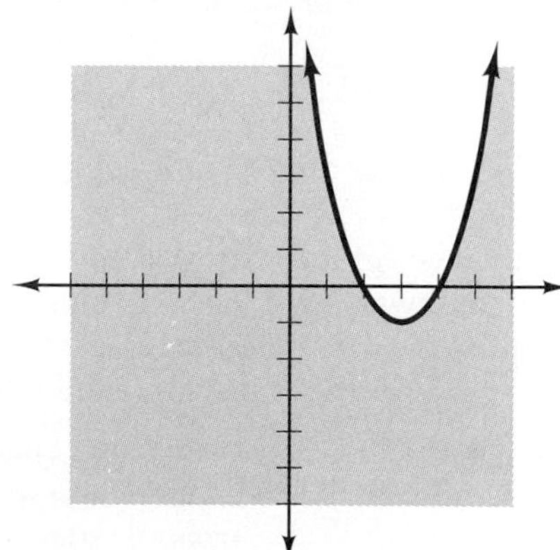

WORDS TO KNOW

rectangular coordinate system (p. 673)
origin (p. 673)
coordinates (p. 673)
quadrants (p. 673)
plot points (p. 674)
ordered pair (p. 679)
linear equation (p. 679)
linear equation in two variables (p. 679)
ordered pair as solution (p. 680)
function notation (p. 682)
dependent variable (p. 682)
independent variable (p. 682)
table of solutions to graph equations (p. 684)
x-intercept (p. 685)
y-intercept (p. 685)
graph equations using the intercepts (p. 688)
slope (p. 690)
graph equations using the slope–intercept method (p. 690)
linear inequality in two variables (p. 695)
boundary (p. 696)
axis of symmetry (p. 701)
vertex (p. 701)

CHAPTER TRIAL TEST

Give the coordinates of each point in Fig. 18-36.

1. *A* **2.** *B* **3.** *C*

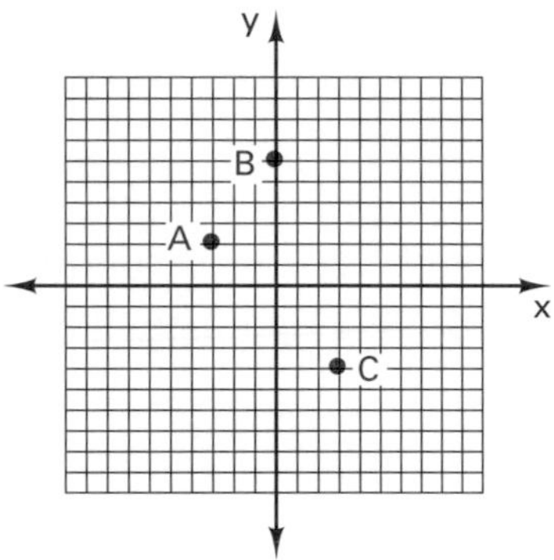

Figure 18-36

Identify the quadrant that contains each of the following points.

4. *D*, $(-2, 3)$ **5.** *E*, $(-3, -1)$ **6.** *F*, $(4, 7)$
7. *G*, $(5, -1)$

Graph the following equations.

8. $2x - y = 4$ **9.** $y = 5 - x$ **10.** $\frac{1}{2}x = y - 1$

11. Find the *x*- and *y*-intercepts of the equation $x - 4y = 2$.

12. Find the *x*- and *y*-intercepts of the equation $y = -8 + x$.

13. Plot the graph for the relationship between horsepower (hp) and revolutions per minute (rpm) in a test engine. Use the same intervals as those in the following data table.

x-axis (rpm)	*y*-axis (hp)
500	30
1000	45
1500	60
2000	75

Graph the following linear inequalities.

14. $x + 2y < 4$ **15.** $2x - y \leq 2$ **16.** $y \leq 2x + 2$
17. $x + y < 1$

Graph the following quadratic equations.

18. $y = x^2 + 2$ **19.** $y = x^2 + 2x + 1$ **20.** $y = x^2 + 4x + 4$

Graph the following quadratic inequalities with a graphing calculator.

21. $y \leq x^2 - 6x + 8$ **22.** $y > 2x^2$ **23.** $y < -\frac{1}{2}x^2$

CHAPTER

19 Slope and Distance

The roof of a house *slopes* to allow runoff of rain, melting snow, and other types of precipitation. Often, a line graph *slopes* to show an increase or decrease in some quantities, such as the relationship between temperature and the effectiveness of film developer. The *slope* of a line, then, is common in many technologies and is useful as a mathematical tool in the hands of trained workers.

19-1 SLOPE

Learning Objectives

L.O.1. Calculate the slope of a line, given two points on the line.
L.O.2. Determine the slope of a horizontal or vertical line.

L.O.1. Calculate the slope of a line, given two points on the line.

Finding the Slope, Given Two Points

A property of lines that is very useful in mathematics and in technical applications is the *slope* of a line.

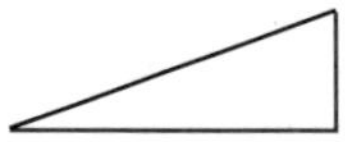

■ DEFINITION 19-1.1 **Slope.** The **slope** of a line is the ratio of the vertical rise of a line to the horizontal run of a line (see Fig. 19-1).

Rise

Run

Figure 19-1

The definition of slope may be expressed as a formula.

Formula 19-1.1 *Slope:*

$$\text{Slope} = \frac{\text{rise}}{\text{run}}$$

$$= \frac{\Delta y}{\Delta x} = \frac{y_2 - y_1}{x_2 - x_1}$$

where Δy is vertical change and Δx is horizontal change.

$$P_1 = (x_1, y_1), \qquad P_2 = (x_2, y_2)$$

P_1 and P_2 are any two points on the line.

To find the slope of a line, determine the coordinates of any two points that are located on the line. Then calculate the change (*difference*) in the y-coordinates to find the vertical rise. Calculate the change (*difference*) in the x-coordinates to find the horizontal run. Make a ratio of the rise to the run and reduce the ratio to lowest terms.

EXAMPLE 19-1.1 Find the slope of a line if the points $(2, -1)$ and $(5, 3)$ are on the line (see Fig. 19-2).

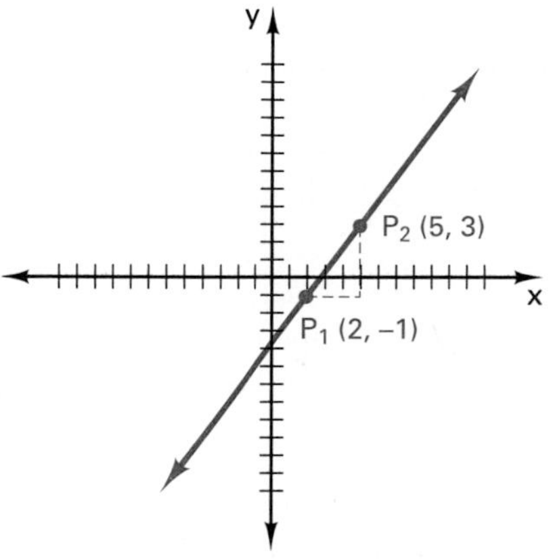

Figure 19-2

Solution: We begin by labeling $(2, -1)$ as point 1 (P_1) and $(5, 3)$ as point 2 (P_2). The coordinates of P_1 are referred to as (x_1, y_1) and the coordinates of P_2 are referred to as (x_2, y_2). We will use the Greek capital letter delta (Δ) to indicate a change and write the slope definition symbolically.

$$\text{Change in } y = \Delta y = \text{rise} = y_2 - y_1 = 3 - (-1) = 4$$

$$\text{Change in } x = \Delta x = \text{run} = x_2 - x_1 = 5 - 2 = 3$$

$$\text{Slope} = \frac{\Delta y}{\Delta x} = \frac{y_2 - y_1}{x_2 - x_1} = \frac{3 - (-1)}{5 - 2} = \frac{4}{3}$$

Thus, the slope of the line through points $(2, -1)$ and $(5, 3)$ is

$$\frac{\Delta y}{\Delta x} = \frac{\text{rise}}{\text{run}} = \frac{4}{3}$$

■

Tips and Traps

It is important to realize that the designation of P_1 and P_2 is not critical. Let's find the slope of this same line designating P_1 as $(5, 3)$ and P_2 as $(2, -1)$.

$$\Delta y = \text{rise} = y_2 - y_1 = -1 - 3 = -4$$

$$\Delta x = \text{run} = x_2 - x_1 = 2 - 5 = -3$$

$$\frac{\Delta y}{\Delta x} = \frac{\text{rise}}{\text{run}} = \frac{-4}{-3} = \frac{4}{3}$$

Notice in Example 19-1.1 that, with either point designated as point 1, the slope is positive. When the slope of a line is positive, the line *rises* from left to right. When the slope of a line is negative, the line *falls* from left to right. Figure 19-3 shows a line with a slope of $\frac{4}{3}$ and a line with a slope of $-\frac{4}{3}$.

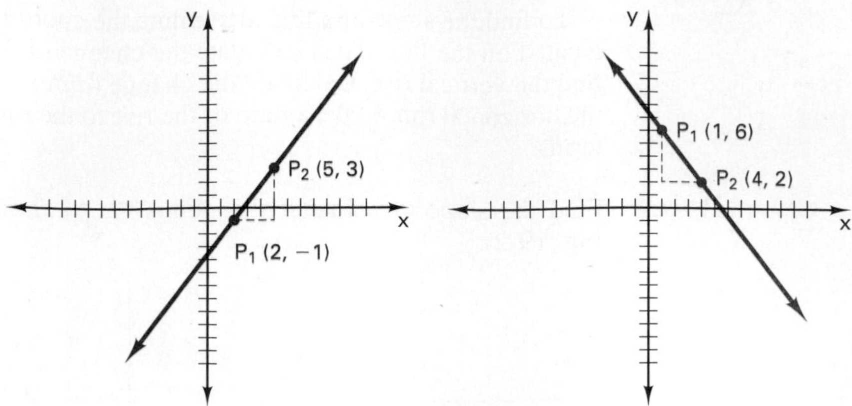

Figure 19-3

EXAMPLE 19-1.2 Find the slope of the line passing through the following points.

(a) (1, 6) and (4, 2) **(b)** (−5, −1) and (3, −2)
(c) (0, 0) and (−5, 3)

Solution

(a) Either point can be designated as P_1. The other point will be P_2.

$$P_1 = (x_1, y_1), \qquad P_2 = (x_2, y_2)$$

Then

$$\text{Slope} = \frac{\Delta y}{\Delta x} = \frac{y_2 - y_1}{x_2 - x_1}$$

Let $P_1 = (1, 6)$ and $P_2 = (4, 2)$. Then

$$\frac{\Delta y}{\Delta x} = \frac{2 - 6}{4 - (1)} = \frac{-4}{3} = -\frac{4}{3}$$

The slope is $-\frac{4}{3}$.

(b) Let $P_1 = (-5, -1)$ and $P_2 = (3, -2)$.

$$\text{Slope} = \frac{\Delta y}{\Delta x} = \frac{-2 - (-1)}{3 - (-5)} = \frac{-2 + 1}{3 + 5} = \frac{-1}{8} = -\frac{1}{8}$$

The slope is $-\frac{1}{8}$.

(c) Let $P_1 = (0, 0)$ and $P_2 = (-5, 3)$.

$$\text{Slope} = \frac{\Delta y}{\Delta x} = \frac{3 - 0}{-5 - 0} = \frac{3}{-5} = -\frac{3}{5}$$

The slope is $-\frac{3}{5}$. ■

L.O.2. Determine the slope of a horizontal or vertical line.

Slope of Horizontal and Vertical Lines

There are two special lines whose slope we want to examine, the horizontal line and the vertical line. Let's first look at the horizontal line that passes through the points (3, 0) and (−2, 0).

EXAMPLE 19-1.3 Find the slope of the horizontal line that passes through (3, 0) and (−2, 0).

Solution: We will let $P_1 = (3, 0)$ and $P_2 = (-2, 0)$.

$$\text{Slope} = \frac{\Delta y}{\Delta x} = \frac{y_2 - y_1}{x_2 - x_1} = \frac{0 - 0}{-2 - 3} = \frac{0}{-5} = 0$$

■

Since any horizontal line has no change in y values or no rise, *the slope of any horizontal line is zero.*

> **Rule 19-1.1** ***Horizontal lines:***
>
> The slope of a horizontal line is zero.
> Two points are on the same horizontal line if their y-coordinates are equal.

Next, let's look at the vertical line that passes through the points (0, 1) and (0, 8).

EXAMPLE 19-1.4 Find the slope of the vertical line that passes through the points (0, 1) and (0, 8).

Solution: We will let $P_1 = (0, 1)$ and $P_2 = (0, 8)$.

$$\text{Slope} = \frac{\Delta y}{\Delta x} = \frac{y_2 - y_1}{x_2 - x_1} = \frac{8 - 1}{0 - 0} = \frac{7}{0}$$

Remember, division by zero is impossible; therefore, the slope of the vertical line passing through (0, 1) and (0, 8) is undefined. ■

Since any vertical line has no change in x values or no run, *the slope of any vertical line is undefined.* We can say that a vertical line has *no slope.*

> **Rule 19-1.2** ***Vertical lines:***
>
> A vertical line has no slope. That is, the slope of a vertical line is undefined.
> Two points are on the same vertical line if their x-coordinates are equal.

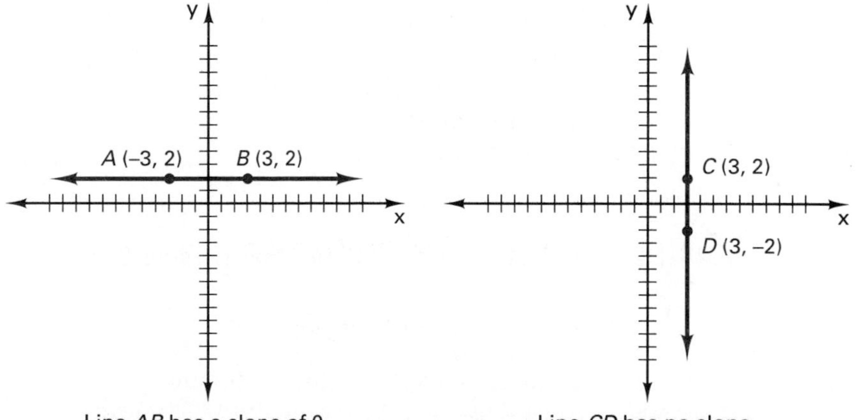

Figure 19-4

Tips and Traps

A slope of zero and no slope are *not* the same.

SELF-STUDY EXERCISES 19-1

L.O.1. Find the slope of the line passing through the following pairs of points.

1. $(-3, 3)$ and $(1, 5)$
2. $(4, -1)$ and $(1, 4)$
3. $(3, 1)$ and $(5, 7)$
4. $(-1, -1)$ and $(3, 3)$
5. $(4, 5)$ and $(-4, -1)$
6. $(7, 2)$ and $(-3, 2)$
7. $(4, -5)$ and $(0, 0)$
8. $(2, -2)$ and $(5, -5)$
9. $(-5, 2)$ and $(-5, -4)$
10. $(4, -4)$ and $(1, 5)$

L.O.2. Identify the horizontal and vertical lines in Exercises 1 through 10 and explain why you identified each as horizontal or vertical.

19-2 POINT–SLOPE FORM OF AN EQUATION

Learning Objectives

L.O.1. Find the equation of a line, given the slope and one point.
L.O.2. Find the equation of a line, given two points on the line.

The slope of a line can be determined from any two points on that line. However, it is sometimes desirable to know the equation of the line so that any point on the line can be determined.

L.O.1. Find the equation of a line, given the slope and one point.

Finding the Equation, Given the Slope and One Point

We can write the equation of a straight line passing through two given points by using one of the points and the slope of the line. We will substitute the slope of the line and the coordinates of one point on the line into the *point–slope form* of the equation of a line.

Formula 19-2.1 *Point–slope form of an equation of a straight line:*

$$y - y_1 = m(x - x_1) \quad \text{or} \quad m = \frac{y - y_1}{x - x_1}$$

where y_1 and x_1 are coordinates of the given point and m is the slope of the line passing through that point. The x and y represent the variables of the equation being sought.

EXAMPLE 19-2.1 Find the equation of the line passing through the point $(3, -2)$ with a slope of $\frac{2}{3}$.

Solution: Substitute into the formula

$$y - y_1 = m(x - x_1)$$

where y_1 and x_1 are the coordinates of the given point and m is the slope of the line passing through that point. x and y represent the variables of the equation being sought.

$$y - (-2) = \frac{2}{3}(x - 3)$$

$$y + 2 = \frac{2}{3}(x - 3) \qquad \textit{(Distribute.)}$$

$$y + 2 = \frac{2}{3}x - 2 \qquad \left(\frac{2}{\not{3}} \cdot \frac{-\not{3}^{\,-1}}{1} = -2\right)$$

$$y = \frac{2}{3}x - 2 - 2 \qquad \textit{(Transpose.)}$$

$$y = \frac{2}{3}x - 4$$

■

Another method for finding the equation is based on a variation of the slope formula. (x_1, y_1) is the known point. (x, y) is the unknown point.

$$m = \frac{y - y_1}{x - x_1}$$

$$\frac{2}{3} = \frac{y - (-2)}{x - 3} \quad \text{or} \quad \frac{2}{3} = \frac{y + 2}{x - 3}$$

We will rearrange the equation or solve the equation for y.

$$3(y + 2) = 2(x - 3) \qquad \textit{(Cross multiply.)}$$

$$3y + 6 = 2x - 6 \qquad \textit{(Distribute.)}$$

$$3y = 2x - 6 - 6 \qquad \textit{(Transpose.)}$$

$$3y = 2x - 12$$

$$\frac{3y}{3} = \frac{2x - 12}{3}$$

$$y = \frac{2x - 12}{3} \quad \text{or} \quad y = \frac{2x}{3} - \frac{12}{3}$$

$$y = \frac{2x}{3} - 4 \quad \text{or} \quad y = \frac{2}{3}x - 4$$

L.O.2. Find the equation of a line, given two points on the line.

Finding the Equation, Given Two Points

To find the equation of a line when two points on the line are known, we will apply Formulas 19-1.1 and 19-2.1.

EXAMPLE 19-2.2 Find the equation of the line that passes through the points (0, 8) and (5, 0).

Solution: We first need to find the slope of the line passing through the points (0, 8) and (5, 0). Let $P_1 = (0, 8)$ and $P_2 = (5, 0)$.

$$\text{Slope} = \frac{\Delta y}{\Delta x} = \frac{y_2 - y_1}{x_2 - x_1} = \frac{0 - 8}{5 - 0} = -\frac{8}{5} \qquad \textit{(Formula 19-1.1)}$$

Then, using the slope and one of the points, we substitute into the point–slope form of an equation. Using P_1 gives us

$$y - y_1 = m(x - x_1) \qquad \textit{(Formula 19-2.1)}$$

$$y - 8 = -\frac{8}{5}(x - 0)$$

$$y - 8 = -\frac{8}{5}x$$

Solving for y, we obtain

$$y = -\frac{8}{5}x + 8$$

Suppose that we had used P_2 (5, 0) instead of P_1 in the point–slope form of the equation. Since there is only one line that passes through (0, 8) and (5, 0), we must have the same equation, no matter which point is used. Let's find the equation of the line passing through (0, 8) and (5, 0) using the point (5, 0) and the slope $-\frac{8}{5}$.

$$y - y_1 = m(x - x_1)$$

$$y - 0 = -\frac{8}{5}(x - 5)$$

Solving for y yields

$$y = -\frac{8}{5}x - \frac{8}{5}(-5) \quad \text{or} \quad y = -\frac{8}{5}x + 8$$

This is the same equation that was found using the point (0, 8). ■

EXAMPLE 19-2.3 Find the equation of the line that passes through the points $(-3, 4)$ and $(7, 4)$.

Solution: We will let $P_1 = (-3, 4)$ and $P_2 = (7, 4)$.

$$\text{Slope} = \frac{\Delta y}{\Delta x} = \frac{y_2 - y_1}{x_2 - x_1} = \frac{4 - 4}{7 - (-3)} = \frac{0}{7 + 3} = \frac{0}{10} = 0$$

Using the point–slope form of an equation and P_1, we have

$$y - y_1 = m(x - x_1)$$

$$y - 4 = 0[x - (-3)]$$

$$y - 4 = 0$$

$$y = 4$$

■

There is one special situation when the point–slope form of an equation *cannot* be used to determine the equation of a line. This is when the slope is undefined. Remember, *any vertical line has an undefined slope*. We can readily identify such a line if the coordinates of the two points have the same x-coordinate. The equation of the line will be $x = k$, where k is the common x-coordinate.

EXAMPLE 19-2.4 Find the equation of the line that passes through the points (5, 3) and (5, 7).

Solution: Since the x-coordinate is the same in both points, we know the slope of the line is undefined and the equation of the line is $x = k$, where k is the common x-coordinate. In this example, $k = 5$; thus, the equation of the line through (5, 3) and (5, 7) is $x = 5$. ■

Tips and Traps

A horizontal line has an equation in the form $y = k$, where k is the y-coordinate of any point on the line.

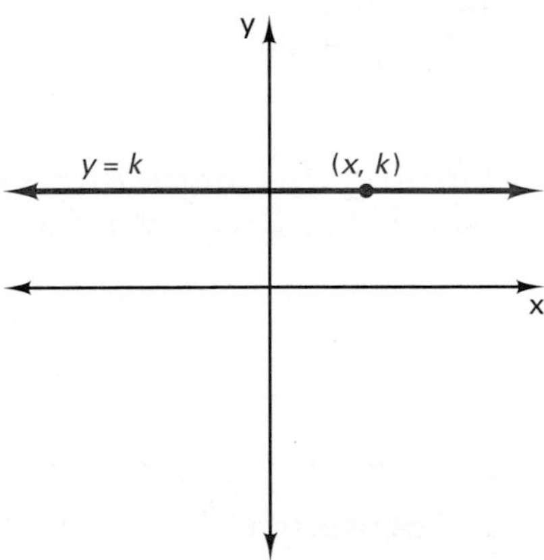

A vertical line has an equation in the form $x = k$, where k is the x-coordinate of any point on the line.

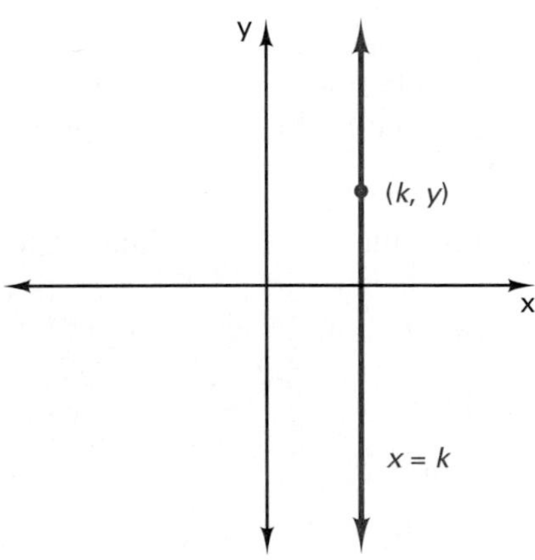

SELF-STUDY EXERCISES 19-2

L.O.1. Find the equation of a line passing through the given point with the given slope. Solve the equation for y when necessary.

1. $(-8, 3), m = \frac{2}{3}$ **2.** $(4, 1), m = -\frac{1}{2}$ **3.** $(-3, -5), m = 2$

4. $(0, -1), m = 1$

L.O.2. Find the equation of a line passing through the given pairs of points. Solve the equation for y.

5. $(4, 6)$ and $(7, 1)$ **6.** $(-1, 6)$ and $(-1, 4)$ **7.** $(-1, -3)$ and $(3, -3)$

8. $(-4, 4)$ and $(-4, -2)$ **9.** $(-2, -4)$ and $(5, 10)$ **10.** $(-4, 0)$ and $(6, 0)$

19-3 SLOPE–INTERCEPT FORM OF AN EQUATION

Learning Objectives

L.O.1. Find the slope and y-intercept of a line, given the equation of the line.
L.O.2. Find the equation of a line, given the slope and y-intercept.

L.O.1. Find the slope and y-intercept of a line, given the equation of the line.

Finding the Slope and y-Intercept of a Line, Given the Equation

So far in our discussion of equations of straight lines, we have always preferred that the final form of the equation be solved for y. When a linear equation is solved for y, we say that it is the *slope–intercept form* of an equation of a straight line.

Formula 19-3.1 *Slope–intercept form of an equation of a straight line:*

$$y = mx + b$$

where m is the slope and b is the y-coordinate of the y-intercept, $(0, b)$.

When an equation is in the form $y = mx + b$, we can determine the slope of the line and the y-intercept by inspection.

EXAMPLE 19-3.1 Find the slope and y-intercept of the lines with the following equations.

(a) $y = 4x + 3$ **(b)** $y = -5x - 3$

(c) $y = -\frac{3}{5}x + \frac{1}{2}$ **(d)** $y = 2.3x - 4.7$

Solution

(a) The slope (m) is the coefficient of x.

$$m = 4$$

The y-coordinate of the y-intercept (b) is the number term.

$$b = 3$$

(b) Slope, $m = -5$; $b = -3$, y-intercept $= (0, -3)$

(c) Slope, $m = -\frac{3}{5}$; $b = \frac{1}{2}$, y-intercept $= \left(0, \frac{1}{2}\right)$

(d) Slope, $m = 2.3$; $b = -4.7$, y-intercept $= (0, -4.7)$ ■

Tips and Traps

Even though b is the y-coordinate of the y-intercept, it is common to refer to b as the y-intercept.

EXAMPLE 19-3.2 Rewrite the following equations in the slope–intercept form. Then find the slope and y-intercept.

(a) $x = y + 5$ **(b)** $2x - 3y = 9$ **(c)** $5y + 2x = 0$
(d) $1.5x - 3y = 4.5$ **(e)** $2y = 4$

Solution: To rewrite an equation in slope–intercept form, we solve the equation for y. Then the coefficient of x is the slope and the number term or constant is the y-intercept.

(a) $x = y + 5$

$x - 5 = y$ or $y = x - 5$

slope, $m = 1$; $b = -5$

(b) $2x - 3y = 9$

$2x - 9 = 3y$

or

$$3y = 2x - 9$$
$$\frac{3y}{3} = \frac{2x - 9}{3}$$
$$y = \frac{2x}{3} - \frac{9}{3}$$
$$y = \frac{2}{3}x - 3$$

slope, $m = \frac{2}{3}$; y-intercept, $b = -3$

(c) $5y + 2x = 0$

$$5y = -2x$$
$$\frac{5y}{5} = -\frac{2x}{5}$$
$$y = -\frac{2}{5}x$$

slope, $m = -\frac{2}{5}$; y-intercept, $b = 0$

Since there is no number term, the value of b is zero.

(d) $1.5x - 3y = 4.5$

$1.5x - 4.5 = 3y$

or

$$3y = 1.5x - 4.5$$
$$\frac{3y}{3} = \frac{1.5x - 4.5}{3}$$
$$y = 0.5x - 1.5$$

slope, $m = 0.5$; y-intercept, $b = -1.5$

(e) $2y = 4$

$$\frac{2y}{2} = \frac{4}{2}$$
$$y = 2$$

slope, $m = 0$; y-intercept, $b = 2$

Since there is no x-term, the coefficient of x or the slope is zero. ■

Tips and Traps

Any vertical line will be an exception in using the slope–intercept form of an equation. In the equation of a vertical line such as $x = 4$, there is no y-term; so the equation cannot be solved for y. We recall that the graph is a vertical line that crosses the x-axis at (4, 0). However, it does not cross the y-axis, so it does not have a y-intercept. We also recall that the slope of a vertical line is undefined. We say that it has no slope.

L.O.2. Find the equation of a line, given the slope and y-intercept.

Writing the Equation of a Line Using the Slope and y-Intercept

The equation of a line can be written by using the slope–intercept form of an equation, $y = mx + b$.

EXAMPLE 19-3.3 Write the equation for a line with a slope of -3 and a y-intercept of 5.

Solution

$$\text{slope} = m = -3$$
$$y\text{-intercept} = b = 5$$
$$y = mx + b$$
$$y = -3x + 5$$

■

The equation of a line can be used to solve applied problems.

EXAMPLE 19-3.4 The following graph shows the cost of producing picture frames.

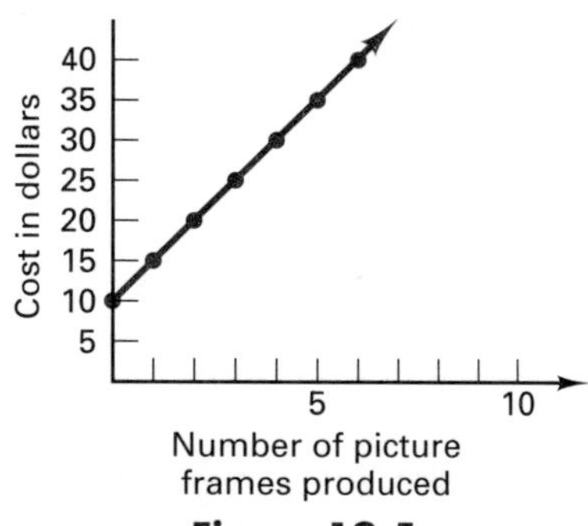

Figure 19-5

(a) Write an equation that represents the graph.
(b) Using the equation, find the cost of producing 20 picture frames.

Solution

(a) The y-intercept represents the fixed cost of producing picture frames. An example of a fixed cost would be the cost of the necessary tools. From the graph, $b = \$10$. The slope m is 5 units vertical change for every 1 unit horizontal change which is 5.

Thus, the equation of the line is

$$y = 5x + 10$$

This type of equation is typically referred to in the business world as a cost function. The notation generally used is $C(x) = 5x + 10$.

(b) Find y when $x = 20$.

$$y = 5x + 10 \qquad \text{or} \qquad C(x) = 5x + 10$$
$$y = 5(20) + 10 \qquad C(20) = 5(20) + 10$$
$$y = 100 + 10 \qquad = 100 + 10$$
$$y = \$110 \qquad = \$110$$

SELF-STUDY EXERCISES 19-3

L.O.1. Determine the slope and y-intercept of the following equations by inspection.

1. $y = 4x + 3$ **2.** $y = -5x + 6$ **3.** $y = -\frac{7}{8}x - 3$

4. $y = 3$

Rewrite the following equations in slope–intercept form and determine the slope and y-intercept.

5. $4x - 2y = 10$ **6.** $2y = 5$ **7.** $\frac{1}{2}y + x = 3$

8. $2.1y - 4.2x = 10.5$ **9.** $2x = 8$ **10.** $x - 5 = 4$

L.O.2. Write the equations of lines with the following slopes and y-intercepts.

11. $m = \frac{1}{4}, b = 7$ **12.** $m = -8, b = -4$

13.

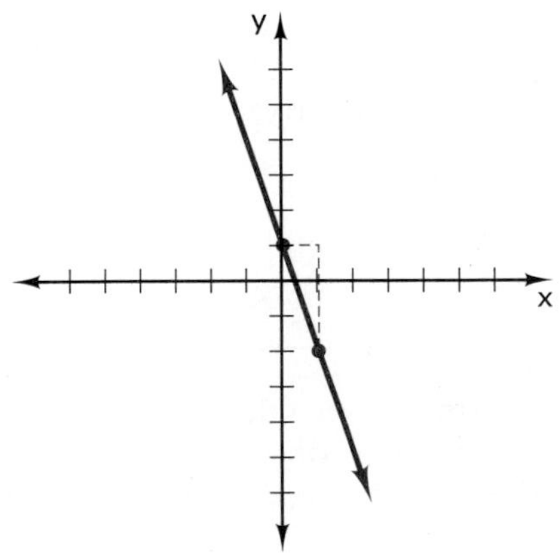

14.

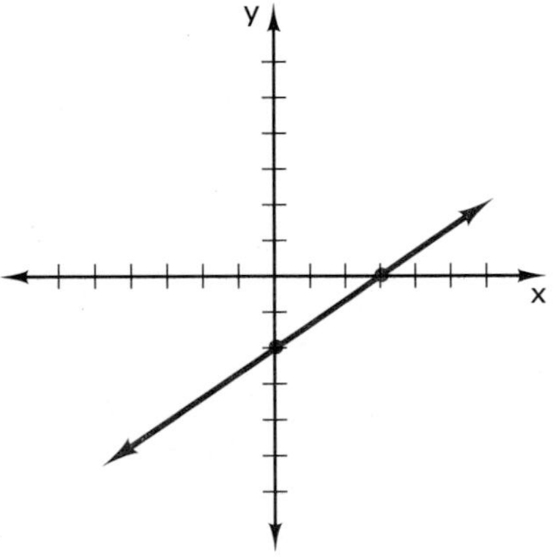

15.

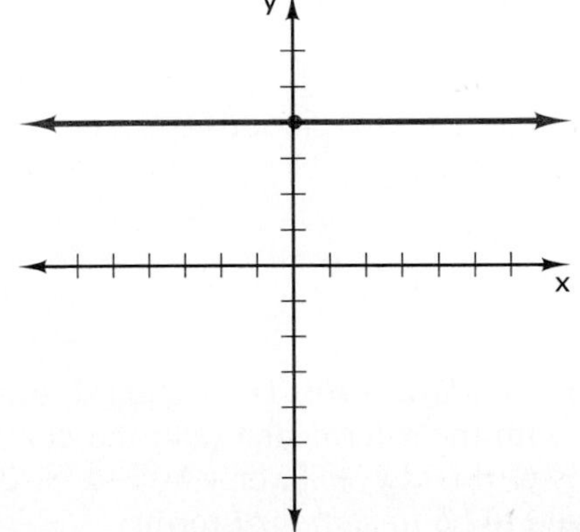

19-4 PARALLEL LINES

Learning Objective

L.O.1. Find the equation of a line, given a point on the line and the equation of a line parallel to a given line.

L.O.1. Find the equation of a line, given a point on the line and the equation of a line parallel to a given line.

Equations of Parallel Lines

Any line can have several lines that are parallel to it. Figure 19-6 illustrates two examples of lines that are parallel to the line $x + y = 5$.

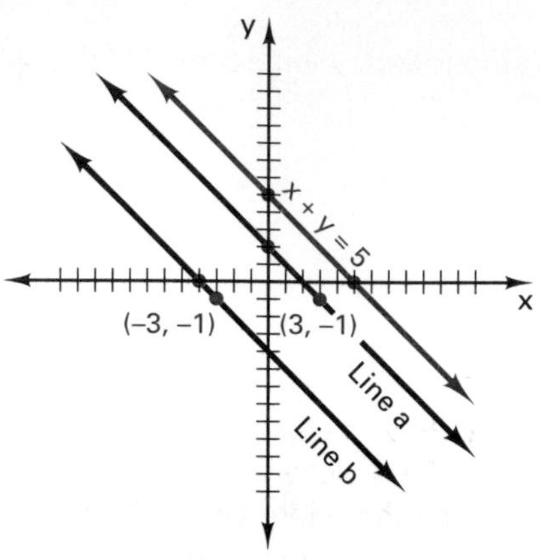

Figure 19-6

■ **DEFINITION 19-4.1 Parallel Lines. Parallel lines** are two or more lines in the same plane that are everywhere the same distance apart.

All lines parallel to a given line have a common property:

> **Rule 19-4.1** *Parallel lines*
>
> The slopes of parallel lines are equal.

If we write $x + y = 5$ in slope–intercept form, we see that $y = -x + 5$, and the slope is -1. Thus, *any* equation that has a slope of -1 is parallel to (or coincides with) the line whose equation is $x + y = 5$. *Coincides* means that one line lies on top of the other; that is, they are the same line.

EXAMPLE 19-4.1 Write the equations for lines a and b in Figure 19-6.

Solution: Line a has a y-intercept of 2. It is parallel to $x + y = 5$. Solved for y, $x + y = 5$ is $y = -x + 5$, with a slope of -1.

Line a: $y = -x + 2$

Line b has a y-intercept of -4.

Line b: $y = -x - 4$ ■

Another form of equations of lines is the *standard form.* The standard form of an equation has the variables on the left, with the x-term first, and the constant on the right. Thus, the equation of line a is $y = -1x + 2$, or $y = -x + 2$, or $x + y = 2$. The equation $x + y = 2$ is said to be in *standard* form.

Formula 19-4.1 *Standard form of an equation of a straight line:*

$$ax + by = c$$

where a, b, and c are integers and $a \geq 0$.

The equation of a parallel line can be found if any point, not just the y-intercept, is known.

In line a of Figure 19-6, one point is $(3, -1)$. We can substitute $x = 3$, $y = -1$, and $m = -1$ into the point–slope form of an equation.

$$\begin{aligned} y - y_1 &= m(x - x_1) \\ y - (-1) &= -1(x - 3) \\ y + 1 &= -x + 3 \\ y &= -x + 3 - 1 \\ y &= -x + 2 \end{aligned}$$

Any point we choose on line a will give the same equation, $y = -x + 2$ or $x + y = 2$.

Now, let's write the equation for line b using the slope and a known point other than the y-intercept. To do so, we will use the following procedure. Line b in Fig. 19-6 is parallel to the line $x + y = 5$ and passes through the point $(-3, -1)$.

Rule 19-4.2 *To find the equation of a line that passes through a given point and is parallel to a given line:*

1. Determine the slope (m) of the *given* line. The slope of the parallel line will be the same as the slope of the given line.
2. Write the equation for the parallel line by substituting the values for m, x_1, and y_1 into the point–slope form of the equation, $y - y_1 = m(x - x_1)$.
3. Write the equation in slope–intercept or standard form.

EXAMPLE 19-4.2 Find the equation of line b in Fig. 19-6.

Solution: The slope of the line $x + y = 5$ is -1 since $x + y = 5$ is equal to $y = -x + 5$. Substitute -1 for m, -3 for x_1, and -1 for y_1 into the point–slope form of the equation.

$$\begin{aligned} y - y_1 &= m(x - x_1) \\ y - (-1) &= -1[x - (-3)] \\ y + 1 &= -1(x + 3) \\ y + 1 &= -x - 3 \\ y &= -x - 3 - 1 \\ y &= -x - 4 \qquad \textit{(slope–intercept form)} \end{aligned}$$

or

$$x + y = -4 \qquad \textit{(standard form)}$$

■

EXAMPLE 19-4.3 Find the equation of a line that is parallel to $2y = 3x + 8$ and passes through the point (2, 1).

Solution: Find the slope of $2y = 3x + 8$.

$$2y = 3x + 8$$

$$\frac{2y}{2} = \frac{3x + 8}{2}$$

$$y = \frac{3}{2}x + 4$$

Thus, the slope of $2y = 3x + 8$ is $\frac{3}{2}$.

Next, find the equation of a line passing through the point (2, 1) that is parallel to $2y = 3x + 8$.

$$y - y_1 = m(x - x_1)$$

$$y - 1 = \frac{3}{2}(x - 2)$$

$$y - 1 = \frac{3}{2}x - 3$$

$$y = \frac{3}{2}x - 3 + 1$$

$$y = \frac{3}{2}x - 2 \qquad \textit{(slope–intercept form)}$$

To write the equation in standard form, we start by clearing fractions.

$$2y = 2\left(\frac{3}{2}x - 2\right)$$

$$2y = 3x - 4$$

or

$$-3x + 2y = -4$$ *(Each term is multiplied by -1 so that the coefficient of x is positive.)*

or

$$3x - 2y = 4$$ *(standard form)*

■

SELF-STUDY EXERCISES 19-4

L.O.1. Write equations for the following in standard form. Verify your answers using a graphing calculator.

1. Find the equation of the line that is parallel to the line $x + y = 6$ and passes through the point $(2, -3)$.
2. Find the equation of the line that is parallel to the line $2x + y = 5$ and passes through the point (1, 7).
3. Find the equation of the line that is parallel to the line $3y = x - 2$ and passes through the point (4, 0).
4. Find the equation of the line that is parallel to the line $3x - y = -2$ and passes through the point $(-3, -2)$.

5. Find the equation of the line that is parallel to the line $x + 3y = 7$ and passes through the point (4, 1).

6. Find the equation of the line that is parallel to the line $3x - y = 4$ and passes through the point (0, 3).

7. Find the equation of the line that is parallel to the line $2x + 3y = 5$ and passes through the point (1, 1).

8. Find the equation of the line that is parallel to the line $3x + 2y = 1$ and passes through the point (2, 0).

9. Find the equation of the line that is parallel to the line $2x - 5y = 0$ and passes through the point (3, −1).

10. Find the equation of the line that is parallel to the line $-3x + 4y = -1$ and passes through the point $\left(\frac{1}{2}, 0\right)$.

19-5 PERPENDICULAR LINES

Learning Objective

L.O.1. Find the equation of a line, given a point on the line and the equation of a line perpendicular to a given line.

L.O.1. Find the equation of a line, given a point on the line and the equation of a line perpendicular to a given line.

Equations of Perpendicular Lines

Any line can have several lines that are perpendicular to it, but there is only one line perpendicular to a *given* line that passes through a *given* point. Figure 19-7 illustrates a line that is perpendicular to the line $2x + y = 7$ and passes through the point (6, 8). (In some cases the given point may lie on the given line.)

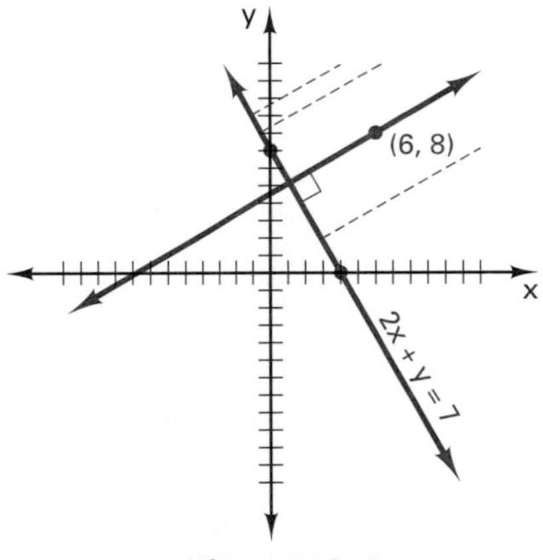

Figure 19-7

■ **Definition 19-5.1 Perpendicular Lines. Perpendicular lines** are two lines that intersect to form right angles (90° angles).

In Fig. 19-7, the perpendicular line that passes through the given point is indicated by the symbol ⊾, which means a right or 90° angle is formed by the lines.

The slope of the equation $2x + y = 7$ is -2, which can be seen if the equation is written as $y = -2x + 7$ in the slope–intercept form. We will determine the equation of a line that passes through (6, 8) and is perpendicular to $2x + y = 7$ by using an important property of perpendicular lines:

> **Rule 19-5.1** *Perpendicular Lines*
>
> The slope of *any* line perpendicular to a given line is the *negative reciprocal* of the slope of the given line.

Thus, if the slope of $2x + y = 7$ or $y = -2x + 7$ is -2, then the slope of *any* line perpendicular to that line is $+\frac{1}{2}$ $\left(\text{the negative reciprocal of } -2 \text{ or } -\frac{2}{1}\right)$.

Now, we know the slope $\left(+\frac{1}{2}\right)$ and one point (6, 8) through which the perpendicular line passes. We can find the equation of the perpendicular line by using the point–slope form of the equation.

$$y - y_1 = m(x - x_1) \qquad \left[m = \frac{1}{2}, P_1 = (6, 8)\right]$$

$$y - 8 = \frac{1}{2}(x - 6)$$

$$y - 8 = \frac{1}{2}x - 3$$

$$y = \frac{1}{2}x - 3 + 8$$

$$y = \frac{1}{2}x + 5$$

Then the equation of the perpendicular line is

$$y = \frac{1}{2}x + 5 \qquad \textit{(slope–intercept form)}$$

or

$$x - 2y = -10 \qquad \textit{(standard form)}$$

Tips and Traps

The negative reciprocal of a number is not necessarily a negative value. It will have the opposite sign.

To find the negative reciprocal:

1. Interchange the numerator and denominator.
2. Give the reciprocal the opposite sign.

The negative reciprocal of -5 is $+\frac{1}{5}$.

The negative reciprocal of $-\frac{3}{4}$ is $+\frac{4}{3}$.

The negative reciprocal of $\frac{4}{5}$ is $-\frac{5}{4}$.

To further illustrate the relationship of the slopes of perpendicular lines, examine Figures 19-8 and 19-9.

The term *normal* is often used in mathematics instead of the term *perpendicular*. The statement ''line *a is normal to* line *b*'' has the same meaning as the statement ''line *a is perpendicular to* line *b*.''

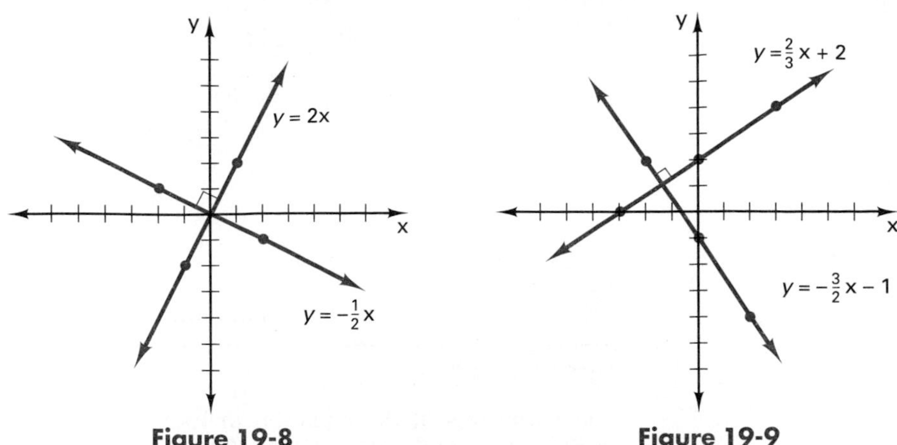

Figure 19-8 **Figure 19-9**

The following rule is used for finding the equation of a line that is perpendicular (normal) to a given line at a given point on the line or through a given point not on the line.

> **Rule 19-5.2** ***To find the equation of a line that is perpendicular (normal) to a given line at a given point on the line or through a given point not on the line:***
>
> 1. Determine the slope of the *given* line.
> 2. Find the *negative reciprocal* of this slope. The negative reciprocal is the slope of the perpendicular line.
> 3. Write the equation for the perpendicular line by substituting the coordinates of the *given* point for x_1 and y_1 and the slope of the *perpendicular* line for m into the point–slope form of the equation $y - y_1 = m(x - x_1)$.
> 4. Write equation in slope–intercept or standard form.

Look at the following example.

EXAMPLE 19-5.1 Find the equation of a line that is perpendicular to the line $y = \frac{2}{3}x + 8$ and passes through the point (3, 10).

Solution: First, find the slope of $y = \frac{2}{3}x + 8$. Since the equation is in slope–intercept form, the slope can be read from the equation. The slope is $\frac{2}{3}$. Next, find the negative reciprocal of $\frac{2}{3}$. The negative reciprocal of $\frac{2}{3}$ is $-\frac{3}{2}$.

Write the equation.

$$y - y_1 = m(x - x_1) \qquad \textit{(point–slope form)}$$

$$y - 10 = -\frac{3}{2}(x - 3) \qquad \textit{(Substitute values.)}$$

$$y - 10 = -\frac{3}{2}x + \frac{9}{2}$$

$$y = -\frac{3}{2}x + \frac{9}{2} + 10$$

$$y = -\frac{3}{2}x + \frac{9}{2} + \frac{20}{2}$$

$$y = -\frac{3}{2}x + \frac{29}{2} \qquad \textit{(slope–intercept form)}$$

or

$$2y = -3x + 29$$

$$3x + 2y = 29 \qquad \textit{(standard form)}$$ ■

Tips and Traps

The variations of the equation in Example 19-5.1 represent the same line. The slope–intercept form is useful for determining properties of a graph by inspection. The standard form is desirable because it contains no fractions.

EXAMPLE 19-5.2 Find the equation of the line normal to $4x + y = -3$ and passing through $(0, -3)$.

Solution

The slope of $4x + y = -3$ is -4. Rearranging the equation $y = -4x - 3$ in slope–intercept form, we can read the slope directly. The negative reciprocal of -4 is $\frac{1}{4}$. Thus, $m = \frac{1}{4}$ (slope of perpendicular).

$$y - y_1 = m(x - x_1)$$

$$y - (-3) = \frac{1}{4}(x - 0) \qquad \left[m = \frac{1}{4},\ p_1 = (0, -3)\right]$$

$$y + 3 = \frac{1}{4}x$$

$$y = \frac{1}{4}x - 3 \qquad \textit{(equation of the normal in slope–intercept form)}$$

or

$$4y = x - 12$$

$$x - 4y = 12 \qquad \textit{(equation of the normal in standard form)}$$ ■

EXAMPLE 19-5.3 Which of the following equations represents the normal of the line $2x - 3y = 1$ passing through $(2, -1)$?

(a) $y = \frac{2}{3}x - \frac{1}{3}$ **(b)** $y = \frac{2}{3}x - \frac{7}{3}$

(c) $y = -\frac{3}{2}x - \frac{1}{3}$ **(d)** $y = -\frac{3}{2}x + 2$

Solution: The slope of $2x - 3y = 1$ is $\frac{2}{3}$. To read the slope, rearrange the equation in slope–intercept form as follows:

$$2x - 3y = 1$$

$$-3y = -2x + 1$$

$$\frac{-3y}{-3} = \frac{-2x}{-3} + \frac{1}{-3}$$

$$y = \frac{2}{3}x - \frac{1}{3} \qquad \textit{(slope–intercept form)}$$

The slope of any normal line would be $-\frac{3}{2}$. Thus, (c) and (d) are the only possible solutions. Since the choices are in slope–intercept form, find the y-intercept of the normal line. The y-intercept of a normal line passing through $(2, -1)$ is found by using the slope-intercept form of an equation.

$$y = mx + b$$

$$-1 = -\frac{3}{2}(2) + b \qquad \textit{(Substitute } m = -\frac{3}{2}, x = 2, \textit{ and } y = -1.\textit{)}$$

$$-1 = -3 + b$$

$$3 - 1 = b \qquad \textit{(Solve for b.)}$$

$$2 = b$$

Thus, the correct equation is

$$y = -\frac{3}{2}x + 2$$

or choice (d). ■

SELF-STUDY EXERCISES 19-5

L.O.1. Write the equations for the following in standard form. Verify your results using a graphing calculator. Be sure the range is set so that the vertical and horizontal increments are equal.

1. Find the equation of the line that is perpendicular to the line $x + y = 6$ and passes through the point $(2, 3)$.
2. Find the equation of the line that is normal to the line $2x + y = 5$ and passes through the point $(1, 7)$.
3. Find the equation of the line that is perpendicular to the line $3y = x - 2$ and passes through the point $(4, 0)$.
4. Find the equation of the line that is normal to the line $3x - y = 2$ and passes through the point $(-3, -2)$.
5. Find the equation of the line that is normal to the line $x + 3y = 7$ and passes through the point $(4, 1)$.
6. Find the equation of the line that is perpendicular to the line $x + 2y = 7$ and passes through the point $(-2, 3)$.
7. Find the equation of the line that is perpendicular to the line $2x + 3y = 4$ and passes through the point $(3, -1)$.
8. Find the equation of the line that is normal to the line $4x + y = 1$ and passes through the point $(0, 0)$.
9. Find the equation of the line that is perpendicular to the line $2x + 2y = 3$ and passes through the point $\left(\frac{1}{2}, 2\right)$.
10. Find the equation of the line that is perpendicular to the line $5x - y = 6$ and passes through the point $\left(5, -\frac{1}{5}\right)$.

19-6 DISTANCE AND MIDPOINTS

Learning Objectives

L.O.1. Find the distance between two points.
L.O.2. Find the midpoint between two points.

L.O.1. Find the distance between two points.

Distance

Many times it is desirable to know the straight-line distance between two points. If we consider two points such as $P_1 = (4, 2)$ and $P_2 = (7, 6)$, we can visualize the straight-line distance between these two points on a pair of coordinate axes as in Fig. 19-10. To find the distance, we extend a vertical line down from P_2 and a horizontal line across from P_1 to form a triangle.

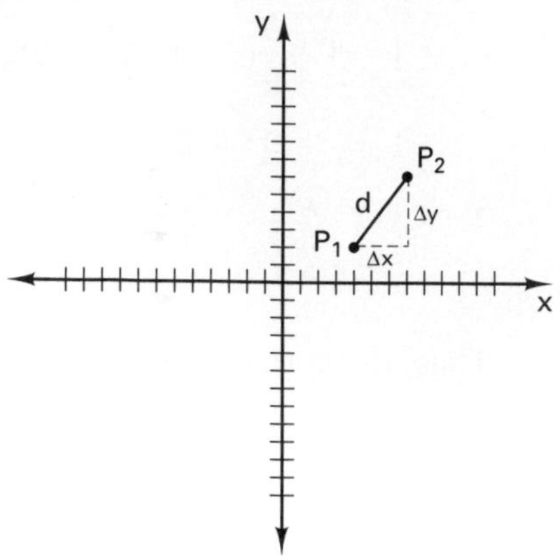

Figure 19-10

We have labeled the straight-line distance from P_1 to P_2 as d, the change in horizontal movement as Δx, and the change in vertical movement as Δy. The triangle formed is a *right triangle*. The *distance formula* is based on the special property of right triangles, the *Pythagorean theorem.*

The distance from P_1 to P_2 is the principal square root of the sum of the squares of the change in horizontal movement and the change in vertical movement. That is, $d^2 = (\Delta x)^2 + (\Delta y)^2$. Solving for d by taking the square root of each side of the equation, we obtain

$$d = \sqrt{(\Delta x)^2 + (\Delta y)^2}$$

If $P_1 = (x_1, y_1)$ and $P_2 = (x_2, y_2)$, the distance formula can be stated in this form:

Formula 19-6.1

Distance: $d = \sqrt{(x_2 - x_1)^2 + (y_2 - y_1)^2}$

EXAMPLE 19-6.1 Find the distance from (4, 2) to (7, 6).

Solution: These points are illustrated in Fig. 19-10. Point (4, 2) is P_1, and point (7, 6) is P_2.

Now, substitute the following values into the distance formula: $x_1 = 4, x_2 = 7, y_1 = 2$, and $y_2 = 6$.

$$d = \sqrt{(7 - 4)^2 + (6 - 2)^2}$$
$$d = \sqrt{3^2 + 4^2}$$
$$d = \sqrt{9 + 16}$$
$$d = \sqrt{25}$$
$$d = 5$$

Thus, the distance from (4, 2) to (7, 6) is 5 units. ■

Tips and Traps

The distance formula can be used to calculate the distance between two points no matter which point we designate as P_1 and which as P_2. Let's rework Example 19-6.1 letting $P_1 = (7, 6)$ and $P_2 = (4, 2)$. Thus, $x_1 = 7$, $x_2 = 4$, $y_1 = 6$, and $y_2 = 2$.

$$d = \sqrt{(4 - 7)^2 + (2 - 6)^2}$$

$$d = \sqrt{(-3)^2 + (-4)^2} \qquad [(-3)^2 = +9;\ (-4)^2 = +16]$$

$$d = \sqrt{9 + 16}$$

$$d = \sqrt{25}$$

$$d = 5$$

EXAMPLE 19-6.2 Find the distance from (5, −2) to (−3, −4).

Solution: Let (5, −2) be P_1 and (−3, −4) be P_2. [We could have designated (−3, −4) as P_1 and (5, −2) as P_2.]

Then $x_1 = 5$, $x_2 = -3$, $y_1 = -2$, and $y_2 = -4$.

$$d = \sqrt{(-3 - 5)^2 + (-4 - (-2))^2}$$

$$d = \sqrt{(-8)^2 + (-2)^2} \qquad [(-4 + 2) = -2]$$

$$d = \sqrt{64 + 4}$$

$$d = \sqrt{68}$$

$$d = 8.246 \qquad \textit{(to the nearest thousandth)}$$

■

L.O.2. Find the midpoint between two points.

Midpoints

Two points determine a line that is infinite in length. However, the two points also determine a *line* segment that has a certain length. The two points are called the *end points* of the line segment.

On a number line or measuring device, the coordinate of the *midpoint* between two points is the average of the coordinates of the points.

> **Formula 19-6.2**
>
> To find the coordinate of the midpoint between two points *on the number line* or linear measuring device, average the end points.
>
> $$\text{Midpoint} = \frac{P_1 + P_2}{2}$$
>
> where P_1 and P_2 are points on the number line or measuring device.

EXAMPLE 19-6.3 Find the midpoint between two points on a metric rule at 2.8 and 5.6.

Solution

$$\begin{aligned}\text{Midpoint} &= \frac{P_1 + P_2}{2}\\ &= \frac{2.8 + 5.6}{2}\\ &= \frac{8.4}{2}\\ &= 4.2\end{aligned}$$

■

The coordinates of the midpoint between two points on a coordinate system can be found by averaging the respective coordinates.

Formula 19-6.3

To find the coordinates of the midpoint of a line segment *on a coordinate system*, average the respective coordinates of the end points of the segment.

$$\text{Midpoint} = \left(\frac{x_1 + x_2}{2}, \frac{y_1 + y_2}{2}\right)$$

where P_1 and P_2 are end points of the segment and $P_1 = (x_1, y_1)$ and $P_2 = (x_2, y_2)$.

EXAMPLE 19-6.4 Find the midpoint of each segment with the given end points.

(a) (2, 4) and (6, 10) **(b)** $(-1, 4)$ and $(2, -3)$
(c) $(3, -5)$ and origin

Solution

(a) (2, 4) and (6, 10)

$$\begin{aligned}\text{Midpoint} &= \left(\frac{x_1 + x_2}{2}, \frac{y_1 + y_2}{2}\right)\\ &= \left(\frac{2 + 6}{2}, \frac{4 + 10}{2}\right) \quad \textit{(Substitute values.)}\\ &= \left(\frac{8}{2}, \frac{14}{2}\right)\\ &= (4, 7)\end{aligned}$$

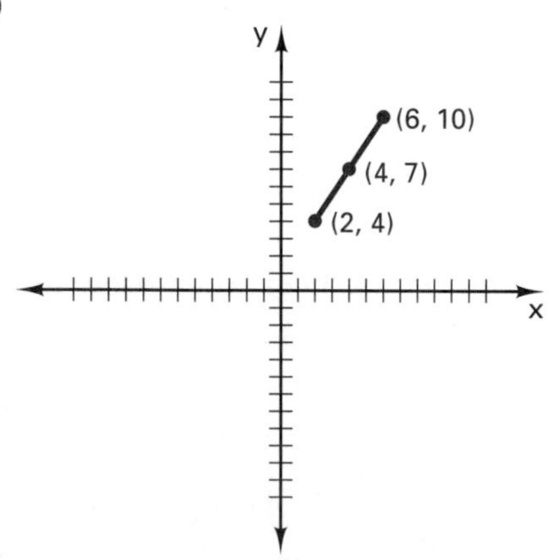

Figure 19-11

(b) $(-1, 4)$ and $(2, -3)$

$$\text{Midpoint} = \left(\frac{x_1 + x_2}{2}, \frac{y_1 + y_2}{2}\right)$$

$$= \left(\frac{-1 + 2}{2}, \frac{4 - 3}{2}\right) \qquad \textit{(Substitute values.)}$$

$$= \left(\frac{1}{2}, \frac{1}{2}\right)$$

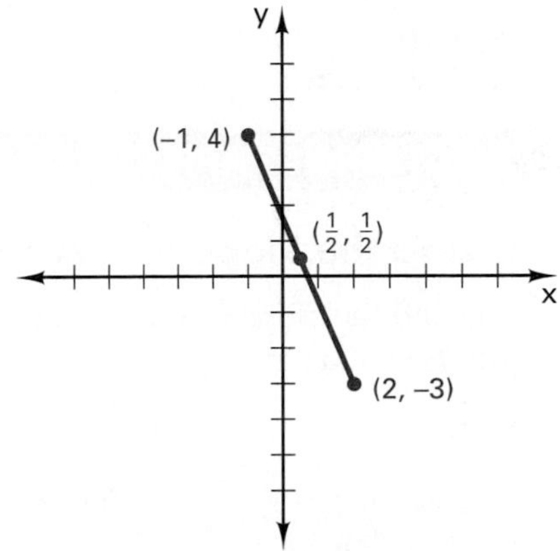

Figure 19-12

(c) $(3, -5)$ and origin

$$\text{Midpoint} = \left(\frac{x_1 + x_2}{2}, \frac{y_1 + y_2}{2}\right)$$

$$= \left(\frac{3 + 0}{2}, \frac{-5 + 0}{2}\right) \qquad \textit{(Substitute values.)}$$

$$= \left(\frac{3}{2}, \frac{-5}{2}\right)$$

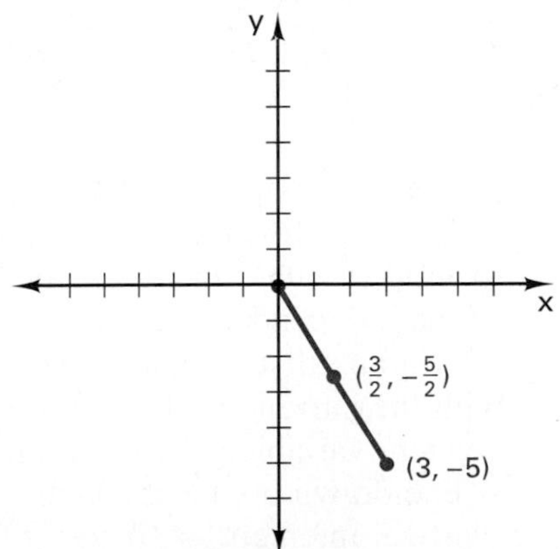

Figure 19-13

Tips and Traps

To find the coordinates of the midpoint of a segment from the origin to any point, halve each coordinate of the point that is not at the origin.

SELF-STUDY EXERCISES 19-6

L.O.1. Graph the line segment determined by the given end points and find the distance between each of the following pairs of points. Express the answer to the nearest thousandth when necessary.

1. (7, 10) and (1, 2)
2. (7, −7) and (2, 5)
3. (5, 0) and (0, 5)
4. (−2, −2) and (3, −4)
5. (5, 7) and (0, −3)
6. (8, −2) and (−4, 3)
7. (−4, 6) and (2, −2)
8. (3, 5) and (0, 1)
9. (5, 4) and (−7, −5)
10. (7, 2) and (−2, −3)

L.O.2. 11–20. Find the coordinates of the midpoint of the segments determined in Exercises 1 through 10.

ELECTRONICS APPLICATION

Using the Slope to Find Resistance

To find the slope of the line that goes through the two points ($x1$, $y1$) and ($x2$, $y2$), the formulas are

$$\text{slope} = \frac{\text{rise}}{\text{run}} = \frac{y2 - y1}{x2 - x1} \quad \text{or} \quad \text{slope} = \frac{y1 - y2}{x1 - x2}$$

Assume that you have a graph with current on the horizontal axis and voltage on the vertical axis. To find the slope of the line through the points (1 mA, 3 V) and (4 mA, 6 V),

$$\text{slope} = \frac{3\text{ V} - 6\text{ V}}{1\text{ mA} - 4\text{ mA}} = \frac{-3\text{ V}}{-3\text{ mA}} = 1\text{ k}\Omega \quad \text{or}$$

$$\text{slope} = \frac{6\text{ V} - 3\text{ V}}{4\text{ mA} - 1\text{ mA}} = \frac{3\text{ V}}{3\text{ mA}} = 1\text{ k}\Omega$$

Notice that volts divided by amperes gives ohms and $1 \div \frac{1}{1000}(m) = 1000(k)$.

To find the slope of the line through (4V, 5 mA) and (6 V, 4 mA),

$$\text{slope} = \frac{5\text{ mA} - 4\text{ mA}}{4\text{ V} - 6\text{ V}} = \frac{1\text{ mA}}{-2\text{ V}} = -1\text{ mS}$$

Notice that amperes divided by volts gives siemens.

Usually, the independent variable is the one that is said to vary from one value to another value, and the dependent variable is calculated based on the given value of the independent variable.

A graph that compares current and voltage for a series circuit is usually drawn with the current on the horizontal axis, since the current is the reference or independent variable, and the voltage is on the vertical axis as the dependent variable. A line drawn on this graph that goes through the center of the graph or the origin (vertical intercept = 0) has

$$\text{slope} = \frac{\text{rise}}{\text{run}} = \frac{\Delta\text{ voltage}}{\Delta\text{ current}} = \text{resistance } (R)$$

So the equation of the line for a series circuit is

$$E = RI \text{ (for dc circuits)} \quad \text{or} \quad E = IZ \text{ (for ac circuits)}$$

A graph that compares current and voltage for a parallel circuit is usually drawn with the voltage on the horizontal axis, since the voltage is the reference or independent variable, and the current is on the vertical axis as the dependent variable. A line drawn on this graph that passes through the origin has

$$\text{slope} = \frac{\text{rise}}{\text{run}} = \frac{\Delta \text{ current}}{\Delta \text{ voltage}} = \text{conductance}$$

So the equation of the line is

$$I = GE \text{ (for dc circuits)} \quad \text{or} \quad I = VY \text{ (for ac circuits)}$$

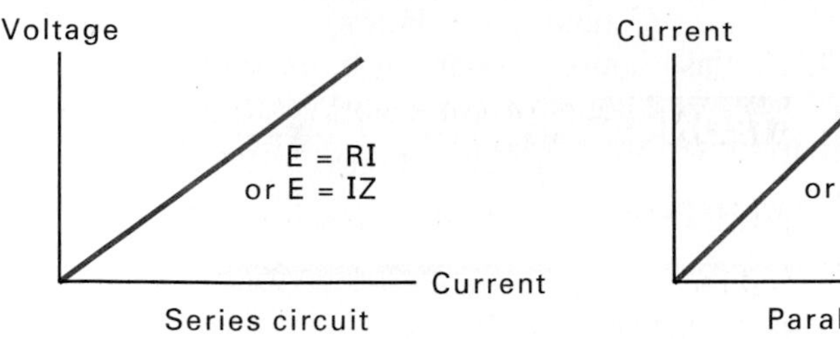

Exercises

Find the slope of the line that passes through each of the following pairs of points. Be sure to include sign, prefix, and units as well as the number in the answer.

1. (4 mA, 12 V) and (7 mA, 15 V)
2. (4 mA, 12 V) and (9 mA, 8 V)
3. (15 V, 7 mA) and (20 V, 9 mA)
4. (25 V, 4 mA) and (37 V, 1 mA)
5. (0.65 V, 1 mA) and (0.75 V, 2 mA)
6. (9 μS, 11 V) and (12 μS, 17 V)
7. (9 μS, 11 V) and (12 μS, 5 V)
8. (4 hours, 200 miles) and (5 hours, 250 miles)
9. (7 hours, 700 kilometers) and (9 hours, 900 kilometers)
10. (2 class hours, 4 hours of homework) and (3 class hours, 6 hours of homework)

Answers for Exercises

Find the slope of the line that passes through each of the following pairs of points. Be sure to include sign, prefix and units, as well as the number in the answer.

1. (4 mA, 12 V) and (7 mA, 15 V) $\frac{(12\text{ V} - 15\text{ V})}{(4\text{ mA} - 7\text{ mA})} = \frac{-3\text{ V}}{(-3\text{ mA})} = +1\text{ k}\Omega$
2. (4 mA, 12 V) and (9 mA, 8 V) $\frac{(12\text{ V} - 8\text{ V})}{(4\text{ mA} - 9\text{ mA})} = \frac{4\text{ V}}{(-5\text{ mA})} = -0.8\text{ k}\Omega$ or $-800\ \Omega$
3. (15 V, 7 mA) and (20 V, 9 mA) $\frac{(7\text{ mA} - 9\text{ mA})}{(15\text{ V} - 20\text{ V})} = \frac{-2\text{ mA}}{(-5\text{ V})} = 0.4\text{ mS}$ or $400\ \mu\text{S}$
4. (25 V, 4 mA) and (37 V, 1 mA) $\frac{(4\text{ mA} - 1\text{ mA})}{(25\text{ V} - 37\text{ V})} = \frac{3\text{ mA}}{(-12\text{ V})} = 0.25\text{ mS}$ or $250\ \mu\text{S}$
5. (0.65 V, 1 mA) and (0.75 V, 2 mA) $\frac{(1\text{ mA} - 2\text{ mA})}{(0.65\text{ V} - 0.75\text{ V})} = \frac{-1\text{ mA}}{(-0.10\text{ V})} =$ 10 mS

6. (9 μS, 11 V) and (12 μS, 17 V) $\frac{(11\text{ V} - 17\text{ V})}{(9\ \mu\text{S} - 12\ \mu\text{S})} = \frac{-6\text{ V}}{(-3\ \mu\text{S})} = 2\text{ M}\Omega$ or 2,000,000 Ω
7. (9 μS, 11 V) and (12 μS, 5 V) $\frac{(11\text{ V} - 5\text{ V})}{(9\ \mu\text{S} - 12\ \mu\text{S})} = \frac{+6\text{ V}}{(-3\ \mu\text{S})} = -2\text{ M}\Omega$ or −2,000,000 Ω
8. (4 hours, 200 miles) and (5 hours, 250 miles) $\frac{(200\text{ miles} - 250\text{ miles})}{(4\text{ hours} - 5\text{ hours})} =$ 50 miles per hour
9. (7 hours, 700 kilometers) and (9 hours, 900 kilometers) $\frac{(700\text{ kilometers} - 900\text{ kilometeres})}{(7\text{ hours} - 9\text{ hours})} =$ 100 kilometers per hour
10. (2 class hours, 4 hours of homework) and (3 class hours, 6 hours of homework) $\frac{(4\text{ hours of homework} - 6\text{ hours of homework})}{(2\text{ class hours} - 3\text{ class hours})} =$ 2 hours of homework per class hour

ASSIGNMENT EXERCISES

19-1

Find the slope of the line passing through the following pairs of points.

1. $(-2, 2)$ and $(1, 3)$
2. $(3, -1)$ and $(1, 3)$
3. $(3, 2)$ and $(5, 6)$
4. $(-1, -1)$ and $(2, 2)$
5. $(4, 3)$ and $(-4, -2)$
6. $(6, 2)$ and $(-3, 2)$
7. $(3, -4)$ and $(0, 0)$
8. $(1, -1)$ and $(5, -5)$
9. $(-4, 1)$ and $(-4, 3)$
10. $(4, -4)$ and $(1, 3)$
11. $(5, 0)$ and $(-2, 4)$
12. $(-2, 1)$ and $(0, 3)$
13. $(-4, -8)$ and $(-2, -1)$
14. $(3, 3)$ and $(3, 0)$
15. $(5, -3)$ and $(-1, -3)$
16. $(-5, -1)$ and $(-7\ -3)$
17. $(-7, 0)$ and $(-7, 5)$
18. $(3, 5)$ and $(2, 5)$
19. $(5, 9)$ and $(7, 11)$
20. $(3, 5)$ and $(5, 3)$
21. Write the coordinates of two points that lie on the same horizontal line.
22. Write the coordinates of two points that lie on the same vertical line.

19-2

Find the equation of a line passing through the given point with the given slope. Solve the equation for y if necessary.

23. $(-6, 2)$, $m = \frac{1}{3}$
24. $(3, 2)$, $m = -\frac{2}{5}$
25. $(4, 0)$, $m = \frac{3}{4}$
26. $(0, -2)$, $m = 2$
27. $(2, 3)$, $m = 4$
28. $(6, 0)$, $m = -1$
29. $(5, -4)$, $m = -\frac{2}{3}$
30. $(-1, -5)$, $m = -3$

Find the equation of a line passing through the given pairs of points. Solve the equation for y if necessary.

31. $(-5, 2)$ and $(6, 1)$
32. $(1, 4)$ and $(-1, 3)$
33. $(-1, -3)$ and $(3, 4)$
34. $(-3, 0)$ and $(4, 0)$
35. $(-2, -3)$ and $(3, 6)$
36. $(2, -4)$ and $(3, -4)$
37. $(5, 2)$ and $(6, 3)$
38. $(4, 6)$ and $(1, -1)$
39. $(-1, -2)$ and $(-3, -4)$
40. $(4, 0)$ and $(4, -3)$
41. $(5, -2)$ and $(3, -2)$
42. $(5, 4)$ and $(0, 4)$

19-3

Determine the slope and y-intercept of the following equations by inspection.

43. $y = 3x + \frac{1}{4}$
44. $y = \frac{2}{3}x - \frac{3}{5}$
45. $y = -5x + 4$

46. $y = 7$

47. $x = 8$

48. $y = \frac{1}{3}x - \frac{5}{8}$

49. $y = \frac{x}{8} - 5$

50. $y = -\frac{x}{5} + 2$

Rewrite the following equations in slope–intercept form and determine the slope and y-intercept.

51. $2x + y = 8$

52. $4x + y = 5$

53. $3x - 2y = 6$

54. $5x - 3y = 15$

55. $\frac{3}{5}x - y = 4$

56. $2.2y - 6.6x = 4.4$

57. $3y = 5$

58. $3x - 6y = 12$

Write the equations using the given slope and y-intercept.

59. $m = 3, b = -2$

60. Slope $= \frac{3}{5}$; y-intercept $= -7$

61.

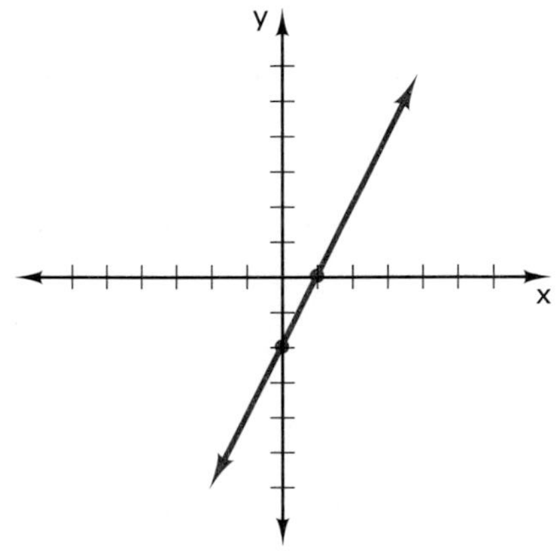

62.

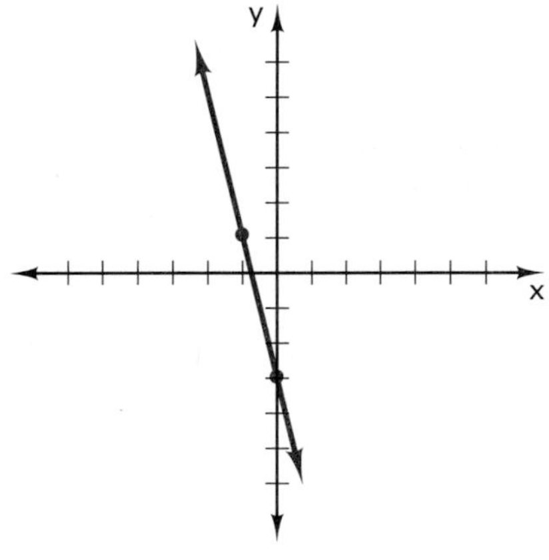

19-4

Write equations for the following in standard form. Graph both equations for each Exercise on a graphing calculator to verify parallelism.

63. Find the equation of the line that is parallel to the line $x + y = 4$ and passes through the point $(2, 5)$.

64. Find the equation of the line that is parallel to the line $3x + y = 6$ and passes through the point $(1, 0)$.

65. Find the equation of the line that is parallel to the line $2y = x - 3$ and passes through the point $(2, -3)$.

66. Find the equation of the line that is parallel to the line $4x - y = -1$ and passes through the point $(0, -2)$.

67. Find the equation of the line that is parallel to the line $x - 3y = 5$ and passes through the point $(5, -5)$.

68. Find the equation of the line that is parallel to the line $3x - 2y = 2$ and passes through the point $(0, -3)$.

69. Find the equation of the line that is parallel to the line $x + 3y = 6$ and passes through the point $(-4, -2)$.

70. Find the equation of the line that is parallel to the line $4x + 3y = 1$ and passes through the point $\left(3, -\frac{1}{2}\right)$

71. Find the equation of the line that is parallel to the line $3x - 4y = 0$ and passes through the point $\left(\frac{1}{3}, 2\right)$

72. Find the equation of the line that is parallel to the line $-2x + 3y = 2$ and passes through the point $(-1, -1)$.

19-5

Write the equations for the following in standard form. Use a graphing calculator to verify that the two lines in each Exercise are perpendicular.

73. Find the equation of the line that is perpendicular to the line $x + y = 4$ and passes through the point $(-3, 1)$.

74. Find the equation of the line that is normal to the line $3x + y = 6$ and passes through the point $(1, 4)$.

75. Find the equation of the line that is perpendicular to the line $x + 2y = 5$ and passes through the point $(-2, 0)$.

76. Find the equation of the line that is normal to the line $2x + 2y = 4$ and passes through the point $(0, 0)$.

77. Find the equation of the line that is normal to the line $5x + y = 8$ and passes through the point $(-1, 2)$.

78. Find the equation of the line that is perpendicular to the line $3y = x - 4$ and passes through the point $(2, 3)$.

79. Find the equation of the line that is perpendicular to the line $5x - y = 10$ and passes through the point $\left(\frac{1}{2}, 3\right)$.

80. Find the equation of the line that is normal to the line $x - 3y = 6$ and passes through the point $(-2, 4)$.

81. Find the equation of the line that is perpendicular to the line $4x - y = 8$ and passes through the point $\left(4, -\frac{1}{2}\right)$.

82. Find the equation of the line that is perpendicular to the line $4y = 2x + 1$ and passes through the point $(3, -1)$.

19-6

Find the distance between each of the following pairs of points. Express the answer to the nearest thousandth when necessary.

83. $(3, 6)$ and $(-1, 4)$
84. $(4, 0)$ and $(0, -1)$
85. $(3, -3)$ and $(0, 7)$
86. $(2, 7)$ and $(-3, -2)$
87. $(0, 0)$ and $(-3, 5)$
88. $(-5, -5)$ and $(4, 0)$
89. $(5, 2)$ and $(-3, -3)$
90. $(1, -5)$ and $(-2, -5)$
91. $(-5, -4)$ and $(2, -2)$
92. $(8, 3)$ and $(3, -4)$

93–102. Calculate the coordinates for the midpoint of each segment whose end points are given in Exercises 83 through 92.

CHALLENGE PROBLEM

103. Holly Holder is introducing a new lipstick in Brightglow's product line. As marketing manager, she estimates the cost of the new lipstick will be \$4.53 per item, plus an additional cost of \$5,000.00.

(a) Make a table of the estimated cost for producing 0 lipsticks, 100 lipsticks, 1000 lipsticks, and 2000 lipsticks.

(b) Represent these costs as ordered pairs (number of lipsticks, cost).

(c) Graph the ordered pairs and sketch a line that fits the ordered pairs.

(d) Write a formula (using function notation) that represents the cost of the new lipstick as a function of the number of lipsticks produced.

(e) Holly projects the selling price of the new lipstick to be \$8.99. How many lipsticks must be sold for the company to recover its cost of producing the new item?

CONCEPTS ANALYSIS

1. (a) Describe the graph of a line whose slope is positive.
   (b) Describe the graph of a line whose slope is negative.
   (c) Describe the graph of a line whose slope is a fraction between 0 and 1.
   (d) Describe the graph of a line whose slope is a number >1.
   (e) Describe the graph of a line whose slope is a fraction between -1 and 0.
   (f) Describe the graph of a line whose slope is a number <-1.
2. What is the slope of a horizontal line? Why?
3. What is the slope of a vertical line? Why?
4. If the standard form of an equation of a horizontal line is $y = k$, what does k represent?
5. If the standard form of an equation of a horizontal line is $x = k$, what does k represent?
6. If an equation is solved for y, what information about the graph can we read from the equation?
7. How do the slopes of parallel lines compare?
8. How do the slopes of perpendicular lines compare?
9. How are the distance formula and the Pythagorean theorem related?
10. List the steps necessary to find the x-intercept of a line if you are given two points on the line.

CHAPTER SUMMARY

Objectives	What to Remember	Examples
Section 19-1		
1. Calculate the slope of a line, given two points on the line.	The slope of a line joining two points is the difference of the y-coordinates divided by the difference of the x-coordinates, that is, rise/run. $m = \frac{y_2 - y_1}{x_2 - x_1}$	Find the slope of the line passing through the points (3, 1) and $(-2, 5)$. $\frac{5 - 1}{-2 - 3} = \frac{4}{-5}$ or $-\frac{4}{5}$
2. Determine the slope of a horizontal or a vertical line.	The slope of a horizontal line is zero, and the coordinates of points on the same horizontal line have the same y-coordinate. The slope of a vertical line is not defined, and points on the same vertical line have the same x-coordinate.	Examine the coordinate pairs and indicate which points lie on the same horizontal line and which lie on the same vertical line. A $(4, -3)$, B $(2, 5)$, C $(4, 5)$, D $(2, -3)$ Points A and D and points B and C lie on the same horizontal line. Points A and C and points B and D lie on the same vertical line.
Section 19-2		
1. Find the equation of a line, given the slope and one point.	Use the point–slope form of an equation to find the equation when given the slope and coordinates of a point on a line. $y - y_1 = m(x - x_1)$	A line has slope 2 and passes through the point (3, 4). Find the equation of the line. $y - y_1 = m(x - x_1)$ $y - 4 = 2(x - 3)$ $y - 4 = 2x - 6$ $y = 2x - 2$

2. Find the equation of a line, given two points on the line.

Find the slope; then use the point–slope form of the equation.

Find the equation of a line that passes through the two points (3, 1), (−3, 2). First, find the slope.

$$m = \frac{2 - 1}{-3 - 3} = \frac{1}{-6} \text{ or } -\frac{1}{6}$$

$$y - y_1 = m(x - x_1)$$

$$y - 1 = -\frac{1}{6}(x - 3)$$

$$y - 1 = -\frac{1}{6}x + \frac{1}{2}$$

$$y = -\frac{1}{6}x + \frac{3}{2}$$

$$\text{or } 6y = -1x + 9$$

$$x + 6y = 9$$

Section 19-3

1. Find the slope and y-intercept of a line, given the equation of the line.

Write the equation in slope–intercept form (solve for y).

$$y = mx + b$$

The coefficient of x is the slope and the constant is the y-coordinate of the y-intercept.

Find the slope and y-intercept for the equation $2x - 4y = 8$. Solve for y.

$$-4y = -2x + 8$$

$$y = \frac{1}{2}x - 2$$

Slope $= \frac{1}{2}$.

y-intercept $= (0, -2)$.

2. Find the equation of a line, given the slope and y-intercept.

Use the slope–intercept form of the equation, $y = mx + b$, and substitute values for m and b, the slope and y-intercept, respectively.

Write the equation of a line that has slope $\frac{2}{3}$ and y-intercept $(0, -1)$.

Use the slope–intercept form of the equation:

$$y = mx + b$$

$$y = \frac{2}{3}x - 1$$

Section 19-4

1. Find the equation of a line, given a point on the line and the equation of a line parallel to a given line.

The slopes of parallel lines are equal; that is, lines that have the same slope are parallel lines. Equations that have equal slopes will have graphs that are parallel lines.

Determine which two of the three given equations will have graphs that are parallel lines:
(a) $y = 3x - 5$;
(b) $3x - 2y = 10$;
(c) $6x - 2y = 8$.

Write each of the three equations in slope–intercept form and compare the slopes (coefficients of x).

(a) $y = 3x - 5$

(b) $y = \frac{3}{2}x - 5$

(c) $y = 3x - 4$

(a) and (c) have the same slope and thus their graphs are parallel.

To find the equation of a line parallel to a given line and passing through a given point, use the point–slope form of the linear equation. Substitute the value of the slope of the given equation and the coordinates of the given point into the point-slope form of an equation.

Find the equation of a line that passes through the point (2, 3) and is parallel to the line whose equation is $y = 4x - 1$. Write the new equation in slope–intercept form.

Use the slope of the given equation, 4, for the slope of the new equation.

$$\begin{aligned} y - y_1 &= m(x - x_1) \\ y - 3 &= 4(x - 2) \\ y - 3 &= 4x - 8 \\ y &= 4x - 5 \end{aligned}$$

Section 19-5

1. Find the equation of a line, given a point on the line and the equation of a line perpendicular to a given line.

The slopes of perpendicular lines are negative reciprocals; that is, lines that have slopes that are negative reciprocals are perpendicular lines. Equations that have slopes that are negative reciprocals have graphs that are perpendicular lines.

Determine which two of the three given equations will have graphs that are perpendicular lines:
(a) $y = 3x - 5$;
(b) $3x + 9y = 10$;
(c) $9x + 3y = 12$

Write each of the three equations in slope–intercept form and compare the slopes (coefficients of x).

(a) $y = 3x - 5$

(b) $y = -\frac{1}{3}x + \frac{10}{9}$

(c) $y = -3x + 4$

(a) and (b) have the slopes that are negative reciprocals and thus their graphs are perpendicular. Note that (a) and (c) are not perpendicular. Their slopes are opposites but *not* reciprocals.

To write the equation of a line perpendicular to a given line, substitute the values of the negative reciprocal of the slope of the given equation and the coordinates of the given point into the point-slope form of an equation.

Find the equation of a line that passes through the point $(-1, 2)$ and is perpendicular to the line whose equation is $y = \frac{1}{3}x - 5$. Write the equation in standard form.

The slope for the new equation is the negative reciprocal of $\frac{1}{3}$, which is -3.

$$\begin{aligned} y - y_1 &= m(x - x_1) \\ y - 2 &= -3(x + 1) \\ y - 2 &= -3x - 3 \\ 3x + y &= -3 + 2 \\ 3x + y &= -1 \quad \text{(standard form)} \end{aligned}$$

Section 19-6

1. Find the distance between two points.	Use the distance formula: $d = \sqrt{(x_2 - x_1)^2 + (y_2 - y_1)^2}$	Find the distance between the points (3, 2) and (−1, 5) $d = \sqrt{(x_2 - x_1)^2 + (y_2 - y_1)^2}$ $d = \sqrt{(-1 - 3)^2 + (5 - 2)^2}$ $d = \sqrt{(-4)^2 + 3^2}$ $d = \sqrt{16 + 9}$ $d = \sqrt{25}$ $d = 5$
2. Find the midpoint between two points.	Use the midpoint formula: $\left(\frac{(x_2 + x_1)}{2}, \frac{(y_2 + y_1)}{2}\right)$	Find the midpoint of the segment joining the points (3, 2) and (−1, 5) $\left(\frac{(x_2 + x_1)}{2}, \frac{(y_2 + y_1)}{2}\right)$ $\left(\frac{-1 + 3}{2}, \frac{5 + 2}{2}\right)$ $\left(\frac{2}{2}, \frac{7}{2}\right)$ $\left(1, \frac{7}{2}\right)$

WORDS TO KNOW

slope (p. 718)
rise (p. 718)
run (p. 718)
horizontal line (p. 721)
vertical line (p. 721)
zero slope (p. 721)
undefined slope (p. 721)
point–slope form of equation (p. 722)
slope–intercept form of equation (p. 726)
cost function (p. 728)
standard form of equation (p. 730)
slope of parallel lines (p. 730)
slope of perpendicular lines (p. 733)
negative reciprocal (p. 733)
normal (p. 734)
distance formula (p. 738)
Pythagorean theorem (p. 738)
coordinates of midpoint of line segment (p. 739)

CHAPTER TRIAL TEST

Find the slope of the line passing through the following pairs of points.

1. (−3, 6) and (3, 2)
2. (0, 4) and (−1, 6)
3. (1, −5) and (3, 0)
4. (5, 3) and (−2, 3)
5. (−1, −1) and (2, 2)
6. (−1, 5) and (−1, 7)

Find the equation of the line passing through the given point with the given slope. Solve the equation for y.

7. (3, −5), $m = \frac{2}{3}$
8. (5, 1), $m = -2$

Find the equation of the line passing through the following pairs of points. Solve the equation for y.

9. (1, 3) and (4, 5)
10. (−1, 1) and (4, −4)
11. (5, 2) and (−1, 2)
12. (7, 4) and (−3, −1)

13. What is the slope and y-intercept of a line whose equation is $y = 3x - 22$?

Rewrite the following equations in slope–intercept form and determine the slope and y-intercept.

14. $x - y = 4$

15. $x = 4y$

16. $2y - x = 3$

17. $\frac{1}{3}y + 2x = 1$

Write equations in slope-intercept form using the given slope and y-intercept or the graph.

18. slope $= -2$, y-intercept $= -3$

19. 

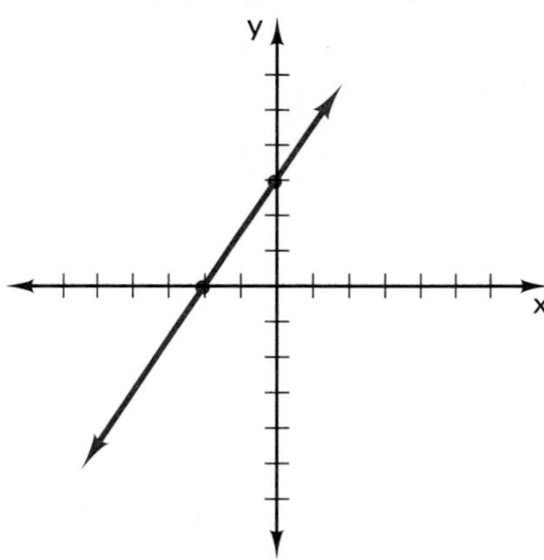

Write the equations for the following in standard form.

20. Find the equation of the line that is parallel to $y - 2x = 3$ and passes through the point $(2, 5)$.

21. Find the equation of the line that is parallel to $2x + y = 4$ and passes through the point $(4, -3)$.

22. Find the equation of the line perpendicular to $y - 2x = 3$ and passing through the point $(2, 5)$.

23. Find the equation of the line perpendicular to $2x + y = 4$ and passing through the point $(4, -3)$.

Find the distance between the following points and find the coordinates of the midpoint of the line segment made by the following pairs of points.

24. $(5, 2)$ and $(-7, -3)$

25. $(-2, 1)$ and $(3, 3)$

CHAPTER

20 Systems of Equations and Inequalities

We have already gained some degree of familiarity with graphing equations and inequalities to arrive at a solution. Now we extend our skills to graphing two equations or inequalities each with two variables and to learning other procedures for solving systems of equations.

20-1 SOLVING SYSTEMS OF EQUATIONS AND INEQUALITIES GRAPHICALLY

Learning Objectives

L.O.1. Solve a system of equations by graphing.
L.O.2. Solve a system of inequalities by graphing.

In Chapter 18 we learned to graph equations and inequalities with two variables. Since an equation with two variables has many ordered pairs of solutions, a graph gives a pictorial view of these solutions. In a similar way, since an inequality with two variables has many values that satisfy the conditions of the inequality, a graph gives a pictorial view of these values as well.

L.O.1. Solve a system of equations by graphing.

Graphing Systems of Equations

A system of two equations with each having two variables is solved when we find the one ordered pair of solutions that satisfies *both* equations. One method of solving systems of two equations is to graph the ordered pairs of solutions of each equation and find the intersection of these graphs. The point where the two graphs intersect represents the ordered pair of solutions that satisfies both equations.

First, let's look at an example that could be described as a system of two equations with two variables.

EXAMPLE 20-1.1 A board is 20 ft long. It needs to be cut so that one piece is 2 ft longer than the other. What should be the length of each piece?

Solution: First, we write equations to describe the conditions of the problem. Since the board is not cut into equal pieces, we let the letter l represent the *longer* piece and the letter s represent the *shorter* piece.

The first condition of the problem states that the total length of the board is 20 ft. Thus, the two pieces (l and s) total 20 ft.

$$l + s = 20$$

The second condition of the problem is that one piece is 2 ft longer than the other. Thus, the shorter piece plus 2 ft equals the longer piece.

$$s + 2 = l$$

The two equations become a *system of equations.*

$$l + s = 20$$
$$s + 2 = l$$

Graph each equation on the same set of axes and examine the intersection of the graphs. To do this, we can make a table of solutions for each equation.

$$l + s = 20 \qquad s + 2 = l$$
$$l = 20 - s \qquad l = s + 2$$

s	l
9	11
10	10
11	9

s	l
9	11
10	12
11	13

Values for s like 1 and 2 would require 19 and 18 units for l on the graph. Therefore, to fit the values easily on a smaller graph, we selected s values near 10. Now we graph the s values on the x-axis and the l values on the y-axis. We then write each equation along its line graph for easy identification (see Fig. 20-1).

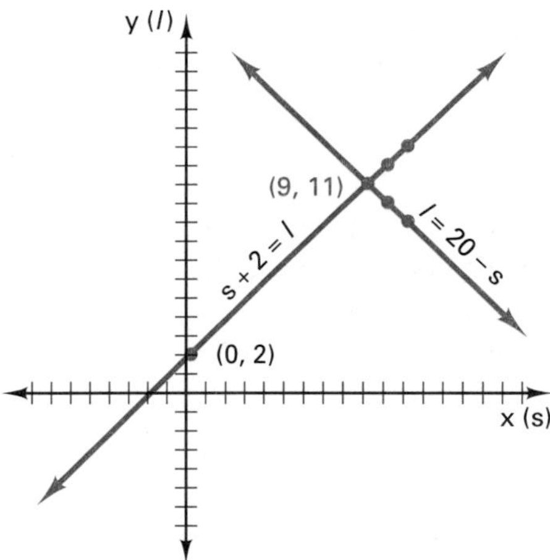

Figure 20-1

The point of intersection is (9, 11), which means that $s = 9, l = 11$. We can check to see if the solution ($s = 9, l = 11$) satisfies both equations.

$$l + s = 20 \qquad s + 2 = l$$
$$11 + 9 = 20 \qquad 9 + 2 = 11$$
$$20 = 20 \qquad 11 = 11$$

The ordered pair checks in both equations. ■

When graphing two straight lines, three possibilities can occur.

1. The two lines can intersect in just one point. This means the system is *independent.*
2. The two lines will not intersect at all. This means that the system is *inconsistent* and there is no pair of values that will satisfy both equations.
3. The two lines will coincide or fall exactly in the same place. This means that the system is *dependent* and the equations are identical or are multiples, and any pair of values that satisfy one equation will satisfy both equations.

The last two possibilities rarely occur in practical applications.

General Tips for Using the Calculator

Systems of linear equations can be graphed on a graphing calculator or computer.

1. Graph the first equation.
2. Graph the second equation without clearing the graph screen.
3. Use the [Trace] feature to determine the approximate coordinates of the intersection of the graphs.
4. Use the [Zoom] or [Box] feature to get a closer view and thus a more accurate approximation of the intersection of the graphs.

L.O.2. Solve a system of inequalities by graphing.

Graphing Systems of Inequalities

In some instances it is desirable to graph two inequalities and find the overlapping or common portion of the two graphs. Look at the next example.

EXAMPLE 20-1.2 Shade the portion on the graph that is represented by the following conditions.

$$y \leq 3x + 5 \quad \textit{and} \quad x + y > 7$$

Solution: First, we will graph the inequality $y \leq 3x + 5$ by graphing the equation $y = 3x + 5$. Using intercepts, when $x = 0$, $y = 5$; and when $y = 0$, $x = -\frac{5}{3}$. To find an additional point, we can let $x = 1$. Then

$$y = 3(1) + 5$$
$$y = 8$$

Graph the ordered pairs (0, 5), $\left(-\frac{5}{3}, 0\right)$, (1, 8). Draw a solid line through the points since the inequality is "$\leq$" rather than "$<$." "Less than or equal to" or "greater than or equal to" means the points on the line are part of the solution set and are indicated with a solid line. (See Fig. 20-2.)

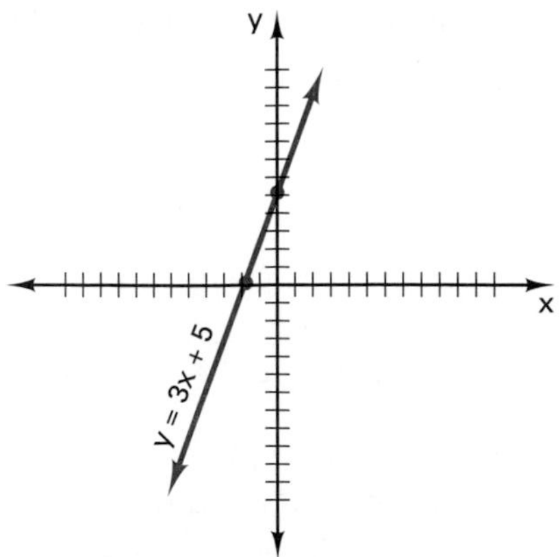

Figure 20-2

Which side of the boundary represented by the equation has values for x and y so that $y \leq 3x + 5$? Select a point on each side of the line, such as (0, 8) and (0, 0). *(Select numbers that will make calculations easy!)*

For (0, 8)	For (0, 0)
$y \leq 3x + 5$	$y \leq 3x + 5$
$8 \leq 3(0) + 5$	$0 \leq 3(0) + 5$
$8 \leq 0 + 5$	$0 \leq 0 + 5$
$8 \leq 5$	$0 \leq 5$
False	True

Shade the side of the line that contains the point (0, 0) because the inequality $0 \leq 5$ makes a true statement (see Fig. 20-3).

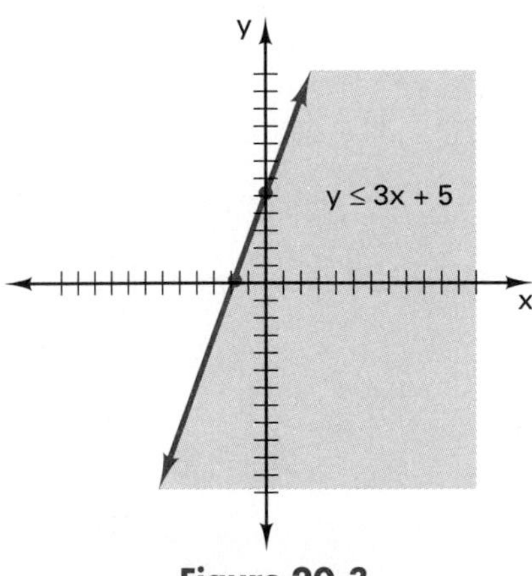

Figure 20-3

Now graph the line $x + y = 7$ and determine the portion of the graph that represents $x + y > 7$. Using intercepts, when $x = 0$, $y = 7$; and when $y = 0$, $x = 7$. For an additional point, when $x = 1$, $y = 6$. Notice that we make the line dashed to indicate the inequality is strictly ''greater than.'' That is, the line is *not* included in the solution set.

We next select a point on each side of the line, such as (0, 0) and (0, 10).

For (0, 0)	For (0, 10)
$x + y > 7$	$x + y > 7$
$0 + 0 > 7$	$0 + 10 > 7$
$0 > 7$	$10 > 7$
False	True

Thus, to show the solution set for $x + y > 7$, shade the portion of the graph that contains the point (0, 10) (see Fig. 20-4).

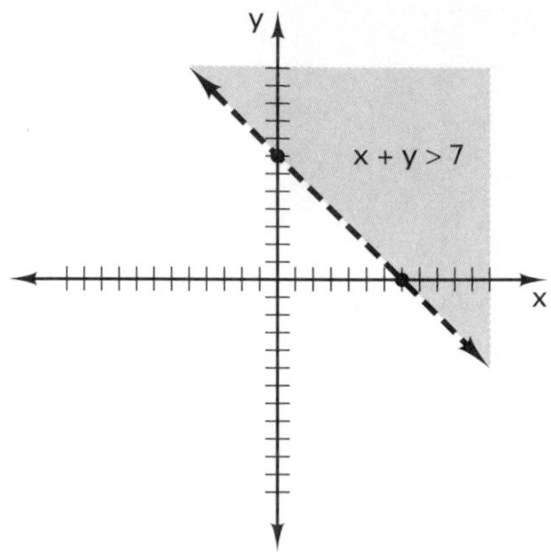

Figure 20-4

Now, visualize both graphs on the same axes. The portion that has overlapping shading represents the points that satisfy *both* conditions and forms the solution set for the system of inequalities (see Fig. 20-5). ■

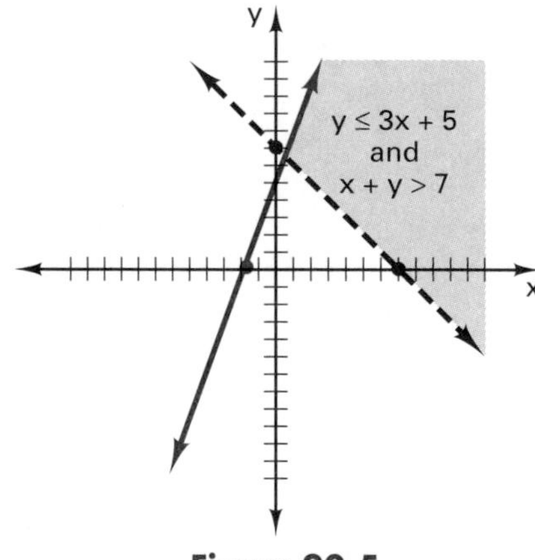

Figure 20-5

General Tips for Using the Calculator

Not all graphing calculators or computer graphing software will shade the intersecting region for the solution of a system of inequalities. However, the calculator can graph the boundaries. Through practice and watching for patterns, you will be able to shade the appropriate region from a mental analysis of the conditions of the problem.

SELF-STUDY EXERCISES 20-1

L.O.1. Solve the following systems of equations by graphing.

1. $x + y = 12$
$x - y = 2$

2. $2x + y = 9$
$3x - y = 6$

3. $2x - y = 5$
$4x - 2y = 8$

4. $x - y = 9$
$3x - 3y = 27$

5. $3x + 2y = 8$
$x + y = 2$

L.O.2. Graph each inequality and shade the portion on the graph that is represented by the following sets of conditions.

6. $y < 2x + 1$ *and* $x + y > 5$

7. $x + y \leq 3$ *and* $x - y \geq 2$

8. $x - 2y < -1$ *and* $x + 2y > 3$

9. $x + y > 4$ *and* $x - y > -3$

10. $3x - 2y < 8$ *and* $2x + y \leq -4$

20-2 SOLVING SYSTEMS OF EQUATIONS USING THE ADDITION METHOD

Learning Objectives

L.O.1. Solve a system of equations containing opposite variable terms by the addition method.

L.O.2. Solve a system of equations not containing opposite variable terms by the addition method.

L.O.3. Apply the addition method to a system of equations with no solution.

We can see that solving systems of equations graphically is time consuming. Also, many solutions to systems of equations are not whole numbers. Graphically, it is difficult to plot fractions. Therefore, we need a convenient algebraic procedure for solving systems of equations. We will introduce only the two methods most commonly used to solve a system of equations with two variables. The first method is the *addition method.*

L.O.1. Solve a system of equations containing opposite variable terms by the addition method.

Solving Systems of Equations Containing Opposite Terms

The addition method incorporates the concept that equals added to equals give equals. Thus, when we add two equations, the result is still an equation.

However, the addition method is effective *only* if one of the variables is *eliminated* in the addition process. The following examples contain two opposite terms that add to zero when the equations are added.

EXAMPLE 20-2.1 Add the following pairs of equations.

(a) $x + y = 7$
$x - y = 5$

(b) $3x + 8y = 7$
$-3x - 3y = 8$

(c) $x - 2y = 4$
$3x + 2y = 8$

Solution

(a)
$$\begin{array}{r} x + y = 7 \\ x - y = 5 \\ \hline 2x + 0y = 12 \\ 2x \quad = 12 \end{array}$$

(b)
$$\begin{array}{r} 3x + 8y = 7 \\ -3x - 3y = 8 \\ \hline 0x + 5y = 15 \\ 5y = 15 \end{array}$$

(c)
$$\begin{array}{r} x - 2y = 4 \\ 3x + 2y = 8 \\ \hline 4x + 0y = 12 \\ 4x \quad = 12 \end{array}$$

■

In each problem in Example 20-2.1, one unknown was eliminated by the addition method. Now the remaining equation that contains only one variable can be solved to find the value of that variable in the system. Then the other variable can be found by substitution. Let's finish solving each problem in Example 20-2.1.

(a)
$$\begin{aligned} x + y &= 7 \\ x - y &= 5 \\ \hline 2x \quad &= 12 \\ x &= 6 \end{aligned}$$

Now substitute 6 in place of x in *either* of the given equations.

$$\begin{aligned} x + y &= 7 \\ 6 + y &= 7 \\ y &= 7 - 6 \\ y &= 1 \end{aligned}$$

The solution is $x = 6$ and $y = 1$, or (6, 1).
Check in the *other* equation:

$$\begin{aligned} x - y &= 5 \\ 6 - 1 &= 5 \\ 5 &= 5 \end{aligned}$$

Checking in the *other* equation helps ensure that no errors have been made in the addition process or in the substitution process.

(b)
$$\begin{aligned} 3x + 8y &= 7 \\ -3x - 3y &= 8 \\ \hline 5y &= 15 \\ y &= 3 \end{aligned}$$

Substitute 3 for y in the first equation.

$$\begin{aligned} 3x + 8y &= 7 \\ 3x + 8(3) &= 7 \\ 3x + 24 &= 7 \\ 3x &= 7 - 24 \\ 3x &= -17 \\ x &= \frac{-17}{3} \end{aligned}$$

The solution is $\left(-\frac{17}{3}, 3\right)$.
Check in the other equation:

$$\begin{aligned} -3x - 3y &= 8 \\ -3\left(\frac{-17}{3}\right) - 3(3) &= 8 \\ 17 - 9 &= 8 \\ 8 &= 8 \end{aligned}$$

(c)

$$\begin{array}{rcl} x - 2y &=& 4 \\ 3x + 2y &=& 8 \\ \hline 4x \quad &=& 12 \\ x &=& 3 \end{array}$$

Substitute 3 for x in the first equation.

$$\begin{aligned} x - 2y &= 4 \\ 3 - 2y &= 4 \\ -2y &= 4 - 3 \\ -2y &= 1 \\ y &= -\frac{1}{2} \end{aligned}$$

The solution is $\left(3, -\frac{1}{2}\right)$.

Check:

$$\begin{aligned} 3x + 2y &= 8 \\ 3(3) + 2\left(-\frac{1}{2}\right) &= 8 \\ 9 + -1 &= 8 \\ 8 &= 8 \end{aligned}$$

■

L.O.2. Solve a system of equations not containing opposite variable terms by the addition method.

Solving Systems of Equations without Opposite Terms

Remember, in the addition method one of the unknowns must be eliminated; that is, the terms must add to 0. We can also use the property that both sides of an equation may be multiplied by the same number and the equality is preserved and the value of the variable remains the same. Thus, if neither pair of variable terms in a system of equations is opposite terms that will add to 0, we multiply one or both of the equations by numbers that will *cause* the terms of one of the variables to add to 0.

Rule 20-2.1 ***To solve a system of equations using the addition method (elimination):***

1. If necessary, multiply one or both equations by numbers that will cause the terms of one of the variables to add to zero.
2. Add the two equations to eliminate a variable.
3. Solve the equation from step 2.
4. Substitute the solution from step 3 in either equation and solve for the remaining variable.
5. Check

Example 20-2.2 Solve the following system of equations.

$$2x + y = 7$$
$$x + y = 3$$

Solution: In this system, neither of the like variable terms will immediately add to 0. Several different strategies can be used. If the x in the second equation had a coefficient of -2, the x terms would add to 0. Or if either y term in the first or second equation had a coefficient of -1, the y terms would add to 0. Thus, we have at least three choices: multiply the first equation by -1, multiply the second equation by -1, or multiply the second equation by -2.

Choice 1.

$$\begin{aligned} -2x - y &= -7 \quad \textit{(Multiply first equation by } -1.) \\ x + y &= 3 \quad \textit{(Add the equations.)} \\ \hline -x \quad &= -4 \end{aligned}$$

Choice 2.

$$\begin{aligned} 2x + y &= 7 \\ -x - y &= -3 \quad \textit{(Multiply second equation by } -1 \textit{ and add the equations.)} \\ \hline x \quad &= 4 \end{aligned}$$

Choice 3.

$$\begin{aligned} 2x + y &= 7 \\ -2x - 2y &= -6 \quad \textit{(Multiply second equation by } -2 \textit{ and add the equations.)} \\ \hline -y &= 1 \end{aligned}$$

Any of the three choices is now appropriate for the addition method. Let's complete choice 1.

$$\begin{aligned} -2x - y &= -7 \\ x + y &= 3 \\ \hline -x \quad &= -4 \\ x &= 4 \end{aligned}$$

Substitute 4 for x in the first equation.

$$\begin{aligned} -2(4) - y &= -7 \\ -8 - y &= -7 \\ -y &= -7 + 8 \\ -y &= 1 \\ y &= -1 \end{aligned}$$

The solution is $(4, -1)$.

Check the solution in the first original equation.

$$\begin{aligned} 2x + y &= 7 \\ 2(4) + (-1) &= 7 \\ 8 + (-1) &= 7 \\ 7 &= 7 \end{aligned}$$

Check the solution in the second original equation.

$$\begin{aligned} x + y &= 3 \\ 4 + (-1) &= 3 \\ 3 &= 3 \end{aligned}$$

■

Tips and Traps

When an original equation is altered, it is very important to check your solution in *both* original equations. This will enable you to identify mistakes such as forgetting to multiply *each* term in the equation by a number.

EXAMPLE 20-2.3 Solve the following system of equations.

$$\begin{aligned} x &= 3y + 7 \\ 2x + 3y &= 2 \end{aligned}$$

Solution: In this system the first equation does not have both letter terms on one side of the equation. Our standard form for solving systems of equations using the addition method requires that the letter terms be on one side and the number term on the other. Also, the letter terms in both equations need to be in the same order so that like terms will be in columns. Thus, we will rearrange the first equation.

$$x = 3y + 7 \longrightarrow x - 3y = 7$$

Now, we will continue to solve the system.

Add to eliminate the y terms:

$$\begin{aligned} x - 3y &= 7 \\ 2x + 3y &= 2 \\ \hline 3x \qquad &= 9 \\ x &= 3 \end{aligned}$$

Substitute:

$$\begin{aligned} x - 3y &= 7 \\ 3 - 3y &= 7 \\ -3y &= 7 - 3 \\ -3y &= 4 \\ y &= -\frac{4}{3} \end{aligned}$$

The solution is $\left(3, -\frac{4}{3}\right)$.

Check the solution in the original equations:

$$\begin{aligned} x &= 3y + 7 & 2x + 3y &= 2 \\ 3 &= 3\left(-\frac{4}{3}\right) + 7 & 2(3) + 3\left(-\frac{4}{3}\right) &= 2 \\ 3 &= -4 + 7 & 6 + (-4) &= 2 \\ 3 &= 3 & 2 &= 2 \end{aligned}$$

■

Example 20-2.4 Solve the following system of equations.

$$2x + 3y = 1$$
$$3x + 4y = 2$$

Solution: In this system there is no number that we can multiply just one equation by so that a letter will be eliminated. Therefore, we need to multiply each equation by some number. There are several possibilities; we will look at just two.

Choice 1. We multiply the first equation by -3 so that the coefficient of x will be -6. Then we multiply the second equation by $+2$ so that the coefficient of x will be $+6$. We then have

$$\begin{array}{r} -6x - 9y = -3 \\ 6x + 8y = 4 \\ \hline -y = 1 \end{array}$$

Choice 2. To eliminate y, we could multiply the first equation by 4 and the second equation by -3. There is no real preference between choices 1 and 2, so we will complete choice 1.

Add:

$$\begin{array}{r} -6x - 9y = -3 \\ 6x + 8y = 4 \\ \hline -y = 1 \\ y = -1 \end{array}$$

Substitute:

$$\begin{aligned} 2x + 3y &= 1 \\ 2x + 3(-1) &= 1 \\ 2x - 3 &= 1 \\ 2x &= 1 + 3 \\ 2x &= 4 \\ x &= 2 \end{aligned}$$

The solution is $(2, -1)$.

Check the solution in the original equations:

$$\begin{aligned} 2x + 3y &= 1 & \qquad 3x + 4y &= 2 \\ 2(2) + 3(-1) &= 1 & 3(2) + 4(-1) &= 2 \\ 4 + (-3) &= 1 & 6 + (-4) &= 2 \\ 1 &= 1 & 2 &= 2 \end{aligned}$$

■

EXAMPLE 20-2.5 Solve the following system of equations.

$$x = 4y$$
$$x - y = 6$$

Solution: We need to rearrange the first equation into our standard form for solving systems of equations by the addition method.

$$x = 4y \longrightarrow x - 4y = 0$$

Now our system is

$$x - 4y = 0$$
$$x - y = 6$$

One way of eliminating a letter term is to multiply the first equation by -1 and add the equations.

$$\begin{array}{r} -x + 4y = 0 \\ x - y = 6 \\ \hline 3y = 6 \\ y = 2 \end{array}$$

Substitute:

$$x = 4y$$
$$x = 4(2)$$
$$x = 8$$

The solution is (8, 2).
Check:

$$\begin{array}{ll} x = 4y & x - y = 6 \\ 8 = 4(2) & 8 - 2 = 6 \\ 8 = 8 & 6 = 6 \end{array}$$

■

L.O.3. Apply the addition method to a system of equations with no solution.

Using Addition in Systems of Equations with No Solutions

In Section 20-1, we saw that there are situations when there is no solution to a system of equations. In one situation, the graphs of the two equations did not intersect and in the other the graphs of the two equations coincided. We will look at examples of these situations again and attempt to solve them using the addition method.

EXAMPLE 20-2.6 Solve the system

$$x + y = 7$$
$$x + y = 5$$

Solution: To eliminate the x terms, let's multiply the second equation by -1 and add the equations.

$$\begin{array}{r}x + y = 7\\ -x - y = -5\\ \hline 0 = 2\end{array}$$

Notice, both variables are eliminated and the resulting equation, $0 = 2$, is *false*. Thus, there are no solutions to this system. ■

Tips and Traps

When solving a system of equations, if both variables are eliminated and the resulting statement is false, then the equations are *inconsistent* and have no solution.

EXAMPLE 20-2.7 Solve the system

$$\begin{aligned}2x - y &= 7\\ 4x - 2y &= 14\end{aligned}$$

Solution: The y's can be eliminated by multiplying the first equation by -2 and adding the equations.

$$\begin{array}{r}-4x + 2y = -14\\ 4x - 2y = 14\\ \hline 0 = 0\end{array}$$

Again, both variables are eliminated; however, this time the result is a *true* statement ($0 = 0$). In this situation all solutions of one equation are also solutions of the other equation. ■

Tips and Traps

When solving a system of equations, if both variables are eliminated and the resulting statement is true, then the equations are *dependent* and have many solutions. The equations are equivalent.

SELF-STUDY EXERCISES 20-2

L.O.1. Solve the following systems of equations using the addition method.

1. $a - 2b = 7$
$3a + 2b = 13$

2. $3m + 4n = 8$
$2m - 4n = 12$

3. $x - 4y = 5$
$-x - 3y = 2$

4. $a - b = 6$
$2a + b = 3$

5. $x + 2y = 5$
$3x - 2y = 3$

L.O.2. Solve the following systems of equations using the addition method.

6. $3x + y = 9$
$x + y = 3$

7. $7x + 2y = 17$
$y = 3x + 2$

8. $a + 6b = 18$
$4a - 3b = 0$

9. $3x + y = -1$
$4x - 2y = -8$

10. $a = 6y$
$2a - y = 11$

L.O.3. Solve the following systems of equations using the addition method.

11. $x + y = 8$
$x + y = 3$

12. $x + 2y = 9$
$x + y = 3$

13. $3x - 2y = 6$
$9x - 6y = 18$

14. $2a + 4b = 10$
$a + 2b = 5$

15. $3a - b = 14$
$a - 3b = 2$

20-3 SOLVING SYSTEMS OF EQUATIONS BY THE SUBSTITUTION METHOD

Learning Objective

L.O.1. Solve a system of equations by the substitution method.

L.O.1. Solve a system of equations by the substitution method.

Using Substitution to Solve Systems of Equations

Another method for solving systems of equations is by substitution. Recall that in formula rearrangement (Section 13-2), whenever more than one variable is used in an equation or formula, we can rearrange the equation or solve for a particular variable. In the *substitution method* for solving systems of equations, we solve one equation for one of the variables and then substitute that solution in place of the variable in the other equation.

> **Rule 20-3.1** *To solve a system of equations by substitution:*
>
> 1. Rearrange either equation to isolate one of the variables.
> 2. Substitute the expression from step 1 into the *other* equation and solve for the variable.
> 3. Substitute the solution from step 2 into the equation from step 1 to find the value of the other variable.
> 4. Check.

EXAMPLE 20-3.1 Solve the following system of equations using the substitution method.

$$x + y = 15$$
$$y = 2x$$

Solution

$x + y = 15$ [*The second equation is already solved for y (step 1).*]

$y = 2x$

$x + y = 15$ [*Substitute 2x for y in first equation and solve (step 2).*]

$x + 2x = 15$

$3x = 15$

$x = 5$

$y = 2x$ [*Substitute the solution for x in second equation to find y (step 3).*]

$y = 2(5)$

$y = 10$

The solution is (5, 10).
Check:

$$x + y = 15 \qquad y = 2x$$
$$5 + 10 = 15 \qquad 10 = 2(5)$$
$$15 = 15 \qquad 10 = 10$$

(The solution checks in both equations.) ■

EXAMPLE 20-3.2 Solve the following system of equations using the substitution method.

$$2x - 3y = -14$$
$$x + 5y = 19$$

Solution: In using the substitution method, either equation can be solved for either unknown. In this example, since the x term in the second equation has a coefficient of 1, the simplest choice would be to solve the second equation for x.

Step 1

$$x + 5y = 19$$
$$x = 19 - 5y$$

Step 2

$$2x - 3y = -14$$
$$2(19 - 5y) - 3y = -14$$
$$38 - 10y - 3y = -14$$
$$38 - 13y = -14$$
$$-13y = -14 - 38$$
$$-13y = -52$$
$$\frac{-13y}{-13} = \frac{-52}{-13}$$
$$y = 4$$

Step 3

$$x = 19 - 5y$$
$$x = 19 - 5(4)$$
$$x = 19 - 20$$
$$x = -1$$

The solution is $(-1, 4)$.
Check the roots $x = -1$, $y = 4$ in both original equations.

$$2x - 3y = -14 \qquad x + 5y = 19$$
$$2(-1) - 3(4) = -14 \qquad -1 + 5(4) = 19$$
$$-2 - 12 = -14 \qquad -1 + 20 = 19$$
$$-14 = -14 \qquad 19 = 19$$

■

SELF-STUDY EXERCISES 20-3

L.O.1. Solve the following systems of equations using the substitution method.

1. $2a + 2b = 60$
$a = 10 + b$

2. $7r + c = 42$
$3r - 8 = c$

3. $x - 35 = -2y$
$3x - 2y = 17$

4. $x + y = 12$
$x = 2 + y$

5. $2p + 3k = 2$
$2p - 3k = 0$

6. $x + 2y = 5$
$x = 3y$

20-4 PROBLEM SOLVING USING SYSTEMS OF EQUATIONS

Learning Objective

L.O.1. Use a system of equations to solve application problems.

L.O.1. Use a system of equations to solve application problems.

Using Systems of Equations to Solve Problems

Many job-related problems can be solved by setting up and solving systems of equations such as in Example 20-1.1. We will look at some additional situations.

EXAMPLE 20-4.1 A radio–television repair person purchased 10 pairs of rabbit-ear antennas and four power supply lines for \$48. Soon after, the repair person purchased another three pairs of rabbit-ear antennas and five power supply lines for \$22. How much did each antenna and each power supply line cost?

Solution: Let a = cost of one antenna. Let l = cost of one power supply line.

$$\text{Total cost} = \text{no. items} \times \text{cost of each item}$$

From this relationship, we derive a system of equations.

$$10a + 4l = 48$$
$$3a + 5l = 22$$

(Multiply each equation by a number to eliminate one variable. The first equation is multiplied by 3. The second equation is multiplied by −10.)

$$30a + 12l = 144$$
$$-30a - 50l = -220$$

(Add equations and solve.)

$$-38l = -76$$
$$l = \frac{-76}{-38}$$
$$l = 2$$

A power supply line costs \$2.00. Substituting, we get

$$10a + 4(2) = 48$$
$$10a + 8 = 48$$
$$10a = 48 - 8$$
$$10a = 40$$
$$a = \frac{40}{10}$$
$$a = 4$$

Rabbit-ear antennas cost \$4.00 each. ■

EXAMPLE 20-4.2 A restaurant ordered three hampers of blue crabs and one hamper of shrimp that together weighed 89 lb. In another order, two hampers of shrimp and five hampers of blue crabs weighed 160 lb. How much did one hamper of crabs weigh and how much did one hamper of shrimp weigh?

Solution: Let c = weight of one hamper of crabs. Let s = weight of one hamper of shrimp.

$$\text{Total weight} = \text{no. of containers} \times \text{weight of each container}$$

From this relationship, we derive our system of equations.

$$\begin{aligned} 3c + 1s &= 89 \\ 5c + 2s &= 160 \end{aligned}$$

(Multiply first equation by -2. Then add both equations.)

$$\begin{aligned} -6c - 2s &= -178 \\ 5c + 2s &= 160 \\ \hline -c \quad &= -18 \\ c &= 18 \end{aligned}$$

$$\begin{aligned} 5c + 2s &= 160 \\ 5(18) + 2s &= 160 \\ 90 + 2s &= 160 \\ 2s &= 160 - 90 \\ 2s &= 70 \\ s &= \frac{70}{2} \\ s &= 35 \end{aligned}$$

(Substitute for c.)

One hamper of blue crabs weighed 18 lb, and one hamper of shrimp weighed 35 lb. ■

EXAMPLE 20-4.3 Two dry cells connected in series have a total internal resistance of 0.09 ohm. The difference between the internal resistance of each dry cell is 0.03 ohm. How much is each internal resistance?

Solution: Let r_1 = larger resistance. Let r_2 = smaller resistance. Total series resistance = internal resistance 1 + internal resistance 2. From this relationship, we derive a system of equations.

$$\begin{aligned} r_1 + r_2 &= 0.09 \\ r_1 - r_2 &= 0.03 \\ \hline 2r_1 \quad &= 0.12 \\ r_1 &= \frac{0.12}{2} \\ r_1 &= 0.06 \end{aligned}$$

(Add and solve.)

$$\begin{aligned} r_1 + r_2 &= 0.09 \\ 0.06 + r_2 &= 0.09 \\ r_2 &= 0.09 - 0.06 \\ r_2 &= 0.03 \end{aligned}$$

(Substitute for r_1.)

Thus, the larger internal resistance is 0.06 ohm and the smaller internal resistance is 0.03 ohm. ■

EXAMPLE 20-4.4 A tank holds a solution that is 10% herbicide. Another tank holds a solution that is 50% herbicide. If a farmer wants to mix the two solutions to get 200 gallons of a solution that is 25% herbicide, how many gallons of each solution should be mixed?

Solution: Let h = number of gallons of 10% solution. Let H = number of gallons of 50% solution.

Total (200 gal) = no. gal 10% solution + no. gal 50% solution

Amount of herbicide: 10%(h) + 50%(H) = 25%(200)

From these relationships, we derive a system of equations. Remember to convert percents to decimals.

$$h + H = 200$$

$$0.10h + 0.50H = 50 \qquad \textit{(0.25 × 200 gal = 50 gal)}$$

$$0.10(200 - H) + 0.50H = 50 \qquad \textit{(Substitute in second equation: } h = 200 - H\textit{)}$$

$$20 - 0.10H + 0.50H = 50$$

$$20 + 0.40H = 50$$

$$0.40H = 50 - 20$$

$$0.40H = 30$$

$$H = \frac{30}{0.40}$$

$$H = 75 \text{ gal}$$

$$h + 75 = 200 \qquad \textit{(Substitute 75 for H in an original equation.)}$$

$$h = 200 - 75$$

$$h = 125 \text{ gal}$$

The farmer must mix 75 gallons of the 10% solution and 125 gallons of the 50% solution to make 200 gallons of a 25% solution. ■

EXAMPLE 20-4.5 A canoeist paddles with the current and completes 21 miles in 2.5 hours. Paddling against the same current for the return trip would take 3.5 hours. Find the canoeist's rate in calm water and the rate of the current.

Solution: Let c = rate of canoeist in calm water. Let w = rate of the water current, rate with current = $c + w$, and rate against current = $c - w$. From these relationships, we set up a system of equations.

$$2.5(c + w) = 21 \qquad \textit{(Time × rate = distance is basis for each equation.)}$$

$$3.5(c - w) = 21$$

$$2.5c + 2.5w = 21 \qquad \textit{(Distribute multiplication in each equation.)}$$

$$3.5c - 3.5w = 21$$

$$3.5c + 3.5w = 29.4 \qquad \textit{(Multiply first equation by 1.4. Then add.)}$$

$$\underline{3.5c - 3.5w = 21}$$

$$7c = 50.4$$

$$\frac{7c}{7} = \frac{50.4}{7}$$

$$c = 7.2 \text{ mph}$$

$$\begin{aligned}
2.5(7.2 + w) &= 21 && \textit{(Substitute 7.2 for c in an original equation.)}\\
18 + 2.5w &= 21 && \textit{(Distribute multiplication.)}\\
2.5w &= 21 - 18\\
2.5w &= 3\\
\frac{2.5w}{2.5} &= \frac{3}{2.5}\\
w &= 1.2 \text{ mph}
\end{aligned}$$

The canoeist's rate is 7.2 mph; the water's rate is 1.2 mph. ■

EXAMPLE 20-4.6 Rosita has $5500 to invest and for tax purposes wants to earn exactly $500 interest for 1 year. She wants to invest part at 10% and the remainder at 5%. How much must she invest at each interest rate to earn exactly $500 interest in 1 year?

Solution: Let x = amount invested at 10%. Let y = amount invested at 5%. Interest for one year = rate × amount invested. Remember to convert percents to decimals. Using these relationships, we derive a system of equations.

$$\begin{aligned}
x + y &= 5500 && \textit{(Sum of parts = total.)}\\
0.10x + 0.05y &= 500 && \textit{(Rate × amount = interest.)}\\
0.10(5500 - y) + 0.05y &= 500 && \textit{(Substitute in second equation: } x = 5500 - y.\textit{)}\\
550 - 0.10y + 0.05y &= 500\\
550 - 0.05y &= 500 && (-0.10 + 0.05 = -0.05)\\
-0.05y &= 500 - 550\\
-0.05y &= -50\\
y &= \frac{-50}{-0.05}\\
y &= \$1000 \text{ at } 5\%\\
x + 1000 &= 5500 && \textit{(Substitute \$1000 for y in an original equation.)}\\
x &= 5500 - 1000\\
x &= \$4500 \text{ at } 10\%
\end{aligned}$$

Rosita must invest $4500 at 10% and $1000 at 5% for 1 year to earn $500 interest. ■

Tips and Traps

To solve an applied problem with a system of equations, first select one variable, such as x, to represent the first unknown quantity and a second variable, such as y, to represent the corresponding unknown quantity. Then express the remaining unknown quantities in terms of the two selected variables and appropriate numerical values. Finally, set up an equation to state each of the two conditions or equalities of the problem. Solve the system by graphing, addition, or substitution.

SELF-STUDY EXERCISES 20-4

L.O.1. Solve the following word problems using systems of equations with two unknowns.

1. Two boards together are 48 in. If one board is 17 in. shorter than the other, find the length of each board.
2. A broker invested $35,000 in two different stocks. One earned dividends at 4% and the other at 5%. If a $1570 dividend was earned on both stocks together, how much was invested in each? (*Reminder:* Change 4% to 0.04 and 5% to 0.05.)
3. A department store buyer ordered 12 shirts and 8 hats for $180 one month and 24 shirts and 10 hats for $324 the following month. What was the cost of each shirt and each hat?
4. A mechanic makes $15 on each 8-cylinder-engine tune-up and $10 on each 4-cylinder-engine tune-up. If the mechanic did 10 tune-ups and made a total of $135, how many 8-cylinder jobs and how many 4-cylinder jobs were completed?
5. Thirty resistors and 15 capacitors cost $12. Ten resistors and 20 capacitors cost $8.50. How much does each capacitor and resistor cost?
6. A private airplane flew 420 miles in 3 hours with the wind. The return trip took 3.5 hours. Find the rate of the plane in calm air and the rate of the wind.
7. In 1 year, Charles Temple earned $660 in interest on two investments totaling $8000. If he received 7% and 9% rates of return, how much did he invest at each rate?
8. A motorboat went 40 miles with the current in 3 hours. The return trip against the current took 4 hours. How fast was the current? What would have been the speed of the boat in calm water?
9. A visitor to south Louisiana purchased 3 pounds of dark-roast pure coffee and 4 pounds of coffee with chicory for $15.10 in a local supermarket. Another visitor at the same store purchased 2 pounds of coffee with chicory and 5 pounds of dark-roast pure coffee for $16.30. How much did each coffee cost per pound?
10. A lawn-care technician wants to spread a 200-pound seed mixture that is 50% bluegrass. If the technician has on hand a mixture that is 75% bluegrass and a mixture that is 10% bluegrass, how many pounds of each mixture is needed to make 200 pounds of the 50% mixture? Round to nearest whole pound.
11. A college bookstore received a partial shipment of 50 scientific calculators and 25 graphing calculators at a total cost of $1300. Later the bookstore received the balance of the calculators: 25 scientific and 50 graphing at a cost of $1775. Find the cost of each calculator.
12. A consumer received two 1-year loans totaling $10,000 at interest rates of 10% and 15%. If the consumer paid $1300 interest, how much money was borrowed at each rate?
13. For the first performance at the Kickin' Kuntry Klub, 40 reserved seats and 80 general admission seats were sold for $2000. For the second performance, 50 reserved seats and 90 general admission seats were sold for $2350. What was the cost for a reserved seat and for a general admission seat?
14. A photographer has a container with a solution of 75% developer and a container with a solution of 25% developer. If she wants to mix the solutions to get 8 pints of solution with 50% developer, how many pints of each solution does she need to mix?

ELECTRONICS APPLICATION

Mesh Currents

Circuits are frequently solved using mesh currents to determine individual currents in a circuit. For instance, the analysis of a circuit with three currents might yield the following system of equations for I_1, I_2, and I_3.

$$\begin{aligned} I_1 - I_2 - I_3 &= 0 \\ 6I_1 + 4I_3 &= 12\text{ A} \\ 3I_2 - 4I_3 &= -3\text{ A} \end{aligned}$$

There are several ways to solve this system. One way is to solve the first equation for I_1 in terms of I_2 and I_3 and substitute that into the second equation. Then the second and third equations will form a pair of equations in two unknowns, which are easy to solve using many methods. The proof is to put the final values all back into the original equation to see if they work.

$$I_1 - I_2 - I_3 = 0 \quad \text{rearranges to give} \quad I_1 = 0 + I_2 + I_3$$

Substituting that value for I_1 into the second equation gives

$$6I_1 + 4I_3 = 12;\ 6(I_2 + I_3) + 4I_3 = 12;\ 6I_2 + 10I_3 = 12$$

By dividing each term by 2, this reduces to

$$3I_2 + 5I_3 = 6$$

This gives the following pair from the modified equation two and the original equation three. (Illustrated is the addition method of solving a pair of equations.)

$$\begin{aligned} 3I_2 + 5I_3 &= 6\text{ A} \\ 3I_2 - 4I_3 &= -3\text{ A} \end{aligned}$$

$$\begin{aligned} 3I_2 + 5I_3 &= 6 \\ -3I_2 + 4I_3 &= 3 \\ \hline 9I_3 &= 9 \\ I_3 &= 1\text{ A} \end{aligned}$$

Substituting into equation 3:

$$\begin{aligned} 3I_2 - 4\,(1\text{ A}) &= -3\text{ A} \\ 3I_2 &= -3\text{ A} + 4\text{ A} \\ 3I_2 &= 1\text{ A} \\ I_2 &= 0.333333\text{ A} \end{aligned}$$

Then

$$I_1 = 0 + I_2 + I_3 = 0 + 0.3333333\text{ A} + 1\text{ A} = +1.333333\text{ A}$$

Proof: $I_1 - I_2 - I_3 = 0 = 1.33333\text{ A} - 0.333333\text{ A} - 1\text{ A} = 0$

$$6I_1 + 4I_3 = 12\text{ A} = 6(1.3333\text{ A}) + 4(1\text{ A}) = 12\text{ A}$$

$$3I_2 - 4I_3 = -3\text{ A} = 3(0.33333\text{ A}) - 4(1\text{ A}) = -3\text{ A} \qquad \text{Yes}$$

Exercises

Solve each of the following systems of equations, which are derived from actual circuits. Prove your answers by going back to the original equations.

1. Set A:
$$\begin{aligned} I_1 - 2I_2 + 3I_3 &= 4\text{ A} \\ 2I_1 + I_2 - 4I_3 &= 3\text{ A} \\ I_1 + 2I_3 &= 8\text{ A} \end{aligned}$$

2. Set B: $I_1 + I_2 + I_3 = -4$ A
$2I_1 + 3I_2 + 4I_3 = 0$
$-I_1 - I_2 + 2I_3 = 8$ A

3. Set C: $I_1 + I_2 + I_3 = +4$ A
$2I_1 + 3I_2 + 4I_3 = 0$
$-I_1 - I_2 + 2I_3 = 8$ A

4. Set D: $3I_1 + 2I_2 + 2I_3 = 3$ A
$2I_1 + 6I_2 + 3I_3 = 0$
$I_1 + 2I_2 + I_3 = 1$ A

5. Set E: $I_1 - I_2 + 3I_3 = 2$ A
$-I_1 + I_2 = 7$ A
$-I_1 + 2I_2 + 6I_3 = 4$ A

Answers for Exercises

Solve each of the following systems of equations, which are derived from actual circuits. Prove your answers by going back to the original equations.

Set	Solutions	Proof
1. Set A: $I_1 - 2I_2 + 3I_3 = 4$ A	$I_1 = 4$ A	$4 - 6 + 6 = 4$
$2I_1 + I_2 - 4I_3 = 3$ A	$I_2 = 3$ A	$8 + 3 - 8 = 3$
$I_1 + 2I_3 = 8$ A	$I_3 = 2$ A	$4 + 4 = 8$
2. Set B: $I_1 + I_2 + I_3 = -4$ A	$I_1 = -10.667$ A	$-10.667 + 5.333 + 1.333 = -4$
$2I_1 + 3I_2 + 4I_3 = 0$	$I_2 = 5.3333$ A	$-21.333 + 16 + 5.3333 = 0$
$-I_1 - I_2 + 2I_3 = 8$ A	$I_3 = 1.3333$ A	$10.667 - 5.3333 + 2.667 = 8$ A
3. Set C: $I_1 + I_2 + I_3 = +4$ A	$I_1 = 16$ A	$16 - 16 + 4 = 4$
$2I_1 + 3I_2 + 4I_3 = 0$	$I_2 = -16$ A	$32 - 48 + 16 = 0$
$-I_1 - I_2 + 2I_3 = 8$ A	$I_3 = 4$ A	$-16 + 16 + 8 = 8$
4. Set D: $3I_1 + 2I_2 + 2I_3 = 3$ A	$I_1 = 3$ A	$9 + 2 - 8 = 3$
$2I_1 + 6I_2 + 3I_3 = 0$	$I_2 = 1$ A	$6 + 6 - 12 = 0$
$I_1 + 2I_2 + I_3 = 1$ A	$I_3 = -4$ A	$3 + 2 - 4 = 1$
5. Set E: $I_1 - I_2 + 3I_3 = 2$ A	$I_1 = -28$ A	$-28 + 21 + 9 = 2$
$-I_1 + I_2 = 7$ A	$I_2 = -21$ A	$28 - 21 = 7$
$-I_1 + 2I_2 + 6I_3 = 4$ A	$I_3 = 3$ A	$28 - 42 + 18 = 4$

ASSIGNMENT EXERCISES

20-1

Solve the following systems of equations by graphing.

1. $x + y = 8$
$x - y = 2$

2. $3x + 2y = 13$
$x - 2y = 7$

3. $2x + 2y = 10$
$3x + 3y = 15$

4. $x + y = 1$
$3x - 4y = 10$

5. $2x - y = 5$
$4x - 2y = 2$

Shade the portion of the graph that is represented by the following sets of conditions.

6. $y \geq x + 3$ *and* $x + y < 4$
7. $2x + y < 6$ *and* $x - y < 1$
8. $x + 2y < -1$ *and* $x + 2y > 3$
9. $2x + y > 3$ *and* $x - y \leq 1$
10. $x + 2y > -4$ *and* $x - 2y > -1$

20-2

Solve the following systems of equations using the addition method.

11. $3x + y = 9$
$2x - y = 6$

12. $2a + 3b = 8$
$a - b = 4$

13. $Q = 2P + 8$
$2Q + 3P = 2$

14. $4j + k = 3$
$8j + 2k = 6$

15. $r = 2y + 6$
$2r + y = 2$

16. $3a + 3b = 3$
$2a - 2b = 6$

17. $c = 2y$
$2c + 3y = 21$

18. $2x + 4y = 9$
$x + 2y = 3$

19. $3R - 2S = 7$
$-14 = -6R + 4S$

20. $x - 3 = -y$
$2y = 9 - x$

21. $c = 2 + 3d$
$3c - 14 = d$

22. $Q - 10 = T$
$T = 2 - 2Q$

23. $x - 18 = -6y$
$4x - 0 = 3y$

24. $R + S = 3$
$S - 9 = -3R$

25. $3a - 2b = 6$
$6a - 12 = b$

26. $a - b = 2$
$a + b = 12$

27. $x + 2y = 7$
$x - y = 1$

28. $2c + 3b = 2$
$2c - 3b = 0$

29. $x + 2r = 5.5$
$2x = 1.5r$

30. $7x + 2y = 6$
$4y = 12 - x$

20-3

Solve the following systems of equations using either the addition or the substitution method.

31. $a + 7b = 32$
$3a - b = 8$

32. $x + y = 1$
$4x + 3y = 0$

33. $c - d = 2$
$c = 12 - d$

34. $3a + 4b = 0$
$a + 3b = 5$

35. $7x - 4 = -4y$
$3x + y = 6$

36. $5Q - 4R = -1$
$R + 3Q = -38$

37. $a = 2b + 11$
$3a + 11 = -5b$

38. $y = 5 - 2x$
$3x - 2y = 4$

39. $c = 2q$
$2c + q = 2$

40. $3x + 2y = 10$
$y = 6 - x$

41. $4x - 2.5y = 2$
$2x - 1.5y = -10$

42. $2a - c = 4$
$a = 2 + c$

43. $4d - 7 = -c$
$3c - 6 = -6d$

44. $x = 10 - y$
$5x + 2y = 11$

45. $3.5a + 2b = 2$
$0.5b = 3 - 1.5a$

46. $c + d = 12$
$c - d = 2$

47. $x + 4y = 20$
$4x + 5y = 58$

48. $a + 5y = 7$
$a + 4y = 8$

49. $3a + 1 = -2b$
$4b + 23 = 15a$

50. $6y + 0 = -5p$
$4y - 3p = 38$

20-4

Solve the following word problems using systems of equations with two unknowns.

51. Three electricians and four apprentices earned a total of $365 on one job. At the same rate of pay, one electrician and two apprentices earned a total of $145. How much pay did each apprentice and electrician receive?

52. Six bushels of bran and 2 bushels of corn weigh 182 lb. If 2 bushels of bran and 4 bushels of corn weigh 154 lb, how much does 1 bushel of bran and 1 bushel of corn each weigh?

53. A painter paid $22.50 for 2 quarts of white shellac and 5 quarts of thinner. If 3 quarts of shellac and 2 quarts of thinner cost the painter's helper $14.50, what is the cost of each quart of shellac and thinner?

54. A main current of electricity is the sum of two smaller currents whose difference is 0.8 A. What are the two smaller currents if the main current is 10 A?

55. The sum of two angles is 175°. Their difference is 63°. What is the measure of each angle?

56. A jonboat traveled 20 miles in 2 hours with the current. The return trip took 3 hours. Find the rate of the boat in calm water and the rate of the current.

57. In 1 year, Sholanda Brown earned $560 on two investments totaling $5000. If she received 10% and 12% rates of return, how much did she invest at each rate?

58. A plane flew 300 miles against the wind in 4 hours. The return trip with the wind took 3 hours. How fast was the wind? What would have been the speed of the plane in calm air?

59. A restaurant purchased 30 pounds of Columbian coffee and 10 pounds of blended coffee for \$190. In a second purchase, the same restaurant paid \$120 for 20 pounds of Columbian coffee and 5 pounds of blended coffee. How much did each coffee cost per pound?

60. A rancher wants to spread a 300-pound grass seed mixture that is 50% tall fescue. If the rancher has on hand a seed mixture that is 80% tall fescue and a mixture that is 20% tall fescue, how many pounds of each mixture is needed to make 300 pounds of the 50% mixture?

61. An automotive service station purchased 25 maps of Ohio and 8 maps of Alaska at a total cost of \$65.55. Later the station purchased 20 maps of Ohio and 5 maps of Alaska for \$49.50. Find the cost of each map.

62. Bev Witonski made two 1-year investments totaling \$7000 at interest rates of 8% and 12%. If she received \$760 in return, how much money was invested at each rate?

63. At the first of the month, store buyer Kathy Miller placed a \$12,525 order for 20 name-brand suits and 35 suits with generic labels. At the end of the month, she placed a \$15,725 order for 30 name-brand suits and 35 generic-label suits. How much did she pay for each type of suit?

64. A taxidermist has a container with a solution of 10% tanning chemical and a container with a solution of 50% tanning chemical. If the taxidermist wants to mix the solutions to get 10 gallons of solution with 25% tanning chemical, how many gallons of each solution should be mixed?

65. Jorge makes 5% commission on telephone sales and 6% commission on showroom sales. If his sales totaled \$40,000 and his commission was \$2250, how much did he sell by telephone? How much did he sell on the showroom floor?

66. Sing-Fong has 60 coins in nickels and quarters. The total value of the coins is \$12. How many coins of each type does she have?

67. How many gallons of 75% fertilizer and 25% fertilizer are needed to make a mixture of 8 gallons of 50% fertilizer?

Examine the following problems to see whether they are worked correctly. If there are errors, make corrections.

68. Solve by addition: $a - b = 6$ and $a + b = 2$.

$$\begin{array}{rcl} a - b &=& 6 \\ a + b &=& 2 \\ \hline 2a &=& 8 \\ a &=& 4 \end{array} \qquad \begin{array}{rcl} 4 - b &=& 6 \\ -b &=& 6 - 4 \\ -b &=& 2 \\ b &=& -2 \end{array} \qquad \text{Check:} \quad \begin{array}{rcl} a - b &=& 6 \\ 4 - (-2) &=& 6 \\ 4 + 2 &=& 6 \\ 6 &=& 6 \end{array}$$

69. Solve by addition: $2c + 3d = 9$ and $3c + d = 10$.

$$\begin{array}{rcl} 2c + 3d &=& 9 \\ 3c + d &=& 10 \quad \text{(Multiply by } -3.) \\ 2c + 3d &=& 9 \\ -6c - 3d &=& -30 \\ \hline -4c &=& -21 \\ \dfrac{-4c}{-4} &=& \dfrac{-21}{-4} \\ c &=& 5.25 \end{array} \qquad \begin{array}{rcl} 2c + 3d &=& 9 \\ 2(5.25) + 3d &=& 9 \\ 10.5 + 3d &=& 9 \\ 3d &=& 9 - 10.5 \\ 3d &=& -1.5 \\ \dfrac{3d}{3} &=& -\dfrac{1.5}{3} \\ d &=& -0.5 \end{array}$$

CHALLENGE PROBLEMS

70. Write an applied mixture problem that can be solved with the given system of equations with two unknowns. Solve the system and check the results.

$$\begin{aligned} 0.5x + 0.3y &= 8 \\ x + y &= 20 \end{aligned}$$

71. Write an applied problem that can be solved with the given system of equations with two unknowns. Solve the system and check the results.

$$\begin{aligned} 3x + 5y &= 28 \\ x - y &= 4 \end{aligned}$$

72. Solve Problem 55 by writing one equation with one variable.

CONCEPTS ANALYSIS

1. What does the graphical solution of a system of two equations with two unknowns represent?
2. What can be said about the solution of a system of two equations if the graphs of the two equations are parallel?
3. What can be said about the solution of a system of two equations if the graphs of the two coincide?
4. What can be said about the solution of a system of two equations if the graphs of the two intersect in exactly one point?
5. How do you determine the solution to a system of inequalities graphically. Can a system of inequalities be solved algebraically?
6. Explain the addition or elimination method of solving a system of two equations with two unknowns.
7. Explain the substitution method of solving a system of two equations with two unknowns.
8. What does it mean when solving a system of two equations with two unknowns if the system produces a statement like $0 = 0$?
9. What does it mean when solving a system of two equations with two unknowns if the system produces a statement like $3 = 0$?
10. Can a system of two linear equations in two unknowns have exactly two ordered pairs as solutions? Explain your answer.

CHAPTER SUMMARY

Objectives	What to Remember	Examples
Section 20-1		
1. Solve a system of equations by graphing.	Graph each equation. The intersection of the two lines is the solution to the system. (Table of solutions, intercepts, or slope-intercept may also be used to graph each equation.)	Graph $x + y = 2$ and $x - y = 4$.

x	y
0	2
1	1
2	0

x	y
0	−4
1	−3
2	−2

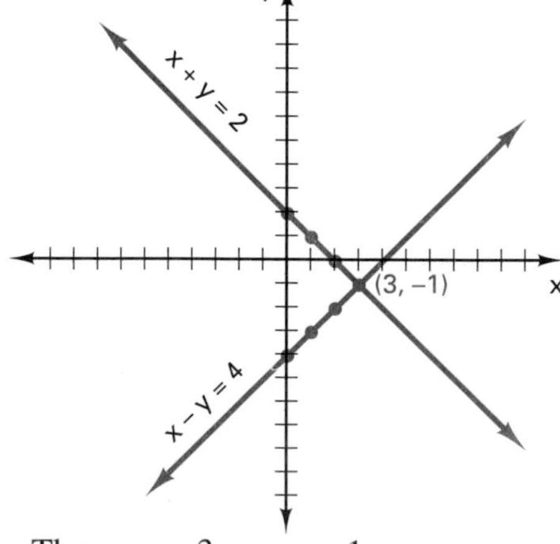

Thus, $x = 3$, $y = -1$.

2. Solve a system of inequalities by graphing.

Rewrite each inequality as an equation. Graph each equation and shade area that satisfies each inequality. Where shaded areas overlap is the set of solutions to the system of inequalities.

Graph $x + y \leq -2$ and $x - y > 3$.

For $x + y = -2$,

when $x = 0, y = -2$

when $y = 0, x = -2$

when $x = 1, y = -3$

For $x - y = 3$,

when $x = 0, y = -3$

when $y = 0, x = 3$

when $x = 1, y = -2$

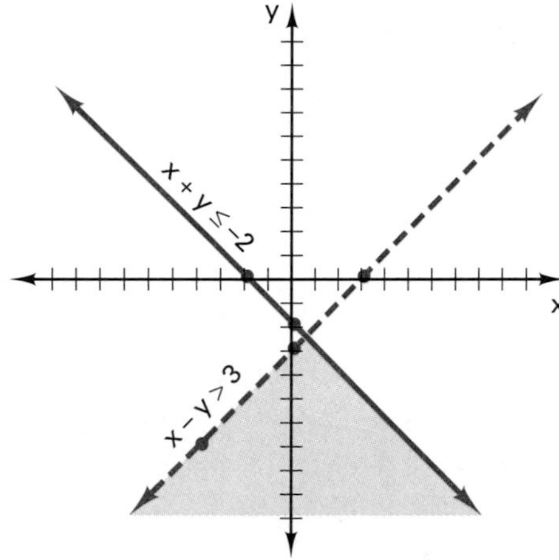

The overlapping shaded area meets both conditions of the system of inequalities.

Section 20-2

1. Solve a system of equations containing opposite variable terms by the addition method.

To solve a system of equations with two opposite variable terms, add the equations so that the opposite terms add to zero. Solve the new equation and substitute the root in an original equation. Check the roots.

Solve $x - y = 6$ and $x + y = 4$. ($-y$ and $+y$ are opposites.)

$$\begin{array}{r} x - y = 6 \\ \underline{x + y = 4} \\ 2x \quad = 10 \end{array}$$

$$\frac{2x}{2} = \frac{10}{2}$$

$$x = 5$$

$$x - y = 6$$

$$5 - y = 6$$

$$-y = 6 - 5$$

$$-y = 1$$

$$y = -1$$

The solution is $(5, -1)$.

2. Solve a system of equations not containing opposite variable terms by the addition method.

To solve a system of equations not containing opposite variables that will add to zero:

1. Multiply one or both equations by signed numbers that will cause one of the variables to add to zero.
2. Add equations.
3. Solve equation from step 2.
4. Substitue the root in an original equation and solve for the remaining variable.
5. Check solutions.

Solve $2x - y = 6$ and $4x + 2y = 4$.
Multiply first equation by 2.

$$\begin{aligned} 4x - 2y &= 12 \\ 4x + 2y &= 4 \\ \hline 8x \quad &= 16 \\ \frac{8x}{8} &= \frac{16}{8} \\ x &= 2 \\ 4x + 2y &= 4 \\ 4(2) + 2y &= 4 \\ 8 + 2y &= 4 \\ 2y &= 4 - 8 \\ 2y &= -4 \\ \frac{2y}{2} &= \frac{-4}{2} \\ y &= -2 \end{aligned}$$

Solution: $(2, -2)$.

3. Apply the addition method to a system of equations with no solution.

Apply the addition rule. But note that *both* variables are eliminated. If the resulting statement is false, there is no solution. If the resulting statement is true, there are many solutions.

Solve $a + b = 2$ and $a + b = 4$.
Multiply first equation by -1.

$$\begin{aligned} -a - b &= -2 \\ a + b &= 4 \\ \hline 0 &= 2 \end{aligned}$$

no solutions

Section 20-3

1. Solve a system of equations by the substitution method.

To solve a system of equations by substitution:

1. Rearrange either equation to isolate one of the variables.
2. Substitute expression from step 1 in the *other* equation and solve.
3. Substitute the root from step 2 in the equation from step 1 and solve for the other variable.
4. Check.

Solve $2x + y = 6$ and $2x - 2y = 8$. Isolate y in first equation:

$$y = 6 - 2x$$

Substitute $6 - 2x$ for y in second equation.

$$\begin{aligned} 2x - 2(6 - 2x) &= 8 \\ 2x - 12 + 4x &= 8 \\ 6x - 12 &= 8 \\ 6x &= 8 + 12 \\ 6x &= 20 \\ \frac{6x}{6} &= \frac{20}{6} \\ x &= \frac{10}{3} \end{aligned}$$

$$\begin{aligned} y &= 6 - 2\left(\frac{10}{3}\right) \\ y &= \frac{18}{3} - \frac{20}{3} \\ y &= -\frac{2}{3} \end{aligned}$$

Solution: $\left(\frac{10}{3}, -\frac{2}{3}\right)$

Section 20-4

1. Use a system of equations to solve application problems.	Let two variables represent the unknown amounts. Use numbers and the variables to represent the conditions of the problem. Set up a system of equations and solve.	A taxidermist bought two pairs of glass deer eyes and five pairs of glass duck eyes for \$15. She later purchased three pairs of deer eyes and five pairs of duck eyes for \$20. Find the price of each type of glass eye.

Let x = price of a pair of deer eyes.
Let y = price of a pair of duck eyes.
Price = number of eyes times price of eye type:

$$\begin{aligned} 2x + 5y &= 15 \\ 3x + 5y &= 20 \end{aligned}$$

Multiply first equation by -1:

$$\begin{aligned} -2x - 5y &= -15 \\ 3x + 5y &= 20 \\ \hline x \quad &= 5 \\ 2x + 5y &= 15 \\ 2(5) + 5y &= 15 \\ 10 + 5y &= 15 \\ 5y &= 15 - 10 \\ 5y &= 5 \\ \frac{5y}{5} &= \frac{5}{5} \\ y &= 1 \end{aligned}$$

The deer eyes cost \$5 a pair and duck eyes cost \$1 a pair.

WORDS TO KNOW

system of equations (p. 753)
system of inequalities (p. 754)
opposites (p. 757)
addition method (p. 757)
inconsistent system (p. 764)
independent system (p. 764)
dependent system (p. 764)
substitution method (p. 765)

CHAPTER TRIAL TEST

Solve the following systems of equations graphically.

1. $2a + b = 10$
$a - b = 5$

2. $x + y = 8$
$3x + 2y = 12$

3. $3x + 4y = 6$
$x + y = 5$

4. $a - 3b = 7$
$a - 5 = b$

5. $2c - 3d = 6$
$c - 12 = 3d$

Shade the portion on the graph that is represented by the conditions in the following problems.

6. $y \leq 2x + 1$ *and* $x + y \geq 5$

7. $x + y < 4$ *and* $y > 3x + 2$

Solve the following systems of equations using the addition method.

8. $x + y = 6$
$x - y = 2$

9. $p + 2m = 0$
$2p = -m$

10. $6p + 5t = -16$
$3p - 3 = 3t$

11. $3x + y = 5$
$2x - y = 0$

12. $7c - 2b = -2$
$c - 4b = -4$

Solve the following systems of equations with two unknowns using the substitution method.

13. $4x + 3y = 14$
$x - y = 0$

14. $a + 2y = 6$
$a + 3y = 3$

15. $7p + r = -6$
$3p + r = 6$

Solve the following systems of equations with two unknowns using either the addition or the substitution method.

16. $4x + 4 = -4y$
$6 + y = -6x$

17. $38 + d = -3a$
$5a + 1 = 4d$

Solve the following word problems using systems of equations with two unknowns.

18. Two lengths of stereo speaker wire total 32.5 ft. One length is 2.9 ft longer than the other. How long is each length of speaker wire?

19. Two currents add up to 35 A and their difference is 5 A. How many amperes are in each current?

20. Six packages of common nails and four packages of finishing nails weigh 6.5 lb. If two packages of common nails and three packages of finishing nails weigh 3.0 lb, how much does one package of each kind of nail weigh?

21. The length of a piece of sheet metal is $1\frac{1}{2}$ times the width. The difference between the length and the width is 17 in. Find the length and the width.

22. A mixture of fieldstone is needed for a construction job and will cost \$378 for the 27 tons of stone. The stone is of two types, one costing \$18 per ton and one costing \$12 per ton. How many tons of each is required?

23. A broker invested \$25,000 in two different stocks. One stock earned dividends at 11% and the other at 12.5%. If a dividend of \$3050 was earned on both stocks together, how much was invested in each stock?

24. A mason purchased 2-in. cold-rolled channels and $\frac{3}{4}$-in. cold-rolled channels whose total weight was 820 lb. The difference in weight between the heavier 2-in. and ligher $\frac{3}{4}$-in. channels was 280 lb. How many pounds of each type of channel did the mason purchase?

25. A total capacitance in parallel is the sum of two capacitances. If the capacitance totals 0.00027 farad (F) and the difference between the two capacitances is 0.00016 F, what is the value of each capacitance in the system?

CHAPTER

21 Selected Concepts of Geometry

Geometry is one of the oldest and most useful of the mathematical sciences. *Geometry* involves the study and measurement of shapes according to their sizes, volumes, and positions. A knowledge of geometry is necessary in many technologies.

Before we can formally study *angles*, we must first examine the basic terms and concepts that form the basis of any study of angles. It is especially important to understand the geometric meanings of the terms used and to distinguish their meanings. In the following section, notice the precise meanings given to some ordinary words when they are used in geometry.

21-1 BASIC TERMINOLOGY AND NOTATION

Learning Objectives

L.O.1. Use a variety of notations to represent points, lines, line segments, rays, and planes.
L.O.2. Distinguish among lines that intersect, that coincide, and that are parallel.
L.O.3. Use a variety of notations to represent angles.
L.O.4. Classify angles according to size.

L.O.1. Use a variety of notations to represent points, lines, line segments, rays, and planes.

Basic Terms and Concepts

Geometry can be described as the study of size, shape, position, and other properties of the objects around us. The basic terms used in geometry are *point*, *line*, and *plane*. Generally, these terms are not defined. Instead, they are usually merely described. Once described, they may be used in definitions of other terms and concepts.

A *point* is a location or position that has no size or dimension. A dot will be used to represent a point, and a capital letter will usually be used to label the point (Fig. 21-1).

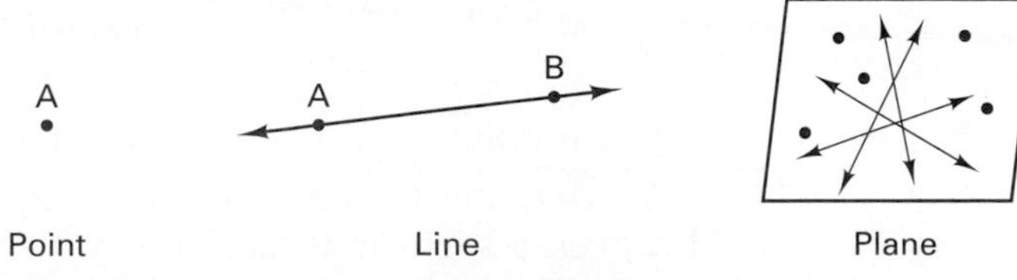

Figure 21-1

A *line* extends indefinitely in both directions. It has length but no width. In our discussions the word *line* will always refer to a straight line unless otherwise specified. In Fig. 21-1, the line is identified by naming any two points on the line (such as A and B).

A *plane* is a flat, smooth surface that extends indefinitely. A plane contains an infinite number of points and lines (Fig. 21-1).

Since a line extends indefinitely in both directions, most geometric applications deal with parts of lines. A part of a line is called a *line segment* or *segment*. A line segment starts and stops at distinct points that we call *end points*. A line segment is defined as follows:

■ **DEFINITION 21-1.1 Line Segment or Segment.** A **line segment**, or **segment**, consists of all points on the line between and including two points which are called end points (Fig. 21-2).

Figure 21-2

The notation for a line that extends through points A and B is $\overleftrightarrow{AB}$ (read "line AB"). The notation for the line segment including points A and B and all the points between is $\overline{AB}$ (read "line segment AB").

Another term used in connection with parts of a line is *ray*. Before the definition of a ray is given, consider the beam of light from a flashlight. The beam is like a ray. It seems to continue indefinitely in one direction.

■ **DEFINITION 21-1.2 Ray.** A **ray** consists of a point on a line and all points of the line on one side of the point (Fig. 21-3).

Figure 21-3

The point from which the ray originates is called the *end point*, and all other points on the ray are called *interior points* of the ray. A ray is named by its end point and any interior point on the ray. In Fig. 21-3, we use the notation $\overrightarrow{RS}$ to denote the ray whose end point is R and that passes through S.

To see the contrast in the notation used for a line, line segment, and ray, look at Fig. 21-4.

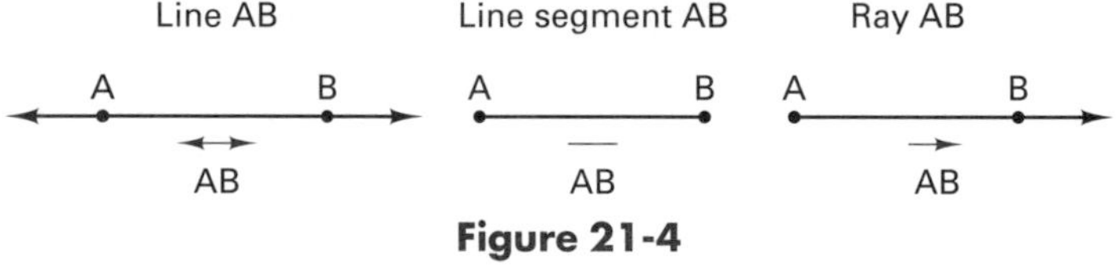

Figure 21-4

To further understand the difference between lines, segments, and rays, consider a line with several points designated on the line (Fig. 21-5).

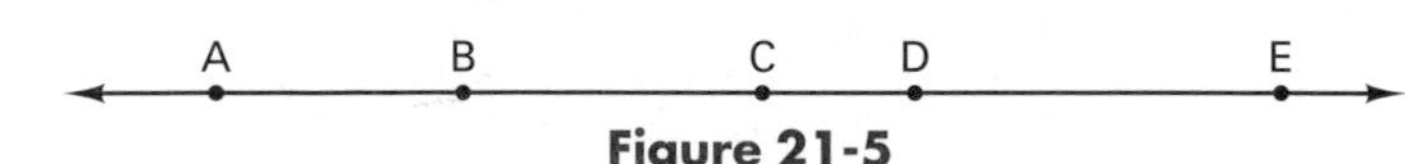

Figure 21-5

Any two points can be used to name the line in Fig. 21-5. For example, $\overleftrightarrow{AB}$, $\overleftrightarrow{AC}$, $\overleftrightarrow{CE}$, $\overleftrightarrow{BD}$, and $\overleftrightarrow{BC}$ are some of the possible ways to name the line.

However, a segment is named *only* by its end points. Thus, in Fig. 21-5, $\overline{AB}$ is not the same segment as $\overline{AC}$, but $\overleftrightarrow{AB}$ and $\overleftrightarrow{AC}$ represent the same line.

Again in Fig. 21-5, $\overrightarrow{BC}$ and $\overrightarrow{BD}$ represent the same ray, but $\overline{BC}$ and $\overline{BD}$ do not represent the same segment.

Notation is important in geometry because it allows us to shorten the amount of writing needed to describe a situation, and it allows us to recognize or recall certain properties at a glance.

L.O.2. Distinguish among lines that intersect, that coincide, and that are parallel.

Intersecting, Coinciding, and Parallel Lines

Since a line can be extended indefinitely in either direction, if two lines are drawn in the same plane, there are three things that can possibly happen:

1. The two lines will *intersect* in *one and only one point*. In Fig. 21-6, $\overleftrightarrow{AB}$ and $\overleftrightarrow{CD}$ intersect at point E.

Figure 21-6

2. The two lines will *coincide*; that is, one line will fit exactly on the other. In Fig. 21-6, $\overleftrightarrow{EF}$ and $\overleftrightarrow{GH}$ coincide.
3. The two lines will never intersect. In Fig. 21-6, $\overleftrightarrow{IJ}$ and $\overleftrightarrow{KL}$ are the same distance from each other along their entire lengths and so never touch.

The relationship described in the third situation has a special name, *parallel* lines. The symbol $\|$ is used for parallel lines. (See Section 19-4.)

L.O.3. Use a variety of notations to represent angles.

Naming Angles

This objective deals with a special type of intersection of straight lines—the angle. Most of us are familiar with angles in everyday life. The corners of a soccer field are angles. The lines between bricks and tiles form angles, and so on. This section and the ones that follow will introduce us formally to the study of angles for technical applications.

When two lines intersect in a point, *angles* are formed, as shown in Fig. 21-7.

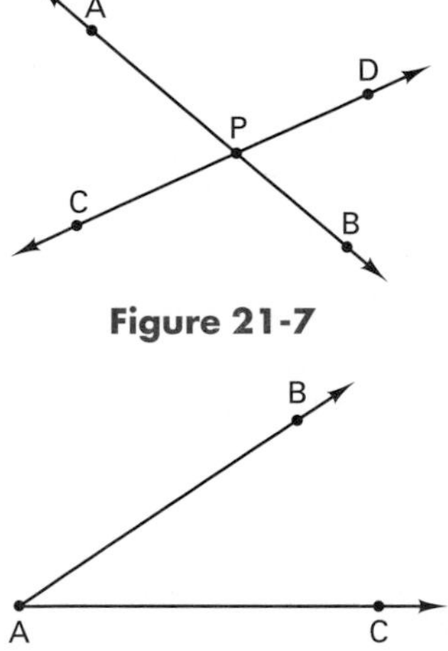

Figure 21-7

Figure 21-8

■ **Definition 21-1.3 Angle.** An **angle** is a geometric figure formed by two rays that interesect in a point, and the point of intersection is the end point of each ray.

In Fig. 21-8, rays $\overrightarrow{AB}$ and $\overrightarrow{AC}$ intersect at point A. Point A is the end point of $\overrightarrow{AB}$ and $\overrightarrow{AC}$. $\overrightarrow{AB}$ and $\overrightarrow{AC}$ are called the *sides* of the angle. Point A is called the *vertex* of the angle.

There are several ways to name an angle. An angle can be named using a number or lowercase letter, or it can be named by the capital letter that names the vertex point, or it can be named by using three capital letters. If three capital letters are used, two of the letters will name interior points of each of the two rays, and the middle letter will name the vertex point of the angle. Using the symbol $\angle$ for angle, the angle in Fig. 21-9 can be named $\angle 1$, $\angle KLM$, $\angle MLK$, or $\angle L$. If one

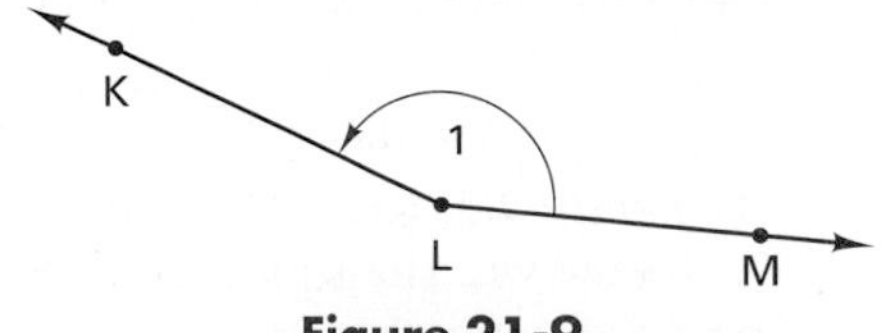

Figure 21-9

capital letter is used to name an angle, this letter is always the vertex letter. One capital letter is used only when it is perfectly clear which angle is designated by this letter. If three letters are used to name an angle, the vertex letter will be the center letter. To name the angle in Fig. 21-10 with three letters, we write $\angle XZY$ or $\angle YZX$. This angle can also be named $\angle a$ or $\angle Z$.

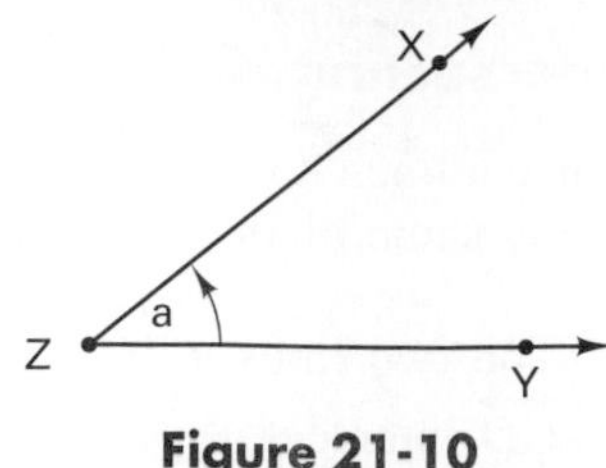

Figure 21-10

Figure 21-11 illustrates how the intersection of two rays actually forms two angles. Throughout this textbook, we will always refer to the smaller of the two angles formed by two rays unless the other angle is specifically indicated.

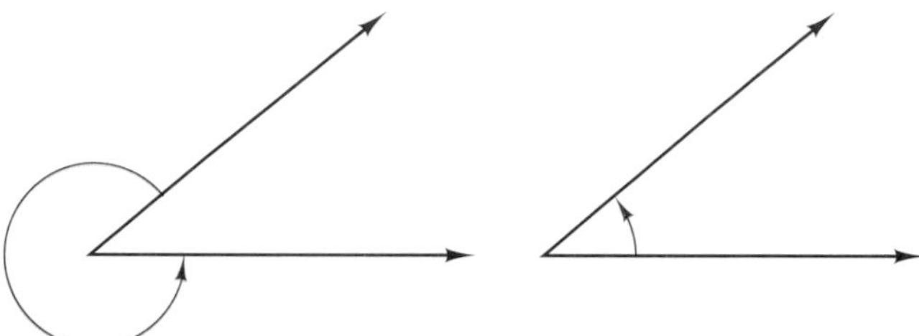

Figure 21-11

L.O.4. Classify angles according to size.

Classification of Angles by Size

The measure of an angle is determined by the amount of opening between the two sides of the angle. The length of the sides does not affect the angle measure. Two measuring units are commonly used to measure angles, *degrees* and *radians*. In this chapter we will use only degrees to measure angles. Radian measures of angles will be discussed later in trigonometry.

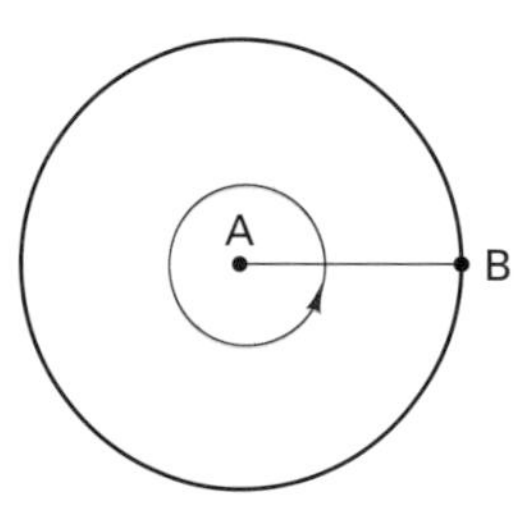

Figure 21-12

Consider the hands of a clock to be the sides of an angle. When the two hands of the clock are both pointing to the same number, the measure of the angle formed is 0 degrees (0°). An angle of 0 degrees is used in trigonometry but is seldom used in geometric applications. During 1 hour the minute hand makes one complete revolution. Ignoring the movement of the hour hand, this revolution of the minute hand contains 360 degrees (360°). Figure 21-12 illustrates a revolution or rotation of 360°. This rotation can be either clockwise or counterclockwise. A complete rotation counterclockwise is $+360°$. A complete rotation clockwise is $-360°$.

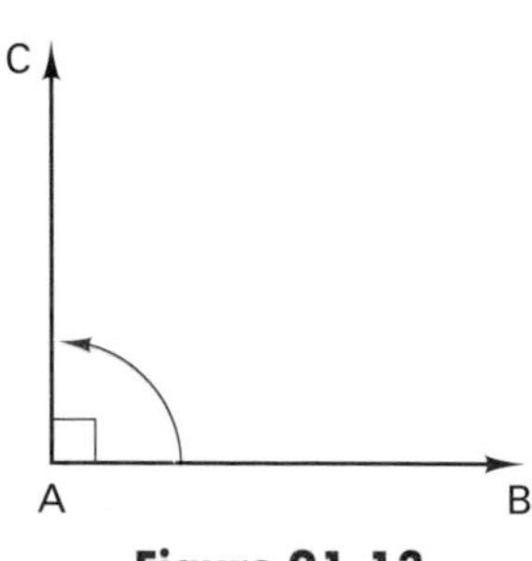

Figure 21-13

■ **Definition 21-1.4 Degree.** A **degree** is a unit for measuring angles. It represents $\frac{1}{360}$ of a complete rotation about the vertex.

Suppose, in Fig. 21-13, that $\overrightarrow{AC}$ rotates from $\overrightarrow{AB}$ through one-fourth of a circle. Then $\overrightarrow{AC}$ and $\overrightarrow{AB}$ form a 90° angle ($\frac{1}{4}$ of $360 = 90$). This angle is called a *right angle*. The symbol for a right angle is ∟.

If two lines intersect so that right angles are formed (90° angles), the lines are *perpendicular* to each other. (See Section 19-5.) The symbol for perpendicular is $\perp$. However, the right-angle symbol implies that the lines forming the angle are perpendicular.

If a string is suspended at one end and weighted at the other, the line it forms is a *vertical line*. A line that is perpendicular to the vertical line is a *horizontal line*. In Fig. 21-14, $\overleftrightarrow{AB}$ is a vertical line, and $\overleftrightarrow{AB}$ and $\overleftrightarrow{CD}$ form right angles. Thus, $\overleftrightarrow{AB} \perp \overleftrightarrow{CD}$ and $\overleftrightarrow{CD}$ is a horizontal line.

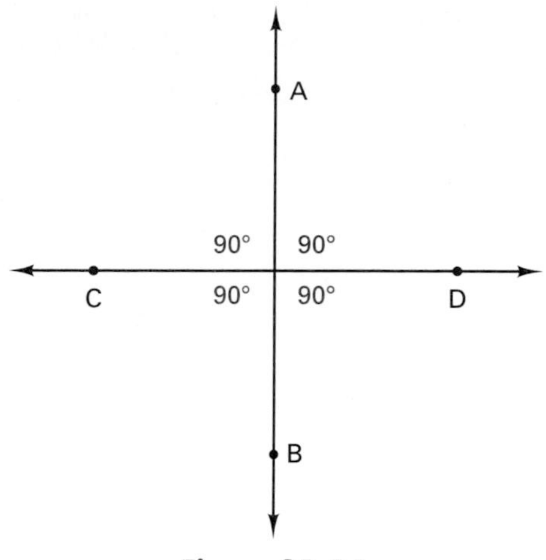

Figure 21-14

Look now at Fig. 21-15. When $\overrightarrow{AC}$ rotates one-half a circle from $\overrightarrow{AB}$, an angle of 180° is formed $\left(\frac{1}{2} \text{ of } 360 = 180\right)$. This angle is called a *straight angle*.

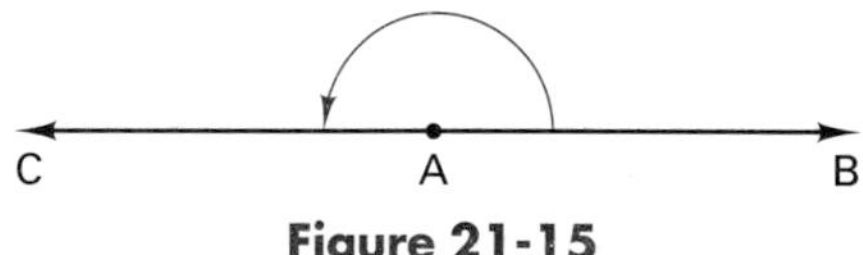

Figure 21-15

These two angles, the right angle (90°) and the straight angle (180°), are used to define two sets of angles that are used often in geometry. An angle that is less than 90° but more than 0° is called an *acute angle*. An angle that is more than 90° but less than 180° is called an *obtuse angle* (see Fig. 21-16).

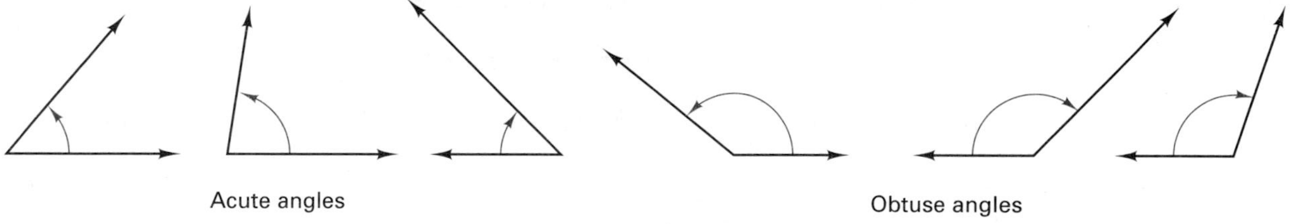

Acute angles

Obtuse angles

Figure 21-16

The following definitions will help us by summarizing the classification of angles.

■ **DEFINITION 21-1.5 Right Angle.** A **right angle** (90°) represents one-fourth of a circle or one-fourth of a complete rotation.

■ **DEFINITION 21-1.6 Straight Angle.** A **straight angle** (180°) represents one-half of a circle or one-half of a complete rotation.

■ **DEFINITION 21-1.7 Acute angle.** An **acute angle** is an angle less than 90° but more than 0°.

■ **DEFINITION 21-1.8 Obtuse Angle.** An **obtuse angle** is an angle more than 90° but less than 180°.

Two other useful terms involving angles are *complementary* and *supplementary* angles.

■ **DEFINITION 21-1.9 Complementary Angles. Complementary angles** are two angles whose sum is one right angle or 90° (Fig. 21-17).

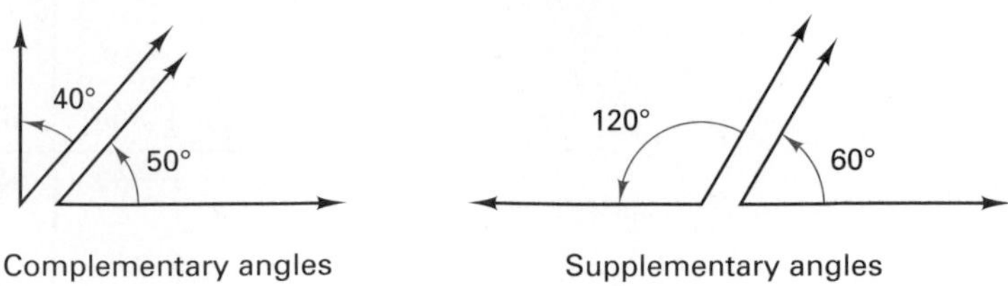

Figure 21-17

■ **DEFINITION 21-1.10 Supplementary Angles. Supplementary angles** are two angles whose sum is one straight angle or 180° (Fig. 21-17).

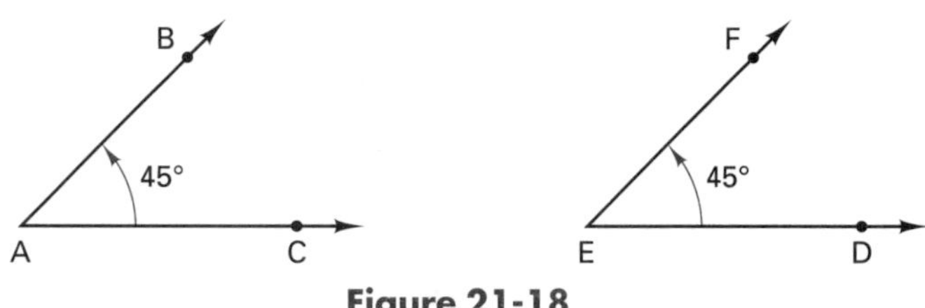

Figure 21-18

Tips and Traps

When the word *equal* is used in describing the relationship between two angles, we are implying that the measures of the angles are equal. Another word that is often used in geometry is *congruent*. When geometric figures are congruent, one figure can be placed on top of the other, and the two figures will match perfectly. If two angles are congruent, the measures of the angles are equal. Also, if the measures of two angles are equal, the angles are congruent. The symbol for congruence is $\cong$. For instance, in Fig. 21-18, the two angles have equal measures, both 45°. Since they have the same measures, they are said to be congruent. That is, $\angle BAC \cong \angle FED$.

SELF-STUDY EXERCISES 21-1

L.O.1. Use Fig. 21-19 for the following exercises.

1. Name the line in three different ways.

2. Name in two different ways the ray whose end point is P and whose interior points are Q and R.

3. Name the segment whose end points are Q and R.

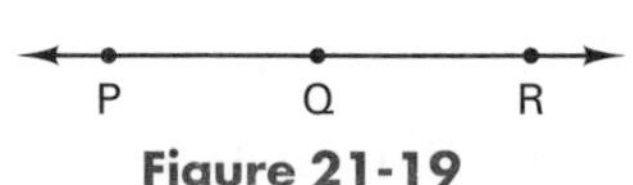

Figure 21-19

Use Fig. 21-20 for the following exercises.

4. Does $\overleftrightarrow{XY}$ represent the same line as $\overleftrightarrow{YZ}$?

5. Does $\overrightarrow{XY}$ represent the same ray as $\overrightarrow{YZ}$?

6. Does $\overline{XY}$ represent the same segment as $\overline{YZ}$?

7. Is $\overleftrightarrow{WX}$ the same as $\overleftrightarrow{WY}$?

8. Is $\overrightarrow{XY}$ the same as $\overrightarrow{XZ}$?

9. Is $\overline{XY}$ the same as $\overline{YX}$?

10. Is $\overrightarrow{XW}$ the same as XY?

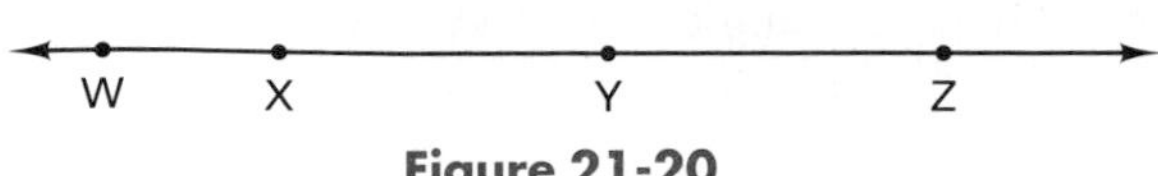

Figure 21-20

L.O.2. Use Fig. 21-21 for the following exercises.

11. $\overleftrightarrow{AB}$ and $\overleftrightarrow{CD}$ are _____ lines.

12. $\overleftrightarrow{EF}$ and $\overleftrightarrow{GH}$ _____ at point O.

13. $\overleftrightarrow{IJ}$ and $\overleftrightarrow{KL}$ _____.

14.–15. $\overleftrightarrow{AB}$ and $\overleftrightarrow{CD}$ will never _____ or _____.

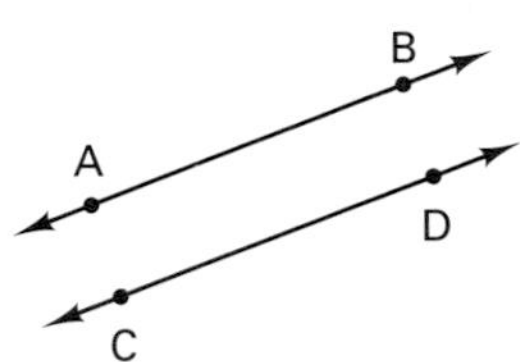

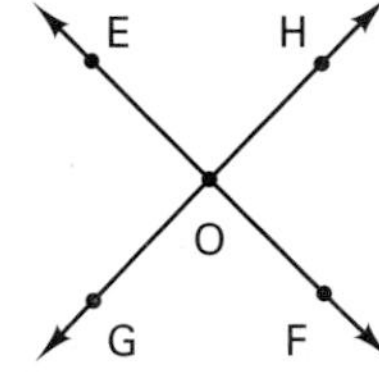

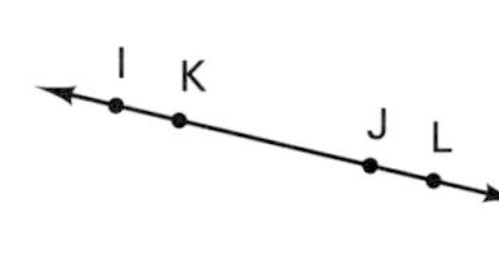

Figure 21-21

L.O.3. Use Fig. 21-22 for the following exercises.

16. Name the angle in two different ways using three capital letters.

17. Name the angle using one capital letter.

18. Name the angle using a number.

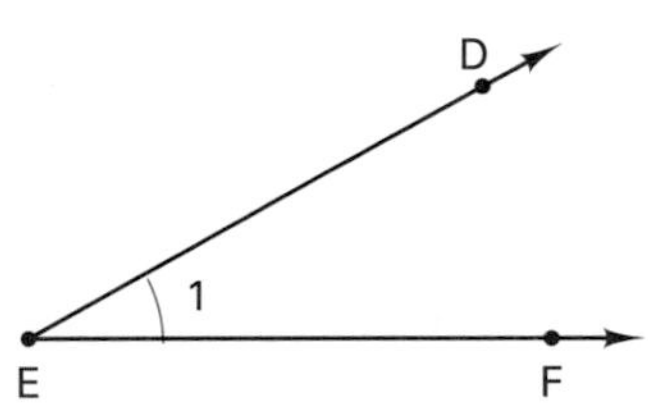

Figure 21-22

Use Fig. 21-23 for the following exercises.

19. Name the angle using one lowercase letter.

20. Name the angle using three capital letters.

21. Name the angle using one capital letter.

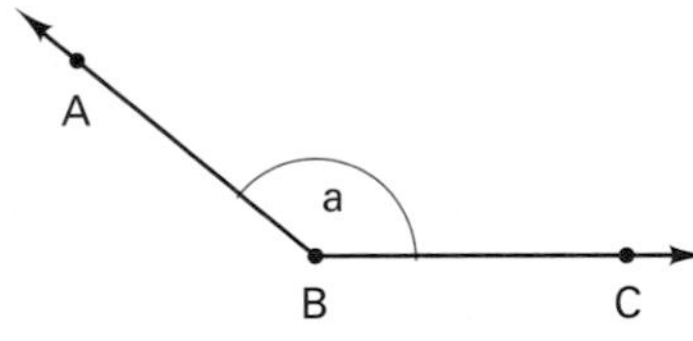

Figure 21-23

L.O.4. Fill in the blanks.

22. One complete rotation is _____°.

23. One-half a complete rotation is _____°.

24. One-fourth a complete rotation is _____°.

Classify the following angle measures using the terms *right*, *straight*, *acute*, or *obtuse*.

25. 38°

26. 95°

27. 90°

28. 153°

29. 10°

30. 180°

31. 60°

32. 120°

Tell whether the following angle pairs are complementary.

33. 17°, 73° **34.** 38°, 142° **35.** 52°, 48°

Tell whether the following angle pairs are supplementary.

36. 60°, 30° **37.** 110°, 70° **38.** 42°, 138°

39. 23°, 67°

40. Two perpendicular lines form a _____° angle. **41.** Two 35° angles are said to be _____angles.

21-2 ANGLE CALCULATIONS

Learning Objectives

L.O.1. Add and subtract angle measures.
L.O.2. Change minutes and seconds to the decimal part of a degree.
L.O.3. Change the decimal part of a degree to minutes and seconds.
L.O.4. Multiply and divide angle measures.

A device used to measure angles is called a *protractor*. The most common type of protractor is a semicircle with two scales from 0° to 180°. An *index mark* is in the middle of the straight edge of the protractor (Fig. 21-24).

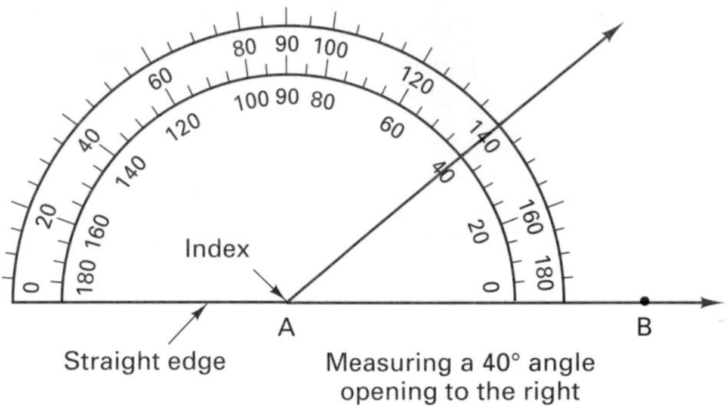

Figure 21-24

To measure an angle that opens to the right, the degree measure is read from the lower scale. Notice in Fig. 21-24 that the lower scale starts with 0 at the right. The index of the protractor is aligned with the vertex of the angle, and the straight edge lies along the lower side of the angle.

To measure an angle that opens to the left, the degree measure is read from the upper scale. Notice in Fig. 21-25 that the upper scale starts with 0 at the left. The index of the protractor is aligned with the vertex of the angle and lies along the lower side of the angle.

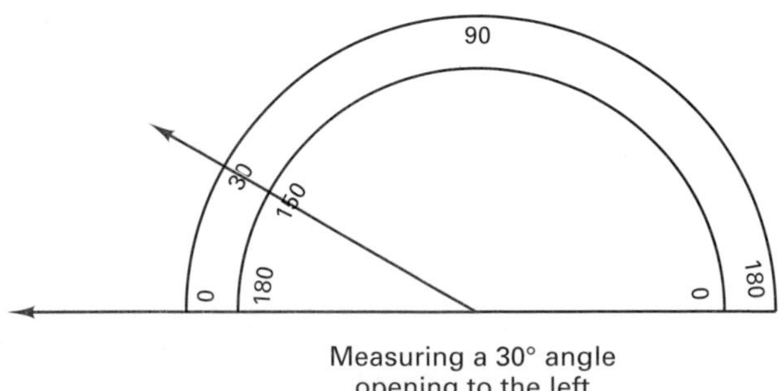

Figure 21-25

Degrees are divided into 60 equal parts. Each of the equal parts is called a *minute.*

$$1 \text{ degree } (1°) = 60 \text{ minutes } (60')$$

The symbol ′ is used for minutes. Similarly, the minute is divided into 60 equal parts called *seconds.*

$$1 \text{ minute } (1') = 60 \text{ seconds } (60'')$$

The symbol ″ is used for seconds. Thus,

$$1° = 60' = 3600''$$

L.O.1. Add and subtract angle measures.

Adding and Subtracting Angle Measures

Angle measures can be added or subtracted. Keep in mind that only like measures can be added or subtracted.

EXAMPLE 21-2.1 Add 12°15′54″ and 82°28′19″.

Solution

$$\begin{array}{r} 12°\,15'\,54'' \\ +\ 82°\,28'\,19'' \\ \hline 94°\,43'\,73'' \end{array}$$

(Arrange the measures in columns of like measures and add.)

Since 73″ = 1′13″, simplify by adding 1′ to the minutes column and subtracting 60″ from the seconds column. Thus, 94°43′73″ = 94°44′13″. ■

EXAMPLE 21-2.2 Add 71°14′ and 82°12″.

Solution

$$\begin{array}{r} 71°\,14' \\ +\ 82°\,12'' \\ \hline 153°\,14'\,12'' \end{array}$$

(Arrange in columns of like measures and add.) ■

EXAMPLE 21-2.3 Subtract 15°32′ from 37°15′.

Solution

$$\begin{array}{rcr} 37°\,15' & = & 36°75' \\ -\ 15°\,32' & = & 15°32' \\ \hline & & 21°43' \end{array}$$

(Borrow 1° from 37°. 1° = 60′ and 60′ + 15′ = 75′.) ■

EXAMPLE 21-2.4 Subtract 3°12′30″ from 15°.

Solution

$$\begin{array}{rcr} \overset{14}{\cancel{15}}\ \overset{59}{\cancel{60}}\ 60 & = & 14°59'60'' \\ -\ 3°\,12'\,30' & = & 3°12'30'' \\ \hline & & 11°47'30'' \end{array}$$

(Borrow 1° from 15°. 1° = 60′. Borrow 1′ from 60′. 1′ = 60″. Then subtract.) ■

EXAMPLE 21-2.5 Find the complement of an angle of 35°25′40″.

Solution: To find the complement of an angle, we subtract the given angle measure from 90°.

$$\begin{array}{rcl} \overset{89}{\cancel{90}}{}^{\circ}\ \overset{59}{}\ \overset{60}{\cancel{60}} & = & 89^\circ 59' 60'' \\ -\ 35^\circ\ 25'\ 40'' & = & \underline{35^\circ 25' 40''} \\ & & 54^\circ 34' 20'' \end{array}$$

■

L.O.2. Change minutes and seconds to the decimal part of a degree.

Changing Minutes and Seconds to Decimal Degrees

With the increased popularity of the calculator, it is sometimes necessary to change minutes or seconds to decimal equivalents. Minutes or seconds are first changed to their fractional part of a degree. Then the fraction is changed to its decimal equivalent by dividing the numerator by the denominator.

Remember, $1' = \frac{1}{60}$ of a degree, and $1'' = \frac{1}{3600}$ of a degree.

> **Rule 21-2.1** *To change minutes to a decimal part of a degree:*
>
> Divide minutes by 60.

> **Rule 21-2.2** *To change seconds to a decimal part of a degree:*
>
> Divide seconds by 3600.

EXAMPLE 21-2.6 Change 15′ to its decimal degree equivalent.

Solution

$$\frac{15}{60} = 0.25^\circ$$

■

EXAMPLE 21-2.7 Change 37″ to its decimal degree equivalent.

Solution

$$\frac{37}{3600} = 0.0103^\circ \qquad \textit{(to the nearest ten-thousandth)}$$

■

EXAMPLE 21-2.8 Change 22′35″ to its decimal degree equivalent.

Solution

$$22' = \frac{22}{60} = 0.3667^\circ \qquad \textit{(to the nearest ten-thousandth)}$$

$$35'' = \frac{35}{3600} = 0.0097^\circ \qquad \textit{(to the nearest ten-thousandth)}$$

$$0.3667^\circ + 0.0097^\circ = 0.3764^\circ$$

■

General Tips for Using the Calculator

Many calculators have a key that will automatically convert between degrees, minutes, and seconds and decimal degrees. The key is often labeled [° ′ ″]. This function key is also called a sexagesimal function key because of its relationships with the number 60. The same key is also used for hours, minutes, and seconds.

Scientific Calculator:

To enter a degree measure in degrees, minutes, and seconds notation:

1. Enter the number of degrees, followed by the [° ′ ″] key.
2. Enter the number of minutes, followed by the [° ′ ″] key.
3. Enter the number of seconds, followed by the [° ′ ″] key. The display on many calculators will automatically show the equivalent measure in decimal degrees. To see the degree, minutes, and seconds notation, press the shift, inverse, or 2nd function key, followed by the [° ′ ″] key.

Calculations can be performed in degrees, minutes, and seconds and the results displayed in either notation.

To enter a degree measure in decimal notation:

1. Enter the measure in decimal degrees.
2. To see the equivalent degrees, minutes, and seconds notation, press the shift, inverse, or 2nd function key, followed by the [° ′ ″] key.

Here are some previously worked examples, performed on the calculator.

- Example 21-2.5

  90° [−] 35 [° ′ ″] 25 [° ′ ″] 40 [° ′ ″] [=] ⇒ 54.57222222
  [INV] [° ′ ″] ⇒ 54°34°20.

  Some calculator displays separate the degrees and minutes and the minutes and seconds with the same notation. The traditional ° ′ ″ symbols should be used in handwritten displays of the results.

- Example 21-2.10

  .43 [INV] [° ′ ″] ⇒ 0°25°48 is written as 0°25′48″.

Graphing Calculator:

On some graphing calculators, the degrees, minutes, and seconds function is found on the [MATH] menu, accessed by pressing [SHIFT] [Graph]. Then select [DMS] (F4). Enter the angle measure using [F1] for degrees, minutes, and seconds. Press [EXE] to display the decimal degrees equivalent.

An angle entered as decimal degrees can be displayed as degrees, minutes, and seconds by pressing the [EXE] key, followed by [F2].

L.O.3. Change the decimal part of a degree to minutes and seconds.

Changing Decimal Part of a Degree to Minutes and Seconds

A decimal part of a degree can be changed to minutes and seconds by reversing the procedure. To change a decimal part of a degree to minutes, multiply by 60. Similarly, to change the decimal part of a minute to seconds, multiply by 60.

> **Rule 21-2.3** ***To change a decimal part of a degree to minutes:***
> Multiply the decimal part of a degree by 60.

> **Rule 21-2.4** ***To change a decimal part of a minute to seconds:***
> Multiply the decimal part of a minute by 60.

If we are changing a decimal part of a *degree* to *seconds*, we can multiply by 3600.

EXAMPLE 21-2.9 Change 0.75° to minutes.

Solution

$$0.75 \times 60 = 45'$$ ■

EXAMPLE 21-2.10 Change 0.43° to minutes and seconds.

Solution

$0.43 \times 60 = 25.8'$ *(degrees to minutes)*

$0.8 \times 60 = 48''$ *(decimal part of minute to seconds)*

Thus, $0.43° = 25'48''$. ■

L.O.4. Multiply and divide angle measures.

Multiplying and Dividing Angle Measures

Angle calculations involving multiplication and division are sometimes necessary. We may multiply or divide angle measures by a number as we do with U.S. Customary measures and then simplify as in Example 21-2.1.

EXAMPLE 21-2.11 An angle of 42°27′32″ needs to be increased to four times its size. Find the measure of the new angle in degrees, minutes, and seconds.

Solution

$$\begin{array}{r} 42°27'32'' \\ \times \qquad\quad 4 \\ \hline 168°108'128'' \end{array} = 169°50'8''$$

(Multiply each part of the measure by 4.)

(Simplify result.) ■

EXAMPLE 21-2.12 An angle of 70°15′16″ is divided into three equal angles. Find the measure of each angle in degrees, minutes, and seconds.

Solution

$23° \quad 25'5\frac{1}{3}''$ or $23°25'5''$ *(to nearest second)*

$$\begin{array}{l} 3\overline{)70° \quad 15' \quad 16''} \\ \quad \underline{69} \\ \quad\; 1° = \underline{60'} \\ \qquad\quad\; 75' \\ \qquad\quad\; \underline{75'} \\ \qquad\qquad\; 16'' \\ \qquad\qquad\; \underline{15''} \\ \qquad\qquad\;\; 1'' \end{array}$$ ■

EXAMPLE 21-2.13 An angle of $57°42'17''$ is divided into four equal parts. Change the measure to its decimal equivalent and find the measure of each part in decimal degrees.

Solution

$$57°42'17'' = 57 + \frac{42}{60} + \frac{17}{3600}$$

$$= 57 + 0.7 + 0.0047 = 57.7047° \quad \textit{(rounded)}$$

$$\frac{57.7047}{4} = 14.4262° \quad \textit{(rounded)}$$

■

SELF-STUDY EXERCISES 21-2

L.O.1. Add or subtract as indicated. Simplify the answers.

1. $15°47'18'' + 38°12'42''$

2. $83°19'54'' - 37°11'36''$

3. $152°28'19'' - 114°35'23''$

4. $45°15'38'' + 28°47'34''$

5. $47°30' - 30°30'15''$

6. $90° - 35°15'48''$

7. Find the supplement of an angle whose measure is $115°35'14''$.

L.O.2. Change to decimal degree equivalents. Express the decimals to the nearest ten-thousandth.

8. $47'$

9. $36''$

10. $5'14''$

11. $10'15''$

12. An angle of $59°24'$ is divided into two equal angles. Change the measure to its decimal equivalent and express the measure of the two equal angles in decimal degrees.

L.O.3. Change to equivalent minutes and seconds. Round to the nearest second when necessary.

13. $0.35°$

14. $0.20°$

15. $0.12°$

16. $0.213°$

17. $0.3149°$

L.O.4. In the following exercises, round angle measures to the nearest second when necessary.

18. An angle of $90°$ is divided into 12 equal parts. Find the measure of each angle in degrees and minutes.

19. An angle of $80°$ is divided into seven equal parts. Find the measure of each angle in degrees, minutes, and seconds.

20. An angle of $12°43'49''$ needs to be three times as large. Find the measure of the new angle in degrees to the nearest ten-thousandth.

21. An angle of $25°17'39''$ is divided into two equal angles. Find the measure of each angle in degrees, minutes, and seconds.

21-3 TRIANGLES

Learning Objectives

L.O.1. Classify triangles by sides.
L.O.2. Relate the sides and angles of a triangle.
L.O.3. Determine if two triangles are congruent using inductive and deductive reasoning.

L.O.4. Use the properties of a 45°, 45°, 90° triangle to find missing parts and to solve applied problems.
L.O.5. Use the properties of a 30°, 60°, 90° triangle to find missing parts and to solve applied problems.

We have examined various relationships that exist among lines and angles in the previous sections. In this section we study one of the most useful mathematical figures for any technician who uses geometry—the *triangle*. The relationships among the sides and angles in the triangle enable us to obtain much information that is implied but not always expressed in certain applications.

Triangles can be classified according to the relationship of their sides or their angles. We classified triangles by angles in Chapter 13. In this section, we will classify triangles by their sides.

L.O.1. Classify triangles by sides.

Classifying Triangles by Sides

There are three possible relationships among the three sides of a triangle, and these relationships result in a special name for each type of triangle.

■ **DEFINITION 21-3.1 Equilateral Triangle.** An **equilateral triangle** is a triangle with three equal sides. The three angles of an equilateral triangle are also equal. Each angle measures 60° (Fig. 21-26).

■ **DEFINITION 21-3.2 Scalene Triangle.** A **scalene triangle** is a triangle with *all* three sides unequal (Fig. 21-27).

■ **DEFINITION 21-3.3 Isosceles Triangle.** An **isosceles triangle** is a triangle with *exactly* two equal sides. The angles opposite these equal sides are also equal (Fig. 21-26).

L.O.2. Relate the sides and angles of a triangle.

Relationship of Triangle Sides to Angles

An equilateral (three sides equal) triangle also has three equal angles. The two equal sides of an isosceles triangle have opposite angles that are equal (Fig. 21-26).

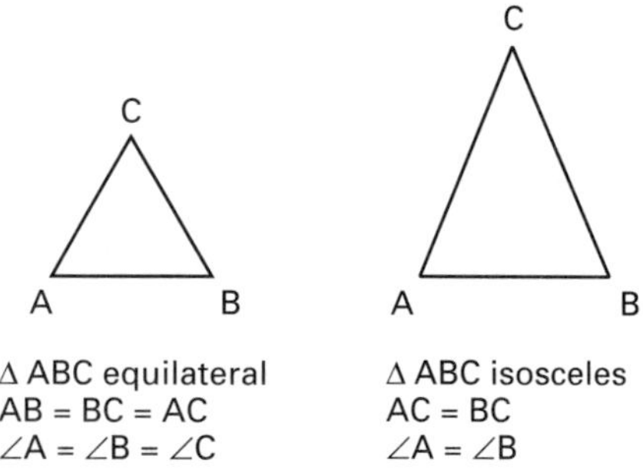

Figure 21-26

In a scalene triangle, where no sides are equal, we can also state an important relationship between the sides and their opposite angles (Fig. 21-27).

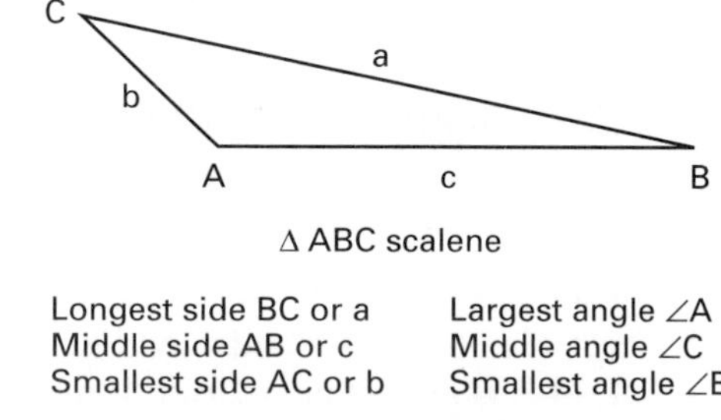

Figure 21-27

Rule 21-3.1

If the three sides of a triangle are unequal, the *largest* angle will be opposite the *longest* side and the *smallest* angle will be opposite the *shortest* side.

EXAMPLE 21-3.1 Use Rule 21-3.1 to identify the longest and shortest sides of the triangle in Fig. 21-28.

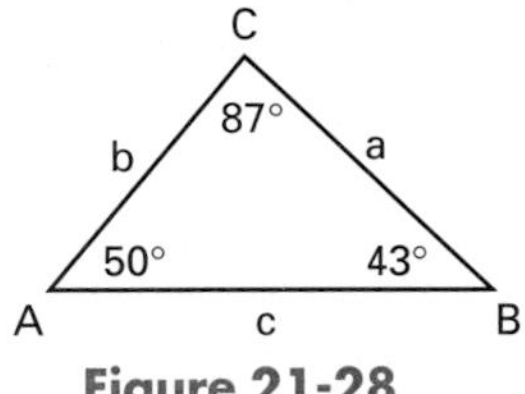

Figure 21-28

Solution: The longest side is AB or c (87° is largest angle). The shortest side is AC or b (43° is smallest angle). ■

EXAMPLE 21-3.2 Use Rule 21-3.1 to identify the largest and smallest angles in the triangle in Fig. 21-29.

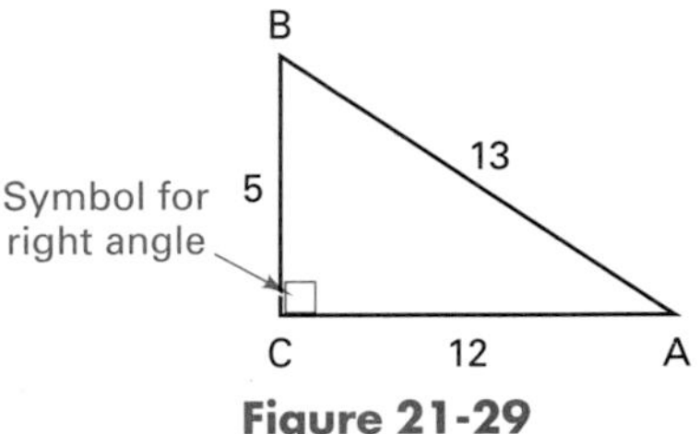

Figure 21-29

Solution: The largest angle is $\angle C$ (13 is the longest side). The smallest angle is $\angle A$ (5 is the shortest side). ■

When applying Rule 21-3.1 to a right triangle, we see that the hypotenuse (side opposite the 90° angle) is always the longest side. Figure 21-29 shows the hypotenuse (side AB) to be the longest side.

L.O.3. Determine if two triangles are congruent using inductive and deductive reasoning.

Congruent Triangles and Inductive and Deductive Reasoning

When we work with triangles that have the same size and shape, we are working with *congruent* triangles. These triangles can be fitted exactly on top of each other.

In Fig. 21-30, $\triangle ABC$ will fit exactly over $\triangle RST$. They are congruent triangles. (The symbol $\cong$ means congruent.) That is, $\triangle ABC \cong \triangle RST$. Each angle in $\triangle ABC$ has an angle in $\triangle RST$ that is its equal. We say these pairs of angles *correspond*. In Fig. 21-30, $\angle A$ corresponds to $\angle R$ because they are equal. Also, $\angle C$ corresponds to $\angle T$, and $\angle B$ corresponds to $\angle S$. The equal sides also correspond. $\overline{AB}$ corresponds to $\overline{RS}$, $\overline{CB}$ corresponds to $\overline{TS}$, $\overline{AC}$ corresponds to $\overline{RT}$.

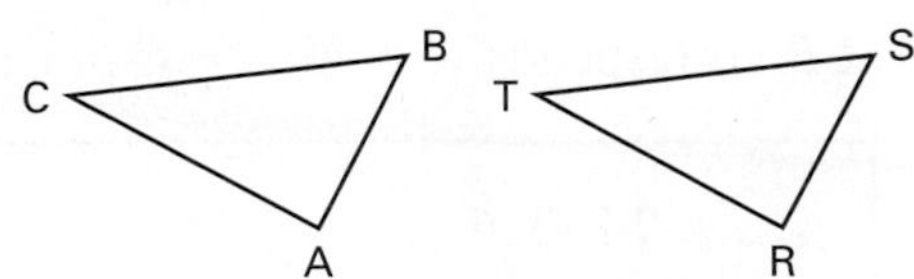

Figure 21-30

■ **DEFINITION 21-3.4 Congruent Triangles. Congruent triangles** are triangles in which the corresponding sides and angles are equal.

We can establish congruent triangles even when we know that only certain angles and sides are equal.

> **Rule 21-3.2**
>
> If the three sides of one triangle are equal to the corresponding three sides of another triangle, the triangles are congruent.

EXAMPLE 21-3.3 Write the corresponding sides of the two triangles in Fig. 21-31.

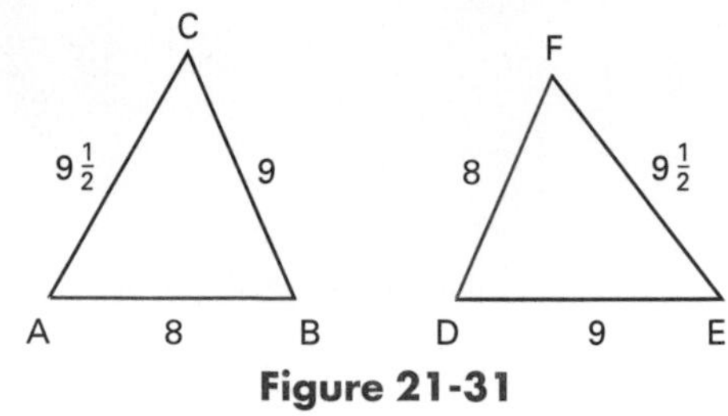

Figure 21-31

Solution

$$\triangle ABC \cong \triangle DEF$$
$$\overline{AB} \text{ corresponds to } \overline{DF}$$
$$\overline{BC} \text{ corresponds to } \overline{DE}$$
$$\overline{AC} \text{ corresponds to } \overline{EF}$$

■

> **Rule 21-3.3**
>
> If two sides and the included angle of one triangle are equal to two sides and the included angle of another triangle, the triangles are congruent.

EXAMPLE 21-3.4 List the corresponding sides and angles of the two triangles in Fig. 21-32.

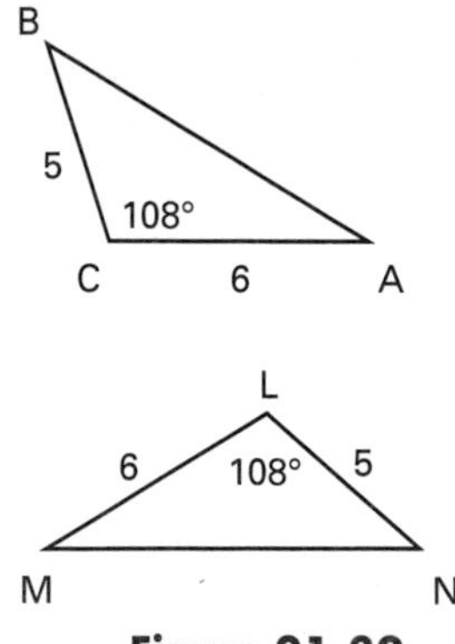

Figure 21-32

Solution

$\overline{BC}$ corresponds to $\overline{LN}$.

$\overline{AC}$ corresponds to $\overline{LM}$.

$\angle C$ corresponds to $\angle L$.

$\overline{AB}$ corresponds to $\overline{MN}$. *(They are opposite equal angles.)*

$\angle A$ corresponds to $\angle M$. *(They are opposite equal sides.)*

$\angle B$ corresponds to $\angle N$. *(They are opposite equal sides.)* ■

> **Rule 21-3.4**
>
> If two angles and the common side of one triangle are equal to two angles and the common side of another triangle, the triangles are congruent.

EXAMPLE 21-3.5 List the corresponding angles and sides of the two triangles in Fig. 21-33.

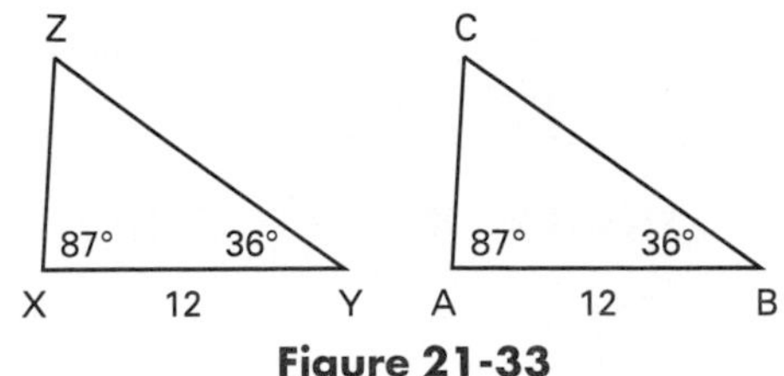

Figure 21-33

Solution

$\angle X$ corresponds to $\angle A$.	$\angle Z$ corresponds to $\angle C$.
$\angle Y$ corresponds to $\angle B$.	$\overline{XZ}$ corresponds to $\overline{AC}$.
$\overline{XY}$ corresponds to $\overline{AB}$.	$\overline{YZ}$ corresponds to $\overline{BC}$.

■

The study of geometry applies two basic types of reasoning, *inductive* and *deductive* reasoning. *Inductive reasoning* starts with investigation and experimentation. Early mathematicians discovered the properties of geometry through inductive reasoning. After extensive investigation, general conclusions are drawn from the results of specific cases.

An example of inductive reasoning would be to examine the results of cutting a square into four parts by cutting along the two diagonals (Figure 21-34). Compare the resulting four triangles. Are they congruent triangles? Make a square of a different size and repeat the exercise (Figure 21-35). Continued repetitions will lead you to conclude that the diagonals of a square divide the square into four congruent triangles.

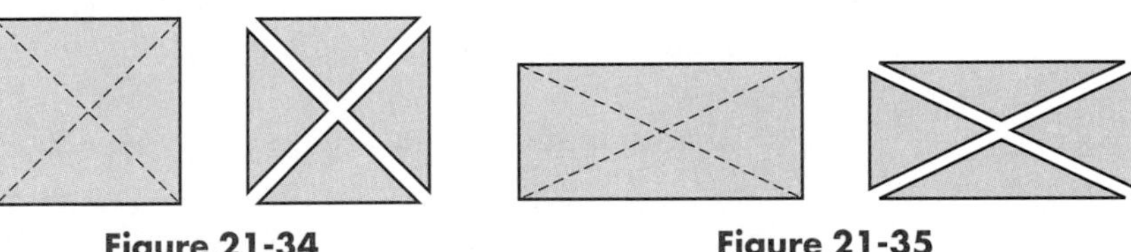

Figure 21-34 **Figure 21-35**

Now, cut a rectangle along the two diagonals. Are the resulting four triangles congruent? No. What conclusions appear to result from this experiment? Repeat the experiment with different sizes of rectangles. What properties seem apparent? There are two pairs of congruent triangles.

Deductive reasoning starts with accepted principles and additional properties are concluded from these accepted principles. Accepting the congruent triangle properties in Rules 21-3.2, 21-3.3, and 21-3.4 and an additional property that the diagonals of a square bisect (cut in half) the angles of the square, we will use deductive reasoning to show that the two diagonals of a square form four congruent triangles.

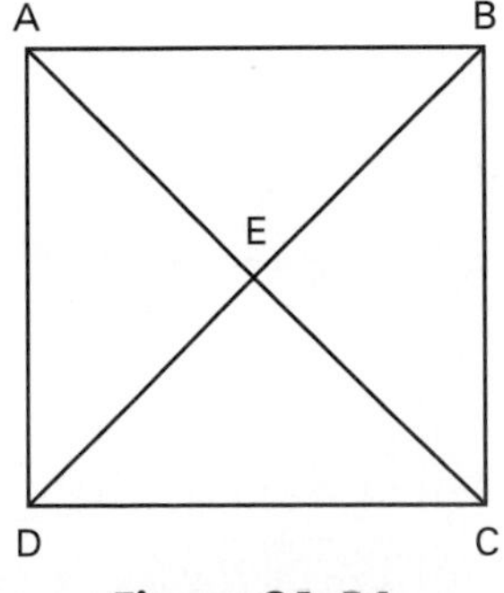

Figure 21-36

Square *ABCD* forms the four triangles $\triangle AED$, $\triangle AEB$, $\triangle BEC$, and $\triangle DEC$ (Figure 21-36). Since the four sides of a square are equal in measure, sides *AD* and *AB* are equal in measure. Since the diagonals of a square bisect the angles, $\angle DAE$, $\angle EAB$, $\angle ABE$, $\angle EBC$, $\angle BCE$, $\angle ECD$, $\angle CDE$, $\angle EDA$ are all 45° angles. Then, in $\triangle AED$ and $\triangle AEB$ we have two angles and a common side of one triangle equal to two angles and a common side of the other. By applying Rule 21-3.4, we can say that $\triangle AED$ and $\triangle AEB$ are congruent. Similarly, we could continue the argument to show that all four triangles are congruent.

In a systematic or axiomatic study of geometric concepts, all properties are developed or proved from a limited number of basic principles. In our study of geometric concepts, we will use an informal approach. While we will apply both inductive and deductive reasoning in our arguments for solving problems, we will not introduce all the concepts necessary to make formal proofs of geometric properties.

L.O.4. Use the properties of a 45°, 45°, 90° triangle to find missing parts and to solve applied problems.

The 45°, 45°, 90° Right Triangle

We saw that an isosceles triangle is a triangle with two equal sides. An isosceles right triangle is a right triangle whose legs are equal. The angles opposite the equal legs are *base angles*. Then the two base angles are also equal since they are the angles opposite the equal sides (Fig. 21-37).

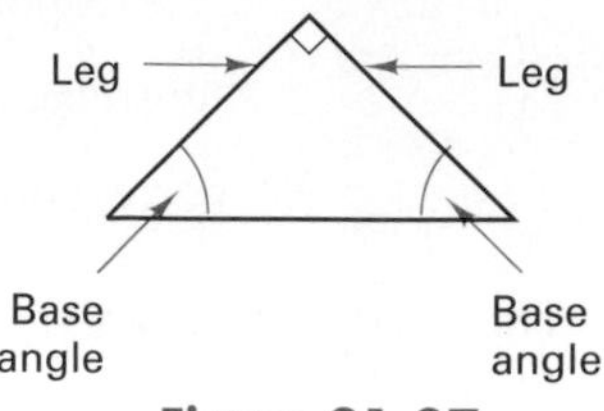

Figure 21-37

Since the sum of the angles of the triangle is 180°, the sum of the two base angles of a right triangle is 180° − 90° or 90°. If both angles are equal, as in the isosceles right triangle, then each angle is $\frac{1}{2}(90°)$ or 45°. Thus, we get the name 45°, 45°, 90° right triangle.

■ **Definition 21-3.5 45°, 45°, 90° Right Triangle.** A **45°, 45°, 90° right triangle** is an isosceles right triangle in which the two base angles are 45° each.

This triangle is frequently used in applications of the Pythagorean theorem.

Example 21-3.6 Find the hypotenuse of an isosceles right triangle, each of whose equal sides is 2 m (Fig. 21-38).

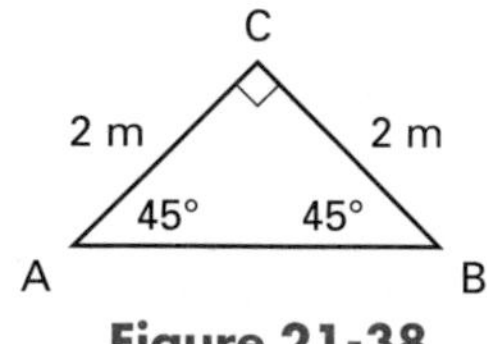

Figure 21-38

Solution: Using the Pythagorean theorem, we have

$$
\begin{aligned}
2^2 + 2^2 &= (AB)^2 \\
4 + 4 &= (AB)^2 \\
8 &= (AB)^2 \\
\sqrt{8} &= AB \\
2.828 &= AB \qquad \textit{(rounded)}
\end{aligned}
$$

Thus, the hypotenuse is 2.828 m.

Notice that we can also calculate $AB = 2.828$ by simplifying $\sqrt{8}$.

$$
\begin{aligned}
AB &= \sqrt{8} = \sqrt{4 \cdot 2} = 2\sqrt{2} \\
AB &= 2(1.414213562) \\
AB &= 2.828 \qquad \textit{(rounded)}
\end{aligned}
$$

The hypotenuse, then, is the product of a leg such as AC and $\sqrt{2}$. ■

Rule 21-3.5 ***To find the hypotenuse of a 45°, 45°, 90° right triangle:***
Multiply the measure of a leg by $\sqrt{2}$ or 1.414213562. That is,

$$\text{hypotenuse} = \text{leg}\sqrt{2}$$

If this relationship is remembered, it can save computational time on the job.

If we need to find a leg of a 45°, 45°, 90° right triangle, we can use another rule based on Rule 21-3.5.

Example 21-3.7 Find AC and BC if $AB = 5$ cm (Fig. 21-39).

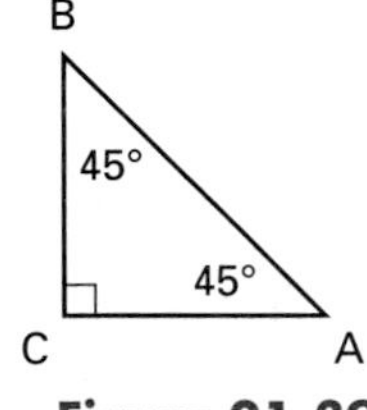

Figure 21-39

Solution: The hypotenuse is equal to the product of a leg and $\sqrt{2}$.

$$5 = AC\sqrt{2}$$

$$\frac{5}{\sqrt{2}} = \frac{AC\sqrt{2}}{\sqrt{2}} \quad \textit{(Solve for AC, a leg.)}$$

$$\frac{5}{\sqrt{2}} = AC$$

$$\frac{5}{\sqrt{2}} \cdot \frac{\sqrt{2}}{\sqrt{2}} = AC \quad \textit{(Rationalize.)}$$

$$\frac{5\sqrt{2}}{2} = AC$$

$$3.536 \text{ cm} = AC \quad \textit{(rounded)}$$

When using a calculator, we can find the decimal value of AC just as easily without rationalizing.

$$\frac{5}{\sqrt{2}} = AC = 5 \div \sqrt{2}$$

$AC = 3.536$ cm, from [5] [÷] [2] [√] [=] ⇒ 3.536 *(rounded)*■

Rule 21-3.6 ***To find a leg of a 45°, 45°, 90° right triangle:***
Divide the product of the hypotenuse and $\sqrt{2}$ by 2 (Fig. 21-40). That is,

$$\text{leg} = \frac{\text{hypotenuse}\ \sqrt{2}}{2}$$

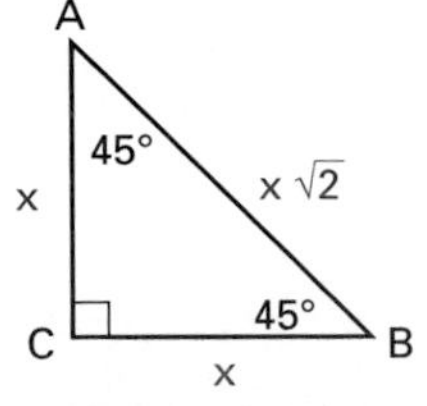

Figure 21-40

The sides of a 45°, 45°, 90° triangle are related by the 1, 1, $\sqrt{2}$ respectively.

L.O.5. Use the properties of a 30°, 60°, 90° triangle to find missing parts and to solve applied problems.

The 30°, 60°, 90° Right Triangle

Another special case of the Pythagorean theorem is the *30°, 60°, 90° right triangle*, which often arises in applications. If we draw the altitude of an *equilateral* triangle, we form two 30°, 60°, 90° right triangles (Fig. 21-41).

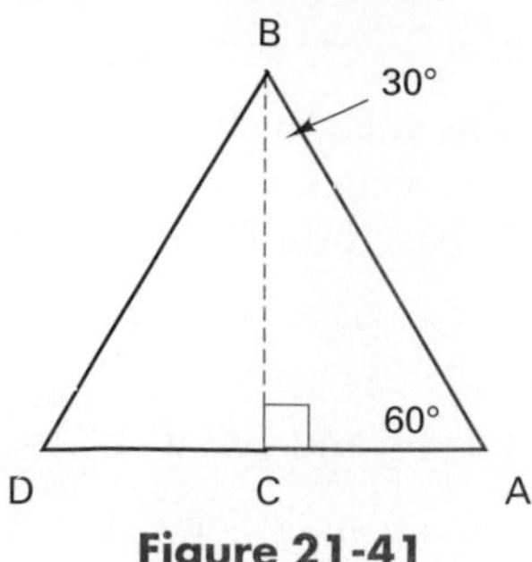

Figure 21-41

■ **DEFINITION 21-3.6 Altitude of Equilateral Triangle.** The **altitude** *of an equilateral triangle* is a line drawn from the midpoint of the base to the opposite vertex, dividing the triangle into two congruent right triangles.

Since the altitude of an equilateral triangle bisects (divides in half) the vertex angle and the base, we know that two 30° angles are formed at B and that $AC = CD$. Since $AD = AB$ and $2AC = AD$, we know that $2AC = AB$, or $AC = \frac{1}{2}AB$ or the side opposite the 30° angle is one-half the hypotenuse.

EXAMPLE 21-3.8 If $AC = 5$ cm, find AB and BC (Fig. 21-42).

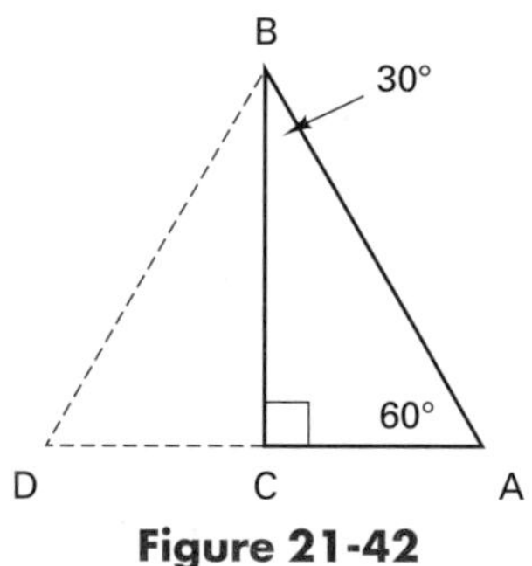

Figure 21-42

Solution: Side AB (the hypotenuse) is twice AC since the altitude divides AD in half, and in an equilateral triangle all sides are equal. Therefore, $AB = 2(5) = 10$ cm. Now that we know two sides of the right triangle, we can use the Pythagorean theorem to find the third side.

$$
\begin{aligned}
(AB)^2 &= (AC)^2 + (BC)^2 \\
10^2 &= 5^2 + (BC)^2 \\
100 &= 25 + (BC)^2 \\
75 &= (BC)^2 \\
\sqrt{75} &= BC \\
8.660 &= BC \qquad \textit{(rounded)}
\end{aligned}
$$

Thus, BC is 8.660 cm.

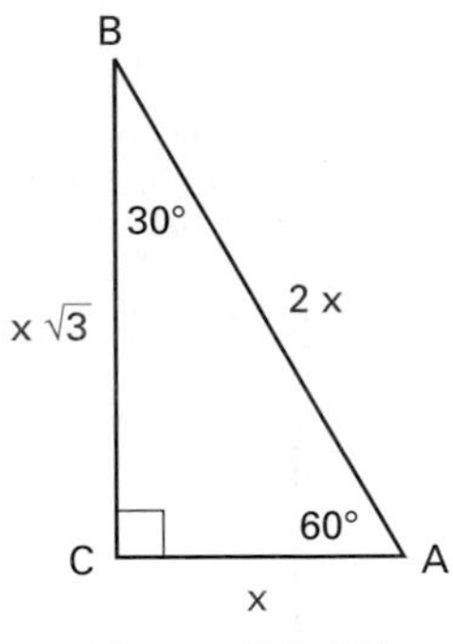

Figure 21-43

Notice that we can also calculate $BC = 8.660$ by simplifying $\sqrt{75}$.

$$BC = \sqrt{75} = \sqrt{25 \cdot 3} = 5\sqrt{3}$$
$$BC = 5(1.7320508)$$
$$BC = 8.660 \qquad (rounded)$$

Then BC (the side opposite the 60° angle) is the product of AC (the side opposite the 30° angle) and $\sqrt{3}$. ■

Once we have learned the relationships of the three sides, we can find two sides of any 30°, 60°, 90° right triangle if we know only one side (Fig. 21-43).

Rule 21-3.7 ***To find the hypotenuse of a 30°, 60°, 90° right triangle:***

Multiply the side opposite the 30° angle by 2.

Rule 21-3.8 ***To find the side opposite the 30° angle in a 30°, 60°, 90° right triangle:***

Divide the hypotenuse by 2.

Rule 21-3.9 ***To find the side opposite the 60° angle in a 30°, 60°, 90° right triangle:***

Multiply the side opposite the 30° angle by $\sqrt{3}$.

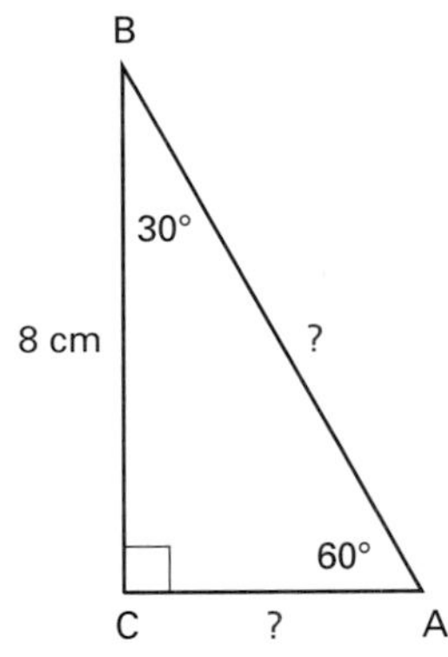

Figure 21-44

Tips and Traps

To summarize the relationships in Fig. 21-43:

$$AB = 2(AC)$$
$$AC = \frac{AB}{2}$$
$$BC = AC\sqrt{3}$$

The sides of a 30°, 60°, 90° triangle are related by the 1, $\sqrt{3}$, 2 respectively.

EXAMPLE 21-3.9 Find AC and AB if $BC = 8$ cm (Fig. 21-44).

Solution

$$BC = x\sqrt{3} = 8 \qquad (x = AC)$$

Solving for x, we have

$$\frac{x\sqrt{3}}{\sqrt{3}} = \frac{8}{\sqrt{3}}$$

Rationalizing, we have

$$x = \frac{8}{\sqrt{3}} \cdot \frac{\sqrt{3}}{\sqrt{3}}$$

$$x = \frac{8\sqrt{3}}{3}$$

$$x = 4.619 \text{ cm} \qquad \textit{(rounded)}$$

$$AB = 2x = 2\left(\frac{8\sqrt{3}}{3}\right) = \frac{16\sqrt{3}}{3} = 9.238 \text{ cm} \qquad \textit{(rounded)}$$

When using a scientific calculator, we can find the decimal value of AC without rationalizing.

$$AC = \frac{8}{\sqrt{3}} = 8 \div \sqrt{3}$$

$AC = 4.619$ cm, from $\boxed{8}$ $\boxed{\div}$ $\boxed{3}$ $\boxed{\sqrt{\ }}$ $\boxed{=}$ $\Rightarrow$ 4.6188021

SELF-STUDY EXERCISES 21-3

L.O.1. Fill in the blanks.

1. A triangle with no equal sides is called a(n) _____ triangle.

2. A triangle with three equal sides is called a(n) _____ triangle.

3. A triangle with only two equal sides is called a(n) _____ triangle.

L.O.2. Identify the longest and shortest sides in the following figures.

4.

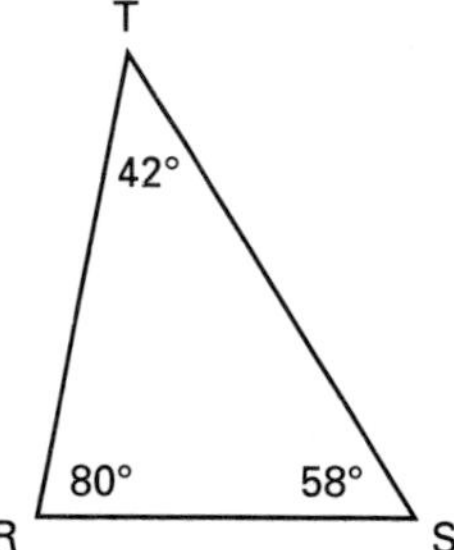

5.

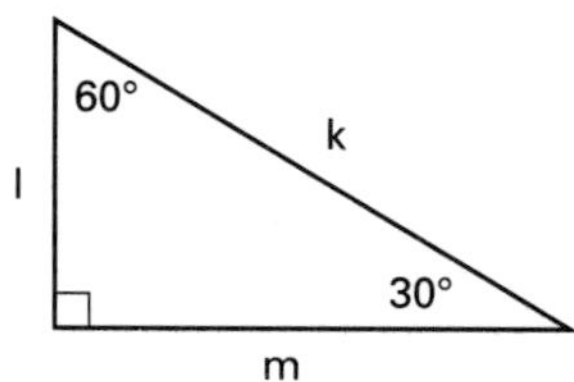

List the angles in order of size from largest to smallest in the following figures.

6.

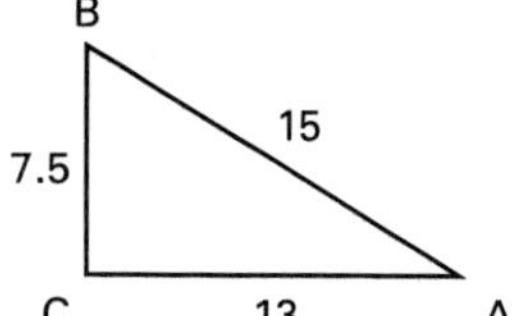

7.

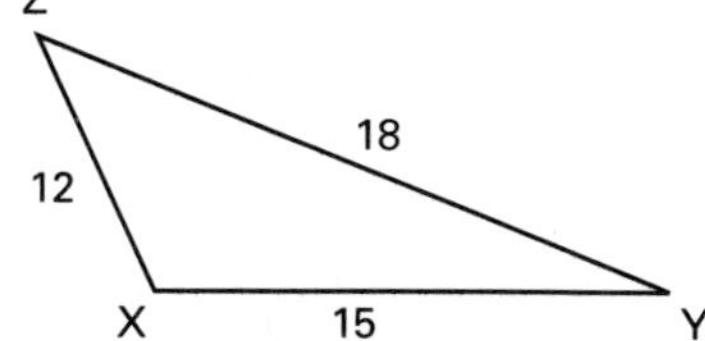

L.O.3. Write the corresponding parts not given for the congruent triangles in the following figures.

8.

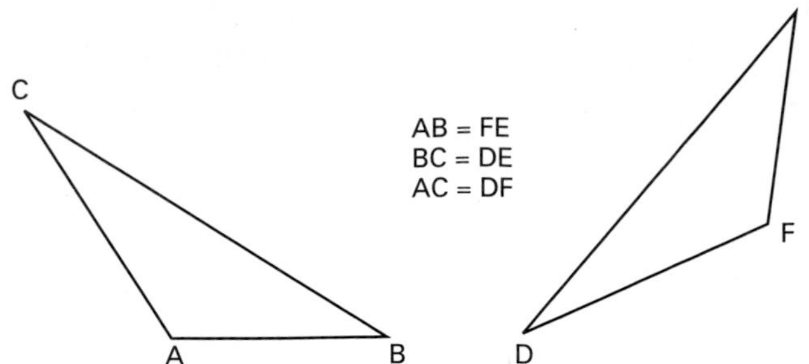

9.

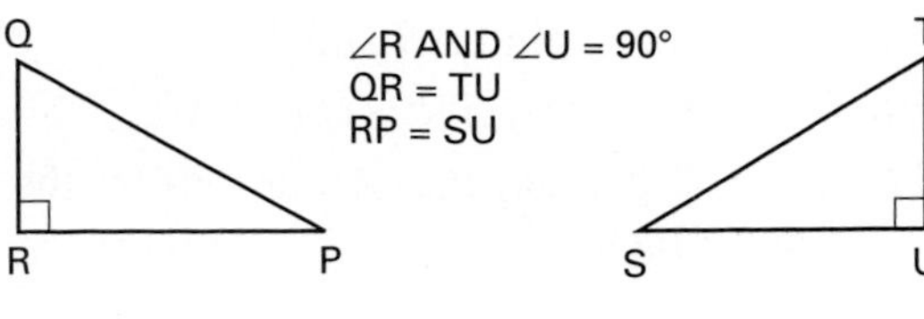

10.

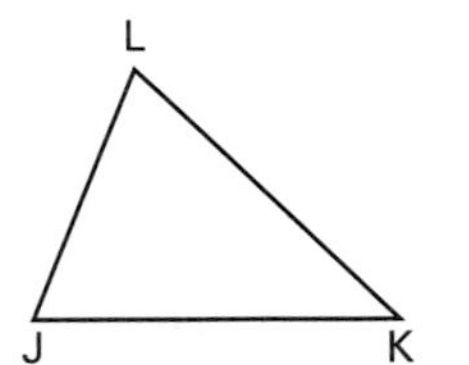

JK = NM
∠J = ∠M
∠K = ∠N

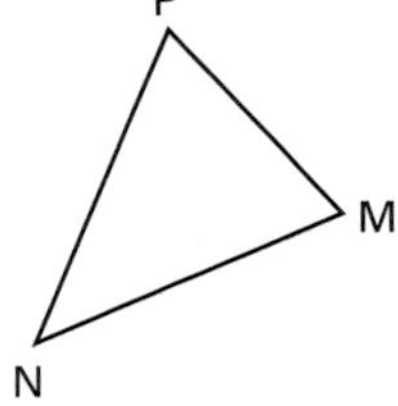

L.O.4. Use Fig. 21-45 to solve the following exercises. Round the final answers to the nearest thousandth if necessary.

11. $AC = 12$ cm; find BC and AB.

12. $AB = 10$ m; find AC and BC.

13. $BC = 7\sqrt{2}$ m; find AC and AB.

14. $AB = 8\sqrt{2}$ m; find AC and BC.

15. $AB = 12\sqrt{3}$ mm; find AC and BC.

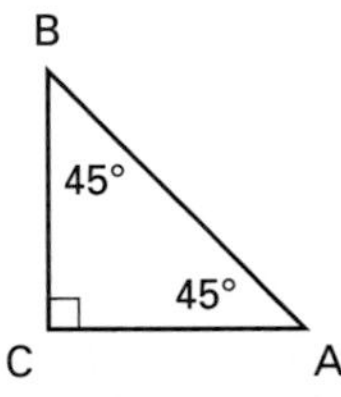

Figure 21-45

16. A rafter 19.5 ft long makes a 45° angle with a joist. If the rafter has an 18-in. overhang, find the length of the joist to the nearest inch (Fig. 21-46).

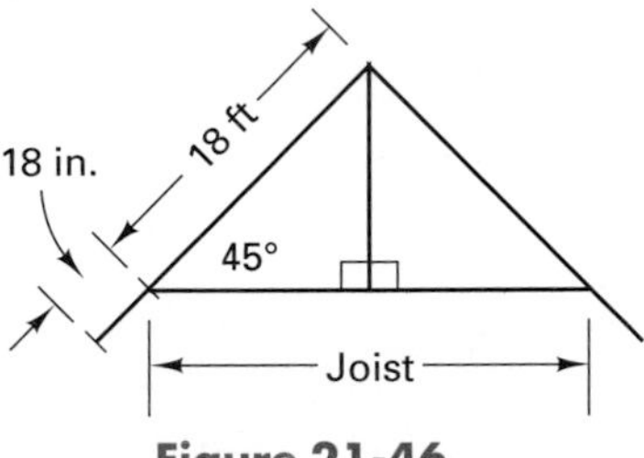

Figure 21-46

17. An elevated tank must be connected to a pipe on the ground. The connecting pipe is 53.5 ft long and will form a 45° angle where it connects to the tank (Fig. 21-47). At what horizontal distance from the tank should the ground pipe stop so that the connection can be made?

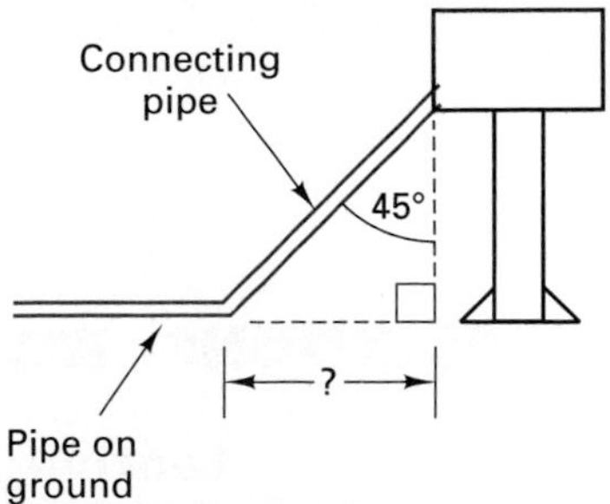

Figure 21-47

L.O.5. Use Fig. 21-48 to solve the following exercises. Round the final answers to the nearest thousandth if necessary.

18. $AC = 6$ cm; find AB and BC.

19. $AB = 18$ mm; find AC and BC.

20. $BC = 8$ inches; find AC and AB.

21. $BC = 7\sqrt{2}$ cm; find AC and AB.

22. $AC = 2$ ft 9 in.; find AB and BC to the nearest inch.

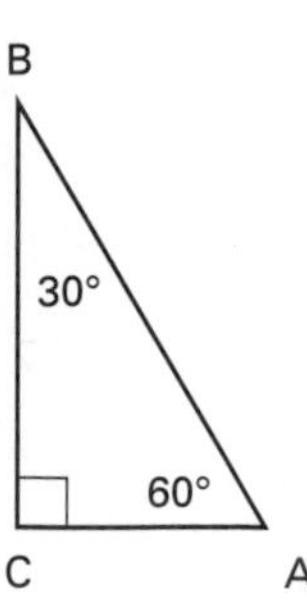

Figure 21-48

Solve the following problems. Round the final answers to the nearest thousandth if necessary.

23. Find the length of the conduit $ABCD$ (Fig. 21-49) if $AB = 18$ ft, $CK = 6$ ft, $CD = 5$ ft, and $\angle KCB = 60°$.

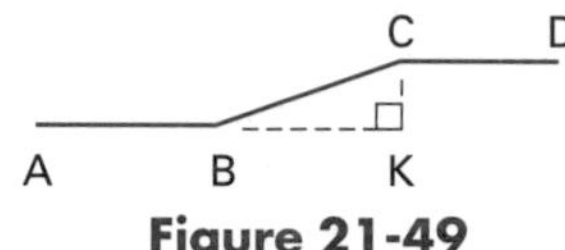

Figure 21-49

24. A rafter makes a 30° angle with the horizontal. If the rise is 9 ft, find the rafter length and the run to the nearest inch (Fig. 21-50).

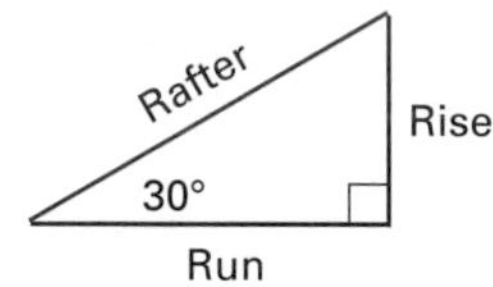

Figure 21-50

25. Find the depth of a V-slot in the form of an equilateral triangle if the cross-sectional opening is 5.2 cm across (Fig. 21-51).

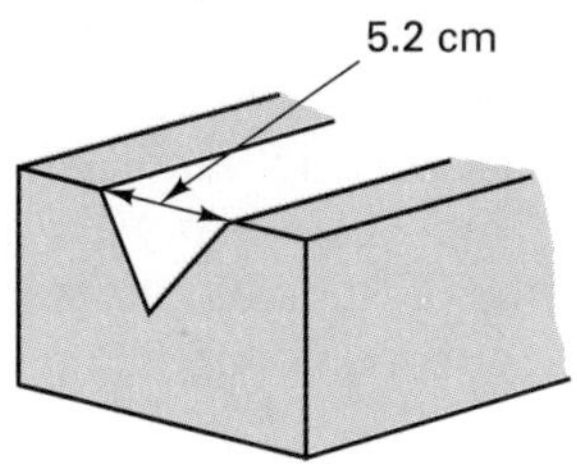

Figure 21-51

21-4 POLYGONS

Learning Objectives

L.O.1. Find the missing dimensions of composite shapes.
L.O.2. Find the perimeter and area of composite shapes.
L.O.3. Find the number of degrees in each angle of a regular polygon.

As we look about us, we see polygons that are not squares, rectangles, parallelograms, trapezoids, or triangles. Yet our work may require us to calculate the perimeter and area of these polygons. In such cases we can use what we already know to find both perimeter and area. Such figures can be considered as *composites.*

■ **Definition 21-4.1 Composite.** A **composite** is a geometric figure made up of two or more geometric figures.

For instance, an L-shaped slab foundation for a building is really a composite of two rectangles. As suggested in Fig. 21-52, the two rectangles forming the composite figure may be considered in more than one way. Here layout I is partitioned vertically, whereas layout II is partitioned horizontally.

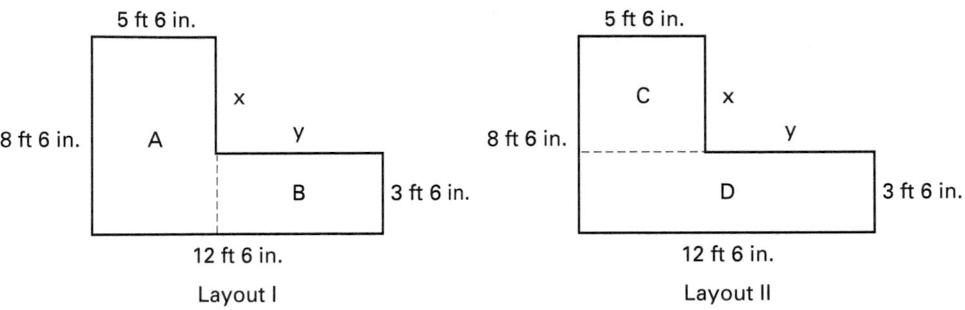

Figure 21-52

L.O.1. Find the missing dimensions of composite shapes.

Finding Missing Dimensions

Observe also that such a layout may not always have all its dimensions indicated. However, the missing dimensions can be inferred or calculated from our knowledge of polygons and the dimensions that are indicated on the layout itself.

Example 21-4.1 Find the missing dimensions x and y on the slab layout in Fig. 21-52.

Solution: In layout I, the side of B opposite its $3'6''$ side is also $3'6''$ because opposite sides of a rectangle are equal. The side of A opposite its $8'6''$ side is, for the same reason, $8'6''$. Dimension x must therefore be the difference between $8'6''$ and $3'6''$.

$$x = 8'6'' - 3'6''$$

$$x = 5'$$

If we think of layout II as two horizontal rectangles, we can find dimension y. The side opposite the $5'6''$ side of rectangle C must be $5'6''$. The side of D opposite the $12'6''$ side must also be $12'6''$. Dimension y must therefore be the difference between $12'6''$ and $5'6''$.

$$y = 12'6'' - 5'6''$$

$$y = 7'$$

The missing dimensions are therefore $x = 5'$ and $y = 7'$. ■

L.O.2. Find the perimeter and area of composite shapes.

Perimeter and Area of Composite Figures

Perimeter is the sum of the sides of a figure or layout. The number and the length of the sides will vary from one composite shape to the next, so no specific formula can be given to cover the variety of composite shapes that can be found. We can use a general formula.

> **Formula 21-4.1** *Perimeter of a composite:*
>
> $$P = a + b + c, \text{ etc.}$$

EXAMPLE 21-4.2 Find the number of feet of 4-in. stock needed for the base plates of a room that has the layout shown in Fig. 21-53. Make no allowances for openings when estimating the linear footage of the base plates that form the perimeter.

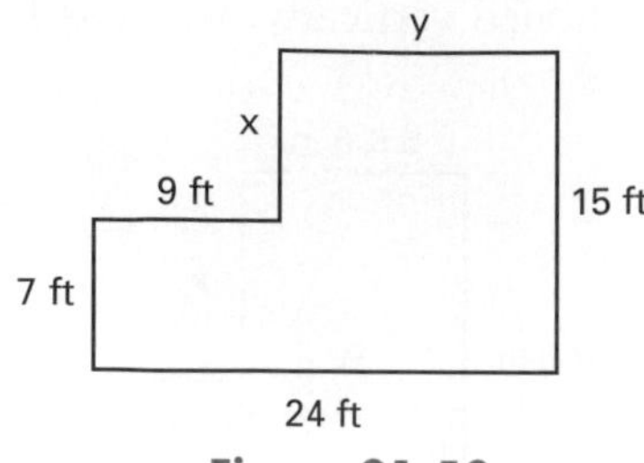

Figure 21-53

Solution

1. Find the missing dimensions.

$$x = 15 \text{ ft} - 7 \text{ ft} \qquad y = 24 \text{ ft} - 9 \text{ ft}$$
$$x = 8 \text{ ft} \qquad y = 15 \text{ ft}$$

2. Apply the general formula for the perimeter of a polygon.

$$P = a + b + c, \text{ etc.}$$
$$P = 7 \text{ ft} + 24 \text{ ft} + 15 \text{ ft} + 15 \text{ ft} + 8 \text{ ft} + 9 \text{ ft} = 78 \text{ ft}$$

3. Count the number of sides on the layout to make sure that each has been substituted into the formula.

The room needs 78 ft of 4-in. stock for the base plates. ■

EXAMPLE 21-4.3 Find the number of square yards of carpeting required for the room in Example 21-4.2. ($9 \text{ ft}^2 = 1 \text{ yd}^2$.)

Solution

1. Divide the composite into two polygons whose areas we can compute (Fig. 21-54). In this case, *A* is a rectangle and *B* is a square.

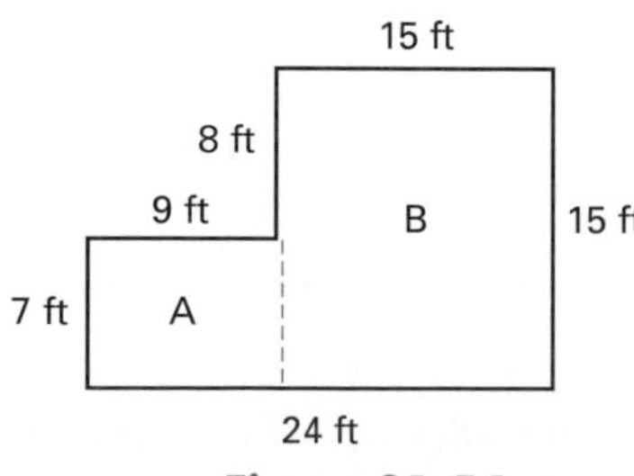

Figure 21-54

2. Find the areas of each smaller polygon and add them together.

Rectangle A	Square B
$A_1 = lw$	$A_2 = s^2$
$A_1 = 9 \times 7$	$A_2 = 15^2$
$A_1 = 63 \text{ ft}^2$	$A_2 = 225 \text{ ft}^2$

$$A_1 + A_2 = \text{total area}$$

$$63 + 225 = 288 \text{ ft}^2$$

3. Convert square feet to square yards using unity ratio.

$$\frac{\overset{32}{\cancel{288}} \cancel{\text{ft}^2}}{1} \times \frac{1 \text{ yd}^2}{\underset{1}{\cancel{9}} \cancel{\text{ft}^2}} = 32 \text{ yd}^2$$

The room requires 32 yd^2 of carpeting. ■

EXAMPLE 21-4.4 Find the area of a gable end of a gambrel roof whose dimensions are shown in Fig. 21-55.

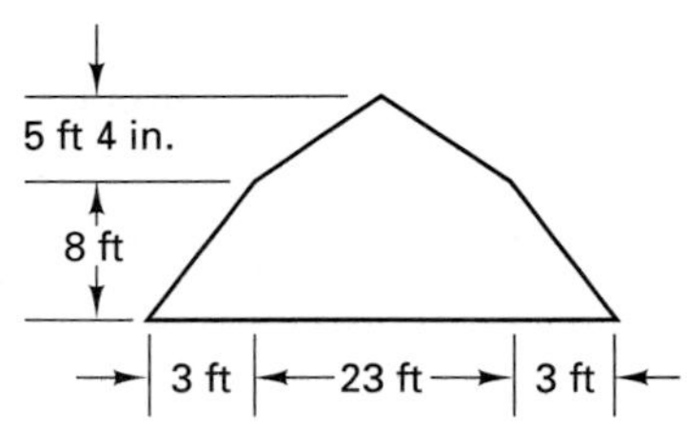

Figure 21-55

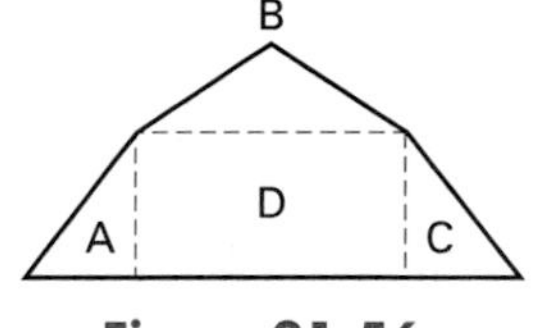

Figure 21-56

Solution

1. Divide the gable end of the gambrel roof into polygons whose areas can be calculated (Fig. 21-56).
2. Calculate the areas of the three triangles and the rectangle. Find the sum of the areas.

Triangle B	Triangles A or C	Rectangle D
Base $= 23'$	Base $= 3'$	Length $= 23'$
Height $= 5'4'' = 5\frac{1}{3}'$	Height $= 8'$	Width $= 8'$
$A_1 = \frac{1}{2}hb$	$A_2 = \frac{1}{2}hb$	$A_3 = lw$
$A_1 = \frac{1}{2}\left(5\frac{1}{3}\right)(23)$	$A_2 = \frac{1}{\underset{1}{\cancel{2}}}(\overset{4}{\cancel{8}})(3)$	$A_3 = 23(8)$
$A_1 = \frac{1}{\underset{1}{\cancel{2}}}\left(\frac{\overset{8}{\cancel{16}}}{3}\right)(23)$	$A_2 = 12 \text{ ft}^2$	$A_3 = 184 \text{ ft}^2$
$A_1 = \frac{184}{3} = 61\frac{1}{3} \text{ ft}^2$		

$$\underbrace{A_1 + 2(A_2) + A_3}_{\text{2 doubled to include } \triangle A \text{ and } \triangle C} = \text{area of gable end}$$

$$61\frac{1}{3} + 2(12) + 184 = 61\frac{1}{3} + 24 + 184 = 269\frac{1}{3} \text{ ft}^2$$

The area of the gable end of the gambrel roof is $269\frac{1}{3}$ ft^2. ■

Sometimes we may be figuring area and find that it is more convenient to calculate an overall area and subtract a smaller area. We saw this for rectangles previously when we figured the area of a room and then subtracted the area of a fireplace hearth that projects into the room. Our purpose was to find the area to be covered with carpet, so we had to exclude the area of the fireplace.

The following example is a similar case that involves composite shapes.

EXAMPLE 21-4.5 Find the area of the flat metal piece shown in Fig. 21-57.

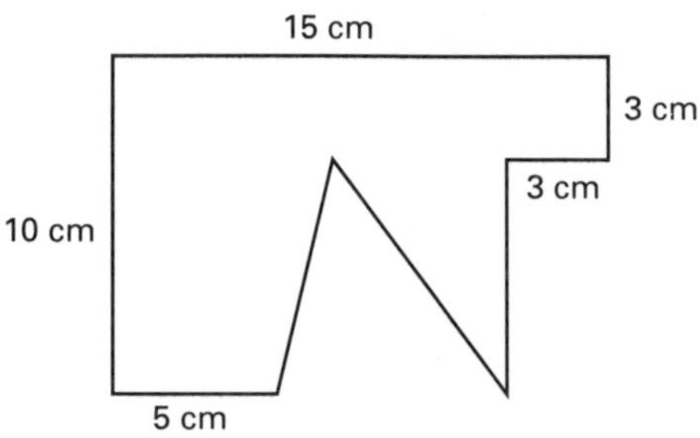

Figure 21-57

Solution

1. Divide the figure into polygons for which we can calculate areas (Fig. 21-58).

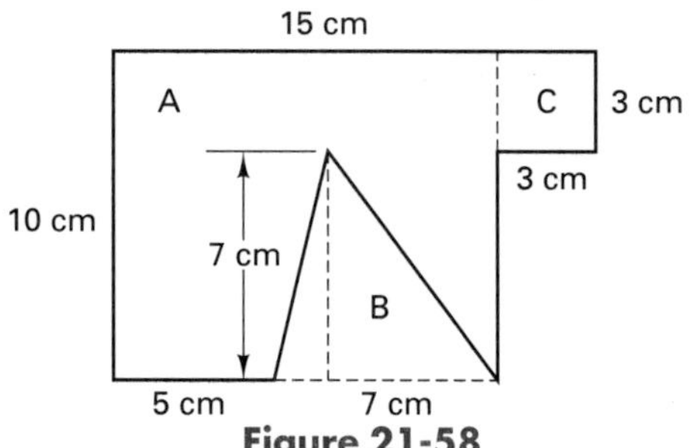

Figure 21-58

2. Find missing dimensions.
3. Find the area of square $C\ (A)_1$, rectangle $A\ (A)_2$, and triangle $B\ (A)_3$. The area of the piece of metal is $A_1 + A_2 - A_3$.

Square	Rectangle	Triangle
$A_1 = s^2$	$A_2 = lw$	$A_3 = \frac{1}{2}bh$
$A_1 = 3^2$	$A_2 = 12(10)$	$A_3 = \frac{1}{2}(7)(7)$
$A_1 = 9 \text{ cm}^2$	$A_2 = 120 \text{ cm}^2$	$A_3 = \frac{1}{2}(49)$
		$A_3 = 24.5 \text{ cm}^2$

$$\text{Area of piece} = A_1 + A_2 - A_3 = 9 + 120 - 24.5 = 104.5 \text{ cm}^2$$

The area of the flat piece of metal is 104.5 cm^2. ■

L.O.3. Find the number of degrees in each angle of a regular polygon.

Degrees in Angles of Regular Polygons

Polygons with equal sides and angles like squares are called *regular polygons.* There are several other regular polygons with definite shapes and specific names that are treated as composite shapes for calculating their areas. Let's examine the regular polygons more closely.

■ **DEFINITION 21-4.2 Regular Polygon.** A **regular polygon** is a polygon with equal sides and equal angles.

> **Rule 21-4.1** ***To find the number of degrees in each angle of a regular polygon:***
>
> Multiply the number of sides less 2 by 180° and divide by the number of sides.
>
> $$\text{Degrees per angle} = \frac{180° \text{ (number of sides } - 2)}{\text{number of sides}}$$

EXAMPLE 21-4.6 Find the number of degrees in each angle of the following regular polygons.

Triangle (3 sides)
Square (4 sides)
Pentagon (5 sides)
Hexagon (6 sides)
Octagon (8 sides)

Solution

$$\text{Triangle:} \quad \frac{180° (3 - 2)}{3} = \frac{180° (1)}{3} = 60°$$

$$\text{Square:} \quad \frac{180° (4 - 2)}{4} = \frac{180° (2)}{4} = \frac{360°}{4} = 90°$$

$$\text{Pentagon:} \quad \frac{180° (5 - 2)}{5} = \frac{180° (3)}{5} = \frac{540°}{5} = 108°$$

$$\text{Hexagon:} \quad \frac{180° (6 - 2)}{6} = \frac{180° (4)}{6} = \frac{720°}{6} = 120°$$

$$\text{Octagon:} \quad \frac{180° (8 - 2)}{8} = \frac{180° (6)}{8} = \frac{1080°}{8} = 135°$$

■

> **Rule 21-4.2**
>
> If lines are drawn from the center of a regular polygon to each vertex, congruent triangles are formed.

Figure 21-59 will show how this property applies to several different regular polygons: the equilateral triangle, square, regular pentagon, and regular hexagon. To inductively accept this property, construct various regular polygons, make cuts from the vertex to the center, and compare the resulting triangles.

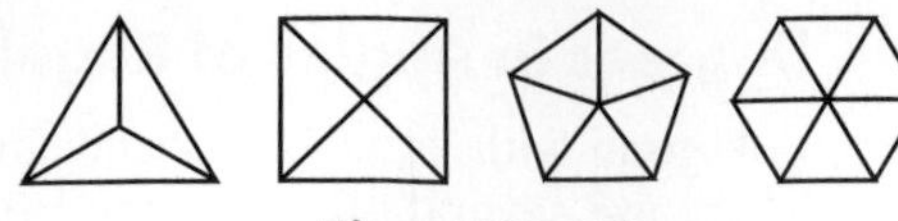

Figure 21-59

EXAMPLE 21-4.7 Find the floor area of a recreational building at a park if it forms a regular hexagon and each side is 20 ft long. The perpendicular distance from one side to the center of the building is 17.3 ft (see Fig. 21-60).

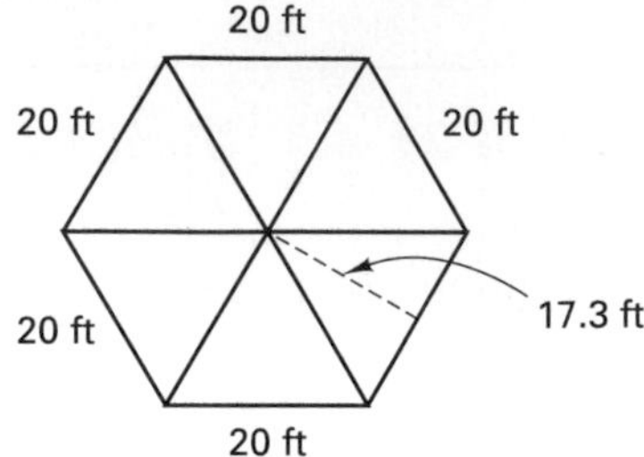

Figure 21-60

Solution

(Divide the regular hexagon into congruent triangles by connecting the center of the hexagon with each vertex of the hexagon.)

$A = \frac{1}{2}hb$ *(Find the area of one triangle.)*

$A = \frac{1}{\cancel{2}_1}(17.3)(\cancel{20}^{10})$ *(We use the known distance from the side of the hexagon to the center for the height of the triangle.)*

$A = 173 \text{ ft}^2$

$173 \times 6 = 1038 \text{ ft}^2$ *(Multiply the area of the one triangle by 6, since the hexagon has been divided into six congruent triangles.)*

The floor area of the recreational building is therefore 1038 ft^2. ■

SELF-STUDY EXERCISES 21-4

L.O.1. Find the missing dimensions x and y of the following figures.

1. **2.**

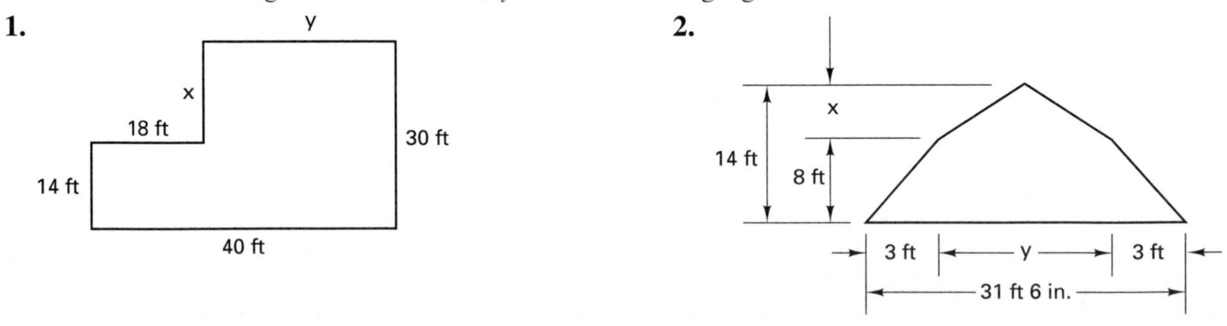

L.O.2. Find the perimeter and the area of the following figures.

3.

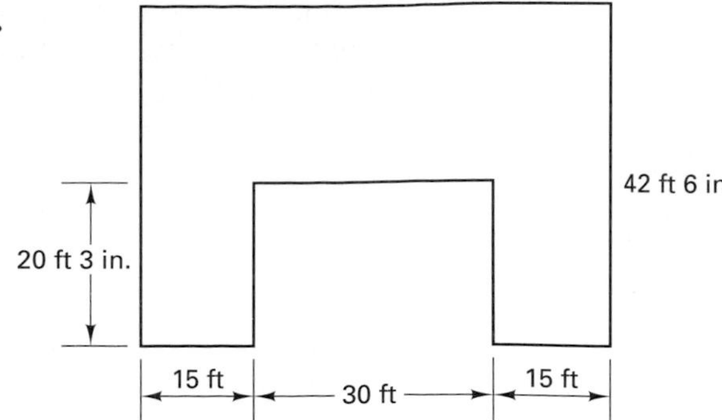

4.

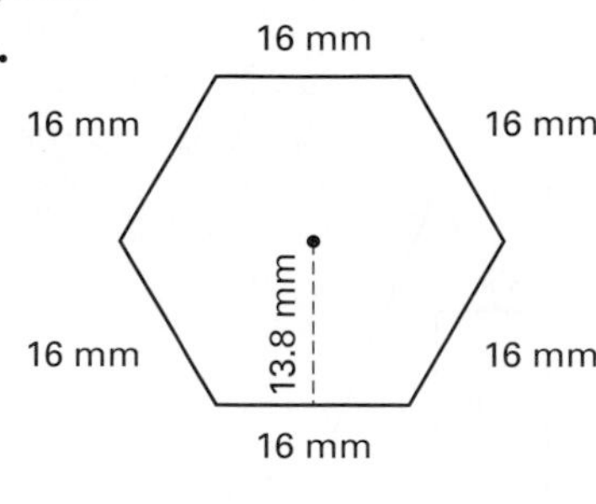

Solve the following problems involving perimeter and area of composite figures.

5. A house features a den with the layout shown in Fig. 21-61. Find the number of feet of 4-in. stock needed for the base plates around the perimeter. Make no allowances for doorways or other openings.

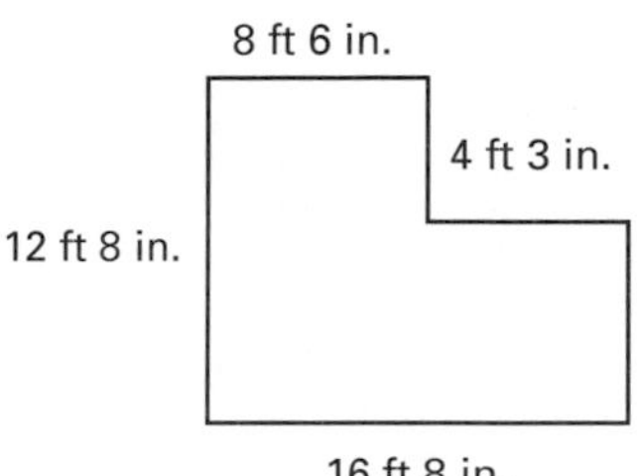

Figure 21-61

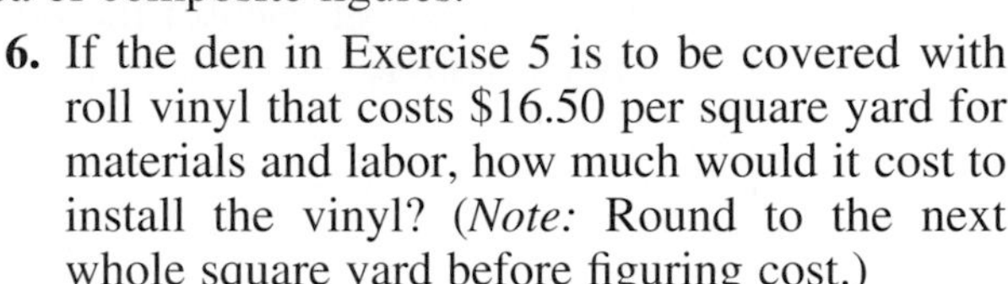

6. If the den in Exercise 5 is to be covered with roll vinyl that costs $16.50 per square yard for materials and labor, how much would it cost to install the vinyl? (*Note:* Round to the next whole square yard before figuring cost.)

7. The metal piece in Fig. 21-62 has the dimensions as indicated. If the 18-gauge steel weighs 2 lb per square foot, how much does the piece weigh to the nearest pound? (144 in.^{2} = 1 ft^2.)

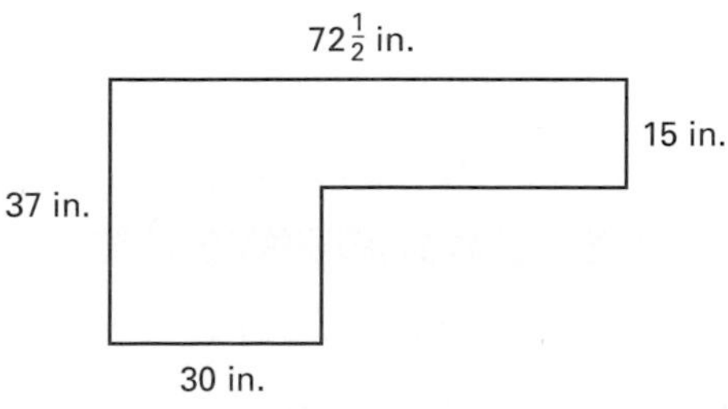

Figure 21-62

8. A swimming pool is in the form of an octagon 15 ft on a side. If the distance from the center of the pool to the midpoint of a side is 18.1 ft, what is the area of the pool? If the area is to be protected by an octagonal cover, what is the measure of each angle of the cover?

9. A No. 5 soccer ball covering is made of 12 colored pentagons 1.75 in. on a side and 20 white hexagons 3.19 in. on a side. If the distance from the midpoint of a side of a colored pentagon to the center of the pentagon is 1.20 in., how many square inches are made of the colored covering?

10. Find the area of the layout shown in Fig. 21-63.

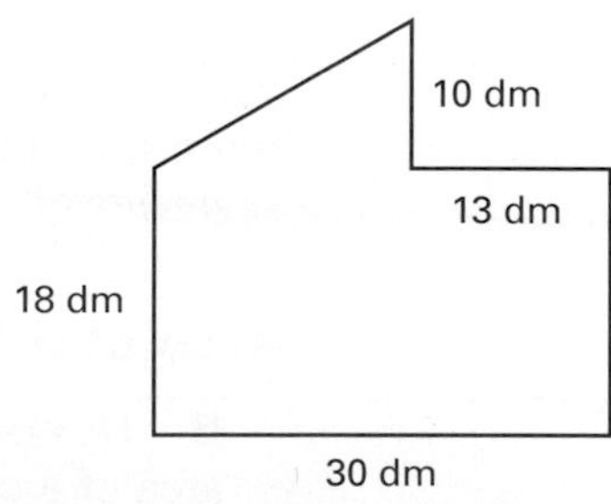

Figure 21-63

11. Find the area of the layout shown in Fig. 21-64.

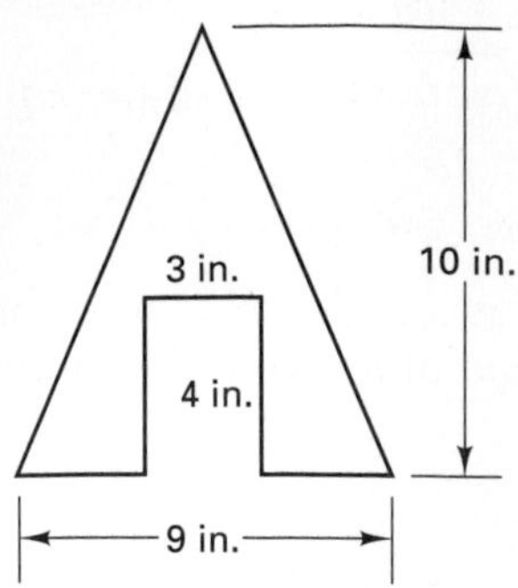

Figure 21-64

L.O.3. Give the specific name of each of the following figures and the number of degrees in each angle.

12.

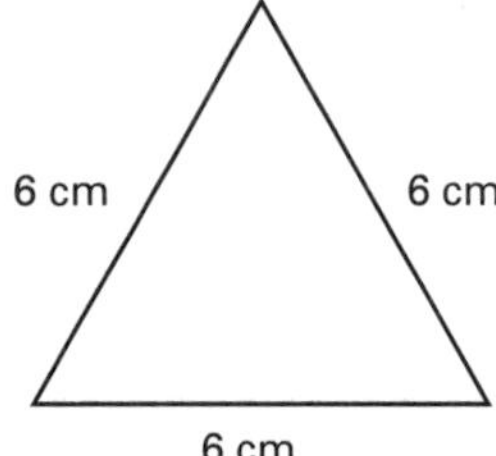

13.

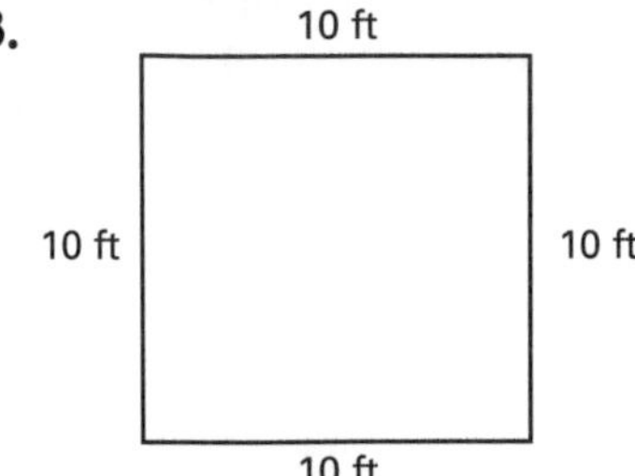

14.

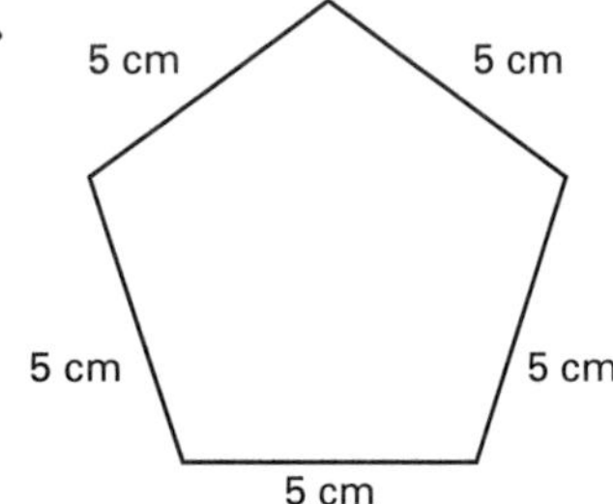

15. Identify the figure in Exercise 4 by giving its specific name and number of degrees in each angle.

21-5 SECTORS AND SEGMENTS OF A CIRCLE

Learning Objectives

L.O.1. Find the area of a sector.
L.O.2. Find the arc length of a sector.
L.O.3. Find the area of a segment.

We have already seen that we sometimes work with figures that are less than a whole circle. We saw previously, for example, the semicircle in several composite figures. In addition to the semicircle, which is half a circle, there are *sectors* and *segments* of a circle.

L.O.1. Find the area of a sector.

Area of a Sector

■ **DEFINITION 21-5.1 Sector.** A **sector** of a circle is the portion of the area of a circle cut off by two radii.

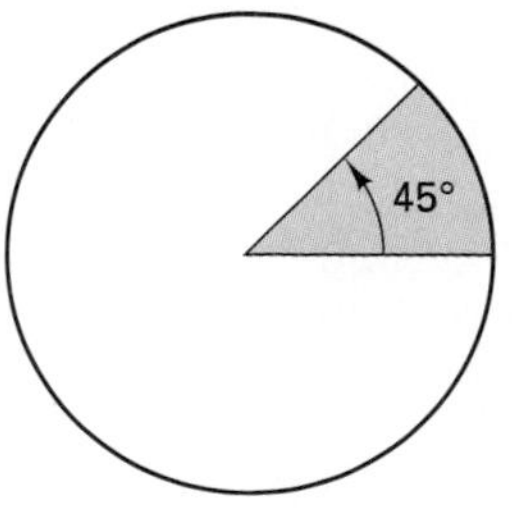

Figure 21-65

To find the area of a sector, we need to calculate how much of the circle is taken up by the sector. In Fig. 21-65, the sector takes up 45° out of the 360° of the whole circle. We can represent this fractional part of the circle as $\frac{45}{360}$. The area of this sector, then, is $\frac{45}{360}$ of the area of the circle.

If we use the Greek letter *theta* (θ) to represent the number of degrees in the angle of the sector, we can express the area of a sector in a formula:

Formula 21-5.1 *Area of a sector:* $\frac{\theta}{360}\pi r^2$

$\frac{\theta}{360}$ = fractional part of circle $\quad \pi r^2$ = area of circle

θ is measured in degrees.

EXAMPLE 21-5.1 Find the area of a sector of 45° in a circle with a radius of 10 in. Round to hundredths.

Solution

$$A = \frac{\theta}{360}\pi r^2$$

$$A = \frac{45}{360}(\pi)(10)^2 \quad \textit{(Substitute for } \theta, \pi, \textit{ and } r.)$$

$$A = 0.125\,(\pi)(100) \quad \textit{(Perform indicated operations.)}$$

$$A = 39.26990817 \text{ in.}^2$$

The area of the sector is 39.27 in.^{2}. ■

EXAMPLE 21-5.2 A cone is to be made from sheet metal. To form a cone, a sector of 40°20′ is cut from a metal circle whose diameter is 20 in. Find the area of the stretchout (portion of the circle) to be formed into the cone to the nearest hundredth (Fig. 21-66).

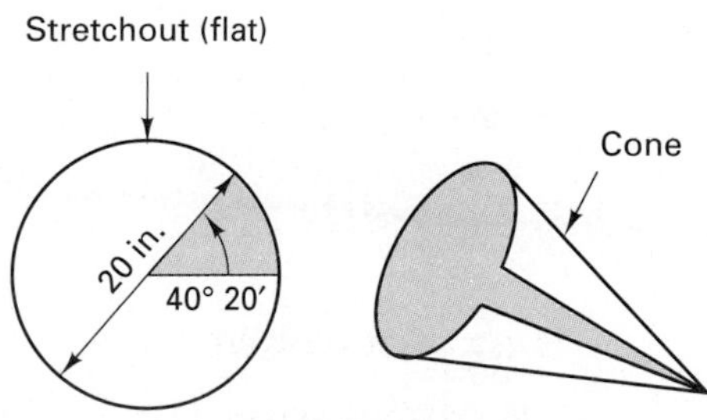

Figure 21-66

Solution

$$\text{Area used for cone} = \text{area of circle} - \text{area of sector}$$

Circle	*Sector*	
$A_1 = \pi r^2$	$A_2 = \dfrac{\theta}{360}\pi r^2$	
$A_1 = \pi(10)^2$	$A_2 = \dfrac{40.3\overline{3}}{360}(314.1592654)$	*(Convert 40° 20′ to 40.333333°. πr^2 has already been figured as 314.1592654 in.².)*
$A_1 = \pi(100)$	$A_2 = 0.112037037(314.1592654)$	
$A_1 = 314.1592654 \text{ in.}^2$	$A_2 = 35.19747324 \text{ in.}^2$	

$\text{Area used for cone} = A_1 - A_2$

$A_3 = 314.1592654 - 35.19747324$

$A_3 = 278.96 \text{ in.}^2$ *(rounded)*

The area of the metal sector used to form the cone is 278.96 in.².

L.O.2. Find the arc length of a sector.

Arc Length of a Sector

Besides finding the area of a sector, we sometimes need to find the arc length of a sector. The *arc length* is the portion of the circumference intercepted by the sides of the sector (See Fig. 21-67). A formal definition of *arc* is given in Definition 21-5.3.

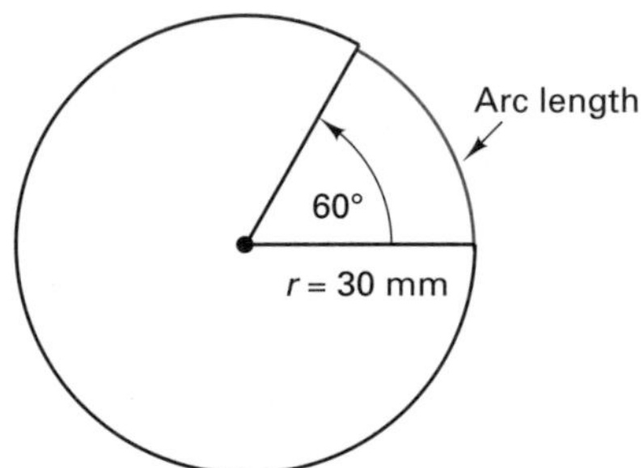

Figure 21-67

> **Formula 21-5.2** ***Arc length:***
>
> $$s = \frac{\theta}{360}(2\pi r) \quad \text{or} \quad s = \frac{\theta}{360}(\pi d)$$
>
> θ is measured in degrees.

EXAMPLE 21-5.3 Find the arc length of the sector formed by a 60′ central angle if the radius is 30 mm (Fig. 21-67).

Solution

$s = \dfrac{\theta}{360}(2\pi r)$

$s = \dfrac{60}{360}(2)(\pi)(30)$ *(Substitute values.)*

$s = 31.41592654$

$s = 31.41 \text{ mm}$ *(rounded)*

L.O.3. Find the area of a segment.

Area of a Segment

If a line segment (called a *chord*) joins the end points of the radii that form a sector, the sector is divided into two figures, a triangle and a *segment* (Fig. 21-68).

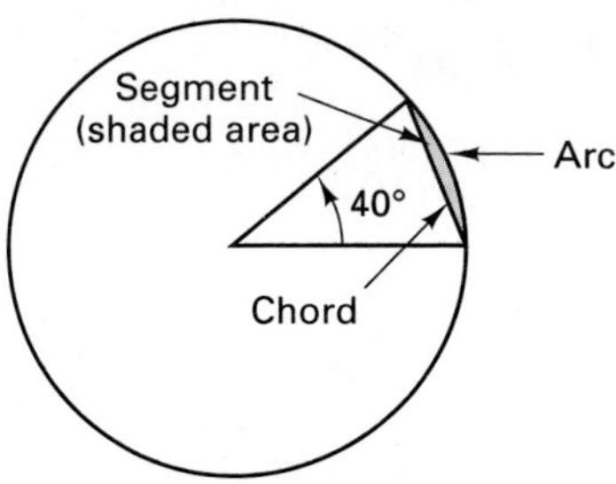

Figure 21-68

■ **Definition 21-5.2 Chord.** A **chord** is a line segment joining two points on the circumference of a circle.

■ **Definition 21-5.3 Arc.** The portion of the circumference cut off by a chord is an **arc**.

■ **Definition 21-5.4 Segment.** A **segment** is the portion of the area of a circle bounded by a chord and an arc.

Since the chord divides the sector into a triangle and a segment, we can calculate the area of the segment by subtracting the area of the triangle from the area of the sector. Expressed symbolically, we have the formula for the area of a segment:

Formula 21-5.3 *Area of a segment:* $A = \frac{\theta}{360}\pi r^2 - \frac{1}{2}bh$

$\frac{\theta}{360}\pi r^2$ = area of sector $\quad \frac{1}{2}bh$ = area of triangle

Example 21-5.4 Find the area of the segment in circle A of Fig. 21-69 to the nearest hundredth.

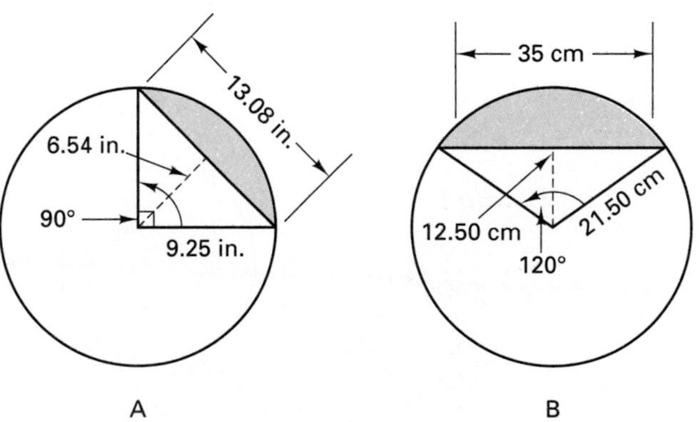

Figure 21-69

Solution

$$A = \frac{\theta}{360}\pi r^2 - \frac{1}{2}bh$$

$$A = \frac{90}{360}(\pi)(9.25)^2 - \frac{1}{2}(13.08)(6.54) \qquad \textit{(Substitute in formula.)}$$

$$A = 67.20063036 - 42.7716$$

$$A = 24.43 \text{ in.}^2 \qquad \textit{(rounded)}$$

The area of the segment is 24.43 in.^{2}. ■

EXAMPLE 21-5.5 A segment of circle B in Fig. 21-69 is removed so that a template for a cam can be made from the rest of the circle. What is the area of the template? Give answer to the nearest hundredth.

Solution

Area of template = area of circle − area of segment

Circle	*Segment*
$A_1 = \pi r^2$	$A_2 = \frac{\theta}{360}\pi r^2 - \frac{1}{2}bh$
$A_1 = \pi(21.50)^2$	$A_2 = \frac{120}{360}(1452.201204) - \frac{1}{2}(35)(12.50)$
$A_1 = 1452.201204 \text{ cm}^2$	$A_2 = 265.317068 \text{ cm}^2$

$$\text{Area of template} = A_1 - A_2$$

$$A_3 = 1452.201204 - 265.317068$$

$$A_3 = 1186.88 \text{ cm}^2 \qquad \textit{(rounded)}$$

The area of the template is 1186.88 cm^2. ■

SELF-STUDY EXERCISES 21-5

L.O.1. Find the area of the sectors of a circle using Fig. 21-70. Round to hundredths.

1. $\angle = 54°$
$r = 16$ cm

2. $\angle = 25°16'$
$r = 30$ mm

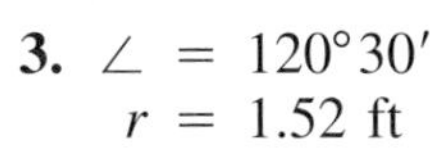

3. $\angle = 120°30'$
$r = 1.52$ ft

4. $\angle = 65°$
$r = 5$ in.

5. $\angle = 150°$
$r = 1.45$ m

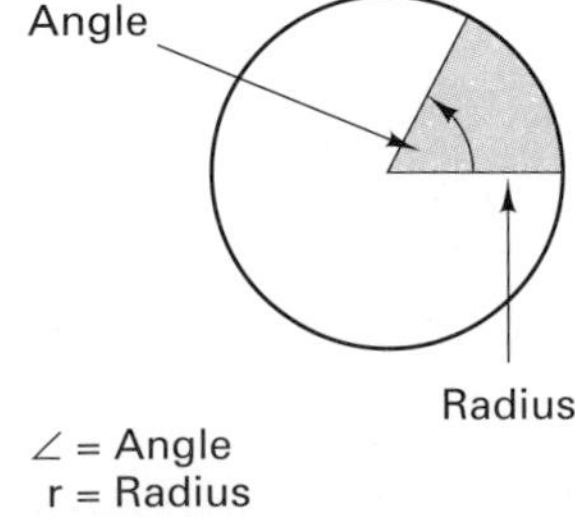

Figure 21-70

6. The library of a contemporary elementary school is circular (Fig. 21-71). The floor plan includes pie-shaped areas reserved for science materials, literary materials, reference materials, and so on. Find the area of the reference section.

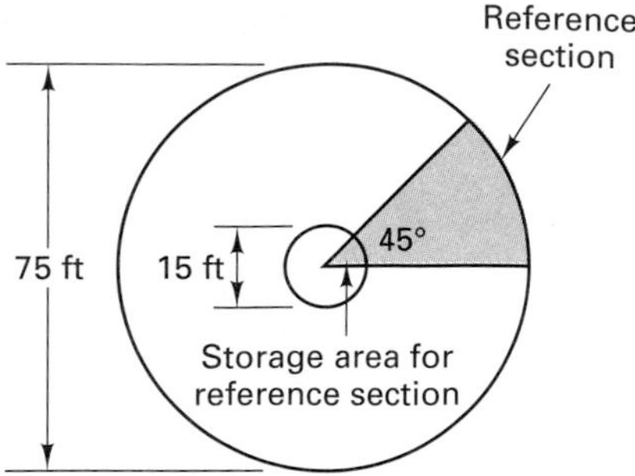

Figure 21-71

7. A mason lays a tile mosaic featuring a four-sector design (Fig. 21-72). What is the area of the design to the nearest hundredth?

60°

6 in.

Figure 21-72

8. If the angle formed by a pendulum swing is 19°30′ and the pendulum is 10 in. long (Fig. 21-73), what area does the pendulum cut as it swings? Express the answer to the nearest tenth. (60′ = 1°.)

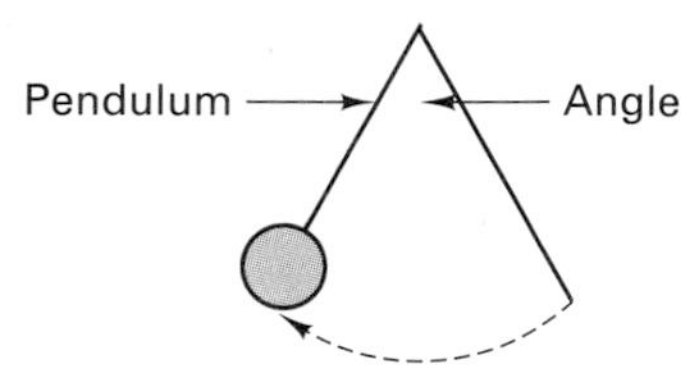

Figure 21-73

L.O.2. Find the arc length of the sectors of a circle using Fig. 21-70. Round to hundredths.

9. $\angle = 54°$
$r = 30$ mm

10. $\angle = 150°$
$r = 5$ in.

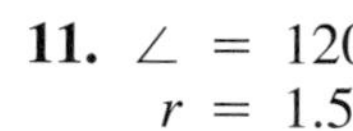

11. $\angle = 120°30'$
$r = 1.52$ ft

12. $\angle = 25°16'$
$r = 16$ cm

13. $\angle = 65°$
$r = 1.45$ m

14. $\angle = 90°$
$r = 10$ in.

L.O.3. Find the area of the segments of a circle using Fig. 21-74. Round to hundredths.

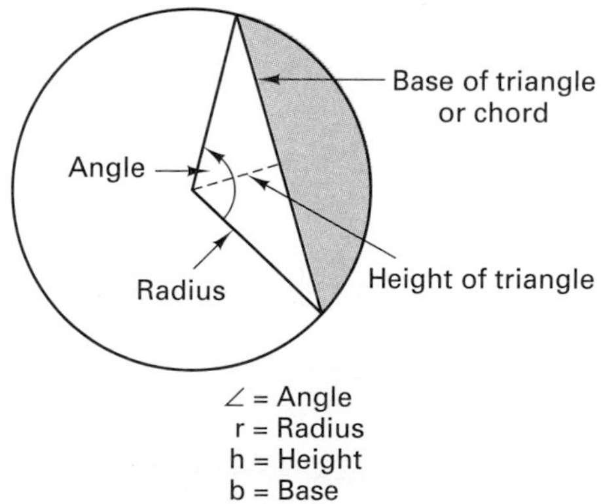

Figure 21-74

15. $\angle = 60°$
$r = 13.3$ cm
$h = 11.52$ cm
$b = 13.3$ cm

16. $\angle = 110°$
$r = 10$ in.
$h = 5.74$ in.
$b = 16.38$ in.

17. $\angle = 105°$
$r = 11.25$ in.
$h = 6.85$ in.
$b = 17.85$ in.

18. $\angle = 60°$
$r = 24$ cm
$h = 20.8$ cm
$b = 24$ cm

19. $\angle = 108°$
$r = 14$ in.
$h = 8.25$ in.
$b = 22.65$ in.

20. A motor shaft has milled on it a flat for a set-screw to rest in order to hold a pulley on the shaft (Fig. 21-75). What is the cross-sectional area of the shaft after being milled? Round to hundredths.

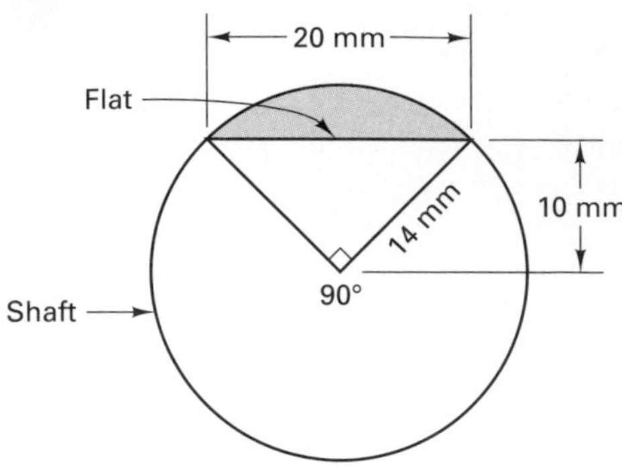

Figure 21-75

21. A contractor pours a concrete patio in the shape of a circle except where the patio touches the exterior wall of the house (Fig. 21-76). What is the area of the patio in square feet? Round to hundredths.

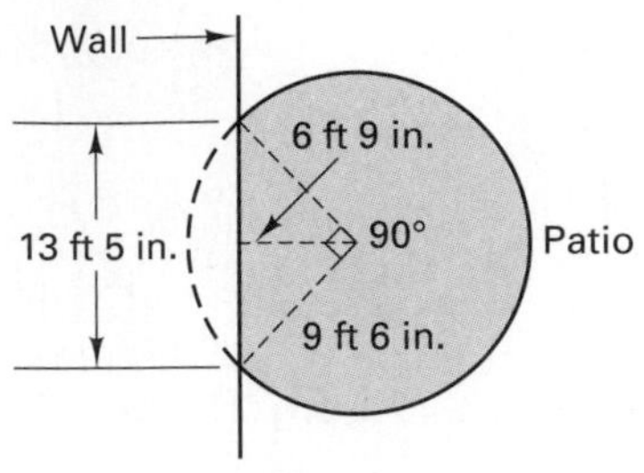

Figure 21-76

21-6 INSCRIBED AND CIRCUMSCRIBED REGULAR POLYGONS AND CIRCLES

Learning Objectives

L.O.1. Use the properties of inscribed and circumscribed equilateral triangles to find missing amounts and to solve applied problems.

L.O.2. Use the properties of inscribed and circumscribed squares to find missing amounts and to solve applied problems.

L.O.3. Use the properties of inscribed and circumscribed hexagons to find missing amounts and to solve applied problems.

Previously, we studied polygons and their areas and perimeters. We also briefly looked at regular polygons, that is, polygons whose sides and angles are equal. In this section we study regular polygons *inscribed* in a circle and *circumscribed* about a circle.

■ **DEFINITION 21-6.1 Inscribed Polygon.** A polygon is **inscribed** *in a circle* when it is inside the circle and all its *vertices* (points where sides of each angle meet) are on the circle. The circle is said to be *circumscribed about the polygon* (Fig. 21-77).

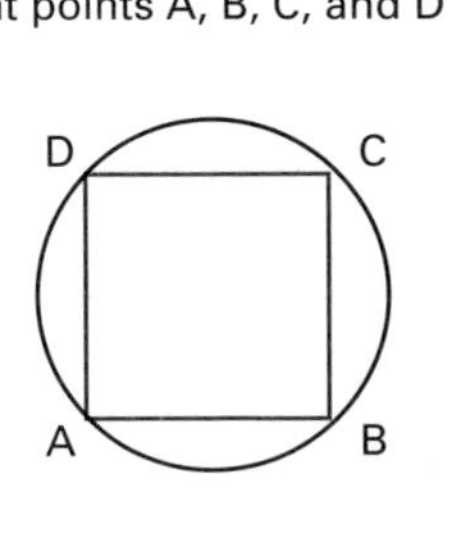

Figure 21-77

■ **Definition 21-6.2 Circumscribed Polygon.** A polygon is **circumscribed** *about a circle* when it is outside the circle and all its sides are *tangent* to (intersecting in exactly one point) the circumference. The circle is said to be *inscribed in the polygon* (Fig. 21-77).

In this section we examine the three most common polygons that are inscribed in or circumscribed about a circle. These are the triangle, the square, and the hexagon.

L.O.1. Use the properties of inscribed and circumscribed equilateral triangles to find missing amounts and to solve applied problems.

Equilateral Triangles

An equilateral triangle can be inscribed in a circle or circumscribed about a circle. If the height (or altitude) is drawn, two congruent right triangles are formed, as shown in Fig. 21-78. Each congruent triangle formed in this way is a 30°, 60°, 90° right triangle in which the hypotenuse is twice the shortest side. The special 30°, 60°, 90° right triangle and its properties may be reviewed in Section 21-3.

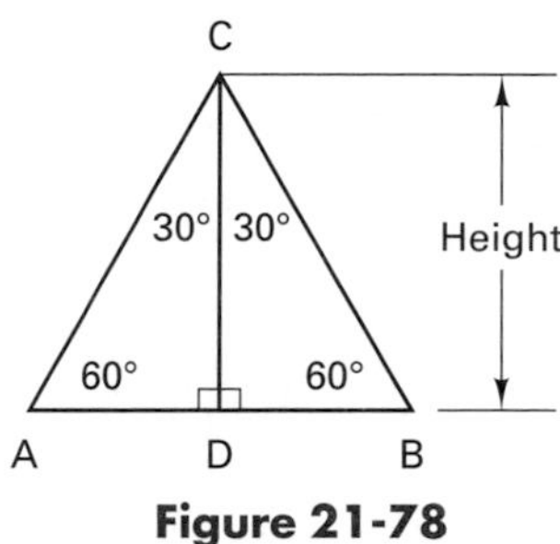

Figure 21-78

Examine Fig. 21-79. When the equilateral triangle is inscribed in a circle, the radius (from the vertex of the triangle to the center of the circle) is two-thirds the height of the triangle. When the equilateral triangle is circumscribed about a circle, the radius (from the center of the circle to the base of the triangle) is one-third the height of the triangle.

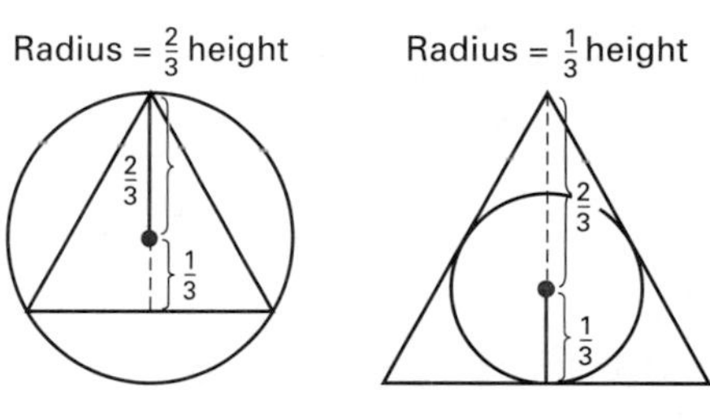

Figure 21-79

Rule 21-6.1

The height of an equilateral triangle forms congruent 30°, 60°, 90° right triangles.

Rule 21-6.2

The radius of a circle circumscribed about an equilateral triangle is two-thirds the height of the triangle or twice the distance from the center of the circle to the base of the triangle.

Rule 21-6.3

The radius of a circle inscribed in an equilateral triangle is one-third the height of the triangle or one-half the distance from the vertex of the triangle to the center of the circle.

Example 21-6.1 Find the following dimensions for the inscribed equilateral triangle in Fig. 21-80, where $CD = 13$ mm and $AO = 15$ mm.

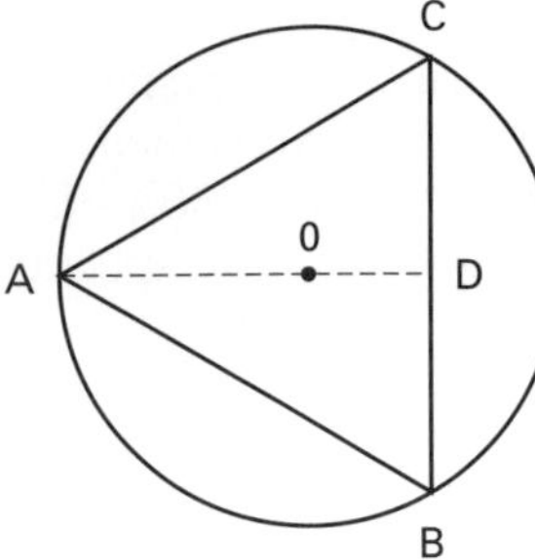

Figure 21-80

(a) Height of $\triangle ABC$ **(b)** BC **(c)** BD **(d)** $\angle CAD$
(e) $\angle ACD$ **(f)** Area of $\triangle ABC$

Solution

(a)
$$AO = \frac{2}{3}AD \qquad \textit{(AO = two thirds of height)}$$
$$15 = \frac{2}{3}AD \qquad \textit{(Substitute.)}$$
$$\frac{3}{2}(15) = \left(\frac{3}{2}\right)\frac{2}{3}AD \qquad \textit{(Solve for AD.)}$$
$$\frac{45}{2} = AD$$
$$22.5 = AD$$

Or, using the formula for finding a leg of a 30°, 60°, 90° triangle (Section 21-3),

$$AD = CD\sqrt{3}$$
$$AD = 13\sqrt{3}$$
$$AD = 22.5 \text{ mm} \qquad \textit{(rounded)}$$

(b) $BC = 2CD$ *(30°, 60°, 90° $\triangle$)*
$BC = 26$ mm

(c) $BD = 13$ mm ($\triangle ADC \cong \triangle ADB$)

(d) $\angle CAD = 30°$

(e) $\angle ACD = 60°$

(f)
$$A = \frac{1}{2}bh$$
$$A = \frac{1}{2}(BC)(AD)$$
$$A = \frac{1}{2}(26)(22.5)$$
$$A = 292.5 \text{ mm}^2$$

■

EXAMPLE 21-6.2 One end of a shaft 35 mm in diameter is milled as shown in Fig. 21-81. Find the radius of the smaller circle.

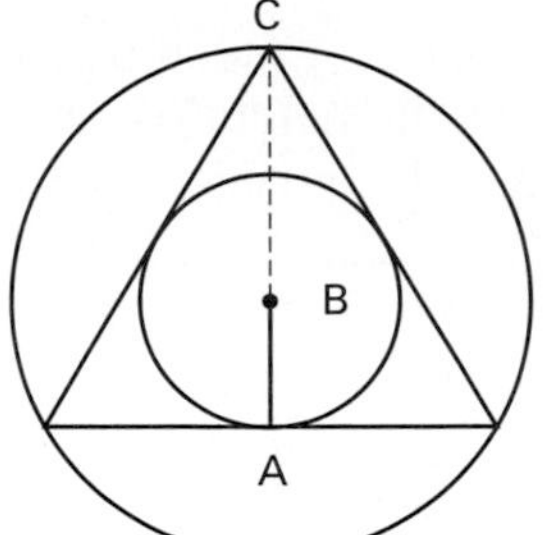

Figure 21-81

Solution

$$BC = \frac{1}{2} \text{ diameter of large circle}$$

$$BC = \frac{1}{2}(35)$$

$$BC = 17.5 \text{ mm}$$

$$AB = \frac{1}{2}BC$$

$$AB = \frac{1}{2}(17.5)$$

$$AB = 8.75 \text{ mm}$$

Thus, the radius of the smaller circle is 8.75 mm. ■

EXAMPLE 21-6.3 Find the area of the larger circle in Fig. 21-81. Round to hundredths.

Solution

$$A = \pi r^2 = \pi(17.5)^2 = 962.11 \text{ mm}^2 \qquad (rounded)$$

Therefore, the area of the larger circle is 962.11 mm^2. ■

L.O.2. Use the properties of inscribed and circumscribed squares to find missing amounts and to solve applied problems.

Squares

A square has four equal sides and four equal angles, each 90°. If a diagonal divides the square, the result is two congruent right triangles whose angles are 45°, 45°, and 90°. A second diagonal results in a division of the square into four 45°, 45°, 90° congruent right triangles as shown in Fig. 21-82. The special 45°, 45°, 90° right triangle and its properties may be reviewed in Section 21-3. Diagonals *AC* and *BD* are equal in length and bisect each other, that is, divide each other equally. Point *O*, where the diagonals intersect, is the center of the square and the center of any inscribed or circumscribed circle (Fig. 21-83).

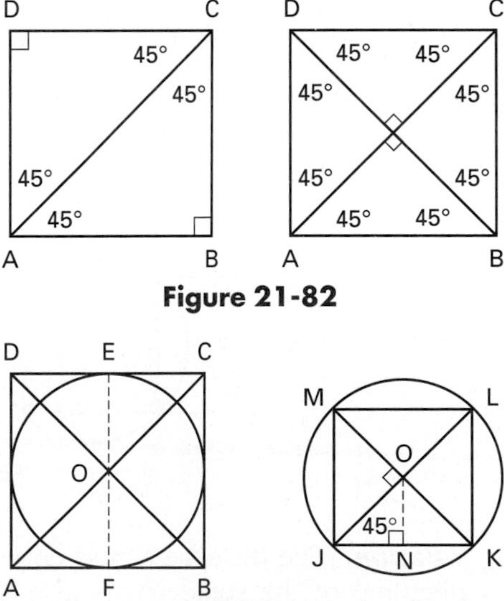

Figure 21-82

Figure 21-83

When the circle is inscribed in the square, the diameter (such as EF) equals the length of a side. The radius (such as FO) equals the height of a 45°, 45°, 90° right triangle formed by the two diagonals.

When the circle is circumscribed about the square, the diameter (such as LJ) equals the diagonal of the square. The radius (such as KO) is the height of a 45°, 45°, 90° right triangle (such as $\triangle JKL$) formed by one diagonal.

Rule 21-6.4

The diagonals of a square inscribed in or circumscribed about a circle form congruent 45°, 45°, 90° right triangles.

Rule 21-6.5

The diameter of a circle inscribed in a square equals a side of the square.

Rule 21-6.6

The diameter of a circle circumscribed about a square equals a diagonal of the square.

Rule 21-6.7

The radius of a circle inscribed in a square equals the height of a 45°, 45°, 90° right triangle formed by two diagonals.

Rule 21-6.8

The radius of a circle circumscribed about a square equals the height of a 45°, 45°, 90° right triangle formed by one diagonal.

EXAMPLE 21-6.4 To mill a square bolt head $1\frac{1}{2}$ in. across flats, round stock of what diameter would be needed? Answer to the nearest hundredth (Fig. 21-84).

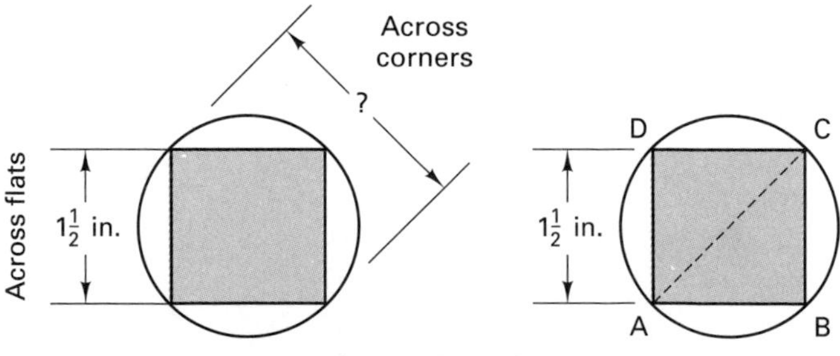

Figure 21-84

Solution: The distance across corners is the diameter of the circle as well as the diagonal of the square. It is also the hypotenuse of the two 45°, 45°, 90° right triangles formed by one diagonal. Use the property of 45°, 45°, 90° right triangles to find the hypotenuse (Rule 21-3.5).

$$\text{Hypotenuse} = \text{leg}\sqrt{2}$$

$$\text{Hypotenuse} = 1.5\sqrt{2} \qquad \left(1\frac{1}{2}\ in. = 1.5\ in.\right)$$

$$\text{Hypotenuse} = 2.12 \text{ in.} \qquad (rounded)$$

Therefore, the diameter of the round stock is 2.12 in. ■

Example 21-6.5 Find the diagonal of a square if the area of the circumscribed circle is 22 in.^{2}. Round to hundredths (Fig. 21-85).

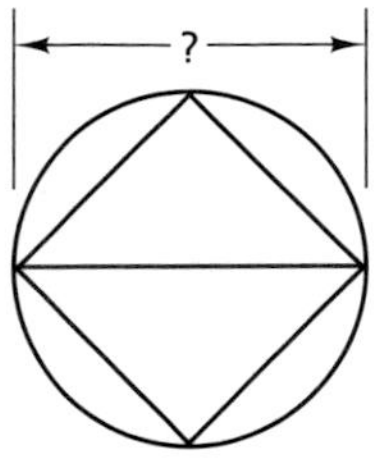

Figure 21-85

Solution

$$A = \pi r^2$$

$$22 = \pi r^2 \qquad (Substitute.)$$

$$\frac{22}{\pi} = r^2$$

$$\sqrt{\frac{22}{\pi}} = r$$

$$2.646283714 = r$$

$$\text{Diagonal} = \text{diameter} = 2 \times \text{radius} = 2(2.646283714) = 5.29 \text{ in.} \qquad (rounded)$$

The diagonal of the inscribed square is 5.29 in. ■

Example 21-6.6 What is the minimum depth of cut required to mill a circle on the end of a square piece of stock whose side is 4 cm? Answer to hundredths (Fig. 21-86).

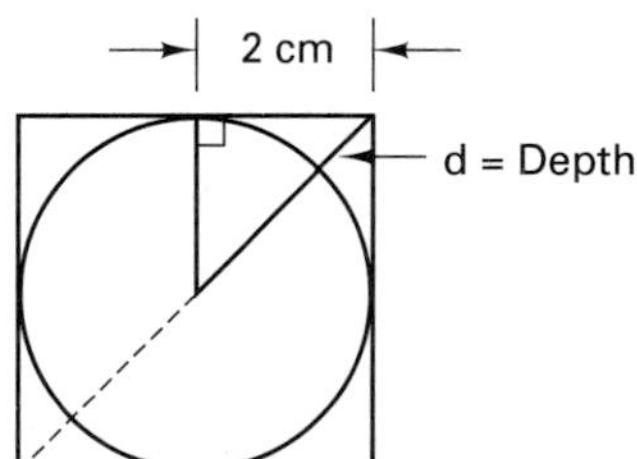

Figure 21-86

Solution: Since the side of the square is 4 cm, the diameter of the circle is 4 cm and the radius is 2 cm. Also, if the radius is drawn perpendicular to the top side, a right triangle is formed as indicated.

Let d = depth of milling, which equals the hypotenuse of the small triangle minus the radius of the circle. Using the Pythagorean theorem, let c = hypotenuse.

$$c^2 = a^2 + b^2$$

$$c^2 = 2^2 + 2^2$$

$$c^2 = 4 + 4$$

$$c^2 = 8$$

$$c = \sqrt{8}$$

$$c = 2.828427125 \text{ cm}$$

Hypotenuse − radius = depth of milling.

$$2.828427125 - 2 = 0.83 \text{ cm} \qquad (rounded)$$

The minimum depth of milling should be 0.83 cm. ■

L.O.3. Use the properties of inscribed and circumscribed hexagons to find missing amounts and to solve applied problems.

Hexagons

A regular hexagon is a figure with six equal sides and six equal angles of 120° each. If three diagonals joining pairs of opposite vertices are drawn, they divide the hexagon into six congruent equilateral triangles (all angles 60°) (see Fig. 21-87).

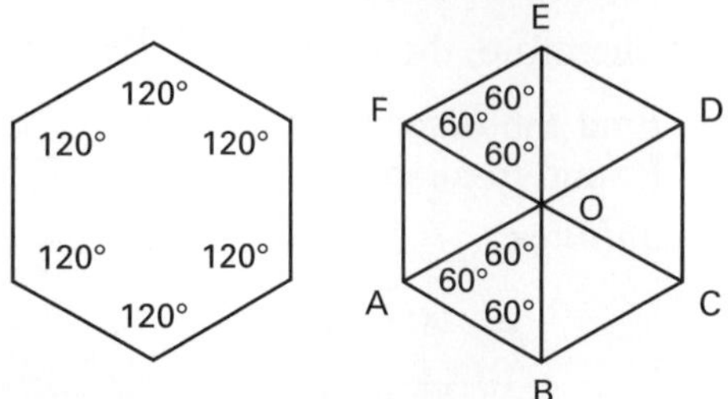

Figure 21-87

In Fig. 21-88, for the inscribed circle the height (such as GO) is the radius, and for the circumscribed circle any side of the equilateral triangles (such as AO) equals the radius. The height divides any of the six triangles into two congruent 30°, 60°, 90° right triangles.

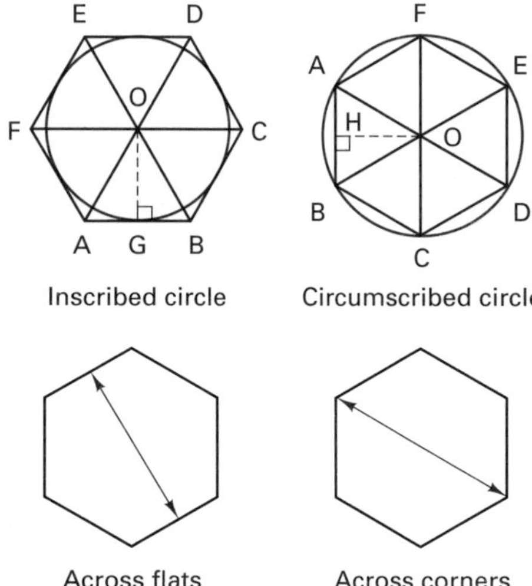

Figure 21-88

> **Rule 21-6.9**
>
> The three diagonals that join opposite vertices of a regular hexagon form six congruent equilateral triangles.

> **Rule 21-6.10**
>
> The height of any of the equilateral triangles of Rule 21-6.9 forms two congruent 30°, 60°, 90° right triangles.

> **Rule 21-6.11**
>
> The diameter of a circle circumscribed about a regular hexagon is a diagonal of the hexagon (**distance across corners**).

Rule 21-6.12

The radius of a circle circumscribed about a regular hexagon is a side of an equilateral triangle formed by the three diagonals or one-half the distance across corners.

Rule 21-6.13

The radius of a circle inscribed in a regular hexagon is the height of an equilateral triangle formed by the three diagonals.

Rule 21-6.14

The diameter of a circle inscribed in a regular hexagon is the **distance across flats** or twice the height of an equilateral triangle formed by the three diagonals.

EXAMPLE 21-6.7 Find the following dimensions for Fig. 21-89. When appropriate, round to hundredths.

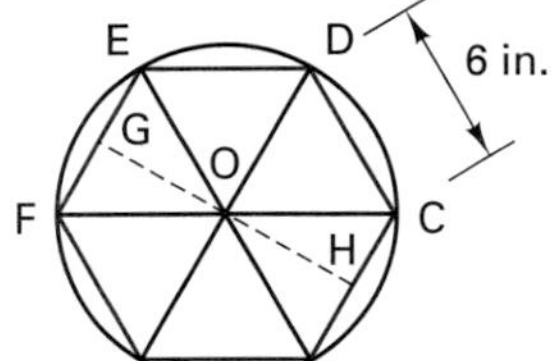

Figure 21-89

(a) $\angle EOD$ **(b)** $\angle COH$ **(c)** $\angle BHO$
(d) Radius **(e)** Distance across corners **(f)** FO

Use Fig. 21-90. Follow the preceding directions.

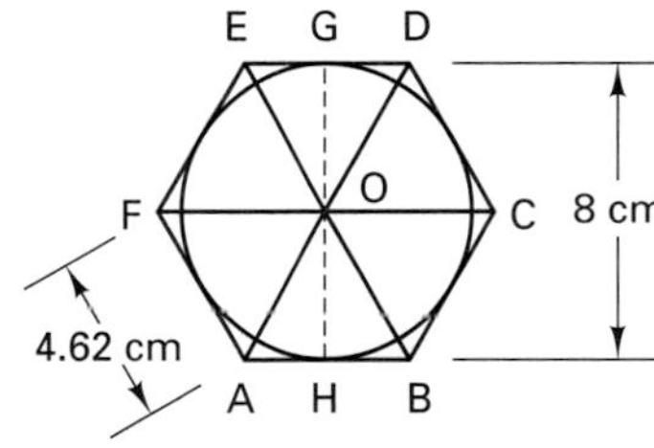

Figure 21-90

(g) Distance across flats **(h)** EO
(i) Diagonal **(j)** $\angle FED$

Solution

(a) $\angle EOD = 60°$ *(This is an angle of an equilateral triangle.)*

(b) $\angle COH = 30°$ *(Height bisects 60° angle to form 30°, 60°, 90° triangle.)*

(c) $\angle BHO = 90°$ *(Height forms right angle with base.)*

(d) Radius = 6 in. *(Radius is a side of triangle.)*

(e) Distance across corners = 12 in. *(Distance across corners is the diameter.)*

(f) FO = 6 in. *(This is a side of an equilateral triangle.)*

(g) Distance across flats = 8 cm *(This is the diameter of the inscribed circle.)*

(h) $EO = 4.62$ cm — *(An equilateral triangle has equal sides.)*

(i) 9.24 cm — *(Diagonal is twice the side.)*

(j) $\angle FED = 120°$ — *(This is the angle at the vertex of a hexagon.)* ■

EXAMPLE 21-6.8 If a regular hexagon is cut from a circle whose radius is 6.45 in., how much of the circle is not used? Round to hundredths (Fig. 21-91).

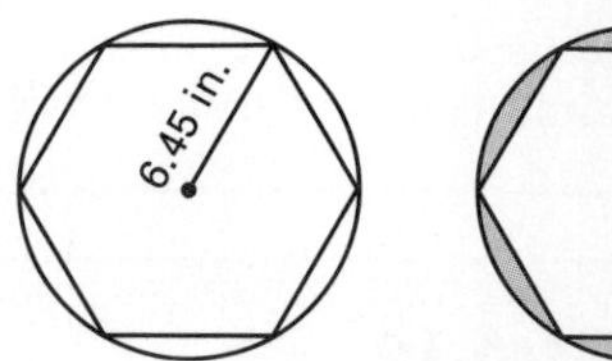

Figure 21-91

Solution: To solve, we find the area of the circle and the area of the hexagon and subtract to find the difference (portion not used).

$$A_1 = \pi r^2$$

$$A_1 = \pi(6.45)^2$$

$$A_1 = \pi(41.6025)$$

$$A_1 = 130.6981084 \text{ in.}^2$$

$$A_2 = 6\left(\frac{1}{2}bh\right)$$

(Since the height of an equilateral triangle forms a 30°, 60°, 90° triangle, $h = \frac{b}{2}\sqrt{3} = 3.225\sqrt{3} = 5.585863854$ *in.)*

$$A_2 = 6\left[\frac{1}{2}(6.45)(5.585863854)\right]$$

$$A_2 = 108.0864656 \text{ in.}^2$$

$$\text{Waste} = A_1 - A_2$$

$$\text{Waste} = 22.61 \text{ in.}^2$$

(rounded)

There would be 22.61 in.^{2} of waste from the circle. ■

EXAMPLE 21-6.9 **(a)** What is the smallest-diameter round stock that a hex-bolt head $\frac{1}{4}$ in. on a side can be milled from (Fig. 21-92)?

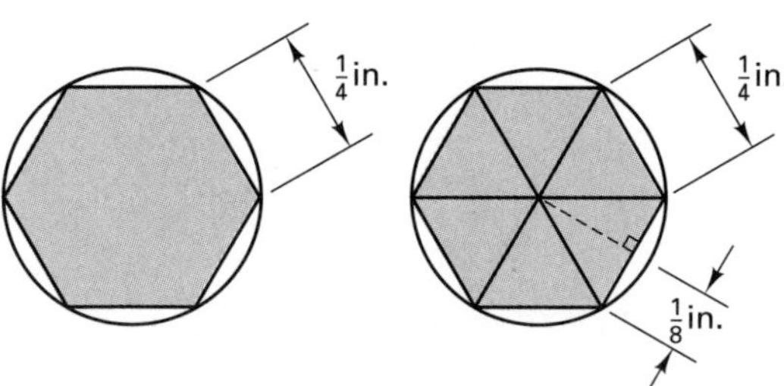

Figure 21-92

(b) What is the distance across the corners of the hex-bolt head?
(c) What is the distance across the flats?

Round answers to hundredths.

Solution

(a) Smallest-diameter round stock $= \frac{1}{2}$ in.

Diagonal $= 2 \times$ side $=$ diameter $\quad \left(2 \times \frac{1}{4} = \frac{1}{2}.\right)$

(b) Distance across corners $= \frac{1}{2}$ in.

Diagonal $=$ distance across corners

(c) Draw diagonals to form equilateral triangles. Insert the height and known dimensions. Use the properties of a 30°, 60°, 90° right triangle.

$a = b\sqrt{3}$ *(a = height of 30°, 60°, 90° triangle; b = base of 30°, 60°, 90° triangle.)*

$a = \frac{1}{8}\sqrt{3}$

$a = 0.21650635$ in.

Distance across flats $= 2 \times$ height $= 2 \times 0.21650635 = 0.43$ in. *(to nearest hundredth)*

The distance across the flats is 0.43 in. ■

SELF-STUDY EXERCISES 21-6

L.O.1. Use Fig. 21-93 to find the following dimensions for an equilateral triangle inscribed in a circle. Round to hundredths.

1. $\angle CAB$ **2.** $\angle BCD$ **3.** $\angle ADC$
4. CB **5.** AC **6.** Diameter
7. Area of $\triangle ABC$ **8.** AD **9.** Radius
10. Circumference

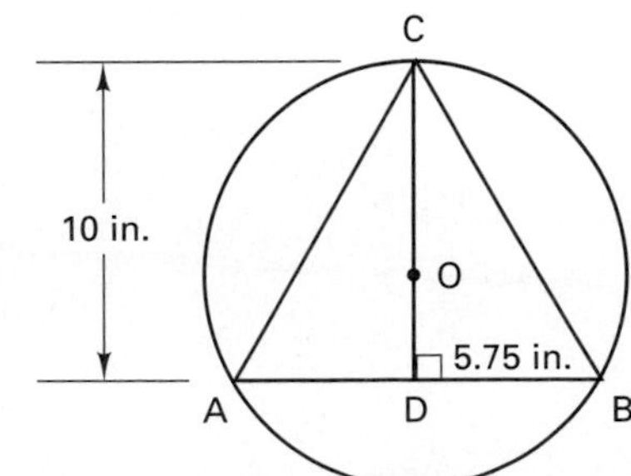

Figure 21-93

Solve the following problems involving inscribed or circumscribed equilateral triangles. Round to hundredths.

11. The area of an equilateral triangle with a base of 10 cm is 43.5 cm^2. The triangle is circumscribed about a circle. What is the radius of the inscribed circle?

12. One end of a circular steel rod is milled into an equilateral triangle whose height is $\frac{3}{4}$ in. What is the diameter of the rod if the smallest diameter possible for the job was used?

L.O.2. Find the measures of the inscribed square in Fig. 21-94.

13. $\angle DOC$ **14.** $\angle BCO$ **15.** BO

16. $\angle BOE$ **17.** CE

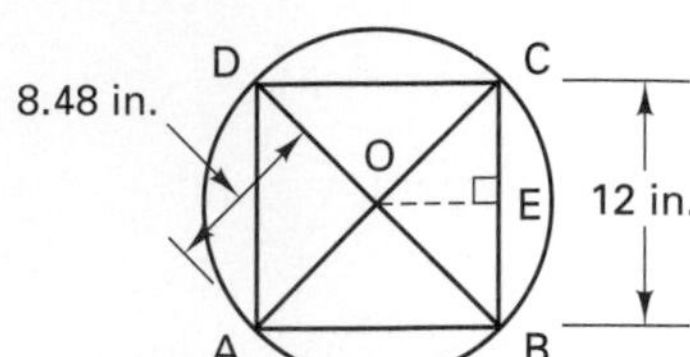

Figure 21-94

Find the measures of the circumscribed square in Fig. 21-95.

18. DO **19.** Radius of circle **20.** $\angle AOB$

21. BE

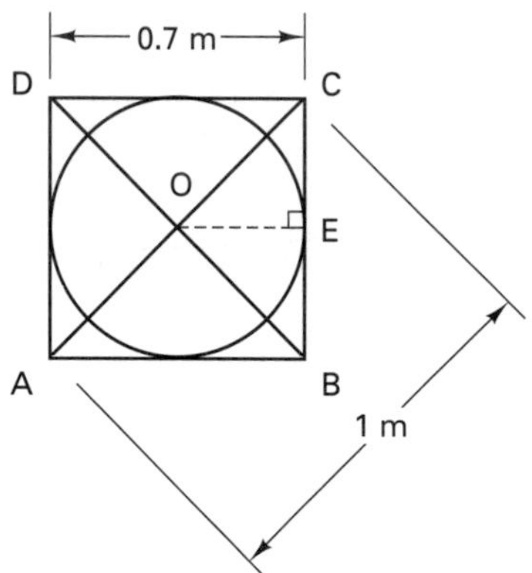

Figure 21-95

Solve the following problems involving squares and circles. Round the answers to hundredths.

22. To what depth must a 2-in. shaft (Fig. 21-96) be milled on an end to form a square 1.414 in. on a side?

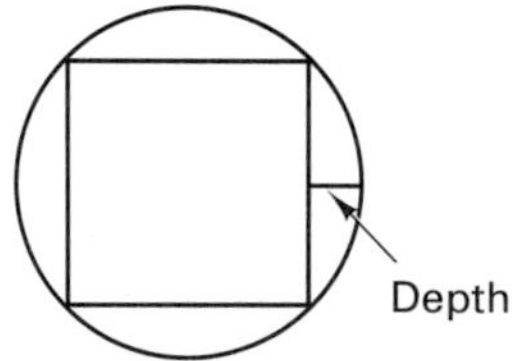

Figure 21-96

23. What is the smallest-diameter round stock (Fig. 21-97) needed to mill a square $\frac{3}{4}$ in. on a side on the end of the stock?

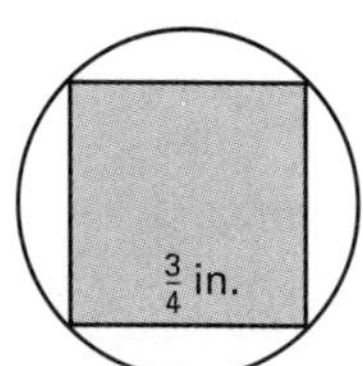

Figure 21-97

24. What is the area of the largest circle that is inscribed in a square with a perimeter of 36 dkm?

L.O.3. Find the following dimensions for Fig. 21-98 if $GO = 17.32$ mm and $FO = 20$ mm.

25. Distance across flats
26. $\angle CDE$
27. $\angle EFO$
28. Distance across corners
29. $\angle FGO$
30. GA
31. AB

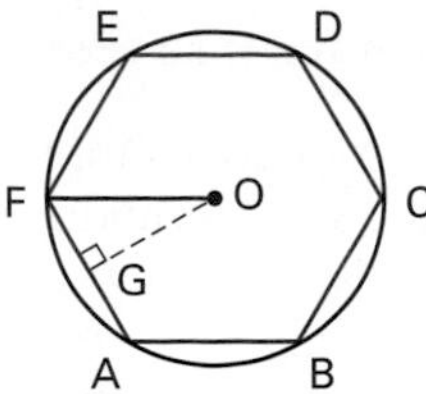

Figure 21-98

Use Fig. 21-99. Follow the preceding directions.

32. $\angle ODC$
33. $\angle EOG$
34. OC

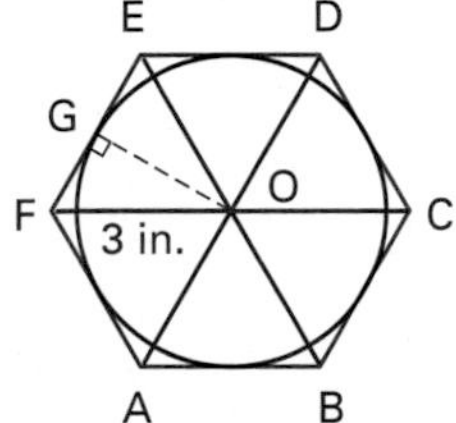

Figure 21-99

35. What is the distance across the flats of a hexagon cut from a circle blank with a circumference of 145 mm (Fig. 21-100)? Round to hundredths.

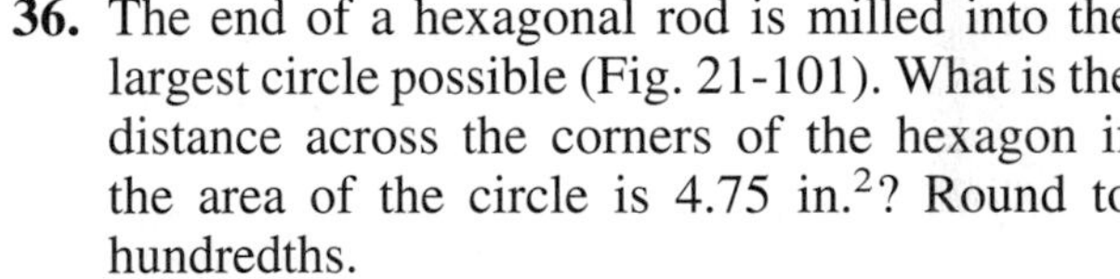

36. The end of a hexagonal rod is milled into the largest circle possible (Fig. 21-101). What is the distance across the corners of the hexagon if the area of the circle is 4.75 in.^{2}? Round to hundredths.

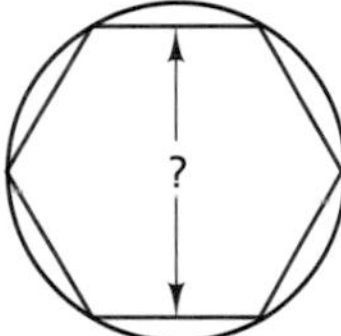

Figure 21-100

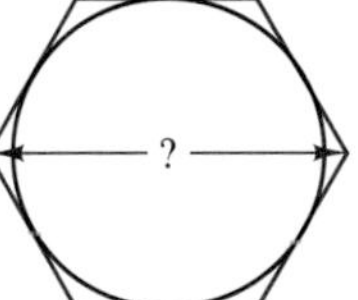

Figure 21-101

ELECTRONICS APPLICATION

Phase Angle

The phase angle is very important in electronics. One application is comparing voltage and current.

An inductive circuit contains an inductor or coil, usually using the symbol L, and the inductance across the coil is measured in henrys (H). If an alternating current is applied (ac), the voltage in that circuit will always lead the current. This can be easily measured on an oscilloscope. You can also say that the current lags the voltage.

Assume that the sine waves that represent the voltage and the current are shown on the oscilloscope. The two sine waves will be slightly offset from each other. The distance of separation will be constant and is called the *phase angle*.

Suppose one full cycle (from zero up to the maximum and then back to zero and to the negative maximum and then back to zero) measures 8 cm on the oscilloscope, and the amount of offset is 1 cm. Then the two sine waves are $\frac{1}{8}$ of a cycle apart. But one full cycle is considered to be 360°. Therefore, 1 cm of offset can be converted to degrees by taking $360° \times \frac{1}{8} = 45° =$ phase angle. If the offset is 0.8 cm, then the phase angle $= 360° \times \frac{0.8}{8} = 36°$.

If the inductive circuit is a simple series circuit with only one inductor and only one resistor in it, and an alternating current is applied, then the voltage measured across the inductor (called V_L) will lead the voltage across the resistor (called V_R) by 90°. This always happens. We can also state this result by saying that the voltage across the resistor lags the voltage across the inductor by 90°. Again, this is easy to measure on an oscilloscope.

If the circuit is a capacitive (rather than inductive) and is a simple series circuit with only one capacitor and only one resistor in it, and an alternating current is applied, then the voltage measured across the capacitor (called V_C) will lag the voltage across the resistor (called V_R) by 90°. This always happens. We can also state this result by saying that the voltage across the capacitor lags the voltage across the resistor by 90°. Again, this is easy to measure on an oscilloscope.

When measuring current and voltage for a capacitive circuit, the current will always lead the voltage, or the voltage will always lag the current. When measuring current and voltage for an inductive circuit, the current will always lag the voltage, or the voltage will always lead the current. We remember this by using

ELI the ICE man

where $E =$ electromotive force (voltage), $I =$ current, $L =$ inductive circuit, and $C =$ capacitive circuit.

Exercises

Fill in all the blanks in the following chart, which shows readings and interpretations from two sine waves on an oscilloscope. Also, for each pair of sine waves, is the circuit inductive or capacitive?

	Sine Wave 1	Sine Wave 2	Length of One Cycle (cm)	Amount of Offset (cm)	Phase Angle (°)	Inductive or Capacitive?
1.	E	I	8	3		
2.	E	I	6	2		
3.	E	I	5	1		
4.	I	E	4	0.5		
5.	I	E	8	0.5		
6.	I	E	6	1		
7.	V_L	V_R	5			
8.	V_R	V_L		1		
9.	V_C	V_R		2		
10.	V_R	V_C		1.5		

Answers for Exercises

Fill in all the blanks in the following chart, which shows readings and interpretations from two sine waves on an oscilloscope. Also, for each pair of sine waves, is the circuit inductive or capacitive?

	Sine Wave 1	Sine Wave 2	Length of One Cycle (cm)	Amount of Offset (cm)	Phase Angle (°)	Inductive or Capacitive?
1.	E	I	8	3	135	Inductive
2.	E	I	6	2	120	Inductive
3.	E	I	5	1	72	Inductive
4.	I	E	4	0.5	45	Capacitive
5.	I	E	8	0.5	22.5	Capacitive
6.	I	E	6	1	60	Capacitive
7.	V_L	V_R	5	1.25	90	Inductive
8.	V_R	V_L	4	1	90	Inductive
9.	V_C	V_R	8	2	90	Capacitive
10.	V_R	V_C	6	1.5	90	Capacitive

ASSIGNMENT EXERCISES

21-1

Give the proper notation for the following.

1. Line AB

2. Segment AB

3. Ray AB

Use Fig. 21-102 for the following exercises.

4. Name the line in two different ways.

5. Name the ray with end point N and with an interior point O.

6. Name the ray with end point M and with an interior point L.

7. Is $\overline{MN}$ the same as $\overline{MO}$?

8. Is $\overleftrightarrow{NO}$ the same as $\overleftrightarrow{NM}$?

9. Is $\overrightarrow{NO}$ the same as $\overrightarrow{NM}$?

10. Is $\overrightarrow{MN}$ the same as $\overrightarrow{MO}$?

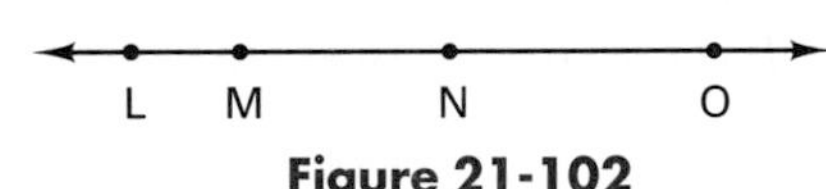

Figure 21-102

Use Fig. 21-103 for the following exercises.

11. Which lines are parallel?

12. Which lines coincide?

13. Which lines intersect?

14. When will $\overleftrightarrow{AB}$ and $\overleftrightarrow{CD}$ meet?

15. Is $\overleftrightarrow{IJ}$ the same as $\overleftrightarrow{KJ}$?

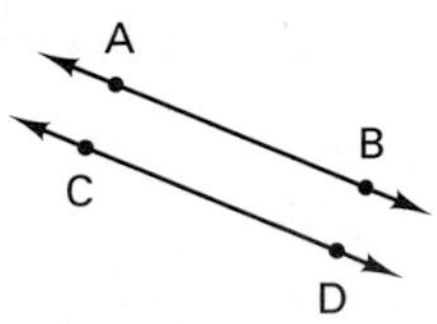

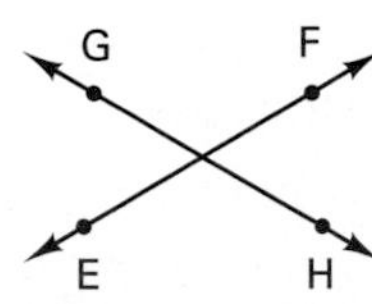

Figure 21-103

Use Fig. 21-104 for the following exercises.

16. Name $\angle a$ using three capital letters.

17. Name $\angle b$ using three capital letters.

18. Name $\angle c$ using three capital letters.

19. Name $\angle a$ using one capital letter.

20. Name $\angle b$ using one capital letter.

21. Name $\angle c$ using one capital letter.

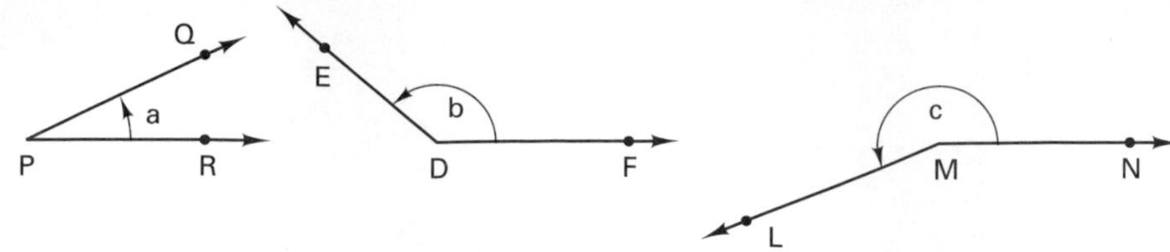

Figure 21-104

Classify the following angle measures using the terms *right*, *straight*, *acute*, or *obtuse*.

22. 50° **23.** 90° **24.** 120°

25. 18° **26.** 75° **27.** 180°

28. 89° **29.** 179° **30.** 30°

State whether the following angle pairs are complementary, supplementary, or neither.

31. 63°, 37° **32.** 98°, 62° **33.** 135°, 45°

34. 45°, 45° **35.** 21°, 79° **36.** 90°, 90°

37. Congruent angles have _____ measures.

38.–39. Perpendicular lines are formed when a _____ line and a _____ line intersect.

21-2

Add or subtract as indicated. Simplify the answers.

40. $\begin{array}{r} 34°28'41'' \\ +\ 18°37'50'' \\ \hline \end{array}$

41. $\begin{array}{r} 115°34'29'' \\ -\ 84°26'18'' \\ \hline \end{array}$

42. $\begin{array}{r} 34°29'35'' \\ +\ 19°30'25'' \\ \hline \end{array}$

43. $\begin{array}{r} 64°15'37'' \\ -\ 29°37'41'' \\ \hline \end{array}$

44. $\begin{array}{r} 80° \\ -\ 28°14'28'' \\ \hline \end{array}$

45. $\begin{array}{r} 74° \\ -\ 13°19'42'' \\ \hline \end{array}$

46. Find the complement of an angle whose measure is 35°29′14″.

Change to decimal degree equivalents. Express the decimals to the nearest ten-thousandth.

47. 29′ **48.** 47″ **49.** 7′34″

50. An angle of 34°36′48″ is divided into two equal angles. Express the measure of each of the equal angles in decimal degrees. Round to the nearest ten-thousandth.

Change to equivalent minutes and seconds. Round to the nearest second when necessary.

51. 0.75° **52.** 0.46° **53.** 0.2176°

In the following exercises, round angle measures to the nearest second when necessary.

54. A right angle is divided into eight equal parts. Find the measure of each angle in degrees and minutes.

55. An angle of 140° is divided into six equal parts. Find the measure of each angle in degrees and minutes.

56. An angle of 18°52′48″ needs to be three times as large. Find the measure of the new angle in decimal degrees.

57. An angle of 37°19′41″ is divided into two equal angles. Find the measure of each equal angle in degrees, minutes, and seconds.

21-3

Classify the triangles in the following figures according to their sides.

58.

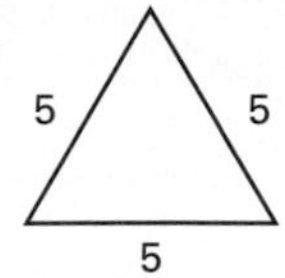

59.

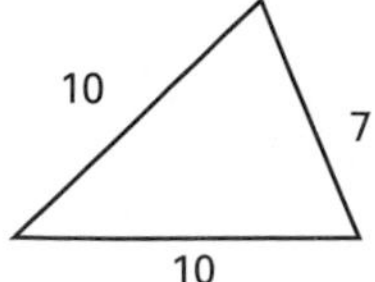

60.

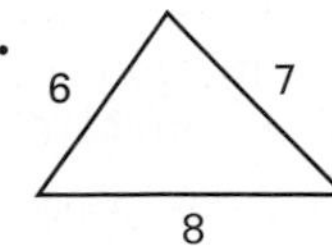

Identify the longest and shortest sides in the following figures.

61.

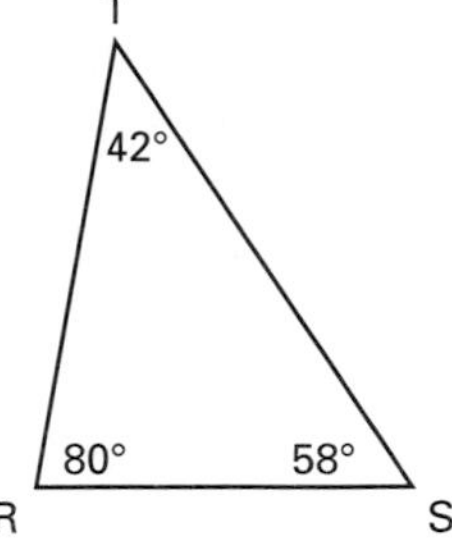

62.

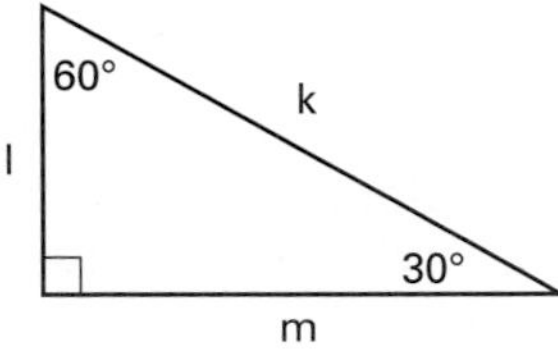

List the angles in order of size from largest to smallest in the following figures.

63.

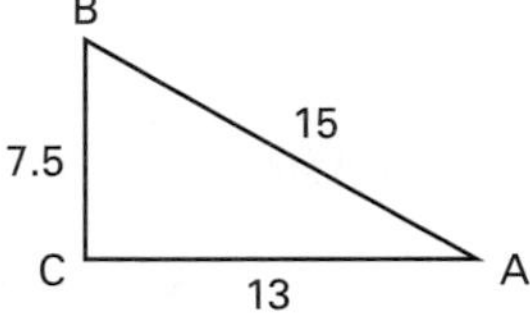

64.

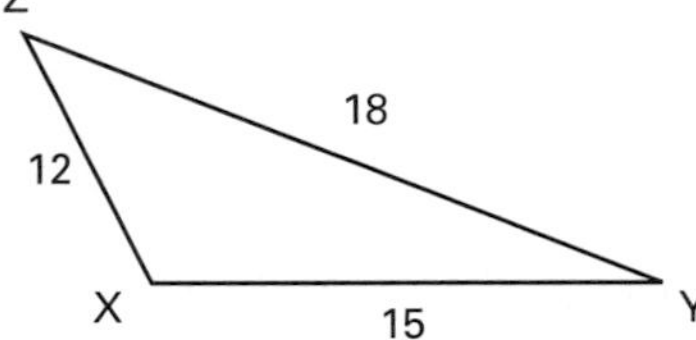

Write the corresponding parts not given for the congruent triangles in the following figures.

65.

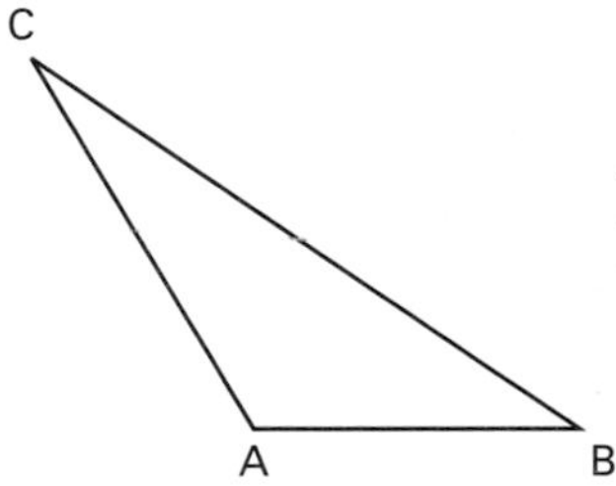

∠A = ∠F
AB = FE
AC = DF

E
F
D

66.

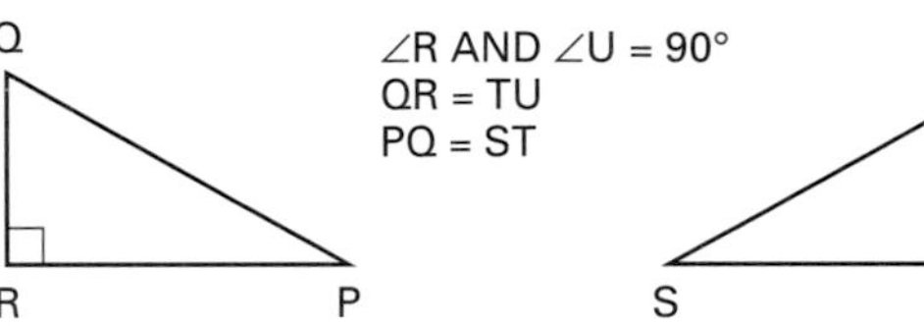

67.

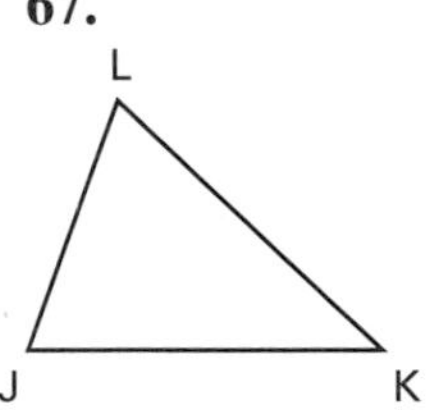

∠K = ∠N
∠L = ∠P
LK = NP

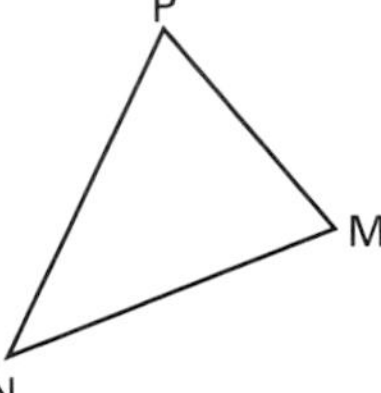

Use Fig. 21-105 to solve the following exercises. Round the final answers to the nearest thousandth if necessary.

68. $RS = 8$ mm; find ST and RT.
69. $RT = 15$ cm; find RS and ST.
70. $ST = 7.2$ ft; find RS and RT.
71. $RT = 9\sqrt{2}$ hm; find RS and ST.
72. $RT = 8\sqrt{5}$ dkm; find RS and ST.

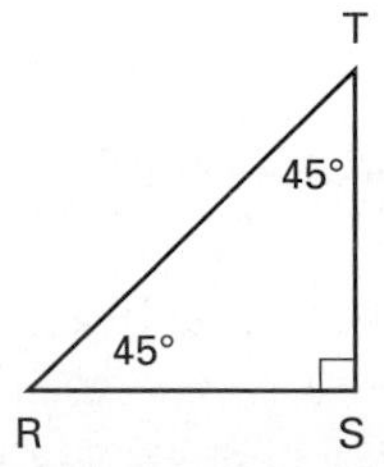

Figure 21-105

Use Fig. 21-106 to solve the following exercises. Round the final answers to the nearest thousandth if necessary.

73. $AC = 12$ dm; find AB and BC.
74. $AB = 15$ km; find AC and BC.
75. $BC = 10$ in.; find AC and AB.
76. $BC = 8\sqrt{2}$ hm; find AC and AB.
77. $AC = 40$ ft 7 in.; find AB and BC to the nearest inch.

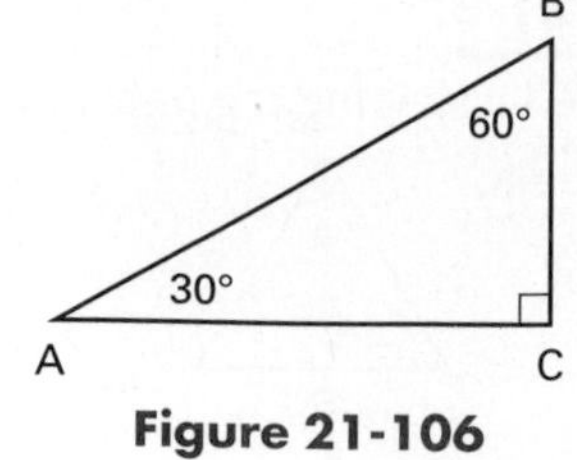

Figure 21-106

Solve the following problems. Round the final answers to the nearest thousandth if necessary.

78. The sides of an equilateral triangle are 4 dm in length. Find the altitude of the triangle. Round the answer to the nearest thousandth of a decimeter.

79. Find the total length to the nearest inch of the conduit $ABCD$ if $XY = 8$ ft, $CE = 14$ in., and angle $CBE = 30°$ (Fig. 21-107).

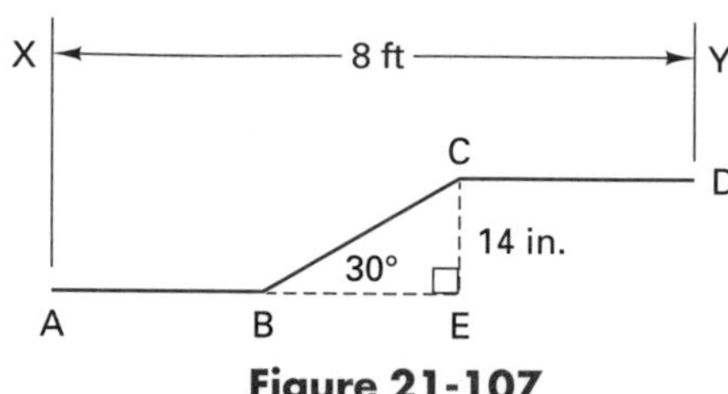

Figure 21-107

80. A piece of round steel is milled to a point at one end. The angle formed by the point is the angle of taper. If the steel has a 16-mm diameter and a 60° angle of taper, find the length c of the taper (Fig. 21-108).

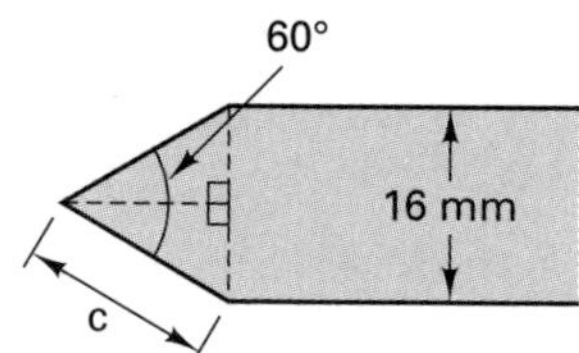

Figure 21-108

81. A V-slot forms a triangle whose angle at the vertex is 60°. The depth of the slot is 17 mm (see Fig. 21-109). Find the width of the V-slot.

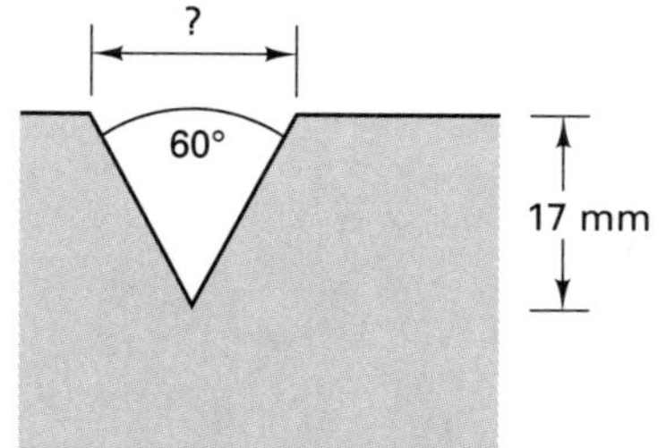

Figure 21-109

82. A manufacturer recommends attaching a guy wire to its 30-ft antenna at a 45° angle. If the antenna is installed on a flat surface, how long must the guy wire be (to the nearest foot) if it is attached to the antenna 4 ft from the top?

21-4

Find the perimeter and the area of the following figures.

83.

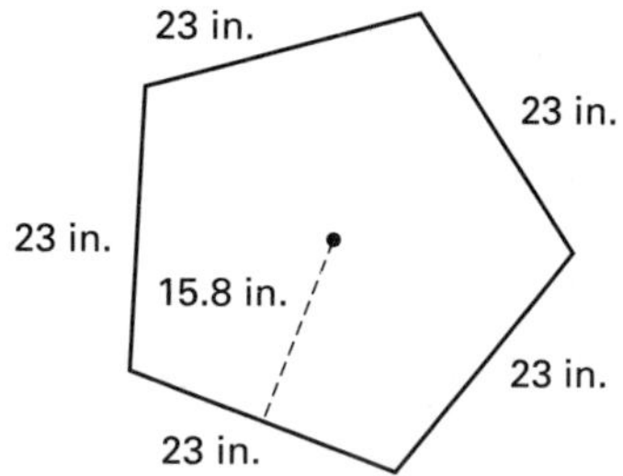

84.

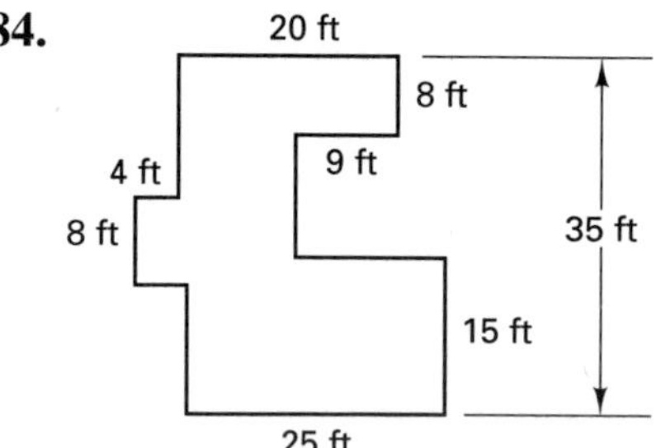

85. Identify the figure in Exercise 83 by giving its specific name and the number of degrees in each angle.

Solve the following problems involving the perimeter and area of composite figures.

86. How many sections of 2- × 4-ft ceiling tiles will be required for the layout in Fig. 21-110?

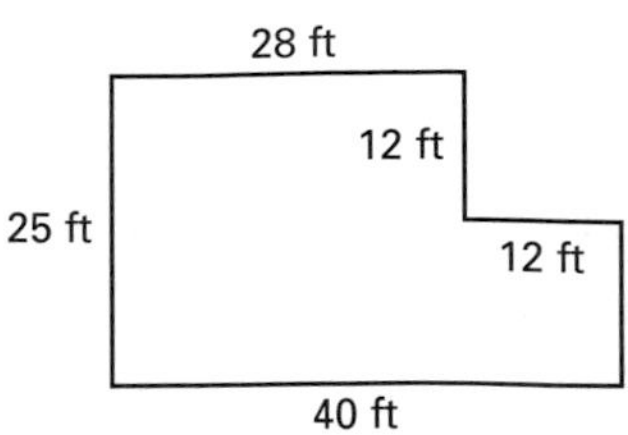

Figure 21-110

87. How many feet of baseboard will be needed to install around the room in Fig. 21-110 if we make no allowance for openings or doorways?

88. A gable end of a gambrel roof (Fig. 21-111) is to be covered with 12-in. bevel siding at an $8\frac{1}{2}$-in. exposure to the weather. Estimate the square footage of siding needed if 18% is allotted for lap and waste. Answer to the nearest foot.

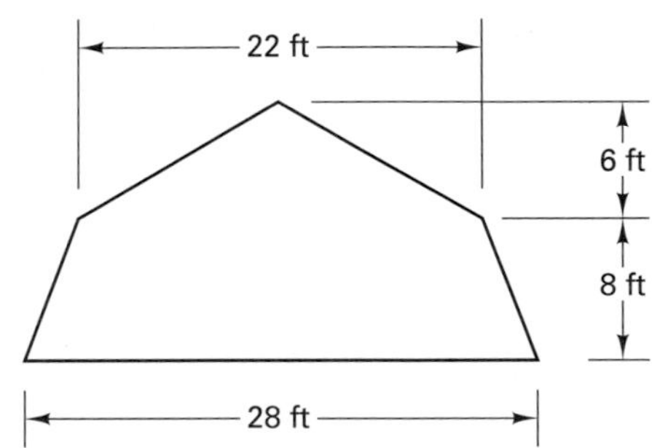

Figure 21-111

89. Find the square feet of floor space inside the walls of the plan of Fig. 21-112. Answer to the nearest square foot.

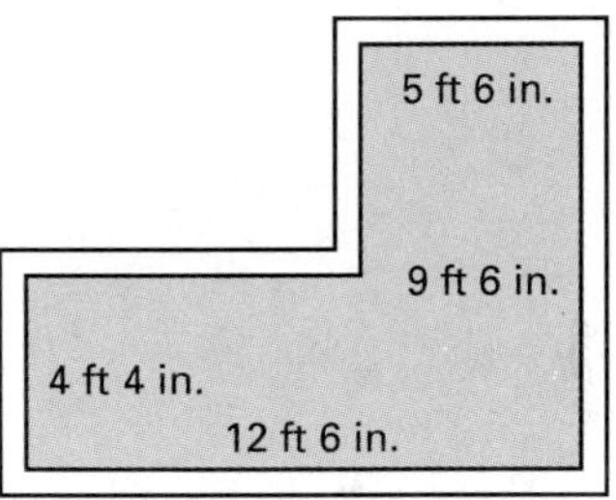

Figure 21-112

90. If one 2- × 4-in. stud is estimated for every linear foot of a wall when studs are 16 in. on center from each other, how many studs are needed for the outside walls of the floor plan in Fig. 21-113?

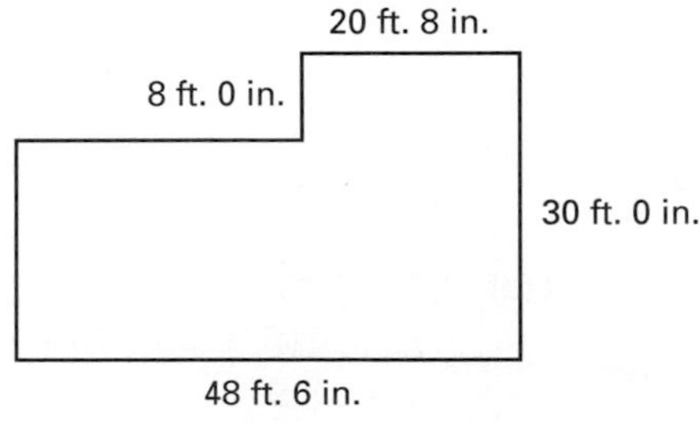

Figure 21-113

91. A home has concrete steps in the rear (Fig. 21-114). If the two sides are covered with a brick facing, how many bricks are needed if $\frac{1}{4}$-in. mortar joints are used? (Six bricks cover 1 square foot). Round any part of a square foot to the next-highest square foot. (144 in.^{2} = 1 ft^2.)

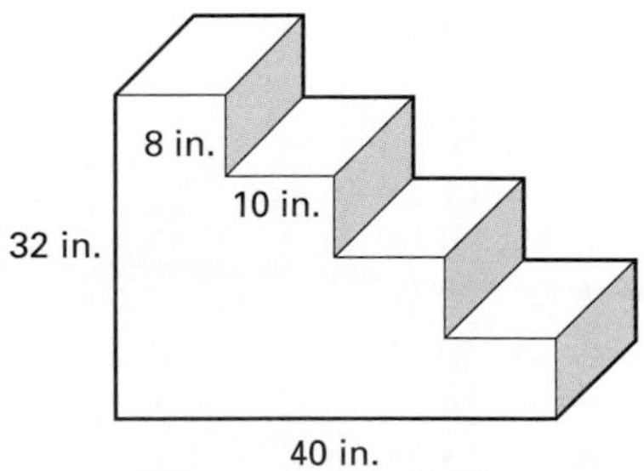

Figure 21-114

92. A mason will lay tile to cover an area in the form of a regular octagon whose sides are each 18 ft. If the distance from the midpoint of a side to the center of the octagon is 21.7 ft, how many tiles will the job require if $7\frac{1}{2}$ tiles will cover 1 square foot and 36 tiles are added for waste?

21-5

Find the area of the sectors of a circle using Fig. 21-115. Round to hundredths.

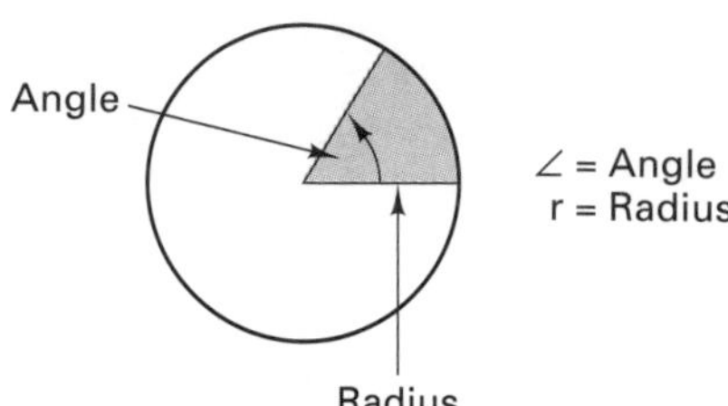

Figure 21-115

93. $\angle = 45°9'$, $r = 2.58$ cm

94. $\angle = 165°$, $r = 15$ in.

95. $\angle = 15°15'$, $r = 110$ mm

96. $\angle = 75°$, $r = 1$ m

97. $\angle = 40°$, $r = 2\frac{1}{2}$ ft

Find the area of the segments of a circle using Fig. 21-116. Round to hundredths.

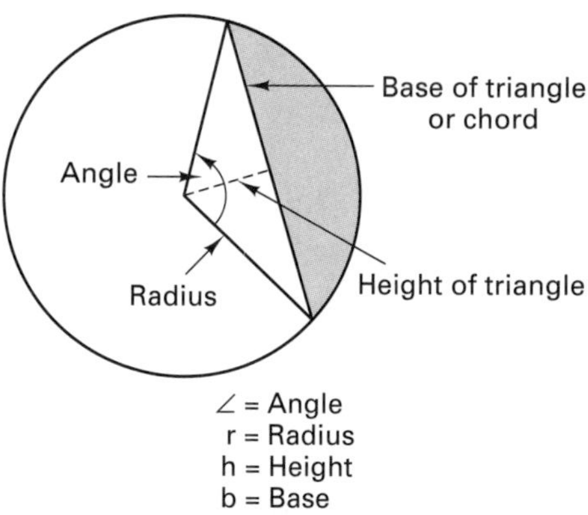

Figure 21-116

98. $\angle = 52°30'$, $r = 11.25$ in., $h = 10.09$ in., $b = 10$ in.

99. $\angle = 45°$, $r = 14$ mm, $h = 12.9$ mm, $b = 10.9$ mm

100. $\angle = 55°$, $r = 10''$, $h = 8.9''$, $b = 9.2''$

101. $\angle = 30°$, $r = 24$ cm, $h = 23.2$ cm, $b = 12.4$ cm

102. $\angle = 60°$, $r = 13.3$ cm, $h = 11.5$ cm, $b = 13.3$ cm

Solve the following problems involving sectors and segments of a circle.

103. What is the area (to nearest tenth) of the flat top bifocal lens shown in Fig. 21-117?

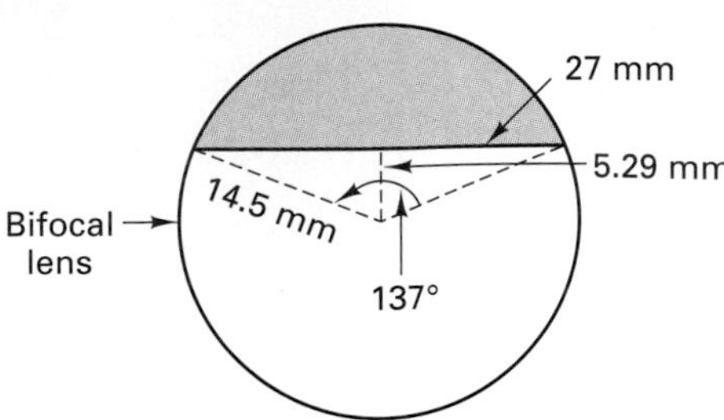

Figure 21-117

104. The minute hand on a grandfather clock cuts an angle of 30° as it moves from 12 to 1 on the clock face. If the minute hand is 27.5 cm long, what is the area of the clock face over which the minute hand moves from 12 to 1? Round to tenths.

105. What is the area of a patio (Fig. 21-118) that has one rounded corner? Round to hundredths.

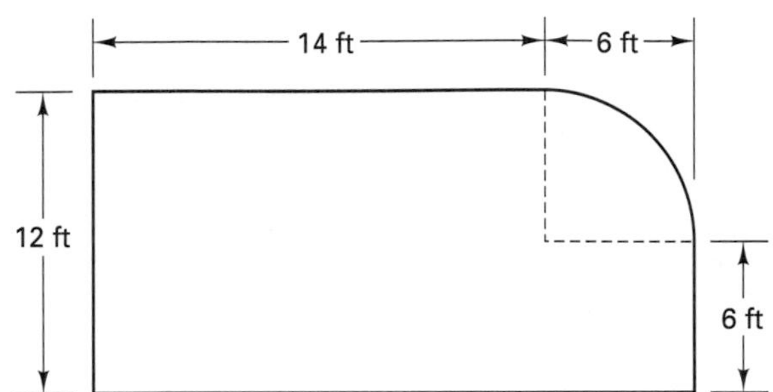

Figure 21-118

106. A machine cuts a 12-in.-diameter frozen pizza into slices whose sides form 72° angles at the center of the pizza. What is the surface area of each slice to the nearest square inch?

107. A drain pipe 20 in. in diameter has 4 in. of water in it (Fig. 21-119). What is the cross-sectional area of the water in the pipe to the nearest hundredth?

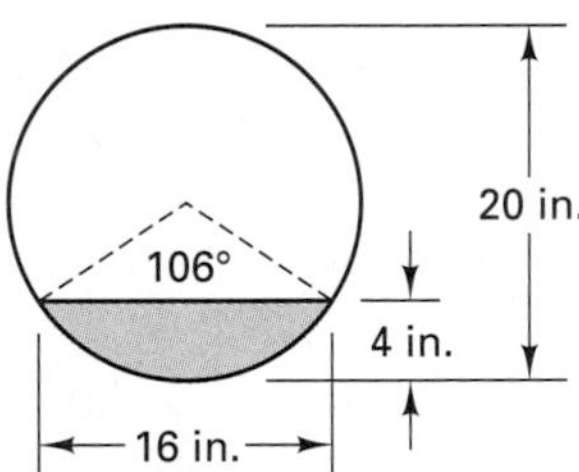

Figure 21-119

Find the arc length of the sectors of a circle using Fig. 21-115. Round to hundredths.

108. $\angle = 40°$
$r = 1.45$ ft.

109. $\angle = 180°$
$r = 10$ in.

110. $\angle = 30°15'$
$r = 20$ cm

111. $\angle = 70°10'$
$r = 30$ mm

21-6

For each figure, supply the missing dimensions. For an inscribed triangle (Fig. 21-120).

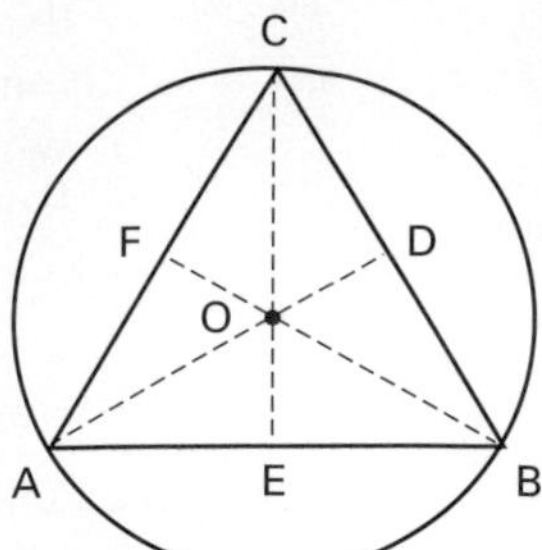

Figure 21-120

112. If $FB = 9$, $FO =$ ____ **113.** If $AB = 10$, $AE =$ ____ **114.** $\angle ABF =$ ____

For a circumscribed square (Fig. 21-121).

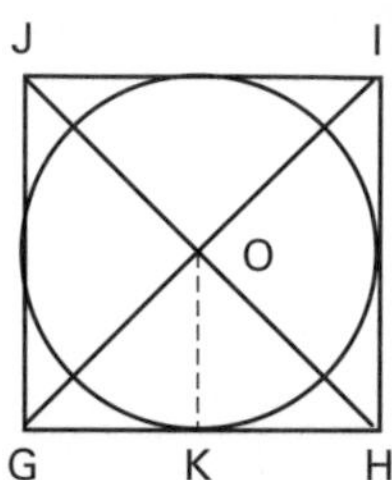

Figure 21-121

115. $\angle GJO =$ ____ **116.** If $GO = 7$, $HO =$ ____ **117.** If $KO = 10$, $IJ =$ ____

For an inscribed hexagon (Fig. 21-122).

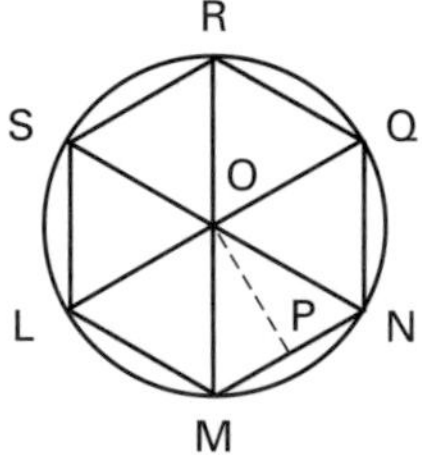

Figure 21-122

118. If $QO = 15$, $MN =$ ____ **119.** $\angle MOP =$ ____ **120.** $\angle MNO =$ ____

Solve the following problems involving polygons and circles. Round the answers to hundredths.

121. A shaft with a 20-mm diameter is milled at one end as illustrated in Fig. 21-123. What is the radius of the inscribed circle?

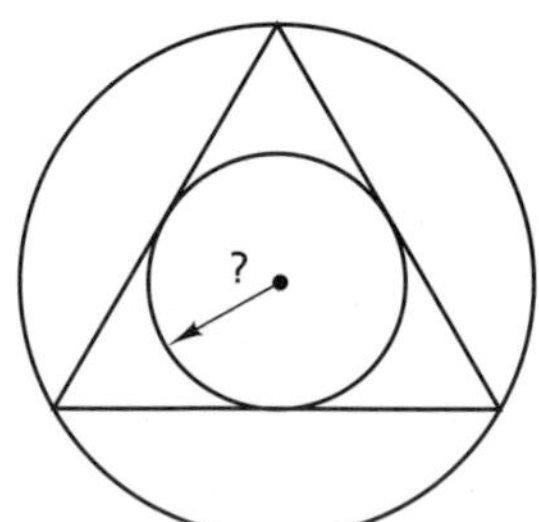

Figure 21-123

122. A hex nut 0.87 in. on a side will be milled from the smallest-diameter round stock possible (Fig. 21-124). What must the diameter of the stock be?

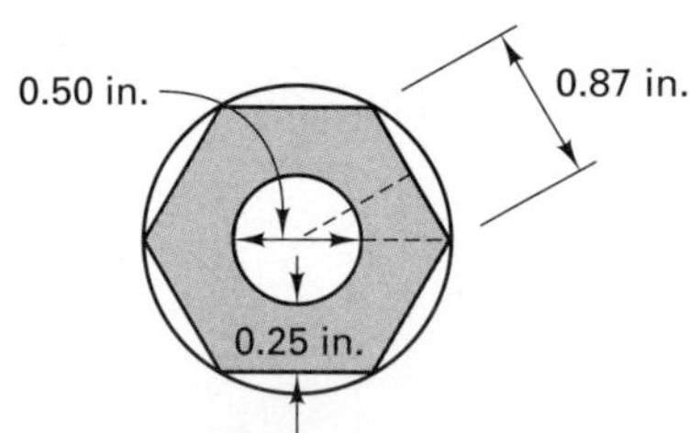

Figure 21-124

123. A square is milled on the end of a 5-cm shaft (Fig. 21-125). If the square is the largest that can be milled on the 5-cm shaft, what is the length of a side to hundredths?

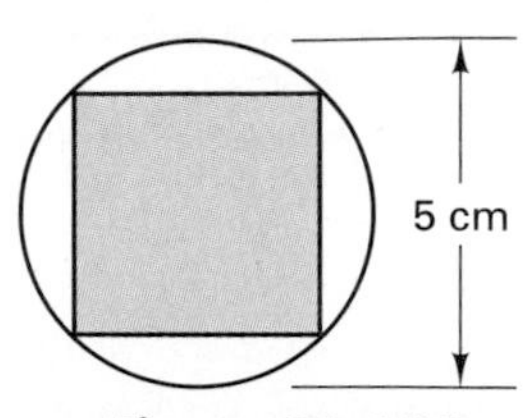

Figure 21-125

CHALLENGE PROBLEMS

124. Compare the area of a circle with a radius of 5 cm to the area of an inscribed equilateral triangle. Compare the area of a circle with a radius of 5 cm to the area of an inscribed square. Compare the area of a circle with a radius of 5 cm to the area of the inscribed polygon in Fig. 21-126.

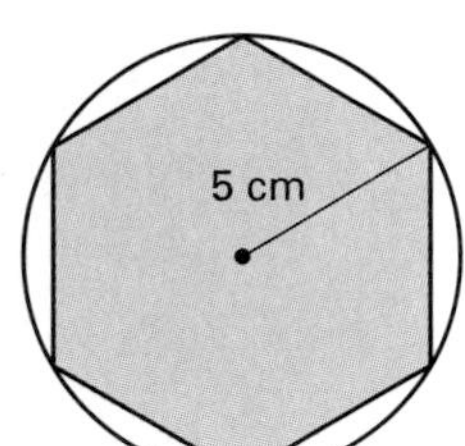

Figure 21-126

125. Estimate the area of the inscribed regular pentagon in Fig. 21-127 by giving two values that the area will be between.

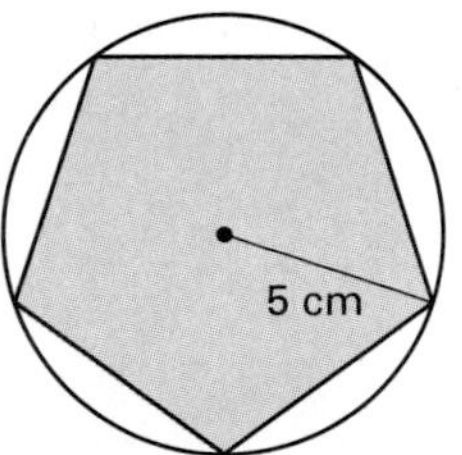

Figure 21-127

Present a convincing argument for your estimate. Similarly, estimate the area of an inscribed regular octagon and give an argument for your estimate.

CONCEPTS ANALYSIS

1. Can the measures 8 cm, 12 cm, and 25 cm represent the sides of a triangle? Illustrate and explain your answer.

2. The Pythagorean theorem $a^2 + b^2 = c^2$ applies to right triangles where a and b are legs and c is the hypotenuse. For what type of triangle is the statement $a^2 + b^2 > c^2$ true? Illustrate your answer.

3. For what type of triangle is the statement $a^2 + b^2 < c^2$ true? Illustrate your answer.

4. A triangle has 6, 3, and $3\sqrt{3}$ as the measures of its sides. Sketch the triangle placing the measures on the appropriate sides and show the measures of each of the three angles of the triangle.

5. Explain the process for finding the perimeter of a composite figure.

6. Explain the process for finding the area of a composite figure.

7. Draw a composite figure for which the area can be found by using either addition or subtraction and explain each approach.

8. Write in words the formula for finding the degrees in each angle of a regular polygon.

9. Describe a short cut for finding the perimeter of an L shaped figure. Illustrate your procedure with a problem.

10. Two pieces of property have the same area (acreage) and equal desirability for development. Both will require expensive fencing. One piece is a square 6000 feet on each side. The other piece is a rectangle 400 feet by 900 feet. Which piece of property requires the least amount of fencing and is thus more desirable?

CHAPTER SUMMARY

Objectives	What to Remember	Examples
Section 21-1		
1. Use a variety of notations to represent points, lines, line segments, rays, and planes.	A dot represents a point and a capital letter names the point. A line extends in both directions and is named by any two points on the line (with $\leftrightarrow$ above the letters, such as $\overleftrightarrow{AB}$). A line segment has a beginning and ending point, which are named by letters (with a $\overline{}$ above the letters, such as $\overline{CD}$). A ray extends from a point on a line and includes all points on one side of the point and is named by the starting point and any other point on the line forming the ray (with a $\rightarrow$ above the letters, such as $\overrightarrow{AC}$). A plane contains an infinite number of points and lines on a flat surface. (Line segments are often informally named just by letters, such as CD.)	Use proper notation for the following: (a) Line GH (b) Line segment OP (c) Ray ST (a) $\overleftrightarrow{GH}$ (b) $\overline{OP}$ (c) $\overrightarrow{ST}$
2. Distinguish among lines that intersect, that coincide, and that are parallel.	Lines that intersect meet at only one point. Lines that coincide will fit exactly on one another. Lines that are parallel are the same distance from one another and will never intersect.	Classify the pairs of lines in Fig. 21-128. (a) 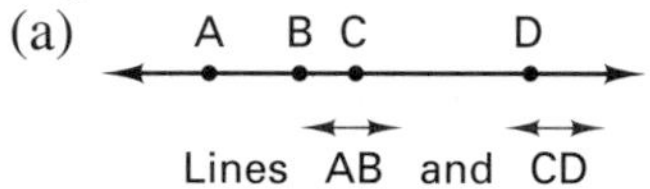 (b) 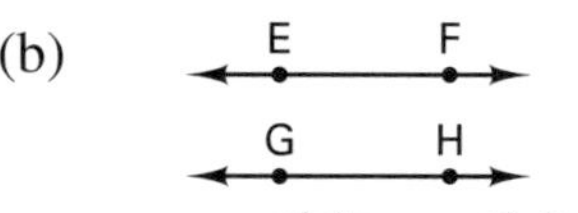 (c) 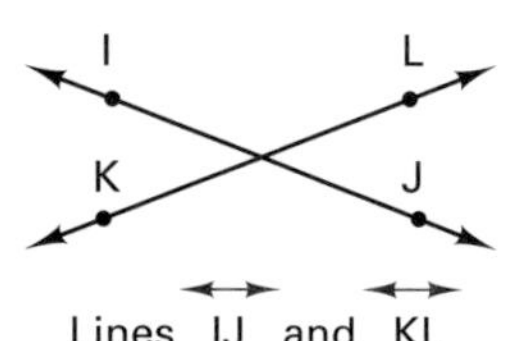 **Figure 21-128** (a) Coinciding (b) Parallel (c) Intersecting
3. Use a variety of notations to represent angles.	When two rays intersect in a point (end point of rays), an angle is formed. Angles may be named with three capital letters (end point in middle and one point on each ray), such as $\angle ABC$. They may be named by only the middle letter (vertex of angle), such as $\angle B$. They may be assigned a number or lowercase letter placed within the vertex, such as $\angle 2$ or $\angle d$.	Name the angle in Fig. 21-129 three ways. 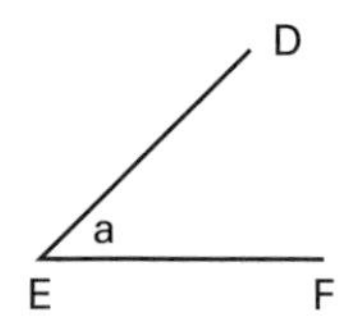 **Figure 21-129** $\angle DEF$, $\angle E$, $\angle a$

4. Classify angles according to size.	A right angle contains 90° or one-fourth a rotation. A straight angle contains 180° or one-half a rotation. An acute angle contains less than 90° but more than 0°. An obtuse angle contains more than 90° but less than 180°.	Identify the following angles: (a) 30° (b) 100° (c) 90° (d) 180° (a) acute (b) obtuse (c) right (d) straight
Section 21-2		
1. Add and subtract angle measures.	Only like measures can be added or subtracted. Arrange measures in columns of like measures. Add or subtract each column separately as for operations with U.S. customary measures (yards, feet, inches). Borrowing and writing answers in standard form are also similar to operations with U.S. customary measures.	Perform the following: (a) $\begin{array}{r} 35°15'10'' \\ +\ 10°55'\ 5'' \\ \hline 45°70'15'' \\ =\ 46°10'15'' \end{array}$ (b) Borrow. $\begin{array}{r} {\scriptstyle 34\ \ 75} \\ 35°15'10'' \\ -\ 10°55'\ 5'' \\ \hline 24°20'\ 5'' \end{array}$
2. Change minutes and seconds to a decimal part of a degree.	To change minutes to a decimal part of a degree: Divide minutes by 60. To change seconds to a decimal part of a degree: Divide seconds by 3600.	Change 15′36″ to a decimal $15 \div 60 = 0.25$ $36 \div 3600 = 0.01$ $0.25 + 0.01 = 0.26°$
3. Change the decimal part of a degree to minutes and seconds.	To change a decimal part of a degree to minutes: Multiply the decimal part of a degree by 60. To change a decimal part of a minute to seconds: Multiply the decimal part of a minute by 60.	Change 0.48° to minutes and seconds. $0.48 \times 60 = 28.8'$ $0.8 \times 60 = 48''$ Thus, 0.48° = 28′48″.
4. Multiply and divide angle measures.	Multiply or divide angle measures as U.S. customary measures are multiplied or divided, and simplify the answers.	Perform the following: (a) $\begin{array}{r} 25°\ 40'10'' \\ \times\ \ \ \ \ \ \ \ 3 \\ \hline 75°120'30'' \\ =\ 77°\ \ \ \ \ \ 30'' \end{array}$ (b) $\begin{array}{r} 8°33'23\frac{1}{3}'' \\ 3\overline{)25°40'10''} \end{array}$
Section 21-3		
1. Classify triangles by sides.	A scalene triangle has all three sides unequal. An isosceles triangle has exactly two equal sides. An equilateral traingle has three equal sides.	Identify the following triangles by the measures of their sides: (a) 6 cm, 6 cm, 6 cm (b) 12 in., 10 in., 12 in. (c) 10 cm, 12 cm, 15 cm (a) Equilateral (b) Isosceles (c) Scalene
2. Relate the sides and angles of a triangle.	An equilateral triangle has three equal angles. The equal sides of an isosceles triangle have opposite angles that are equal. If the three sides of a triangle are unequal, the largest angle will be opposite the longest side, and the smallest angle will be opposite the shortest side. In a right triangle, the hypotenuse is always the longest side.	Identify the largest and the smallest angles in triangles with the following sides: 12 in., 10 in., 9 in. Largest angle is opposite 12-in. side; smallest angle is opposite 9-in. side.

3. Determine if two triangles are congruent using inductive and deductive reasoning.

Triangles are congruent (one can fit exactly over the other) (a) if the three sides of one equal the corresponding sides of the other, (b) if two sides and the included angle are equal to two sides and the included angle of the other triangle, or (c) if two angles and the common side are equal to two angles and the common side on the other triangle.

Determine whether the pairs of triangles in Figure 21-130 are congruent.

(a) Congruent
(b) Not congruent
(c) Congruent
(d) Congruent

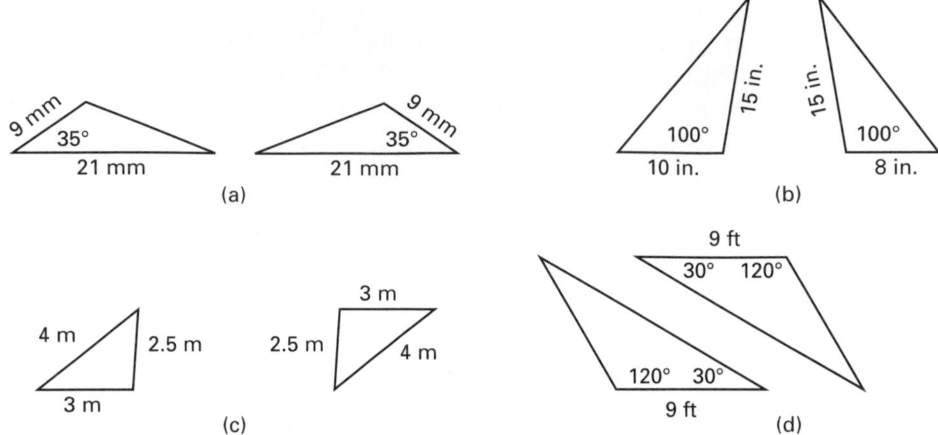

Figure 21-130

4. Use the properties of a 45°, 45°, 90° triangle to find missing parts and to solve applied problems.

To find the hypotenuse of a 45°, 45°, 90° triangle: Multiply a leg by $\sqrt{2}$. To find a leg of a 45°, 45°, 90° triangle: Divide the product of the hypotenuse and $\sqrt{2}$ by 2.

Given a 45°, 45°, 90° triangle, find:
(a) the hypotenuse if a leg is 10 cm
(b) a leg if the hypotenuse is 15 in.

(a) $10(\sqrt{2}) = 14.14213562$ or 14.1 cm (rounded)

(b) $\frac{15(\sqrt{2})}{2} = 10.60660172$ or 10.6 in. (rounded)

5. Use the properties of a 30°, 60°, 90° triangle to find missing parts and to solve applied problems.

To find the hypotenuse of a 30°, 60°, 90° triangle: Multiply the side opposite the 30° angle by 2. To find the side opposite the 30° angle in a 30°, 60°, 90° triangle: Divide the hypotenuse by 2. To find the side opposite the 60° angle of a 30°, 60°, 90° triangle: Multiply the side opposite the 30° angle by $\sqrt{3}$.

Given a 30°, 60°, 90° triangle, find: (a) the hypotenuse if the side opposite 30° angle = 4.5 cm; (b) the side opposite the 30° angle if hypotenuse is 20 in.; (c) the side opposite the 60° angle if side opposite 30° angle is 35 mm.

(a) $4.5(2) = 9$ cm

(b) $\frac{20}{2} = 10$ in.

(c) $35(\sqrt{3}) = 60.62177826$ or 60.6 in. (rounded)

Section 21-4

1. Find missing dimensions of composite shapes.

If a composite shape does not have all the dimensions marked, the missing dimensions can be inferred or calculated from our knowledge of polygons and the dimensions on the layout itself.

Find the dimensions x and y of the layout in Fig. 21-131.

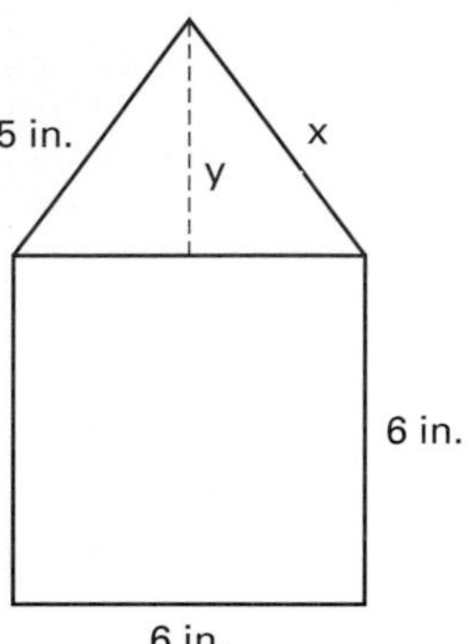

Figure 21-131

The layout is a square topped by an isosceles triangle. $x = 5$ in., since legs of isosceles triangle are equal. $y =$ height of triangle. From apex of triangle, drop a perpendicular line to the base, bisecting the base and forming two right triangles whose bases are each 3 in. (half of 6 in.). Use the Pythagorean theorem to find the altitude.

$$\begin{aligned} x^2 &= y^2 + \left(\frac{1}{2}\text{ base}\right)^2 \\ 25 &= y^2 + 3^2 \\ 25 &= y^2 + 9 \\ y^2 &= 25 - 9 \\ y^2 &= 16 \\ y &= 4 \end{aligned}$$

2. Find the perimeter and area of composite shapes.

Perimeter of a composite: $P = a + b + c$, etc.

Area of a composite: Use area formulas for separate polygons.

Find the perimeter and area of the layout in Fig. 21-131.

$$P = 6 + 6 + 6 + 5 + 5$$
$$P = 28 \text{ in.}$$

$$A_{\text{square}} = 6^2 = 36 \text{ in.}^2$$

$$\begin{aligned} A_{\text{triangle}} &= \frac{1}{2}bh \\ &= \frac{1}{2}(3)(5.83) \\ &= 8.745 \text{ in.}^2 \end{aligned}$$

$$\begin{aligned} A_{\text{layout}} &= 36 + 8.745 \\ &= 44.745 \text{ in.}^2 \end{aligned}$$

3. Find the number of degrees in each angle of a regular polygon.

To find the number of degrees in each angle of a regular polygon: Multiply the number of sides less 2 by 180° and divide the product by the number of sides.

Find the number of degrees in each angle of a pentagon.

A pentagon has five sides.

$$\text{Degrees per angle} = \frac{180(5 - 2)}{5} = \frac{180(3)}{5} = \frac{540}{5} = 108°$$

Section 21-5

1. Find the area of a sector.

A sector is a portion of a circle cut off by two radii.

$$\text{Area of a sector} = \frac{\theta}{360}\pi r^2$$

where θ is the degrees of the angle formed by the radii.

Find the area of a sector whose central angle is 50° and whose radius is 20 cm (Fig. 21-132).

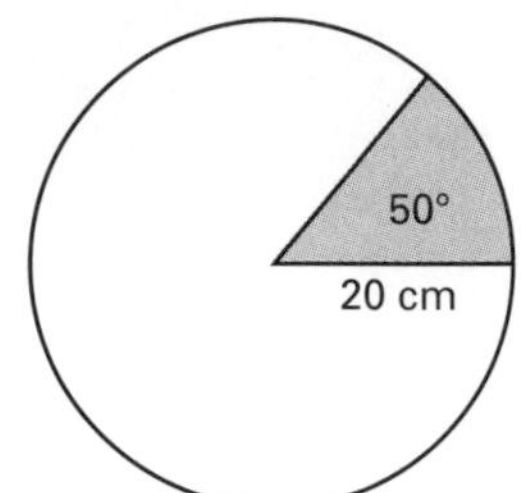

Figure 21-132

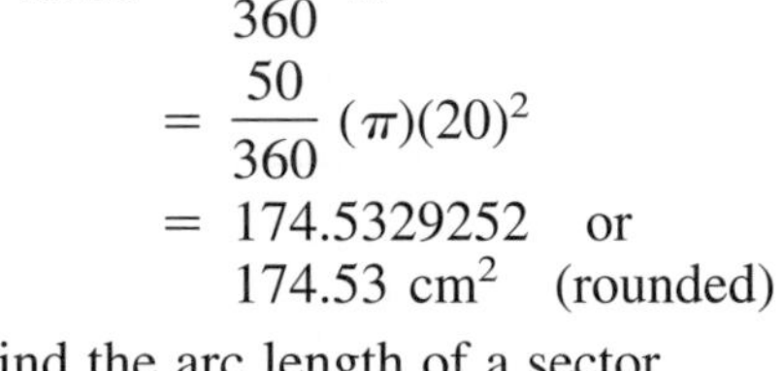

$$\text{Area} = \frac{\theta}{360}\pi r^2 = \frac{50}{360}(\pi)(20)^2 = 174.5329252 \quad \text{or} \quad 174.53 \text{ cm}^2 \quad \text{(rounded)}$$

2. Find the arc length of a sector.

Arc length is the portion of the circumference cut off by the sides of a sector.

$$\text{Arc length } (s) = \frac{\theta}{360}(2\pi r)$$

Find the arc length of a sector formed by a 50° central angle if the radius is 20 cm (Fig. 21-133).

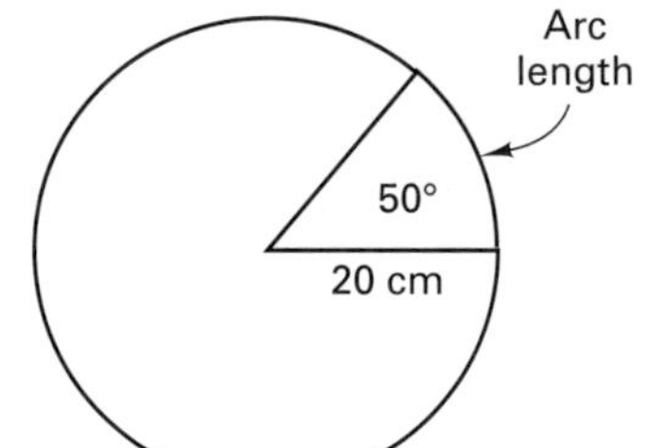

Figure 21-133

$$s = \frac{\theta}{360}(2\pi r)$$

$$s = \frac{50}{360}(2)(\pi)(20)$$

$$s = 17.45329252 \quad \text{or} \quad 17.45 \text{ cm}$$

3. Find the area of a segment.

A chord is a line segment joining two points on a circle. An arc is the portion of the circumference cut off by a chord. A segment is the portion of a circle bounded by a chord and an arc.

The area of a segment is the area of the sector less the area of the triangle formed by the chord and the central angle of the sector:

$$A = \frac{\theta}{360}\pi r^2 - \frac{1}{2}bh$$

Find the area of the segment formed by a 12-cm chord if the height of the triangle part of the sector is 6 cm. The central angle is 90° (Fig. 21-134).

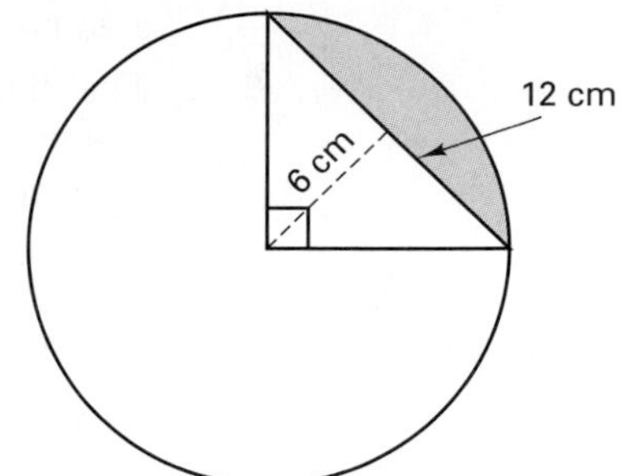

Figure 21-134

$$A = \frac{\theta}{360}\pi r^2 - \frac{1}{2}bh$$

$$A = \frac{90}{360}(\pi)(8.48)^2 - \frac{1}{2}(12)(6)$$

$$A = 0.25(\pi)(71.9104) - 0.5(72)$$

$A = 20.47829609$ or 20.48 cm^2 (rounded)

Section 21-6

1. Use the properties of inscribed and circumscribed equilateral triangles to find missing amounts and to solve applied problems.

The height of an equilateral triangle forms congruent 30°, 60°, 90° right triangles.

The radius of a circle circumscribed about an equilateral triangle is two-thirds the height of the triangle or twice the distance from the center of the circle to the base of the triangle.

The radius of a circle inscribed in an equilateral triangle is one-third the height of the triangle or one-half the distance from the vertex of the triangle to the center of the circle.

One end of a shaft 30 mm in diameter is milled as shown in Fig. 21-135. Find the diameter of the smaller circle.

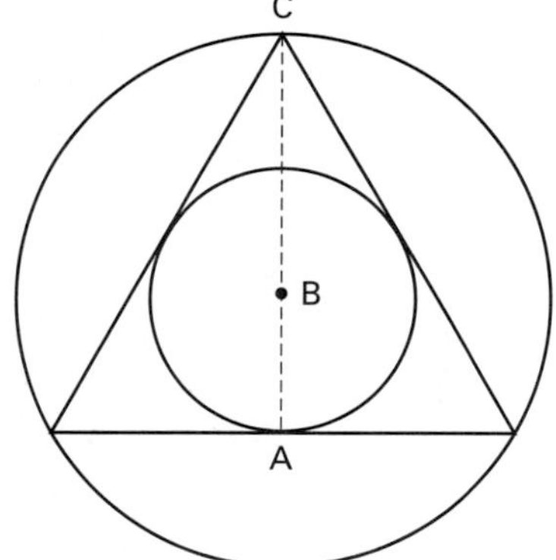

Figure 21-135

BC is half the diameter of large circle.

$$BC = \frac{1}{2}(30)$$

$$BC = 15 \text{ mm}$$

$$AB = \frac{1}{2}BC$$

$$AB = \frac{1}{2}(15)$$

$$AB = 7.5 \text{ mm}$$

(radius of small circle)

Diameter of small circle $= 2(7.5) =$ 15 mm.

2. Use the properties of inscribed and circumscribed squares to find missing amounts and to solve applied problems.

The diagonal of a square inscribed in or circumscribed about a circle form congruent 45°, 45°, 90° right triangles.

The diameter of a circle inscribed in a square equals a side of the square.

The diameter of a circle circumscribed about a square equals a diagonal of the square.

The radius of a circle inscribed in a square equals the height of a 45°, 45°, 90° right triangle formed by two diagonals.

The radius of a circle circumscribed about a square equals the height of a 45°, 45°, 90° right triangle formed by one diagonal.

The end of a 20-mm round rod is milled as shown in Fig. 21-136. Find the area of the square.

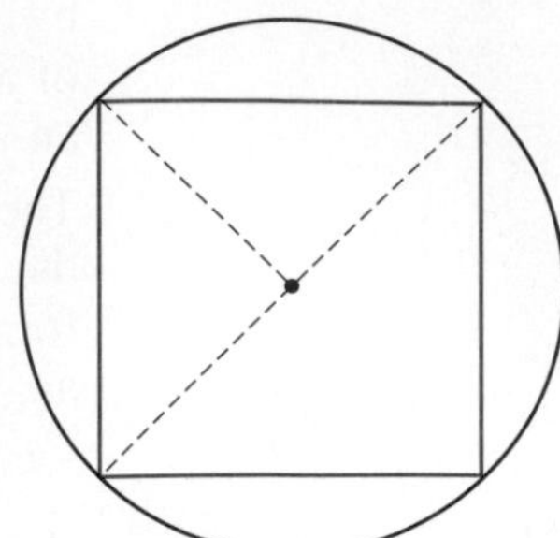

Figure 21-136

The radius equals the height of a right triangle formed by one diagonal. The diameter is the base. Find the area of the triangle and double it.

$$\begin{aligned} A_{\text{triangle}} &= \frac{1}{2}bh \\ &= \frac{1}{2}(20)(10) \\ &= 100 \text{ mm}^2 \end{aligned}$$

Area of square end: $100 \times 2 = 200 \text{ mm}^2$.

3. Use the properties of inscribed and circumscribed hexagons to find missing amounts and to solve applied problems.

The three diagonals that join opposite vertices of a regular hexagon form six congruent equilateral triangles.

The height of any of the equilateral triangles described above forms two congruent 30°, 60°, 90° right triangles. The diameter of a circle circumscribed about a regular hexagon is a diagonal of the hexagon (distance across corners).

The radius of a circle circumscribed about a regular hexagon is a side of an equilateral triangle formed by the three diagonals or one-half the distance across corners.

The radius of a circle inscribed in a regular hexagon is the height of an equilateral triangle formed by the three diagonals.

The diameter of a circle inscribed in a regular hexagon is the distance across flats or twice the height of an equilateral triangle formed by the three diagonals.

Find the following dimensions for Fig. 21-137 if the diameter is 10 cm.

(a) $\angle AOF$
(b) Radius
(c) Distance across corners

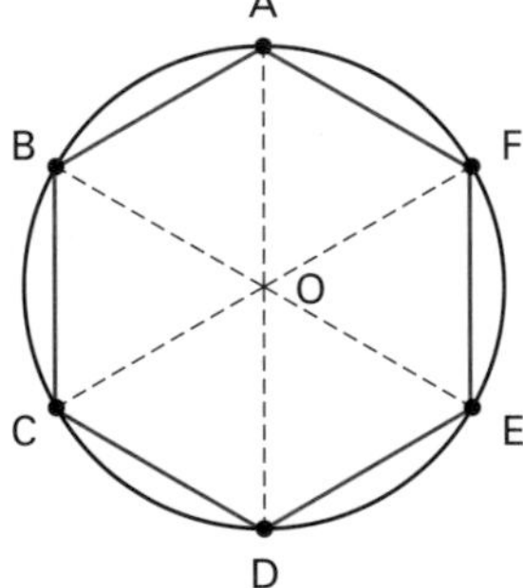

Figure 21-137

(a) $\angle AOF = 60°$ (Angle of equilateral triangle.)
(b) 5 cm (Radius of circumscribed circle is a side of equilateral triangle.)
(c) 10 cm (Diameter of circumscribed circle is distance across corners.)

WORDS TO KNOW

geometry (p. 781)
plane (p. 781)
point (p. 781)
line (p. 781)
line segment (p. 782)
ray (p. 782)
end point (p. 782)
interior point (p. 782)
intersect (p. 783)
coincide (p. 783)
parallel (p. 783)
angle (p. 783)
side of an angle (p. 783)
vertex (p. 783)
degree (p. 784)
minute (p. 784)
second (p. 784)
perpendicular (p. 784)
vertical (p. 785)
straight angle (p. 785)
acute angle (p. 785)
right angle (p. 785)
obtuse angle (p. 786)
complementary angles (p. 786)
supplementary angles (p. 786)
congruent (p. 786)
protractor (p. 788)
index mark on protractor (p. 788)
scalene triangle (p. 794)
isosceles triangle (p. 794)
equilateral triangle (p. 794)
corresponding sides (p. 795)
included angle (p. 796)
inductive reasoning (p. 797)
deductive reasoning (p. 797)
45°, 45°, 90° right triangle (p. 798)
30°, 60°, 90° right triangle (p. 800)
hypotenuse (p. 800)
leg (p. 800)
altitude (p. 800)
composite (p. 805)
perimeter (p. 805)
regular polygon (p. 809)
sector (p. 812)
segment (p. 812)
area (p. 813)
arc (p. 814)
arc length (p. 814)
chord (p. 815)
inscribed (p. 818)
circumscribed (p. 818)
tangent (p. 819)
distance across flats (p. 822)
distance across corners (p. 822)

CHAPTER TRIAL TEST

1. Show the proper notation to name the line segment from *B* to *C* in Fig. 21-138.

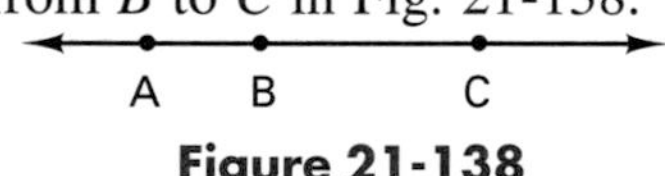

Figure 21-138

2. Name the angle in Fig. 21-139 in four different ways.

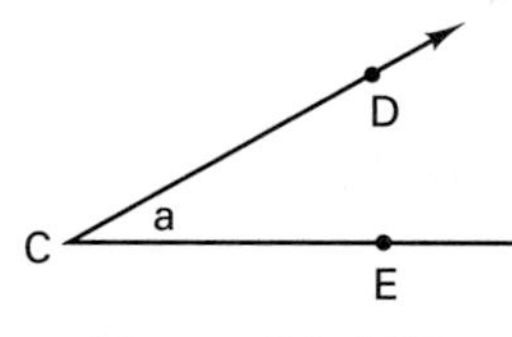

Figure 21-139

Classify the following angles as right, straight, acute, or obtuse.

3. 42°

4. 158°

Identify the lines in the following figures as intersecting, parallel, or perpendicular.

5.

6.

7.

8. Subtract 40°37′26″ from 75°.

9. An angle of 47°16′28″ is divided into three equal angles. Find the measure of each angle in degrees, minutes, and seconds (to the nearest second).

10. Change 15′32″ to degrees to the nearest ten-thousandth.

11. Change 0.3125° to minutes and seconds.

12. Name the following triangles: (a) all sides unequal, (b) exactly two sides equal, and (c) all sides equal.

13. Identify the largest and the smallest angles in Fig. 21-140.

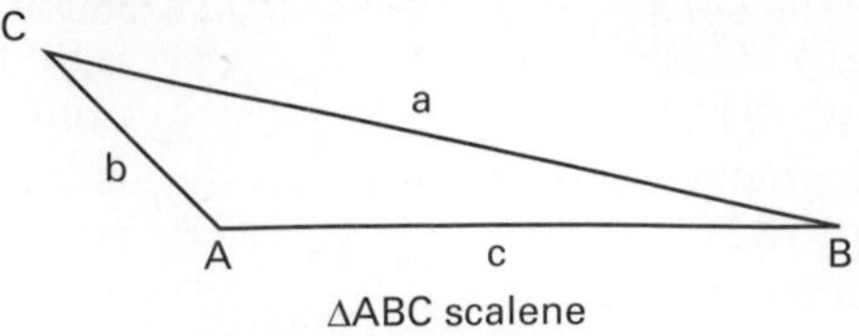

ΔABC scalene

Figure 21-140

14. In Fig. 21-141, identify (a) the angle corresponding to ∠*DEF* and (b) the side corresponding to *HI*.

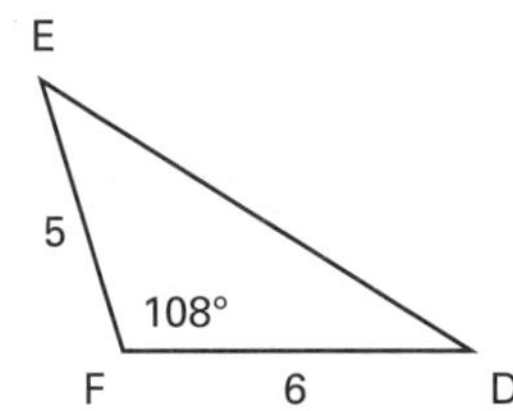

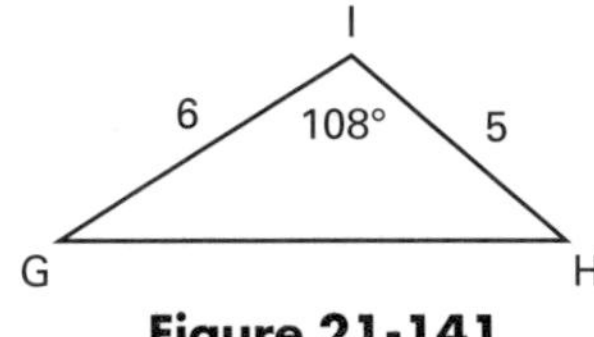

Figure 21-141

15. A conduit *ABCD* must be made so that ∠*CBE* is 45° (Fig. 21-142). If *BE* = 4 cm and *AK* = 12 cm, find the length of the conduit to the nearest thousandth centimeter.

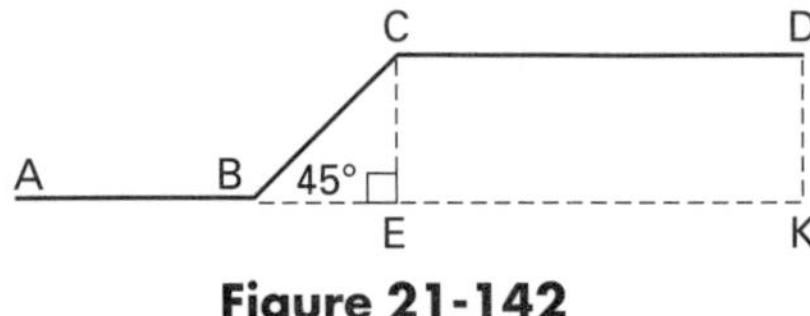

Figure 21-142

16. The rafters of a house make a 30° angle with the joists (Fig. 21-143). If the rafters have an 18-in. overhang and the center of the roof is 10 ft above the joists, how long must the rafters be cut?

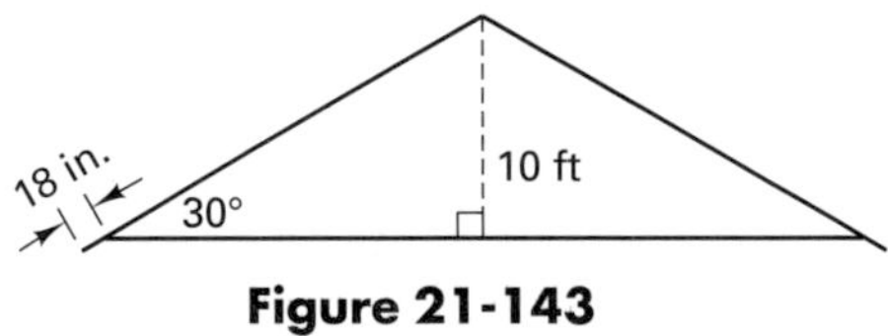

Figure 21-143

17. Find the perimeter of the composite in Fig. 21-144.

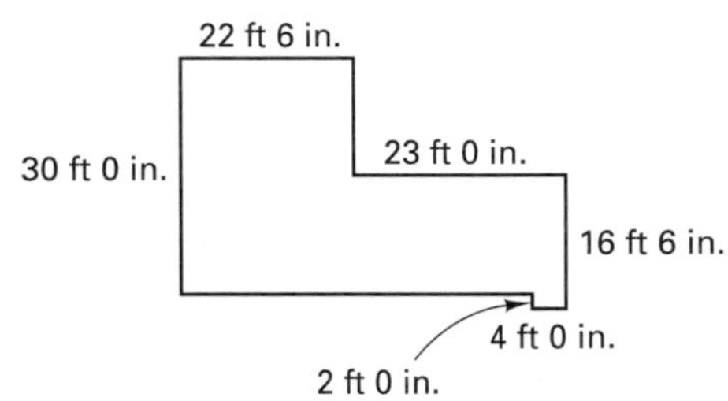

Figure 21-144

18. A section of a hip roof is a trapezoid measuring 35 ft at the bottom, 15 ft at the top, and 10 ft high. Find the area of this section of the roof in square feet.

19. Find the number of degrees in each angle of a regular octagon (eight sides).

20. Find the area of the sector in Fig. 21-145 if the central angle is 55° and the radius is 45 mm. Round to hundredths.

21. Find the arc length cut off by the sector in Fig. 21-145. Round to hundredths.

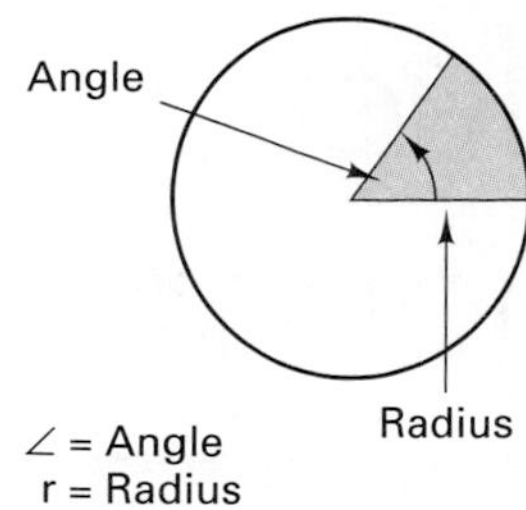

Figure 21-145

22. A segment is removed from a flat metal circle so that the piece will rest on a horizontal base (Fig. 21-146). Find the area of the segment that was removed.

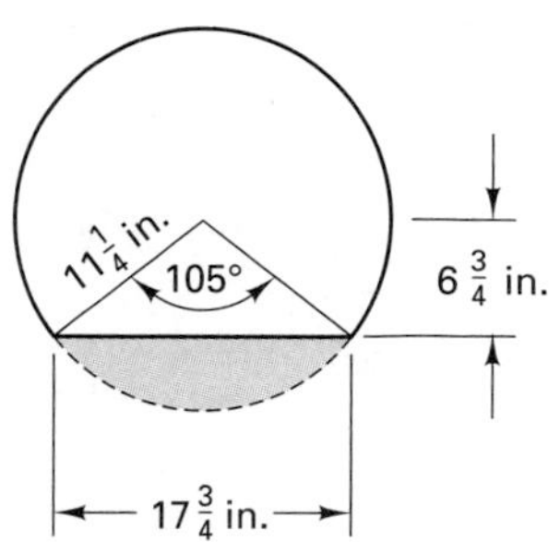

Figure 21-146

23. Find the circumference of a circle inscribed in an equilateral triangle if the height of the triangle is 12 in. (Fig. 21-147).

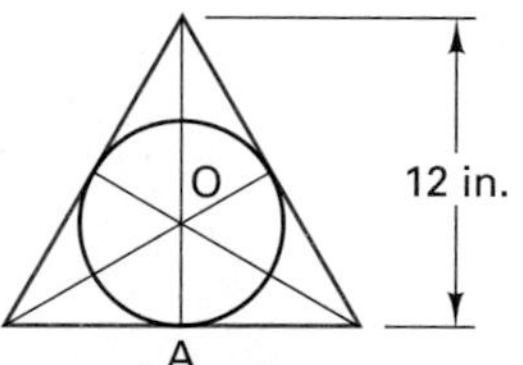

Figure 21-147

24. A square metal rod is milled so that a circle (Fig. 21-148) is formed at the end. If the square cross section is 1.8 in. across the corners, what is the diameter of the circle?

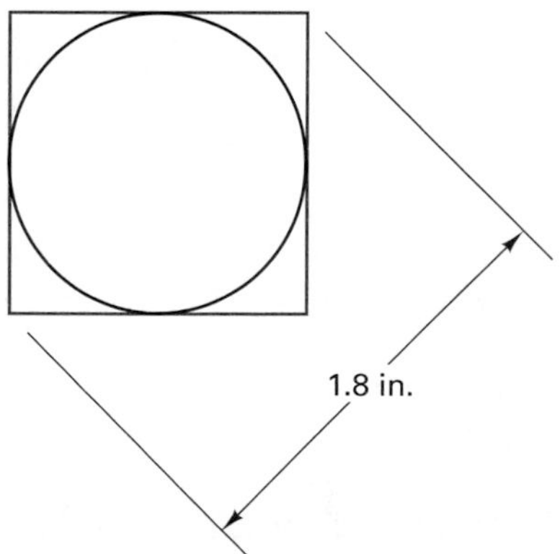

Figure 21-148

25. A circle is milled on the cross-sectional end of a hexagonal rod (Fig. 21-149). If the distance across the flats of the cross section is $\frac{1}{2}$ in., what is the circumference of the circle?

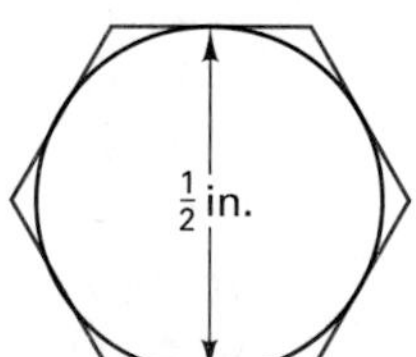

Figure 21-149

CHAPTER

22 Introduction to Trigonometry

One of the most important uses of *trigonometry* is to allow us to find by indirect measurement lengths or distances that are difficult or impossible to measure directly. Although trigonometry, which means "triangle measurement," involves more than a study of just the triangle, we will limit our study to the trigonometry of the triangle.

In our study of trigonometry, we will build on much of our knowledge of angles and triangles from geometry. We will also see that certain calculations are made more quickly and in fewer steps than they can be made using geometry alone. Also, we will see that certain calculations that cannot be made using geometry alone can be made using trigonometry.

22-1 RADIANS AND DEGREES

Learning Objectives

L.O.1. Find arc length, given a central angle in radian measure.
L.O.2. Find the area of a sector, given a central angle in radian measure.
L.O.3. Convert angles in degree measure to angles in radian measure.
L.O.4. Convert angles in radian measure to angles in degree measure.

In this section we study the angles formed by radii of a circle and how such angles are measured. We also use that measure to find additional information about the circle.

In geometry, we learned that angles can be measured in units called degrees. Another unit for measuring angles is the *radian*.

■ DEFINITION 22-1.1 **Radian.** A **radian** is the measure of a central angle of a circle whose intercepted arc is equal in length to the radius of the circle (Fig. 22-1).

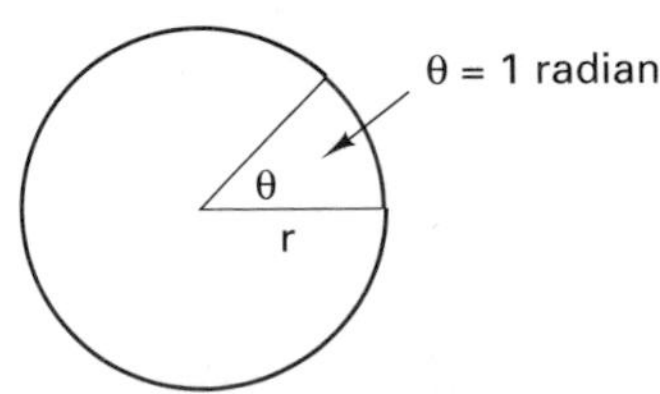

Figure 22-1

The circumference of a circle is related to the radius by the formula $C = 2\pi r$. Thus, the ratio of the circumference to the radius of any circle is $\frac{C}{r} = 2\pi$. That is, the length of the radius could be measured off 2π times (about 6.28 times) along the circumference. A complete rotation is 2π radians (see Fig. 22-2).

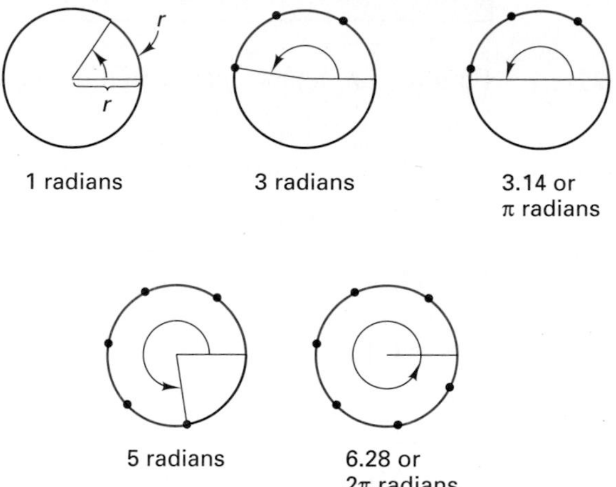

Figure 22-2

How are radians and degrees related? Since a central angle measuring 1 radian makes an arc length equal to the radius, we will use the arc-length formula that uses degrees to find the equivalent angle measure in degrees for 1 radian.

$$\text{Arc length} = \frac{\theta}{360}2\pi r$$

If the radius and arc length are equal to 1,

$$1 = \frac{\theta}{360}2\pi\,(1)$$

$$360 = \theta\,(2\pi) \qquad \textit{(Multiply both sides by 360.)}$$

$$\frac{360}{2\pi} = \frac{\theta\,(2\pi)}{2\pi} \qquad \textit{(Divide both sides by } 2\pi.\textit{)}$$

$$57.29577951 = \theta$$

One radian is approximately 57.3°.

L.O.1. Find arc length given a central angle in radian measure.

Arc Length

Radian measure is used very often in physics and engineering mechanics. One application of radian measure is in determining *arc length*, which is the length of the arc intercepted by a central angle.

The circumference of a circle is $2\pi r$, where r is the radius of the circle. 2π is the radian measure of a complete rotation. To find the arc length, s, of an arc, we can multiply the radian measure of the central angle formed by the end points of the arc and the center of the circle times the radius of the circle.

$$\text{Arc length} = (\text{radian measure of angle})(\text{radius})$$

We will use the Greek lowercase letter theta (θ) to represent the radian measure of the central angle in the formula for arc length.

Formula 22-1.1 *Arc length:*

$$s = \theta r$$

θ is measured in radians.

EXAMPLE 22-1.1 Find the arc length intercepted on the circumference of the circle in Fig. 22-3 by a central angle of $\frac{\pi}{3}$ radians (rad) if the radius of the circle is 10 cm.

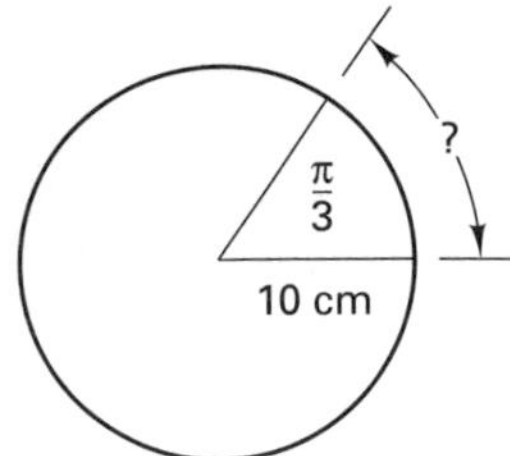

Figure 22-3

Solution

$$s = \theta r$$

$$s = \frac{\pi}{3}(10)$$

$$s = \frac{\pi(10)}{3}$$

$$s = 10.47 \text{ cm} \qquad \textit{(to the nearest hundredth)}$$

Thus, the arc length of the intercepted arc is 10.47 cm. ■

EXAMPLE 22-1.2 Find the radian measure of an angle at the center of a circle of radius 5 m. The angle intercepts an arc length of 12.5 m (Fig. 22-4).

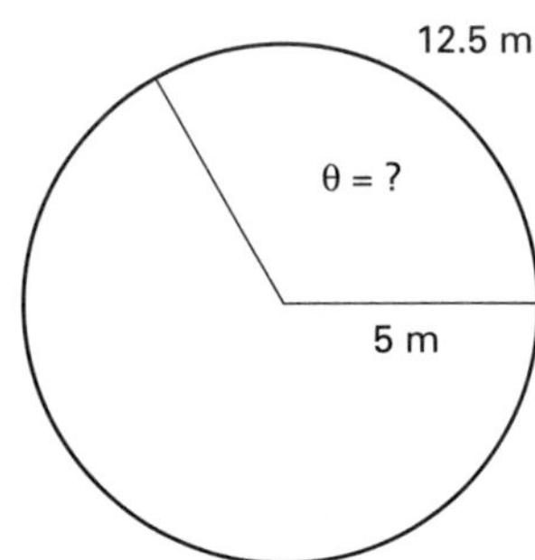

Figure 22-4

Solution: Since $s = \theta r$,

$$\theta = \frac{s}{r} \qquad \textit{(Formula is rearranged for } \theta.\textit{)}$$

$$\theta = \frac{12.5}{5}$$

$$\theta = 2.5 \text{ rad}$$

Thus, the angle is 2.5 rad. ■

EXAMPLE 22-1.3 Find the radius of an arc if the length of the arc is 8.22 cm and the intercepted central angle is 3 rad (Fig. 22-5).

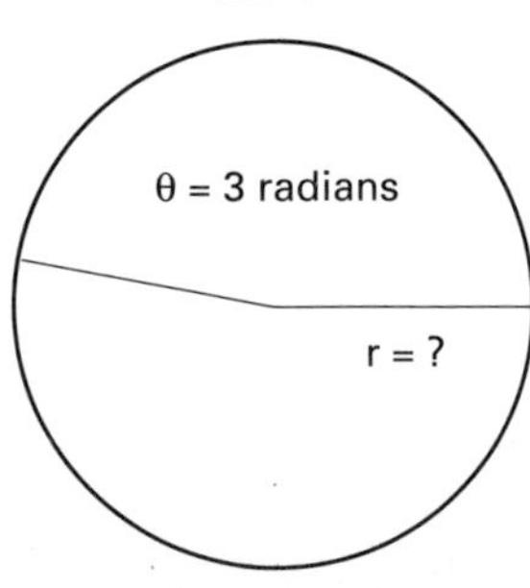

Figure 22-5

Solution: Since $s = \theta r$,

$$r = \frac{s}{\theta} \qquad \textit{(Formula is rearranged for } r.\textit{)}$$

$$r = \frac{8.22}{3}$$

$$r = 2.74 \text{ cm}$$

Therefore, the radius of the arc is 2.74 cm. ■

L.O.2. Find the area of a sector given a central angle in radian measure.

Area of Sector

The formula for the area of a circle is $A = \pi r^2$. To see how this relates to radian measures, we can rewrite the formula as $A = \frac{1}{2}(2\pi)r^2$. (2π rad is a complete rotation.) In Section 21-5 the formula for the area of a *sector* was given as $A = \frac{\theta}{360}\pi r^2$, where θ is expressed in degrees. An alternative formula using the angle measure expressed in radians may be used. To derive this formula for the area of part of a circle or the area of a sector, we substitute θ (the radian measure of the central angle that determines the sector) for 2π in the formula $A = \frac{1}{2}(2\pi)r^2$.

> **Formula 22-1.2** *Area of sector:*
>
> $$A = \frac{1}{2}\theta r^2$$
>
> θ is measured in radians.

EXAMPLE 22-1.4 Find the area of a sector whose central angle is 5 rad and whose radius is 7.2 in.

Solution

$$A = \frac{1}{2}\theta r^2$$

$$A = \frac{1}{2}(5)(7.2)^2$$

$$A = 129.6 \text{ in.}^2$$

Thus, the area of the sector is 129.6 in.^{2}. ■

L.O.3. Convert angles in degree measure to angles in radian measure.

Converting Degrees to Radians

Many job applications will require us to convert radians to degrees for certain calculations and convert degrees to radians for others.

A complete rotation is 360° or 2π rad. To convert from one type of angle measure to another, we multiply by a *unity ratio* that relates degrees and radians. Since $360° = 2\pi$ rad, we can simplify the relationship to $180° = \pi$ rad. $\left(\frac{360°}{2} = \frac{2\pi}{2}\text{ rad}\right)$

> **Rule 22-1.1**
>
> To convert degrees to radians, multiply degrees by the unity ratio $\frac{\pi \text{ rad}}{180°}$.

EXAMPLE 22-1.5 Convert the following degree measures to radians. Use the calculator value for π to change to the decimal answer rounded to the nearest hundredth.

(a) 20° **(b)** 50° **(c)** 175° **(d)** 270° **(e)** 67°

Solution

(a) Since $360° = 2\pi$ rad or $180° = \pi$ rad, we use this relationship in a unity ratio to convert degrees to radians.

$$20° \times \frac{\pi \text{ rad}}{180°} = \frac{20(\pi)}{180} = 0.35 \text{ rad}$$

$$\textbf{(b) } 50° \times \frac{\pi \text{ rad}}{180°} = \frac{50(\pi)}{180} = 0.87 \text{ rad}$$

$$\textbf{(c) } 175° \times \frac{\pi \text{ rad}}{180°} = \frac{175(\pi)}{180} = 3.05 \text{ rad}$$

$$\textbf{(d) } 270° \times \frac{\pi \text{ rad}}{180°} = \frac{270(\pi)}{180} = 4.71 \text{ rad}$$

$$\textbf{(e) } 67° \times \frac{\pi \text{ rad}}{180°} = \frac{67(\pi)}{180} = 1.17 \text{ rad}$$

■

When parts of a degree are expressed in decimals, we use the same procedure as before to convert to radians. When angle measures are expressed in degrees, minutes, and seconds, we must first convert the minutes and seconds to a decimal part of a degree before converting to radians. (This procedure was discussed in detail in Section 21-2).

EXAMPLE 22-1.6 Convert the following measures to radians. Round the degree measure to the nearest ten-thousandth; then round the radian measure to the nearest hundredth.

(a) 15.25° **(b)** 25°20′ **(c)** 110°25′30″

Solution

$$\textbf{(a) } 15.25° \times \frac{\pi \text{ rad}}{180°} = \frac{15.25(\pi)}{180} = 0.27 \text{ rad} \quad \textit{(rounded)}$$

$$\textbf{(b) } 25°20' = 25.3333° \quad \left(20' \times \frac{1°}{60'} = \frac{20°}{60} = 0.3333°\right)$$

$$25.3333° \times \frac{\pi \text{ rad}}{180°} = \frac{25.3333(\pi)}{180} = 0.44 \text{ rad} \quad \textit{(rounded)}$$

$$\textbf{(c) } 110°25'30'' = 110.425°$$

$$\left(110 + \frac{25}{60} + \frac{30}{3600} = 110.425\right)$$

$$110.425° \times \frac{\pi \text{ rad}}{180°} = \frac{110.425(\pi)}{180} = 1.93 \text{ rad} \quad \textit{(rounded)}$$

■

L.O.4. Convert angles in radian measure to angles in degree measure.

Converting Radians to Degrees

Rule 22-1.2

To convert radians to degrees, multiply radians by the unity ratio $\frac{180°}{\pi \text{ rad}}$.

EXAMPLE 22-1.7 Convert the following to degrees. Round to the nearest ten-thousandth of a degree.

(a) 2 rad **(b)** $\frac{\pi}{2}$ rad **(c)** 0.25 rad

Solution

$$\textbf{(a) } 2 \cancel{\text{rad}} \times \frac{180°}{\pi \cancel{\text{rad}}} = \frac{360}{\pi} = 114.5916° \quad \textit{(rounded)}$$

(b) $\frac{\overset{1}{\cancel{\pi}}}{2}\cancel{\text{rad}} \times \frac{180°}{\underset{1}{\cancel{\pi}\cancel{\text{rad}}}} = \frac{180}{2} = 90°$ *(Since π canceled, we did not have to substitute for π.)*

(c) $0.25\ \cancel{\text{rad}} \times \frac{180°}{\pi\ \cancel{\text{rad}}} = \frac{(0.25)(180)}{\pi} = 14.3239°$ *(rounded)*

■

Example 22-1.8 Convert the following to degrees, minutes, and seconds. Round to the nearest second.

(a) 1 rad (b) 3.2 rad

Solution

(a) $1\ \cancel{\text{rad}} \times \frac{180°}{\pi\ \cancel{\text{rad}}} = \frac{180°}{\pi} = 57.29577951°$

$0.29577951° \times \frac{60'}{1°} = 17.74677077'$ *(Convert decimal part of degree to minutes.)*

$0.74677077' \times \frac{60''}{1'} = 45''$ *(Convert decimal part of minute to seconds, to the nearest second.)*

Thus, 1 rad $= 57°17'45''$.

(b) $3.2\ \cancel{\text{rad}} \times \frac{180°}{\pi\ \cancel{\text{rad}}} = \frac{3.2(180)}{\pi} = 183.3464944°$

$0.3464944° \times \frac{60'}{1°} = 20.7896664'$ *(Convert decimal part of degree to minutes.)*

$0.7896664' \times \frac{60''}{1'} = 47''$ *(Convert decimal part of minute to seconds, to the nearest second.)*

Thus, 3.2 rad $= 183°20'47''$.

■

Whenever the central angle measure is expressed in degrees, we may use either Formula 21-5.1 or Formula 22-1.2. To use Formula 22-1.2, we must convert the central angle measure from degrees to radians. In the following example, the area is found using both formulas.

Example 22-1.9 Find the area of a sector whose central angle is 135° and whose radius is 2.7 cm.

Solution: Using Formula 21-5.1, $A = \frac{\theta}{360}\pi r^2$, where θ is given in degrees:

$A = \left(\frac{135}{360}\right)(\pi)(2.7)^2 = 8.59\text{ cm}^2$ *(to nearest hundredth)*

Using Formula 22-1.2, 135° must first be changed to radians.

$135° = 135° \times \frac{\pi\text{ rad}}{180°} = \frac{135(\pi)}{180} = 2.35619449\text{ rad}$

Then $A = \frac{1}{2}\theta r^2$, where θ is given in radian measure.

$$A = \frac{1}{2}(2.35619449)(2.7)^2 = 8.59 \text{ cm}^2 \qquad \textit{(to nearest hundredth)}$$

Therefore, the area of the sector is approximately 8.59 cm^2. ■

SELF-STUDY EXERCISES 22-1

L.O.1. Solve the following problems. Round answers to hundredths if necessary.

1. Find the arc length intercepted on the circumference of a circle by a central angle of 2.15 rad if the radius of the circle is 3 in.
2. Find the arc length intercepted on the circumference of a circle by a central angle of 4 rad if the radius of the circle is 3.5 cm.
3. Find the radian measure of an angle at the center of a circle of radius 2 in. if the angle intercepts an arc length of 8.5 in.
4. Find the radian measure of an angle at the center of a circle of radius 4.3 cm if the angle intercepts an arc length of 15 cm.
5. Find the radius of an arc if the length of the arc is 14.7 cm and the intercepted central angle is 2.1 rad.
6. Find the radius of an arc if the length of the arc is 12.375 in. and the intercepted central angle is 2.75 rad.

L.O.2.

7. Find the area of a sector whose central angle is 2.14 rad and whose radius is 4 in.
8. Find the area of a sector whose central angle is 6 rad and whose radius is 1.2 cm.
9. Find the radius of a sector if the area of the sector is 7.5 cm^2 and the central angle is 3 rad.
10. How many radians does the central angle of a sector measure if its area is 1.7 $in.^2$ and its radius is 2 in.?

L.O.3. Convert the following degree measures to radians rounded to the nearest hundredth.

11. 45° **12.** 56° **13.** 78°
14. 140°

Convert the following measures to degrees rounded to the nearest ten-thousandth. Then convert to radians to the nearest hundredth.

15. 21°45′ **16.** 177°33′ **17.** 44°54′12″
18. 10°31′15″

L.O.4. Convert the following radian measures to degrees. Round to the nearest ten-thousandth of a degree.

19. $\frac{\pi}{4}$ rad **20.** $\frac{\pi}{6}$ rad **21.** 2.5 rad
22. 1.4 rad

Convert the following radian measures to degrees, minutes, and seconds. Round to the nearest second.

23. 0.5 rad **24.** $\frac{\pi}{8}$ rad **25.** 0.75 rad
26. 1.1 rad

27. Find the arc length of an arc whose intercepted angle is 38° and whose radius is 2.3 cm. Round to hundredths.
28. Find the number of degrees in a central angle whose arc length is 3.2 in. and whose radius is 3 in. Round to the nearest whole degree.
29. Find the area of a sector whose central angle is 105° and whose radius is 7.2 cm. Round to hundredths.
30. Find the number of degrees to the nearest ten-thousandth of a central angle of a sector whose area is 5.6 cm^2 and whose radius is 4 cm.

22-2 TRIGONOMETRIC FUNCTIONS

Learning Objectives

L.O.1. Find the sine, cosine, and tangent of right triangles, given the measures of at least two sides.

L.O.2. Find the cosecant, secant, and cotangent of right triangles, given the measures of at least two sides.

L.O.1. Find the sine, cosine, and tangent of right triangles, given the measures of at least two sides.

Sine, Cosine, Tangent

In geometry, we studied some basic properties of similar triangles and right triangles. In trigonometry, we can determine the measure of either acute angle of a right triangle if we know the measure of at least two sides of the right triangle. We will define several functions that are ratios of various sides of a triangle, and these ratios will be used later in determining the angles of a triangle.

Figure 22-6 shows a right triangle, ABC, with the sides of the triangle labeled according to their relationship to angle A. The *hypotenuse* is the side opposite the right angle of the triangle, and the hypotenuse forms one of the sides of angle A. The other side that forms angle A will be called the *adjacent side* of angle A. The third side of the triangle is called the *opposite side* of angle A.

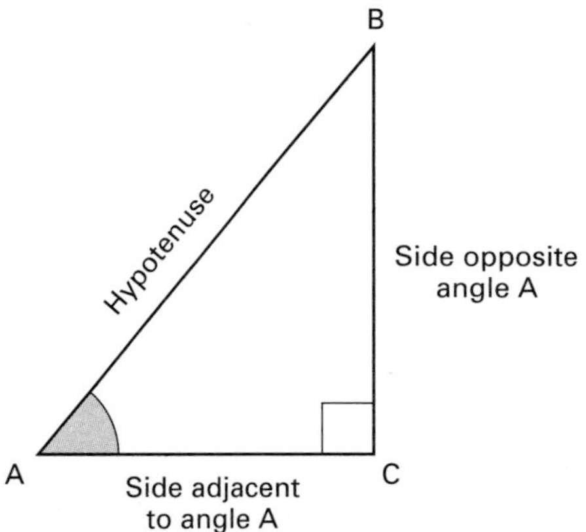

Figure 22-6

In Fig. 22-7, the sides of the right triangle ABC are labeled according to their relationship to angle B. The hypotenuse forms one of the sides of angle B, the other side that forms angle B is the adjacent side of angle B, and the third side is the opposite side of angle B. We used in Section 13-3 the term *hypotenuse*, but the terms *adjacent side* and *opposite side* are also very important in understanding trigonometric functions.

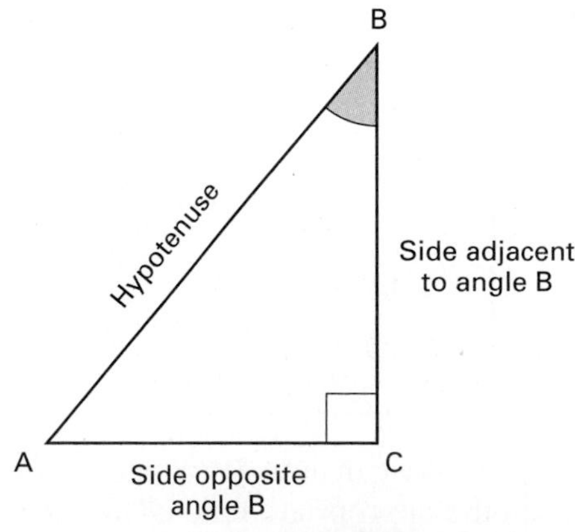

Figure 22-7

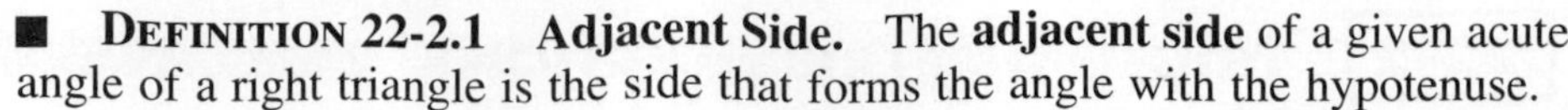

■ **Definition 22-2.1 Adjacent Side.** The **adjacent side** of a given acute angle of a right triangle is the side that forms the angle with the hypotenuse.

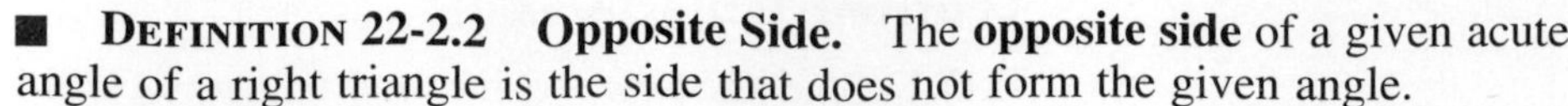

■ **Definition 22-2.2 Opposite Side.** The **opposite side** of a given acute angle of a right triangle is the side that does not form the given angle.

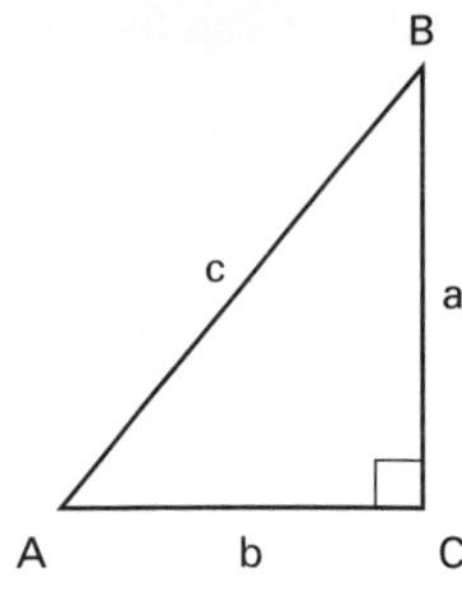

Standard right triangle

Figure 22-8

The three most commonly used trigonometric functions are the *sine*, *cosine*, and *tangent*. The sine, cosine, and tangent of angle A (Fig. 22-8) are defined to be the following ratios of the sides of the right triangle.

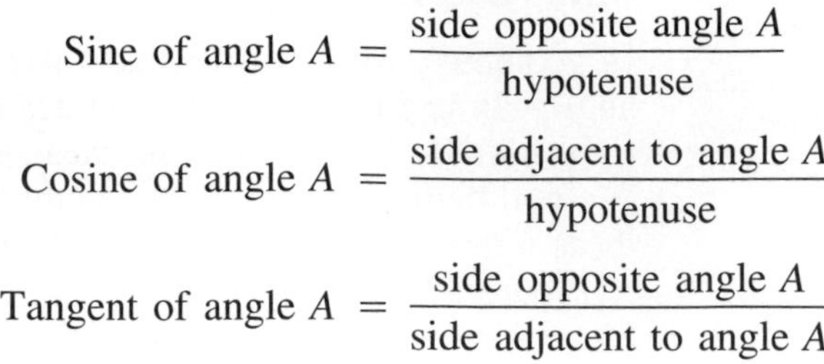

$$\text{Sine of angle } A = \frac{\text{side opposite angle } A}{\text{hypotenuse}}$$

$$\text{Cosine of angle } A = \frac{\text{side adjacent to angle } A}{\text{hypotenuse}}$$

$$\text{Tangent of angle } A = \frac{\text{side opposite angle } A}{\text{side adjacent to angle } A}$$

For convenience, we will abbreviate the sine, cosine, and tangent functions as *sin*, *cos*, and *tan*, respectively.

Furthermore, we will abbreviate the names of the ratios and use the letters that designate the sides of the *standard right triangle* in Fig. 22-8. In the standard right triangle, side a is opposite angle A, side b is opposite angle B, and side c (hypotenuse) is opposite angle C (right angle).

Thus, we can identify the functions of angle A as follows.

Formula 22-2.1 ***Trigonometric functions of angle A:***

$$\sin A = \frac{\text{side opposite } \angle A}{\text{hypotenuse}} = \frac{a}{c}$$

$$\cos A = \frac{\text{side adjacent to } \angle A}{\text{hypotenuse}} = \frac{b}{c}$$

$$\tan A = \frac{\text{side opposite } \angle A}{\text{side adjacent to } \angle A} = \frac{a}{b}$$

Similarly, we can identify the sine, cosine, and tangent of the other acute angle, angle B.

Formula 22-2.2 ***Trigonometric functions of angle B:***

$$\sin B = \frac{\text{side opposite } \angle B}{\text{hypotenuse}} = \frac{b}{c}$$

$$\cos B = \frac{\text{side adjacent to } \angle B}{\text{hypotenuse}} = \frac{a}{c}$$

$$\tan B = \frac{\text{side opposite } \angle B}{\text{side adjacent to } \angle B} = \frac{b}{a}$$

We can now determine the sine, cosine, and tangent of a right triangle by writing the appropriate ratio of the sides of the triangle and expressing the ratio in lowest terms or as a decimal equivalent.

EXAMPLE 22-2.1 Find the sine, cosine, and tangent of angles A and B. Leave the answers as fractions in lowest terms (Fig. 22-9).

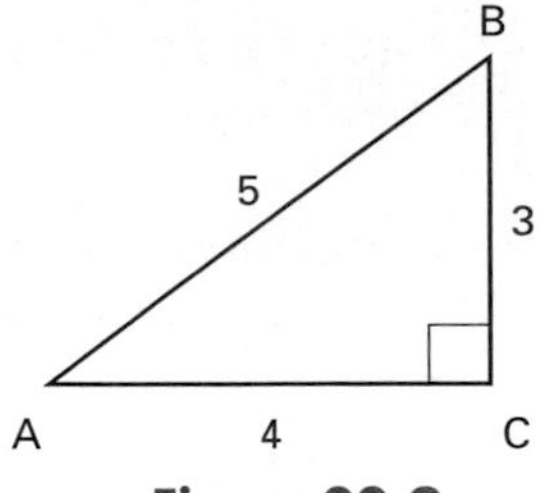

Figure 22-9

Solution

$$\sin A = \frac{\text{opposite}}{\text{hypotenuse}} = \frac{3}{5}$$

$$\cos A = \frac{\text{adjacent}}{\text{hypotenuse}} = \frac{4}{5}$$

$$\tan A = \frac{\text{opposite}}{\text{adjacent}} = \frac{3}{4}$$

$$\sin B = \frac{\text{opposite}}{\text{hypotenuse}} = \frac{4}{5}$$

$$\cos B = \frac{\text{adjacent}}{\text{hypotenuse}} = \frac{3}{5}$$

$$\tan B = \frac{\text{opposite}}{\text{adjacent}} = \frac{4}{3}$$

■

EXAMPLE 22-2.2 Find the sine, cosine, and tangent of angles A and B. Write the answers as decimals. Round to the nearest ten-thousandth (Fig. 22-10).

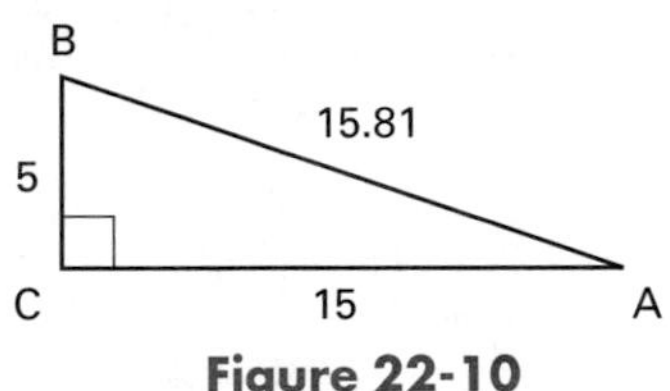

Figure 22-10

Solution

$$\sin A = \frac{\text{opposite}}{\text{hypotenuse}} = \frac{5}{15.81} = 0.3163$$

$$\cos A = \frac{\text{adjacent}}{\text{hypotenuse}} = \frac{15}{15.81} = 0.9488$$

$$\tan A = \frac{\text{opposite}}{\text{adjacent}} = \frac{5}{15} = 0.3333$$

$$\sin B = \frac{\text{opposite}}{\text{hypotenuse}} = \frac{15}{15.81} = 0.9488$$

$$\cos B = \frac{\text{adjacent}}{\text{hypotenuse}} = \frac{5}{15.81} = 0.3163$$

$$\tan B = \frac{\text{opposite}}{\text{adjacent}} = \frac{15}{5} = 3$$

■

Tips and Traps

When both terms of a ratio are expressed in the same unit of measure, that is, *like* units, then the ratio itself is unitless because the common units cancel. Thus, values of trigonometric ratios are numerical values with no unit of measure.

Example 22-2.3 Find the sine, cosine, and tangent of angles A and B. Leave the answers as fractions in lowest terms (Fig. 22-11).

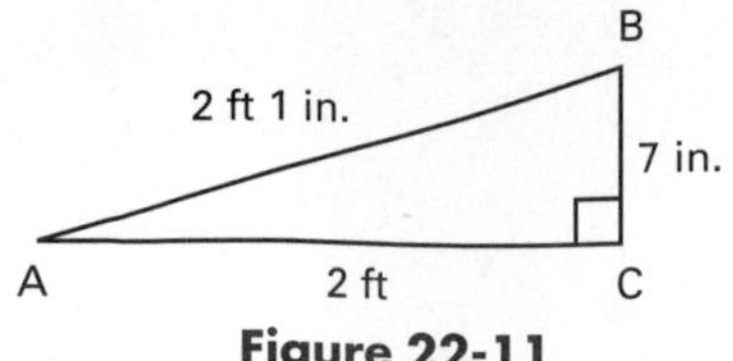

Figure 22-11

Solution

$$\sin A = \frac{\text{opposite}}{\text{hypotenuse}} = \frac{7 \text{ in.}}{2 \text{ ft } 1 \text{ in.}} = \frac{7 \text{ in.}}{25 \text{ in.}} = \frac{7}{25}$$

$$\cos A = \frac{\text{adjacent}}{\text{hypotenuse}} = \frac{2 \text{ ft}}{2 \text{ ft } 1 \text{ in.}} = \frac{24 \text{ in.}}{25 \text{ in.}} = \frac{24}{25}$$

$$\tan A = \frac{\text{opposite}}{\text{adjacent}} = \frac{7 \text{ in.}}{2 \text{ ft}} = \frac{7 \text{ in.}}{24 \text{ in.}} = \frac{7}{24}$$

$$\sin B = \frac{\text{opposite}}{\text{hypotenuse}} = \frac{2 \text{ ft}}{2 \text{ ft } 1 \text{ in.}} = \frac{24 \text{ in.}}{25 \text{ in.}} = \frac{24}{25}$$

$$\cos B = \frac{\text{adjacent}}{\text{hypotenuse}} = \frac{7 \text{ in.}}{2 \text{ ft } 1 \text{ in.}} = \frac{7 \text{ in.}}{25 \text{ in.}} = \frac{7}{25}$$

$$\tan B = \frac{\text{opposite}}{\text{adjacent}} = \frac{2 \text{ ft}}{7 \text{ in.}} = \frac{24 \text{ in.}}{7 \text{ in.}} = \frac{24}{7}$$

■

L.O.2. Find the cosecant, secant, and cotangent of right triangles, given the measures of at least two sides.

Cosecant, Secant, Cotangent

There are three other trigonometric functions. These functions are not used as often as the sine, cosine, and tangent funtions, but because of the relationships between these new functions and the previously learned functions, these new functions are useful to know.

Each of the three basic trigonometric functions (sine, cosine, and tangent) has a *reciprocal function.* The *cosecant* (*csc*) function is the reciprocal of the sine function, the *secant* (*sec*) function is the reciprocal of the cosine function, and the *cotangent* (*cot*) function is the reciprocal of the tangent function.

Using Fig. 22-12, we can write the reciprocal trigonometric functions of angle A.

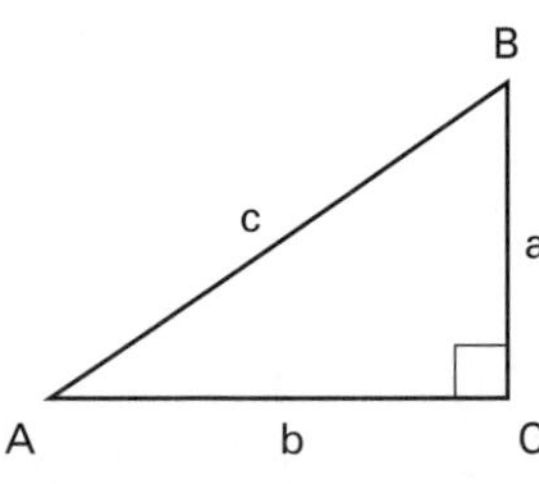

Standard right triangle

Figure 22-12

Formula 22-2.3 ***Reciprocal trigonometric functions of angle A:***

$$\csc A = \frac{\text{hypotenuse}}{\text{opposite}} = \frac{c}{a}$$

$$\sec A = \frac{\text{hypotenuse}}{\text{adjacent}} = \frac{c}{b}$$

$$\cot A = \frac{\text{adjacent}}{\text{opposite}} = \frac{b}{a}$$

Similarly, we can write the reciprocal trigonometric functions of angle B in Fig. 22-12.

Formula 22-2.4 ***Reciprocal trigonometric functions of angle B:***

$$\csc B = \frac{\text{hypotenuse}}{\text{opposite}} = \frac{c}{b}$$

$$\sec B = \frac{\text{hypotenuse}}{\text{adjacent}} = \frac{c}{a}$$

$$\cot B = \frac{\text{adjacent}}{\text{opposite}} = \frac{a}{b}$$

EXAMPLE 22-2.4 Write the six trigonometric ratios for angles A and B in Fig. 22-13.

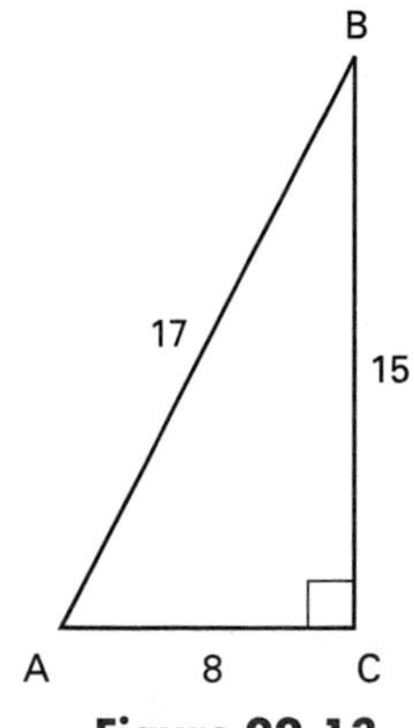

Figure 22-13

Solution

$$\sin A = \frac{\text{opp}}{\text{hyp}} = \frac{15}{17} \qquad \sin B = \frac{\text{opp}}{\text{hyp}} = \frac{8}{17}$$

$$\cos A = \frac{\text{adj}}{\text{hyp}} = \frac{8}{17} \qquad \cos B = \frac{\text{adj}}{\text{hyp}} = \frac{15}{17}$$

$$\tan A = \frac{\text{opp}}{\text{adj}} = \frac{15}{8} \qquad \tan B = \frac{\text{opp}}{\text{adj}} = \frac{8}{15}$$

$$\csc A = \frac{\text{hyp}}{\text{opp}} = \frac{17}{15} \qquad \csc B = \frac{\text{hyp}}{\text{opp}} = \frac{17}{8}$$

$$\sec A = \frac{\text{hyp}}{\text{adj}} = \frac{17}{8} \qquad \sec B = \frac{\text{hyp}}{\text{adj}} = \frac{17}{15}$$

$$\cot A = \frac{\text{adj}}{\text{opp}} = \frac{8}{15} \qquad \cot B = \frac{\text{adj}}{\text{opp}} = \frac{15}{8}$$

(For convenience, opposite, hypotenuse, *and* adjacent *are abbreviated here and elsewhere.)*

■

EXAMPLE 22-2.5 Find the six trigonometric functions for angles A and B in Fig. 22-14. Express ratios to the nearest ten-thousandth.

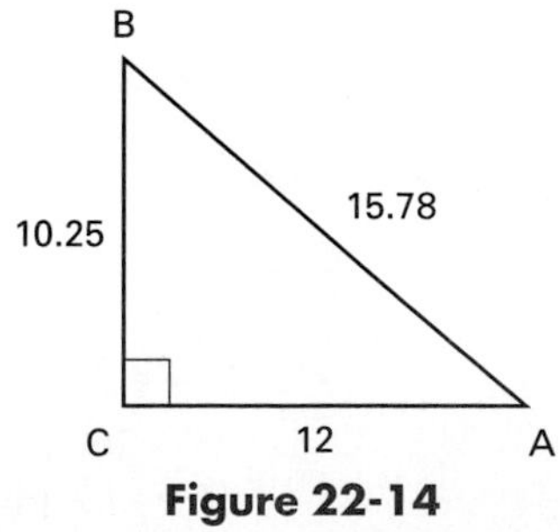

Figure 22-14

Solution

$$\sin A = \frac{10.25}{15.78} = 0.6496 \qquad \sin B = \frac{12}{15.78} = 0.7605$$

$$\cos A = \frac{12}{15.78} = 0.7605 \qquad \cos B = \frac{10.25}{15.78} = 0.6496$$

$$\tan A = \frac{10.25}{12} = 0.8542 \qquad \tan B = \frac{12}{10.25} = 1.1707$$

$$\csc A = \frac{15.78}{10.25} = 1.5395 \qquad \csc B = \frac{15.78}{12} = 1.3150$$

$$\sec A = \frac{15.78}{12} = 1.3150 \qquad \sec B = \frac{15.78}{10.25} = 1.5395$$

$$\cot A = \frac{12}{10.25} = 1.1707 \qquad \cot B = \frac{10.25}{12} = 0.8542$$

■

SELF-STUDY EXERCISES 22-2

L.O.1. Using Fig. 22-15, find the indicated trigonometric ratios for the following exercises.

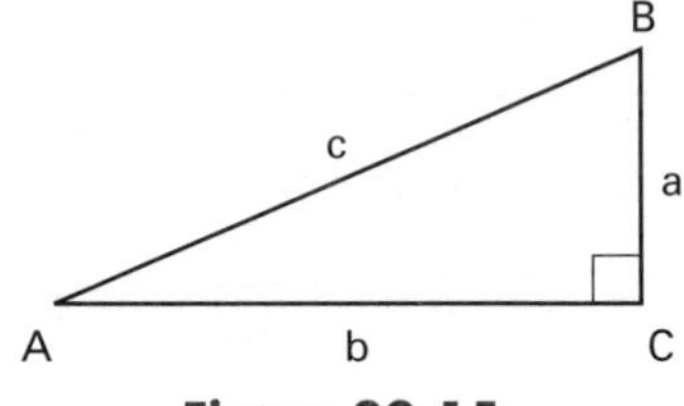

Figure 22-15

$a = 5$, $b = 12$, and $c = 13$. Express the ratios as fractions in lowest terms.

1. $\sin A$ **2.** $\cos A$ **3.** $\tan A$ **4.** $\sin B$ **5.** $\cos B$ **6.** $\tan B$

$a = 9$, $b = 12$, and $c = 15$. Express the ratios as fractions in lowest terms.

7. $\sin A$ **8.** $\cos A$ **9.** $\tan A$ **10.** $\sin B$ **11.** $\cos B$ **12.** $\tan B$

$a = 16$, $b = 30$, and $c = 34$. Express the ratios as fractions in lowest terms.

13. $\sin A$ **14.** $\cos A$ **15.** $\tan A$ **16.** $\sin B$ **17.** $\cos B$ **18.** $\tan B$

$a = 9$, $b = 14$, and $c = 16.64$. Express the ratios as decimals to the nearest ten-thousandth.

19. $\sin A$ **20.** $\cos A$ **21.** $\tan A$ **22.** $\sin B$ **23.** $\cos B$ **24.** $\tan B$

L.O.2. Determine the six trigonometric ratios for angles A and B in Fig. 22-16.

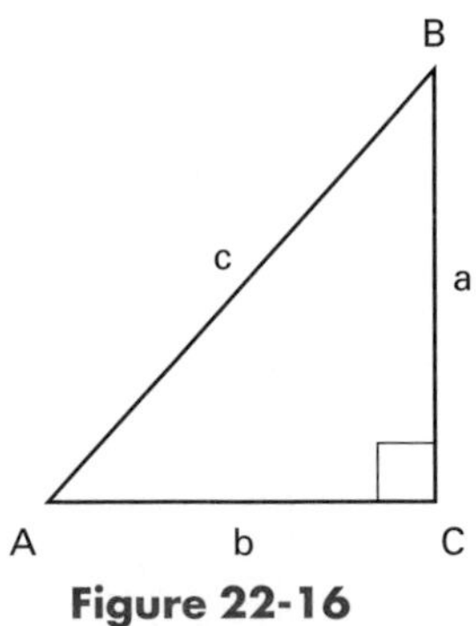

Figure 22-16

25. $a = 12$, $b = 16$, $c = 20$

26. $a = 8.15$, $b = 5.32$, $c = 9.73$. Express the ratios as decimals to the nearest ten-thousandth.

22-3 USING A CALCULATOR TO FIND TRIGONOMETRIC VALUES

Learning Objectives

L.O.1. Find trigonometric values for sine, cosine, and tangent using a calculator.
L.O.2. Find the angle measure given a trigonometric value.
L.O.3. Find the trigonometric values for cosecant, secant, and cotangent using a calculator.

A number of trigonometric ratios are used so frequently that mathematicians have prepared tables that give the various calculations. These tables of trigonometric functions were used extensively before the increased availability of the scientific calculator.

In studying similar triangles we found that corresponding angles of similar triangles are equal and that corresponding sides are proportional. These properties of similar triangles lead us to a very important property of trigonometric functions:

Every angle of a specified measure has a specific set of values for its trigonometric functions.

L.O.1. Find trigonometric values for sine, cosine, and tangent using a calculator.

Finding the Sine, Cosine, or Tangent

It is recommended that a calculator be used to find trigonometric values. To find trigonometric values on a calculator, the keys *sin*, *cos*, and *tan* are used. The angle measure can be entered as a degree measure or a radian measure, depending on the selected mode. Most calculators are preset to use degree mode unless another mode is selected.

General Tips for Using the Calculator

To find trigonometric values for sine, cosine, and tangent using a scientific calculator:

1. Set the calculator to the desired mode of angle measure. Calculators generally have three angle modes: degrees, radians, and gradients.
2. Enter the angle measure and then press the appropriate trigonometric function key (sin, cos, tan).

To find trigonometric values for sine, cosine, and tangent using a graphing calculator:

1. Set the calculator to the desired mode.

Casio-fx 7700: [Shift] [1] (DRG) [F1] [EXE] for degrees

[Shift] [1] (DRG) [F2] [EXE] for radians

To check mode, press [M Disp]

TI-81 or 82 [Mode] (move arrow keys to highlight Rad or Deg) [ENTER] [CLEAR]

To check mode, press [Mode].

2. Press the appropriate trigonometric key, then the angle measure, followed by [EXE] or [ENTER].

EXAMPLE 22-3.1 Using a calculator, find the following trigonometric values.

(a) sin 27° **(b)** cos 52° **(c)** tan 85° **(d)** sin 20°30′

(e) cos 1.34 **(f)** $\tan \frac{\pi}{4}$

Solution

(a) Be sure the calculator is in degree mode.
Scientific: 27 [SIN] ⇒ 0.453990499
Graphing: [SIN] 27 [EXE] ⇒ 0.4539904997

(b) Be sure the calculator is in degree mode.
Scientific: 52 [COS] ⇒ 0.615661475
Graphing: [COS] 52 [EXE] ⇒ 0.6156614753

(c) Be sure the calculator is in degree mode.
Scientific: 85 [TAN] ⇒ 11.4300523
Graphing: [TAN] 85 [EXE] ⇒ 11.4300523

(d) Be sure the calculator is in degree mode.
Scientific: 20.5 [SIN] ⇒ 0.350207381
(Some scientific calculators allow an angle measure to be entered using degrees, minutes, and seconds *or* using the decimal degree equivalent.)
Graphing: [SIN] 20.5 [EXE] ⇒ 0.3502073813 ■

Tips and Traps

To simplify notation, we will agree that angle measures in degrees must be accompanied by the degree symbol (°). Angle measures in radians will have no symbol or abbreviation indicated. For example, sin 1° means sine of one degree. Sin 1 means sine of one radian.

(e) Reset the calculator to radian mode.
Scientific: 1.34 [COS] ⇒ 0.2287528078
Graphing: [COS] 1.34 [EXE] ⇒ 0.2287528078

(f) Be sure the calculator is in radian mode.
Scientific: [(] [EXP] (π) [÷] 4 [)] [TAN] ⇒ 1
Graphing: [TAN] [(] [Shift] [EXP] (π) [÷] 4 [EXE] ⇒ 1

L.O.2. Find the angle measure, given a trigonometric value.

Finding Angle Measures from Trigonometric Values

On many occasions it is necessary to determine the angle measure from a trigonometric value.

EXAMPLE 22-3.2 Determine the degree measure of an angle of the following trigonometric value using a scientific or graphing calculator. θ represents the unknown angle measure. Round to the nearest tenth of a degree.

(a) $\sin \theta = 0.6561$ **(b)** $\cos \theta = 0.4226$ **(c)** $\tan \theta = 0.3327$

Determine the radian measure of the following to the nearest thousandth.

(d) $\sin \theta = 0.3287$ **(e)** $\tan \theta = 2.825$

General Tips for Using the Calculator

To find angle measures from trigonometric values using a calculator:

Scientific:

1. Set the calculator to the desired mode of angle measure (degrees or radians).
2. Enter the trigonometric value and press the *inverse* function key [INV] or *shift* function key [Shift] and then the appropriate trigonometric function key (sin, cos, tan). Calculators may have $\sin^{-1}$, $\cos^{-1}$, and $\tan^{-1}$ as inverse function keys.

Graphing:

1. Set the calculator to the desired mode of angle measure (degrees or radians).
2. Enter the appropriate inverse trigonometric function key ($\sin^{-1}$, $\cos^{-1}$, or $\tan^{-1}$). Enter the trigonometic value and [EXE] or [ENTER].

Solution

(a) Be sure the calculator is in degree mode.

Scientific: .6561 [INV] [SIN] (SIN^{-1}) ⇒ 41.0031105
= 41.0°

Graphing: [INV] [SIN] (SIN^{-1}) .6561 [EXE] ⇒ 41.0031105
= 41.0°

(b) Be sure the calculator is in degree mode.

Scientific: .4226 [INV] [COS] (COS^{-1}) ⇒ 65.00115448
= 65.0°

Graphing: [SHIFT] [COS] (COS^{-1}) .4226 [EXE] ⇒ 65.00115448
= 65.0°

(c) Be sure the calculator is in degree mode.

Scientific: .3327 [INV] [TAN] (TAN^{-1}) ⇒ 18.40228403
= 18.4°

Graphing: [SHIFT] [TAN] (TAN^{-1}) .3327 [EXE] ⇒ 18.40228403
= 18.4°

(d) Be sure the calculator is in radian mode.

Scientific: .3287 [INV] [SIN] (SIN^{-1}) ⇒ 0.334926759
= 0.335 rad

Graphing: [SHIFT] [SIN] (SIN^{-1}) .3287 [EXE] ⇒ 0.3349267596
= 0.335 rad

(e) Be sure the calculator is in radian mode.

Scientific: 2.825 [INV] [TAN] (TAN^{-1}) $\Rightarrow$ 1.230578215
$= 1.231$ rad

Graphing: [SHIFT] [TAN] (TAN^{-1}) 2.825 [EXE] $\Rightarrow$ 1.230578215
$= 1.231$ rad ■

A calculator can be used to investigate relationships among trigonometric functions. Examine the following values found on a calculator.

$\sin 40° = 0.6428$	$\cos 50° = 0.6428$
$\sin 30° = 0.5$	$\cos 60° = 0.5$
$\sin 80° = 0.9848$	$\cos 10° = 0.9848$
$\sin 90° = 1$	$\cos 0° = 1$

We can generalize our observations.

Rule 22-3.1 ***The relationship of the function of an angle and the function of its complement:***

The sine of an angle equals the cosine of its complement.

L.O.3. Find the trigonometric values for cosecant, secant, and cotangent using a calculator.

Reciprocal Functions

In using a calculator, the reciprocal key, [1/x], can be used to find the value of reciprocal trigonometric functions. For instance, after the sine of an angle is found, the cosecant can be determined by pressing the reciprocal key.

The cotangent, secant, and cosecant functions are not used as often as the other functions.

The following identities indicate the reciprocal relationships of the tangent and cotangent, the sine and cosecant, and the cosine and secant functions.

$$\cot \theta = \frac{1}{\tan \theta} \qquad \csc \theta = \frac{1}{\sin \theta} \qquad \sec \theta = \frac{1}{\cos \theta}$$

EXAMPLE 22-3.3 Determine the value of the indicated trigonometric functions.

(a) csc 18.5° **(b)** sec 1.0821

Solution

(a) $\csc 18.5° = \dfrac{1}{\sin 18.5°} = \dfrac{1}{0.317304656} = 3.1515$ *(to nearest ten-thousandth)*

Scientific (in degree mode):
18.5 [SIN] [1/x] $\Rightarrow$ 3.151545305
Graphing (in degree mode):
[SIN] 18.5 [EXE] [Shift] [x^{-1}] [EXE] $\Rightarrow$ 3.151545305

(b) $\sec 1.0821 = \frac{1}{\cos 1.0821} = \frac{1}{0.469475214} = 2.1300$ *(to nearest ten-thousandth)*

Scientific (in radian mode):

1.0821 [COS] [1/x] ⇒ 2.130037898

Graphing (in radian mode):

[COS] 1.0821 [EXE] [x^{-1}] [EXE] ⇒ 2.130037898 ■

Examine the following trigonometric values, looking for patterns.

$\tan 30° = 0.5774$	$\cot 60° = 0.5774$
$\tan 10° = 0.1763$	$\cot 80° = 0.1763$
$\tan 45° = 1$	$\cot 45° = 1$
$\tan 90° =$ error *(undefined)*	$\cot 0° =$ error *(undefined)*
$\sec 30° = 1.1547$	$\csc 60° = 1.1547$
$\sec 10° = 1.0154$	$\csc 80° = 1.0154$
$\sec 45° = 1.4142$	$\csc 45° = 1.4142$
$\sec 90° =$ error	$\csc 0° =$ error

Rule 22-3.2 ***The relationship of the function of an angle and the function of its complement.***

The tangent of an angle equals the cotangent of its complement.
The secant of an angle equals the cosecant of its complement.

Tips and Traps

Trigonometric functions can be paired as cofunctions or as reciprocal functions.

Cofunctions

sine θ	cosine θ
secant θ	cosecant θ
tangent θ	cotangent θ

The value of a function of an angle and the value of the cofunction of the complement of the angle are equal.

For degrees:	For radians:
$\sin \theta = \cos (90° - \theta)$	$\sin \theta = \cos \left(\frac{\pi}{2} - \theta\right)$
$\sec \theta = \csc (90° - \theta)$	$\sec \theta = \csc \left(\frac{\pi}{2} - \theta\right)$
$\tan \theta = \cot (90° - \theta)$	$\tan \theta = \cot \left(\frac{\pi}{2} - \theta\right)$

Reciprocal functions

$$\text{sine } \theta = \frac{1}{\text{cosecant } \theta}$$

$$\text{cosine } \theta = \frac{1}{\text{secant } \theta}$$

$$\text{tangent } \theta = \frac{1}{\text{cotangent } \theta}$$

EXAMPLE 22-3.4 Find the degree measure of an angle whose trigonometric value is given. Express the measure to the nearest tenth of a degree.

(a) sec θ = 2.7320 **(b)** csc θ = 5.9137

Find the radian measure to the nearest hundredth radian.

(c) cot θ = 0.2167

Solution: First, find the reciprocal of the given value. Then find the inverse of the reciprocal function.

(a) *Scientific* (in degree mode): *Graphing:*

2.732 [1/x] [INV] [COS] (COS^{-1}) 2.732 [Shift] [)] (x^{-1}) [EXE] [Shift] [COS] (COS^{-1}) [Ans] [EXE]

$\theta = 68.5°$

(b) *Scientific* (in degree mode): *Graphing:*

5.9137 [1/x] [INV] [SIN] (SIN^{-1}) 5.9137 [Shift] [)] (x^{-1}) [EXE] [Shift] [SIN] (SIN^{-1}) [Ans] [EXE]

$\theta = 9.7°$

(c) *Scientific* (in radian mode): *Graphing:*

.2167 [1/x] [INV] [TAN] (TAN^{-1}) .2167 [Shift] [)] (x^{-1}) [EXE] [Shift] [TAN] (TAN^{-1}) [Ans] [EXE]

$\theta = 1.36$ rad ■

SELF-STUDY EXERCISES 22-3

L.O.1. Use a calculator to find the following trigonometric values. Express answers in ten-thousandths.

1. sin 21° **2.** cos 3.5° **3.** tan 47°
4. cos 52.5° **5.** sin 0.5498 **6.** cos 21°30′
7. cos 1.1519 **8.** cos 0.3665 **9.** sin 53°30′
10. tan 42.5° **11.** tan 47.7° **12.** sin 62°10′
13. cos 12°40′ **14.** cos 1.0530 **15.** tan 73°14′
16. sin 1.2363 **17.** cos 46.8° **18.** cos 0.3549
19. cos 1.1636 **20.** tan 12.4°

L.O.2. Determine the degree measure of an angle of the following trigonometric values. θ represents the unknown angle measure. Express answer to the nearest tenth of a degree.

21. sin θ = 0.3420 **22.** cos θ = 0.9239 **23.** tan θ = 2.356
24. cos θ = 0.4617 **25.** cos θ = 0.540 **26.** sin θ = 0.5712

27. $\tan \theta = 1.265$
28. $\cos \theta = 0.137$
29. $\sin \theta = 0.6298$
30. $\cos \theta = 0.9325$

Determine the radian measure of an angle of the following trigonometric values. Express answer to the nearest ten-thousandth.

31. $\sin \theta = 0.5299$
32. $\tan \theta = 0.8098$
33. $\cos \theta = 0.6947$
34. $\cos \theta = 0.3907$
35. $\cos \theta = 0.968$
36. $\sin \theta = 0.9959$
37. $\tan \theta = 0.3160$
38. $\tan \theta = 2.430$
39. $\cos \theta = 0.9610$
40. $\cos \theta = 0.4210$

L.O.3. Find the indicated trig values using a calculator. Round to the nearest ten-thousandth.

41. cot 24.5°
42. csc 42°
43. sec 0.2443
44. csc 1.3788
45. sec 1.2165
46. cot 28.6°
47. cot 87°
48. sec 42°30′
49. csc 0.2136
50. sec 1.0372

Find θ to the nearest tenth of a degree.

51. $\cot \theta = 0.4238$
52. $\sec \theta = 1.8291$
53. $\csc \theta = 3.7129$
54. $\sec \theta = 8.2156$
55. $\cot \theta = 1.7318$

ASSIGNMENT EXERCISES

22-1

Convert the following degree measures to radians rounded to the nearest hundredth.

1. 60°
2. 212°
3. 300°

Convert the following measures to degrees rounded to the nearest ten-thousandth. Then convert to radians to the nearest hundredth.

4. 25°30′
5. 99°45′
6. 120°20′40″

Convert the following radian measures to degrees. Round to the nearest ten-thousandth of a degree.

7. $\frac{5\pi}{6}$ rad
8. 2.4 rad
9. 1.7 rad

Convert the following radian measures to degrees, minutes, and seconds. Round to the nearest second.

10. 0.9 rad
11. $\frac{3\pi}{8}$ rad
12. 1.2 rad

Find the arc length, radius, or central angle (in radians) for each of the following. Round to hundredths when necessary.

13. Find s if $\theta = 0.7$ rad and $r = 2.3$ cm.
14. Find θ if $s = 6.2$ cm and $r = 5$ cm.
15. Find r if $\theta = 2.1$ rad and $s = 3.6$ ft.

Find the area of the sector, the radius, or the central angle (in radians) of the sector for each of the following. Round to hundredths when necessary.

16. Find A if $\theta = 0.88$ rad and $r = 1.5$ m.
17. Find r if $\theta = 4.2$ rad and $A = 24$ in.^{2}.
18. Find θ if $r = 4$ cm and $A = 12$ cm^2.
19. A pendulum 6 in. long swings through an angle of 20°. Find the arc length the pendulum swings over from one extreme position to the other. Round to hundredths.
20. A movable part on a machine swings through an angle of 40° with an arc length of 8 in. What is the length of the part? Round to hundredths.

21. Find the area of a sector whose central angle is 85° and whose radius is 4.6 cm. ($A = \frac{1}{2}\theta r^2$, where θ is in radians.) Round to hundredths.

22. Find the number of degrees to the nearest ten-thousandth of a central angle of a sector whose area is 8.4 cm^2 and whose radius is 6 cm.

22-2

Find the indicated trigonometric ratios using Fig. 22-17.

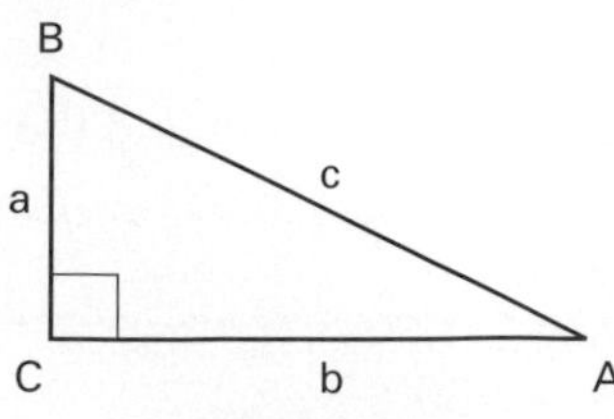

Figure 22-17

$a = 15$, $b = 20$, $c = 25$. Express the ratios as fractions in lowest terms.

23. sin A **24.** cos A **25.** tan A **26.** csc A **27.** sec A **28.** cot A
29. sin B **30.** cos B **31.** tan B **32.** csc B **33.** sec B **34.** cot B

$a = 2$ ft, $b = 10$ in., $c = 2$ ft 2 in. Express the ratios as fractions in lowest terms.

35. sin A **36.** sec B **37.** cot B **38.** csc A **39.** tan B **40.** cos A

$a = 7$, $b = 10.5$, $c = 12.62$. Express the ratios in decimals to the nearest ten-thousandth.

41. cos B **42.** csc A **43.** tan A **44.** sin B **45.** sec A **46.** sin A
47. cot A

22-3

Use a calculator to find the following trigonometric values. Round to the nearest ten-thousandth.

48. cos 42.5° **49.** sin 0.4712 **50.** sin 65.5°
51. cot 73° **52.** tan 1.0210 **53.** tan 47°
54. tan 15.6° **55.** sin 0.8610 **56.** tan 25° 40′
57. cos 32°50′ **58.** cot 0.7510 **59.** cos 80°10′

Determine the degree measure of an angle having the following trigonometric values. θ represents the unknown angle measure. Express answers to the nearest tenth of a degree.

60. $\sin\theta = 0.5446$ **61.** $\cos\theta = 0.6088$ **62.** $\tan\theta = 0.8720$
63. $\cot\theta = 0.9884$ **64.** $\cos\theta = 0.8897$ **65.** $\cot\theta = 3.340$

Determine the radian measure of an angle having the following trigonometric values. Round to the nearest ten-thousandth.

66. $\sin\theta = 0.9205$ **67.** $\tan\theta = 2.723$ **68.** $\cos\theta = 0.9450$
69. $\cot\theta = 0.3772$ **70.** $\sin\theta = 0.2896$ **71.** $\tan\theta = 0.3440$

Find the indicated trigonometric values. Round to the nearest ten-thousandth.

72. sec 15.5° **73.** csc 71° **74.** sec 0.4363
75. csc 1.0821 **76.** sec 1.2886 **77.** csc 0.4829

CHALLENGE PROBLEMS

78. Is there an angle between 0° and 360° for which the sine and cosine of the angle are equal? Justify your answer by making a table of values for the sine and cosine of angles between 0° and 360° and by graphing $\sin x$ and $\cos x$ for values between 0° and 360°.

79. Investigate the relationship described in problem 78 for the sine and tangent functions and justify your answer.

CONCEPTS ANALYSIS

1. Explain how the sine function is a ratio.

2. Use a calculator to find the value of the sine of several angles in each of the four quadrants and make a general statement about the greatest and least values the sine function can have.

3. Draw three right triangles. The acute angles of the first triangle are 30° and 60°. The angles of the second are 25° and 65°. The angles of the third are 80° and 10°. For each triangle, find the sine of the first angle and the cosine of the second angle. Compare the sine and cosine values for each triangle. What generalization can you make from these three examples?

4. Draw three right triangles of your choice and verify the generalization you made in question 3.

5. Use a calculator to find the sine of the given angles and make a table with the given angle in the first column and the sine of the angle in the second column. What is the measure of the angle at which the sine of the angle begins to repeat?
30°, 60°, 90°, 120°, 150°, 180°, 210°, 240°, 270°, 300°, 330°, 360°, 390°, 420°.

6. Find the sine of several angles larger than those given in question 5 to determine where, if any, the sine begins to repeat again.

7. Make a general statement about the sine function as it repeats. Compare your general statement with the graph of the sine function. (You may need to translate angle measures to radian measures and evaluate π as approximately 3.1.)

8. Use a graphing calculator to graph the following functions on the same grid. $\sin x$; $2 \sin x$; $3 \sin x$; $\frac{1}{2} \sin x$; $\frac{1}{3} \sin x$. Describe the impact the coefficient of the sine function has on the graph of the function. This coefficient is called the amplitude. Check a dictionary for the definition of the word amplitude. Does the dictionary definition make sense in view of your description?

Set the calculator range to 0, 360, 1, −3, 3, 0.1, 0, 360, 3.6 before graphing. Calculator steps for first graph: [Graph] [Sin] [X,θ,T] [EXE]

9. Graph $\sin x$ and $\sin 2x$ on the same graph. Compare these graphs. How do they differ and how are they alike?

10. Graph $\sin x$ and $\sin \left(\frac{1}{2}\right)x$ on the same graph. Compare these two graphs to the graphs for question 9. Describe the effect the coefficient of x has on the graph of the sine function. Graph other sine functions in which the coefficient of x is different to verify your description.

CHAPTER SUMMARY

Objectives	What to Remember	Examples
Section 22-1		
1. Find arc length given a central angle in radian measure.	To find the length of an arc of a circle, multiply the radian measure of the central angle by the radius of the circle. Use $s = \theta r$, where θ is given in radians.	Find the arc length intercepted on the circumference of a circle by a central angle of $\frac{\pi}{4}$ if the radius of the circle is 24 meters. $s = \theta r$; $s = \frac{\pi}{4}(24)$; $s = \pi(6)$; $s = 6\pi$; $s = 18.850$ m
2. Find the area of a sector, given a central angle in radian measure.	The area of a sector is $\frac{1}{2}\theta r^2$, where θ is given in radians.	Find the area of a sector that has a central angle of 1.4 rad and a radius of 5.2 ft. $A = \frac{1}{2}\theta r^2$; $A = 0.5(1.4)(5.2)^2$; $A = 18.928 \text{ ft}^2$
3. Convert angles in degree measure to angles in radian measure.	Degree angle measures are converted to radian measures by multiplying by the unity ratio $\frac{\pi \text{ rad}}{180°}$.	Write 82° in radian measure and round to the nearest thousandth. $82 \times \frac{\pi}{180} = 1.431$ rad
4. Convert angles in radian measure to angles in degree measure.	Radian angle measures are converted to degree measures by multiplying by the unity ratio $\frac{180°}{\pi \text{ rad}}$.	Write 0.45 radians in degree measure and round to the nearest tenth of a degree. $0.45 \times \frac{180}{\pi} = 25.8°$
Section 22-2		
1. Find the sine, cosine, and tangent of right triangles, given the measures of at least two sides.	Use the ratios for each of the three trigonometric functions to calculate the value of the function when the measures of appropriate sides are given. The Pythagorean theorem may be needed to find the length of a third side in some instances. $\text{sine } A = \frac{\text{opposite}}{\text{hypotenuse}}$; $\text{cosine } A = \frac{\text{adjacent}}{\text{hypotenuse}}$; $\text{tangent } A = \frac{\text{opposite}}{\text{adjacent}}$.	Find the sine, cosine, and tangent of angle A in the triangle illustrated and round to the nearest ten-thousandth. $\sin A = \frac{15}{17} = 0.8824$; $\cos A = \frac{8}{17} = 0.4706$; $\tan A = \frac{15}{8} = 1.875$ 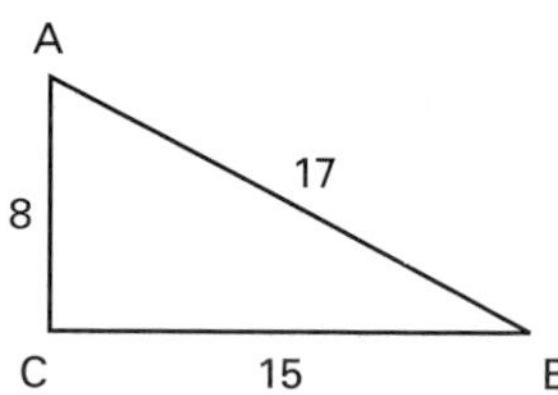

2. Find the cosecant, secant, and cotangent of right triangles, given the measures of at least two sides.	The cosecant, secant, and cotangent are reciprocal functions of the sine, cosine, and tangent, respectively. $\text{cosecant } A = \frac{\text{hypotenuse}}{\text{opposite}};$ $\text{secant } A = \frac{\text{hypotenuse}}{\text{adjacent}};$ $\text{cotangent } A = \frac{\text{adjacent}}{\text{opposite}}.$	Find the cosecant, secant, and cotangent of angle A in the triangle illustrated in objective 1 and round to the nearest ten-thousandth. $\csc A = \frac{17}{15} = 1.1333$ $\sec A = \frac{17}{8} = 2.125$ $\cot A = \frac{8}{15} = 0.5333$

Section 22-3

1. Find trigonometric values for sine, cosine, and tangent using a calculator.	Use the [SIN], [COS], and [TAN] keys to find the trigonometric value of a specified angle. Be sure the calculator is set in the appropriate mode, degree or radian.	Use a calculator to find the following: sin 35° cos 198° tan 125° sin 2.8 cos 2.98 tan 4.2 *Scientific:* 35 [SIN] ⇒ 0.573576436 198 [COS] ⇒ −0.951056516 125 [TAN] ⇒ −1.428148007 Change calculator to radian mode: 2.8 [SIN] ⇒ 0.33498815 2.98 [COS] ⇒ −0.986972292 4.2 [TAN] ⇒ 1.777779774 *Graphing:* [SIN] 35 [EXE] ⇒ 0.5735764364 [COS] 198 [EXE] ⇒ −0.9510565163 [TAN] 125 [EXE] ⇒ −1.428148007 Change to radian mode: [Shift] 1 [F2] [EXE] (Casio) [SIN] 2.8 [F5] [EXE] ⇒ 0.3349881502 [COS] 2.98 [F5] [EXE] ⇒ −0.9869722927 [TAN] 4.2 [F5] [EXE] ⇒ 1.777779775 TI-81 or 82 does not require F5 key.
2. Find the angle measure, given a trigonometric value.	To find the angle measure when the trigonometric value is given, use the inverse trigonometric function key, which is usually accessed by using the [Shift] key.	Find the value of x: $0.906307787 = \cos x$ or $x = \cos^{-1} 0.906307787$. $x = 25°$ Keystrokes for both scientific and graphing calculators are relatively similar for this process.

3. Find the trigonometric values for cosecant, secant, and cotangent using a calculator.	Use the [SIN], [COS], [TAN] calculator keys and the inverse function key [1/x] to find values of the inverse functions.	Use a calculator to find the following: csc 25° sec 1 *Scientific:* 25 [SIN] [1/x] ⇒ 2.366201583 *Graphing:* [SIN] 25 [x^{-1}] [EXE] ⇒ 2.366201583 *Scientific* (set mode to radians): 1 [COS] [1/x] ⇒ 1.850815718 *Graphing* (select radian mode): [COS] 1 [F5] [EXE] ⇒ 0.5403023059 [x^{-1}] [EXE] ⇒ 1.850815718

WORDS TO KNOW

radian (p. 850)
degree (p. 850)
arc length (p. 851)
sector (p. 852)
unity ratio (p. 853)
hypotenuse (p. 857)
adjacent side (p. 857)
opposite side (p. 857)
sine (p. 858)
cosine (p. 858)
tangent (p. 858)
standard right triangle (p. 858)
cosecant (p. 860)
cotangent (p. 860)
secant (p. 860)

CHAPTER TRIAL TEST

Convert the following degree measures to radians rounded to the nearest hundredth.

1. 35° **2.** 122° **3.** 315°
4. 240°

Convert the following measures to degrees rounded to the nearest ten-thousandth. Then convert to radians to the nearest hundredth.

5. 15°25′ **6.** 142°32′15″ **7.** 16°12′
8. 32°18′37″

Convert the following radian measures to degrees. Round to the nearest ten-thousandth of a degree.

9. $\frac{5\pi}{8}$ rad **10.** 3.1 rad

Convert the following radian measures to degrees, minutes, and seconds. Round to the nearest second.

11. 1.2 rad **12.** $\frac{\pi}{6}$ rad

Use the relationships of arc length, area, central angle, and radius to solve the following problems relating to sectors. Round to hundredths if necessary.

13. Find s if $\theta = 0.5$ and $r = 2$ in.
14. Find θ if $s = 5.3$ m and $r = 7$ m.
15. Find r if $\theta = 1.7$ and $s = 2.9$ m.
16. Find A if $\theta = 35°$ and $r = 7.3$ cm.
17. Find r if $\theta = 1.6$ and $A = 7.2\ m^2$.
18. Find θ if $r = 7$ m and $A = 14\ m^2$.

Write the ratios as fractions in lowest terms for the trigonometric functions using triangle ABC in Fig. 22-18. $a = 10$, $b = 24$, $c = 26$.

19. $\sin A$

20. $\tan B$

21. $\csc A$

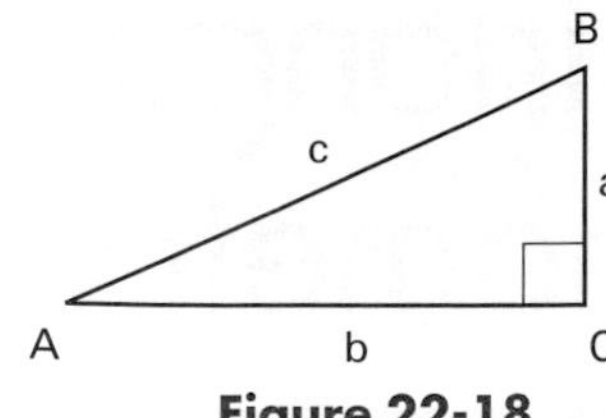

Figure 22-18

Express the following trigonometric values (Fig. 22-18) in decimals to the nearest ten-thousandth.
$a = 5$, $b = 11.5$, $c = 12.54$

22. $\cot A$

23. $\cos A$

24. $\sin B$

Use a calculator to find the following trigonometric values.

25. $\sin 53°$

26. $\cos 68.5°$

27. $\tan 42°30'$

28. $\sin 61°10'$

29. $\sin 1.1519$

30. $\cos 1.0297$

Determine the degree measure of an angle having the following trigonometric values. θ represents the unknown angle measure. Express answers to the nearest tenth of a degree.

31. $\sin \theta = 0.2756$

32. $\tan \theta = 1.280$

33. $\cos \theta = 0.9426$

34. $\cot \theta = 1.540$

35. $\sin \theta = 0.5900$

36. $\tan \theta = 0.0275$

Determine the radian measure of an angle having the following trigonometric values. Round to the nearest ten-thousandth.

37. $\sin \theta = 0.7660$

38. $\cos \theta = 0.8387$

39. $\tan \theta = 0.3259$

Find the indicated trigonometric values. Round to the nearest ten-thousandth.

40. $\sec 25.5°$

41. $\csc 47°$

42. $\sec 0.3316$

CHAPTER

23 Right-Triangle Trigonometry

Right triangles are used extensively. Thus, it is important for us to gain a working knowledge of solving right triangles, that is, finding the measures of all sides and angles. We also need to become skilled in determining how to use our knowledge to solve job-related applications of right triangles.

By using the trigonometric relationships we learned in Chapter 22, we can find all the angles and sides of a right triangle if we know a side and any other part.

23-1 SINE, COSINE, AND TANGENT FUNCTIONS

Learning Objectives

L.O.1. Find missing parts of a right triangle using the sine function.
L.O.2. Find missing parts of a right triangle using the cosine function.
L.O.3. Find missing parts of a right triangle using the tangent function.

The sine, cosine, and tangent functions can be used to find certain parts of a right triangle. Formula rearrangement and other algebraic principles may be needed also, depending on what information is given in the problem and what information needs to be found.

L.O.1. Find missing parts of a right triangle using the sine function.

Using the Sine Function

We will abbreviate the formula to read

$$\sin \theta = \frac{\text{opp}}{\text{hyp}}$$

Using this formula, we can find parts of right triangles even when the triangles are other than the standard right triangle.

Rule 23-1.1

Use the sine function to find the measure of an angle of a right triangle, the side opposite the angle, or the hypotenuse, when we are given two of these three parts of the triangle.

Tips and Traps

For consistency, we will round all trigonometric ratios to four significant digits and all angle values to the nearest 0.1° throughout this chapter.

The *significant digits* of a whole number or integer are the digits beginning with the first nonzero digit on the left and ending with the last nonzero digit on the right. The significant digits of a decimal number are the digits beginning with the first nonzero digit on the left and ending with the last digit on the *right of the decimal point.*

5000	1 significant digit	5 is first *and* last nonzero digit.
250	2 significant digits	2 is first and 5 is last nonzero digit.
205	3 significant digits	2 is first and 5 is last nonzero digit.
0.004	1 significant digit	4 is first nonzero *and* last digit.
3.05	3 significant digits	3 is first nonzero and 5 is last digit.
2.070	4 significant digits	2 is first nonzero digit and 0 is last digit.

To round a number to a certain number of significant digits:

1. From the left, count the number of significant digits desired. Notice the last of these significant digits.
2. If the next digit to the right is less than 5, do not change the last significant digit. If the next digit to the right is 5 or greater, add 1 to the last significant digit.
3. Then, if the last significant digit is to the *left* of a decimal point, replace all digits after the last significant digit up to the decimal point with zeros and drop all digits after the decimal point. If the last significant digit is to the *right* of a decimal point, drop all digits after the last significant digit.

Let's examine a few instances of rounding to four significant digits when a calculator answer is obtained.

$$\sin A = \frac{7}{21} = 0.3333333333$$

$$= 0.3333 \quad \textit{(rounded to four significant digits)}$$

$$a = 15(0.8039) = 12.0585$$

$$= 12.06 \quad \textit{(rounded to four significant digits)}$$

$$t = \frac{25}{0.5299} = 47.17871296$$

$$= 47.18 \quad \textit{(rounded to four significant digits)}$$

$$f = \frac{24}{0.8434} = 28.45624852$$

$$f = 28.46 \quad \textit{(rounded to four significant digits)}$$

EXAMPLE 23-1.1 Find angle A if $a = 7$ and $c = 21$ (Fig. 23-1).

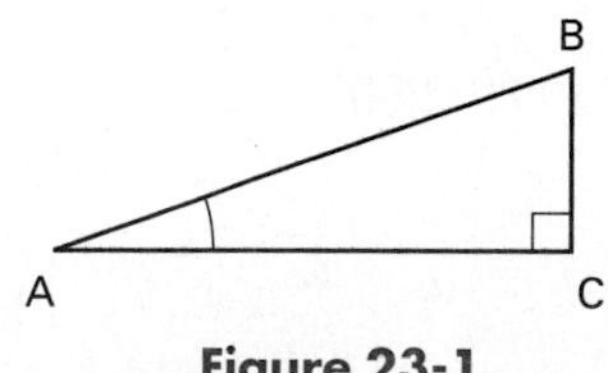

Figure 23-1

Solution: To find angle A we use the sine ratio because we are given the side *opposite* $\angle A$ and the hypotenuse.

$$\sin A = \frac{\text{opp}}{\text{hyp}}$$

$$\sin A = \frac{7}{21}$$

$$\sin A = 0.3333333333$$

$$A = 19.5^\circ \qquad \textit{(using } \boxed{\sin^{-1}} \textit{ key and rounding)}$$

Note: We did not round until the angle measure was found. ■

EXAMPLE 23-1.2 Find side a in triangle ABC (Fig. 23-2).

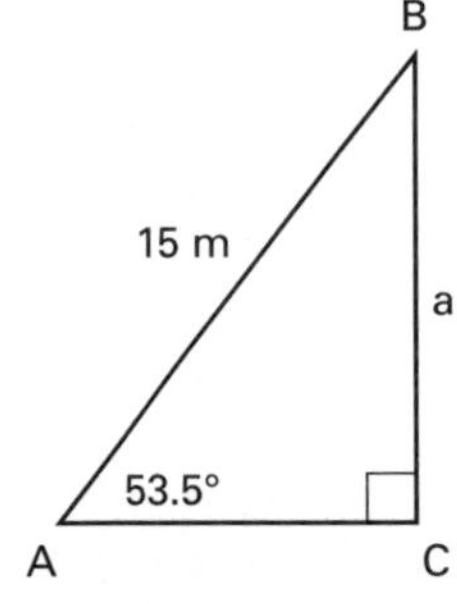

Figure 23-2

Solution: Since we are given $\angle A$ and the hypotenuse and need to find the side opposite $\angle A$, we again use the sine function.

$$\sin A = \frac{\text{opp}}{\text{hyp}}$$

$$\sin 53.5^\circ = \frac{a}{15} \qquad \textit{(}\sin 53.5^\circ = 0.80385686\textit{)}$$

$$0.80385686 = \frac{a}{15}$$

$$15(0.80385686) = a$$

$$a = 12.06 \text{ m} \qquad \textit{(rounded to four significant digits)}$$ ■

EXAMPLE 23-1.3 Find the hypotenuse in triangle RST (Fig. 23-3).

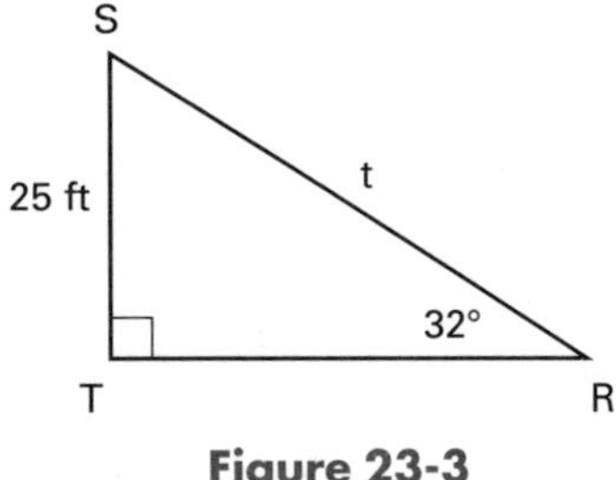

Figure 23-3

Solution: Since we are given an angle and the side opposite the angle and need to find the hypotenuse, the sine function is used.

$$\sin R = \frac{\text{opp}}{\text{hyp}}$$

$$\sin 32^\circ = \frac{25}{t}$$

$$0.529919264 = \frac{25}{t} \qquad \textit{(}\sin 32^\circ = 0.529919264\textit{)}$$

$$0.529919264t = 25$$

$$t = \frac{25}{0.529919264}$$

$$t = 47.18 \text{ ft} \qquad \textit{(rounded to four significant digits)}$$ ■

When we find the measures of all sides and angles of a triangle, we say that we are *solving* the triangle. In Example 23-1.4, we will find values of all sides and angles of the triangle using the sine function.

Tips and Traps

In most problems when a triangle is being solved, there is a choice of which missing part can be found first. Since the rounded value for that missing part is sometimes used to find other missing parts, final answers may vary slightly due to rounding discrepancies. For instance, the angles of a triangle may add to as little as 179° or as much as 181°. Or the length of a side may be slightly different in the last significant digit. If the full calculator value is used in finding other missing parts, the rounding discrepancy will be reduced.

EXAMPLE 23-1.4 Solve triangle *DEF* (Fig. 23-4).

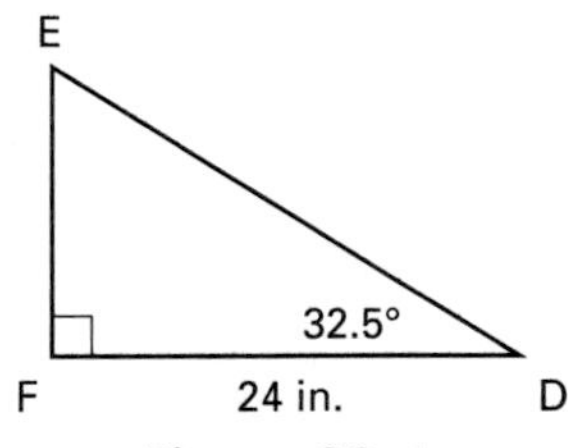

Figure 23-4

Solution: To use the sine function, we must have *two* of the following parts: an angle, its opposite side, or the hypotenuse. Since we only have one angle and one side (not opposite the angle), we must find the third angle of the triangle. Angles *D* and *E* are complementary (add up to 90°), so

$$\angle E = 90° - 32.5° = 57.5°$$

Now we can find the hypotenuse because we have an angle, *E*, and its opposite side, *e*.

$$\sin E = \frac{\text{opp}}{\text{hyp}}$$

$$\sin 57.5° = \frac{24}{f}$$

$$0.843391445 = \frac{24}{f} \qquad \textit{(sin 57.5° = 0.843391445)}$$

$$0.843391445f = 24$$

$$f = \frac{24}{0.843391445}$$

$$f = 28.46 \text{ in.} \qquad \textit{(rounded to four significant digits)}$$

To find side *d*, we have

$$\sin D = \frac{\text{opp}}{\text{hyp}}$$

$$\sin 32.5° = \frac{d}{28.45653714} \qquad \textit{(using full calculator value for f)}$$

$$0.537299608 = \frac{d}{28.45653714} \qquad \textit{(sin 32.5° = 0.537299608)}$$

$$0.537299608(28.45653714) = d$$

$$15.29 \text{ in.} = d \qquad \textit{(rounded to four significant digits)}$$

The solved triangle looks like Fig. 23-5. ■

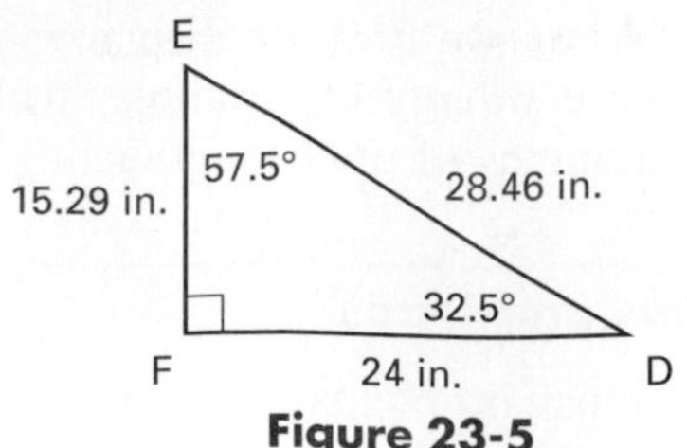

Figure 23-5

Tips and Traps

In this and other problems involving right triangles, our computations of the sides can be checked by using the Pythagorean theorem. We will check the solution to Example 23-1.4 as an illustration.

$$(\text{hyp})^2 = (\text{leg})^2 + (\text{leg})^2$$

$$(28.46)^2 \doteq (15.29)^2 + (24)^2$$

$$809.9716 \doteq 233.7841 + 576$$

$$809.9716 \doteq 809.7841$$

(Difference is caused by rounding discrepancy. $\doteq$ means "is approximately equal to.")

L.O.2. Find missing parts of a right triangle using the cosine function.

Using the Cosine Function

In some of the previous problems, when we had to find a side not opposite the given angle, we had to first find an acute angle by subtracting the given acute angle from 90°. If we use the cosine function, however, we can find the same side by using the given angle, rather than its complement.

The abbreviated cosine ratio is

$$\cos \theta = \frac{\text{adj}}{\text{hyp}}$$

Keep in mind that angle θ is made up of two sides of the right triangle. One of these sides is the hypotenuse and the other side is the side *adjacent* to angle θ.

> **Rule 23-1.2**
>
> Use the cosine function to find the measure of an angle of a right triangle, its *adjacent* side, or the hypotenuse if we are given two of these three parts of the triangle.

Although a check of the solution will not be shown, if we wish to check the sides, the Pythagorean theorem may be used (See Tips and Traps following Example 23-1.4.).

EXAMPLE 23-1.5 Find angle A (Fig. 23-6).

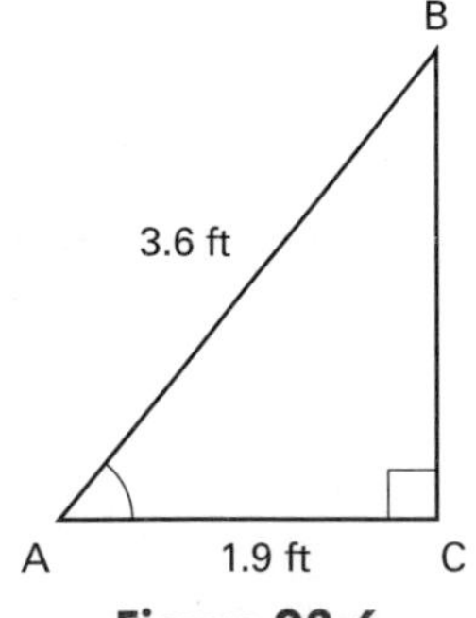

Figure 23-6

Solution: Since we are asked to find an angle and are given the hypotenuse and the side adjacent to the angle, the cosine function is used.

$$\cos A = \frac{\text{adj}}{\text{hyp}}$$

$$\cos A = \frac{1.9}{3.6}$$

$$\cos A = 0.527777777$$

$$A = 58.1°$$

(using $\boxed{\cos^{-1}}$ *key and rounding)* ■

EXAMPLE 23-1.6 Find side b (Fig. 23-7).

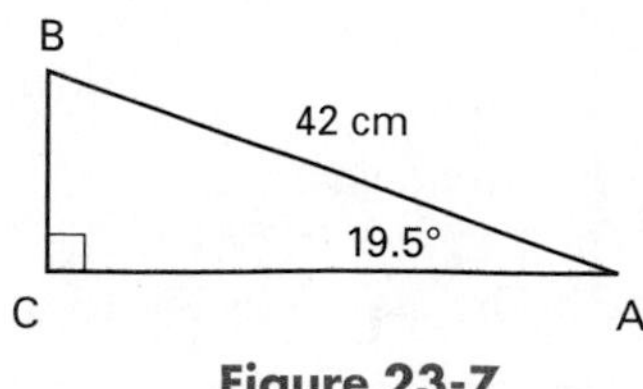

Figure 23-7

Solution: We use the cosine function because we are given the hypotenuse and an angle to find its adjacent side.

$$\cos A = \frac{\text{adj}}{\text{hyp}}$$

$$\cos 19.5° = \frac{b}{42}$$

$$(0.942641491)(42) = b \qquad (\cos 19.5° = 0.942641491)$$

$$39.59 \text{ cm} = b \qquad \text{(four significant digits)}$$

EXAMPLE 23-1.7 Find side t (Fig. 23-8).

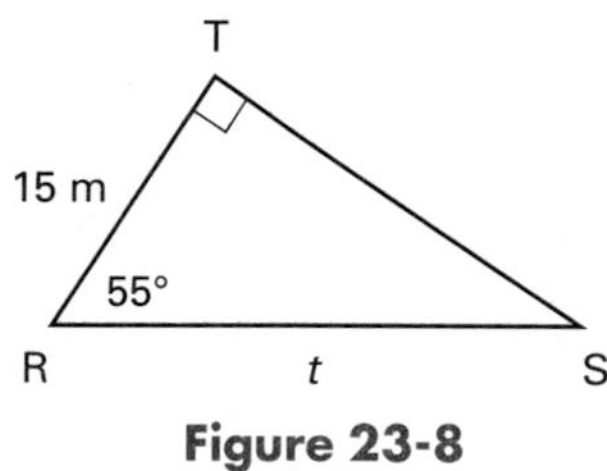

Figure 23-8

Solution: Again the cosine function is used because we must find the hypotenuse if given an angle and its adjacent side.

$$\cos R = \frac{\text{adj}}{\text{hyp}}$$

$$\cos 55° = \frac{15}{t}$$

$$0.573576436 = \frac{15}{t} \qquad (\cos 55° = 0.573576436)$$

$$t(0.573576436) = 15$$

$$t = \frac{15}{0.573576436}$$

$$t = 26.15 \text{ m} \qquad \text{(four significant digits)}$$

L.O.3. Find missing parts of a right triangle using the tangent function.

Using the Tangent Function

If we have a right triangle in which we know only the length of the two legs, we cannot use the sine or cosine function to solve the triangle unless we use the Pythagorean theorem to find the length of the hypotenuse. We can use the tangent function directly.

$$\tan \theta = \frac{\text{opp}}{\text{adj}}$$

> **Rule 23-1.3** *Use the tangent function if we are given any* **two** *of the following three parts of a right triangle:*
>
> 1. An acute angle
> 2. The side opposite the acute angle
> 3. The side adjacent to the acute angle

Whenever we wish to check our calculation of the sides, we may use the Pythagorean theorem as in the Tips and Traps following Example 23-1.4.

EXAMPLE 23-1.8 Find angle A (Fig. 23-9).

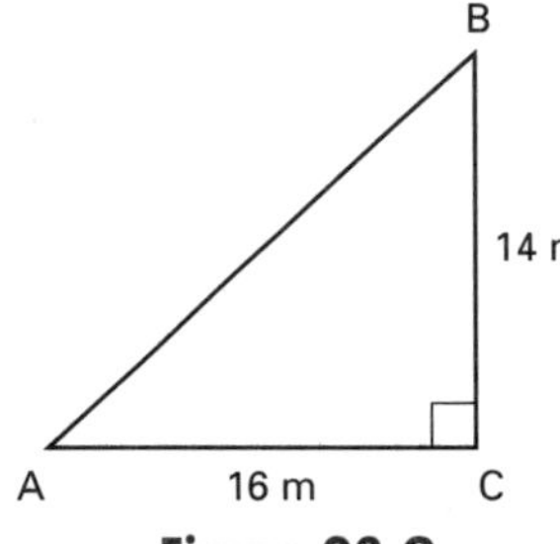

Figure 23-9

Solution: Since we are looking for an angle and are given the side opposite and the side adjacent to the angle, we use the tangent function.

$$\tan A = \frac{\text{opp}}{\text{adj}}$$

$$\tan A = \frac{14}{16}$$

$$\tan A = 0.875$$

$$A = 41.2° \qquad \textit{(using } \boxed{\tan^{-1}} \textit{ key and rounding)}$$ ■

EXAMPLE 23-1.9 Use the tangent function to find a (Fig. 23-10).

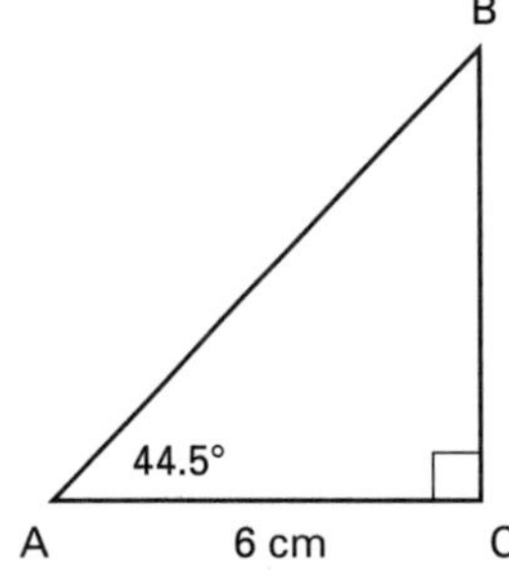

Figure 23-10

Solution: In this example, we are given an acute angle and the side adjacent to the acute angle. We are looking for the opposite side and thus use the tangent function.

$$\tan A = \frac{\text{opp}}{\text{adj}}$$

$$\tan 44.5° = \frac{a}{6}$$

$$0.982697263 = \frac{a}{6} \qquad \textit{(tan 44.5° = 0.982697263)}$$

$$0.982697263(6) = a$$

$$5.896 \text{ m} = a \qquad \textit{(four significant digits)}$$ ■

EXAMPLE 23-1.10 Find side b (Fig. 23-11).

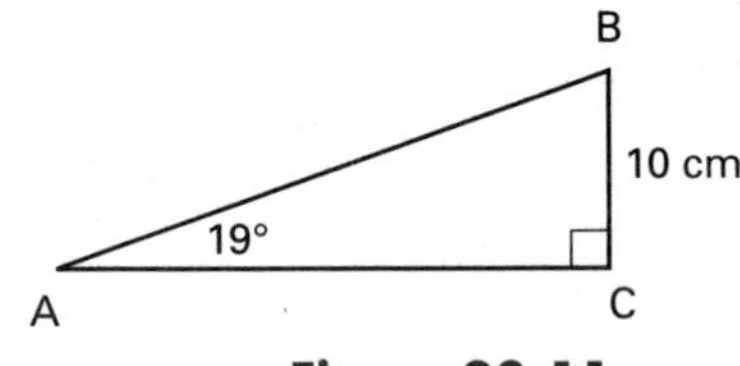

Figure 23-11

Solution: Here we know an acute angle and its opposite side. We are looking for the side adjacent to the acute angle.

$$\tan A = \frac{\text{opp}}{\text{adj}}$$

$$\tan 19° = \frac{10}{b}$$

$$0.344327613 = \frac{10}{b} \qquad (\tan 19° = 0.344327613)$$

$$0.344327613b = 10$$

$$b = \frac{10}{0.344327613}$$

$$b = 29.04 \text{ cm} \qquad \text{(four significant digits)}$$

■

SELF-STUDY EXERCISES 23-1

L.O.1. Use the sine function to find the indicated parts of the triangle *LMN* (Fig. 23-14). Round length of sides to four significant digits and angles to the nearest 0.1°.

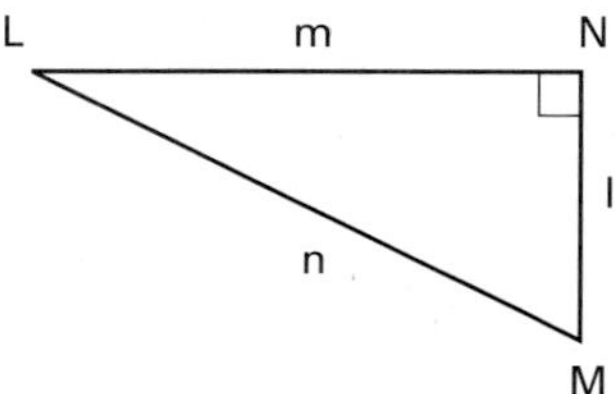

Figure 23-14

1. Find M if $n = 15$ m and $m = 7$ m.

2. Find l if $n = 13$ in. and $L = 32°$.

3. Find m if $l = 15$ m and $L = 28°$.

4. Find n if $l = 12$ ft and $M = 42°$.

5. Find M if $m = 13$ cm and $n = 19$ cm.

6. Find M if $n = 3.7$ in. and $l = 2.4$ in.

Solve triangle *STU* (Fig. 23-15) using the sine function. Check the measures of the sides by using the Pythagorean theorem. Round as above.

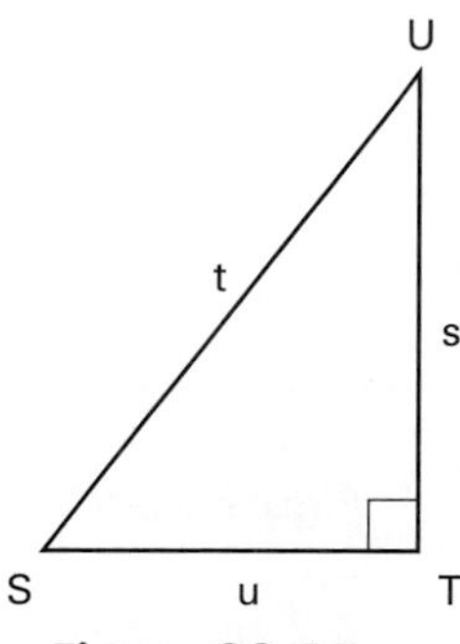

Figure 23-15

7. Solve if $t = 18$ yd and $s = 14$ yd.

8. Solve if $U = 45°$ and $u = 4.7$ m.

9. Solve if $S = 34.5°$ and $t = 8.5$ mm.

10. Solve if $S = 16°$ and $s = 14$ m.

L.O.2. Use the cosine function to find the indicated parts of triangle KLM (Fig. 23-16). Round sides to four significant digits and angles to the nearest 0.1°.

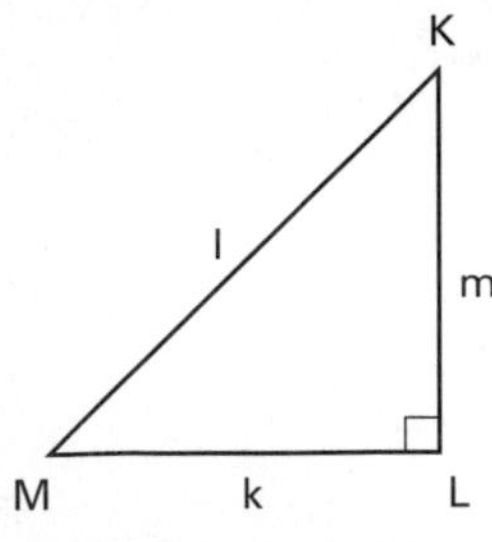

Figure 23-16

11. Find M if $k = 13$ m and $l = 16$ m.

12. Find k if $l = 11$ cm and $M = 24°$.

13. Find l if $M = 31°$ and $k = 27$ ft.

14. Find l if $m = 15$ dm and $M = 25°$.

15. Find k if $K = 72°$ and $l = 16.7$ mm.

16. Find l if $K = 67°$ and $k = 13$ yd.

Use the sine *or* cosine function to solve triangle QRS (Fig. 23-17). Round as above.

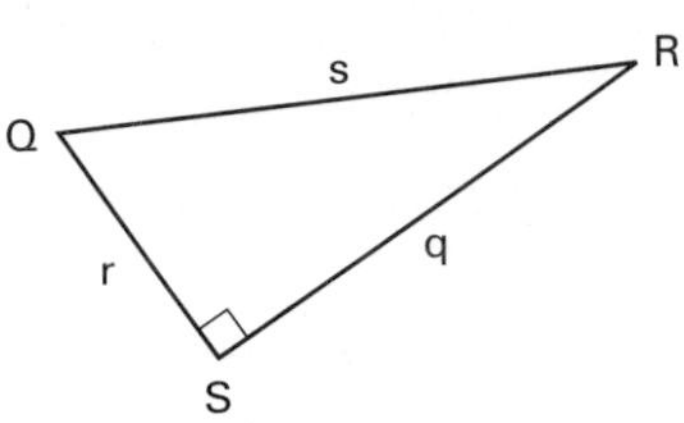

Figure 23-17

17. Solve if $s = 23$ ft and $q = 16$ ft.

18. Solve if $s = 17$ cm and $R = 46°$.

19. Solve if $q = 14$ dkm and $Q = 73.5°$.

20. Solve if $R = 59.5°$ and $q = 8$ m.

L.O.3. Use the tangent function to find the indicated parts of triangle ABC (Fig. 23-12). Round sides to four significant digits and angles to the nearest 0.1°.

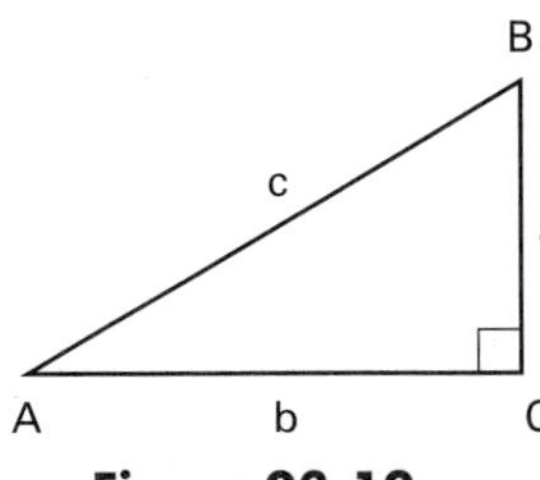

Figure 23-12

21. Find A if $b = 11$ cm and $a = 6$ cm.

22. Find b if $a = 1.9$ m and $A = 25°$.

23. Find a if $A = 40.5°$ and $b = 7$ ft.

24. Find A if $b = 10.8$ m and $a = 4.7$ m.

25. Find a if $A = 43°$ and $b = 0.05$ cm.

26. Find a if $B = 68°$ and $b = 0.03$ m.

Solve triangle *DEF* (Fig. 23-13). Round sides to four significant digits and angles to the nearest 0.1°.

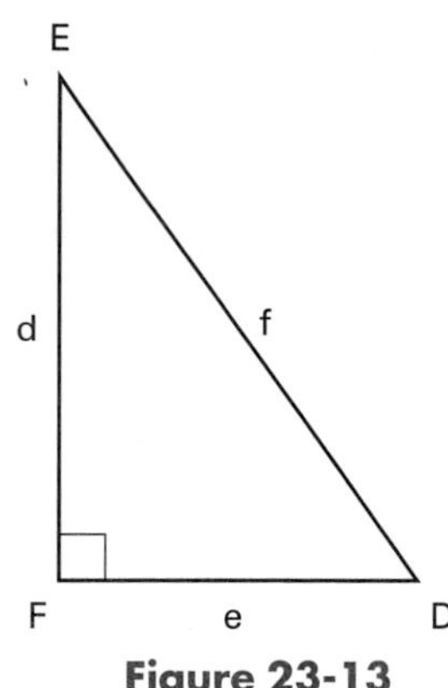

Figure 23-13

27. Solve if $e = 4.6$ m and $d = 3.2$ m.

28. Solve if $D = 42°$ and $e = 7$ ft.

29. Solve if $E = 73.5°$ and $e = 20.13$ in.

30. Solve if $d = 11$ ft and $e = 8$ ft.

23-2 APPLIED PROBLEMS USING RIGHT-TRIANGLE TRIGONOMETRY

Learning Objectives

L.O.1. Select the most direct method for solving right triangles.
L.O.2. Solve applied problems using right-triangle trigonometry.

In this section, we learn how to select the function (theorem, property, or the like) that minimizes the number of steps needed to solve a problem, in order to solve applied problems more efficiently and quickly.

L.O.1. Select the most direct method for solving right triangles.

Selecting the Most Direct Method to Solve Right Triangles

Our first task in solving any problem involving right triangles, especially problems on the job, is to *select the most convenient function* to use in each particular problem. Following are two basic rules or guidelines to use.

Rule 23-2.1

Where possible, choose the function that uses parts that are given in the problem, rather than parts that are not given and must be calculated.

Rule 23-2.2

Where possible, choose the function that gives the desired part directly, that is, without having to find other parts first.

Example 23-2.1 In $\triangle ABC$, find c (Fig. 23-18).

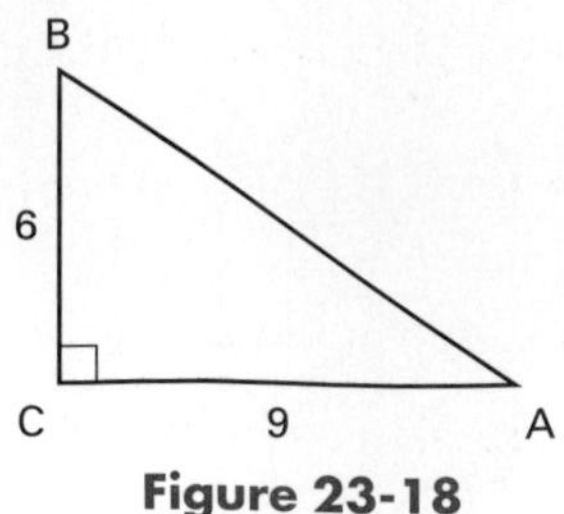

Figure 23-18

Solution: The most direct method of finding side c is the Pythagorean theorem.

$$c^2 = a^2 + b^2$$
$$c^2 = 6^2 + 9^2$$
$$c^2 = 36 + 81$$
$$c^2 = 117$$
$$c = 10.82 \quad \textit{(four significant digits)}$$

For illustrative purposes only, we will use the trigonometric functions to find c.

$$\tan A = \frac{\text{opp}}{\text{adj}} \quad \textit{(values given for this function)}$$
$$\tan A = \frac{6}{9}$$
$$\tan A = 0.666666666$$
$$A = 33.7° \quad \textit{(using } \boxed{tan^{-1}} \textit{ key and rounding)}$$
$$90° - A = B \quad \textit{(complementary angles)}$$
$$90° - 33.69006753° = 56.30993247° \quad \textit{(Use full calculator value of A.)}$$
$$\cos B = \frac{\text{adj}}{\text{hyp}} \quad \textit{(values now available for cosine function)}$$
$$\cos 56.30993247° = \frac{6}{c}$$
$$0.554700196 = \frac{6}{c}$$
$$0.554700196c = 6$$
$$c = \frac{6}{0.554700196}$$
$$c = 10.82 \quad \textit{(four significant digits)}$$

■

Example 23-2.2 In $\triangle ABC$ (Fig. 23-19), find angle B.

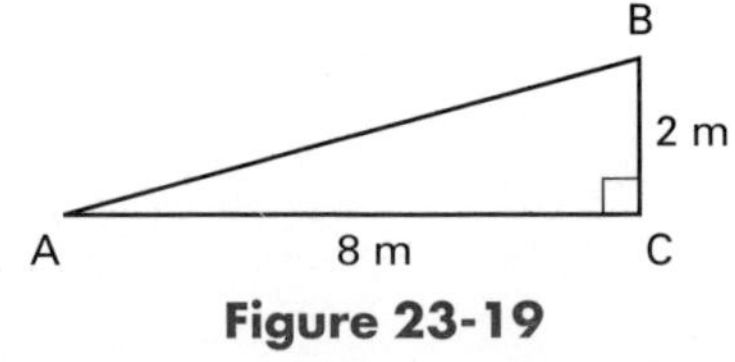

Figure 23-19

Solution: Here the Pythagorean theorem would not give us any angle measure. It would give us the hypotenuse, and this would enable us to use $\sin\theta = \dfrac{\text{opp}}{\text{hyp}}$ or $\cos\theta = \dfrac{\text{adj}}{\text{hyp}}$ to find angle B. However, since we are given the sides adjacent to and opposite angle B, we should use the tangent function, $\tan\theta = \dfrac{\text{opp}}{\text{adj}}$, for a quick, direct solution.

$$\tan B = \frac{8}{2} \qquad \textit{(8 is opposite } \angle B.)$$

$$\tan B = 4$$

$$B = 76.0° \qquad \textit{(using } \boxed{tan^{-1}} \textit{ key and rounding to nearest tenth degree)}$$ ■

L.O.2. Solve applied problems using right-triangle trigonometry.

Solving Technical Problems Using Right-triangle Trigonometry

Many technical applications can be solved by the use of right triangles. The following examples show some work-related applications of right triangles.

In solving technical problems, we should draw a diagram or picture to help us visualize the various relationships.

EXAMPLE 23-2.3 A jet takes off at a 30° angle. If the runway (from takeoff) is 875 ft long, find the altitude of the airplane as it flies over the end of the runway (Fig. 23-20).

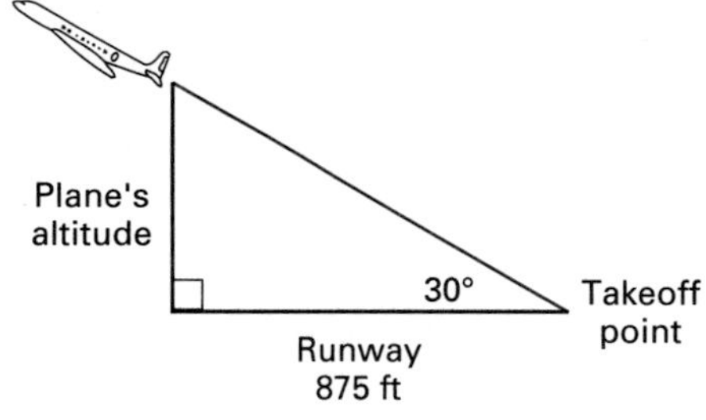

Figure 23-20

Solution

$$\tan 30° = \frac{\text{opp}}{\text{adj}}$$

$$0.577350269 = \frac{\text{opp}}{875} \qquad \textit{(tan 30° = 0.577350269)}$$

$$0.577350269(875) = \text{opp}$$

$$505.2 \text{ ft} = \text{plane's altitude} \qquad \textit{(four significant digits)}$$ ■

Many right-triangle applications make use of the terminology *angle of elevation* and *angle of depression* (Fig. 23-21). The angle of elevation is generally used

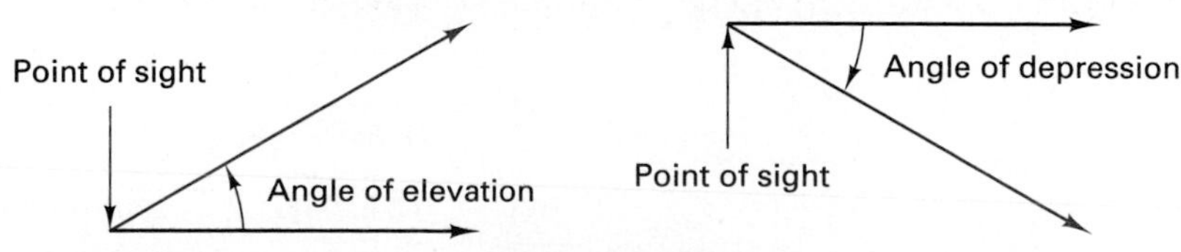

Figure 23-21

when we are looking *up* at an object. The angle of depression is used to describe the location of an objective *below* eye level. *Both* angles are the angles formed by a line of sight and a horizontal line from the point of sight.

EXAMPLE 23-2.4 A stretch of roadway drops 30 ft for every 300 ft of road. Find the *angle of declination* of the road (Fig. 23-22).

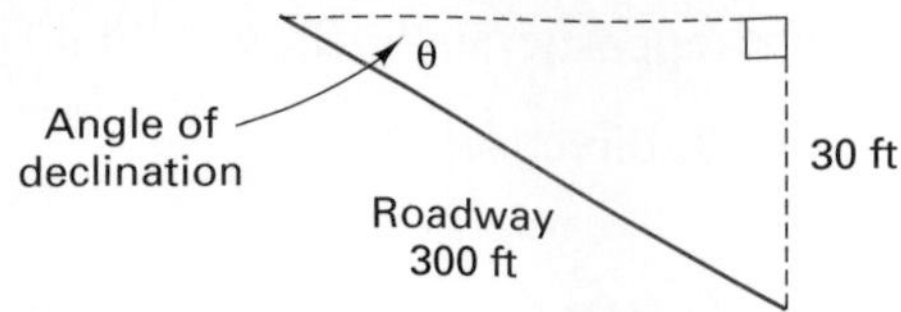

Figure 23-22

Solution: The angle of declination is the angle of depression.

$$\sin \theta = \frac{\text{opp}}{\text{hyp}}$$

$$\sin \theta = \frac{30}{300}$$

$$\sin \theta = 0.1$$

$$\theta = 5.7° \quad \textit{(using } \boxed{\sin^{-1}} \textit{ key and rounding)}$$

The angle of declination of the road is 5.7°. ■

EXAMPLE 23-2.5 A surveyor locates two points on a steel column so that it can be set plumb (perpendicular to the horizon). If the angle of elevation is 15° and the surveyor's transit is 175 ft from the column (Fig. 23-23), find the distance from the transit to the upper point on the column. (A transit is a surveying instrument for measuring angles.)

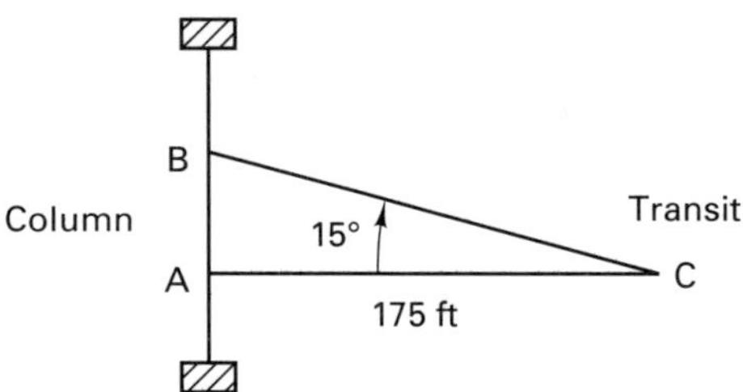

Figure 23-23

Solution: To find the distance from the transit to point *B* on the column, we use the cosine function

$$\cos \theta = \frac{\text{adj}}{\text{hyp}}$$

$$\cos 15° = \frac{175}{\text{hyp}}$$

$$0.965925826 = \frac{175}{\text{hyp}} \qquad \textit{(}\cos 15° = 0.965925826\textit{)}$$

$$(0.965925826)\ \text{hyp} = 175$$

$$\text{hyp} = \frac{175}{0.965925826}$$

$$\text{hyp} = 181.2 \text{ ft} \qquad \textit{(four significant digits)}$$

Point *B* is 181.2 ft from the transit. ■

Example 23-2.6 Find the angle a rafter makes with a joist of a house if the rise is 12 ft and the span is 30 ft (Fig. 23-24). Find the length of the rafter.

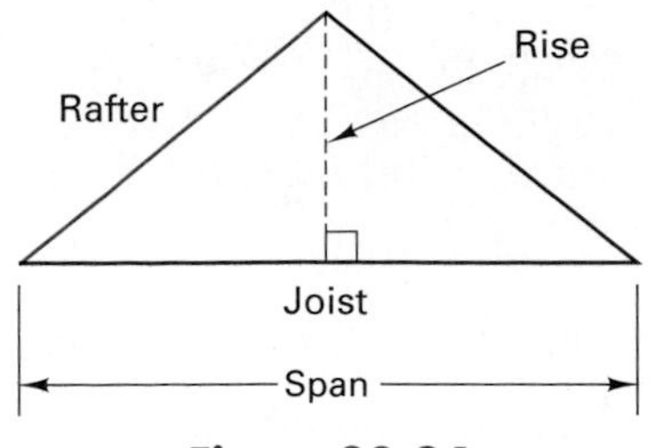

Figure 23-24

Solution: The span is twice the distance from the outside stud to the center point of the joist. Therefore, to solve the right triangle for the desired angle, we draw the triangle shown in Fig. 23-25.

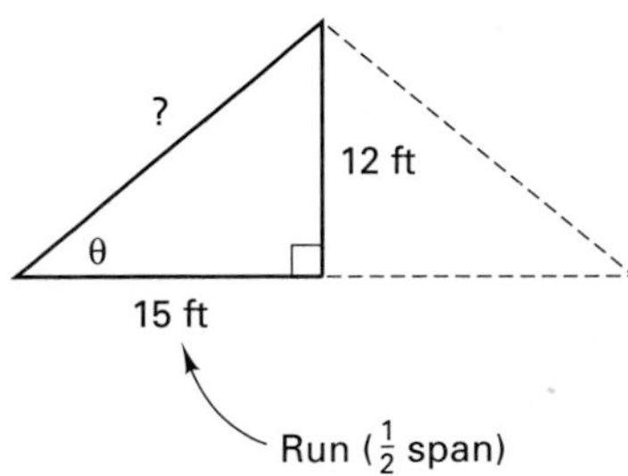

Figure 23-25

$$\tan\theta = \frac{\text{opp}}{\text{adj}}$$

$$\tan\theta = \frac{12}{15}$$

$$\tan\theta = 0.8$$

$$\theta = 38.7° \qquad \textit{(using } \boxed{\tan^{-1}} \textit{ key and rounding)}$$

We can find the length of the rafter directly by using the Pythagorean theorem

$$(\text{hyp})^2 = 12^2 + 15^2$$

$$(\text{hyp})^2 = 144 + 225$$

$$(\text{hyp})^2 = 369$$

$$\text{hyp} = 19.21 \text{ ft} \qquad \textit{(rounded)}$$

The angle the rafter makes with the joist is 38.7° and the rafter is 19.21 ft. ■

Example 23-2.7 Find the angle formed by the connecting rod in the mechanical assembly shown in Fig. 23-26.

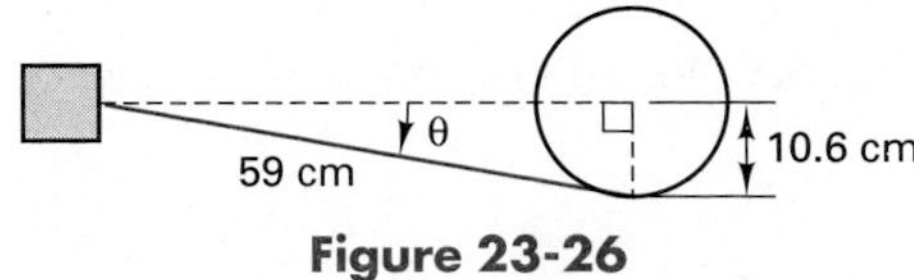

Figure 23-26

Solution: Given the hypotenuse and the side opposite the desired angle, we use the sine function.

$$\sin\theta = \frac{\text{opp}}{\text{hyp}}$$

$$\sin\theta = \frac{10.6}{59}$$

$$\sin\theta = 0.179661016$$

$$\theta = 10.4° \qquad \textit{(using } \boxed{\sin^{-1}} \textit{ key and rounding)}$$

The angle formed by the connecting rod is 10.4° ■

Example 23-2.8 Find the impedance Z of a circuit with 20 ohms of reactance X_L represented by the vector diagram in Fig. 23-27.

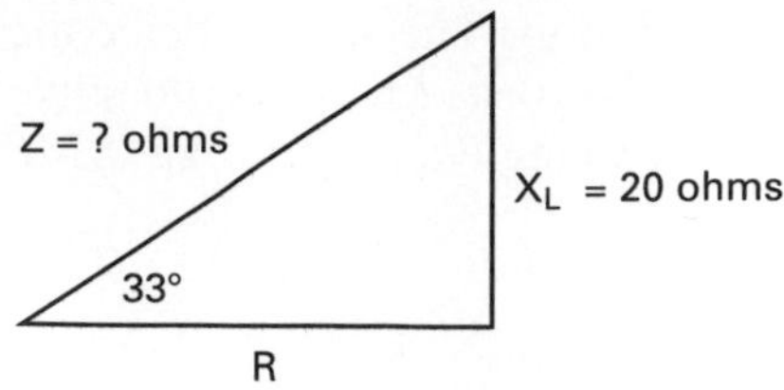

Figure 23-27

Solution: Since we are looking for the hypotenuse Z and are given an angle and the opposite side, we use the sine function.

$$\sin = \frac{\text{opp}}{\text{hyp}}$$

$$\sin 33° = \frac{X_L}{Z}$$

$$0.544639035 = \frac{20}{Z} \qquad (\sin 33° = 0.544639035)$$

$$0.544639035Z = 20$$

$$Z = \frac{20}{0.544639035}$$

$$Z = 36.72 \text{ ohms}$$

The impedance Z is 36.72 ohms. ■

SELF-STUDY EXERCISES 23-2

L.O.1. Find the indicated part of the following right triangles by the most direct method.

1.

x

40°

115 m

2.

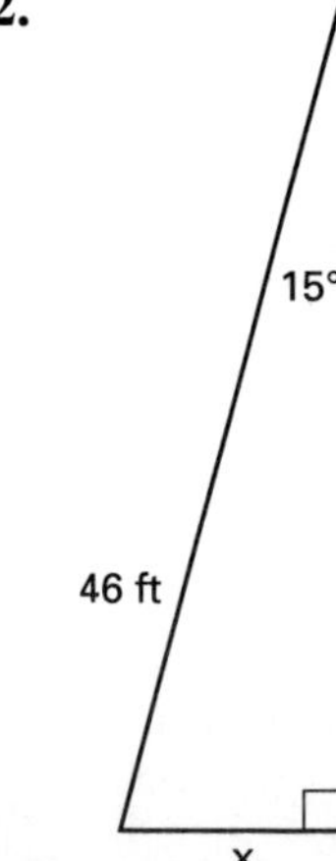

3.

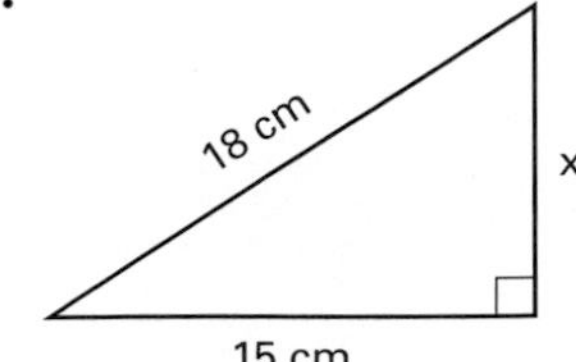

4. 17 cm 70° x

5.

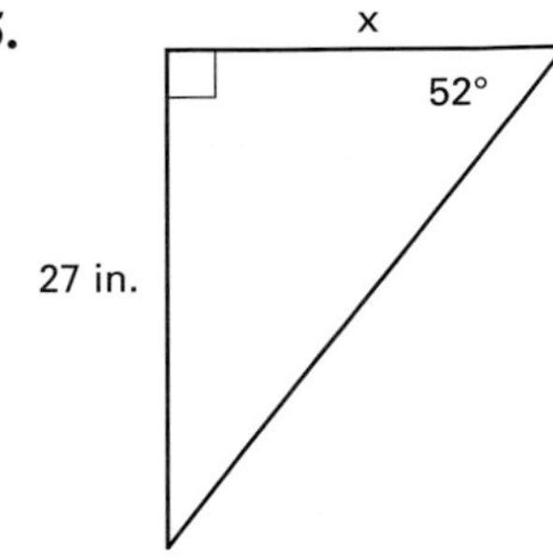

L.O.2. Use the trigonometric functions to solve the following technical problems. Round side lengths to four significant digits and angles to 0.1°.

6. A sign is attached to a building by a triangular brace. If the horizontal length of the brace is 48 in. and the angle at the sign is 25° (Fig. 23-28), what is the length of the wall support piece?

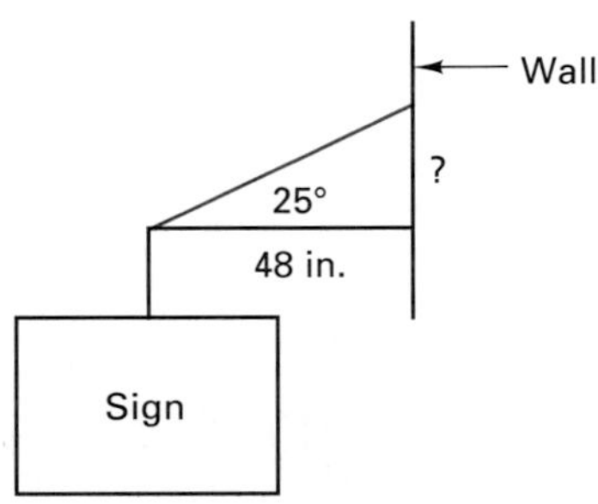

Figure 23-28

7. A surveyor uses right triangles to measure inaccessible property lines. To measure a property line that crosses a pond, a surveyor sights to a point across the pond, then makes a right angle, measures 50 ft, and sights the point across the pond with a 47° angle (Fig. 23-29). Find the distance across the pond from the initial point.

8. At what angle must a jet descend if it is 900 ft above the end of the runway and must touch down 1500 ft from the runway's end?

9. A 50-ft wire is used to brace a utility pole. If the wire is attached 4 ft from the top of the 35-ft pole, how far from the base of the pole will the wire be attached to the ground?

10. A roadway rises 4 ft for every 15 ft along the road. What is the angle of inclination of the roadway?

11. A shadow cast by a tree is 32 ft long when the angle of inclination of the sun is 36°. How tall is the tree?

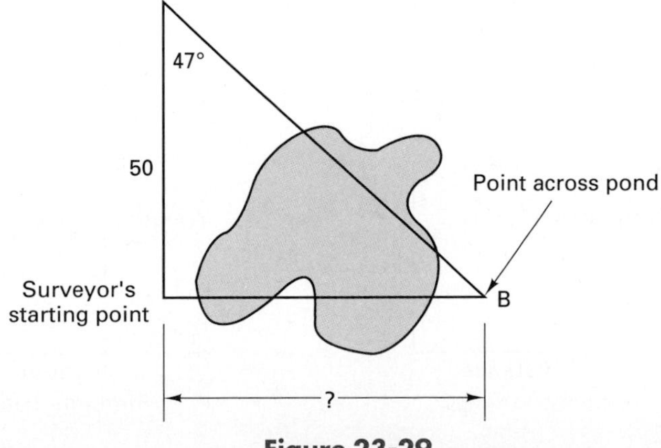

Figure 23-29

12. The vector diagram of the circuit in Fig. 23-30 has a known impedance Z. Find the reactance X_L. All units are in ohms.

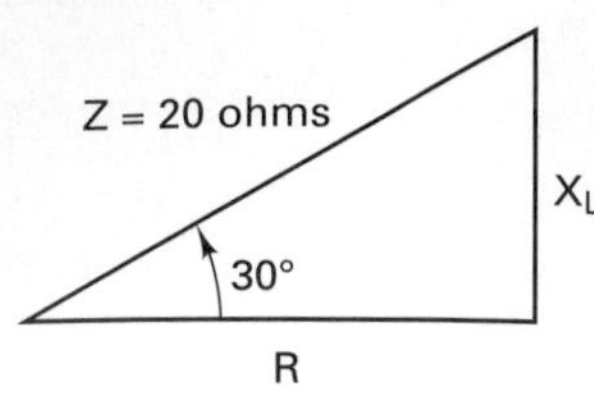

Figure 23-30

13. Using Fig. 23-30, find resistance R.

14. A piston assembly at the midpoint of its stroke forms a right triangle (Fig. 23-31). Find the length of rod R.

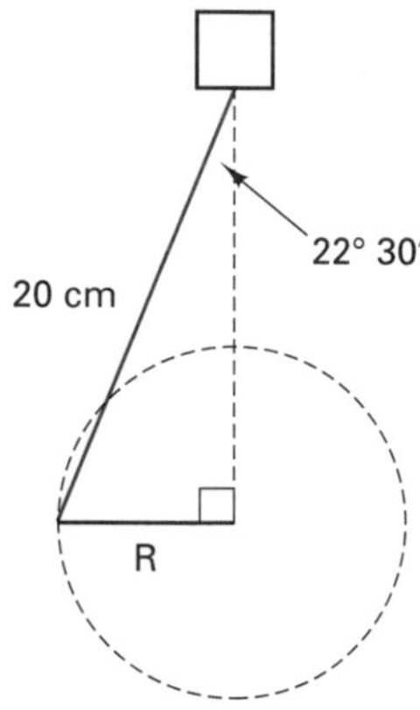

Figure 23-31

15. Find the angle a rafter makes with a joist of a house if the rise is 18 ft and the span is 50 ft. Refer to Fig. 23-24 on page 889.

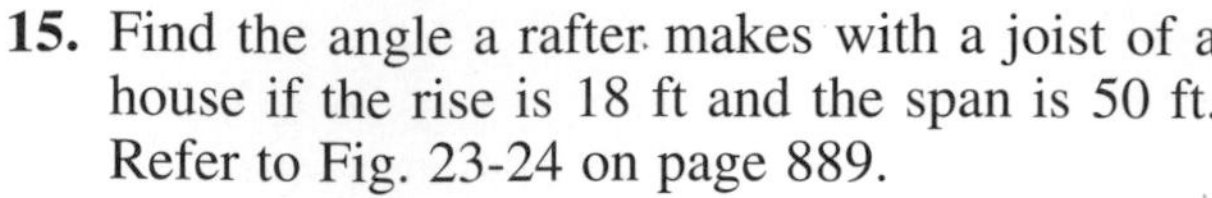

ELECTRONICS APPLICATION

Electronics Triangles

Technicians use many ac triangles, but the two main ones are the ohms triangle and the siemens triangle. Remember that ohms and siemens are reciprocals of each other. Series circuits use the ohms triangle and parallel circuits use the siemens triangle.

For each triangle, the units are the same on all three legs. The hypotenuse and the angle combine to give the polar number, and the horizontal side and the vertical side combine to give the rectangular number. The polar number shows the location of a point in a way different from the rectangular coordinate system.

Most analysis is done with rectangular numbers. All measurements are done with polar numbers. The two difficult parts are the vocabulary and how to find the missing sides and angles. All the triangles are handled in the same way.

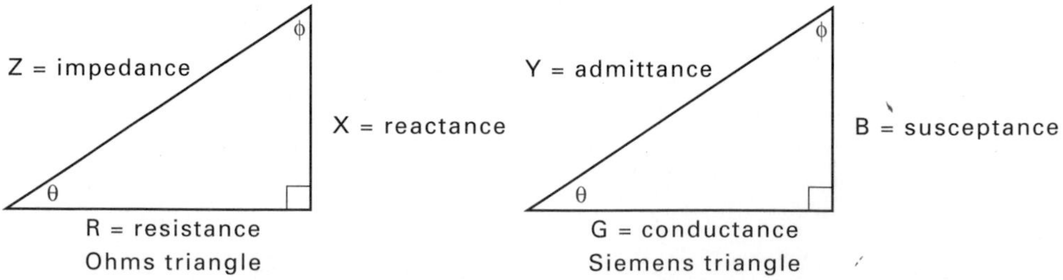

All three sides in the ohms triangle are measured in ohms. All three sides in the siemens triangle are measured in siemens, the reciprocal of ohms. If a component has a high resistance, it has a low conductance, and vice versa.

R and G are used for both dc (direct current) and ac (alternating current) analysis. X, Z, B, and Y are used only for ac analysis.

The Pythagorean theorem still holds. The formula can be rearranged to solve for any missing side if you know the other two sides.

$$Z^2 = R^2 + X^2 \qquad Y^2 = G^2 + B^2$$

$$R^2 = Z^2 - X^2 \qquad G^2 = Y^2 - B^2$$

$$X^2 = Z^2 - R^2 \qquad B^2 = Y^2 - G^2$$

Since each of these equations has a squared term on the left, that unknown can be solved for by taking the square root of both sides. Use parentheses, as that makes it easier on your calculator. Most calculators are designed for you to use the square-root key *after* the sum of squares is in the display. The calculator sequence for solving for Z on most calculators is

[(] R [x^2] [+] X [x^2] [)] [$\sqrt{\ }$]

or

R [x^2] [+] X [x^2] [=] [$\sqrt{\ }$]

The algebraic solutions for Z, R, X, Y, G, B are

$$Z = \sqrt{(R^2 + X^2)} \qquad Y = \sqrt{(G^2 + B^2)}$$

$$R = \sqrt{(Z^2 - X^2)} \qquad G = \sqrt{(Y^2 - B^2)}$$

$$X = \sqrt{(Z^2 - R^2)} \qquad B = \sqrt{(Y^2 - G^2)}$$

In electronics, the sides for R and G are always positive and rest on the horizontal axis. X and B are always perpendicular to the horizontal axis, and they may be positive or negative depending on whether the circuit is inductive or capacitive. The hypotenuse, Z or Y, is always considered to be positive since it is the square root of sum of squares. The angle θ is also called the *phase angle* and is positive or negative, depending upon the slope of the hypotenuse. The angle ϕ is the angle between the hypotenuse and the vertical side.

The angles θ and ϕ can be calculated in several different ways, depending on which sides or angles you were originally given. Another standard notation for $\sin^{-1}$, $\cos^{-1}$, and $\tan^{-1}$ is arcsin, arccos, and arctan, respectively.

$$\theta = \arctan\ (\text{rise/run}) = \arctan\ (X/R) = \tan^{-1}\ (X/R)$$

$$\theta = \arctan\ (\text{rise/run}) = \arctan\ (B/G) = \tan^{-1}\ (B/G)$$

$$\theta = \arcsin\ (\text{opp}\ \theta/\text{hyp}) = \arcsin\ (X/Z) = \sin^{-1}\ (X/Z)$$

$$\theta = \arcsin\ (\text{opp}\ \theta/\text{hyp}) = \arcsin\ (B/Y) = \sin^{-1}\ (B/Y)$$

$$\theta = \arccos\ (\text{adj}\ \theta/\text{hyp}) = \arccos\ (R/Z) = \cos^{-1}\ (R/Z)$$

$$\theta = \arccos\ (\text{adj}\ \theta/\text{hyp}) = \arccos\ (G/Y) = \cos^{-1}\ (G/Y)$$

$$\phi = 90^\circ - \theta$$

$$\theta = 90^\circ - \phi$$

The calculator sequence on calculators for finding an arcfunction is the same as the inverse fraction.

For arcsin, use:

x [÷] z [=] [sin] ($\sin^{-1}$) or [(] x [÷] z [)] [sin] ($\sin^{-1}$)

How do you find the missing sides and angles if two sides are given or if one side and one angle are given?

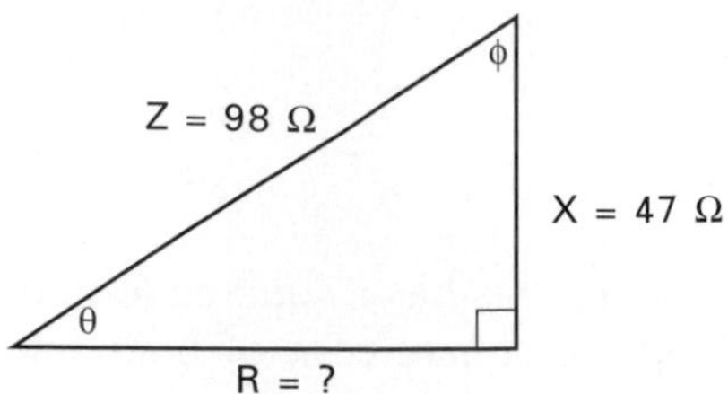

Example of two sides given

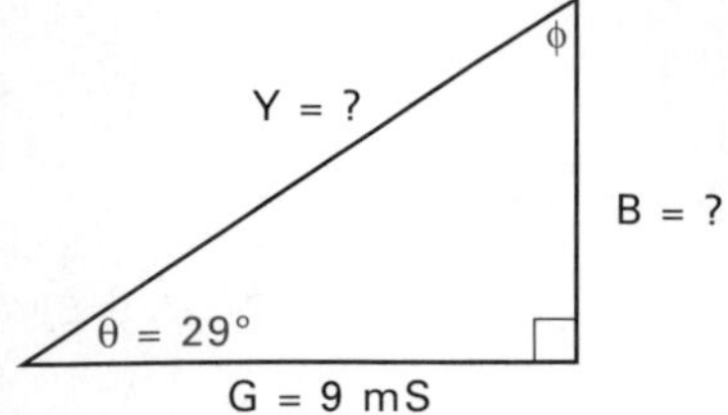

Example of side and angle given

$$Z^2 = R^2 + X^2$$
$$R^2 = Z^2 - X^2$$
$$R = \sqrt{(Z^2 - X^2)}$$
$$R = \sqrt{(98^2 - 47^2)}$$
$$R = 85.99\ \Omega$$
$$\theta = \arcsin (X/Z) = \sin^{-1} (X/Z)$$
$$\theta = \arcsin (47/98) = \sin^{-1} (47/98)$$
$$\theta = 28.66°$$
$$\phi = 90° - 28.66° = 61.34°$$

Proof: $\tan 28.66° \stackrel{?}{=} 47/85.99$, True

$$\cos \theta = G/Y \qquad \tan \theta = B/G$$
$$Y = G/\cos \theta \qquad B = G \tan \theta$$
$$Y = 9/\cos 29° \qquad B = 9 \tan 29°$$
$$Y = 10.29\text{ mS} \qquad B = 4.99\text{ mS}$$

Proof: $Y^2 = G^2 + B^2$

$$Y^2 - G^2 - B^2 = 0$$
$$10.29^2 - 9^2 - 4.99^2 \stackrel{?}{=} 0, \text{ True}$$
$$\phi = 90° - 29° = 61°$$

A complete calculator sequence including proof for both is

98 [x^2] [−] 47 [x^2] [=] [√] [MIN] 47 [÷] 98 [=] [SIN^{-1}] [TAN] [−] 47 [÷] [MR] [=]

⇒ −2.E-12

(*Note:* calculate B first and store it, and then calculate Y.)

9 [×] 29 [TAN] [=] [MIN] 9 [/] 29 cos [÷] [x^2] [−] 9 [x^2] [−] [MR] [x^2] [=] ⇒ −1E − 09

Fill in all answers right on the triangle with the correct units. Always give a proof.

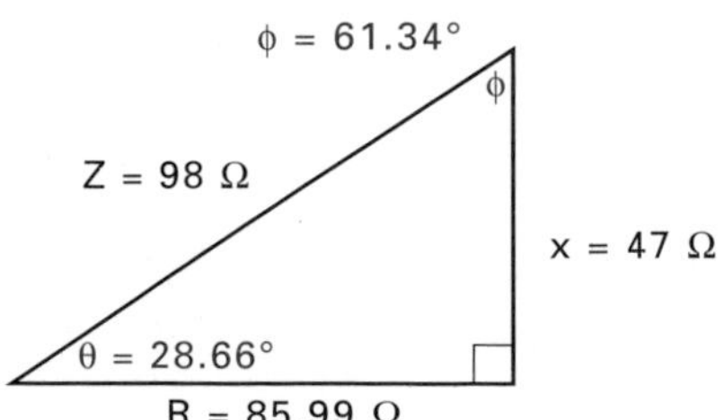

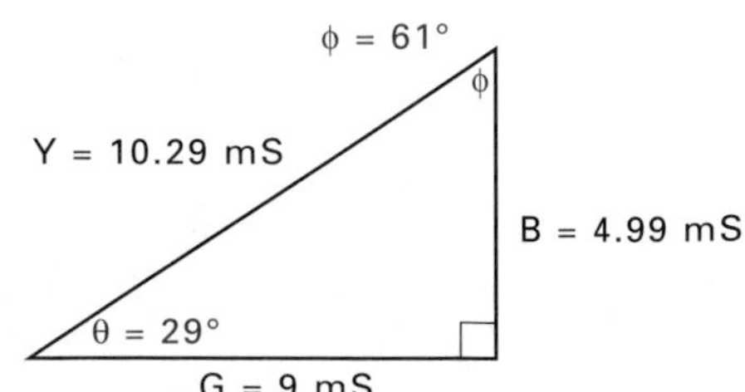

Exercises

Redraw each of the following triangles. Fill in all missing sides and angles. Show your proof. Include correct units.

1.

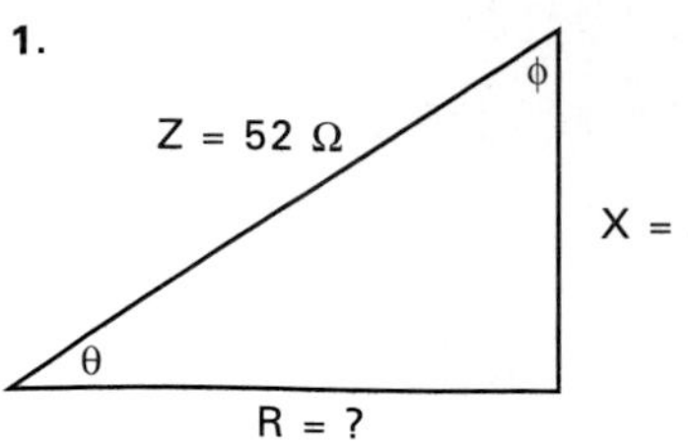

2.

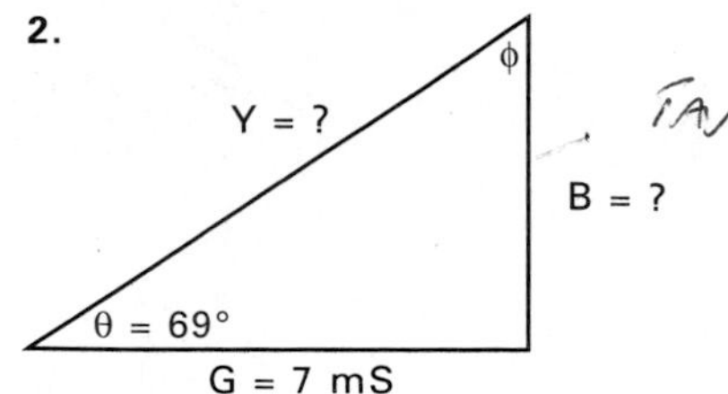

3.

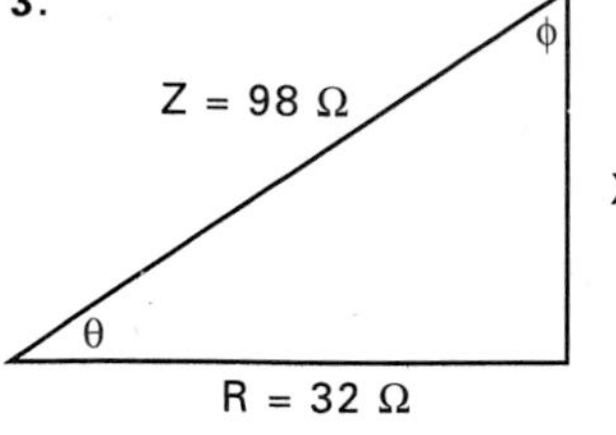

4.

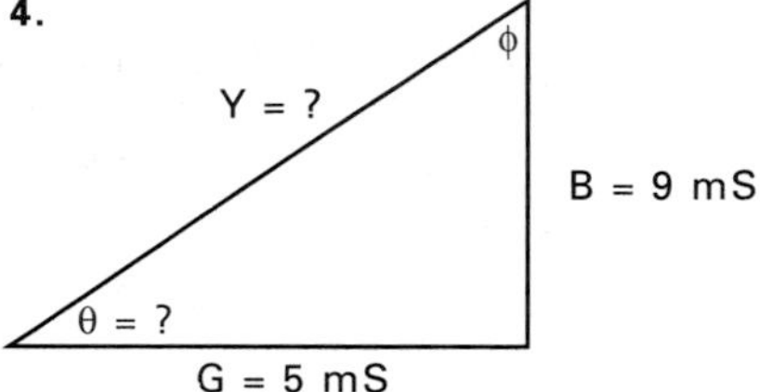

5.

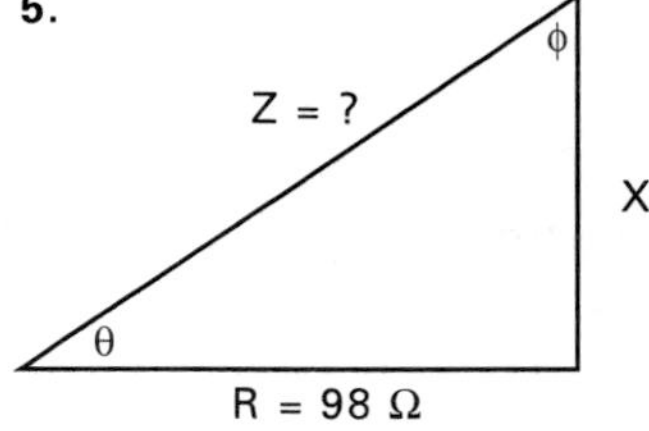

6.

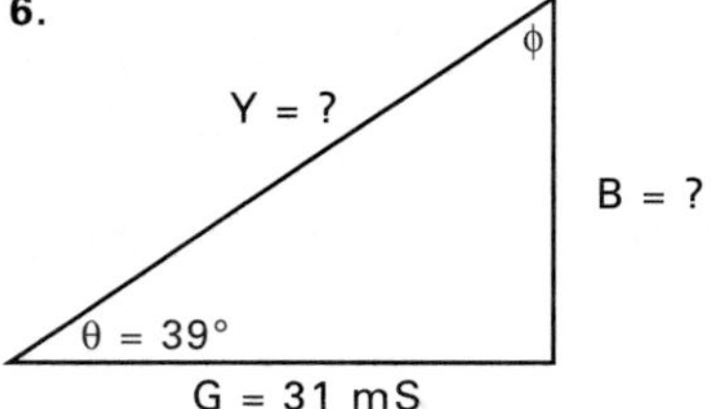

7.

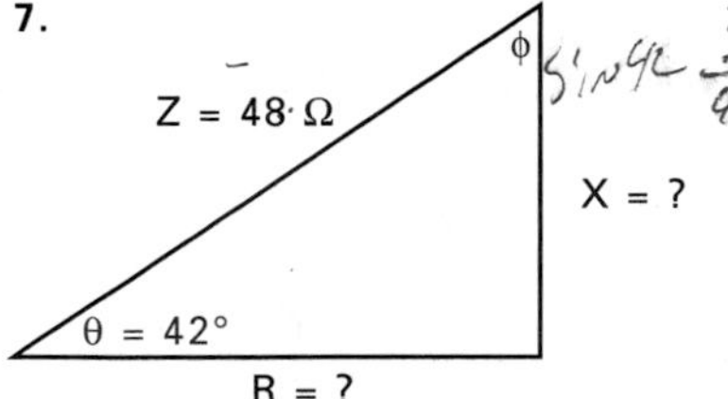

8.

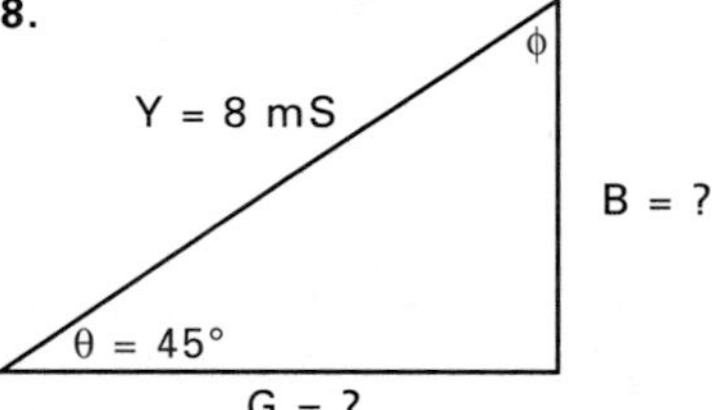

Answers for Exercises

Redraw each of the following triangles. Fill in all missing sides and angles. Show your proof. Include correct units.

1. ϕ = 58.72°

Z = 52 Ω

X = 27 Ω

θ = 31.28°

R = 44.44 Ω

2. ϕ = 21°

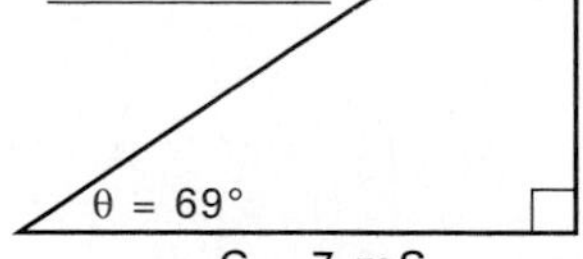

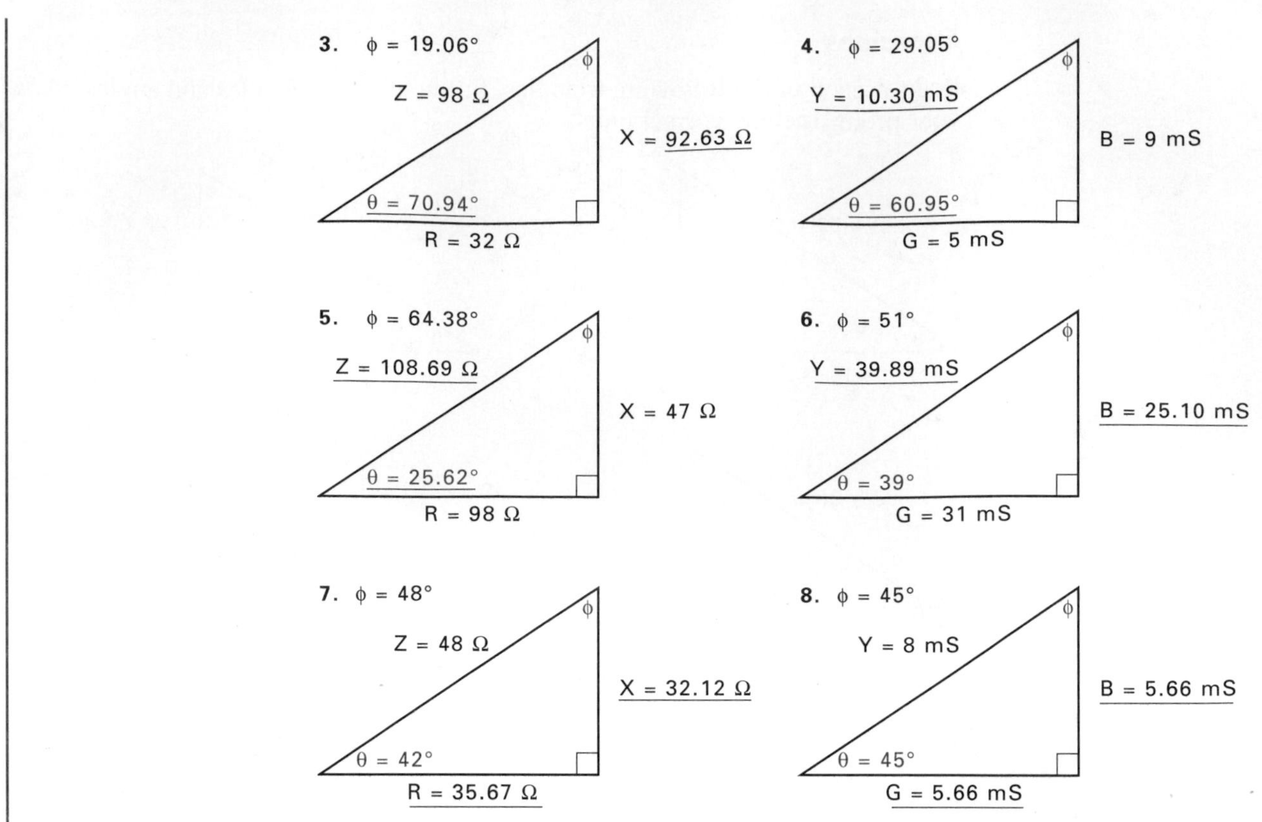

ASSIGNMENT EXERCISES

23-1

Solve the triangles in the following figures. Round sides to four significant digits and round angles to the nearest 0.1°.

1.

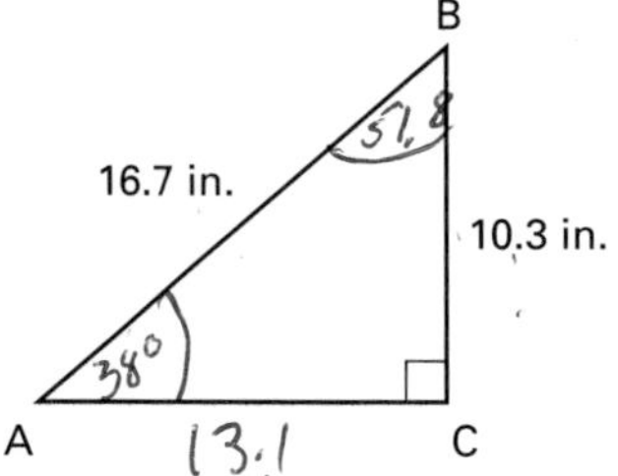

2.

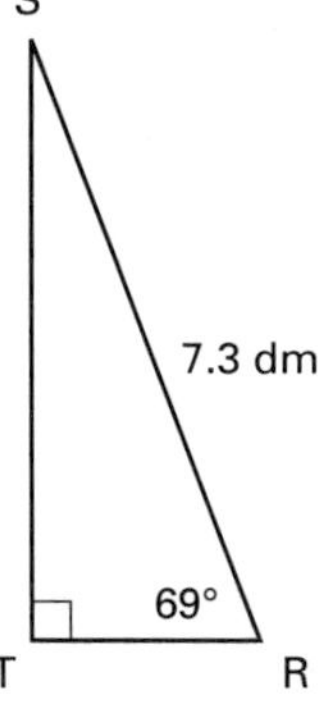

3.

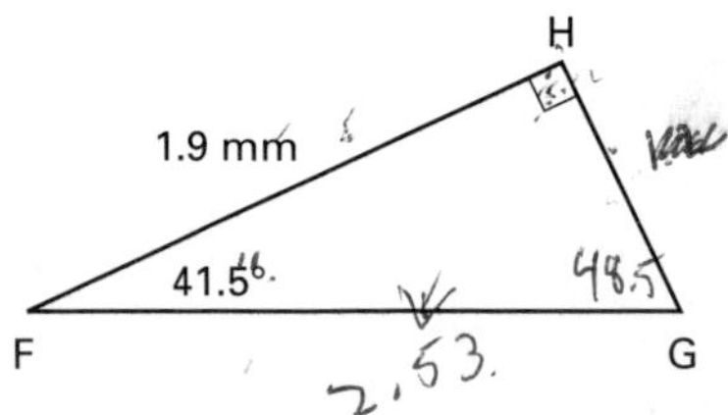

4.

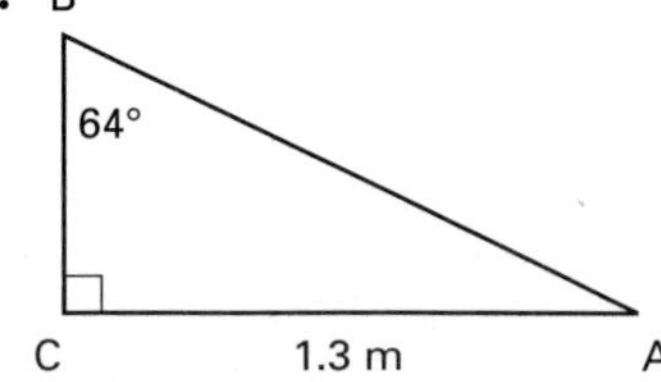

5.

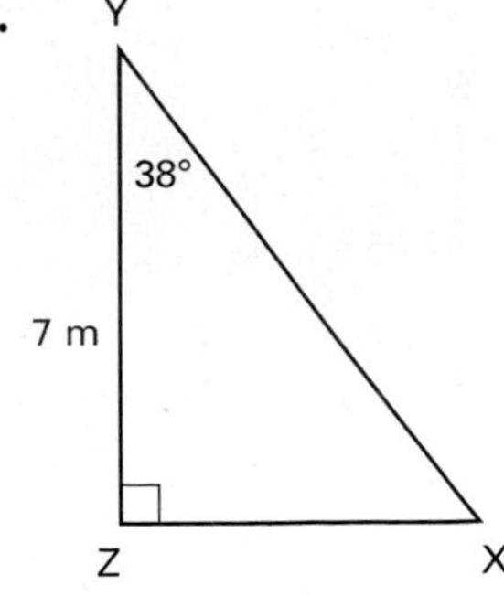

6.

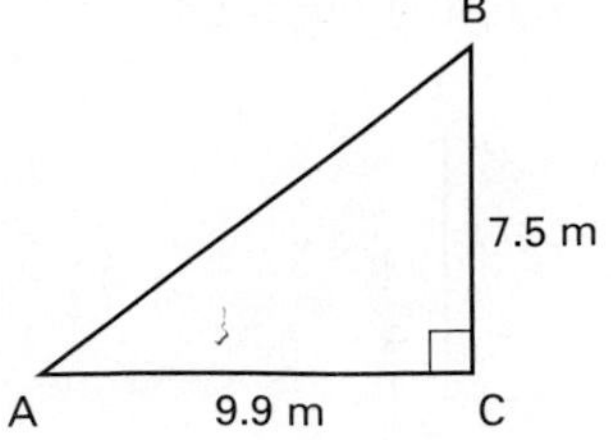

7.

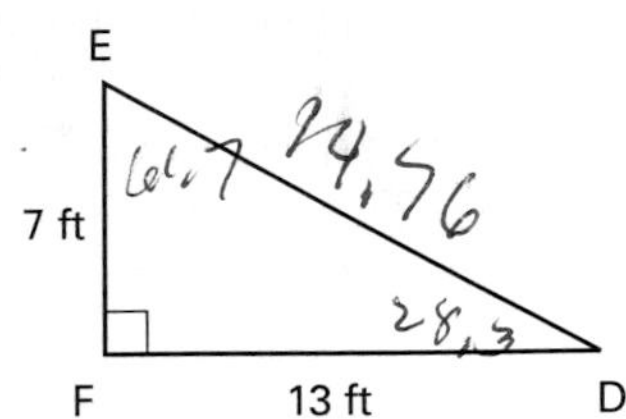

8.

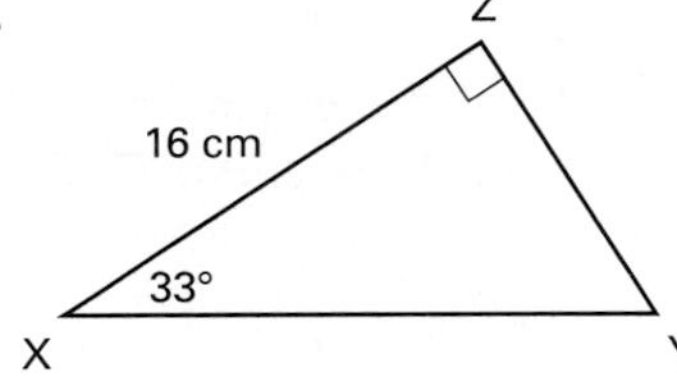

9.

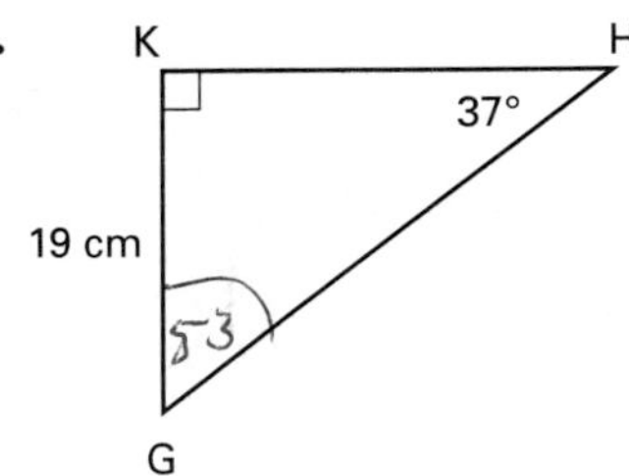

10.

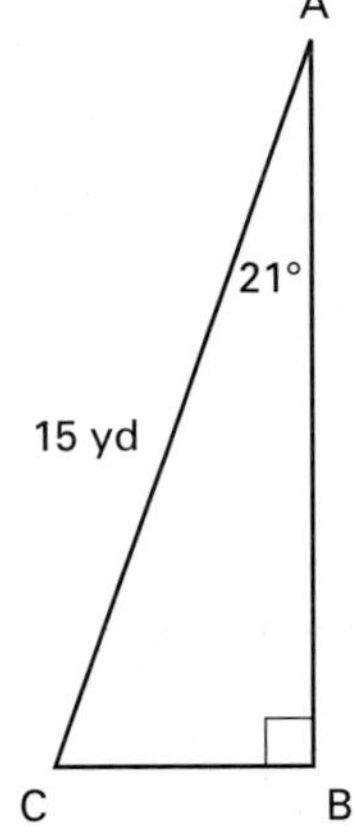

23-2

Find the indicated parts of the triangles in the following figures. Round side lengths to four significant digits and angles to the nearest 0.1°.

11. Find A.

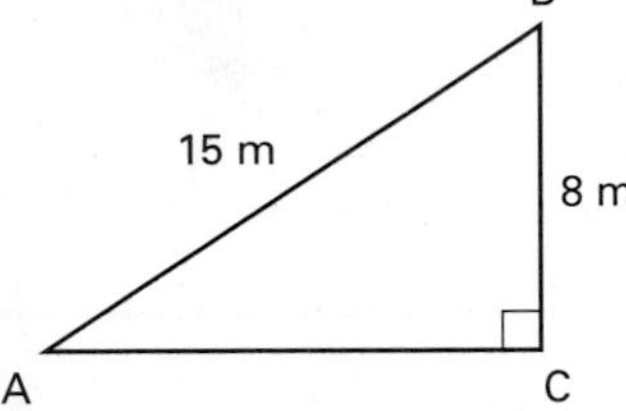

12. Find A.

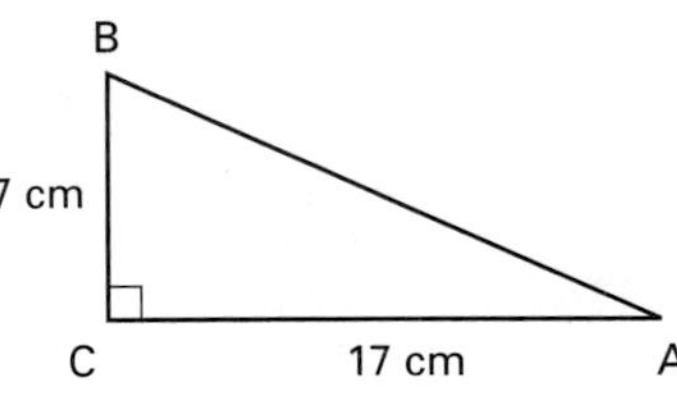

13. Find a.

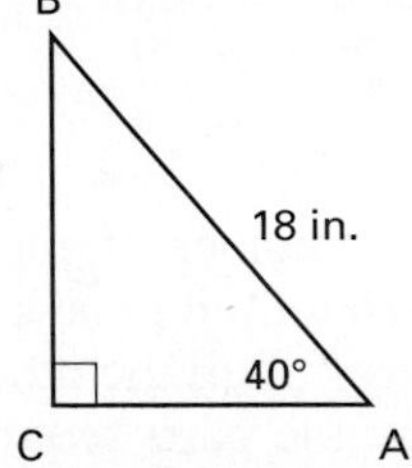

Solve the triangles in the following figures.

14.

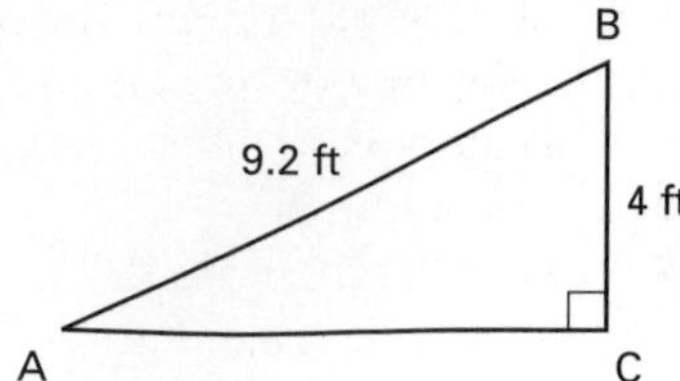

15.

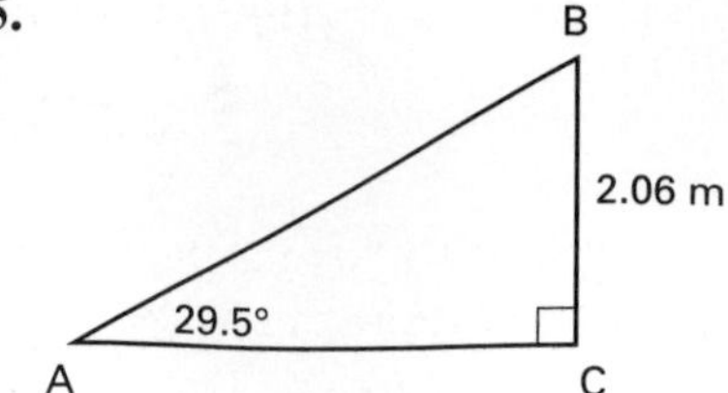

16. A railway inclines 14°. How many feet of track must be laid if the hill is 15 ft high?

17. A corner shelf is cut so that the sides placed on the wall are 37 in. and 42 in. What are the measures of the acute angles?

18. From a point 5 ft above the ground and 20 ft from the base of a building, a surveyor uses a transit to sight an angle of 38° to the top of the building. Find the height of the building.

19. A surveyor makes the measures indicated in Fig. 23-32. Solve the triangle. All parts of the triangle should be included in a report.

20. What length of rafter is needed for a roof if the rafters will form an angle of 35.5° with a joist and the rise is 8 ft?

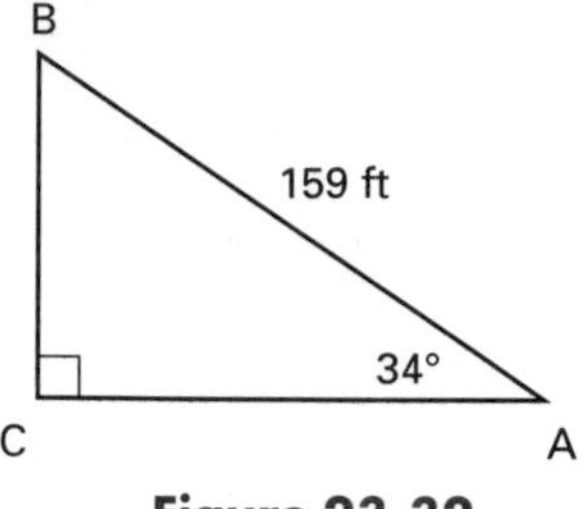

Figure 23-32

CHALLENGE PROBLEMS

21. Draw triangle ABC so that angle C is 90° and angle A is 70°. Point E is on side $\overline{AB}$ and is 10 cm from A. Point D is on $\overline{BC}$ and $\overline{DE}$ is parallel to $\overline{CA}$. If $\overline{CA}$ is 24 cm, find the length of $\overline{BD}$.

22. You are designing a stairway for a home and you need to determine how many steps the stairway must have. If the stairway is to be 8 feet high and the floorspace allotted is 14 feet, determine how many steps are needed and determine the optimum measure (rise and run) for each step.

CONCEPTS ANALYSIS

1. Use a calculator to find sin 45° and cos 45°. Using your knowledge of geometry, explain the relationship between the sin 45° and cos 45°. Does it follow that the sine and cosine of the same angle are always equal? Are there other angles for which the two functions are equal?

2. If you are given an angle and an adjacent side of a right triangle, explain how you would find the hypotenuse of the triangle.

3. You know the measure of both legs of a right triangle and the measure of one of the acute angles. Describe two different ways you could use to find the length of the hypotenuse.

4. What is the least number of measures of the parts of a triangle required to be known in order to find the measures of all the other parts of the triangle? What are the parts required?

5. Explain the differences and similarities between angle of inclination and angle of declination.

6. Using a calculator compare the sine of an acute angle with the sine of its complement for several angles. Compare the sine of an acute angle with the cosine of its complement for several angles. Generalize your findings.

7. Devise a right triangle that has the measure of an angle (other than the right angle) and a side given. Then use trigonometric functions or other methods to find the measures of all the other sides and angles.

8. Describe the calculator steps required to find the measure of an angle whose sine function is given.

9. Use your calculator to discover which of the three trigonometric functions-sine, cosine, tangent-can have values greater than 1. For what range of angles is the value of this function greater than 1?

10. Use your graphing calculator to graph the trigonometric functions sine and cosine. Tell how the graphs are similar and how they are different.

CHAPTER SUMMARY

Objectives	What to Remember	Examples
Section 23-1		
1. Find missing parts of a right triangle using the sine function.	$\sin\theta = \dfrac{\text{opposite side}}{\text{hypotenuse}}$	Use the sine function to find the indicated part of the triangle. 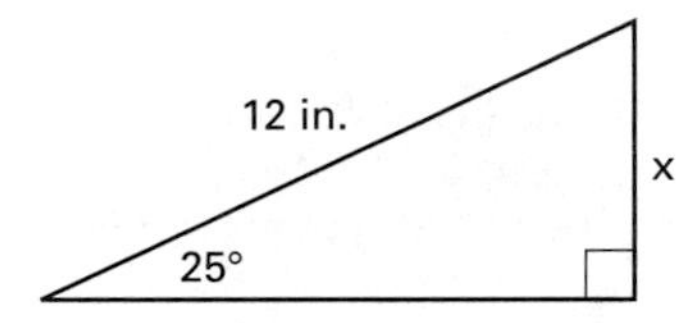 $\sin 25° = \dfrac{x}{12}$ $x = 12(0.422618261)$ $x = 5.07$ in.
2. Find missing parts of a right triangle using the cosine function.	$\cos\theta = \dfrac{\text{adjacent side}}{\text{hypotenuse}}$	Use the cosine function to find the indicated part of the triangle. 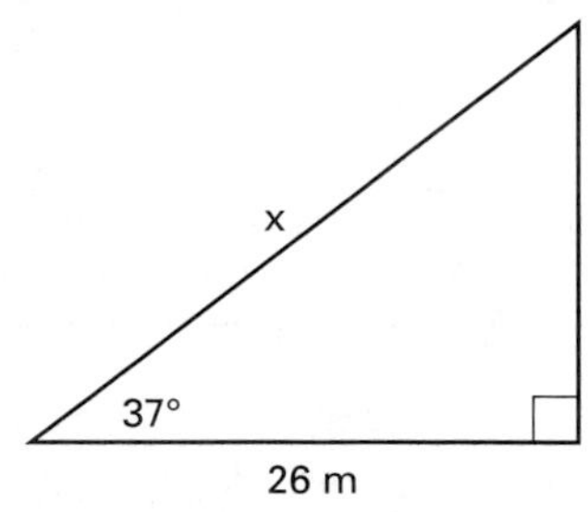 $\cos 37° = \dfrac{26}{x}$ $x = \dfrac{26}{\cos 37°}$ $x = \dfrac{26}{0.79863551}$ $x = 32.56$ m

3. Find missing parts of a right triangle using the tangent function.

$$\tan \theta = \frac{\text{opposite side}}{\text{adjacent side}}$$

Use the tangent function to find the indicated part of the triangle.

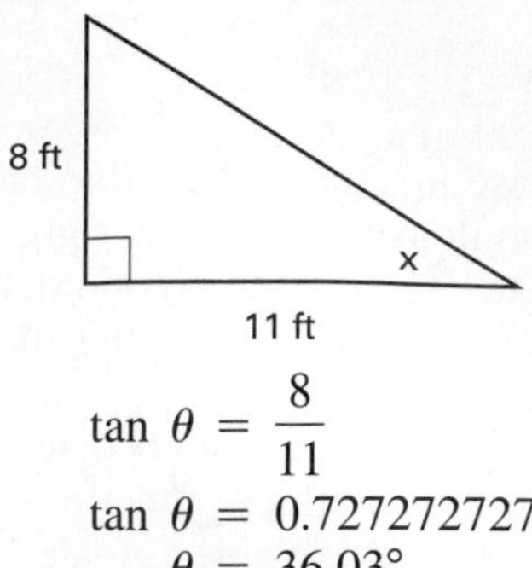

$$\tan \theta = \frac{8}{11}$$
$$\tan \theta = 0.727272727$$
$$\theta = 36.03°$$

Section 23-2

1. Select the most direct method for solving right triangles.

Where possible, choose the function that uses parts that are given in the problem, rather than parts that are not given and must be calculated.

Where possible, choose the function that gives the desired part directly, that is, without having to find other parts first.

Find the indicated parts.

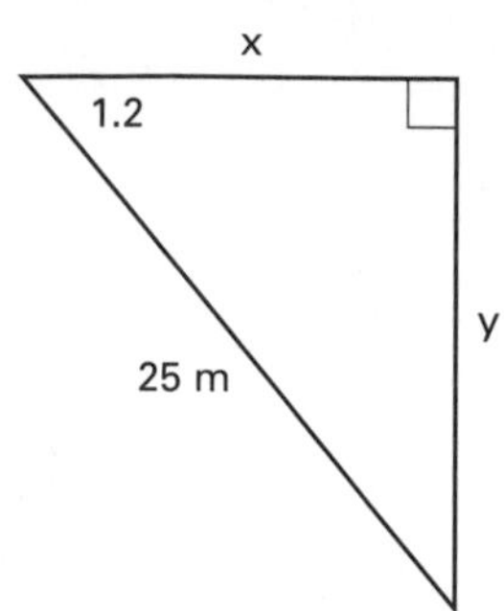

$$\cos 1.2 = \frac{x}{25}$$
$$x = 25(\cos 1.2)$$
$$x = 25(0.362357754)$$
$$x = 9.06 \text{ m}$$
$$\sin 1.2 = \frac{y}{25}$$
$$y = 25(\sin 1.2)$$
$$y = 25(0.932039086)$$
$$y = 23.30 \text{ m}$$

2. Solve applied problems using right-triangle trigonometry.

To solve a problem using the trigonometric functions, first determine the given parts and the missing part or parts; then identify the trigonometric function that relates the given and missing parts. Use this function and the given information to find the missing information.

A jet takes off at a 25° angle. Find the distance traveled by the plane from the takeoff point to the end of the runway if the runway is 950 feet long.

$$\cos 25° = \frac{950}{x}$$
$$x = \frac{950}{\cos 25°}$$
$$x = \frac{950}{0.991202811}$$
$$x = 958.4 \text{ ft}$$

WORDS TO KNOW

significant digits (p. 887)
angle of elevation (p. 887)
angle of depression (p. 887)
angle of declination (p. 888)

CHAPTER TRIAL TEST

The following measures are indicated for a standard right triangle, ABC, where C is the right angle. Draw the figures and use the sine, cosine, or tangent function to find the indicated parts of the triangle. Round side lengths to four significant digits and angles to the nearest 0.1°.

1. $a = 16$ m, $b = 14$ m, find A.

2. $a = 7$ in., $A = 33°$, find c.

3. $c = 17$ ft, $B = 25°$, find a.

4. $a = 21$ m, $A = 48.5°$, find b.

5. $a = 32$ cm, $c = 47$ cm, find A.

6. $c = 12$ m, $A = 35°$, find a.

7. $b = 21$ cm, $A = 17°$, find a.

8. $b = 1$ cm, $A = 87°$, find c.

9. $b = 3.1$ m, $c = 6.8$ m, find A.

10. $a = 0.15$ m, $c = 0.46$ m, find A.

11. Solve triangle ABC in Fig. 23-33. Round as above.

12. Solve triangle DEF in Fig. 23-34. Round as above.

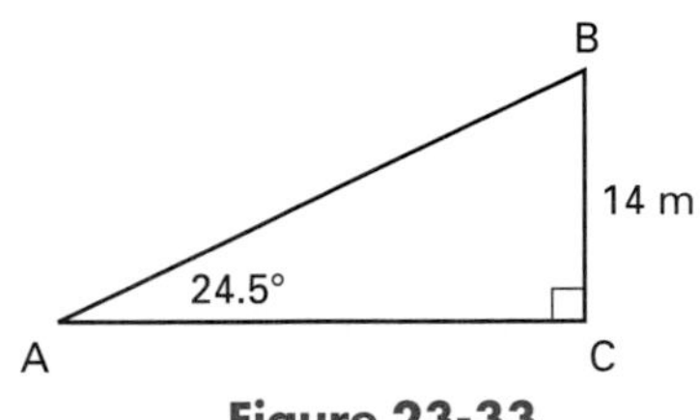

Figure 23-33

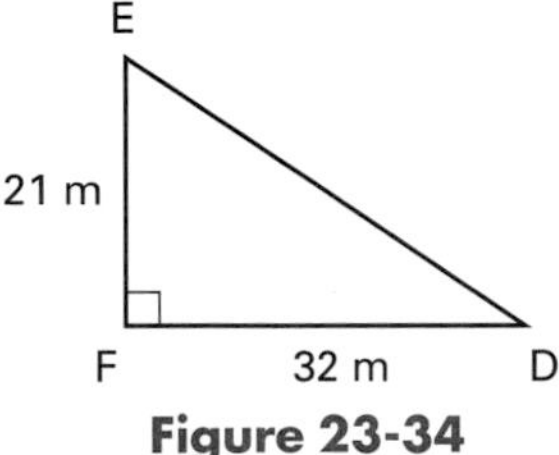

Figure 23-34

Solve the following problems. Round as above unless otherwise indicated.

13. A stair has a rise of 4 in. for every 5 in. of run. What is the angle of inclination of the stair?

14. A surveyor uses indirect measurement to find the width of a river at a certain point. The surveyor marks off 50 ft along the river bank at right angles with the river and then sights an angle of 43° to point A across the river (Fig. 23-35). Find the width of the river.

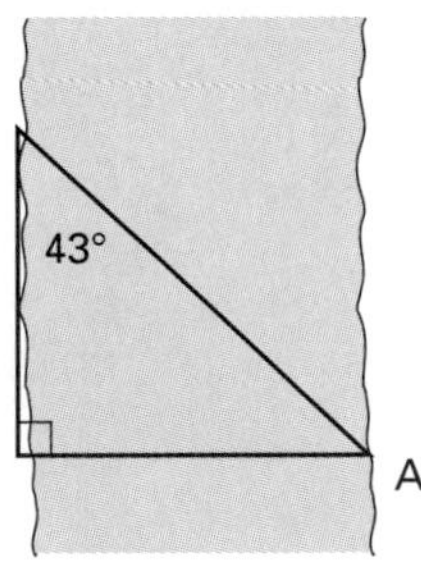

Figure 23-35

15. Steel girders are reinforced by placing steel supports between two runners so that right triangles are formed (Fig. 23-36). If the runners are 24 in. apart and a 30° angle is desired, find the length of the support that will be placed at a 30° angle.

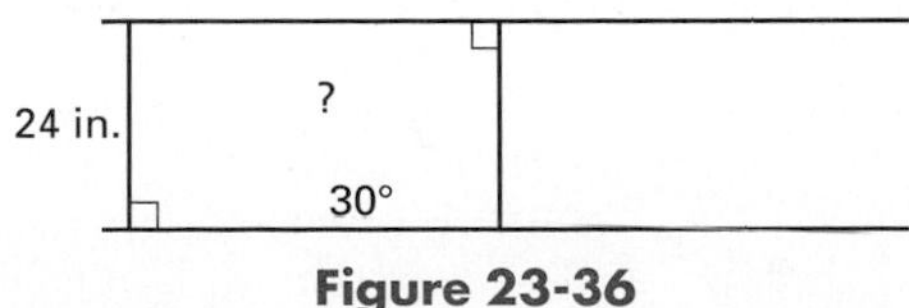

Figure 23-36

16. In the circuit represented by the diagram of Fig. 23-37, find the total current I_t. All units are in amps.

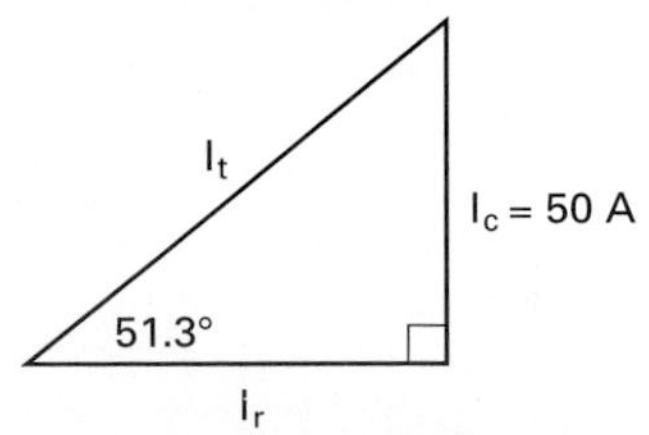

Figure 23-37

17. Find the current in the resistance branch I_r of the circuit in Problem 16.

18. The minimum clearances for the installation of a metal chimney pipe are shown in Fig. 23-38. How far from the ridge should the hole be cut for the pipe to pass through the roof?

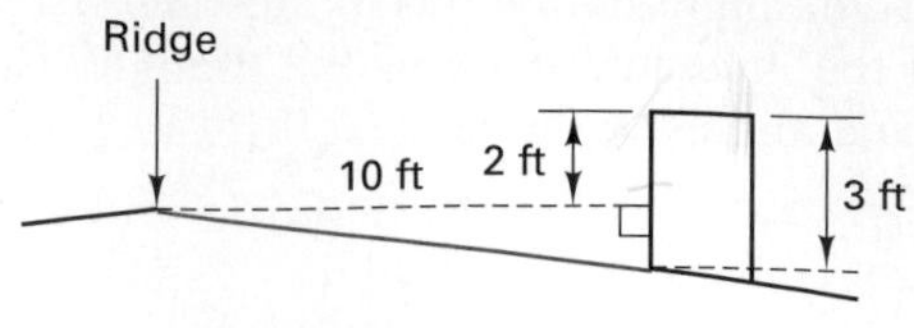

Figure 23-38

19. What angle is formed by the chimney pipe and the roof where the pipe passes through the roof (Fig. 23-38)?

20. Elbows are used to form bends in rigid pipe and are measured in degrees. What is the angle of bend of the elbow in the installation shown in Fig. 23-39 to the nearest whole degree?

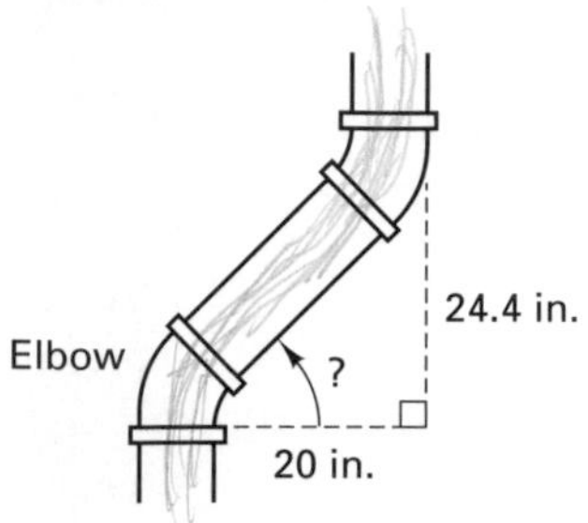

Figure 23-39

21. Find the angle formed by the connecting rod and the horizontal in the mechanical assembly shown in Fig. 23-40.

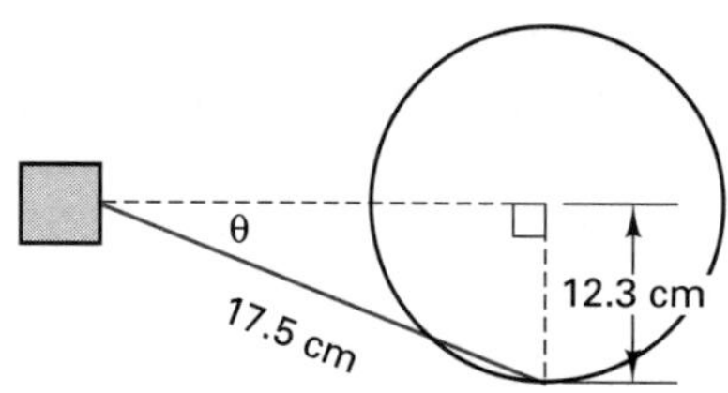

Figure 23-40

22. A utility pole is 40 ft above ground level. A guy wire must be attached to the pole 3 ft from the top to give it support. If the guy wire forms a 20° angle with the ground, how long must the guy wire be? Disregard length needed for attaching the wire to the pole or ground. Round to hundredths.

23. Solve for reactance X_L and resistance R in Fig. 23-41. All units are in ohms. Round to hundredths.

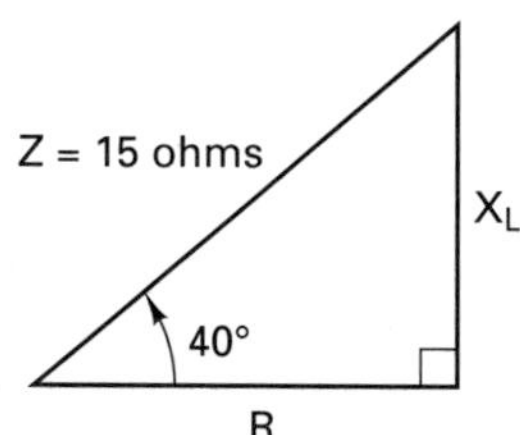

Figure 23-41

CHAPTER

24 Oblique Triangles

Although right triangles are perhaps the most frequently used triangles, we often work with triangles that are not right triangles. In this chapter, we will apply the laws of sines and cosines to oblique triangles, find the area of oblique triangles when only selected parts are given, and use *vectors* for solving triangles that have an angle greater than 90°.

24-1 VECTORS

Learning Objectives

L.O.1. Find the magnitude of a vector in standard position, given the coordinates of the end point.
L.O.2. Find the direction of a vector in standard position, given the coordinates of the end point.
L.O.3. Find the sum of vectors.

L.O.1. Find the magnitude of a vector in standard position, given the coordinates of the end point.

Magnitude of Vectors

Quantities that we have discussed so far in this text have been described by specifying their size or magnitude. Quantities such as area, volume, length, and temperature, which are characterized by magnitude only, are called *scalars*. There are many other quantities such as electrical current, force, velocity, and acceleration that are called *vectors*.

■ **DEFINITION 24-1.1 Vector.** A quantity described by magnitude (or length) and direction is a *vector*. A vector represents a shift from one point to another.

When we consider the speed of a plane to be 500 mph, we are considering a scalar quantity. However, when we consider the speed of a plane *traveling northeast from a given location*, we are concerned with both the distance traveled and the direction. This is a vector quantity. We had previously seen such quantities in the vector diagrams used to solve electronic problems by means of right triangles in Chapter 23 and elsewhere.

Vectors are represented by straight arrows. The length of the arrow represents the *magnitude* of the vector (Fig. 24-1). The curved arrow shows the counterclockwise *direction* of the vector (Fig. 24-2), with the end of the arrow being the beginning point and the arrowhead being the end point.

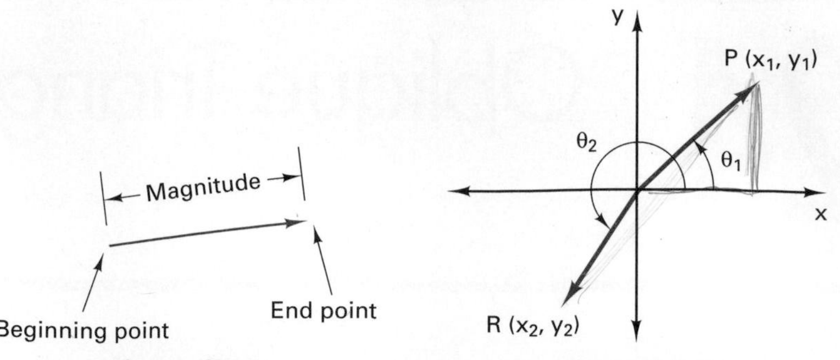

Figure 24-1 **Figure 24-2**

In relating trigonometric functions to vector quantities, we will place the vectors in *standard position*. Standard position places the beginning point of the vector at the origin of a rectangular coordinate system. *The direction of the vector is the counterclockwise angle measured from the positive x-axis (horizontal axis).*

In Fig. 24-2, vector P is a first quadrant vector that has a direction of θ_1 and an endpoint at the point P. The magnitude of a vector can be determined if the vector is in standard position and the x- and y-coordinates of the end point are known. Vector R is a third-quadrant vector that has a direction of θ_2 and an end point at point R. Its magnitude may be similarly determined.

Example 24-1.1 Find the magnitude of vector P in Fig. 24-2 if the coordinates of P are (4, 3).

Solution: From P, we can draw a vertical line to the x-axis, forming a right triangle with the x-axis and the vector P. The length of the side of the triangle along the x-axis is the x-coordinate of the point P, or 4. The length of the vertical side of the triangle is the y-coordinate of the point P, or 3. The length of the hypotenuse of the triangle is the magnitude of the vector. We will use the Pythagorean theorem to determine the magnitude of vector P.

$$p^2 = x^2 + y^2$$

$$p^2 = 4^2 + 3^2$$

$$p^2 = 16 + 9$$

$$p^2 = 25$$

$$p = \pm\sqrt{25}$$

$$p = \pm 5 \qquad \textit{(or +5 since length or magnitude is positive)}$$

The magnitude of vector P is 5. ■

Example 24-1.2 Find the magnitude of vector R in Fig. 24-2 if the coordinates of point R are $(-2, -5)$.

Solution: The magnitude of a vector is always positive; however, the x- and y-coordinates can be negative.

$$p^2 = x^2 + y^2$$

$$p^2 = (-2)^2 + (-5)^2$$

$$p^2 = 4 + 25$$

$$p^2 = 29$$

$$p = \sqrt{29}$$

$$p = 5.385 \qquad \textit{(rounded)}$$

■

L.O.2. Find the direction of a vector in standard position, given the coordinates of the end point.

Direction of Vectors

If the vector in standard position falls in quadrant I, we can use right-triangle trigonometry to find the direction of the vector. For a vector in standard position with an end point in quadrant I, the tangent function can be used to find the angle or direction of the vector.

EXAMPLE 24-1.3 Find the direction of vector P in Fig. 24-2 if the coordinates of P are (4, 3).

Solution: The right triangle is formed by the vector, the x-axis, and a line from the end point of the vector to the x-axis. The angle at the origin has an opposite-side value of 3 and an adjacent-side value of 4.

$$\tan \theta = \frac{3}{4}$$

$$\tan \theta = 0.75$$

$$\theta = 36.87°$$

■

L.O.3. Find the sum of vectors.

Sum of Vectors

Two vectors with the same direction can be added by aligning the beginning point of one vector with the end point of the other. The vector represented by the sum is called the *resultant* vector, or resultant. The resultant has the same direction as the vectors being added and a magnitude that is the sum of the two magnitudes.

EXAMPLE 24-1.4 Add two vectors with a direction of 35° if the magnitudes of the vectors are 4 and 5, respectively (Fig. 24-3).

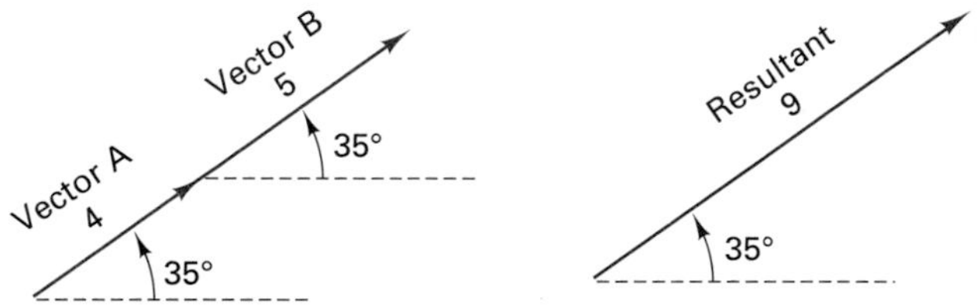

Figure 24-3

Solution

The resultant vector has a magnitude of 9 and a direction of 35°. ■

Two vectors with opposite directions can be added by aligning the beginning point of the vector with the smaller magnitude with the end point of the vector with the larger magnitude. The resultant has the same direction as the vector with the larger magnitude and a magnitude that is the difference of the two magnitudes.

EXAMPLE 24-1.5 Add a vector with a direction of 60° and a magnitude of 6 to a vector with a direction of 240° and a magnitude of 4 (Figure 24-4).

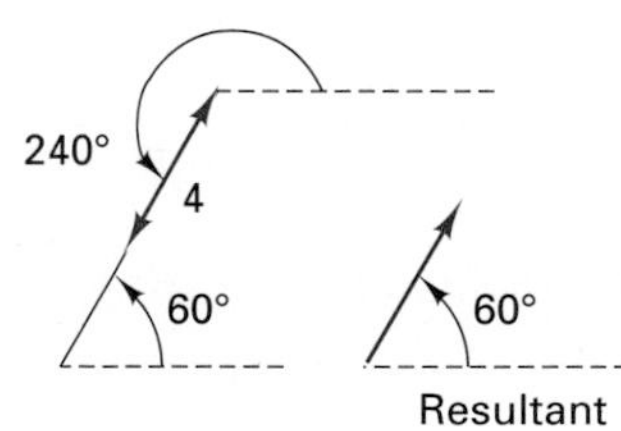

Figure 24-4

Solution

Subtract the magnitudes: $6 - 4 = 2$. The resultant has a direction of 60° and a magnitude of 2. ■

Adding any two vectors that have different directions is accomplished by performing the shifts of each vector in succession.

Example 24-1.6 Show a graphical representation of the sum of a 45° vector with a magnitude of 5 (vector A) and a 60° vector with a magnitude of 6 (vector B) (see Fig. 24-5).

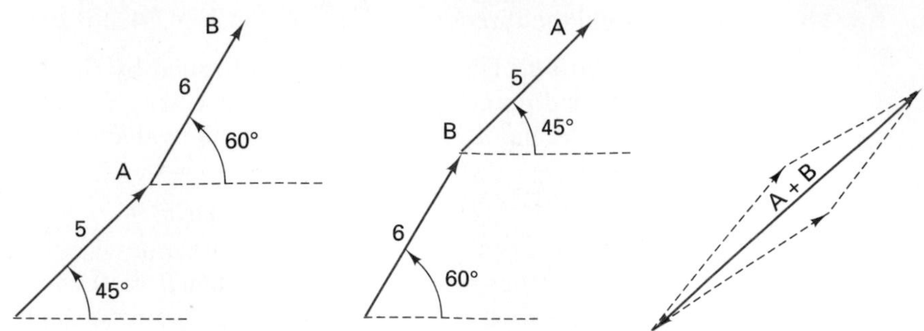

A +B or B + A = resultant A + B

Figure 24-5

Solution ■

Complex numbers are sometimes used to represent vectors. The x-coordinate of a vector in standard position is represented as the real component. The y-coordinate of a vector in standard position is represented as the imaginary component. Vector A could be written in complex notation by first finding the x- and y-coordinates of the end point.

x-coordinate	y-coordinate
$\cos 45° = \frac{x}{5}$	$\sin 45° = \frac{y}{5}$
$5(\cos 45°) = x$	$5(\sin 45°) = y$
$3.535533906 = x$	$3.535533906 = y$

Vector A in complex form: $3.535533906 + 3.535533906i$

x-coordinate	y-coordinate
$\cos 60° = \frac{x}{6}$	$\sin 60° = \frac{y}{6}$
$6(\cos 60°) = x$	$6(\sin 60°) = y$
$3 = x$	$5.196152423 = y$

Vector B in complex form: $3 + 5.196152423i$

$$\text{Resultant } A + B = (3.535533906 + 3.535533906i) + (3 + 5.196152423i)$$
$$= 6.535533906 + 8.731686329i \text{ or } 6.54 + 8.74i$$

Example 24-1.7 Find the magnitude and direction of the resultant $A + B$ in Example 24-1.6.

Solution: The resultant in complex form is $6.535533906 + 8.731686329i$; thus, the x-coordinate of the end point is 6.535533906 and the y-coordinate is 8.731686329.

$$\text{Magnitude} = \sqrt{(6.535533906)^2 + (8.731686329)^2}$$
$$\text{Magnitude} = \sqrt{118.9555496}$$
$$\text{Magnitude} = 10.92$$
$$\tan\theta = \frac{8.731686329}{6.535533906}$$
$$\tan\theta = 1.336032596$$
$$\theta = 53.19^\circ \quad \textit{(rounded)}$$

Tips and Traps

Vectors written in complex form are often used in electronics. However, the imaginary part is referred to as the j-factor.

Resultant vector A + B from Example 24-1.6 would be written as 6.54 + j8.74. The customary notation is for j to be followed by the coefficient.

SELF-STUDY EXERCISES 24-1

L.O.1. Find the magnitude of the vectors in standard position with end points at the following points. Round to the nearest thousandth.

1. (5, 12) **2.** (−12, 9) **3.** (2, −7)
4. (−8, −3) **5.** (1.5, 2.3)

L.O.2. Find the direction of the vectors in standard position with end points at the following points. Round to the nearest hundredth.

6. (5, 12) **7.** (6, 8) **8.** (8, 3)
9. (2, 5) **10.** (1, 4)

11. Find the resultant vector of two vectors that have a direction of 42° and magnitude of 7 and 12 respectively.

12. Two vectors have a direction of 72°. Find the sum of the vectors if their magnitudes are 1 and 7 respectively.

13. Find the sum of two vectors if one has a direction of 45° and a magnitude of 7 and the other has a direction of 225° and a magnitude of 8.

14. Find the sum of two vectors if one has a direction of 75° and a magnitude of 15 and the other has a direction of 255° and a magnitude of 9.

L.O.3. Find the magnitude and direction of the resultant of the sum of the two given vectors in complex notation.

15. $5 + 3i$ and $7 + 2i$ **16.** $1 + 2i$ and $5 + 2i$

24-2 TRIGONOMETRIC FUNCTIONS FOR ANY ANGLE

Learning Objectives

L.O.1. Find related, acute angles for angles or vectors in quadrants II, III, and IV.
L.O.2. Determine the signs of trigonometric values of angles more than 90°.
L.O.3. Find the trigonometric values of angles of more than 90° using a calculator.

L.O.1. Find related, acute angles for angles or vectors in quadrants II, III, and IV.

Related Angles

The direction of any vector in standard position can be determined, if the coordinates of the end point are known, by applying our knowledge of trigonometric functions. A right triangle that we will refer to as our *reference triangle* can be formed by drawing a vertical line from the end point of the vector to the x-axis. See the example of a reference triangle in Fig. 24-6. The angle θ will be called the *related angle*.

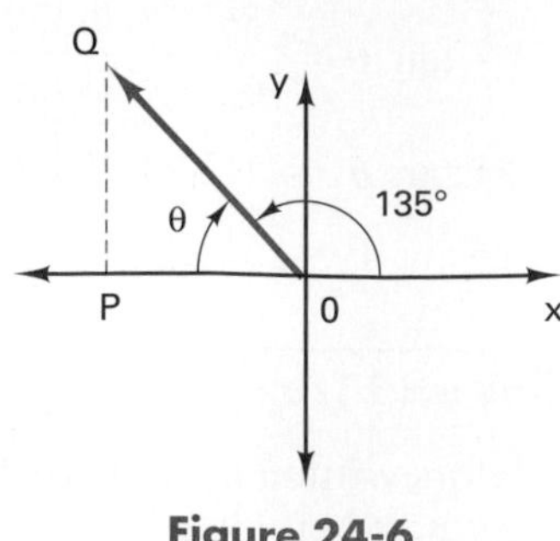

Figure 24-6

■ **DEFINITION 24-2.1 Related Angle.** The **related angle** is the acute angle formed by the x-axis and the vector.

In quadrant I, the related angle is the same as the direction of the vector. Therefore, the direction of the vector in quadrant I is always less than 90°. For vectors in quadrants II, III, and IV, see Figs. 24-6, 24-7, and 24-8.

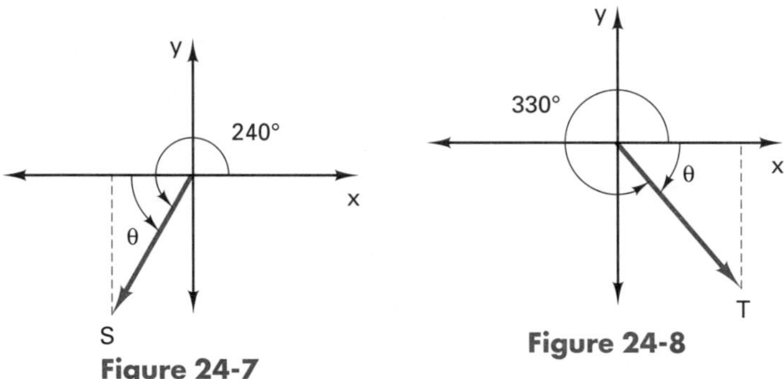

Figure 24-7

Figure 24-8

In Fig. 24-6, $\overline{PQ}$, the x-axis, and vector Q form a right triangle. The direction of the vector is 135°; therefore, the related angle is 45° (180° − 135°). 135° is a second quadrant angle. Second quadrant angles are more than 90° and less than 180°, or more than $\frac{\pi}{2}$ radians (1.57) and less than π radians (3.14). The related angle for any second-quadrant vector can be found by subtracting the direction of the vector from 180° (or π radians).

Third-quadrant angles are more than 180° and less than 270° or more than π radians (3.14) and less than $\frac{3\pi}{2}$ radians (4.71). In Fig. 24-7, vector S is a third-quadrant vector. The related angle θ is 60° (240° − 180°). The related angle for any third-quadrant vector is found by subtracting 180° (or π radians) from the direction of the vector.

Vector T in Fig. 24-8 is a fourth-quadrant vector. Fourth-quadrant angles are more than 270° and less than 360°, or more than $\frac{3\pi}{2}$ radians (4.71) and less than 2π radians (6.28). The related angle θ is 30° (360° − 330°). The related angle for any fourth-quadrant vector is found by subtracting the direction of the vector from 360° (or 2π radians).

Rule 24-2.1

The related angle for quadrant I angles and vectors is equal to the direction (angle) of the vector. The related angle for angles and vectors more than 90° can be found as follows:

Quadrant II angle: $180° - \theta_2$ or $\pi - \theta_2$

Quadrant III angle: $\theta_3 - 180°$ or $\theta_3 - \pi$

Quadrant IV angle: $360° - \theta_4$ or $2\pi - \theta_4$

Related angles will always be angles less than 90° or $\frac{\pi}{2}$ radians.

EXAMPLE 24-2.1 Find the related angle for the following angles.

(a) 210° **(b)** 1.93 rad

Solution

(a) 210° is between 180° and 270° and is a quadrant III angle. Then $210° - 180° = 30°$. The related angle is 30°.

(b) 1.93 rad is between π and $\frac{\pi}{2}$ rad and is a quadrant II angle. Then $\pi - 1.93 = 1.211592654$ rad. The related angle is 1.21 rad to the nearest hundredth. ■

L.O.2. Determine the signs of trigonometric values of angles more than 90°.

Signs of Trigonometric Functions for Angles or Vectors Greater than 90°

To determine the appropriate sign of trigonometric functions for angles more than 90°, we will examine the trigonometric function in each quadrant. The sign of the function indicates the direction or quadrant of the vector.

The vector in Fig. 24-9 has a magnitude of r and direction of θ. To relate our trigonometric functions to vectors, the magnitude of the vector is the hypotenuse, the side opposite θ is y, and the side adjacent to θ is x. The magnitude of a vector (r) is always positive. In quadrant I, the x-value and the y-value are both positive. See Example 23-2.8 for an example of an electrical problem with a vector in quadrant I.

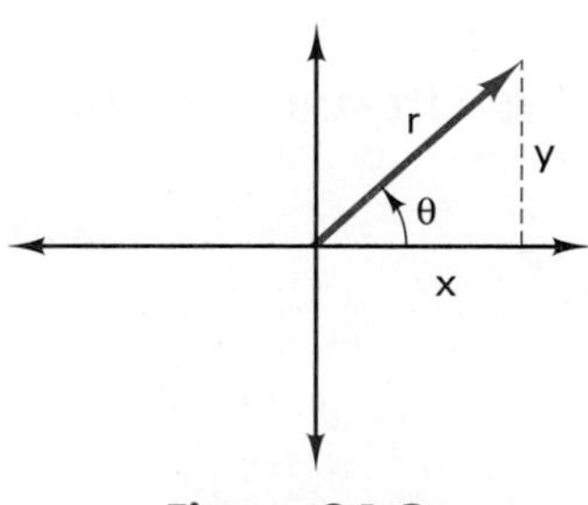

Figure 24-9

■ **DEFINITION 24-2.2 Functions in quadrant I:**

$$\sin\,\theta_1 = \frac{y}{r} \qquad \csc\,\theta_1 = \frac{r}{y}$$

$$\cos\,\theta_1 = \frac{x}{r} \qquad \sec\,\theta_1 = \frac{r}{x}$$

$$\tan\,\theta_1 = \frac{y}{x} \qquad \cot\,\theta_1 = \frac{x}{y}$$

In quadrant I, the sign of all six trigonometric functions is positive.

For quadrant II vectors (Fig. 24-10), again the magnitude (r) is positive and the y value is positive, but the x value is negative.

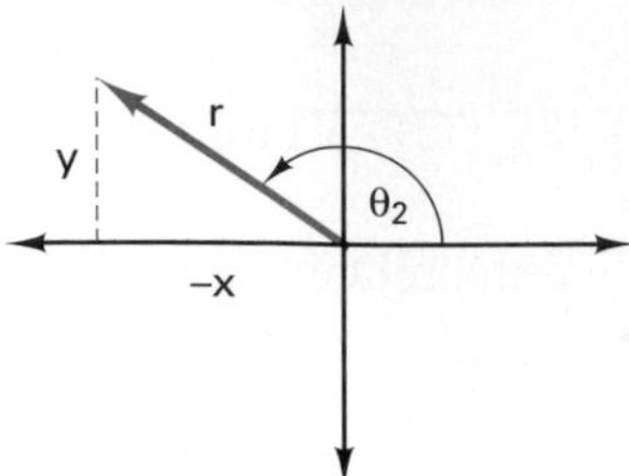

Figure 24-10

■ **DEFINITION 24-2.3 Functions in quadrant II:**

$$\sin\,\theta_1 = \frac{y}{r} \qquad \csc\,\theta_2 = \frac{r}{y}$$

$$\cos\,\theta_2 = \frac{-x}{r} = -\frac{x}{r} \qquad \sec\,\theta_2 = -\frac{r}{x}$$

$$\tan\,\theta_2 = \frac{y}{-x} = -\frac{y}{x} \qquad \cot\,\theta_2 = -\frac{x}{y}$$

Thus, in quadrant II the sine and cosecant functions are positive, and the remaining functions are negative.

Quadrant III vectors (Fig. 24-11) have a positive magnitude (r), negative x value, and negative y value.

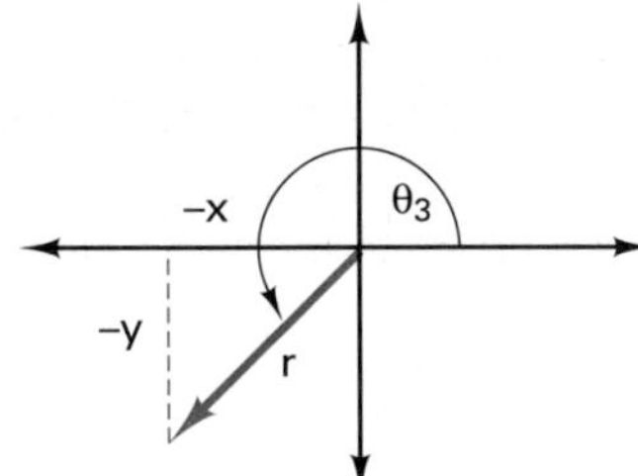

Figure 24-11

■ **DEFINITION 24-2.4 Functions in quadrant III:**

$$\sin\,\theta_3 = \frac{-y}{r} = -\frac{y}{r} \qquad \csc\,\theta_3 = -\frac{r}{y}$$

$$\cos\,\theta_3 = \frac{-x}{r} = -\frac{x}{r} \qquad \sec\,\theta_3 = -\frac{r}{x}$$

$$\tan\,\theta_3 = \frac{-y}{-x} = \frac{y}{x} \qquad \cot\,\theta_3 = \frac{x}{y}$$

Thus, in quadrant III the tangent and cotangent functions are positive and the remaining functions are negative.

Quadrant IV vectors (Fig. 24-12) have a positive magnitude (r), positive x value, and negative y value.

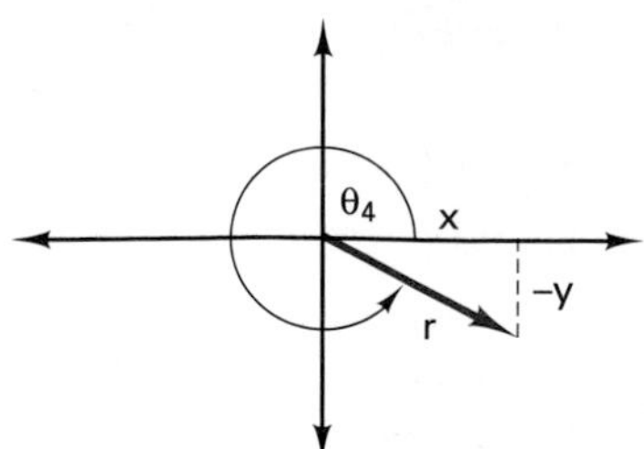

Figure 24-12

■ DEFINITION 24-2.5 **Functions in quadrant IV:**

$$\sin \theta_4 = \frac{-y}{r} = -\frac{y}{r} \qquad \csc \theta_4 = -\frac{r}{y}$$

$$\cos \theta_4 = \frac{x}{r} \qquad \sec \theta_4 = \frac{r}{x}$$

$$\tan \theta_4 = \frac{-y}{x} = -\frac{y}{x} \qquad \cot \theta_4 = -\frac{x}{y}$$

Thus, in quadrant IV the cosine and secant functions are positive and the remaining functions are negative.

Tips and Traps

To help remember the signs of the trigonometric functions in the various quadrants, we may find the following reminder helpful.

ALL-SIN-TAN-COS

This gives the positive functions in the quadrants I to IV, respectively. Also, the functions cosecant, contangent, and secant have the same signs as their respective reciprocal functions.

L.O.3. Find the trigonometric values of angles of more than 90° using a calculator.

Values of Trigonometric Functions of Angles or Vectors Greater Than 90°

General Tips for Using the Calculator

To find the value of the trigonometric function of an angle more than 90° using a calculator:

1. Set the calculator in degree or radian mode as desired.
2. Enter the degrees or radians.
3. Enter the trigonometric function.

EXAMPLE 24-2.2 Using a calculator and the π key, find the values of the following trigonometric functions. Round final answer to four significant digits.

(a) sin 155° **(b)** cos 3 **(c)** tan 208° **(d)** sin 4.2

(e) cos 304.5° **(f)** $\tan \frac{5\pi}{3}$

Solution

(**a**) $\sin 155° = 0.4226$ (**b**) $\cos 3 = -0.9900$
(**c**) $\tan 208° = 0.5317$ (**d**) $\sin 4.2 = -0.8716$
(**e**) $\cos 304.5° = 0.5664$ (**f**) $\tan \frac{5\pi}{3} = -1.732$ ■

EXAMPLE 24-2.3 Find the direction in degrees of a vector in standard position if the coordinates of its end point are (6, −8).

Solution: From the coordinates of the end point, we can determine the quadrant of the vector. The vector is a quadrant IV vector. Also, from the coordinates of the end point we know that the x-coordinate is 6 and the y-coordinate is −8. Then $\tan \theta_4 = \frac{-8}{6} = -1.333333333$. From the calculator, we see that $\theta = -53.1°$ to the nearest 0.1°. The related angle for a quadrant IV angle of −53.1° is 53.1°. If $360° - \theta_4 = 53.1°$, $\theta_4 = 306.9°$. Thus, the direction of the vector is 306.9°. ■

SELF-STUDY EXERCISES 24-2

L.O.1. Find the related angle for the following angles. (Use the calculator value for π and round to hundredths.)

1. 120° **2.** 195° **3.** 290°
4. 345° **5.** 148° **6.** 250°
7. 212° **8.** 118° **9.** 2.18 rad
10. 5.84 rad

L.O.2. Give the signs of all six trigonometric functions of vectors with the following end points.

11. (3, 5) **12.** (−2, 6) **13.** (−4, −2)
14. (5, −3) **15.** (5, 0)

L.O.3. Using a calculator and the π key, find the value of the following functions. Round to four significant digits.

16. sin 210° **17.** tan 140° **18.** cos 2.5
19. cos 4 **20.** sin 300° **21.** tan 6
22. cos 100° **23.** $\sin \frac{5\pi}{6}$

24. Find the direction in degrees of a vector in standard position if the coordinates of its end points are (−3, 2). Round to the nearest 0.1°.

25. Find the direction in radians of a vector in standard position if the coordinates of its end points are (−2, −1). Round to the nearest hundredth.

24-3 LAW OF SINES

Learning Objectives

L.O.1. Find missing parts of an oblique triangle, given two angles and an opposite side.
L.O.2. Find missing parts of an oblique triangle, given two sides and an angle opposite one of them.
L.O.3. Solve applied problems using the law of sines.

In Chapter 21, we defined an oblique triangle as a triangle that does not contain a right angle. Since these triangles do not have right angles, we cannot use the trigonometric functions directly as we did previously to find sides and angles of right triangles. However, two formulas based on the trigonometric functions of right triangles can be used to solve oblique triangles.

In this section we will study one of these formulas, the *law of sines.*

Formula 24-3.1

The **law of sines** states that the ratios of the sides of a triangle to the sines of the angles opposite these respective sides are equal (Fig. 24-13).

Law of Sines

$$\frac{a}{\sin A} = \frac{b}{\sin B} = \frac{c}{\sin C}$$

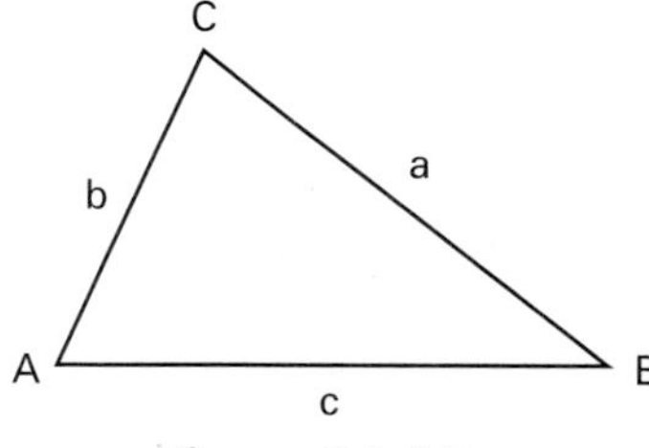

Figure 24-13

Tips and Traps

The law of sines is used to solve triangles when either of the following conditions exists:

1. Two angles and the side opposite one of them are known.
2. Two sides and the angle opposite one of them are known.

If we know two angles of a triangle, we can always find the third angle. Therefore, in 1 above, if we know two angles and any side, we can solve the triangle.

L.O.1. Find missing parts of an oblique triangle, given two angles and an opposite side.

Solving Triangles Using Two Angles and an Opposite Side

Application problems often require solving triangles when we know two angles and a side of a triangle. In the examples that follow, all digits of the calculated value that show in the display will be given. Rounding should be done after the last calculation has been made.

EXAMPLE 24-3.1 Solve triangle ABC if $A = 50°$, $B = 75°$, and $b = 12$ ft.

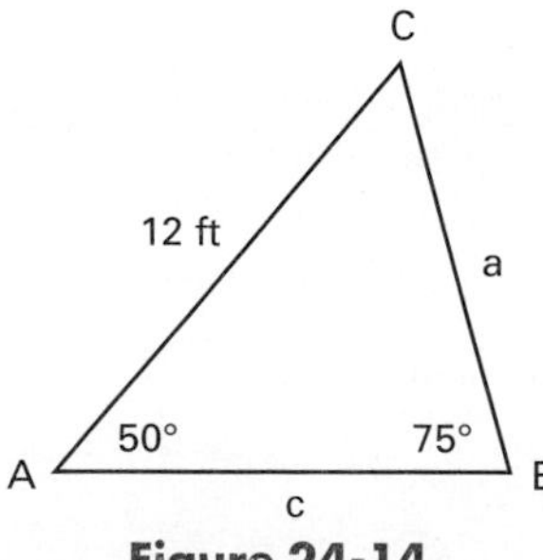

Figure 24-14

Solution: The first step in solving the triangle is to sketch it and label the parts, as shown in Fig. 24-14. To solve the triangle, we need to find angle C, side a, and side c.

$$A + B + C = 180°$$

$$50° + 75° + C = 180° \quad \textit{(Transpose.)}$$

$$C = 180° - (50° + 75°)$$

$$C = 55°$$

To find a and c, we use the law of sines. First, we find a. To find a, we choose the proportion that contains a (the unknown quantity) and three known quantities.

$$\frac{a}{\sin A} = \frac{b}{\sin B} \qquad \textit{(We know b, A, and B.)}$$

$$\frac{a}{\sin 50^\circ} = \frac{12}{\sin 75^\circ} \qquad \textit{(Substitute.)}$$

$$\frac{a}{0.7660444431} = \frac{12}{0.9659258263} \qquad \textit{(Cross multiply.)}$$

$$0.9659258263a = 12(0.7660444431)$$

$$0.9659258263a = 9.192309916$$

Scientific calculator steps:

50 [SIN] [×] 12 [=] [÷] 75 [sin] [=] ⇒ 9.51681078

$$a = 9.5 \qquad \textit{(rounded to nearest tenth)}$$

To find c, we choose a proportion that contains the unknown quantity c and three known quantities.

$$\frac{b}{\sin B} = \frac{c}{\sin C} \qquad \textit{(We found angle C earlier.)}$$

$$\frac{12}{\sin 75^\circ} = \frac{c}{\sin 55^\circ}$$

$$\frac{12}{0.9659258263} = \frac{c}{0.8191520443} \qquad \textit{(Cross multiply.)}$$

$$0.9659258263c = 12(0.8191520443)$$

$$c = 10.17658319$$

Scientific calculator steps:

55 [SIN] [×] 12 [÷] 75 [SIN] [=] ⇒ 10.17658319

$$c = 10.2 \qquad \textit{(rounded to nearest tenth)}$$

As a check for our work, the longest side should be opposite the largest angle, and the shortest side should be opposite the smallest angle (see Fig. 24-15). ■

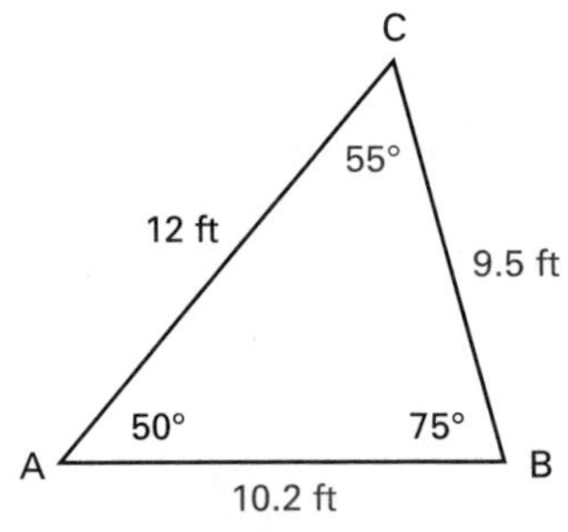

Figure 24-15

EXAMPLE 24-3.2 Solve triangle ABC if $A = 35°$, $a = 7$ cm, and $B = 40°$.

Solution: We first sketch the triangle and label its parts (Fig. 24-16). To solve the triangle, we need to find angle C, side b, and side c.

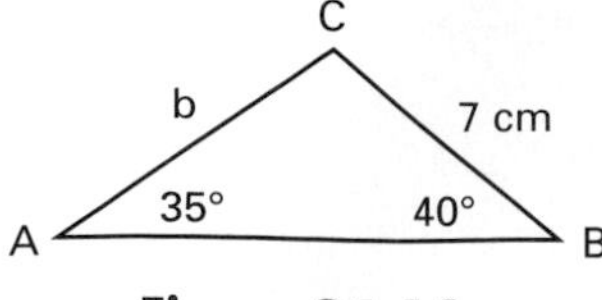

Figure 24-16

$$\begin{aligned} A + B + C &= 180° \\ 35° + 40° + C &= 180° \\ C &= 180° - (35° + 40°) \\ C &= 105° \end{aligned}$$

To find b and c, we use the law of sines. First, we find b.

$$\frac{a}{\sin A} = \frac{b}{\sin B}$$

When choosing the proportion, such as the one above, be sure to choose one in which substitutes can be made for three terms.

$$\begin{aligned} \frac{7}{\sin 35°} &= \frac{b}{\sin 40°} \\ \frac{7}{0.5735764364} &= \frac{b}{0.6427876097} \\ 0.5735764364b &= 7(0.6427876097) \\ 0.5735764364b &= 4.499513268 \\ b &= 7.8 \qquad \textit{(rounded to nearest tenth)} \end{aligned}$$

To find c, we choose the proportion that allows substitution for as many of the *originally* given sides and angles as possible.

$$\begin{aligned} \frac{a}{\sin A} &= \frac{c}{\sin C} \\ \frac{7}{\sin 35°} &= \frac{c}{\sin 105°} \\ \frac{7}{0.5735764364} &= \frac{c}{0.9659258263} \\ 0.5735764364c &= 7(0.9659258263) \\ 0.5735764364c &= 6.761703505 \\ c &= 11.8 \qquad \textit{(rounded to nearest tenth)} \end{aligned}$$

As a check for our work, the longest side should be opposite the largest angle, and the shortest side should be opposite the smallest angle (Fig. 24-17). ■

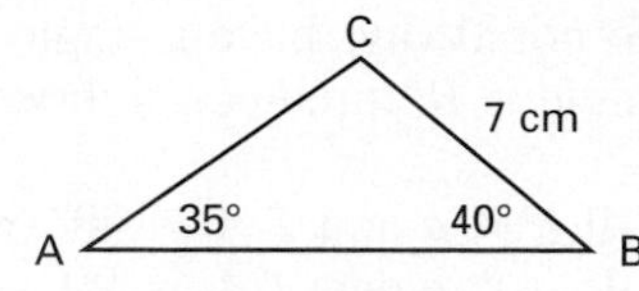

Figure 24-17

EXAMPLE 24-3.3 Determine the unknown angles and sides of a piece of land described by the triangle in Fig. 24-18.

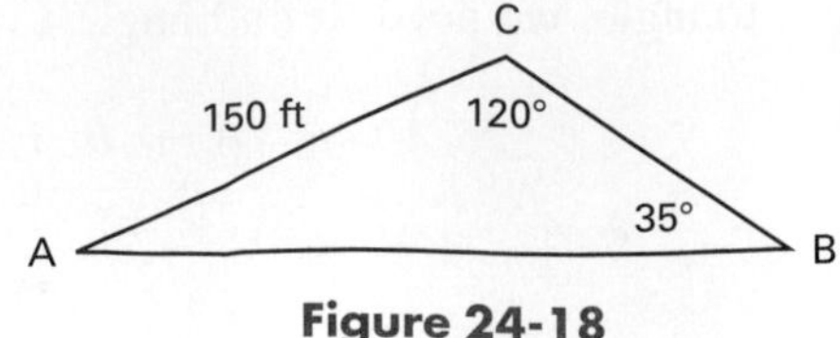

Figure 24-18

Solution

$$A + B + C = 180°$$
$$A + 35° + 120° = 180°$$
$$A = 180 - (35° + 120°)$$
$$A = 25°$$

Next, we find side a.

$$\frac{b}{\sin B} = \frac{a}{\sin A}$$
$$\frac{150}{\sin 35°} = \frac{a}{\sin 25°}$$
$$\frac{150}{0.5735764364} = \frac{a}{0.4226182617}$$
$$0.5735764364a = 150(0.4226182617)$$
$$0.5735764364a = 63.39273926$$
$$a = 110.5 \qquad \textit{(rounded to nearest tenth)}$$

Next, we find c.

$$\frac{150}{\sin 35°} = \frac{c}{\sin 120°}$$
$$\frac{150}{0.5735764364} = \frac{c}{0.8660254038}$$
$$0.5735764364c = 150(0.8660254038)$$
$$0.5735764364 = 129.9038106$$
$$c = 226.5 \qquad \textit{(rounded to nearest tenth)}$$

■

L.O.2. Find missing parts of an oblique triangle, given two sides and an angle opposite one of them.

Solving Triangles Using Two Sides and an Opposite Angle

When two sides of a triangle and an angle opposite one of them are known, we do not always have a single triangle. If the given sides are a and b and the given angle is B, three possibilities may exist (Fig. 24-19).

- If $b < a$ and $\angle A \neq 90°$, we have two possible solutions.
- If $b < a$ and $\angle A = 90°$, we have one solution.
- If $b \geq a$, we have one solution.

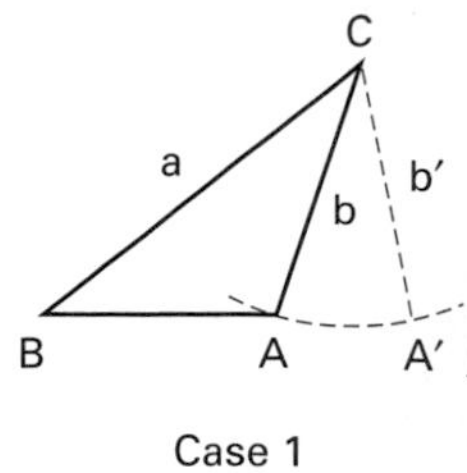

Case 1

Two solutions for A, C, and c since b can meet side c in two points, A and A′.

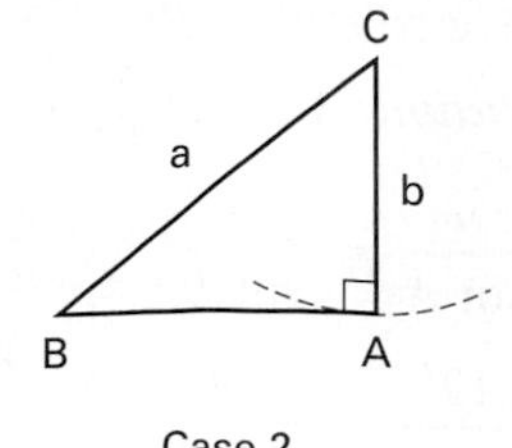

Case 2

One solution since b meets side c in exactly one point.

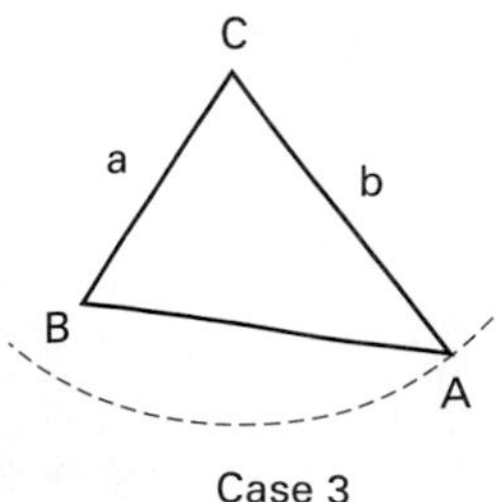

Case 3

One solution since b meets side c in only one point.

Figure 24-19

Note that side b above could be any side of the triangle that is opposite the given angle. Side a is then the other given side. Because of the lack of clarity when two sides and an angle opposite one of them are given, the situation is called the *ambiguous case.* (''Ambiguous'' means that the given information can be interpreted in more than one way.)

EXAMPLE 24-3.4 Solve triangle ABC if $A = 35°$, $a = 8$, and $b = 11.426$ (Fig. 24-20).

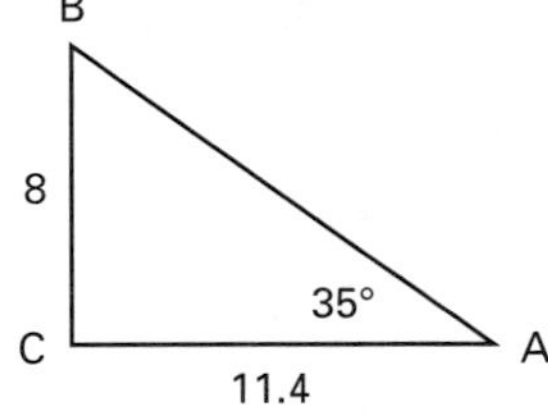

Figure 24-20

Solution

$$\frac{a}{\sin A} = \frac{b}{\sin B}$$

$$\frac{8}{\sin 35°} = \frac{11.426}{\sin B}$$

$$8 \sin B = 11.426 (\sin 35°)$$

$$8 \sin B = 6.553684362$$

$$\sin B = 0.8192105453$$

$$B = 55.0° \qquad \textit{(nearest 0.1°)}$$

$$C = 180° - (55° + 35°)$$

$$C = 90° \qquad \textit{(ABC is a right triangle.)}$$

Since this is a right triangle, we have only one possible solution. To find the third side of this right triangle, we can use the Pythagorean theorem or the law of sines.

Using the law of sines, we have

$$\frac{a}{\sin A} = \frac{c}{\sin C}$$

$$\frac{8}{\sin 35°} = \frac{c}{\sin 90°}$$

$$0.5735764364c = 8(1)$$

$$0.5735764364c = 8$$

$$c = 13.9 \qquad \textit{(rounded to nearest tenth)}$$

Figure 24-21

The solved triangle is shown in Fig. 24-21. ■

Example 24-3.5 Solve triangle ABC if $a = 12$, $b = 8$, and $B = 40°$ (Fig. 24-22).

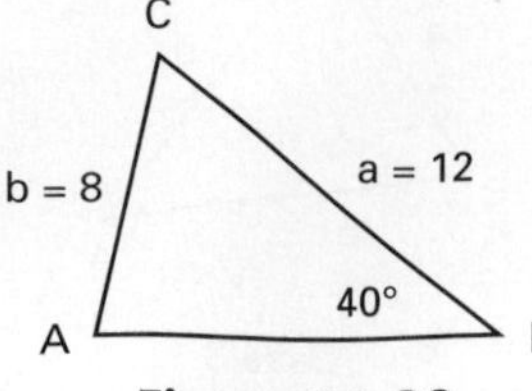

Figure 24-22

Solution: To find A, we have

$$\frac{a}{\sin A} = \frac{b}{\sin B}$$

$$\frac{12}{\sin A} = \frac{8}{\sin 40°}$$

$$8 \sin A = 12(0.6427876097)$$

$$8 \sin A = 7.713451316$$

$$\sin A = 0.9641814145$$

$$A = 74.61856831 \text{ or } 105.3814317°$$

There are two angles less than 180° that have a sine of approximately 0.9641814145. These angles are 74.61856831° and 105.3814317°. The angle 74.61856831° is in Quadrant I. The other angle is in Quadrant II and has 74.61856831° as its reference angle. Therefore, $180° - 74.61856831° = 105.3814317°$ is the measure of the other angle. You can verify this by comparing the sine of the two angles. This situation occurs when the side opposite the given angle is less than the other given side. As a result, we have *two* possible solutions.

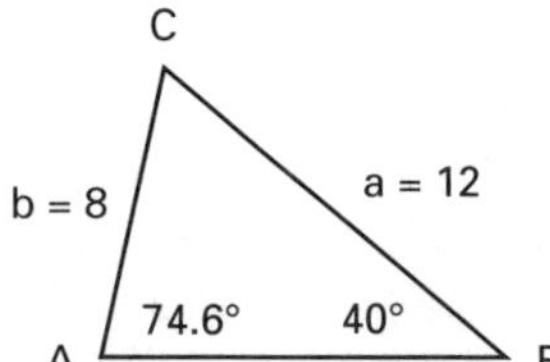

Figure 24-23

Possible solution 1: $A = 74.61856831°$ (Fig. 24-23).

$$C = 180° - (74.61856831° + 40°)$$

$$C = 65.3814317°$$

Now we find side c.

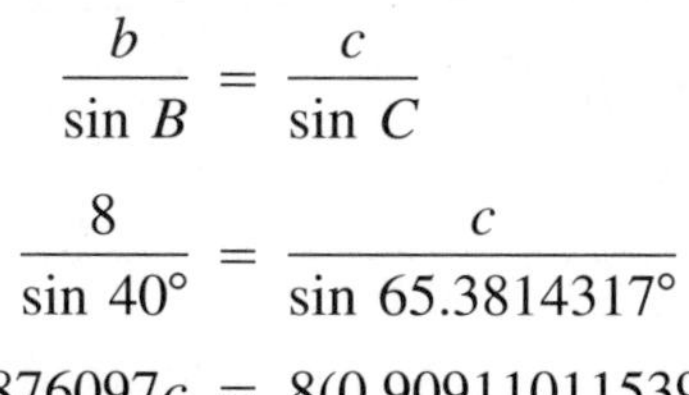

$$\frac{b}{\sin B} = \frac{c}{\sin C}$$

$$\frac{8}{\sin 40°} = \frac{c}{\sin 65.3814317°}$$

$$0.6427876097c = 8(0.90911011539)$$

$$c = 11.3 \qquad \textit{(rounded to tenths)}$$

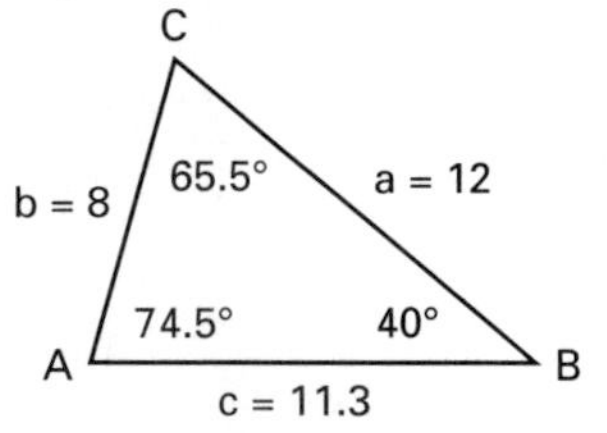

Figure 24-24

The solved triangle is shown in Fig. 24-24.

Possible solution 2: $A = 105.3814317°$ (Fig. 24-25).

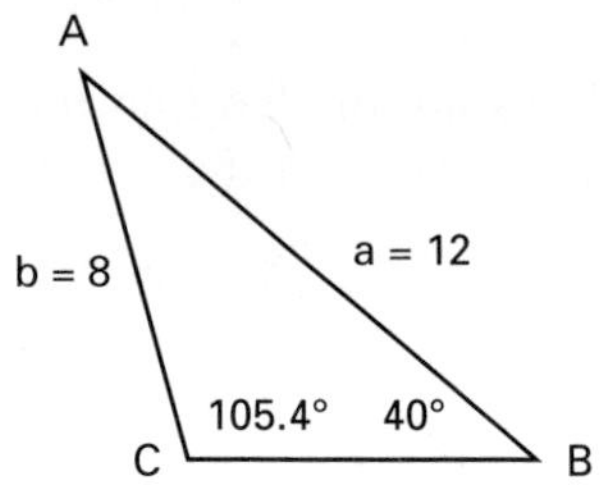

$$C = 180° - (105.3814317° + 40°)$$

$$C = 34.6185683°$$

Now we find side c.

Figure 24-25

$$\frac{b}{\sin B} = \frac{c}{\sin C}$$

$$\frac{8}{\sin 40°} = \frac{c}{\sin 34.6185683°}$$

$$0.6427876097c = 8(0.5681104756)$$

$$0.6427876097c = 4.544883805$$

$$c = 7.1 \qquad \textit{(rounded to tenths)}$$

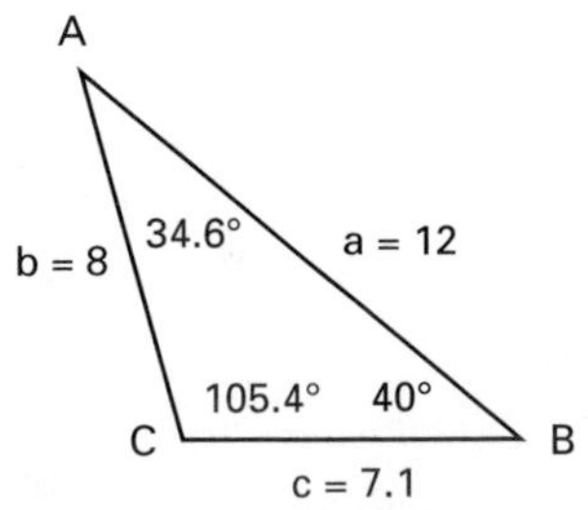

Figure 24-26

The solved triangle is shown in Fig. 24-26. ■

EXAMPLE 24-3.6 Solve triangle ABC if $b = 10$, $c = 12$, and $C = 40°$ (Fig. 24-27).

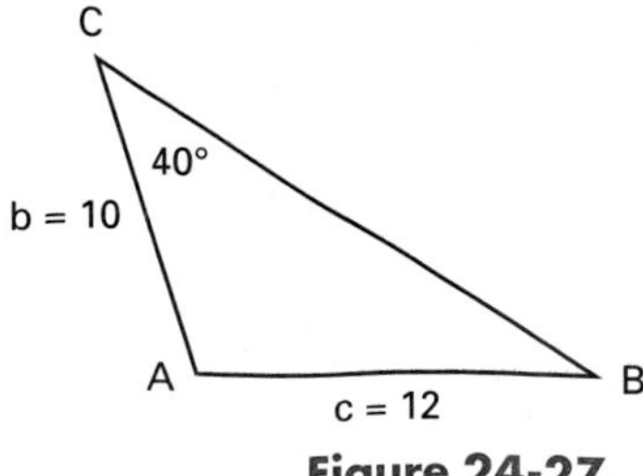

Figure 24-27

Solution: Solve for B first. Since the side opposite the given angle is *longer* than the other side, we expect *one* solution.

$$\frac{c}{\sin C} = \frac{b}{\sin B}$$

$$\frac{12}{\sin 40°} = \frac{10}{\sin B}$$

$$12 \sin B = 10 \sin 40°$$

$$12 \sin B = 10(0.6427876097)$$

$$\sin B = 0.5356563414$$

$$B = 32.38843382°$$

Both 32.38843382° and 147.6115662° have a sine of approximately 0.5356563414; however, we cannot have an angle of 147.115662° in this triangle because the triangle already has a 40° angle. 147.6115662° + 40° = 187.6115662°. The *three* angles of a triangle total only 180°. Therefore, we have only *one* solution. Finding A, we have

$$A = 180° - (40° + 32.38843382°)$$

$$A = 107.6115662°$$

Solving for a, we have

$$\frac{a}{\sin A} = \frac{c}{\sin C}$$

$$\frac{a}{\sin 107.6115662°} = \frac{12}{\sin 40°}$$

$$a \sin 40° = 12 \sin 107.6115662°$$

$$0.6427876097a = 12(0.9531296095)$$

$$0.60427876097a = 11.43755531$$

$$a = 17.8 \qquad \textit{(rounded to nearest tenth)}$$

The solved triangle is shown in Fig. 24-28. ■

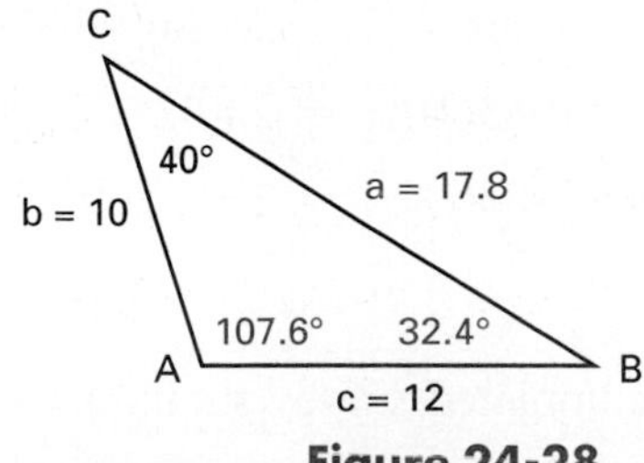

Figure 24-28

L.O.3. Solve applied problems using the law of sines.

Applied Problems

Since real-world applications many times do not involve right triangles, the law of sines is very useful in solving applied problems.

EXAMPLE 24-3.7 A technician checking a surveyor's report is given the information shown in Fig. 24-29. Calculate the missing information.

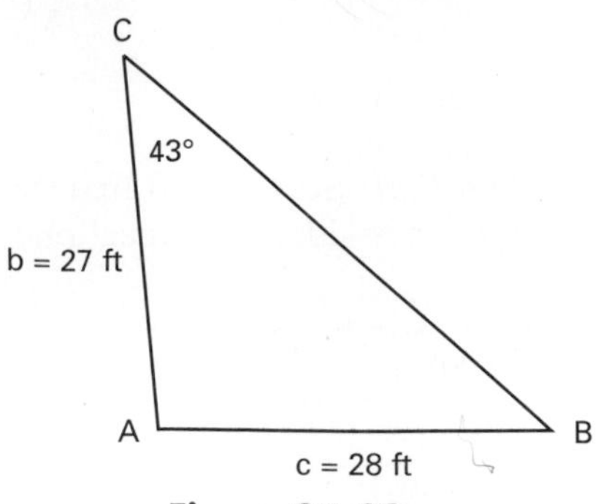

Figure 24-29

Solution: Find B.

$$\frac{c}{\sin C} = \frac{b}{\sin B}$$

$$\frac{28}{\sin 43°} = \frac{27}{\sin B}$$

$$28 \sin B = 27 \sin 43°$$

$$28 \sin B = 27(0.6819983601)$$

$$\sin B = 0.6576412757$$

$$B = 41.12023021°$$

There is only one case here because we are given the triangle. Also, the other angle whose sine is 0.6576412757 is 138.8797696°, which we exclude because 138.8797696° + 43° = 181.8797696°.

To find angle A, we have

$$A = 180° - (43 + 41.12023021)$$

$$A = 95.87976979°$$

$$\frac{a}{\sin A} = \frac{c}{\sin C}$$

$$\frac{a}{\sin 95.87976979°} = \frac{28}{\sin 43°}$$

$$a \sin 43° = 28 \sin 95.87976979°$$

$$0.6819983601a = 28(0.9947390495)$$

$$0.6819983601a = 27.85269339$$

$$a = 40.8 \text{ ft} \qquad \textit{(rounded to nearest tenth)}$$

The completed survey should have the measures shown in Fig. 24-30. ■

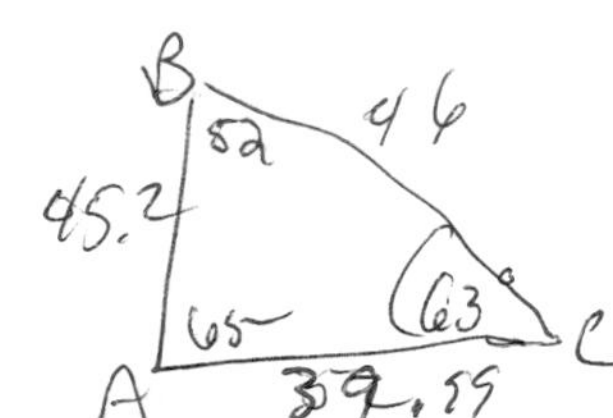

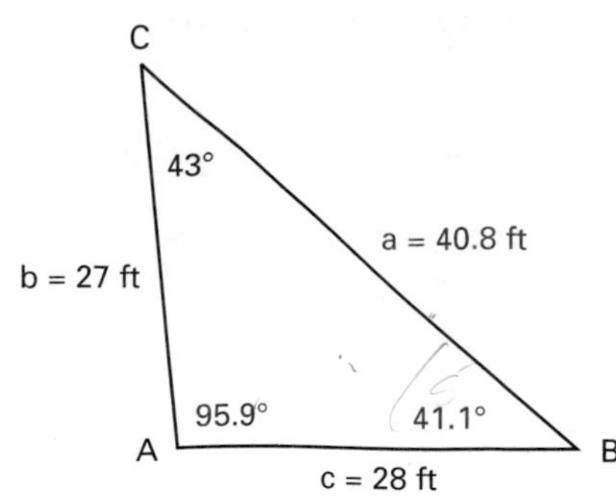

Figure 24-30

SELF-STUDY EXERCISES 24-3

L.O.1. Solve each of the following oblique triangles using the law of sines. Round the final answer for sides to tenths and angles to the nearest 0.1°.

1. $a = 46, A = 65°, B = 52°$

2. $b = 7.2, B = 58°, C = 72°$

3. $a = 65, B = 60°, A = 87°$

4. $c = 3.2, A = 120°, C = 30°$

5. $b = 12, A = 95°, B = 35°$

L.O.2. Use the law of sines to solve the following triangles. If a triangle has two possibilities, find both solutions. Round sides to tenths and angles to the nearest 0.1°.

6. $a = 42, b = 24, A = 40°$

7. $b = 15, c = 3, B = 70°$

8. $a = 18, c = 9, C = 20°$

9. $a = 8, b = 4, A = 30°$

L.O.3.

10. Find the missing angle and sides of the plot of land described by Fig. 24-31.

11. Find the distance from A to B on the surveyed plot shown in Fig. 24-32.

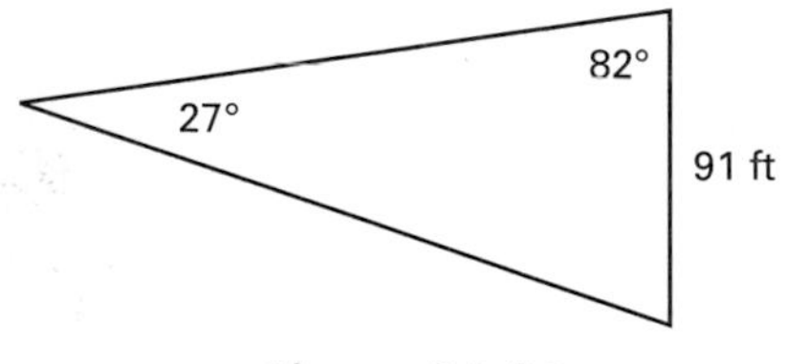

Figure 24-31

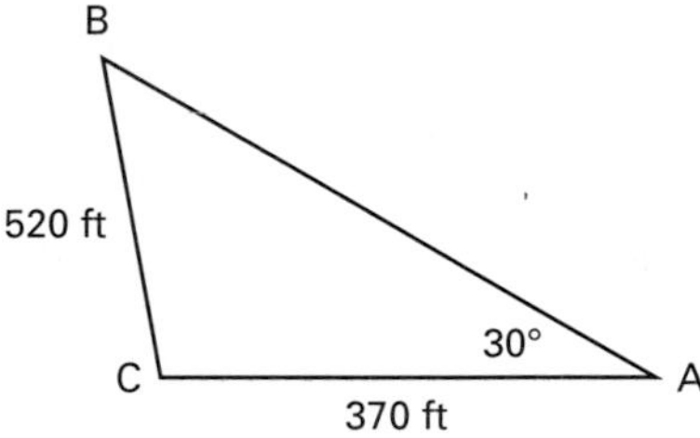

Figure 24-32

24-4 LAW OF COSINES

Learning Objectives

L.O.1. Find the missing parts of an oblique triangle, given three sides of the triangle.

L.O.2. Find the missing parts of an oblique triangle, given two sides and the included angle of the triangle.

L.O.3. Solve applied problems using the law of cosines and the law of sines.

In some cases, our given information will not allow us to use the law of sines. For example, if we know all three sides of a triangle, we cannot use the law of sines to find the angles. However, in a case such as this, we can use the *law of cosines*. This law is based on the trigonometric functions just as the law of sines is.

Formula 24-4.1

The **law of cosines** states that the square of any side of a triangle equals the sum of the squares of the other sides minus twice the product of the other two sides and the cosine of the angle opposite the first side. For triangle ABC:

Law of Cosines

$$a^2 = b^2 + c^2 - 2bc \cos A$$

$$b^2 = a^2 + c^2 - 2ac \cos B$$

$$c^2 = a^2 + b^2 - 2ab \cos C$$

Tips and Traps

To use the law of cosines efficiently, we are given

1. Three sides of a triangle
   or
2. Two sides and the included angle of a triangle.

We can sometimes use *either* the law of sines or the law of cosines to solve certain triangles. However, whenever possible, the law of sines should be used because it is easier to evaluate and is therefore preferred.

L.O.1. Find the missing parts of an oblique triangle, given three sides of the triangle.

Solving Triangles Given Three Sides

EXAMPLE 24-4.1 Find the angles in triangle ABC (Fig. 24-33).

Solution: We may find any one of the angles first. If we choose to find angle A first, we must use the formula that contains $\cos A$: $a^2 = b^2 + c^2 - 2bc \cos A$. We rearrange the formula to solve for $\cos A$.

$$a^2 - b^2 - c^2 = -2bc \cos A$$

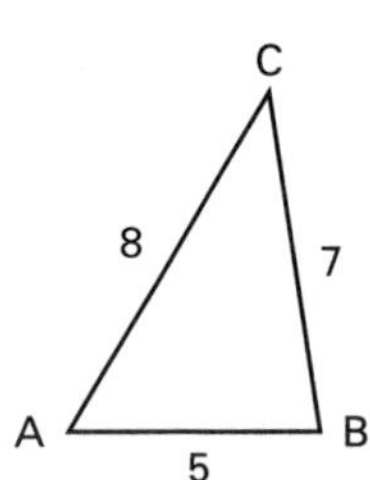

Figure 24-33

Now we multiply each term on both sides by -1 to reduce the number of negative signs.

$$-a^2 + b^2 + c^2 = 2bc \cos A$$

$$\frac{-a^2 + b^2 + c^2}{2bc} = \cos A$$

In our problem, $a = 7$, $b = 8$, and $c = 5$. Substituting, we have

$$\frac{-(7)^2 + 8^2 + 5^2}{2(8)(5)} = \cos A$$

$$\frac{-49 + 64 + 25}{80} = \cos A$$

$$\frac{40}{80} = \cos A$$

$$0.5 = \cos A$$

$$60° = A$$

Scientific calculator steps:

7 [x^2] [+/−] [+] 8 [x^2] [+] 5 [x^2] [=] [÷] 2 [÷] 8 [÷] 5 [=] [INV] [COS] ⇒ 60

To find angle B, we use the law of cosines so that we can use only given values. The law of sines could be used but would involve using a calculated value. (See Example 24-4.2 for use of calculated value and law of sines.)

$$b^2 = a^2 + c^2 - 2\,ac \cos B$$

Rearranging the formula as above, we have

$$\frac{-b^2 + a^2 + c^2}{2ac} = \cos B$$

Now we substitute $a = 7$, $b = 8$, and $c = 5$.

$$\frac{-(8)^2 + 7^2 + 5^2}{2(7)(5)} = \cos B$$

$$\frac{-64 + 49 + 25}{70} = \cos B$$

$$\frac{10}{70} = \cos B$$

$$0.142857142 = \cos B$$

$$81.7867893° = B$$

The law of cosines or the law of sines can be used to find the third angle. However, the quickest way to find this angle is to subtract the sum of A and B from 180°.

$$C = 180° - (A + B)$$

$$C = 180° - (60° + 81.7867893°)$$

$$C = 180° - (141.7867893°)$$

$$C = 38.2132107°$$

■

EXAMPLE 24-4.2 Find the three angles of triangle JKM (Fig. 24-34).

Figure 24-34

Solution: To find angle J, we write the law of cosines for $\cos J$.

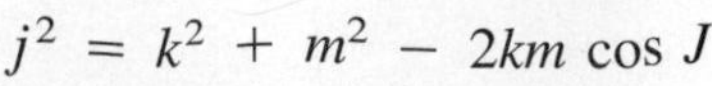

$$j^2 = k^2 + m^2 - 2km \cos J$$

Rearrange for $\cos J$.

$$\cos J = \frac{k^2 + m^2 - j^2}{2km}$$

$$j = 6.8,\ k = 8.2,\ m = 7.1$$

$$\cos J = \frac{(8.2)^2 + (7.1)^2 - (6.8)^2}{2(8.2)(7.1)}$$

$$\cos J = \frac{67.24 + 50.41 - 46.24}{116.44}$$

$$\cos J = \frac{71.41}{116.44}$$

$$\cos J = 0.613277224$$

$$J = 52.2° \qquad \textit{(rounded to nearest 0.1°)}$$

We may now use the law of cosines or the law of sines to find the second angle. Using the law of sines requires us to use the angle that we just calculated, so we should use full calculator values if we use the law of sines. In this Example we will use the law of sines instead of the law of cosines.

$$\frac{j}{\sin J} = \frac{k}{\sin K}$$

$$\frac{6.8}{\sin 52.17315328°} = \frac{8.2}{\sin K}$$

$$6.8 \sin K = 8.2(0.789867739)$$

$$6.8 \sin K = 6.476915465$$

$$\sin K = 0.952487568$$

$$K = 72.3° \qquad \textit{(rounded to nearest 0.1°)}$$

$$M = 180° - (52.2° + 72.3°)$$

$$M = 180° - 124.5°$$

$$M = 55.5°$$

■

EXAMPLE 24-4.3 Find the angles in triangle ABC (Fig. 24-35).

Solution: We will find A first.

$$a^2 = b^2 + c^2 - 2bc \cos A$$

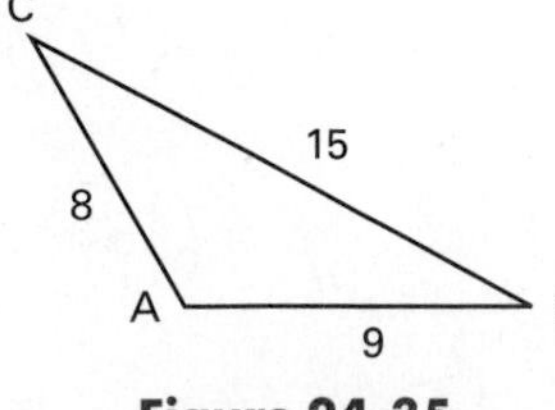

Figure 24-35

Solve for $\cos A$.

$$\cos A = \frac{b^2 + c^2 - a^2}{2bc}$$

$$\cos A = \frac{8^2 + 9^2 - 15^2}{2(8)(9)}$$

$$\cos A = \frac{64 + 81 - 225}{144}$$

$$\cos A = \frac{-80}{144}$$

$$\cos A = -0.555555955$$

$$A = 123.7 \qquad \textit{(rounded to nearest 0.1°)}$$

Recall from Section 24-2 that the cosine is *negative* in the second and third quadrants. Since A is either acute ($<90°$) or obtuse ($>90°$ but $<180°$), it must be in the second quadrant since its cosine is negative (Fig. 24-36). If a calculator is used and -0.555555555 is entered, 123.7489886° will appear in the display. As in the preceding example, we can now use the law of sines to find the second angle.

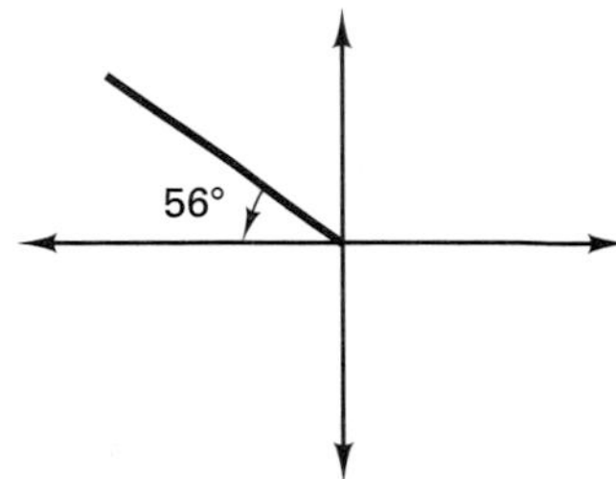

Figure 24-36

$$\frac{a}{\sin A} = \frac{b}{\sin B}$$

$$\frac{15}{\sin 123.7489886°} = \frac{8}{\sin B}$$

$$15 \sin B = 8 \sin 123.7°$$

$$15 \sin B = 8(0.831479419)$$

$$15 \sin B = 6.651835354$$

$$\sin B = 0.44345569$$

$$B = 26.3° \qquad \textit{(rounded to 0.1°)}$$

$$C = 180° - (A + B)$$

$$C = 180° - (123.7° + 26.3°)$$

$$C = 30.0°$$

■

L.O.2. Find the missing parts of an oblique triangle, given two sides and the included angle of the triangle.

Solving Triangles Given Two Sides and the Included Angle

The law of cosines is needed to solve oblique triangles if two sides and the included angle are given.

EXAMPLE 24-4.4 Solve triangle ABC (Fig. 24-37).

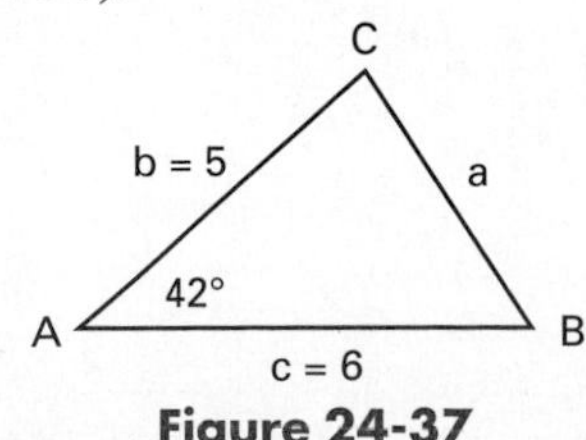

Figure 24-37

Solution: We can find a by using the law of cosines.

$$a^2 = b^2 + c^2 - 2bc \cos A$$
$$a^2 = 5^2 + 6^2 - 2(5)(6) \cos 42°$$
$$a^2 = 25 + 36 - 60 \cos 42° \qquad \textit{(cos 42° = 0.743144825)}$$
$$a^2 = 25 + 36 - 44.58868953$$
$$a^2 = 16.41131047$$
$$a = 4.051 \qquad \textit{(rounded to four significant digits from 4.051087566)}$$

Use the law of sines to find the two angles.

$$\frac{a}{\sin A} = \frac{b}{\sin B}$$
$$\frac{4.051087566}{\sin 42°} = \frac{5}{\sin B}$$
$$4.051087566 \sin B = 5 \sin 42° \qquad \textit{(sin 42° = 0.669130606)}$$
$$\sin B = 0.825865394$$
$$B = 55.7° \qquad \textit{(rounded to nearest 0.1°)}$$
$$\frac{a}{\sin A} = \frac{c}{\sin C}$$
$$\frac{4.051087566}{\sin 42°} = \frac{6}{\sin C}$$
$$4.051087566 \sin C = 6 \sin 42°$$
$$\sin C = 0.9910384739$$
$$C = 82.3° \qquad \textit{(rounded to nearest 0.1°)}$$
$$42° + 55.7° + 82.3° = 180°$$

L.O.3. Solve applied problems using the law of cosines and the law of sines.

The sum of the angles should add to 180°, the longest side should be opposite the largest angle, and the shortest side should be opposite the smallest angle. ■

Solving Applied Problems

Example 24-4.5 A vertical 45-ft pole is placed on a hill that is inclined 17° to the horizontal (Fig. 24-38). How long a guy wire is needed if the guy wire is placed 5 ft from the top of the pole and attached to the ground at a point 32 ft uphill from the base of the pole?

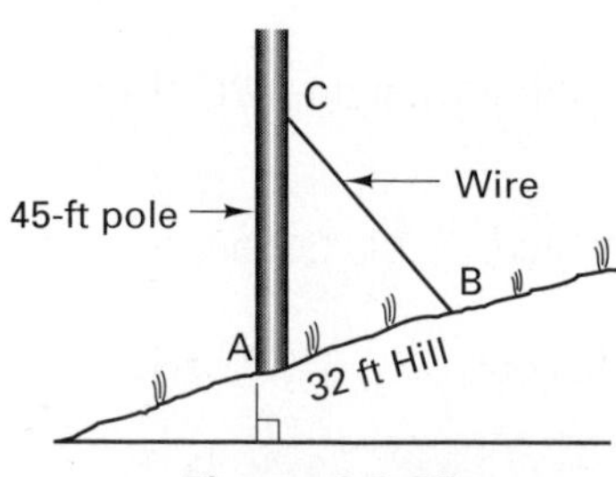

Figure 24-38

Solution: Point A is where the pole enters the ground. Point B is where the wire is attached to the ground. Point C is where the wire is attached to the pole (5 ft from the top of the pole). The length of the wire is side a in the triangle. Since AC makes a 90° angle with the horizontal, we know that angle A is $90 - 17$ or 73°. Using the law of cosines, we have

$$a^2 = b^2 + c^2 - 2bc \cos A$$

$$a^2 = 40^2 + 32^2 - 2(40)(32) \cos 73°$$

$$a^2 = 1600 + 1024 - 2560(0.2923717047)$$

$$a^2 = 1600 + 1024 - 748.471564$$

$$a^2 = 1875.528436$$

$$a = 43.31 \text{ ft} \qquad \textit{(rounded to four significant digits)}$$

■

SELF STUDY EXERCISES 24-4

L.O.1. Solve the triangles in the following figures. Round side lengths to four significant digits and angles to the nearest 0.1°.

1.

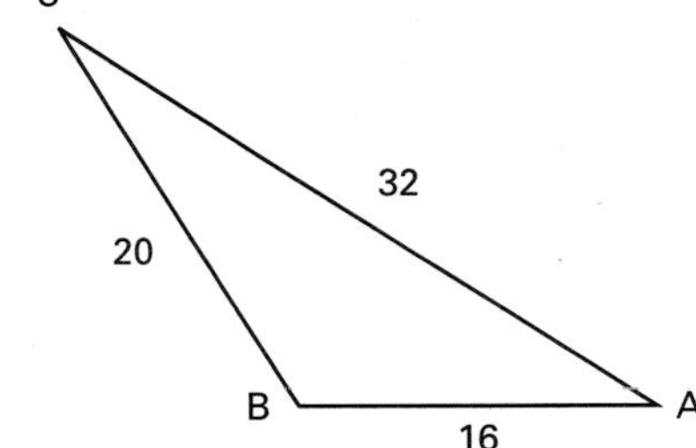

2.

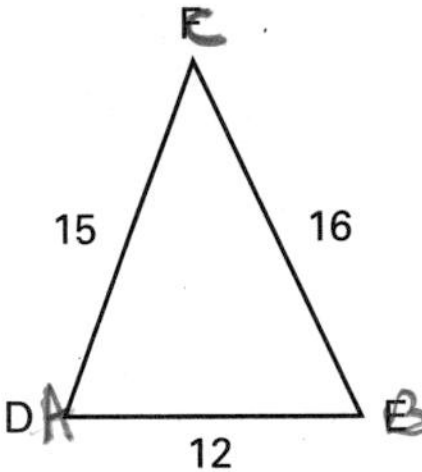

L.O.2.

3.

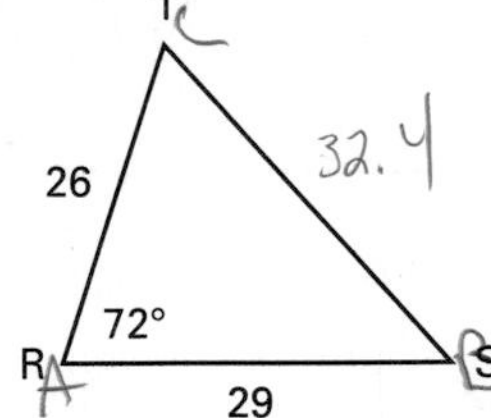

4.

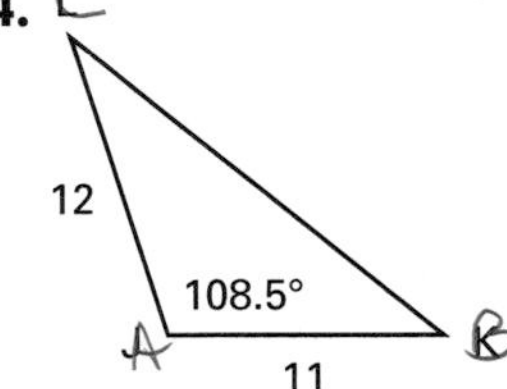

L.O.3. Solve the following problems. Round as above.

5. A hill is inclined 20° to the horizontal. A pole stands vertically on the side of the hill with 35 ft above the ground. How much wire will it take to reach from a point 2 ft from the top of the pole to a point on the ground 27 ft downhill from the base of the pole?

6. A triangular tabletop is to be 8.4 ft by 6.7 ft by 9.3 ft. What angles must be cut?

24-5 AREA OF TRIANGLES

Learning Objectives

L.O.1. Find the area of a triangle when the height is unknown and at least three parts of the triangle are known.

L.O.2. Find the area of a triangle using Heron's formula when the three sides of the triangle are known.
L.O.3. Solve applied problems involving the area of a triangle.

In Chapter 13, we learned to find the area of right triangles and other triangles when the height was given. However, we can now use trigonometry to find the area of triangles even if we do not know the height. We can also find the area of a triangle if we know only the lengths of each of the three sides.

L.O.1. Find the area of a triangle when the height is unknown and at least three parts of the triangle are known.

Finding the Area of a Triangle When the Height Is Unknown

We can use trigonometry to find the area of *any* triangle if we know the measure of any two sides and the included angle.

In Fig. 24-39, the area of triangle ABC can be found by using one of three formulas.

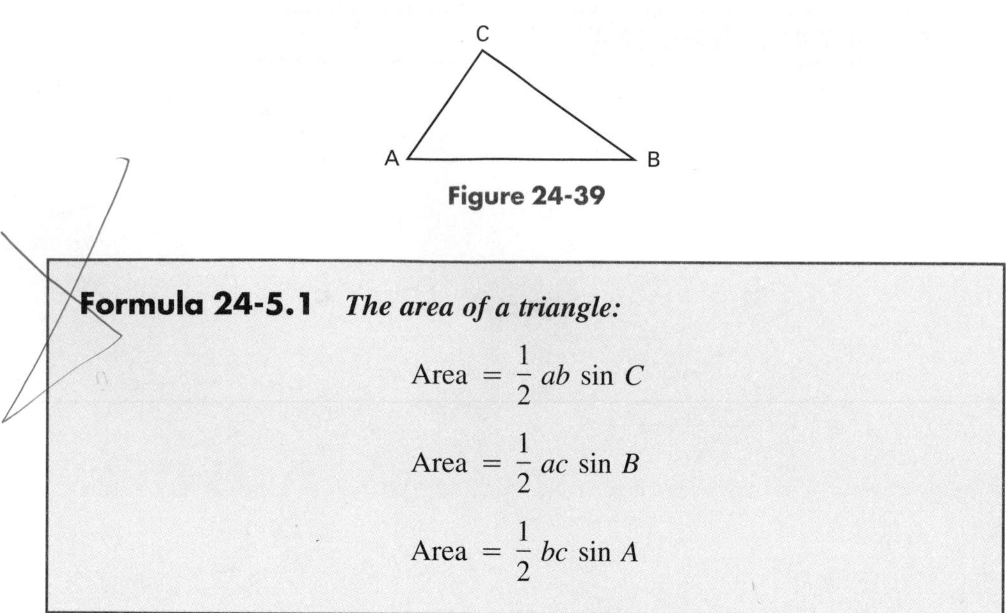

Figure 24-39

Formula 24-5.1 *The area of a triangle:*

$$\text{Area} = \frac{1}{2}ab \sin C$$

$$\text{Area} = \frac{1}{2}ac \sin B$$

$$\text{Area} = \frac{1}{2}bc \sin A$$

These three formulas may be stated as a rule.

Rule 24-5.1 *To find the area of a triangle without the height:*

The area of a triangle equals one-half the product of two sides times the sine of the included angle.

EXAMPLE 24-5.1 Find the area of triangle ABC (Fig 24-40).

Solution

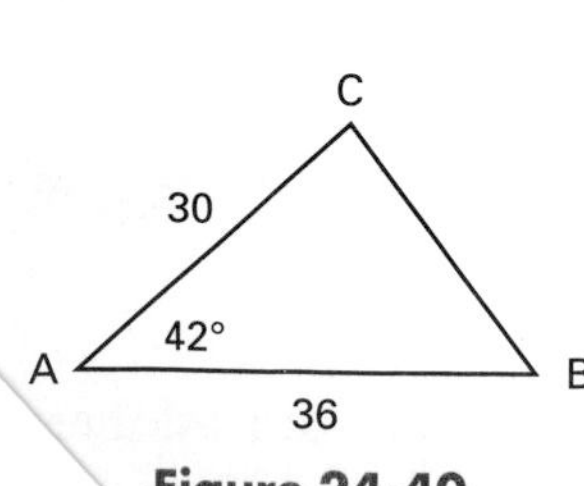

Figure 24-40

$$\text{Area} = \frac{1}{2}bc \sin A$$

$$\text{Area} = \frac{1}{2}(30)(36)(\sin 42°)$$

$$\text{Area} = 540(0.6691306064)$$

$$\text{Area} = 361.3 \text{ square units}$$ *(rounded to four significant digits)* ■

EXAMPLE 24-5.2 Find the area of triangle DEF (Fig. 24-41).

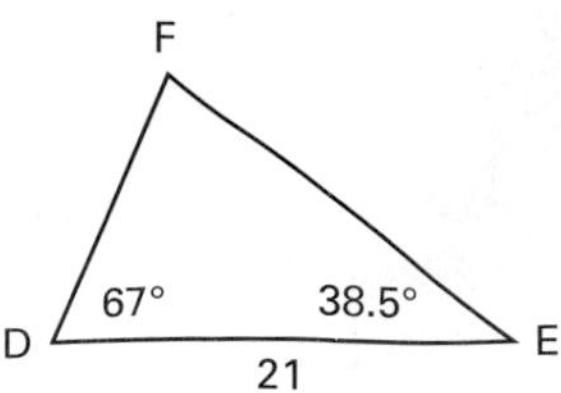

Figure 24-41

Solution: In this example we are not given two sides and the included angle. However, we can use the law of sines to find side d or side e. This would give us two sides and the included angle. To use the law of sines, we first need to find $\angle F$.

$$F = 180° - (67° + 38.5°)$$

$$F = 74.5°$$

We then find side d.

$$\frac{f}{\sin F} = \frac{d}{\sin D}$$

$$\frac{21}{\sin 74.5°} = \frac{d}{\sin 67°}$$

$$0.9636304532d = 21(0.9205048535)$$

$$d = 20.06 \qquad \textit{(from 20.06018164)}$$

Using the formula for area, we have

$$\text{Area} = \frac{1}{2}(21)(20.06018164)\sin 38.5° \qquad \textit{(Angle E is included between d and f.)}$$

$$\text{Area} = (210.6319072)(0.6225146366)$$

$$\text{Area} = 131.1 \text{ square units} \qquad \textit{(rounded to four significant digits)}$$

■

EXAMPLE 24-5.3 Find the area of triangle ABC (Fig. 24-42).

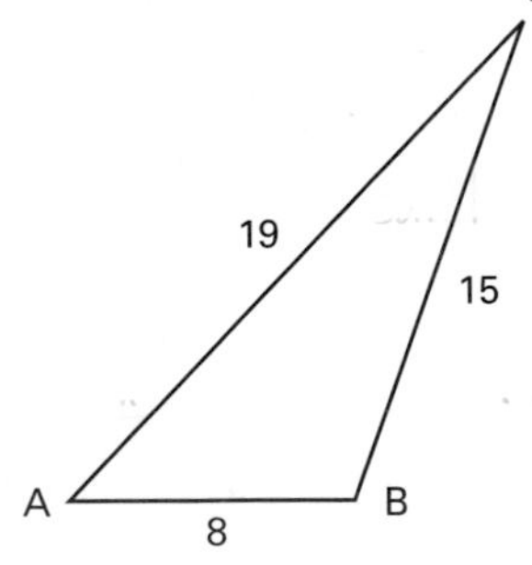

Figure 24-42

Solution: To find the area of $\triangle ABC$, we must find the measure of an angle. The law of cosines must be used here.

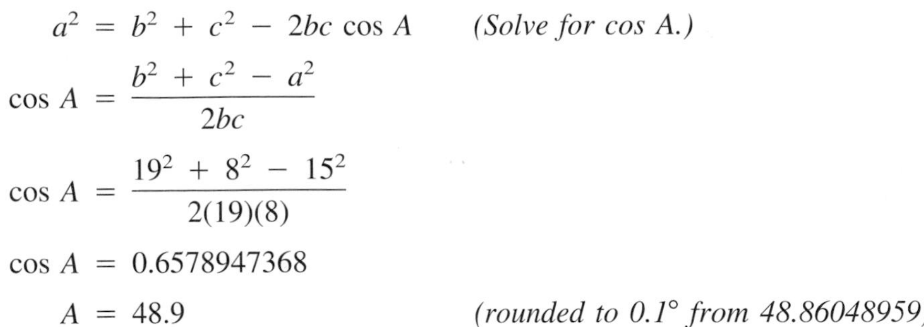

$$a^2 = b^2 + c^2 - 2bc\cos A \qquad \textit{(Solve for cos A.)}$$

$$\cos A = \frac{b^2 + c^2 - a^2}{2bc}$$

$$\cos A = \frac{19^2 + 8^2 - 15^2}{2(19)(8)}$$

$$\cos A = 0.6578947368$$

$$A = 48.9 \qquad \textit{(rounded to 0.1° from 48.86048959)}$$

Using the area formula, we have

$$\text{Area} = \frac{1}{2}(19)(8)\sin 48.86048958°$$

$$\text{Area} = 76(0.7531098958)$$

$$\text{Area} = 57.24 \text{ square units} \qquad \textit{(rounded to four significant digits)}$$

■

L.O.2. Find the area of a triangle using Heron's formula when the three sides of the triangle are known.

Heron's Formula for Area

There is another way to calculate the area of a triangle when all three sides are known. The formula is known as *Heron's formula.*

Formula 24-5.2 **Heron's formula** ***for the area of a triangle:***

$$\text{Area} = \sqrt{s(s - a)(s - b)(s - c)}$$

where $s = \frac{1}{2}(a + b + c)$ and a, b, and c are the lengths of the three sides.

Using this formula to find the area of the triangle in Example 24-5.3, we first find s; then we substitute in the formula.

$$s = \frac{1}{2}(a + b + c)$$

$$s = \frac{1}{2}(19 + 15 + 8)$$

$$s = 21$$

$$\text{Area} = \sqrt{s(s - a)(s - b)(s - c)}$$

$$\text{Area} = \sqrt{21(21 - 19)(21 - 15)(21 - 8)}$$

$$\text{Area} = \sqrt{21(2)(6)(13)}$$

$$\text{Area} = \sqrt{3276}$$

57.24 square units *(rounded to four significant digits)*

L.O.3. Solve applied problems involving the area of a triangle.

The two calculated areas are the same. ■

Solving Applied Problems

Example 24-5.4 A triangular piece of property has sides that measure 120 ft long, 150 ft long, and 100 ft long. Find the area of the lot.

Solution: Using the formula

$$\text{Area} = \sqrt{s(s - a)(s - b)(s - c)}$$

we have

$$s = \frac{370}{2} = 185$$

$$\text{Area} = \sqrt{185(185 - 120)(185 - 150)(185 - 100)}$$

$$\text{Area} = \sqrt{185(65)(35)(85)}$$

$$\text{Area} = \sqrt{35{,}774{,}375}$$

$$\text{Area} = 5981 \text{ ft}^2$$

(to nearest square foot)

SELF-STUDY EXERCISES 24-5

L.O.1. Find the area of the triangles shown in the following figures. Use the law of sines or the law of cosines when necessary. Round answers to three significant digits.

1.

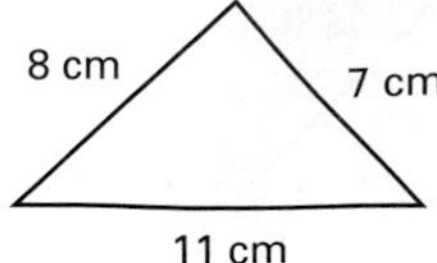

2.

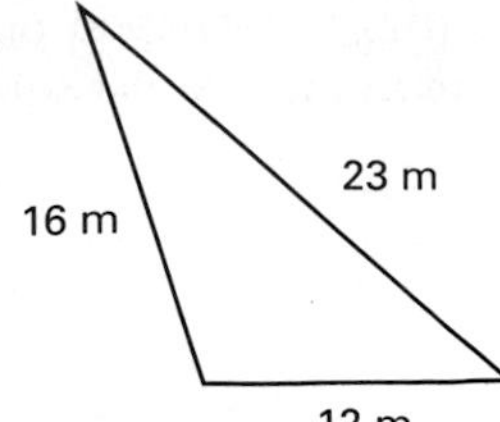

3.

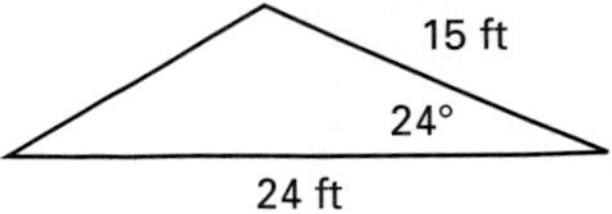

4.

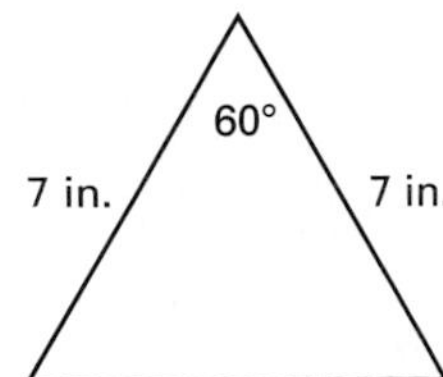

5.

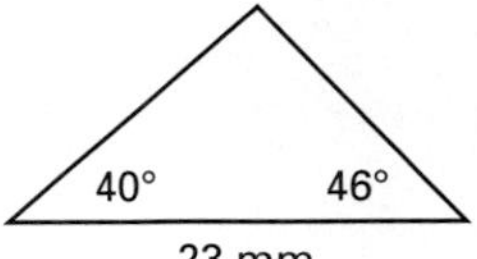

6.

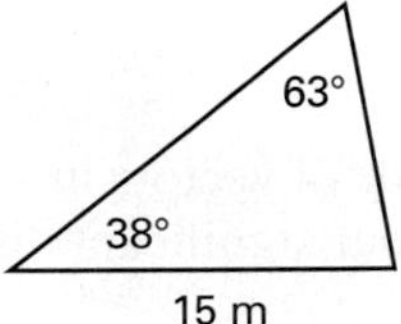

7.

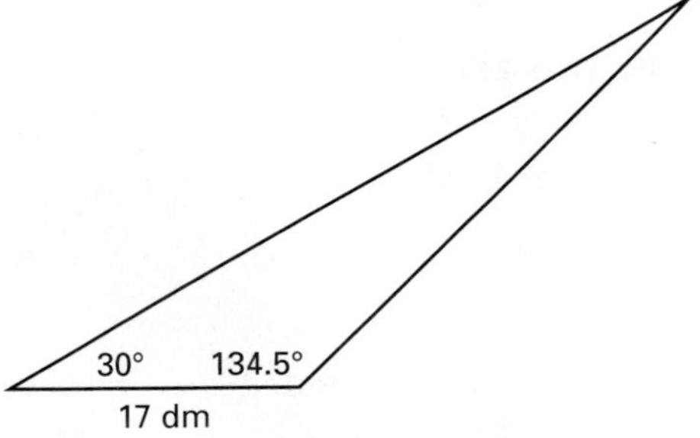

8.

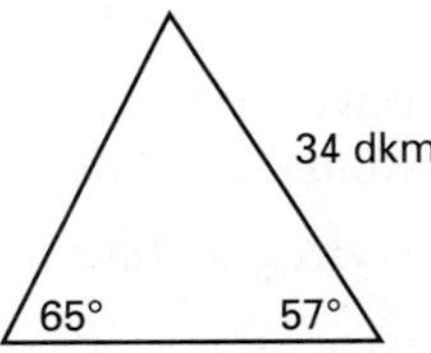

L.O.2.

9.

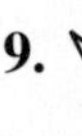

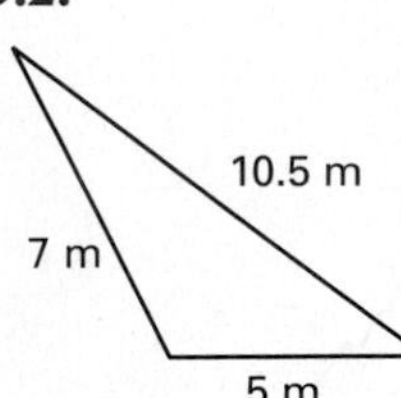

10.

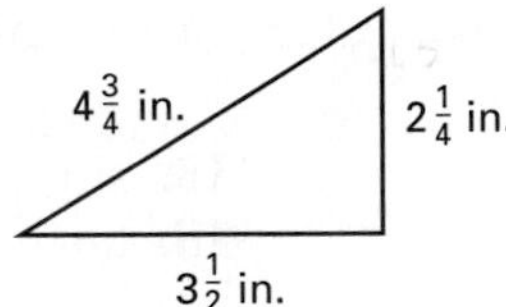

11.

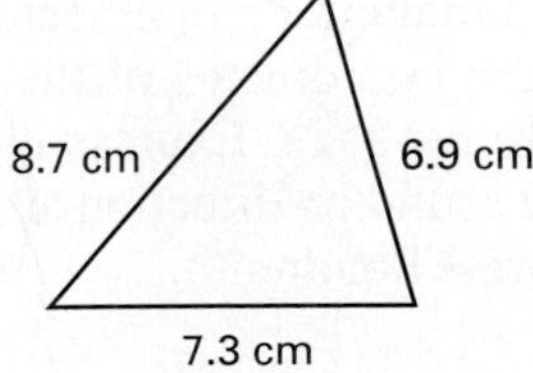

12.

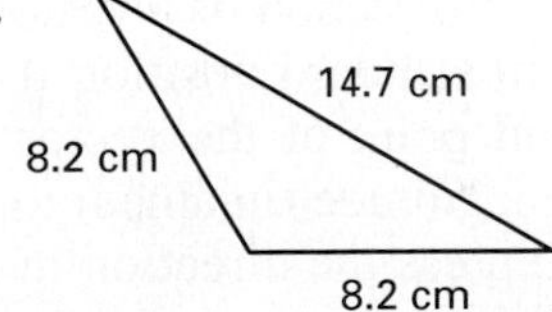

L.O.3.

13. A tool and die maker for a medical implant manufacturer must calculate the area of a triangle whose sides are 2.06 cm, 3.27 cm, and 4.16 cm in order to make a die. What is the area of the triangle to three significant digits?

14. What is the area of a piece of carpeting that has sides measuring 14.3 ft, 12.8 ft, and 11.6 ft? Round to the nearest tenth of a square foot.

15. What is the area of the irregular-shaped property in Fig. 24-43? Round to the nearest square foot.

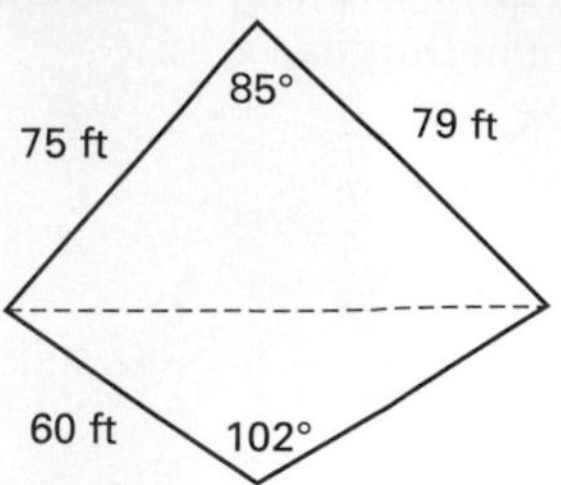

Figure 24-43

ASSIGNMENT EXERCISES

24-1

Find the direction and magnitude of vectors in standard position with the following end points. Round lengths to four significant digits and angles to the nearest tenth of a degree.

1. (4, 6)
2. (2, 8)
3. (6, 1)

Find the magnitude and direction of the resultant of the sum of the two given vectors in complex notation. Round as above.

4. $3 + 5i$ and $2 + 3i$
5. $1 + 4i$ and $6 + 8i$

24-2

Find the related angle for the following angles.

6. 115°
7. 3.04 rad
8. 4.75 rad
9. 221°
10. 305°
11. 5.4 rad
12. 138.5°
13. 212°15′10″

Using the calculator, find the values of the following functions. Round to four significant digits.

14. cos 250°
15. sin 2.1
16. tan 175°
17. sin 340°
18. tan 4.5
19. cos 290°
20. $\sin \frac{3\pi}{4}$
21. $\tan \frac{5\pi}{4}$

22. Find the magnitude and direction of a vector of an electrical current in standard position if the coordinates of the end point of the vector are (2, −3). Round the magnitude (in amps) to the nearest hundredth. Express the direction in degrees to the nearest 0.1°.

23. Find the direction and magnitude of a vector in standard position if the coordinates of the end point of the vector are (−2, 2). Express its direction in radians and round its direction and its magnitude to the nearest hundredth.

24-3

Solve the following triangles. If a triangle has two possibilities, find both solutions. Round sides to tenths and angles to the nearest 0.1°.

24. $A = 60°, B = 40°, b = 20$
25. $B = 120°, C = 20°, a = 8$
26. $A = 60°, B = 60°, a = 10$
27. $a = 5, c = 7, C = 45°$

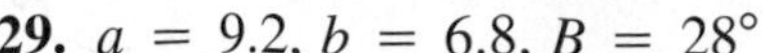

28. $b = 10$, $c = 8$, $B = 52°$

29. $a = 9.2$, $b = 6.8$, $B = 28°$

30. A surveyor needs the measure of JK in Fig. 24-44. Find JK to the nearest foot.

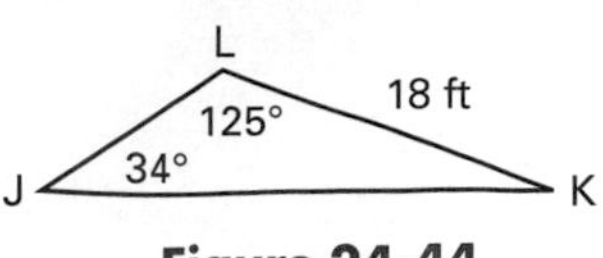

Figure 24-44

31. In Fig. 24-45, find RS.

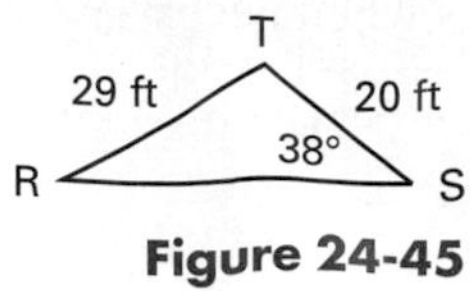

Figure 24-45

24-4

Solve the triangles in the following figures. Round sides to four significant digits and angles to nearest 0.1°.

32.

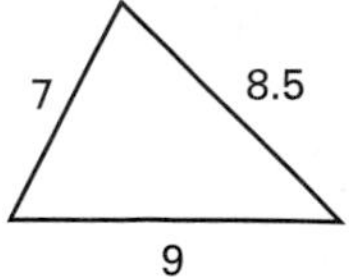

33.

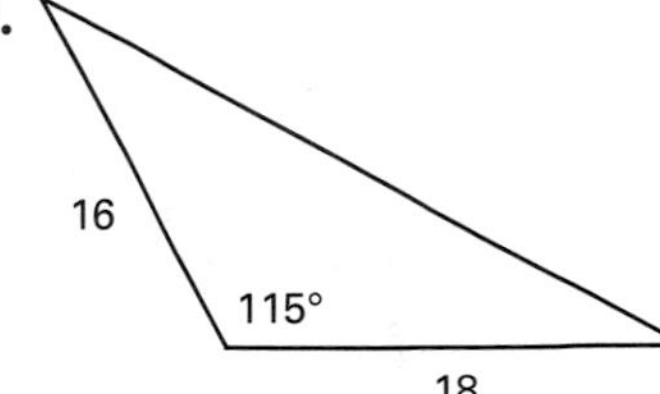

34.

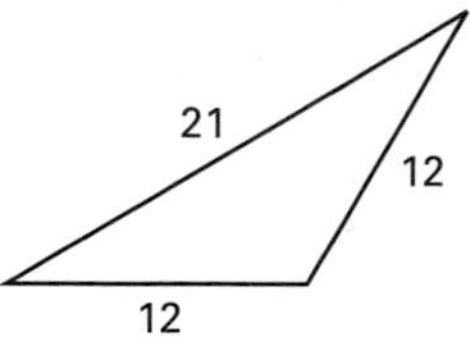

35.

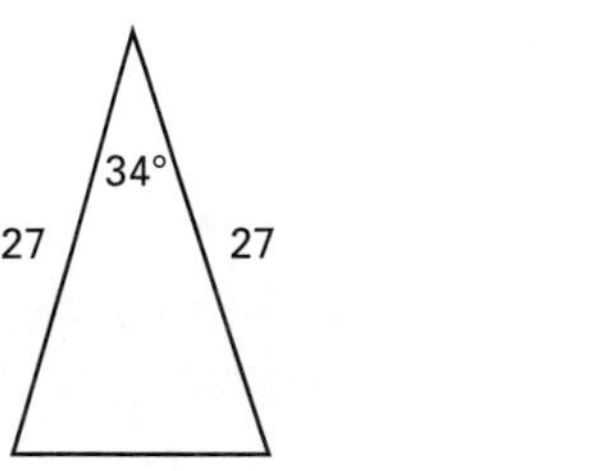

36.

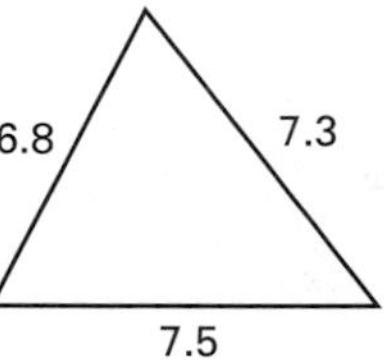

37.

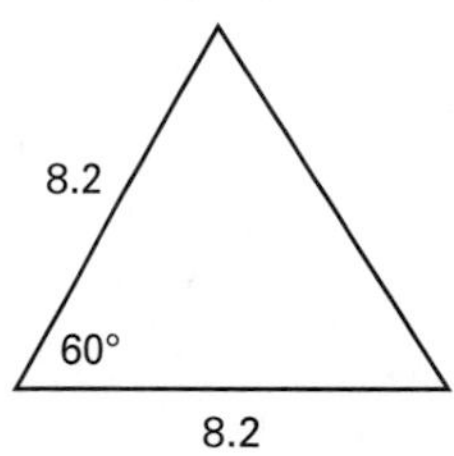

Solve the following problems. Round as in Problems 32-37.

38. A triangular lot has sides 180 ft long, 160 ft long, and 123.5 ft long. Find the angles of the lot.

39. A vertical 50-ft pole stands on top of a hill inclined 18° to the horizontal. What length of wire is needed to reach from a point 6 ft from the pole's top to a point 75 ft downhill from the base of the pole?

40. A ship sails from a harbor 35 nautical miles east then 42 nautical miles in a direction 32° south of east (Fig. 24-46). How far is the ship from the harbor?

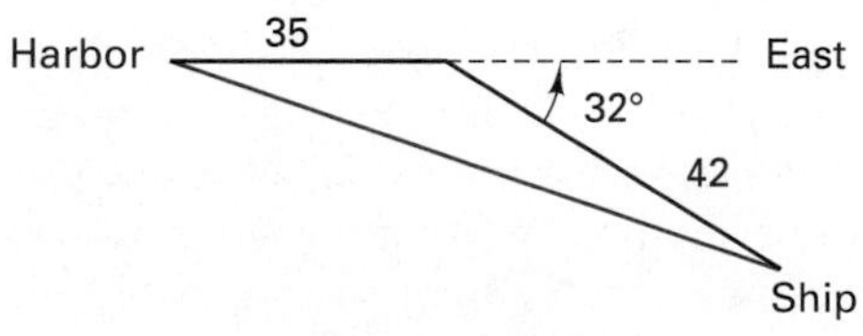

Figure 24-46

41. A hill with a 35° grade (inclined to the horizontal) is cut down for a roadbed to a 10° grade. If the distance from the base to the top of the original hill is 800 ft (Fig. 24-47), how many vertical feet will be removed from the top of the hill, and what is the distance from the bottom to the top of the hill for the roadbed?

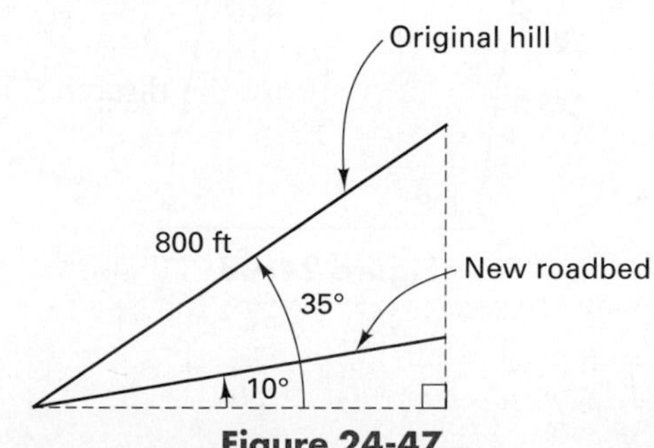

Figure 24-47

24-5

Find the area of the triangles shown in the following figures. Use the law of sines or the law of cosines when necessary. Round the answers to three significant digits.

42.

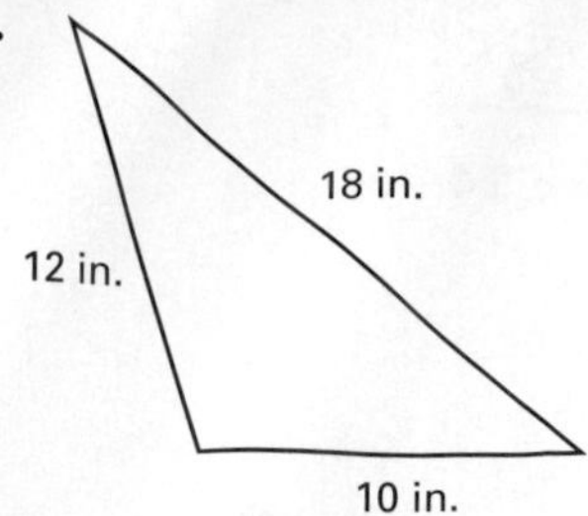

43.

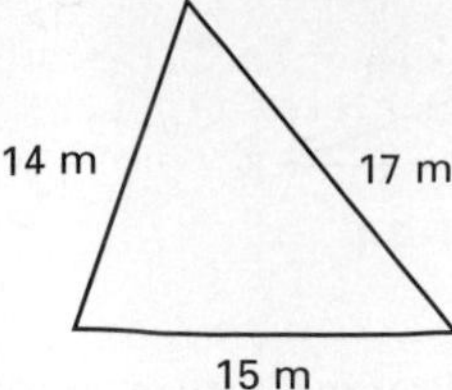

44.

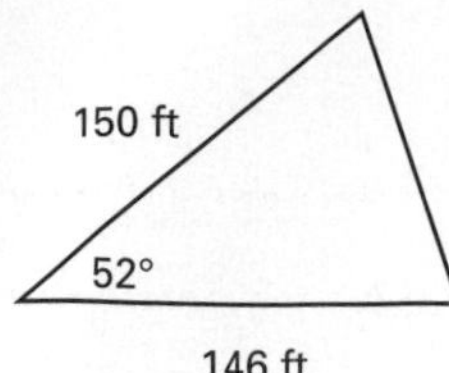

45.

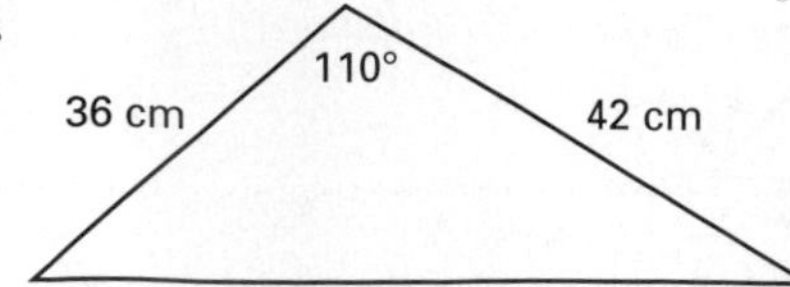

46.

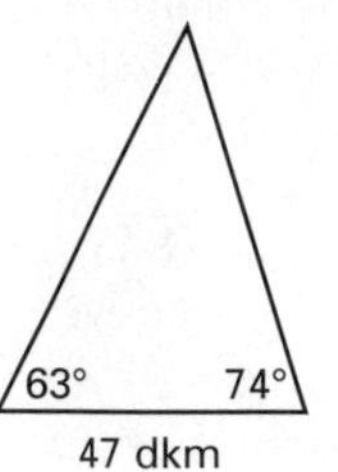

47.

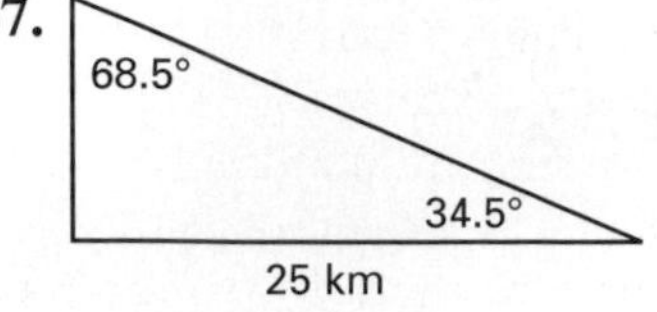

48.

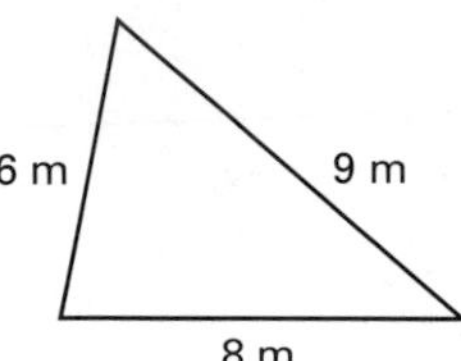

49.

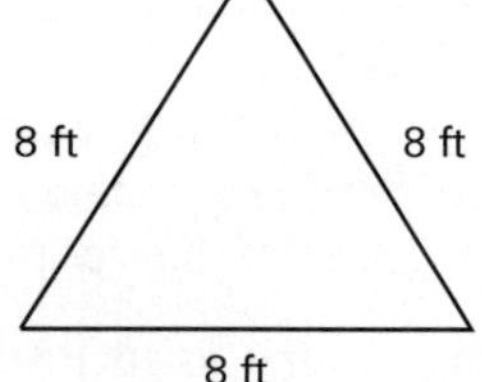

50.

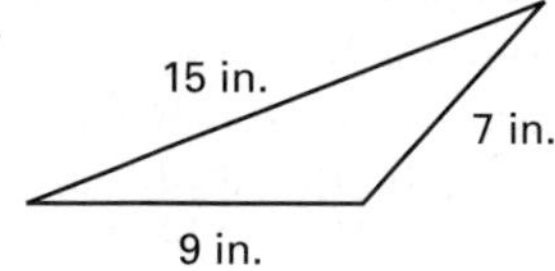

51. Find the area of a triangular tabletop whose sides are 46 in., 37 in., and 40 in.

52. What is the area of a triangular plot of ground with sides that measure 146 ft, 85 ft, and 195 ft?

53. Find the area of the irregular-shaped property in Fig. 24-48.

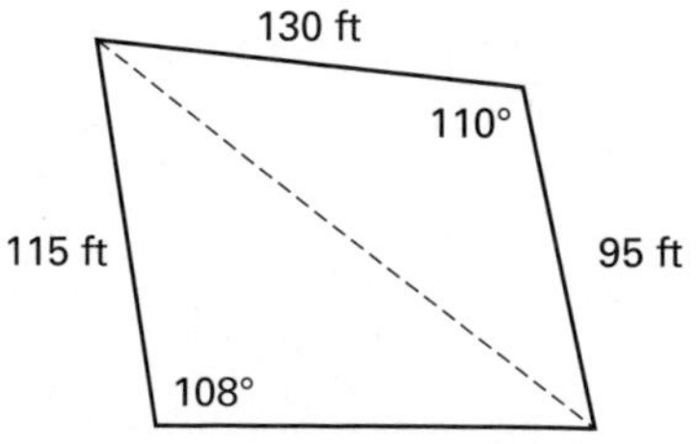

Figure 24-48

CHALLENGE PROBLEM

54. Cy Pipkin needs to finish his surveying report by finding the area of the plot of triangular land he has surveyed. Read his survey report and finish his report. Be sure to show all calculations so he can check them.

    Survey report

    From a corner of the property, one property line measures 196 feet and the other property line measures 216 feet. The angle formed by these two property lines measures 42°.

CONCEPTS ANALYSIS

1. Speed is a scalar measure, whereas velocity is a vector measure. What are the differences and similarities in these two measures?
2. Explain in your words what a related angle is, and explain how to find the related angle for an angle located in the third quadrant.
3. Locate a point in the second quadrant by a set of coordinates. Find the magnitude and direction of the vector determined by the point, and find the related angle for the vector.
4. Give the sign of the sine function in each of the four quadrants. Use your calculator to find the value of an angle in each of the four quadrants to validate your answer.
5. Give the sign of the tangent function in each of the four quadrants. Use your calculator to find the value of an angle in each of the four quadrants to validate your answer.
6. What parts of a triangle must be given in order to use the law of sines?
7. Draw a triangle and assign values to three parts of the triangle; then use the law of sines and other methods to find the remaining angles and sides.
8. Write the law of cosines in words.
9. Write the law of sines in words.
10. How would you know whether to select the law of sines or the law of cosines to find missing angles or sides of a triangle?

CHAPTER SUMMARY

Objectives	What to Remember	Examples
Section 24-1		
1. Find the magnitude of a vector in standard position, given the coordinates of the end point.	Use the Pythagorean theorem. $P = \sqrt{x^2 + y^2}$	Find the magnitude of a vector in standard position with the end point at (4, 2). $P = \sqrt{x^2 + y^2}$ $P = \sqrt{4^2 + 2^2}$ $P = \sqrt{16 + 4}$ $P = \sqrt{20}$ magnitude = 4.472 *(rounded)*

2. Find the direction of a vector in standard position, given the coordinates of the end point.

Use the tangent function.

$$\tan \theta = \frac{y}{x}$$

Find the direction of a vector in standard position with the end point at (2, 4).

$$\tan \theta = \frac{4}{2}$$
$$\tan \theta = 2$$
$$\theta = 63.4 \quad \textit{(rounded)}$$

direction = 63.4°

3. Find the sum of vectors.

1. Represent the vectors in the form of complex numbers, where the x-coordinate of the end point of a vector in standard position is the real component and the y-coordinate of the end point of a vector is the imaginary component.
2. Add the like components.

Add vectors A and B.

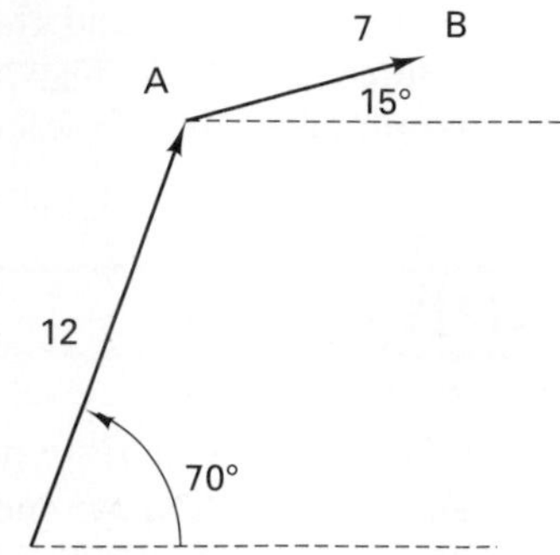

Vector A:

x-coordinate

$$\cos 70° = \frac{x}{12}$$
$$12(0.3420201433) = x$$
$$x = 4.1 \quad \textit{(rounded)}$$

y-coordinate

$$\sin 70° = \frac{y}{12}$$
$$12(0.9396926208) = y$$
$$y = 11.3 \quad \textit{(rounded)}$$

Vector $A = 4.1 + 11.3i$

Vector B:

x-coordinate

$$\cos 15° = \frac{x}{7}$$
$$7(0.9659258263) = x$$
$$x = 6.8 \quad \textit{(rounded)}$$

y-coordinate

$$\sin 15° = \frac{y}{7}$$
$$7(0.258819051) = y$$
$$y = 1.8 \quad \textit{(rounded)}$$

Vector $B = 6.8 + 1.8i$

Vector $A + B = 10.9 + 13.1i$

Section 24-2

1. Find related, acute angles for angles in quadrants II, III, and IV.

To find a related angle for an angle in any quadrant, draw the angle and form the third side of a triangle by drawing a line perpendicular to the x-axis from the line forming the angle. The related angle is the angle formed by the ray forming the angle and the x-axis.

Quadrant II angle:

$$180° - \theta \quad \text{or} \quad \pi - \theta$$

Quadrant III angle:

$$\theta - 180° \quad \text{or} \quad \theta - \pi$$

Quadrant IV angle:

$$360° - \theta \quad \text{or} \quad 2\pi - \theta$$

Find the related angle for an angle of 156°. The angle is in quadrant II; thus, $180° - 156° = 24°$.

2. Determine the signs of trigonometric values of angles more than 90°.

The mnemonic device ALL-SIN-TAN-COS can be used to remember the signs of the trigonometric functions in the various quadrants. The signs of *all* trigonometric functions are positive in the first quadrant. The sign of the SIN function is positive in the second quadrant (COS and TAN are negative). The sign of the TAN function is positive in the third quadrant. The sign of the COS function is positive in the fourth quadrant.

What is the sign of cos 145°? 145° is a quadrant II angle and the cosine function is negative in quadrant II. Thus, the sign of cos 145° is negative.

3. Find the trigonometric values of angles more than 90° using a calculator.

Most scientific calculators will now give the trigonometric value of an angle regardless of the quadrant in which it is found.

Use a scientific or graphing calculator to find the value of the following trigonometric functions. Round to 5 significant digits.

$$\sin 125° = 0.81915$$
$$\cos 215° = -0.81915$$
$$\tan 335° = -0.46631$$

Section 24-3

1. Find missing parts of an oblique triangle, given two angles and an opposite side.

Use the law of sines to find the missing part of an oblique triangle when two angles and a side opposite one of them are given.

$$\frac{a}{\sin A} = \frac{b}{\sin B} = \frac{c}{\sin C}$$

Find side AB in the given triangle.

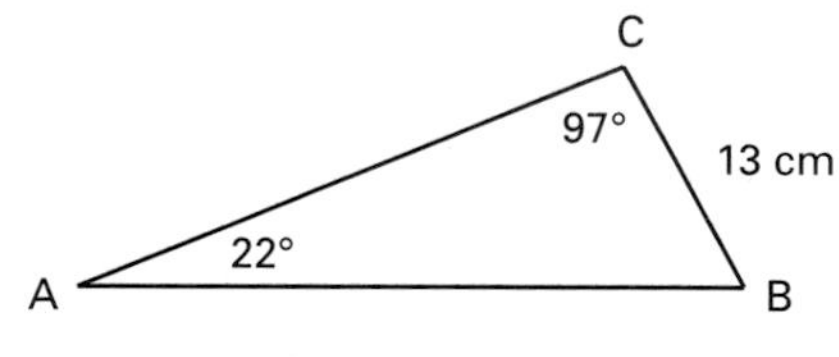

$$\frac{a}{\sin A} = \frac{c}{\sin C}$$
$$\frac{13}{\sin 22} = \frac{c}{\sin 97}$$
$$c \sin 22 = 13 \sin 97$$
$$c = \frac{13 \sin 97}{\sin 22}$$
$$c = 34.4 \text{ cm} \quad \textit{(rounded)}$$
$$AB = c = 34.4 \text{ cm}$$

2. Find missing parts of an oblique triangle, given two sides and an angle opposite one of them.

Use the law of sines to find the missing part of an oblique triangle when two sides and an angle opposite one of them are given.

Find the measure of angle A in the given triangle.

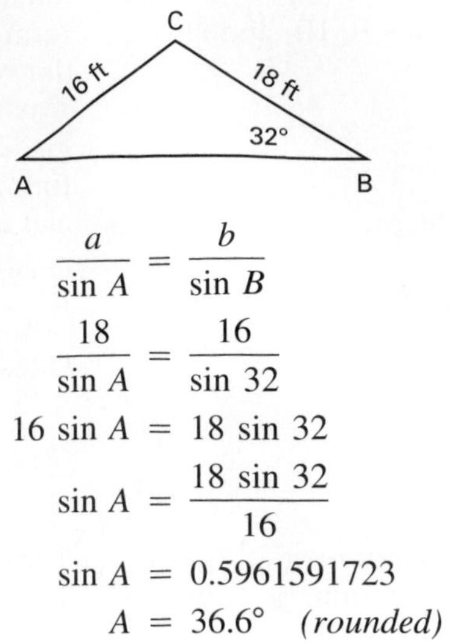

$$\frac{a}{\sin A} = \frac{b}{\sin B}$$

$$\frac{18}{\sin A} = \frac{16}{\sin 32}$$

$$16 \sin A = 18 \sin 32$$

$$\sin A = \frac{18 \sin 32}{16}$$

$$\sin A = 0.5961591723$$

$$A = 36.6° \quad \textit{(rounded)}$$

Given $\triangle ABC$, if $b < a$ and $\angle A \neq 90°$, we have two possible solutions. If $b < a$ and $\angle A = 90°$, we have one solution. If $b > a$, we have one solution.

Determine how many solutions exist for the following triangle based on the given facts. Explain your response.

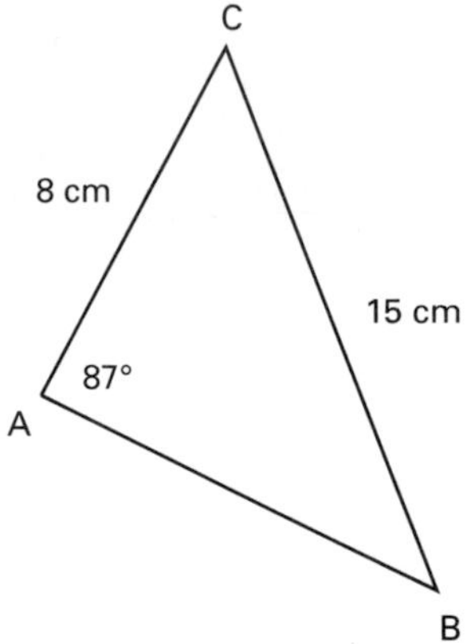

Since $b < a$ and $\angle A < 90°$, there are two possible solutions.

3. Solve applied problems using the law of sines.

Identify the given facts in the problem and represent these facts in graphic form. Use the law of sines to find the necessary parts of the problem.

A surveyor took the following measurements for a triangular plot of land. Use the law of sines to find the missing angles and side of the plot.

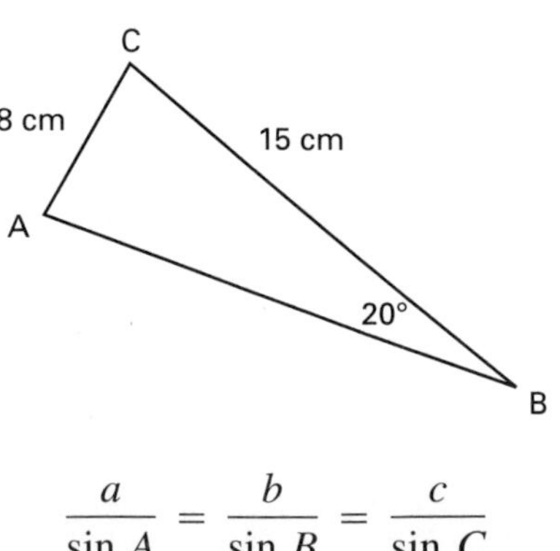

$$\frac{a}{\sin A} = \frac{b}{\sin B} = \frac{c}{\sin C}$$

Find $\angle A$.

$$\frac{15}{\sin A} = \frac{8}{\sin 20}$$
$$8 \sin A = 15 \sin 20$$
$$\sin A = \frac{15 \sin 20}{8}$$
$$\sin A = 0.6412877617$$
$$A = 39.9° \quad \textit{(rounded)}$$

Find $\angle C$.

$$\angle C = 180 - 20 - 39.9$$
$$\angle C = 120.1°$$

Find AB.

$$\frac{c}{\sin C} = \frac{b}{\sin B}$$
$$\frac{c}{\sin 120.1°} = \frac{8}{\sin 20°}$$
$$c \sin 20° = 8 \sin 120.1°$$
$$c = \frac{8 \sin 120.1°}{\sin 20}$$
$$c = 20.24 \text{ cm} \quad \textit{(rounded)}$$

Section 24-4

1. Find the missing parts of an oblique triangle, given three sides of the triangle.

To find an angle when three sides are given, we need to use the law of cosines, but first we have to rearrange the formula to solve for cos A. The rearranged formula is:

$$\cos A = \frac{-a^2 + b^2 + c^2}{2bc}$$

Find the measure of angle A in the given triangle.

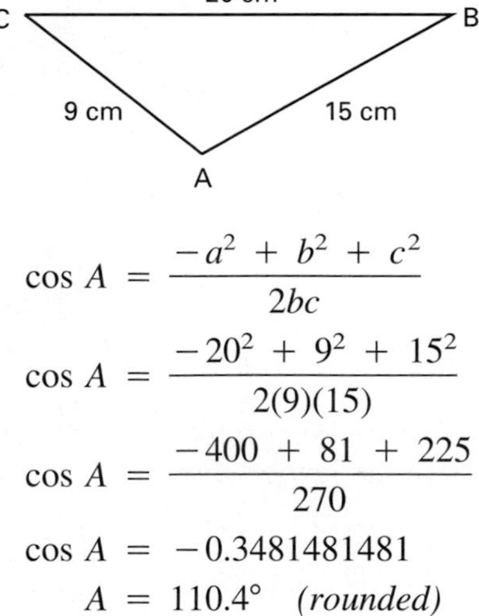

$$\cos A = \frac{-a^2 + b^2 + c^2}{2bc}$$
$$\cos A = \frac{-20^2 + 9^2 + 15^2}{2(9)(15)}$$
$$\cos A = \frac{-400 + 81 + 225}{270}$$
$$\cos A = -0.3481481481$$
$$A = 110.4° \quad \textit{(rounded)}$$

2. Find the missing parts of an oblique triangle, given two sides and the included angle of the triangle.

Use the law of cosines to find the missing parts of an oblique triangle when given three sides of the triangle.

$$a = \sqrt{b^2 + c^2 - 2bc \cos A}$$

In words, the side opposite the given angle equals the square root of the quantity found by the square of one given side plus the square of the other given side minus twice the product of the two given sides and the cosine of the given angle.

Find the third side of the given triangle.

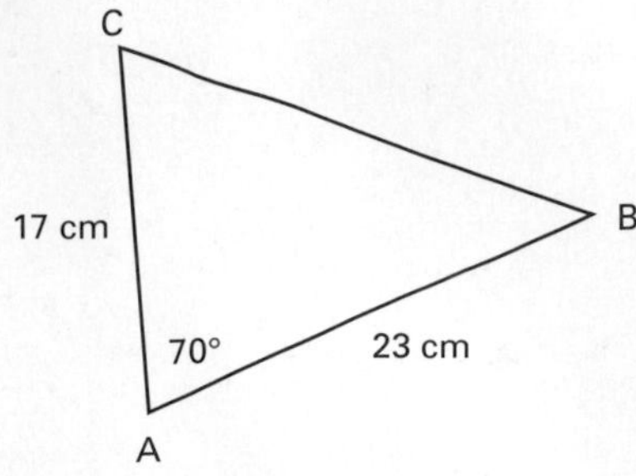

$$a = \sqrt{b^2 + c^2 - 2bc \cos A}$$
$$a = \sqrt{17^2 + 23^2 - 2(17)(23) \cos 70°}$$
$$a = \sqrt{289 + 529 - 267.4597521}$$
$$a = \sqrt{550.5402479}$$
$$a = 23.5 \text{ cm} \quad \textit{(rounded)}$$

Find as many parts as possible without using information that you have calculated.

Find the measures of all the angles and sides of the given triangle.

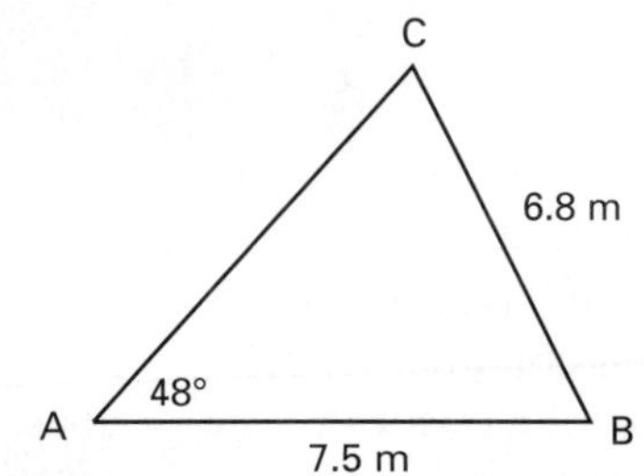

$$\frac{a}{\sin A} = \frac{c}{\sin C}$$
$$\frac{6.8}{\sin 48°} = \frac{7.5}{\sin C}$$
$$\sin C = \frac{7.5 \sin 48°}{6.8}$$
$$\sin C = 0.8196450281$$
$$C = 55.04927548°$$
$$\angle B = 180 - \angle A - \angle C$$
$$\angle B = 180 - 48 - 55.04927548$$
$$\angle B = 76.95072452°$$

Use the law of sines or law of cosines to find b or AC.

$$\frac{a}{\sin A} = \frac{b}{\sin B}$$
$$b = \frac{a \sin B}{\sin A}$$
$$b = \frac{6.8 \sin 76.95072452°}{\sin 48°}$$
$$b = 8.9 \text{ m}$$

3. Solve applied problems using the law of cosines and the law of sines.	Draw a sketch to represent the data given in the problem. Then identify which law is needed to solve the problem based on the given data.	A vertical utility pole is placed on a hill. The pole is 40 ft and a 35-ft guy wire is placed at the top of the pole to be attached to a ground anchor. If the angle formed by the hill and the pole is 43°, how far up the hill should the guy wire be placed if no allowance is given for the amount of wire required to attach the guy wire to its anchor on the pole.

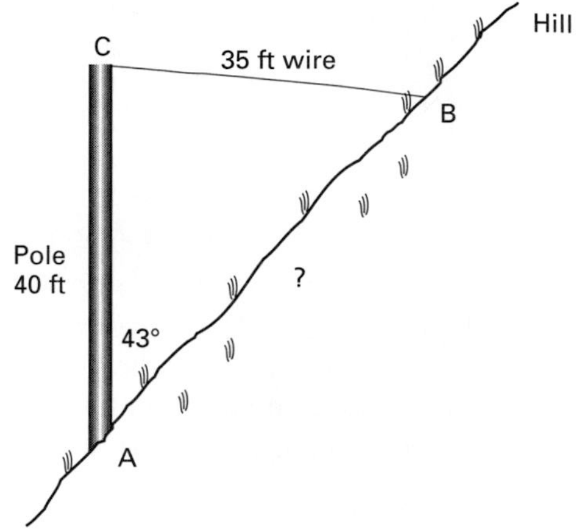

$$\frac{a}{\sin A} = \frac{b}{\sin B}$$

$$\frac{35}{\sin 43°} = \frac{40}{\sin B}$$

$$\sin B = \frac{40 \sin 43°}{35}$$

$$\sin B = 0.7794266972$$

$$B = 51.208114270°$$

$$\angle C = 180 - 43 - 51.208114270$$

$$\angle C = 85.79188573°$$

$$c = \sqrt{a^2 + b^2 - 2ab \cos C}$$

$$c = \sqrt{35^2 + 40^2 - 2(35)(40) \cos 85.79188573°}$$

$$c = \sqrt{1225 + 1600 - 205.462423}$$

$$c = \sqrt{2619.537577}$$

$$c = 51.2 \text{ ft} \quad \textit{(rounded)}$$

Section 24-5

1. Find the area of a triangle when the height is unknown and at least three parts of the triangle are known.

Area $= \frac{1}{2}ab \sin C$, where a and b are two adjacent sides of a triangle and angle C is the angle between sides a and b.

Find the area of triangle ABC.

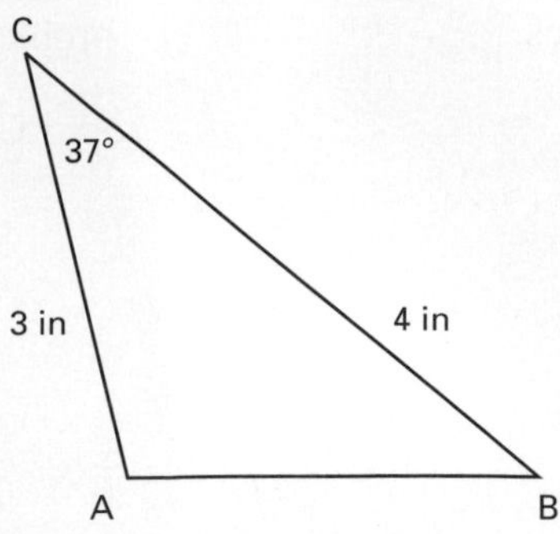

$$\text{Area} = \frac{1}{2}ab \sin C$$
$$\text{Area} = 0.5(4)(3)(\sin 37°)$$
$$\text{Area} = 3.6 \text{ in.}^2$$

2. Find the area of a triangle using Heron's formula when the three sides of the triangle are known.

Area $= \sqrt{s(s - a)(s - b)(s - c)}$, where a, b, and c are the lengths of the three sides and s is one-half the perimeter of the triangle.

$$\left(s = \frac{1}{2}(a + b + c)\right)$$

Find the area of triangle ABC.

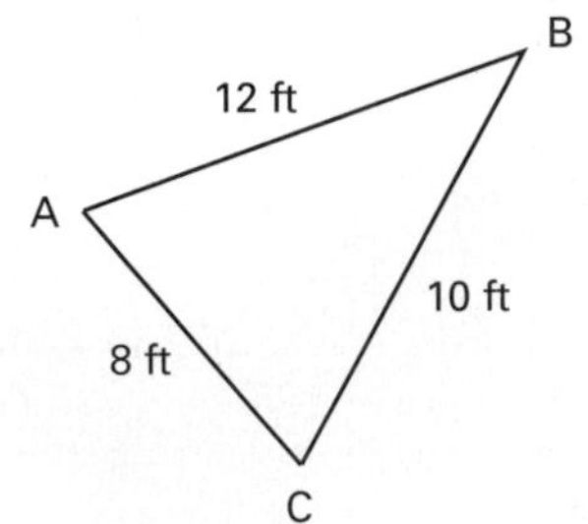

$$s = \frac{1}{2}(a + b + c)$$
$$s = \frac{1}{2}(8 + 10 + 12)$$
$$s = \frac{1}{2}(30)$$
$$s = 15$$
$$\text{Area} = \sqrt{s(s - a)(s - b)(s - c)}$$
$$\text{Area} = \sqrt{15(15 - 10)(15 - 8)(15 - 12)}$$
$$\text{Area} = \sqrt{15(5)(7)(3)}$$
$$\text{Area} = \sqrt{1575}$$
$$\text{Area} = 39.7 \text{ ft}^2$$

3. Solve applied problems involving the area of a triangle.

To solve applied problems, illustrate the data with a figure and use the formula that is appropriate for the given data.

Find the square footage in a triangular lot that measures 108 ft, 125 ft, and 97 ft on each of the sides, respectively.

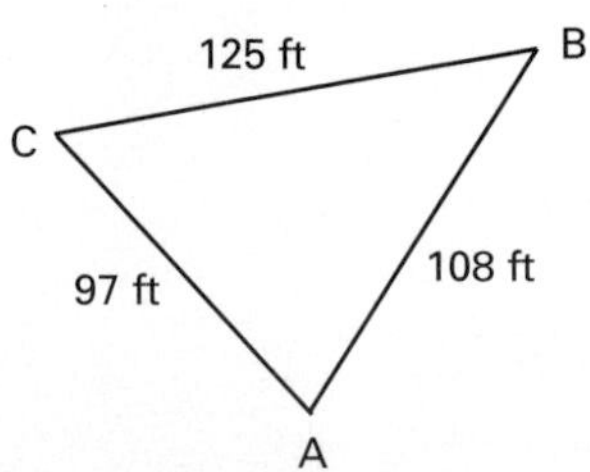

$$s = \frac{1}{2}(a + b + c)$$

$$s = \frac{1}{2}(125 + 97 + 108)$$

$$s = \frac{1}{2}(330)$$

$$s = 165$$

$$A = \sqrt{s(s - a)(s - b)(s - c)}$$

$$A = \sqrt{165(165 - 125)(165 - 97)(165 - 108)}$$

$$A = \sqrt{165(40)(68)(57)}$$

$$A = \sqrt{25{,}581{,}600}$$

$$A = 5057.8 \text{ ft}^2$$

WORDS TO KNOW

vector (p. 903)
magnitude (p. 903)
direction (p. 904)
related angle (p. 907)
law of sines (p. 913)
ambiguous case (p. 917)
law of cosines (p. 922)
Heron's formula (p. 929)

CHAPTER TRIAL TEST

Find the values of the following. Round to the nearest ten-thousandth.

1. sin 125°

2. tan 140°

3. cos 160°

4. Find the length of the vector whose end point coordinates are (8, 15).

5. Find the angle of the vector whose end point coordinates are $(-8, 8)$.

Use the law of sines or the law of cosines to find the side or angle indicated in the following figures. Round sides to four significant digits and angles to the nearest 0.1°.

6.

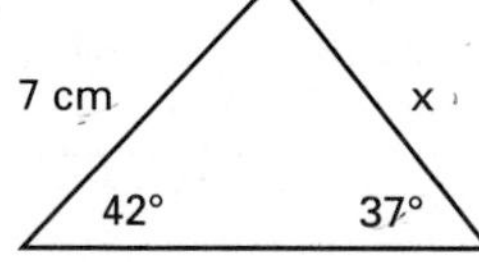

7.

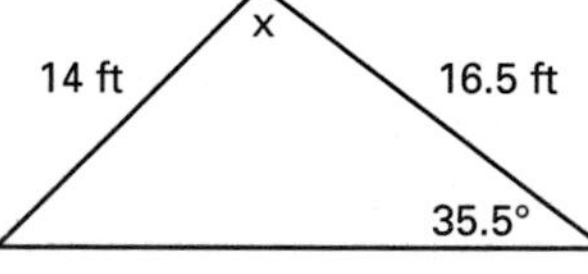

8.

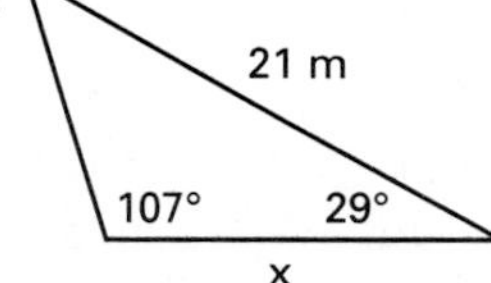

9.

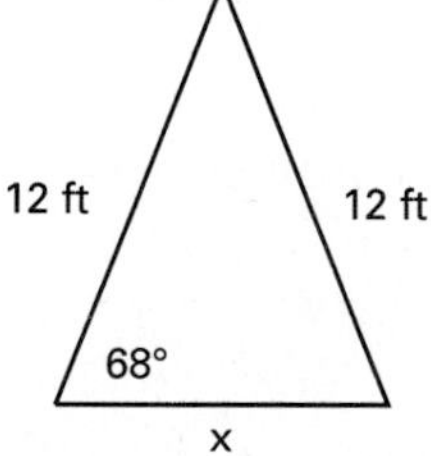

10.

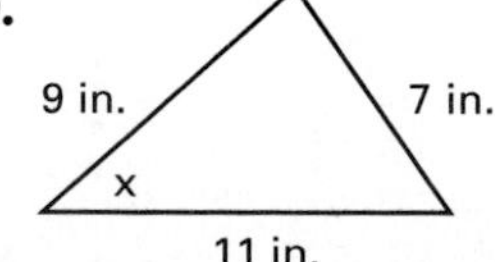

11.

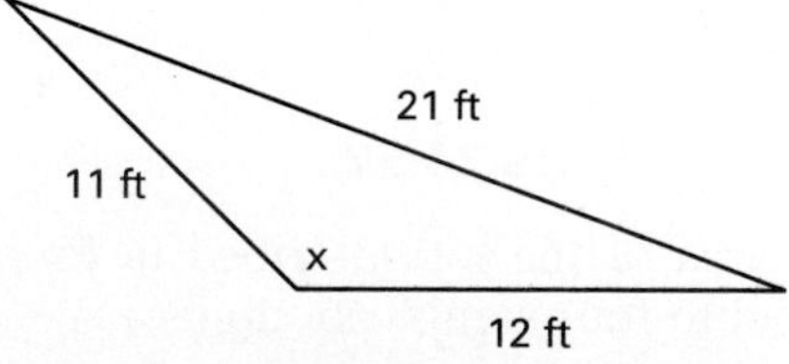

12.

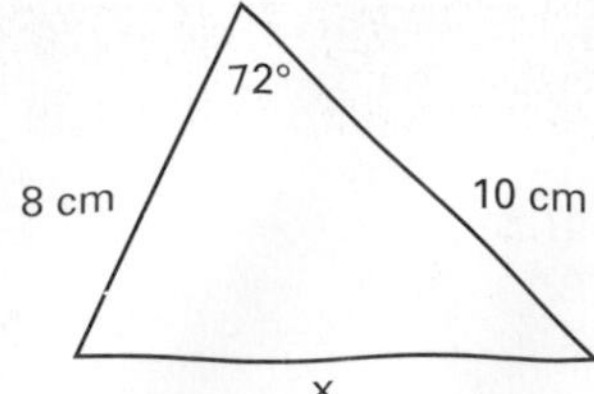

13.

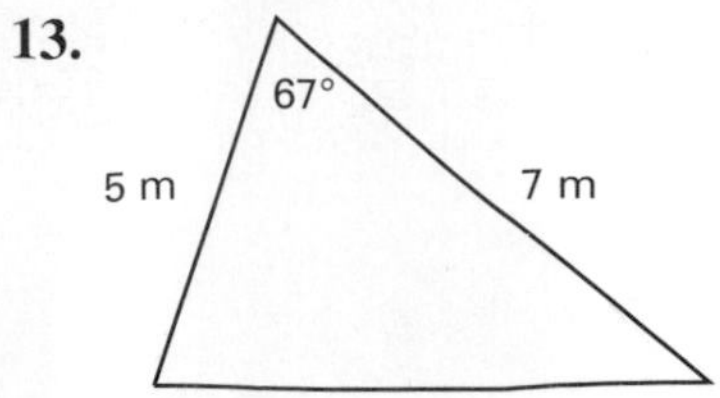

14. Find the area of triangle ABC in Fig. 24-49. Round to four significant digits.

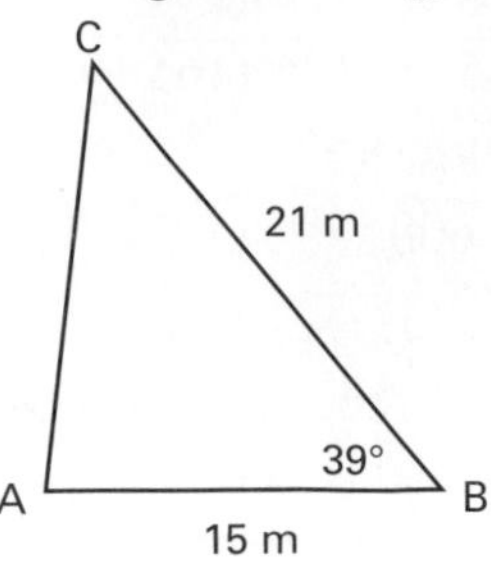

Figure 24-49

15. Find the area of triangle RST in Fig. 24-50. Round to four significant digits.

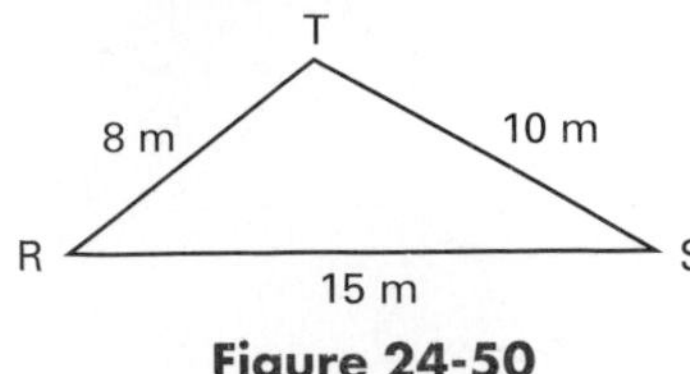

Figure 24-50

16. Find the direction in degrees (to nearest 0.1°) of the vector I_t (total current) in Fig. 24-51.

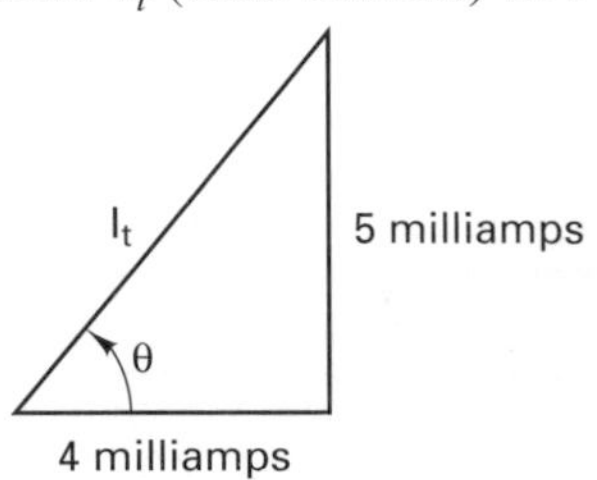

Figure 24-51

17. A surveyor measures two sides of a triangular lot and the angle formed by these two sides. What is the length of the third side if the two measured sides are 76 ft and 110 ft and the angle formed by these two sides is 107°?

18. What is the area of a triangular cast if the sides measure 31 mm, 42 mm, and 27 mm?

19. Find the area of a triangular flower bed if two sides measure 12 ft and 15 ft and the included angle measures 48°.

20. A plane is flown from an airport due east for 32 mi, and then turns 15° north of east and travels 72 mi. How far is the plane from the airport?

21. Find the magnitude of the vector I_t in Problem 16 to the nearest tenth.

22. A connecting rod 30 cm long joins a crank 20 cm long to form a triangle with a third, imaginary line (Fig. 24-52). If the crank and the imaginary line form an angle of 150°, find the angle formed by the connecting rod and the imaginary line.

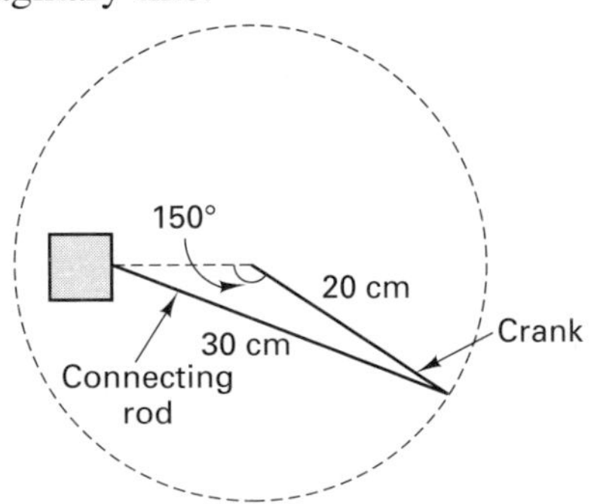

Figure 24-52

23. Find the magnitude of vector Z in Fig. 24-53 if vectors R and X_L form a right angle. All units are in ohms.

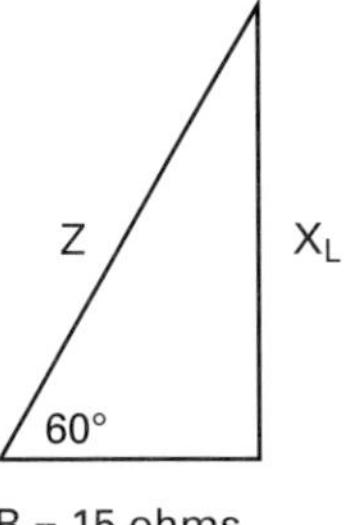

Figure 24-53

24. Find the area of the lot described in Problem 17. (Round to four significant digits.)

Selected Answers to Student Exercise Material

Answers to all Self-Study Exercises and answers to odd-numbered Assignment Exercises and odd-numbered problems in the Chapter Trial Tests are included here. Answers to all even-numbered exercises and problems may be found in the Instructor's Manual.

Self-Study Exercises 1-1 (pp. 5–6)

L.O.1.

1. hundred millions **2.** ten thousands **3.** hundreds **4.** ten millions **5.** billions **6.** hundred thousands **7.** tens **8.** ten millions **9.** billions **10.** ten thousands

L.O.2.

11. six thousand, seven hundred, four **12.** eighty-nine thousand, twenty-one **13.** six hundred sixty-two million, nine hundred thousand, seven hundred fourteen **14.** three million, one hundred one **15.** fifteen billion, four hundred seven million, two hundred ninety-four thousand, three hundred seventy-six. **16.** one hundred fifty **17.** 7,000,000,400 **18.** 1,627,106 **19.** 58,201 **20.** $1,006

L.O.3.

21. 590 **22.** 8,900 **23.** 9,250 **24.** 83,000 **25.** 200,000 **26.** 50 **27.** 900 **28.** 8,000 **29.** 4,000 **30.** 500

Self-Study Exercises 1-2 (pp. 10–11)

L.O.1.

1. 15 **2.** 27 **3.** 22 **4.** 36 **5.** 37 bolts

L.O.2.

6. 9,192 **7.** 16,956 **8.** 106,285 **9.** 2310 **10.** 97,614 **11.** $456 **12.** 24,349 lb **13.** 298 screws **14.** 4553 bricks **15.** 50 gal

L.O.3.

16. 17,100; 17,017 **17.** 84,200; 84,213 **18.** 402,300; 402,199 **19.** $3,440; $3,442 **20.** $2,300; $2,260 **21.** 8,900; 9,239 **22.** 600; 624 **23.** 9,000; 8,449 **24.** 800; 801 **25.** 60,000; 52,801 **26.** 50,000; 49,241 **27.** 159 lb **28.** No, total capacity needed is 115 gal. **29.** Yes, 479 pages needed **30.** 350 ft

L.O.4.

31. 190,000; 190,786 **32.** 1,900,000; 1,859,867 **33.** 50,000; 54,359
34. 51,000,000; 47,520,014 **35.** $17,034

Self-Study Exercises 1-3 (pp. 15–16)

L.O.1.

1. 5 **2.** 1 **3.** 7 **4.** 2 **5.** 3 **6.** 7 **7.** 5 **8.** 2 **9.** 4
10. 1

L.O.2.

11. 24 **12.** 401 **13.** 1020 **14.** 115 **15.** 53,036 **16.** 22 bags
17. 341 boxes **18.** 321 ft

L.O.3.

19. 45 **20.** 608 **21.** 1,573 **22.** 22,205 **23.** 100,000; 91,034
24. 2000; 2089 **25.** 0; 174 **26.** 200; 599 **27.** 70,000; 70,382
28. 1300; 1097

L.O.4.

29. 4,000; 4,397 **30.** 1000; 1187 **31.** 500,000; 508,275 **32.** 17 L
33. 186 bricks **34.** 213 in. **35.** 125 in. **36.** 45 in.

Self-Study Exercises 1-4 (pp. 24–25)

L.O.1.

1. (a) 15 (b) 56 (c) 63 (d) 24 **2.** $42 **3.** commutative property of multiplication
4. Numbers may be grouped in any way for multiplication.

$$\begin{aligned} 3[(5)(2)] &= [3(5)]2 \\ 3(10) &= (15)2 \\ 30 &= 30 \end{aligned}$$

Answers may vary.

5. 0 **6.** 378 **7.** 84 **8.** 0

L.O.2.

9. 224 **10.** 581 **11.** 630 **12.** 102,612 **13.** 864 pieces
14. 4096 washers **15.** 84 books **16.** 700 tickets **17.** $288
18. 249 students

L.O.3.

19. $5 \cdot 7 + 5 \cdot 4 \quad 5(11)$
$35 + 20 \quad 55$
55

20. $3(9) + 3(11) \quad 3(20)$
$27 + 33 \quad 60$
60

21. $14(3) + 14(11) \quad 14(14)$
$42 + 154 \quad 196$
196

22. $4(5) - 4(2) \quad 4(3)$
$20 - 8 \quad 12$
12

23. $12(16) - 12(3) \quad 12(13)$
$192 - 36 \quad 156$
156

24. $7(18) - 7(1) \quad 7(17)$
$126 - 7 \quad 119$
119

25. $P = 40$ ft **26.** $P = 60$ in. **27.** 56 ft **28.** 170 ft

L.O.4.

29. $768 **30.** 12,354 ft^2 **31.** 1,164 cm^2 **32.** 1,344 ft^2 **33.** 96 ft^2

L.O.5.

34. 33,465,696 **35.** 233,790,450 **36.** 561,500,160 **37.** 0
38. $36,180 **39.** $21,996 **40.** 147,000,000,000 **41.** 16,755,200,000
42. 420,000,000

Self-Study Exercises 1-5 (pp. 33–34)

L.O.1.

1. $8 \div 4$ **2.** $9 \div 3$ **3.** $24 \div 6$ **4.** $30 \div 7$ **5.** $6 \div 2$
6. $8 \div 4$ **7.** $9\overline{)45}$ **8.** $6\overline{)24}$ **9.** $3\overline{)21}$ **10.** $8\overline{)56}$ **11.** $2\overline{)8}$
12. $6\overline{)12}$ **13.** $\frac{14}{2}$ **14.** $\frac{16}{2}$

L.O.2.

15. 75 **16.** 23 **17.** 20 **18.** 3 **19.** 16 **20.** 12 **21.** 43
22. 47 **23.** 12 **24.** 23R1 **25.** 124R30 **26.** 56R80 **27.** 32 ft
28. $1245 **29.** 6 in.

L.O.3.

30. 100; 128R24 **31.** 4000; 3151R90 **32.** 4000; 3731R53
33. 800; 843 **34.** 128 loads **35.** 127 cords

L.O.4.

36. 1471 **37.** 301 **38.** 4003 **39.** 89,500 **40.** 40,200 **41.** 6201
42. 1070 bricks **43.** $10,500 **44.** $105 **45.** $1,950 **46.** $29,580
47. $4 **48.** 24 parts **49.** 10 lengths **50.** approximately 16 ft
51. 10 in.

Self-Study Exercises 1-6 (pp. 40–41)

L.O.1.

1. 4; 3 **2.** 9; 4 **3.** 2; 9 **4.** 1 **5.** 49 **6.** 1000 **7.** 16
8. 9 **9.** 15 **10.** 8 **11.** 1 **12.** 8^1 **13.** 145^1 **14.** 12^1
15. 23^1 **16.** leaves base unchanged **17.** changes value to 1; 0^0 is undefined

L.O.2.

18. 64 **19.** 4 **20.** 625 **21.** 196 **22.** 169 **23.** 1 **24.** 10,000
25. 64 **26.** 81 **27.** 289 **28.** 324 **29.** 10,201 **30.** 484
31. 5 **32.** 7 **33.** 9 **34.** 14 **35.** Use it as a factor two times.
36. Find a number that is used as a factor two times to give the desired number.

L.O.3.

37. 225 **38.** 343 **39.** 78,125 **40.** 18 **41.** 28 **42.** 33

L.O.4.

43. $10^3 = 1000$ **44.** 32,000 **45.** 30,000 **46.** 20,000 **47.** 10,200
48. 22,000 **49.** 25 **50.** 21 **51.** 3 **52.** 9 **53.** 250 **54.** 12
55. Shift the decimal to the right in the number being multiplied by the power of ten as indicated by the exponent. **56.** Shift the decimal to the left in the number being multiplied by the power of ten as indicated by the exponent.

Self-Study Exercises 1-7 (pp. 43–44)

L.O.1.

1. 26 **2.** 18 **3.** 9 **4.** 8 **5.** 20 **6.** 32 **7.** 30 **8.** 15
9. 156 **10.** 109 **11.** 23 **12.** 45 **13.** 16 **14.** 3 **15.** 9
16. 5 **17.** 7 **18.** 15 **19.** 160 **20.** 1458

L.O.2.

21. 483 **22.** 443 **23.** 70 **24.** 13 **25.** 11 **26.** 29 **27.** 5
28. 39 **29.** 7 **30.** 14 **31.** 20 **32.** 30 **33.** parentheses
34. addition or subtraction

Self-Study Exercises 1-8 (pp. 46–47)

L.O.1.

1. 1, 2, 3, 4, 6, 8, 12, 24 **2.** 1, 2, 3, 4, 6, 9, 12, 18, 36 **3.** 1, 3, 5, 9, 15, 45 **4.** 1, 2, 4, 8, 16, 32 **5.** 1, 2, 4, 8, 16 **6.** 1, 3, 9, 27 **7.** 1, 2, 4, 5, 10, 20 **8.** 1, 2, 3, 5, 6, 10, 15, 30 **9.** 1, 2, 3, 4, 6, 12 **10.** 1, 2, 4, 8 **11.** 1, 2, 4 **12.** 1, 3, 5, 15 **13.** 1, 3, 9, 27, 81 **14.** 1, 2, 4, 8, 16, 32, 64 **15.** 1, 2, 19, 38 **16.** 1, 2, 23, 46

L.O.2.

17. $1 \cdot 5, 2 \cdot 5, 3 \cdot 5, 4 \cdot 5, 5 \cdot 5, 6 \cdot 5$
18. $1 \cdot 6, 2 \cdot 6, 3 \cdot 6, 4 \cdot 6, 5 \cdot 6, 6 \cdot 6$
19. $1 \cdot 8, 2 \cdot 8, 3 \cdot 8, 4 \cdot 8, 5 \cdot 8, 6 \cdot 8$
20. $1 \cdot 9, 2 \cdot 9, 3 \cdot 9, 4 \cdot 9, 5 \cdot 9, 6 \cdot 9$
21. $1 \cdot 10, 2 \cdot 10, 3 \cdot 10, 4 \cdot 10, 5 \cdot 10, 6 \cdot 10$
22. $1 \cdot 30, 2 \cdot 30, 3 \cdot 30, 4 \cdot 30, 5 \cdot 30, 6 \cdot 30$

23. $1 \cdot 5 = 5$, $2 \cdot 5 = 10$, $3 \cdot 5 = 15$, $4 \cdot 5 = 20$, $5 \cdot 5 = 25$
24. $1 \cdot 12 = 12$, $2 \cdot 12 = 24$, $3 \cdot 12 = 36$, $4 \cdot 12 = 48$, $5 \cdot 12 = 60$
25. $1 \cdot 7 = 7$, $2 \cdot 7 = 14$, $3 \cdot 7 = 21$, $4 \cdot 7 = 28$, $5 \cdot 7 = 35$
26. $1 \cdot 3 = 3$, $2 \cdot 3 = 6$, $3 \cdot 3 = 9$, $4 \cdot 3 = 12$, $5 \cdot 3 = 15$
27. $1 \cdot 50 = 50$, $2 \cdot 50 = 100$, $3 \cdot 50 = 150$, $4 \cdot 50 = 200$, $5 \cdot 50 = 250$
28. $1 \cdot 4 = 4$, $2 \cdot 4 = 8$, $3 \cdot 4 = 12$, $4 \cdot 4 = 16$, $5 \cdot 4 = 20$

L.O.3.

29. No. The sum of the digits is not divisible by 3. **30.** Yes. The last digit of the number is 0. **31.** No. The last 2 digits of the number is not divisible by 4. **32.** Yes. The sum of the digits is divisible by 3. **33.** Yes. The division has no remainder. **34.** Yes. The three digits of the number are divisible by 8. **35.** Yes. The sum of the digits is divisible by 3. **36.** Yes. The sum of the digits is divisible by 3 and the last digit is an even number.

Self-Study Exercises 1-9 (p. 48)

L.O.1.

1. $1 \cdot 14, 2 \cdot 7$; 1, 2, 7, 14 **2.** $1 \cdot 22, 2 \cdot 11$; 1, 2, 11, 22 **3.** $1 \cdot 11$; 1, 11 **4.** $1 \cdot 17$; 1, 17 **5.** $1 \cdot 18, 2 \cdot 9, 3 \cdot 6$; 1, 2, 3, 6, 9, 18 **6.** $1 \cdot 24, 2 \cdot 12, 3 \cdot 8, 4 \cdot 6$; 1, 2, 3, 4, 6, 8, 12, 24

L.O.2.

7. $2 \cdot 5 \cdot 5$ **8.** $2 \cdot 2 \cdot 13$ **9.** $3 \cdot 3 \cdot 5 \cdot 5$ **10.** $5 \cdot 5 \cdot 5$ **11.** $2 \cdot 2 \cdot 5 \cdot 5$ **12.** $2 \cdot 2 \cdot 2 \cdot 5 \cdot 5$ **13.** $5 \cdot 13$ **14.** $3 \cdot 5 \cdot 5$ **15.** $11 \cdot 11$ **16.** $2 \cdot 2 \cdot 2 \cdot 2 \cdot 3 \cdot 3$ **17.** $2^3 \cdot 71$ **18.** $2^4 \cdot 7$ **19.** $2^2 \cdot 31$ **20.** $2^2 \cdot 41$ **21.** $2^3 \cdot 3^2$ **22.** $2^2 \cdot 3^2 \cdot 5^2$

Self-Study Exercises 1-10 (pp. 50–51)

L.O.1.

1. 6 **2.** 30 **3.** 56 **4.** 12 **5.** 180 **6.** 60 **7.** 24 **8.** 18 **9.** 24 **10.** 700 **11.** 27 **12.** 16 **13.** 90 **14.** 120 **15.** 180 **16.** 60 **17.** 96 **18.** 72 **19.** 60 **20.** 300 **21.** 66 **22.** 312

L.O.2.

23. 6 **24.** 5 **25.** 1 **26.** 1 **27.** 2 **28.** 2 **29.** 15 **30.** 5
31. 12 **32.** 6

Assignment Exercises, Chapter 1 (pp. 51–54)

1. (a) tens (b) ten thousands (c) millions **3.** Fifty-six million, one hundred nine thousand, one hundred ten **5.** 1,256,401 **7.** (a) 40 (b) 70 (c) 20 **9.** (a) 30,000 (b) 4900 (c) 10,000 (d) 49,000 (e) 190,000 (f) 9000 **11.** (a) 23 (b) 18 (c) 28 (d) 25 **13.** 20 cans **15.** 20 points **17.** (a) 9,000; \$29,091 (b) 36,000; 36,048 **19.** (a) 3 (b) 2 (c) 2 (d) 5 (e) 0 (f) 2 (g) 144 (h) 12,140 **21.** 17 sets **23.** 200 miles; 190 miles **25.** 84 **27.** 1,143 **29.** 13,725 **31.** 3,349,890 **33.** 394,254,080 **35.** \$1,407 **37.** \$43,920 **39.** 20 **41.** 60 **43.** 80 **45.** \$328 **47.** \$14,800; \$13,140 **49.** 1,140,000 ft^2; 1,202,800 ft^2

51. (a) $5 \div 3$ (b) $3\overline{)5}$ (c) $\frac{5}{3}$ **53.** 1 **55.** not possible or undefined

57. 52R11 **59.** 13 **61.** 2008R6 **63.** 23 with 11 left over **65.** 58R3 **67.** 5; 5R45 **69.** 100; 126R95 **71.** 40 each; 48 each **73.** (a) 5, 6 (b) 12, 2 (c) 10, 6 **75.** (a) 904^1 (b) 76^1 (c) 3^1 **77.** (a) 4 (b) 169 (c) 49 (d) 144 **79.** (a) 50 (b) 12 (c) 17 (d) 9 **81.** (a) $3 \times 10^2 = 300$ (b) $75 \times 10^4 = 750{,}000$ (c) $22 \times 10^3 = 22{,}000$ (d) $5 \times 10^2 = 500$ **83.** 5 **85.** 54 **87.** 12 **89.** 28 **91.** 5 **93.** 1, 2, 3, 4, 6, 8, 9, 12, 18, 24, 36, 72 **95.** 1, 3, 7, 21 **97.** 15, 30, 45, 60, 75 **99.** Yes, the sum of the digits is divisible by 3. **101.** $2 \cdot 2 \cdot 11 = 2^2 \cdot 11$ **103.** $2 \cdot 2 \cdot 2 \cdot 3 \cdot 3 \cdot 3 = 2^3 \cdot 3^3$

105. $1 \cdot 48$
$2 \cdot 24$
$3 \cdot 16$
$4 \cdot 12$
$6 \cdot 8$
1, 2, 3, 4, 6, 8, 12, 16, 24, 48

107. $1 \cdot 51$
$3 \cdot 17$
1, 3, 17, 51

109. 360

111. 180 **113.** 2 **115.** 6 **117.** 15 yd × 144 (Answers may vary.)

Trial Test, Chapter 1 (p. 59)

1. Five million, thirty thousand, one hundred two **3.** 2,700 **5.** 1007 **7.** \$9,271,314 **9.** 134 **11.** 106 **13.** \$800; \$764 **15.** \$15,000; \$15,075 **17.** 182 books

19. $1 \cdot 42$
$2 \cdot 21$
$3 \cdot 14$
$6 \cdot 7$

21. $2 \cdot 2 \cdot 3 \cdot 7$ or $2^2 \cdot 3 \cdot 7$

23. 420 **25.** Commutative means we can add numbers in any order. $2 + 4 = 6$; $4 + 2 = 6$. Associative means we can group numbers differently. $(2 + 1) + 3 = 6$; $2 + (1 + 3) = 6$.

Self-Study Exercises 2-1 (p. 62)

L.O.1.

14. 4. 9. 11. 10. 7. 2. 1. 5. 3. 6. 12. 8. 13. 15.

−11 −10 −9 −8 −7 −6 −5 −4 −3 −2 −1 0 1 2 3 4 5 6 7 8 9 10 11 12

L.O.2.

16. < **17.** < **18.** > **19.** > **20.** > **21.** > **22.** >
23. < **24.** < **25.** <

L.O.3.

26. 8 **27.** 0 **28.** 86 **29.** 32 **30.** 74 **31.** −1 **32.** 32
33. −74 **34.** 2 **35.** 49

L.O.4.

36. 12 **37.** −58 **38.** no opposite of 0, or zero is its own opposite.
39. −529 **40.** 3,825

Self-Study Exercises 2-2 (pp. 66–67)

L.O.1.

1. 17 **2.** −13 **3.** 99 **4.** −59 **5.** −48 **6.** −161

L.O.2.

7. −2 **8.** 4 **9.** −2 **10.** 2 **11.** −10 **12.** −40 **13.** 7
14. 15 **15.** −8 **16.** −33 **17.** 9 **18.** −11 **19.** 13 **20.** 9
21. 3

L.O.3.

22. 0 **23.** 0 **24.** 0 **25.** 0 **26.** 0

L.O.4.

27. 9 **28.** −3 **29.** −7 **30.** 18 **31.** −8 **32.** −28

Self-Study Exercises 2-3 (p. 71)

L.O.1.

1. −12 **2.** 6 **3.** −6 **4.** 4 **5.** −25 **6.** −3 **7.** 8
8. −3 **9.** −9 **10.** 13 **11.** −1 **12.** −8

L.O.2.

13. −4 **14.** 1 **15.** 14 **16.** −3 **17.** −9 **18.** −5 **19.** 13
20. 2

L.O.3.

21. 15 **22.** −8 **23.** −12 **24.** 8 **25.** 7 **26.** 10

Self-Study Exercises 2-4 (p. 74)

L.O.1.

1. 40 **2.** 12 **3.** 35 **4.** 21 **5.** 24 **6.** 6

L.O.2.

7. −15 **8.** −10 **9.** −32 **10.** −12 **11.** −56 **12.** −24

L.O.3.

13. 0 **14.** 0 **15.** 0 **16.** 0 **17.** 0 **18.** 0

L.O.4.

19. −60 **20.** 36 **21.** 0 **22.** 0 **23.** −42 **24.** 54

L.O.5.

25. 9 **26.** −8 **27.** 25 **28.** 1 **29.** 625 **30.** 81

Self-Study Exercises 2-5 (p. 76)

L.O.1.

1. 2 **2.** −3 **3.** −4 **4.** −4 **5.** −4 **6.** 8 **7.** 5 **8.** −6
9. 8 **10.** 5

L.O.2.

11. not defined **12.** not defined **13.** 0 **14.** 0 **15.** 0
16. not defined **17.** not defined

Self-Study Exercises 2-6 (p. 80)

L.O.1.

1. -4 **2.** -30 **3.** -2 **4.** 17 **5.** -2 **6.** 24 **7.** 8 **8.** 24
9. -5 **10.** -7 **11.** -3 **12.** 14

L.O.2.

13. -12 **14.** -1260 **15.** 27 **16.** -14 **17.** 28 **18.** -35
19. -29 **20.** -132

Assignment Exercises, Chapter 2 (pp. 81–82)

1. G **3.** F **5.** E **7.** A **9.** L **11.** M **13.** $>$ **15.** $>$
17. $<$ **19.** 5 **21.** 3 **23.** 8 **25.** 12 **27.** -15 **29.** -11
31. 5 **33.** -4 **35.** 0 **37.** 7 **39.** 3 **41.** -5 **43.** 2 **45.** 8
47. -3 **49.** 40 **51.** -14 **53.** -12 **55.** 0 **57.** 0
59. -168 **61.** 343 **63.** 2 **65.** 2 **67.** -4 **69.** 0 **71.** 56
73. 1 **75.** 36 **77.** 73 **79.** -2 **81.** 391,464
83.

9:00 A.M.:	−1°C	10:00 A.M.:	0°C
11:00 A.M.:	0°C	12:00 P.M.:	1°C
1:00 P.M.:	1°C	2:00 P.M.:	4°C
3:00 P.M.:	0°C	4:00 P.M.:	−7°C
5:00 P.M.:	−15°C	6:00 P.M.:	−27°C

Trial Test, Chapter 2 (p. 85)

1. $<$ **3.** $<$ **5.** -8 **7.** 4 **9.** -48 **11.** 10 **13.** 3 **15.** 8
17. -10 **19.** 0 **21.** -7 **23.** not defined **25.** -3 **27.** -2
29. 1 **31.** 33 **33.** 22 **35.** -5

Self-Study Exercises 3-1 (pp. 90–91)

L.O.1.

1. $\frac{1}{3}$ **2.** $\frac{2}{5}$ **3.** $\frac{1}{2}$ **4.** $\frac{5}{6}$ **5.** $\frac{3}{4}$ **6.** $\frac{3}{8}$ **7.** $\frac{7}{12}$ **8.** $\frac{1}{16}$ **9.** $\frac{3}{3}$
10. $\frac{4}{6}$ or $\frac{2}{3}$ **11.** (a) 6 (b) 5 (c) 5 (d) 6 (e) 5, 6 (f) 6 (g) 5 (h) proper (i) 5, 6 (j) less than 1 **12.** (a) 8 (b) 8 (c) 8 (d) 8 (e) 8, 8 (f) 8 (g) 8 (h) improper (i) 8, 8 (j) equal to 1 **13.** (a) 5 (b) 11 (c) 11 (d) 5 (e) 11, 5 (f) 5 (g) 11 (h) improper (i) 11, 5 (j) more than 1 **14.** (a) 3 (b) 12 (c) 12 (d) 3 (e) 12, 3 (f) 3 (g) 12 (h) improper (i) 12, 3 (j) more than 1
15. b **16.** f **17.** d **18.** a **19.** c **20.** c **21.** a **22.** b
23. d **24.** e **25.** c **26.** $6\frac{27}{100}$ **27.** $\frac{3}{10}$ **28.** $\frac{689}{1000}$ **29.** $3\frac{7}{10}$
30. $1\frac{53}{100}$

Self-Study Exercises 3-2 (pp. 95–96)

L.O.1.

1. $\frac{4}{5}, \frac{8}{10}, \frac{12}{15}, \frac{16}{20}, \frac{20}{25}, \frac{24}{30}$ **2.** $\frac{7}{10}, \frac{14}{20}, \frac{21}{30}, \frac{28}{40}, \frac{35}{50}, \frac{42}{60}$ **3.** $\frac{3}{4} = \frac{18}{24}$ **4.** 6
5. 12 **6.** 36 **7.** 5 **8.** 20 **9.** 21 **10.** 12

L.O.2.

11. $\frac{1}{2}$ **12.** $\frac{3}{5}$ **13.** $\frac{3}{4}$ **14.** $\frac{5}{16}$ **15.** $\frac{1}{2}$ **16.** $\frac{7}{8}$ **17.** $\frac{5}{16}$ **18.** $\frac{1}{4}$
19. $\frac{1}{4}$ **20.** $\frac{6}{25}$ **21.** $\frac{5}{8}$ **22.** $\frac{1}{4}$ **23.** $\frac{3}{4}$ **24.** $\frac{3}{16}$ **25.** $\frac{7}{32}$

Self-Study Exercises 3-3 (pp. 98–99)

L.O.1.

1. $2\frac{2}{5}$ **2.** $1\frac{3}{7}$ **3.** 1 **4.** $4\frac{4}{7}$ **5.** 4 **6.** $2\frac{1}{7}$ **7.** $2\frac{5}{9}$ **8.** $9\frac{2}{5}$
9. $9\frac{5}{9}$ **10.** $1\frac{17}{21}$ **11.** $3\frac{4}{5}$ **12.** 16 **13.** $7\frac{1}{5}$ **14.** $9\frac{1}{2}$ **15.** 9

L.O.2.

16. $\frac{7}{3}$ **17.** $\frac{25}{8}$ **18.** $\frac{15}{8}$ **19.** $\frac{77}{12}$ **20.** $\frac{9}{1}$ **21.** $\frac{31}{8}$ **22.** $\frac{89}{12}$
23. $\frac{103}{16}$ **24.** $\frac{257}{32}$ **25.** $\frac{69}{64}$ **26.** $\frac{73}{10}$ **27.** $\frac{26}{3}$ **28.** $\frac{100}{3}$ **29.** $\frac{200}{3}$
30. $\frac{25}{2}$ **31.** 15 **32.** 18 **33.** 56 **34.** 32 **35.** 48

Self-Study Exercises 3-4 (pp. 101–102)

L.O.1.

1. 72 **2.** 30 **3.** 50 **4.** 48 **5.** 24

L.O.2.

6. $\frac{2}{3}$ **7.** $\frac{7}{16}$ **8.** $\frac{8}{9}$ **9.** $\frac{11}{16}$ **10.** $\frac{15}{32}$ **11.** $\frac{7}{12}$ **12.** $\frac{4}{5}$ **13.** $\frac{9}{10}$
14. $\frac{4}{15}$ **15.** $\frac{1}{2}$ **16.** No, $\frac{15}{64}$ is greater **17.** Yes, $\frac{3}{8}$ is greater
18. Yes, $\frac{7}{16}$ is greater **19.** Yes **20.** $\frac{5}{8}$ **21.** No **22.** No **23.** Yes
24. No **25.** too small

Self-Study Exercises 3-5 (pp. 107–108)

L.O.1.

1. $\frac{3}{8}$ **2.** $1\frac{3}{8}$ **3.** $\frac{5}{8}$ **4.** $\frac{25}{32}$ **5.** $\frac{9}{16}$ **6.** $1\frac{7}{16}$ **7.** $\frac{11}{64}$ **8.** $1\frac{19}{40}$
9. $1\frac{23}{36}$ **10.** $1\frac{1}{12}$ **11.** $\frac{15}{16}$ in. **12.** $1\frac{27}{32}$ in. **13.** $1\frac{15}{16}$ in.
14. $1\frac{7}{8}$ in. **15.** $1\frac{1}{4}$ in.

L.O.2.

16. $6\frac{4}{5}$ **17.** $4\frac{1}{8}$ **18.** $16\frac{13}{16}$ **19.** $1\frac{11}{18}$ **20.** $4\frac{13}{16}$ **21.** $5\frac{17}{32}$
22. $4\frac{11}{16}$ **23.** $13\frac{1}{2}$ **24.** $4\frac{7}{15}$ **25.** $12\frac{11}{16}$ **26.** $8\frac{13}{16}$ in. **27.** $3\frac{15}{16}$ in.
28. $11\frac{5}{8}$ gal **29.** $24\frac{5}{32}$ in. **30.** $22\frac{7}{8}$ in.

Self-Study Exercises 3-6 (p. 113)

L.O.1.

1. $\frac{1}{4}$ **2.** $\frac{3}{16}$ **3.** $\frac{1}{16}$ **4.** $\frac{1}{8}$ **5.** $\frac{9}{64}$ **6.** $\frac{1}{8}$

L.O.2.

7. $4\frac{11}{16}$ **8.** $17\frac{3}{4}$ **9.** $4\frac{15}{16}$ **10.** $5\frac{21}{32}$ **11.** $8\frac{13}{16}$ in. **12.** $10\frac{1}{8}$ in. **13.** $3\frac{1}{10}$ lb **14.** $\frac{3}{16}$ in. **15.** $\frac{29}{32}$

Self-Study Exercises 3-7 (p. 118)

L.O.1.

1. $\frac{3}{32}$ **2.** $\frac{7}{32}$ **3.** $\frac{7}{16}$ **4.** $\frac{7}{12}$ **5.** $\frac{1}{3}$ **6.** $\frac{5}{32}$ **7.** $\frac{1}{15}$ **8.** $\frac{11}{25}$ **9.** $\frac{2}{25}$ **10.** $\frac{21}{32}$

L.O.2.

11. $21\frac{7}{8}$ **12.** 75 **13.** $4\frac{1}{8}$ **14.** $36\frac{1}{10}$ **15.** $1\frac{21}{40}$ **16.** $2\frac{1}{6}$ **17.** $18\frac{3}{4}$ L **18.** 90 in. **19.** $15\frac{5}{8}$ in. **20.** 264 kg copper, 84 kg tin, 36 kg zinc

Self-Study Exercises 3-8 (pp. 123–124)

L.O.1.

1. $\frac{7}{3}$ **2.** $\frac{1}{8}$ **3.** $\frac{5}{11}$ **4.** $\frac{5}{1}$ **5.** $\frac{1}{7}$

L.O.2.

6. $\frac{9}{10}$ **7.** $\frac{11}{12}$ **8.** 2 **9.** $1\frac{1}{20}$ **10.** $2\frac{1}{12}$ **11.** $4\frac{2}{3}$ **12.** $\frac{2}{5}$ **13.** 8

L.O.3.

14. $13\frac{1}{3}$ **15.** 32 **16.** $\frac{5}{8}$ **17.** $14\frac{3}{4}$ **18.** 7 **19.** 18 2 × 4's

20. 10 lengths **21.** 12 shovels **22.** $23\frac{11}{12}$ in. **23.** 20 ft × 15 ft

24. 4 whole pieces **25.** $7\frac{25}{47}$ in. **26.** $23\frac{15}{37}$ in. **27.** 8 pieces

Self-Study Exercises 3-9 (p. 126)

L.O.1.

1. $\frac{1}{10}$ **2.** 3 **3.** $5\frac{3}{5}$ **4.** $\frac{12}{25}$ **5.** $\frac{1}{3}$ **6.** 5 **7.** 28 **8.** $\frac{2}{9}$ **9.** 3 **10.** $\frac{1}{3}$

Self-Study Exercises 3-10 (p. 128)

L.O.1.

1. $-\frac{-5}{8}, -\frac{5}{-8}, \frac{-5}{-8}$ **2.** $-\frac{-3}{-4}, \frac{-3}{4}, \frac{3}{-4}$ **3.** $-\frac{2}{-5}, -\frac{-2}{5}, \frac{2}{5}$

4. $-\frac{7}{8}, \frac{-7}{8}, \frac{7}{-8}$ **5.** $-\frac{-7}{8}, -\frac{7}{-8}, \frac{-7}{-8}$

L.O.2.

6. $-\frac{2}{8}$ or $-\frac{1}{4}$ **7.** $-1\frac{1}{10}$ **8.** $1\frac{1}{10}$ **9.** $\frac{6}{11}$ **10.** $-\frac{25}{32}$

Self-Study Exercises 3-11 (p. 130)

L.O.1.

1. $1\frac{5}{24}$ **2.** $-2\frac{16}{35}$ **3.** $14\frac{59}{96}$ **4.** $1\frac{3}{4}$ **5.** $\frac{4}{5}$ **6.** $\frac{1}{2}$ **7.** $-\frac{1}{24}$

8. $282\frac{1}{10}$

Assignment Exercises, Chapter 3 (pp. 131–137)

1. $\frac{3}{8}$ **3.** $\frac{1}{3}$ **5.** $\frac{13}{24}$ **7.** (a) 9 (b) 4 (c) 4 (d) 9 (e) 9, 4 (f) 4 (g) 9 (h) improper (i) greater than 1 (j) 9, 4 **9.** f **11.** d **13.** e **15.** d

17. $\frac{87}{100}$ **19.** $2\frac{3}{100}$ **21.** 15 **23.** 25 **25.** 10 **27.** 24 **29.** 11

31. $\frac{1}{2}$ **33.** $\frac{1}{8}$ **35.** $\frac{1}{4}$ **37.** $\frac{17}{32}$ **39.** $\frac{3}{8}$ **41.** $\frac{3}{4}$ **43.** $4\frac{1}{2}$

45. $3\frac{3}{5}$ **47.** $4\frac{7}{8}$ **49.** $5\frac{3}{8}$ **51.** $87\frac{1}{2}$ **53.** $\frac{8}{1}$ **55.** $\frac{57}{8}$ **57.** $\frac{147}{16}$

59. $\frac{23}{5}$ **61.** $\frac{12}{1}$ **63.** 20 **65.** 33 **67.** $\frac{3}{8}$ **69.** $\frac{3}{16}$ **71.** $\frac{27}{32}$

73. $\frac{9}{19}$ **75.** 60 **77.** 16 **79.** 60 **81.** larger **83.** no

85. smaller **87.** $\frac{21}{64}$ **89.** $1\frac{13}{30}$ **91.** $8\frac{19}{32}$ **93.** $10\frac{9}{32}$ **95.** $18\frac{1}{16}$ in.

97. $12\frac{19}{32}$ in. **99.** $15\frac{25}{32}$ in. **101.** $4\frac{3}{4}$ in. **103.** $\frac{7}{8}$ in. **105.** $\frac{1}{3}$

107. $1\frac{5}{8}$ **109.** $5\frac{31}{32}$ **111.** $7\frac{11}{16}$ **113.** $\frac{7}{16}$ in. **115.** $1\frac{29}{64}$ in.

117. $\frac{7}{24}$ **119.** $\frac{7}{24}$ **121.** $\frac{1}{2}$ **123.** $7\frac{7}{8}$ **125.** $1\frac{1}{5}$ **127.** $100\frac{1}{2}$ in.

129. $2\frac{3}{4}$ cups **131.** $1\frac{1}{6}$ **133.** $9\frac{1}{3}$ **135.** 24 **137.** 2 **139.** $\frac{3}{5}$

141. $2\frac{55}{64}$ in. **143.** $\frac{3}{16}$ yd. **145.** $\frac{1}{18}$ **147.** $5\frac{1}{3}$ **149.** $\frac{1}{4}$ **151.** $\frac{1}{8}$

153. $-\frac{-3}{-8}, +\frac{-3}{8}, \frac{3}{-8}$ **155.** $\frac{7}{8}, -\frac{-7}{8}, -\frac{7}{-8}$ **157.** $\frac{62}{63}$ **159.** $1\frac{5}{7}$

161. $-\frac{64}{305}$

Trial Test, Chapter 3 (pp. 142–143)

1. $\frac{3}{4}$ **3.** 3 **5.** $\frac{34}{7}$ **7.** $\frac{1}{4}$ **9.** $3\frac{8}{9}$ **11.** $\frac{1}{2}$ **13.** $13\frac{1}{2}$ **15.** $1\frac{5}{12}$ **17.** $7\frac{13}{14}$ **19.** $\frac{1}{9}$ **21.** $\frac{1}{6}$ **23.** $\frac{5}{16}$ **25.** $\frac{2}{7}$ **27.** $3\frac{5}{6}$ cups

Self-Study Exercises 4-1 (p. 151)

1. thousandths **2.** ones **3.** hundred-thousandths **4.** millionths **5.** ten-thousandths **6.** 4 **7.** 0 **8.** 6 **9.** 3 **10.** 7 **11.** twenty-one and three hundred eighty-seven thousandths **12.** four hundred twenty and fifty-nine thousandths **13.** eighty-nine hundredths **14.** five hundred sixty-eight ten-thousandths **15.** thirty and two thousand three hundred seventy-nine hundred thousandths **16.** twenty-one and two hundred five thousand eighty-five millionths **17.** 3.42 **18.** 78.195 **19.** 500.0005 **20.** 0.75034 **21.** 0.5 **22.** 0.23 **23.** 0.07 **24.** 6.83 **25.** 0.079 **26.** 0.468 **27.** 5.87 **28.** 0.108 **29.** 6.03 **30.** 4.00 **31.** 3.72 **32.** 7.08 **33.** 0.3 **34.** 0.56 **35.** 2.75 **36.** 0.2 **37.** 8.88 **38.** 0.25 **39.** 0.913 **40.** 0.76 **41.** 5.983 **42.** 1.972 **43.** 0.179, 0.23, 0.314 **44.** 1.87, 1.9, 1.92 **45.** 72.07, 72.1, 73 **46.** 0.837 in. **47.** the reading **48.** yes **49.** 0.04 in. **50.** No. 10 wire

Self-Study Exercises 4-2 (pp. 155–156)

1. 500 **2.** 6200 **3.** 8000 **4.** 430,000 **5.** 40,000 **6.** 40,000 **7.** 285,000,000 **8.** 470 **9.** 83,000,000,000 **10.** 300,000,000 **11.** 43 **12.** 367 **13.** 8 **14.** 103 **15.** 3 **16.** 8.1 **17.** 12.9 **18.** 42.6 **19.** 83.2 **20.** 6.0 **21.** 7.04 **22.** 42.07 **23.** 0.79 **24.** 3.20 **25.** 7.77 **26.** 0.217 **27.** 0.020 **28.** 1.509 **29.** 4.238 **30.** 7.004 **31.** \$219 **32.** \$83 **33.** \$507 **34.** \$3 **35.** \$6 **36.** \$8.24 **37.** \$0.29 **38.** \$0.53 **39.** \$5.80 **40.** \$238.92 **41.** 0.784 **42.** 3.82 **43.** 2.8 **44.** 500 **45.** 8 **46.** 60 **47.** 0.5 **48.** 0.009 **49.** 0.1 **50.** 3 **51.** 50 **52.** 80 **53.** 50 **54.** \$3 **55.** 20 **56.** 0.4, 2, 0.5, 1, 0.07, 2 **57.** 0.03, 0.03, 0.02, 0.03, 0.04

Self-Study Exercises 4-3 (p. 159)

1. 15.7 **2.** 34.18 **3.** 8.13 **4.** 87.4 **5.** 129.97 **6.** 26.03 **7.** 78.2 **8.** 45.7 **9.** 261.335 **10.** 356.612 **11.** 6.984 **12.** 9.525 **13.** 1126.6 **14.** 413.6 **15.** 18.1 **16.** 18.8 **17.** 0.81805 **18.** 1.17642 **19.** 55.513 **20.** 85.411 **21.** 3.077 in. **22.** 15.503 A **23.** \$181.25 **24.** 391.4 ft **25.** \$19.93 **26.** \$5.04 **27.** 6.93 **28.** 15.834 **29.** 803.693 **30.** 56.14 **31.** 4.094 **32.** 3.9 **33.** 291.82 **34.** 7.5 **35.** 310.8 **36.** 4.4 **37.** 15.1 lb **38.** 12.08 in., 12.10 in. **39.** 59.83 cm **40.** 4.189 in., 4.201 in. **41.** 0.22 dm **42.** 11.55 A **43.** \$8.75 **44.** \$2.25 **45.** \$4.75

Self-Study Exercises 4-4 (pp. 162–163)

1. 15.486 **2.** 56.55 **3.** 3.2445 **4.** 0.05805 **5.** 0.08672 **6.** 7.141 **7.** 0.0834 **8.** 170.12283 **9.** 0.38381871 **10.** 4.9386274 **11.** 30.66 **12.** 596.97 **13.** 50.7357 **14.** 38.6232 **15.** 339.04 **16.** 540.27 **17.** 254 **18.** 184.2 **19.** 0.9307 **20.** 0.7602 **21.** 0.58635 **22.** 0.73265 **23.** 9.21702 **24.** 11.42356 **25.** 0.0176 **26.** 0.027045 **27.** 0.915371 **28.** 0.390483 **29.** 0.00015

30. 0.00056 **31.** 3.957 in. **32.** 5.25 in. **33.** \$27.48 **34.** 151.2 in. **35.** \$64.20 **36.** \$10,728 **37.** 0.375 in. **38.** \$1960.50 **39.** 15 A **40.** \$78 **41.** 423 **42.** 783.6 **43.** 480 **44.** 5236 **45.** 486.7 **46.** 290 **47.** 42,385 **48.** 600 **49.** 29,730 **50.** 42,300 **51.** 8346 **52.** 872,300 **53.** 5930 **54.** 6872.5 **55.** 42.35 **56.** \$697 **57.** \$550 **58.** \$12,700 **59.** 350 watts **60.** yes

Self-Study Exercises 4-5 (pp. 167–168)

1. 1.26 **2.** 3.09 **3.** 0.063 **4.** 285 **5.** 5.9 **6.** 45 **7.** 10.9 **8.** 0.19 **9.** 1.06 **10.** 0.33 **11.** 25 **12.** 20,700 **13.** 90,200 **14.** 10,700 **15.** 1.8 **16.** 6 lb **17.** 23 rolls **18.** 600 revolutions **19.** 0.7 ft **20.** 93.75 volts **21.** $19.84\frac{8}{13}$ **22.** $1.82\frac{1}{12}$ **23.** $2601.16\frac{2}{3}$ **24.** $1.98\frac{4}{7}$ **25.** $0.07\frac{27}{29}$ **26.** 169.3 **27.** 0.16 **28.** 9 **29.** \$0.96 **30.** \$1 **31.** \$575 **32.** \$1.98 **33.** 0.0812 in. **34.** 0.2 in. **35.** 0.319 **36.** 0.4027 **37.** 0.5 **38.** 0.04917 **39.** 0.02081 **40.** 0.000921 **41.** 0.00835 **42.** 0.0002003 **43.** 0.00004 **44.** 0.008 **45.** 0.4592 **46.** 2.396 **47.** 0.0084 **48.** 0.00394 **49.** 0.0000000005 **50.** \$0.126 or \$0.13 (rounded) **51.** 0.02175 lb **52.** \$0.258 or \$0.26 (rounded) **53.** 11.458 ohms **54.** 0.75 in.

Self-Study Exercises 4-6 (pp. 171–172)

1. 80.4° **2.** 85 **3.** 454 bales **4.** \$765.32 **5.** 1.69 in. **6.** 3.5 A **7.** 187 **8.** \$14 **9.** \$500 **10.** 459 **11.** 290 **12.** \$6 **13.** \$600 **14.** 36,000 **15.** 300 **16.** 0.8 **17.** 20 **18.** 5 or between 5 and 6 **19.** 10 **20.** 70 **21.** 20 or between 20 and 30 **22.** \$220 **23.** 174,000 bricks **24.** 150 ft **25.** 360,000 lb **26.** 30 switches

Self-Study Exercises 4-7 (pp. 175–176)

1. $\frac{1}{2}$ **2.** $\frac{1}{10}$ **3.** $\frac{1}{5}$ **4.** $\frac{7}{10}$ **5.** $\frac{1}{4}$ **6.** $\frac{1}{40}$ **7.** $3\frac{9}{10}$ **8.** $4\frac{4}{5}$ **9.** $\frac{2}{3}$ **10.** $\frac{7}{8}$ **11.** $\frac{3}{8}$ **12.** $\frac{1}{8}$ **13.** $\frac{1}{6}$ **14.** $\frac{3}{8}$ **15.** $\frac{5}{8}$ **16.** $\frac{3}{4}$ **17.** $\frac{3}{16}$ **18.** $2\frac{3}{8}$ **19.** $\frac{5}{6}$ **20.** $\frac{5}{16}$ **21.** $3\frac{1}{8}$ **22.** 0.4 **23.** 0.3 **24.** 0.875 **25.** 0.625 **26.** 0.45 **27.** 0.98 **28.** 0.21 **29.** 3.875 **30.** 1.4375 **31.** 4.5625 **32.** $0.\overline{6}$ or 0.66 . . . **33.** $0.\overline{27}$ or 0.2727 . . . **34.** $0.\overline{7}$ or 0.77 . . . **35.** $0.83\frac{1}{3}$ **36.** $0.58\frac{1}{3}$ **37.** 2.046875 in. **38.** 1.03125 in. **39.** 4.5% **40.** 3.75% **41.** 0.125 in. **42.** 0.4375 in.

Self-Study Exercises 4-8 (p. 177)

1. −26.297 **2.** −1.11 **3.** −91.44 **4.** −110.72 **5.** −59.04 **6.** 340.71 **7.** $-27.7\overline{3}$ **8.** 0.413 **9.** −26.6

Assignment Exercises, Chapter 4 (pp. 178–181)

1. 0.3 **3.** 0.04 **5.** 2.1 **7.** thousandths **9.** 7 **11.** six and eight hundred three thousandths **13.** 0.625 **15.** 4.79 **17.** 0.02, 0.021, 0.0216 **19.** $\frac{7}{8}$ **21.** 320 **23.** 7000 **25.** 500 **27.** 50,000

29. 27,000,000,000 **31.** 41.4 **33.** 6.90 **35.** 23.4610 **37.** 24
39. \$43 **41.** 80 **43.** 10 **45.** 0.07 **47.** \$3 **49.** 1.4
51. 68.21 **53.** 90.247 **55.** 171.22 **57.** 3.127 **59.** 350.313
61. 77.00 **63.** 48 **65.** 389 **67.** 8.291, 8.301 in. **69.** 0.43 in.
71. 2.888 in. **73.** 8.93, 8.94 in. **75.** 2.1 in. **77.** 2.09
79. 0.02133 **81.** 362.1 **83.** 0.002204 **85.** 23.3828 **87.** 4082
89. 85,000 **91.** \$850 **93.** 0.12096 in. **95.** \$250.04
97. \$5735.52 **99.** 14.6700072 lb/in.^{2} **101.** 0.06 **103.** 23
105. 12 **107.** 36.26 **109.** \$2 **111.** 48.23 **113.** 5.938
115. \$8.97 **117.** 0.575 gal **119.** 150 **121.** 38 **123.** 360
125. 10 or between 10 and 20 **127.** 51.2 **129.** 48.79 ft **131.** \$72.25
133. \$12 **135.** $\frac{2}{5}$ **137.** $\frac{3}{4}$ **139.** $\frac{5}{6}$ **141.** $\frac{7}{8}$ **143.** 0.875
145. 5.25 **147.** 0.73 **149.** $0.\overline{3}$ or 0.33 . . . **151.** $0.\overline{5}$ or 0.55 . . .
153. $0.41\frac{2}{3}$ **155.** 0.015625 **157.** 0.046875 **159.** 0.078125
161. 0.109375 **163.** $\frac{11}{16}$ **165.** $\frac{7}{16}$ **167.** $\frac{3}{8}$ **169.** $\frac{5}{8}$
171. 0.875 in. **173.** $\frac{11}{32}$ in. **175.** -3.5 **177.** 53.5
179. (a) 0.8 (b) 0.3 (c) 0.03 (d) 0.25 (e) 0.004 (f) 0.6325 (rounded)

Trial Test, Chapter 4 (pp. 184–185)

1. 7.027 **3.** 7.2 **5.** 2.88 **7.** \$4.83 **9.** \$22 **11.** 4.02
13. 14.96 **15.** 3.04 **17.** 9.5 **19.** 1.8331 **21.** 0.204 **23.** 42,730
25. 1.11 **27.** 1410 **29.** 11.6 **31.** 0.005238 **33.** 83 **35.** 11
37. 0.6 **39.** 1200 **41.** 70 or between 70 and 80 **43.** 0.8 **45.** 0.71
47. $\frac{7}{40}$ **49.** $\frac{1}{3}$ **51.** \$17,500 **53.** 1.1 **55.** -10.81

Self-Study Exercises 5-1 (pp. 188–189)

L.O.1.

1. 40% **2.** 70% **3.** $62\frac{1}{2}\%$ **4.** $77\frac{7}{9}\%$ **5.** $\frac{7}{10}\%$ **6.** $\frac{2}{7}\%$
7. 20% **8.** 14% **9.** 0.7% **10.** 1.25% **11.** 500% **12.** 800%
13. $133\frac{1}{3}\%$ **14.** 350% **15.** 430% **16.** 220% **17.** 305%
18. 720% **19.** 1510% **20.** 3625%

Self-Study Exercises 5-2 (p. 194)

L.O.1.

1. $\frac{9}{25}$, 0.36 **2.** $\frac{9}{20}$, 0.45 **3.** $\frac{1}{5}$, 0.20 **4.** $\frac{3}{4}$, 0.75 **5.** $\frac{1}{16}$, 0.0625
6. $\frac{5}{8}$, 0.625 **7.** $\frac{2}{3}$, $0.66\frac{2}{3}$ **8.** $\frac{3}{500}$, 0.006 **9.** $\frac{1}{500}$, 0.002
10. $\frac{1}{2000}$, 0.0005 **11.** $\frac{1}{12}$, $0.08\frac{1}{3}$ **12.** $\frac{3}{16}$, 0.1875

L.O.2.

13. 8 **14.** 4 **15.** $2\frac{1}{5}$, 2.5 **16.** $4\frac{1}{4}$, 4.25 **17.** $1\frac{19}{25}$, 1.76

18. $3\frac{4}{5}$, 3.8 **19.** $1\frac{3}{8}$, 1.375 **20.** $3\frac{7}{8}$, 3.875 **21.** $1\frac{2}{3}$ **22.** $3\frac{1}{6}$
23. 1.153 **24.** 2.125 **25.** 1.0625

Self-Study Exercises 5-3 (p. 196)

L.O.1.

1. $\frac{1}{10}$, 0.1 **2.** 25%, 0.25 **3.** 20%, $\frac{1}{5}$ **4.** $33\frac{1}{3}$%, $0.33\frac{1}{3}$ **5.** $\frac{1}{2}$, 0.5

6. 80%, 0.8 **7.** 75%, $\frac{3}{4}$ **8.** $\frac{2}{3}$, $0.66\frac{2}{3}$ **9.** 100%, $\frac{1}{1}$ **10.** 30%, 0.3

11. $\frac{2}{5}$, 0.4 **12.** 70%, $\frac{7}{10}$ **13.** 90%, 0.9 **14.** $\frac{3}{5}$, 0.6 **15.** 100%, 1

16. 25%, $\frac{1}{4}$ **17.** $66\frac{2}{3}$%, $0.66\frac{2}{3}$ **18.** $\frac{7}{10}$, 0.7 **19.** 50%, $\frac{1}{2}$

20. 60%, 0.6 **21.** 10%, $\frac{1}{10}$ **22.** $\frac{1}{5}$, 0.2 **23.** $33\frac{1}{3}$%, $\frac{1}{3}$ **24.** $66\frac{2}{3}$%, $\frac{2}{3}$

25. 75%, 0.75 **26.** $\frac{3}{10}$, 0.3 **27.** 90%, $\frac{9}{10}$ **28.** 20%, 0.2

29. 60%, $\frac{3}{5}$ **30.** $\frac{1}{1}$, 1

Self-Study Exercises 5-4 (pp. 209–210)

L.O.1.

1. 3.75 **2.** 0.625 **3.** $66\frac{2}{3}$% **4.** $37\frac{1}{2}$% **5.** 14.25 **6.** 350
7. 9.375 **8.** 220 **9.** 20% **10.** 200

L.O.2.

11. 75 **12.** 63 **13.** 206 **14.** 115.92 **15.** 0.675 **16.** 0.94
17. 155.25 **18.** 231 **19.** 924 **20.** 345

L.O.3.

21. 25% **22.** 30% **23.** $33\frac{1}{3}$% **24.** 32.9% **25.** 15.75%

26. 16% **27.** 0.8% **28.** $0.66\frac{2}{3}$% **29.** 20% **30.** 111.25%

L.O.4.

31. 72 **32.** 50 **33.** 344 **34.** 46 **35.** 360 **36.** 275 **37.** 75
38. 250 **39.** 18.4 **40.** 261

L.O.5.

41. 1.0625 lb **42.** 3% **43.** 200 hp **44.** 0.021 lb **45.** 5%
46. $575 **47.** 1200 lb **48.** 11% **49.** 9,625 parts
50. 100,880 welds

Self-Study Exercises 5-5 (pp. 215–216)

L.O.1.

1. 108 **2.** 109.2 **3.** 10.2 **4.** 33.75

L.O.2.

5. 27 in. **6.** 1815 board feet **7.** 2393 board feet **8.** 25,908 bricks
9. $16,802.50 **10.** 1366.4 yd^3

L.O.3.

11. 7% **12.** 45% **13.** 20% **14.** 5% **15.** 350 hp **16.** 10.5 yd^3 **17.** 3019 board feet **18.** 28.5 lb **19.** 40 in. **20.** 20%

Self-Study Exercises 5-6 (pp. 222–223)

L.O.1.

1. \$4.55, \$80.38 **2.** \$829.06 **3.** \$23.30 **4.** 25% **5.** 23%

L.O.2.

6. \$17.52 **7.** \$3,866.00 **8.** \$944.50 **9.** \$434.84 **10.** \$3205.75

L.O.3.

11. \$9.75 **12.** \$300 **13.** 1.75% **14.** \$367.50 **15.** \$1752

Assignment Exercises, Chapter 5 (pp. 225–230)

1. $33\frac{1}{3}\%$ **3.** 75% **5.** 200% **7.** 6% **9.** 370% **11.** 0.04%

13. $\frac{2}{15}\%$ **15.** 1000% **17.** $12\frac{1}{2}\%$ **19.** $45\frac{5}{11}\%$ **21.** 60%

23. $16\frac{2}{3}\%$ **25.** 45% **27.** 60% **29.** 30% **31.** 0.5% **33.** 80%

35. $\frac{13}{20}$, 0.65 **37.** $\frac{1}{4}$, 0.25 **39.** $\frac{3}{8}$, 0.375 **41.** $\frac{1}{500}$, 0.002

43. $\frac{1}{1250}$, 0.0008 **45.** $\frac{5}{16}$, 0.3125 **47.** 2 **49.** $4\frac{1}{2}$, 4.5

51. $2\frac{3}{5}$, 2.6 **53.** $4\frac{5}{8}$, 4.625 **55.** $3\frac{1}{3}$ **57.** 3.1875 **59.** 0.215

61. 0.0775 **63.** 0.19525 **65.** 1.25 **67.** $1\frac{2}{5}$ **69.** 20%, 0.2

71. 40%, $\frac{2}{5}$ **73.** 60%, 0.6 **75.** 10%, 0.1 **77.** 100%, 1

79. $\frac{3}{10}$, 0.3 **81.** 90%, 0.9 **83.** $\frac{1}{8}$, 0.125 **85.** $\frac{1}{20}$, 0.05 **87.** $\frac{5}{8}$, 0.625 **89.** 24 **91.** 0.4375 **93.** 60% **95.** 250% **97.** 83 **99.** 152 **101.** 15.3% **103.** 500% **105.** 84 **107.** 37.5% **109.** 1.14% **111.** \$266 **113.** 4% **115.** 200 students **117.** 10% **119.** 7% **121.** 14% **123.** 142.1 kg **125.** 62 cm, 63 cm

127. \$1.69 **129.** 12.5% **131.** 57.5 lb **133.** $33\frac{1}{3}\%$ **135.** 4.8%

137. 289 hp **139.** \$439.84 **141.** \$365.66 **143.** \$1,707.60 **145.** 1.75% **147.** \$11,019.75 **149.** \$0.89 **151.** 8.5% **153.** \$369.69 **155.** \$355 **157.** 26.6% **159.** 28.6%

Trial Test, Chapter 5 (pp. 236–237)

1. $43\frac{3}{4}\%$ **3.** 320% **5.** $\frac{1}{400}$ **7.** 0.005 **9.** 0.8 **11.** 9

13. 305 **15.** 115 **17.** 67.10 **19.** \$525 **21.** \$2500 **23.** 21.14% **25.** 13.17% **27.** 5% **29.** \$104.87

Self-Study Exercises 6-1 (pp. 249–250)

L.O.1.

1. lb **2.** oz **3.** lb or oz **4.** gal **5.** qt **6.** T **7.** lb **8.** qt **9.** ft and in. **10.** c **11.** in. **12.** yd **13.** mi **14.** oz **15.** oz **16.** mi **17.** lb or oz **18.** lb **19.** lb and oz **20.** oz

L.O.2.

21. 48 in. **22.** 21 ft **23.** 4400 yd **24.** $9\frac{1}{3}$ yd **25.** 2 mi

26. 48 oz **27.** $2\frac{3}{10}$ or 2.3 lb **28.** 732.8 oz **29.** 19.2 oz

30. 408 oz, 25.5 lb **31.** 20 qt **32.** 13 pt **33.** $1\frac{1}{2}$ gal **34.** 60 pt

35. 9 gal **36.** 256 oz **37.** $1\frac{1}{2}$ or 1.5 yd **38.** 108 in. **39.** 7040 ft

40. 24 c

L.O.3.

41. 3 ft 8 in. **42.** 2 mi 1095 ft **43.** 3 lb $3\frac{1}{2}$ oz **44.** 2 gal 1 qt

45. 2 gal **46.** 2 T 500 lb **47.** 2 yd 2 ft 11 in. **48.** 2 qt 2 c 2 oz **49.** 2 ft 10 in. **50.** 6 lb 9 oz **51.** 4 gal 2 qt 16 oz **52.** 6 qt 20 oz **53.** 2 mi 5 yd 2 ft 1 in. **54.** 5 ft 4 in. **55.** 3 T 600 lb 15 oz **56.** 4 lb 5 oz **57.** 2 lb 5 oz **58.** 2 yd 2 ft 4 in. **59.** 2 yd 2 ft 11 in. **60.** 1 mi 875 ft 6 in.

Self-Study Exercises 6-2 (pp. 252–253)

L.O.1.

1. 2 lb 5 oz **2.** 4 ft 7 in. **3.** 15 lb 11 oz **4.** 18 ft 3 in.

5. 8 qt $\frac{1}{2}$ pt **6.** 14 gal 1 qt **7.** 9 yd 1 ft **8.** 5 c 1 oz **9.** 5 ft 4 in.

10. 7 yd 1 ft 5 in. **11.** 2 ft 7 in. **12.** 11 ft **13.** 5 lb 15 oz **14.** 12 lb

L.O.2.

15. 6 in. **16.** 1 pt **17.** 2 ft **18.** 4 lb 14 oz **19.** 1 lb 6 oz **20.** 6 lb 11 oz **21.** 11 in. **22.** 10 lb 13 oz **23.** 2 in.

24. 3 gal 3 qt $1\frac{1}{2}$ pt **25.** 4 ft 2 in. **26.** 2 ft 3 in. **27.** 45 sec

28. 39 sec **29.** 69 lb 7 oz **30.** 16 lb 14 oz

Self-Study Exercises 6-3 (pp. 259–260)

L.O.1.

1. 60 mi **2.** 108 gal **3.** 252 lb **4.** 13 qt **5.** 57 lb 8 oz **6.** 38 gal 2 qt **7.** 43 gal 3 qt **8.** 58 ft **9.** 36 lb **10.** 7 qt 1 pt

L.O.2.

11. 35 in.^{2} **12.** 108 ft^2 **13.** 180 yd^2 **14.** 108 mi^2 **15.** 378 tiles **16.** 47 gal 2 oz **17.** 7 qt 1 pt **18.** 6 lb 4 oz **19.** 7 lb 8 oz

L.O.3.

20. 6 gal **21.** 1 day 15 hr **22.** 10 yd 1 ft 3 in. **23.** 1 yd 1 ft 7 in.

24. 3 qt $\frac{1}{2}$ pt **25.** $6\frac{2}{3}$ gal **26.** 1 hr 40 min **27.** $5\frac{1}{4}$ qt
28. 7 ft 6 in. **29.** 3 gal 2 qt **30.** 9 pieces **31.** 5 ft
32. 3 gal 1 qt 5 oz **33.** 24 lb 3 oz

L.O.4.

34. 3 **35.** 17 **36.** 8 **37.** 18 **38.** 3 **39.** 8 pieces
40. 9 boxes **41.** 24 cans **42.** 9 tickets **43.** 9 pieces

L.O.5.

44. $15\frac{\text{gal}}{\text{sec}}$ **45.** $\frac{3}{4}\frac{\text{lb}}{\text{min}}$ **46.** $15{,}840\frac{\text{ft}}{\text{hr}}$ **47.** $2{,}304\frac{\text{oz}}{\text{min}}$ **48.** $2\frac{\text{qt}}{\text{sec}}$
49. $20\frac{\text{oz}}{\text{hr}}$ **50.** $40\frac{\text{ft}}{\text{min}}$ **51.** $440\frac{\text{ft}}{\text{sec}}$ **52.** $44\frac{\text{ft}}{\text{sec}}$ **53.** $3\frac{\text{qt}}{\text{min}}$
54. $53\frac{1}{3}\frac{\text{lb}}{\text{min}}$ **55.** $\frac{5}{6}\frac{\text{gal}}{\text{sec}}$

Self-Study Exercises 6-4 (pp. 272–274)

L.O.1.

1. (a) 1000 meters (b) 10 liters (c) $\frac{1}{10}$ of a gram (d) $\frac{1}{1000}$ of a meter
(e) 100 grams (f) $\frac{1}{100}$ of a liter **2.** b **3.** b **4.** c **5.** b **6.** a
7. d **8.** b **9.** c **10.** a **11.** b **12.** c **13.** b **14.** a
15. b **16.** b

L.O.2.

17. 40 **18.** 70 **19.** 580 **20.** 80 **21.** 2.5 **22.** 210 **23.** 85
24. 142 **25.** 153 mL **26.** 460 m **27.** 75 dkg **28.** 160 mm
29. 400 **30.** 8,000 **31.** 58,000 **32.** 800 **33.** 250 **34.** 2,100
35. 102,500 **36.** 8,330 **37.** 2,000,000 **38.** 70 **39.** 236 L
40. 467 cm **41.** 38,000 dg **42.** 13,000 cm **43.** 2.8 **44.** 23.8
45. 10.1 **46.** 6 **47.** 2.9 **48.** 19.25 **49.** 1.7 **50.** 438.9 dm
51. 4.7 g **52.** 0.225 dL **53.** 2.743 **54.** 0.385 **55.** 0.15
56. 0.08 **57.** 2964.84 **58.** 0.2983 **59.** 0.0003 **60.** 0.004
61. 0.002857 **62.** 15.285 **63.** 0.0297 hm **64.** 0.00003 L

L.O.3.

65. 11 m **66.** 12 hL **67.** 6 cg **68.** 2.4 dm or 24 cm **69.** 5.9 cL or 59 mL **70.** cannot add **71.** 10.1 kL or 101 hL **72.** 0.55 g or 55 cg
73. cannot subtract **74.** 7.002 km or 7,002 m **75.** 45.5 m
76. 1.47 kL or 147 dkL **77.** 516 m **78.** 40.8 m **79.** 150.96 dm
80. 969.5 m **81.** 13 m **82.** 9 cL **83.** 163 g **84.** 0.4 m or 4 dm
85. 5 **86.** 30 **87.** 16 prescriptions **88.** 80 containers **89.** 5.74 dL or 57.4 cL **90.** 2.3 dkm or 23 m **91.** 165.7 hm or 16.57 km
92. 9.1 kL or 91 hL **93.** 1.25 cL **94.** 70 mm **95.** 7.5 dm
96. 0.8 m or 8 dm **97.** 21,250 containers **98.** 19 vials

Self-Study Exercises 6-5 (pp. 276–277)

L.O.1.

1. 354.33 in. **2.** 130.8 yd **3.** 26.04 mi **4.** 6.36 liq qt **5.** 11 L
6. 59.4 lb **7.** 22.5 kg **8.** 17.78 cm **9.** 5.4 m **10.** 1.32288 oz
11. 21.85 L **12.** 30 m **13.** 27 kg **14.** 241.5 km **15.** 32.7 yd
16. 235.6 mi **17.** 7.44 mi **18.** 13.2 L **19.** 328 ft

Self-Study Exercises 6-6 (pp. 281–282)

L.O.1.

1. $4\frac{9}{16}$ in. **2.** $4\frac{1}{16}$ in. **3.** $3\frac{13}{16}$ in. **4.** $3\frac{3}{8}$ in. **5.** $2\frac{1}{4}$ in. **6.** 2 in.
7. $1\frac{3}{4}$ in. **8.** $1\frac{3}{16}$ in. **9.** $\frac{3}{4}$ in. **10.** $\frac{3}{8}$ in.

L.O.2.

11. 115 mm or 11.5 cm **12.** 102 mm or 1.02 cm **13.** 96 mm or 9.6 cm
14. 85 mm or 8.5 cm **15.** 57 mm or 5.7 cm **16.** 50 mm or 5 cm
17. 44 mm or 4.4 cm **18.** 30 mm or 3 cm **19.** 19 mm or 1.9 cm
20. 10 mm or 1 cm

Assignment Exercises, Chapter 6 (pp. 282–286)

1. lb or oz **3.** qt **5.** lb or oz **7.** T **9.** qt **11.** c **13.** yd
15. oz **17.** 4 yd **19.** 6,336 ft **21.** 80 oz **23.** $42\frac{1}{2}$ lb
25. 304 oz or 19 lb **27.** 15 pt **29.** 24 pt **31.** 384,000 oz
33. 6,600 ft **35.** 7 ft 5 in. **37.** 13 lb $1\frac{1}{2}$ oz **39.** 2 gal 1 pt
41. 4 yd 4 in. **43.** 4 ft 1 in. **45.** 2 gal 16 oz **47.** 3 ft 11 in.
49. 10 ft 11 in. **51.** 8 gal 2 qt **53.** 14 oz **55.** 1 ft 9 in.
57. 1 ft 5 in. **59.** 504 ft^2 **61.** 63 $in.^2$ **63.** 10 yd 1 ft 3 in.
65. $5\frac{5}{12}$ ft **67.** 4 lb 8 oz **69.** 3.5 **71.** $4\frac{4}{9}$ **73.** $300\frac{mi}{hr}$
75. $60\frac{mi}{hr}$ **77.** $1.25\frac{gal}{min}$ **79.** $1\frac{1}{2}$ qt **81.** kilo- **83.** milli-
85. centi- **87.** 10 times **89.** $\frac{1}{1000}$ of **91.** 1000 times **93.** a
95. a **97.** c **99.** b **101.** 6.71 dkm **103.** 2,300 mm
105. 12,300 mm **107.** 230,00 mm **109.** 413.27 km **111.** 3.945 hg
113. 30.00974 kg **115.** Cannot add unlike measures. **117.** 748 cg or 7.48 g **119.** 61.47 cg **121.** 15 **123.** 8.5 hL **125.** 18.9 m
127. 245 mL **129.** 6 m **131.** 100 servings **133.** 40 shirts
135. 234.35 yd **137.** 15.9 liq qt **139.** 70.4 lb **141.** 22.86 cm
143. 156.88 qt **145.** 60 m **147.** 281.75 km **149.** $5\frac{1}{4}$ in.
151. $4\frac{7}{16}$ in. **153.** $3\frac{15}{16}$ in. **155.** $3\frac{9}{16}$ in. **157.** $2\frac{3}{4}$ in.
159. 117 mm or 118 mm **161.** 99 mm **163.** 60 mm **165.** 45 mm
167. 20 mm **169.** Answers will vary.

Trial Test, Chapter 6 (pp. 289–290)

1. 36 in. **3.** 8 gal **5.** $4\frac{qt}{sec}$ **7.** 165 yd **9.** $80\frac{2}{3}\frac{ft}{sec}$ **11.** $1\frac{1}{4}$ in.
13. 2 gal 1 qt 4 oz **15.** deci- **17.** 0.298 km **19.** 9.48 L or 94.8 dL
21. 120.75 km **23.** 8.48 pt **25.** $735\frac{m}{sec}$

Self-Study Exercises 7-1 (pp. 299–301)

L.O.1.

1. $P = 12$ cm
$A = 9\text{ cm}^2$ **2.** square **3.** 193 ft^2 **4.** 590 ft **5.** 900 yd^2

6. $\frac{1}{16}\text{ mi}^2$ **7.** 5268 ft **8.** 81 tiles **9.** 36 tiles

10. perimeter = 80 in.
area = 400 in.^2

L.O.2.

11. $P = 10$ ft
$A = 6\text{ ft}^2$ **12.** rectangle **13.** 42,500 ft^2 **14.** 180 ft^2

15. 241 ft^2 **16.** 59 ft **17.** 142.5 board feet **18.** 156 ft **19.** 17 ft
20. 130 in.

L.O.3.

21. $P = 38$ in.
$A = 72\text{ in.}^2$ **22.** parallelogram **23.** 156 ft **24.** 9000 ft^2

25. 124 in. **26.** 18 tiles **27.** 1,111 parallelograms **28.** 160 in.
29. 108 in. **30.** 9 signs

Self-Study Exercises 7-2 (pp. 307–309)

L.O.1.

1. 25.1 cm **2.** 18.8 in. **3.** 9.4 ft **4.** 17.3 m

L.O.2.

5. 0.6 m^2 **6.** 2.1 ft^2 **7.** 14.6 cm^2 **8.** 1.0 in.^2

L.O.3.

9. 592.7 ft^2 **10.** 268.27 in. **11.** 1178.1 mm^2 **12.** Yes, the cross-section area of the third pipe is larger than the combined area of the other two pipes. **13.** 168.5 mm^2 **14.** Yes, the combined cross-sectional area of the two pipes is 25.1 in.^2 which is greater than 20 in.^2—the area of the large pipe. **15.** 2827.4 ft/min **16.** 1570.8 ft/min **17.** 78.5 ft^2 **18.** 0.448 in. **19.** 4.4 m

Assignment Exercises, Chapter 7 (pp. 309–311)

1. $P = 57$ cm
$A = 162$ cm **3.** $P = 210$ mm
$A = 2450\text{ mm}^2$ **5.** 836 ft **7.** 33 yd^2

9. \$14.25 **11.** 3 rolls **13.** 235 ft^2 **15.** $C = 25.13$ m
$A = 50.27\text{ m}^2$

17. $C = 17.42$ cm
$A = 12\text{ cm}^2$ **19.** 5.8 m^2 **21.** 2.4 in. **23.** 33 ft/min

25. 29.5 in. **27.** 1.4 yd^2 **29.** 29.5 in. **31.** 66 in. diameter

Trial Test, Chapter 7 (pp. 314–315)

1. $P = 91$ ft
$A = 514.5\text{ ft}^2$ **3.** $P = 12.2$ in.
$A = 6.08\text{ in.}^2$ **5.** \$52 **7.** 282 ft^2 **9.** center

11. radius **13.** 37.70 in. **15.** 385.62 ft **17.** 0.29 in.^2 **19.** 0.15 in.^2

Self-Study Exercises 8-1 (pp. 319–320)

L.O.1.

1. 8.6% **2.** 35.5% **3.** 56.3%

L.O.2.

4. Debt retirement **5.** Misc. expenses and General government
6. Social projects and Education costs

L.O.3.

7. 5 Amps **8.** 50 volts **9.** 35 volts **10.** 25 ohms

Self-Study Exercises 8-2 (pp. 325–326)

L.O.1.

1. 10 **2.** 2 **3.** $\frac{1}{3}$ **4.** $\frac{1}{7}$ **5.** 12% **6.** 28% **7.** 35-37 and 38-40
8. 20-22 and 23-25 **9.** 7 **10.** 16

	Midpoint	Tally	Class Frequency
11.	14	||	2
12.	11	||||	4
13.	8	卌 ||	7
14.	5	卌 卌 卌 卌	20

	Midpoint	Tally	Class Frequency
15.	93	||	2
16.	88	卌	5
17.	83	卌 |||	8
18.	78	卌 ||||	9
19.	73	|||	3
20.	68	卌 |	6
21.	63	||	2
22.	58	卌	5

L.O.2.

23.

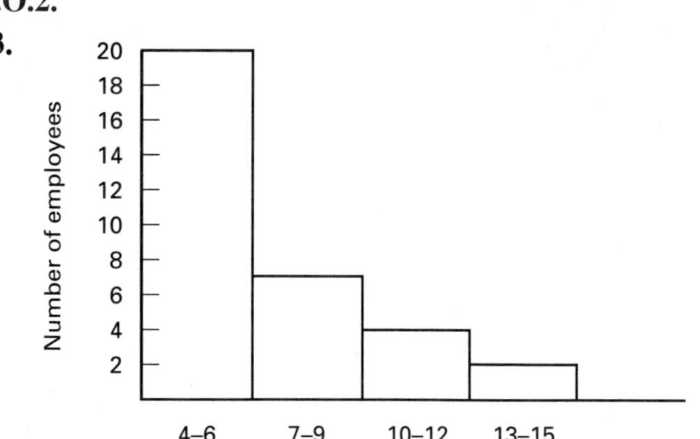

24.

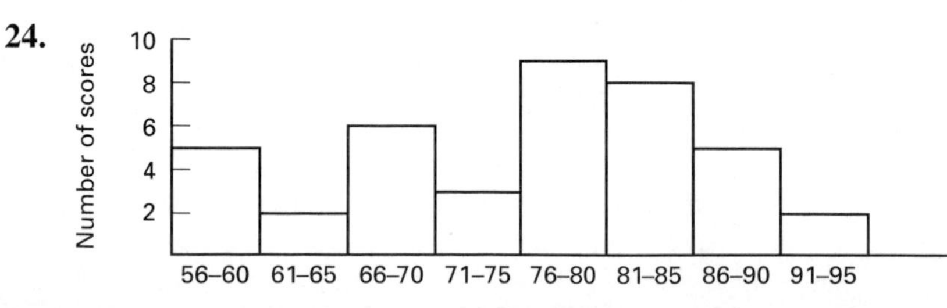

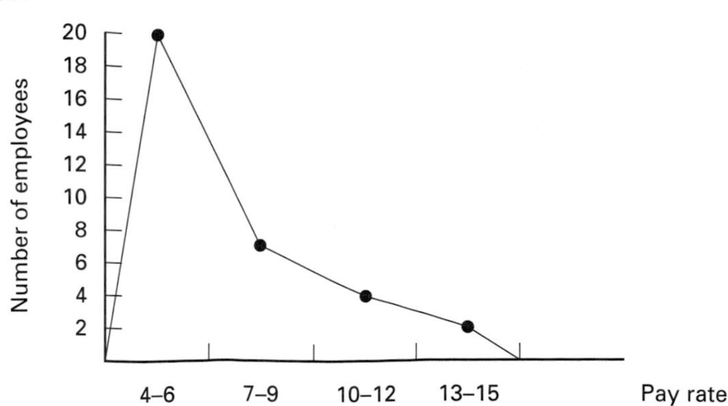

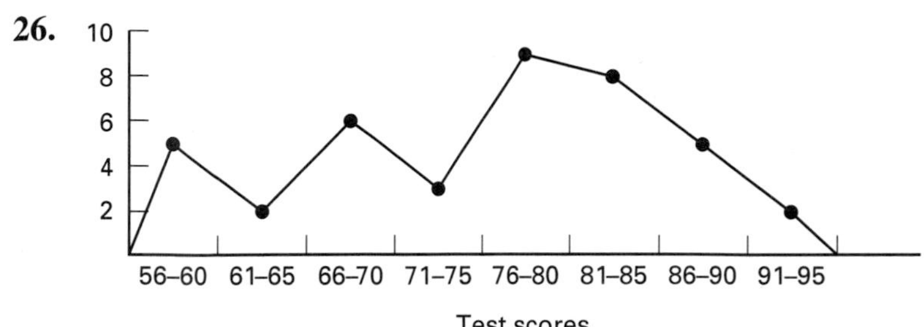

Self-Study Exercises 8-3 (p. 330)

L.O.1.

1. 16 **2.** 17 **3.** 66 **4.** 73.75 **5.** 33.7 **6.** 66.6 **7.** 42.33°F **8.** 12.67°C **9.** $34.80 **10.** $39.20 **11.** 14.67 in. **12.** 8 in. **13.** 67.33 **14.** 66 **15.** 17.3 runs **16.** 27.2 cars **17.** 24.6 mpg **18.** 10.2 mpg

L.O.2.

19. 44 **20.** 43 **21.** 15.5 **22.** 26.5 **23.** $30 **24.** $66 **25.** $4.85 **26.** $5.85 **27.** 2 **28.** 5 **29.** no mode **30.** no mode **31.** $67 **32.** $32 **33.** 4 **34.** $1.97 **35.** 24 **36.** 23 **37.** 13 **38.** 18 **39.** $25 **40.** $33 **41.** 48°F

Self-Study Exercises 8-4 (pp. 333–334)

L.O.1.

1. Keaton Brienne Renee
Keaton Renee Brienne
Brienne Keaton Renee
Brienne Renee Keaton
Renee Keaton Brienne
Renee Brienne Keaton

2. ABCD BACD
ABDC BADC
ACBD BCAD
ADBC BDAC
ACDB BCDA
ADCB BDCA
CABD DABC
CADB DACB
CBAD DBAC
CDAB DCAB
CBDA DBCA
CDBA DCBA
24 ways
$4 \cdot 3 \cdot 2 \cdot 1 = 24$ ways

3. $4 \cdot 3 \cdot 2 \cdot 1 = 24$ ways

4. $5 \cdot 4 \cdot 3 \cdot 2 \cdot 1 = 120$ ways

5. $5 \cdot 4 \cdot 3 \cdot 2 \cdot 1 = 120$ ways

L.O.2.

6. $\frac{1}{24}, \frac{1}{23}$ **7.** $\frac{1}{4}$ **8.** $\frac{11}{48}$ **9.** $\frac{2}{5}$ **10.** $\frac{1}{3}$

Assignment Exercises, Chapter 8 (pp. 334–338)

1. 1989, 1991, 1992 **3.** 1990, 1993, 1994 **5.** 12.9% **7.** 17.4%

9. 70 mph **11.** 50 **13.** 110 **15.** $\frac{1}{2}$

	Midpoint	Tally	Class Frequency
17.	60.5	~~||||~~ ~~||||~~	10
19.	40.5	~~||||~~ ~~||||~~ ||	12
21.	20.5	~~||||~~ ||	7

23.

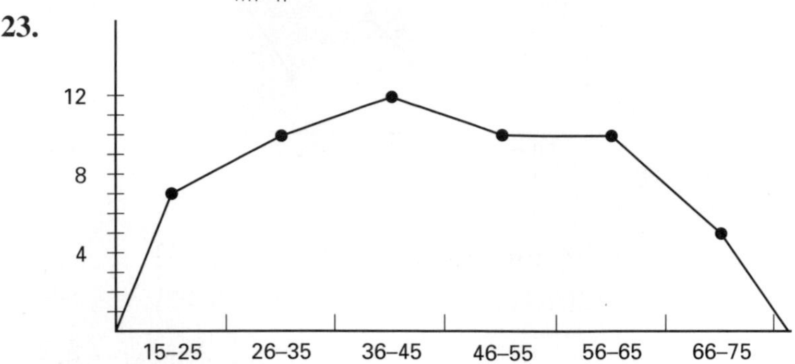

25. 36-45 **27.** 15 **29.** $\frac{1}{2}$ **31.** $\frac{10}{54} = 18.5\%$

33.

Class Interval	Midpoint	Tally	Class Frequency
20–24	22	~~||||~~ ||	7
25–29	27	~~||||~~		6
30–34	32	||||	4

35.

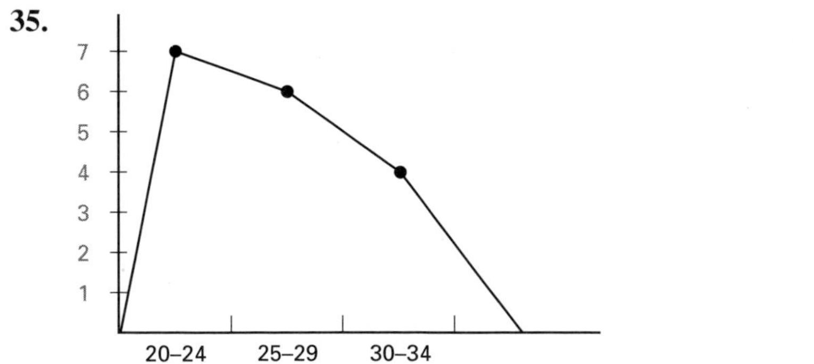

37. 12.6 cars **39.** 13.8 mpg **41.** \$6.03 **43.** \$1.85 **45.** 3.67

47. 63 **49.** 24 **51.** $\frac{10}{13}, \frac{3}{4}$ **53.** 8 **55.** $\frac{1}{5}$ **57.** $\frac{1}{4}$

Trial Test, Chapter 8 (pp. 344–346)

1. bar **3.** line **5.** 2 degrees **7.** \$65 **9.** 25% **11.** $\frac{15}{200} = \frac{3}{40}$

13. English dept., Electronic dept. **15.** 108.3% **17.** $\frac{18}{20} = \frac{9}{10}$ **19.** 6

21. $\frac{1}{2}$

	Midpoint	Tally	Class Frequency
23.	5	𝍸 ||||	9

25. 14 **27.** cannot be determined **29.** mean: 77.9 median: 78 mode: 81 **31.** 6 **33.** $\frac{3}{5}$

Self-Study Exercises 9-1 (p. 349)

L.O.1.

1. -15 **2.** -1 **3.** -2 **4.** -10 **5.** -33 **6.** 42 **7.** -5
8. 8 **9.** $\frac{8}{35}$ **10.** -2.3

L.O.2.

11. 6 **12.** 0 **13.** -17 **14.** -19 **15.** 4 **16.** -6 **17.** -8
18. -34 **19.** -17 **20.** 16

Self-Study Exercises 9-2 (p. 355)

L.O.1.

1. $n = 11$ **2.** $m = 6$ **3.** $y = -4$ **4.** $x = 6$ **5.** $p = 9$
6. $b = -9$ **7.** $\boxed{7} + \boxed{c}$ **8.** $\boxed{4a} - \boxed{7}$ **9.** $\boxed{3x} - \boxed{2(x + 3)}$ **10.** $\boxed{\frac{a}{3}}$
11. $\boxed{7xy} + \boxed{3x} - \boxed{4} + \boxed{2(x + y)}$ **12.** $\boxed{14x} + \boxed{3}$ **13.** $\boxed{\frac{7}{a + 5}}$
14. $\boxed{\frac{4x}{7}} + \boxed{5}$ **15.** 5 **16.** -4 **17.** $\frac{1}{5}$ **18.** $\frac{2}{7}$ **19.** 6 **20.** $-\frac{4}{5}$

L.O.2.

21. $-3a$ **22.** $-8x - 2y$ **23.** $7x - 2y$ **24.** $-3a + 6$ **25.** 10
26. $3a + 5b + 8c + 1$

L.O.3.

27. $10x - 20$ **28.** $6y + 13$ **29.** $-4x + 5$ **30.** $-12a + 3$

Self-Study Exercises 9-3 (p. 360)

L.O.1.

1. $x = 8$ **2.** $x = 9$ **3.** $x = 5$ **4.** $n = 30$ **5.** $a = -9$
6. $b = -2$ **7.** $c = 7$ **8.** $x = -9$ **9.** $x = \frac{32}{5}$ **10.** $y = \frac{9}{2}$
11. $n = -\frac{5}{2}$ **12.** $b = \frac{28}{3}$ **13.** $x = -7$ **14.** $y = 2$ **15.** $x = 15$

16. $y = 8$ **17.** $\frac{10}{3}$ **18.** $-\frac{20}{3}$ **19.** $x = 12$ **20.** $n = 56$

L.O.2.

21.–40. Each solution should check.

Self-Study Exercises 9-4 (pp. 365–366)

L.O.1.

1. $x = 6$ **2.** $m = 7$ **3.** $a = -3$ **4.** $m = -\frac{1}{2}$ **5.** $y = 8$
6. $x = 0$

L.O.2.

7. $b = -1$ **8.** $x = 8$ **9.** $t = 6$ **10.** $x = 3$ **11.** $y = 5$
12. $x = -\frac{7}{2}$ **13.** $a = -2$ **14.** $x = -9$ **15.** $x = \frac{1}{4}$ **16.** $x = -8$
17. $t = 7$ **18.** $x = -1$ **19.** $y = 11$ **20.** $y = -2$ **21.** $x = 27$
22. $y = 13$ **23.** $R = -68$ **24.** $P = \frac{5}{6}$ **25.** $x = 14$
26. $x = -\frac{4}{15}$ **27.** $m = \frac{49}{18}$ **28.** $s = \frac{8}{5}$ **29.** $m = \frac{8}{3}$ **30.** $T = \frac{68}{9}$
31. $x = 2$ **32.** $R = 0.8$ **33.** $x = 6.03$ **34.** $x = 0.08$
35. $R = 0.34$

Self-Study Exercises 9-5 (p. 368)

L.O.1.

1. $x = 3$ **2.** $y = -1$ **3.** $x = 1$ **4.** $t = -7$ **5.** $x = \frac{1}{3}$
6. $x = 6$ **7.** $a = -3$ **8.** $x = 2$ **9.** $x = \frac{1}{2}$ **10.** $b = \frac{20}{3}$

Self-Study Exercises 9-6 (p. 375)

L.O.1.

1. $x + 3$ **2.** $x - y$ **3.** $x + 4$ **4.** $x - 7$ **5.** $Y - 10$ **6.** $2a$
7. $2(7m)$ **8.** $a - 15$ **9.** $x + 0.07$ **10.** $x - 0.25$ **11.** $x + 50$
12. $x - 0.025$ **13.** $4(30x + 0.8)$ **14.** $x(y - 5)$

L.O.2.

15. $x + 4 = 12; x = 8$ **16.** $2x - 4 = 6; 5$
17. $x + x + x - 3 = 27;$
Two parts weigh 10 lb.
Third part weighs 7 lb.
18. $24 + x = 60;$
36 gal
19. $4.03 - 3.97 = x;$
0.06 kg
20. $x + 3x = 400$
100 ft of solid pipe
300 ft of perforated pipe

Assignment Exercises, Chapter 9 (pp. 375–382)

1. 9 **3.** 13 **5.** -77 **7.** -19 **9.** -34 **11.** $y = 8$
13. $x = 7$ **15.** $x = 13$ **17.** $\boxed{5} + \boxed{k}$ **19.** $\boxed{6a} - \boxed{4(a - 7)}$
21. $\boxed{9mx} - \boxed{2m} + \boxed{8} + \boxed{6(m + x)}$ **23.** $\boxed{\frac{x + 5}{a + 3}}$ **25.** 7 **27.** $\frac{1}{8}$
29. 12 **31.** $9x$ **33.** $-3x - 2y + 2z$ **35.** $10x + 7y + 11$
37. $21x + 14$ **39.** $-7x + 3$ **41.** $x = 7$ **43.** $b = -\frac{15}{2}$
45. $y = 7$ **47.** $x = -8$ **49.** $x = 15$ **51.** $x = 64$ **53.** $x = -49$
55. $x = 84$ **57.** $x = 5$ **59.** $b = -2$ **61.** $x = 7$ **63.** $x = -4$
65. $x = 3$ **67.** $x = -1$ **69.** $a = 5$ **71.** $x = 5$ **73.** $x = 7$

75. $x = 4$ **77.** $x = 2$ **79.** $x = -3$ **81.** $x = \frac{1}{3}$ **83.** $y = -12$
85. $y = -3$ **87.** $x = \frac{9}{2}$ **89.** $y = 7$ **91.** $x = 9$ **93.** $x = \frac{20}{13}$
95. $x = \frac{7}{20}$ **97.** $c = -\frac{9}{32}$ **99.** $y = 1.5$ **101.** $R = 0.61$
103. $y = -1$ **105.** $x = -9$ **107.** $x = 1$ **109.** $x = 3$
111. $x = 2$ **113.** $x = -2$ **115.** $x = 2$ **117.** $x = 0$ **119.** $x = 3$
121. $x = 3$ **123.** $x = -1$ **125.** $x = -\frac{1}{8}$ **127.** $5 + x$
129. $y - 7$ **131.** $x + 5$ **133.** $m(xy)$ **135.** $\frac{8}{y}$ **137.** $3x$ **139.** $\frac{8}{x}$
141. $x - 6 = 8; x = 14$ **143.** $5(x + 6) = 42 + x; x = 3$
145. $x + x - 3 = 51$; 27 hr, 24 hr **147.** Answers will vary. $x = 20$

Trial Test, Chapter 9 (pp. 385–386)

1. -13 **3.** 4 **5.** 18 **7.** $\boxed{4ab} - \boxed{\frac{2}{x}} + \boxed{3(x + 7)}$ **9.** $\frac{1}{7}$ **11.** $5m$
13. $-7x - 20$ **15.** $2y - 9$ **17.** $k + 8$ **19.** $y = -9$
21. $m = 10$ **23.** $x = 7$ **25.** $x = 7$ **27.** $x = \frac{7}{4}$ **29.** $x = 2$
31. $x = 7$ **33.** $5 + x = 16; x = 11$
35. $x + 4x = 2580$; 516 gal, 2064 gal

Self-Study Exercises 10-1 (pp. 396–397)

1. 28 **2.** -11 **3.** $\frac{18}{7}$ **4.** 0 **5.** $-\frac{27}{5}$ **6.** $-\frac{5}{27}$ **7.** $\frac{8}{5}$
8. $\frac{11}{15}$ **9.** -350 **10.** 48 **11.** 27 **12.** $\frac{1}{3}$ **13.** 13 **14.** $\frac{25}{8}$
15. 0 **16.** $\frac{1}{9}$ **17.** 576 **18.** 4 **19.** 28 **20.** -68 **21.** $\frac{27}{35}$
22. $\frac{108}{5}$ **23.** $\frac{217}{24}$ **24.** 8 **25.** -5 **26.** $\frac{3}{10}$ **27.** 70 **28.** 0
29. $\frac{32}{63}$ **30.** $\frac{32}{5}$ **31.** $-\frac{36}{11}$ **32.** $-\frac{1}{108}$ **33.** 21 **34.** $\frac{4}{25}$ **35.** $\frac{2}{21}$
36. $\frac{9}{7}$ **37.–42.** Answers to 12, 14, 16, 18, 25, and 32 should be checked.
43. 80 fixtures **44.** $1\frac{1}{3}$ hours **45.** $3\frac{1}{13}$ hr **46.** $3\frac{3}{7}$ days
47. $2\frac{1}{3}$ min

Self-Study Exercises 10-2 (pp. 402–403)

L.O.1.

1. 2 **2.** 0.8 **3.** 6.03 **4.** 16 **5.** 4.7 **6.** 0.08 **7.** 0.035
8. 0.34 **9.** -3.3 **10.** 38.8 **11.** 3.3 **12.** -0.2 **13.** 0.6
14. 1.128 **15.** 16

L.O.2.

16. 87.5 lb **17.** 4.71 in. **18.** 12 hr **19.** 6.5 hr **20.** 8.3 Ω
21. $5.30 **22.** 156.25 V **23.** $264 **24.** $136 **25.** $1925

26. 12.25% **27.** 12% **28.** 18.5% **29.** $2777.78 **30.** $\frac{1}{2}$ yr or 6 mo

Self-Study Exercises 10-3 (pp. 413–415)

L.O.1.

1. 3 **2.** 2 **3.** 29 **4.** $\frac{5}{8}$ **5.** $\frac{34}{5}$ **6.** $\frac{17}{14}$ **7.** 3 **8.** 21 **9.** 4

10. $-\frac{4}{5}$

L.O.2.

11. $5.90 **12.** $74.13 **13.** 48 engines **14.** 15 headpieces
15. 1425 mi **16.** 3360 mi **17.** 550 lb **18.** 1.7 gal

L.O.3.

19. 900 rpm **20.** 200 in.^{3} **21.** 20 in. **22.** 18 painters **23.** 10 hr
24. 50 rpm **25.** 5 in. **26.** 4 hr **27.** 5 machines **28.** 120 rpm
29. 4 helpers (5 workers total) **30.** 9 in.

L.O.4.

31. $\angle A = \angle F, \angle B = \angle E, \angle C = \angle D$
32. $PQ = ST, \angle P = \angle S, \angle Q = \angle T$
33. $JL = MP, LK = NP, \angle L = \angle P$ **34.** $a = 20, d = 9$

35. $DC = 6, DE = 5\frac{1}{3}$ **36.** 65 ft

Assignment Exercises, Chapter 10 (pp. 417–420)

1. $\frac{5}{6}$ **3.** $-\frac{4}{15}$ **5.** $\frac{49}{18}$ **7.** $-\frac{60}{13}$ **9.** $\frac{8}{3}$ **11.–15.** Answers to 2, 6, and 10 should be checked. **17.** $-\frac{7}{4}$ **19.** $\frac{16}{7}$ **21.** $\frac{10}{7}$ **23.** 6

25. $\frac{31}{39}$ **27.** $2\frac{1}{10}$ hr **29.** 25 fixtures **31.** 1.9 **33.** 2 **35.** 77.2

37. −11.8 **39.** −0.8 **41.** 1.5 **43.** 0.44 **45.** 0.02 **47.** 3.4

49. 17 Ω **51.** 110 V **53.** 375 lb **55.** 2 **57.** $\frac{7}{6}$ **59.** $\frac{1}{2}$ **61.** $\frac{49}{12}$

63. $-\frac{21}{2}$ **65.** $\frac{3}{14}$ **67.** $-\frac{21}{23}$ **69.** 23 machines **71.** $2\frac{1}{2}$ hr

73. 4500 women **75.** 1500 rpm **77.** 3 days **79.** $4\frac{1}{5}$ ft

81. 81.8 gal **83.** 5.0 hr **85.** 1351 ft **87.** 60 rpm **89.** 3 machines

91. $\frac{AB}{RT} = \frac{BC}{TS} = \frac{AC}{RS}$ **93.** 10 **95.** 100 teeth

97. $2\frac{6}{7}$ gal water, $7\frac{1}{7}$ gal antifreeze

Trial Test, Chapter 10 (pp. 425–426)

1. 16 **3.** 21 **5.** $-\frac{11}{5}$ **7.** $-\frac{22}{7}$ **9.** $-\frac{56}{3}$ **11.** $\frac{7}{25}$ **13.** $\frac{10}{21}$

15. 6.17 **17.** 0.30 **19.** 254.24 **21.** 2.5 **23.** −0.15 **25.** 168.75
27. 1.33 psi **29.** 54.7 L **31.** 11.429 Ω **33.** 107 lb **35.** 5

Self-Study Exercises 11-1 (pp. 433–434)

L.O.1.

1. 125 **2.** $\frac{4}{25}$ **3.** 81 **4.** 0.09 **5.** −1.728 **6.** −512

L.O.2.

7. x^7 **8.** m^4 **9.** a^2 **10.** x^{10} **11.** y^6 **12.** a^2b **13.** y^5
14. x^4 **15.** $\frac{1}{a}$ **16.** b **17.** 1 **18.** $\frac{1}{x^2}$ **19.** y^4 **20.** n^7 **21.** $\frac{1}{x^{10}}$
22. $\frac{1}{n}$ **23.** x^7 **24.** $\frac{1}{x^5}$ **25.** $\frac{1}{x^2}$

L.O.4.

26. x^6 **27.** 1 **28.** $-\frac{1}{8}$ **29.** $\frac{4}{49}$ **30.** $\frac{a^4}{b^4}$ **31.** $8m^6n^3$ **32.** $\frac{x^6}{y^3}$
33. $\frac{9}{25}$ **34.** $4a^2$ **35.** x^6y^3

Self-Study Exercises 11-2 (p. 438)

L.O.1.

1. $7a^2$ **2.** $3x^3$ **3.** $-2a^2 + 3b^2$ **4.** $2a - 2b$ **5.** $-5x^2 - 2x$
6. $5a^2$ **7.** $-x^2 - 2y$ **8.** $2m^2 + n^2$ **9.** $9a + 2b + 10c$
10. $-2x + 3y - 2z$

L.O.2.

11. $14x^3$ **12.** $2m^3$ **13.** $-21m^2$ **14.** $-2y^6$ **15.** $2x^2$ **16.** $-\frac{a}{2}$
17. $\frac{1}{2x^2}$ **18.** $-\frac{3x^3}{4}$ **19.** $3x^2 - 18x$ **20.** $12x^3 - 28x^2 + 32x$
21. $-8x^2 + 12x$ **22.** $10x^2 + 4x^3$ **23.** $3x - 2$ **24.** $4x^3 - 2x - 1$
25. $7x - \frac{1}{x}$ **26.** $4x^2 + 3x$

Self-Study Exercises 11-3 (pp. 447–448)

L.O.1.

1. 37 **2.** 1820 **3.** 0.56 **4.** 1.42 **5.** 780,000 **6.** 62
7. 0.00046 **8.** 0.61 **9.** 0.72 **10.** 42 **11.** 10^{10}
12. 10^{-7} or $\frac{1}{10^7}$ **13.** 10^{-3} or $\frac{1}{10^3}$ **14.** 10 **15.** 10^3 **16.** 10^2
17. $\frac{1}{10^3}$ **18.** 1 **19.** $\frac{1}{10^5}$ **20.** $\frac{1}{10}$

L.O.2.

21. 430 **22.** 0.0065 **23.** 2.2 **24.** 73 **25.** 0.093 **26.** 83,000
27. 0.0058 **28.** 80,000 **29.** 6.732 **30.** 0.00589

L.O.3.

31. 3.92×10^2 **32.** 2×10^{-2} **33.** 7.03×10^0 **34.** 4.2×10^4
35. 8.1×10^{-2} **36.** 2.1×10^{-3} **37.** 2.392×10^1

38. 1.01×10^{-1} **39.** 1.002×10^{0} **40.** 7.21×10^{2} **41.** 4.2×10^{5}
42. 3.26×10^{4} **43.** 2.13×10^{1} **44.** 6.2×10^{-6} **45.** 5.6×10^{1}

L.O.4.

46. 2.144×10^{7} **47.** 5.6×10^{3} **48.** 2.36×10^{-2} **49.** 4.73×10^{-3}
50. 7×10^{2} **51.** 6.5×10^{-5} **52.** 7×10^{2} **53.** 8×10^{-3}
54. 3.2285×10^{13} **55.** 4.2×10^{2} or 420

Self-Study Exercises 11-4 (p. 451)

L.O.1.

1. binomial **2.** binomial **3.** monomial **4.** monomial
5. monomial **6.** monomial **7.** trinomial **8.** trinomial
9. monomial **10.** binomial **11.** binomial **12.** binomial

L.O.2.

13. 1 **14.** 2 **15.** 2, 1, 0 **16.** 3, 1, 0 **17.** 1, 0 **18.** 2, 0
19. 0 **20.** 0 **21.** 1, 0 **22.** 2, 0 **23.** 2 **24.** 3 **25.** 6 **26.** 5
27. 2 **28.** 1

L.O.3.

29. $-3x^2 + 5x$; 2; $-3x^2$; -3 **30.** $-x^3 + 7$; 3; $-x^3$; -1
31. $9x^2 + 4x - 8$; 2; $9x^2$; 9 **32.** $5x^2 - 3x + 8$; 2; $5x^2$; 5
33. $7x^3 + 8x^2 - x - 12$; 3; $7x^3$; 7
34. $-15x^4 + 12x + 7$; 4; $-15x^4$; -15 **35.** $8x^6 - 7x^3 - 7x$; 6; $8x^6$; 8
36. $-14x^8 + x + 15$; 8; $-14x^8$; -14

Self-Study Exercises 11-5 (pp. 455–456)

L.O.1.

1. \$2415.77 **2.** \$2050.40 **3.** (a) 122.14 (b) 149.18 (c) 164.87 (d) 182.21
4. (a) \$1197.22 (b) \$1576.91 **5.** 64 **6.** 0.0041 **7.** 9,765,625
8. 0.0020 **9.** 243 **10.** 0.0032

L.O.2.

11. 2.08×10^{-87} **12.** 4.28×10^{-96} **13.** 5.58×10^{-44} **14.** 7.39
15. 0.05 **16.** 1.23 **17.** 0.03

L.O.3.

18. $x = 7$ **19.** $x = -3$ **20.** $x = 2$ **21.** $x = 10$ **22.** $x = 7$
23. $x = 3$ **24.** $x = 6$ **25.** $x = -4$ **26.** $x = -6$ **27.** $x = \frac{7}{6}$
28. $x = \frac{5}{2}$ **29.** $x = 5$ **30.** $x = 8$

Self-Study Exercises 11-6 (pp. 462–463)

L.O.1.

1. $\log_3 9 = 2$ **2.** $\log_2 32 = 5$ **3.** $\log_9 3 = \frac{1}{2}$ **4.** $\log_{16} 2 = \frac{1}{4}$
5. $\log_4 \frac{1}{16} = -2$ **6.** $\log_3 \frac{1}{81} = -4$

L.O.2.

7. $3^4 = 81$ **8.** $12^2 = 144$ **9.** $2^{-3} = \frac{1}{8}$ **10.** $5^{-2} = \frac{1}{25}$

11. $25^{-0.5} = \frac{1}{5}$ **12.** $4^{-0.5} = \frac{1}{2}$ **13.** $x = 3$ **14.** $x = \frac{1}{81}$
15. $x = 2$ **16.** $x = -2$ **17.** $x = 625$ **18.** $x = -4$

L.O.3.

19. 0.4771 **20.** 0.7782 **21.** 0.3802 **22.** 0.6232 **23.** 2.1761
24. -2.9208 **25.** 1.3863 **26.** 0.9163 **27.** -1.8971 **28.** 5.6168
29. 4.6052

L.O.4.

30. 3 **31.** 6 **32.** 5 **33.** 2 **34.** 1.9358

L.O.5.

35. 3 **36.** 5 **37.** 8

L.O.6.

38. 2.096 **39.** 1.893 **40.** 1.631

Assignment Exercises, Chapter 11 (pp. 465–467)

1. 81 **3.** 244,140,625 **5.** 0.0001 **7.** 429,981,696 **9.** 0.1296
11. x^{10} **13.** x^6 **15.** x^3 **17.** x^{12} **19.** x^{15}
21. $-2x^4 - 4x^3 + 13x$ **23.** $11x^2 - 11y^2$ **25.** $21x^6$ **27.** $-\frac{2x^3}{3}$
29. $10x^3 + 15x^2 - 20x$ **31.** $2x^2 - 4x + 7$
33. 10^{12} or 1,000,000,000,000 **35.** 10^{-3} or $\frac{1}{1000}$ **37.** 8730
39. 5.2×10^4 **41.** 1.7×10^{-4} **43.** 4.368×10^{126} **45.** 3.38×10^{10}
47. No. The term $-3x^{-2}$ has a negative exponent. **49.** 1.25×10^{-6}
51. 2.71×10^{-35} **53.** (a) \$2542.50 (b) \$547.09 **55.** $x = -2$
57. $x = 4$ **59.** $x = -\frac{5}{3}$ **61.** $x = 4$ **63.** $x = -5$ **65.** $x = 1$
67. $x = 4$ **69.** 0.01831563889 **71.** 0.00004539992976
73. $\log_3 81 = 4$ **75.** $\log_{27} 3 = \frac{1}{3}$ **77.** $\log_4 \frac{1}{64} = -3$
79. $\log_9 \frac{1}{3} = -\frac{1}{2}$ **81.** $\log_{12} \frac{1}{144} = -2$ **83.** $11^2 = 121$
85. $15^\circ = 1$ **87.** $7^1 = 7$ **89.** $4^{-2} = \frac{1}{16}$ **91.** $9^{-0.5} = \frac{1}{3}$
93. 0.6990 **95.** 2.2553 **97.** -0.3979 **99.** 5.5984 **101.** -0.2231
103. 2.1133 **105.** 2 **107.** 343 **109.** $x = -2$
111. a. 2 b. 4 c. 8.1761 **113.** 3.17
115. a. $C_t = C_i(0.65)^x$
b. 30,420,000 units per cubic meter after 60 days
8,354,092.5 units per cubic meter after 150 days
c. approximately 500 days or 16 to 17 months

Trial Test, Chapter 11 (pp. 472–473)

1. x^5 **3.** x^3 **5.** $\frac{16}{49}$ **7.** $\frac{x^4}{y^2}$ **9.** $-6a^6$ **11.** $12a^3 - 8a^2 + 20a$
13. 10^6 **15.** 42,000 **17.** 4.2 **19.** 0.042 **21.** 2.4×10^2
23. 8.6×10^{-4} **25.** 7.83×10^{-3} **27.** 2.952×10^0 **29.** 3.5×10^2
31. 1.5×10^6 ohms **33.** 58.09475019 **35.** 248,832 **37.** $x = 9$
39. $x = 2$ **41.** $\log_4 \frac{1}{2} = -\frac{1}{2}$ **43.** $3^{-3} = \frac{1}{27}$ **45.** 3.4657
47. $x = 3$ **49.** 2.5850 **51.** 1.4641 **53.** $S = \$224.94$ thousands

Self-Study Exercises 12-1 (p. 478)

L.O.1.

1. $\frac{1}{5}$ **2.** $\frac{7}{9}$ **3.** $\frac{11}{12}$ **4.** 0.2 **5.** 0.5 **6.** 0.04 **7.** 4.359
8. -2.295 **9.** ± 1.863 **10.** -0.933

L.O.2.

11. x^8 **12.** x^3 **13.** $-x^{15}$ **14.** $\pm x^4$ **15.** $\frac{a^6}{b^{11}}$

Self-Study Exercises 12-2 (p. 481)

L.O.1.

1. 7 **2.** 19 **3.** x **4.** $\frac{1}{5}$ **5.** j **6.** 5.3 **7.** $2\sqrt{6}$ **8.** $7\sqrt{2}$
9. $4\sqrt{3}$ **10.** $x^4\sqrt{x}$ **11.** $y^7\sqrt{y}$ **12.** $2x\sqrt{3x}$ **13.** $2a^2\sqrt{14a}$
14. $6ax^2\sqrt{2a}$ **15.** $2x^2yz^3\sqrt{11xz}$ **16.** Answers will vary.

Self-Study Exercises 12-3 (pp. 487–488)

L.O.1.
1. $12\sqrt{3}$ **2.** $-4\sqrt{5}$ **3.** $2\sqrt{7}$ **4.** $9\sqrt{3} - 8\sqrt{5}$
5. $-3\sqrt{11} + \sqrt{6}$ **6.** $43\sqrt{2} + 6\sqrt{3}$ **7.** $12\sqrt{3}$ **8.** $13\sqrt{5}$
9. $11\sqrt{7}$ **10.** $-3\sqrt{6}$ **11.** $8\sqrt{2} - 6\sqrt{7}$ **12.** $6\sqrt{3}$ **13.** $27\sqrt{5}$
14. $47\sqrt{2}$ **15.** $6\sqrt{10}$ **16.** $-\sqrt{3}$

L.O.2.
17. $15\sqrt{10}$ **18.** 240 **19.** $20x\sqrt{15x}$ **20.** $28x^5\sqrt{6x}$ **21.** $x\sqrt{3x}$
22. $\frac{2\sqrt{2}}{\sqrt{15y}}$ or $\frac{2\sqrt{30y}}{15y}$ **23.** $\frac{2\sqrt{2}}{x\sqrt{7}}$ or $\frac{2\sqrt{14}}{7x}$ **24.** $\frac{2}{3y\sqrt{2y}}$ or $\frac{\sqrt{2y}}{3y^2}$
25. $\frac{4x\sqrt{x}}{3}$ **26.** $x\sqrt{7}$ **27.** $4\sqrt{3}$ **28.** $\frac{x}{\sqrt{6}}$ or $\frac{x\sqrt{6}}{6}$ **29.** $\frac{y}{\sqrt{7}}$ or $\frac{y\sqrt{7}}{7}$
30. $2x\sqrt{2x}$

L.O.3.

31. $2\sqrt{2}$ **32.** $\frac{10\sqrt{3}}{3}$ **33.** $\frac{8}{5\sqrt{3}}$ or $\frac{8\sqrt{3}}{15}$ **34.** $\frac{15x^3\sqrt{x}}{4}$

L.O.4.
35. $\frac{5\sqrt{3}}{3}$ **36.** $\frac{6\sqrt{5}}{5}$ **37.** $\frac{\sqrt{2}}{4}$ **38.** $\frac{\sqrt{55}}{11}$ **39.** $\frac{\sqrt{6}}{3}$ **40.** $\frac{5\sqrt{3x}}{3}$
41. $\frac{2x\sqrt{7}}{7}$ **42.** $\frac{2x^3\sqrt{3x}}{3}$ **43.** $\frac{3}{4\sqrt{3}}$ **44.** $\frac{5}{\sqrt{35}}$ **45.** $\frac{11}{\sqrt{77}}$
46. $\frac{3}{\sqrt{3}}$ **47.** $\frac{3}{2\sqrt{3}}$

Self-Study Exercises 12-4 (p. 490)

L.O.1.

1. $a^{\frac{1}{4}}$ **2.** $m^{\frac{1}{3}}$ **3.** $(3x)^{\frac{1}{5}}$ **4.** $(7a)^{\frac{1}{4}}$ **5.** $(xy)^2$ **6.** $3a^2$ **7.** $32^{\frac{1}{3}}x^{\frac{7}{3}}$
8. $1296x^6y^8$ **9.** $16^{\frac{2}{3}}a^{\frac{8}{3}}b^{\frac{14}{3}}$

L.O.2.

10. a^3 **11.** $b^{\frac{4}{5}}$ **12.** $64xy^{12}$ **13.** $32mn^{10}$ **14.** $a^{\frac{2}{5}}$ **15.** $\frac{1}{b^{\frac{2}{5}}}$

16. $m^{\frac{1}{8}}$ **17.** $x^{\frac{7}{6}}$ **18.** $a^{\frac{5}{3}}$ **19.** $2a^{\frac{17}{5}}$ **20.** $\frac{1}{3a^{\frac{13}{3}}}$ **21.** $\frac{2}{9a^{\frac{9}{8}}}$

Self-Study Exercises 12-5 (p. 494)

L.O.1.

1. $5i$ **2.** $6i$ **3.** $8xi$ **4.** $4y^2i\sqrt{2y}$

L.O.2.

5. i **6.** 1 **7.** 1 **8.** 1

L.O.3.

9. $15 + 0i$ **10.** $0 + 33i$ **11.** $5 + 2i$ **12.** $8 + 4i\sqrt{2}$
13. $7 - i\sqrt{3}$ **14.** $0 + 7i$ **15.** $-4 + 0i$

L.O.4.

16. $11 + 5i$ **17.** $2\sqrt{3} + 2\sqrt{2} - 8i\sqrt{3}$ **18.** $4 + i$ **19.** $12 + 20i$
20. $1 + 9i$

Self-Study Exercises 12-6 (p. 498)

L.O.1.

1. ± 5 **2.** ± 3 **3.** ± 3.464 **4.** ± 0.667 **5.** ± 3 **6.** ± 0.926
7. ± 1.155 **8.** ± 4 **9.** ± 2.449 **10.** ± 2.683

L.O.2.

11. $x = 81$ **12.** $y = 192$ **13.** $y = 66$ **14.** $x = \pm 5.292$
15. $x = -1.979$ **16.** $x = \pm 2$ **17.** $x = 50$ **18.** $x = \pm 4.899$
19. $y = \pm 2.105$ **20.** $y = \pm 2.9$

Assignment Exercises, Chapter 12 (pp. 498–501)

1. $\frac{4}{5}$ **3.** $\frac{1}{7}$ **5.** 0.7 **7.** 5.9 **9.** ± 2.7 **11.** y^6 **13.** $-b^9$
15. 5 **17.** x **19.** $3P\sqrt{P}$ **21.** $3a\sqrt{2b}$ **23.** $4x^2y\sqrt{2x}$
25. $5x^5y^4\sqrt{3y}$ **27.** $-\sqrt{7}$ **29.** $11\sqrt{6}$ **31.** $-28\sqrt{3}$
33. $-5\sqrt{2}$ **35.** $-17\sqrt{2}$ **37.** $24\sqrt{3}$ **39.** $40\sqrt{21}$
41. $-160\sqrt{6}$ **43.** $\frac{3}{4}$ **45.** $\frac{3\sqrt{6}}{8}$ **47.** $\frac{\sqrt{15}}{10}$ **49.** $\sqrt{3}$ **51.** $\frac{9}{16}$
53. $\frac{5\sqrt{17}}{17}$ **55.** $\frac{\sqrt{21}}{6}$ **57.** $\frac{3}{8\sqrt{3}}$ **59.** $\frac{12}{5\sqrt{6}}$ **61.** $x^{\frac{1}{2}}$ **63.** $x^{\frac{4}{5}}$

65. $\sqrt[5]{y^3}$ **67.** $x^{\frac{1}{3}}$ **69.** $(4y)^{\frac{1}{5}}$ **71.** $2b^4$ **73.** a^2 **75.** y

77. $27x^{\frac{3}{4}}y^6$ **79.** $64a^3x^{\frac{3}{2}}$ **81.** $x^{\frac{1}{2}}$ **83.** $a^{\frac{7}{6}}$ **85.** $\frac{1}{x^{\frac{1}{8}}}$ **87.** $a^{\frac{8}{3}}$

89. $2a^{\frac{7}{2}}$ **91.** $\frac{3}{2a^{\frac{22}{5}}}$ **93.** $10i$ **95.** $\pm 2y^3i\sqrt{6y}$ **97.** -1 **99.** i

101. $0 + 15i$ **103.** $0 - 12i$ **105.** $7 - 4i$ **107.** $11 + i$ **109.** ± 6

111. ± 2 **113.** ± 5 **115.** ± 3 **117.** $\frac{3\sqrt{6}}{2}$ or 3.674 **119.** 46

121. ± 10.262 **123.** ± 8.888 **125.** ± 1 **127.** ± 7.874
129. $20 + 15i$

Trial Test, Chapter 12 (pp. 504–505)

1. $6\sqrt{14}$ **3.** $6\sqrt{3}$ **5.** 0 **7.** $4x^2\sqrt{3}$ **9.** $2\sqrt{6}$ **11.** $\frac{7\sqrt{15x}}{2x}$

13. $\frac{\sqrt{30x}}{x}$ **15.** ± 1 **17.** 49 **19.** 9 **21.** ± 7 **23.** ± 6

25. 9.980 **27.** $3x^5$ **29.** $x^2y^3z^6$ **31.** a **33.** $5x^{\frac{1}{6}}y^2$ **35.** $\frac{1}{k^{\frac{2}{5}}}$

37. $2x$ **39.** $\frac{s^2}{r^{\frac{4}{5}}}$ **41.** $-i$ **43.** $-3 + 5i$ **45.** $15 + 4i$

Self-Study Exercises 13-1 (pp. 510–511)

L.O.1.

1. 25 **2.** 15 **3.** 1565 **4.** 9 **5.** 4.5 **6.** $310 **7.** 14.25% **8.** 3 years **9.** $2600 **10.** 90 **11.** 8 **12.** 3378.38 **13.** 13.6

14. 17 **15.** 66 in. **16.** 9.0 cm **17.** $10.50 **18.** $\frac{1}{4}$ mi

19. 153.9 in.^{2} **20.** 6.25 km^2 **21.** $48.75

Self-Study Exercises 13-2 (p. 514)

L.O.1.

1. $X = E - I$ **2.** $K = M - A$ **3.** $r = \frac{S}{2\pi h}$ **4.** $b = \frac{m}{x + y}$

5. $y = \frac{A - mx}{2m}$ **6.** $T_2 = \frac{T_1V_2}{V_1}$ **7.** $r = \sqrt{\frac{V}{\pi h}}$ **8.** $X = \frac{m^2}{Y}$

9. $R = \frac{100P}{B}$ **10.** $b = \frac{P - 2s}{2}$ **11.** $r = \frac{C}{2\pi}$ **12.** $l = \frac{A}{w}$

13. $C = \frac{R}{A - B}$ **14.** $s = \sqrt{A}$ **15.** $R = \frac{D}{T}$ **16.** $D = P - S$

17. $r = \sqrt{\frac{A}{\pi}}$ **18.** $C = \sqrt{a^2 + b^2}$ **19.** $I = A - P$ **20.** $T = \frac{I}{PR}$

Self-Study Exercises 13-3 (p. 519)

L.O.1.

1. 355 **2.** 165 **3.** 290 **4.** 0 **5.** -175 **6.** 344 **7.** 333 **8.** -81

L.O.2.

9. 920 **10.** 558 **11.** 250 **12.** 460 **13.** 640 **14.** 672 **15.** -110 **16.** 140 **17.** -460 **18.** 492

L.O.3.

19. 35 **20.** 0 **21.** 45 **22.** 5 **23.** 15 **24.** 10 **25.** 65 **26.** 50 **27.** 80 **28.** 120

L.O.4.

29. 158 **30.** 59 **31.** 113 **32.** 122 **33.** 68 **34.** 419 **35.** 590 **36.** 770 **37.** 365 **38.** 32

Self-Study Exercises 13-4 (pp. 534–538)

L.O.1.

1. $P = 309.3$ mm, $A = 5053.05$ mm^2 **2.** $P = 52.9$ ft, $A = 162$ ft^2 **3.** 348 in.^{2} **4.** 260 ft^2
5. 8220 ft^2 **6.** 98 ft

L.O.2.

7. $P = 27\frac{1}{2}$ in., $A = 27$ in.^{2} **8.** $P = 48$ cm, $A = 96$ cm^2 **9.** 15.75 cm^2 **10.** 72 ft; 216 ft^2
11. 9.75 ft^2 **12.** 30 ft^2 **13.** 24.75 in.^{2} **14.** $175.50 **15.** 8 ft
16. 10 ft 8 in.

L.O.3.

17. $AC = 24$ cm **18.** $BC = 10$ mm **19.** $AB = 17$ yd **20.** 6.403 cm
21. 6.325 m **22.** 12 ft **23.** 30 ft **24.** 39.783 cm
25. 26 ft 10 in. **26.** 12.806 ft **27.** 13 cm **28.** 45 dm
29. 10.607 mm **30.** 61.083 ft **31.** $E_a = 170.294$

L.O.4.

32. $LSA = 28$ units^2, $TSA = 432$ units^2 **33.** $LSA = 848.23$ units^2, $TSA = 1357.17$ units^2 **34.** 3244 ft^2
35. 39.4 in.^{2} **36.** 14,700 mm^2 **37.** 37.306 **38.** 600 in.^{2} **39.** 69.1

L.O.5.

40. $V = 576$ units^3 **41.** $V = 23.3$ units^3 **42.** 23.3 in.^{3} **43.** 18,850 ft^3
44. 320 cm^3 **45.** 55.36 in.^{3} **46.** 236 ft^3

L.O.6.

47. 314.2 cm^2 **48.** 904.8 in.^{3} **49.** 6,361.7 ft^2 **50.** 356,892.8 gal
51. 50.3 ft^2 **52.** 225.6 gal

L.O.7.

53. $LSA = 188.5$ ft^2; $TSA = 301.6$ ft^2; $V = 301.6$ ft^3 **54.** 4,712.4 ft^3
55. 589.0 cm^2 **56.** 50.3 L **57.** 1,649.3 ft^2 **58.** 10.0 ft **59.** $347

Self-Study Exercises 13-5 (pp. 541–542)

L.O.1.

1. 4.8 ohms **2.** 61.6% **3.** 35 miles per hour **4.** 1.5 amperes
5. 6 ft^3 **6.** 3 amperes **7.** 2 cylinders **8.** 1600 rpm **9.** 71.71 in.
10. 4.4 ohms

Assignment Exercises, Chapter 13 (pp. 544–550)

1. 12.5 **3.** 3 **5.** 1.25 **7.** 9.6% **9.** 10.8 **11.** $193.60
13. 2 years **15.** 50 in. **17.** $89.50 **19.** 95.0 in.^{2}

21. $F = D + m + n$ **23.** $w = \frac{V}{lh}$ **25.** $a = \frac{c}{h + b}$ **27.** $r = \sqrt{\frac{B}{cx}}$

29. $B = \frac{A}{P}$ **31.** $r = \frac{I}{Pt}$ **33.** $r = s + d$ **35.** $t = \frac{V_0 - V}{32}$

37. $t = \frac{A - P}{Pr}$ **39.** $P = S + D$ **41.** 351 **43.** 472 **45.** 203

47. 212 **49.** 185 **51.** $P = 12$ in., $A = 6$ in.^{2} **53.** $P = 339$ mm, $A = 5899.5$ mm^2

55. $A = 8.125 \text{ ft}^2$ **57.** 11.25 ft **59.** 19.5 ft^2 **61.** Two sheets **63.** 48 ft **65.** 199.5 ft^2 **67.** 15 in. **69.** 7.141 ft **71.** 8 yd **73.** 20.248 mi **75.** 30 cm **77.** 21.633 in. **79.** 4.243 in. **81.** 600 cm^3 **83.** 616 cm^2 **85.** 2733.186 cm^2 **87.** 102.315 yd^3 **89.** 60 ft **91.** 348,962 gal **93.** 1017.9 m^2 **95.** 7238.2 ft^2 **97.** 169.6 cm^3 **99.** 377.0 in.^3 **101.** 2 gal **103.** 116.6 ft^2 **105.** 243.3 in.^2 **107.** 2094 ft^3 **109.** 2.75 amperes **111.** 10 ft^3 **113.** 3.2 amperes **115.** 1400 rpm **117.** 2 cylinders **119.** Answers will vary.

Trial Test, Chapter 13 (pp. 555–556)

1. $L = \frac{RA}{P}$ **3.** $r = \sqrt{\frac{d}{\pi sn}}$ **5.** 361 **7.** 383 **9.** 250 ft^2 **11.** 10 dm **13.** 50 in.^2 **15.** 6768 gal **17.** 3848.5 ft^3 **19.** 2080.7 cm^2 **21.** 10.9 ft^3 **23.** 1497.05 lb **25.** 12,000 calories

Self-Study Exercises 14-1 (p. 558)

L.O.1.

1. $7(a + b)$ **2.** $m(m + 2)$ **3.** $3x(2x + 1)$ **4.** $5(ab + 2a + 4b)$ **5.** $2ax(2x + 3a)$ **6.** $a(5 - 7b)$ **7.** $3(4a^2 - 5a + 2)$ **8.** $3x(x^2 - 3x - 2)$ **9.** $2ab(4a + 7b^2)$ **10.** $3m^2(1 - 2m)$

Self-Study Exercises 14-2 (pp. 563–564)

L.O.1.

1. $a^2 + 11a + 24$ **2.** $x^2 + x - 20$ **3.** $y^2 - 10y + 21$ **4.** $2a^2 + 5a - 12$ **5.** $3a^2 - 8ab + 4b^2$ **6.** $5cx - 25xy - cy + 5y^2$ **7.** $6x^2 - 17x + 12$ **8.** $2a^2 - 7ab + 5b^2$ **9.** $21 - 52m + 7m^2$ **10.** $40 - 21x + 2x^2$ **11.** $x^2 + 11x + 28$ **12.** $y^2 - 12y + 35$ **13.** $m^2 - 4m - 21$ **14.** $3bx + 18b - 2x - 12$ **15.** $12r^2 - 7r - 10$ **16.** $35 - 22x + 3x^2$ **17.** $4 - 14m + 6m^2$ **18.** $6 + 13x + 6x^2$ **19.** $2x^2 + x - 15$ **20.** $20x^2 - 13xy - 21y^2$ **21.** $14a^2 + 19ab - 3b^2$ **22.** $30a^2 - 13ab - 10b^2$ **23.** $27x^2 + 30xy - 8y^2$ **24.** $20x^2 - 47xy + 24y^2$ **25.** $21m^2 + 29mn - 10n^2$

L.O.2.

26. $a^2 - 9$ **27.** $4x^2 - 9$ **28.** $a^2 - y^2$ **29.** $16r^2 - 25$ **30.** $25x^2 - 4$ **31.** $49 - m^2$ **32.** $4 - 12x + 9x^2$ **33.** $9x^2 + 24x + 16$ **34.** $Q^2 + 2QL + L^2$ **35.** $a^4 + 2a^2 + 1$ **36.** $4d^2 - 20d + 25$ **37.** $9a^2 + 12ax + 4x^2$ **38.** $9x^2 - 49$ **39.** $36 + 12Q + Q^2$ **40.** $y^2 - 25x^2$ **41.** $16 - 24j + 9j^2$ **42.** $9m^2 - 4p^2$ **43.** $m^4 + 2m^2p^2 + p^4$ **44.** $4a^2 - 49c^2$ **45.** $81 - 234a + 169a^2$ **46.** $x^3 + p^3$ **47.** $Q^3 - L^3$ **48.** $27 + a^3$ **49.** $8x^3 - 64p^3$ **50.** $27m^3 - 8$ **51.** $216 + p^3$ **52.** $125y^3 - p^3$ **53.** $x^3 + 8y^3$ **54.** $Q^3 - 216$ **55.** $27T^3 + 8$

Self-Study Exercises 14-3 (pp. 567–568)

L.O.1.

1. difference **2.** not difference **3.** difference **4.** difference **5.** not difference **6.** difference **7.** $(y + 7)(y - 7)$ **8.** $(4x + 1)(4x - 1)$ **9.** $(3a + 10)(3a - 10)$ **10.** $(2m + 9n)(2m - 9n)$ **11.** $(3x + 8y)(3x - 8y)$ **12.** $(5x + 8)(5x - 8)$ **13.** $(10 + 7x)(10 - 7x)$

14. $(2x + 7y)(2x - 7y)$ **15.** $(11m + 7n)(11m - 7n)$
16. $(9x + 13)(9x - 13)$

L.O.2.

17. not perfect square **18.** perfect square **19.** not perfect square
20. not perfect square **21.** perfect square **22.** perfect square
23. $(x + 3)^2$ **24.** $(x + 7)^2$ **25.** $(x - 6)^2$ **26.** $(x - 8)^2$
27. $(2a + 1)^2$ **28.** $(5x - 1)^2$ **29.** $(3m - 8)^2$ **30.** $(2x - 9)^2$
31. $(x - 6y)^2$ **32.** $(2a - 5b)^2$

L.O.3.

33. difference of two cubes **34.** not a special sum or difference of cubes
35. sum of two cubes **36.** not a special sum or difference of two cubes
37. difference of two cubes **38.** sum of two cubes
39. $(2m - 7)(4m^2 + 14m + 49)$ **40.** $(y - 5)(y^2 + 5y + 25)$
41. $(Q + 3)(Q^2 - 3Q + 9)$ **42.** $(c + 1)(c^2 - c + 1)$
43. $(5d - 2p)(25d^2 + 10dp + 4p^2)$ **44.** $(a + 4)(a^2 - 4a + 16)$
45. $(6a - b)(36a^2 + 6ab + b^2)$ **46.** $(x + Q)(x^2 - xQ + Q^2)$
47. $(2p - 5)(4p^2 + 10p + 25)$ **48.** $(3 - 2y)(9 + 6y + 4y^2)$

Self-Study Exercises 14-4 (pp. 579–580)

L.O.1.

1. $(x + 3)(x + 2)$ **2.** $(x - 7)(x - 4)$ **3.** $(x + 6)(x + 2)$
4. $(x - 3)(x - 1)$ **5.** $(x + 7)(x + 1)$ **6.** $(x + 5)(x + 2)$
7. $(x - 4)(x + 3)$ **8.** $(y - 5)(y + 2)$ **9.** $(y - 3)(y + 2)$
10. $(a + 4)(a - 3)$ **11.** $(b + 3)(b - 1)$ **12.** $(7 + b)(2 - b)$
13. $(x - 4)(x - 3)$ **14.** $(x - 6)(x + 5)$ **15.** $(x + 9)(x + 2)$
16. $(x - 6)(x - 3)$ **17.** $(x - 9)(x + 2)$ **18.** $(x + 18)(x - 1)$
19. $(x + 5)(x + 4)$ **20.** $(x - 10)(x - 2)$ **21.** $(x - 8)(x - 2)$
22. $(x - 16)(x - 1)$ **23.** $(x - 14)(x + 1)$ **24.** $(x - 7)(x + 2)$

L.O.2.

25. $(x + y)(x + 4)$ **26.** $(3x + 2)(2x - y)$ **27.** $(3x + 5)(m - 2n)$
28. $(6x - 7)(5y - 6)$ **29.** $(x - 2)(x + 8)$ **30.** $(3x - 1)(2x - 7)$
31. $(x - 4)(x + 1)$ **32.** $(2x - 1)(4x + 3)$ **33.** $(x - 5)(x + 4)$
34. $(x - 2)(3x + 5)$

L.O.3.

35. $(3x + 1)(x + 2)$ **36.** $(3x + 2)(x + 4)$ **37.** $(3x + 2)(2x + 3)$
38. $(x - 2)(x - 16)$ **39.** $(3x - 4)(2x - 3)$ **40.** $(2x - 5)(x - 2)$
41. $(3x - 5)(2x - 1)$ **42.** $(p + 9)(p - 4)$ **43.** $(6x - 5)(x - 1)$
44. $(4x + 3)(2x + 5)$ **45.** $(3x - 5)(5x + 1)$ **46.** $(y - 12)(y - 3)$
47. $(2x - 7)(x + 1)$ **48.** $(6x - 5)(2x + 3)$ **49.** $(5x + 3)(2x - 1)$
50. $(Q - 11)(Q + 4)$

L.O.4.

51. $(6x - 5y)(x + 2y)$ **52.** $(3a + 2b)(2a - 7b)$ **53.** $(6x - 5)(3x + 2)$
54. $(5x - 4y)(4x + 3y)$ **55.** $(x + 8)(x - 7)$ **56.** $(a + 19)(a + 2)$

L.O.5.

57. $4(x - 1)$ **58.** $(x + 3)(x - 2)$ **59.** $(2x + 3)(x - 1)$
60. $(x + 3)(x - 3)$ **61.** $4(x + 2)(x - 2)$ **62.** $(m + 5)(m - 3)$
63. $2(a + 2)(a + 1)$ **64.** $(b + 3)^2$ **65.** $(4m - 1)^2$
66. $(x + 7)(x + 1)$ **67.** $(2m + 1)(m + 2)$ **68.** $(2m + 1)(m - 3)$
69. $(2a - 5)(a + 1)$ **70.** $(3x - 2)(x + 4)$ **71.** $(3x + 5)(2x - 3)$
72. $(4x - 1)(2x + 3)$

Assignment Exercises, Chapter 14 (pp. 580–581)

1. $5(x + y)$ **3.** $4(3m^2 - 2n^2)$ **5.** $2a(a^2 - 7a - 1)$ **7.** $5x(3x - 4)(x + 1)$ **9.** $6a^2(3a + 2)$ **11.** $x^2 + 11x + 28$ **13.** $m^2 - 4m - 21$ **15.** $12r^2 - 7r - 10$ **17.** $4 - 14m + 6m^2$ **19.** $2x^2 + x - 15$ **21.** $14a^2 + 19ab - 3b^2$ **23.** $27x^2 + 30xy - 8y^2$ **25.** $21m^2 + 29mn - 10n^2$ **27.** $36x^2 - 25$ **29.** $49y^2 - 121$ **31.** $64a^2 - 25b^2$ **33.** $64x^2 - y^2$ **35.** $9y^2 - 16z^2$ **37.** $x^2 + 18x + 81$ **39.** $x^2 - 6x + 9$ **41.** $16x^2 - 120x + 225$ **43.** $64 + 112m + 49m^2$ **45.** $16x^2 - 88x + 121$ **47.** $g^3 - h^3$ **49.** $8H^3 - 27T^3$ **51.** $216 + i^3$ **53.** $z^3 + 8t^3$ **55.** $343T^3 + 8$ **57.** not difference **59.** difference **61.** difference **63.** not perfect-square trinomial **65.** perfect-square trinomial **67.** perfect-square trinomial **69.** difference of two cubes **71.** difference of two cubes **73.** sum of two cubes **75.** $(5y - 2)(5y + 2)$ **77.** *NSP*, this is a sum of two squares, not a difference. **79.** $(a + 1)^2$ **81.** $(4c - 3b)^2$ **83.** $(n - 13)^2$ **85.** $(6a + 7b)^2$ **87.** $(7 - x)^2$ **89.** *NSP*, this is a sum of two squares, not a difference. **91.** *NSP*, the middle term needs a y factor. **93.** $(7 + 9y)(7 - 9y)$ **95.** $(3x + 10y)(3x - 10y)$ **97.** $(3x - y)^2$ **99.** $(3xy + z)(3xy - z)$ **101.** $(x + 2)^2$

103. $\left(\frac{2}{5}x + \frac{1}{4}y\right)\left(\frac{2}{5}x - \frac{1}{4}y\right)$ **105.** $(T - 2)(T^2 + 2T + 4)$

107. $(d + 9)(d^2 - 9d + 81)$ **109.** $(3K + 4)(9K - 12K + 16)$ **111.** $(x + 8)(x + 3)$ **113.** $(x + 10)(x + 3)$ **115.** $(x - 8)(x - 1)$ **117.** $(x - 13)(x + 2)$ **119.** $(x + 8)(x - 3)$ **121.** $(6x + 1)(x + 4)$ **123.** $(5x - 4)(x - 6)$ **125.** $(3x + 7)(2x - 5)$ **127.** $(7x + 8)(x - 3)$ **129.** $(3a + 10)(3a - 10)$ **131.** $(2x + 1)(x - 2)$ **133.** $(a + 9)(a - 9)$ **135.** $(y - 7)^2$ **137.** $(b + 5)(b + 3)$ **139.** $(13 + m)(13 - m)$ **141.** $(x - 8)(x + 4)$ **143.** $(x + 20)(x - 1)$ **145.** $2(x - 4)(x + 2)$ **147.** $2x(x - 6)(x + 1)$ **149.** This enables us to factor more rapidly and easily.

Trial Test, Chapter 14 (p. 585)

1. $3x + 6y$ **3.** $14x^3 + 21x^2 - 35x$ **5.** $m^2 - 49$ **7.** $a^2 + 6a + 9$ **9.** $2x^2 - 11x + 15$ **11.** $x^3 - 8$ **13.** $125a^3 - 27$ **15.** $x(7x + 8)$ **17.** $7ab(a - 2)$ **19.** $(x + 8)(x + 1)$ **21.** $(x - 9)(x + 4)$ **23.** $(3x + 2)(2x - 3)$ **25.** $(3x + 5)(x + 6)$ **27.** $(a + 8b)^2$ **29.** $(b - 5)(b + 2)$ **31.** $(3x + 4)(x - 1)$ **33.** $(3m - 2)(m - 1)$ **35.** $(3a - 2)(9a^2 + 6a + 4)$

Self-Study Exercises 15-1 (pp. 588–589)

L.O.1.

1. $7x^2 - 4x + 5 = 0$ **2.** $8x^2 - 6x - 3 = 0$ **3.** $7x^2 - 5 = 0$ **4.** $x^2 - 6x + 8 = 0$ **5.** $x^2 - 9x + 8 = 0$ **6.** $x^2 - 4x - 8 = 0$ **7.** $3x^2 - 6x + 5 = 0$ **8.** $x^2 - 6x - 5 = 0$ **9.** $x^2 - 16 = 0$ **10.** $8x^2 - 7x - 8 = 0$ **11.** $8x^2 + 8x - 10 = 0$ **12.** $0.3x^2 - 0.4x - 3 = 0$

L.O.2.

13. $x^2 - 5x = 0$
$a = 1, b = -5, c = 0$

14. $3x^2 - 7x + 5 = 0$
$a = 3, b = -7, c = 5$

15. $7x^2 - 4x = 0$
$a = 7, b = -4, c = 0$

16. $3x^2 - 5x + 8 = 0$
$a = 3, b = -5, c = 8$

17. $x^2 - 5x + 6 = 0$
$a = 1, b = -5, c = 6$

18. $11x^2 - 8x = 0$
$a = 11, b = -8, c = 0$

19. $x^2 - x = 0$
$a = 1, b = -1, c = 0$

20. $9x^2 - 7x - 12 = 0$
$a = 9, b = -7, c = -12$

21. $x^2 - 5 = 0$
$a = 1, b = 0, c = -5$

22. $x^2 + 6x - 3 = 0$
$a = 1, b = 6, c = -3$

23. $5x^2 - 0.2x + 1.4 = 0$
$a = 5, b = -0.2, c = 1.4$

24. $\frac{2}{3}x^2 - \frac{5}{6}x - \frac{1}{2} = 0$
$a = \frac{2}{3}, b = -\frac{5}{6}, c = -\frac{1}{2}$

25. $1.3x^2 - 8 = 0$
$a = 1.3, b = 0, c = -8$

26. $\sqrt{3}\,x^2 + \sqrt{5}\,x - 2 = 0$
$a = \sqrt{3}, b = \sqrt{5}, c = -2$

27. $8x^2 - 2x - 3 = 0$ **28.** $5x^2 + 2x - 7 = 0$ **29.** $x^2 + 3x = 0$

Self-Study Exercises 15-2 (p. 591)

L.O.1.

1. $x = \pm 3$ **2.** $x \pm 7$ **3.** $x = \pm\frac{8}{3}$ **4.** $x = \pm\frac{7}{4}$ **5.** $y = \pm 3$ **6.** $x = \pm\frac{9}{2}$ **7.** $x = \pm\sqrt{5}$ or ± 2.236 **8.** $x = \pm 1$ **9.** $x = \pm 2$ **10.** $x = \pm 1.732$ **11.** Isolate the squared letter, then take the square root of both sides. **12.** opposites

Self-Study Exercises 15-3 (p. 593)

L.O.1.

1. $x = 3, 0$ **2.** $x = 0, \frac{7}{3}$ **3.** $x = 0, 2$ **4.** $x = 0, -\frac{1}{2}$ **5.** $x = 0, \frac{1}{2}$ **6.** $x = 0, 3$ **7.** $x = 0, 6$ **8.** $y = 0, -4$ **9.** $x = 0, -\frac{2}{3}$ **10.** $x = 0, \frac{4}{3}$ **11.** An incomplete quadratic equation is missing the number term while a pure quadratic equation is missing the linear term.
12. Yes, the common factor of x will be set equal to zero.

Self-Study Exercises 15-4 (p. 595)

L.O.1.

1. $-3, -2$ **2.** $3, 3$ **3.** $7, -2$ **4.** $3, -6$ **5.** $-4, -3$ **6.** $5, 3$ **7.** $14, -1$ **8.** $3, 6$ **9.** $\frac{1}{2}, 3$ **10.** $-\frac{1}{3}, -4$ **11.** $\frac{3}{5}, -\frac{1}{2}$ **12.** $-\frac{1}{3}, -\frac{3}{2}$ **13.** $-\frac{3}{2}, -5$ **14.** $\frac{4}{3}, 2$ **15.** $\frac{1}{6}, -3$

Self-Study Exercises 15-5 (p. 600)

L.O.1.

1. $3, -\frac{2}{3}$ **2.** $3, -4$ **3.** $\frac{11}{5}, -1$ **4.** $3, 3$ **5.** $3, -2$ **6.** $\frac{3}{4}, -\frac{1}{2}$ **7.** $1.84, -10.84$ **8.** $-0.18, -1.82$ **9.** $3.14, -0.64$ **10.** $3.58, 0.42$

L.O.2.

11. width = 5.55 cm, length = 8.55 cm
12. width = 4 in., length = 10 in.
13. 4 m **14.** width = 11 ft, length = 16 ft
15. length = 110 ft, width = 70 ft (nearest ft) **16.** 111.5 kg

Self-Study Exercises 15-6 (p. 602)

L.O.1.

1. $x = \frac{2}{3}, -1$ **2.** $x = \frac{3 \pm \sqrt{5}}{2}$ or 2.62, 0.38 **3.** $x = \frac{-1 \pm \sqrt{17}}{4}$ or 0.78, −1.28 **4.** no real solutions **5.** $x = \frac{3 \pm \sqrt{37}}{2}$ or 4.54, −1.54 **6.** $x = \frac{-5 \pm \sqrt{97}}{6}$ or 0.81 or −2.47 **7.** If a is positive and c is negative the equation will have real roots. **8.** If $b^2 - 4ac$ is a perfect square and is positive, the roots will be real, rational, and unequal.

Self-Study Exercises 15-7 (p. 604)

L.O.1.

1. 2 **2.** 1 **3.** 4 **4.** 1 **5.** 4 **6.** 7 **7.** $x = 0, 2, -3$
8. $x = 0, \frac{1}{2}, -3$ **9.** $x = 0, \frac{5}{2}, \frac{2}{3}$ **10.** $x = 0, 2, 5$ **11.** $x = 0, 3, -2$
12. $x = 0, -\frac{1}{2}, -2$ **13.** $x = 0, 3, -3$ **14.** $x = 0, \frac{1}{2}, -\frac{1}{2}$
15. $x = 0, \frac{3}{4}, -\frac{3}{4}$

Assignment Exercises, Chapter 15 (pp. 604–606)

1. pure **3.** pure **5.** incomplete **7.** pure **9.** complete
11. $x = \pm 10$ **13.** $x = \pm\frac{3}{2}$ **15.** $y = \pm 1.740$ **17.** $x = \pm 2.828$
19. $x = \pm 2.236$ **21.** $x = \pm 2$ **23.** $x = \pm 4.123$ **25.** $y = \pm 3.055$
27. $x = \pm 4$ **29.** $x = \pm 8$ **31.** 0, 5 **33.** 0, 2 **35.** $0, -\frac{1}{2}$
37. 0, 5 **39.** $0, -\frac{2}{3}$ **41.** 0, −3 **43.** 0, 9 **45.** 0, −8 **47.** $0, \frac{5}{3}$
49. $0, \frac{1}{2}$ **51.** 3, 1 **53.** −5, 2 **55.** −6, −1 **57.** 4, 2
59. $-\frac{2}{3}, \frac{3}{2}$ **61.** $-\frac{2}{5}, \frac{5}{2}$ **63.** $-\frac{3}{4}, -1$ **65.** $\frac{3}{4}, -\frac{1}{3}$ **67.** −21, 2
69. $\frac{2}{3}, -1$ **71.** 3, 2 **73.** 6, −3 **75.** −6, −5 **77.** $a = 1$, $b = -2$, $c = -8$
79. $a = 1$, $b = 3$, $c = -4$ **81.** $a = 1$, $b = -3$, $c = 2$ **83.** 9, 1 **85.** 2, −4 **87.** $2, -\frac{1}{2}$
89. 1.78, −0.28 **91.** −0.23, −1.4 **93.** 22, 11 **95.** 42, 14
97. real, rational, equal **99.** real, irrational, unequal

101. real, rational, unequal **103.** 3 **105.** 1 **107.** 8 **109.** $0, 2, \frac{2}{3}$

111. $0, -2, -3$ **113.** $0, -5, \frac{1}{2}$ **115.** $0, \frac{2}{3}, -3$ **117.** $0, -2, -4$

119. $0, 5, -4$ **121.** $0, 9, -3$ **123.** $0, 8, -5$ **125.** $0, -2, -4$

127. length = 25.9 ft, height = 16 feet

Trial Test, Chapter 15 (pp. 609–610)

1. pure **3.** incomplete **5.** ± 9 **7.** $\pm\frac{4}{3}$ **9.** ± 2.33 **11.** 0, 2

13. 3, 2 **15.** $-\frac{3}{2}, -4$ **17.** 169.79 mils **19.** 1.98 amps

21. $0.37, -5.36$ **23.** 1.7 ft, 4.7 ft **25.** 6 **27.** $0, \frac{6}{5}, \frac{2}{3}$

29. $0, \frac{3}{2}, -5$

Self-Study Exercises 16-1 (p. 614)

L.O.1.

1. $\frac{4}{9}$ **2.** $\frac{3}{8}$ **3.** $2AB^2$ **4.** $\frac{x}{x + 2}$ **5.** $\frac{1}{2}$ **6.** $\frac{3(3x + 2)}{x(2x - 1)}$ or $\frac{9x + 6}{2x^2 - x}$

7. $\frac{1}{A - B}$ **8.** $\frac{x^2 - 3x - 10}{x^2 + 3x - 10}$ **9.** $\frac{x - 2}{x + 2}$ **10.** $\frac{2x - 4}{3}$ or $\frac{2(x - 2)}{3}$

Self-Study Exercises 16-2 (pp. 614–617)

L.O.1.

1. $\frac{10x^3}{9y^2}$ **2.** $3(A - B)$ or $3A - 3B$ **3.** 4 **4.** $\frac{A - 7}{B + 5}$

5. $\frac{3(x + 1)}{x - 1}$ or $\frac{3x + 3}{x - 1}$ **6.** $\frac{1}{12A}$ **7.** $\frac{1}{2}$ **8.** $\frac{x}{x - 2}$

9. $\frac{50(A - B)}{AB}$ or $\frac{50A - 50B}{AB}$ **10.** 1

Self-Study Exercises 16-3 (p. 620)

L.O.1.

1. $\frac{5}{7}$ **2.** $\frac{13}{16}$ **3.** $\frac{3x}{2}$ **4.** $\frac{23x}{24}$ **5.** $\frac{4 + 3x}{2x}$ **6.** $\frac{55}{6x}$

7. $\frac{9x - 1}{(x + 1)(x - 1)}$ or $\frac{9x - 1}{x^2 - 1}$ **8.** $\frac{10x + 6}{(x + 3)(x - 1)}$ or $\frac{10x + 6}{x^2 + 2x - 3}$

9. $\frac{x - 7}{(2x + 1)(x - 1)}$ or $\frac{x - 7}{2x^2 - x - 1}$

10. $\frac{x + 1}{(x - 6)(x - 5)}$ or $\frac{x + 1}{x^2 - 11x + 30}$

Self-Study Exercises 16-4 (pp. 623–624)

L.O.1.

1. $\frac{25}{27}$ **2.** $1\frac{1}{20}$ **3.** $\frac{x - 5}{12x}$ **4.** $\frac{2}{3(x^2 - 2x)}$ or $\frac{2}{3x(x - 2)}$ **5.** $\frac{x + 3}{2}$

6. $\frac{x - 3}{3}$ **7.** $3(x - 6)$ or $3x - 18$ **8.** $\frac{3x^2}{4}$ **9.** $\frac{5(2x + 1)}{3x}$

10. $\frac{8(3x + 2)}{5x}$ or $\frac{24x + 16}{5x}$ **11.** $-\frac{28}{9}$ or $-3\frac{1}{9}$ **12.** $-\frac{1}{7}$

Self-Study Exercises 16-5 (p. 629)

L.O.1.

1. 0 **2.** 0 **3.** 5, 0 **4.** 0, −9 **5.** −8, 8 **6.** 2, −2 **7.** $\frac{3}{4}$, 0
8. 0, −3

L.O.2.

9. 35 **10.** 5 **11.** 1 **12.** $\frac{1}{4}$ **13.** 4 **14.** $-\frac{4}{5}$ **15.** $1\frac{5}{7}$ days
16. 8 investors **17.** 25 mph going, 40 mph returning **18.** 3 min
19. 50 lb dark-roast, 75 lb medium-roast, $2 per pound **20.** 6 ohms

Assignment Exercises, Chapter 16 (pp. 629–631)

1. $\frac{3}{4}$ **3.** $\frac{B^2}{2AC}$ **5.** $\frac{x - 3}{2(x + 3)}$ **7.** 1 **9.** $\frac{M^2 - N^2}{M^2 + N^2}$ **11.** $\frac{1}{1 + y}$
13. 5 **15.** $y + 1$ **17.** $\frac{2}{x + 6}$ **19.** 3 **21.** $\frac{5x^3}{4y^2}$ **23.** 15
25. $\frac{4(2y - 1)}{y - 1}$ **27.** 1 **29.** $\frac{x + 3}{x + 2}$ **31.** $\frac{1}{4}$ **33.** $\frac{(y - 1)^3}{y}$
35. $\frac{y + 3}{y + 2}$ **37.** $\frac{3x^2}{2}$ **39.** $(y + 4)(y - 3)$ **41.** $\frac{2}{3}$ **43.** $\frac{4x}{7}$
45. $\frac{19x}{12}$ **47.** $\frac{15 - 7x}{3x}$ **49.** $\frac{13}{4x}$ **51.** $\frac{10x + 5}{(x - 3)(x + 2)}$
53. $\frac{6x - 38}{(x + 3)(x - 4)}$ **55.** $\frac{x + 3}{x - 5}$ **57.** $\frac{5}{4x - 12}$ **59.** $\frac{4x^2}{3}$ **61.** $\frac{51}{28}$
63. $\frac{3x^2 - 30}{2x^2 + 12}$ **65.** 0, 2 **67.** 0, $\frac{1}{2}$ **69.** $\frac{-20}{3}$ **71.** $\frac{1}{7}$
73. 3 students **75.** $7\frac{1}{2}$ hr

Trial Test, Chapter 16 (pp. 635–636)

1. $\frac{1}{2}$ **3.** $\frac{3x - 4}{x + 3}$ **5.** $\frac{x - 2}{x + 2}$ **7.** $\frac{3A}{y}$ **9.** $\frac{1}{x^2(x + 2y)}$ **11.** $\frac{2x + 1}{x}$
13. $-\frac{5}{(x + 2)(x - 3)}$ **15.** $\frac{12 + x}{4x}$ **17.** $\frac{26x - 4}{(3x - 2)(x + 2)}$
19. $\frac{1}{x + 2y}$ **21.** $\frac{2x^2}{x - 3}$ **23.** 0, −3 **25.** $\frac{11}{7}$ **27.** $-\frac{4(y - 2)}{y + 1}$
29. 5 persons

Self-Study Exercises 17-1 (p. 640)

L.O.1.

1. F **2.** F **3.** T **4.** T

L.O.2.

5. [8, 12]

6. $(5, \infty)$

5

7. $(-\infty, -2)$

−2 0

8. $(-6, \infty)$

−6

9. (843,000 , 1,000,000)

Self-Study Exercises 17-2 (p. 645)

L.O.1.

1. $x < 6$; $(-\infty, 6)$

$x < 6$

4 5 6 7 8

2. $a \geq -3$; $[-3, \infty)$

$a \geq -3$

−4 −3 −2 −1 0 1

3. $y \leq 8$; $(-\infty, 8]$

$y \leq 8$

6 7 8 9 10

4. $b > -1$; $(-1, \infty)$

$b > -1$

−3 −2 −1 0 1 2

5. $t < 6$; $(-\infty, 6)$

$t < 6$

2 3 4 5 6

6. $y < 5$; $(-\infty, 5)$

$y < 5$

3 4 5 6 7

7. $a \leq -2$; $(-\infty, -2]$

$a \leq -2$

−4 −3 −2 −1 0 1

8. $x \leq \frac{1}{4}$; $\left(-\infty, \frac{1}{4}\right]$

$x \leq \frac{1}{4}$

−3 −2 −1 0 $\frac{1}{4}$ 1 2

9. $t \leq 7$; $(-\infty, 7]$

$t \leq 7$

5 6 7 8 9 10

10. $y > 11$; $(11, \infty)$

$y > 11$

9 10 11 12 13 14

11. $x \geq 3$; $[3, \infty)$

$x \geq 3$

0 1 2 3 4 5

12. $x < 1$; $(-\infty, 1)$

$x < 1$

−1 0 1 2 3 4 5

13. $x > \frac{1}{3}$; $\left(\frac{1}{3}, \infty\right)$

$x > \frac{1}{3}$

−1 0 $\frac{1}{3}$ 1 2 3

14. $a \leq -3$; $(-\infty, -3]$

$a \leq -3$

−4 −3 −2 −1 0 1

15. $x < \frac{1}{2}$; $\left(-\infty, \frac{1}{2}\right)$

$x < \frac{1}{2}$

−2 −1 0 $\frac{1}{2}$ 1 2 3

16. $2x + 196 \geq 52{,}800$
$x \geq 26{,}302$

Self-Study Exercises 17-3 (pp. 652–653)

L.O.1.

1. T; all elements in U **2.** F; 2 is not in A **3.** T; $\emptyset$ is a subset of all sets.
4. F; No element of A is in B **5.** $\{\ \}$ or $\emptyset$ **6.** $\{2, 0, -2\}$
7. $\{-4, -2, 0, 1, 2, 3, 4, 5\}$ **8.** $\{-5, -3, -1, 1, 2, 3, 4, 5\}$
9. $\{-5, -4, -3, -2, -1, 1, 2, 3, 4, 5\}$ **10.** $\{-5, -4, -3, -1, 0\}$

L.O.2.

11. $2 < x < 6$; $(2, 6)$

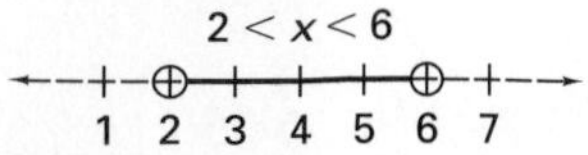

$2 < x < 6$

1 2 3 4 5 6 7

12. $2 < x < 7$; $(2, 7)$

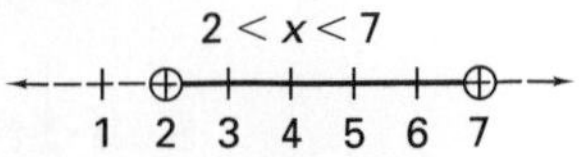

$2 < x < 7$

1 2 3 4 5 6 7

13. No solution; $\emptyset$

14. $x = 5$; $[5]$

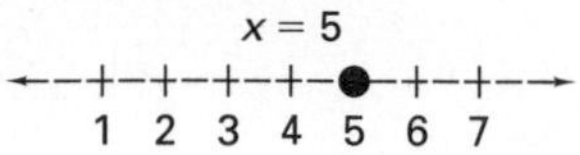

$x = 5$

1 2 3 4 5 6 7

15. No solution; $\emptyset$

16. $3 < x < 7$; $(3, 7)$

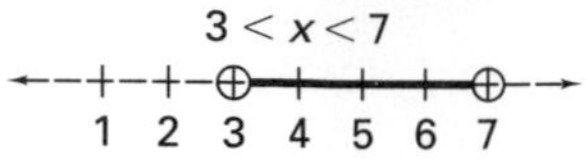

$3 < x < 7$

1 2 3 4 5 6 7

17. $0 < x < 3$; $(0, 3)$

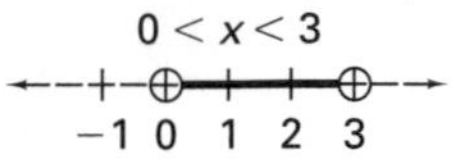

$0 < x < 3$

−1 0 1 2 3

18. $-2 \le x \le 2$; $[-2, 2]$

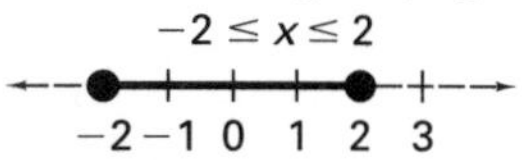

$-2 \le x \le 2$

−2 −1 0 1 2 3

19. $-6 < x < -3$; $(-6, -3)$

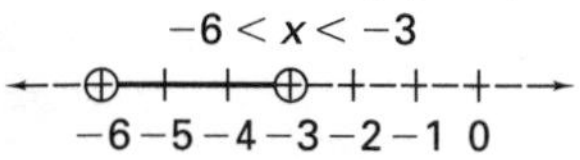

$-6 < x < -3$

−6 −5 −4 −3 −2 −1 0

20. $3 < x < 5$; $(3, 5)$

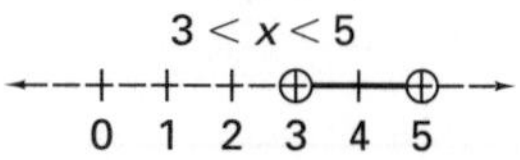

$3 < x < 5$

0 1 2 3 4 5

21. $x < 2$ or $x > 9$

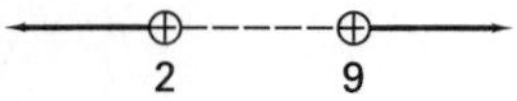

2 9

$(-\infty, 2)$ or $(9, \infty)$

22. $x \le 3$ or $x \ge 7$

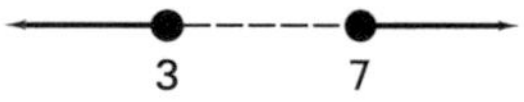

3 7

$(-\infty, 3]$ or $[7, \infty)$

23. $x < 4$ or $x \ge 5$

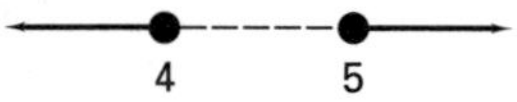

4 5

$(-\infty, 4]$ or $[5, \infty)$

24. $x \le 1$ or $x \ge 3$

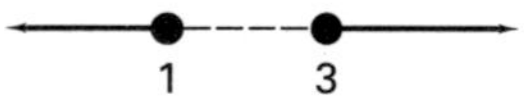

1 3

$(-\infty, 1]$ or $[3, \infty)$

25. $x \le 1\frac{1}{2}$ or $x \ge 3\frac{1}{2}$

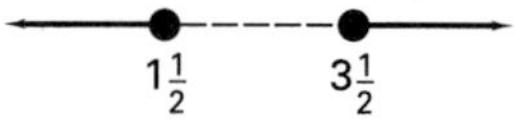

$1\frac{1}{2}$ $3\frac{1}{2}$

$\left(-\infty, 1\frac{1}{2}\right]$ or $\left[3\frac{1}{2}, \infty\right)$

26. $5.22 \le 5.27 \le 5.32$;
$x - 0.05 \le x \le x + 0.05$

Self-Study Exercises 17-4 (p. 660)

L.O.1.

1. $-3 < x < 2$

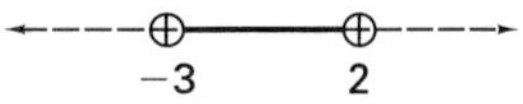

−3 2

2. $x < -3$ or $x > \frac{1}{2}$

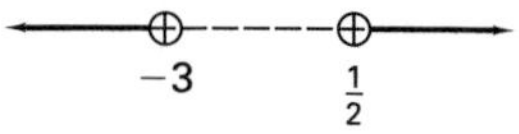

−3 $\frac{1}{2}$

3. $\frac{2}{3} < x < \frac{5}{2}$

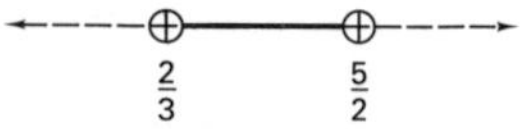

$\frac{2}{3}$ $\frac{5}{2}$

4. $y \le -6$ or $y \ge 2$

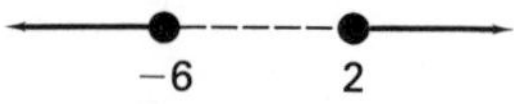

−6 2

5. $-4 < a < 6$

6. $2 \le x \le 5$

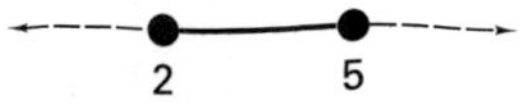

7. $x \le -2$ or $x \ge 3$

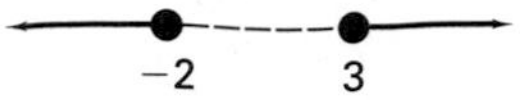

8. $1 \le x \le 3$

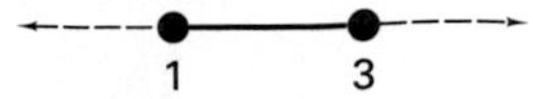

9. $\frac{1}{4} \le x \le 5$

10. $x < -\frac{1}{2}$ or $x > \frac{1}{3}$

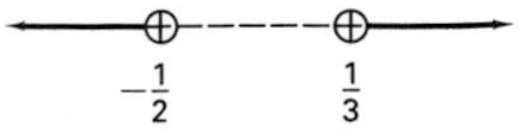

L.O.2.

11. $-7 < x < 3$

12. $x < -8$ or $x > 3$

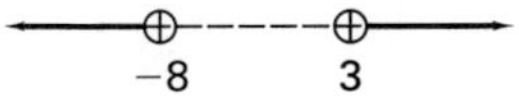

13. $x < \frac{1}{3}$ or $x > 3$

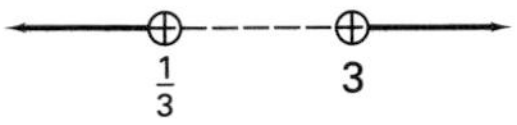

14. $-7 < x < 7$

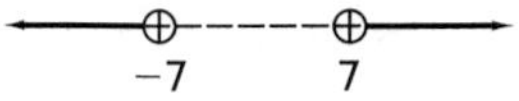

15. $0 < x < 1$

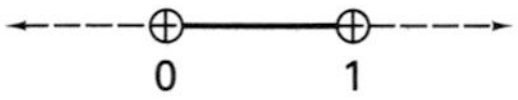

Self-Study Exercises 17-5 (p. 663)

L.O.1.

1. ± 8 **2.** $8, -2$ **3.** $9, 5$ **4.** $6, -3$ **5.** $3, \frac{5}{3}$ **6.** ± 11

7. $10, -6$ **8.** $7, 1$ **9.** $4, -3$ **10.** $\frac{7}{3}, -1$

Self-Study Exercises 17-6 (p. 666)

L.O.1.

1. $-5 < x < 5$

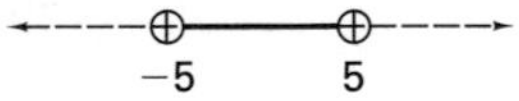

2. $-1 < x < 1$

3. $-2 < x < 2$

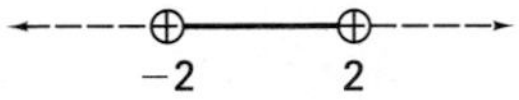

4. $-7 \le x \le 7$

5. $-5 < x < -1$

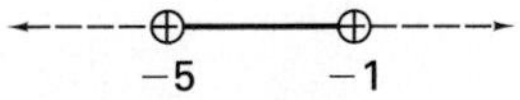

6. $-7 < x < -1$

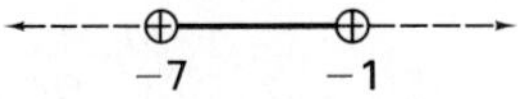

7. $-1 < x < 7$

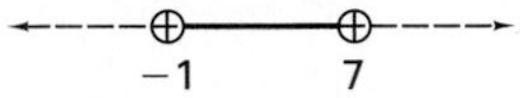

8. $-1 < x < 5$

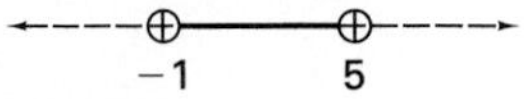

9. $-1 \le x \le 11$

10. $-6 \le x \le 8$

11. $-\frac{2}{3} < x < 4$

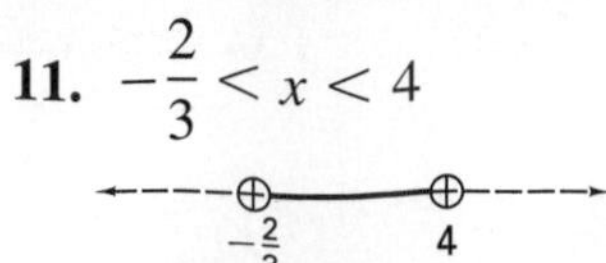

12. $\frac{1}{2} < x < \frac{7}{2}$

13. $-1 \leq x \leq \frac{1}{5}$

14. $-3 \leq x \leq -\frac{1}{2}$

L.O.2.

15. $x < -2$ or $x > 2$

16. $x < -3$ or $x > 3$

17. $x \leq -5$ or $x \geq 5$

18. $x \leq -6$ or $x \geq 6$

19. $x < 3$ or $x > 7$

20. $x < -2$ or $x > 10$

21. $x \leq -8$ or $x \geq 2$

22. $x \leq -10$ or $x \geq 2$

23. $x \leq \frac{5}{2}$ or $x \geq \frac{5}{2}$
or all real numbers

24. $x \leq -\frac{2}{3}$ or $x \geq 2$

25. $x \leq -2$ or $x \geq -\frac{6}{5}$

26. $x \leq -\frac{3}{2}$ or $x \geq 3$

27. $\left\{x \,\middle|\, \frac{1}{3} \leq x \leq 1\right\}$

28. $\{x \mid 0 \leq x \leq 5\}$

29. $\left\{x \,\middle|\, x \leq -\frac{4}{5} \text{ or } x \geq 2\right\}$

30. $\left\{x \,\middle|\, x \leq 1\frac{1}{4} \text{ or } x \geq 1\frac{3}{4}\right\}$

31. no solution
{ } or $\emptyset$

32. $\{x \mid x \in R\}$ all reals

33. $x \leq 0$ or $x \geq 6$ **34.** $x < -\frac{1}{2}$ or $x > 0$ **35.** $1 \leq x \leq 7$
36. no solution **37.** $-7 < x < 7$ **38.** $-11 < x < -5$
39. $x < -12$ or $x > 12$ **40.** $x < -10$ or $x > -6$

Assignment Exercises, Chapter 17, (pp. 666–667)

L.O.1.

1. a set with no elements { } or $\emptyset$ **3.** $5 \in W$

5. $(-7, \infty)$

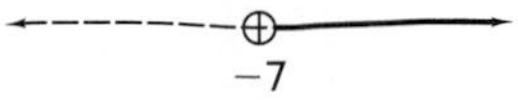

-7

7. $[-4, 2)$

-4 2

9. $(-2, \infty)$

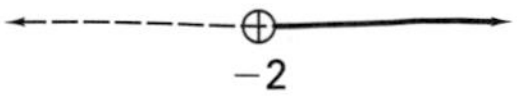

-2

11. $m < 7$

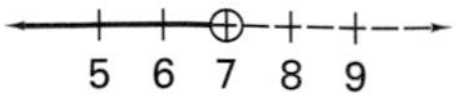

5 6 7 8 9

13. $x > 0$

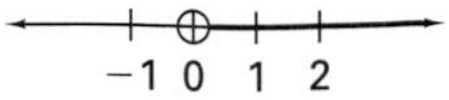

−1 0 1 2

15. $x \leq 3$

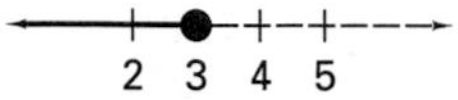

2 3 4 5

17. $x > -9$

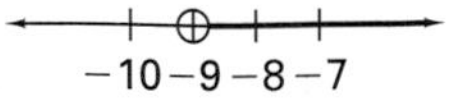

−10 −9 −8 −7

19. $x > -1$

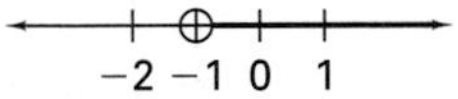

−2 −1 0 1

21. $y \geq -1$

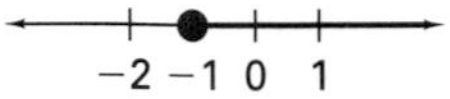

−2 −1 0 1

23. $x \leq 6$

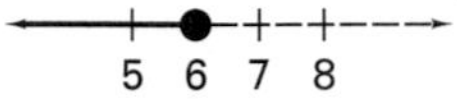

5 6 7 8

25. $b > 7$

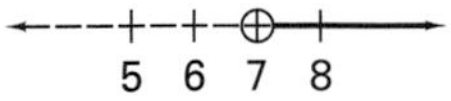

5 6 7 8

27. { } or $\emptyset$ **29.** $\{-1\}$ **31.** $\{5\}$

33. $(-2, \infty)$

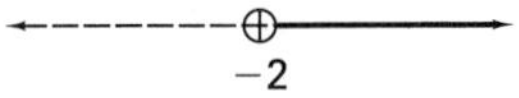

-2

35. $[-3, \infty)$

-3

37. $\left(\frac{1}{2}, \frac{9}{2}\right)$

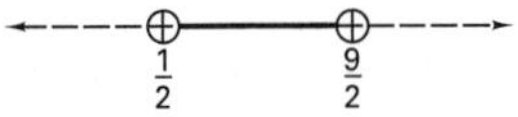

$\frac{1}{2}$ $\frac{9}{2}$

39. $2 < x < 4$; $(2, 4)$

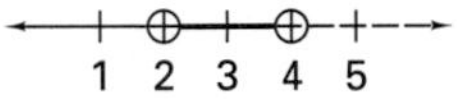

1 2 3 4 5

41. no solution; $\emptyset$

43. $2 < x < 6$; $(2, 6)$

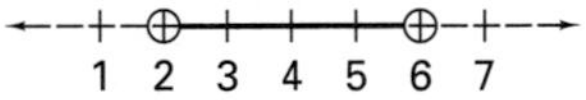

1 2 3 4 5 6 7

45. $-3 < x \leq 2$; $(-3, 2]$

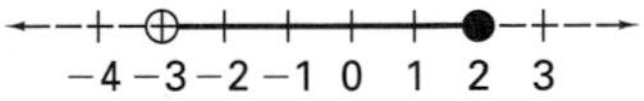

−4 −3 −2 −1 0 1 2 3

47. $-2 \leq x \leq -\frac{3}{4}$; $[-2, \frac{3}{4}]$

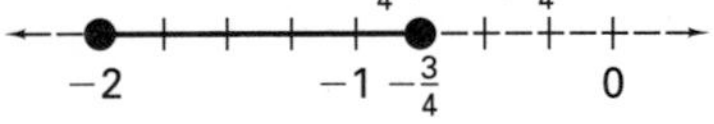

−2 −1 $-\frac{3}{4}$ 0

49. $x < 2$ or $x > 8$

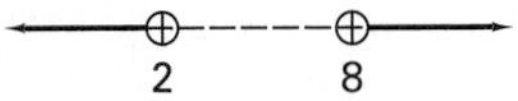

2 8

$(-\infty, 2)$ or $(8, \infty)$

51. $x < -9$ or $x > 8$

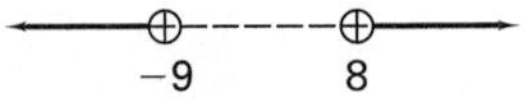

−9 8

$(-\infty, -9)$ or $(8, \infty)$

53. $x < 2$ or $x > 5$

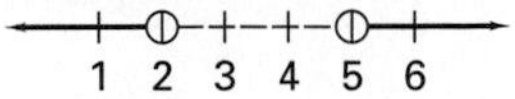

1 2 3 4 5 6

55. $-\frac{1}{3} < x < \frac{3}{2}$

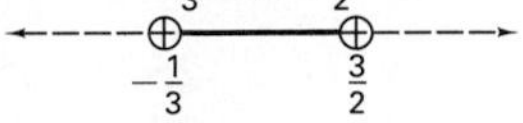

$-\frac{1}{3}$ $\frac{3}{2}$

57. $-1 \leq x \leq 2$

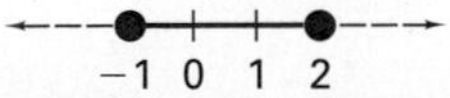

−1 0 1 2

59. $-\frac{1}{2} \leq x \leq 3$

$-\frac{1}{2}$ 3

61. $-5 < x < \frac{3}{2}$

63. $-1 < x < 7$

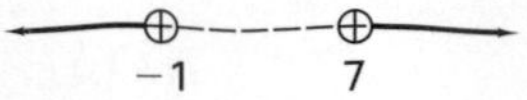

65. $x < -8$ or $x > 0$

67. ± 12 69. $4, -10$ 71. $20, -4$ 73. $6, -\frac{5}{2}$ 75. $1, -\frac{23}{7}$

77. $3, -\frac{13}{7}$ 79. $\frac{11}{3}, \frac{7}{3}$ 81. ± 7 83. ± 16 85. $10, -4$

87. $2, -\frac{1}{2}$ 89. $-1, -3$

91. $-1 < x < 7$

93. $-4 < x < 12$

95. no solution 97. $\$108{,}000 < I < \$250{,}000$

Trial Test, Chapter 17 (pp. 671–672)

L.O.1.

1. $x \geq -12$

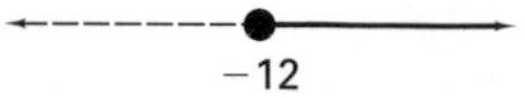

$[-12, \infty)$

3. $x > 3$

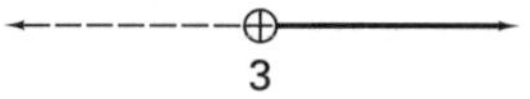

$(3, \infty)$

5. $x > 2$

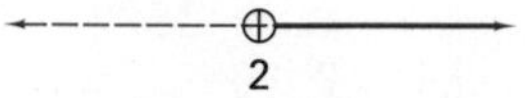

$(2, \infty)$

7. $x \leq -6$

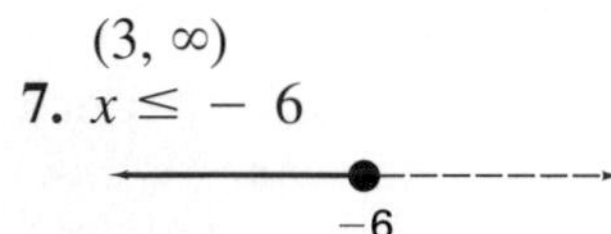

$(-\infty, -6]$

9. $x > -2$

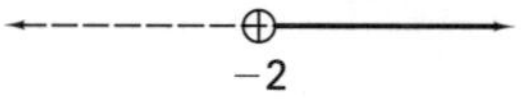

$(-2, \infty)$

11. $-8 < x < 4$

$(-8, 4)$

13. no solution

15. $x < -\frac{3}{2}$ or $x > 1$

$\left(-\infty, -\frac{3}{2}\right)$ or $(1, \infty)$

17. $x < 2$ or $x > 6$

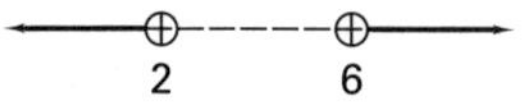

$(-\infty, 2)$ or $(6, \infty)$

19. $A \cup B = \{1, 2, 3, 4, 5, 6, 7, 8, 9\}$

21. $A \cap B' = \{9\}$

23. $-5 < x < 2$

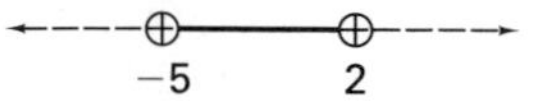

$(-5, 2)$

25. ± 15 27. ± 2

29. $x < -2$ or $x > 2$

$(-\infty, -2)$ or $(2, \infty)$

Self-Study Exercises 18-1 (pp. 678–679)

1.–5.

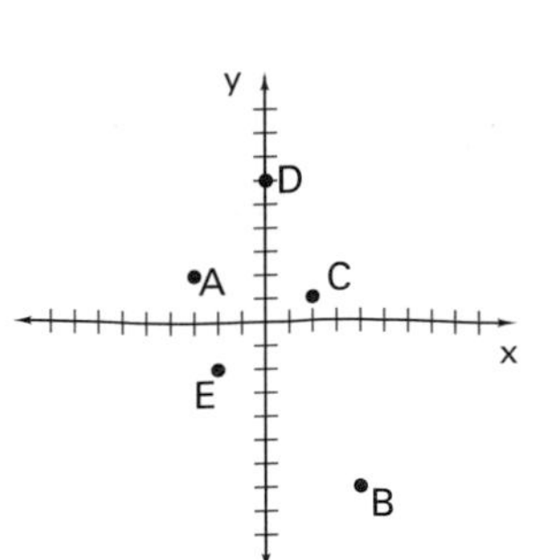

6.–10.

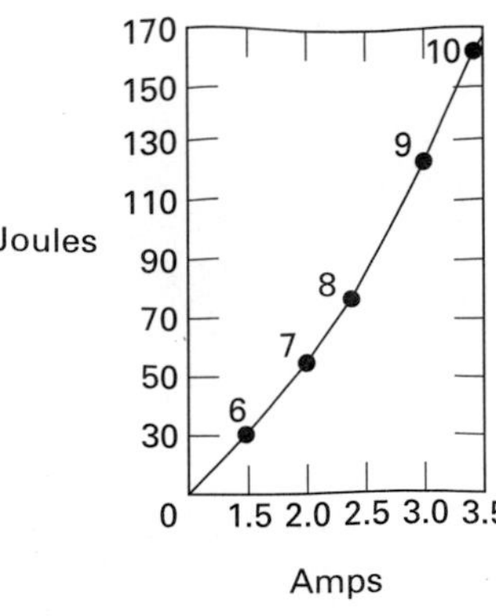

11. II **12.** IV **13.** III **14.** I **15.** II **16.** (9, 1) **17.** (0, −7) **18.** (−3, −3) **19.** (−1, 2) **20.** (0, 9) **21.** (0, 0) **22.** (−7, 0) **23.** (−4, 8) **24.** (8, −6) **25.** (2, 5)

Self-Study Exercises 18-2 (pp. 683–684)

L.O.1.

1. no **2.** yes **3.** yes **4.** no **5.** yes **6.** yes **7.** yes **8.** no **9.** yes **10.** no **11.** yes **12.** yes **13.** yes **14.** no **15.** no **16.** no **17.** yes **18.** no

L.O.2.

19. $y = 5$ **20.** $y = 5$ **21.** $x = -6$ **22.** $x = 21$ **23.** $y = 15$ **24.** $y = 0$ **25.** $x = 8$ **26.** $x = 3$ **27.** $y = 2$ **28.** $y = 1$ **29.** $x = 12$ **30.** $y = -12$ **31.** $x = 18$ **32.** $y = 12$ **33.** (1, 3); (1, −2); (1, 0); (1, 1) **34.** (−1, 4); (3, 4); (0, 4); (2, 4) **35.** (2, −7); (−3, −7); (0, −7); (5, −7) **36.** (9, −2); (9, 3); (9, 0); (9, 2) **37.** \$12

38. \$2 **39.** \$136 **40.** \$13,800 **41.** $f(3) = 9$, $f(-2) = -11$, $f(0) = -3$

42. $f(3) = 13$, $f(-2) = 1$, $f(0) = 7$ **43.** $f(3) = -13$, $f(-2) = 7$, $f(0) = -1$ **44.** $f(3) = -16$, $f(-2) = 14$, $f(0) = 2$

45. $f(3) = 2\frac{1}{2}$, $f(-2) = -\frac{5}{6}$, $f(0) = \frac{1}{2}$ **46.** $f(3) = 1\frac{3}{8}$, $f(-2) = -\frac{1}{2}$, $f(0) = \frac{1}{4}$ **47.** $f(3) = 1\frac{2}{5}$, $f(-2) = 2\frac{2}{5}$, $f(0) = 2$

48. $f(3) = -1\frac{2}{5}$, $f(-2) = 2\frac{3}{5}$, $f(0) = 1$ **49.** $f(3) = 1.87$, $f(-2) = -1.63$, $f(0) = -0.23$ **50.** $f(3) = 6.8$, $f(-2) = -3.7$, $f(0) = 0.5$

51. $f(3) = -12.2$, $f(-2) = -0.2$, $f(0) = -5$ **52.** $f(3) = -3.3$, $f(-2) = -1.8$, $f(0) = -2.4$ **53.** $f(-2) = -3$

54. $f(-1) = 12$ **55.** $f(0) = -4$ **56.** $f(-5) = 31$

Self-Study Exercises 18-3 (pp. 694–695)

Since plotted solutions may vary, check solutions by comparing the x- and y-intercepts.

1.

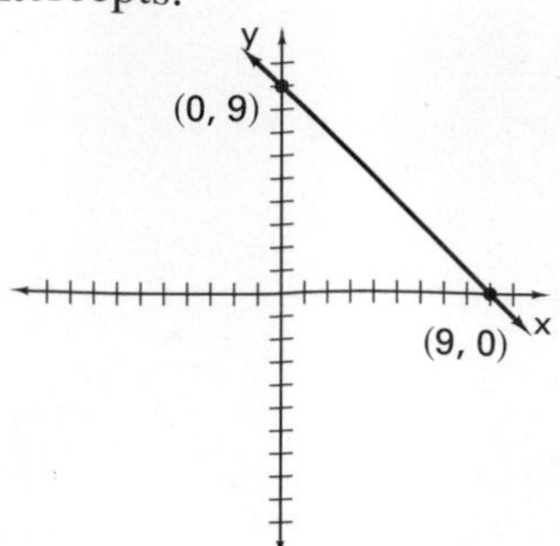

2.

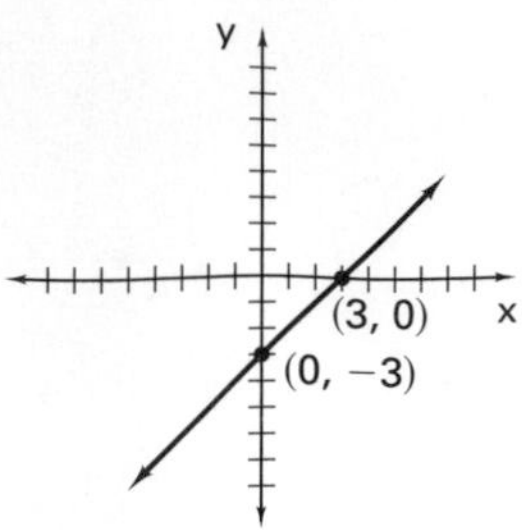

3.

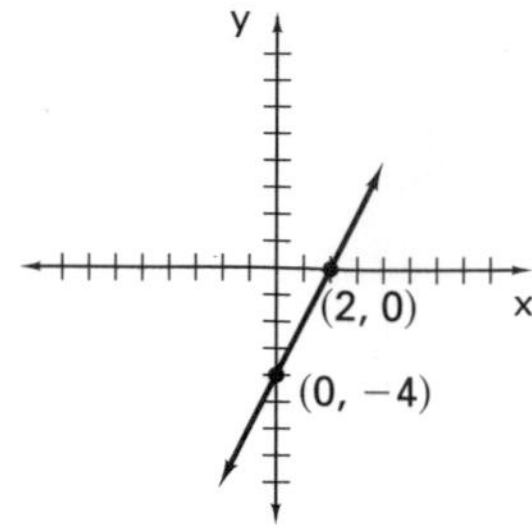

4.

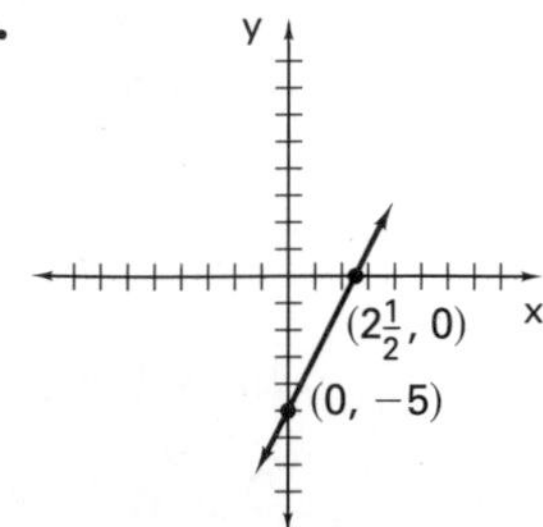

5.

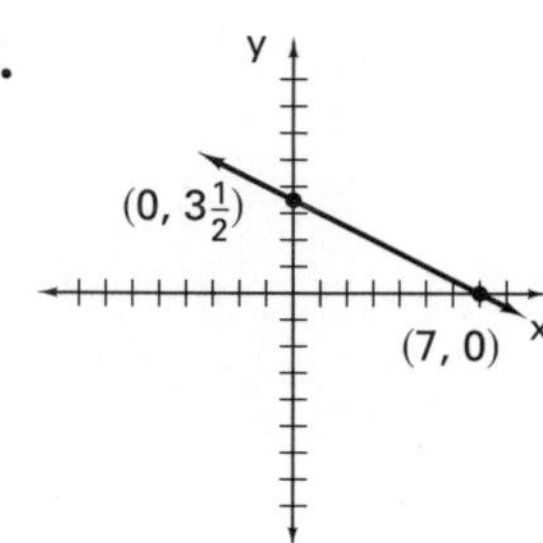

6. x-intercept, (5, 0)

y-intercept, $\left(0, \frac{5}{3}\right)$

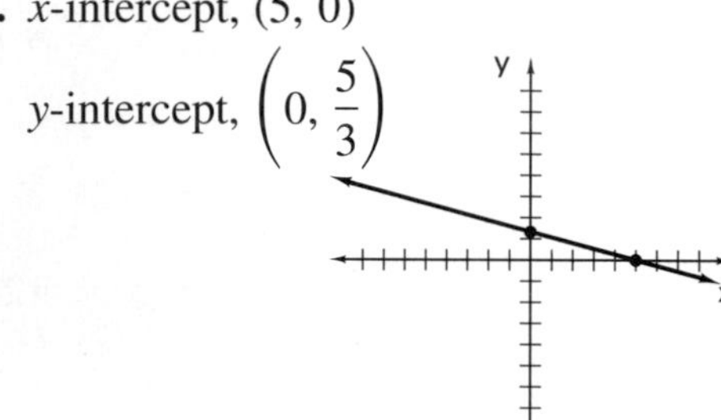

7. x-intercept, $(-4, 0)$

y-intercept, (0, 8)

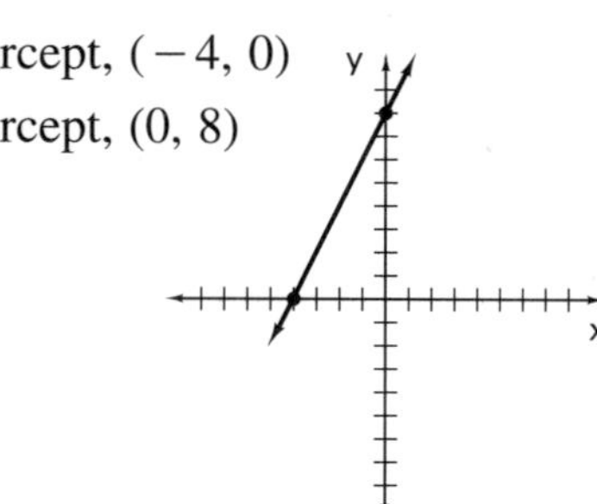

8. x-intercept, $\left(\frac{1}{3}, 0\right)$

y-intercept, $(0, -1)$

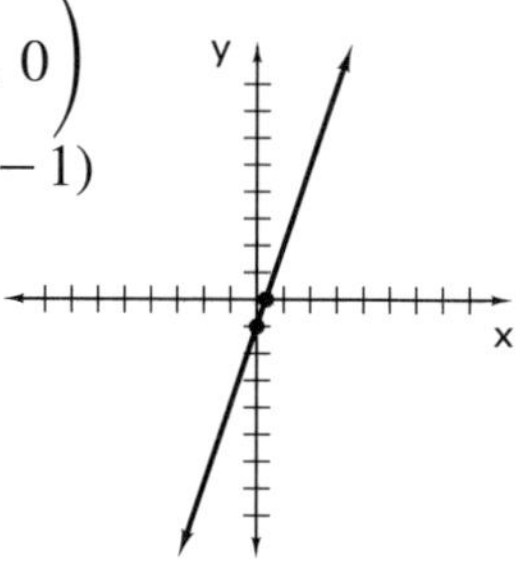

9. x-intercept, $\left(\frac{2}{5}, 0\right)$

y-intercept, $(0, -2)$

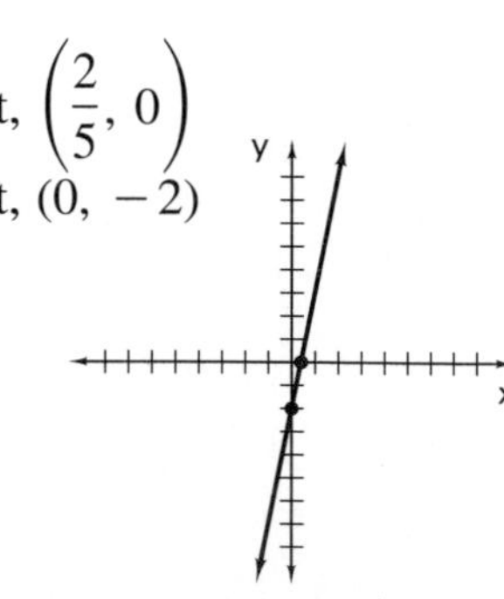

10. x-intercept, (3, 0)

y-intercept, $\left(0, -\frac{3}{2}\right)$

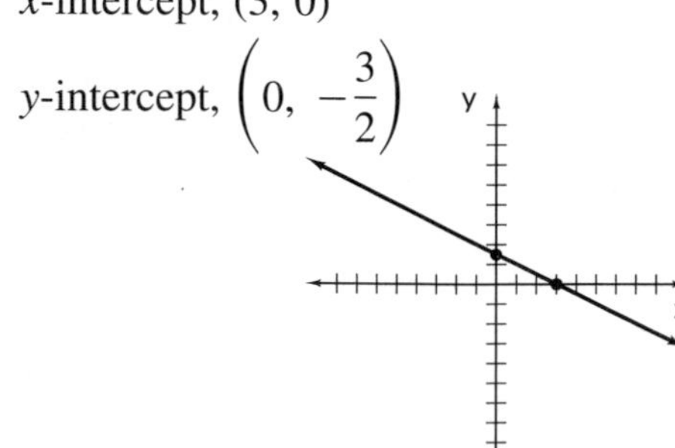

11.

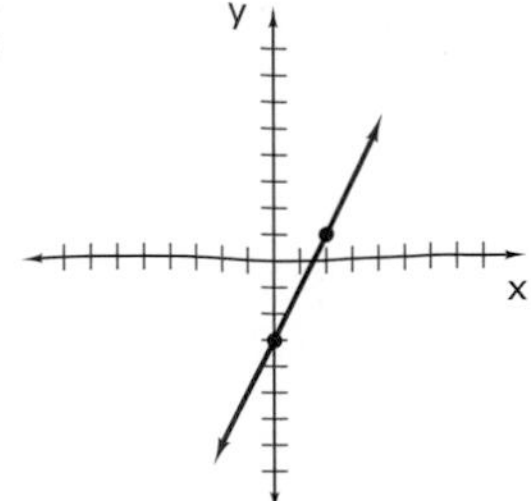

12.

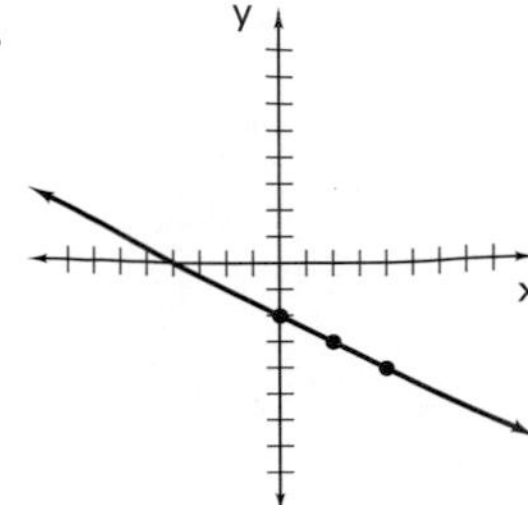

13.

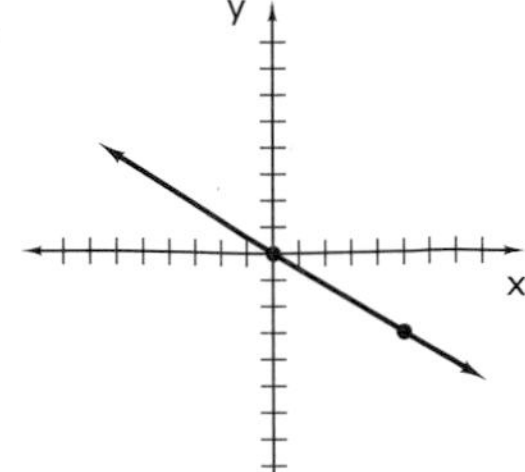

14.

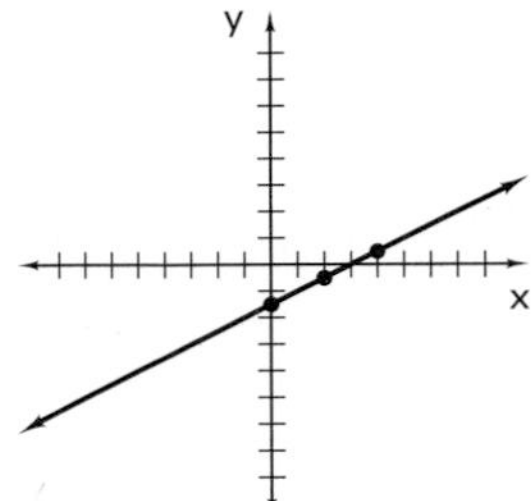

15.

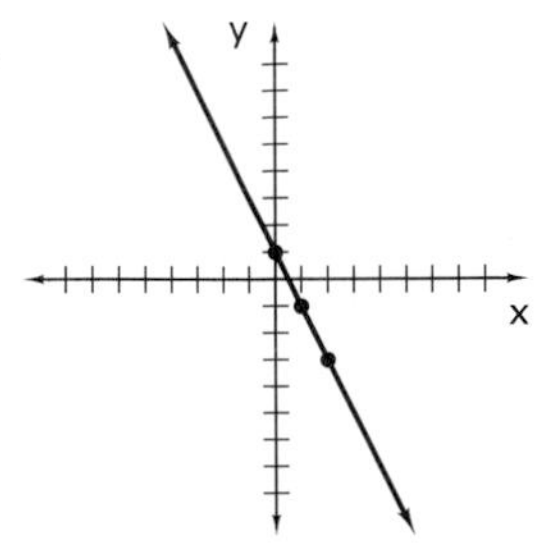

Self-Study Exercises 18-4 (p. 700)

1.

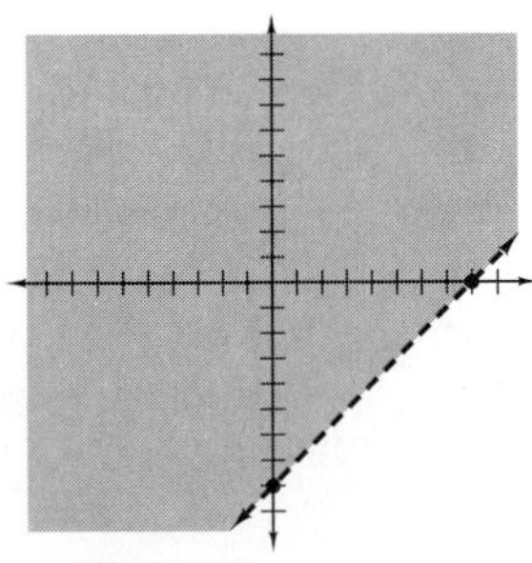

2.

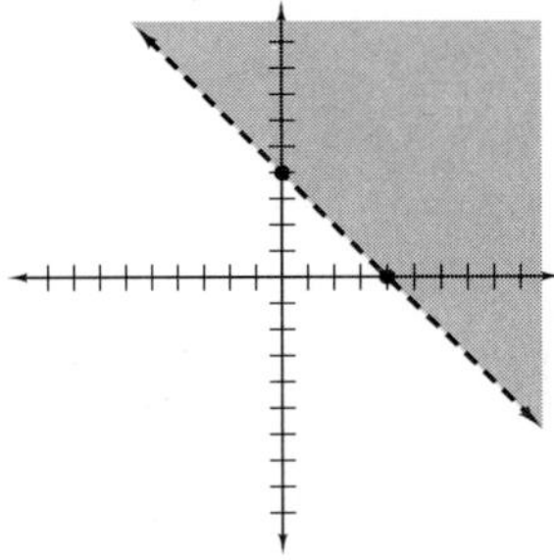

3.

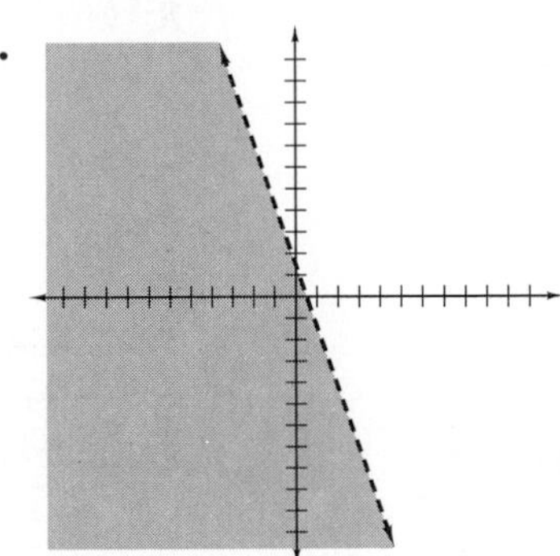

4.

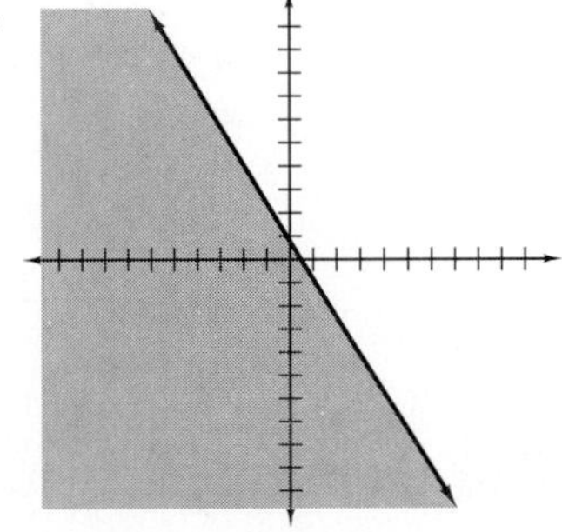

5.

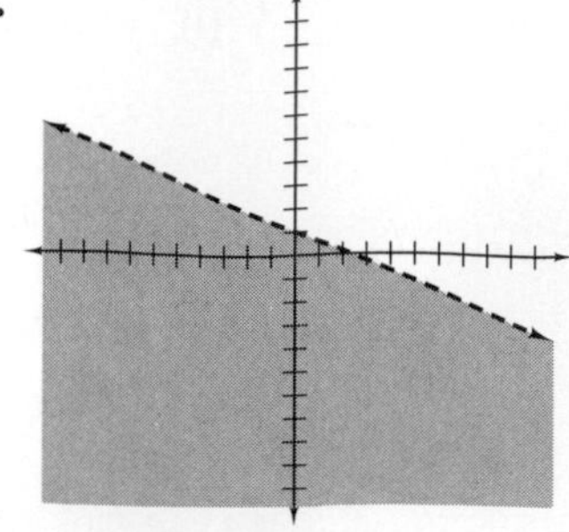

6.

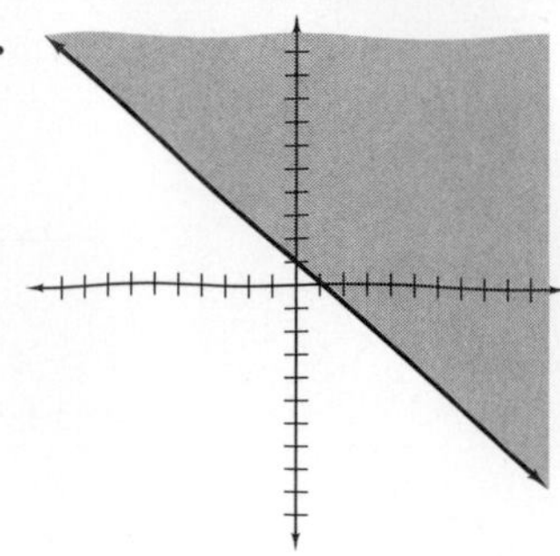

7.

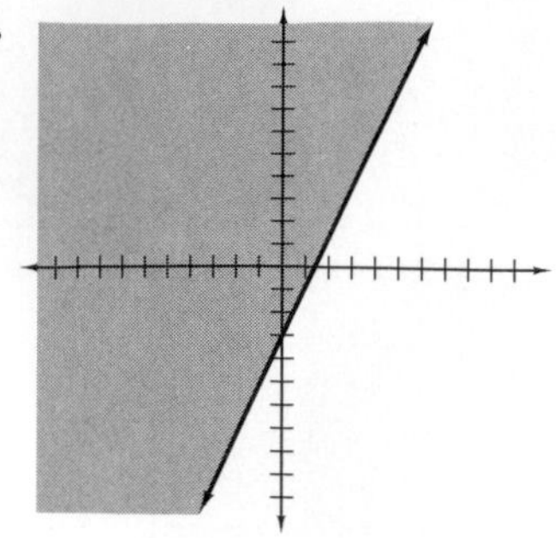

8.

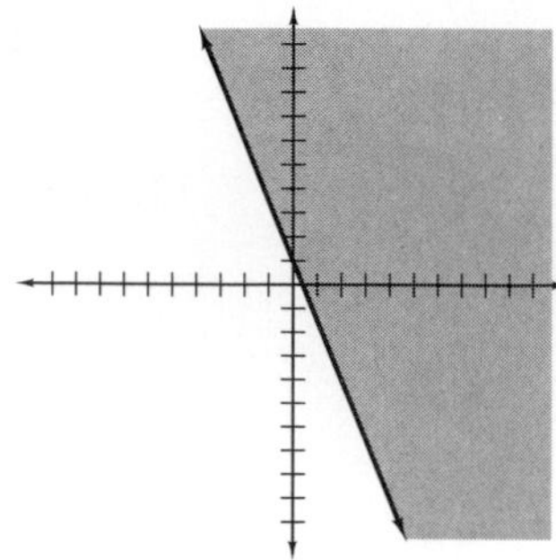

9.

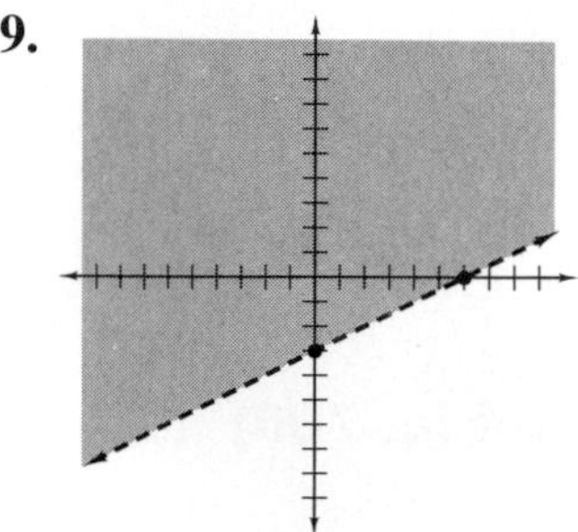

10.

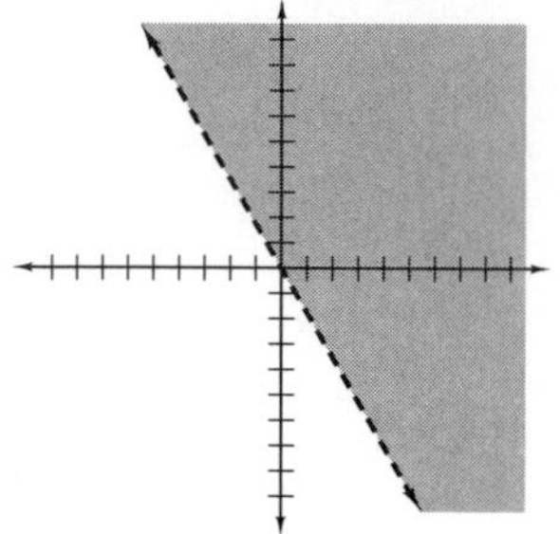

Self-Study Exercises 18-5 (p. 707)

1. linear **2.** quadratic **3.** quadratic **4.** other **5.** linear **6.** other

7. $y = x^2$

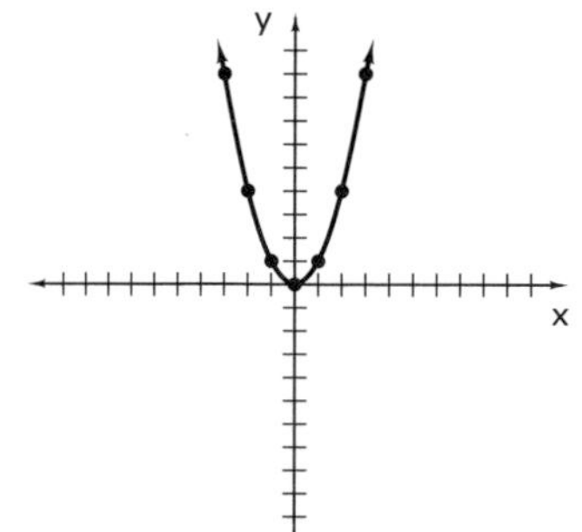

8. $y = 3x^2$

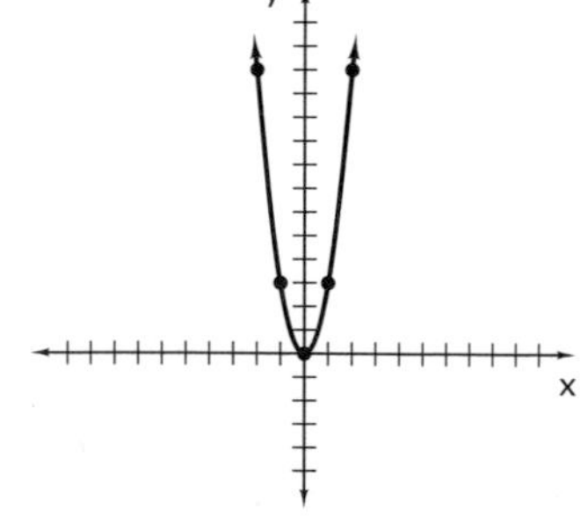

9. $y = \frac{1}{3}x^2$

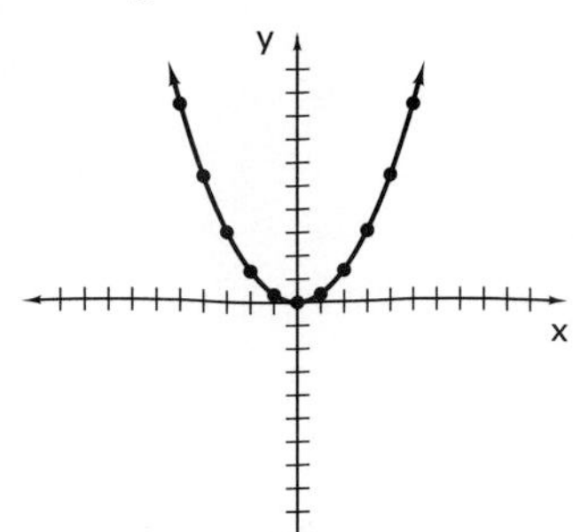

10. $y = -4x^2$

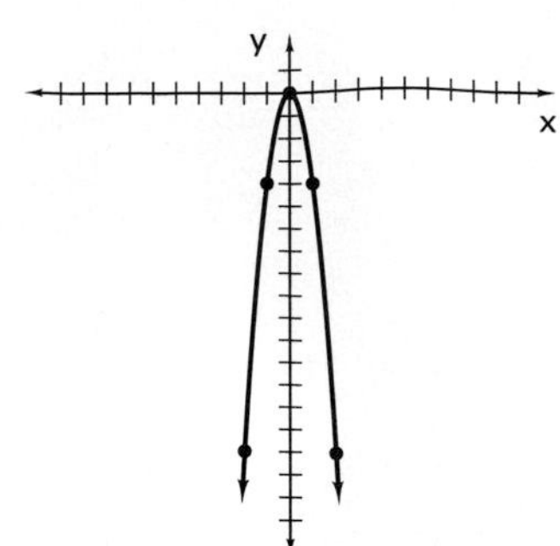

11. $y = -\frac{1}{4}x^2$

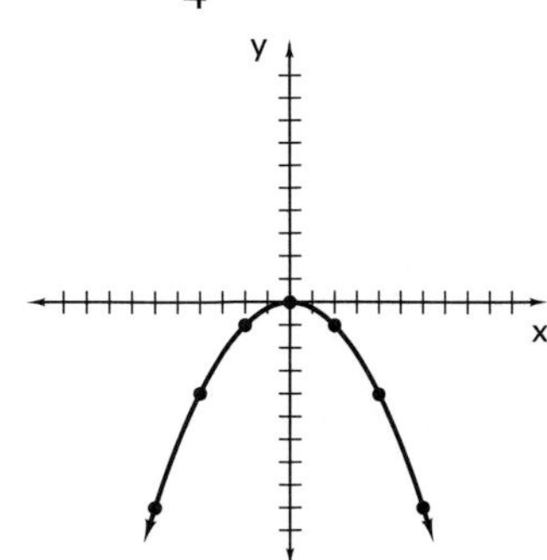

12. $y = x^2 - 4$

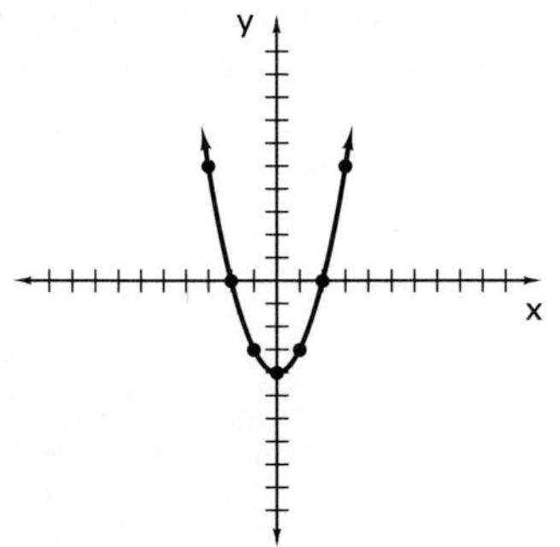

13. $y = x^2 + 4$

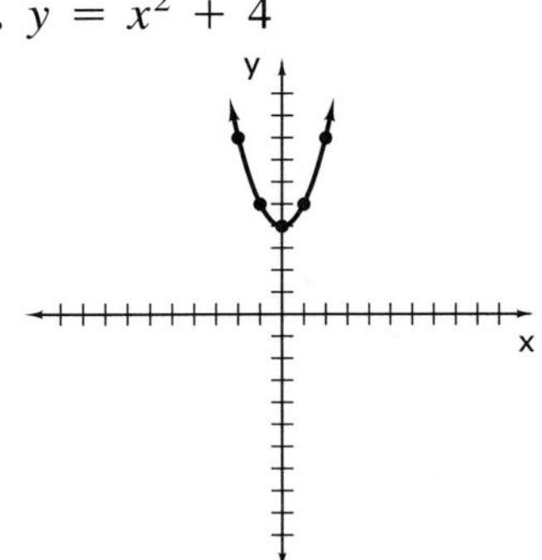

14. $y = x^2 - 6x + 9$

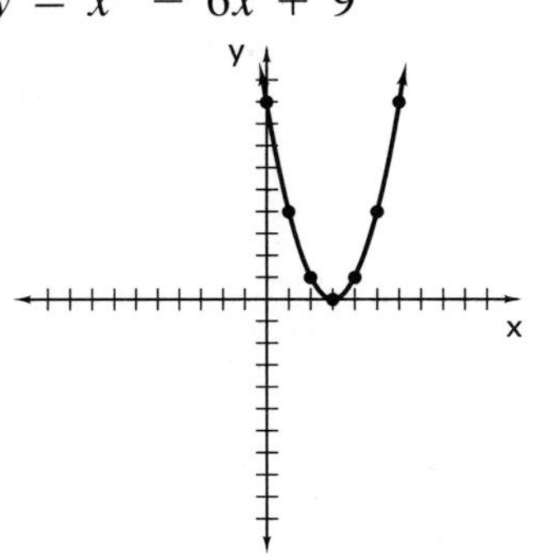

15. $y = -x^2 + 6x - 9 = -1(x^2 - 6x + 9)$
$-1(x - 3)^2$

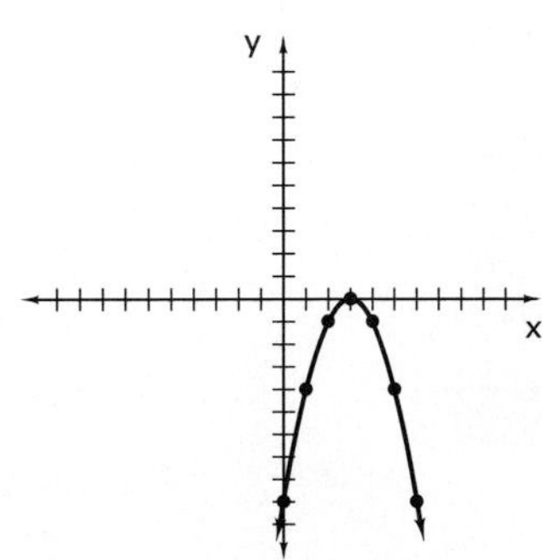

16. $y = 3x^2 + 5x - 2$

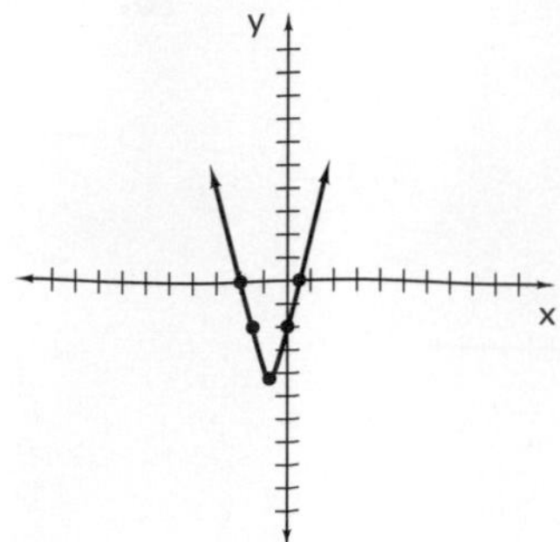

17. $y = (2x - 3)(x - 1)$

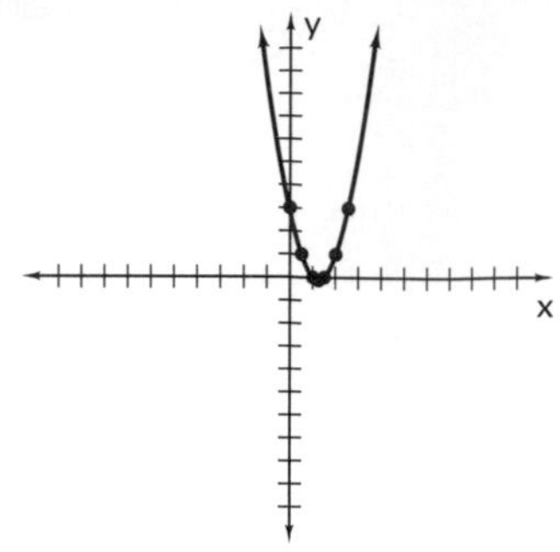

18. $y > 2x^2 - 9x - 5$

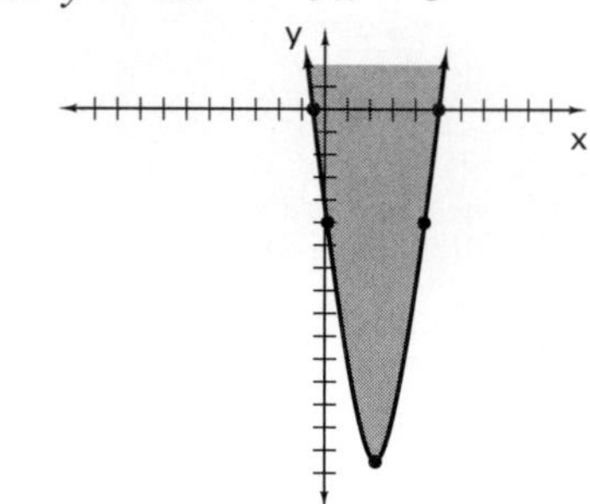

19. $y > -2x^2 + 9x + 5$

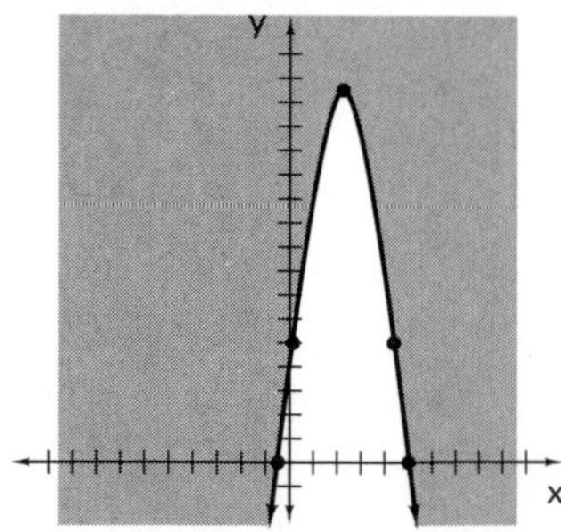

Assignment Exercises, Chapter 18 (pp. 707–710)

1. (2, 4) **3.** $(-4, -4)$ **5.** (0, 0) **7.** (4, 1) **9.** (0, 7)

11.–15.

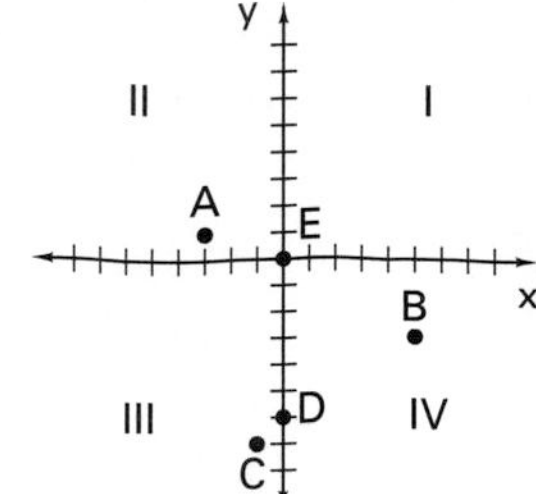

17. c) She will *not* reach her goal.

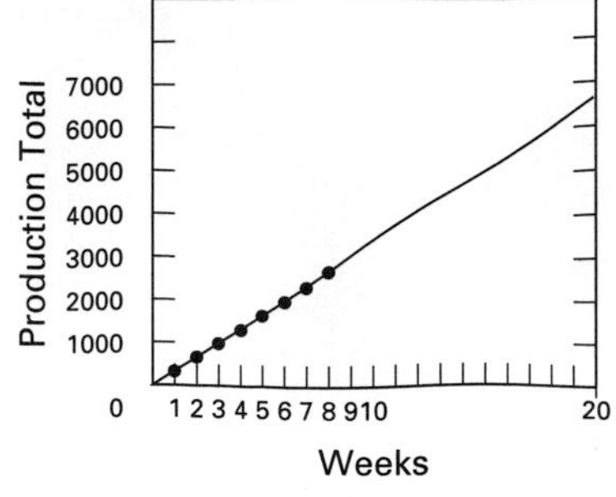

19. yes **21.** yes **23.** $y = 1$

25. x-intercept, (3, 0); y-intercept, (0, 3)

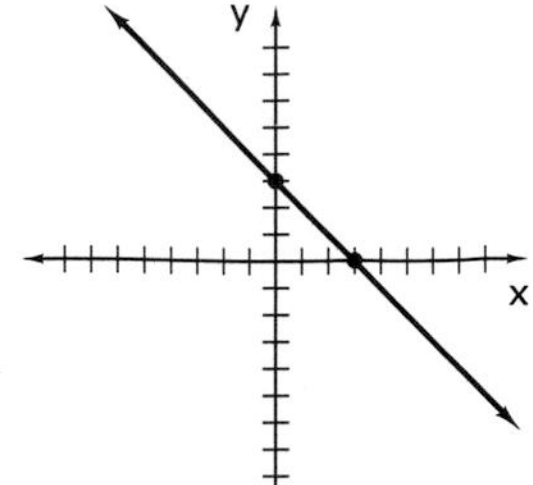

27. x-intercept, $\left(\frac{1}{2}, 0\right)$; y-intercept, (0, 1)

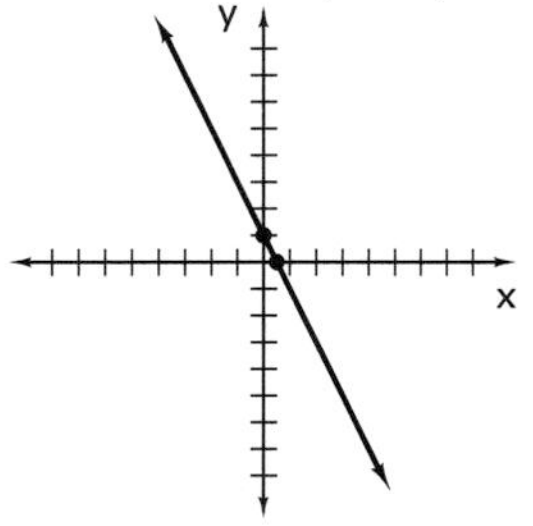

29. x-intercept, $\left(\frac{4}{3}, 0\right)$; y-intercept, $(0, -4)$

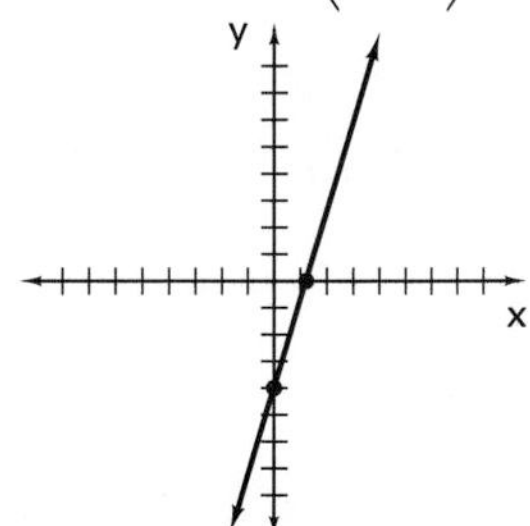

31. x-intercept, (6, 0); y-intercept, (0, 2)

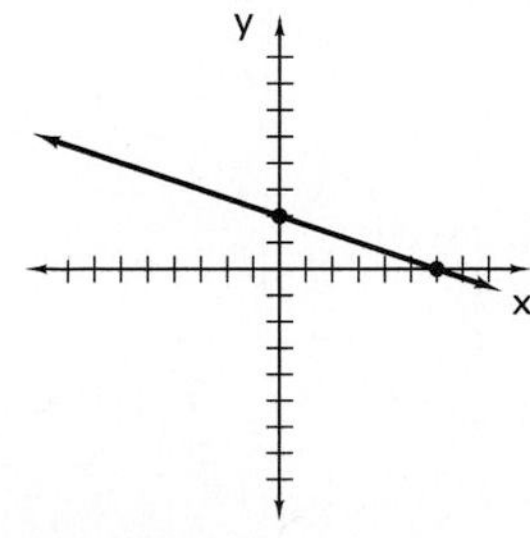

33. x-intercept, $\left(\frac{1}{3}, 0\right)$; y-intercept, $(0, -1)$

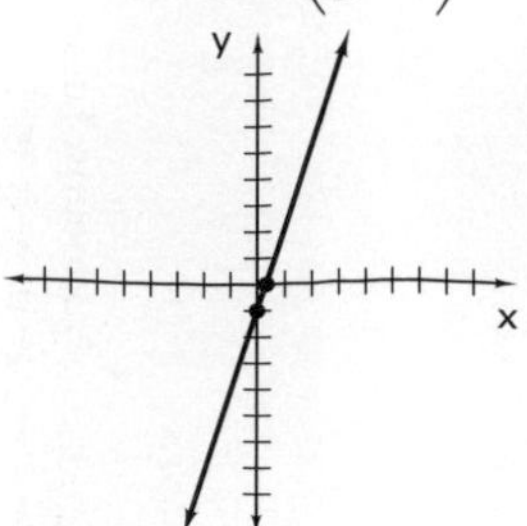

31. $3y = 6 - x$

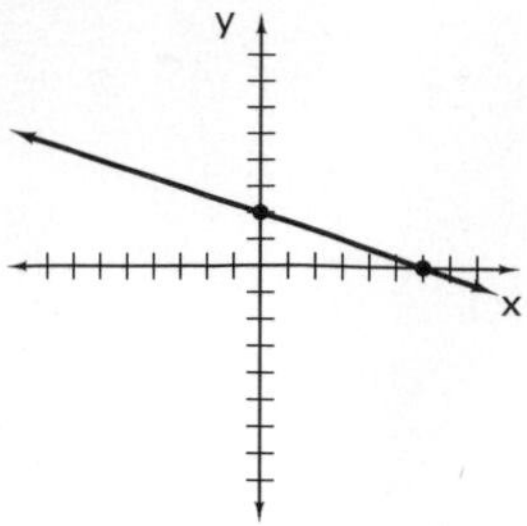

33. $3x - y = 1$

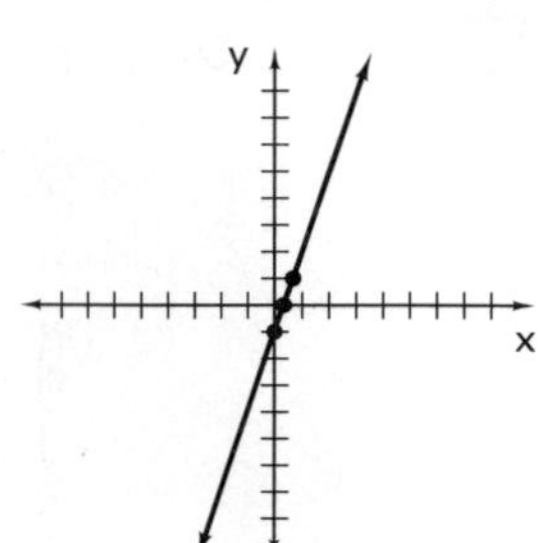

35. $y = 5x - 2, \quad m = \frac{5}{1}, \quad b = -2$

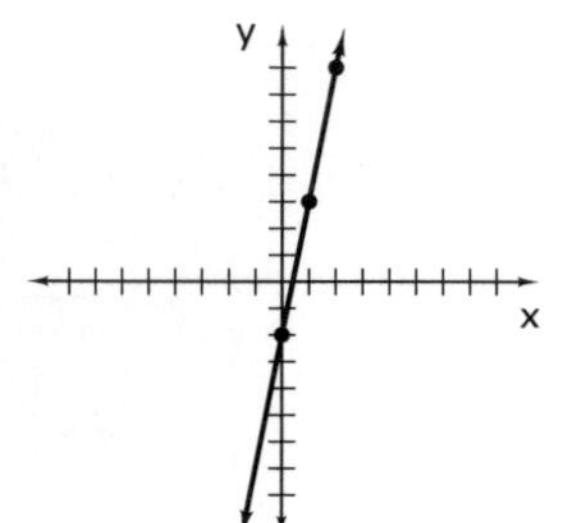

37. $y = -3x - 1, \quad m = \frac{-3}{1}, \quad b = -1$

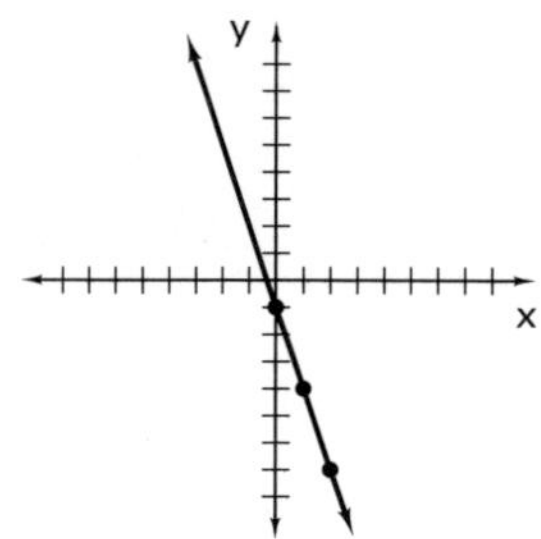

39. $x - y = 4, \quad y = x - 4, \quad m = \frac{1}{1}, \quad b = -4$

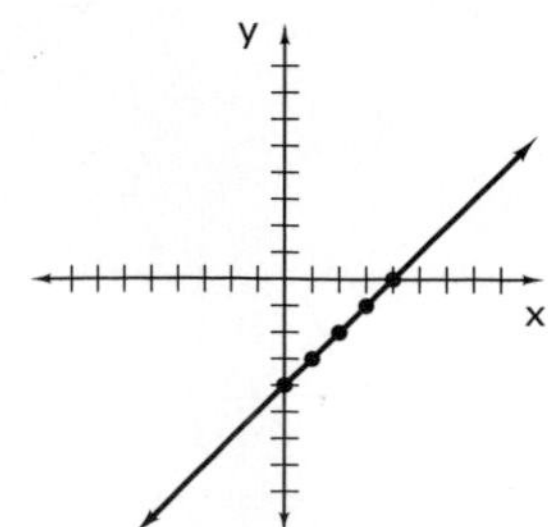

41. $x - 2y = -1$

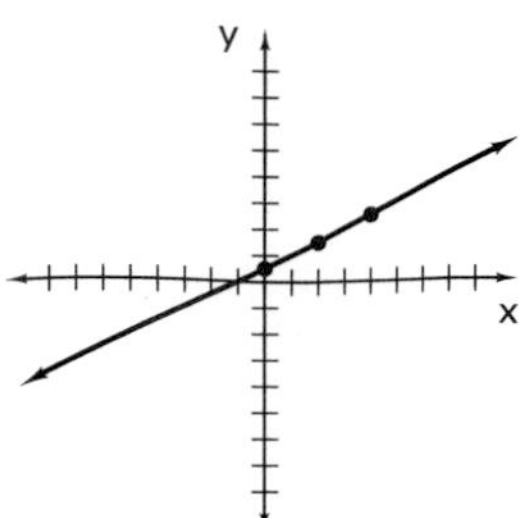

43. $4x + y < 2 \quad \left(\frac{1}{2}, 0\right) \; (0, 2)$

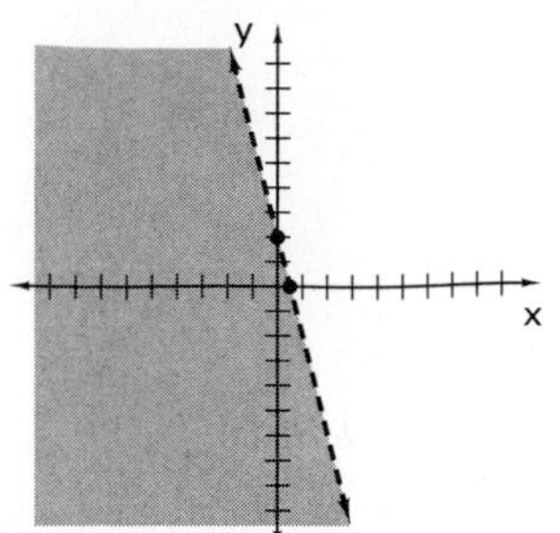

45. $3x + y \leq 2 \quad \left(\frac{2}{3}, 0\right) \; (0, 2)$

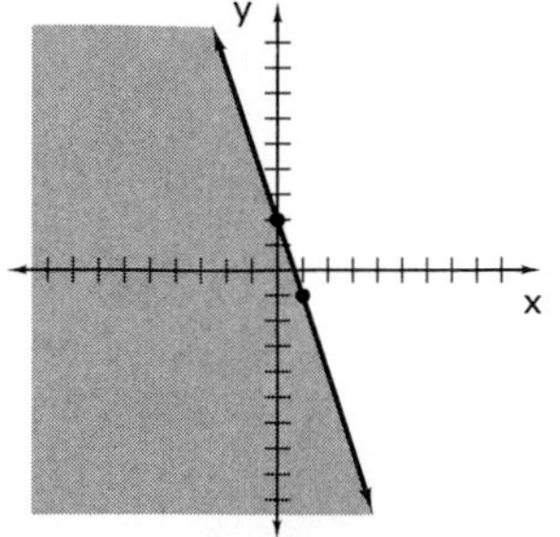

47. $x - 2y < 8 \quad (8, 0) \; (0, -4)$

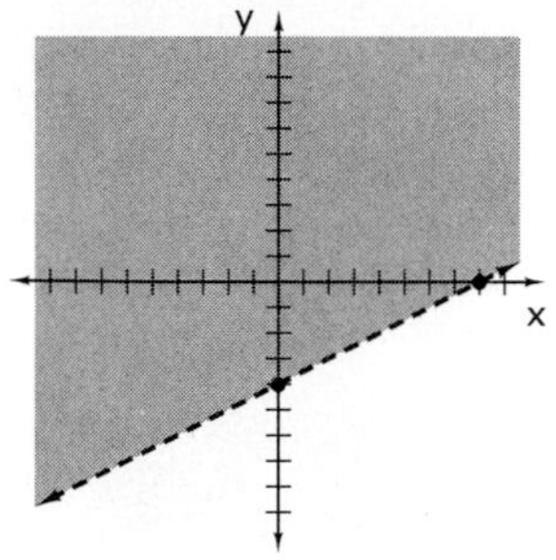

49. $y \geq 3x - 2 \quad \left(\frac{2}{3}, 0\right) \; (0, -2)$

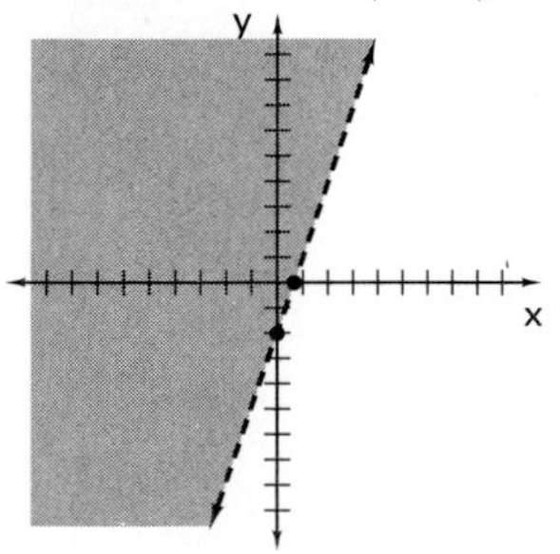

51. $y > \frac{2}{3}x - 2 \quad (3, 0) \; (0, -2)$

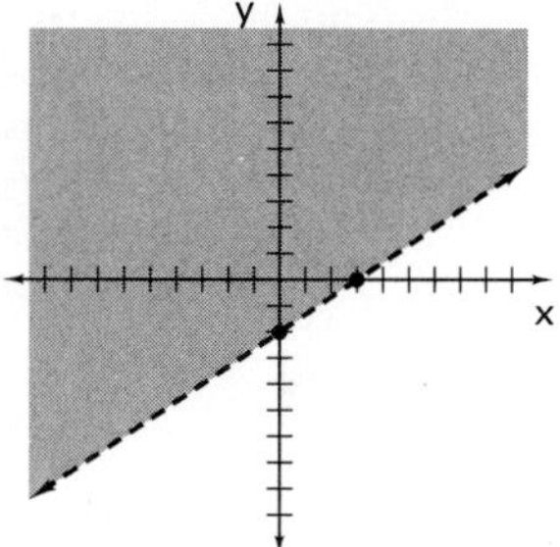

53. $y = x^2 + 1$

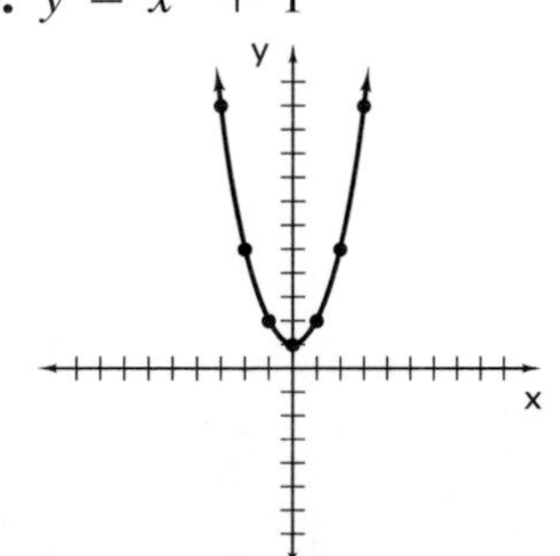

55. $y = -x^2 - 1$

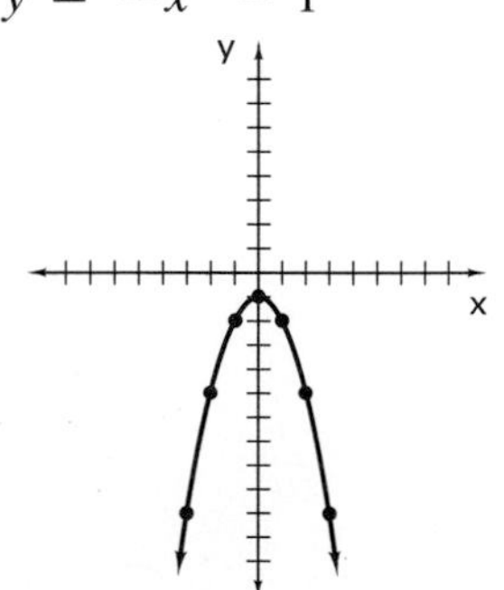

57. $y = 3x^2 - 5$

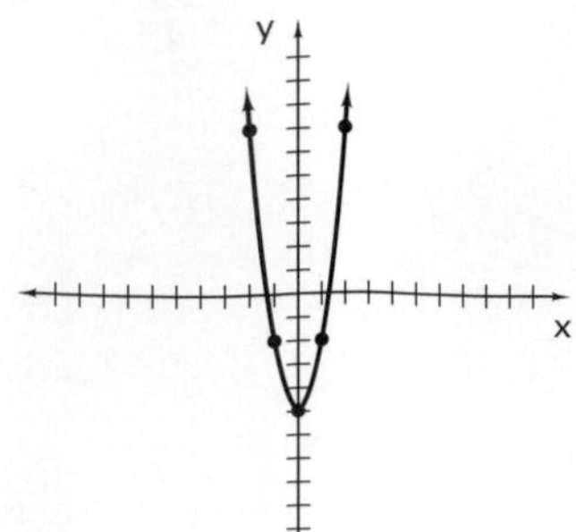

59. $y = -x^2 + 2x - 1$

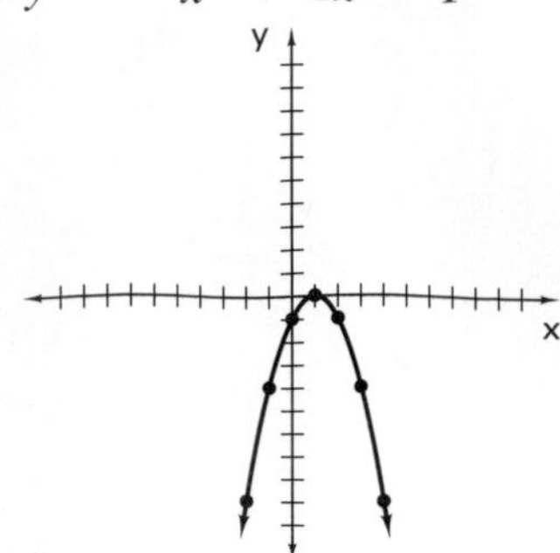

61. $y = -x^2 + 4x - 4$

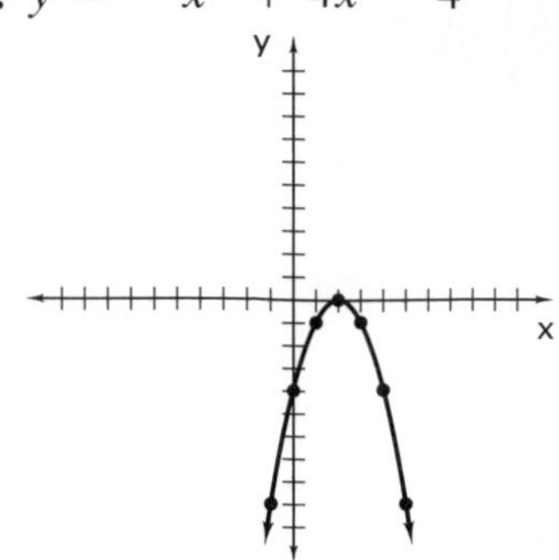

63. $y \geq -2x^2$

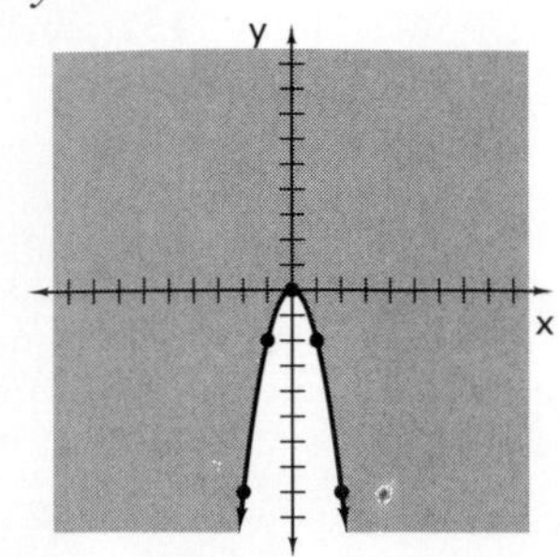

65.

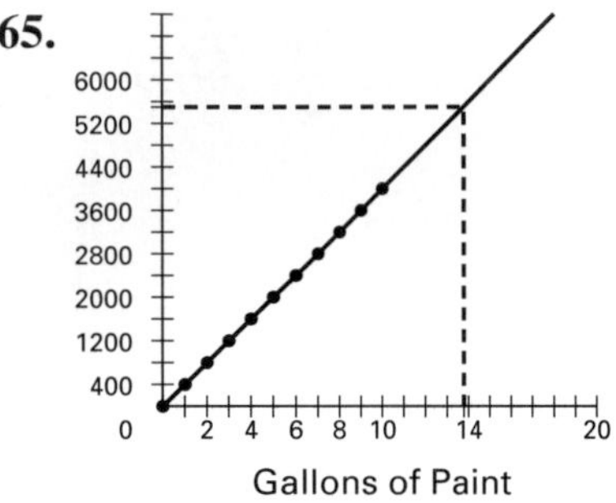

65. a)

gal	ft^2
1	400
2	800
3	1200
4	1600
5	2000
6	2400
7	2800
8	3200
9	3600
10	4000

b) see graph
c) 14 gal
d) \$279.30
e) \$302.34

67. amount applied to interest

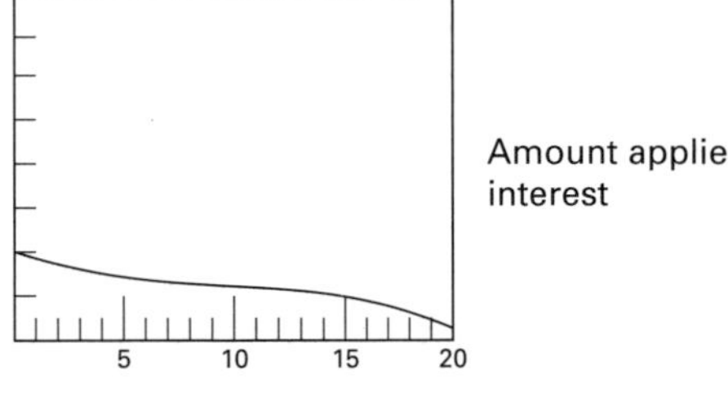

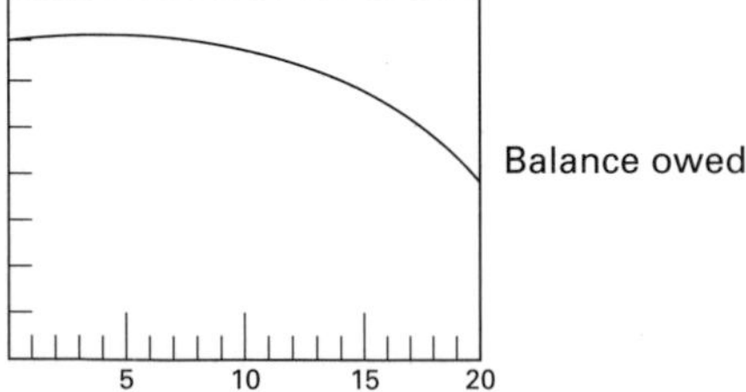

Trial Test, Chapter 18 (pp. 716–717)

1. $(-3, 2)$ **3.** $(3, -4)$ **5.** III **7.** IV

9. $y = 5 - x$

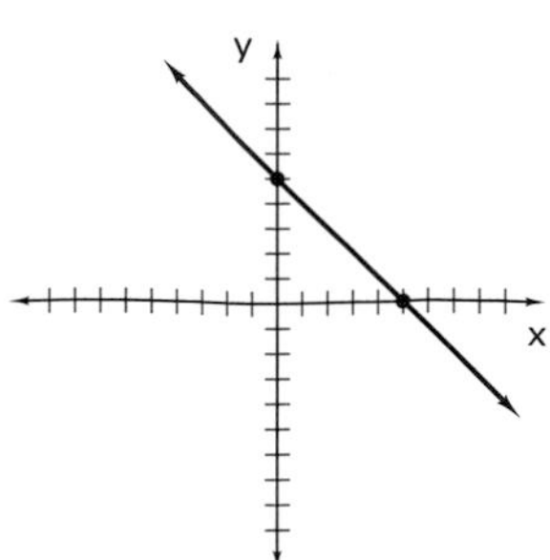

11. x-intercept: $(2, 0)$

y-intercept: $\left(0, -\frac{1}{2}\right)$

13.

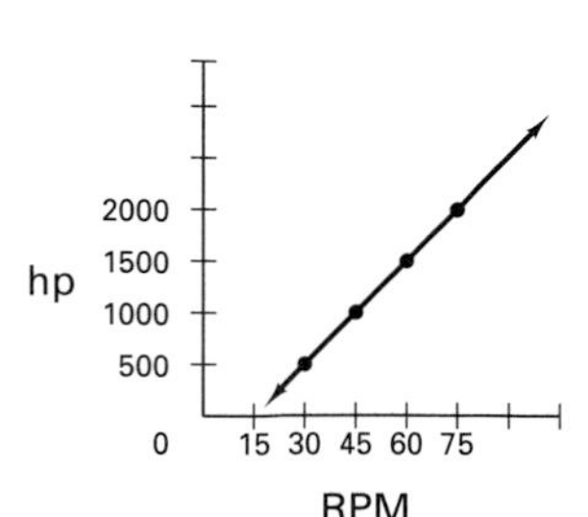

15. $2x - y \leq 2$

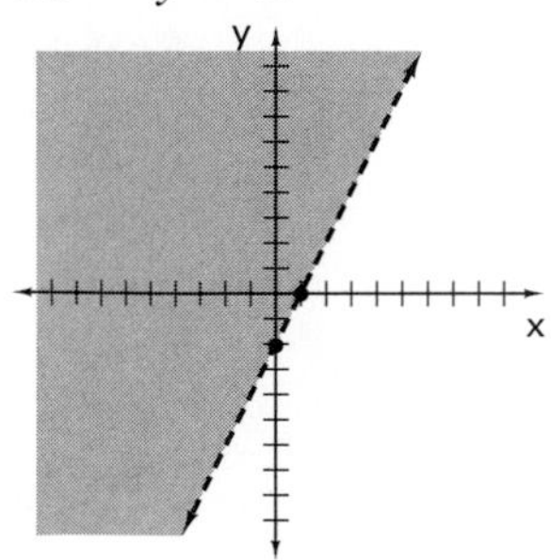

$2x - y = 2$

17. $x + y < 1$

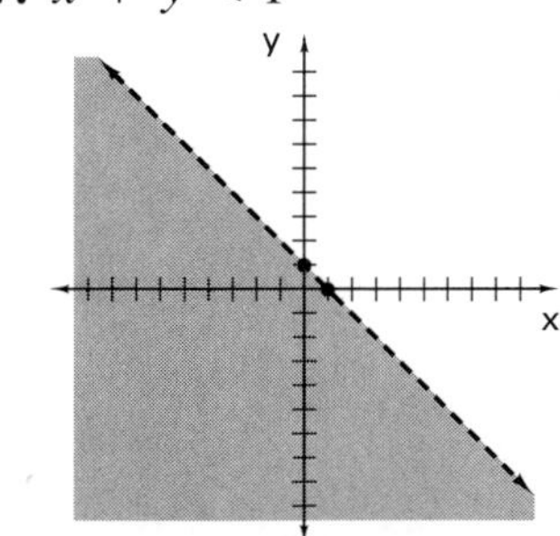

19. $y = x^2 + 2x + 1$

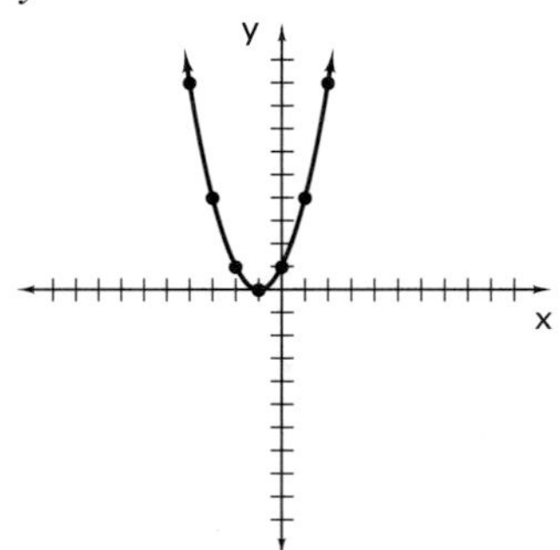

21. $y \leq x^2 - 6x + 8$

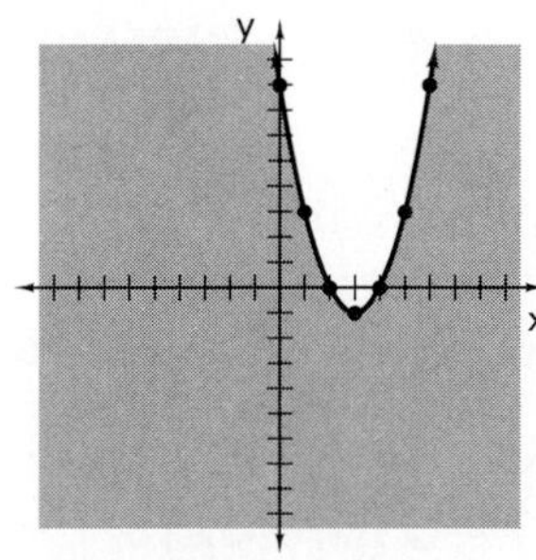

23. $y < -\frac{1}{2}x^2$

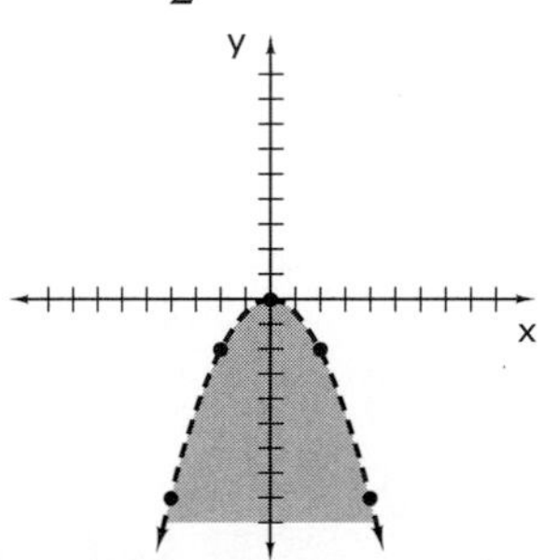

Self-Study Exercises 19-1 (p. 722)

L.O.1.

1. $\frac{1}{2}$ **2.** $-\frac{5}{3}$ **3.** 3 **4.** 1 **5.** $\frac{3}{4}$ **6.** 0 **7.** $-\frac{5}{4}$ **8.** -1

9. no slope **10.** -3

L.O.1.

Problem 6 is horizontal, slope is zero.
Problem 9 is vertical, slope not defined.

Self-Study Exercises 19-2 (pp. 725–726)

L.O.1.

1. $y = \frac{2}{3}x + \frac{25}{3}$ **2.** $y = -\frac{1}{2}x + 3$ **3.** $y = 2x + 1$ **4.** $y = x - 1$

L.O.2.

5. $y = -\frac{5}{3}x + \frac{38}{3}$ **6.** $x = -1$ **7.** $y = -3$ **8.** $x = -4$

9. $y = 2x$ **10.** $y = 0$

Self-Study Exercises 19-3 (p. 729)

L.O.1.

1. slope $= 4$
y-intercept $= 3$

2. slope $= -5$
y-intercept $= 6$

3. slope $= -\frac{7}{8}$
y-intercept $= -3$

4. slope $= 0$
y-intercept $= 3$

5. $y = 2x - 5$
slope $= 2$ y-intercept $= -5$

6. $y = \frac{5}{2}$
slope $= 0$ y-intercept $= \frac{5}{2}$

7. $y = -2x + 6$
slope $= -2$ y-intercept $= 6$

8. $y = 2x + 5$
slope $= 2$ y-intercept $= 5$

9. $0 = -2x + 8$
slope $= \infty$ y-intercept $=$ none

10. $0 = -x + 9$
slope $= \infty$ y-intercept $=$ none

L.O.2.

11. $y = \frac{1}{4}x + 7$ **12.** $y = -8x - 4$ **13.** $y = -3x + 1$

14. $y = \frac{2}{3}x - 2$ **15.** $y = 4$

Self-Study Exercises 19-4 (pp. 732–733)

L.O.1.

1. $x + y = -1$ **2.** $2x + y = 9$ **3.** $x - 3y = 4$ **4.** $3x - y = -7$
5. $x + 3y = 7$ **6.** $3x - y = -3$ **7.** $2x + 3y = 5$
8. $3x + 2y = 6$ **9.** $2x - 5y = 11$ **10.** $6x - 8y = 3$

Self-Study Exercises 19-5 (p. 737)

L.O.1.

1. $x - y = -1$ **2.** $x - 2y = -13$ **3.** $3x + y = 12$
4. $x + 3y = -9$ **5.** $3x - y = 11$ **6.** $2x - y = -7$
7. $3x - 2y = 11$ **8.** $x - 4y = 0$ **9.** $2x - 2y = -3$
10. $x + 5y = 4$

Self-Study Exercises 19-6 (p. 742)

L.O.1.

1. 10 **2.** 13 **3.** 7.071 **4.** 5.385 **5.** 11.180 **6.** 13 **7.** 10
8. 5 **9.** 15 **10.** 10.296

L.O.2.

11. (4, 6) **12.** $\left(\frac{9}{2}, -1\right)$ **13.** $\left(\frac{5}{2}, \frac{5}{2}\right)$ **14.** $\left(\frac{1}{2}, -3\right)$ **15.** $\left(\frac{5}{2}, 2\right)$
16. $\left(2, \frac{1}{2}\right)$ **17.** $(-1, 2)$ **18.** $\left(\frac{3}{2}, 3\right)$ **19.** $\left(-1, -\frac{1}{2}\right)$
20. $\left(\frac{5}{2}, -\frac{1}{2}\right)$

Assignment Exercises, Chapter 19 (pp. 744–746)

1. $\frac{1}{3}$ **3.** 2 **5.** $\frac{5}{8}$ **7.** $-\frac{4}{3}$ **9.** no slope **11.** $-\frac{4}{7}$ **13.** $\frac{7}{2}$
15. 0 **17.** no slope **19.** 1 **21.** $(-1, 2)(3, 2)$ Answers will vary.
23. $y = \frac{1}{3}x + 4$ **25.** $y = \frac{3}{4}x - 3$ **27.** $y = 4x - 5$
29. $y = -\frac{2}{3}x - \frac{2}{3}$ **31.** $y = -\frac{1}{11}x + \frac{17}{11}$ **33.** $y = \frac{7}{4}x - \frac{5}{4}$
35. $y = \frac{9}{5}x + \frac{3}{5}$ **37.** $y = x - 3$ **39.** $y = x - 1$ **41.** $y = -2$

43. $m = 3$, $b = \frac{1}{4}$ **45.** $m = -5$, $b = 4$ **47.** no slope, no y-intercept **49.** $m = \frac{1}{8}$, $b = -5$

51. $y = -2x + 8$, $m = -2, b = 8$ **53.** $y = \frac{3}{2}x - 3$, $m = \frac{3}{2}, b = -3$ **55.** $y = \frac{3}{5}x - 4$, $m = \frac{3}{5}, b = -4$

57. $y = \frac{5}{3}$, $m = 0,\ b = \frac{5}{3}$ **59.** $y = 3x - 2$ **61.** $y = 2x - 2$

63. $x + y = 7$ **65.** $x - 2y = 8$ **67.** $x - 3y = 20$
69. $x + 3y = -10$ **71.** $3x - 4y = -7$ **73.** $x - y = -4$
75. $2x - y = -4$ **77.** $x - 5y = -11$ **79.** $2x + 10y = 31$
81. $x + 4y = 2$ **83.** 4.472 **85.** 10.440 **87.** 5.831 **89.** 9.434
91. 7.280 **93.** (1, 5) **95.** $\left(1\frac{1}{2}, 3\right)$ **97.** $\left(-1\frac{1}{2}, 2\frac{1}{2}\right)$
99. $\left(1, -\frac{1}{2}\right)$ **101.** $\left(-1\frac{1}{2}, -3\right)$

103. a)

Lipsticks	Cost
x	y
0	5000
100	5453
1000	9530
2000	14,060

b) (0, 5000); (100, 5453); (1000, 9530); (2000, 14,060)

c)

20,000
15,000
10,000
5000
500 1000 1500 2000

d) $f(x) = 4.53x + 5000$

e) $4.53x + 5000 = 8.99x$
$5000 = 4.46x$
$x = 1121.076$
or 1122 lipsticks

Trial Test, Chapter 19 (pp. 750–751)

1. $-\frac{2}{3}$ **3.** $\frac{5}{2}$ **5.** 1 **7.** $y = \frac{2}{3}x - 7$ **9.** $y = \frac{2}{3}x + \frac{7}{3}$

11. $y = 2$ **13.** $m = 3$, $b = -22$ **15.** $y = \frac{1}{4}x$, $m = \frac{1}{4}, b = 0$

17. $y = -6x + 3$, $m = -6, b = 3$ **19.** $y = \frac{3}{2}x + 3$ **21.** $y = -2x + 5$

23. $y = \frac{1}{2}x - 5$ **25.** 5.385, $\left(\frac{1}{2}, 2\right)$

Self-Study Exercises 20-1 (pp. 757)

L.O.1.

1. $x = 7, y = 5$

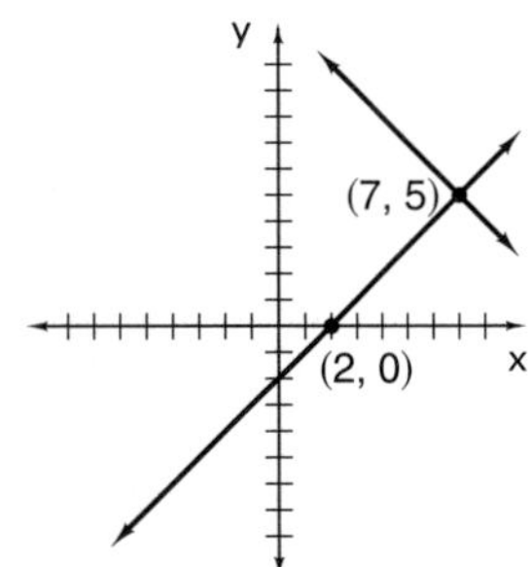

2. $x = 3, y = 3$

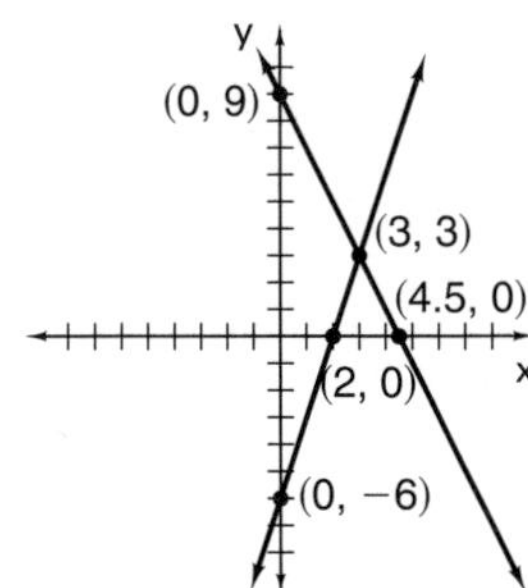

3. No solution, no intersection.

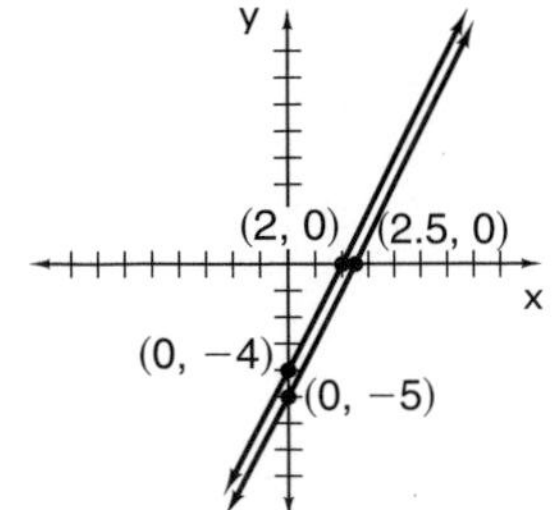

4. Many solutions, lines coincide.

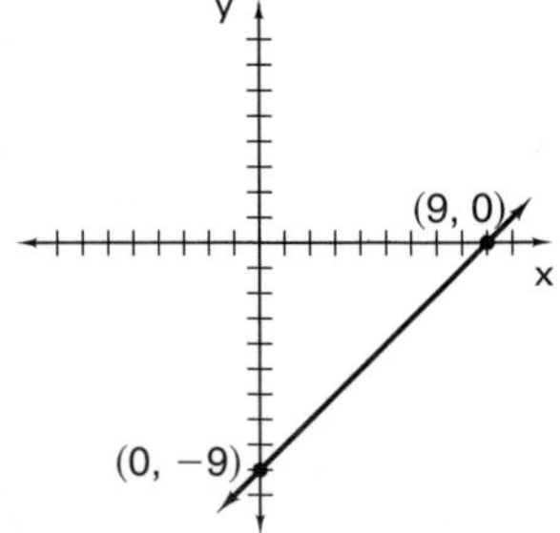

5. $x = 4, y = -2$

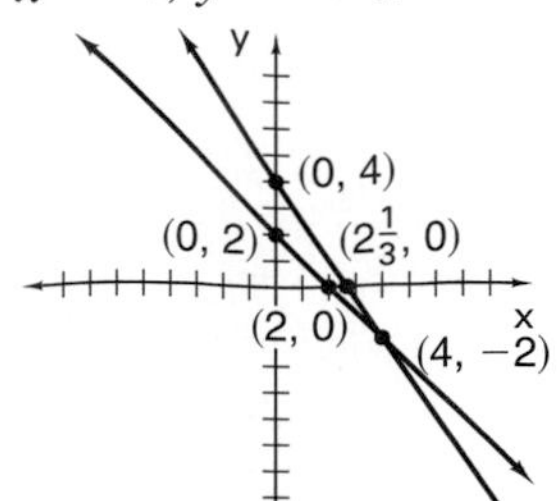

6.

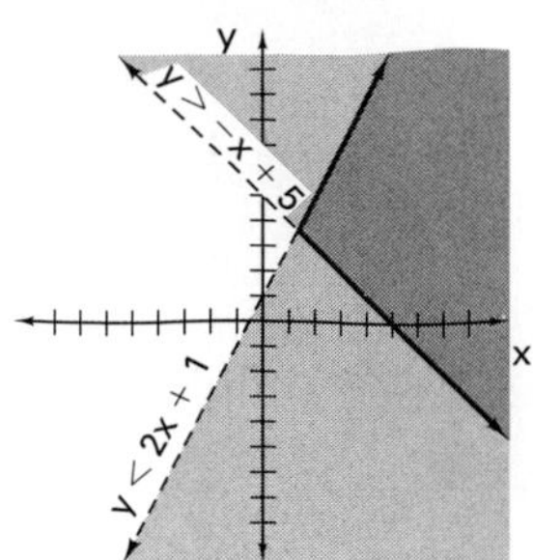

7.

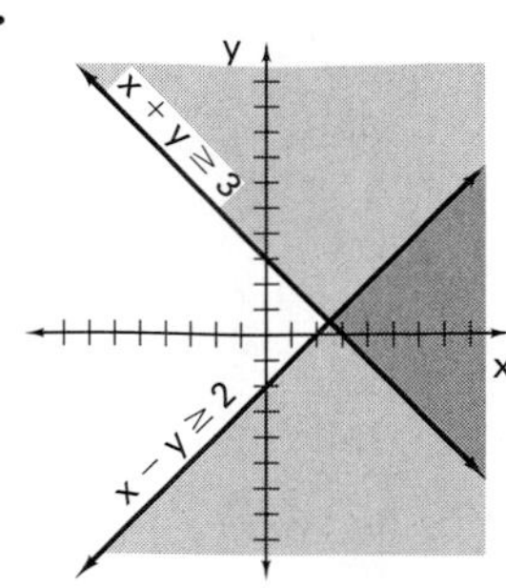

8. $x - 2y < -1$ $\qquad$ $x + 2y > 3$

$$y > \frac{1}{2}x + \frac{1}{2} \qquad y > -\frac{1}{2}x + 3$$

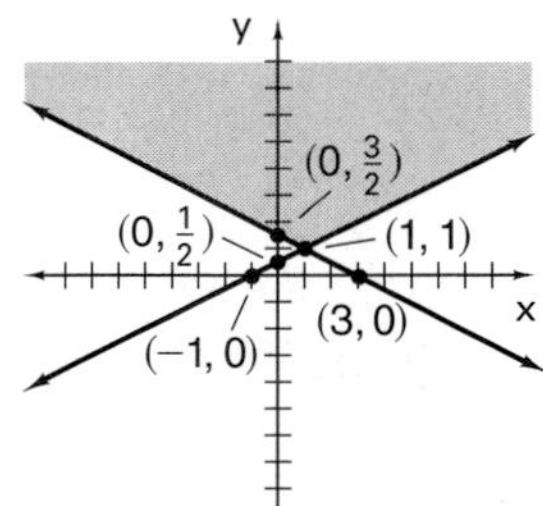

9. $x + y > 4$ $\qquad$ $x - y > -3$

$$y > -x + 4 \qquad y < x + 3$$

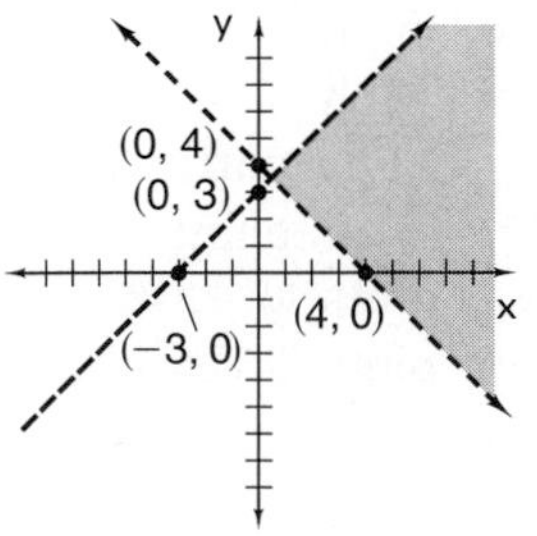

10. $3x - 2y < 8$ $\qquad$ $2x + y \leq -4$

$$y > \frac{3}{2}x - 4 \qquad y \leq -2x - 4$$

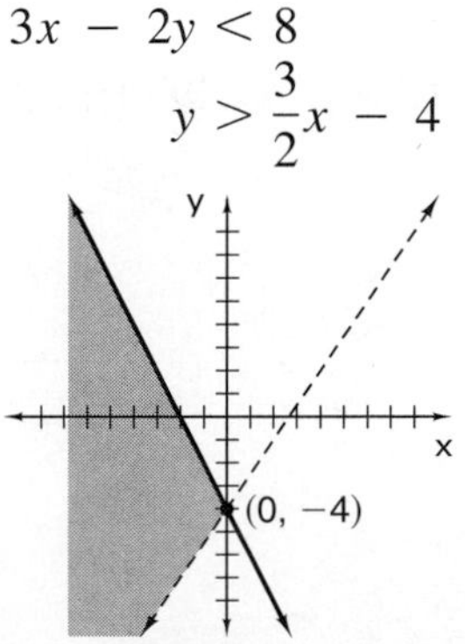

Self-Study Exercises 20-2 (pp. 764–765)

L.O.1.

1. $a = 5, b = -1$ **2.** $m = 4, n = -1$ **3.** $x = 1, y = -1$

4. $a = 3, b = -3$ **5.** $x = 2, y = \frac{3}{2}$

L.O.2.

6. $x = 3, y = 0$ **7.** $x = 1, y = 5$ **8.** $a = 2, b = \frac{8}{3}$

9. $x = -1, y = 2$ **10.** $a = 6, y = 1$

L.O.3.

11. inconsistent; no solution **12.** $x = -3, y = 6$ **13.** dependent; many solutions **14.** dependent; many solutions **15.** $a = 5, b = 1$

Self-Study Exercises 20-3 (p. 766)

L.O.1.

1. $a = 20, b = 10$ **2.** $r = 5, c = 7$ **3.** $x = 13, y = 11$

4. $x = 7, y = 5$ **5.** $p = \frac{1}{2}, k = \frac{1}{3}$ **6.** $x = 3, y = 1$

Self-Study Exercises 20-4 (p. 771)

L.O.1.

1. short 15.5 in.
long 32.5 in.

2. \$18,000 @ 4%
\$17,000 @ 5%

3. \$11 per shirt
\$ 6 per hat

4. (7) 8-cylinder jobs
(3) 4-cylinder jobs

5. resistor \$0.25
capacitor \$0.30

6. rate of plane = 130 mph
rate of wind = 10 mph

7. \$3000 @ 7%
\$5000 @ 9%

8. 1.67 mph
11.67 mph

9. dark roast, \$2.50
with chicory, \$1.90

10. 75% mixture, 123 lb
10% mixture 77 lb

11. scientific \$11.00
graphing \$30.00

12. \$4000 @ 10%
\$6000 @ 15%

13. reserved \$20.00
general \$15.00

14. 4 pints @ 75%
4 pints @ 25%

Assignment Exercises, Chapter 20 (pp. 773–776)

1.

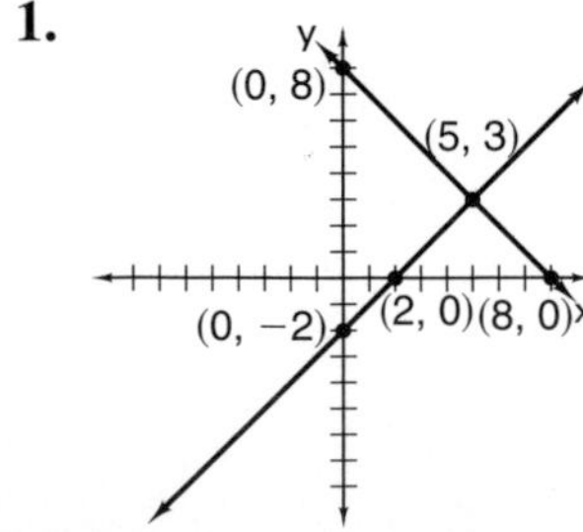

3.

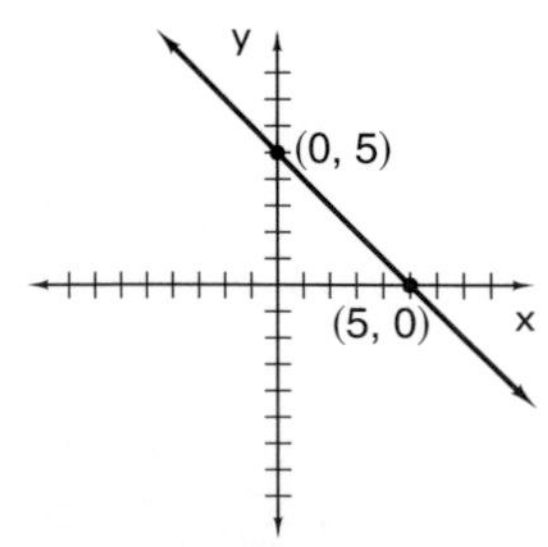

many solutions, lines coincide

5.

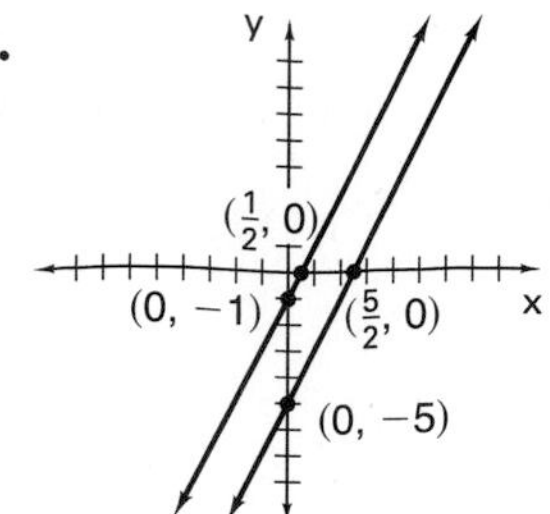

7.

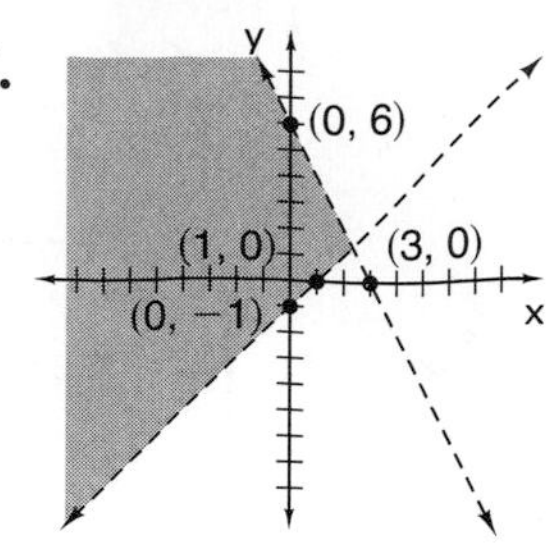

no solutions, lines parallel

9.

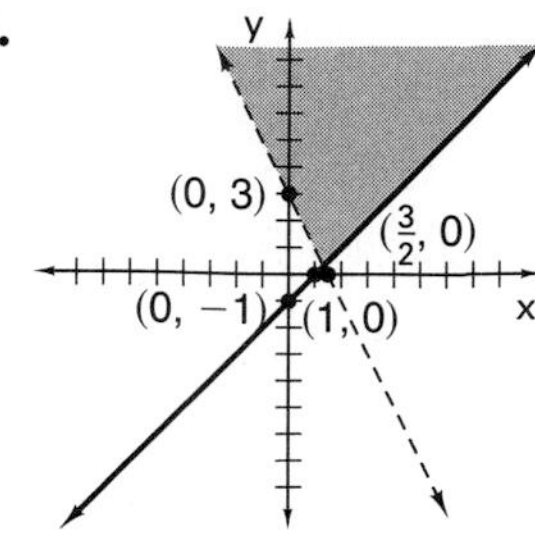

11. 3, 0 **13.** $-2, 4$ **15.** $2, -2$ **17.** 6, 3 **19.** dependent; many solutions **21.** 5, 1 **23.** $2, \frac{8}{3}$ **25.** 2, 0 **27.** 3, 2 **29.** 2, 1.5 **31.** 4, 4 **33.** 7, 5 **35.** $4, -6$ **37.** $3, -4$ **39.** $\frac{4}{5}, \frac{2}{5}$ **41.** 28, 44 **43.** $-3, \frac{5}{2}$ **45.** $4, -6$ **47.** 12, 2 **49.** $1, -2$ **51.** electrician \$75; apprentice \$35 **53.** shellac \$2.50; thinner \$3.50 **55.** 119°, 56° **57.** \$2000 @ 10%; \$3000 @ 12% **59.** columbian = \$5; blended = \$4 **61.** Ohio = \$1.95; Alaska = \$2.10 **63.** name brand = \$320; generic = \$175 **65.** telephone = \$15000; showroom = \$25000 **67.** 4 gal. 75%; 4 gal. 25% **69.** In line 4, $-6c$ should be $-9c$. $c = 3$, $d = 1$ **71.** Answers will vary.

Trial Test, Chapter 20 (pp. 779–780)

1. (5, 0)

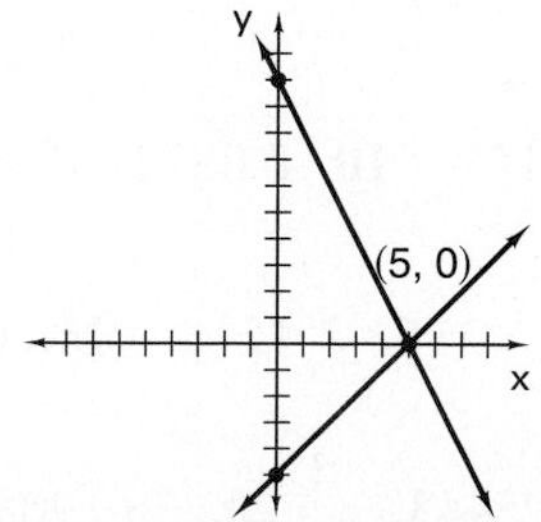

3. (14, −9)

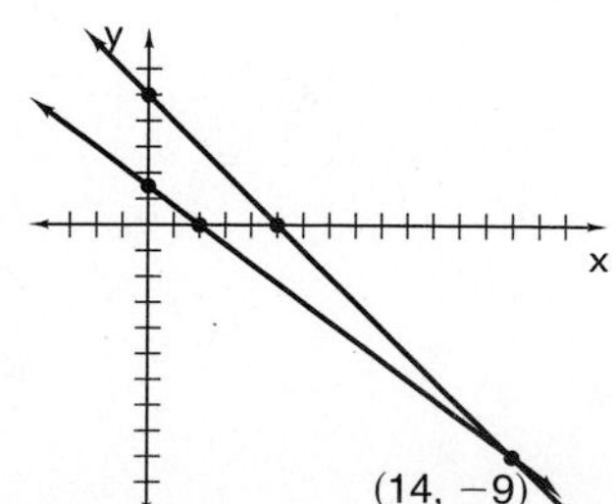

5. $(-6, -6)$ **7.** $\left(\frac{1}{2}, \frac{7}{2}\right)$

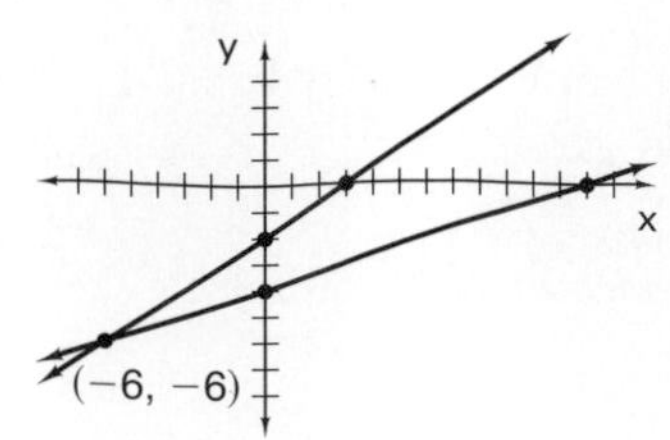

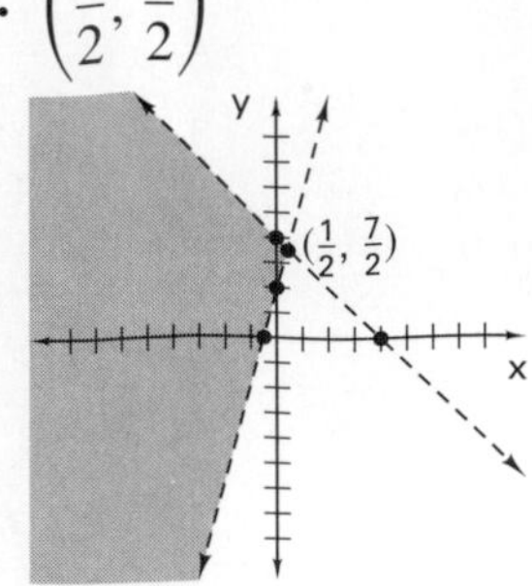

9. 0, 0 **11.** 1, 2 **13.** 2, 2 **15.** −3, 15 **17.** −9, −11
19. 20*A*, 15*A* **21.** $L = 51$ in. $W = 34$ in. **23.** $20000 @ 12.5% $ 5000 @ 11%
25. 0.000215 *F* 0.000055 *F*

Self-Study Exercises 21-1 (pp. 786–788)

L.O.1.

1. $\overleftrightarrow{PQ}$, $\overleftrightarrow{QR}$, & $\overleftrightarrow{PR}$ **2.** $\overrightarrow{PQ}$, $\overrightarrow{PR}$ **3.** $\overline{QR}$ **4.** Yes
5. No. End points are different. **6.** No **7.** Yes **8.** Yes
9. Yes. $\overline{XY}$ and $\overline{YX}$ contain same points. **10.** No

L.O.2.

11. parallel **12.** intersect **13.** coincide **14–15.** coincide or intersect

L.O.3.

16. $\angle DEF$, $\angle FED$ **17.** $\angle E$ **18.** $\angle 1$ **19.** $\angle a$ **20.** $\angle ABC$
21. $\angle B$

L.O.4.

22. 360° **23.** 180° **24.** 90° **25.** acute **26.** obtuse **27.** right
28. obtuse **29.** acute **30.** straight **31.** acute **32.** obtuse
33. Yes **34.** No **35.** No **36.** No **37.** Yes **38.** Yes **39.** No
40. 90° **41.** congruent

Self-Study Exercises 21-2 (p. 793)

L.O.1.

1. 54° **2.** 46°8′18″ **3.** 37°52′56″ **4.** 74°3′12″ **5.** 16°59′45″
6. 54°44′12″ **7.** 64°24′46″

L.O.2.

8. 0.7833° **9.** 0.01° **10.** 0.0872° **11.** 0.1708° **12.** 29.7°

L.O.3.

13. 21′ **14.** 0°12′ **15.** 7′12″ **16.** 0°12′47″ **17.** 18′54″

L.O.4.

18. 7°30′ **19.** 11°25′43″ **20.** 38.1908° **21.** 12°38′50″

Self-Study Exercises 21-3 (pp. 802–804)

L.O.1.

1. scalene **2.** equilateral **3.** isosceles

L.O.2.

4. longest $\overline{TS}$, shortest $\overline{RS}$ **5.** longest k, shortest l **6.** $\angle C, \angle B, \angle A$
7. $\angle X, \angle Z, \angle Y$

L.O.3.

8. $\angle A = \angle F, \angle C = \angle D, \angle B = \angle E$
9. $\angle P = \angle S, \angle Q = \angle T, QP = TS$
10. $\angle L = \angle P, PM = LJ, LK = PN$

L.O.4.

11. $BC = 12$ cm, $AB = 16.971$ cm **12.** $AC = 7.071$ m, $BC = 7.071$ m
13. $AC = 9.899$ m, $AB = 14.0$ m **14.** $AC = 8$ m, $BC = 8$ m
15. $AC = 14.697$ mm, $BC = 14.697$ mm **16.** 25′5″ **17.** 37.83 ft

L.O.5.

18. $AB = 12$ cm, $BC = 10.392$ cm **19.** $AC = 9.0$ mm, $BC = 15.588$ mm
20. $AC = 4.619$ in, $AB = 9.238$ in. **21.** $AC = 5.715$ cm, $AB = 11.431$ cm
22. $AB = 5'6''$, $BC = 4'9''$ **23.** 35′
24. rafter = 18′0″, run = 15′7″ **25.** 4.5 cm

Self-Study Exercises 21-4 (pp. 810–812)

L.O.1.

1. $x = 16'$, $y = 22'$ **2.** $x = 6'$, $y = 25'6''$

L.O.2.

3. 245′6″ = P **4.** 96 mm = P **5.** $p = 58'8''$
1942.5 ft^2 = A 662.4 mm^2 = A
6. 20 yd^2 @ \$16.50 = \$330 **7.** 24 lb **8.** 1086 ft^2, $\angle = 135°$
9. 5.25 in.^{2} **10.** 625 dm^2

L.O.3.

11. 33 in.^{2} **12.** triangle, $\angle 60°$ **13.** square, $\angle 90°$
14. pentagon, $\angle 108°$ **15.** hexagon, $\angle 120°$

Self-Study Exercises 21-5 (pp. 816–818)

L.O.1.

1. 120.64 cm^2 **2.** 198.44 cm^2 **3.** 2.43 ft^2 **4.** 14.18 in.^{2}
5. 2.75 m^2 **6.** 530.14 ft^2 **7.** 75.40 in.^{2} **8.** 17.0 in.^{2}

L.O.2.

9. 28.27 mm **10.** 13.09 in. **11.** 3.20 ft **12.** 7.06 cm
13. 1.64 m **14.** 15.71 in.

L.O.3.

15. 16.01 cm^2 **16.** 48.98 in.^{2} **17.** 54.83 in.^{2} **18.** 51.99 cm^2
19. 91.29 in.^{2} **20.** 561.81 mm^2 **21.** 257.93 ft^2

Self-Study Exercises 21-6 (pp. 827–829)

L.O.1.

1. 60° **2.** 30° **3.** 90° **4.** 11.5 in. **5.** 11.5 in. **6.** 13.33 in. **7.** 57.5 in.^{2} **8.** 5.75 in. **9.** 6.67 in. **10.** 41.88 in. **11.** 2.9 cm **12.** 1 in.

L.O.2.

13. 90° **14.** 45° **15.** 8.49 in. **16.** 45° **17.** 6 in. **18.** 0.5 m **19.** 0.35 m **20.** 90° **21.** 0.35 m **22.** 0.29 in. **23.** 1.06 in. **24.** 63.62 dkm^2

L.O.3.

25. 34.64 mm **26.** 120° **27.** 60° **28.** 40 mm **29.** 90° **30.** 10 mm **31.** 20 mm **32.** 60° **33.** 30° **34.** 3 in. **35.** 39.97 mm **36.** 2.84 in.

Assignment Exercises, Chapter 21 (pp. 831–839)

1. $\overleftrightarrow{AB}$ **3.** $\overrightarrow{AB}$ **5.** $\overrightarrow{NO}$ **7.** No **9.** No **11.** $\overleftrightarrow{AB}$ & $\overleftrightarrow{CD}$; $\overleftrightarrow{IJ}$ & $\overleftrightarrow{KL}$ **13.** $\overleftrightarrow{EF}$ & $\overleftrightarrow{GH}$ **15.** Yes **17.** $\angle EDF$ **19.** $\angle P$ **21.** $\angle M$ **23.** right **25.** acute **27.** straight **29.** obtuse **31.** neither **33.** supplementary **35.** neither **37.** equal **39.** vertical **41.** 31°08′11″ **43.** 34°37′56″ **45.** 60°40′18″ **47.** 0.4833° **49.** 0.1261° **51.** 0°45′0″ **53.** 0°13′3″ **55.** 23°20′ **57.** 18°39′50.5″ **59.** isosceles **61.** longest *TS*, shortest *RS* **63.** $\angle C$, $\angle B$, $\angle A$ **65.** $\angle C = \angle D$, $\angle B = \angle E$, $CB = DE$ **67.** $\angle M = \angle J$, $JL = PM$, $JK = MN$ **69.** RS = 10.607 cm, ST = 10.607 cm **71.** RS = 9.0 hm, ST = 9.0 hm **73.** AB = 13.856 dm, BC = 6.928 dm **75.** AC = 17.321 in., AB = 20.0 in. **77.** AB = 46′10″, BC = 23′5″ **79.** 8′4″ **81.** 19.63 mm **83.** 115 in., 908.5 in.^{2} **85.** pentagon, 108° **87.** 130 ft **89.** 83 ft^2 **91.** 72 bricks **93.** 2.62 cm^2 **95.** 1610.28 mm^2 **97.** 2.18 ft^2 **99.** 6.66 mm^2 **101.** 6.96 cm^2 **103.** 179.95 mm^2 **105.** 232.27 ft^2 **107.** 44.50 in.^{2} **109.** 31.42 in. **111.** 36.74 mm **113.** $AE = 5$ **115.** $\angle GJO = 45°$ **117.** $IJ = 20$ **119.** $\angle MOP = 30°$ **121.** 5 mm **123.** 3.54 cm **125.** area pentagon < area circle $\pi(5)^2 = 78.54$ cm^2
area pentagon > area square 50.00 cm^2

Trial Test, Chapter 21 (pp. 847–849)

1. $\overline{BC}$ **3.** acute **5.** parallel **7.** perpendicular **9.** 15°45′29″ **11.** 0°18′45″ **13.** largest $\angle A$, smallest $\angle B$ **15.** 13.657 cm **17.** 155′ **19.** 135° **21.** 43.20 mm **23.** 25.13″ **25.** 1.57 in.

Self-Study Exercises 22-1 (p. 856)

L.O.1.

1. 6.45 in. **2.** 14 cm **3.** 4.25 rad **4.** 3.49 rad **5.** 7 cm **6.** 4.5 in.

L.O.2.

7. 17.12 in.^{2} **8.** 4.32 cm^2 **9.** 2.24 cm **10.** 0.85 rad

L.O.3.

11. 0.79 rad **12.** 0.98 rad **13.** 1.36 rad **14.** 2.44 rad **15.** 0.38 rad **16.** 3.10 rad **17.** 0.78 rad **18.** 0.18 rad

L.O.4.

19. 45° **20.** 30° **21.** 143.2394° **22.** 80.2141° **23.** 28°38′52″
24. 22°30′ **25.** 42°58′19″ **26.** 63°1′31″ **27.** 1.53 cm **28.** 61°
29. 47.50 cm^2 **30.** 40.1070°

Self-Study Exercises 22-2 (p. 862)

L.O.1.

1. $\frac{5}{13}$ **2.** $\frac{12}{13}$ **3.** $\frac{5}{12}$ **4.** $\frac{12}{13}$ **5.** $\frac{5}{13}$ **6.** $\frac{12}{5}$ **7.** $\frac{9}{15} = \frac{3}{5}$
8. $\frac{12}{15} = \frac{4}{5}$ **9.** $\frac{9}{12} = \frac{3}{4}$ **10.** $\frac{12}{15} = \frac{4}{5}$ **11.** $\frac{9}{15} = \frac{3}{5}$ **12.** $\frac{12}{9} = \frac{4}{3}$
13. $\frac{16}{34} = \frac{8}{17}$ **14.** $\frac{30}{34} = \frac{15}{17}$ **15.** $\frac{16}{30} = \frac{8}{15}$ **16.** $\frac{30}{34} = \frac{15}{17}$
17. $\frac{16}{34} = \frac{8}{17}$ **18.** $\frac{30}{16} = \frac{15}{8}$ **19.** $\frac{9}{16.64} = 0.5409$
20. $\frac{14}{16.64} = 0.8413$ **21.** $\frac{9}{14} = 0.6429$ **22.** $\frac{14}{16.64} = 0.8413$
23. $\frac{9}{16.64} = 0.5409$ **24.** $\frac{14}{9} = 1.5556$

L.O.2.

25. $\sin A = \frac{12}{20} = \frac{3}{5}$ $\sin B = \frac{16}{20} = \frac{4}{5}$

$\cos A = \frac{16}{20} = \frac{4}{5}$ $\cos B = \frac{12}{20} = \frac{3}{5}$

$\tan A = \frac{12}{16} = \frac{3}{4}$ $\tan B = \frac{16}{12} = \frac{4}{3}$

$\csc A = \frac{20}{12} = \frac{5}{3}$ $\csc B = \frac{20}{16} = \frac{5}{4}$

$\sec A = \frac{20}{16} = \frac{5}{4}$ $\sec B = \frac{20}{12} = \frac{5}{3}$

$\cot A = \frac{16}{12} = \frac{4}{3}$ $\cot B = \frac{12}{16} = \frac{3}{4}$

26. $\sin A = \frac{8.15}{9.73} = 0.8376$ $\sin B = \frac{5.32}{9.73} = 0.5468$

$\cos A = \frac{5.32}{9.73} = 0.5468$ $\cos B = \frac{8.15}{9.73} = 0.8376$

$\tan A = \frac{8.15}{5.32} = 1.5320$ $\tan B = \frac{5.32}{8.15} = 0.6528$

$\csc A = \frac{9.73}{8.15} = 1.1939$ $\csc B = \frac{9.73}{5.32} = 1.8289$

$\sec A = \frac{9.73}{5.32} = 1.8289$ $\sec B = \frac{9.73}{8.15} = 1.1939$

$\cot A = \frac{5.32}{8.15} = 0.6528$ $\cot B = \frac{8.15}{5.32} = 1.5320$

Self-Study Exercises 22-3 (pp. 868–869)

L.O.1.

1. 0.3584 **2.** 0.9981 **3.** 1.0724 **4.** 0.6088 **5.** 0.5225
6. 0.9304 **7.** 0.4068 **8.** 0.9336 **9.** 0.8039 **10.** 0.9163
11. 1.0990 **12.** 0.8843 **13.** 0.9757 **14.** 0.4950 **15.** 3.3191
16. 0.9446 **17.** 0.6845 **18.** 0.9377 **19.** 0.3960 **20.** 0.2199

L.O.2.

21. 20.0° **22.** 22.5° **23.** 67° **24.** 62.5° **25.** 57.3° **26.** 34.8°
27. 51.7° **28.** 82.1° **29.** 39.0° **30.** 21.2° **31.** 0.5585 rad
32. 0.6807 rad **33.** 0.8028 rad **34.** 1.1694 rad **35.** 0.3527 rad
36. 1.4802 rad **37.** 0.3061 rad **38.** 1.1804 rad **39.** 0.2802 rad
40. 1.1362 rad

L.O.3.

41. 2.1943 **42.** 1.4945 **43.** 1.0306 **44.** 1.0187 **45.** 2.8824
46. 1.8341 **47.** 0.0524 **48.** 1.3563 **49.** 4.7174 **50.** 1.9661
51. 67.0° **52.** 56.9° **53.** 15.6° **54.** 83.0° **55.** 30.0°

Assignment Exercises, Chapter 22 (pp. 869–871)

1. 1.05 radians **3.** 5.24 radians **5.** 1.74 radians **7.** 150°
9. 97.4028° **11.** 67°30′ **13.** 1.61 cm **15.** 1.71 ft **17.** 3.38 in.
19. 2.09 in. **21.** 15.70 cm^2 **23.** $\frac{15}{25} = \frac{3}{5}$ **25.** $\frac{15}{20} = \frac{3}{4}$ **27.** $\frac{25}{20} = \frac{5}{4}$
29. $\frac{20}{25} = \frac{4}{5}$ **31.** $\frac{20}{15} = \frac{4}{3}$ **33.** $\frac{25}{15} = \frac{5}{3}$ **35.** $\frac{2 \text{ ft}}{2 \text{ ft } 2 \text{ in.}} = \frac{24 \text{ in.}}{26 \text{ in.}} = \frac{12}{13}$
37. $\frac{2 \text{ ft}}{10 \text{ in.}} = \frac{24 \text{ in.}}{10 \text{ in.}} = \frac{12}{5}$ **39.** $\frac{10 \text{ in.}}{2 \text{ ft}} = \frac{10 \text{ in.}}{24 \text{ in.}} = \frac{5}{12}$
41. $\frac{7}{12.62} = 0.5547$ **43.** $\frac{7}{10.5} = 0.6667$ **45.** $\frac{12.62}{10.5} = 1.2019$
47. $\frac{10.5}{7} = 1.5000$ **49.** 0.4540 **51.** 0.3057 **53.** 1.0724
55. 0.7085 **57.** 0.8403 **59.** 0.1708 **61.** 52.5° **63.** 45.3°
65. 16.7° **67.** 1.2188 rad **69.** 1.2101 rad **71.** 0.3313 rad
73. 1.0576 **75.** 1.1326 **77.** 2.1536 **79.** Answers will vary.

Trial Test, Chapter 22 (pp. 874–875)

1. 0.61 radians **3.** 5.50 radians **5.** 0.27 radian **7.** 0.28 radian
9. 112.5° **11.** 68°45′18″ **13.** 1 in. **15.** 1.71 m **17.** 3 m
19. $\frac{10}{26} = \frac{5}{13}$ **21.** $\frac{26}{10} = \frac{13}{5}$ **23.** $\frac{11.5}{12.54} = 0.9171$ **25.** 0.7986
27. 0.9163 **29.** 0.9135 **31.** 16.0° **33.** 19.5° **35.** 36.2° rad
37. 0.8726 rad **39.** 0.3150 rad **41.** 1.3673

Self-Study Exercises 23-1 (pp. 883–885)

L.O.1.

1. 27.8° **2.** 6.889 in. **3.** 28.21 m **4.** 16.15 ft **5.** 43.2°
6. 49.6°

7.
$$\begin{aligned} S &= 51.1° \\ U &= 38.9° \\ u &= 11.31 \text{ yd} \\ 18^2 &= 14^2 + (11.31)^2 \\ 324 &\doteq 323.9161 \end{aligned}$$

8.
$$\begin{aligned} S &= 45° \\ t &= 6.647 \text{ m} \\ s &= 4.700 \text{ m} \\ (6.647)^2 &= (4.7)^2 + (4.7)^2 \\ 44.18 &= 44.18 \end{aligned}$$

9. $U = 55.5°$
$s = 4.814$ mm
$u = 7.005$ mm
$(8.5)^2 \doteq (4.814)^2 + (7.005)^2$
$72.25 \doteq 23.17 + 49.07$
$72.25 \doteq 72.24$

10. $U = 74°$
$t = 50.79$ m
$u = 48.82$ m
$(50.79)^2 \doteq (48.82)^2 + (14)^2$
$2579.62 \doteq 2383.39 + 196$
$2579.62 \doteq 2579.39$

L.O.2.

11. 35.7° **12.** 10.05 cm **13.** 31.50 ft **14.** 35.49 dm
15. 15.88 mm **16.** 14.12 yd **17.** $R = 45.9°$, $Q = 44.1°$, $r = 16.52$ ft **18.** $Q = 44°$, $r = 12.23$ cm, $q = 11.81$ cm
19. $R = 16.5°$, $r = 4.147$ dkm, $s = 14.60$ dkm **20.** $Q = 30.5°$, $r = 13.58$ m, $s = 15.76$ m

L.O.3.

21. 28.6° **22.** 4.075 m **23.** 5.979 ft **24.** 23.5° **25.** 0.04663 cm
26. 0.01212 m **27.** $D = 34.8°$, $E = 55.2°$, $f = 5.604$ m **28.** $E = 48°$, $d = 6.303$ ft, $f = 9.419$ ft
29. $D = 16.5°$, $d = 5.963$ in., $f = 20.99$ in. **30.** $D = 54.0°$, $E = 36.0°$, $f = 13.60$ ft

Self-Study Exercises 23-2 (pp. 890–892)

L.O.1.

1. 150.1 m **2.** 11.91 ft **3.** 9.950 cm **4.** 46.71 cm **5.** 21.09 in.

L.O.2.

6. 22.38 in. **7.** 53.62 ft **8.** 59.0° **9.** 39.23 ft **10.** 14.9°
11. 23.25 ft **12.** 10 ohms **13.** 17.32 ohms **14.** 7.654 cm
15. 35.8°

Assignment Exercises, Chapter 23 (pp. 896–898)

1. $A = 38.1°$, $B = 51.9°$, $b = 13.15$ in.
3. $G = 48.5°$, $f = 1.681$ mm, $h = 2.537$ mm
5. $X = 52°$, $y = 5.469$ m, $z = 8.883$ m
7. $D = 28.3°$, $E = 61.7°$, $f = 14.76$ ft
9. $G = 53°$, $g = 25.21$ m, $k = 31.57$ m
11. 32.2° **13.** 11.57 in. **15.** $B = 60.5°$, $b = 3.641$ m, $c = 4.183$ m
17. 48.6° and 41.4° **19.** $B = 56°$, $a = 88.91$ ft, $b = 131.8$ ft
21. 56.54 cm

Trial Test, Chapter 23 (pp. 901–902)

1. 48.8° **3.** 15.41 ft **5.** 42.9° **7.** 6.420 cm **9.** 62.9°
11. $B = 65.5°$, $b = 30.72$ m, $c = 33.76$ m **13.** 38.7° **15.** 48 in.
17. 40.06 A **19.** 84.3° **21.** 44.7° **23.** $X_L = 9.64$ ohms; $R = 11.49$ ohms

Self-Study Exercises 24-1 (p. 907)

L.O.1.

1. 13 **2.** 15 **3.** 7.280 **4.** 8.544 **5.** 2.746

L.O.2.

6. 67.38° **7.** 53.13° **8.** 20.56° **9.** 68.20° **10.** 75.96°
11. 19; 42° **12.** 8; 72° **13.** 1; 225° **14.** 6; 75°

L.O.3.

15. $12 + 5i$; 13; 22.62° **16.** $6 + 4i$; 7.211; 33.69°

Self-Study Exercises 24-2 (p. 912)

L.O.1.

1. 60° **2.** 15° **3.** 70° **4.** 15° **5.** 32° **6.** 70° **7.** 32°
8. 62° **9.** 0.96 **10.** 0.44

L.O.2.

11. all positive **12.** sin, csc positive; others negative
13. tan, cot positive; others negative **14.** cos, sec positive; others negative
15. cos, sec positive; tan, sin zero; cot, csc undefined

L.O.3.

16. -0.5000 **17.** -0.8391 **18.** -0.8011 **19.** -0.6536
20. -0.8660 **21.** -0.2910 **22.** -0.1736 **23.** 0.5000 **24.** 146.3°
25. 2.68° rad

Self-Study Exercises 24-3 (p. 921)

L.O.1.

1. $C = 63°$; $b = 40.0$; $c = 45.2$ **2.** $A = 50°$; $a = 6.5$; $c = 8.1$
3. $C = 33°$; $b = 56.4$; $c = 35.5$ **4.** $B = 30°$; $a = 5.5$; $b = 3.2$
5. $C = 50°$; $a = 20.8$; $c = 16.0$

L.O.2.

6. $B = 21.5°$; $C = 118.5°$; $c = 57.4$ **7.** $C = 10.8°$; $A = 99.2°$; $a = 15.8$
8. $A = 43.2°$; $B = 116.8°$; $b = 23.5$ or $A = 136.8°$; $B = 23.2°$; $b = 10.4$
9. $B = 14.5°$; $C = 135.5°$; $c = 11.2$

L.O.3.

10.

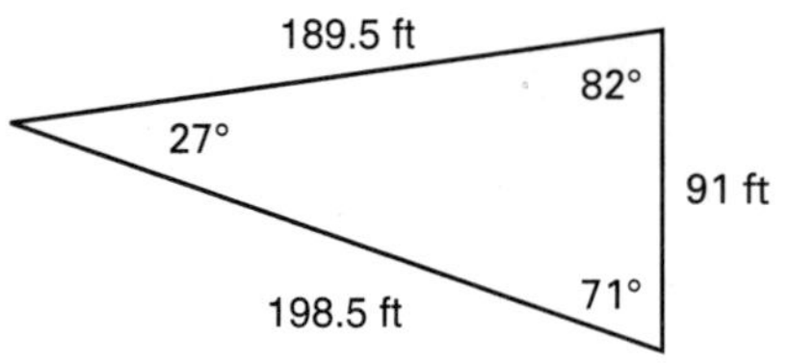

189.5 ft
82°
27°
91 ft
198.5 ft
71°

11. 806.4 ft

Self-Study Exercises 24-4 (p. 927)

L.O.1.

1. $A = 30.8°$ **2.** $D = 71.7°$
$B = 125.1°$ $E = 62.9°$
$C = 24.1°$ $F = 45.4°$

L.O.2.

3. $r = 32.42$ **4.** $j = 18.68$
$S = 49.7°$ $K = 37.5°$
$T = 58.3°$ $L = 34°$

L.O.3.

5. 49.27 ft **6.** 60.8°
44.1°
75.1°

Self-Study Exercises 24-5 (pp. 931–932)

L.O.1.

1. 27.9 cm^2 **2.** 90.4 m^2 **3.** 73.2 ft^2 **4.** 21.2 $in.^2$ **5.** 123 mm^2
6. 76.3 m^2 **7.** 193 dm^2 **8.** 454 dkm^2

L.O.2.

9. 15.0 m^2 **10.** 3.71 $in.^2$ **11.** 24.4 cm^2 **12.** 26.7 cm^2

L.O.3.

13. 3.32 cm^2 **14.** 70.4 ft^2 **15.** 5108 ft^2

Assignment Exercises, Chapter 24 (pp. 932–935)

1. 7.211; 56.3° **3.** 6.083; 9.5° **5.** 13.89; 59.7° **7.** 0.1016 **9.** 41°
11. 0.8832 **13.** 32°15′10″ **15.** 0.8632 **17.** −0.3420 **19.** 0.3420
21. 1 **23.** 2.83; 2.36 rad **25.** $A = 40°$; $b = 10.8$; $c = 4.3$
27. $A = 30.3°$; $B = 104.7°$; $b = 9.6$
29. 1st − $A = 39.4°$; $C = 112.6°$; $c = 13.4$ **31.** 42.0 ft
2nd − $A = 140.6°$; $C = 11.4°$; $c = 2.9$

33.

34.6°
28.70
16
115°
30.4°
18

35.

34°
27
27
73°
73°
15.79

37.

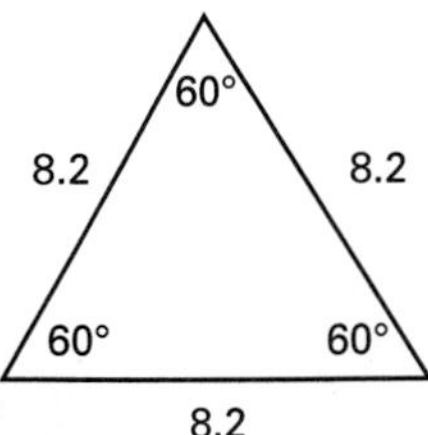

39. 97.98 ft

41. 343.3 ft; 665.4 ft **43.** 99.7 m^2 **45.** 710 cm^2 **47.** 185 km^2
49. 27.7 ft^2 **51.** 709 $in.^2$ **53.** 12,000 ft^2

Trial Test, Chapter 24 (pp. 943–944)

1. 0.8192 **3.** −0.9397 **5.** 135° **7.** 101.3° **9.** 8.991 ft
11. 131.8° **13.** 6.830 m **15.** 36.98 m^2 **17.** 150.9 ft **19.** 66.88 ft^2
21. 6.4 milliamps **23.** 30 ohms

Index